SOILS UNDER CYCLIC AND TRANSIENT LOADING

PROCEEDINGS OF THE INTERNATIONAL SYMPOSIUM ON
SOILS UNDER CYCLIC AND TRANSIENT LOADING / SWANSEA / 7–11 JANUARY 1980

Soils under Cyclic and Transient Loading

Edited by
G.N. PANDE & O.C. ZIENKIEWICZ
Dept. Civil Engineering, University College of Swansea, Swansea

VOLUME ONE
1 Modelling cyclic and transient behaviour in laboratory
2 Constitutive relations for soils

A.A.BALKEMA / ROTTERDAM / 1980

The text of the various papers in this volume were set individually
by typists under the supervision of each of the authors concerned.

For the complete set of two volumes, ISBN 90 6191 076 5
For volume 1, ISBN 90 6191 084 6
For volume 2, ISBN 90 6191 085 4

Printed in the Netherlands

Preface

The recent activity in the design and construction of off-shore, nuclear and large earth structures has given a considerable impetus to research on behaviour of soils subjected to static cyclic and transient loading. Computer methods are now available for the solution of a variety of soil structure interaction problems. The validity of such solutions hinges on the soil model used and here the state of art is less advanced but rapidly developing. A Symposium was planned at the Department of Civil Engineering, University College of Swansea, Swansea, from 7th-11th January, 1980 to provide a forum for discussion and exchange of views between researchers and designers on relevant topics such as constitutive relations of soils, cyclic degradation, shakedown, liquefaction, calibration of numerical models, field measurements and prediction of cyclic and dynamic response. More than eighty contributions were offered from as many as twenty countries. These papers are published in this volume of Proceedings. In addition a number of distinguished scholars and experts were invited to deliver 'keynote' lectures at the Symposium. Long abstracts of their lectures are also included. Unfortunately a number of contributions could not arrive on time and hence are not included in this volume. The papers are roughly classified according to the following five major themes of the Symposium.

1. Modelling cyclic and transient behaviour in laboratory
2. Constitutive relations for soils
3. Liquefaction and soil dynamics
4. Numerical and analytical methods of analysis
5. Applications, case histories, field measurement and calibration of models

We would like to thank all authors and those colleagues chairing the various sessions of the Symposium for their manifold efforts. We are also thankful to the members of the "Advisory Panel" viz.

Dr. H.A.M. van Eekelen
Shell Research Centre,
Rijswijk
THE NETHERLANDS

Professor D.L. Finn
British Columbia University,
Vancouver,
CANADA

Professor G. Gudehus
Institut fur Bodenmechanik und Felmechanik,
Karlesruhe,
WEST GERMANY

Dr. I.M. Smith
Simon Engineering Laboratories,
Manchester,
U.K.

Professor K. Hoeg
Norwegian Geotechnical Institute,
Oslo,
NORWAY

Professor A. Verruijt
Delft University of Technology,
THE NETHERLANDS

Professor Z. Mrőz
Institute of Fundamental Research,
Warsaw,
POLAND

Professor C.P. Wroth,
University of Oxford,
Oxford,
U.K.

Professor H.B. Seed
University of California,
Berkeley,
U.S.A.

We had considerable support from them during planning and conducting of the Symposium.

October, 1979

G. N. PANDE
O. C. ZIENKIEWICZ

Contents

VOLUME ONE

1 Modelling cyclic and transient behaviour in laboratory

2 Constitutive relations for soils

4 Numerical and analytical methods of analysis

5 Applications, case histories, field measurement and calibration of models

1. Modelling cyclic and transient behaviour in laboratory

Estimation of shakedown displacement in sand bodies with the aid of model tests

A.HETTLER & G.GUDEHUS
University of Karlsruhe, Karlsruhe, Germany

INTRODUCTION

The German Federal Minister for Research and Technology intensely supports the research into wheel-on-rail systems. Among these activities two groups in Karlsruhe (including the authors) and in Berlin (including Th. Dietrich) are dealing with the long-term plastic behaviour of foundations under repeated loads.

Preliminary studies on related subjects have been made, dealing with piles (Dietrich, 1977), shallow foundations (Holzlöhner, 1977), and offshore structures (Gudehus, 1978) under repeated loads. The second author is of opinion that the behaviour of sand bodies under cyclic loads can as yet scarcely be dealt with by finite element methods. On the other hand, Dietrich (Dissertation, 1973) has pointed out in several papers that model tests can yield quantitatively reliable solutions in certain cases. Thus it was decided to concentrate upon model tests in the research project introduced above. The present paper outlines the model theory with a few examples in order to demonstrate that this concept is by no means restricted to wheel-on-rail systems.

For reasons of brevity and simplicity, the following restrictions have been made within this paper, viz.
-the sand bodies are dry, so that pore pressure production is excluded;
-the external load is quasistatic (i.e. inertia effects can be neglected), and equal for all cycles;
-the loads are low enough to exclude incremental collapse (Goldscheider, 1977).
It may only be mentioned that the model theory can also be extended to cover some of the cases excluded here.

THE MODEL LAW

According to the rules of dimensional analysis one can write the following formula (Dietrich, 1980) for the shakedown displacement s, viz.

$$s/B = f(N; \frac{P}{\gamma B}; \frac{L}{B}; G_i; M_i) \quad ; \qquad (1)$$

with the symbols

B, a characteristic length of the structure;

N, the number of cycles;

p, a characteristic external pressure (possible further loading quantities assumed to be proportional to p);

γ, the unit weight of soil;

$L := \sqrt[4]{\frac{EI}{\gamma B}}$, the elastic length of the flexible structure having a flexural stiffness EI (this variable disappears in Equ.1 in case of rigid structures);

G_i, set of length ratios characterizing the dimensionless geometry;

M_i, set of dimensionless material parameters.

Equation 1 implies the decisive assumption that the only dimensional material parameter of the soil is γ. This is a basic property of rigid-granular materials (Dietrich, 1976), and also of certain materials of the rate type (Gudehus and Kolymbas, 1979). As a consequence, the angles of friction and dilatancy are independent of stress level, and the incremental stiffness is proportional to stress level. This statement is at variance with the widely accepted opinion concerning stress level influence. The discrepancy can partly be explained by bifurcation (Vardoulakis, 1979) and progressive failure (Gudehus, 1978). It may suffice here to justify our model law by succesful examples of application. The restriction

3

to cases without bifurcation and/or progressive failure has to be kept in mind,
however.

On principle the function f of Equ.1 can
be determined by model tests with the same
sand as in situ. Because of the big number
of variables, however, this procedure would
not be practicable. Dietrich (1977) has
proposed a representation of f by factors,
viz.

$$f = f_G(G_i) \; f_p \; (\frac{p}{\gamma B}) f_N(N) \quad . \qquad (2)$$

The load factor f_p can be further specified
by a power law

$$f_p = (\frac{p}{\gamma B})^\alpha \quad , \qquad (3)$$

with an empirical exponent α. Equ. 3 reflects the dimensionless stress-strain law
from triaxial tests, e.g., and is typical
of a rigid-granular material (Dietrich,
1976). It appears to be justified to consider α as density independent (see next
section). Due to Equ. 3 geometrically similar systems under loads of different intensities are self-similar in the sense of
Dietrich (1980).

The factor f_N can often be approximated
by the empirical function

$$f_N = c_1(1 + c_0 \ln N), \qquad (4)$$

with two empirical constants c_0 and c_1.
Equ. 4 may even be taken as defining equation for shakedown of sand bodies (Goldscheider, 1977). Note that Equ. 4 implies
equal loading in each cycle. Frequency
and load-time distribution (zig-zag or
sinusoidal, e.g.) do not enter.

Thus the semi-empirical shakedown displacement formula reads

$$s/B = A(\frac{p}{\gamma B})^\alpha \; (1 + c_0 \ln N) \qquad (5)$$

The shape factor A and the constants α and
c_0 have to be determined for one set of
length ratios in a single model test. Thus
the number of model tests required is drastically reduced so that the amount of work
is justified for practical applications.
In the following sections we demonstrate
by a sequence of examples that Equ. 5 is
sufficiently accurate.

INFLUENCE OF LOAD INTENSITY FOR SINGLE
LOADING

A series of model tests by Görner (1932)
may first elucidate that there is no scale
effect due to stress level. Circular stiff
foundations of r=2,5 cm to 49 cm have been

punched into carefully prepared sand beds.
One can use r as characteristic length and
neglect the elastic length L. Geometrical
similarity is given as the foundations were
not embedded.

Fig. 1 shows the dependance of dimensionless settlement, s/r on the dimensionless
load, p/ r, in a log-log plot. The points

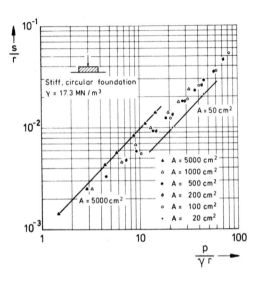

Fig. 1a: high density, heavily compacted

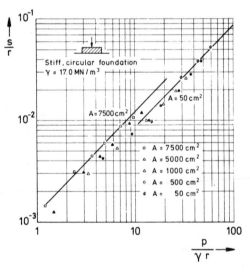

Fig. 1b: high density

Fig. 1a, b: Static model tests by Görner
varying area from 20cm² up to 7500 cm², curves
in dimensionless log-log representation for
different densities

4

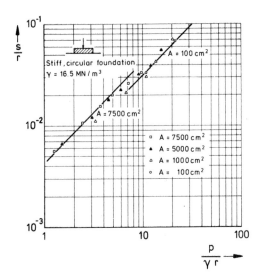

Fig. 1c: medium density

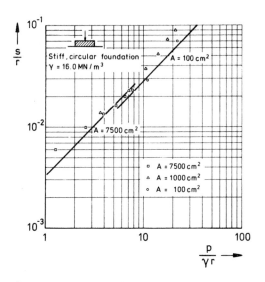

Fig. 1d: loose density

Fig. 1c, d: Static model tests by Görner
varying area from 20cm^2 up to 7500cm^2 curves
in dimensionless log-log-representation for
different densities

for different loads can be approximated by
straight lines. This dependance is covered
by Equ. 3 with an exponent of $\alpha \simeq 1$. Fig. 1a,
derived from the tests with heavily compac-
ted sand, shows parallel lines indicating
a scale factor of about 2 for the settlement when increasing the foundation radius
by a factor of 16. The scale factor is

lower for normal compaction (Fig. 1b) and
reaches one for medium (Fig. 1c) and loose density (Fig. 1d). It appears from
Fig. 1 that our model law, Equ. 1 or 5,
should be improved by a scale factor for
high densities. This would be a wrong inference, however. Actually, a 'hidden' model law was violated for the dense sand
tests: heavy compaction by tamping has produced a zone of increased horizontal stress
the depth of which was not scaled according
to the foundation radius. Thus the subsoil
had an increased stiffness up to a bigger
relative depth for the smaller foundations.
As the incremental stiffness is strongly
dependent on stress ratio (Gudehus and Kolymbas, 1979) the sand bed in a model test
has to be prepared such that the stress
ratios are the same at equivalent points
of model and prototype. Overlooking this
'hidden' model law can heavily deteriorate
the results as can frequently be seen in
the literature. Görner's (1932) tests with
medium density comply fairly well with this
model law and it can be seen from Fig. 1c
that there is in fact no scale effect due
to stress level.

We next consider the variability of the
exponent α. Comparing Figs. 1a, b, c, and
d indicates that α does not depend on density. This statement is part of Dietrich's
theory (1980). Fig. 2, obtained from a se-

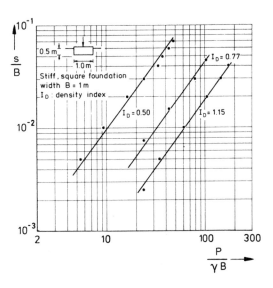

Fig. 2: Static model-tests (Karlsruhe 1966)
varying density, curves in dimensionless
log-log-representation

5

ries of tests with square foundations (Leu-sink et al., 1966), gives the same information. Fig. 3 represents α values evaluated

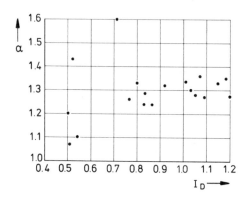

Fig. 3: Static model-tests (Karlsruhe 1966), exponent α in function of density index

from further tests made in Karlsruhe (Leu-sink et al., 1966). Apart from an increase scatter for loose states there is no systematic influence of relative density upon α. The scatter may be attributed to non-uniformities of the sand bed. It appears that α is the same for different shapes of foundation, but variable with the granulometry.

FOUNDATIONS UNDER REPEATED LOADS

A series of field tests with different piles under horizontal loads by Alizadeh and Davisson (1970) may serve to underline our model formula, Equ. 5. Piles of different materials and diameters have been loaded once and repeatedly. Fig. 4 represents dimensionless horizontal displacements for single horizontal load. Division of Q by L^2 yields p as a kind of earth pressure, and $p = Q/L^2$ is the quantity that enters Equ. 5. The log-log plot of Fig. 4 again supports the power law. It is remarkable that the exponent $\alpha \simeq 1,40$ is independent of the type of pile.

Fig. 5 is a log-log plot of dimensionless shakedown displacement versus the number of cycles. It can be seen that the s/B values for two different relative loads $Q/\gamma BL^2$ differ by a constant factor for any number of cycles. This observation supports the representation of Equ. 2, i.e. the factorization with respect to the number of cycles N.

The factor f_N was calculated for different relative loads with the aid of Equ.2 and is plotted versus the logarithm of N in Fig. 6. One obtains a single straight line

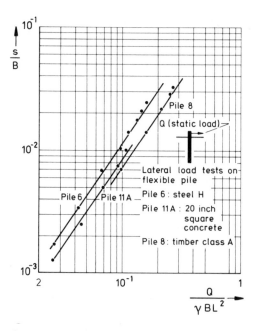

Fig. 4: Static lateral load tests on piles by Alizadeh and Davisson (1970) with different kind of piles, load displacement curves in dimensionless log-log-representation

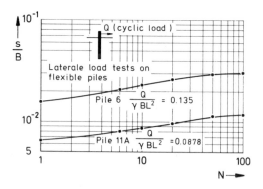

Fig. 5: Cyclic lateral load tests on piles by Alizadeh and Davisson (1970), displacement versus number of cycles in log-log representation

for two different piles which supports our Equ. 4.
Finally we again consider circular shallow foundations, but now under cyclic vertical loads. Holzlöhner (1977) has carried out model tests with up to 0.36 m diameter slabs and evaluated them along the lines of Dietrich (1980). This study went beyond the scope of the present paper as the influ-

6

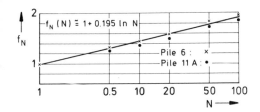

Fig. 6.: Cyclic lateral load tests on piles (1970), f_N versus the logarithm of N

ence of a simultaneously acting permanent load and of the frequency of load cycles has also been considered. Here we only deal with the influence of load intensity and number of cycles.

Fig. 7 shows the dimensionless shakedown settlement versus lnN. It can clearly be seen that Equ. 4 is valid for different loads. A further evaluation towards our funktion f_p is not given here as Holzlöhner (1977) has worked with different permanent

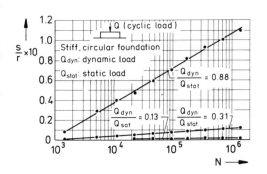

Fig. 7: Cyclic model tests on stiff foundations by Holzlöhner (1977) varying load intensity, dimensionless settlement versus the logarithm of N

loads such that the model law for forces is more complicated than Equ. 3.

CONCLUDING REMARKS

Having in mind the typical scatter of shakedown displacements in sand bodies our model law, Equ. 5, is of sufficient accuracy for foundation practice. This has certain consequences some of which may be briefly outlined here.

For cases of immediate practical interest it is recommendable to work with model setups. Of course, a classification of typical geometry and load conditions has to be made. Thus it was decided to construct two model

setups for the railway research project mentioned above: one representing the infinite flexible strip under transient load (by Th. Dietrich), and the other one for the flexible grid under cyclic loads (in Karlsruhe). Tests in the same setups will also serve to establish sufficient conditions to avoid incremental collapse. For some more details of the same research project see Gudehus (1979).

For further validation and possible extension of Dietrich's model law some more fundamental studies are needed. It is intended to study the influence of stress level and stress ratio upon incremental stiffness in a newly designed triaxial apparatus. The authors are of opinion that most of the presently used triaxial apparatuses cannot serve for this purpose, as a strictly uniform deformation of samples is not secured with them. More complicated loading conditions are also studied in suitably chosen principal model tests. Such tests will lead to improved load factors. The theoretical basis for the factorized model law is further established by Dietrich (1980).

In the long run it can be hoped that model tests can partly be replaced by finite element calculations. Certainly one has to start with monotonous loading. We are still rather far from an acceptable solution for this comparably simpler case, as
- the constitutive laws presently used for sand (Gudehus, 1979a) mostly do not comply with the model law;
- convergence of calculation and uniqueness of solutions are not secured (these mathematical difficulties are partly due to inadequate constitutive laws);
- the results are mostly not evaluated according to laws of similarity (as was done by Winter, 1979, e.g.), and thus the required amount of work for an individual practical example is too big.

It will take a long time before finite element solutions for cases of cyclic loading can also meet these requirements.

REFERENCES

Alizadeh, M. and M.T.Davisson 1970, Lateral load tests on piles - Arkansas River project, Journal of the Soil Mechanics and Foundations Division, ASCE, Vol.96, No. SM 5, Sept. 1970, 1583-1603.
Dietrich, Th. und Dolling, H.J., 1973, Die Möglichkeiten experimenteller Untersuchungen im Grundbau bei besonderer Berücksichtigung der Modelltechnik, Wissenschaftliche Berichte aus der Arbeit der Bundesanstalt für Materialprüfung.

Dietrich, Th., 1976, Der psammische Stoff
   als mechanisches Modell des Sandes,
   Dissertation, Universität Karlsruhe.
Dietrich, Th., 1977, Experimental study
   of flexible piles in sand cyclically
   displaced at low frequency, Proceedings
   of International Symposium on testing
   in situ of concrete structures, Sept.
   1977.
Dietrich, Th., 1980, On structures in sand
   permitting of factorization of the influence of repetition of loading, International Symposium on soils under cyclic
   and transient loading, Swansea.
Görner, E.W., 1932, Über den Einfluß der
   Flächengröße auf die Einsenkung von
   Gründungskörpern, Geologie und Bauwesen,
   Band 4, Heft 1-4.
Goldscheider, M., 1977, Shakedown and incremental collapse in dry sand bodies,
   Proceedings of International Symposium
   on Dynamical Methods in Soil and Rock
   Mechanics, Karlsruhe, Volume 2.
Gudehus, G., 1978, Engineering approxima-
   tion for some stability problems in geo-
   mechanics, Advances in analysis of geo-
   technical instabilities, University of
   Waterloo, Waterloo Press, SM Study Nr.
   13, Paper 1.
Gudehus, G., 1978, Stability of saturated
   granular bodies under cyclic load, Proceedings of International Symposium on
   Dynamical Methods in Soil and Rock Mechanics, Karlsruhe, Volume 2.
Gudehus, G. and Kolymbas, D., 1979, A constitutive law of the rate type for soils,
   Proceedings of the Third International
   Conference on Numerical Methods in Geomechanics, Aachen.
Gudehus, G., 1979, Foundation and soil
   mechanics problems for wheel-on-rail
   systems, Int.Symp. on traffic and transportation technologies, Hamburg.
Holzlöhner, U., 1977, Residual settlements
   in sand due to repeated loading, Proceedings of International Symposium on Dynamical Methods in Soil and Rock Mechanics, Karlsruhe, Volume 2.
Leussink, Blinde und Abel, 1966, Versuche
   über die Sohldruckverteilung unter star-
   ren Gründungskörpern auf kohäsionslosem
   Sand, Veröffentlichungen des Instituts
   für Bodenmechanik und Felsmechanik der
   Universität Karlsruhe, Heft 22.
Vardoulakis, I., 1979, Bifurcation analysis of the triaxial test on sand samples,
   Acta Mechanica 32, 35-54.
Winter, H., 1979, Fließen von Tonböden:
   Eine mathematische Theorie und ihre An-
   wendung auf den Fließwiderstand von Pfäh-
   len, Dissertation, Universität Karlsruhe.

The behaviour of Leighton Buzzard Sand
in cyclic simple shear tests

D.M.WOOD
Cambridge University Engineering Department, UK

M.BUDHU
University of Guyana, Georgetown

INTRODUCTION

Rotation of principal axes of stress and
strain is unavoidable in most practical
problems in which soil masses are subjected
to changing loads. The amount of data
that has been produced from laboratory
tests in which controlled rotation of principal axes is possible is, however, limited.
The primary reason for this shortage of
data lies with mechanical difficulties of
devising an apparatus in which soil specimens can be given complete freedom to
deform with completely independent rotations of the principal axes of stress and
strain.

   The simple shear apparatus is one device
in which rotation of axes does occur, but,
as will become apparent, this apparatus is
not without its shortcomings.

THE BEHAVIOUR OF THE SIMPLE SHEAR TEST

The initial development of the Cambridge
simple shear apparatus by Roscoe (1953)
sprang from the earlier work of Hvorslev
(1936) testing samples of clay in a direct
shear box. Hvorslev appreciated that if
the aim was to understand the failure
states of soils then it was very necessary
to obtain information about the water content of the soil (which may be regarded
with void ratio or specific volume as a
"state parameter") not averaged over the
whole sample at failure, but specific to
the narrow zone, forced by the form of
the apparatus, in which the deformation of
the soil was in fact concentrated (Fig.1).

   Roscoe's apparatus was intended to make
the sample deform uniformly (Fig.2), and
make the failure zone occupy the whole
sample. However, although the Cambridge
simple shear apparatus has been developed
to a stage at which, by an arrangement of

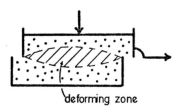

Fig.1. Direct shear box

of hinges and slides (Fig.3) the whole
sample can plausibly deform uniformly
(Bassett, 1967; Cole, 1967), there remains
a basic mechanical problem. Even if soil
were not a material exhibiting dilatancy,
the lengths of the ends AB and CD of the
sample (Fig.2) would have to change, and
sliding of soil particles on these ends
would have to occur as the sample was
deformed. The need to allow the soil to
expand or contract under shearing increases
the problem. The need to permit free change of length is provided by the mechanism
sketched in Fig.3, but in the Cambridge
apparatus the ends are made smooth in order
to allow free movement between the ends
and the soil particles. Consequently the
complementary shear stresses required can
not be provided and a resulting non-uniform distribution of shear stress on the
top and bottom boundaries of the sample is
to be expected (Fig.4).

   Roscoe (1953) reported an analysis of an
elastic material subjected to these, non-
ideal, simple shear boundary conditions and
showed that the distribution of shear stress
shown qualitatively in Fig.4 was accompanied by variations of normal stress across
these boundaries, with tensile and compressive stress singularities produced at the
leading (X) and trailing (Y) edges (Fig.5)
respectively. A soil/metal interface can
not withstand a normal tensile stress:

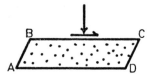

Fig.2. Simple shear

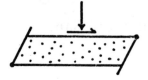

Fig.3. Mechanism of Cambridge simple shear apparatus

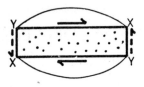

Fig.4. Expected distribution of boundary shear stresses in simple shear apparatus

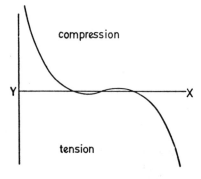

Fig.5. Distribution of normal stresses in elastic analysis (after Roscoe, 1953)

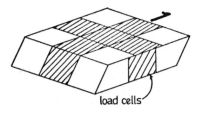

Fig.6. Load cells in Cambridge simple shear apparatus

Roscoe showed some simple tests on samples of plasticine where separation of the samples from the apparatus occurred at the leading edges.

Roscoe's elastic analysis indicated that the internal strains might be more uniform than the stresses, but that the stresses on the top and bottom boundaries of the central third of the sample were reasonably uniform. (Prévost and Höeg (1976) showed, again for an elastic material, that slip at the top and bottom boundaries produced a marked worsening of the uniformity over the central third; Finn, Pickering and Bransby (1971) demonstrated the importance of having sufficiently rough boundaries in permitting sand samples to show their full potential for resistance to liquefaction.) The philosophy at Cambridge, then, has been to accept that the stresses are not uniformly distributed over the boundaries, but to surround the sample with load cells, or contact stress transducers (Fig.6)(as originally described by Arthur and Roscoe (1961)) so that the stress state in the central third of the sample can be deduced by a procedure such as that described by Wood, Drescher and Budhu (1979). Stress conditions in this central third are then assumed to be representative for an element of soil subjected to true simple shear.

That the boundary stresses are non-uniform implies, even though those boundaries are rigid, that the internal deformations are likely also to be non-uniform. Duncan and Dunlop (1969) have analysed the simple shear deformation of a soil sample and show zones of progressive failure developing within the sample as the boundary movements are increased. It is likely however, that for small deformations internal non-uniformities may be negligible – Duncan and Dunlop show zones of progressive failure (evidence of major inhomogeneity) only for shear strains in excess of 5%.

In presenting results of simple shear tests conducted in the Cambridge simple shear apparatus it is tacitly assumed that the non-uniformities of deformation may be neglected. However, the changing distributions of internal strains - both shear and volumetric - have been monitored by means of radiography (Roscoe, Arthur and James, 1963). An array of lead shot markers is placed in the sample at the start of the test and the positions of these markers measured from x-ray film at appropriate stages during the test. When testing samples of dense sand, x-rays can be used to show up any zones of preferential dilation (rupture bands) that may develop - in which a marked reduction in absorption of x-rays occurs.

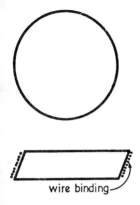

wire binding

Fig.7. Circular simple shear apparatus

The Cambridge simple shear apparatus, then, in summary, tests a sample which is square in cross section, and is surrounded by an array of load cells measuring the distribution of stresses at the boundary of the sample, and permits the investigation of internal deformations by means of x-rays.

By contrast, the circular simple shear apparatus that are, as a result of their simpler construction and operation, in more widespread use, test samples which are circular in cross section, and contained within a rubber membrane surrounded by metal rings (Kjellman, 1951) or reinforced with a helical wire binding (Bjerrum and Landva, 1966) in order to keep the cross-sectional area of the sample constant (Fig.7). In the apparatus generally available only the average stresses applied to the horizontal boundaries of the sample are measured, no measurement is made of the lateral stresses, and the deformation

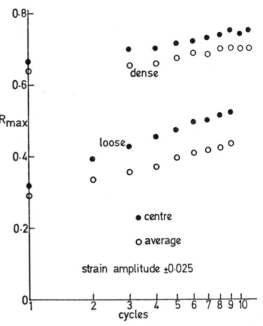

Fig.9. Comparison of central and average stress ratios in cyclic loading

of the sample is assumed uniform. The theoretical analysis of Lucks, Christian, Brandow and Höeg (1972) showed that approximately 70% of the sample in the circular simple shear apparatus was uniformly stressed.

The assumption of uniformity of deformations may not be seriously in error provided only small deformations (monotonic or cyclic) of the sample are being considered. The use only of the average stress conditions may produce a conservative estimate of the strength of the sample: Figs. 8 and 9 show comparisons of development of stress ratio R in monotonic and cyclic tests on dense Leighton Buzzard sand, between results computed using the average stress conditions, and using the stress conditions obtaining over the central third of the sample. The absence of measurements of lateral stresses makes it difficult to establish reliable comparisons between results obtained in the simple shear apparatus and results obtained in other apparatus - such, for example, as the triaxial apparatus.

Tests have been conducted at Cambridge in an apparatus described by Budhu (1979) in which a sample circular in section has been contained between horizontal boundaries made up of arrays of load cells, and within a rubber membrane reinforced with

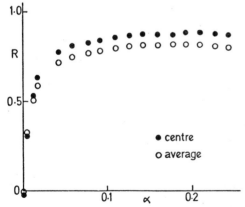

Fig.8. Comparison of central and average stress ratios in monotonic loading of dense sand

11

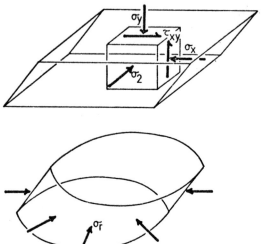

Fig.10. Stress components in square and circular simple shear apparatus

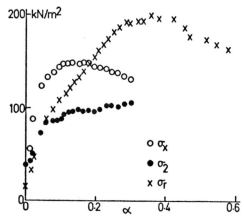

Fig.11. Lateral stresses in monotonic loading of dense sand in square and circular simple shear apparatus

strain gauge wire so that some indication of the lateral radial stresses (σ_r in Fig.10) acting on the sample in this apparatus can be deduced from the hoop tension measured in the wire. Moussa (1974) and Youd and Craven (1975) have deduced lateral stresses for the circular simple shear apparatus using the same procedure.

In analysing results of simple shear tests it has been assumed that the lateral normal stresses on vertical planes in and perpendicular to the plane of shearing (σ_x and σ_2 in Fig.10) are equal (for example, Prévost and Höeg, 1976; van Eekelen and Potts, 1978) - experimental evidence from monotonic loading tests on dense sand (Fig.11) suggests that this is certainly not always correct.

The flexibility of the membrane in the circular apparatus permits some local changes of curvature round the sample - but these are unlikely to be sufficient to produce the differences in lateral stress which the soil manages to generate in the square apparatus. (Although the purpose of the reinforcement is nominally to maintain a plane strain condition of deformation, it actually more closely maintains a constant sample perimeter.) Deduced lateral stresses from a monotonic loading test on dense sand at the same stress level are shown in Fig.11. Because of the flexibility of the boundaries in the circular apparatus, preferential dilation is able to occur across an inclined zone (sketched from a radiograph in Fig.12) and the lateral stresses consequently increase

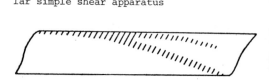

Fig.12. Rupture zone in circular apparatus (sketch of radiograph). Leighton Buzzard sand

rapidly. This radiograph also shows the curved deformation of the ends of the sample, with the membrane developing a shape like that recorded by Youd (1972) (Fig.13) - a non-ideal consequence of the flexibility of the boundaries.

Zones of preferential dilation can be observed in tests in the square apparatus (Fig.14 and 15). When the sand being tested is 14/25 Leighton Buzzard sand - with an average grain size of about 1mm - these zones are roughly horizontal. The sample is 100mm long and initially about 20mm

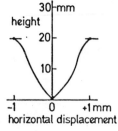

Fig.13. Profiles of edge of circular sample at shear maxima (after Youd, 1972)

Fig.14. Rupture zone in Cambridge apparatus (sketch of radiograph). Leighton Buzzard sand

Fig.15. Rupture zones in Cambridge apparatus (sketch of radiograph). Fine sand

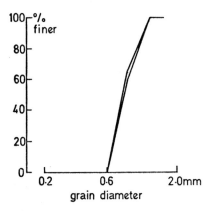

Fig.16. Grain size distribution for Leighton Buzzard sand (after Cole, 1967)

high; Roscoe (1970) talks of rupture zones with a thickness of about ten grain diameters, so that these would be expected to occupy roughly half of the sample. It is to be expected that the formation of such zones will be influenced both by the proportions of the apparatus - an increased ratio of length to height of sample tending to increase the degree of uniformity (Bassett, 1967) - and by the ratio of height of sample to grain size. Fig.15 shows a sketch of dilatant zones for a fine sand (grain diameter about 0.2mm) with the same initial sample dimensions. It is clear that this sand has found much more freedom to develop non-uniformities - and that rupture has become possible on several of the potential surfaces predicted by Arthur, Dunstan, Al-Ani and Assadi (1977).

Non-uniformities of water content were shown to exist by Casagrande (1975) who reported tests on a saturated medium loose sand in a circular simple shear apparatus : the samples were frozen and cut up after a test so that the distribution of water content could be determined. Where such internal volume changes are occurring the reliability of average boundary measurements of deformations becomes doubtful. In particular, a test in which no overall volume change of the sample is permitted is unlikely to be truly undrained for any elements within the sample.

THE BEHAVIOUR OF LEIGHTON BUZZARD SAND

The tests to be described here have all

been conducted in the square simple shear apparatus designed by Stroud (1971), on dry 14/25 Leighton Buzzard sand. This is a coarse sand, with mean grain size about 1mm, and typical grading shown in Fig.16. Samples have all been formed only by pouring : the maximum void ratio being 0.78 for loose samples, and the minimum void ratio being 0.53 for dense samples. The tests have all been performed with a constant mean applied vertical stress of 98.1 kN/m^2 (though this does not guarantee that the vertical stress applied to the central third of the sample will be constant). The tests were run in this way, rather than with constant sample volume, because the large pressures generated by dense sands when dilation is prevented are destructive of load cells. The cyclic tests were run with constant rates of shear deformation between present limits - giving a triangular waveform.

Typical results for a monotonic test on dense sand are shown in Figs.17 and 18. Fig.17 shows the development of stress ratio R (= τ_{yx}/σ_y) for the centre of the sample with shear strain α ; and Fig.18 shows the change in void ratio, calculated for the sample as a whole.

In a cyclic test, in which the shear strain is varied between limits on each side of the initial position, the normal stress distribution (shown in Fig.5 for the elastic material) will fluctuate during each cycle. Distributions for a cyclic test with shear strain amplitude $\alpha = \pm 0.025$ are shown in Fig.19. The stress distribution is shown as a smooth curve although the load cells in the simple shear apparatus yield discontinuous information. The curves have been fitted in such a way that the average stress and

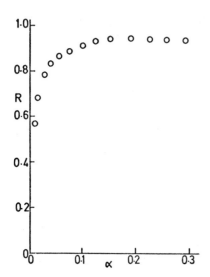

Fig.17. Monotonic test on dense sand, stress ratio:shear strain

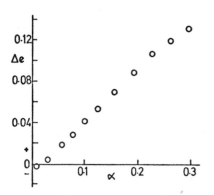

Fig.18. Void ratio changes in monotonic test on dense sand

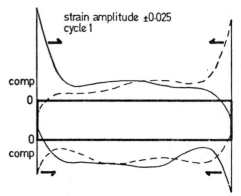

Fig.19. Distributions of normal stress on horizontal boundaries at shear maxima during cyclic test on dense sand

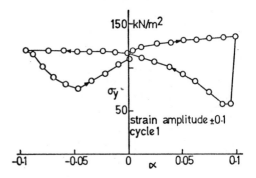

Fig.20. Variation of vertical normal stress at centre of sample in cyclic test on dense sand

the eccentricity of loading for each of the load cells are correctly matched. Curves are shown for the peak forward and peak reverse conditions. Although the average stress at the centre remains reasonably constant the ends of the sample are subjected to very severe variations of stress, which must have some influence on the behaviour of the samples.

The procedure that is used to calculate the stress state in the centre of the sample makes use of all the available load cell information - including the measured eccentricities - and produces an average stress tensor which is as nearly as possible compatible with this information (Wood, Drescher and Budhu, 1979). Although

the tests are conducted with constant applied average vertical stress, the vertical stress, σ_y, computed for the centre of the sample can vary significantly from the average. Fig.20 shows variations of σ_y (for an applied average stress of 98.1 kN/m^2) for a test with shear strain amplitude ± 0.1. Typically, the stress drops rapidly when the direction of straining is reversed and then builds up again gradually as the opposite peak of the cycle is approached.

The loops of stress ratio R against shear strain α show typically hysteretic behaviour (Fig.21a,b). The variation of maximum stress ratio with number of cycles for loose and dense sands is shown in Fig.22a, b. The solid curves come from a smoothed interpolation of the experimental contours of equal maximum stress ratio in a (number of cycles):(strain amplitude) space (Fig. 23a,b) and provide a reasonable fit to the data points in Fig.22.

14

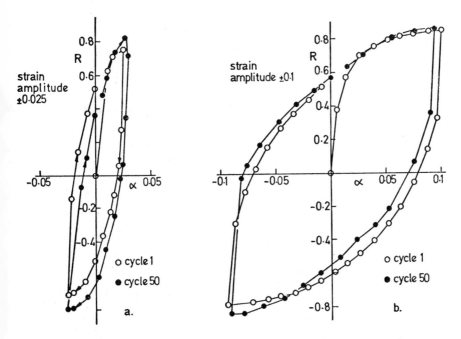

Fig.21. Stress ratio : strain loops for cyclic tests on dense sand

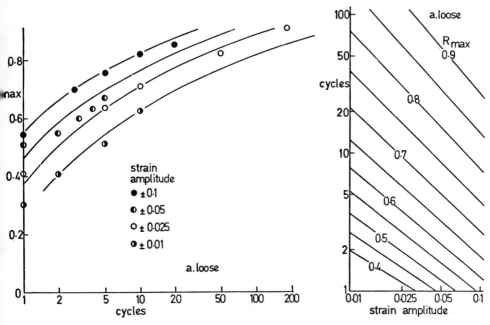

Fig.22a. Variation of stress ratio with number of cycles for tests on loose sand

Fig.23a. Contours of maximum stress ratio from cyclic tests on loose sand

15

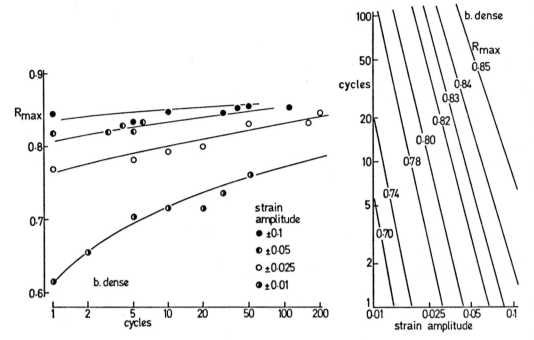

Fig.22b. Variation of stress ratio with number of cyclès for tests on dense sand

Fig.23b. Contours of maximum stress ratio from cyclic tests on dense sand

The observed behaviour of the sand must be very much bound up with the changes of void ratio that are occurring within each cycle. Subjected to a moderately small cyclic shear strain of amplitude ±0.025 the initially dense sand becomes denser (Fig.24a) - confirming Youd's (1971) finding that cyclic simple shear straining can produce lower void ratios than can be produced by other simple means. Although within each cycle the dominant effect is one of dilation, the overall result is a net contraction (a reduction in void ratio of about 0.018 after 200 cycles). The main volumetric decrease occurs during the unloading phase of each cycle - a feature incorporated by Finn, Lee and Martin (1975) in their model of the behaviour of sand in simple shear.

With a larger strain amplitude of ±0.1 the net volume change is small but expansive (Fig.24b) - the dilatant tendency dominates (and internal inhomogeneities can perhaps be anticipated).

In tests on initially loose samples the overall effect for all strain amplitudes is volumetric compression (though some dilation is shown towards peak strain in each cycle). The variation of void ratio at the end of each cycle with number of

cycles is shown for loose sand and dense sand in Figs.25a,b.

Complete presentation of the stress-strain behaviour of soils in the simple shear apparatus is complicated by the fact that the stress state traces a path in a four-dimensional space - the four dimensions may be, for example, the three principal stresses and the direction of the major principal stress. The extra dimensions must never be forgotten in studying two dimensional plots.

One simple relationship between stress and strain parameters, which was originally established by Roscoe, Bassett and Cole (1967) for monotonic loading, concerns the orientations of the principal axes of stress, stress increment and strain increment. Three angles, ψ, χ and ξ respectively may be defined as shown in Fig.26. Roscoe, Bassett and Cole (1967) showed that aftèr a small initial strain the axes of strain increment and stress coincide ($\psi = \xi$). For cyclic tests (Fig.27) each time the direction of straining is reversed a major deviation of ψ and ξ occurs (in fact, initially axes of strain increment and stress increment coincide, with $\chi = \xi$, as would be expected for an elastic material) but after a small strain ψ and ξ again

16

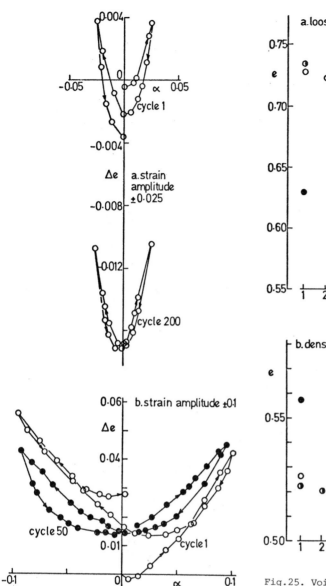

Fig.24. Volume changes within cycles for cyclic tests on dense sand

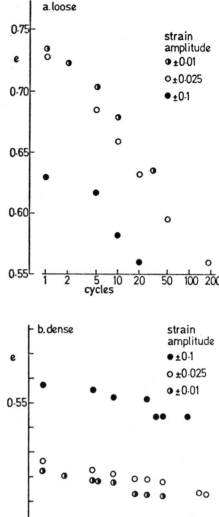

Fig.25. Void ratio at end of each cycle in cyclic tests on (a) loose , and (b) dense sand

become equal - an equality that is expected for an isotropic plastic material.

The stress path in a two dimensional principal stress space for the plane of shearing, with $t = \frac{1}{2}(\sigma_1 - \sigma_3)$ and $s = \frac{1}{2}(\sigma_1 + \sigma_3)$, is shown in Fig.28 for a test with an amplitude ±0.12 . The rotations of principal axes that occur during a cycle are not shown in this space - it gives perhaps a false impression to plot both halves of the cycle on top of each other. The form of the cycle plotted is typical for other amplitudes. It should be noted that at the end of the cycle, with the sample unstrained, the stress state is already very close to failure.

Finally, the variation of the principal stress normal to the plane of shearing during a cyclic test with shear strain amplitude ±0.1 is shown in Fig.29. Considerations of isotropic plasticity suggest

17

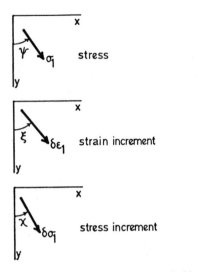

Fig.26. Angular parameters defining
orientations of principal axes

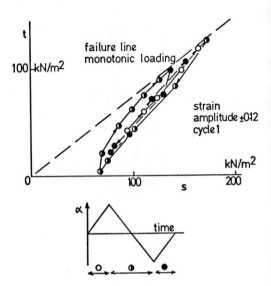

Fig.28. Stress path for one cycle of test
on dense sand

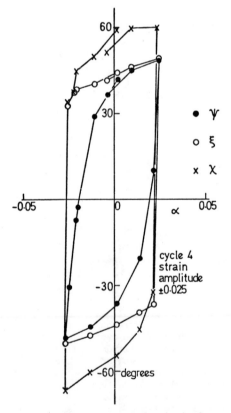

Fig.27. Orientations of principal axes
during cyclic test on dense sand

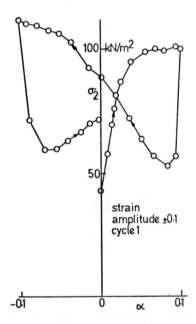

Fig.29. Variation of intermediate principal stress in cyclic test on dense sand

that σ_2 should equal s once plastic flow
has been developed. If other, anisotropic
(in the sense of Lensky, 1960) plasticity
relations are assumed (for example, van
Eekelen and Potts, 1978) then other ratios
of σ_2 to s are expected for plastic flow :

Stroud (1971) showed that the ratio σ_2/s tends to a constant value of about 0.74 in monotonic straining of Leighton Buzzard sand. Results from cyclic tests show that while this value may be reached in each cycle at the peaks of strain, the reduction in σ_2 that occurs immediately after the reversal is smaller than the drop in s so that the ratio may briefly rise nearly to 1.0.

DISCUSSION

A discussion of the mechanics of the cyclic simple shear test might begin by suggesting that the dilatant structure, that has emerged from developed plastic flow in one direction, then looks only like a loose structure for the reversed flow. Immediately on reversal the movement of the end flaps back into the sand - a passive process - tends to increase the normal stress at what was the leading edge, some redistribution of normal stress over the horizontal surfaces occurs with the stress at the centre, σ_y, dropping temporarily. Because the sand structure appears loose to the reversed deformation, there is a marked tendency for contraction which is reflected in the reduction of the lateral stresses - since no deformation can occur in these directions. As the reversed deformation continues, particle rearrangement continues until plastic flow develops in the reverse direction, and the tendency to dilation produces a build up of lateral stresses once more.

Youd (1970) and Arthur, Dunstan, Al-Ani and Assadi (1977) have discussed models for the behaviour of sand in which two opposing mechanisms operate simultaneously. Youd considers combinations of densification and dilatation, and suggests that there is a "low strain densification bias" - which has been confirmed in the present experimental programme. Though the tests here have all been conducted on dry sand, it might be inferred that saturated dense samples subjected to low strains without drainage would show a tendency to build up positive pore pressures. Evidently in these dry, drained tests the tendency for continued contraction must become gradually exhausted as the structure of the sand becomes denser: in the saturated, undrained tests the structure will remain approximately at its initial density.

Arthur et al (1977) suggest that a contribution to strain softening comes from local failures at inherent discontinuities within the granular material, while strain hardening occurs as a result

of the rearrangement of particle geometry and particle contacts during plastic strain following the local failures. This idea of opposing mechanisms links up with Finn, Bransby and Pickering's (1970) observation that resistance to liquefaction of sand can not be explained only by void ratio - other factors, related to the strain induced anisotropy of particle contacts are also relevant. This can be observed in the variation of maximum stress ratio R_{max} with number of cycles shown for dense sand in Fig.22b. At low strain amplitudes some permanent adjustment of particle positions is possible, with improved ability to resist deformation - and increased stress ratio. At higher strain amplitudes, however, the benefit of particle rearrangement is lost in the dilational flow needed to meet the imposed strain and little change in stress ratio occurs.

CONCLUSION

The intention in this work has been as much to study the mechanics of the simple shear test as to investigate the behaviour of Leighton Buzzard sand. The amplitudes of strains that have been applied are perhaps larger than may be relevant to most practical situations, but they have been chosen in order to show up, perhaps in an unduly pessimistic light, the main features of the material and test behaviour. The amplitudes are certainly well outside the range studied by Seed and Silver (1972).

The disadvantage of the simple shear apparatus lies in its end conditions. We might conclude that, on Saint-Venant's (1855) principle, a longer apparatus would give us more uniform conditions in the centre - and indeed a large shaking table such as that used by De Alba, Seed and Chan (1976) is really a long simple shear apparatus. It is interesting, however, that De Alba et al concluded that"test data from properly conducted cyclic simple shear tests in Roscoe-type simple shear devices ... closely approximate the results obtained from the large scale shaking table tests, in terms of the relationship between cyclic stress ratio and the number of stress cycles required to cause initial liquefaction." The advantage of the Cambridge simple shear apparatus over the shaking table is that the surrounding array of load cells does permit a complete evaluation of the stress state in the centre of the sample, which is important if more general aspects of stress-strain response than potential for liquefaction are to be studied.

It is perhaps unduly simplistic to say that the results presented, plausible though they are in general, in detail tell more about the simple shear apparatus than about the sand. It may perhaps be more accurate to say that the results tell more about the behaviour of sand in the simple shear apparatus than about the behaviour of sand in simple shear.

ACKNOWLEDGEMENTS

The experimental work described in this paper was financed through a contract with the Building Research Establishment. M.Budhu is grateful for a Commonwealth Scholarship enabling him to carry out research at Cambridge.

DEFINITION OF SYMBOLS

e   void ratio
$R = \tau_{yx}/\sigma_y$ stress ratio
$s = \frac{1}{2}(\sigma_1 + \sigma_3)$ } plane strain
$t = \frac{1}{2}(\sigma_1 - \sigma_3)$ } stress parameters
α shear strain
ε_1 major principal strain
ξ orientation of principal strain increment
σ_1, σ_3 principal stresses in plane of shearing
σ_2 intermediate principal stress
σ_r radial stress
σ_x, σ_y normal stresses
τ_{yx} shear stress
χ orientation of principal stress
ψ orientation of principal stress increment.

REFERENCES

Arthur,J.R.F., Dunstan,R., Al-Ani,Q.A.J.L. and Assadi,A.(1977) Plastic deformation and failure in granular media. Geotechnique 27 1, 53-74.

Arthur,J.R.F. and Roscoe,K.H.(1961) An earth pressure cell for the measurement of normal and shear stresses. Civ. Eng. Publ. Wks. Rev., 56 659, 765-770.

Bassett,R.H.(1967) The behaviour of granular materials in the simple shear apparatus, Ph.D. Thesis, Cambridge University.

Bjerrum,L. and Landva,A.(1966) Direct simple-shear tests on a Norwegian quick clay. Geotechnique 16 1, 1-20.

Budhu,M.(1979) Simple shear deformation of sands. Ph.D. Thesis, Cambridge University.

Casagrande,A.(1975) Liquefaction and cyclic deformation of sands: a critical review. Proc. 5th Panam. Conf. Soil Mechs, Buenos Aires(published as Harvard Soil Mechanics Series No.88, January 1976).

Cole,E.R.L.(1967) The behaviour of soils in the simple-shear apparatus. Ph.D. Thesis, Cambridge University.

De Alba,P., Seed,H.B. and Chan,C.K.(1976), Sand liquefaction in large-scale simple shear tests. Proc. ASCE 102 GT9, 909-927.

Duncan,J.M. and Dunlop,P.(1969) Behaviour of soils in simple shear tests. Proc. 7th Int. Conf. Soil Mechs., Mexico, 1 101-109.

Finn,W.D.L., Bransby,P.L. and Pickering, D.J.(1970) Effect of strain history on liquefaction of sand. Proc. ASCE 96 SM6, 1917-1934.

Finn,W.D.L., Lee, K.W. and Martin,G.R.(1975) Stress-strain relations for sand in simple shear. Dept. Civ. Eng., Univ. of British Columbia, Vancouver, Canada. Soil Mechs. Series No.26.

Finn,W.D.L., Pickering,D.J. and Bransby, P.L.(1971) Sand liquefaction in triaxial and simple shear tests. Proc. ASCE 97 SM4, 639-659.

Hvorslev,M.J.(1937) Über die Festigkeitseigenschaften gestörter bindiger Böden. Ingeniørvidenskabelige Skrifter A, No.45, København.

Kjellman,W.(1951) Testing the shear strength of clay in Sweden. Geotechnique 2 3, 225-232.

Lensky,V.S.(1960) Analysis of plastic behaviour of metals under complex loading. Proc. 2nd Symp. on Naval Structural Mechanics, Brown University, in: Plasticity, ed. E.H.Lee and P.S.Symonds, Pergamon Press, New York, 259-278.

Lucks,A.S., Christian,J.T., Brandow,G.E. and Höeg,K.(1972) Stress conditions in NGI simple shear test. Proc. ASCE 98 SM1, 155-160.

Moussa,A.A.(1974) Radial stresses in sand in constant volume static and cyclic simple shear tests. Norwegian Geotechnical Institute, Internal report 51505-10.

Prévost,J-H. and Höeg,K.(1976) Reanalysis of simple shear soil testing. Canadian Geotechnical Journal 13 4, 418-429.

Roscoe,K.H.(1953) An apparatus for the application of simple shear to soil samples. Proc. 3rd Int. Conf. Soil Mech., Zurich 1 2, 186-191.

Roscoe,K.H.(1970) Tenth Rankine Lecture: The influence of strains in soil mechanics. Geotechnique 20 2, 129-170.

Roscoe,K.H., Arthur,J.R.F, and James,R.G. (1963) The determination of strains in soils by an X-ray method. Civ. Eng. publ. Wks. Rev., 58 684, 873-876 & 685, 1009-1012.

Roscoe,K.H., Bassett,R.H. and Cole,E.R.L.
(1967) Principal axes observed during
simple shear of a sand. Proc. Geotechnical Conf., Oslo, $\underline{1}$, 231-237.

Saint-Venant,B.de (1855) Mémoire sur la
torsion des prismes, avec des considéra-
tions sur leur flexion, ainsi sur l'équi-
libre intérieur des solides élastiques
en général, et des formules pratiques
pour le càlcul de leur résistance à
divers efforts s'exerçant simultanément.
Mémoires presentées par divers savants
étrangers à l'Académie des Sciences, $\underline{14}$
233-560.

Seed,H.B. and Silver,M.L.(1972) Settlement of dry sands during earthquakes.
Proc. ASCE $\underline{98}$ SM4, 381-397.

Stroud,M.A.(1971) The behaviour of sand
at low stress levels in the simple shear
apparatus. Ph.D. Thesis, Cambridge
University.

van Eekelen,H.A.M. and Potts,D.M.(1978)
The behaviour of Drammen clay under cyclic loading. Geotechnique $\underline{28}$ 2, 173-
196.

Wood,D.M., Drescher,A. and Budhu,M.(1979)
On the determination of the stress state
in the simple shear apparatus. Submitted
for publication.

Youd,T.L.(1970) Densification and shear
of sand during vibration. Proc. ASCE
$\underline{96}$ SM3, 863-880.

Youd,T.L.(1971) Maximum density of sand
by repeated straining in simple shear.
Highway Research Record, $\underline{374}$ 1-6.

Youd,T.L.(1972) Compaction of sands by
repeated shear straining. Proc. ASCE
$\underline{98}$ SM7, 709-725.

Youd,T.L. and Craven,T.N.(1975) Lateral
stress in sands during cyclic loading.
Proc. ASCE $\underline{101}$ GT2, 217-221.

An application of cyclic triaxial testing
to field model tests

STEPHEN J. HAIN
University of New South Wales, Sydney, Australia

1 SUMMARY

The prediction of prototype displacements
from the results of 1 g field model tests
requires some knowledge of the relationship between modulus and the confining
stress level. In this paper the triaxial
test is used to examine under cyclic
loading conditions the relationship between the secant modulus associated with
permanent deformation and the confining
stress level. The tests have been carried
out on dry samples of the fine Oosterschelde
sand at varying porosities.

A relationship of the form $E_s \propto \sigma_3^y$ is
proposed and the value of the exponent y
is shown to be dependent on the stress
ratio ($R = \sigma_1/\sigma_3$) and the sample porosity.
For stress ratios less than 2.0 values of
y less than 0.2 are predicted. This is
outside the normally assumed range of
values of 0.5 to 0.6. In this investigation measured displacements were corrected
for end effects and at low stress ratios
these form a significant proportion of the
measured values. At higher stress ratios
the importance of the end effects correction diminshes and in these circumstances
y values in the range 0.5 to 0.6 are
predicted.

2 INTRODUCTION

The geotechnical engineer is continually
expanding the scope of his operations and
in this environment the available analytical models can often be found wanting. In
such a situation the construction and testing of a scale model can provide extremely
valuable information to develop understanding and thereby assist with the basic predictive task. Scale models can either be
laboratory or field models and although

both involve significant assumptions the
considerable financial commitment involved
in the latter would seem to have limited
its use. However, as the magnitude of
civil engineering structures increases the
need for more sophisticated prediction
would seem to call for an increased use of
field model tests. A recent example of a
situation where field testing was considered justified was the testing of a one
third scale model caisson for the Dutch
Delta Works, (Heins and De Leeuw 1977).

With the increased use of large scale
centrifuges for model testing (Rowe 1975)
one of the major deficiencies of laboratory model tests, unrepresentative stress
similarity, has largely been overcome.
However, the problem of preparing a representative supporting soil mass still
remains. It would seem to be in this
area that the field model tests has many
advantages and therefore some attention
should be given to the prediction of prototype behaviour from the results of a
1 g field model test.

Consider a pseudo elastic approach to
the prediction of permanent displacement
of a cyclicly loaded foundation,

$$S = B.q.\frac{1}{E}.I \tag{1}$$

where
S = foundation displacement
B = some representative dimension of the
    foundation
q = average applied stress
E = pseudo elastic soil modulus for a
    particular stress range
I = influence factor dependent on founda-
    tion geometry and soil layer thickness

Using the subscripts p for prototye and m
for model the ratio of displacement for

a given loading condition is

$$\frac{S_p}{S_m} = \left(\frac{B_p}{B_m}\right) \cdot \left(\frac{q_p}{q_m}\right) \cdot \left(\frac{E_m}{E_p}\right) \qquad (2)$$

The present analysis is restricted to sands and thus,

$$q \propto \gamma.D.N_q + \tfrac{1}{2}.\gamma.B.N_\gamma \qquad (3)$$

where

γ = unit weight of soil
D,B = relevant dimensions
N_q, N_γ = bearing capacity factors

Further, it is assumed (Hardin and Black 1966) that,

$$E \propto \sigma_o^y \qquad (4)$$

where

σ_o = isotropic component of the initial static stress
y = an exponent

Introducing the quantities,

$N_s = \dfrac{x_m}{x_p}$ = linear scale factor

$N_g = \dfrac{g_m}{g_p}$ = centrifugal acceleration factor

and using Eqs. (3) & (4), Eq. (2) becomes

$$\frac{S_p}{S_m} = \left(\frac{1}{N_s}\right) \cdot \left(\frac{1}{N_s N_g}\right) \cdot (N_s N_g)^y \qquad (5)$$

thus,

$$\frac{S_p}{S_m} = \frac{(N_s N_g)^y}{N_s^2 N_g} \qquad (6)$$

For a model test performed in a centrifuge under increased gravity true stress similarity can be achieved and hence $N_s N_g = 1$ and $S_p/S_m = 1/N_s$. However, for a field test which is performed under normal gravity $N_g = 1$ and hence $S_p/S_m = 1/N_s^{2-y}$. In this latter situation some knowledge of the exponent y is required if prototype displacements are to be predicted from the model test results.

In this paper the triaxial test is used to examine under cyclic loading conditions the relationship between the secant modulus association with permanent displacement and the confining stress. The material used for the investigation was fine sand from the Oosterschelde region in Holland. This material had been used in previous investigations (Rowe and Craig 1976, 1978).

3 EXPERIMENTAL PROCEDURE

3.1 Test Equipment

The test equipment used was similar to

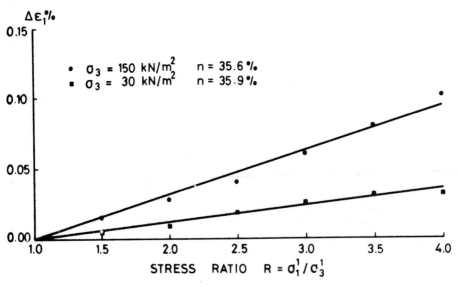

Fig. 1. Recoverable strain ($\Delta\varepsilon\%$) versus stress ratio (R) for two samples, with approximately equal porosities, at different confining pressures.

that used at the University of Manchester for cyclic loading investigations (Khaffaf 1975) and (Ellis 1975). This uses a standard 100 mm triaxial cell with end platens modified for lubricated ends (Rowe and Barden 1964). Cyclic deviator loads were applied using a double acting air jack. The loading rod incorporated a load cell near the top platen thus avoiding friction losses in the '0' ring sleeve at the top of the cell.

Air pressure to the jack was controlled by a regulator valve and two solenoid valves. The solenoid valves, controlled by a timer, alternately connect the jack to the pressure source or the atmosphere. A range of frequencies less than 0.5 Hz was possible with this system; however 0.2 Hz was used for all tests. The cell pressure remained constant during any particular test.

Vertical deformation of the sample was measured using an LVDT and a dial gauge mounted externally on the top of the cell and monitoring movement of the loading rod. Volumetric deformations were not measured as all tests were performed using the dry material.

3.2 Sample Preparation

Initially the lubricated ends were pre-

pared and subjected to a stress of 25 kN/m² for 1 hr to ensure consistent properties. A 100 mm dia. by 100 mm high mould and extension collar were attached and the dry sand was poured rapidly into the mould. This produced a very loose sample of porosity approximately 45%. A small vibrator was then applied to the side of the mould and the sample vibrated for different lengths of time to achieve a porosity in the range 38% to 34%. A similar method had been used with considerable success (Ellis 1975) for saturated samples and a comparable level of consistency was obtained for the dry fine sand used in this investigation.

3.3 Test Programme

Initially a series of calibration tests were performed to determine the corrections which resulted from end effects. There are three end effects which can influence the measured values of axial displacement for a given deviator stress.
(i)   explusion of the silicone grease from between the end platen and the latex disc.
(ii)   compression of the latex disc
(iii) penetration of the sand particles into the latex disc

In these tests the lubricated ends were prepared in the standard manner and then

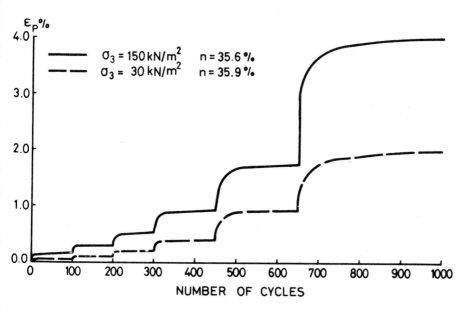

Fig. 2. Permanent strain ($\varepsilon p\%$) versus number of cycles of deviator stress application for two samples, with approximately equal porosities, at different confining pressure. Samples exhibit shakedown with increase in deviator stress.

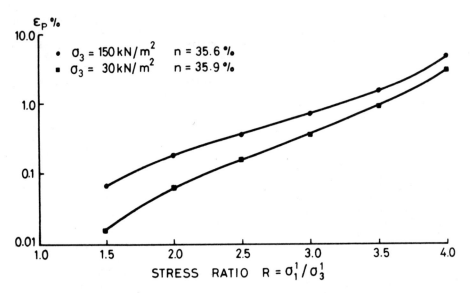

Fig. 3. Permanent strain ($\varepsilon p\%$) at equilibrium versus stress ratio (R) for two samples with approximately equal porosities, at different confining pressures.

covered with a very thin layer of sand particles. A dummy aluminium sample was then inserted and subjected to the same test conditions as for the sand samples. Several tests were performed at the various cell pressures and in each case the displacements due to the end effects reached an equilibrium value after approximately 50 cycles of each deviator stress.

Tests were performed at constant cell pressures of 30 kN/m^2 and 150 kN/m^2 as this difference was considered large enough to represent possible differences between model and prototype stress levels. During the test the deviator stress was cycled between zero and a maximum value which was increased throughout the test to give increments of stress ratio ($R = \sigma_1'/\sigma_3'$) equal to 0.5 up to a maximum stress ratio of 4. The sample was subjected to a particular deviator stress until the increase in axial displacement was less than 0.002 mm per 10 cycles.

4 INTERPRETATION OF RESULTS

4.1 Recoverable Deformations

Recoverable deformation was not of primary interest in this investigation however readings were taken for each test and two typical results are shown in Figure 1. The values of recoverable strain plotted

are the final equilibrium values minus the end effects correction. When the deviator stress was increased during a test the recoverable strain increased but very quickly attained an equilibrium value. This was in contrast to the permanent deformation which required significantly more cycles to reach an equilibrium value.

These results can appropriately be expressed in terms of a resilient modulus (M_R) defined as the amplitude of the deviator stress divided by the recoverable strain. For a given confining pressure and sample porosity a constant value of resilient modulus was observed regardless of the deviator stress level (Figure 1).

This series of tests did not allow a thorough investigation of resilient modulus. However, it would seem that a relationship of the form,

$$M_R = K \cdot \sigma_3^n \qquad (7)$$

where
K,n = constants for the material depending on density and moisture content could provide a basis for future evaluation.

4.2 Permanent Deformations

During a test when the deviator stress was increased there was an immediate increase in permanent deformation which eventually

26

attained an equilibrium value. This shakedown behaviour is shown in Figure 2 for
two typical test results. The equilibrium
value of permanent deformation chosen to
represent the test result was the permanent
deformation when the rate of increase was
less than 0.002 mm per 10 cycles. Using a
2.5 sec duration deviator stress pulse
applied at a frequency of 0.2 Hz this
required 100 to 150 cycles for most deviator
stress increments. However, for the
higher stress ratios, approaching failure,
a significantly greater number of cycles
was required to reach equilibrium.

For a 0.65 sec duration deviator stress
pulse applied at a frequency of 0.83 Hz
permanent deformations were reported to be
increasing slightly after 2×10^6 cycles
(Morgan 1966). For a 0.1 sec duration
deviator stress pulse applied at a frequency
of 0.5 Hz permanent deformations were
still increasing significantly after 10^5
cycles (Barksdale 1972). While for a
sinusoidal 0.5 sec duration deviator stress
pulse applied at a frequency of 1 Hz an
equilibrium situation was obtained after
10^4 cycles (Brown 1974). The above information suggests an inverse relationship

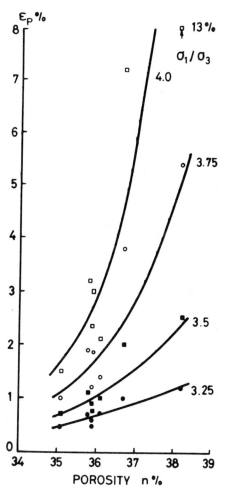

Fig. 5. Permanent strain ($\varepsilon p\%$) versus
porosity (n%) for samples tested at a
confining pressure of 30 kN/m^2. Stress
ratios 3.25, 3.5, 3.75 and 4.0 shown.

between length of the stress pulse and
the number of cycles required to give
an equilibrium situation. Little work
appears to have been conducted regarding
the number of cycles to equilibrium for
stress pulses of various durations yet
there is significant variation in the
published results.

To enable accurate resolution of the
permanent strain values the test results
were plotted on a graph of stress ratio
versus the logarithm of permanent strain.
Two typical graphs are shown in Figure 3.
Selecting from these graphs values of
permanent strain for given stress ratios,

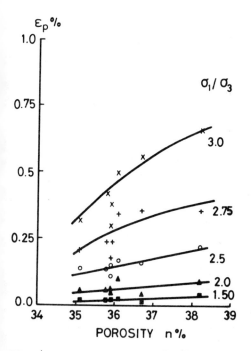

Fig. 4. Permanent strain ($\varepsilon p\%$) versus
porosity (n%) for samples tested at a
confining pressure of 30 kN/m^2. Stress
ratios 1.5, 2.0, 2.5, 2.75 and 3.0 shown.

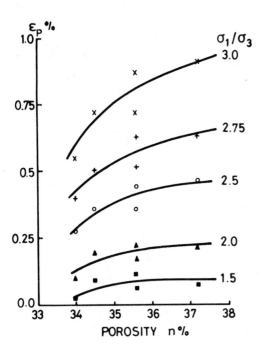

Fig. 6. Permanent strain (ε%) versus porosity (n%) for samples tested at a confining pressure of 150 kN/m². Stress ratios 1.5, 2.0, 2.5, 2.75 and 3.0 shown.

Table 1. Permanent Strains for Sample Porosity of 36%.

$R = \dfrac{\sigma_1}{\sigma_3}$	Ep% σ_3=150 kN/m²	Ep% σ_3=30 kN/m²
1.50	.090	.025
1.75	.140	.035
2.00	.210	.060
2.25	.295	.090
2.50	.430	.145
2.75	.595	.280
3.00	.835	.455
3.25	1.22	.67
3.50	1.70	1.05
3.75	3.05	1.73
4.00	5.10	2.83

graphs showing the relationship between permanent strain and sample porosity can be obtained. To allow for a clearer presentation values for stress ratios up to 3 have been plotted to a larger scale. These graphs are shown in Figures 4 to 7 for the confining pressures considered. The results shown some scatter, but, definite trends emerge depending on the stress ratio and a general behaviour pattern can be established. At low stress ratios porosity has only a small influence on the permanent strain. However, at larger stress ratios, and particularly as failure conditions are approached, there is a dramatic increase in permanent strain. By establishing a curve of best fit to the data given in Figures 4 to 7, it is possible to predict permanent strains for given sample porosities. Table 1 shows the results of this analysis for a sample porosity of 36%.

Using these permanent strains a secant modulus associated with permanent deformation may be calculated as:

$$E_s = \frac{\sigma_1 - \sigma_3}{\varepsilon_p} = \frac{\sigma_3(R-1)}{\varepsilon_p} \qquad (8)$$

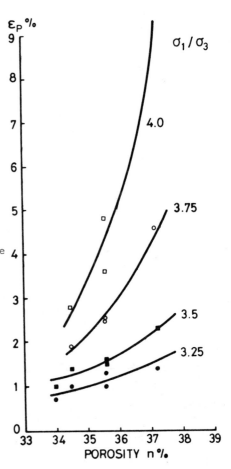

Fig. 7. Permanent strain (εp%) versus porosity (n%) for samples tested at a confining pressure of 150 kN/m². Stress ratios 3.25, 3.5, 3.75 and 4.0.

28

where E_s = secant modulus for permanent deformation

ε_p = permanent strain

Furthermore, if this modulus is assumed to be a function of the initial confining stress (Eq.4) then for two tests performed at different confining pressures,

$$\frac{E_{S1}}{E_{S2}} = \{\frac{\sigma_{31}}{\sigma_{32}}\}^y \qquad (9)$$

From equation (8),

$$\frac{E_{S1}}{E_{S2}} = \frac{\sigma_{31}(R-1)\,\varepsilon_{p_2}}{\sigma_{32}(R-1)\,\varepsilon_{p_1}} \qquad (10)$$

Comparing modulii values at the same stress ratios for the two tests gives:

$$\frac{E_{S1}}{E_{S2}} = \{\frac{\sigma_{31} \cdot \varepsilon_{p_2}}{\sigma_{32} \cdot \varepsilon_{p_1}}\} \qquad (11)$$

and combining equations (9) and (11) and taking logarithms yields:

$$y = 1 + \frac{\log(\varepsilon_{p_2}/\varepsilon_{p_1})}{\log(\sigma_{31}/\sigma_{32})} \qquad (12)$$

Using Eq. 12 and a table of permanent strain values for a given porosity the value of the exponent y in Eq. 4 can be calculated for any given stress ratio. Figure 8 shows a graph of y versus stress ratio for three values of sample porosity. It is apparent that the stress ratio has an important influence on the relationship between soil modulus and confining pressure in the present circumstances. It should be noted that the value of the exponent y of Eq. 4 is always less than 1 and that this results from the increased particle crushing which accompanies an increase in confining pressure.

It is generally recognised Biarez (1961) that for the cyclic modulus of sands y can be assumed to lie between 0.5 and 0.6. Reference to Figure 8 shows that at stress ratios less than 2.5 the value of y is considerably smaller than the normally assumed values. However, the writer believes that most published information relating to the value of y has been obtained from triaxial tests where no steps were taken to correct for end effects. At low stress ratios the correction for end effects forms a signficant proportion of the total measured displacement and when taken into consideration leads to a considerably lower value of y. The variation of y with stress ratio, shown

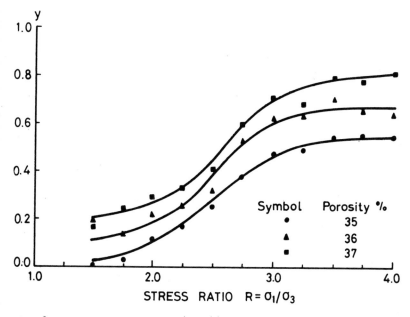

Fig. 8. Value of exponent y (Eq. 4) versus stress ratio (R) for sample porosities of 35%, 36% and 37%.

in Figure 8, is consistent with an increased influence of grain crushing at low values of stress ratio. It is also interesting to note that the most rapid change in the value of y occurs at approximately the same stress ratio as the onset of dilation in a static test. However more research is required to give a definite reason for the variation of y with stress level.

5 APPLICATION TO FIELD MODEL TESTS

The results given in Figure 8 can be used to predict the full scale field behaviour of a cyclically loaded foundation from the results of a 1 g model test. It is assumed that the triaxial stress conditions used to evaluate y are a reasonable approximation to the stress conditions of a representative element in the supporting soil. Taking the safety factor on ultimate load as 2.5 then, Safety Factor =

$$\frac{\tan \phi_f}{\tan \phi_m} = \frac{\tan[2(\tan^{-1} \sqrt{R_f} - 45)]}{\tan[2(\tan^{-1} \sqrt{R_m} - 45)]}$$

where
ϕ_f = value of ϕ at failure
ϕ_m = value of ϕ mobilized
R_f = stress ratio corresponding to ϕ_f
R_m = stress ratio corresponding to ϕ_m

For R_f = 4.5, R_m = 1.9 and from Figure 8, y is equal to 0.15 for a porosity of 36%. Thus from Eq. 6.

$$S_p = \frac{1}{N_S^{1.85}} \cdot S_m \tag{13}$$

For the case N_s=0.33, Eq. 13 indicates that the prototype displacements will be 7.8 times the model displacements. If the normally assumed value for y of 0.5 had been taken the ratio between prototype and model displacements is given as 5.3. Rowe and Craig (1976) carried out a series of centrifuge tests to simulate the field model test of Heins and De Leeuw (1977) and to predict the prototype field behaviour. In this case the field model was not truely scaled in stress and the ratio between prototype and model displacements was about 9.

6 CONCLUSIONS

The prediction of prototype displacements from the results of a 1 g field model test requires some knowledge of the relationship between modulus and the confining stress level. For cyclic load conditions this relationship can conveniently be examined by using the triaxial test and by assuming that the relationship between modulus and confining pressure is given by Eq. 4. The present series of tests have been carried out on dry samples of the fine Oosterschelde sand at varying porosities.
The most important finding of the present study is the relationship between the exponent y of Eq. 4 and the stress ratio (Figure 8). This graph shows that at stress ratios less than 2.5 the value of y is considerably smaller than the normally assumed range of values of 0.5 to 0.6. The present study differs from previous studies in that careful steps were taken to correct the measured displacements for end effects. At low stress ratios these corrections form a significant proportion of the measured total displacement. At larger stress ratios, greater than 3, the importance of the end effects is much smaller and this study predicts y values in the range 0.5 to 0.6 in accordance with previously published work.

ACKNOWLEDGEMENTS

This work was performed while the writer was on Study Leave at the University of Manchester. The writer is particularly indebted to Professor P.W. Rowe for formulating the objectives of the study and for many helpful discussions.

REFERENCES

Barksdale, R.D. 1972, Laboratory evaluation of ruting in base course materials, 3rd Int. Conf. on Structural Design of Asphalt Pavements, Vol. 1, London, pp. 161-174.
Biarez, J. 1961, Contribution a l'étude des properties mécanique sols et des materiaux pulverulents, D.Sc Thesis, University of Grenoble.
Brown, S.F. 1974, Repeated load testing of a granular material, Jnl. Geotechnical Engineering Division, A.S.C.E., Vol. 100, GT7, pp. 825-841.
Ellis, G.P. 1975, The behaviour of north sea sand under cyclic loading conditions, M.Sc. Thesis, University of Manchester.
Hardin, B.O. and W.L. Black, 1966, Sand stiffness under various triaxial stresses Jnl. Soil Mechanics and Foundations Division, A.S.C.E., Vol. 92, SM2, pp. 27-42.

Heins, W.F. and E.H. De Leeuw, 1977, Large
scale cyclic loading tests 8th Int. Conf.
Soil Mech. and Fndn. Eng., Tokyo, Vol. 2,
pp. 541-544.

Khaffaf, J.H. 1975, The behaviour of
saturated undrained clay under cyclic
loading conditions, M.S.C. Thesis, University of Manchester.

Morgan, J.R. 1966, The response of granular
materials to repeated loading, Proc. 3rd
ARRB Conf., Vol. 3, No. 2, pp. 1178-1192.

Rowe, P.W. 1975, Application of centrifugal
models to geotechnical structures, Soil
Mechanics - Recent Developments, 1975
pp. 1-26, W.H. Sellen, N.S.W., Australia.

Rowe, P.W. and L. Barden, 1964, Importance
of free ends in triaxial testing, Jnl.
Soil Mechanics and Foundations Division,
A.S.C.E., Vol. 90, SM1, p. 1.

Rowe, P.W. and W.H. Craig, 1976, Studies
of offshore caissons founded on
Oosterschelde sand, Design and construction of offshore structures. Institution of Civil Engineers, London, pp.
39-45.

Rowe, P.W. and W.H. Craig, 1978, Predictions of caisson and pier performance by
dynamically loaded centrifugal models.
Symposium on Foundation Aspects of
Coastal Structures, L.G.M. Delft.

Development of pore water pressures in a dense calcareous sand under repeated compressive stress cycles

MANOJ DATTA, G. VENKATAPPA RAO & SHASHI K. GULHATI
Indian Institute of Technology, New Delhi, India

1 SYNOPSIS

This paper presents the results of a laboratory study undertaken to evaluate the factors which control the development of pore water pressure during pile driving in calcareous sands so as to better understand the observed phenomena of low resistance during pile driving and increased resistance to redriving after stoppages. Consolidated undrained tests were conducted under repeated compressive stress cycles in a triaxial apparatus on a dense calcareous sand so as to simulate conditions during pile driving. The results show that pore water pressures developed under cyclic loading are strongly related to the static pore water pressure response of the sand and they increase linearly with log of number of cycles. Positive pore water pressures develop in calcareous sands which are highly susceptible to crushing, or which may exist in a loose state due to early diagenesis of the sediment.

2 INTRODUCTION

Calcareous sands consisting of skeletal remains of marine organisms are found in abundance on the continental shelves of marine areas, lying between latitudes 30°N and 30°S. These sands tend to crush easily during shear as compared to terrigeneous sands (Datta et. al. 1979). Piles driven in calcareous sands encounter low resistance to driving (McClelland 1974) and stoppages during driving of piles is found to cause an increase in the resistance upon redriving due to soil 'set up' (Aggarwal et. al.1977). Similar behaviour has been observed for piles driven in chalk (Higginbottom 1965), which is essentially a cemented calcareous silt. Low resistance to driving in chalk has been explained in terms of remoulding of chalk around the pile during driving producing 'putty chalk' (Higginbottom 1965) and inducing development of positive pore water pressures. Soil 'set up' in chalk is thought to occur on account of dissipation of pore water pressures during stoppages, which cause an increase in the strength of the strata around the pile thus exhibiting increased resistance to redriving (Vijayvergia et. al.1977). In calcareous sands crushing of grains is thought to reduce the skin friction and end bearing capacity (Angemeer et. al. 1973 and 1975). No explanation has been advanced for the phenomenon of soil set up in calcareous sands.

That positive pore water pressures develop in sands subjected to cyclic shear stresses has been observed in laboratory studies and the liquefaction of sands during earthquakes which is a consequence of these positive pore water pressures, is well documented (Seed and Lee 1966). More recently, development of positive pore water pressures during ocean storm loading in a dense sand strata were reported on the basis of laboratory studies conducted in relation to the design of the Ekofisk Tank in the North Sea (Bjerrum 1973) and positive pore water pressures were recorded during a severe storm after Ekofisk Tank was installed(Lee 1976a).

This paper presents the results of a study conducted on a dense calcareous sand which was subjected to repeated compressive stress cycles under undrained triaxial condition so as to simulate conditions under pile driving. The study was designed to isolate the influence of number of cycles, stress level and confining pressure on the pore water pressure response, and also on the magnitude of strains developed, magnitude of crushing and on subsequent static behaviour of the sand.

3 LABORATORY SIMULATION

During an earthquake, a soil element is subjected to deformations resulting from the upward propogation of shear waves from underlying layers causing the soil element to experience a series of cyclic shear strains which reverse direction many times during an earthquake. This is usually simulated in the laboratory by testing samples in triaxial shear, simple shear or torsional shear under undrained conditions by applying alternating cyclic stresses of a constant amplitude and frequency (usually 120 cycles per minute) which approximately reproduces the complex and erratic sequence of ground motions during an earthquake (Seed and Lee 1966, Peacock and Seed 1968, Ishibashi and Sherif 1974).

During ocean storms, as a wave moves across an offshore gravity platform it exerts a force on the structure, first along the direction of its movement and then in the opposite direction. This causes the foundation soil strata to experience a series of stress cycles (in the horizontal direction due to the force and in the vertical direction due to the overturning moment induced by the wave), which reverse in direction depending upon the frequency of the waves. The stress conditions are comparable to those under an earthquake but the frequency of cycles is much lower (3 to 12 cycles per minute) and the duration of ocean storms is longer than that of earthquakes. The laboratory simulation is similar to that for earthquake loading with a lower frequency of cycling(Bjerrum 1973, Anderson 1976). In the cases where frequency of cycles does not effect the soil behaviour, higher frequencies may be used to shorten the total testing time(Lee and Focht 1975).

Pile driving essentially consists of a series of stress waves which travel along the length of the pile, causing the soil to shear and the pile to penetrate. A soil element at the pile-soil interface is repeatedly subjected to shear stresses in the direction of movement of the pile corresponding to the blows of the hammer. The soil may experience as many as 35 to 65 blows per minute at full stroke of the hammer for different hammers. This can be simulated in the laboratory by subjecting a soil sample in a triaxial apparatus to repeated compressive stress cycles under undrained condition such that the failure plane experiences shear stress cycles similar to those experienced by a soil element at the pile-soil interface. The field parameters which have to be simulated in the laboratory are (a) energy input of the hammer, (b) rate of driving i.e. number of blows per minute, (c) total number of blows experienced by the soil & (d) depth below mudline. The last three parameters can be duplicated in the laboratory by controlling the frequency & total no. of cycles and the confining pressure. The rated energy of the hammer is a critical parameter which determines the penetration per blow, during pile driving. The stress level induced by a blow of the hammer is so high that the soil is sheared beyond failure resulting in large displacement (i.e. large plastic strains) of the pile. Ideally, any simulation in the laboratory would require the stress levels of the cyclic loading to be greater than the undrained strength. However, in the present study stress levels upto 75 percent of the undrained strength have been used to evaluate the influence of increase in stress level on the soil behaviour.

Tests conducted in the laboratory to simulate earthquake loads and ocean storm wave loads are usually performed under alternating stress cycles and at low stress levels. These results are, as such, of little relevance to the pile driving problem. However, certain studies have been performed to understand the influence of stress levels on the pore water pressure and strain development response of soils. Their results are discussed in the following text.

Bjerrum (1973), studied the influence of increase in stress level on the pore pressure development under alternating cyclic stresses using a simple shear device. He demonstrated that the rate of increase of pore water pressure per cycle increased with increasing stress level. However, the range of shear stresses used in his study was small compared to the failure shear stress. Anderson (1976) investigated the effect of cyclic shear stress level on the strains and pore water pressures induced during two-way cyclic loading in simple shear tests conducted on clay and showed that cyclic shear stress levels below 33 percent of the static shear stress at failure had negligible effect on soil behaviour. The test results indicated that cyclic strains increased with increase in stress level. Negative pore water pressures were recorded during the initial cycles at stress levels greater than 60 percent of the static failure shear stress. Herman and Houston (1976) reported results of cyclic tests conducted on two deep ocean soils - a turbidite and a calcareous ooze, in a triaxial apparatus. Both soils were sheared under static conditions to a certain stress level and then cyclic stresses were applied without causing stress reversal. It is interesting to note that the application of cyclic loads after initially shearing to high stress levels caused the pore water pressures to become increasingly negative during the initial cycles. After a certain number of cycles the pore pressures gradually increased and became positive.

From this review it may be concluded that the stress level plays an important role in the development of pore water pressure and strain under cyclic loading and this aspect needs further investigation.

The effect of alternating cyclic stresses on subsequent static behaviour was studied by Lee (1976). He showed that soil samples after being subjected to cyclic loading developed static undrained strength of the same magnitude as that of an undisturbed sample, provided sufficient strains were allowed to develop. This aspect has also been studied in the present investigation.

4 EXPERIMENTAL INVESTIGATION

4.1 Sand Tested

A calcareous sand obtained from the west coast of India was chosen for the present study because of its high susceptibility to crushing in comparison to terrigeneous sands (Datta et. al. 1979). It consisted of a mixture of shell fragments and skeletal remains of marine organisms with large intraparticle voids having a carbonate content of greater than 85 percent. The physical properties of the sand are given in Table 1.

Table 1 Characteristics of calcareous sand.

Particle size	Coarse
Coefficient of Uniformity	1.50
Maximum Void Ratio	1.39
Minimum Void Ratio	0.93
Specific Gravity	2.81

4.2 Testing programme

The number of blows delivered per minute by a hammer depends upon its size and make and usually vary between 35 to 65 for hammers being currently used to install deep penetration piles in the offshore regions. Previous studies have shown that cyclic shear behaviour of sands is not affected by the frequency of cycles lying between 10 to 240 cycles per minute (Peacock and Seed 1968). For the present study, a frequency of 30 cycles per minute was chosen as a representative value. The number of blows experienced by a soil element depend upon its depth below the mudline. The soil lying near the mudline is subjected to the maximum number of stress cycles whereas that near the bottom of the pile experiences the least number of stress cycles. In this study each soil sample was subjected to 400 cycles of compressive stress..

Two series of consolidated undrained triaxial tests were performed under confining pressures ($\overline{\sigma}_{3C}$) of 2 and 4 kg/cm^2 on sand samples prepared at their minimum void ratio i.e. a relative density of 100 percent. In Test Series I, the samples were sheared under static conditions upto 20 percent strain. In Test Series II, the samples were first subjected to repeated compressive stress cycles at a frequency of 30 cycles per minute. During each cycle the deviator stress varied from zero to a given stress level (which was constant for each test) and then back to zero. Stress levels upto as much as 75 percent of the static undrained strength were used. The term stress level has been used in this paper to indicate the ratio of cyclic shear stress (τ_{cy}) to static undrained shear strength (S_u) i.e.

$$\text{Stress level} = \frac{\tau_{cy}}{S_u} =$$

$$\frac{1/2(\text{Cyclic deviator stress},(\sigma_1 - \sigma_3)_{cy})}{1/2(\text{Static deviator stress at failure, } (\sigma_1 - \sigma_3)_f)}$$

$$= \frac{(\sigma_1 - \sigma_3)_{cy}}{(\sigma_1 - \sigma_3)_f}$$

Each sample was thus subjected to 400 cycles after which it was statically sheared upto 20 percent strain. The magnitude of crushing for each sample was evaluated after the completion of testing, for samples of both the test series.

4.3 Testing procedure

Cylindrical sand samples 3.81 cm in diameter and 7.62 cm high were prepared by placing deaired sand under water in a split former. The minimum void ratio was obtained by tamping the sand and simultaneously tapping the former from outside to produce small vibrations. Back pressure was applied to ensure saturation and only those samples which had a B factor of greater than 0.95 were tested. All tests were performed using INSTRON 1195 Universal Testing machine, a strain controlled machine having provision for applying cyclic loads between fixed load limits. The load limits were ascertained from the stress level to be used. The strain rates were chosen so as to yield a frequency of 30 cycles/min. However, since only discrete strain rates could be used the frequencies of cycling varied between 23 to 37 cycles per minute for different tests. The variations of stress, strain and pore water pressure versus time were recorded on a strip chart recorder by amplifying the output signals from a load cell, an extensometer and a pressure transducer respectively. Deformation rate of 0.5 mm/min was adopted for static shearing. Sieve analysis was done after each test to obtain the grain size distribution curve and evaluate the magnitude of crushing.

5 RESULTS AND DISCUSSION

Fig. 1 shows a typical load, displacement and pore water pressure versus time record of a test conducted on a sample consolidated under confining pressure of 4.0 kg/cm^2 and subjected to a cyclic stress level of 0.5. It is evident from the figure that the first cycle causes development of large permanent strain and residual pore water pressure where as subsequent cycles cause smaller increments in their magnitude.

5.1 Pore water pressures

Figs. 2(a) and (b) depict the variation of induced (residual) pore water pressures expressed as a fraction of the confining pressure ($\Delta u / \overline{\sigma}_{3c}$) with the number of cycles (N) on a log scale for confining pressures of 2 and 4 kg/cm^2 respectively. Each figure presents the results of tests conducted at different stress levels. The plots

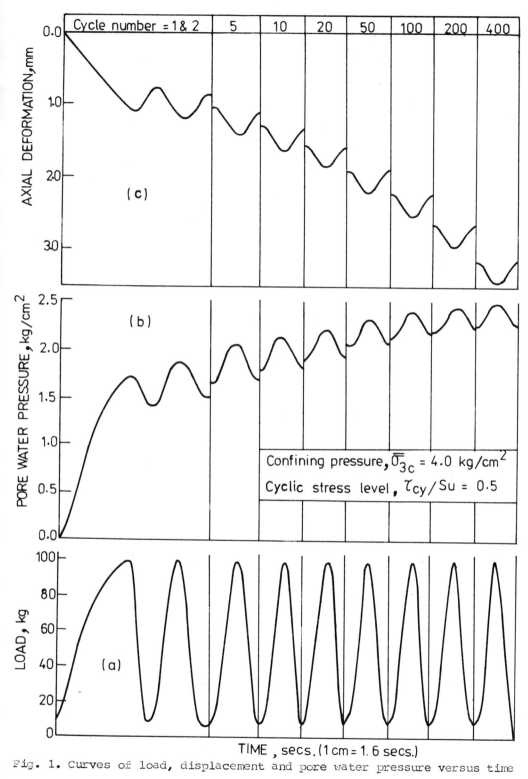

Fig. 1. Curves of load, displacement and pore water pressure versus time

37

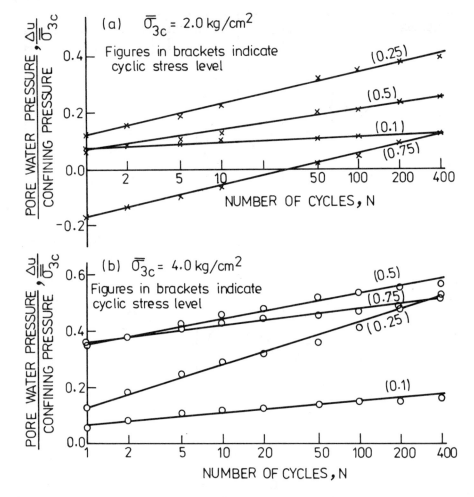

Fig. 2. Variation of induced (residual) pore water pressures with number of cycles

reveal that the pore pressures increase linearly with log N and can be expressed as follows:

$$\Delta u / \bar{\sigma}_{3c} = C + K \log N \quad \ldots(1)$$

where C is the value of $\Delta u / \bar{\sigma}_{3c}$ after the first cycle and K is the slope of the line.
Both C and K vary with the stress level and confining pressure. Janbu (1976) used a cause-effect relation and derived a relationship similar to Equation (1), which he suggested was applicable for over-consolidated clays. He, however, did not include the constant C in his equation.

To assess the influence of the stress level on the pore pressure development, the results presented in Figs. 2(a) and (b) have been re-plotted in Figs. 3 (a) and (b) with the stress level as abscissa. The lines of N = 1 represent values of C as a function of the stress level. Also plotted in the two figures are the $\Delta u / \bar{\sigma}_{3c}$ versus τ/S_u curves obtained from static consolidated undrained tests of Test Series I (τ/S_u for static test $= (\sigma_1 - \sigma_3)/(\sigma_1 - \sigma_3)_f$). It is of interest to note that the C values essentially follow the same trend as the curve obtained from static tests. A plot of K vs τ/S_u is shown in Fig. 4 which also

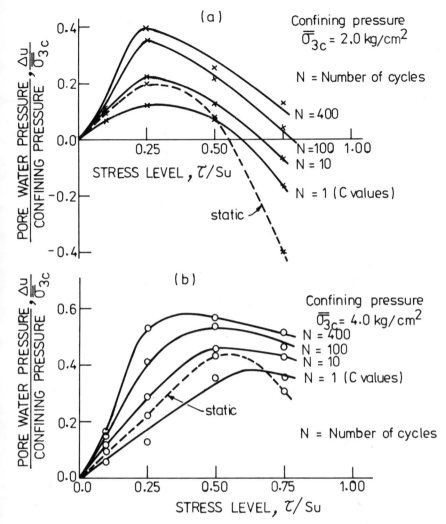

Fig. 3. Variation of induced (residual) pore water pressures
with stress level

presents the $\Delta u / \bar{\sigma}_{3c}$ versus τ/S_u curves obtained from the static tests for $\bar{\sigma}_{3c}$ equal to 2 and 4 kg/cm². It is not possible to arrive at any conclusion regarding the variation of K with stress level.

If one refers to the curves of $\Delta u / \bar{\sigma}_{3c}$ vs τ/S_u in Figs.3(a) and (b) for N = 400 cycles it is observed that they still possess the same shape as that of $\Delta u / \bar{\sigma}_{3c}$ vs τ/S_u determined from static tests. This indicates that the pore water pressure response under cyclic loading is strongly related to the static

behaviour of the soil. Since an increase in the confining pressure causes higher static pore water pressures to develop, the pore water pressures induced under cyclic loading also show a similar trend.

Bjerrum (1973) showed that an increase in the stress level resulted in a higher rate of pore water pressure increase per cycle leading to the development of higher pore water pressures. The range of stress levels used in his study was low. The present results indicate that, initially, pore water pressure increases with stress level but as

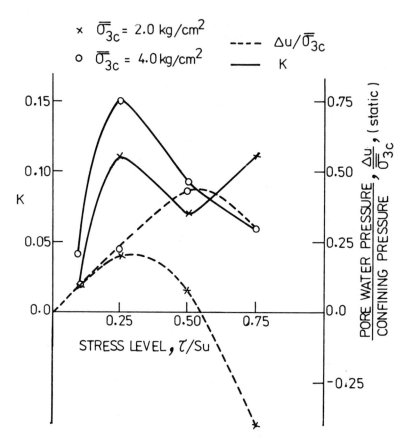

Fig. 4. Variation of K with stress level

the stress level crosses a certain magnitude the induced pore water pressures tend to decrease with increase in stress level and may also become negative (refer Fig. 3(a) and (b)).

It is of interest to compare the results of the present study with those of alternating stress cyclic loading to assess the influence of type of cyclic loading on the pore water pressure development. Table 2 compares the magnitude of parameters used in the studies conducted by Seed and Lee (1966) and Bjerrum (1973) along with those of the present study. As mentioned earlier the stress levels used (indicated by cyclic stress/confining stress in Table 2) in the present study are much larger than those used for earthquake and ocean wave loading problems. Fig. 5 presents a typical result of pore water pressure deve-

lopment with the number of cycles for each of the three studies from which one notes that pore water pressures increase very rapidly under alternating stress cycles in comparison to the increase under repeated compressive stress cycles. This is probably so because in alternating stress cycles the extension part of the cycle plays a dominant role in the development of positive pore water pressures (Lee 1976).

5.2 Strains

The variation of permanent strains (ϵ_p) with the number of cycles on log scale is shown in Fig. 6 for confining pressures of 2.0 and 4.0 kg/cm^2. Results of tests conducted at various stress levels are presented in the figure. At low stress levels ($\leqslant 0.25$) strains increase

Table 2. A Comparison of range of parameters used in the present study with those of other studies.

Parameter	Seed and Lee (1966)	Bjerrum (1973)	Present study
Loading	Alternating stress cycles	Alternating stress cycles	Compressive stress cycles
Apparatus	Triaxial	Direct Shear	Triaxial
Sample state	Loose to dense	Medium dense to dense	Dense
Confining stress* (Kg/cm^2)	0.5 to 12.0	1.54	2.0 and 4.0
Cyclic stress** (Kg/cm^2)	$\pm$0.2 to $\pm$2.0	$\pm$0.15 to $\pm$0.37	0$\leftrightarrow$0.77 to 0$\leftrightarrow$5.85
Frequency (cycles/min)	120	5	30
$\dfrac{\text{Cyclic stress}}{\text{Confining stress}}$	$\pm$0.25 to $\pm$0.40	$\pm$0.1 to $\pm$0.24	0$\leftrightarrow$0.19 to 0$\leftrightarrow$2.93

* Normal stress for direct shear tests; cell pressure for triaxial tests.
** Cyclic shear stress for direct shear tests; 1/2 (cyclic deviator stress) for triaxial tests.

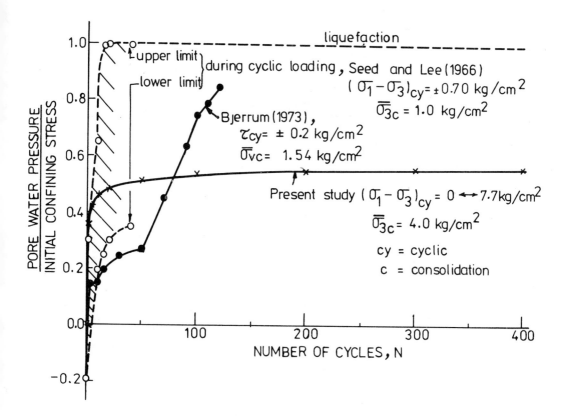

Fig. 5. Comparison of development of pore water pressures

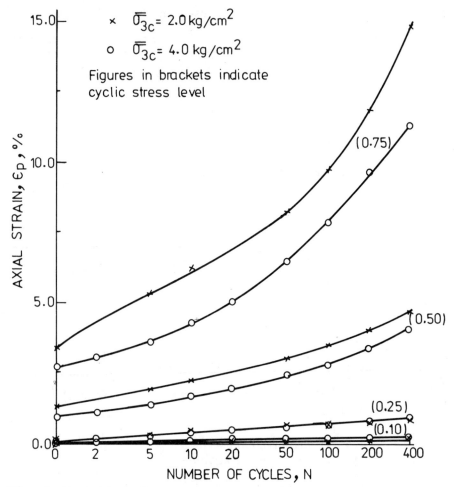

Fig. 6. Variation of permanent strains with number of cycles

essentially linearly with log N. This however does not hold good at higher stress levels.

To assess the influence of stress level on the strain development, the results have been replotted in Figs. 7(a) and (b) with τ/S_u as abscissa. At low stress levels ($\leqslant 0.25$) the strains developed are less than 1 percent. There is a marked increase in the magnitude of strains developed at stress levels greater than 0.25. For the same stress level, larger strains are developed under a confining pressure of 2 kg/cm² in comparison to those developed under a confining pressure of 4 kg/cm². This is in accordance with the ϵ_p versus τ/S_u

relationships determined from static tests which are also shown in Figs. 7(a) and (b). It must be noted that for stress levels higher than 0.50, large strains develop even though the stresses are still lower than the undrained strength. Further the strains developed under repeated compressive stress cycles are much larger as compared to those under alternating stress cyclic loading in which the residual strains tend to be zero or low until liquefaction occurs (Seed and Lee 1966).

5.3 Static behaviour after cyclic loading:

42

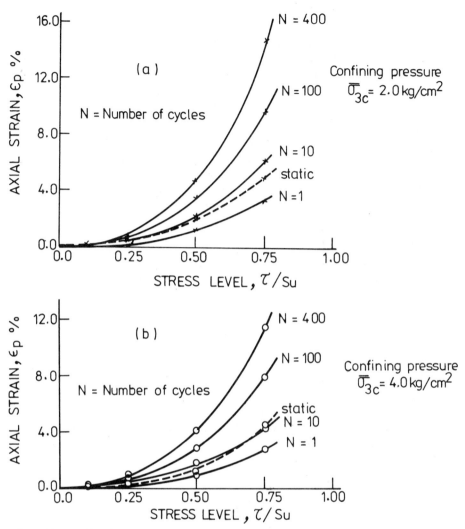

Fig. 7. Variation of permanent strains with stress level

The results of static tests (Test Series I) and of static tests after cyclic loading (Test Series II) are presented in Fig. 8 for $\overline{\sigma}_{3c} = 4.0$ kg/cm². Similar behaviour was also observed for $\overline{\sigma}_{3c} = 2.0$ kg/cm². It is noted that the peak deviator stresses are essentially the same for both Test Series I and II. The static pore water pressures which develop in samples initially subjected to cyclic loading become equal to those observed in static tests. These observations are consistent with those of Lee (1976) who noted similar behaviour in soils subjec-

ted to alternating stress cycles. This indicates that cyclic loading has no significant influence upon subsequent static behaviour.

5.4 Crushing

The magnitude of crushing was evaluated in terms of a crushing coefficient, Cc, a dimensionless parameter defined earlier by the authors (Datta et. al. 1979). It was found that the magnitude of crushing after shearing was the same both for Test series I and II for each confining

43

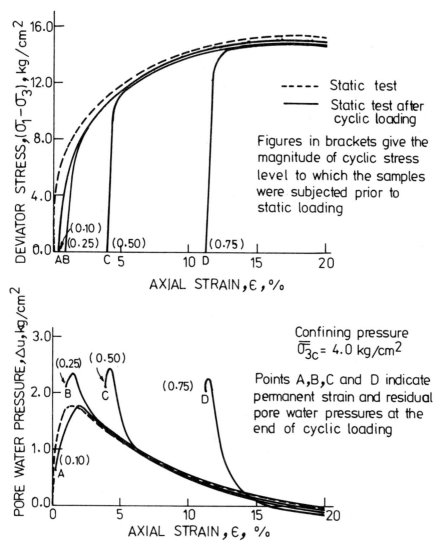

Fig. 8. Static stress-strain-pore water pressure behaviour after cyclic loading

pressure (Cc was equal to 1.9 at $\bar{\sigma}_{3c}$ = 2 kg/cm² and equal to 2.1 at $\bar{\sigma}_{3c}$ = 4.0 kg/cm²). This suggests that crushing essentially depends upon the permanent strain (equal to 20 percent for this study) and is not influenced by the type of loading i.e. whether cyclic or static. This observation is of significance because the magnitude of crushing has been shown to influence the static pore water pressure behaviour of calcareous sands (Datta et.al. 1979a). Dense sands which crush easily, develop positive pore water pressures at low confining pressures where as sands which are resistant to crushing exhibit high negative pore water pressures at low confining pressures. This is shown in Fig. 9 which compares the static $\Delta u / \bar{\sigma}_{3c}$ versus τ / S_u curve for the calcareous sand under study (sand A) with that of another calcareous sand which shows lower susceptibility to crushing (sand B) at $\bar{\sigma}_{3c}$ = 4 kg/cm² (both have a relative density of 100 percent). Since the

44

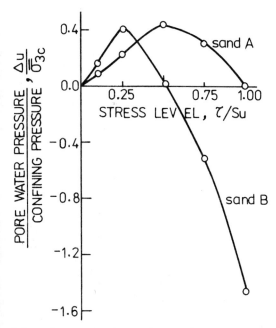

Fig. 9. Variation of static pore water pressures with stress level

pore pressure response under cyclic loading has been shown to be dependant on static behaviour, it is obvious that sand B will develop negative pore water pressure at high stress levels where as sand A will and has been shown to develop positive pore water pressure. This brings out the fact that sands which crush readily will tend to develop positive pore pressures during cyclic loading.

It must be noted that the above discussion has been confined to the sands in the densest state. For higher porosities, static pore water pressures would be more positive (Bishop and Eldin 1953, Bjerrum et. al. 1961) leading to higher values of C, thus causing larger positive pore water pressures to develop during cyclic loading. This observation is of significance because calcareous sands can exist in the loose state as a result of early diagenesis which is known to occur in calcareous sediments preventing any densification of the sediment.

5.5 Reliability of Test Results

It is often pointed out that redistribution of water in dense sand samples makes the top portion of the sample soft, causing 'cyclic mobility' in the laboratory specimens thus complicating the results, which may be unreliable for application in the field (Green and Ferguson 1971, Castro 1975, Hoeg 1976). It is felt that the redistribution of water may only become of significance in tests conducted under alternating cyclic stresses in which the extension cycle is responsible for the deterioration of the sample. If any softening at the top of the sample had occurred in the present study, it would have reflected as an initial flat portion in the stress-strain curves of subsequent static tests. Such a characteristic is observed in the shape of the stress-strain curves presented by Lee (1976)(a typical curve is shown in Fig. 10). Also presented in Fig.10 is the shape of a typical static stress-strain curve after cyclic loading in the present tests which shows that the initial portion of the curve is very stiff, thus ruling out the possibility of occurrence of softening of the sample by redistribution of water.

6 CONCLUSIONS

The process of driving a pile repeatedly subjects the surrounding soil to large compressive stress

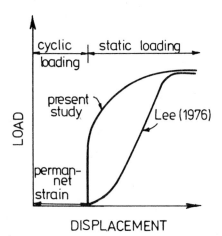

Fig. 10. Static stress-strain behaviour after cyclic loading

45

cycles which have a frequency corresponding to the number of blows delivered per minute by the hammer. During the installation of deep penetration piles for ocean structures, the rate of pile driving usually lies between 35 to 65 blows per minute. This rapid rate of stress application causes the surrounding strata to shear under essentially undrained conditions. The present study shows that the pore water pressures induced in a sand under repeated compressive stress cycles (of different magnitudes and under different confining pressures) are strongly related to the static pore water pressure response of the sand. They increase linearly with log of number of cycles after the first cycle. Increase in susceptibility to crushing and increase in the porosity of the sand cause higher pore water pressures to develop under large compressive cyclic stresses.

Calcareous sands often consist of skeletal remains of marine organisms which crush easily. They may also exist in a loose state due to diagenesis of the sediment. In both cases, positive pore water pressures are likely to be induced during pile driving. This conclusion reinforces the view that the phenomenon of low resistance during driving and increased resistance to redriving after stoppages in piles driven in calcareous sands is on account of the development of positive pore water pressures during driving and their subsequent dissipation during stoppages.

This study also shows that under compressive stress cycles large strains may develop even when the stresses are lower than the static failure stress of the sand. Cyclic loading does not significantly effect the subsequent static undrained behaviour. A comparison of the results of the present study with those of tests conducted under alternating stress cycles reveal that the pore water pressure development is more rapid under alternating stress cycles than under compressive stress cycles.

7 REFERENCES

Aggarwal, S.L., A.K. Malhotra and R. Banerjee 1977, Engineering properties of calcareous soils affecting the design of deep penetration piles for offshore structures, Proceedings Offshore Technology Conference, Houston, Texas, 3:503-512.

Anderson, K.H. 1976, Behaviour of clay subjected to undrained cyclic loading, Proceedings BOSS'76, Conference on Behaviour of Offshore Structures, Trondheim, Norway, 1:392-403.

Angemeer, J., E.D. Carlson and J.H. Klick 1973, Techniques and results of offshore piles load testing in calcareous soils, Proceedings Offshore Technology Conference, Houston, Texas, 2:677-692.

Angemeer, J., E.D. Carlson, S.Stroud and H. Kurzeme 1975, Pile load tests in calcareous soils conducted in 400 ft of water from a semi-submersible exploration rig, Proceedings Offshore Technology Conference, Houston, Texas, 2: 657-670.

Bishop, A.W. and A.K.G. Eldin 1953, The effect of stress history on the relation between $\emptyset$ and porosity of sand, Proceedings Third International Conference on Soil Mechanics and Foundation Engineering, 1 : 100-105.

Bjerrum, L. 1973, Geotechnical problems involved in foundation of structures in the North Sea, Geotechnique,23:3:319-358.

Bjerrum, L., S. Krimstan and D. Kummereje 1961, The shear strength of a fine sand, Proceedings Fifth International Conference on Soil Mechanics and Foundation Engineering, 1:29-37.

Castro, G. 1975, Liquefaction and cyclic mobility of saturated sands, Journal of Geotechnical Engineering Division, ASCE, 101: GT6:551-569.

Datta, M., S.K. Gulhati and G. Venkatappa Rao 1979, Crushing of calcareous sands during shear, Proceedings Offshore Technology Conference, Houston, Texas, 1459-1467.

Datta, M., S.K. Gulhati and G. Venkatappa Rao 1979a, Undrained shear behaviour of calcareous sands, Indian Geotechnical Journal, Communicated.

Green and Ferguson 1971, On Liquefaction of sand: Report of

Lecture by A. Cassagrande, Geotechnique, 21:3:197-202.

Herman, H.G. and W.N. Houston 1976, Response of sea floor soils to combined static and cyclic loading, Proceedings Offshore Technology Conference, Houston, Texas, 1:53-60.

Higginbottom, I.E. 1965, The Engineering Geology of chalk, Proceedings Symposium on Chalk in Earthworks and Foundations, London, 1-13.

Hoeg, K. 1976, Foundation Engineering for fixed offshore structures, Proceedings BOSS'76, Conference on Behaviour of Offshore Structures, Trondheim, Norway, 1:39-69.

Ishibashi, I. and M.A. Sherif, 1974, Soil liquefaction by torsional simple shear device, Journal of Geotechnical Engineering Division, ASCE, 100:GT8:871-881.

Janbu, N. 1976, Chairman's Report, Soils session one, Soils Under Cyclic Loading, Proceedings BOSS' 76, Conference on Behaviour of Offshore Structures, Trondheim, Norway, 2:373-383.

Lee, K.L. 1976, Fundamental considerations for cyclic triaxial tests on saturated sand, Proceedings BOSS'76, Conference on Behaviour of Offshore Structures, Trondheim, Norway, 1:355-373.

Lee, K.L. 1976a, Predicted and measured pore pressures in the Ekofisk Tank foundation, Discussion, Proceedings BOSS'76, Conference on Behaviour of Offshore Structures, Trondheim, Norway, 2:384-398.

Lee, K.L. and J.A. Focht 1975, Liquefaction potential at Ekofisk Tank in the North Sea, Journal of Geotechnical Engineering Division, ASCE, 101:GT1:1-18.

McClelland, B. 1974, Design of deep penetration piles for ocean structures, Journal of Geotechnical Engineering Division, ASCE, 100:GT7:705-747.

Peacock, W.H. and H.B. Seed 1968, Sand liquefaction under cyclic loading simple shear conditions, Journal of Soil Mechanics and Foundations Division, ASCE, 94:SM3:689-708.

Seed, H.B. and K.L. Lee 1966, Liquefaction of saturated sands during cyclic loading, Journal of Soil Mechanics and Foundations Division, ASCE, 92:SM6:105-134.

Vijayvergia, V.N., A.P. Cheng and H.J. Kolk 1977, Effect of soil set up on pile drivability in chalk, Journal of Geotechnical Engineering Division, ASCE, 103:GT10:1069-1082.

Partially drained cyclic storm tests
in a direct simple shear device

BERND SCHUPPENER
Federal Institute for Waterway Engineering, Hamburg, Germany

1 INTRODUCTION

Wave loads of storms induce cyclic shear and normal stresses in the soil beneath foundations of offshore-structures. Under these cyclic stress changes even dense sand will behave contractive (Goldscheider 1975). Depending on the drainage conditions there will either be a volume decrease or a generation of excess pore water pressure if the volume changes are delayed or prevented, due to long drainage paths or cyclic shears stresses at high frequencies. Compared to earthquake loading wave loads have low frequencies (o.o5 to o.25 Hz). But beneath gravity type offshore structures with foundation areas of about 1o.ooo m², the drainage paths for excess pore pressure dissipation are comparatively long. Therefore considerable excess pore water pressure may develop in the subsoil during a 1oo-year-storm. When excess pore water pressures are taken into account for stability analyses of foundations, it is normally conservatively assumed, that the volume changes in the soil during loading are zero, i.e. that there is no pore pressure dissipation. As storms last one or two days, this assumption will yield far too high excess pore water pressures. Therefore a realistic prediction of the excess pore water pressures generated in cohesionless soil must take into account the pore pressure dissipation during the storm.

2 EXISTING CONCEPTS FOR THE DETERMINATION OF EXCESS PORE WATER PRESSURE

Bjerrum (1973) was the first to present a procedure for the determination of the excess pore water pressure, which develops beneath a gravity type offshore structure due to cyclic wave loads. Based on results of undrained cyclic simple shear tests, he calculated the excess pore pressure increment of every single wave of a storm and determined the excess pore water at the end of the storm by summing up the pore pressure increments induced by the waves of the 1oo-year-storm. The procedure works as follows: The irregular wave loading during a storm is simplified by

Table 1. Determination of excess pore water pressure due to wave loads of a 1oo-year-storm after Bjerrum (1973)

Height of waves	Number of waves	$\frac{\tau_{hp}}{\bar{\sigma}_{vc}}$	$\tan \beta = \frac{\Delta u}{\sigma_{vc} \cdot N}$	$\frac{\Delta u}{\sigma_{vc}}$
m	–	–	%	%
4 – 8	485	0,07	0,006	2,9
8 – 12	471	0,12	0,013	6,1
12 – 16	282	0,17	0,030	8,5
16 – 20	121	0,22	0,065	7,9
20 – 24	32	0,26	0,150	4,8
24 – 26	3	0,30	0,300	0,9
Σ	1394		Σ	31,1

arranging the waves according to their height (Tab.1, column 1 and 2). Next the cyclic horizontal shear stresses τ_{hp} are calculated (index "p" for pulsating), which are induced in the subsoil beneath the foundation. These cyclic shear stresses τ_{hp} are normalized by dividing them by the average vertical normal stress σ_{vc} due to the dead weight of the structure (column 3, Tab. 1). This normalized cyclic shear stress is called cyclic shear stress ratio. To calculate the pore water pressure of one range of wave heights the value of the normalized pore pressure increase per load cycle:

$$\tan \beta = \Delta u / (\sigma_{vc} \cdot N) \qquad (1)$$

is needed. (column 4, Tab. 1). This tan ß-value Bjerrum derived by undrained cyclic test, which were run at different shear stress ratios τ_{hp}/σ_{vc} in the NGI direct simple shear apparatus. Multipling the number of waves N_i in column 2 with the tan ß-value of column 4, the normalized excess pore water pressure of this range of wave heights is obtained (see column 5). Summing up the values of column 5 Bjerrum yields the normalized excess pore water pressure at the end of the storm.

It is obvious that this procedure will produce too high excess pore water pressures because the pore pressure dissipation due to the drainage during the storm has not been taken into account.

Franke and Schuppener (1976) have incorporated the drainage effect into Bjerrum's concept. By an iterative numerical procedure the pore pressure decrease is calculated by means of the conventional theory of consolidation, which is then superimposed on the excess pore pressure of column 5 in Tab. 1.

Simultaneously Goldscheider and Gudehus (1976) and See et al. (1976) proposed to combine the differential equation, which is used in the theory of consolidation, with a pumping term in ψ, which describes the pore pressure increase per time due to cyclic loading. For one dimensional conditions this equation reads:

$$\frac{\partial u}{\partial t} = c_v \cdot \frac{\partial^2 u}{\partial z^2} + \psi \qquad (2)$$

where t is the time, u the pore pressure, z the depth, and c_v the coefficient of consolidation. Using Bjerrum's tan ß-value of eq (1) the pumping term ψ becomes:

$$\psi = \frac{\Delta u}{t} = \frac{\tan\beta \cdot \sigma_{vc}}{T} \qquad (3)$$

with T as wave period. For a constant pumping term Gudehus (1977) gives solutions in closed form for u (t,z) for several types of boundary conditions of the drainage system. Rahman et al. (1977) also use this combination of pumping term and theory of consolidation in a finite element programme, by which they compute the excess pore pressure distribution beneath a gravity type offshore foundation during a storm.

All of these concepts are based on the assumption that for a given cyclic shear stress ratio the pumping term ψ or Bjerrums tan ß - value is constant, which means that, it is not influenced by partial drainage. Results of partially drained cyclic test demonstrate however, that the pumping term ψ is strongly affected by the partial drainage. In Fig. 1 two cyclic simpl shear tests with saturated sand are compared. One of the tests was run undrained and the specimen failed after 9 cycles when a normalized pore pressure of $\Delta u/\sigma_{vc} = 0,43$ was attained. In the second test the specimen was prepared and loaded identically, but a pore pressure dissipation of 20 % per 5 cycles was allowed, the experimental technique will be described in the next paragraph. It can be seen that the effect of pore pressure dissipation is most striking: due to partial drainage the specimen gains so much strength that after reaching a maximum at 20 cycles the pore pressure decreases again and would have reached zero, if the test had been continued.

Fig. 1 therefore shows quite clearly, that a partial drainage during a storm will affect the excess pore pressure in two aspects:
- firstly the drainage will reduce the absolute value of the pore pressure,
- secondly the pumping term ψ or tan ß-value will be reduced, which is the most important effect of

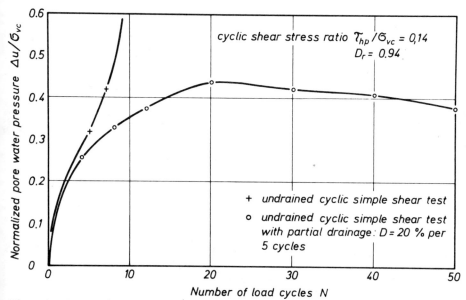

Fig. 1. Comparison between an undrained and partially drained cyclic simple shear test.

partial drainage.

This reduction of the pumping term may be explained by a reorientation of the grain structure, which is induced by small volume reductions during partial drainage. By this reorientation of the grain structure, the soil gains a higher shearing resistance, so that the pumping term ψ will decrease.

It is evident therefore, that all excess pore pressure computations, which do not account for the decrease of the pumping term due to the partial drainage, must yield unrealistically high pore pressures. This results in an over conservative and uneconomical design.

Smits (1978) was the first to propose a procedure taking into account the decreasing pumping term ψ due to partial drainage. He also uses the differential equation of the theory of consolidation extended by the pumping term. But, he applies the closed form solutions for pore pressure step wise, only for short time intervals, where a constant pumping term may be assumed. During the next time interval a corrected pumping term is then introduduced, which is determined by the change in porosity Δn of the preceding time interval. The procedure

therefore requires the knowledge of how at a given shear stress ratio τ_{hp}/σ_{vc} the pumping term ψ changes with porosity n. This relation can only be determined experimentally, of which the most difficult part is the experimental determination of the change in porosity Δn, which requires very accurate measurements of the volume changes during partial drainage. With sufficient accuracy these measurements can only be obtained with loose to medium dense sands. Approaching dense to very dense sands, the volume changes due to partial drainage are so small, that the precision of most devices measuring volume changes are inadequate. Moreover compliance effects like the membrane penetration (Kiekbusch and Schuppener 1977) tend to make the interpretation of volume change measurements difficult.

3. A NEW PROCEDURE TO DETERMINE EXCESS PORE PRESSURES DURING A STORM

To avoid the experimental problems of a sufficiently accurate determination of the pumping ψ in dependence on changes in porosity n, a different procedure of excess pore pressure determination has been developed. Basically the new proce-

51

dure uses the stress path method
(Lambe 1967, Lambe and Marr 1979):
The complete stress history,
and the drainage conditions of a
characteristic soil element below
a gravity type offshore structure
are simulated with a saturated sand
specimen in a cyclic simple shear
device, while the excess pore water
pressure is measured. The stress
history of this so called partially
drained storm test includes:
- the stresses and stress changes
  due to a possible overconsolidation,
- the stresses by the dead weight
  of the construction and
- the cyclic shear loads of the
  waves of a 1oo-year-storm.
The procedure consists of four main
steps. Firstly, an investigation
of the subsoil of the location has to
be performed. With regard to the partially drained storm tests this soil
investigation should supply upper
and lower limits for the coefficient
of permeability k of the sand, the
Young's Modulus E_s and or relative
density I_D. Moreover it should clarify, whether the sand has been
subjected to an overconsolidation
during its geological history.

Secondly, the pore pressure dissipation per time due to the drainage
must be theoretically determined.
For a given geometry of the foundation it can be estimated by means
of the theory of consolidation
assuming conservative values for
the coefficients of permeability k
and of the Young's Modulus E_s. In
this calculation charts as those
published by Davis and Poulos (1972)
were found to be very expedient.

Thirdly, the irregular loading of
the offshore structure due to waves
of a 1oo-year-storm has to be converted into a loading programme for
the partially drained storm test.
This loading programme should on
the one hand be as representative
of the design storm as possible; on
the other hand the testing procedure
should be as simple as possible.
Above all, it must be guaranteed that
the loading programme will yield
values of excess pore water pressure,
which are on the safe side of the
stability analyses. To ensure a conservative value for the excess pore
water pressure the following method
was chosen: Similar as in Bjerrum's
concept the waves of a storm or a

Table 2. Loading programm for a partially drained cyclic storm test

Height of waves	Number of waves N_i	Number of load cycle N	Stress ratio τ_{hp}/σ_{vc}
m	-	-	-
4 - 8	25o	1 - 25o	0.07
8 - 12	225	251 - 475	0.12
12 - 16	150	476 - 625	0.17
16 - 20	60	626 - 685	0.22
20 - 24	15	686 - 700	0.26
24 - 26	2	701 - 702	0.30

storm period are arranged according
to their wave height (see column 1
and 2 in Tab. 2). The cyclic loading of the storm test begins with
the shear stress level τ_{hp}/σ_{vc}
of the lowest wave height (column 3
and 4 in Tab. 2). The number of
load cycles the specimen is subjected to, equals the number of waves
of the corresponding range of wave
heights. When this value is reached,
the cyclic shear stress level
τ_{hp}/σ_{vc} is increased to the next
value etc. till the highest shear
stress level corresponding to the
highest wave of the design storm
is reached.

The fourth and last step is the
partially drained cyclic storm test.
To begin with a sand specimen is
prepared with the relative density
I_D or Young's Modulus E_s determined
in step one. The specimen is then
saturated using backpressure and
consolidated to the stresses, which
act on a representative soil element in the base of the foundation
due to the dead weight of the structure. In case the sand has experienced an overconsolidation in its
geological history , the reconstituted sand specimen may also be
overconsolidated in the laboratory.
After the consolidation the undrained cyclic loading is performed according to a loading programme modelling the wave loads of
a 1oo-year-storm. To account for
pore pressure dissipation due to
drainage an experimental procedure
is used, which was first described
by Lee and Focht (1975) which they
called partial drainage. In this
procedure the undrained cyclic

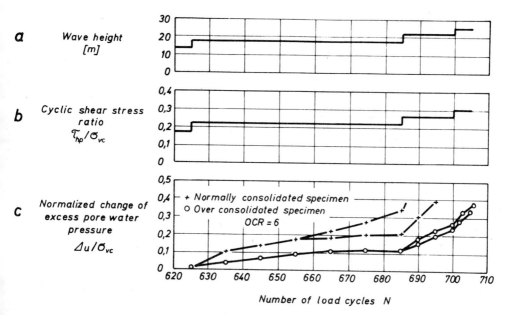

a   Wave height [m]

b   Cyclic shear stress ratio τ_{hp}/σ_{vc}

c   Normalized change of excess pore water pressure $\Delta u/\sigma_{vc}$

+ Normally consolidated specimen
o Over consolidated specimen
OCR = 6

Number of load cycles N

Fig. 2. Results of the partially drained storm tests for the Research Platform "Nordsee" for a 1oo-year-storm

loading is stopped after a preselected number of load cycles. The measured excess pore water pressure is then reduced by that amount, which was calculated for the pore pressure dissipation in step two. This reduction of the excess pore pressure to its nominal value is done by increasing the backpressure to this nominal value, whereafter the drainage valve is opened to allow the specimen to consolidate by the reduced excess pore pressure. The drainage line is then closed and the undrained cyclic loading is continued. After the preselected number of load cycles it is interrupted to repeat the procedure.

The excess pore pressure measured at the end of the cyclic loading programme may then be introduced in the stability analyses for the offshore foundation of a 1oo-year-storm to determine the safety factor e.g. against horizontal sliding under the horizontal load of a wave of 3o m height.

4. RESULTS OF PARTIALLY DRAINED CYCLIC STORM TESTS

4.1 Tests for the Research Platt-form "NORDSEE"

The Research Platform "NORDSEE" is founded on dense fine to medium sand. The foundation body rests on the sea floor by 8 single contact areas of about 1oo m^2. To estimate the pore pressure dissipation due to drainage a Young's modulus E_s = 2oo MN/m^2 and a coefficient of permeability of k = 4 . 10^{-5} m/s for the sand was assumed. With the charts of Davis and Poulos (1972) a time of about t_{95} = 25o s is needed for a degree of consolidation of U = 95 %. If the average wave period is T = 10 s and the pore pressure dissipation is in a first approximation assumed to be linear with consolidation time, there will be 20 % of consolidation during 5 wave periods. In a partially drained cyclic storm test modelling these conditions the excess pore pressure therefore has to be reduced by 20 % after 5 load cycles.

Partially drained cyclic storm tests with normally- and overconsolidated sand were performed in a simple shear apparatus (Franke and Schuppener 1976, Schuppener 1978) using the loading programme of Tab.2. In Fig. 2 the number of load cycles N is plotted against a) the wave height, b) the corresponding cyclic shear stress ratio and c) the normalized excess pore pressure $\Delta u/\sigma_{vc}$. The figure begins with the 625th load cycle where for the first time a small excess pore pressure is generated, when reaching a shear stress level of $\tau_{hp}/\sigma_{vc} = 0.22$ corresponding to a wave height of 18 m. A significant difference in the behaviour of normally and overconsolidated specimen can be seen. (The overconsolidated specimen were subjected to vertical and horizontal normal stresses of $\sigma_{vm} = 1.200$ kN/m^2 and $\sigma_{hm} = 720$ kN/m^2 respectively and then unloaded to almost zero again. Thereafter they were subjected to the same consolidation stresses as the normally consolidated samples, which were $\sigma_{vc} = 200$ kN/m^2 and $\sigma_{hc} = 120$ kN/m^2 for the vertical and the horizontal normal stress.) The normally consolidated specimen liquefy at a normalized pore pressure of $\Delta u/\sigma_{vc} = 0.4$ before the highest shear stress ratio $\tau_{hp}/\sigma_{vc} = 0.3$ is reached. In contrast to this the overconsolidated specimen even sustained the two cycles at maximum shear stress ratio without failure.

4.2 A test modelling the Nov. 1973 storm for the Ekofisk Tank

The fine to medium sand below the Ekofisk Tank (EFT) is very dense and most likely overconsolidated (Bjerrum 1973, Lee and Focht 1975). The foundation is near circular and has an area of about 8000 m^2. With a Young's modulus of $E_s = 200$ MN/m^2 and a coefficient of permeability of $k = 4 \cdot 10^{-5}$ m/s the time for 95 % consolidation is about $t_{95} = 5000$ s. To approximate the pore pressure dissipation below the EFT again a linear relation is assumed between consolidation time and excess pore water pressure. With an average wave period of T = 10 s the excess pore pressure in the storm tests has to be reduced by 10 % after 50 undrained load cycles.

Since its construction comprehensive measurements of the behaviour of the EFT have been performed, which contained the observation of excess pore waters in the sand below the base of the tank. During the first major storm in Nov. 1973 normalized excess pore pressures of $\Delta u/\sigma_{vc}$ from 0.01 to 0.15 were measured (Clausen et al. 1975). The maximum wave height during the storm was 16.0 m, which is about two thirds of the storm shown in

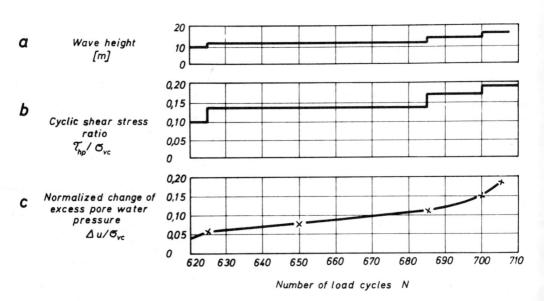

Fig.3. Results of the partially drained storm test for the Ekofisk Tank simulating the Nov. 1973 storm.

Tab. 2. Incidentally equal wave heights produce an equal average cyclic shear stress ratio below the base of the research platform "NORDSEE" and the Ekofisk Tank. In the loading programme modelling the storm of November 1973 the wave heights and cyclic shear stress ratios of the 1oo-year-storm of Tab. 2 were reduced by about 30 %. The results of the partially drained cyclic storm test are shown in Fig. 3, beginning with the 625th cycle, where for the first time a significant excess pore water pressure was measured. At a shear stress ratio of $\tau_{hp}/\sigma_{vc} = $ o.19 which corresponds to the maximum wave of 16.o m, the normalized excess pore water pressure reached a value of $\Delta u/\sigma_{vc} = $ 0.15. Considering that the gradual increase of the shear stress level in the test will tend to produce higher excess pore water pressures compared to the irregular sequence of wave heights during a storm, the agreement between the result of the partially drained cyclic storm test and the in situ measurement is good.

5 SUMMARY AND CONCLUSION

After discussing existing concepts a new and simple procedure is presented by which the excess pore water pressures generated below gravity type offshore structures due to wave loads of a storm may be determined. The effect of drainage is taken into account by a special experimental technique.

The results of a partially drained storm test modelling the stress history of a soil element below the Ekofisk Tank during the storm of Nov. 1973 showed good agreement with in situ measurements. As the proposed procedure is necessarily based on a number of simplifications the author is fully aware that this good agreement between test and in situ measurement can only be a first argument to substantiate the proposed method.

6 REFERENCES

Bjerrum, L. 1973, Geotechnical problems involved in foundations of structures in the North Sea, Geotechnique 16,No.3: 319-358

Clausen, C.J.F. & Di Biagio, E. & Duncan, J.M. & Andersen, K.H. 1975, Observed behaviour of the Ekofisk oil storage tank foundation, Norweg. Geotechn. Instit. No.1o8.

Davis, E.M. & Poulos, H.G. 1972, Rate of settlement under two- and three dimensional conditions, Geotechnique 22, No. 1:95-114.

Franke, E. & Schuppener, B. 1976, Offshore-Flachgründung der Forschungsplattform NORDSEE, Besonderheiten der Baugrunderkundung und der Gründungsberechnung, Vorträge der Baugrundtagung:551-573.

Goldscheider, M. 1975, Dilatanzverhalten von Sand bei geknickten Verformungswegen, Mech. Res. Comm. 2: 143-148.

Goldscheider, M. & Gudehus, G. 1976, Einige bodenmechanische Probleme bei Küsten- und Offshore-Bauwerken, Vorträge der Baugrundtagung: 5o7-522.

Gudehus, G. 1977, Stability of saturated granular bodies under cyclic load, Proc. Dynamical Methods in Soil and Rock Mechanics, Vol. 2:195-214

Kiekbusch, M. & Schuppener, B. 1977, Membrane penetration and its effect on pore pressures. J.Geot. Eng.Div. ASCE, 1o3, GT 11:1267-1279.

Lambe, T.W. & Marr, W.A. 1979, Stress path method:second edition, J.Geot.Eng.Div. ASCE 1o5, GT 6: 727-738.

Lee, K.L. & Focht, J.A. 1975, Liquefaction potential at Ekofisk Tank in North Sea, J.Geot.Eng.Div. ASCE 1o1, GT 1:1-18.

Rahman, M.S., Seed, H.B. & Booker, J.R. 1977, Pore pressure development under offshore gravity structures, J.Geot.Eng.Div. ASCE 1o3, GT 12: 1419-1436.

Schuppener, B. 1978, Some aspects of the testing procedures for cyclic loading of sand samples, Proc. Dynamical Methods in Soil on Rock Mechanics, Vol. 2:131-147.

Smits, F.P. 1978, Excess pore pressures and displacements due to wave induced loading of a caisson foundation as predicted by plasticity analysis, Proc. Foundation Aspects of Coastal Structures, Delft, Vol. 1.

Cyclic strengths of undisturbed cohesive soils
of western Tokyo

KENJI ISHIHARA
University of Tokyo, Tokyo, Japan

SUSUMU YASUDA
Kisojiban Consultants Co. Ltd., Tokyo, Japan

SYNOPSIS

Cyclic triaxial tests were performed on un-disturbed samples of two cohesive soils, alluvial clay and loam, obtained from Tama City in the western part of Tokyo. The iso-tropically consolidated specimens were first subjected to an initial static axial stress under drained condition and then to se-quences of cyclic axial stress with each sequence applying 30 cycles of constant load and the amplitude of each sequence increasing successively until the specimen deformed to a pre-determined failure strain. The specimens were saturated and the se-quences of shear stress were applied in un-drained conditions. The cyclic strength was defined as the initial static shear stress plus the amplitude of cyclic shear stress required to cause failure in the specimen in a given number of cycles. The result of the tests employing different proportions of initial static and cyclic shear stress showed that without initial static shear stress the cyclic strength at 30 cycles of loading was smaller than the static strength. However, the cyclic strength increased by about 20% over the static strength when an initial shear stress as much as half the static strength was applied to the specimen. The test results also showed that soils with the lower plasticity index reduced their strength more remarkably under cyclic loading than soils with the higher plasticity index.

INTRODUCTION

In making stability analysis of man-made fills or natural slopes during earthquakes, the strength of cohesive soils that is mobilized in the field under cyclic loading conditions must be known. Cyclic loading is viewed as repetition of rapid loads.

The effect of rapid loading on the strength of cohesive soils was studied by Casagrande and Wilson (1951), Whitman (1957), Kawakami (1960) and Olson and Parola (1967). These studies showed that the strength of cohe-sive soils under monotonous loading condi-tions increased as the speed of loading increased. The effect of load repetition on the cohesive soils was investigated by Seed and Chan (1966), Ellis and Hartman (1967), and Lee (1979). The result of these studies consistently revealed that the strength of cohesive soils tended to decrease as the number of load repetition increased.

Since the cyclic loading is considered, as mentioned above, as the repetition of rapid loads, the tendency of increasing the strength due to rapidity of load application appears concurrently with the tendency of decreasing the strength due to cyclic nature of load application. Therefore, a proper assessment of the soil strength under dynamic loading conditions during earthquakes requires a better understanding of the extent of the interaction of these two mutually conflicting processes acting together to mobilize actual strength of cohesive soils.

The extent to which these two factors govern the clay strength has been shown further by Seed and Chan (1966) to depend on the magnitude of initial shear stress to which a soil has been subjected before the soil is made to fail by cyclic load appli-cation. Thus, it is also necessary to consider the effect of initial shear stress in understanding the clay strength under cyclic loading condition.

The primary objectives of this paper were, therefore, to find out possible influence of the aforementioned factors more defini-tively for two cohesive clays existing in the western part of Tokyo.

LOADING SCHEME AND DEFINITION OF CYCLIC
STRESS-STRAIN BEHAVIOR

The loading scheme employed in this investigation to determine the cyclic strength
is schematically illustrated in Fig. 1 in
terms of stress versus strain plot. A
specimen was first consolidated under an
isotropic pressure $\sigma_o' = 100 KN/m^2$ as illustrated by point A in Fig. 1. An axial
stress, σ_s, was then applied statically to
the specimen under drained condition as
indicated by point B in the figure. The
static axial stress, σ_s, was applied to
simulate the in situ shear stress which may
exist in a soil element beneath a sloping
surface or adjacent to structures. Several
sequences of uniform cyclic axial stresses
with stepwise increasing amplitudes were
further superimposed, 30 times each, on
the specimen under undrained conditions.
For instance, a cyclic stress with a small
amplitude, σ_{d1}, was applied until the
specimen deformed to a strain, ε_1, as
illustrated by point C in Fig. 1. In this
sequence of cyclic stress application, the
axial strains corresponding to 10 and 30
cycles were noted as points B' and B". In
the second sequence of cyclic stress application, the points such as C' and C" indicating the axial strains corresponding to
10 and 30 cycles were also noted. Similarly, the points such as D' and D" were
marked in the third sequence of cyclic

stress application in which the amplitude
of cyclic loading was increased to σ_{d3}.
On the basis of the test results such as
those shown Fig. 1, it was possible to construct a stress versus strain curve for a
given number of cyclic loading. For
instance, for 10 cycles of loading, a curved
line was drawn through the points B', C'
and D', starting from the point B in Fig.1.
This curve is rewritten in Fig. 2. Likewise, when a curve was drawn through the
points B", C" and D" starting from the
point B, the stress versus strain curve was
obtained for the loading involving 30 cycles
of axial stress. The curve obtained in
this manner is also shown in Fig. 2. In
the same fashion, a stress-strain curve can
be constructed for any given number of
cycles. Thus, for a given value of the
initial shear stress, a set of stress-strain
curves for different numbers of cycles can
be obtained as schematically illustrated in
Fig. 2
In the above loading scheme, cyclic loads
with stepwise increasing amplitudes are
applied to a specimen in sequence. Therefore, it is likely that the response of the
specimen to one sequence of cyclic loading
is affected by other preceding sequences
with lesser amplitudes. Since the preceding
sequences act toward increasing strains in
the specimen, their effect appears in the
stress-strain curve as if the specimen were
less stiff than a fresh specimen that had

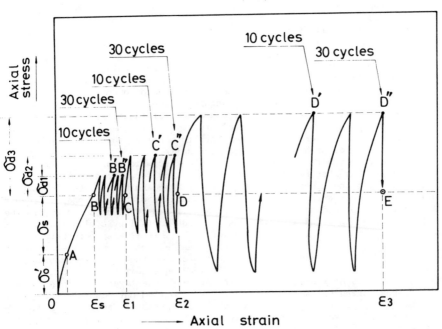

Fig. 1 Construction of a stress-strain curve for cyclic loading

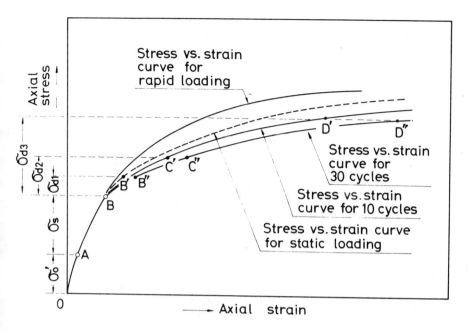

Fig. 2 A set of stress-strain curves for different number of cycles

not been subjected previously to cyclic loads with lesser amplitudes. In order to avoid the stress history effect as above, it would be preferable to use fresh specimens for each stress amplitude in an entire cyclic loading scheme and to determine the points such as B', C' and D' in Fig. 1 which are obtained after a specified number of cycles has been applied to the fresh specimens.

Although the cyclic test program as above was considered more desirable, it was not adopted in the present study because of the limited number of identical undisturbed specimens that were made available for the tests. In the case where the available number of test samples is limited, the loading scheme as explained in Fig. 1 seems to be a reasonable alternative to the most idealistic test program in which fresh specimens are always used for each load amplitude.

Fig. 2 shows schematically the stress-strain curve that is obtained in the static test with a monotonously increasing load. A stress versus strain curve obtained for a transient load in which a load is increased monotonously but much more rapidly than in the static test is also shown. A number of test results obtained thus far indicates that the stress-strain curve is stiffest in the transient loading condition, giving the greatest strength. The stress-strain curves under cyclic loading conditions generally tend to become more gently sloped, and tend to give smaller strength than the stress versus strain curve under the transient loading conditions. In addition, the cyclic loading tests tend to decrease the stiffness and the strength of the specimen as the specified number of cycles is increased, as illustrated in Fig. 2. The stress-strain curve for the static loading lies somewhere between the curve for the transient loading and the curve for a cyclic loading with a large number of cycles.

The above general trend regarding the change in the stress-strain and strength characteristics has been observed in many laboratory tests, particularly in the case of cohesive soils as well as in cohesiveless soils tested under undrained conditions.

SITE AND SOIL DESCRIPTION

The present study was performed at a proposed site for a large-scale land development project for the construction of a new residential district in the western part of Tokyo Metropolis called Tama. This area was covered generally with volcanic ash with its thickness ranging from a several meters to approximately 10 meters. Below this there exist cemented sands and hard clay or silt deposits of diluvial origin. The area consists of small hills 10 to 20

59

meters high. Some sections of the hills were eroded by many small streams, forming alluvial deposits in low-lying portions in the middle to lower reaches of the rivers.

To obtain broad acres of flat lands, the hilly sections were excavated and borrow materials were dumped over the low-lying alluvial portions. One of the geotechnical engineering problems facing such large-scale earth works was the stability of the man-made fills during future earthquakes. The dumped soils formed slopes with 1 vertical to 1.8 to 2.0 horizontal ratio near the edges of the fills, and the concern arose for possible failure induced by earthquakes.

The type of soils in this area most critical to the slope stability were the borrow material "loam" in the man-made fill consisting mainly of volcanic ashes and the clays in the alluvial deposit underlying the man-made fill.

The alluvial clay is a fluvial deposit with varying properties including sometimes interbeds or lenses of silt or sand. The clay was normally consolidated and it was impossible to obtain samples in blocks, because of its large consistency and the ground water table being almost as high as the exposed ground surface.

The man-made fill had been constructed by dumping excavated soils down into valleys with bulldozers. No special compaction method was employed, however, the previously dumped underlying soils were compacted to some extent due to the weight of the bulldozer moving over them.

SAMPLING PROCEDURES

All samples were obtained on the ground surface. A small section approximately 200 m by 200 m was chosen on a flat alluvial deposit. Within this section, four adjacent sites approximately 20 m apart from each other were selected for sampling locations. The clays in this area were soft and the ground water table was only about 20 cm below the ground surface. At the sampling spot, sampling brass tubes 1.2 mm thick, 7.5 cm in diameter and 15 cm in length, were pushed upright into the soil being sampled. When the tube penetrated half of its length, the soil surrounding the tube was removed to facilitate pushing the tube further to its full length. The tube containing undisturbed specimen was dug out carefully by means of a scoop. After levelling off both ends of the specimen, it was waxed and capped in the field. The specimen was transported to the soil testing laboratory with foam rubber protection. To distinguish samples

from four nearby sites, the samples were labelled as site 1, 2, 3 and 4.

Samples from the dumped fill were obtained from a small selected area also approximately 200 m by 200 m and four sites about 20 m distant from each other in this area were chosen as sampling site. Sampler from each of these four sites were labelled again as site 1, 2, 3 and 4. The ground water table in this sampling site was located about 1.5 m below the ground surface. In order to obtain as much saturated specimens as possible, the surface soil was removed as far as the level of the ground water table. The sampling procedure carried out on this exposed surface was exactly the same as the sampling of clays from the alluvial deposit.

LABORATORY TESTING

In the laboratory, the specimen was extruded carefully from the brass tube by means of a small piston. The specimen was trimmed with the use of a sharp knife for shaping a final test specimen 5 cm in diameter and 10 cm in length. Cyclic triaxial tests were performed with porous stone at the top and bottom of the specimen for the drainage. In all tests, specimens were consolidated under an effective confining pressure of $\sigma_o' = 100 KN/m^2$ with a back pressure of $300 KN/m^2$. Then the specimens' volume were determined to define void ratios. Additional vertical stresses, σ_s, were applied to the specimens under drained condition, and maintained for approximately one hour until the associated axial deformation ceased to take place. This was done to reproduce the static stress condition existing in the sloping ground surface. After the consolidation, cyclic axial stresses with stepwise increasing amplitudes, σ_d, 30 cycles each were applied to the specimens, according to the loading scheme as illustrated in Fig. 1. All cyclic triaxial tests were performed with a 1 Hz cyclic frequency.

In addition, the conventional type of consolidated undrained tests were performed using a consolidation pressure of $\sigma_o' = 100$ KN/m^2 to determine the static strength, σ_f, for comparison with the cyclic strength. Both static and cyclic strengths were determined from a peak deviator stress or from maximum deviator stress recorded at an axial strain of 15%.

PROPERTIES OF SOILS

Grain size distributions of the soils from the alluvial clay deposit are shown in

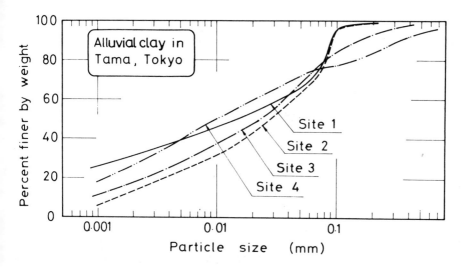

Fig. 3　Grain size distribution curves of alluvial clay

Table 1　Properties of test samples of alluvial clays

Sampling Site	Specific Gravity Gs	Void Ratio e*	Liquid Limit W_L	Plastic Limit Wp	Plasticity Index Ip
1	2.65	1.97	80	39	41
2	2.39	2.54	111	64	47
3	2.44	2.35	106	60	46
4	2.48	1.94	91	44	47

* Average value among test specimens

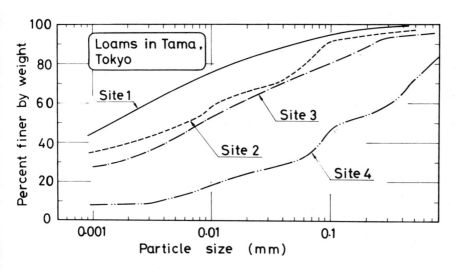

Fig. 4　Grain size distribution curves of loam

Table 2 Properties of test samples of loams

Sampling Site	Specific Gravity Gs	Void Ratio e*	Liquid Limit W_L	Plastic Limit Wp	Plasticity Index Ip
1	2.69	1.89	100	60	40
2	2.57	1.64	72	45	27
3	2.51	1.94	81	48	33
4	2.65	1.08	47	29	18

* Average value among test specimens

Fig. 3 for each of the sampling sites. The grain size characteristics of the soils did not differ appreciably from one sampling site to another. The sand fraction remaining by #200 sieve accounted for approximately 25% of the total weight and clay fraction, finer than 0.005 mm, accounted for 30%. Physical properties of this soil listed in Table 1 showed that the soil was medium plastic with a plasticity value of about 45.

Grain size characteristics of the soils from the man-made fills are shown in Fig. 4. The soils had almost similar grain size distributions with the sand fraction amounting only to about 20% except for the loam from the sampling site 4. The list of physical properties shown in Table 2 indicates that the loam had relatively small plasticity index values ranging between 18 and 40.

TEST RESULTS

The results of two static consolidated undrained tests performed on the alluvial clay and the loam are shown in Fig. 5 where the deviator stress versus the axial strain are plotted. These are the test results on the specimens from the sampling sites 4 both for the alluvial clay and the loam. The tests on the specimens from the other sites showed similar stress-strain

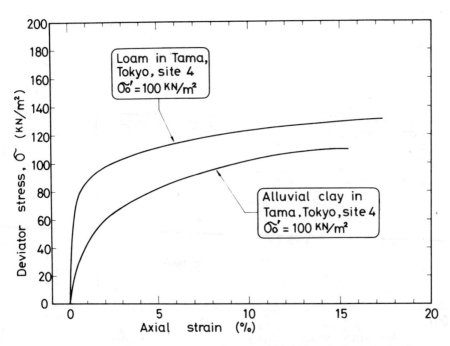

Fig. 5 Stress versus strain curves in consolidated undrained tests for alluvial clay and loam

behavior and strength characteristics. Fig. 5 shows that the alluvial clay had a static strength on the order of 110KN/m^2 in terms of deviator stress and the loam had a static strength amounting to 130KN/m^2 which was larger than that of the alluvial clay. These strengths were used in the following data reduction process to normalize the cyclic strengths of the two soils obtained at different cycles with different initial static stress.

Typical results of cyclic triaxial tests on the specimens from the alluvial clay deposit are presented in Fig. 6 in which the deviator stress, $\sigma_s + \sigma_d$, (initial static stress plus amplitude of cyclic stress) is plotted versus the axial strain measured after the application of the initial static stress. Fig. 6 indicates that the cyclic strength at the cycles of 10, 20 and 30 decreased generally in this order as the number of cycles increased, but that the difference in the cyclic strength between 10, 20 and 30 cycles was small. The test results in Fig. 6 also show that the cyclic strength increased substantially as the initial static stress was increased. For instance, the specimen subjected to cyclic loading without initial shear stress exhibits a strength of 75 KN/m^2, whereas the specimen first loaded to one-third of the way to static failure, $\sigma_s = 31.6$KN/m^2, and then loaded cyclically shows a strength of 107KN/m^2.

Representative results of cyclic triaxial tests on the specimens from the man-made fills are demonstrated in Fig. 7 where the deviator stress is plotted against the axial strain. Note that the strain in the abscissa pertains only to the strain that developed during the cyclic loading. It may be seen from the figure that the cyclic strength of the loam subjected directly to cyclic stress without initial static shear stress was considerably lower than the loam specimen having been subjected to larger initial shear stress.

DISCUSSION OF THE TEST RESULTS

From the stress-strain curves corresponding to 30 cycles of loading as exemplified in Figs. 6 and 7, the maximum deviator stress, $\sigma_s + \sigma_d$, was read off to determine the cyclic strength at different initial shear stresses. The cyclic strengths thus determined were normalized to the static strength, σ_f, and plotted in the ordinate of Figs. 8 and 9 versus the initial shear stress divided by the static strength. It may be seen from these figures that the cyclic strengths at 30 cycles of loading could increase generally by zero to 30 percent over the static strength, when the initial shear stress, about 30 to 60 percent of the static strength, has been applied to the specimen before it is made

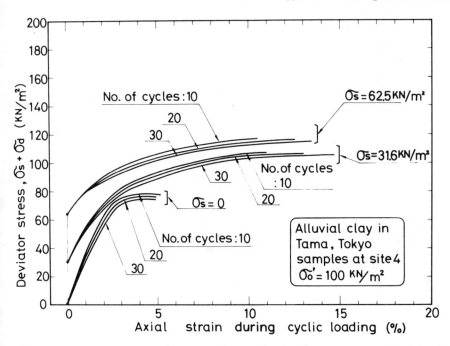

Fig. 6 Stress versus strain curves in cyclic loading tests on alluvial clay

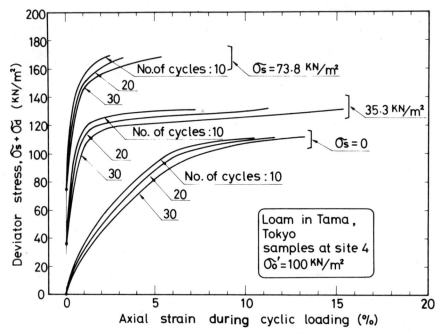

Fig. 7   Stress versus strain curves in cyclic loading tests on loam

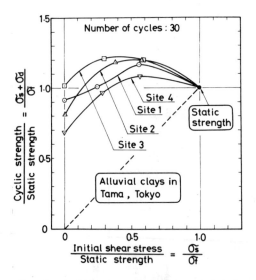

Fig. 8 Cyclic strengths versus initial shear stress (alluvial clay)

to fail under cyclic loading condition. The figures also show that when there is no initial sustained stress, that is, $\sigma_s/\sigma_f = 0$, the cyclic strength is considerably lower than the static strength. The tendency of cyclic strength varying as above depending upon the initial shear stress may be most logically interpreted as follows.

It was indicated above that the strength of cohesive soils under rapid loading conditions is greater than the strength obtained in static loading conditions in which the speed of loading is slow. This has been known as the rate effect. In the cyclic loading as executed in this investigation with a frequency of 1 Hz, the speed of loading at any instance of cycling is far greater than the speed at which the specimen is loaded in the normal static test. Therefore, even in the cyclic loading, the rate effect still shows up in the specimen's behavior when the number of shear stress application is small. It is also well-known that cyclic stress applied to cohesive soils makes soils soft and weak in both stiffness and strength. This may be called the deteriorating effect. This effect is suspected to be the result of a buildup of pore water pressure or from breakdown of inherent structure existing

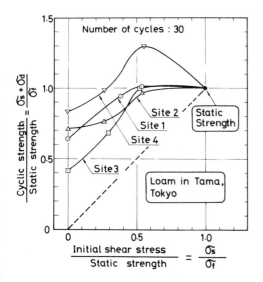

Fig. 9 Cyclic strengths versus initial shear stress (loam)

applied initially to the specimen. In a case where cyclic stress amplitude is greater than the initial static stress, a specimen is alternately subjected to tri-axial compression and triaxial extension. In other words, the specimen is subjected to stress changes reversing its direction between the triaxial compression and tri-axial extension. When the initial static stress is equal to zero, the cyclic stress application becomes such that it involves a complete reversal of direction. There-fore, it may be mentioned that the strength deteriorating effect becomes more pronounc-ed as the degree of stress reversal between the triaxial compression and extension becomes larger.

In the cyclic loading test as executed in this study with a frequency of 1 Hz, it may thus be speculated that the rate effect towards increasing strength must be acting together with the deteriorating effect of cyclic loading, with the combined effect of the two tending to reduce the strength. With this fact in mind, it is possible to offer a logical interpretation to the observed test results shown in Figs. 8 and 9. When the initial shear stress is as much as half the static strength, the chance of the specimen being subjected to a high degree of stress reversal is small. The deteriorating effect is, therefore, outweighed by the rate effect, hence the

in cohesive soils. It has also been shown by Seed and Chan (1966) that the strength deteriorating effect due to cyclic loading becomes more pronounced as the proportion of the cyclic stress amplitude increases relative to the static component of stress

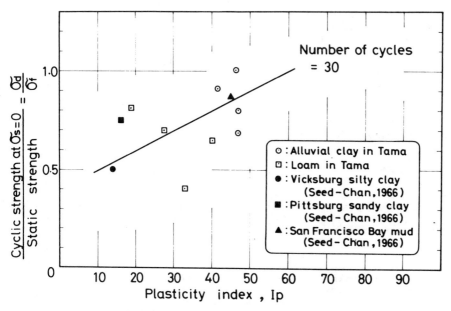

Fig. 10 Cyclic strength in reversing cyclic loading as functions of plasticity index of soils

cyclic strength becomes greater than the static strength. On the contrary, when the cyclic loading is in a fully reversing condition with a zero value of initial shear stress, the deteriorating effect becomes predominant and a remarkable reduction in cyclic strength is observed in the test results shown in Figs. 8 and 9.

The reduction in cyclic strength at the zero value of initial shear stress may be viewed as representing the inherent characteristics of a soil losing its strength under cyclic loading conditions. Therefore, the ratio between cyclic strength and static strength at zero initial shear stress will be taken up as an index parameter to express the extent of strength loss of a given soil due to cyclic loading.

There are good reasons to believe that the cyclic strength deterioration is related closely to the sensitivity of soil as the strength deterioration due to cyclic loading can be viewed as being similar in nature to the remolding effect. Bjerrum (1954) has shown that the sensitivity of clay increases as the plasticity index of the clay decreases. Therefore, it may as well be assumed that the strength deterioration caused by cyclic loading will be more pronounced in soils with a lower plasticity index than in soils having a higher plasticity index value. In view of this hypothesis, the ratio between cyclic strength to static strength at zero initial shear stress was read off from Figs. 8 and 9, and plotted in Fig. 10 versus the plasticity index of each soil. It may be seen from Fig. 10 that the loam with lower plasticity index shows a larger reduction in cyclic strength than the alluvial clay having higher plasticity index. Three test data obtained by Seed and Chan (1966) from the similar tests are also plotted in Fig. 10. Seed's data plot near the average line drawn in the figure in support of the increasing loss of cyclic strength with decreasing plasticity index of soil.

CONCLUSIONS

Cyclic triaxial tests on undistudbed samples of two kinds of cohesive soils have shown that the cyclic strength (defined by the initial static shear stress plus the amplitude of superimposed cyclic shear stress required to cause failure in a given number of cycles) decreased as the number of cycles being considered increased. It has also been shown that the cyclic strength at 30 cycles of loading increased by approximately 20% over the static strength when the initial static stress

approximately half the static strength was applied to the soil specimen. When the cyclic stress was applied without the initial static stress, the cyclic strength, however, was observed as falling below the static strength. This reduction in cyclic strength below the static strength was shown to be more pronounced in soils with lower plasticity index than in soils having higher plasticity index.

ACKNOWLEDGEMENTS

The cyclic test program described in this paper was conducted as part of the activities of the committee organized for studying seismic stability of man-made fills in the Tama land development area. The committee was chaired by Professor G. Kuno of Chuo University and supervised by Mr. K. Kiga of the Japan Housing Corporation. The overall support of these persons is gratefully acknowledged. The authors also wish to thank Dr. Kenji Mori for editing the original draft.

REFERENCES

Bjerrum, L. (1954), "Geotechnical Properties of Norwegian Marine Clays, "Geotechnique, Vol. IV, p. 49.
Casagrande, A. and Wilson, S.D. (1951), "Effect of Rate of Loading on the Strength of Clay and Shales at Constant Water Content," Geotechnique, Vol. 2, pp. 251-263.
Ellis, W. and Hartman, V.B. (1967), "Dynamic Soil Strength and Slope Stability," Proc. ASCE, SM. 4, pp. 355-373.
Houston, W.N. and Mitchell, J.K. (1969), "Property Interrelationships in Sensitive Clays," Proc. ASCE, Vol. 95, SM, 4, pp. 1037-1062.
Kawakami, F. (1960), "Properties of Compacted Soils under Transient Loads," Soil and Foundation, Vol. 1, No. 2, pp. 23-29.
Lee, K.L. (1979), "Cyclic Strength of a Sensitive Clay of Eastern Canada," Canadian Geotechnical Journal, Vol. 16, No. 1, pp. 163-176.
Olson, R.E. and Parola, J.F. (1967), Dynamic Shearing Properties of Compacted Clay," Proc. International Symposium on Wave Propergation and Dynamic Properties of Earth Materials, University of New Mexico, pp. 173-181.
Seed, H.B. and Chan, C.K. (1966), "Clay Strength under Earthquake Loading Conditions," Proc. ASCE, SM. 2, pp. 53-78.
Whitman, R.V. (1957), "The Behavior of Soils under Transient Loading," Proc. 4th International Conference on Soil Mechanics and Foundation Engineering, Vol. 1, pp.207-210.

The mechanical properties of cement stabilised soils in the conditions of load repetitions

HENRYK KOBA & BOGDAN STYPULKOWSKI
Wroclaw Technical University, Wroclaw, Poland

1 INTRODUCTION

In spite of many years' development of analytic methods of pavements' design and possibilities of the numerical methods use, the problem of optimization in the field of pavement thickness has not been solved yet. The use of analytic methods in the design of the pavement thickness is in the first place limited by difficulties connected with determination of materials' properties built in the pavements. Conditions of materials' cooperation in the pavement structure are also unknown. It gives occasion to use in most of cases conventional rates e.g. CBR in the design of flexible pavements thickness. But acceptance of thickness and system of each courses is determined by technological reasons. Nevertheless it does not ensure planned durability of the pavement.

The authors' research of soil-cement indicates that a knowledge of appropriate mechanical properties of material allows to predict in advance durability of pavement. Results of numerous research works demonstrate that pavements with stabilised courses are very economical but in most of cases they don't hold to the end of planned exploitation period. Mostly it is caused by the cracks of stabilised bases and bituminous surface layers sometimes. Insufficient thickness of stabilised cement courses is one of essential reasons of these cracks. Up to the present in many countries thickness of stabilised layers have not been large (15-25 centimetres). They have mostly been determined by technological conditions. At the moment tendency to thickness increasing of stabilised layers is occured. According to Keil's theory stabilised layers 40-60 cm thick ensure 50 years period of pavements durability (Keil, 1975). All these established facts are resulted from cxperience only but without theoretical base. Optimum thickness of the stabilised layer can be fixed basing on the specific character of the traffic, mechanical properties of the applied materials and features of a subgrade.

2 DESIGN METHOD OF PAVEMENT BASING ON FATIGUE STRENGTH

In the pavement design the estimation of load-carrying ability and durability of the pavement constitutes a research question to solve. Of the quality of the traffic volume arises the following fact. The resistance to carrying of load repetitions is the most objective rate to estimate the pavement work. This resistance can be determined by mechanical properties of the applied materials together with their fatigue strength.

In a case of the pavement containing the carrying layer of the cement stabilised soil, extensible stresses in lower zones of the courze constitute basic strength criterion.

Determination of the stresses' value and resulting from them permissible thickness of the layer demands:

1. Appropriately directed ana-

lysis of the pavement's traffic
loads.

2. Analysis of the stresses state
in the structure system of the pavement.

3. Estimation of the pavement's
durability in fatigue expression.

2.1 Analysis of the pavement's traffic loads

Fixing of the proper traffic volumes in the designed pavement demands to pay particular attention
to their quantity and repeatability.

Determination of the traffic quantity can be carried out basing on:

1. Average annual day intensity
of existing or designed traffic
(real vehicles per a day).

2. Type structure of road traffic
(percentage share of each vehicles'
groups).

3. Forecast of traffic increase.

Basing on these data and load
distribution of axles in each kinds
of vehicles we can determine a number of single axles that will cross
a given stretch of road in the period of its exploitation. Load of
each axles is accepted as the statical influence (loaded vehicles)
increased by a dynamical addition
for axles with load:

< 80 kM - 30%
$\geqslant 80$ kM - 20%

2.2 Analysis of the stresses state in the structure of the pavement

In theoretical considerations describing the work of the pavement's
structure, modeles of pavement based on the theory of granulated
mediums, creeping visco-elastic
bodies and the theory of elasticity
(Borkowski, 1973). Difficulties
connected with the fixing of materials' real properties and conditions of the pavement work has made
application, in practice, of methods based above all on the elasticity theory. In these solutions the
structure of the pavement is considered as a complex of elastic layers
resting upon elastic half space
(Hanuška, 1973). Of more interesting solutions based on the theory
of elasticity we can mention 2-
-layered system (Fig.1), which was

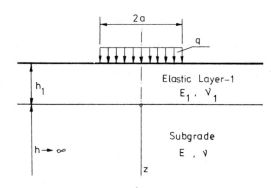

Fig.1. Two-layered elastic system

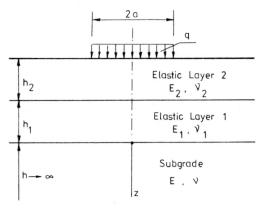

Fig.2. Three-layered elastic system

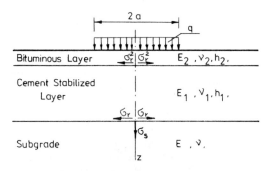

Fig.3. Static scheme of pavement

also engaged in the research work
by Burmister, Fox, Odemark, Kogan
and 3-layered system (Fig.2) that
was developed by Jones, Kirk, Odemark, Fox and others.

To the pavemet's construction
containing the carrying course made of cement stabilised soil we
can take the system presented in
Figure 3.

In the solution we accept:

1. The construction of the pavement works in elastic strains' range.
2. The destruction of the pavement is the consequence of fatigue changes in the carrying courses.
3. Extensible stresses in the lower zone of the stabilised course constitute the principal strength criterion.

The complete analysis of stresses state in the layer's system of the pavement we can carry out basing on the numerical methods.

2.3 Estimation of pavement durability

The estimation of the pavement durability with the carrying course containing cement stabilised soil can be related to the fatigue durability of the stabilised course.

Thickness of bituminous layer is accepted technologically within about 8-12 cm. Calculation resolve themselves into fixing of the stabilised course's proper thickness.

In the road traffic vehicles with different load of the wheels – P_i take part. Their effect on fatigue durability of the pavement is various and should be analysed separately. To determine the total effect of different loads on the course's durability we can make use of one of the fatigue hypothesis, for example Miner's theory (Kohler, 1974). According to this hypothesis a structure exposed to the action of the loads; $P_1, \ldots, P_i, \ldots, P_m$, developing stresses: $\sigma_1, \ldots, \sigma_i, \ldots, \sigma_m$, related to the number of loads cycles: $n_1, \ldots, n_i, \ldots, n_m$ suffers destruction when:

$$\sum_{i=1}^{m} \frac{n_i}{N_i} = 1$$

where:

n_i – a number of loads' cycles at the data level of stresses.

N_i – a limiting number of loads' cycles.

Graphic interpretation of Miner's hypothesis is presented in Figure 4. Basing on this theory wanted thickness of the stabilised course we can calculate according to the scheme showed in Figure 5.

To solve questions connected with the pavement design according

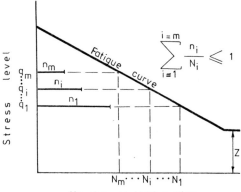

$$\sum_{i=1}^{i=m} \frac{n_i}{N_i} \leqslant 1$$

Fig.4. Graphic interpretation of Miner's hypothesis

to the proposed method, the knowledge of the following material's mechanical properties of the layer is necessary:
1. Elasticity modulus – E
2. Poisson's ratio –ν and
3. Tensile strength of the material in the conditions of load repetitions.

2.4 The mechanical properties of cement stabilised soils

Elasticity modulus and Poisson's ratio of cement stabilised soils we can fix (Koba, 1979) basing on the strains' analysis of:
1. A cylindrical specimen in compression
2. A bended beam.

Fixing of tensile strength requires information on the resistance of a material to carrying load repetitions thus fatigue strength. It predominates opinion that load repetitions cause at first formation of material's microcacks, which – as a number of load cycles increase – enlarge and then unite making destruction of the structure.

Material's fatigue occurs at considerably smaller stresses than static strength. And the fatigue scrap is the brittle scrap and is not preceded by perceptible plastic strains. Till now particularly a wide range of fatigue research has been realized in relation to typical structural materials – metals and their alloys. At present it is more and more

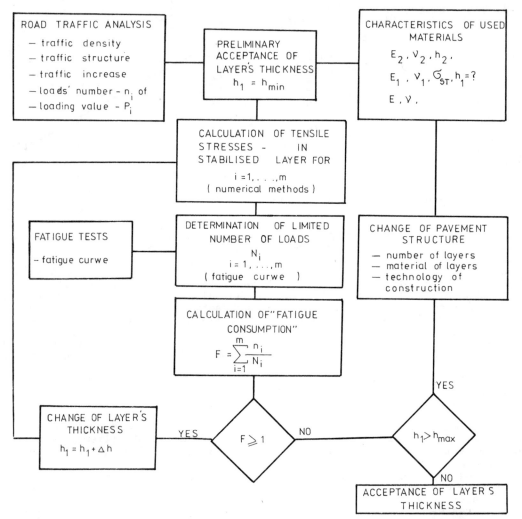

Fig.5. The calculation scheme of layer's thickness of cement stabilised soil

frequently paid attention to the phenomenon of material's fatigue in the pavements' structures, too. Fatigue effects are suspected mainly in bituminous mixes, cement concrete, as well as cement stabilised soils. Attempts on fatigue tests of cement stabilised soils have been carried out in the USA (Sargious, 1976) and German Democratic Republic (Rossberg, 1967). Published results of these works are fragmentary. They refer to individual kinds of soil. And it is difficult to put these results into practice. Taking into account existing in this range needs, the article represents a research method and results on fati-

gue tests of cement stabilised soils.

3 LABORATORY TESTS

In fatigue examinations of metals and cement concretes it is usually made use of a dynamic pulsator (e.g. Amsler's pulsator) or special strength machines. These plants as a general rule are characterized by large frequency of loading and they are rather destined for materials testing, which strength is considerably larger than strength of cement stabilised soils. These factors have decided on the construction's need of prototype machine for fatigue tests of cement stabilised soils.

3.1 Fatigue testing machine

The machine constructed at the Institute of Civil Engineering in
Wroclaw Technical University allows
simple and rapid determination of
fatigue strength of cement stabilised soils. Testing is carried out
on the specimens shaped like beams
of 5x5x30 cm. Loads are forced mechanically by rotary motion of the
transmission shaft with seated eccentric on it. The machine enables
testing of 12 specimens at the same
time. The statical scheme of testing
is represented in Figure 6. Each of
12 measuring positions are equiped
with a counter of loading cycles
switched off automaticaly as soon
as the beam breaks. They are equiped
with the checking strains devices,
too. Loads are forced with frequency of 0,55 Hz (33cycles per minute).
Time of loading can be regulated
from 0 to 1,84 sec. The course scheme of loading changes of the specimen in fatigue tests is shown in
Figure 7.

3.2 Results of carried out tests

Laboratory tests have generally been
limited to fixing of the determination methodology of fatigue strength
in relation to cement stabilised

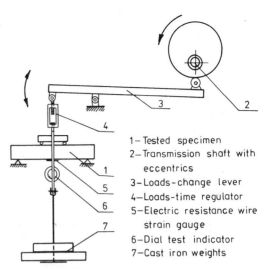

1— Tested specimen
2— Transmission shaft with
   eccentrics
3— Loads-change lever
4— Loads-time regulator
5— Electric resistance wire
   strain gauge
6— Dial test indicator
7— Cast iron weights

Fig.6.Static scheme of fatigue
tests

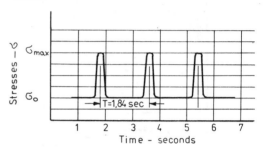

Fig.7.Course of loading changes of
specimen in fatigue tests

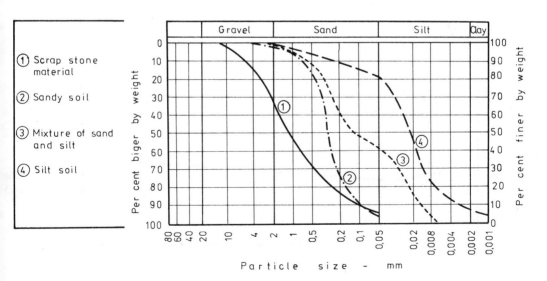

Fig.8 Grain-size distribution of soils

soils. For láck of explicity as to determination and influence of some factors on strength characteristics of cement stabilised soils the authors have been carried out series of distinctive tests. In the first place the tests have regarded a choice of the soil's kind and influence of specimens "ripening period" on strength characteristics.

The result has been that the tests have been carried out on scrap stone material, sandy soil, silt soil and sand and silt mixture. Granulation curves of tested soils are given in Figure 8.

The influence of cement setting period on strength characteristics of cement stabilised soils has been analysed basing on:
1. Compressive strength (axial)
2. Tensile strength at bending
3. Elasticity modulus
4. Poisson's ratio
Examplary change of mechanical properties for sandy soil with 10 per cent cement content is shown in Figure 9. Basing on the represented in Figure 9 tests' results it has been accepted a period of 28 days as an authoritative period of cement setting regarding specimens destined

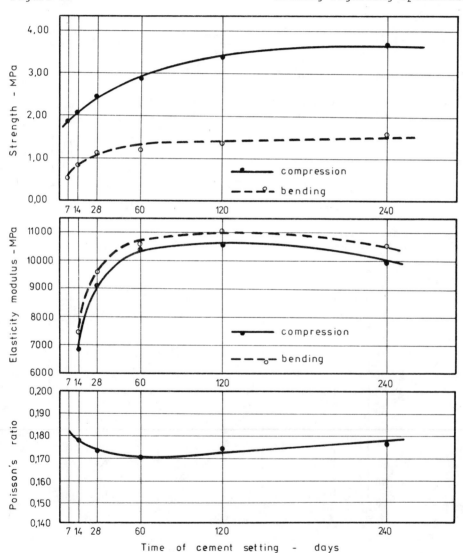

Fig.9. Change of mechanical properties of sandy soil with 10 per cent cement content

72

for fatigue tests. Each series of fatigue tests have included 55 specimens. Specimens have been loaded of: 0.90; 0.80; 0.70; 0.65; 0.60; 0.55 and 0.50 static value of destructive force - R_{STAT}. They have been loaded with frequency of 0.55 Hz. Time of loading - 0.2 sec. The testing result of each specimen has been a number of loads cycles to the moment of its breakage. Specimens, which have withstood a limited number of loading cycles (a measure base) - $N_G = 2 \cdot 10^6$, according to the fatigue test methodology have been treated as not destructive.

The test results have been elaborated basing on the mathematical statistics separately for limited and permanent fatigue strength. The examplary interpretation of the tests' results of sandy soil with 10 per cent cement content is represented in Figure 10.

In the range of limited fatigue strength obtained fests' results we can circumscribe by equation:

1. Sandy soil with:
   - 8 per cent cement content

$$logN = 10.7130 - 10.7816 \frac{\sigma}{\sigma_{ST}}$$

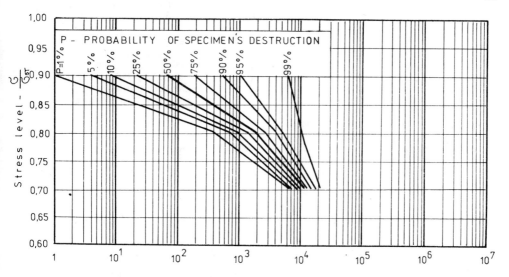

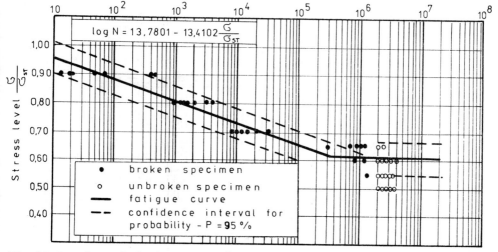

Fig.10 Results of fatigue tests' of sandy soil with 10 per cent cement content

73

- 10 per cent cement content:

$$\log N = 13.7801 - 13.4102 \frac{\sigma}{\sigma_{ST}}$$

- 12 per cent cement content:

$$\log N = 10.7439 - 9.4913 \frac{\sigma}{\sigma_{ST}} \ ,$$

2. Silt soil with 12 per cent cement content:

$$\log N = 13.0459 - 12.4533 \frac{\sigma}{\sigma_{ST}} .$$

3. Mixture of sand and silt (1:1) with 12 per cent cement content:

$$\log N = 16.0450 - 16.0179 \frac{\sigma}{\sigma_{ST}}$$

4. Scrap stone material with 4 per cent cement content:

$$\log N = 11.7774 - 11.423 \frac{\sigma}{\sigma_{ST}} .$$

The graphic illustration of fatigue curves (Wöhler's curve) for all tested soils is given in Figure 11. All tested materials prove the limiting resistance to load repetitions in the range of stresses: σ 0.35σ_{ST}. In the range of fatigue strength obtained results have approximate character. It results from accepted "measure base" of $N_G = 2 \cdot 10^6$ loading cycles. One should judge that for larger value of the limiting number of loads cycles, smaller value of destructive stresses should has been obtained. Great accordance in the course of fatigue curves for different kinds of soil allows to judge that it is possible to fix a typical fatigue curve independent of soil's kind and quantity of cement content.

In the course of fatigue tests, the influence of load repetitions on change of elasticity modulus of cement stabilised soils has also been analysed. The influence of these loads was analysed basing on the specimens strains as the number of loading cycles had been increased. The examplary graphic illustration of the testing results of sandy soil with 10 per cent cement content is given in Figure 12. Constant value of elastic strains, in the whole period of tests, points to lack of the influence of load repetitons on change of elasticity modulus of tested materials. Similar testing results have been obtained for the other kinds of soils.

4 CONCLUSIONS

Represented in short considerations and results of authors' research are an excerpt of works which they conduct on the development of pavement design method on the basis of mater-

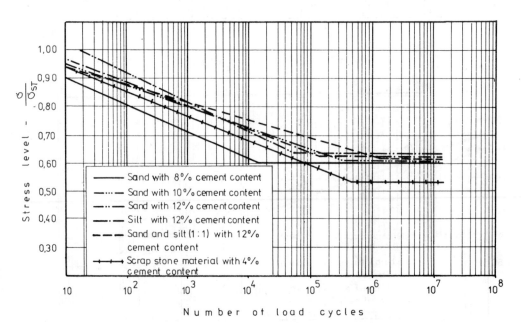

Fig.11 Fatigue curves of cement stabilised soils

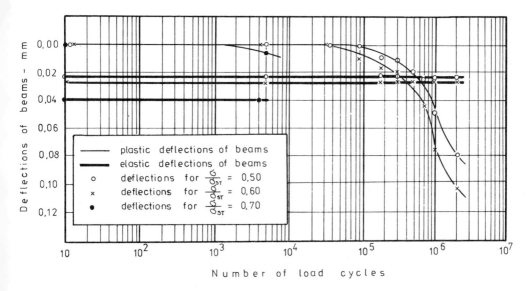

Fig.12 Changes course of specimens strains in conditions of load repetitions

ial fatigue strength. Obtained till now test results indicates application's possibility of fatigue strength of cement stabilised soils in pavement design to planned its life. More detailed recognition of these questions will facilitate to make more complete analysis of pavement work conditions on the basis of characteristics of cement modified soils taking into consideration natural subgrade.

5 REFERENCES

Borkowski, H. 1973, Teoretyczne modele konstrukcji nawierzchni drogowych, Drogownictwo. 3 i 4.

Hanuska, A. & Novotný, B. 1973, Použitie rovnic teórie pružnosti na posúdenie mnohovrstvových vozoviek, Inženýrske Stavby. 11: 502--505.

Keil, K. 1975, Stabilitat von Tragkorper und Decke in Strassenbau, Strasse, Brucke, Tunnel 2.

Koba, H. 1979, Cechy mechaniczne gruntów stabilizowanych cementem w warunkach obciążeń powtarzalnych, Wroclaw, Report of the Institute of Civil Engineering of Wroclaw Technical University - - 2/79 PRE 4/79, p. 55-70.

Kohler, G. 1974, Dickenbemessung-flexibler Fahrbahndecken auf Ermüdungsverhalten, Strassen und Tiefbau. 28: 13-21.

Rossberg, K.1967, Prüfung von zement-verfestigen Erdstoffen auf Dauer-biegezugfestigkeit, Die Strasse, 12: 541-546.

Sargious, M. 1976, Pavements and surfacings for higways and airports. London, Applied Science Publishers LTD.

Mechanical behaviour of clays under cyclic loading

Y. MEIMON
Institut Francais du Pétrole, Rueil-Malmaison, France

P. Y. HICHER
Ecole Centrale de Paris, Chatenay-Malabry, France

1 SUMMARY

A programme of triaxial tests has been conducted on clays to determine their mechanical behaviour under cyclic loading. In a typical loading test (stress deviator $q = q_m \pm q_c$, $q_{min} = q_m - q_c$, $q_{min} \geqslant 0$), for which frequency of .1 HZ corresponds to an average wave, q_m is applied during a given time under drained or undrained conditions before cyclic loading begins.

According to the clay (kaolinite - bentonite) and to its overconsolidation ratio, the tests have resulted in stabilization or failure. The fundamental role played by the mineralogy and more generally the structure of clays under cyclic loading is shown. The tests point out the influence of experimental parameters such as q_m, q_c, drainage and the effect of the number of cycles on the strain and pore pressure evolution. By storm loading tests different aspects of the strain hardening of the clays have been isolated : volumetric strain hardening due to the drainage and deviatoric strain hardening due to the maximum shear stress applied to the sample.

Also discussed in the paper are the bases of an experimental methodology useful for the determination of the properties of the cohesive soils under cyclic loading, for the calculations of marine foundations. Finally stress strain functions, extrapolating the clay behaviour from one to a given number of cycles are presented. These are useful for computations using the finite element method.

2 INTRODUCTION

To evaluate the safety against collapse of large marine foundations, which are submitted to cyclic loading, the prediction of the stress strain behaviour of the soil under non monotonic loading is necessary. Previous works show that stress strain elastoplastic laws describe quite well the behaviour of clays under monotonic loading. One can refer to the models initiated by the Cam-clay law (Schoefield and Wroth, 1968) or by the incremental law (Darve, 1974). A great advantage of these laws is to point out the dependence of the clay behaviour on the stress path loading and the stress history. When considering cyclic loading, progress have been accomplished with non associate plasticity models, anisotropic hardening stress strain laws (Mroz and all 1979), incremental laws (Darve and all 1979). However, the knowledge seems to be less advanced to take into account the strain hardening parameters or the effect of the viscosity.

Moreover, presently a computation by the finite element method (F.E.M.) cycle by cycle and using a fundamental law is unrealistic due to its high cost. A current approach is to reproduce in laboratory the same stress paths as those created by the foundation in the subsoil and to obtain stress strain functions extrapolating the behaviour of the soil from one to a given number of cycles (Andersen, 1976 ; Bonin and all 1976) Often, the definition of these stress paths is difficult or cover many cyclic loading types (piles, shallow foundations). According to the type of cyclic loading the strain response can be completely different (Hicher, 1979). Also, parameters such as drainage conditions overconsolidation ratio and clay mineralogy may influence greatly the stress strain evolution during the cyclic loading. The reduction of the undrained shear properties of clays after cyclic loading is well established.

The study presented below concerns one way cyclic loading triaxial tests carried out at the Soil Mechanics Laboratory of the

Ecole Centrale de Paris for a research programme conducted by the Association de Recherches en Mécanique des Sols Marins. Two clays were tested, each of them for two overconsolidation ratios (OCR. 1 and OCR.4). This paper is a synthesis of the results obtained in this experimentation. The bases of an experimental methodology useful for the determination of the properties of the cohesive soils under cyclic loading are discussed. Some stress strain extrapolation functions are also presented.

3 CHARACTERISTICS OF THE TRIAXIAL TEST EXPERIMENTATION

3.1 Usual clay properties

The two clays are a bentonite (smectite vermiculite) and a kaolinite. They were prepared in laboratory from a liquid mixture, first consolidated to 100 kPa in a wide consolidometer during three weeks. The normally consolidated samples were consolidated in the triaxial cell to 200 kPa following the classic method to have a good saturation. The overconsolidated samples were first consolidated in the cell to 800 kPa ; then, the confining pressure was lowered to 200 kPa (drainage being allowed) to obtain an O.C.R. of 4. All the samples were 35 mm in diameter and 70-75mm long. As the final confining effective pressure was equal for all the tests, a direct comparison between normally and overconsolidated soils is possible. In these conditions, the usual clay properties are summarized in table 1

Table 1. Usual properties of the clays

	Kaolinite	Bentonite
Percentage $<2\mu$	64	69
Kaolinite (mineral)	76	0
Illite	24	0
Smectite	0	100
W_L %	70	105
W_p %	40	51
$\emptyset$ (degrees)	25	22
M	1	.87
W (OCR1) %	45	65
N (OCR4) %	39	55
q_f (OCR1) kPa	60	68
q_f (OCR4) kPa	125	135

3.2 Test apparatus and procedures

In a typical one way cyclic loading test stresses are controlled. The triaxial M.T.S. press is programmed by a function generator.

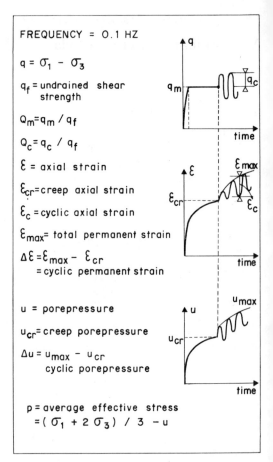

FREQUENCY = 0.1 HZ

$q = \sigma_1 - \sigma_3$

q_f = undrained shear strength

$Q_m = q_m / q_f$

$Q_c = q_c / q_f$

ε = axial strain

ε_{cr} = creep axial strain

ε_c = cyclic axial strain

ε_{max} = total permanent strain

$\Delta\varepsilon = \varepsilon_{max} - \varepsilon_{cr}$ = cyclic permanent strain

u = porepressure

u_{cr} = creep porepressure

$\Delta u = u_{max} - u_{cr}$ cyclic porepressure

p = average effective stress = $(\sigma_1 + 2\sigma_3) / 3 - u$

Fig.1. Notations and test procedure.

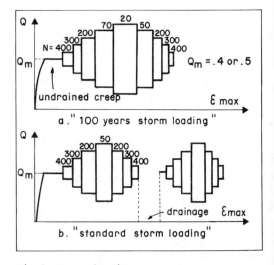

a." 100 years storm loading"

b. "standard storm loading"

Fig.2. Storm loading test programmes.

78

The chosen signal is sinusoïdal with a frequency of ·1 Hz corresponding to an average wave. The axial displacement and the pore pressure are continuously plotted during the test. The test is composed of two stages :

i) First, a stress deviator qm is quickly applied and at a given time the strains are practically stabilized. This is the creep sequence of the test which may represent the setting of a structure on the soil. The creep is undrained to correspond to imme- diate effect of the wave on the foundation. It is drained if the wave effect is delayed. By convention, undrained creep time is fixed to one hour. For drained creep, 24h to 48h may elapse until the pore pressure dissipates.

ii) The second stage is the undrained cy- clic loading : a cyclic stress deviator of amplitude q_c is applied around q_m during 1500 to 8000 cycles. The amplitude q_c is constant in a standard test but it varies in a storm loading test. Figures 1 and 2 precises all the notations and the test procedures. A storm loading test is general- ly composed of two identical loading programmes separated by a drainage sequence.

In a test the data parameters are : q_m (or Q_m), q_c (or Q_c), the number of cycles N, the drainage conditions, the clay type. The measured parameters are : pore pressure (u_{max} or Δu), permanent axial deformation (ε_{max} or $\Delta\varepsilon$), cyclic axial deformation (ε_c), volumetric deformation (ε_v).

3.3 Triaxial test programme

Most of the 75 tests were undrained creep cyclic loading tests. They were generally stopped at 2000 cycles. A typical range of values of Q_m was : .4, .5, .6. In the constant amplitude cyclic tests, Q_c took three or four values from .05 to .3. In fact, some tests were doubled to analyse the scatter of the results. The drained creep cyclic tests were carried out to characte- rize the effect of the modification in testing procedure before the beginning of the cyclic loading. An equal number of tests was performed on each clay.

4 EFFECT OF THE STRESS PATH FOLLOWED BEFORE THE CYCLIC LOADING.

4.1. Effect of the undrained creep

The main effect of the undrained creep stage is a degradation of the initial structure greater than in a standard monotonic loading triaxial test. This is shown in fig. 3 where the axial strain obtained at the end of the creep is plotted with that of a

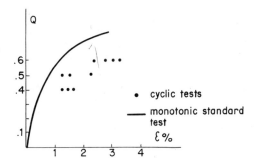

Fig.3. Axial strain at the end of the creep sequence. OCR 4 kaolinite.

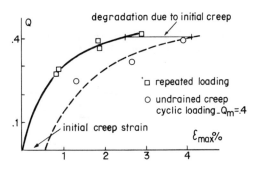

Fig.4. Degradation due to initial undrained creep. OCR 1 Bentonite.

standard compression test. This degrada- tion is amplified when the cyclic loading is applied : at stabilization, the perma- nent axial strain of an undrained creep cyclic loading test is greater than that of a similar repeated loading test ($Q_m=Q_c$, no creep sequence) (fig. 4). Though the undrained creep disturbs the initial structure of the clay, the scatter of the results after the cyclic loading remains acceptable. This fact is probably due to the choice of a conventional creep time at the end of which strains are nearly stabilized. However, it can be useful to analyse only the evolution of the cyclic part of the axial strain $\Delta\varepsilon$ in place of ε_{max}.

In conclusion, when considering that the soil under a foundation is partially drained, the cyclic loading test procedure including an undrained creep stage cer- tainly underestimates the strength proper- ties of clays.

4.2 Effect of the drained creep sequence

When drainage is allowed during the creep, the pore pressure due to the application of Q_m dissipates. However, the stress path followed is different from that of an anisotropic consolidation (fig. 5). The main difference is that the q/p ratio, characteristic of the shear level applied to the sample vary during the creep first exceeding the constant shear level of the similar anisotropic consolidation, then decreasing to it. Therefore, there is no evolution of the permanent strain and pore pressure when the q/p ratio during the cyclic loading remains lower than the maximum reached during the creep sequence (fig. 6 , Table 2) So, in addition to the volumetric strain hardening due to the drainage (which is reliable to the volumetric strain ε_v), the drained creep induces a deviatoric strain hardening.

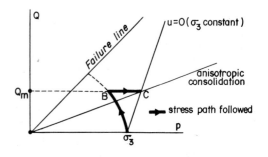

Fig.5. Stress path followed during drained creep.

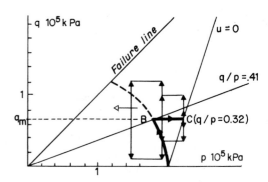

Fig.6. Stress paths for different drained creep cyclic tests on OCR 1. kaolinite $Q_m=.6$

Table 2. q/p ratio evolution during drained creep cyclic tests on OCR1 kaolinite. $Q_m=.6$

Q_c	$\Delta\varepsilon$	q/p ratio	
%		First cycle	Last cycle
6.7	.04	.36	.36
23.3	1.13	.47	.53
40.0	3.00	.63	.83

The comparison between the results of undrained creep cyclic tests and drained creep cyclic tests leads to the same conclusion : the increase of the shear strength q_f due to consolidation during the drained creep cannot completely explain the stronger behaviour of the clay (see fig. 7 where Q_m and Q_c are reduced for drained creep cyclic tests to take in account the increase of q_f).

4.3 Consequences for the testing procedure

The preceding results point out the importance of the stress path followed before the cyclic loading begins. As the initial stress state in a natural soil is anisotropic (K_o state) it seems necessary to perform an anisotropic consolidation before applying cyclic loads to the sample. It is certain that the quantitative results will be modified by this new procedure. This fact is already well known for monotonic loading of soft clays (Bjerrum 1973, Meimon 1975).

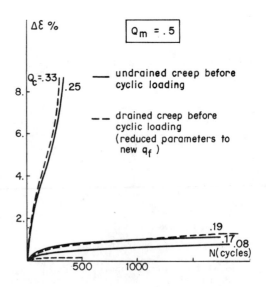

Fig.7. Comparison between undrained creep and drained creep cyclic tests. OCR 1. kaolinite

5 PHENOMENOLOGY OF CYCLIC LOADING

5.1 Stabilization criterion

In the experimentation, it was found that a cyclic test resulted in the failure of the sample or the stabilization of the strains or the continuous evolution of the strain without failure. The pore pressure and the axial strain did not stabilize at the same time (fig. 8). Generally, stabilization occured later for the overconsolidated samples than for the normally consolidated samples (fig. 9). The prolongation to 8 000 cycles of the cyclic loading on samples already stabilized at 1 000 cycles did not modify the strains and the effective stresses.

Hence it is thought that :

i) a good criterion of stabilization is to wait the stabilization of both the pore pressure and axial strain.

ii) there are only two possible issues to a cyclic test : stabilization or failure

iii) a time test of 1 500-2 000 cycles is sufficient for normally consolidated materials but it must be prolonged for overconsolidated materials to about 5 000 cycles.

5.2 Failure mode in the constant amplitude tests

The effective stress paths from the first cycle to the end of the undrained creep cyclic loading tests are plotted on figure 10 for both the normally consolidated and the overconsolidated (OCR 4) kaolinite. It can be seen that :

i) all the failure points are located very near the classic failure line q = M p.

ii) for the normally consolidated samples, failure occurs by reduction of the effective stresses.

iii) for the oversonsolidated samples, there is practically no evolution of the pore pressure during cyclic loading even at failure.

These results are independent of the drainage condition during the creep sequence Hence, when O.C.R. is 4, the failure seems to be solely the result of progressive destruction of the clay interparticle contacts without any change in effective stresses.

Now extrapolating to in situ behaviour, if two identical cyclic wave loadings are applied separated by a calm period some predictions can be made :

i) for an OCR 1 clayey sediment, there will be drainage and pore pressure dissi-

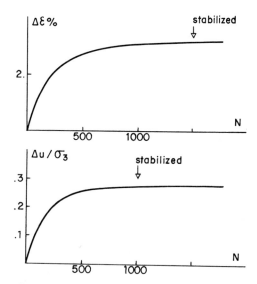

Fig.8. Stabilization for cyclic strain and pore pressure. OCR 1 bentonite. $Q_m=.4$

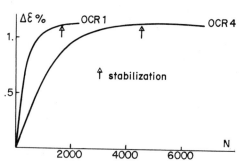

Fig.9. Stabilization times of OCR 1 and OCR 4 materials. (kaolinite).

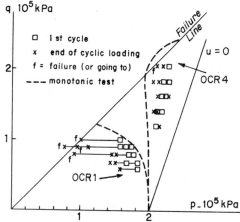

Fig.10. Stress path during the cyclic loading in undrained creep cyclic tests. OCR 1 and OCR 4 kaolinite.

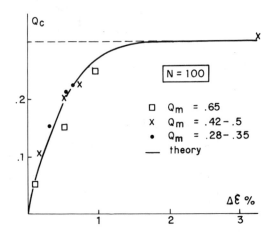 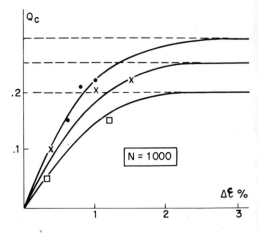

Fig.11.(Q_c,$\Delta\varepsilon$) relationships (OCR 4 bentonite) for undrained creep cyclic tests.

pation leading to an increase of the clay skeleton resistance against the second wave, but also to volumetric strain and settlement during the consolidation.

ii) for an OCR 4 sediment, the permanent strains due to each wave will be nearly equal

Return now to figure 7 : for each value of Q_m, it exists a limit amplitude Q_{cl} from which failure is obtained. Q_{cl} depends also on the followed procedure (drained or undrained creep) and on the overconsolidation ratio. Some estimations of Q_{cl} are summarized in table 3. The accuracy of the Q_{cl} determination is directly function of the number of cyclic tests : generally, at least 4 on 5 tests are necessary to have a good precision. Q_{cl} is independent of the number of applied cycles and can easily be used in preliminary studies.

Table 3. Estimation of Q_{cl} for undrained creep cyclic tests on OCR 1 kaolinite

Q_m	$< Q_{cl} <$	
.4	.18,	.27
.5	.17,	.26
.6	.15,	.23

5.3 Influence of Q_m and Q_c in the constant amplitude tests

In fig. 10, a plotted point corresponds to a value of $Q_m + Q_c$. A cyclic loading test at a given value $Q_m + Q_c$ can result in stabilization or failure : failure is obtained for the upper value of Q_c, stabilization

for the lower. Hence, the amplitude Q_c (and not $Q_m + Q_c$) is a fundamental parameter governing the stress strain behaviour during the cyclic loading. The relationship between Q_c and $\Delta\varepsilon$ at different values of Q_m and N is shown in fig. 11 for the OCR4 bentonite :

i) At a constant number of cycles N, the curves are graded following the increase in Q_m

ii) When N is small it is difficult to differenciate these curves

iii) All the curves present a limit level Q_{cl} at which the permanent strain increases till the failure. This time, Q_{cl} depends on the number of cycles N.

A typical relationship between Q_c and the cyclic strain ε_c is plotted on fig. 12. It seems to be independent of Q_m that is pro-

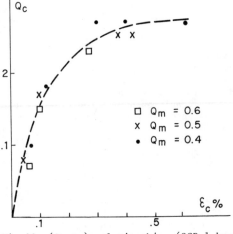

Fig.12. (Q_c,ε_c) relationship (OCR 1 kaolinite) for undrained creep cyclic tests.

bably due to the small variation of this parameter in the tests.
The cyclic strain modulus E_c ($E_c = 2 Q_c/\varepsilon_c$) decreases when the number of cycles increases except for the low Q_c - levels. (fig. 13). For the normally consolidated materials, the relationship between Q_c and $\Delta u/\sigma_{3c}$ has the same shape as the (Q_c, $\Delta\varepsilon$) curves leading to the same definition of Q_{c1} (fig.14).

For all these relationships, a mathematical formulation is possible as it is shown in section 7

5.4 Strain hardening

The storm loading tests allow to precise the role played by the strain hardening which was pointed out in section 4. A typical test can be divided in four parts (fig.15).
 (a) the loading by increasing Q_c amplitudes steps up to the maximum step : the permanent strains reached at the end of each step are practically the same as those in the similar constant amplitude tests at the same number of cycles. Therefore the virgin material does not remember the cyclic loading when Q_c is increasing.
 (b) the unloading by decreasing Q_c steps. The influence of the maximum shear level applied to the sample appears clearly in fig. 16 : the highest shear level corresponds to the lowest permanent strains at the inferior steps. Moreover the cyclic strain ε_c is higher in an unloading step than in a loading step (Table 4). This is the effect of the deviatoric strain hardening which is characterized here by the reduction of the permanent axial strain but also the reduction of the cyclic strain modulus E_c when the sample is unloaded .
 (c) the drainage time after the complete unloading of the sample to the confining pressure (200 kPa). During this sequence the permanent pore pressure dissipates (if it exists) and the sample consolidates. It is the well known effect of the volumetric strain hardening.
 (d) the second loading programme which is identical with the first. If volumetric strain hardening had taken place, the strains are considerably reduced (in a ratio of 10 in fig. 16). This is almost exclusively due to the increase of q_f and in fact, the applied loads are lower in relative value than in the first loading programme.
 In the case of the overconsolidated materials, the pore pressure does not vary during the cyclic loading (as in the constant amplitude tests) : there is no consolidation during the stage (c). It is shown in fig. 17 that :

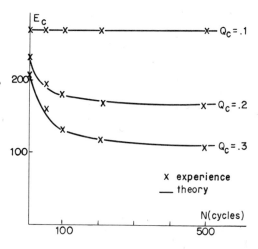

Fig.13. Cyclic strain modulus evolution. Undrained creep cyclic test on OCR 1 Bentonite.

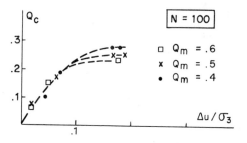

Fig.14. (Q_c,$\Delta u/\sigma_3$) relationship for undrained creep cyclic tests.(OCR 1 kaolinite)

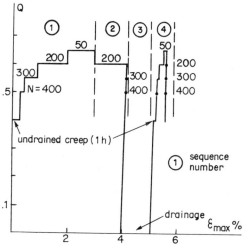

Fig.15. Typical storm loading test. OCR 1 kaolinite.

83

i)the strains obtained at the two maxima steps of cyclic loading are nearly the same and the failure of the sample can occur.

ii) the strains obtained at the inferior steps during the 2nd programme are lower than those of the similar steps in the first programme.

This is the result of a deviatoric strain hardening. For this O.C.R.4. storms of equal intensity could cumulate similar strains even if they are separated by a calm period.

Table 4. Cyclic strain in a storm loading test on OCR1 kaolinite

Q_c	N	ε_c loading	ε_c unloading
.1	400	.05	.08
.15	300	.075	.14
.2	200	.13	.20
.25	50-70	.19	.27

Concerning the experimental methodology, the storm loading tests allow to qualify the capacity of resistance of the material subjected to variable amplitude cyclic loads. It seems that the effect of strain hardening can be quantified only by means of a fundamental stress strain law. However, storm loading tests, where the in situ boundary and loading conditions would be precisely reproduced, could give an estimation of safety regarding the determination of the cyclic clay properties in constant amplitude tests. Moreover, they would permit to verify the usual methods of strain calculations for variable cyclic stress amplitudes from the constant strain curves obtained in constant cyclic stress amplitude tests (Andersen, 1976)

5.5 Change in the shear behaviour of the clays after the cyclic loading.

After the cyclic loading, a standard compression triaxial test was performed on most of the samples. The reduction of the undrained shear strength q_f depends on the disturbance of the clay structure created by the cyclic loading. It can be measured in relation to ε_{max}, the permanent axial strain at the end of the cyclic test (fig. 18). It is found that :

i) the reduction is lower than 8% if ε_{max} does not exceed 5 %. On the contrary, it can reach up to 40 %.

ii) the reduction does not depend on the overconsolidation ratio

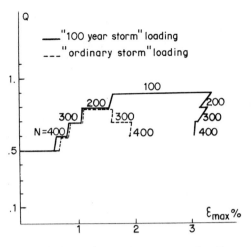

Fig.16. Comparison of two storm loading tests on OCR 1 Bentonite

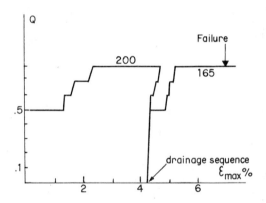

Fig.17. Storm loading test on OCR 4 Bentonite

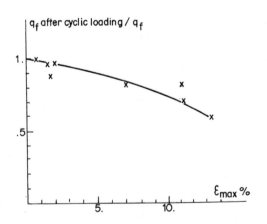

Fig.18. Reduction of shear strength corresponding to ε_{max}. OCR 1 - OCR 4 kaolinite.

84

The effective stress path followed after the cyclic loading corresponds to that of an overconsolidated sample for the OCR 1 clays while it does not vary for the OCR 4 clays (fig. 19).

6 INFLUENCE OF THE MINERALOGY

The physical chemical history of a clay depends on the mineralogy, the concentration of hydration cations, the geological history of the sediment. As the two clays were prepared in laboratory, it is only possible to analyse the effect of the mineralogy. The structure of the kaolinite is floculated : the particles are randomly distributed, separated at the contact points by a thin adsorbed water layer ; the particles arrangement can be easily modified The structure of the bentonite is oriented: it is more stable, the particles are parallel with a few contact points, the adsorbed water layers are thick. Bentonite particles are smaller than kaolinite particles.

Hence, the difference between the clays structures can explain differences of their behaviour during the cyclic loading tests :

i) permanent strains are greater for the kaolinite than for the bentonite in the constant amplitude tests. For example, a (Q_m=.5, Q_c=.25) test on kaolinite went to failure while a (Q_m=.5, Q_m=.29) test on kaolinite went to stabilization

ii) in the storm loading tests, when the sample is unloaded, the kaolinite continues to strain while the bentonite recovers a part of the strain (fig. 15-16)

iii) the bentonite strain stabilizes less rapidly during the creep step than the kaolinite strain.

7 COMPUTATION METHODS AND EXTRAPOLATION FUNCTIONS.

7.1 Brief review of computation methods.

Most of the computation methods use extrapolation functions to describe the behaviour of clays under cyclic loading. The well known finite element method (F.E.M.) must be used to calculate large marine foundations. (Zienkiewicz, 1979). Industrial applications have been performed with the F.E.M. on bidimensional models. The computation steps are : (Bonin and all 1976)

i) monotonic loading of the foundation subsoil corresponding to the setting of the structure.

ii) description of a part of the first cycle.

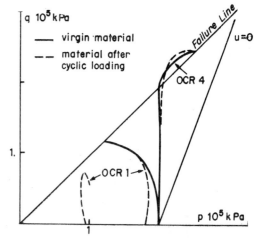

Fig.19. Comparison between the stress paths in monotonic triaxial tests before and after cyclic loading (kaolinite).

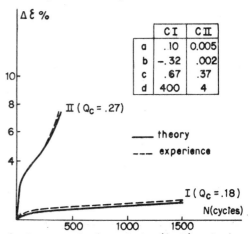

	CI	CII
a	.10	0.005
b	-.32	.002
c	.67	.37
d	400	4

Fig.20. Formulation of the ($\Delta\varepsilon$,N) relationship OCR 1 kaolinite

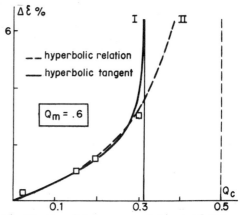

Fig.21. Possible representations of the ($\Delta\varepsilon$,Q_c) relationship OCR 1 Bentonite.

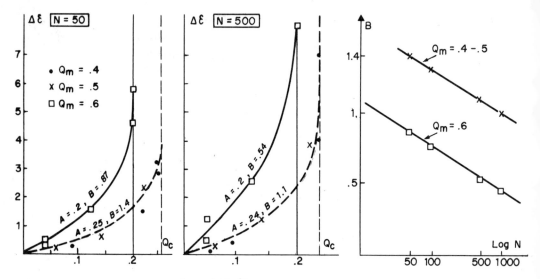

Fig.22. Representation of $\Delta\varepsilon = \frac{1}{B}\log\frac{1+Q_c/A}{1-Q_c/A}$ for OCR 4 kaolinite

iii) extrapolation to a given number of cycles of the permanent strain and corresponding reduction of mechanical properties.

iv) treatment of an another cyclic loading.

It is thought that to start correctly the extrapolation it is necessary to use a sophisticated law which would permit to compute at regular intervals a complete cycle using an effective stress analysis. The extrapolation functions have to cover all the stress paths encountered under the foundation : two way cyclic loading, one way cyclic loading

Some limitations of this analysis have to be pointed out :

i) strain hardening, which is a fundamental parameter when variable loading is applied, is not taken in account.

ii) during the cyclic loading, implicit hypothesis is made that in every point of the subsoil, the average stress tensor (Q_m in the above tests) remains constant which is probably inexact.

7.2 Extrapolation functions

In spite of the small number of tests for each type of clay, an attempt to obtain strain stress extrapolation functions has been made for the undrained creep constant amplitude cyclic tests. The cyclic part of axial strain $\Delta\varepsilon$ can be related to Q or to N:

(1) $\Delta\varepsilon = (|f(N)|_{Q_c}) Q_m$ or

(2) $\Delta\varepsilon = (|g(Q_c)|_{Q_m})N$

The first relationship can be represented by the function :

(3) $\Delta\varepsilon = aNe^{bN} + \frac{N}{cN+d}$ (see fig. 20)

a,b,c,d depend on Q_m and Q_c and it is not easy to calculate them. So the second relationship is preferred. An hyperbolic function leads generally to an unrealistic asymptote (fig. 21). A possible function is:

(4) $\Delta\varepsilon = \frac{1}{B}$ Log $\frac{1+Q_c/A}{1-Q_c/A}$

or (4') $Q_c = A\,\text{th}(\frac{B}{2}\Delta\varepsilon)$

A represents the limit cyclic amplitude Q_{c1} for a given value of N and Q_m

B is reliable to the deformability of the material and theoretically depends on N and Q_m Relations (4) or (4') are plotted respectively for the OCR4 kaolinite (fig. 22) and for the OCR4 bentonite (fig. 11). To use the relation (4) in a FEM analysis it is necessary to transform Q_m and Q_c into deviatoric stress invariants and to apply a strain tensor with a null volumetric strain condition. A similar relation can be used for extrapolating the pore pressure for the normally consolidated materials :

(5) $(\frac{\Delta u}{\sigma_{3c}}) = \frac{1}{B'}$ log $\frac{1+Q_c/A}{1-Q_c/A}$

Finally, the cyclic strain modulus E_c is well represented by an hyperbolic function (fig. 13) :

$E_c = E_\infty + \frac{C}{N+D}$ where E_∞, C and

D depend on Q_c.

8 CONCLUSIONS

The experimentation leads to the main following conclusions :

i) though the clays were prepared in laboratory, the qualitative results seem to be general.

ii) the stress path followed before the beginning of the cyclic loading governs the cyclic behaviour of the clays.

iii) samples have to be consolidated to Ko state to precisely correspond to in situ conditions.

iv) the mechanical history (i.e. overconsolidation ratio) of the clay influences the failure mode : reduction of effective stresses for OCR1, progressive destruction of particles contacts without variation of the pore pressure for OCR4

v) the strain hardening is a fundamental parameter to take in account when variable amplitude cyclic loading is applied. A stress strain law must be developed including both deviatoric strain hardening and volumetric strain hardening as independant parameters

vi) mineralogy can influence very greatly the quantitative results of cyclic loading tests.

vii) neglecting the strain hardening effect, functions extrapolating the cyclic behaviour of the materials can be determined using an unique relationship. It may be integrated in a finite element method calculation.

Further research is already undertaken in the way of numerical simulation of the clay behaviour under cyclic loading by means of a general stress strain law taking in account all the results of the experimentation and leading to general calculations of marine foundations.

AKNOWLEDGMENTS

We gratefully aknowledge the members of the Association de Recherche en Mécanique des Sols Marins (I.F.P., CNEXO, CFP TOTAL, SNEA(P), CFEM, CG DORIS, ETPM, SEA TANK CO BOUYGUES OFFSHORE) for granting permission to publish the paper.

REFERENCES

Andersen 1976, Behaviour of clay subjected to undrained cyclic loading, Proc. Conf. Behaviour of Offshore structures, 1 : 92-403, Trondheim.

Bjerrum, L. 1973, Problems of soil mechanics and construction on soft clays and structurally instable soils, Proc. 8e Int. Conf. Soils Mechanics and Found. Engineering, 3 : 111-159, Moscow, USSR.

Bonin, Deleuil and Zaleski-Zamenhof 1976, Foundation analysis of marine gravity structures submitted to cyclic loading, Offshore Technology Conf., OTC 2475, Dallas, Texas.

Darve 1974, Contribution à la détermination de la loi rhéologique incrementale des sols, These de Docteur Ingénieur, Grenoble.

Darve, Flavigny and Vuaillat 1979, Une loi rhéologique complète pour matériaux argileux, 7e Europ. Conf. Soils Mechanics and Found. Engineering, 1 : 119-125, Brighton.

Hicher 1979, Contribution à l'étude de la fatigue des argiles, These de Docteur Ingénieur, Ecole Centrale de Paris

Meimon 1975, Lois de comportement des sols mous et application au calcul d'ouvrages en terre, These de Docteur Ingénieur, Paris 6.

Mröz, Norris and Zienkiewicz 1979, Application of an anisotropic hardening model, Geotechnique, 29 (1) : 1-34.

Schoefield and Wroth 1968, Critical State soil mechanics, London, Mc Graw Hill

Voyiatzoglou, Propriétés mécaniques des argiles, These de Docteur Ingénieur (to be published) Ecole Centrale de Paris.

Zienkiewicz 1979, Constitutive laws and numerical analysis for soil under static, transient or cyclic loads, Proc. Conf. Behaviour of Offshore Structures, 1 : 391-406, London.

Simplified procedure to characterize permanent strain in sand subjected to cyclic loading

R. W. LENTZ
University of Missouri, Rolla, USA

G. Y. BALADI
Michigan State University, East Lansing, USA

1 ABSTRACT

The trend toward ever increasing axle loads on highway and airport pavements has led to the development of numerous domestic and international methods of pavement design and/or rehabilitation. All of these methods agree that, for flexible pavements, subgrades should not undergo significant volume change and/or permanent deformation under the application of traffic loads. Thus, parameters for characterizing permanent deformation of subgrade materials are required in each of these methods. Currently, most design agencies use empirical design nomographs coupled with empirical subgrade strength parameters based on individual experience and local environmental conditions. Further, the lack of standard laboratory tests to evaluate subgrade materials adds significantly to the variability between design outcomes when applying different methods to a single pavement.

This paper introduces a very simple and economical test procedure to evaluate parameters of cohesionless subgrade soils. By this procedure, permanent strain in sand subjected to cyclic loading can be characterized using stress and strain parameters from the universally accepted static triaxial test. To develop the procedure duplicate samples were tested using both static triaxial apparatus and a closed-loop electro-hydraulically actuated triaxial system. The dynamic test results were normalized with respect to parameters obtained from the corresponding static triaxial test. The normalized cyclic principal stress difference showed a unique relationship to the normalized accumulated permanent strain. This relationship was found to be independent of moisture content, density and confining pressure.

Benefits to be gained by use of such a simplified procedure include significant savings of laboratory time and energy as well as reduced equipment and personnel costs. Also, practicing engineers are more likely to accept the use of rational design methods if they have available a simple test procedure to characterize material behavior.

2 INTRODUCTION

The trend towards ever increasing axle loads on highway and airport pavements has revealed the inadequacy of the current empirical design methods for flexible pavements. Empirical methods are based on correlations of pavement performance with some empirical test, such as CBR or stabilometer which categorizes material strength, or use of limiting subgrade strain criteria from elastic layer theory (Yoder, et al. 1975). These methods lack the ability to predict the amount of deformation anticipated after a given number of load applications. When pavement loads exceed the range for which performance data is available empirical methods fail. Since soil is known to behave in a non-linear fashion, performance under higher axle loads cannot be extrapolated from performance at lower load levels.

Rational methods of pavement design have been proposed to overcome this deficiency. These are usually quasi-elastic (elastic theory to predict stresses coupled with permanent strains determined by repeated load laboratory tests) (Yoder, et al. 1975). Some methods also use viscoelastic theory coupled with laboratory testing (Kenis 1977), (Hufferd, et al. 1978). To be useful these methods must have the capability of predicting cumulative permanent deformations which will occur as a consequence of traffic loading. This requires the development of an adequate method for characterization of permanent strain (Pell, et al. 1972),

(Hufferd, et al. 1978). Further, these methods should be simple, economical and do not require new complicated expensive equipments and/or testing procedure. Such method is presented in this paper.

3 BACKGROUND INFORMATION

Parameters which affect the accumulation of permanent strain in cohesionless material have been reported to be number of load repetitions, stress history, confining pressure, stress level, and density (Yoder, et al. 1975), (Hufferd, et al 1978), (Brown 1974), (Morgan 1966), (Barksdale 1972), (Chou 1977), (Kalcheff, et al. 1973).

The effect of number of load repetitions on permanent strain has been reported by several investigators with some indicating that the relationship is a straight line on a semilogarithmic plot (Barksdale 1972) and others that it is a straight line on log-log plots (Yoder, et al. 1975). The effect of previous load applications has been found to cause a significant reduction in the amount of permanent strain experienced under subsequent loading (Kalcheff, et al. 1973). It has been reported that for a given cyclic principal stress difference, increasing the confining pressure results in a decrease in permanent strain (Brown 1974), (Morgan 1966), (Barksdale 1972). For a constant confining pressure the permanent strain after any number of load cycles has been found to depend directly on the magnitude of the principal stress difference (Morgan 1966), (Barksdale 1972). Cyclic stress versus permanent strain curves have been shown to be analogous to static stress-strain curves (Barksdale 1972) and describable using hyperbolic functions developed for static test results (Duncan, et al. 1970), (Kondner, et al. 1963). Reduction in density has been shown to cause an increase in permanent strain accumulation (Barksdale 1972), (Kalcheff 1976).

4 TESTING PROCEDURE AND EQUIPMENT

The material used in the testing program was a uniform, medium sand typical of the north half of the State of Michigan. For verification purposes a few tests were also conducted on samples of fine stamp sand (crushed rock from a stamp-mill). Grain size distribution curves for both materials are shown in Figure 1. For more information, the reader is referred to (Lentz 1979).

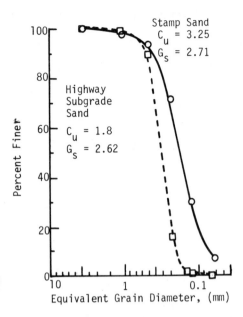

FIGURE 1   GRAIN SIZE DISTRIBUTION CURVES FOR HIGHWAY SUBGRADE AND STAMP SANDS.

The research program involved running drained cyclic triaxial tests on 5.08 cm. (2 inch) diameter x 13.716 cm. (5.4 inch) long samples compacted moist. Identical samples were tested under drained static triaxial conditions to obtain stress-strain curves to use for normalizing the dynamic test results. For both static and dynamic tests, loads were measured using a load cell mounted directly beneath the sample base. Deformation was measured using a linear variable differential transformer (LVDT) mounted across the length of the sample. The loading system consisted of a closed-loop electrohydraulic actuator operated in the load controlled mode. All cyclic triaxial tests were conducted to at least 10,000 cycles using a sinusoidal wave form with a frequency of one hertz. Three levels of confining pressure (σ_3) and two levels of density were used. For each combination of these variables, several levels of cyclic principal stress difference, σ_d, were used. Since stress history has a large influence on permanent strain, each combination of variables required a new sample.

The static triaxial tests were performed using the same triaxial cell used for dynamic tests. Loads were applied gradually in increments using the electrohydraulic actuator. Increments of load were approximately ten percent of the estimated sample strength as suggested by Bishop and Henkel (Bishop, et al. 1962). The size of the load increment was reduced as the failure stress was approached to allow for a reliable determination of strength. Each load increment was maintained until the rate of strain had become very small before the deformation reading was recorded. This procedure was expected to produce the same stress-strain curve as would conventional constant strain rate triaxial equipment.

5 TEST RESULTS

Typical results for a series of cyclic triaxial tests are shown in Figure 2. The samples were compacted moist to 99 percent of the maximum dry density determined by AASHTO method T-180 and tested at a confining pressure, σ_3, of 34.47 kPa (5 psi). Similar plots were obtained for samples tested at 172.37 kPa (25 psi) and 344.74 kPa (50 psi) confining pressures.

The change in permanent strain is large during the first few cycles and then gradually decreases after a large number of load repetitions. Thus, the data can conveniently be presented on a plot of permanent strain versus logarithm of number of load cycles. It can be seen from Figure 2 that the data could be approximated by a straight line. Least squares technique was used to pass a best fit straight line through each set of data. The equation of the line is of the form

$$\varepsilon_p = a + b \ln N$$

where ε = accumulated permanent strain
N = number of load repetitions
a,b = regression constants from lease squares best fit.

The constant a represents the permanent strain occurring during the first cycle of load and the constant b represents the rate of change in permanent strain with increasing number of load repetitions.
Typical results of a static triaxial test is presented in Figure 3. It should be noted that the sample was a duplicate to that tested under dynamic conditions.

6 DISCUSSION

The results of the cyclic tests can be presented in the form of cyclic stress (σ_d)

versus permanent strain (ε_p) plots at any given number of load repetitions (N). This has been done for three different confining pressures at N = 10,000 cycles and is shown in Figure 4. From this figure the strong effect of confining pressure (σ_3) is obvious. Since the strength of granular material is so dependent on confining pressure it seems reasonable to study the effect of σ_3 and/or σ_d as a percentage of the static strength. To accomplish this the value of σ_d for each cyclic test was normalized by dividing it by the peak strength, S_d, of an identical sample tested at the same confining pressure under static triaxial conditions. The data shown in Figure 4 has been normalized and replotted in Figure 5. It can be seen that this normalizing procedure draws the curves closer together and reduces, but does not eliminate, the total effect of confining pressure. At this point it was thought that normalizing the permanent strain by some reference strain obtained in the static triaxial test could eliminate the effect of confining pressure. The criteria for selecting this normalizing strain value is that: a) it should contain the plastic deformation characteristics of the sand under the given test conditions and b) must be a well defined value that is reproducible by different operators. Based on these criteria the static strain at 95% of peak strength ($\varepsilon_{.95 S_d}$) was selected as the normalizing value. At this load a large amount of the total strain is permanent, thus representing the plastic characteristics of the material. However, the curve is still rising steeply enough so that the strain value is well defined. The determination of this normalizing strain, $\varepsilon_{0.95 S_d}$, can be seen by the dashed lines in Figure 3. Each combination of confining pressure, moisture, and sample density requires a separate static stress-strain curve to obtain normalizing parameters, S_d, and $\varepsilon_{.95 S_d}$.

When the cyclic permanent strains in Figure 5 were normalized by dividing by the appropriate static reference strain, $\varepsilon_{.95 S_d}$, the curves collapse together to produce one single curve as shown in Figure 6. Also shown in Figure 6 are additional normalized results for samples at a lower density. Note that the points plotted in the figure represent samples tested at three different confining pressures and two densities, yet the data can be reasonably represented by a single curve. The significance of this is that with this curve and only the results of a static triaxial stress-strain test the permanent strain after 10,000 cycles at any level of cyclic principal stress difference can be predicted.

91

Since hyperbolic functions have been shown by others (Barksdale 1972), (Monismith, et al. 1975) to apply to cyclic stress-permanent strain curves, it is appropriate to apply the same type function to the normalized data in Figure 6. Least squares procedure was used to obtain the best fit hyperbolic curve shown in the figure.

To verify that this curve applies to other material in addition to the subgrade sand used in the testing program, several additional tests were performed on a second material. This material was a crushed stamp sand having a finer gradation than the subgrade sand as shown in Figure 1. The stamp sand also had a different mineralogical composition and much more angular particle shape. Due to the particle shape the stamp sand compacted to a much lower density than the subgrade sand when compacted with the same effort. Cyclic and static triaxial tests were performed on samples of stamp sand at confining pressures of 34.47 kPa (5 psi) and 172.37 kPa (25 psi). When the normalizing procedure described above was applied to the stamp sand, the normalized results plot right among the results of the subgrade sand. This indicates the procedure may be applicable to a range of cohesionless materials.

7 BENEFITS TO PRACTICING ENGINEERS

In practice, using the material characterization procedure presented will result in significant saving of laboratory time and will obviate the need for expensive testing equipment. Also, rational pavement design methods which require characterization of permanent strain behavior will be more likely to gain quick acceptance by practicing engineers if they have available a simple test method.

Work is continuing on the development of a general constitutive equation which will require only the stress-strain results from static triaxial tests in order to predict accumulated permanent strain after any number of load cycles. Also, applicability to a wider range of subgrade soils, including cohesive, is being tested.

8 CONCLUSION

This paper has presented a simple procedure for characterizing the permanent strain behavior of cohesionless subgrade material by using stress-strain curves obtained from static triaxial tests. More research is needed to extend the procedure to a wider variety of subgrade materials and to develop a general constitutive equation for predict-ing permanent strain.

The adoption of this procedure in practice will save much laboratory time and money in meeting material characterization needs.

9 REFERENCES

Barksdale, R.D. 1972, "Laboratory Evaluation of Rutting in Base Course Materials," Proc. of the 3rd Int'l Conf. - Structural Design of Asphalt Pavements, London, England, 161-174.

Bishop, A.W. & D.J. Henkel 1962, THE MEASUREMENT OF SOIL PROPERTIES IN THE TRI-AXIAL TEST, 2nd Edition, Edward Arnold (Publishers) Ltd., London.

Brown, S.F. 1974, "Repeated Load Testing of a Granular Material," Journal of the Geotechnical Engineering Division, ASCE, Vol 100, No. GT7, Proc. Paper 10684, 825-841.

Chou, Y.T. 1977, "Engineering Behavior of Pavement Materials: State of the Art," U. S. Army Engineers Waterways Experiment Station, CE, Vicksburg, Miss., Technical Report S-77-9.

Duncan, J.M. & C.Y. Chan 1970, "Nonlinear Analysis of Stress and Strain in Soils," Journal of Soil Mechanics and Foundations Division, ASCE, Vol. 96 SM5, 1629-1653.

Hufferd, W.L. & J.S. Lai 1978, "Analysis of N-layered Ciscoelastic Pavement Systems," Final Report to Federal Highway Administration, Report No. FHWA-RD-78-22, 220.

Kalcheff, I.V. & R. G. Hicks 1973, "A Test Procedure for Determining the Resilient Properties of Granular Materials," Journal of Testing and Evaluation, JTEVA, Vol. 1, No. 6, 472-479.

Kalcheff, I.V. 1976, "Characteristics of Graded Aggregate as Related to Their Behavior Under Varying Loads and Environments," Presented at Conference of Graded Aggregate Base Materials in Flexible Pavements, Oak Brook, Ill. Sponsored by National Crushed Stone Association, Wash., D.C.

Kenis, W.J. 1977, "Predictive Design Procedure - A Design Method for Flexible Pavements Using the VESYS Structural Subsystem," Proceedings of the 4th International Conference - Structural Design of Asphalt Pavements, Vo. I, Ann Arbor, Michigan, 101-130.

Kondner, R.L. & J.S. Zelasko 1963, "A Hyperbolic Stress-Strain Formulation for Sands," Proceedings, Second International Pan-American Conference of Soil Mechanics and Foundation Engineering, Brazil, Vol. I, 289-324.

Lentz, R.W. 1979, "Permanent Deformation of Cohesionless Subgrade Material Under

Cyclic Loading," Michigan State University, East Lansing, Michigan. Ph.D. Thesis.

Monismith, C.L., N. Ogawa & C.R. Freeme 1975, "Permanent Deformation Characteristics of Subgrade Soils Due to Repeated Loading," Transportation Research Record 537, 1-17.

Morgan, J.R. 1966, "The Response of Granular Materials to Repeated Loading," Proc. of the 3rd Conference of the Australian Road Research Board, Volume 3, Part 2, Sydney, Australia, 1178-1191

Pell, P.S. & S.F. Brown 1972, "The Characteristics of Materials for the Design of Flexible Pavement Structures," Proceedings of the 3rd International Conference - Structural Design of Asphalt Pavements, London, England, 326-342.

Yoder, E.J. & M.W. Witczak 1975, "Principles of Pavement Design," 2nd Edition, John Wiley and Sons, Inc., New York.

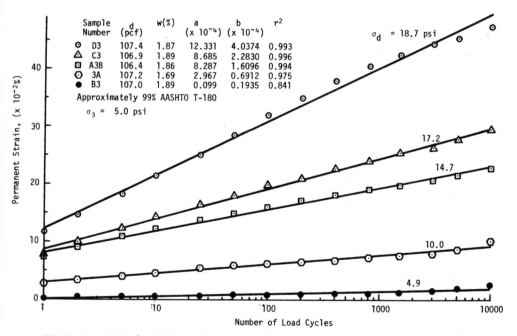

FIGURE 2 EFFECT OF CYCLIC PRINCIPAL STRESS DIFFERENCE AND NUMBER OF LOAD CYCLES ON PERMANENT STRAIN AT CONSTANT CONFINING PRESSURE FOR HIGHWAY SUBGRADE SAND MATERIALS.

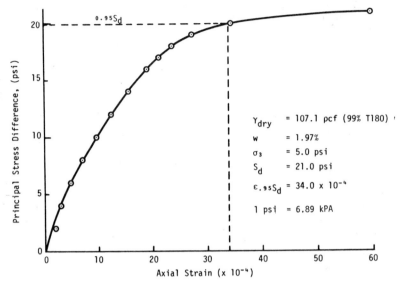

FIGURE   3   STATIC STRESS-STRAIN FOR HIGHWAY SUBGRADE SAND.

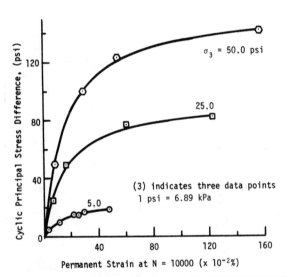

FIGURE   4   CYCLIC PRINCIPAL STRESS DIFFERENCE VERSUS
PERMANENT STRAIN AT 10,000 LOAD APPLICATIONS.

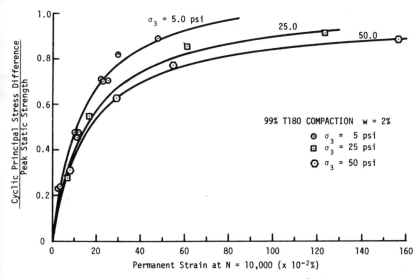

FIGURE   5   NORMALIZED CYCLIC PRINCIPAL STRESS DIFFERENCE VERSUS PERMANENT STRAIN.

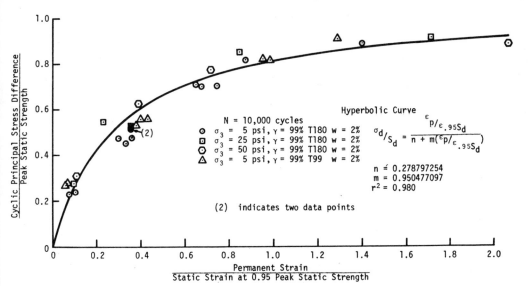

FIGURE   6   NORMALIZED CYCLIC PRINCIPAL STRESS DIFFERENCE VERSUS NORMALIZED PERMANENT STRAIN.

A test equipment for large scale vertical cyclic loading tests
on saturated sand

W.F. VAN IMPE
Ghent State University, Ghent, Belgium

1 INTRODUCTION AND SHORT REVIEW

1.1 The problem of liquefaction of saturated sands due to cyclic loading has lead
to widespread use of laboratory cyclic
loading tests to assess the liquefaction
characteristics of sands.

In these tests, watersaturated sand samples are subjected to cyclic loading mostly under undrained conditions. Thus laboratory tests of the basic principles generally include simulation of cyclic loading in
the field. Further due to the widespread
damage caused by the liquefaction (of
sand) during earthquakes, it is convenient
to proceed with laboratory tests by stimulating cyclic loading in the field during
such phenomena. From this point of view
the triaxial equipment has become most
commonly used and to a lesser degree, the
simple shear test on account of the increased difficulty to perform it. As shown in
table 1, under simulated earthquake conditions, the laboratory test equipment of
the shear or torsion type gives the most
appropriate results. Nonetheless due to
the ease of handling of the equipment the
most convenient test usually is the consolidated undrained cyclic triaxial test.

While executing cyclic and undrained
loading tests, the progressive increase in
pore-water pressure is observed until the
critical value is reached (when the pore-
water pressure becomes equal to the confining pressure in the cell). At that moment
(the so called initial liquefaction moment)
the effective confining stress is reduced
to zero and cyclic strain amplitudes may
become suddenly very large, depending on
the initial density of the sand sample.
Starting from a very loose saturated sand,
a liquefaction with unlimited liquefaction
potential will occur, while starting form
a dense sand, only cyclic strain deforma-

tions with a limited deformation potential
will be observed : "liquefaction" is self
neutralising in the case of dense packed
sands. Some authors use the expression
"cyclic mobility" for this type of "liquefaction".

1.2 The various factors influencing the
number of load cycles causing the onset of
the so called "initial liquefaction" have
been extensively described in literature.
The most reliable are :
1. parameters in connection with the
type of soil
- the mean grain size and grain size
distribution
- the shape of the sand particles
- the fabric of the soil skeleton
- the density.
2. parameters in connection with the
initial stress conditions of the soil
- consolidating stress on the sample
- type and magnitude of some possible
precompression.
3. parameters of the test equipment and
test procedure
- the type of test equipment itself
- maximum and minimum stress levels during the test
- duration of the cyclic loading on the
sample
- shape of the cyclic loading diagram
- dimensions of the sample
- measuring devices for pore pressure
measurement
- end bearing plates on the sample (lubricated or not)
- type and dimensions of the membrane
around the sample.
The implicit assumption in all liquefaction tests performed on saturated sands
under undrained condition is that no volume
changes can occur during the test. Only in

Table 1.

Apparatus used	Stress condition		Mohr – diagram	Principal stresses at failure	Remarks
	During consolidation	At failure			
Stresses in situ induced by earthquakes				$\sigma_1 = \frac{\lambda_0 + 1}{2}\,\sigma_v + \sqrt{\tau_{vh}^2 + (\frac{1-\lambda_0}{2})^2\,\sigma_v}$ $\sigma_2 = \lambda_0 \sigma_v$ $\sigma_3 = \frac{\lambda_0 - 1}{2}\,\sigma_v - \sqrt{\tau_{vh}^2 + (\frac{1-\lambda_0}{2})^2\,\sigma_v}$	
Conventional triaxial test equipment				$\begin{cases}\sigma_1 = \sigma_3 + \Delta\sigma \\ \sigma_2 = \sigma_3 \\ \sigma_3 = \sigma_3 \end{cases}$ $\begin{cases}\sigma_1 = \sigma_3 \\ \sigma_2 = \sigma_3 \\ \sigma_3 = \sigma_3 - \Delta\sigma \end{cases}$	Tests by Castro and by Seed and Lee
Conventional triaxial test equipment with constant medium stress				$\begin{cases}\sigma_1 = \sigma_3 - \Delta\sigma \\ \sigma_2 = \sigma_3 - \Delta\sigma \\ \sigma_3 = \sigma_3 - \Delta\sigma \end{cases}$ $\begin{cases}\sigma_1 = \sigma_3 + \Delta\sigma \\ \sigma_2 = \sigma_3 + \Delta\sigma \\ \sigma_3 = \sigma_3 - \Delta\sigma \end{cases}$	
Conventional simple shear test equipment				equal to the in situ-stresses during earthquakes	Value of $\lambda_0 \sigma_v$ on the A or B plane are unknown
Shear–torsion equipment				equal to the in situ-stresses during earthquakes	Value of $\lambda_0 \sigma_v$ on the A plane is unknown
Triaxial torsion apparatus				equal to the in situ-stresses during earthquakes	Shear stress and shear strain are not uniformely distributed

this case the measured pore-water pressure would be the correct one. However, every pore pressure measurement on a saturated soil is affected by the measuring system itself. If for example the volume of the system confining the sand sample, increases as the pore-water pressure rises, the apparent bulk modulus of the water phase in the sample is decreased and consequently the measured pore-water pressures become less than those that would have occurred with a confining system of zero compliance. In the case of a simple shear test, such compliance may occur due to changes in membrane thickness, additional membrane stretch in corners and slight expansion of the confining frame, and compliance in the pore pressure measurement system. In respect of the membrane penetration effects in triaxial tests, a lot of research has been done, all concluding that the major source of compliance, due to the variations in membrane penetration into the peripheral voids between the grains, with increasing or decreasing confining pressure, is very important (fig. 1). Studies made by (Newland & Allely 1959), (Frydman, Zeitlen & Alpan 1973) and (Raju & Sadasivan 1974) have shown that the reduction of the rate of change in the pore-water pressure due to the membrane penetration in undrained triaxial tests is very important. Thus the effective stress-path, the stress-strain relations and the undrained strength are influenced accordingly in the unconservative sense. Some authors have found some underestimations of the pore-pressure values by membrane penetration during undrained triaxial tests of 30 % tot 60 %. Such deviations of the pore-pressures developed actually are unacceptable. Analogous problems of unconservative deviations result from the end-bearing plates in the triaxial equipment (Vernese & Lee 1977).

Even when one should succeed in neutralizing all of these parameters, influencing the results of undrained triaxial tests in the unconservative sense, there still remains a very important inconvenience inherent to the normal triaxial equipment. Indeed, at the moment of liquefying of a loose saturated sand in reality, the structure of the flowing sand becomes different from the static one, the grains are constantly rotating in relation to the surrounding grains so as to offer minimum frictional resistance. Prof. Casagrande termed this structure of the sand : the "flow structure". As suggested by Prof. Casagrande, this flow structure can only develop in a dynamic loaded sand when, as

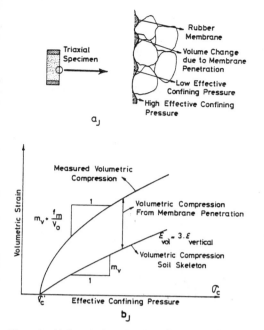

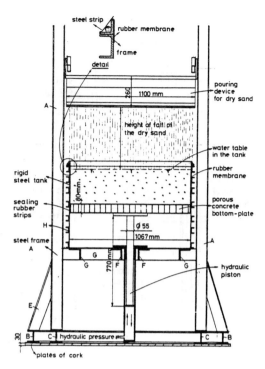

Fig. 1. Volumetric strain of triaxial sample as a function of the effective confining pressure (after Lade & Hernandez 1977).

Fig. 2. Testing tank with pouring device.

in nature, the driving force of the cyclic loading on the sand, being maintained constant during the flow. So the cyclic triaxial tests should be executed "load-controlled" at any moment. Even during the liquefaction phenomenon itself, when for example up to 25 % vertical strain of the sample in the triaxial cell occurs in less than 0,2 or 0,1 seconds, the vertical driving load should remain constant. If such driving load cannot follow up the progressing settlements, the applied cyclic stress suddenly can be reduced due to any quick settlement. In these very short time intervals of reduced driving forces, discharging the saturated sandmass necessarily leads to a change of structure, making the sand sample more resistant to further cyclic loading.

The normal triaxial equipment for cyclic loading of sand, simply is uncapable of keeping the driving forces constant during such a short time collapse of the sample. Due to (Casagrande 1976) and (Castro 1969) a special loading device for undrained cyclic triaxial tests was developed to overcome the above mentioned inconvenience of the normal triaxial equipment. The loading device consists of a loading yoke with a hanger for the weights with which

the load is applied. Counterweights to balance this dead weight of the yoke and hanger are applied through a pulley arrangement of a steel wire passing over a wheel.

Analogously, starting from the idea of a loading device with a dead weight to keep the vertical loads constant during liquefaction, the author has developed, at the laboratory of soil mechanics of Ghent University, equipment for the cyclic vertical loading of large scale samples without any membrane supply.

2. EQUIPMENT AND PREPARATION OF THE SAMPLES IN THE TEST TECHNIQUE USED BY THE AUTHOR

2.1 Sample preparation

In a rigid steel tank of 1,067 m by 1,067 m and 0,7 m depth, with stiffening steel profiles at the outside, a device consisting of a porous concrete bottom plate of 1,065 m by 1,065 m was made in such way that it could move inside the steel tank over a height of 0,50 m. The concrete plate is set in a steel frame to which the upward or downward force is transfered by a hydraulic piston control-

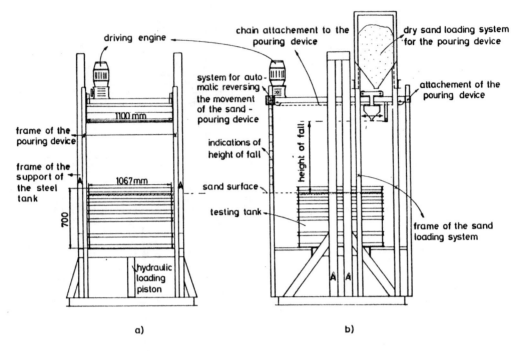

Fig. 3. Pouring the sand.

ling the movement (fig. 2).

The porous concrete constituents are : aggregate 4/8 : 16 kN/m^3 ; cement P40 : 3 kN/m^3 ; water 95 ℓ/m^3. A permeability test showed the concrete bottom plate to have a permeability of 5.10^{-5} m/sec.

At the four sides of the movable bottom plate and covering each of the four inside walls of the tank, special permeable tissues are stretched, in such way that they are able to follow the movements of the concrete bottom plate, in order to avoid any relative movement between the poured sand sample and the fixed stiff walls of the tank (fig. 2). The movable bottom plate is sealed up against the tank walls with a very flexible rubberstrip. Above the test tank a sand pouring device can be fastened at different heights. In this way the free fall height of the oven-dried sand may be changed, (fig. 3). The sand pouring system, designed as shown on fig. 3-4 is able to move back- and forwards over a distance of 1,45 m with a speed of 6×10^{-2} m/sec which may be assumed to be constant over the surface of the test tank, the turning point being situated sufficiently far out of the tank walls. At a given height of the sand pouring system above the tank, the height of fall of the sand is kept constant by means of

the movable concrete bottom plate. The changes of the relative density or the dry unit weight of the sand, with the height of fall is given on figures 7a and 7b and the entire characteristics of the Molsand are shown in figures 5 and 6. The relative density of the prepared sand sample of about 0,5 m^3 was calculated by weighing carefully at the end of the loading tests the poured sandmass (after drying it completely at 105°C) and by measuring with great accuracy the volume of this sandmass in the test tank.

Because of the purpose of this research work, it was necessary to saturate the big sand sample. Several procedures were investigated before a satisfactory degree of saturation (> 99 %) was reached. It was discovered very fast that no procedure could be invented to saturate fully starting from a dry sand sample 1,07 m x 1,07 m x 0,5 m. Even using CO_2-gaz replacing the air in the sandpores combined with depression systems, a saturation degree of at maximum 82 % was reached. Raining the sand in a watertable adjusted at the top of level of the tank resulted in a very small variation of the relative density of the sand regardless the height of fall.

The most appropriate procedure was found to be the following : the dry sand was pou-

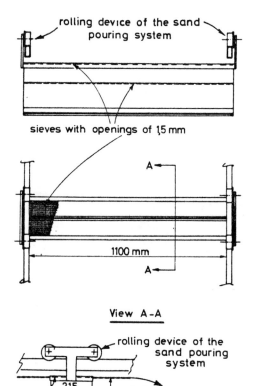

rolling device of the sand pouring system

sieves with openings of 1,5 mm

A◄—

1100 mm

A◄—

View A-A

rolling device of the sand pouring system

215

260

chain for transport of the sand pouring device

sieves

2

Fig. 4. Sand pouring system.

red in layers of 3 mm of thickness. After pouring each layer, the mechanism was stopped and water was filled up very slowly through the porous bottom plate. When the water table reached a height of about 10 mm above the new sand surface, the waterlevel was lowered again very slowly, exactly up to the level of the new sand surface. Subsequently, another layer of 3 mm of thickness was poured and again saturation was realised in the same way. This procedure was stopped when the bottom plate reached the tank bottom, so forming the saturated sand sample of 0,5 m of thickness. The variation of the relative density with the height of fall of the saturated sand sample prepared in the mentioned way, is shown on figure 7a. The degree of saturation, which is most determining in the research on liquefaction problems (Chaney 1978), varied from $v = 97,8$ % up to $v = 98,5$ %. In order to improve the degree of saturation, the mentioned pouring process was adjusted as

follows. Each time the water level was rised up to 5 mm above the sand surface, a kind of steel fork having a width of 1 m stretched to about 3 mm into the water, was moved uniformly and with a speed of about 20 mm/sec twice to and from over the entire tank length. Most of the air bubbles locked up between the poured sand grains were freed increasing the degree of saturation up to 99,2 % or even 99,5 %.

2.2 The consolidation of the saturated sample

Once the saturated sample prepared as described above, a rubber bag with a central opening and a square outside shape is placed on the sandsurface (fig. 8). Subsequently the water table was raised until the tank was entirely filled and then closed hermetically by means of rubberstrips and a very stiff top plate, anchored to the tank. The mentioned top plate is provided with three openings, one of them being a central opening O_1 of 20 mm of diameter with a ball bushing and loading piston. About 0,20 m above the top plate, the loading piston is provided with a load-cell type Philips P.R. 6200/13, capable of measuring the load on the sand with an accuracy of < 0,2 %. Calibration tests provide a measure for the friction resistance on the loading piston at the ball bushing level. The dead weight is resting on the mentioned load cell.

The rubber bag underneath the top plate rests on the sand surface and is surrounded by the water up to the top level of the tank. Through the second opening O_2, the rubber bag can be filled with compressed air at any chosen pressure. While pressurizing it, the valve at O_3 in the top plate is opened, allowing excess water in the zone between the sandsurface and the topplate to dissipate. Once the rubberbag reached the chosen pressure σ_c and its shape became as imposed by the tank walls and a central steel gullet, the valve O_3 was closed. At the desired consolidating pressure σ_c, the sample was drained through the drainage tube d_1 by manipulating the valve C_1 (fig. 8).

Relating to the permeabilities of the sand and the porous concrete bottom plate, we considered a consolidation period to be over, when no more water expulsion through the drainage valve was measured in a time interval of 2 hours.

The values of the relative density after the mentioned consolidation period (and before any cyclic loading took place) are calculated from the calibration curves shown

101

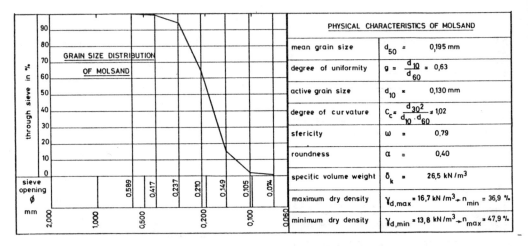

GRAIN SIZE DISTRIBUTION OF MOLSAND	PHYSICAL CHARACTERISTICS OF MOLSAND		
	mean grain size	d_{50} =	0,195 mm
	degree of uniformity	$g = \dfrac{d_{10}}{d_{60}}$ =	0,63
	active grain size	d_{10} =	0,130 mm
	degree of curvature	$C_c = \dfrac{d_{30}^2}{d_{10} \cdot d_{60}}$ =	1,02
	sfericity	ω =	0,79
	roundness	α =	0,40
	specific volume weight	δ_k =	26,5 kN/m³
	maximum dry density	$\gamma_{d,max}$ = 16,7 kN/m³ → n_{min} =	36,9 %
	minimum dry density	$\gamma_{d,min}$ = 13,8 kN/m³ → n_{max} =	47,9 %

Fig. 5. Physical characteristics of the Molsand.

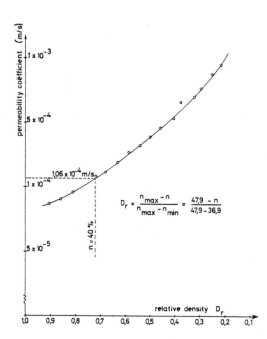

$$D_r = \frac{n_{max} - n}{n_{max} - n_{min}} = \frac{47,9 - n}{47,9 - 36,9}$$

Fig. 6. Measured permeability of Molsand as a function of relative density.

in fig. 7a and 7b giving the variation of the dry unit volume weight $\gamma_{d,o}$ at zero consolidation stress with the drop height of the poured sand and the variation of γ_d with applied consolidation stress σ_c.

Each of these curves makes it possible to derive the value of the relative density D_r of the saturated and consolidated sand volume at the beginning of the cyclic loading test, starting from the imposed drop height of the sand and from the applied vertical consolidation stress.

3. VERTICAL CYCLIC LOADING AND PORE PRESSURE MEASUREMENT

As indicated on fig. 8, a special air pressure device is used for dynamic vertical loading. This dynamic loading system is connected rigidly with the dead weight frame. The outline of operation of the dynamic air pressure cylinder is shown on fig. 9a. From the air pressure system available in the Laboratory of Soil Mechanics, a pressure tank of about 1,5 m³ is filled up continuously with compressed air at 10³ kN/m². From the pressure tank, a reducing valve (3) takes care of the desired final pressure on the piston of the cylinder. The value of the mentioned air pressure can be read off on the pressure-gauge (4). By a dynamic working valve (2), the compressed air is admitted either in the upper or in the lower part of the cylinder, at a frequency of 2 Hz. The dynamic force may be added to or substracted from the static force as shown on fig. 9b. A special regulation device (1) takes care of the form of the loading wave indicated on fig. 9b. Such device seemed to be necessary to compensate the normally asymmetric form of an air pressure induced wave form.

As we wanted our dynamic working load to vary between about 200 N and 2,5 kN, two double working air pressure cylinders are used. One of them in the load-interval from 150 N up to 700 N, the second for the dynamic loading tests from 600 kN up to

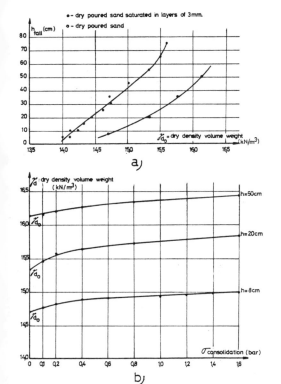

• — dry poured sand saturated in layers of 3mm.
○ — dry poured sand

a)

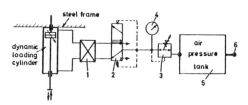

Fig. 9a. Outline of operation of dynamic air pressure cylinder.

b)

Fig. 7. Dry volume weight as a function of h_{fall} and σ_{consol}.

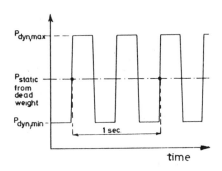

Fig. 9b. Dynamic wave form.

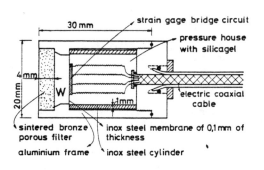

Fig. 10. Electrical pore-water pressure cell.

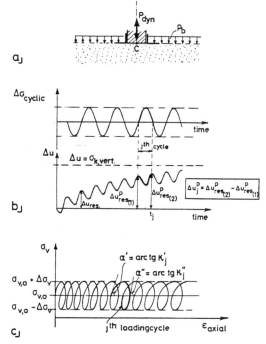

Fig. 11.

103

about 3 kN. Each of the mentioned double working cylinders has a maximum length of free travel of 200 mm.

The effective pressure regulated by the reducing valve (3) and the diameter and length of all pressure tubes ensure that the dynamic loading cycle remains constant at any moment, even if very quick settlements or liquefaction phenomenae should occur.

The recording of the dynamic and static loading on the sand, of the vertical movements of the loading plate on the sand and of the pore pressure measurements is obtained by three H.P.-7404 A oscillographic recorders with twelf amplifier systems in accordance with each of the measuring cells.

On fig. 10, a shematic view of a pore-water pressure cell is given. Each of this cells is installed, during the sand pouring, in the sand mass at known co-ordinates. On the stainless steel membrane (0,1 mm of thickness) of each cell, a strain gage bridge is attached of the type Philips P.R. 9849 k/09F with a specific resistance of 120,4 Ω/bridge, a transverse sensitivity factor of $\pm$ 1 % and a linearity deviation < 1 %. The thin stainless steel membrane can be deformed by the water pressure transfered from the pore water in the sand to the water in the space W (fig. 10). For measuring the water pressure at the thin steel membrane, a sintered bronze porous filter with uniform pore holes of about 0,12 mm separates the sand mass from the measuring membrane. The little space W in each cell always is filled up with de-aired water entirely before installing the pore pressure cell inside the sandmass.

4. INTERPRETATION OF THE MEASURED PORE PRESSURE VALUES DURING THE VERTICAL CYCLIC LOADING TESTS

Due to the outer cyclic loading, a unit volume of completely saturated undrained sand will show a redistribution of the grains. In this way it happens that the outer loading is transferred from the contact points of the grains to the pore water, developing the pore water pressure. Respecting the argumentation given by (Martin et al. 1975), we hereafter make a proposal for interpretation of the measured waterpressure in our testing procedure. After one more loading cycle (fig. 11b) the residual pore water pressure Δu^P_j rises up to the value $\Delta u^P_j = \Delta u^P_{res(2)} - \Delta u^P_{res(1)}$. This change of the residual pore pressure results in a decrease of the pore volume $\Delta n_{o,j}$ existing at the start of the jth

loading cycle. The value of $\Delta n_{o,j}$ can be calculated as :

$$- \Delta n_{o,j} = n_{o,j} \cdot \Delta u^P_j \cdot C_w + (1 - n_{o,j}) \times$$
$$\times \Delta u^P_j \cdot C_s \qquad (1)$$

in which
$C_w \cdot \Delta u^P_j$ = the volume change (per unit volume of completely saturated sand) of the pore water due to the loading Δu^P_j (C_w = $5 \cdot 10^{-7}$ m^2/kN).

$C_s \cdot \Delta u^P_j$ = the volume change (per unit volume of completely saturated sand) of the sand particles due to the loading Δu^P_j (we took $C_s = 0,2 \cdot 10^{-7}$ m^2/kN).

From the other point of view, there is a change $\Delta \varepsilon^P_{j,dev}$ in volume of the entire soil skeleton due to the rearrangements of the grains. Moreover, the rise of the residual pore pressure results, for the saturated and undrained unit volume of sand, in an equivalent decrease of the interparticle contact pressure (the effective stress in the soil skeleton). This reduction of the effective stress leads to a certain release of the volumetric strain of the soil skeleton itself. One can calculate a value for this release of volumetric strain of the soil skeleton as :

$$\Delta \varepsilon_{vol,j} = \left| \frac{\Delta u^P_j}{K_j} \right| \qquad (2)$$

with K_j = the bulk modulus of the saturated sand skeleton.

The value of K_j, we derive from the measured pore pressure curves (fig. 11b) as :

$$K_j = \frac{K'_j + K''_j}{2} \qquad (3)$$

Assuming compatibility of the volumes, the change of the pore volume should be equal to the net change of volume of the soil skeleton. So, one can state :

$$\Delta n_{o,j} = \Delta \varepsilon^P_{j,dev} - \left| \frac{\Delta u^P_j}{K_j} \right| \qquad (4)$$

$$- \left[\overline{n}_{o,j} \cdot \Delta u^P_j \cdot C_w + (1 - n_{o,j}) \times \right.$$
$$\left. \times \Delta u^P_j \cdot C_s \right] = \Delta \varepsilon^P_{j,dev} - \left| \frac{\Delta u^P_j}{K_j} \right| \qquad (5)$$

From this equation one can calculate the change of the volume of the saturated ske-

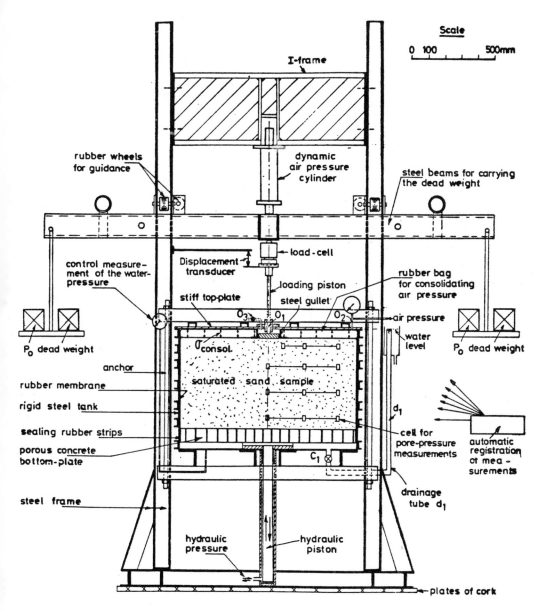

Fig. 8. Consolidating the saturated sand sample.

105

leton due to the cyclic loading in the jth loading cycle :

$$\Delta\epsilon^p_{j,dev} = \frac{\Delta u^p_j}{K_j} - \left[n_{o,j} \cdot C_w + \right.$$

$$\left. + (1 - n_{o,j}) C_s \right] \cdot \Delta u^p_j \qquad (6)$$

The volume changes of the measuring electronic equipment itself, due to the cyclic loading at the top of the sandlayer, appeared in our case to be too small to be taken into account. So we don't bring in any correction to the measured pore pressure values.

It must be emphasized ones more that complete saturation on the sand sample must be obtained as good as possible, to justify the whole reasoning. Otherwise, the unique value of C_w, depending very much on the degree of saturation, is no longer applicable ; neither is valid then any longer the entire relationship (4).

Finally, in accordance with the definitions of a pore pressure generating parameter β given by (Bjerrum 1973) and (Gudehus 1977), in our testing procedure one calculates a β_{j_w} -value as :

$$\beta_{j_w} = \frac{\Delta u^p_j}{\sigma_c \cdot N_j} \qquad (7)$$

Formulae (4) and (6) only are applicable if one assumes the entire volume of the sample to be constant and if the stresses are uniformly distributed inside the sample. Otherwise, migration of water from one place in the sample to another will take place. So, in our testing procedure we should write the mentioned equations as a sommation over the whole sample volume inside a totally undeformable testing tank. In each little part-volume of the mentioned sommation, it so is assumed the stresses are uniformly distributed and in consequence no water migration takes place.

5. CONCLUSION

In the authors opinion one should try to reform the classical way of cyclic loading testing into large scale tests with load control without any membrane supply. An attempt was made to develop such equipment for vertical cyclic loading on a large scale sand sample, measuring the pore pressure developing inside the sample.

6. REFERENCES

Bjerrum, L. 1973, Geotechnical problems involved in foundations of structures in the North Sea, Geotechnique, Vol. 23, N° 3, pp. 319-358.

Casagrande, A. 1976, Liquefaction and cyclic deformation of sands - A critical review, Harvard Soil Mechanics Series, N°88, pp. 1-52.

Castro, G. 1969, Liquefaction of sands, Harvard Soil Mechanics Series, N° 81, pp. 1-112.

Chaney, R.C. 1978, Deformation of earth-dams under earthquake loading, Ph. D.-Dissertation, UCLA.

De Alba, P., Seed, H.B. & Chan, C.K. 1976, Sand liquefaction in large scale simple shear tests, Journal of the Geotechnical Engineering Division, ASCE, Vol. 102, n° GT9, pp. 909-927.

Frydman, S., Zeitlen, J.G. & Alpan, I. 1973, The membrane effect in triaxial testing of granular soils, Journal of testing and evaluation, Vol. 1, n° 1, pp. 37-41.

Gudehus, G. 1977, Stability of saturated granular bodies under cyclic load, Proc. Dynamical methods in soil and rock mechanics, Karlsruhe, Vol. 2, pp. 195-212.

Kiekbusch, M. & Schuppener, B. 1977, Membrane penetration and its effect on pore pressures, Journal of the Geotechnical Engineering Division, ASCE, Vol. 103, N° GT11, pp. 1267-1279.

Lade, P.V. & Hernandez, S.B. 1977, Membrane penetration effects in undrained tests, Journal of the Geotechnical Engineering Division, ASCE, Vol. 103, n° GT2, pp. 109-125.

Lee, K.L., Morrison, R.A. & Haley, S.C. 1969, A note on the pore pressure parameter B, Proc. 7th International Conference on Soil Mechanics and Foundation Engineering, Vol. 1, Mexico, pp. 231-238.

Martin, G.R., Finn, W.D.L. & Seed, H.B. 1975, Fundamentals of liquefaction under cyclic loading, Journal of the Geotechnical Engineering Division, ASCE, Vol. 101, N° GT5, pp. 423-436.

Mulilis, J.P., Townsend, F.C. & Horz, R.C. 1978, Triaxial testing techniques and sand liquefaction, Dynamic Geotechnical Testing, ASTM STP654, American Society for Testing and Materials, pp. 265-279.

Newland, P.L. & Allely, B.H. 1959, Volume changes during undrained triaxial tests on saturated dilatant granular materials, Geotechnique, Vol. 9, N°4, pp. 174-182.

Park, T.K. & Silver, M.L. 1975, Dynamic triaxial and simple shear behavior of sand, Journal of the Geotechnical En-

gineering Division, ASCE, Vol. 101, N°
GT6, pp. 513-529.

Poorooshasb, H.B. et al. 1966, Yielding and
flow of sand in triaxial compression :
part 1, Canadian Geotechnical Journal,
Vol. 3, N° 4, pp. 179-190.

Raju, V.S. & Sadasivan, S.K. 1974, Membrane penetration in triaxial tests on
sands, Journal of the Geotechnical Engineering Division, ASCE, Vol. 100, N° GT4,
pp. 482-489.

Seed, H.B. & Lee, K.L. 1966, Liquefaction
of saturated sands during cyclic loading,
Journal of the Soil Mechanics and Foundations Division, ASCE, Vol. 92, N° SM6,
pp. 105-134.

Seed, H.B., Martin, Ph.P. & Lysmer, J.
1976, Pore-water pressure changes during
soil liquefaction, Journal of the Geotechnical Engineering Division, ASCE,
Vol. 102, N° GT4, pp. 323-346.

Seed, H.B. & Peacock, W.H. 1971, Test procedures for measuring soil liquefaction
characteristics, Journal of the Soil
Mechanics and Foundations Division, ASCE,
Vol. 97, N°SM8, pp. 1099-1119.

Silver, M.L. & Seed, H.B. 1971, Volume
changes in sands during cyclic loading,
Journal of the Soil Mechanics and Foundations Division, ASCE, Vol. 97, N°SM9,
pp. 1171-1182.

Tinoco, F.H. 1977, Pore pressure parameters
and sand liquefaction, Proc. 9th International Conference on Soil Mechanics
and Foundation Engineering, Vol. 2,
Tokyo, pp. 409-418.

Van Impe, W. 1977, De vrije wringingsproef;
een belangrijke bijdrage tot de studie
van het gedrag van funderingen onder in-
vloed van grondtrillingen, Tijdschrift
der Openbare Werken van België, nr 6,
pp. 501-519.

Vernese, F.J. & Lee, K.L. 1977, Effect of
frictionless caps and bases in the cyclic triaxial test, Contract Report S-77
-1, U.S. Army Engineer Waterways Experiment Station, Vicksburg, pp. 1-112.

Stress-path dependent stress strain-volume change behaviour of a granular soil

A. VARADARAJAN & S.S. MISHRA
Indian Institute of Technology, New Delhi, India

1 INTRODUCTION

Solutions of Geotechnical Engineering problems require predictions of deformations and stresses during various stages of loading. Powerful numerical methods are available to make such predictions even for complicated problems. To get accurate results, realistic stress-strain relationships of soil should be incorporated in the analyses. It is now well known that the stress-strain-volume change relationships of soils are dependent on a number of factors such as soil type, density, stress level and stress-path (Duncan and Chang, 1970, Yudhbir and Varadarajan, 1975, Lade and Duncan, 1976). Attempts are continuously being made to develop analytical models for soils incorporating all such factors (Desai, 1977).

A programme of research is under progress for the formulation of an analytical model for the predictions of stress-strain-volume change relationship of granular soils. This paper presents part of the results of the investigations of a granular soil.

2 TESTING PROGRAMME

2.1 Tests conducted

Two test series were carried out using standard triaxial testing equipment on a locally available sand. In the first series, the samples were consolidated isotropically and sheared to failure using three stress-paths (Fig 1a), i) vertical stress, σ_1 increasing and lateral stress, σ_3 constant, ii) σ_3 decreasing and σ_1 increasing such that the average mean stress ($\sigma_1 + 2\sigma_3$)/3, p_a constant and iii) σ_1 constant and σ_3 decreasing. Four consolidation stresses, σ_c, 1, 2, 3 and 4 kg/cm^2

were used.

In the second series, the samples were anisotropically consolidated with a stress ratio, K = 0.36 (1-sin $\emptyset'$, $\emptyset'$=40° is the angle of shearing resistance) to approximate the insitu-state of stress condition. The samples were then sheared using eight stress-paths (Fig 1b), i) σ_1 constant and σ_3 increasing, ii) σ_1 and σ_3 increasing such that σ_3/σ_1 = 0.36, iii) σ_3 constant and σ_1 increasing, iv) σ_3 decreasing and σ_1 increasing such that ($\sigma_1+\sigma_3$)/2, p constant v) σ_1 constant and σ_3 decreasing, vi) σ_1 and σ_3 decreasing such that σ_3/σ_1 = 0.36, vii) σ_3 constant and σ_1 decreasing and viii) σ_1 decreasing and σ_3 increasing such that p was constant. Four stress levels at p_c=1.85, 3.70, 5.55 and 7.40 kg/cm^2 were adopted.

2.2 Soil used

A locally procured sand from Badarpur near Delhi was used for the tests. It is uniformly graded with 80 per cent of the soil having grain sizes between 0.25 mm-0.60 mm. The sand is composed of angular grains of mainly quartz. A single batch of sand was used for all the tests. The specific gravity of the sand material was 2.66.

2.3 Experimental procedure

Standard triaxial compression testing equipment was used for the tests. 3.81 cm diameter and 7.62 cm long cylindrical samples were adopted for all the tests. The samples were prepared under saturated condition (Bishop and Henkel, 1957). A uniform procedure of tamping the mould was used to achieve the same density of the samples. The samples were towards dense condition. Cell pressure was applied by the self compensating mercury pot system and the vertical loading was applied through the loading frame, proving ring being used for load measurement. For tests with σ_3 constant, strain controlled loading was applied. For other stress-path tests cell pressure and the vertical loading were applied by manual control in small increments. The loading to be applied was estimated at every increments. The loading to be applied was estimated at every increment after using area corrections for the sample. For speedy calculations during experiment a standardised programme was used with the aid of a 200 step programmable calculator. Vertical displacement and volume change readings were taken after steady condition was reached for each increment of loading. Smaller increments of loading were used at loadings near failure. The vertical displacements and volume changes were measured to the accuracy of 0.000254 cm and 0.01 cc respectively.

3 EXPERIMENTAL RESULTS

3.1 Consolidation and rebound

Fig.2 shows consolidation and rebound curves for isotropic and anisotropic stress conditions. The relationships for loading for both the cases show similar behaviour at low stress levels, but at high stress levels, the anisotropic consolidation curve shows higher volume change. The unloading curves indicate more irrecoverable volumetric strain in the case of anisotropic unloading.

3.2 Stress-strain-volume change relationships

Isotropically consolidated samples

Figs.3,4 and 5 show stress-strain-volume change relationships of σ_3 constant, p_a constant and σ_1 constant stress-path tests for only three stress levels. All the tests indicate increase in vertical failure strain with higher consolidation pressures. For a consolidation pressure, σ_3 constant stress-path test shows largest vertical deformation and σ_1 constant stress-path tests indicate lowest value.

The volumetric strains observed are more compressive in the case of σ_3 constant stress-path tests than the other two cases. In general, volume expansion is noted for all the tests near failure state.

Anisotropically consolidated samples

Fig.1c shows the stress-paths used in σ_3-σ_1

stress-space. The stress-paths are designated as ASP0, ASP70.2, ASP90, ASP135, ASP180, ASP250.2, ASP270 and ASP315, the numbers indicating the angles with respect to horizontal in the counter clockwise direction. ASP indicates anisotropically consolidated and SP refers to the stress-path. Following similar notations ISP90, ISP116.5 and ISP180 refer to stress-paths of isotropically consolidated samples discussed earlier.

In Figs. 6,7 and 8 are shown stress-strain-volume change relationships for ASP90, ASP135 and ASP180 tests. These tests were carried out till failure was reached. In all these cases higher vertical failure strains are observed at higher preshear stress levels, p_c (= $\sigma_1 + \sigma_3$ /2). For a p_c value, ASP90 test shows highest value of vertical strain and ASP180 indicates the lowest value. This trend is similar to that observed in ISP tests also. In all the tests volume expansion is generally observed from the very beginning of the test. ASP90 and ASP135 tests show comparable volume expansion while ASP180 test indicates less volume expansion. This trend is not same as for ISP tests.

Figs. 9 and 10 show stress-strain-volume change relationships for ASP70.2 and ASP 250.2 tests. These are loading and unloading tests along the K-line. The relationships are nearly linear. As would be expected ASP70.2 shows compression and ASP 250.2 indicates expansion.

Stress-strain-volume change relationships for ASP315 and ASP0 are shown in Figs. 11 and 12. In these tests shear stress was decreased upto zero value only (due to the limitation of standard triaxial cell). In these stress-paths p(= $\sigma_1 + \sigma_3$ /2) was increased. In the initial stages of loading small increase in vertical strains (<0.1 per cent) and subsequent reductions in the values are observed. All these tests show volumetric compression during all stages of loading.

Fig. 13 shows stress-strain-volume change relationships for ASP270. In this test p and q (= $\sigma_1 - \sigma_3$/2) were decreased. These relationships are slightly nonlinear. In this stress range very small volume expansion is observed.

4 ELASTIC STRESS-STRAIN THEORY

4.1 Isotropically consolidated samples

To study the effect of stress-path on the stress-strain volume change relationships the Young's moduli and Poisson's ratios were calculated at initial stress levels using the relationships based on theory of elasticity (Lambe and Whitman, 1968 and Yudhbir and Varadarajan, 1975). For standard triaxial stress conditions the Young's modulus, E and Poisson's ratio, ν are given by

$$E = \frac{(\Delta \sigma_1 + 2 \Delta \sigma_3)(\Delta \sigma_1 - \Delta \sigma_3)}{\Delta \sigma_3 (\Delta \epsilon_1 - 2 \Delta \epsilon_3) + \Delta \sigma_1 \Delta \epsilon_3} \quad (1)$$

$$\nu = \frac{\Delta \sigma_3 \Delta \epsilon_1 - \Delta \sigma_1 \Delta \epsilon_3}{\Delta \sigma_3 (\Delta \epsilon_1 - 2 \Delta \epsilon_3) + \Delta \sigma_1 \Delta \epsilon_3} \quad (2)$$

where, $\Delta \sigma_1$ and $\Delta \sigma_3$ = change in σ_1 and σ_3 respectively and, $\Delta \epsilon_1$ and $\Delta \epsilon_3$ = corresponding changes in vertical and lateral strains. $\Delta \epsilon_3$ was calculated from vertical strain and volumetric strain values.

The effect of consolidation stress on E value was expressed by the relationships

$$E_i = K_{ur} p_a \left(\frac{\sigma_3}{p_a}\right)^n \quad (3)$$

where

E_i = the initial tangent modulus

K_{ur} = modulus number

p_a = atmospheric pressure in the same pressure units as E

σ_3 = consolidation pressure and

n = exponent determining the rate of variation of E with σ_3.

The K_{ur}, n and average ν values are shown in Table 1. The parameters are function of stress-path as also observed in the case of clays (Yudhbir and Varadarajan, 1975). The effect is not as significant as it is for clays. Similar variations for a sand is also noted by Desai (1977).

Table 1. Elastic parameters

Stress-path	K_{ur}	n	Average ν_i
ISP90	270	0.726	0.35
ISP116.5	347	0.560	0.00
ISP180	285	1.070	0.50
ASP0	4500	0.370	0.71
ASP70.2	420	0.430	0.33
ASP90	150	1.470	0.32
ASP135	200	1.240	0.34
ASP180	3550	0.490	0.57
ASP250.2	610	0.510	0.09
ASP270	3100	0.51	0.21
ASP315	530	1.66	1.31

4.2 Anisotropically consolidated samples

In the case of the stress-paths used in these series, it was not possible to adopt the same procedure of prediction of E and values since in some of the stress-paths, failure could not be reached. For evaluation of E and values comparable stress changes the stress-probe level system used by Lewis and Burland (1970) was adopted as follows

$$\sigma_n = \sqrt{\sigma_1^2 + \sigma_2^2 + \sigma_3^2} \quad (4)$$

$$\Delta \sigma_n = 0.2 \sigma_n = \sqrt{\Delta \sigma_1^2 + \Delta \sigma_2^2 + \Delta \sigma_3^2} \quad (5)$$

$\Delta \sigma_1$ and $\Delta \sigma_3$ values were worked out for this stress level for different stress-paths and corresponding values of $\Delta \epsilon_1$ and $\Delta \epsilon_3$ were obtained from the experimental results. E and ν values and K_{ur} and n values were then calculated. In evaluating K_{ur} and n values, p_c values were used in place of σ_3 values. The results are tabulated in Table 1. For quantifying the stress-paths for use in the Finite Element analysis, the gradient of the stress-path in the p-q stress-space was adopted earlier (Yudhbir and Varadarajan, 1975). Due to the limitations of this procedure (for example p constant tests, the gradient will be infinity), here the gradient of the stress-paths in terms of angles, θ (as defined earlier) in the σ_1-σ_3 stress space is used for quantifying the stress-paths. Accordingly plots have been made between θ and K_{ur} and, θ and n (Fig.14). It can be clearly seen that the parameters are very much a function of stress-path. The variations of modulus values between the lowest and the highest value is in the order of 30. For the analysis of problems at higher factors of safety, these relationships may be readily used for incorporating stress-path dependent parameters.

5 ELASTOPLASTIC STRESS-STRAIN THEORY

For investigating the effect of stress-path in the stress-strain-volume change relationships the elastoplatic stress-strain theory, given by Lade and Duncan (1975) was also used. The relevant expressions are given in Appendix I.

The parameters were evaluated from ISP90, ISP116.5 and ISP180 test results. They are presented in Table 2. It may be observed that the parameters for the prediction of plastic strains vary with different stress-paths. The dimensionless parameter M for

ISP90, ISP116.5 and ISP180 are 4.6×10^{-4} 1×10^{-4} and 1×10^{-5} respectively, which indicate the order of magnitude of the effect of stress-path. The constant K_2 and stress level f relationship are shown in Fig.15. This relationship is interesting to note, is fairly independent of not only the consolidation pressure as observed by Lade and Duncan (1975) but also the stress-path.

Table 2 Plastic parameters

Stress path parameters	ISP90	ISP116.5	ISP180
K_1	63.10	63.10	63.10
f_t	33.00	29.00	28.00
A	0.4475	0.4475	0.4475
r_f	0.95	0.696	0.795
M	4.6×10^{-4}	1×10^{-4}	1×10^{-5}
l	1.76	2.322	3.322

To evaluate the influence of these parameters, the prediction of stress-strain-volume charge relationships was done for ISP90, ISP116.5 and ISP180 stress-paths at a consolidation pressure of 2 kg/cm^2.

ISP90 parameters were used for the prediction of stress-strain-volume change relationships for ISP90, ISP116.5 and ISP180 as indicated in first column sketches in Fig. 16. The predictions of the relationships for ISP116.5 and ISP180 are indicated in the second and third column sketches of Fig.16.

Fig.17 shows the experimental and predicted stress-strain-volume change relationships. It may be observed that the predictions by ISP90 parameters as would be expected are very close to the experimental results. However, the predictions by ISP 180 parameters are far away from the experimental values. ISP116.5 parameters predict the stress-strain relationships rather very closely, but the strain-volume change prediction is not satisfactory.

In Fig.18 are shown stress-strain-volume change relationships for ISP116.5 tests. The prediction by ISP116.5 parameters in better than those by the parameters of ISP 90 and ISP180.

Fig.19 shows the prediction of stress-strain volume change relationships by ISP 180, ISP90 and ISP116.5 parameters. ISP180 parameters only predicts the relationships close to the experimental results.

These predicted results indicate that the parameters evaluated by one stress-path tests do not predict the stress-strain-volume change relationships accurately for other stress-paths. With this much of limited data, it appears, for realistic predictions, relevant stress-path test parameters should be evaluated and used. Apart from the inherent short comings of the theory in predicting the stress-strain-volume change relationships for stress-paths involving decreasing stress ratios ($df < 0$), (Lade and Duncan, 1975) appears to require modifications to incorporate the effect of stress-paths also for accurate predictions of stress-strain-volume change relationships. Further work is in progress for the predictions of anisotropically consolidated samples and other types of sand.

6 CONCLUSIONS

From the stress-strain-volume change relations of Badarpur sand, the following conclusions are drawn.

Anisotropic unloading gives more irrecoverable strain than the isotropic unloading.

Both isotropically consolidated and anisotropically consolidated samples show that the stress-strain volume change relationships are very much influenced by stress-paths.

Elastic parameters E, K_{ur}, n and ν evaluated at initial stress level are shown to be functions of stress-paths. Quantitative relationships between stress-paths and K_{ur} and n values have been established.

The parameters evaluated using elasto-plastic theory from isotropically consolidated samples are shown to be stress-path dependent. However, the stress level dependent plastic potential function is found to be independent of the stress-path.

7 REFERENCES

Bishop, A.W. and D.J. Hankel 1957, The measurement of soil properties in the triaxial test, Edward Arnold (Publishers) Ltd., London.

Desai, C.S. 1977a, Constitutive law for geologic media. In C.S. Desai and J.T. Christian (ed.), Numerical methods in geotechnical engineering, p.65-115, McGraw-Hill Book Company, New York.

Desai, C.S. 1977b, Deep foundations.In C.S. Desai and J.T. Christian (ed.), Numerical methods in geotechnical engineering, p.235-271. McGraw-Hill Book Company, New York.

Duncan, J.M. and C.Y. Chang 1970, Nonlinear analysis of stress and strain in soil, Jnl. of the soil mechanics and foundation division, ASCE 96: 1629-1653.

Lade, P.V. and J.M. Duncan 1975, Elasto-plastic stress-strain theory for cohesionless soil, Jnl. of the geotechnical engng. divn : Proc. ASCE 101:1037-1054.

Lade, P.V. and J.M. Duncan 1976, Stress-path dependent behaviour of cohesionless soil, Jnl. of the geotechnical engng. divn. Proc. ASCE 102:51-68.

Lambe, T.W. and R.V. Whitman 1969, Soil mechanics, John Wiley & Sons, Inc. New York.

Lewin, P.I. and J.B. Burland 1970, Stress probe experiments on saturated normally consolidated clay, Geotechnique, 20(1): 38-56.

Yudhbir and A. Varadarajan 1975, Stress-path dependent deformation moduli of clay, Jnl. of the geotechnical engn.divn. Proc. ASCE 101:315-327.

APPENDIX I

Lade and Duncan,1975,theory

$$\{d\epsilon_{ij}\} = \{d\epsilon_{ij}^e\} + \{d\epsilon_{ij}^p\} \tag{1}$$

where

$d\epsilon_{ij}$ = total strain increment

$d\epsilon_{ij}^e$ = elastic component of the strain increment, and

$d\epsilon_{ij}^p$ = plastic component of the strain increment

Failure criterion is given by,

$$I_1^3 - K_1 I_3 = 0 \tag{2}$$

where,

$$I_1 = \sigma_1 + \sigma_2 + \sigma_3$$

and, $I_3 = \sigma_1 \sigma_2 \sigma_3$

and K_1 = constant whose value depend on the density of the soil.

$$f = \frac{I_1^3}{I_3} \tag{3}$$

f = stress level
 = K_1 at failure $\tag{4}$

Plastic potential function $g(\sigma_{ij})$ is given as

$$g = I_1^3 - K_2 I_3 \tag{5}$$

where,

K_2 = constant for any given value of f

$$\Delta\epsilon_{ij}^p = \Delta\lambda\frac{\partial g}{\partial \sigma_{ij}} \tag{6}$$

where,

$\Delta\lambda$ = a constant

From Eq.6 plastic stress-strain relationship is expressed as

$$\Delta\epsilon_1^p = \Delta\lambda K_2(\frac{3}{K_2} I_1^2 - \sigma_2 \sigma_3)$$

$$\Delta\epsilon_2^p = \Delta\lambda K_2(\frac{3}{K_2} I_1^2 - \sigma_1 \sigma_3) \tag{7}$$

$$\Delta\epsilon_3^p = \Delta\lambda K_2(\frac{3}{K_2} I_1^2 - \sigma_1 \sigma_2)$$

Applying the first and third of Eq.7 the ratio of $\Delta\epsilon_1^p$ and $\Delta\epsilon_3^p$ for triaxial compression

$$-\nu^p = \frac{\Delta\epsilon_3^p}{\Delta\epsilon_1^p} \tag{8a}$$

$$-\nu^p = \frac{3\Delta\lambda K_2 (\frac{3}{K_2} I_1^2 - \sigma_1 \sigma_3)}{\Delta\lambda K_2 (\frac{3}{K_2} I_1^2 - \sigma_3^2)} \tag{8b}$$

and solving for K_2

$$K_2 = \frac{3 I_1^2 (1 + \nu^p)}{\sigma_3(\sigma_1 + \nu^p \sigma_3)} \tag{9}$$

Determination of K_2 - from the plot of K_2 vs $f(= I_1^3/I_3)$ the variation of K_2 with f is related by a simple linear equation.

$$K_2 = Af + 27 (1-A) \tag{10}$$

where A is the inclination of straight line in the plot of K_2 vs f.
Plastic work done at each stage of test is calculated from

$$w_p = \int \{\sigma_{ij}\}^T \{d\epsilon_{ij}^p\} \tag{11}$$

in which

$\{\sigma_{ij}\}^T \{d\epsilon_{ij}^p\}$ is the plastic work

113

done per unit volume during strain increment $\{d\epsilon^P_{ij}\}$

Determination of : Proportionality constant $\Delta\lambda$ is given by

$$\Delta\lambda = \frac{dW_p}{3g} \tag{12}$$

The $f-W_p$ plot give rise a concept of threshold stress level, f_t and assume that values of f between 27 and f_t no plastic strains occur and no plastic work is done.

Relationship between W_p and $(f - f_t)$ is approximated by hyperbolae as

$$(f-f_t) = \frac{W_p}{a+d\,W_p} \tag{13}$$

The reciprocal of parameter 'a' is the initial slope of $W_p - (f-f_t)$ relationship. The value of 'a' increases with confining pressure and the variation is expressed as

$$a = M\,p_a\left(\frac{\sigma_3}{p_a}\right)^1 \tag{14}$$

where,

p_a = atmospheric pressure expressed in the same unit as 'a' and σ_3

M and l are dimensionless numbers.

Reciprocal of parameter d in Eq. 13 is the ultimate value of $(f-f_t)$. The failure ratio

$$r_f = \frac{K_1 - f_t}{(f-f_t)_{ult}} \tag{15}$$

in which $(f-f_t)_{ult} = \frac{1}{d}$

From Eq.13 the increment in plastic work is expressed as

$$dW_p = \frac{a\,df}{\left(1 - r_f\,\dfrac{f-f_t}{K_1-f_t}\right)^2} \tag{16}$$

in which 'a' is given by Eq.14; r_f by Eq. 15; f = the current value of stress level; and $df(\gtrless 0)$ = difference in f between two successive stress states.

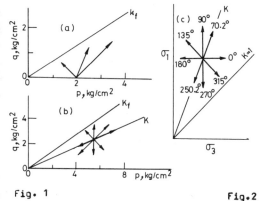

Fig. 1

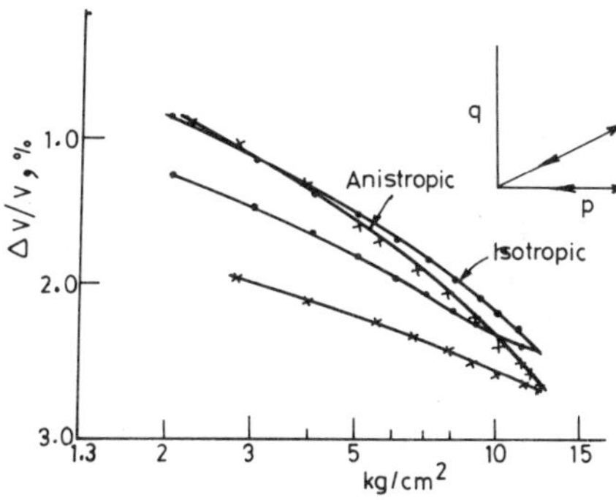

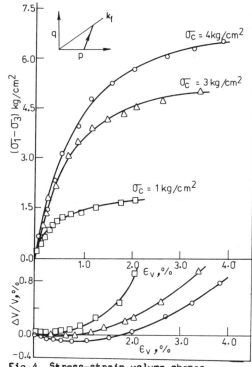

Fig.2 Consolidation and rebound curves for isotropic
and anisotropic stress conditions.

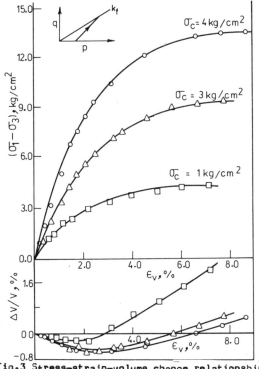

Fig.3 Stress-strain-volume change relationship
$\overline{\sigma_3}$ constant test (ISP90)

Fig.4 Stress-strain volume change
relationship p_a constant test (ISP1165.)

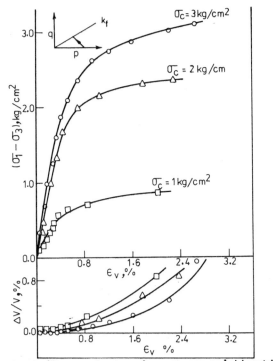

Fig.5 Stress-strain volume change relationship σ_1 constant test (ISP180)

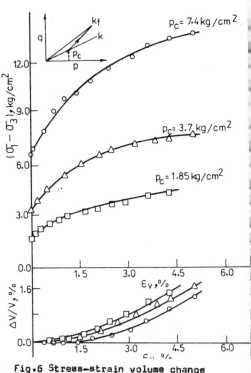

Fig.6 Stress-strain volume change relationship (ASP90)

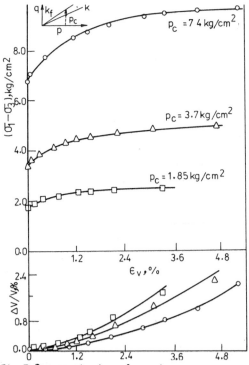

Fig.7 Stress-strain volume change relationship (ASP135)

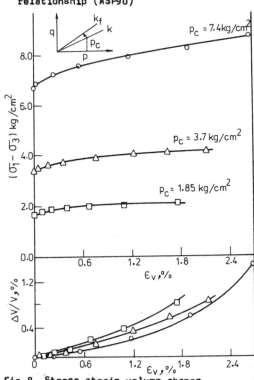

Fig.8 Stress-strain volume change relationship (ASP180)

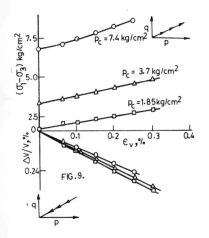

FIG.9.

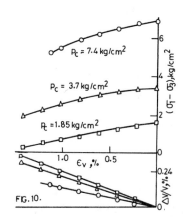

FIG.10.

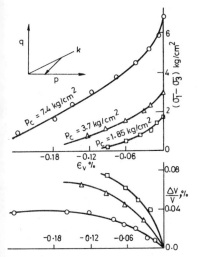

Fig.11 Stress-strain volume change relationship (AS315)

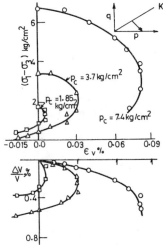

Fig.12 Stress-strain volume change relationship (ASPO)

117

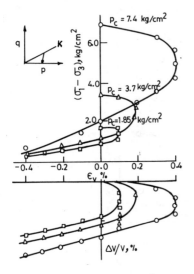

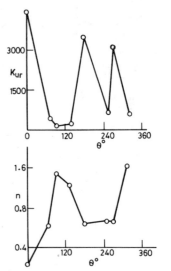

Fig.13 Stress-strain volume change rela-
       tionship (ASP270)

Fig.14 K_{ur}, n, θ relationship

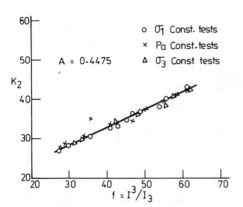

Fig.15 K_2, f relationship

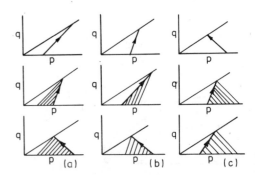

Fig.16   Method of prediction of stress-strain-v
         change by different stress-path paramet

118

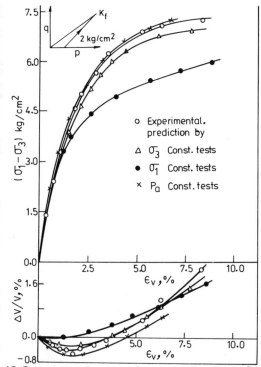

Fig.17 Experimental and predicted stress-strain volume change relationship by ISP90.

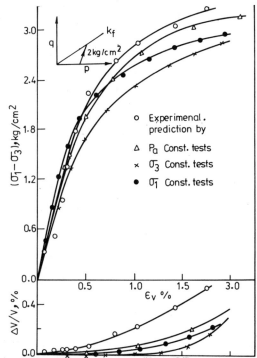

Fig.18 Experimental and predicted stress-strain volume change relationship by ISP116.5

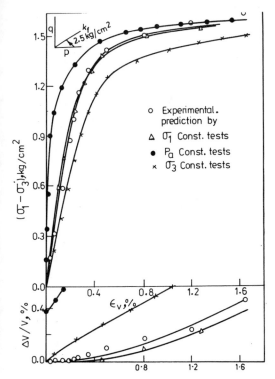

Fig.19 Experimental and predicted stress-strain volume change relationship by ISP180

119

Dynamic properties of sand subjected to initial shear stress

K. UCHIDA, T. SAWADA & T. HASEGAWA
Kyoto University, Kyoto, Japan

1 INTRODUCTION

Since soil under a sloping surface is subjected to shear stress prior to applying
dynamic stress, dynamic behavior of soil
structures with slope, such as fill dams,
may be affected by initial shear stress.
Most previous works of evaluating dynamic
deformation characteristics do not take account of the effects of initial shear stress
because they are mainly interested in horizontal ground. There are, on the other hand,
a few works investigating the effects of
initial shear stress: Hardin and Drnevich
(1972a) and Kuribayashi, Iwasaki and Tatsuoka (1974) showed that shear moduli of sand
were hardly affected by small initial shear
stress; Timmerman and Wu (1969) indicated
that shear strain developed under cyclic
loading was affected by initial shear
stress; Toki and Kitago (1974) showed that
the anisotropic structure of soil particles
due to initial shear stress affected the
increase of axial strain and the volume
change under cyclic loading.

The purpose of this study is to determine
quantitatively shear moduli G and damping
ratios D used in the equivalent linear method, taking account of the effects of initial
shear stress. The dynamic triaxial tests of
saturated Toyoura Sand are performed applying static shear stresses prior to the application of dynamic stresses.

It is pointed out from the test results
that shear strain and pore pressure develop
in a different way from those under the isotropic stress condition. In order to clarify
these effects of initial shear stress, the
quantitative relationships between pore
pressure and the stress ratio (q_a/p') and
between shear strain amplitude and the
stress ratio (q_a/p') are established. Furthermore, to determine the plastic shear
strain developed in the dynamic tests, the
relationships are investigated among the

plastic shear strain γ_p, stress p' and q,
and the state parameter S_s, which was proposed by Moroto (1976).

There are many factors which affect shear
moduli such as those suggested by Hardin and
Black (1966). Their influence on shear moduli is examined herein. And, using these factors, the quantitative determination of
shear moduli G and damping ratios D is investigated. Finally, an illustrative example
of earthquake response analysis is conducted
using the results obtained in this study.

2 TEST MATERIAL AND PROCEDURES

The material used in the present investigation is saturated Toyoura Sand, properties
of which are listed in Table 1.

The dynamic triaxial test apparatus used
is an electrohydraulic servo controlled type
in which dynamic load can be applied only in
the axial direction by the lower piston.
Soil samples 5 cm in diameter and 10 cm in
height are prepared by pouring de-aired saturated sand with a spoon into a mold which
is filled with de-aired water in advance.
The specified initial void ratio e_o=0.69 to
0.71 is obtained by tapping the mold with
the spoon three times in each three directions for every five spoons, which make an
angle of about 120° in plane.

Table 1   Physical properties of sand

Specific gravity	G_s	2.65
Limit density (in saturated condition)	e_{min}	0.588
	e_{max}	0.910
Grading	D_{10}	0.165 mm
	D_{50}	0.120 mm
	U_c	1.48

The test procedures are as follows:

1. Isotropic confining pressure ($\sigma_3 = 0.5$ kg/cm^2 and 1.0 kg/cm^2) is applied to the sample and the pore pressure coefficient B is confirmed to be above 0.95.

2. The sample is isotropically consolidated for about 30 minutes.

3. The static shear stress determined by the static mobilized stress ratio MSR (defined as MSR=$(q/p')/(q/p')_{fail}$, in which $q=\sigma_1'-\sigma_3'$ and $p'=(\sigma_1'+2\sigma_3')/3$) is applied in the load-controlled and drained condition.

4. The sinusoidal dynamic load is applied 30 cycles in the undrained condition.

The test programs are listed in Table 2.

Analogue data of load, deformation and pore pressure are recorded by a data recorder, are converted to digital data and are analysed by a digital computer.

Table 2   Test programs

Consolidation pressure σ_3 (kg/cm^2)	0.5, 1.0
Initial void ratio e_o	0.69 to 0.71
Static mobilized stress ratio MSR	0.40, 0.50, 0.57, 0.67
Frequency of vibration f (Hz)	1, 2, 4

3 EXPERIMENTAL RESULTS

It was shown from the results of static triaxial compression tests that the stress ratio at failure $(q/p')_{fail}$. was 1.656 for the initial void ratio e_o=0.70. From this value, the intensities of initial shear stress q_s were determined corresponding to MSR=0.40, 0.50, 0.57, 0.67 and σ_3=0.5 kg/cm^2 and 1.0 kg/cm^2. In the process of applying initial shear stress, the volume change which occured was entirely on the compression side irrespective of σ_3 and MSR. On the other hand, the characteristic phenomena in the process of the dynamic tests were the significant increase of plastic shear strain and the development of negative pore pressure for MSR above 0.57.

First, the significant increase of plastic shear strain is examined. The hysteretic curve in Fig. 1 represents an illustrative relationship between shear stress q and shear strain γ which is obtained by applying dynamic stress after the application of initial shear stress. It is seen in Fig. 1 that the plastic shear strain γ_p increases significantly in the first 10 cycles. This phenomenon seems to occur due to the anisotropic structure of soil by the application of

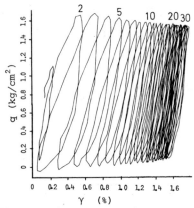

Fig. 1   Typical hysteretic curve   q ~ γ

initial shear stress. In more than 10 cycles, however, the increase of γ_p becomes less significant because of the rearrangement of soil particles by repeated loading. The most adequate way to investigate the quantitative determination of γ_p may be to investigate shearing deformation taking account of the plastic work. From this point of view, the parameter S_s is introduced, which was established by Moroto (1976). It is the state parameter independent of stress paths for granular materials in static triaxial compression tests. The increment of that parameter S_s is defined by the ratio of the increment of plastic work done due to shear deformation dW_s to effective mean principal stress p'

$$dS_s = dW_s/p' = d\varepsilon_{vd} + \frac{2}{3}\cdot(q/p')\cdot d\gamma_p \quad (1)$$

in which $d\varepsilon_{vd}$ is the increment of plastic volumetric strain due to dilatancy. From the experimental facts of γ_p=G(η) and ε_{vd}=D(η), Eq. (1) can be rewritten as

$$dS_s = D'(\eta)\cdot d\eta + \frac{2}{3}\cdot\eta\cdot G'(\eta)\cdot d\eta \quad (2)$$

in which η is q/p'. Therefore, it is clearly shown that S_s is independent of stress path from the theoretical consideration mentioned above. Eq. (1) can be rewritten with use of $d\varepsilon_p$=(2/3)·$d\gamma_p$ as follows:

$$dS_s = (\eta + d\varepsilon_{vd}/d\varepsilon_p)\cdot d\varepsilon_p = \eta_w\cdot d\varepsilon_p. \quad (3)$$

It is seen from Eq. (3) that the gradient of the curve of the relationship S_s ~ γ_p, that is η_w, should be constant or the function of only γ_p so S_s may be the state parameter.

In this study, the parameter S_s is calculated in initial shear process and dynamic process. The relationship S_s ~ γ_p obtained in both process is shown in Fig. 2. It is seen in this figure that the parameter S_s satis-

122

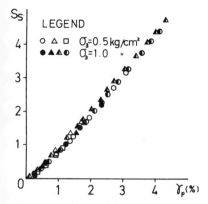

Fig. 2   Relationship $S_S \sim \gamma_p$

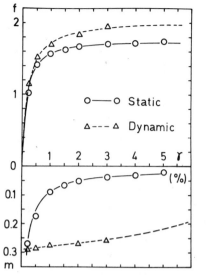

Fig. 4   Relationships $\gamma_p \sim f$ and $\gamma_p \sim m$

fies the condition of the state parameter, not only in the static process but also in the dynamic process, and also that γ_p is directly related to S_S and therefore can be determined by the current stress state irrespective of stress paths. This fact coincides with the concept of equi-γ-lines proposed by Ishihara, Tatsuoka and Yasuda (1975). So, the values of p' and q/p' corresponding to constant γ_p are plotted on semilog graph paper as shown in Fig. 3. This figure shows that the equi-γ-lines can be expressed exactly by the yield function of Poorooshasb (1971),

$$f = q/p' + m \cdot \ln p' . \qquad (4)$$

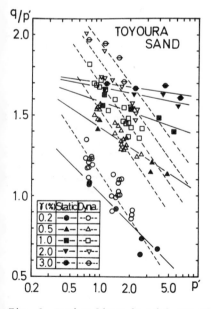

Fig. 3   Equi-γ-lines in q/p' and p'

Then, the slope of lines in Fig. 3, that is m, and the value of (q/p') at p'=1, that is f, are plotted with γ_p as shown in Fig. 4. This figure shows that the two parameters in static condition m_s and f_s, and those in dynamic condition m_d and f_d are expressed properly by the following equations and the coefficients listed in Table 3.

$$f_s = \gamma_p / (a_s + b_s \cdot \gamma_p) \qquad (5)$$
$$m_s = \alpha_s \cdot (\gamma_p)^{\beta_s} \qquad (6)$$
$$f_d = \gamma_p / (a_d + b_d \cdot \gamma_p) \qquad (7)$$
$$m_d = (0.1 - \gamma_p)/(\alpha_d + \beta_d \cdot (0.1 - \gamma_p)). \qquad (8)$$

Secondly, with regard to pore pressure developed in the dynamic test, it is interesting to note the following: In the case of MSR=0.40 and 0.50, since there was much volume left to compress in the static test, the positive pore pressure developed even after the sample reached the void ratio at which dilatancy occured in the static test. The negative pore pressure, on the other hand, developed in the case of MSR=0.57 and 0.67, in which the volumetric strain at the end of initial shear process was near the maximum compression point. Judging from these facts it may be concluded that the occurrence of positive or negative pore pressure is determined by the anisotropic structure of soil

Table 3   Values of coefficients a, b, α, β

	a	b	α	β
Static	0.0007	0.5663	0.0028	-0.7516
Dynamic	0.0009	0.4898	0.0582	1.9839

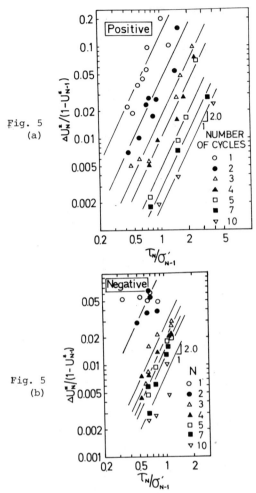

Fig. 5 (a)

Fig. 5 (b)

Fig. 5 Increment of pore pressure rise in each cycle

Figs. 5 (a) and (b) show the relationships in each cycle between the modified pore pressure rise $\Delta U_N^* / (1-U_{N-1}^*)$ and stress ratio (τ_N / σ_{N-1}') for the positive and negative pore pressure data respectively. It is seen in these figures that the lines are almost parallel and they have the same slope α of 2.0. The other three parameters C_1, C_2 and C_3 can be determined by curve fitting into the experimental data of $(\Delta U_N^* / (1-U_{N-1}^*)) / (\tau_N / \sigma_{N-1}')^\alpha$ and N. The values of those parameters are listed in Table 4. Thus, it may be concluded that the pore pressure rise in each cycle can be determined uniquely by MSR, N and (τ_N / σ_{N-1}') and that the effective stresses can be obtained at any time in the dynamic condition.

Table 4   Values of parameters C_1. C_2, C_3, α

		α	C_1	C_2	C_3
Positive	(MSR<0.57)	2.0	0.20	3.20	-0.10
Negative	(N<2.198)	2.0	1.00	4.42	-1.50
(MSR≥0.57)	(N≥2.198)	2.0	0.033	1.62	2.46

4 QUANTITATIVE REPRESENTATION OF DYNAMIC PROPERTIES

In general, the representation of a nonlinear hysteretic curve in the equivalent linear method is realized by the two parameters; one is shear modulus G which is practically defined as the slope of the line connecting the point on the curve corresponding to the maximum stress (or the maximum strain) with the origin, the other is damping ratio D which is practically defined as the ratio of the dissipation energy per cyclic loading ΔW to the strain energy stored in the material per cyclic loading W.

As mentioned above, in the dynamic condition applying initial shear stress, the significant increase of plastic shear strain occurs particularly in the first 10 cycles and the energy is dissipated more than in the isotropic stress condition due to the plastic work. Thus, the starting and the final points on the hysteretic curve in each cycle do not coincide. Therefore the two parameters must be evaluated according to the actual hysteretic curve as shown in Fig. 6. That is, shear modulus G is defined as

$$G = q_a / \gamma_a \tag{10}$$

and damping ratio D is defined as

$$D = \frac{1}{4\pi} \cdot \frac{\Delta W}{W} . \tag{11}$$

particles which is constructed by applying initial shear stress. Then, applying the equation proposed by Sherif, Ishibashi and Tsuchiya (1978) to the test results, it is shown that the change of pore pressure can be expressed properly by the equation,

$$\Delta U_N^* = (1-U_{N-1}^*) \cdot \frac{C_1 \cdot N}{N^{C_2-C_3}} \cdot (\frac{\tau_N}{\sigma_{N-1}'})^\alpha \tag{9}$$

in which ΔU_N^* = the increment of normalized pore pressure during Nth cycle; U_{N-1}^* = the normalized residual pore pressure at (N-1)th cycle (the normalized pore pressure equals the actual pore pressure divided by the initial effective confining pressure); N = number of cycles; τ_N = the shear stress at Nth cycle; σ_{N-1}' = the effective confining pressure at the end of (N-1)th cycle; and C_1, C_2, C_3 and α are material parameters.

124

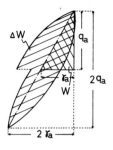

Fig. 6 Definition of
shear modulus G
and damping ratio D

4.1 Shear modulus G

To neglect the effect of void ratio on
shear moduli, shear modulus G is divided by
the shear modulus G_o at very small shear
strain amplitude (Hardin and Richart 1963)
which is practically used by many investigators including Hardin and Drnevich.

$$G_o = 700 \cdot \frac{(2.17-e)^2}{1+e} \cdot (p_S')^{0.5} \qquad (12)$$

in which p_S' is initial effective confining
pressure, and p_S' and G_o are represented in
kg/cm^2. As a result of selecting from many
factors indicated by Hardin and Black
(1966), major factors affecting G/G_o are
considered to be consolidation pressure σ_3,
static mobilized stress ratio MSR, initial
effective confining pressure p_S', effective
mean principal stress p', shear stress amplitude q_a, shear strain amplitude γ_a, number of cycles N and frequency of vibration
f. Figs. 7 (a) and (b) show the relationship
$G/G_o \sim \gamma_a$ at N=10 for σ_3=0.5 kg/cm^2 and 1.0
kg/cm^2 respectively. It can be considered
in these figures that frequency of vibration
at least from 1 Hz to 4 Hz has almost no effect on G/G_o. Furthermore, it is interesting
to note that G/G_o at any γ_a increases with
increasing of MSR and σ_3 and that the curve
fitted into the experimental data are almost
parallel to each other. Then, it is examined
whether the change of G/G_o can be determined
quantitatively by shear strain amplitude γ_a
and effective mean principal stress p' according to the following equation which is
originally established by Silver and Seed
(1971),

$$G/G_o = K(\gamma_a) \cdot (p')^{m' (\gamma_a)} \qquad (13)$$

in which $K(\gamma_a)$ equals G/G_o at p'=1 kg/cm^2,
$[G/G_o]_{p'=1}$. However since the dynamic tests
at p'=1 kg/cm^2 were not carried out in this
study, $K(\gamma_a)$ was obtained from the calculation with use of the experimental data and
the various curves of the relationship
$m' (\gamma_a) \sim \gamma_a$ established by Iwasaki, Tatsuoka

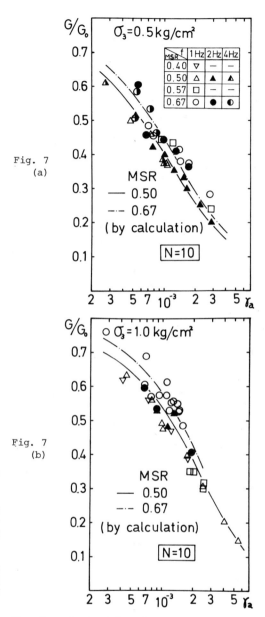

Fig. 7
(a)

Fig. 7
(b)

Fig. 7 Relationship $G/G_o \sim \gamma_a$ at N=10

and Takagi (1978), which are shown in Fig. 8.
As a result of the calculation, it is found
that almost the same relationship
$[G/G_o]_{p'=1} \sim \gamma_a$ is obtained to any $m' (\gamma_a) \sim \gamma_a$
curve. Fig. 9 shows $[G/G_o]_{p'=1} \sim \gamma_a$ relation-
ship at N=10. The curves in Figs. 7 (a) and
(b) are obtained by use of the curve in
Fig. 9. Good agreement is found between
those curves and the experimental data in
the figures. Fig. 10 shows a comparison of
the relationship $[G/G_o]_{p'=1} \sim \gamma_a$ among the

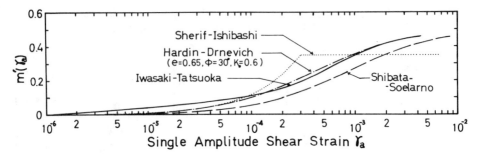

Fig. 8 Relationship $m'(\gamma_a) \sim \gamma_a$ (Iwasaki, Tatsuoka and Takagi 1978)

curve by the authors, the average curve by Seed and Idriss (1970) and the curves by the other investigators listed in Table 5. It is seen that the difference among those curves is rather small and that the curve by the authors is located midway between the curve by Shibata and Soelarno (1975) and that by Iwasaki, Tatsuoka and Takagi (1978).

4.2 Damping ratio D

The relationship between damping ratio D and shear modulus ratio G/G_o at N=10 is examined as shown in Fig. 11. It can be considered that frequency of vibration has almost no effect on damping ratio. This result agrees with the previous result (for example, Hardin 1965) that the major part of internal damping in soils is primarily caused by hysteretic damping independent of frequency of vibration. Moreover, it is seen in Fig. 11 that the relationship $D \sim G/G_o$ is determined irrespective of MSR and p'. This slightly downward convex curve in Fig. 11 can be ex-

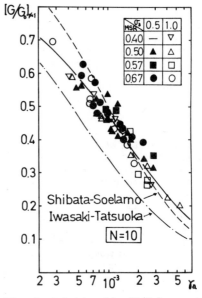

Fig. 9 Relationship $[G/G_o]_{p'=1} \sim \gamma_a$ at N=10

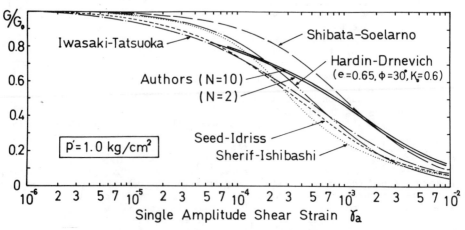

Fig. 10 Comparison among relationships $[G/G_o]_{p'=1} \sim \gamma_a$

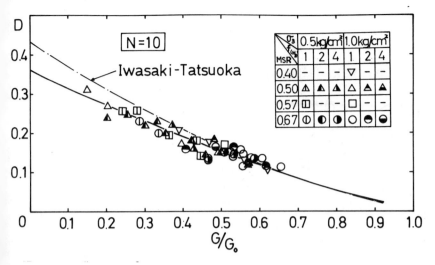

Fig. 11 Relationship $D \sim G/G_o$ at N=10

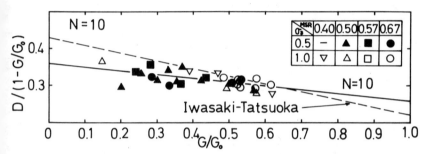

Fig. 12 Relationship $D/(1-G/G_o) \sim G/G_o$ at N=10

pressed by the following equation proposed by Lazan (1968),

$$D = \frac{H_o}{2\pi} \cdot (1 - G/G_o). \qquad (14)$$

This equation is similar to that proposed by Hardin and Drnevich (1972b). The difference between them is that the coefficient H_o of the former varies with strain amplitude but the coefficient of the latter is constant. Then the plots of $D/(1-G/G_o)$ versus G/G_o are shown in Fig.12. It is seen that the coefficient H_o is not constant but has a linear relationship with G/G_o. Accordingly, the relationship between D and G/G_o can be expressed exactly by the following equation,

$$D = (a + b \cdot G/G_o) \cdot (1 - G/G_o). \qquad (15)$$

Fig. 13 shows a comparison of the relationship $D \sim G/G_o$ among the curve by the authors, the average curve by Seed and Idriss and the curves by the other investigators listed in Table 5. It is found that the curve by the

authors is near the curve by Tatsuoka, Iwasaki and Takagi (1978) and that comparing the curve by the authors with the curves by the other investigators, the curve by the authors shows rather higher values of D.

5 CONSIDERATION OF THE EFFECTS DUE TO INITIAL SHEAR STRESS

The anisotropic structure of soil particles due to initial shear stress has effects on not only plastic shear strain and pore pressure as mentioned above but also shear strain amplitude. Figs. 14 (a), (b) and (c) show the relationship between shear strain amplitude γ_a and the ratio of shear stress amplitude q_a to effective mean principal stress p', (q_a/p'), plotting on logarithmic graph paper for $\sigma_3 = 1.0$ kg/cm^2 and N=2, 5 and 10 respectively. It is seen that γ_a is related to (q_a/p') by the following equation,

$$\gamma_a = a \cdot (q_a/p')^b. \qquad (16)$$

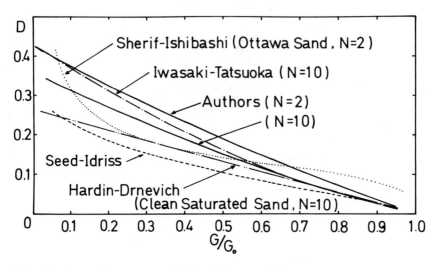

Fig. 13 Comparison among relationships $D \sim G/G_o$

Table 5 Equations established by other investigators

Reference	Relationship G/G_o (or G) $\sim \gamma_a$	Relationship $D \sim G/G_o$ (or γ_a)
Hardin-Drnevich (Hardin & Drnevich 1972b)	$G/G_o = 1 / (1+\gamma_a/\gamma_r)$ γ_r: the reference strain ($=\tau_f/G_o$) $\tau_f=[(\frac{1+K_o}{2} \cdot \sigma_y' \cdot \sin\phi + C \cdot \cos\phi)^2$ $\quad - (\frac{1-K_o}{2} \cdot \sigma_y')^2]^{0.5}$ $G_o = 700 \cdot \frac{(2.17-e)^2}{1+e} \cdot (p')^{0.5}$	$D = D_o \cdot (1 - G/G_o)$ $D_o = (28 - 1.5 \cdot \log N)/100$
Sherif-Ishibashi (Sherif & Ishibashi 1976) (Sherif, Ishibashi & Gaddah 1977)	$(\gamma_a < 3\times10^{-4})$ $G = 2.8\phi \cdot (\frac{p'}{0.0703})^{1167\gamma_a+0.5} \cdot 40$ $\quad \cdot (0.205)^{\gamma_a/0.0003} \cdot 0.0703$ $(\gamma_a \geq 3\times10^{-4})$ $G = 2.8\phi \cdot (\frac{p'}{0.0703})^{0.85} \cdot (100 \cdot \gamma_a)^{-0.6}$ $\quad \cdot 0.0703$	$D = 0.1648 \cdot (5.86-p') \cdot (\gamma_a)^{0.3}$ (Ottawa Sand)
Iwasaki-Tatsuoka (Iwasaki, Tatsuoka & Takagi 1978) (Tatsuoka, Iwasaki & Takagi 1978)	$G/G_o = [G/G_o]_{p'=1} \cdot (p')^{m'(\gamma_a)}$ $G_o = 900 \cdot \frac{(2.17-e)^2}{1+e} \cdot (p')^{0.4}$	$D = D_{max} \cdot (1 - G/G_o)$ (D_{max} is indicated by the curve in their paper, which can be expressed by the following equation.) $D_{max} = 0.4295 - 0.2078 \cdot G/G_o$
Shibata-Soelarno (Shibata & Soelarno 1975)	$G/G_o = 1 / (1+1000 \cdot \gamma_a/(p')^{0.5})$	D : not specified

* p', G and G_o are represented in kg/cm^2; ϕ is represented in degree.

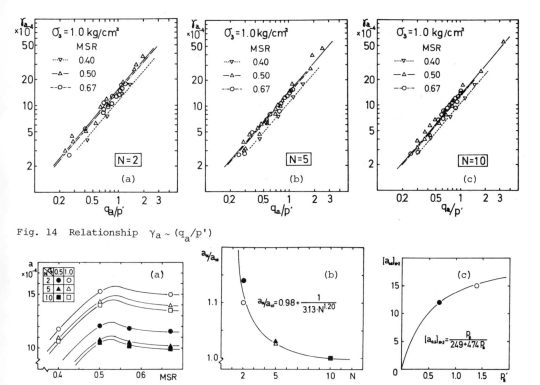

Fig. 14   Relationship $\gamma_a \sim (q_a/p')$

Fig. 15   Variations of coefficient 'a' in Eq. (16)

in which the coefficient 'b', which is the slope of the line in those figures, is constant irrespective of MSR, σ_3 and N, so 'b' is equal to 1.15. On the other hand, the coefficient 'a' varies with MSR, σ_3 and N. The relationship 'a' $\sim$ MSR is shown in Fig. 15 (a), the relationship $(a_{0.5})_N/(a_{0.5})_{10} \sim N$ in Fig. 15 (b), which is the ratio of the value of 'a' at any N and for MSR=0.50 to that at N=10 and for MSR=0.50, and the relationship $(a_{0.5})_2 \sim$ initial confining pressure p_s' in Fig. 15 (c), which is the value of 'a' at N=2 and for MSR=0.50. As mentioned in Section 3, the intensity of initial shear stress which is determined by MSR and σ_3 influences the anisotropic structure of soil and pore pressure. This is the reason γ_a varies with MSR and σ_3. Consequently, it may be concluded that initial shear stress considerably influences shear strain amplitude and thus the dynamic deformation characteristics which have strong dependency on shear strain amplitude.

Next, it seems that the anisotropic structure of soil due to initial shear stress is gradually changed by repeated loading into a rather isotropic structure. So, the effects of intial shear stress clearly appear in the dynamic deformation characteristics at a small number of cycles. Fig. 16

and 17 show the relationships $G/G_0 \sim \gamma_a$ and $[G/G_0]_{p'=1} \sim \gamma_a$ at N=2 and 5 respectively. It may be seen from the comparison among Figs. 9, 17 (a) and 17 (b) that number of cycles has small effect on the relationship $[G/G_0]_{p'=1} \sim \gamma_a$. Fig. 18 shows the relationship $D \sim G/G_0$ at N=2 and Fig. 19 shows the relationship $D/(1-G/G_0) \sim G/G_0$ at N=2 and 5. It may be seen in Fig. 19 that the straight lines for different number of cycles are almost parallel ('b' in Eq. (16) is about -0.10.) and that the maximum damping ratio $(D)_{G/G_0=0}$ decreases with increasing number of cycles as shown in Fig. 20. Accordingly, it may be concluded that the change of soil structure with increasing number of cycles has a small effect on shear moduli but a greater effect on damping ratio.

Finally, the comparison is made between the relationship $[G/G_0]_{p'=1} \sim \gamma_a$ by the authors and that by Shibata and Soelarno which is obtained in the isotropic stress condition with use of the same material and the same type apparatus as used in this study. It is seen in Fig. 10 that $[G/G_0]_{p'=1}$ applied initial shear stress is smaller than that in the isotropic condition for γ_a less than about 10^{-3}, but both $[G/G_0]p'=1$ almost agree for γ_a larger than 10^{-3}. This may be caused by the following: The anisotropic

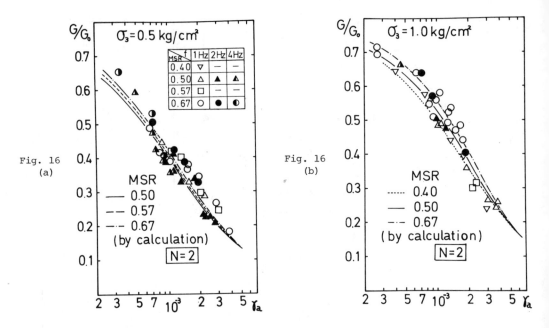

Fig. 16 (a)

Fig. 16 (b)

Fig. 16 Relationship $G/G_o \sim \gamma_a$ at N=2   (a) for σ_3=0.5 kg/cm^2 (b) for σ_3=1.0 kg/cm^2

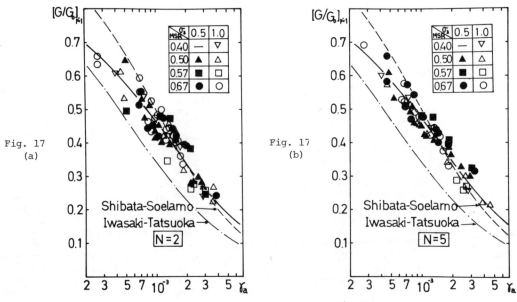

Fig. 17 (a)

Fig. 17 (b)

Fig. 17 Relationship $[G/G_o]_{p'=1} \sim \gamma_a$ (a) at N=2 (b) at N=5

structure of soil due to initial shear stress reduces G/G_o for smaller shear strain amplitude, but that effect disappears as the anisotropic structure approaches isotropic structure for larger shear strain amplitude.

Accordingly, it may be concluded that the change of soil structure with increasing shear strain amplitude has significant effects on shear moduli and that the effects of initial shear stress on shear moduli

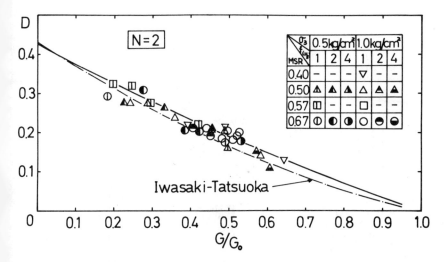

Fig. 18 Relationship $D \sim G/G_o$ at N=2

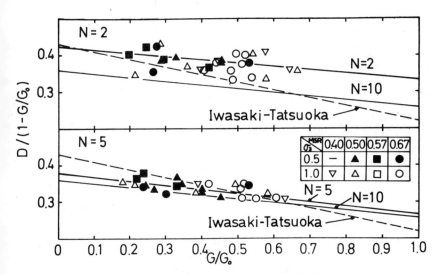

Fig. 19 Relationship $D/(1-G/G_o) \sim G/G_o$ at N=2 and 5

disappears for γ_a larger than about 10^{-3}.

6 APPLICATION TO DYNAMIC RESPONSE ANALYSIS

As a result of this study, the dynamic deformation characteristics of saturated Toyoura Sand subjected to initial shear stress
can be determined quantitatively according
to the following procedures: First, when the
values of static mobilized stress ratio MSR,
initial confining pressure p'_s, shear stress
amplitude q_a and number of cycles N (for
example, q_a equal to 0.6 times the peak

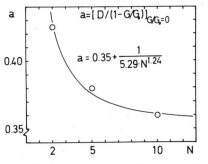

Fig. 20 Relationship $(D)_{G/G_o=0}$ ('a') $\sim$ N

response shear stress amplitude and N equal
to the number of cycles of q_a specified
above) are known, shear strain amplitude γ_a
can be determined by Eq. (16) and Figs. 19
(a), (b) and (c), and pore pressure U can be
determined by Eq. (9) and Table 4. Secondly,
with use of γ_a and effective mean principal
stress p' ($= p_s' - U$), shear moduli G can
be determined by Eqs. (12) and (13) and
Figs. 8 and 9, and damping ratio D can be
determined by Eq. (15) and Fig. 20. Thirdly,
the change of the state parameter S_S and
plastic shear strain γ_p according to response stress history can be determined by
Eqs. (4) through (8) and Fig. 2. These procedures are shown schematically in Fig. 21.

As mentioned above, shear moduli G and
damping ratios D for dynamic response analysis can be easily determined according to
response stress history using the effective
stresses. It is easy to incorporate those
equations into computer programs for the
dynamic response analysis. Figs. 22 (a) and
(b) show the calculated results of an example of the earthquake response analysis
using the results obtained in this study.
In the analysis, the model dam of Toyoura
Sand is 37.5 m in height and the acceleration data of the Taft earthquake (1952) is
applied.

7 CONCLUSIONS

From an experimental study on the dynamic
properties of saturated Toyoura Sand subjected to initial shear stress under a dynamic triaxial test, the following conclusions are obtained.
1. Initial shear stress influences both
shear strain amplitude and pore pressure
developed by repeated loading. Judging from
the strain-dependency in shear moduli and
damping ratios, and the dependency of shear
moduli on effective mean principal stress,
initial shear stress considerably influences
the dynamic deformation characteristics.
2. Due to the anisotropic structure in-
curred by applying initial shear stress,
plastic shear strain significantly increases , particularly in the first 10 cycles. Plastic shear strain is directly related to the state parameter S_S by Moroto
and so is related to stress p' and q when
the yield function established by Poorooshasb is introduced.
3. The ratio of shear modulus G to that
at very small shear strain amplitude G_o,
G/G_o, is determined as a function of G/G_o
for p'=1.0 kg/cm^2, $[G/G_o]_{p'=1}$, effective
mean principal stress p' and the power
m'(γ_a) such as Eq. (13).
4. Damping ratio D is related to G/G_o by
Eq. (15) irrespective of static mobilized

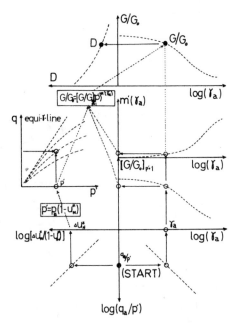

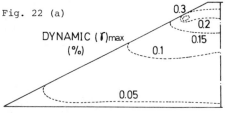

Fig. 21 Schematic presentation of procedure
determining the dynamic properties

Fig. 22 (a)

DYNAMIC $(\gamma)_{max}$
(%)

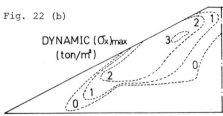

Fig. 22 (b)

DYNAMIC $(\sigma_x)_{max}$
(ton/m^2)

Fig. 22 Typical results of earthquake
response analysis

stress ratio MSR and effective mean principal stress p', while the relationship D ∼
G/G_o is influenced by number of cycles N.
5. The investigation of G/G_o and D at
small number of cycles shows that the change
of structure of soil with increasing number
of cycles has a great effect on damping ratios, while it has a rather small effect on
shear moduli.

6. The comparison between the relationship $[G/G_o]_{p'=1} \sim \gamma_a$ by the authors obtained by applying initial shear stress and that by Shibata and Soelarno in the isotropic stress condition makes it clear that the anisotropic structure of soil due to initial shear stress reduces the value of G/G_o for small shear strain amplitude but with the change of the anisotropic structure of soil with increasing shear strain amplitude, this effect on G/G_o decreases.

7. Little difference can be found between the results of G/G_o and D by the authors and those by other investigators. The curve expressing the relationship $[G/G_o]_{p'=1} \sim \gamma_a$ by the authors is located midway between the curve by Shibata and Soelarno and that by Iwasaki, Tatsuoka and Takagi. As for damping ratio D, the result by the authors gives relatively higher values than those by the other investigators.

8. On the basis of the results in this study, shear moduli G and damping ratios D for the dynamic response analyses can be determined quantitatively.

8' REFERENCES

Hardin, B.O. 1965, The nature of damping in sands. Proc. ASCE, Vol. 91, No. SM1: 63-97.

Hardin, B.O. & Black, W.L. 1966, Sand stiffness under various triaxial stress. Proc. ASCE, Vol. 92, No. SM2: 27-42.

Hardin, B.O. & Drnevich, V.P. 1972a, Shear modulus and damping in soils: Measurement and parameter effects. Proc. ASCE, Vol. 98, No. SM6: 603-624.

Hardin, B.O. & Drnevich, V.P. 1972b, Shear modulus and damping in soils: Design equations and curves. Proc. ASCE, Vol. 98, No. SM7: 667-692.

Hardin, B.O. & Richart,F.E.,Jr. 1963, Elastic wave velocities in granular soils. Proc. ASCE, Vol. 89, No. SM1: 33-65.

Ishihara, K., Tatsuoka, F. & Yasuda, S. 1975, Undrained deformation and liquefaction of sand under cyclic stresses. Soils and Foundations, Vol. 15, No. 1: 29-44.

Iwasaki, T., Tatsuoka, F. & Takagi, Y. 1978, Shear moduli of sands under cyclic torsional shear loading. Soils and Foundations, Vol. 18, No. 1: 39-56.

Kuribayashi, E., Iwasaki, T. & Tatsuoka, F. 1974, Effects of stress conditions on dynamic properties of sands. Bull. Int. Inst. Seismology and Earthquake Eng., Vol. 12: 117-130.

Lazan, B.J. 1968, Damping of materials and members in structural mechanics. Pergamon Press Ltd., London.

Moroto, N. 1976, A new parameter to measure degree of shear deformation of granular material in triaxial compression tests. Soils and Foundations, Vol.16, No. 4: 1-9.

Poorooshasb, H.B. 1971, Deformation of sand in triaxial compression. Proc. 4th Asian Regional Conf. on Soil Mech. and Found. Eng., Bangkok, Vol. 1: 63-66.

Seed, H.B. & Idriss, I.M. 1970, Soil moduli and damping factors for dynamic response analyses. EERC Report, No. 70-10.

Sherif, M.A. & Ishibashi, I. 1976, Dynamic shear moduli for dry sands. Proc. ASCE, Vol. 102, No. GT11: 1171-1184.

Sherif, M.A., Ishibashi, I. & Gaddah, A.H. 1977, Damping ratio for dry sands. Proc. ASCE, Vol. 103, No. GT7: 743-756.

Sherif, M.A., Ishibashi, I. & Tsuchiya, C. 1978, Pore pressure prediction during earthquake loading. Soils and Foundations, Vol. 18, No. 4: 19-30.

Shibata, T. & Soelarno, D.S. 1975, Stress-Strain characteristics of sands under cyclic loading. Proc. Japanese Soc. Civil Engineers, No. 239: 57-65 (in Japanese).

Silver, M.L. & Seed, H.B. 1971, Deformation characteristics of sands under cyclic loading. Proc. ASCE, Vol. 97, No. SM8: 1081-1098.

Tatsuoka, F., Iwasaki, T. & Takagi, Y. 1978, Hysteretic damping of sands under cyclic loading and its relation to shear modulus. Soils and Foundations, Vol. 18, No. 2: 25-40.

Timmerman, D.H. & Wu, T.H. 1969, Behavior of dry sands under cyclic loading. Proc. ASCE, Vol. 95, No. SM4: 1092-1112.

Toki, S. & Kitago, S. 1974, Effects of repeated loading on deformation behavior of dry sand. J. Japanese Soc. Soil Mech. and Found. Eng., Vol.14, No. 1: 95-103 (in Japanese).

Cyclic strength and shear modulus as a function of time

JOHN VRYMOED & WILLIAM BENNETT
Department of Water Resources, Sacramento, USA

SIAMAK JAFROUDI & C.K.SHEN
University of California, Davis, USA

1 SYNOPSIS

Recent laboratory investigations have identified differences in cyclic strengths and shear moduli between undisturbed and remolded soil samples. An increase in shear modulus was found by Afifi, Anderson, Richart, Woods, and others, when samples were consolidated for a period of time. Cyclic strengths also increased when samples were similarly subjected to a period of sustained pressure (Seed, 1976). The increases in shear modulus and cyclic strength, however, have not been investigated simultaneously using the same soil.

This paper summarizes a laboratory investigation on undrained, cyclically-loaded, triaxial test samples consisting of a silty-clayey sand. The authors compare the variations in shear modulus and cyclic strength as a function of time. These variations are then combined in an analysis to determine the overall effect on dynamic stability. The tested samples consist of undisturbed specimens recovered from a 4-year-old compacted embankment and remolded samples of the same material consolidated for one hour to 61 days.

2 INTRODUCTION

A complete examination of the dynamic stability of an embankment is usually performed in four phases:
1) review and analysis of existing information, 2) subsurface exploration and laboratory testing, 3)

geologic study and seismic investigation, and 4) static and dynamic stress analyses and interpretation of results.

The determination of the shear modulus and cyclic strength of the different soils of an embankment plays a crucial role in the final determination of the embankment's overall stability. The induced stress as determined in phase 4, is in many cases almost proportional to the assigned shear modulus. Recent studies have shown that the dynamic properties of a soil may be significantly influenced by the duration of confinement. To date, investigations have concentrated exclusively on either the change in a soil's shear modulus or the change in a soil's cyclic strength with duration of confinement. Since the induced stress which is influenced by the shear modulus must ultimately be compared to the cyclic strength for an indication of stability, it is important to study the effect of time on both shear modulus and cyclic strength simultaneously.

3 PREVIOUS STUDIES

Results of earlier studies show that the shear modulus of soils is dependent upon the duration of confinement. This dependency is significant for the more fine-grained soils. Afifi and Richart (1973) concluded that the duration of confinement had little effect on the shear modulus of soils with a mean grain size greater than 0.04 mm. This effect was significant for

135

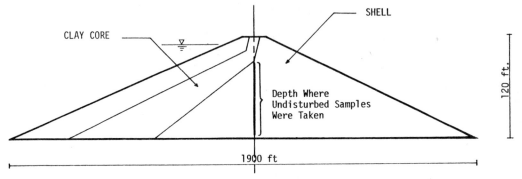

Figure No. 1   Perris Dam   Maximum Section (1 ft. = .3048 m)

finer grained soils.

Anderson and Stokoe (1978) indicated that the greatest increase in shear modulus occurred during primary consolidation. Subsequent increases in shear modulus were identified as long-term time effects. The increases in shear modulus were found to occur for low and high amplitude shear strains, the latter approaching 0.1 percent.

Other investigators (Anderson, 1975; Stokoe, 1975) obtained good agreement between laboratory and field determined shear moduli when careful consideration was given to time effects. If time effects were not included, the laboratory determined shear moduli were less than the in situ determined shear moduli.

The literature contains limited data on the effect of duration of confinement on the cyclic strength of soils. It is generally accepted, however, that the cyclic strength of reconstituted samples underestimates the cyclic strength of undisturbed samples of the same material. The lower cyclic strength of reconstituted samples can be attributed to the absence of any bonding at the particle contacts due to cementation. Coupled with effects of cementation is the effect of secondary consolidation which likewise contributed to the underestimation of the cyclic strength of reconsituted samples. Mori et al. (1978) indicated that the difference in cyclic strength between reconstituted and undisturbed samples due to these effects can be on the order of 75 percent.

Duration of confinement affects the shear modulus and cyclic strength

of laboratory samples as previous investigators have indicated. The reasons for an increase in shear modulus and an increase in cyclic strength are the same. Data coupling both of these increases, however, for a given soil was not found in the literature survey conducted.

4 LABORATORY INVESTIGATION

4.1 Materials

Undisturbed samples were obtained from Perris Dam, a 36.6 m (120 ft.) high zoned earth embankment located in Riverside County, California. The dam is owned and operated by the California Department of Water Resources (DWR) and was completed in October of 1972. A cross-section of the dam is shown on Figure No. 1. A 15.2 centimetre (6-in.) diameter hole was drilled through the crest to a depth of 36.6 m at the dam's maximum section. Undisturbed samples, 7.1 cm in diameter, were taken in the downstream shell zone using a pitcher barrel sampler.

The sampled material is classified as a silty-clayey sand. Table 1 gives the overall average soil characteristics as determined from construction records. The in situ density corresponds to 99-100 percent of the DWR compaction standard 957 kN-m/m^3 (20,000 ft-lbs/ft^3).

The material for the reconstituted samples were obtained from the borrow area used for the shell zone of the dam. When this material was compacted to a 100 percent of the DWR

136

standard, a density of 2098 kN-m/m^3 (131) (pcf) was obtained with a corresponding moisture content of 9.1 percent. This density was higher than the overall average density obtained during construction. Rather than matching the density, however, it was decided to match the percent compaction as well as the overall average gradation. The diameter and height of the reconstituted samples was 7.1 and 14.2 cm respectively.

Table 1. Soil characteristics.

% passing No. 4 sieve (4.76 mm)	100
% passing No. 200 sieve (0.074 mm)	50
% passing 5 micron	20
Liquid limit	23
Plastic index	7
Specific gravity	2.75
Average density (Kg/m^3)	2011.9
(lbs/ft^3)	125.6
% water content	10.8

4.2 Testing procedures

An MTS hydraulic actuator was used to apply the loads and strains to the sample. The MTS can be used in both a constant load or displacement mode. Loads and displacements were applied in the form of a sinusoidal wave pattern. An electronic feedback control kept the load and displacement constant.

Load and displacement were measured using a 1500 Kg load cell and a 25.4 cm linear variable differential transformer respectively. The digital voltmeter made it possible to accurately apply the static load. Hysteretic load-deformation loops were recorded by a Houston X-Y recorder.

The reconstituted and undisturbed samples were placed in triaxial cells and back pressure saturated. The confining pressure under which the undisturbed samples were aged was determined as follows:

$$\sigma_{3c}' = \left[\gamma H + 2\left(\frac{\mu}{1+\mu}\right)\gamma H\right] \Big/ 3 \qquad \text{eq. 1}$$

where:

σ_{3c}' = consolidation pressure applied to samples
γ = unit weight of soil
H = depth where sample was taken
μ = poisson's ratio

The value of poisson's ratio was determined using the following relationship:

$$\mu = G - F \log\left(\frac{\sigma_h}{P_a}\right) \qquad \text{eq. 2}$$

Where G and F are parameters determined in the laboratory (Wong, 1974) and P_a and σ_h are the atmospheric and vertical pressures respectively. Testing of the Perris Dam shell material (Woodward-Clyde, 1975) indicated values of G and F of 0.38 and 0.12 respectively.

Only one effective confining pressure was used for the study, therefore, only undisturbed samples from approximately the same depth were tested. This depth corresponds to mean effective confining pressure of 430 kN/m^2 (62.5 psi) with the use of eq. 1.

All samples were tested at this confining pressure in both the cyclic strength and shear modulus phases of the study. The tests were carried out under isotropic stress conditions.

In the determination of shear modulus, the MTS machine was used in the constant displacement mode. The induced shear strain of 5.5×10^{-2} percent was kept constant for all samples tested. This shear strain was selected because it is representative of shear strains induced in well compacted soil embankments by moderate ground shaking.

After the shear modulus determination, the samples were tested to determine their cyclic strengths. A cyclic stress ratio (deviatoric shear stress normalized with the confining pressure) of 0.45 was applied to all samples. This stress ratio resulted in a double amplitude strain of 10 percent in a number of cycles characteristic of moderate ground shaking.

Tests were carried out to determine what effect the prior shear modulus testing had on the cyclic strength of the samples. Results indicated that there was no measurable influence on the cyclic strengths determined for this material compacted to the aforementioned density. This maximized the use of samples which is important in a study involving time dependency.

Twenty-one reconstituted samples and four undisturbed samples obtained from the embankment were tested in the above manner. The aging of the reconstituted samples varied from several hours to 61 days. The undisturbed samples represent samples aged for more than 1,000 days. Some of the undisturbed samples were also aged within the testing apparatus for an additional period of 4 hours and 10 days to determine what effect that procedure had on the shear modulus and cyclic strength.

5 FINDINGS

5.1 Dynamic properties

Figures 2 and 3 present the variance of dynamic properties as a function of time of confinement.

In Figure 2, the shear modulus recorded for the first five stress cycles is plotted with the logarithm of time in days. The shear modulus was determined from the slope of the chord passing through the peaks of the axial load-deformation hysteresis loop. The figure indicates nearly a straight line relationship within the scattering of data. The modulus increases with increasing time of confinement. For the undisturbed samples aged in situ for more than 1,000 days, the modulus is 15 percent greater than that of the average one-day remolded sample.

Figure 3 relates the change in damping ratio with time of confinement. Damping was calculated from the area of the hysteresis loop. The plot reveals the damping generally decreases with increasing duration of confinement. The damping for the undisturbed samples is approximately 35 percent lower than the damping recorded from a one-day remolded sample.

In Figure 3, the undisturbed sample subjected to additional aging in the testing apparatus (10 days) appears to have a somewhat higher modulus than the sample aged only an additional four hours prior to testing. However, the damping ratios for those particular samples, as seen in Figure 3, are not significantly different and are within the scattering of the data. Although the quantity of data is limited, the test results suggest additional duration of confinement or aging of undisturbed samples prior to testing may increase the stiffness of samples by roughly 5 percent. This increase would tend to balance the loss in stiffness due to the disturbance caused by the sampling, extrusion and other subsequent handling of the soil samples.

5.2 Cyclic strength

Figure 4 relates the change in cyclic strength with time. Each triaxial sample was subjected to the same cyclic deviator stress and the number of cycles to 2, 5, and 10 percent peak-to-peak axial strain was recorded. The figure displays the logarithm of number of cycles to these strains as a function of duration of confinement. As one would predict, the number of cycles to a given strain generally increases with time, indicating the cyclic strength of the soil improves with age. For the range of one day to four years, there is a four-fold increase in the number of cycles.

Additional aging of the undisturbed samples generally produces a small increase in the sample's ability to resist deformation under cyclic load. In terms of cycles, the increase is approximately 12 percent for the larger strain criteria.

6 ANALYSIS

6.1 Procedure

A dynamic analysis was carried out to determine the overall effect on stability having taken into account the variations of modulus, damping and cyclic strength with time. A vertical homogeneous soil column, representing the area of Perris Dam where the undisturbed samples were taken, was mathematically modeled using computer program SHAKE

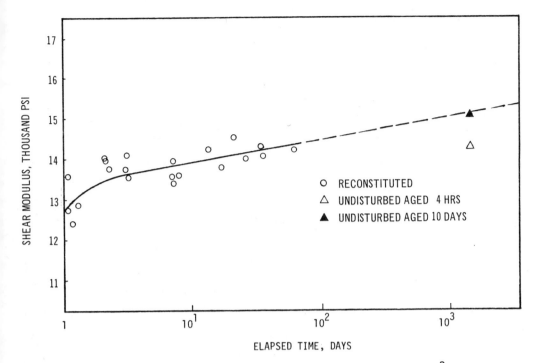

Figure No. 2   Shear Modulus Versus Time   1 psi = 6.89 kN/m^2

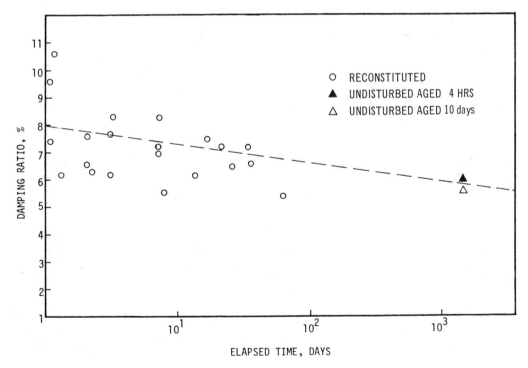

Figure No. 3   Damping Ratio Versus Time

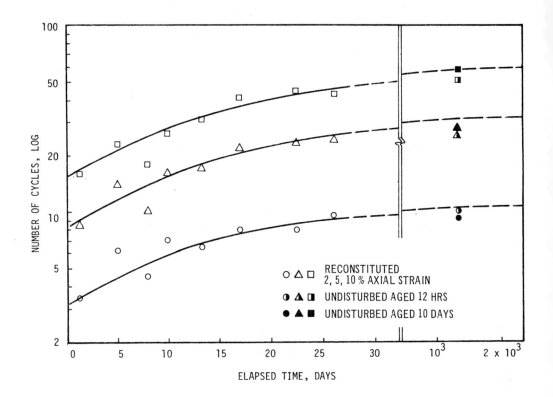

Figure No. 4   Number of Cycles Versus Time

(Schnabel et al., 1972). SHAKE uses a vertically propagating wave theory to compute the dynamic shear stresses of a horizontally layered soil deposit. Studies (Vrymoed, 1978) have shown that a sloped embankment can be adequately represented by such a model. The earthquake accelerogram used for the model was the South 69° East component of the Taft record, 1952 Kern County, California Earthquake, with a scaled peak acceleration of 0.2g.

Separate computations were made incorporating both the newly remolded (one day) and the undisturbed soil properties. The undisturbed soil properties were derived from the undisturbed sample aged an additional 10 days. The irregular shear stress time histories produced by SHAKE were converted to an uniform series of 10 stress cycles as outlined by Seed et al. (1975). The resulting average stress as a function of depth for both the new and aged soil profiles is represented by the solid lines in Figure 5a.

The shear stress required to cause 10 percent axial strain in 10 cycles represents the cyclic strength and is similarly shown as a function of depth by the broken lines in Figure 5a. The variation of cyclic strength with depth was assumed to be consistent with previous cyclic strength testing on the same material (Woodward-Clyde, 1975).

6.2 Results

Figure 5b presents the results of the dynamic analysis. Factor of safety, defined as the ratio of cyclic strength to induced stress, is given as a function of depth for both cases. The factor of safety, in general, is greater for the aged deposit. The gain in cyclic strength with time appears to out-weigh the increase in shear stress. This increase in shear stress is due to the increase shear modulus for the aged deposit.

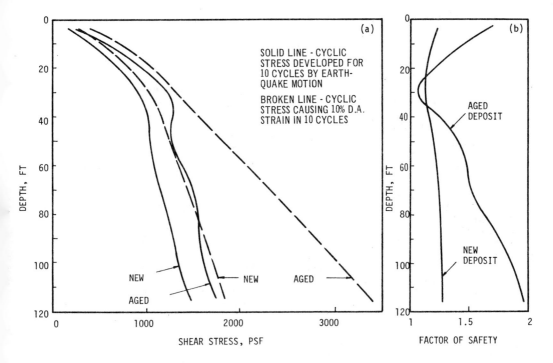

Figures Nos. 5a, 5b Induced Shear Stress and Strength Versus Depth, Factor of Safety Versus Depth (1 psf = 47.9 N/m², 1 ft. = .3048 metre)

However, Figure 5b also shows that the factor of safety is less for the aged deposit than for the new deposit at a depth of 9 m (30 ft.). It would appear that other factors such as depth of a deposit and strong motion characteristics may cause a decrease in the factor of safety at this particular depth.

7 CONCLUSIONS

This investigation indicates that for the soil tested: 1) the shear modulus at high shear strains increases and the damping ratio decreases with the duration of confinement; 2) aging under a constant isotropic confining pressure increases the soil's resistance to cyclic stress-induced deformation; 3) the disturbance introduced to field samples may be compensated for, in part, by a longer period of confinement; and 4) using aged samples to determine both dynamic properties and dynamic strength generally increases the soil deposit's dynamic

stability, however, lower safety factors may result in specific portions of the deposit.

The views expressed in this paper are those of the authors and not the State of California.

8 REFERENCES

Afifi, S.S. and Richart, F.E., Jr. 1973, Stress-History Effects on Shear Modulus of Soils, Soils and Foundations Vol. 13, No. 1.

Afifi, S.S. and Woods, R.D. 1971, Long-Term Pressure Effects on Shear Modulus of Soils, Journal of Soil Mechanics and Foundations Division, ASCE Vol. 97, No. SM10.

Anderson, D.G., Espana, C., and McLamore, V.R. 1978, Estimating Insitu Shear Modulus at Competent Sites, Proceedings Earthquake Engineering and Soil Dynamics, Pasadena, California, ASCE Vol. 1: 181-197.

Anderson, D.G. and Richart, E.F., Jr. 1976, Effects of Straining on Shear Modulus of Clays, Journal of the Geotechnical Engineering Division, ASCE Vol. 102, GT9.

Anderson, D.G. and Stokoe, K.H., II 1978, Shear Modulus: A Time-Dependent Material Property, Special Technical Publication, American Society for Testing Material, ASTM, STP #654.

Anderson, D.G. and Woods, R.D. 1975, Comparison of Field and Laboratory Shear Moduli, Proceedings In Situ Measurement of Soil Properties, Raleigh, North Carolina, ASCE Vol. 1.

Anderson, D.G. and Woods 1976, Time-Dependent Increase in Shear Modulus of Clay, Journal of the Geotechnical Engineering Division, ASCE Vol. 102, GT5.

Arango, I., Moriwaki, Y., and Brown, F. 1978, In Situ and Laboratory Shear Velocity and Modulus, Proceedings, Earthquake Engineering and Soil Dynamics, Pasadena, California, ASCE, Vol. 1.

Marcuson, W.F., III and Townsend, F.C. 1976, Effects of Specimen Reconstitution on Cyclic Triaxial Results, U.S. Army Engineer Waterways Experiment Station Miscellaneour Paper S-76-5.

Marcuson, W.F., III and Wahls, H.E. 1972, Time Effects on Dynamic Shear Modulus of Clays, Journal of Soil Mechanics and Foundations Division, ASCE Vol. 98, No. SM12.

Mori, K., Seed, H. Bolton and Chan, C.K. 1978, Influence of Sample Disturbance on Sand Response to Cyclic Loading, Journal of Geotechnical Engineering Division, ASCE Vol. 104, No. GT3.

Schnabel, P.B., Lysmer, J. and Seed, H.B. 1972, SHAKE, A Computer Program for Earthquake Response Analysis of Horizontally Layered Sites, Earthquake Engineering Research Center, Report No. EERC 72-12, University of California, Berkeley.

Seed, H. Bolton 1976, Evaluation of Soil Liquefaction Effects on Level Ground During Earthquakes, State-of-the-Art Paper presented at Symposium on Soil Liquefaction, ASCE National Convention, Philadelphia.

Seed, H. Bolton, Arango, Ignacio and Chan, Clarence K. 1975, Evaluation of Soil Liquefaction Potential During Earthquake, Earthquake Engineering Research Center, Report No. EERC 75-28, University of California, Berkeley.

Seed, H. Bolton, Idriss, I.M., Makdisi, F., and Banerjee, N. 1975, Representation of Irregular Stress Time Histories by Equivalent Uniform Stress Series in Liquefaction Analysis, Earthquake Engineering Research Center, Report No. EERC 75-29, University of California, Berkeley.

Sherif, M.A. and Ishibashi, I. 1976, Dynamic Shear Moduli for Dry Sands, Journal of the Geotechnical Engineering Division, ASCE Vol. 102, No. GT11.

Stokoe, K.H., Abdel-razzak, K.G. 1975, Shear Moduli of Two Compacted Fills, Proceedings, In Situ Measurement of Soil Properties, Raleigh, North Carolina, ASCE Vol 1:422-447.

Stokoe, K.H. II, and Lodde, P.F. 1978, Dynamic Response of San Francisco Bay Mud, Proceedings, Earthquake Engineering and Soil Dynamics, Pasadena, California, ASCE Vol II:940-959.

Vrymoed, J.L., and Calzascia, E.R. 1978, Simplified Determination of Dynamic Stresses in Earth Dams, Proceedings, Earthquake Engineering and Soil Dynamics, Pasadena, California, ASCE Vol II:991-1006.

Wong, K.S., and Duncan, J.M. 1974, Hyperbolic Stress-Strain Parameters for Nonlinear Finite Element Analyses of Stresses and Movements in Soil Masses, Department of Civil Engineering Institute of Transportation and Traffic Engineering, University of California, Berkeley, Report No. TE-74-3.

Woodward-Clyde Consultants 1975, Evaluation of Seismic Stability of Perris Dam, A Report Prepared for the State of California Department of Water Resources, Division of Design and Construction.

Strength of cohesionless soil under vibratory load

T. CISEK
Silesian Technical University, Gliwice, Poland

1 INTRODUCTION

The soil medium, as well as other structural materials, may be subjected to all kinds of load met in practice. The work of the soil under statical load is considerably well recognized and is far ahead of the knowledge of the effects of dynamic load, both impact and vibratory one. It is proved that all of the above mentioned kinds of loads produce in structural materials, among them also in soil, various effects in the area of their strength and deformations.

Practically, the variety of load effects is the reason for the presently used division of those loads. A large group of subsoil load is of the vibration character. In practice answers for the following questions are searched for:

- does there exist an interrelation between the soil strength and the parameters of vibratory load,
- what their values in given geological conditions are optimal for compaction of soil and sinking various elements in it,
- what parameters of vibratory load

determine the greatest risk for the stability of the existing slopes, road surfaces and engineering foundations.

The presented investigations concerning to cohesionless soil constitute some attempt of the answering those questions.

2 THE CHARACTERISTICS OF TESTED MATERIALS

In the tests there were used soil media with granular compositions as shown on Figure 1. The initial material (medium I) was soil with grains in the interval $0 \div 5.0$mm. The remaining media (II,III and IV) were built through the elimination from medium I of appriopriate fractions.

Investigating the mineralogical composition of the tested media one has distinguished in them over 80 % (volumetrically) quartz, about 5 % orthoclase and small amounts of iron oxides and heavy minerals.

The degree of roundness of the grains was evaluated by the RUCHIN's (1961) method. The results of those tests are placed in table 1. The average specific weight value (γ_s)

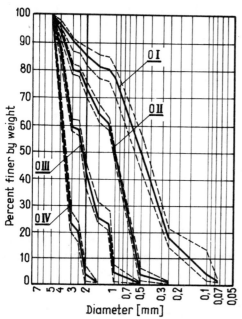

Fig.1. Particle size distribution curve

Table 1. Degree of roundness

Me-dium	Roundness classification					Degree of roundness S^o, %
	0°	1°	2°	3°	4°	
I	8	8	9	22	53	76.0
II	11	9	14	11	45	62.5
III	24	23	11	24	18	47.2
IV	27	25	27	18	9	42.2

of the media was 26.1 kN/m³. The average values of volumetrical weight (γ) depending on the state of density are given in table 2.

Standard strength tests were carried out in a triaxial compression test apparatus. For the tests there were used samples of the following dimensions: d=0.065m; h=0.127m. For medium II there were carried out additional tests on samples with the following dimensions: d=0.038m; h=0.085m.

Table 2. Average values of unit weight γ, kN/m³

Me-dium	Standard density		Density during test
	max	min	
I	17.81	16.76	17.68
II	17.27	16.37	17.07
III	16.23	15.65	16.01
IV	15.99	14.86	15.25

The speed of the protruding of the pivot in the apparatus was $2.66 \cdot 10^5$ m/sec. Drained tests were carried out for two kinds of moisture content states
- dry air humidity condition (w1),
- total saturation (w2).
The standard strength was tested at four levels of stress (σ_3). That strength was defined as the maximum value of the stress deviator $(\sigma_1-\sigma_3)_{f=0}$. The average values of the standard friction angle ($\Phi_{u_{f=0}}$) is listed in table 3.

Table 3. Standard total stress strength parameter $\Phi_{u(f=0)}$, rad

Me-dium	Samples d=0.065m		Samples d=0.038m
	w1	w2	w1
I	0.66123	0.64616	—
II	0.64215	0.63834	0.62223
III	0.64523	0.63651	—
IV	0.64015	0.63501	—

As shown in the introduced results,

in the tests there were used media which are often applied in practice for the loadbearing fills under the fundations for machines. They are also used in draining layers of road subsoil. These are situations, where this type of material works under vibratory load. Another reason for accepting those media for tests is the meagre knowledge of the effects of subjecting them to vibratory load.

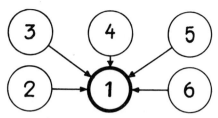

1 – Triaxial test apparatus
2 – Generator of water pulsation
3 – Measuring and testing equipment
4 – Equipment for water pressure stabilisation
5 – Stand for forming and compaction of specimens
6 – Some additional parts of testing equipment

Fig.2.Ideal scheme of the test stand

3 THE METHOD AND THE TECHNIQUE OF THE TESTS

The testing of the strength of the soil against shear pressure stress in the conditions of vibratory load was carried out with the application of the phenomenological method of research.

3.1 The apparatus used for the tests

Taking as basis the considerations contained in the works by Seed and Lee (1966), Stewarski (1972), Stawnicer and Karpushin (1973) and by Chain (1975) a triaxial compression test apparatus was used in the tests. The modification of that apparatus allowed the tests with periodically changable water pressure in the test cell. Besides the typical equipment for triaxial compression test apparatus there was introduced some additional equipment, which is shown on Figure 2.(The details of the test stand in paper by Cisek – 1979).

3.2 The test conditions and techniques

Conditions of tests were following:
- they were carried out for laboratory chosen granular compositions,
- as a content value for all tested samples of a given medium there was accepted its initial porosity,
- in the tests there were determined: standard strength – $(\sigma_1 - \sigma_3)_{f=0}$, post-vibratory strength – $(\sigma_1 - \sigma_3)^*$, vibratory strength – $(\sigma_1 - \sigma_3)^w$,
- as a measure of strength in all the cases there was accepted the maximum value $\sigma_1 - \sigma_3$ in the function

$$\sigma_1 - \sigma_3 = F(\varepsilon_1) \qquad (1)$$

Standard strength was tested after the conditions stated in pt 2.

Post-vibratory strength was tested after the completed process of loading a sample with vibratory water pressure in the test cell of the apparatus. The tests consisted of:
- the producing of a given level of hydrostatic stress,
- applying of minimum 2000 cycles

of additional dynamic water pressure with programmed frequency and amplitude. The fixing of 2000 cycles of vibratory load was based on the results of the diagnostic tests that were carried out. It was noted, that after that number of cycles there was a practical disapearance of the deformations of the cohesionless medium (similarly to Timmerman and Wu – 1970),

– applying of static vertical load $(\Delta\sigma_1)$ leading to the collapse of the sample in an identical way as in standard tests.

In the tests of vibratory strength the collapse loads were applied to the sample during the time of water pulsating in the cell. The application of loads was commenced after there passed enough time (depending on the frequency of the vibrations) allowing the subjecting of the sample to minimum two thousand cycles.

3.3 The programme of tests

Practically all the characteristics of the ground are of multi-factorial nature. Own observations and informations in literature point out to the possibility of a dependence of the strength of vibratory loaded soil on three groups of factors concerning:

– load,
– material,
– geometry.

In the tests there were taken into account factors as follows:

a) load factors:
  – the level of static stress in the test cell (σ_{st}),

  – the amplitude σ_a (contained in the parameter $\varkappa$) and the frequency of vibratory load (f),

b) material factors:
  – the variability of grain compositions (Rg-OI,II,III,IV),
  – extreme moisture contents of the media (w1, w2),

c) geometrical factors:
  – the sizes of the tested samples (d_1, h_1 and d_2, h_2).

Vibratory stresses of water in the testing cell of the triaxial apparatus were changing in a harmonic way from σ_{st} to σ_{max}. The function of the change of stresses can be described with the following formula:

$$\sigma \cong \sigma_{st} + \frac{\sigma_{max} - \sigma_{st}}{2}(1 + \sin\omega t) \qquad (2)$$

or

$$\sigma \cong \sigma_{st} + \varkappa\,\sigma_{st}(1 + \sin\omega t) \qquad (3)$$

The description of functions (2) and (3) is given in Figure 3.

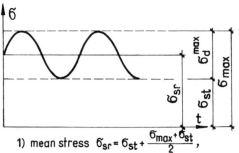

1) mean stress $\sigma_{sr} = \sigma_{st} + \dfrac{\sigma_{max} + \sigma_{st}}{2}$,

2) amplitude $\sigma_a = \dfrac{\sigma_{max} - \sigma_{st}}{2}$ or $\sigma_a = \sigma_{sr} - \sigma_{st}$,

3) $\varkappa = \dfrac{\sigma_a}{\sigma_{st}}$.

Fig.3. Function of water pressure changes

The searched function of answer for the tested factors has the following form:

$$\sigma_1 - \sigma_3 = F(Rg; w; f; \varkappa; \sigma_{st}; d) \qquad (4)$$

The hyper-surface of answer is placed in six-dimensional space. The carrying out of a complete experiment for all sets of factors values was not possible. In the introductory tests there were built orthogonal plans of first order (Nalimow and Czernowa - 1967). The achieved results pointed out to the necessity of plans building of higher orders. For reasons of both principal and technical nature such plans were resigned of.

In the main tests for the matrix of complete planning of the experiment there was applied selection of the Galilean type. The plans of tests for the appriopriate central data were as follows:
- for the function:

$$(\sigma_1 - \sigma_3)^w = F(Rg; \varkappa) \qquad (5)$$

where $Rg = OI, OII, OIII, OIV$
$\varkappa \in [0 \div 0.4]$ with change every 0.05

for central values: $w = w1$, $f = 8.85Hz$, $\sigma_{st} = 0.225MPa$, $d = 0.065m$,
- for the function:

$$(\sigma_1 - \sigma_3)^{w;*} = F(Rg; \sigma_{st}; f; w; d) \qquad (6)$$

where $Rg = OI, OII, OIII$
$\sigma_{st} = 0.338; 0.282; 0.225;$ and $0.169MPa$

$$f = \begin{cases} \text{for } (\sigma_1 - \sigma_3)^w - 3.90; 8.85; 11.96; \\ \qquad 15.91; 20.38; 27.54; \\ \qquad \text{and } 36.62Hz \\ \text{for } (\sigma_1 - \sigma_3)^* - 3.90/0; 8.85/0; \\ \qquad 11.96/0; 15.91/0; \\ \qquad 20.38/0; 27.54/0Hz \\ \text{for} (\sigma_1 - \sigma_3)_{f=0} - 0Hz \end{cases}$$

4 THE RESULTS OF THE EXPERIMENTS

The results in the majority of cases are shown as lines determined by the a mean values of the parallel repetitions (minimum three ones).

4.1 Vibratory strength as stress amplitude function

In the tests the stress amplitude (σ_a) was considered against its value in relation with static stresses (σ_{st}). Results achieved in the test of function (5) are shown on Figure 4.

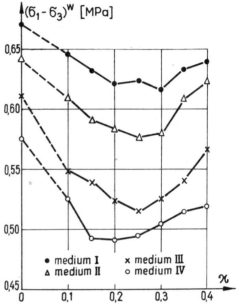

Fig.4. Results for the partial function (5). (w1, d=0.065m, f=8.85Hz, $\sigma_{st} = 0.225MPa$)

Results allowed to formulate the following conclusions:
- vibratory strength of the tested media clearly depends on the value of $\varkappa$,
- the results point out to a non-linear dependence $(\sigma_1 - \sigma_3)^w = F(\varkappa)$ which is nearing a paraboloical one,
- minimal strengths are achieved at the value of $\varkappa$ between $0.15 \div 0.30$,
- the range of strength changes is influenced by the kind of the medium.

147

4.2 Strength as vibratory stress ferquency function

The results of the 634 experiments connected with the analysis of function (6) are contained in Figures 5,6,7,8,9,10,11.

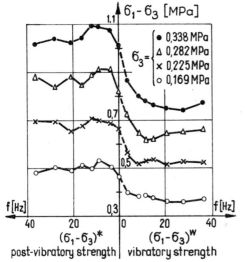

Fig.5.Results for the partial function (6). (OI, w1, d=0.065m)

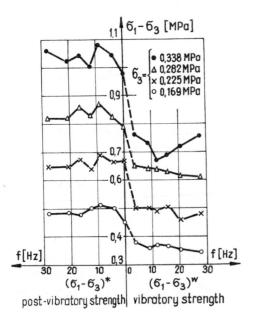

Fig.6.Results for the partial function (6). (OI, w2, d=0.065m)

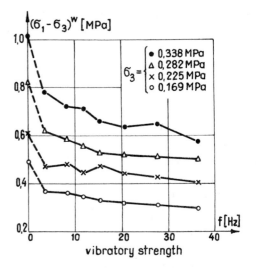

Fig.7.Results for the partial function (6). (OII, w1, d=0.065m)

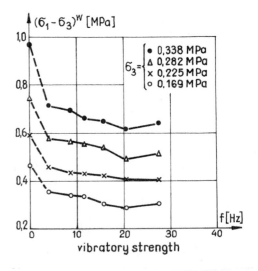

Fig.8.Results for the partial function (6). (OII, w2, d=0.065m)

148

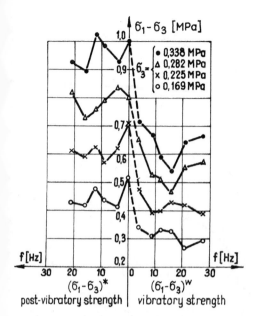

Fig.9.Results for the partial function (6). (OII, w1, d=0.038m)

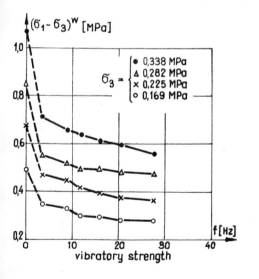

Fig.10.Results for the partial function (6). (OIII, w1, d=0.065m)

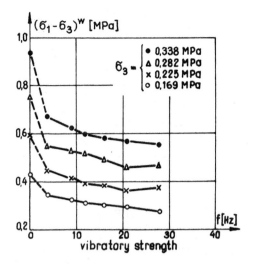

Fig.11.Results for the partial function (6). (OIII, w2, d=0.065m)

In order to determine the percentage changes of post-vibratory and vibratory strength in relation to standard strength one defined degrees and intervals of its changes. The degree of post-vibratory strength changes $(K_{\sigma_i}^*)$ is

$$K_{\sigma_i}^* = \frac{(\sigma_1 - \sigma_3)^* - (\sigma_1 - \sigma_3)_{f=0}}{(\sigma_1 - \sigma_3)_{f=0}} \cdot 100\% \qquad (7)$$

The degree of vibratory strength changes $(K_{\sigma_i}^w)$ is

$$K_{\sigma_i}^w = \frac{(\sigma_1 - \sigma_3)^w - (\sigma_1 - \sigma_3)_{f=0}}{(\sigma_1 - \sigma_3)_{f=0}} \cdot 100\% \qquad (8)$$

The interval of strength changes $(K_{\sigma_i}^r)$ is

$$K_{\sigma_i}^r = K_{\sigma_i}^* - K_{\sigma_i}^w \qquad (9)$$

The interval of strength changes $(K_{\sigma_i}^r)$ is equivalent to the vibratory resistance of the soil in the area of strength. Under the term of soil vibratory resistance in the area of its strength one should understand the possible percentage soil strength changes after the vibration and at

the time of its duration. If intervals of strength changes ($K_{\sigma i}^r$) are
smaller the more resistant against
vibration in the area of strength
is a given soil.

Exemplary evaluation of the deg-
rees of strength changes $K_{\sigma i}^*$ and
$K_{\sigma i}^w$ and vibratory resistance of the
tested soils contain Figures 12,13
14.

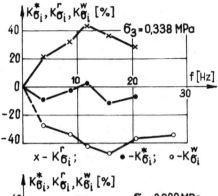

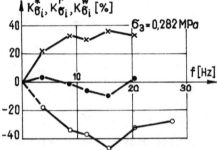

Fig.14.Degrees of vibratory and post-vibratory strength, and of vibratory resistance changes. (OII, w1, d=0.038m)

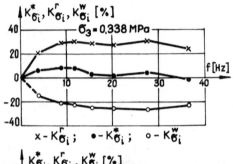

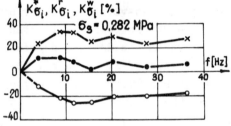

Fig.12.Degrees of vibratory and
post-vibratory strength, and of vibratory resistance changes. (OI,
w1, d=0.065m)

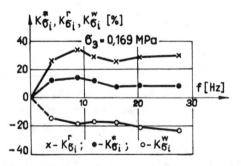

Fig.13.Degrees of vibratory and post-
vibratory strength, and of vibratory
resistance changes.(OI, w2, d=0.065m)

The introduced results lead to the
following conclusions:
- vibratory loads bring about changes of standard strength of the
  tested soils. The changes concern
  both to the vibratory and post-vibratory strength,
- the changes of strength depend
  substantially on the frequency
  of vibration,
- the range and character of the
  changes of strength are also affected by: the level of stress in
  the test cell, the kind of tested
  material and the size of the model,
- vibration is the cause of lose-
  ning or compaction of the soil
  which shows up in way of the decrease or increase of its post-
  vibratery strength,

150

- in each of the tested cases the
  vibratory strength of soils was
  lower than standard strength and
  post-vibratory strength (even up
  to 55%),
- with the growth of vibratory frequency decreases vibratory strength.
  Such character of changes prevails
  in the received results. Nevertheless in some instances function
  (6) has its minimal values,
- together with the increase of the
  homogeneity of the granular composition increase appriopriatelly
  the maximal decreases of vibratory strength of soil,
- vibratory resistance of the tested media depends to a small extent on the frequency of vibration
  (after excess an appriopriate frequency value) and stabilizes on
  30% level approximately.

4.3 Strength parameters

The results presented in the previous chapter allow to describe partial functions for the dependence

$$(\Phi_u; c_u)^{*;w} = F(Rg; f; w; d) \qquad (10)$$

One has used for that purpose the
linear hypothesis of Coulomb-Mohr.
The application of the principles
of the static soil failure theory
for the evaluation of vibratory
strength parameters Φ_u^w, c_u^w has required replacing of the principal
stresses which are, in fact, time
functions by a static load. Principal stresses were namely calculated
as mean values of vibratory ones
according to formula

$$\sigma_{2,3} = \sigma_{st} + \frac{\sigma_{max} - \sigma_{st}}{2} \qquad (11)$$

Having a sufficient number of scattered results one has approximated
for partial functions included in
a general dependence

$$\Phi_u^w = F(Rg; f; w; d) \qquad (12)$$

The regression function was determined by help of least squares method, taking into account a certain
set of polinomials. The scheme of
this passage is shown on Figure 15.

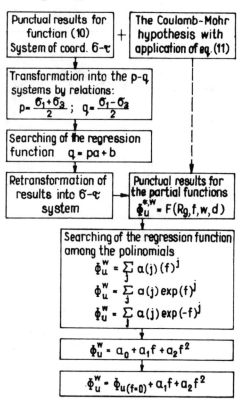

Fig.15.Scheme of procedure for
finding the regression function

The best approximation in all the
tested cases was reached for the
polinomial model

$$\Phi_u^w = a_0 + a_1 f + a_2 f^2 \qquad (13)$$

The regression coefficiens for this
function are given in table 4.

Table 4. Regression ceofficients for the polinomial (13)

Medium, Moisture content	The number of measuring points	Regression coefficients			Coefficient of multiple correlation	Residual variance .10⁻²
		a_o	$a_1 \cdot 10^{-3}$	$a_2 \cdot 10^{-4}$		
OI,w1	8	0.65193	9.04	1.95	0.961	1.49
OI,w2	7	0.62737	9.54	2.90	0.870	3.88
OII,w1	8	0.63339	10.61	1.68	0.991	0.89
OII,w2	7	0.61379	10.83	2.70	0.889	6.27
OII,w1 d=0.038m	7	0.62464	16.31	5.38	0.869	10.60
OIII,w1	7	0.61588	11.99	2.61	0.904	8.70
OIII,w2	7	0.60938	15.53	3.83	0.938	6.63

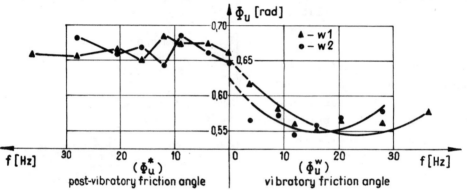

Fig.16.Total stress strength parameter Φ_u as function of frequency. (OI, w1 and w2, d=0.065m)

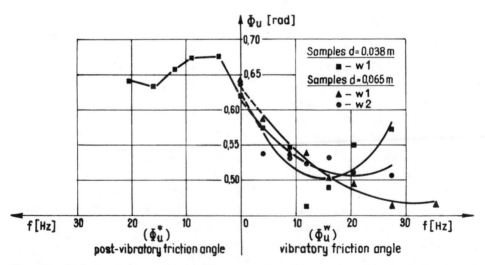

Fig.17. Total stress strength parameter Φ_u as function of frequency. (OII, w1 and w2, d=0.038m and d=0.065m)

152

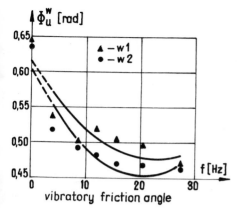

Fig.18.Total stress strength parameter Φ_u^w as function of frequency.
(OIII, w1 and w2, d=0.065m)

From the physical evaluation of the
obtained coefficients it follows
that a_0 is the friction angle determined in standard tests ($\Phi_{u(f=0)}$). So,
the formula (13) takes on the following form:

$$\Phi_u^w = \Phi_{u(f=0)} + a_1 f + a_2 f^2 \qquad (14)$$

The appriopriate regression curves
are given in Figures 16,17,18.

One could speak in an identical
way as in case of strength (formulas 7,8,9) about the degree of the
changes of post-vibratory friction
angle ($K_{\Phi_u}^*$), about the degree of
changes of the vibratory friction
angle ($K_{\Phi_u}^w$) and about the vibratory
resistance of soil in the area of
friction angle $K_{\Phi_u}^r$.

The introduced results of the research concerning the influence of
vibration on the friction angle of
cohesionless media allow to formulate the following conclusions:
- vibratory loads bring about changes of the post-vibratory (Φ_u^*) and
  vibratory (Φ_u^w) friction angle,
- the range of the changes is affec-

ted by the kind of soil, its moisture content and the size of samples,
- those changes are   non-linear,
- function $\Phi_u^w = F(f)$ is of parabolic
  character,
- the values of vibratory friction
  angle (Φ_u^w) are always smaller than
  the values Φ_u^* and $\Phi_{u(f=0)}$,
- together with the increase of the
  homogeneity of granular composition there grow up the maximal
  decreases of Φ_u^w ,
- the values of maximal decreases
  of Φ_u^w and of frequences at which
  they are achieved depend clearly
  on the size of the samples,
- the saturation of the medium influences slightly the range of
  the changes of post-vibratory
  friction angle Φ_u^*, and causes decreasing the values of maximal
  drops Φ_u^w ,
- in the tests there was founded
  the occurence of such frequency
  (f) at which the given medium
  has its smallest vibratory resistance in the area of Φ_u (it corresponds with the maximal values
  of $K_{\Phi_u}^r$).

5 SUMMARY AND FINAL CONCLUSIONS

In spite of the phenomenological
character of the tests there come
to mind some hypothetical justifi-
cations for the received dependen-
ces. They could be motivated on
the basis of different models which
treat soil as continuum or granular
medium. In the case of the tested
soils, analysis of the received effects could well be based on the

model of granular medium.

Shear causes the rebuilding of the structure in the areas of the reached limit state. In the case of shear with the vibratory action at the same time, the rebuilding of the structure in the area of shear takes place simultanously with the rebuilding of the structure of the whole volume of the sample. With changing states of stresses in the sample, due to vibration, there takes also place an appriopriate change of inter granular resistances. It makes moves and turns easier in the areas of shear. In effect one gets appriopriate smaller vibratory strength and values of vibratory friction angle smaller than the standard ones.

As an effect of vibration the rebuilding of the soil structure leads to changes of the initial porosity of soil. Therefore the post-vibratory porosity (which differs from the initial one) influences the values of post-vibratory strength and post-vibratory friction angle.

One may also analyse the problem of reaching minimal values of vibratory strength at certain frequencies of vibration. The tested media consist of grains of considerable masses. Those grains have free vibration frequencies conditioned by a number of factors connected with the position of a grain in the medium. In the case of media of highly uniform particle sizes there exist a greater number of grains of similar free vibration frequencies. Therefore if in such cases the fre-

quency of vibratory load is appriopriate, there may takes place the phenomenon of intergranular resonance. It may be the cause of maximal decrease in vibratory strength. This confirms the noticed decrease of vibratory resistance of soil with the increase of homogeneity of its granular composition.

The tests carried out proved the existence of a substantial influence of the size of the samples on the results. Soil is characterized by a high ability to absorb part of energy which produces the vibration of grains and ability to dissipate it as heat energy. This is probably the reason for lower resistance against vibration of soil samples that are of smaller sizes. These problems require further, wider tests.

The application of the means of static theory of limit equilibrium for the interpretation and evaluation of the results of vibratory tests causes the rising of many reservations. It seems however, that on the present stage of knowledge, this kind of procedure creates suitable conditions for comparative analysis of the results of tests of static and vibratory nature. Realizing the shortcomings of the assumed method of the interpretation of the results, we may solve however some practical problems introducing into the existing statical solutions of some boundary problems the obtained vibratory characteristics. There remains open the matter of practical verification of the received solutions and the need of further search after most right methods of evaluating the soil strength under periodically changing state of stresses.

Timmerman, D.Ch. and T.G.Wu. 1970, Testing of stress and displacements of cohesionless soil under cyclic load (in Russian), Put' i Str-vo Ž.Dorog, No 1.

REFERENCES

Chain, W.Ja.1975, On resistance of sand against vibration (in Russian), Geotechnical Problem, No 24, Dniepropietrovsk.

Cisek, T.1979, The strength of vibratory loaded soil (in Polish). Doct.Thesis (unpublished). Silesian Techn.University, Gliwice.

Nalimov, V.V. and N.A.Černova 1965, Statistical methods of extreme planing researches in Russian . Nauka, Moskva.

Ruchin, L.B.1961, Principles of lithology (in Russian). Leningrad.

Seed, H.B. and K.L.Lee 1966, Liquefaction of saturated sands during cyclic loading, J.Soil Mech.Foun.Div.ASCE. 92, SM6: 105-134.

Stavnicer, L.P. and W.P.Karpenko 1977, Laboratory investigations of sand foundation stability under vibration (in Russian). Osnov. Fund. i Mechan.Gruntov, No 2.

Stevarski, E.1972, Influence of dynamic load on shear resistance change on the ground of laboratory test (in Polish). Doct. Thesis. Min.-Metall.Academy, Kraków.

International Symposium on Soils under Cyclic and Transient Loading / Swansea / 7-11 January 1980

Cyclic behaviour of clay as measured in laboratory

T. J. KVALSTAD & R. DAHLBERG
Det norske Veritas, Hoevik, Norway

1 INTRODUCTION

During the last six years 13 concrete
gravity platforms have been installed in
the North Sea. The geotechnical problems
related to the foundation design of such
structures were critically examined by
Bjerrum (1973), Young, Kraft and Focht
(1976), Høeg (1976) and more recently by
Foss and Dahlberg (1979). The principal
design problems are the stability of the
structure during extreme loading and its
deformations due to ordinary loading, both
types of loading being a combination of
static and cyclic loading.

With the aim to improve our understanding of the basic mechanisms governing the
behaviour of clay, a joint research project financed by 13 sponsors from the oil
industry was carried out on Drammen clay
in 1974-1975. The results obtained in
this study (NGI, 1975) are among the best
and most comprehensive available at present (Van Eekelen and Potts, 1978). To
further enlighten the basic mechanisms
that govern the behaviour of clay under
cyclic loading the final report recommended future testing to concentrate on:

a) Additional tests to study the effect
of storms followed by drainage.

b) Additional tests to study the effect
of anisotropic consolidation.

c) Tests with simultaneous variation in
vertical and horizontal stresses.

d) Tests to study improved preparation
procedures for simple shear tests on overconsolidated clay.

e) Tests with random loading history.

f) Tests with higher load periods to
achieve more reliable cyclic pore water
pressure measurements.

g) Tests to study the effect of magnitude of consolidation stresses.

h) Tests to establish a critical stress

level beneath which cyclic loading has no
effect.

i) Tests with longer consolidation time.

This paper presents the results of cyclic triaxial tests on Drammen clay covering a few of the topics listed above,
namely items b), c) and e). Item g) has
been touched upon by Foss, Dahlberg and
Kvalstad (1978). All tests in this study
have been run on clay with an overconsolidation ratio OCR = 4 which allows for a
good control of the initial soil properties and reflects pretty good the average
degree of overconsolidation of a typical
North Sea clay.

2 DESIGN AGAINST FAILURE IN CYCLIC LOADING

In the early foundation design for gravity
structures the effects of cyclic loading
were considered simply by increasing the
safety factors enough to be on the conservative side. In the NPD, DnV and FIP
regulations of 1977 the safety considerations are introduced as partial safety
coefficients to characteristic loads and
shear strengths. The characteristic shear
strength of the soil related to the ultimate limit state (ULS) must account for
the effects of cyclic loading.

One method to account for these effects
was described by Andersen (1976). According to this method a soil specimen consolidated to stresses representative of
the actual stress consitions in the foundation underneath the gravity structure
is exposed to a "storm loading" before
loaded statically to failure. The static
shear strength thus obtained is less than
the one obtained in a static test without
previous cyclic loading. This reduced
strength is introduced as the characteristic shear strength of the clay in a quasi-

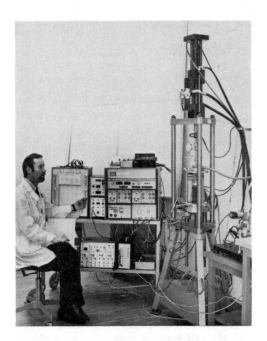

Figure 1 DnV Triaxial Equipment with connected data aquisition system.

static stability analysis.

Another method which accounts for the effects of cyclic loading was described by Foss, Dahlberg and Kvalstad (1978). This method is used to assess the risk of "Failure in cyclic loading", which failure mode should be distinguished from and analysed in addition to "Quasi-static stability failure" which is the traditional failure mode considered for gravity structures. Failure in cyclic loading can be characterized as the accumulated effect of all the waves in a storm which may introduce a complete foundation failure even if all load amplitudes are considerably below the undrained shear strength of the clay. The relevance of this failure mode is demonstrated in the abovementioned paper, which points out the duration of the storm as another factor of importance. Further evidence is obtained from centrifuge model tests on gravity structures on clay reported by Rowe, Craig and Procter (1976) and recently by Andersen, Selnes, Rowe and Craig (1979). The present paper includes some additional information regarding this failure mode.

3 DESCRIPTION OF TRIAXIAL EQUIPMENT

The triaxial testing (Figure 1) is carried out using a closed-loop servocontrolled electro-hydraulic testing system, type MTS, which contains two independently operated hydraulic actuators allowing for a large spectrum of static and dynamic load conditions to be simulated, e.g.

- Stepless anisotropic consolidation at constant principle stress ratio, σ_1/σ_3 and under K_o conditions.
- Advanced dynamic testing with simultaneous independent variation of axial and radial stress as well as static testing at constant axial strain rate.
- Random oscillations simulating a storm or an earthquake.

Internal load transducers are used which eliminate the effects of piston friction.

Automatic operation allows day and night testing. A timing unit turns the tape recorder on and off automatically at preset intervals reducing the amount of stored data to a rational niveau while a 4-channel plotter gives a continuous record of the test. Increased accuracy of the stored analog signals is achieved by use of a high resolution 13-bit Pulse Code Modulation (PCM) system. Since the accuracy of the recording system is higher than the accuracy in the measured values the data acquisition system does not introduce any extra errors.

4 STORM LOADING TESTS

4.1 Simulation of storm loading

Wave forces represent the dominating environmental load on a gravity structure. The wave loads are best described by means of the spectral analysis method. The most important assumption in this method is the linear relationship between wave height and the response for a given wave period. The response of interest in this connection is the wave induced shear stress variation in the soil elements of the foundation.

An irregular sea state is expressed by means of a wave spectrum. By establishing the transfer function for base shear stress vs wave frequency, the response spectrum of the shear stress can be obtained by multiplying the wave spectrum by the square of the transfer function.

The response spectrum for shear stress in a soil element has to be found for a wave spectrum representing the design sea state which is described by means of a significant wave height H_s, a mean zero up-crossing period T_z and the shape of the spectrum.

For North Sea conditions and gravity structures the wave and response spectra

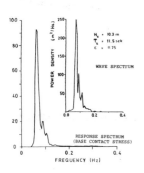

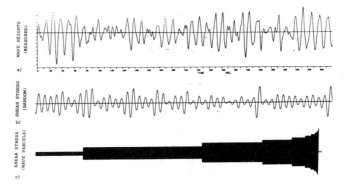

Figure 2   Typical North Sea
wave and response spectra

Figure 3   Time histories: a) measured wave heights
b) generated random signal for storm-loading test
c) generated wave parcel signal (compressed time axis)

are typically narrow (Figure 2). The
Rayleigh distribution may be used to describe the distribution of shear stress
amplitudes for the soil elements. The
most probable highest shear stress $\tau_{c,max}$
may then be expressed as

$$\tau_{c,max} = \tau_{c,s}\sqrt{\frac{\ln N}{2}} \qquad (1)$$

where $\tau_{c,s}$ is the significant shear stress
and N the number of cycles during the
period of stationary conditions (H_s and T_z
constant).

The i'th highest amplitude τ_i may be
expressed as

$$\tau_i = \tau_{c,max}\sqrt{\frac{\log(N/i)}{\log N}} \qquad (2)$$

In a storm the occurrence of individual
waves is random with small and large waves
interspersed (Figure 3a). For practical
reasons, however, cyclic tests on soil
materials are normally carried out either
as constant amplitude or as multi-stage
cyclic tests where, in the latter case,
the shear stress amplitude is increased
after a given number of constant amplitudes. "Storm-loading" tests may be carried out as "wave parcel" tests where the
cycles are arranged in order of increasing
amplitude (Eq. (2) in reverse order) and
sorted in a number of blocks or parcéls
with constant amplitude (Figure 3c) or as
"random" tests (Figure 3b).

It is obvious that random loading is a
more realistic way of testing than multi-
stage or wave-parcel cyclic loading. The
execution of the tests and the analysis
and interpretation of the results are,
however, more complicated and an advanced

testing and data acquisition system is
required.

To compare the development of permanent
pore water pressure and stiffness reduction due to random vs wave parcel type of
loading a series of storm loading tests
was carried out.

Both the random and the wave parcel
cycles are sines with a period of 10.0
seconds. The number of waves of the simulated storm was restricted to 1250 (length
of magnetic tape and tape recorder speed).
The control signal for the electrohydraulic load actuator was generated by an
HP-9125 desk computer with a random number
generator and digital/analog converters.
The analog signals were recorded on magnetic tape.

Equation (2) was used for generation of
the random signal, and the cycle numbers
i were picked randomly in the range 1 to
1250.

In the wave parcel loading the cyclic
shear stress amplitude and the number of
waves defining the different wave parcels
were chosen to fit closely the Rayleigh
distribution given by Eq. (2). The wave
parcel loading used in the present tests
is described in Table 1.

4.2 Test procedures

All samples were consolidated isotropically for one hour at σ = 50 kPa. Then backpressure was applied (u_B = 600 kPa) and
the vertical and radial stresses were increased continuously up to σ'_{vpc} = 400 kPa,
σ'_{hpc} = 200 kPa overnight. Thereafter the
samples were unloaded to isotropic stress
conditions ($\sigma'_{vc} = \sigma'_{hc}$ = 100 kPa). During

159

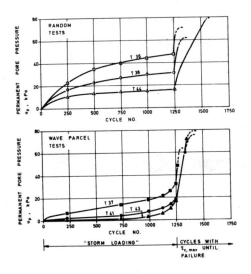

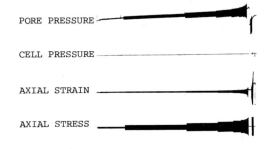

Figure 5 Wave parcel test followed by cyclic test at $\tau_{c,max}$ until failure.

Figure 4 Storm loading tests. Development of permanent pore pressure vs number of cycles.

Table 1. Wave parcel loading

Parcel No.	Cycle No.	Number of cycles ΔN	Relative amplitude $\tau_{c,i}/\tau_{c,max}$
1	1- 226	226	0.083
2	227- 738	512	0.260
3	739- 994	256	0.412
4	995-1122	128	0.518
5	1123-1186	64	0.605
6	1187-1218	32	0.680
7	1219-1234	16	0.747
8	1235-1242	8	0.807
9	1243-1246	4	0.861
10	1247-1248	2	0.908
11	1249	1	0.950
12	1250	1	1.00

Table 2. "Storm loading" tests

Test No.	Load type	Max cyclic shear stress $\tau_{c,max}$, kPa
T36	Random	40.9
T37	Wave parcels	40.4
T38	Random	35.7
T41	Wave parcels	34.2
T43	Wave parcels	31.1
T44	Random	30.5

To keep the effective stress level about equal on the extension and the compression side during the 2-way cycling, a permanent shear stress of about 10 kPa was applied on the compression side. Failure in permanent extension was thus avoided.

In Figure 4 the permanent pore water pressure, u_p, is plotted against number of cycles, N, for the two types of storm loading.

The shear stress levels were kept sufficiently low to avoid failure within the 1250-cycles of storm loading. Thus a direct comparison of permanent pore pressure and cyclic shear strain due to $\tau_{c,max}$ at the end of the storm was possible. (Tests T36 and T37 were, however, quite close to failure after 1250 cycles.)

The tests were continued by applying the associated maximum shear stress, $\tau_{c,max}$, the number of cycles required to reach failure, i.e. a cyclic vertical strain $\varepsilon_v > \pm 2\%$ corresponding to a cyclic shear strain $\gamma_c > \pm 3\%$. The development of strain and pore water pressure as well as the load configuration is always recorded continuously on a 4-channel strip chart

consolidation and unloading the samples were drained through the top and bottom filters.

Storm loading was applied by cycling the axial load at constant cell pressure. Since the object of this test series was to investigate the soil response to random loading and wave parcel loading respectively, both these types of loading were related to about the same storm intensity defined in terms of $\tau_{c,max}$ as explained in 4.1. The storm loading tests cover three different storm intensities i.e. three "random" tests and three "wave parcel" tests at comparable stress levels, Table 2.

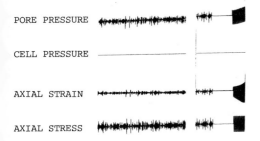

PORE PRESSURE

CELL PRESSURE

AXIAL STRAIN

AXIAL STRESS

Figure 6 Section of random test followed
by cyclic test at $\tau_{c,max}$ until failure.

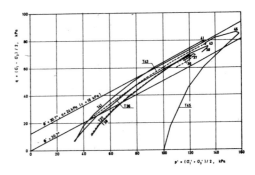

Figure 7 Effective stress paths. Static
tests after cyclic loading.

recorder, see example output in Figures 5
and 6 for the actual storm loading tests.

After cyclic loading the tests were continued by shearing the sample statically
at a constant strain rate of 3% per hour
to a vertical strain of $\epsilon_v > 15\%$. The
undrained shear strength, the failure
strain and corresponding pore water pressure is given in Table 3.

Table 3. Static tests after storm loading

Test No.	Undrained shear strength, s_u kPa	Vertical strain at failure, $\epsilon_{v,f}$ %	Pore water pressure at failure, u_f kPa
T36	67.8	14.0	45.5
T37	68.4	11.8	42.1
T38	74.2	12.1	39.5
T41	77.6	12.5	46.5
T42	68.5	19.3	43.8
T43	75.8	13.6	39.8
T44	-	-	-
T45	83.5	4.1	24.5

The effective stress paths of these static tests (with the exception for test No.
44) are shown in Figure 7. For comparison
the stress path for an ordinary static
test (No. 45), without previous cyclic
loading, is plotted. The failure envelopes indicated in the figure were determined from such ordinary tests. The
effective shear strength parameters seem
to be unaffected by cyclic loading while
the undrained shear strength is reduced,
the reduction increasing with the intensity and duration of the storm.

5 INTERPRETATION AND DISCUSSION OF STORM LOADING TESTS

5.1 "Cyclic shear strength"

In Figure 8 the cyclic shear strain, γ_c,
is plotted vs the corresponding cyclic
shear stress $\tau_{c,max}$ applied at the end of
the "storm" (i.e. after 1250 cycles). No
large differences can be seen between the
random and the wave parcel tests. Compared with the broken line which indicates
the stress-strain curve for the first
cycle in constant amplitude tests, a sharp
break can be observed for $\tau_{c,max}$ between
30 and 35 kPa. The shape of the curve indicates that an increase in $\tau_{c,max}$ to more
than 40 to 45 kPa will lead to very
large cyclic shear strains. A "cyclic
shear strength" may thus be defined by
extrapolation of the curve up to the acceptable cyclic shear strain or an asymptotic value of $\tau_{c,max}$. In Figure 8 this
value may be assumed approximately equal
to 45 kPa.
   It is thus possible for a given design
storm (number of cycles, period, amplitude
distribution) to establish an upper limit
of the maximum response shear stress in a
soil element beyond which very large and
unacceptable strains will occur i.e. the
cyclic shear strength. In the authors'
opinion the corresponding failure mode
must be looked upon as a collapse which
may have similar consequences as a total
failure and should thus be analysed in the
ultimate limit state.

5.2 Application of the "strain-accumula-
tion method"

The "strain accumulation method" described

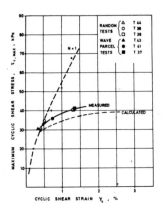

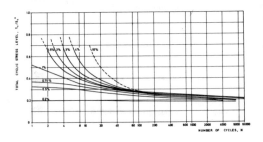

Figure 8 Storm loading: Cyclic shear strain at the end of the storm due to $\tau_{c,max}$. Triaxial test results and calculated curve.

Figure 9 Strain contour diagram based on two-way cyclic triaxial tests (Drammen clay, OCR = 4).

by Andersen (1976) has been shown feasable for prediction of cyclic shear strains due to storm-loading simulated in simple shear (NGI 1975). To allow a check of this method for triaxial testing four two-way cyclic tests with constant amplitude were carried out. Additionally the results from the joint research project (NGI 1975) were used. The results are plotted as a strain contour diagram showing the increase in cyclic shear strain with increasing number of cycles as a function of the total cyclic stress level τ_c/s_u. It should be noted that the s_u value used was s_u = 100 kPa which was the average undrained shear strength in triaxial compression based on previous static tests on Drammen clay. The samples used for constant amplitude tests were from a borehole located 50 to 100 m from the boreholes of the storm loading samples and had an initial water content 3 to 5% higher. These samples might thus have a slightly lower undrained shear strength than the storm loading samples.

A computer program has been developed for the analysis of storm effects based on the "strain accumulation method". The strain contour diagram is given as input. A series of storms with the same content as the wave parcel tests (see Table 1) and variable intensity (i.e. $\tau_{c,max}$) were applied and the resulting cyclic shear strain at the end of each storm due to $\tau_{c,max}$ was evaluated. The procedure, described by Foss, Dahlberg and Kvalstad (1978), yields data for plotting of the stress-strain curve $\tau_{c,max}$ versus cyclic shear strain at the end of the storm.

In Figure 8 which shows the results of

the storm loading tests the calculated curve based on the strain contour method is shown as well. The calculated curve falls slightly below the measured values. This is, in the authors' opinion, mainly due to the lower undrained shear strength of the samples used for determination of the contour diagram.

5.3 Comparison of simple shear and triaxial test results

As mentioned in 4.2 a certain permanent shear stress (approximately 10 kPa) has to be applied to avoid failure in permanent extension in the two-way cyclic triaxial tests. It is obvious that a comparison on a total stress basis should take account of the difference in undrained shear strength in extension and compression. For Drammen clay with OCR = 4 the extension strength $s_{u,e}$ is only about 50% of the compression strength, $s_{u,c}$, while the strength in simple shear, $\tau_{h,f}$ is approximately 70 to 75% of $s_{u,c}$.

In our previous work (Foss, Dahlberg and Kvalstad, 1978) the "cyclic shear strength" was determined to lie in the range 0.65 to 0.75 of the undrained shear strength determined in simple shear. If we now for the triaxial tests instead of using the undrained strength in compression use the average of compression and extension strength $s_{u,av} = \frac{1}{2}(s_{u,c} + s_{u,e})$ the cyclic triaxial shear strength determined in this test series will be about 0.6 of the average strength $s_{u,av}$ and comparable to the experience from the simple shear tests.

162

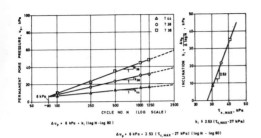

Figure 10 Reinterpretation of Figure 4 in semi-logarithmic scale.

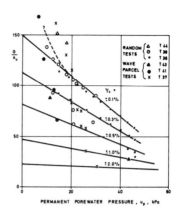

Figure 11 Reduction in stiffness as a function of permanent pore pressure and cyclic shear strain.

5.4 Random loading effective stress model

The stress-strain behaviour of the clay during storm-loading is governed by the effective stresses. As the permanent pore pressure increases due to cyclic shear stress variations the effective stress path moves towards the failure envelope. It was shown in the NGI Report No. 74037 that the cyclic stress-strain behaviour of the clay expressed in terms of effective cyclic stress level and cyclic shear strain is uniquely described by a narrow range of hyperbolic curves comprising all types of tests.

If the pore pressure can be predicted it should thus be possible to predict the shear strain due to cyclic loading at a given time and for a given shear stress during a storm.

In Figure 10a the permanent pore pressure curves from Figure 4 have been re-plotted versus the logarithm of the cycle number. Plotted this way straight line relationships are obtained for the three storm intensities. Furthermore, the inclination of these lines was found to vary linearly with $\tau_{c,max}$ which is a measure of the storm intensity, see Figure 10b. Based on the three "random" storm loading tests the following relationships were established from Figure 10a.

$$\Delta u_p = 6\text{kPa} + k_i \, (\log N - \log 80) \quad (3)$$

From Figure 10b follows

$$k_i \simeq 2.53 \, (\tau_{c,max} - 27\text{kPa}) \quad (4)$$

Inserting Eq. (4) into Eq. (3) yields

$$\Delta u_p = 6\text{kPa} + 2.53 \, (\tau_{c,max} - 27\text{kPa})$$
$$(\log N - \log 80) \quad (5)$$

which closely describes the development of permanent pore pressure under random loading. Figure 11 shows the reduction in stiffness as a function of permanent pore pressure and cyclic shear strain based on results from the storm loading tests. It is obvious that the mean effective normal stress as well as the shear stress are the governing parameters for the soil behaviour. By applying a hyperbolic effective stress-strain relationship similar to the Duncan-Chang model for static loading (Duncan and Chang 1970) a model may be derived which describes the development of cyclic shear stress and strain under "design storm loading".

The comments given above are restricted to symmetric or nearly symmetric loading. If high permanent shear stresses are combined with cyclic shear stress variations permanent shear strains will take place and to some extent influence the build-up of permanent pore pressure. Some comments on these effects will be given in Section 7.

6 EFFECT OF CYCLIC OCTAHEDRAL NORMAL STRESS

Cyclic triaxial tests are normally carried out with constant cell pressure. This leads to cyclic variation in the octahedral normal stress, σ_{oct}, defined as

$$\sigma_{oct} = \frac{1}{3} (\sigma_1 + 2\sigma_3) \qquad (6)$$

where σ_1 is the total vertical stress and σ_3 is the cell pressure. Theoretically, the undrained soil response to a cyclic variation in the octahedral normal stress should be elastic in nature having no effect on the cyclic behaviour of the tested soil. Experimental evidence for this is found in the results reported herein from a test series run on aniso-topically consolidated samples.

The consolidation procedure was identical to that described in 4.2 but the unloading branch ended at a certain residual or per-manent shear stress, τ_p, upon which the cyclic shear stress, τ_c, is superimposed. Table 4 gives the stress conditions in the different tests.

Table 4. Stress conditions in anisotropi-cally consolidated tests.

Test No.	τ_p kPa	τ_c kPa	τ_{max} kPa	τ_{min} kPa	$\Delta\sigma_{oct}$, kPa
T10*	29.5	±28.5	58.0	1.0	±19.0
T18	32.2	±29.5	61.7	2.7	0.0
T9B*	40.0	±20.5	60.5	19.5	±13.7
T17	41.2	±20.5	61.7	20.7	0.0
T8 *	40.0	±28.6	68.6	11.4	±19.0
T16	39.4	±30.0	69.4	9.4	0.0
T7 *	54.3	±20.4	74.7	33.9	±13.6
T15	53.8	±20.1	73.9	33.7	0.0

* Tests run with constant cell pressure, i.e. cyclic octahedral normal stress. Other tests run with σ_{oct} = const.

The effect of cyclic octahedral normal stress was investigated by running two parallel test series, one series with con-stant cell pressure allowing for cyclic variation in the octahedral normal stress and another series where the cell pressure was cycled out of the phase with the axial load to keep the octahedral normal stress constant during cycling, see Figure 12a. The small pore water pressure variation follows very closely the small changes in octahedral normal stress.
Figure 12b shows the pore water pressure response in an ordinary test with constant cell pressure. Also in this case the fluctuation in octahedral normal stress govern the cyclic pore water pressures (i.e. nearly elastic response).

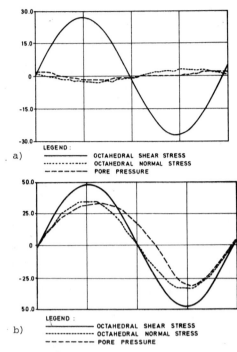

a)
LEGEND:
——————— OCTAHEDRAL SHEAR STRESS
·············· OCTAHEDRAL NORMAL STRESS
– – – – – PORE PRESSURE

b)
LEGEND:
——————— OCTAHEDRAL SHEAR STRESS
·············· OCTAHEDRAL NORMAL STRESS
– – – – – PORE PRESSURE

Figure 12 Pore pressure response to variation in normal octahedral stress a) $\Delta\sigma_{oct} \approx 0$, b) σ_3 = constant.

The results are presented in Figure 13, which gives the ratio of permanent to cyc-lic shear strain, γ_p/γ_c, versus permanent shear strain, γ_p, for all tests in this test series. For comparison the results of four anisotropically consolidated tests with constant cell pressure run in the abovementioned industry sponsored research project are included in Figure 13. These data which are taken from NGI Report No. 74037 fit very well into the picture, see also Table 5.

Table 5. One-way anisotropically consoli-dated triaxial tests reported by NGI (Report No. 74037).

Test No.	τ_{cons}^* kPa	τ_{max} kPa	τ_{min} kPa	τ_c kPa	τ_p kPa
TB26.1	35	65	35	±17.5	52.5
TB27.1	35	75	35	±20	55.0
TB28.3	35	75	-5	±40	35.0
TB28.4	35	65	5	±30	35.0

* τ_{cons} = consolidation shear stress

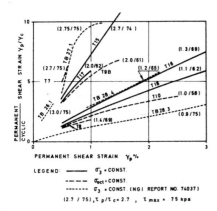

Figure 13 Permanent/cyclic shear strain γ_p/γ_c versus permanent shear strain γ_p.

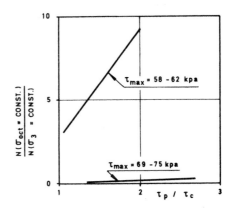

Figure 14 Relative increase in number of cycles, N to reach a permanent shear strain $\gamma_p = 0.5\%$ when σ_{oct} = constant instead of variable.

The only test which is questionable is T8 which lies too low compared with T16 which it should be compared with. With this exception comparable tests with and without cyclic octahedral normal stress indicate that the effect of the octahedral normal stress on the overall cyclic behaviour of the clay is insignificant.

On the other hand Figure 14 shows the effect of cyclic octahedral normal stress on the number of cycles to reach a certain permanent shear strain, here $\gamma_p = 0.5\%$. For maximum shear stresses $\tau_{max} = 58-62$ kPa during cyclic loading the clay becomes more resistant against cyclic loading if the octahedral normal stress, σ_{oct}, is constant, the resistance increasing with increasing ratio τ_p/τ_c. In the tests with $\tau_{max} = 69-75$ kPa the reverse behaviour is observed, namely keeping σ_{oct} constant makes the clay less resistant to cyclic loading. Due to the non-conclusive nature of these results this behaviour should be studied further before any definite conclusions are drawn.

7 EFFECT OF ANISOTROPIC CONSOLIDATION

As a total 8 tests were run on anisotropically consolidated samples, see Table 4. Since the effect of cyclic octahedral normal stress is insignificant judging from Figure 13 all test results have been used to evaluate the effect of anisotropic consolidation on the cyclic behaviour of the clay. Figure 15 is an interpretation and simplification of Figure 13 containing both the DnV test results (with the exception of test T8) and the anisotropically consolidated tests reported by NGI. In addition the results of four isotropically

consolidated triaxial tests with OCR = 4 from the same NGI report are interpreted and included in Figure 15, see also Table 6.

Table 6. One-way cyclic, isotropically consolidated, triaxial tests reported by NGI (Report No. 74037).

Test No.	τ_{cons} kPa	τ_{max} kPa	τ_{min} kPa	τ_c kPa	τ_p kPa
TB2F	0	68.2	0	±34.1	34.1
TB5	0	72.9	0	±36.4	36.4
TB22	0	75.1	0	±37.5	37.5
TB23	0	75.3	0	±37.6	37.6

The results in Figure 15 show that isotropically consolidated samples with consolidation shear stress $\tau_{cons} = 0$ cycled one-way between $\tau_{max} = 2\tau_c$ and $\tau_{min} = \tau_{cons} = 0$ (i.e. $\tau_p = \tau_c$) behaves very much the same as anisotropically consolidated samples with $\tau_p = \tau_{cons}$ which are cycled symmetrically around τ_p with $\tau_c = \tau_p$, i.e. $\tau_p/\tau_c = 1.0$. This means that the definition of $\tau_p = \tau_c$ in isotropically consolidated tests is reasonable.

As the ratio τ_p/τ_c increases the effect of anisotropic consolidation is evident, leading to less cyclic shear strains for a given permanent shear strain level. As the permanent shear strain increases the ratio permanent to cyclic shear strain increases as well. This ratio is greater than one for permanent shear strains less than 0.5% and τ_p/τ_c greater than 1.0

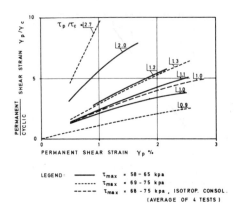

LEGEND:
——— $\tau_{max} = 58 - 65$ kpa
------- $\tau_{max} = 69 - 75$ kpa
- - - - $\tau_{max} = 68 - 75$ kpa, ISOTROP. CONSOL.
(AVERAGE OF 4 TESTS)

Figure 15 Permanent/cyclic shear strain, τ_p/τ_c, versus permanent shear strain for different values of τ_p/τ_c.

indicating that permanent rather than cyclic shear strains are critical. This tendency becomes more pronounced as the ratio τ_p/τ_c increases. When the ratio τ_p/τ_c decreases the tendency to develop permanent shear strains decreases and for τ_p/τ_c equal to about 0.2 to 0.3 a true two-way cyclic loading is obtained without significant permanent shear strains. As the ratio τ_p/τ_c becomes less than 0.2 there is a tendency for the sample to fail in cyclic loading on the extension side, the permanent shear strains increasing in magnitude as the ratio τ_p/τ_c becomes less than zero.

8 CONCLUSIONS

Evaluation of the stability of gravity platforms during severe storms requires modelling of soil behaviour under random cyclic loading. For foundations on clay the cumulative generation of pore pressure results in reduced effective normal stresses and thus reduced stiffness. Laboratory tests (triaxial and simple shear) as well as centrifuge model tests show that unacceptable and uncontrolled cyclic deformations may develop in the course of a storm if the "cyclic shear strength" of the soil is exceeded. This condition which is termed "failure in cyclic loading" should, in the authors' opinion, be analysed in the ultimate limit state.

The "cyclic shear strength" as defined in this paper is a function of the soil material, the initial stress conditions and the stress history as well as the storm duration, and will also be influenced by the wave periods.

The determination of the "cyclic shear strength" has been done experimentally in three different ways; Storm loading test with cycles of different amplitude in random order, storm loading with wave parcel loading, and constant shear stress amplitude tests evaluated by the strain accumulation method described by Andersen (1976). All three methods gave similar results. Rearranging the random storm cycles in wave parcels of increasing amplitude seems to have no significant effect on the cyclic shear strength of the investigated soil material (Drammen clay, OCR $\cong$ 4).

The stress-strain relationship is governed by the effective stresses and an effective stress model may be developed based on random loading tests. For cases with non-symmetrical loading, i.e. high permanent stress levels, permanent strains will develop. This will affect the development of pore pressure as well as cyclic shear strain. The treatment of permanent strains is complex, and based on the test results presented in this paper only some general tendencies have been described. A complete soil model must incorporate these effects as well. However, if tendencies can be modelled within reasonable accuracy and applied to a relevant analysis method like the FEM, an adequate tool for prediction of failure modes and the corresponding capacities of the foundation may be developed.

9 ACKNOWLEDGEMENT

Support for the research work was provided in part by the Royal Norwegian Council for Scientific and Industrial Research (NTNF), which is acknowledged with great appreciation.

10 REFERENCES

DnV 1977, Rules for the design, construction and inspection of offshore structures, Det norske Veritas, Oslo.
FIP 1977, Recommendations for the design and construction of concrete sea structures, Federation Internationale de la Précontrainte, Slough.
NPD 1977, Regulations for the structural design of fixed structures on the Norwegian continental shelf, Norwegian Petroleum Directorate, Stavanger.
Andersen, K.H. 1976, Behaviour of clay subjected to undrained cyclic loading, BOSS'76, 1st Int. Conf. on Behaviour of Offshore Structures, 1: 392-403, Trondheim.

Andersen, K.H., P. Selnes, R.W. Rowe &
  W.H. Craig 1979, Prediction and obser-
  vation of a model gravity platform on
  Drammenclay, BOSS'79, 2nd Int. Conf. on
  Behaviour of Offshore Structures, Paper
  34, London.
Bjerrum, L. 1973, Geotechnical problems
  involved in foundations of structures in
  the North Sea, Geotechnique, 23(3): 319-
  358.
Duncan, J.M. & C.-Y. Chang 1970, Nonlinear
  analysis of stress and strain in soils,
  Journal of the Soil Mechanichs and
  Foundations Division, ASCE, Vol. 96, No.
  SM5, Sept. 1970.
Foss, I & R. Dahlberg 1979, Foundation
  design methods for gravity structures,
  FIP State of the art Report: Foundations
  of concrete gravity structures in the
  North Sea, FIP 6/2: 33-59.
Foss, I., R. Dahlberg & T.J. Kvalstad 1978,
  Design of foundations of gravity struc-
  tures against failure in cyclic loading,
  10th Annual Offshore Techn. Conf., Paper
  No. OTC 3114, Houston.
Høeg, K. 1976, Foundation engineering for
  fixed offshore structures, BOSS'76, 1st
  Int. Conf. on Behaviour of Offshore
  Structures, 1: 36-69, Trondheim.
Norwegian Geotechnical Institute 1975,
  Research project. Repeated loading on
  clay, NGI Report No. 74037.
Van Eekelen, H.A.M. & D.M. Potts 1978,
  The behaviour of Drammen clay under
  cyclic loading, Geotechnique, 28(2):
  173-196.
Young, A.G., L.M. Kraft & J.A. Focht 1976,
  Geotechnical considerations in founda-
  tion design of offshore gravity struc-
  tures, J. Petroleum Techn.: 925-937.

Resilient stress-strain behaviour of a crushed rock

J. W. PAPPIN & S. F. BROWN
University of Nottingham, Nottingham, UK

1 INTRODUCTION

The development of improved design procedures for flexible pavement structures
based on the use of analysis requires a
knowledge of the mechanical characteristics
of the component materials.   Granular
material, usually in the form of crushed
rock, often forms the main structural
layer of the pavement (the road base) and
is also included as the sub-base.

Previous research on this material,
using the repeated load triaxial test, has
produced extensive data (Hicks and
Monismith 1971) showing that the resilient
response is markedly non-linear.   Most
experiments have involved the application
of a repeated deviator stress (q_r) and a
constant confining stress (σ_c).   The
recorded resilient (recoverable) axial
strain (ε_{ar}) has then been used to
calculate a resilient modulus (Mr) as:

$$M_r = \frac{q_r}{\varepsilon_{ar}} \tag{1}$$

The non-linearity has been expressed as a
stress-dependent Mr in an equation of the
form:

$$M_r = K_1 \theta^{K_2} \tag{2}$$

where θ = sum of the principal stresses
when the peak deviator stress is applied
and K_1, K_2 are constants which depend on
the material type and condition.

Fig. 1 shows the type of total stress
path used in these tests in p-q stress
space where:

$$p = \text{mean normal stress} = \frac{1}{3}(\sigma_a + 2\sigma_c) \tag{3}$$

and

$$q = \text{deviator stress} = (\sigma_a - \sigma_c) \tag{4}$$

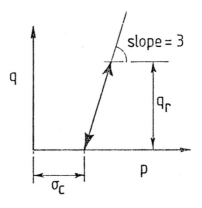

Fig. 1 Typical repeated stress path

The value of θ in Equation 2 is hence
determined from:

$$\theta = 3(\sigma_c + \frac{1}{3}q_r) \tag{5}$$

It will be noted from Fig. 1 that the
deviator stress was pulsed from zero to q_r.
However, some other investigations (Brown
1974, Brown & Hyde 1975) using more general
stress paths showed that Equation 2 was not
entirely adequate for the likely range of
in situ conditions.

2 THE EXPERIMENTS

The recent research at Nottingham has been
concerned with a detailed, accurate
examination of the resilient response of a
particular crushed rock.   The initial
stage of this work was reported by Boyce

et al (1976) who described the experimental procedures and the first set of results. The work presented herein has used essentially the same experimental techniques and has produced much additional data.

The material involved was a crushed limestone of 40 mm maximum particle size having the grading shown in Fig. 2. It falls within the limits set by the Department of Transport (DOE) (1976) for wet-mix macadam road bases and type 1 granular sub-bases. A uniform, high density of 2270 kg/m^3 was achieved by vibratory compaction. The material was tested in the dry state so that effective stresses could be directly determined. Much of the earlier research involved partially saturated materials and effective stresses were not known.

A servo-controlled triaxial apparatus capable of cycling both the cell pressure and axial load was used to test 150 mm diameter by 300 mm high specimens of the material. A wide range of repeated stress paths was used, each one requiring four parameters for its specification. These are shown in Fig. 3 and consist of the mean (p_m and q_m) and repeated (p_r and q_r) components of each parameter.

For each stress path the resulting resilient axial and radial strains were measured directly on the specimen (see Boyce and Brown 1976) and then used to calculate:

resilient volumetric strain:

$$v_r = \varepsilon_{ar} + 2\varepsilon_{rr} \qquad (6)$$

and

resilient shear strain:

$$\varepsilon_r = \frac{2}{3}(\varepsilon_{ar} - \varepsilon_{rr}) \qquad (7)$$

where ε_{ar} and ε_{rr} are the resilient, axial and radial strains respectively, obtained after about 10 cycles of loading. The extensive number and complexity of the repeated stress paths used by Boyce et al (1976) gave rise to resilient strains that could not easily be described in terms of a stress dependent resilient modulus and Poisson's ratio. Hence they derived a set of equations that can be written in the form:

$$2\varepsilon_{ir} + \frac{v_r}{3} = \left(\frac{p_m}{k}\right)^n \sinh \frac{2\sigma_{ir} - p_r}{2\sigma_{im}} \qquad (8)$$

where subscript i indicates either the axial or radial value and k = 1.03x10^{13}kPa and n = 0.33. This relationship is clearly much more complex than the earlier model (Equation 2) and attempts at using it in numerical analysis have proved difficult.

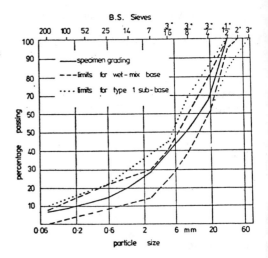

Fig. 2 Grading curve for the material

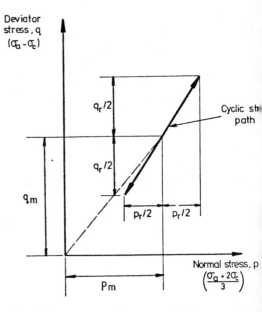

Fig. 3 Components of stress in the repeated load triaxial test

The recent work has involved comprehensive testing of three specimens using sinusoidal wave forms and a frequency of 1 Hz. The repeated stress paths illustrated in Fig. 4 were applied to the specimens at mean normal stress (p_m) values of 48, 96 and 192 kPa. A feature of this arrangement of stress paths is that there are a number having mirror image stress paths above or below the normal stress axis, thus making it possible for direct comparisons to be made between the response

170

of the material in triaxial compression and extension. To remove the possibility of an excessive build-up in permanent strain, no stress paths reached the static failure condition (see Fig. 4). A reduced number of stress paths that only included q_r/p_r values of 0, 3 and ∞ were applied at p_m values of 12 and 24 kPa and tests in the compression region only were applied at a p_m value of 384 kPa. The recorded resilient strains from the three specimens were averaged for each stress path.

Inspection of the results revealed that for those stress paths where $q_m \neq 0$ and having no change in normal stress (i.e. $p_r = 0$), some resilient volumetric strain was observed and similarly for those stress paths where $q_m \neq 0$ and having no change in deviator stress (i.e. $q_r = 0$) some resilient shear strain was observed. This means that the material, when under isotropic stress conditions ($q_m = 0$)

exhibited an isotropic response, but in other stress situations, a complex stress-induced anisotropic behaviour was observed. It follows from this that to use the conventional elastic properties (i.e. tangential stiffness modulus and Poisson's ratio) to describe resilient behaviour would be difficult for this material. Other researchers (Domaschuck and Wade 1969, Brown and Bush 1972) have indicated that bulk and shear modulus are suitable parameters for describing this type of material and from these and suggestions by Thrower (1976) the possibility of using secant bulk and shear modulus was considered. This means that both stresses and strains are referred to a common origin, and to achieve this, body forces within the pavement must be considered, and will give rise to stresses and strains before any wheel load is applied. Using this approach, an attempt was made to describe the resilient strain results. The strains arising from repeated stress paths wholly within the triaxial compression region were initially considered and, due to the nature of the model, volumetric and shear strain were considered separately.

3 THE COMPRESSION STRESS REGION ($\sigma_a > \sigma_c$)

The resilient volumetric strain results were found to be commutable if plotted in stress space. In other words, the resilient volumetric strain resulting from the stress path between point (p_1,q_1) and point (p_2,q_2) when added to the strain from point (p_2,q_2) to (p_3,q_3) equalled the strain produced by going directly from (p_1,q_1) to (p_3,q_3).

Assuming that zero strain occurs at zero stress, a plot of resilient volumetric strain contours (v_c) was established in the p-q stress region as shown in Fig. 5. The equation describing these contours is:

$$v_c = \left(\frac{p}{k}\right)^{0.33}\{1 - 0.08(q/p)^2\} \qquad (9)$$

where $k = 1.9 \times 10^{11}$ kPa. This equation gives rise to a corresponding secant bulk modulus:

$$K = p\frac{(k/p)^{0.33}}{1 - 0.08(q/p)^2} \qquad (10)$$

Assuming the stress path moved from (p_1,q_1) to (p_2,q_2) the resilient volumetric strain can be calculated from:

$$v_r = \left(\frac{p_2}{k}\right)^{0.33}\{1 - 0.08(q_2/p_2)^2\}$$
$$- \left(\frac{p_1}{k}\right)^{0.33}\{1 - 0.08(q_1/p_1)^2\} \qquad (11)$$

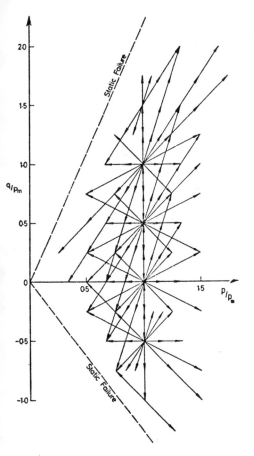

Fig. 4 Stress paths applied to sample at one value of mean normal stress

171

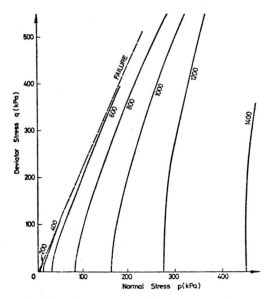

Fig. 5 Contours of resilient volumetric strain (με) in p-q stress space

The results for the resilient shear strains differed fundamentally from the volumetric strains in that they were not commutable in stress space. This shows that the shear strain is stress path dependent and hence the procedure used above for volumetric strain is not directly applicable.

It was found, however, that at any constant mean stress (p_m and q_m) and constant stress path slope (q_r/p_r), the shear strain varied with stress path length (l_r) where:

$$l_r = \{p_r^2 + q_r^2\}^{0.5} \tag{12}$$

in the manner shown in Fig. 6. This behaviour was observed to be true for all tested directions of q_r/p_r and can be described by the relationship:

$$\varepsilon_r = \varepsilon_o\left(\frac{l_r}{l_o}\right)^{1.4} \tag{13}$$

where ε_o and ε_r = shear strains produced by stress paths length l_o and l_r respectively. The measured shear strains were then converted to normalised values. (ε_n)

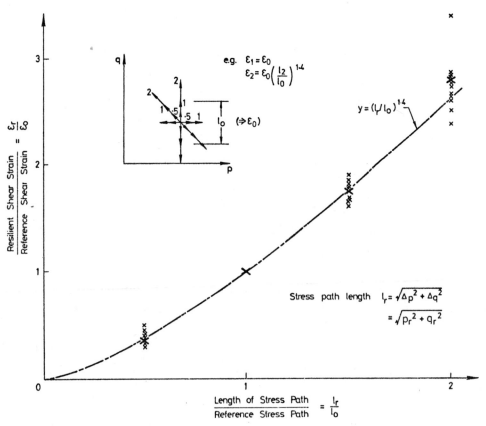

Fig. 6 Effect of stress path length on resilient shear strain

in order to make them commutable in stress space. The normalised values can be expressed as:

$$\varepsilon_n = \varepsilon_o \times \frac{l_r}{l_o} = \varepsilon_r \times \frac{l_r}{l_o} \times \left(\frac{l_o}{l_r}\right)^{1.4}$$

$$= \varepsilon_r \times \left(\frac{l_o}{l_r}\right)^{0.4} \tag{14}$$

For the normalised strain values to be generally commutable it was found that l_o in Equations 13 and 14 should be equated to p_m.

The contours of normalised resilient shear strain are shown in Fig. 7 and can be expressed by the equation:

$$\varepsilon_n = \frac{0.00024 \ q}{p+b} \tag{15}$$

where b is a constant equal to 13 kPa.
The resilient shear strain produced by a stress path from (p_1,q_1) to (p_2,q_2) is:

$$\varepsilon_r = 0.00024\left\{\frac{q_2}{p_2+b} - \frac{q_1}{p_1+b}\right\}\left\{\frac{(p_r^2+q_r^2)^{0.5}}{p_m}\right\}^{0.4} \tag{16}$$

3.1 Stress Path Dependence

To confirm the observations that shear strain was stress path dependent whilst volumetric strain was stress path independent a series of stress paths forming closed loops (see Fig. 8) were applied to two specimens at reference p values (p_H) of 96 and 192 kPa; p_H referring to the value of p at the point H in Fig. 8. Fig. 9 shows comparisons between the resilient strains measured for a single stress path and the sum of the resilient strains measured for a set of stress paths which, when combined start and finish at the same stress points as the single path. The legend on the figure shows the sets of multiple stress paths used for comparison with the single stress paths. It is clear from this figure that comparison of the resilient volumetric strains for differing stress paths is generally very close but, for the resilient shear strains, large discrepancies occur.

A possible explanation for the resilient shear strain response was considered to be provided by hysteresis in the shear stress-strain behaviour as shown in Fig. 10. It is clear from this figure that shear stress varying from point a to b and from b to c will give rise to shear strains x_1 and x_2 respectively. However, a stress path going directly from a to c gives strain y which is clearly greater than $x_1 + x_2$. To investigate this hypothesis

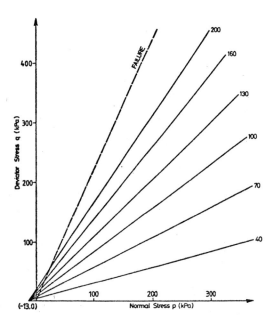

Fig. 7 Normalised resilient shear strain contours ($\mu\varepsilon$) in p-q stress space

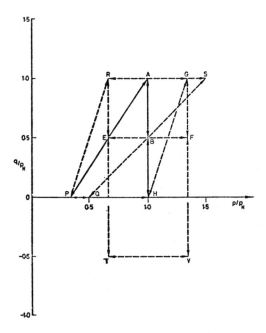

Fig. 8 Stress paths used in 'closed path' testing

173

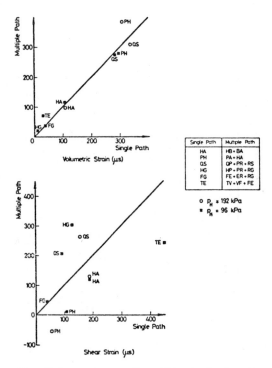

Fig. 9 Comparisons of resilient strains for single and multiple stress paths

Single Path	Multiple Path
HA	HB + BA
PH	PA + HA
QS	QP + PR + RS
HG	HP + PR + RG
FG	FE + ER + RG
TE	TV + VF + FE

○ P_H = 192 kPa
■ P_H = 96 kPa

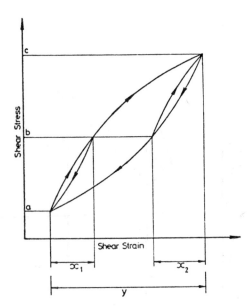

Fig. 10 Possible relationship between repeated shear stress and resilient shear strain

an analogue computer was used to convert the readings taken from the test specimen (i.e. deviator stress, confining stress, axial and radial strain) into normal stress deviator stress, volumetric strain and shear strain. This enabled the resilient volumetric or shear strain to be directly plotted against normal or deviator stress using an X-Y plotter. This process was carried out for a wide range of stress paths and, whilst for stress paths where p was constant (i.e. p_r = 0) the above hypothesis was found to work for the shear strain, hysteresis was also observed in the volumetric strain response. For other stress paths the hysteresis characteristics were found to be quite complex for both volumetric and shear strains. From these results it was concluded that the stress path dependent behaviour cannot be simply explained by considering hysteresis effects alone.

4 THE EXTENSION STRESS REGION ($\sigma_c > \sigma_a$)

From Fig. 4 it can be seen that a number of applied stress paths had mirror images about the normal stress axis. Analysis of the results for these paths indicated that the volumetric strains were not significantl affected by the transition from triaxial compression except at high q/p values. The resilient shear strains, however, were noticeably affected, particularly as q/p increased, i.e. for stress paths closer to failure. After determining the normalised shear strain values in the extension region it was found that the strain contours could be correlated with the compression contours using an expression based on the ratio of major to minor principal stress (σ_1/σ_3). This implies that σ_1/σ_3 and the mean normal stress (p) are the controlling parameters. To apply this theory to the triaxial extension stress condition, any extension stress point, $\sigma_a < \sigma_c$, with its correspondir p and q, must be converted to a compression condition $\bar\sigma_a > \bar\sigma_c$ with its corresponding $\bar p$ and $\bar q$ such that $\bar\sigma_a/\bar\sigma_c$ is equal to σ_c/σ_a and $\bar p$ = p. This gives rise to:

$$\bar q = \frac{-q}{1 + q/3p} \qquad (17)$$

This transformed value of q was used in Equations 11 and 16 (not used for the q_r term) to predict the resilient volumetric and shear strains in the triaxial extension regions respectively. Comparisons between the observed and predicted values using both the transformed and untransformed deviator stress are shown in Figs 11 and 12 for the volumetric and shear resilient strains. The predictions using the stress

174

correction (Equation 17) give consistently better predictions for the resilient shear strains but are seen to have only a minimal effect for the volumetric strains.

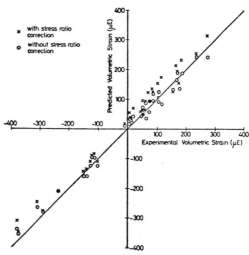

Fig. 11 Comparison of predicted and experimental resilient volumetric strains in triaxial compression-extension

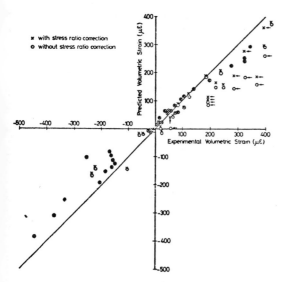

Fig.13 Comparison of predicted and experimental resilient volumetric strains in triaxial extension

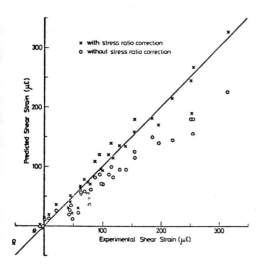

Fig. 12 Comparison of predicted and experimental resilient shear strains in triaxial extension

5 EXTENSION-COMPRESSION STRESS PATHS

To apply the predictive model to the resilient volumetric strain response for those stress paths going from triaxial extension to triaxial compression (see Fig. 4), the transformed values of deviator stress were used, when appropriate, in Equation 11. Fig. 13 shows a comparison between the predicted resilient volumetric strains (predicted with and without transformed q values) and the experimental strains produced by the stress paths applicable to this stress regime. The predictions are seen to be adequate but the effect of the stress ratio correction on the predicted strain is minimal on all stress paths except for those which reach points at or near $q/p = 1$ (these points are arrowed in the figure).

Predictions of resilient shear strain for stress paths of this type are more difficult, because of the stress path dependent nature of the response. In addition, Equation 16 does not apply directly in this stress situation as the resilient shear strain is the sum of, instead of the difference between, the initial and final strain value, since they act in opposite directions (one axial compression and one axial extension). Various attempts have been made to expand the shear strain model to cater for this situation, but unfortunately, no satisfactory solution has been obtained without making several assumptions.

6 APPLICATION OF BEHAVIOUR MODEL

The non-linear resilient behaviour model described above can be directly applied to most non-linear numerical analysis methods, since it readily produced values of secant bulk and shear modulus. These parameters are isotropic and directly describe the complex behaviour of this type of material, despite the stress induced anisotropic characteristics noted above. Great care would have to be exercised if the stress paths applied in an analysis were equivalent to the triaxial compression-extension conditions. However, in most pavement situations it is unlikely that this would arise.

The model is only based on triaxial test results and, therefore, assumptions would have to be made for extending its use to the general three-dimensional stress situations occurring in many analytical problems. The main assumption required is the expansion of the stress ratio correction term (Equation 17) to general three-dimensional stresses. The same logic can be used to convert any stress point $(\sigma_1 > \sigma_2 > \sigma_3)$ to a corresponding triaxial compression condition $\bar{\sigma}_z > \bar{\sigma}_x = \bar{\sigma}_y$ with its corresponding $\bar{p}$ and $\bar{q}$ such that $\bar{\sigma}_z/\bar{\sigma}_x$ is equal to σ_1/σ_3 and $\bar{p} = p$. This gives rise to:

$$\frac{\bar{q}}{p} = \frac{3(\sigma_1 - \sigma_3)}{\sigma_1 + 2\sigma_3} \qquad (18)$$

Other assumptions required would be the definition of stress path length and the incorporation of some means of dealing with principal stress rotation. There is clearly a need for three-dimensional testing of this type of material to clarify these assumptions.

Generally, in pavement applications crushed rock is placed in a partially saturated condition. Further experimental work has been performed at Nottingham University (Pappin 1979) to investigate the resilient behaviour of the material under both saturated and partially saturated conditions. This research has shown that, if effective stresses are used, the prediction model developed above is suitable.

6 CONCLUSIONS

The results of repeated load triaxial tests carried out on specimens of a well graded crushed limestone at a single high density and in the dry state have led to the following conclusions.

1. The complex resilient behaviour of the material could be described in terms of isotropic secant bulk and shear moduli.
2. To achieve this the resilient volumetric and shear strains had to be expressed as contours in p-q stress space, the resilient strain being derived as the change in contour value from the initial to the final stress state.
3. Volumetric resilient strain contours were independent of stress path and were basically dependent on p with q/p becoming significant towards failure. The final expression for resilient volumetric strain was:

$$v_r = \left(\frac{p_2}{k}\right)^{0.33} \{1 - 0.08(q_2/p_2)^2\}$$

$$- \left(\frac{p_1}{k}\right)^{0.33} \{1 - 0.08(q_1/p_1)^2\} \qquad (19)$$

where $k = 1.9 \times 10^{11}$ kPa and p_1, q_1 and p_2, q_2 define the beginning and end points of the repeated stress path.
4. Shear strain was stress path dependent but after this effect was normalised, contours dependent on (q/p+constant) were established. The final expression for resilient shear strain was:

$$\varepsilon_r = 2.4 \times 10^{-4} \left[\frac{q_2}{p_2+b} - \frac{q_1}{p_1+b}\right]$$

$$\left[\frac{(p_r^2 + q_r^2)^{0.5}}{p_m}\right]^{0.4} \qquad (20)$$

where $b = 13$ kPa and p_1, q_1, p_2, and q_2 are defined as above.
5. This fundamental difference between shear and volumetric strain response with respect to stress path effects could not be explained by hysteresis effects alone.
6. A stress ratio correction, based on the principal stress ratio, σ_1/σ_3, provided a satisfactory means of transposing stress conditions in the triaxial extension stress region to the compression region.

ACKNOWLEDGEMENTS

The research described in this paper was carried out in the Department of Civil Engineering, University of Nottingham under contract to the Transport and Road Research Laboratory. The Authors are grateful for the support provided by their Head of Department, Professor R.C. Coates, the advice of Professor P.S. Pell, and the assistance of various colleagues.

REFERENCES

Boyce, J.R. and S.F. Brown 1976, Measure-
ment of elastic strain in granular
materials, Geotechnique, 26(4): 637-640.
Boyce, J.R., S.F. Brown and P.S. Pell 1976,
The resilient behaviour of a granular
material under repeated loading, Proc.
Australian Road Research Board, 8: 8-19.
Brown, S.F. 1974, Repeated load testing of
a granular material, Journ. Geot. Eng.
Div., ASCE, 100, GT7, Proc. paper 10684,
July: 825-841.
Brown, S.F. and D.I. Bush 1972, Dynamic
response of model pavement structure,
Journ. Trans. Eng., ASCE, 98, TE4, Nov.:
1005-1022.
Brown, S.F. and A.F.L. Hyde 1975, The
significance of cyclic confining stress
in repeated load triaxial testing of
granular material, Transportation Research
Record 537: 49-58.
Department of Transport 1976, Specification
for road and bridge works, London,
HMSO.
Domaschuk, L. and N.H. Wade 1969, A study
of bulk and shear moduli for a sand,
Journ. S.Mech. and Found. Div., ASCE,
SM2: 561-581.
Hicks, R.G. and C.L. Monismith 1971,
Factors influencing the resilient
response of granular materials, Highway
Research Record 345: 15-31.
Pappin, J.W. 1979, Characteristics of a
granular material for pavement analysis,
thesis submitted for the degree of Ph.D.,
University of Nottingham.
Thrower, E.N. 1976, Discussion on 'The
resilient behaviour of a granular material
under repeated loading' by J.R. Boyce,
S.F. Brown and P.S. Pell, Proc.
Australian Road Research Board, 8: 8-19.

International Symposium on Soils under Cyclic and Transient Loading / Swansea / 7-11 January 1980

Response of Kaolin to reversals of strain path in undrained triaxial tests

K.KUNTSCHE
Universität Karlsruhe, Karlsruhe, Germany

1 INTRODUCTION

Material constants appearing in constitutive equations can only be determined by
experiments. In addition, only experiments
can yield the inevitable limitations for the
formulation of such equations. Referring to
cohesive soils in particular there is further need of suitable and reliable results
of fundamental tests in this continuum-
mechanics approach.

In continuation of former experimental
work on clay in Karlsruhe, which clarified
some aspects of the time-dependent behaviour
(Leinenkugel, 1976), a series of tests is
in progress now with the intention of studying time-independent characteristics. Viscous properties of clay are widely known
and they cannot - as is often done - be
neglected.

The time-independent material response
is investigated here with strain-controlled
tests with constant strain-rate. For clays
there are two reasons for considering isochoric deformations: first, they are appearing at quick deformation problems and, secondly, the void ratio becomes an independent variable. It is assumed that the water-
saturation of the samples provides the constant volume so that triaxial tests with
these restrictions can be conducted with
monotonous or reversed strain-path. After
a few comments on the material properties
of clays published already (Sec.2) and the
procedure of the reported tests (Sec.3), this
treatise presents some features of the results. Hereby the following points are
stressed out (Sec.4):
- How far do samples deform homogeneously?
- Can the influence of different void ratios be eliminated?
- How do various strain histories influence
  stress and pore-water pressure response?

2 RESPONSE OF NORMALLY CONSOLIDATED CLAY

In studying published test data two uncer-
tainties often arise: First, given results
could be influenced by the quoted viscous
properties. For example, the belief in
elastic-plastic models has misled many ex-
perimentators to perform stress-controlled
tests with arbitrary loading program. The
other shortcoming of many reported results
is the unknown deformation of the sample.
It is obvious that only a homogeneous strain
field allows the determination of stress in
the sample. This fact, however, is often
neglected. Strain-softening, e.g., can frequently be explained by the loss of homogeneity (Vardoulakis, 1979). With these reservations the following is summarized:

2.1 Monotonous paths

Triaxial tests (Rendulic, 1973; Henkel, 1960)
show that in the effective principal stress
plane the contours of equal void ratio, as
evaluated from drained tests, agree with the
effective stress paths of undrained tests.
It is asserted that there is a functional
dependence between effective stresses and
void ratio.

This fact is the origin of the so-called
generalized effective stress principle
(Schofield and Wroth, 1968), stating that
the effective stress-increments determine
the strain-increments. It is indicated
here that this functional dependence is
only shown for monotonous paths.

There are countless strength data expressed by the effective stresses in terms
of the angle of friction φ' or/and by the
undrained strength c_u. As known, the finding of Hvorslev (1937) and Henkel (1960)
concerning c_u states that c_u depends uniquely on the void ratio. They show that an

exponential function can describe this dependence quite well. For brevity questions concerning φ' are not discussed here.

2.2 Non-monotonous paths

Almost all experimental investigations on clay covering stress- or strain-reversals are concerned with cyclic tests adapted to earthquake or wave loading conditions. The work of Thiers and Seed (1968), Castro and Christian (1976), Koutsoftas (1978), Idriss et al.(1978) and Anderson et al.(1978) e.g., yields the following essential features:

There exists a threshold stress or strain value in cyclic loading, above which the undrained shear strength and stiffness decrease. The loss of undrained shear strength however, is remarkably smaller than that of stiffness. Moreover the cumulative generation of pore-water pressures is observed. Below this threshold no such effects occur. Only few results of tests with conventional slow strain rates including stress- or strain-path reversals are available.

In his review on triaxial tests Whitman (1960) reports that in spite of high excess pore-pressure due to cyclic stressing the effective stress path joins the path obtained with monotonous loading,as failure is approached. Lo (1961) and Rao and Gulhati (1973) observe the same path independence of failure in terms of effective stresses. Rao and Gulhati moreover show the important feature of similar shape of the stress-strain curves after reversals. This similar shape is also obtained by Wood and Wroth (1977). So, it can be concluded that, even considering strength only, observations are often contradictory and there is further need of clarification.

3 MATERIAL AND TEST PROCEDURE

The basic requirement of tests is their reproducibility. In particular, tests with clay need accurately defined sample preparation and test procedure. For reasons of comparison to other data the procedure of preparation and test has to be given.

Kaolin clay (from Goldhausen, W.Germany) is used with LL=52, PL=23, 70 % clay fraction, and γ_s = 26,8 kN/m^3.

The material is remoulded with a water content of 100 % in a mixer and transferred to perforated brass moulds with a filter envelope, both under vacuum conditions. After first isotropic consolidation under -80 kN/m^2 (pressure is taken negative) samples of 50 mm dia. are extracted by thin wall tubing and trimmed to a height of 50 mm. Thin circular rubber membranes are put on top and bottom of the sample. The end platens are then lubricated. Axial filter strips are installed on the sample surface to provide further drainage to the bottom. After isotropic consolidation under various cell pressures the undrained samples are compressed or/and extended with constant speed of piston. During the deformation the following quantities are electronically measured: The axial force by a strain gauge system inside the cell, the pore-pressure by a pressure transducer connected to a hypodermic needle, which is inserted axially in the sample, and the axial displacement of the piston by a LVDT arrangement. All data are registrated and evaluated automatically. The axial stress is evaluated with regard to the actual size of the cylindrical sample; other corrections are not taken into consideration. After the test the water content is determined. The various water contents after remoulding, preconsolidation and shear test scatter max. 1%.

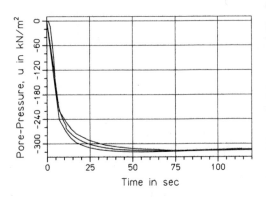

Fig. 1 a) Response of pore-water pressure due to $\Delta p = -\,320$ kN/m^2 of undrained samples

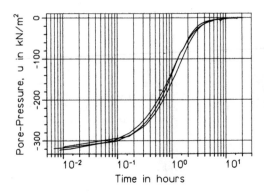

Fig. 1 b) Consolidation behaviour

Other checks of reproducibility are the developments of pore-water pressure u. Fig.1a shows the typical performance of u of three undrained samples (preconsolidated by -80 kN/m^2) which are instantenously subjectet to $\Delta p = -320$kN/m^2. The response occurs almost simultaneously with loading and verifies the intended water-saturation. The subsequent consolidation is represented by Fig. 1b and shows the typical features.

4 TEST RESULTS AND INTERPRETATION

4.1 Sample deformation

The use of smooth lubricated end platens is necessary to prevent shear stresses at the ends, such that the axial principal stress can be obtained. Barden and McDermott (1965) show experimentally that in this case pore-pressure differences within the sample are markedly reduced and barrelling is minimized. As regards barrelling, they recognize it as a stability problem of the triaxial sample and recommend samples with a height to diameter ratio in the range between 1.0 and 1.5. How far lubricated ends can support homogeneous deformation is shown by Balasubramaniam (1976) through X-ray techniques. He observes substantially uniform strains up to about 75 % of the peak-strain. Vardoulakis (1979) analyzes the bifurcation of triaxial sand samples and emphasizes the use of relatively short samples. The chosen lubrication and geometry ($h_o/d_o = 1.0$) of the samples confirm the published data:

The diameters of the samples at top, bottom and in the middle region do not differ more than 3 % of mean diameter within the strain-range of $-0.15 < \varepsilon_1 < 0.10$ ($\varepsilon_1 := \ln h/h_o$, h_o = initial height, h = momentary height).

Thereby the samples are always more restrained at the bottom due to the inserted hypodermic needle. It is noted that with further straining slip band formation or necking occurs only occasionally. Shear stress reaches its plateau at $|\varepsilon_1| = 0,08$-0.1 in compression or extension respectively. From the measurements of diameters and the visual sample observations doubts remain, whether the samples really reach the maximum value of shear stress homogeneously. The scatter of the strength results (s. Fig. 2) might be due to bifurcation.

4.2 Normalization

For the chosen type of testing, stress and strain are fully described by the shear stress $\tau := 1/2 (\sigma_1 - \sigma_3)$ and axial strain ε_1.

Note that the change of mean stress $\Delta p = 1/3 \, \Delta\sigma_1 = 1/3 \, (\sigma_1 - \sigma_3)$ is proportional to τ.

The actual stress response of a sample generally depends on strain increment, strain history and void ratio. (The change of strain rate due to the constant speed of the piston is negligible). Small differences in the void ratios e of two samples may cause high differences in stress response – see the above mentioned exponential relationship between strength and e. In spite of elaborate techniques it was not possible to produce samples with identical void ratios. In order to compare the behaviour of two samples the dependence on void ratio has to be eliminated. The normalization of the stress response by a suitable equivalent stress, which depends uniquely on void ratio, is the simplest attempt to do this. This normalization is only possible when for a given strain-path the influence of void ratio on the stress response might be separated multiplicatively.

Ladd and Foott (1974) or Leinenkugel (1976), e.g., define the equivalent stress as the consolidation pressure, which means that strength in particular is proportional to consolidation pressure. In other words, in a log σ_c vs. e diagram the consolidation line is parallel to the strength line. As Hvorslev (1960) has shown, this is not generally true. Further, as there is always a considerable scatter of void ratios for same σ_c, this definition of equivalent stress seams difficult. In defining an equivalent stress here it is argued that it is more reasonable to use the strength as a function of void ratio for samples with identical strain history. Fig. 2 shows the measured shear strength values of all tests in a log scale vs. void ratio. The equivalent stress is defined as the strength obtained with monotonous compression tests:

$$\log\sigma_e := \log|c_{uc}| = -1.92 \cdot e + 3.41$$

The usefulness of this equivalent stress is depicted by the Fig. 2, 3a and 3b by two reasons: First, normalized strength should be a constant depending on strain history. This implies that in Fig. 2 the c_u vs. e lines of tests with identical strain history are parallel to each other. Showing this, such test results are connected by a straight line. Secondly, normalized stress response of identically strained samples should coincide over the whole range of straining. This is depicted in Fig. 3a for monotonous, and in Fig. 3b for non-monotonous tests.

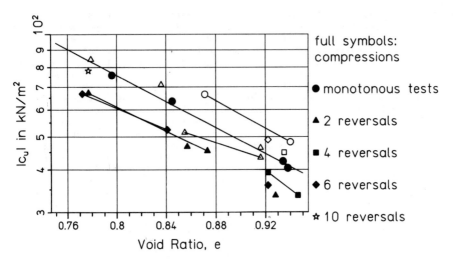

Fig. 2 Undrained shear strength - void ratio relationship

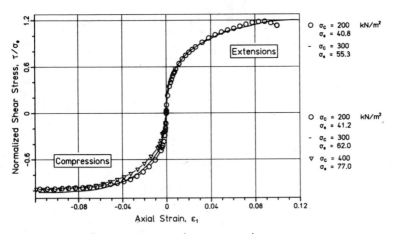

Fig. 3 Normalized stress-strain curves a) monotonous tests

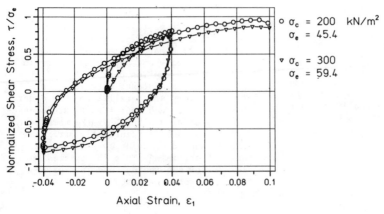

Fig. 3 Normalized stress-strain curves b) tests with two reversals

4.3 Response of stress and pore-water pressure

In Fig. 5 the response of normalized shear stress (a_1–a_4) and corresponding pore-water pressure (b_1–b_4) of 10 representative tests of this series is presented. The following features are inherent in the tests (see also Fig. 3b): First, dealing with monotonous tests only, the normalized stress-strain curves yield the well known shape reaching the strength plateau at $|\varepsilon_1|$ = 0.08 – 0.1. As expected an appreciable loss of strength with further straining occurs only in the extension tests. The shear strength values in extension tests are about 20 % higher compared with those of compression tests (Fig. 3a). Pore-water pressure also develops conventionally: In both types of tests $|u|$ increases to different values which are mainly depending on the different changes of the mean stress Δp. Thereby u increases monotonously even when strength is reached. This means that a definition of an angle φ' is arbitrary to some extent. Here φ' is defined, taking u at that strain where strength-plateau is firstly reached. Two different values of φ' are obtained: $17.9°$ and $16.2°$ in extensions and compressions (Fig. 4), which is inconsistent with the Mohr-Coulomb law.

Secondly, the response of samples subjected to strain path reversals reveal several essential characteristics. The most evident one is the symmetry of stress path after reversal. It follows same curved shape as from the unstrained state. This is true in particular for the stress path of the cyclicly strained test (Fig. 5a_4) which is truly cyclic, i.e. it coincides within each cycle. The pore-water pressure, however, develops totally different: Each reversal into compression – combined with the increase of mean stress $|\Delta p|$ – induces a further accumulation of u. Thereby the value of the residual $|\Delta u|$ depends mainly on the length of the strain-path in the direction of compression (Fig. 5b_1). Hence tests with strain path reversals ending in extension (Fig. 5b_1, b_4) show considerably higher amounts of pore pressure than monotonous tests. In tests ending in compression this behaviour is also observed but not as pronounced.(Fig. 5b_2). The reason is that reversals into extension induce accumulative Δu only in the first stages of straining.

The undrained shear strength is lowered by previous strain-path reversals. The loss increases with the length of the reversed path or of the path preceding the reversal, but not more than about 20 % of the monotonous tests. It is noted, that the strength loss is of the same order of magnitude independent of the type of test, i.e.if the test ends in compression or extension. Therefore it must be concluded that the strength loss is not directly caused by the actual value of the pore-water pressure. This fact may be demonstrated by Fig. 4: The strength of extended samples is certainly reduced but obviously not according to the Mohr-Coulomb law.

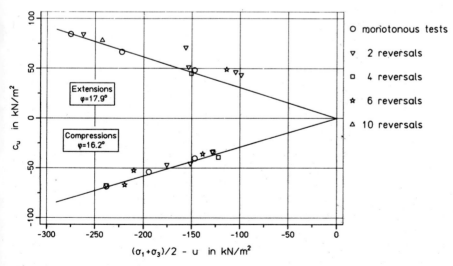

Fig. 4 Mohr-Coulomb diagram

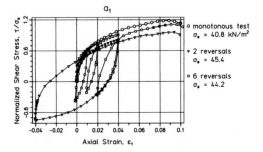

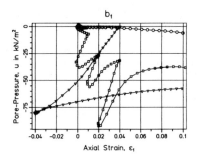

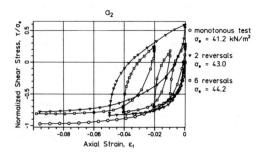

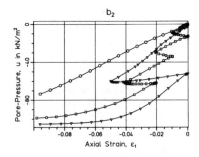

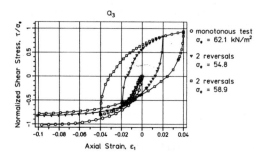

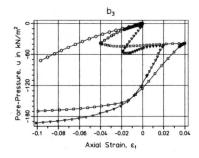

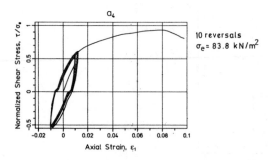

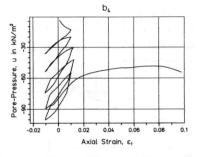

Fig. 5 Response of normalized shear stress (a) and corresponding pore-water pressure (b) ($\sigma_3 = \sigma_c$ = const.)

184

5 CONCLUSIONS

Some typical features of strain-controlled, undrained triaxial tests covering strain path reversals on isotropically consolidated and remoulded Kaolin can be listed:

- Within a wide range of strain, samples remain cylindrical due to the applied lubrication of end-platens and due to the low h_o/d_o - ratio.

- The total-stress response is shown to be proportional to an equivalent stress. For this stress the strength of monotonously compressed samples has been taken.

- Tests with strain path reversals reveal three essential features: first, the stress path after reversal follows the same curve as that from the unstrained state. Secondly, particularly every reversal into compression induces pore-water pressure and, thirdly, there is no obvious relationship between strength loss and developed pore-pressure.

The reported test-results agree with published data with regard to the relative small strength loss after reversals, which is found in all cited papers. The observations further agree with the results of Rao and Gulhati (1973) and Wood and Wroth (1977) referring to the symmetry of the stress-strain curve. The behaviour of identical effective stresses at failure with or without previous reversals, as reported by Whitman (1960) and others, could not be confirmed. Moreover the occuring strength loss cannot be explained by the developed pore-water pressure. This means that the concept of the fatigue-parameter (Van Eekelen and Potts, 1978), for which the pore-pressure is used, cannot be verified.

It is concluded:

The total stress-rate induced by a certain strain-rate depends on the stress state only and not on the stress history or on the developed pore pressure; elastic parts of response are not detectable. Omitting the generalized effective stress principle, this simplifies the formulation of constitutive equations considerably. Cyclic tests with strain amplitudes greater than $\varepsilon_1 = \pm 0.01$ are in progress so that the limits of the above mentioned behaviours can be checked.

ACKNOWLEDGEMENTS

This paper is a partial result of a research project supported by the Deutsche Forschungsgemeinschaft. The tests described in this paper were carried out in the Institut für Bodenmechanik und Felsmechanik under the supervision of Professor G.Gudehus. Many tests were conducted in co-operation with Dr. T.S. Nagaraj, who worked as a guest professor in the institute. His advice and assistance is gratefully acknowledged. The author is also indebted to Dr. D. Kolymbas and Dr. H. Winter for their comments on the paper.

6 REFERENCES

Anderson, K.H., Hansteen, O.E., Høeg, K. and Prévost, J.H., 1978, Soil deformations due to cyclic loads on offshore structures, Numerical Methods in Offshore Engineering, Ed. Zienkiewicz et.al, Wiley, p. 1-40.

Balasubramaniam, A.S., 1976, Local strains and displacement patterns in triaxial specimens of a saturated clay, Soils and Foundations, Vol. 16, No. 1: 101-114.

Barden, L. and McDermott, R.I.W., 1965, Use of free ends in triaxial testing of clays, Journal of the Soil Mech. and Found. Div., Proc. ASCE, Vol. 91, No. SM6: 1-23.

Castro, G. and Christian, I. T., 1976, Shear strength of soils and cyclic loading, Journ. Geotechn. Engng. Div. Proc. ASCE, Vol. 102, No. GT9:887-894.

Henkel, D.J., 1960, The shear strength of saturated remoulded clays, Proc. of the ASCE Res. Conf. on Shear Strength of Cohesive Soils, Boulder, Colo., p 533-554.

Hvorslev, M.J., 1937, Über die Festigkeitseigenschaften gestörter bindiger Böden, Thesis, Kopenhagen.

Hvorslev, M.J., 1960, Physical components of the shear strength of saturated clays, Proc.of the ASCE Res.Conf.on Shear Strength of Coh.Soils,Bould.Co.p 169-273.

Idriss, I.M., Dobry, R. and Singh, R.D., 1978, Nonlinear behaviour of soft clays during cyclic loading, Journ. Geotechn. Engng. Div., Proc. ASCE, Vol, 104, No. GT 12: 1427-1447.

Koutsoftas, D.C., 1978, Effect of cyclic loads on undrained strength of two marine clays. Journ. Geotechn. Engng., Div. Proc. ASCE, Vol. 104, No. GT5: 609-620.

Ladd, C.C. and Foott, R., 1974, New designed procedure for stability of soft clays, Journ. Geotechn. Engng. Div., Proc. ASCE, Vol. 100, No. GT 7: 763-786.

Leinenkugel, H.-J., 1976, Deformations- und Festigkeitsverhalten bindiger Erdstoffe. Experimentelle Ergebnisse und ihre physikalische Deutung, Veröffentlichungen Inst.f.Bodenmechanik und Felsmechanik, Universität Karlsruhe, Heft Nr. 66.

Lo, K.Y., 1961, Stress-strain relationship
and pore-water pressure characteristics
of a normally consolidated clay, Proc.
5th Int. Conf. Soil Mech. Found. Engng.
Paris, 1: 219-224.

Rao, E.S. and Gulhati, S.K., 1978, Behaviour of pore water pressure with strain
under cyclic loading, Symp. Earth & Earth
Struct. under Earthquakes & Dynamic Loads,
Roorkee, India, p. 87-98.

Rendulic, L., 1937, Ein Grundgesetz der
Tonmechanik und sein experimenteller
Beweis, Der Bauingenieur 31/32: 459-467.

Schofield, A. and Wroth, P., 1968, Critical
State Soil Mechanics, London, McGraw-Hill.

Thiers, G.R. and Seed, H.B., 1968, Cyclic
stress-strain-characteristics of clay,
Journ. of the Soil Mech. and Found. Div.
Proc. ASCE, Vol. 94, No. SM2: 555-569.

Van Eekelen, H.A.M. and Potts, D.M., 1978,
The behaviour of drammen clay under
cyclic loading, Géotechnique 28, No. 2
173-196.

Vardoulakis, I., 1979, Bifurcation analysis
of the triaxial test on sand samples.
Acta Mechanica 32: 35-54.

Whitman, R.V., 1960, Some considerations
and data regarding the shear strength
of clays. Proc.of the ASCE Res.Conf.on
Shear..., Boulder, Colorado, p 581-614.

Wood, D.M. and Wroth, C.P., 1977, Some laboratory experiments related to the results of pressuremeter tests, Géotechnique 27, No. 2: 181-201.

Mechanical response of cement stabilized soil under repeated triaxial loading

C.P. NAG, T. RAMAMURTHY & G. VENKATAPPA RAO
Indian Institute of Technology, Delhi, India

1 SYNOPSIS

In this paper the results of axisymmetric triaxial tests under repeated load are reported for soil-cement. The tests were conducted at low stress levels to correspond to those anticipated in highway pavements. Material characteristics in the form of modulus of resilient deformation, resilient and total strains and modulus of elasticity are presented and discussed as a function of load repetition, confining pressure and cyclic deviator stress. The repetitive loading has been found to improve significantly the elastic behaviour of soil-cement.

2. INTRODUCTION

In the past few years increasing attention has been directed by highway engineers to devise rational methods of flexible pavement design. Linear elastic, non-linear elastic and viscoelastic theories for layered system are now being adopted to predict the pavement response. Also some of the recent studies have indicated close resemblance of flexible pavement performance to the stresses, strains and displacements calculated from layered theories. The elastic theories for layered system require material characteristics as inputs in terms of modulus of elasticity, E, and Poisson's ratio, μ, and an intimate knowledge of the stress-strain behaviour of the paving materials and the underlying subgrade soil.

Behaviour of paving materials very much depends on the stress-field, temperature, moisture content etc. and it is therefore important that material characterisation be carried out under simulated field conditions. Whereas other field conditions are easy to simulate in a laboratory test, it is difficult to represent the complex stress conditions that occur within a pavement layer. The pavements are subjected to an applied surface loading which is repetitive in nature with varying frequency and magnitude. An element in the pavement structure is acted upon by a system of generalised stress conditions (a system of normal and shear stresses). Some researchers have adopted axisymmetric triaxial compression test under repeated load as an approximation to the stress condition in the field.

The stress levels reached in a soil-cement base course under design wheel load are sufficiently low as compared to its fatigue limit. Most of the work reported was studied at axial stress about 60 to 70 percent of failure stress. The fatigue behaviour of soil-cement mixes under low stress conditions has been rarely reported. In this paper the test results of repeated load axisymmetric triaxial compression tests on partially saturated cement stabilised Delhi silt under consolidated undrained conditions (CU) are presented. The variation of modulus of elasticity, modulus of resilient deformation, inelastic and recoverable elastic strains have been shown with respect to number of load repetitions.

3 REVIEW

The derivation of proper constitutive relationship for subgrade soils and other paving materials is an important aspect in the solution of several outstanding problems in pavement design. Lazen (1968) has identified three approaches to the problem of defining constitutive relationship. These are (i) the micromechanistic approach, (ii) the phenomenological or solid mechanics approach and (iii) the simulated service evaluation approach involving adhoc tests designed to gather data under particular combination of stress and environmental conditions. Most work in highway engineering has adopted the simulation service approach, by employing variety of dynamic and repeated load test techniques. Both laboratory and field tests were employed to understand the stress-strain behaviour of the pavement materials. For isotropic linear-elastic conditions, however, it is well known that a material can be completely characterised by two elastic constants analogous to modulus of elasticity, E, and Poisson's ratio, μ. To evaluate these parameters under generalised stress conditions a test apparatus has been developed at IIT Delhi (Nag et.al 1979) as part of the comprehensive research programme in progress on material characterization of stabilised soils under simulated field conditions. The present study is part of this programme.

To define material characteristics of various paving materials and soil subgrade many researchers (Mitchell and Shen 1967; Barksdale,1972,Brown 1974, Morris et.al 1974 and Mcleans and Monismith 1975) have adopted stressfield represented by axisymmetric triaxial compression test under repeated loading as an approximation to the in-situ conditions. Both constant and cyclic confining pressure (σ_3) were used. In general their test data are reported in the form of resilient modulus (M_R) defined as the ratio of deviator stress ($\sigma_1 - \sigma_3$) to the resilient strain (ϵ_r), and the axial plastic strain (ϵ_p). For granular materials a logarithmic relationship between M_R and σ_3 has been established as

given by (1).

$$M_R = K\sigma_3^n , \text{---------------(1)}$$

where K and n are experimental constants. Allen and Thompson (1974) have shown from a priliminary series of tests on various granular materials, that there is a difference in the resilient properties determined under constant and under variable confining stress conditions. Brown and Hyde (1976) while verifying the conclusions drawn by Allen and Thompson, have suggested the use of constant confining pressure equal to the mean of cyclic value, since no quantifiable difference in the values of M_R and ϵ_r was observed in these cases. However, this test equivalency was not valid for evaluation of Poisson's ratio which was found to vary considerably under the two test conditions. It was also concluded that the results from constant and cyclic stress tests were compati..ble when interpreted in terms of volumetric and shear stress-strain relationship. Bulk and shear moduli, as functions of applied stress were adjudged better than M_R and ϵ_r, for characterising non-linear materials. For soil subgrades Frazin et.al (1975) have derived relationship between Young's modulus of elasticity, E, and the slope of axial stress-strain curve in axisymmetric triaxial compression test, E_T, as given by equation (2),

$$E = A . E_T \text{--------------(2)}$$

where A is a function of Poisson's ratio, μ, and slenderness ratio, S.

Mitchell & Shen (1967), following the general approach adopted by Dorman and Metcalf (1965) have tested soil-cement specimens under repeated flexure and triaxial compression tests. Their test results indicate decrease of M_R in compression with increase in applied stress; and increase in ultimate compressive strength with increasing stress applications. For specimens tested in flexure, M_R was found to remain unaffected by applied stress level. Wang and Mitchell (1971), while confirming the conclusions of Mitchell and Shen, have indicated dependency of M_R values on both the confining pressure (σ_3) and deviator stress ($\sigma_1 - \sigma_3$), and advanced a relationship relating M_R, I_1 and ($\sigma_1 - \sigma_3$) as given by (3)

$$M_R = K_1(K_2 - \log_e(\sigma_1 - \sigma_3))I_1^{K_3} \text{----(3)}$$

where K_1, K_2, K_3 are constants and I_1 the first stress invariant. It may be noted that the stress levels adopted in these investigations were of the order of 60 percent of failure stress and more, and are much higher than those anticipated in the pavements. Applying Griffith's failure theory on the results obtained by Mitchell and Shen and others, Radd et.al.(1977) have developed two different failure criteria for curing periods upto 4 weeks and 10 weeks and more to describe the behaviour of cement treated pavement materials subjected to repeated multiaxial stress applications. In these criteria number of repetitions to failure are expressed in terms of the maximum stress level applied, and the resulting relationship was independent of pulse duration, frequency and shape.

Only few researchers have reported data on Poisson's ratio and have found it to vary between 0.1 and 0.2 (Fossberg et.al. 1972). Kollias and Williams (1978) have studied Poisson's ratio for various cemented materials ranging from stabilised silty clay to lean concrete, and have found the dynamic value to be considerably higher than the static value. Static value of Poisson's ratio was found to lie around 0.15 for all types of cement bound materials.

In the foregoing it is observed that contradictory results were reported regarding the influence of confining stress and deviator stress on the values of M_R at higher stress levels, hence this aspect has been investigated at low stress levels which are anticipated in the highway pavements. Further change in the strength and modulus values after repetitive loading has also been studied, to have a more comprehensive understanding of the material behaviour.

4 EXPERIMENTAL INVESTIGATIONS

Delhi silt (sand=42%, silt=46%, clay -12%, LL=24.5% and PI=5.5%) was mixed with ordinary portland cement in 4 and 8 percentages by weight of soil to form soil-cement mixes. Cylindrical specimen 38 mm diameter and 76 mm high were moulded by static compaction at optimum conditions of density and moisture content corresponding to Proctor compaction (1860 kg/m^3 and 11%). The compaction device was specially designed to achieve uniform density by compacting in stages from both ends. The specimens were then cured at moulding moisture contents for 28 days in desiccators maintained at 100 percent relative humidity. At the end of curing the specimens were checked for any loss of weight, on account of moisture evaporation. No appreciable change was observed. The specimen was then set for CU test in a triaxial cell and was allowed to consolidate for 30 minutes beyond which consolidation process was found to cease. The specimens were subjected to repeated compression in a 10 tonne Instron Universal testing machine. The specimens were tested under constant confining pressures of 0.049, 0.098 and 0.147 MN/m^2 to correspond to the range of stress generally occuring within the pavement layers. The machine was set for continuous cyclic axial loading so as to produce cyclic deviator stresses of the order of 0.57, 0.38 and 0.19 MN/m^2, at a crosshead speed of 0.33 mm/sec. Depending upon the stiffness of the specimen, the frequency of load cycling was found to lie between 0.5 to 1.0 HZ. The load deformation charts duly magnified were recorded on 2 pen autographic recorder attached to the machine. The permanent deformation of the specimen was recorded by a deflection dial gauge having least count of 0.002 mm.

Tests on Delhi Silt and Soil + 4 percent cement mixes were conducted for 10,000 load applications and then tested to failure under static conditions. For soil + 8 percent cement mix, five identical specimens were subjected to repetitive axial compression to 1,10, 100, 1,000 and 8,000 repetitions and observations for inelastic and recoverable elastic deformations recorded. In all cases, after repetitive loading, the specimens were allowed to rest under isotropic stress condition ($\sigma_1 = \sigma_3$)

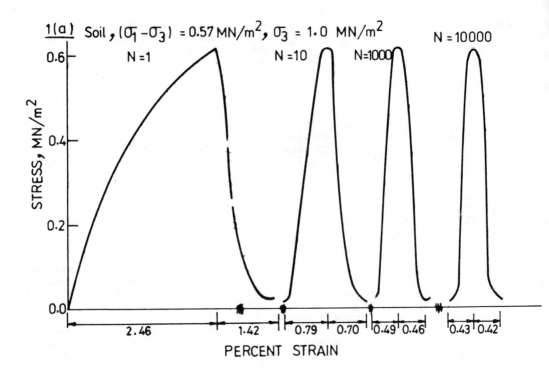

1(a) Soil, $(\sigma_1 - \sigma_3) = 0.57 \, MN/m^2$, $\sigma_3 = 1.0 \, MN/m^2$

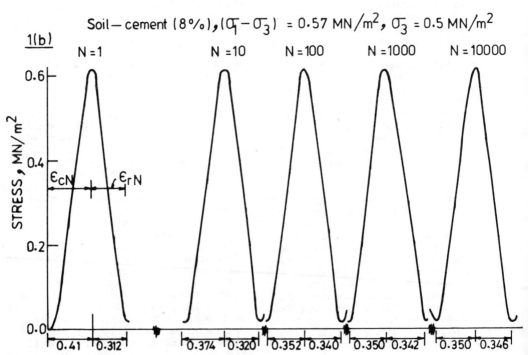

Figure 1 Typical stress-strain curves under repeated axisymmetric triaxial load test for (a) Soil and (b) Soil+8% cement

190

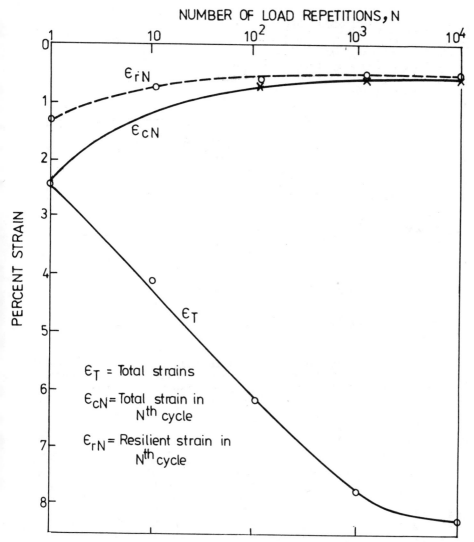

Figure 2 Variation of strains with N, for soil

for 10 minutes. These were then tested under static triaxial loading to failure with crosshead speed of 8.3x10^{-3} mm/sec to ascertain changes in material properties which might have occurred due to repetitive loading. A few tests were also conducted at 80 percent of initial failure load to investigate the change in material properties at near fatigue conditions.

5 RESULTS AND DISCUSSIONS

The typical stress strain curves for soil and soil+4% cement specimen tested under repeated axisymmetric triaxial test are shown in Figure 1(a) and (b). Table 1 presents strength and deformation data of soil and soil-cement specimen tested in axisymmetric triaxial test under static conditions.

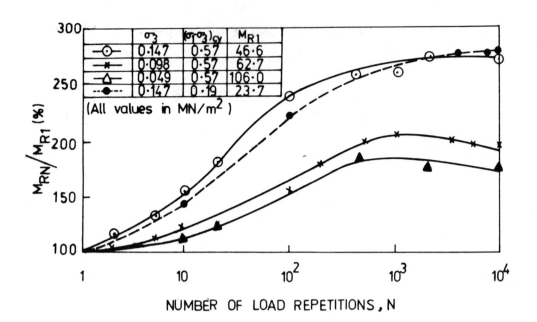

Figure 3 Variation of M_{RN}/M_{R1} with N, for soil

Table 1: Strength and deformation parameters of soil and soil-cement under static triaxial compression test.

Cement content (%)	σ_3 MN/m^2	(σ_1/σ_3) max	E MN/m^2	ϵ max (%)
0	.049	10.2	11.9	8.0
	.098	5.6	17.8	7.2
	.147	3.7	20.0	6.5
4	.049	41.9	132	2.34
	.098	22.5	156	2.28
	.147	19.0	183	2.20
8	.049	78.6	295	2.16
	.098	40.8	335	2.13
	.147	25.5	356	2.09

Test results of the present investigations in duly normalised form are presented with respect to number of load repetitions (N) in Figure 2 to 11 as variation of total strain(ϵ_T), axial cyclic strain in N^{th} cycle (ϵ_{CN}), ratio of resilient strain during N^{th} cycle (ϵ_{rN}) to ϵ_{CN}, axial irrecoverable strains (ϵ_p), modulus of resilient deformation (M_R), modulus of elasticity (E) and strength at failure ($\sigma_1-\sigma_3)_f$. The term cyclic stress level has been used in this paper to indicate the ratio of cyclic deviator stress $(\sigma_1 - \sigma_3)_{CY}$ to deviator stress at failure $(\sigma_1 - \sigma_3)_f$

5.1 SILT

For cyclic deviator stress ($\sigma_1 - \sigma_3)_{CY}$ = 0.57 MN/m^2 (cyclic stress level = 0.6) and σ_3 = 0.098 MN/m^2 total strains (ϵ_T) of silt specimen were found to increase almost linearly upto N = 100 (Fig.2) beyond which the curve between ϵ_T versus N becomes curvilinear with rate of increase of total strain falling gradually. Most of the total strain occured in the first few cycles. The total strain during a cycle (ϵ_{CN}) decreases gradually with number of load repetitions, at N=1, ϵ_{CN} equals 2.48 percent and at N= 10,000; ϵ_{CN}=0.50 percent. The resilient deformation during N^{th} cycle (ϵ_{rN}) decreases gradually from 1.42 percent at N=1 to 0.48 percent at N=10,000, which corresponds to an increase of the ratio $\epsilon_{rN}/\epsilon_{CN}$ from 57.3 percent to 98.7 percent. The ratio of $\epsilon_{rN}/\epsilon_{CN}$ for silt has been plotted alongwith the data of stabilized soil in Figure 6. A study of the above curves reflects

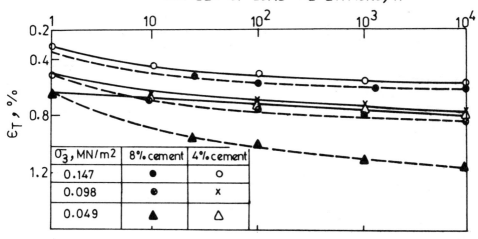

Figure 4 Variation of total strains (ϵ_T) with N, for soil-cement

on the fast accumulation of irrecoverable strains (ϵ_p) initially
amounting to 5.75 percent at N=100
and then steadily increasing to 7.7
percent at N=10,000. The trend of
the variation of parameters discussed above also indicates stiffening of the silt specimen under repetitive loading.

Figure 3 presents variation of the
ratio of modulus of resilient deformation at N^{th} cycle to that at
first cycle (M_{RN}/M_{R1}) for silt tested under different combinations of
$(\sigma_1 - \sigma_3)CY$ and σ_3. The ratio M_{RN}/M_{R1}, initially increases rapidly
for first 1000 repetitions and then
tend to decrease. The ratio of M_{RN}/M_{R1} have recorded an increase of
180 and 280 percent at confining
pressures of 0.049 and 0.147 MN/m²
respectively, at N=1000, and $(\sigma_1-\sigma_3)_{cy}$ = 0.57 MN/m². For a confining
pressure of 0.147 MN/m² when cyclic
deviator stress is increased from
0.19 MN/m² to 0.57 MN/m², the ratio
of M_{RN}/M_{R1} for silt shows a margi-
nal increase at all values of N. A
study of M_R values given in the inset table of Figure 3 indicate that
the M_R values are significantly influenced by both cyclic deviator
stress and confining pressure.
Values of M_{R1} are higher for higher

deviator stress and for lower confining pressures. Ahmed & Larew(1962),
Seed et al. (1962), and Monismith
et al. (1967) have also reported
dependency of M_R values on the axial
and confining pressures.

5.2 Soil-cement

Figures 4,5 and 6 present variation
of ϵ_T, ϵ_{CN} and $\epsilon_{RN}/\epsilon_{CN}$ respectively
with number of load repetitions, N,
for soil-cement specimens (of cement
contents, 4 and 8%) tested under different conditions of $(\sigma_1 - \sigma_3)_{cy}$
and σ_3. The cyclic stress levels
for these tests ranged between 0.10
and 0.30 depending upon cement content and confining pressures. Total
axial strains steadily increased
with increasing number of load repetitions (Figure 4). The effect of
confining pressure is found to be
predominant for soil+4% cement specimens tested under $(\sigma_1 - \sigma_3)_{cy}$ =
0.57 MN/m² to the extent that for
N=1000, ϵ_T value at σ_3=0.049 MN/m²
was about 140 percent more than that
at σ_3=0.147 MN/m². For soil+8% cement specimen this increase was comparatively very low. In all cases
ϵ_{CN} was found to decrease gradually
with N (Figure 5). For $(\sigma_1 - \sigma_3)_{cy}$ = 0.57 MN/m² and σ_3 = 0.049 MN/m²,
soil+4% cement specimens recorded a

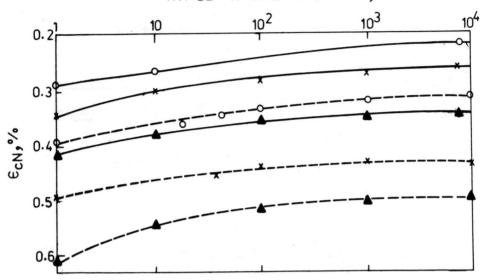

Figure 5 Variation of ϵ_{CN} with N, for soil-cement

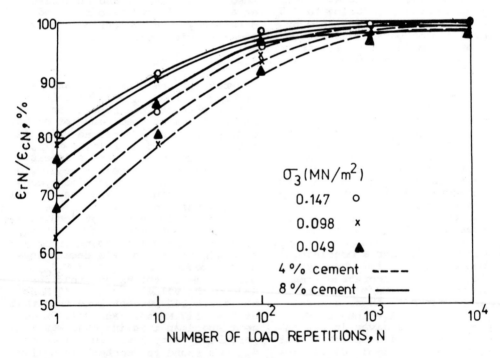

Figure 6 Variation of $\epsilon_{rN}/\epsilon_{CN}$ with N, for soil-cement

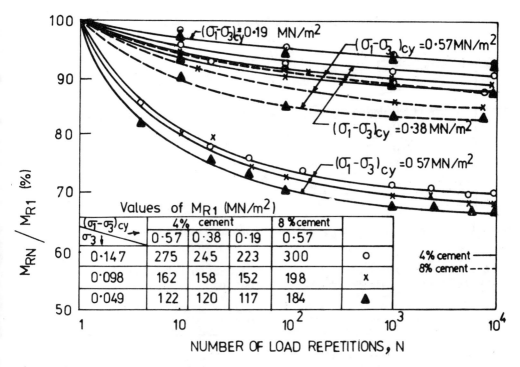

Figure 7 Variation of M_{RN}/M_{R1} with N, for soil-cement

decrease of ϵ_{CN} from 0.67 percent at N=1 to 0.50 percent at N=10,000. The corresponding values at σ_3 = 0.147 MN/m^2 were 0.39 percent and 0.32 percent. Soil+8% cement specimens recorded ϵ_{CN} values which are 60-70 percent of those observed for soil+4% cement specimens under identical test conditions. The ratio of ϵ_{rN} to ϵ_{CN} increased with increase in N for all specimens (Figure 6). Depending upon the stiffness of the specimen and applied confining pressure the ratio of $\epsilon_{rN}/\epsilon_{CN}$ ranged between 92 - 98 percent at N=100, and tends towards 100 percent with further increase in N. For $(\sigma_1 - \sigma_3)_{cy}$ = 0.57 MN/m^2, the effect of increase in σ_3 values from 0.049 MN/m^2 to 0.147 MN/m^2 was found to increase the $\epsilon_{rN}/\epsilon_{CN}$ ratio for soil+8% cement specimen from 96 to 98 percent at N=100 and from 92-95 percent for soil+4% cement specimen. A further study of these curves indicates that the total and resilient strains of soil-cement specimens are sngnificantly effected by cement content and confining pressure.

Figure 7 presents M_{RN}/M_{R1} values plotted against N for soil-cement specimen tested under different combinations of cyclic deviator and confining pressures. A reference to the table given as inset to Figure 7 (which records M_{R1} values) reveals that for all values of confining pressure M_{R1} values marginally increase with increasing cyclic deviator stress. The influence of confining pressure is more significant to the extent that with change of σ_3 from 0.049 to 0.147 MN/m^2, M_{R1} values at all cyclic deviator stresses increase by about 100 percent. Study of the curves plotted in Figure 7 shows that M_{RN}/M_{R1} values record a decreasing trend for all values of $(\sigma_1 - \sigma_3)_{cy}$ and σ_3 with increasing number of load repetitions. This decrease is upto 65-70 percent for $(\sigma_1-\sigma_3)_{cy}$ = 0.57 MN/m^2 and only to 92 percent for $(\sigma_1 - \sigma_3)_{cy}$=0.19 MN/m^2, at N = 10,000. For the same value of $(\sigma_1 - \sigma_3)_{cy}$ the effect of increase in confining pressure is to increase M_{RN}/M_{R1} values marginally. The

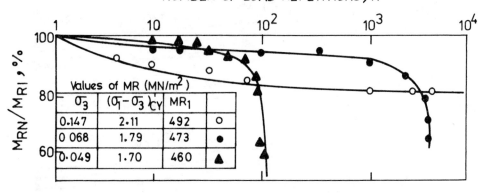

Figure 8 Variation of M_{RN}/M_{R1} with N, for soil-cement

effect is almost insignificant at low cyclic deviator stress equal to 0.19 MN/m². A comparison of the curves of M_{RN}/M_{R1} v/s N, for soil + 8% cement specimens tested under identical conditions of cyclic deviator and confining pressures, reveals that the increased stiffness of soil+8% cement specimens has caused decrease of M_R values by 15 to 18 percent at N=10,000. A few tests on soil+4% cement specimens were conducted with cyclic stress level as 0.80, under different confining pressures. The curves for M_{RN}/M_{R1} v/s N for these tests are plotted in Figure 8, where from it can be deducted that fatigue life of soil cement is greatly influenced by the confining pressures. For 3 equal to 0.049 and 0.068 MN/m² complete failure was observed at 102 and 3034 cycles of load repetitions respectively, and for σ_3=0.147 MN/m², the failure could not be achieved even for 5000 repetitions. A study of these curves reveals that soil-cement specimens at higher stress levels also behave initially in the same manner as discussed above. The rapid decrease of M_{RN}/M_{R1} values after certain number of load repetitions very near to failure suggest crack initiation and its rapid propagation. The crack initiation takes place earlier for low confining pressures and higher confining pressures seems to resist the formation of crack at an early stage.

5.3 Static behaviour after repeated loading

Soil+8% cement specimens subjected to 1,10,100,1000 and 8,000 load repetitions corresponding to $(\sigma_1 - \sigma_3)_{CY}$ = 0.57 MN/m², were then tested under static test conditions to assess changes in the material characteristics. Figure 9 shows the variation of the ratio of failure stress at N repetitions, $(\sigma_1 - \sigma_3)_f$ N, to failure stress at static conditions $(\sigma_1 - \sigma_3)_f$ with respect to N. For all confining pressures it was found that failure strength of soil-cement decreased with increasing number of repetitions. The strength was found to be varying from 100 percent at N=1 to 99 percent at N=8000 for σ_3=0.049 MN/m² and from 99.2 percent at N=1 to 95 percent at N=8000 for σ_3 = 0.147 MN/m². The decrease in strength was found to be more at higher confining pressure.

In Figure 10, the variation of the ratio of modulus of elasticity, E_N/E_0 (E_N = modulus of elasticity after N load repetitions, and E_o = modulus of elasticity for N=0) are plotted against N. E values have shown considerable improvement due to repetitive loading, and the ratio of E_N/E_o have been found to increase from 102% at N = 1 to 118 percent at N = 8000 for 3=0.049 MN/m² and 105 percent to 128 percent for corresponding values of N, at σ_3 = 0.147 MN/m². This effect of increased stiffness due to repetitive loading was

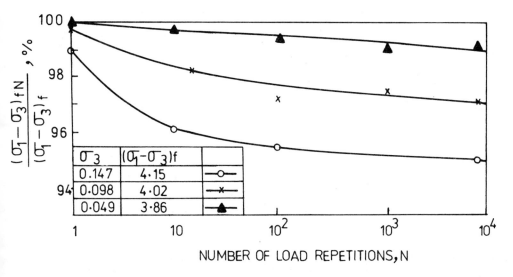

Figure 9 Variation of $(\sigma_1 - \sigma_3)_{fN}/(\sigma_1 - \sigma_3)_f$ with N, for soil+8% cement

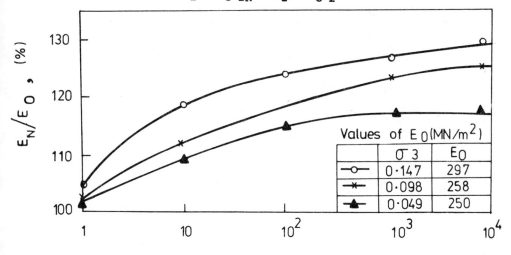

Figure 10 Variation of E_N/E_O with N, for soil+8% cement

also marked in the variation of M_R values with respect to N. The values of modulus of elasticity have shown comparatively large change at higher confining pressures, whereas the M_{RN}/M_{R1} ratio was found marginally more for lower confining pressure.

Figure 11 presents variation of irrecoverable deformations (ϵ_p) for soil+8% cement specimen with number of applied loadrepetitions(N).

ϵ_p was found to be higher for lower confining pressures and higher value of N. The ϵ_p values show large increase during first 100 cycles, beyond which the rate of increase gradually slows down.

Above discussion of the test results of various strength parameters observed for soil-cement describe the typical behaviour of soil-cement subjected to repetitive loading and

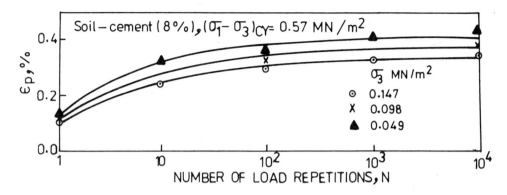

Figure 11 Variation of ϵ_p with N, for soil+8% cement

suggest improvement in elastic properties due to repetition. Reduction in M_R values with increased number of load repetitions was also observed by Mitchell and Shen (1967) for Vicksburg silty clay-cement tested at stress levels of 0.60. The cement content employed by them was 13 percent and one day soaking followed the period of curing before the samples were put to cyclic test. Bofinger (1965) have reported slight decrease in strength with load repetitions. Mitchell and Shen (1967) have also reported increase of strength of soil-cement specimen upto 35 percent of initial failure strength on account of cyclic loading. The difference in the strength results may be attributed to the difference in testing standards employed and the type of soil used. In general the results observed under present investigation conforms to those reported by Mitchell and Shen (1965) Ahmed and Larew (1962) and Wang and Mitchell (1972).

6 CONCLUSIONS

The preceeding discussions on the variation of various material characteristics of soil and soil-cement during and after repetitive loading leads to the following conclusions:-

(i) The partially saturated soil specimen initially show increase in M_R values with increasing value or N suggesting the soil undergoing further compaction under repetitive loading.

(ii) M_R values are greatly influenced by the confining pressure higher M_R values were observed for higher confining pressures.

(iii) For soil-cements, values of M_R decreases with increasing number of load repetitions indicating towards increased stiffness. M_R value is a function of stiffness of the specimen, the applied cyclic deviator stress and the confining pressure. The effect of confining pressure is more predominant at higher cyclic deviator stress.

(iv) Resilient strains increase with increasing number of load repetitions. For N=10,000, the elastic recovery of soil-cement specimens were found to be about 100 percent suggesting completely elastic behaviour under repetitive loading.

(v) For the same cyclic stress level, failure of soil-cement specimen was observed at larger number of load repetitions for higher confining pressures.

(vi) The static modulus of elasticity (E) was found to increase significantly on account of repetitive loading, suggesting significant improvement in the elastic properties of soil-cement. The increment in the values of modulus of elasticity is a function of confining pressure - the increase is higher for higher confining pressure.

The observations made above indicate that material characteristics of soil-cement are significantly improved by repetitive loading. The behaviour of soil-cement under low stress levels remain same as observed at higher axial stresses by other research workers. The experimental data at low stress level presented here is of relevance and use in the design of cement stabilized base courses.

7 REFERENCES

Allen, J.J. and M.R. Thompson, 1974, Resilient response of granular materials subjected to time dependent lateral stresses, TRR 510.

Barksdale, R.D. 1972, Laboratory evaluation of rutting in base course materials, Proc. 3rd Conf. on Structural Design of Asphalt Pavement, 161-174.

Bofinger, H.E. 1965, The fatigue behaviour of soil-cement Australian Road Research 12-20.

Brown, S.F. 1974, Repeated load-testing of a granular material, Journal of Geotechnical Engineering, ASCE, 100:GT 7:825-841.

Brown, S.F. and A.F.L. Hyde, 1975, Significance of cyclic confining stress in repeated load triaxial testing of granular material, TRR 537.

Dorman, G.M. and C.T. Metcalf, 1965, Design curves for flexible pavements based on layered system theory, HRR: 71: 69-84.

Fossberg, P.E., J.K. Mitchell and C.L. Monismith 1972, Load deformation characteristics of a pavement with cement-stabilized base and asphalt concrete surfacing, Proc. 3rd Conf. on Structural Design of Asphalt Pavements, 795-811.

Frazin, M.H., J.K. Raymond and R.B. Corotis, 1975, Evaluation of modulus and Poisson's ratio from triaxial tests, TRR 537: 69-80.

Kolias, S. and R.I.T. Williams, 1978, TRRL Supl. report 344, crowthorne.

Lazen, B.J. 1968, Damping of materials and members in structural mechanics, Pergamon Press.

Mcleans, D.B. and C.L. Monismith 1974, Estimation of permanent deformation in asphalt concrete layers due to repeated traffic loading, TRR 510: 14-31.

Mitchell, J.K. and C.K. Shen 1967, Soil-cement properties determined by repeated loading in relation to bases for flexible pavements, Proc. 2nd Conf. on the Structural Design of Asphalt Pavement: 427-451.

Morris, J., R.C.G. Hass, P. Reilly, and E. Hignell 1974, Permanent deformation in asphalt pavements can be predicted, Proc. Association of Asphalt Paving Technologist.

Nag, C.P., T. Ramamurthy and G. Venkatappa Rao, 1979, Development of universal triaxial test apparatus for evaluation of pavement material characteristics, Int. Symp. on Pavement Evaluation and Overlay Design, Rio-de-janeiro.

Radd, L., C.L. Monismith, and J.K. Mitchell 1977, Fatigue behaviour of cement treated materials, TRR 641.

Wang, M.C. and J.K. Mitchell 1971, Stress and deformation prediction in cement treated soil pavements, HRR 351 : 93-111.

Effective stress changes observed
during undrained cyclic triaxial tests on clay

M. TAKAHASHI, D. W. HIGHT & P. R. VAUGHAN
Imperial College, London, UK

1 SUMMARY

The reliable measurement of pore water pressures in clays under cyclic loading has been restricted to tests with relatively long period cycles by the slow response time of conventional pore pressure transducers.   A fast response time piezometer probe, mounted at the mid-height of triaxial specimens, has been used in an investigation of effective stress changes during compression-extension loading of a sandy clay in cycles of differing frequency.

Observations from this work are presented. The migration of the effective stress path towards the origin of stress space is demonstrated and the rate of migration, in terms of number of cycles, is shown to be dependent on, inter alia, the applied stress level, stress history and cycle period.   Reduction in effective stress with continued compression-extension cycling leads to the development of large plastic strains when the effective stress path approaches the effective stress failure envelope, defined in slow monotonic loading tests.   With continued cycling the observed effective stress path for the extension phase migrates well beyond this failure envelope.   This phenomenon has been explored and appears to be related to differences in the rate of shearing in the cyclic and monotonic loading tests.

2 INTRODUCTION

An investigation is being carried out at Imperial college into the monotonic and cyclic loading behaviour of a low plasticity sandy clay, the Lower Cromer Till from Happisburgh, Norfolk.   The work forms part of a research project related to the per-formance of offshore structures subjected to wave loading in the North Sea.

The nature of the loading offshore is such that soil elements affected by the structures experience significant cyclic rotation of their principal stress directions as well as cyclic variations in the magnitude of the principal stresses.   Such stress changes are particularly difficult to apply in the laboratory.   At present the cyclic load tests have been restricted to the triaxial apparatus and have involved varying the magnitudes of the principal stresses while either keeping their directions constant or causing a $90°$ step rotation of their directions (i.e. in compression-extension tests).   It remains to be seen what the relevance of the latter type of test is to a study of the behaviour of soils undergoing continuous cyclic rotation of principal stress directions.

Observations of effective stress changes during undrained cyclic loading of clays are limited, particularly at rates of cycling appropriate to the field situation, yet they are an essential requirement for explaining behaviour and for developing generalised soil models.   Reliable measurement of pore water pressures throughout each cycle has, therefore, been a major objective in the investigation.   Some of the effective stress phenomena which have been observed in undrained cyclic triaxial tests are now reported and important features which should be embodied in numerical soil models are highlighted.

3 THE SOIL, TEST PROGRAMME AND EQUIPMENT

Soil from the Lower Cromer Till was chosen as representing one of the North Sea sediment types.   Its sampling, composition and natural variability have been described by Hight et al (1979). To achieve consistent and repeatable behaviour, reconstituted samples, 38 mm diameter by 76 mm high, have been used; these were cut from soil blocks which had been consolidated either isotropically or anisotropically in an oedometer from a slurry. The samples were either reconsolidated in the triaxial cell isotropically to an effective stress of 300 kPa (overconsolidated samples were swelled back from this stress) or were tested without further consolidation.

Reference is made in this paper to three forms of undrained triaxial test, namely cyclic compression-extension (two-way cyclic), cyclic compression (one-way cyclic), and monotonic extension tests, conducted at different rates of loading. A summary of the tests and the sample history and loading is given in Table 1.

All the tests were carried out in the hydraulic triaxial cell (Bishop and Wesley, 1975) which was controlled using a closed-loop servo system, incorporating a PDP-11 computer (Hight, 1979). The cyclic tests were stress-controlled, using a sinusoidal waveform for the two-way tests and a triangular waveform for the one-way tests. The monotonic extension tests were either strain-controlled (0.006 mm/min) or load-controlled.

The pore water pressures used to derive effective stress changes were measured at the mid-height of the triaxial samples using a piezometer probe mounted flush with the cylindrical surface of the sample. A description of the probe, its mounting, response time and suitability for pore pressure measurement during cyclic loading has been given by Hight and Takahashi (1979). For the sandy clay under investigation, with a C_v of 1-2 m^2/year, response times of less than one second can be achieved. This enables pore pressures to be followed closely in cycles of 10 seconds duration.

In the work which is reported, the shortest cycle period was 10 minutes; although this and the longer periods are well outside typical cycle periods for offshore loadings, their use ensured that the pore pressures at the mid-height were correctly measured throughout the cycles. Lubricated sample ends were not used and at all the rates of loading a pore pressure difference existed between the centre of the sample and its ends; the magnitude of the difference was larger in compression than in extension. Clearly some equilibration of these pore pressures will have occurred and the degree of equilibration will have increased with increasing cycle period. Evidence from pore pressure measurements at both the base and mid-height indicate that equilibration has little influence on the pore pressure at the centre of the sample in half cycles of compression but tends to depress the central pore pressure in half cycles of extension.

4 EFFECTIVE STRESS PATHS IN UNDRAINED CYCLIC TRIAXIAL TESTS

In Fig. 1(a) an example is given of the effective stress paths followed during cyclic compression-extension loading of a normally consolidated sample. In this example the cyclic stress ratio, defined as the ratio of the applied shear stress to the monotonic undrained shear strength in compression, was 0.75 and the cycle period was 30 minutes. The corresponding stress-strain and pore pressure-strain data for cycles 1, 5 and 8 are presented in Fig. 1(b). In Fig. 1(c) the axial strains and pore pressures measured at the maximum shear stress in compression and at the maximum shear stress in extension are plotted against cycle number.

In terms of total stresses the sample exhibits features normally associated with the cyclic compression-extension loading of clays, e.g. a reduction in stiffness and an increase in hysteresis with number of cycles.

Several other important features are revealed by the effective stress changes plotted in Fig. 1:-

(i) The effective stress paths migrate towards the origin of the stress space; this is a natural consequence of the pore pressures generated during loading being different in magnitude to those generated during unloading. This difference is greater in the half cycle of compression than in the half cycle of extension, giving rise to a lack of symmetry in the effective stress cycles.

(ii) The shape of the effective stress cycles alters as migration takes place and loops develop in the half cycle on the extension side.

202

Table 1. Test programme and sample details

I   Undrained cyclic compression-extension tests.
(Samples trimmed from soil block isotropically consolidated to p'_{max} of 160 kPa and swelled back to p' of 40 kPa.)

Test No.	OCR after isotropic reconsolidation (p'_{max} = 300 kPa)	Water content after consolidation (%)	Cyclic stress* ratio	Cycle period (mins)
C-101	1	15.70	0.56	30
C-98	1	15.51	0.63	30
C-75	1	15.61	0.75	30
C-92	1	15.70	0.75	240/10
C-96	1	15.64	0.75	480
C-80	4	16.10	0.75	30
C-85	7	16.33	0.65	30

II   Undrained cyclic compression tests.
(Samples trimmed from soil block anisotropically consolidated to $\sigma'_{v\ max}$ of 120 kPa and swelled back to σ'_v of 30 kPa.)

C-11	1	15.64	0.55	15
C-12	1	15.75	0.80	15
C-13	1	15.53	0.55	1m 40s
C-14	1	15.71	0.80	1m 40s

* cyclic stress ratio = cyclic shear stress/undrained shear strength in compression

III   Unconsolidated undrained monotonic extension tests.
(Samples trimmed from soil block anisotropically consolidated to $\sigma'_{v\ max}$ of 200 kPa (220 kPa for samples marked *) and swelled back to σ'_v of 50 kPa.)

Test No	Water content during shear (%)	Type of loading
A	First time shearing:-	
C-18*	17.44	Strain-controlled (0.006 mm/min)
C-33	18.12	Strain-controlled (0.006 mm/min)
C-26	17.33	Load controlled (32 secs to apply max shear stress)
B	After cyclic compression-extension:-	
C-20*	16.99	Strain-controlled (0.006 mm/min)
C-23*	17.51	Strain-controlled (0.006 mm/min)
C-58	17.82	Load-controlled (26 secs to apply max shear stress)
C-25	17.61	Maintain deviator load after cyclic compression-extension

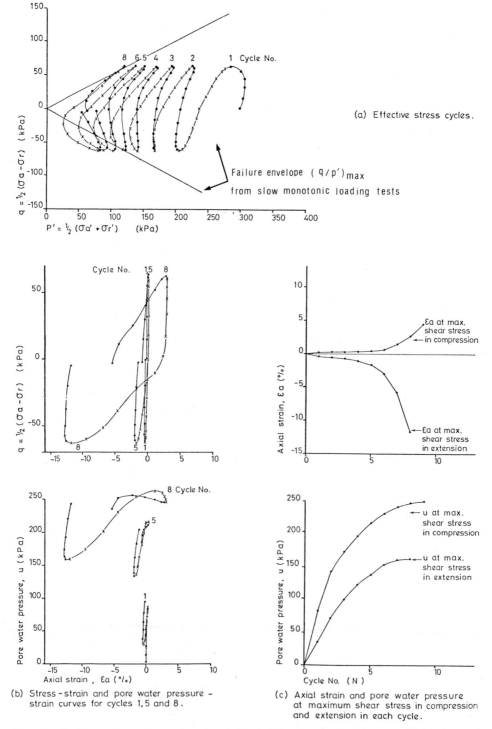

(a) Effective stress cycles.

Failure envelope $(q/p')_{max}$
from slow monotonic loading tests

(b) Stress-strain and pore water pressure - strain curves for cycles 1,5 and 8.

(c) Axial strain and pore water pressure at maximum shear stress in compression and extension in each cycle.

Fig. 1. Cyclic compression-extension triaxial test on normally consolidated sample. Test No. C-75. Cyclic stress ratio = 0.75. Effective confining pressure = 300 kPa.

(iii) Large axial strains develop in the sample when the effective stress path approaches the failure envelope defined by slow strain-controlled monotonic loading tests. (The failure envelope is used herein to mean the envelope to the maximum ratios of q/p', i.e. $(\sigma_1'-\sigma_3')(\sigma_1'+\sigma_3')$, reached in these slow monotonic tests.) The extension cycle leads in the approach to the failure envelope and, correspondingly, large strains occur first in extension.

(iv) When low values of the mean effective stress, p' $(=\frac{1}{2}(\sigma_a'+\sigma_r'))$ are reached the effective stress path in extension travels beyond the failure envelope.

(v) While axial strain development accelerates with the number of cycles, pore pressure generation reduces. In fact, a more-or-less stable effective stress cycle is eventually achieved.

Two of these features are examined more closely, namely the rate of migration of the effective stress path and the large strain behaviour in compression-extension cycles at low effective stress.

5 FACTORS CONTROLLING THE RATE OF MIGRATION OF EFFECTIVE STRESS CYCLES

Three factors have been identified so far as influencing the rate at which the effective stress path migrates towards the origin of stress space during cyclic compression-extension loading: these are the cyclic stress ratio, the consolidation history and the frequency of cyclic loading.

5.1 Cyclic stress ratio

In Fig. 2 the behaviour of three isotropically normally consolidated samples which were cyclically loaded in compression and extension with cyclic stress ratios of 0.56, 0.63 and 0.75 can be compared. The rate of migration can be judged from Fig. 2(a) in terms of the effective stress, p', in the sample at the end of each cycle, N; also indicated by an arrow for each test is the cycle number at which the effective stress path in extension first reached the failure envelope. It is immediately apparent, both from the slope of the p' - N line and the cycle in which

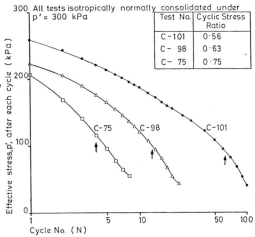

Test No.	Cyclic Stress Ratio
C-101	0·56
C-98	0·63
C-75	0·75

(a) Effective stress, p', after each cycle

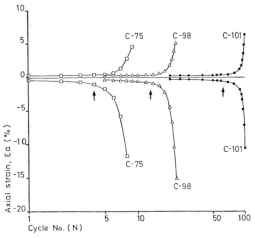

(b) Axial strain at maximum shear stress in compression and extension in each cycle

Fig. 2. Effect of cyclic stress ratio on the migration of the effective stress path of normally consolidated samples.

the failure envelope is reached, that increasing the cyclic stress ratio increases the rate of migration of the effective stress cycles. This merely confirms the observations made previously in terms of total stresses that increasing cyclic stress ratio produces a more rapid deterioration in the sample, see, for example, Andersen (1976).

In Fig. 2(b) the corresponding axial strain development in the three samples can

205

be seen. The pattern is similar in each,
namely a relatively slow build up in strain
until a value of approximately 1% is
reached, after which axial strains increase
dramatically.   In test C-75 (cyclic stress
ratio of 0.75) the development of an axial
strain in extension of 1% and the onset of
large strain behaviour coincide with the
effective stress path in extension reaching the failure envelope.   In the tests
with lower cyclic stress ratios, an axial
strain of 1% does not develop, apparently,
until several cycles after the effective
stress paths in extension reach the failure
envelope.   In fact, the onset of large
strain behaviour coincides more closely
with the effective stress path in compression reaching the failure envelope.
Although this might imply that the onset
of large strain behaviour requires the
effective stress path to reach the failure
envelope in both extension and compression,
it seems more likely that the location of
the failure envelope in extension is incorrect, as discussed subsequently.

It is of interest that all three samples
eventually experienced large cyclic deformations, even with a cyclic stress ratio
as low as 0.56.   Clearly this level is
above the critical level of repeated
loading as defined by Sangrey et al (1969).

5.2 Consolidation history

For samples swelled back by different
amounts from the same maximum consolidation
pressure, there are bound to be differences
in the rate of migration of the effective
stress paths under cyclic loading.   In
fact, for samples with a sufficiently high
overconsolidation ratio, OCR, i.e. a low
initial value of p', then under a high
cyclic stress ratio, the first effective
stress cycle lies close to the failure
envelope and further migration of the
cycles is restricted.   This is demonstrated
in Fig. 3 where the rate of migration
produced by compression-extension loading
of 3 samples, of OCR 1, 4 and 7, can be
compared.   In the samples of OCR 4 and 7
the direction of migration is, in fact,
initially away from the origin of stress
space but subsequently reverses.   When
all samples are at similar low values of
p' their effective stress cycles are of
approximately the same shape.

5.3 Loading frequency

Direct evidence on the effect of cyclic
frequency on the rate of migration of the
effective stress path is not available at

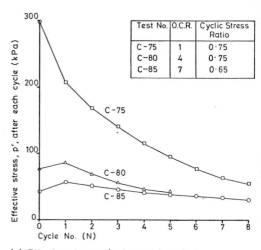

Test No.	O.C.R.	Cyclic Stress Ratio
C-75	1	0·75
C-80	4	0·75
C-85	7	0·65

(a) Effective stress, p', after each cycle.

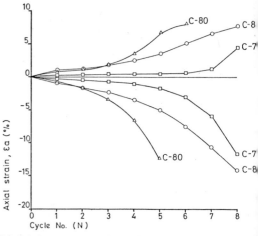

(b) Axial strain at maximum shear stress in
compression and extension in each cycle.

Fig. 3. Effect of consolidation history on
the migration of the effective stress path.

this stage of the investigation from cyclic
compression-extension tests.   However, a
series of cyclic compression tests at
different frequencies with a triangular
waveform was run on samples which were consolidated initially in an oedometer to a
vertical effective stress of 120 kPa and
subsequently consolidated isotropically in
the triaxial cell to an effective stress of
300 kPa.   Some of the conclusions from
this test series are summarised in Fig. 4
in which the effect of cyclic period on
the rate of migration at two cyclic stress
ratios can be seen.   Samples loaded more

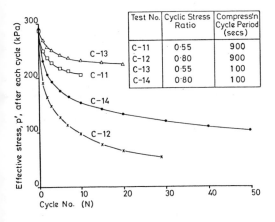

Test No.	Cyclic Stress Ratio	Compress'n Cycle Period (secs)
C-11	0·55	900
C-12	0·80	900
C-13	0·55	1·00
C-14	0·80	1·00

Fig. 4   Effect of cycle period on the migration of the effective stress path in compression cycles on normally consolidated samples.

slowly, 450 seconds to the maximum shear stress, generated higher pore pressures and so migrated more rapidly than samples loaded to the maximum shear stress in 50 seconds.   The rate of axial strain development   showed a corresponding dependence on cycle period.

It seems likely that these findings would apply to compression-extension loading of similar normally consolidated samples.

6 COMPRESSION-EXTENSION BEHAVIOUR AT LOW EFFECTIVE STRESS

The observation in Fig. 1 of the effective stress path in extension migrating beyond any failure envelope anticipated from slow monotonic tests is of considerable interest, particularly since stress paths cross a 45° line from the origin implying negative effective stress in the sample.   Such a situation is not inconsistent with the conditions of the test, since the shape of the sample and the lower pore pressures existing at its ends would allow this state to exist.   A similar observation has been made in cyclic compression-extension tests on reconstituted London Clay.

In order to understand this phenomenon, monotonic extension tests were carried out on samples which had either been previously subject   to cyclic compression-extension loading or were previously unsheared. The details of the tests are given in Table 1 and the results are plotted in Fig.5. Two tests, C-20 and C-23, were sheared at a constant rate of strain after cyclic loading;   for C-20 shearing began from an isotropic stress state, i.e. after completion of a compression-extension cycle, while for C-23 shearing began from the maximum shear stress in compression.   Their behaviour, in terms of effective stress path, was consistent with the behaviour observed in tests C-18 and C-33 on samples which were sheared at the same rate and which had not previously been cyclically loaded.   However, the failure envelope

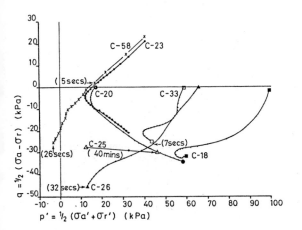

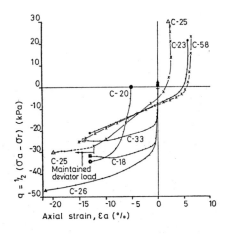

Fig. 5. Effective stress paths and stress-strain curves for monotonic extension tests.

207

based on samples C-20 and C-23 differs from that based on C-18 and C-33. This difference may arise from a change in structure in the material, brought about by cyclic loading. The shift in failure surface may help to account for the apparent non-coincidence between the onset of large strain behaviour and the arrival of the effective stress path in extension at the failure envelope; this was referred to previously. However, the difference is not of major significance and the effective stress paths in extension during cyclic loading travel well outside the steeper failure envelope.

Two samples, C-58 and C-26, were sheared in fast load-controlled extension tests; C-58 had been subject to compression-extension loading which was halted at the maximum shear stress in compression, while C-26 had not been sheared before. The effective stress paths for C-58 and C-26 can be seen in Fig. 5 to travel well beyond any failure envelope which might be based on the results of tests C-20, C-23, C-18 and C-33. This marked difference in effective stress path between the fast load-controlled tests and the slow strain-controlled tests, with their much longer times to failure suggests that it is the rate of loading which is largely responsible for the apparent migration of the effective stress path beyond the failure envelope.

Two additional pieces of evidence confirm that the rate of loading affects the location of the effective stress path in extension. The first relates to test C-25 in which compression-extension loading was halted at the maximum shear stress in extension; the deviator load on the sample was then maintained constan and the sample's behaviour was monitored. Axial elongation occurred, giving rise to an increase in axial stress, as shown in Fig. 5. This was accompanied by a decrease in pore water pressure measured at the mid-height of the sample. After 40 minutes the point representing the effective stress state in the sample had migrated to the location shown in Fig. 5, i.e. to the failure envelope based on C-23, the slow strain-controlled test.

The second confirmatory piece of evidence concerns a series of cyclic compression-extension tests in which the cycle period was varied after low mean effective stresses had developed in the samples. The cycles shown in Fig. 6 for the different periods demonstrate that, with increasing period,

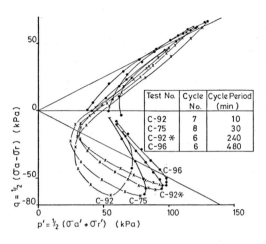

Test No.	Cycle No.	Cycle Period (min)
C-92	7	10
C-75	8	30
C-92 *	6	240
C-96	6	480

Fig. 6. Effect of cycle period on effective stress path at low effective stress.

the effective stress paths conform more closely with the pattern suggested by slow monotonic loading tests.

These observations obviously call into question the validity of the pore water pressures measured at the mid-height of the sample, particularly at the fast rates of loading. The extremely fast response time of the piezometer probe has been discussed and is not in doubt. The authors believe, therefore, that the pore pressure at the periphery of the sample is correctly measured. There are no indications of a significant change in pore pressure caused by stress disturbance at the probe. There may be radial non-uniformity of pore pressure in the sample, but this is unlikely to be large. At fast rates of loading there may be pore pressure irregularities locally within the sample. As stated previously, equilibration of pore pressure gradients between the ends and centre of the sample may affect the measured central pore pressure during slow extension loading.

The conclusions at this stage of the investigation are that the generation of pore pressure is rate dependent and that effective stress paths based on a global measurement of pore pressure in fast monotonic or cyclic loading tests may travel outside a failure envelope defined by slow monotonic loading tests.

208

7 DISCUSSION

In the low plasticity sandy clay which has been investigated, features associated with the undrained cyclic compression-extension loading of other, more plastic, clays can be recognised, e.g.

(i) a reduction in stiffness with increasing number of cycles, and

(ii) an increase in hysteresis and damping with number of cycles.

These features are associated with migration of the effective stress path towards the origin of stress space. The rate at which migration takes place, and hence the rate at which stiffness reduces and hysteresis increases, has been found to depend on:-

(i) cyclic stress ratio,

(ii) the consolidation history, and

(iii) the frequency of the cyclic loading.

When axial strains of approximately 1% have developed, the effective stress path has migrated close to the failure envelope defined by slow monotonic loading tests. At this stage the build up of axial strains accelerates and a more-or-less stable effective stress cycle is established. The shape of the cycle implies a tendency to strongly dilatant behaviour and is independent of initial consolidation history. Although there is a limit to pore pressure generation, axial strains continue to accumulate as a consequence of the non-symmetric nature of the compression-extension loading.

The clay is of low sensitivity and, as might be expected, there is no evidence of a reduction in ultimate undrained compression strength as a result of cyclic loading. However, after the onset of the large strain behaviour described above, this strength can only be mobilised at strain levels which would be unacceptably high in most engineering situations.

It would appear that both the onset of large strain behaviour under cyclic compression-extension loading and the changes in soil response prior to this can be quantified conveniently in terms of the effective stress changes which take place in the soil. Numerical modelling of soil behaviour under cyclic loading should make use of this (see, for example, Van Eekelen and Potts, 1978). If modelling is extended to large strain behaviour then difficulties may be encountered in defining the relevant failure envelope. There is evidence that effective stress paths in extension travel beyond the failure envelope determined in slow monotonic loading tests; this appears to be related to differences in rate of loading and to changes in soil structure brought about by cyclic loading. It may be necessary, then, to consider transient migration of the failure envelope, both under cyclic loading and under rapid monotonic loading to failure.

ACKNOWLEDGEMENTS

Part of the work reported has been supported by grants from the Science Research Council to the London Centre for Marine Technology.

REFERENCES

Andersen, K. H. (1976) Behaviour of clay subjected to undrained cyclic loading. Proc. First Int. Conf. on Behaviour of Offshore Structures, Vol 1, 392-403.

Bishop, A. W. & Wesley, L. D. (1975) A hydraulic triaxial apparatus for stress path testing. Geotechnique, 25, 4, 657-670.

Hight, D. W. (1979) A computer control system for soil testing. In preparation.

Hight, D. W. & Takahashi, M. (1979) A note on pore pressure measurement during cyclic loading of clays. In preparation.

Hight, D. W., El-Ghamrawy, M. K. & Gens, A. (1979) Some results from a laboratory study of a sandy clay and implications regarding its in situ behaviour. Proc. Second Int. Conf. on Behaviour of Offshore Structures, Vol 1, 133-150.

Sangrey, D. A., Henkel, D. J. & Esrig, M.I. (1969) The effective stress response of a saturated clay soil to repeated loading. Canadian Geo. Jnl, 6, 3, 241-252.

Van Eekelen, H. A. M. & Potts, D. M. (1978) The behaviour of Drammen Clay under cyclic loading. Geotechnique, 28, 2, 173-196.

Some results of impact experiments on nearly saturated and dry sand

P. A. RUYGROK & H. VAN DER KOGEL
Soil Mechanics Laboratory, Delft, Netherlands

SUMMARY

Laboratory-scale impact tests on a two-
layer system consisting of a hard-plastic
slab on both a dry and a nearly saturated
sandbed were performed, to investigate dynamic phenomena that may occur in asphalt
covered sand-dikes under wave impact. Test
set-up, instrumentation and calibration
aspects are briefly described. Some general
results are presented, together with a discussion on specific aspects related to
liquefaction and typical phenomena in the
pore pressure records.

1 INTRODUCTION

During the years 1975 thru 1978 a series
of impact experiments has been performed
by the Delft Soil Mechanics Laboratory.
The objective was to obtain insight into
the behaviour of saturated and dry sand
under impact loading. This objective
emerged from a discussion on the response
of a dike under seawave loading. One of the
hypothesis which was put forward was, that
the impact of a seawave could liquefy the
sometimes loose sand layers beneath the
asphalt protection. This could be the start
of a subsequent deterioration of the dike.
Therefore it was decided to model the
asphalt-sand system in the laboratory in
order to study its behaviour.
    In this paper we restrict ourselves to
the design of the experiment, instrumentation and a first presentation of some results.

2 STATEMENT OF THE PROBLEM

The question is to establish criteria in
case of extreme dynamic loading (impact of
breaking waves following storm surge) on
dike slopes with asphalt-concrete cover at
several distributions of saturation degree
of the sand underneath the attacked zone.

Two lines of approach were originally proposed (1971): development of a theoretical
model and test-verification or performance
of model-tests.
    At that time a realistic theoretical
approach towards some urgent questions was
considered impossible at short notice. From
the point of strict model considerations,
the lack of prototype-information and technical problems resulting from geometric
scaling, the possibilities of realistic
modelling were also very restricted.
    To obtain at least additional preliminary
directives, the approach eventually chosen
was to perform experiments with partial
scaling and simplified geometry. The results
(slab deflections, stresses, pressures and
strains) could serve the purpose to check
the merits of a numerical program, to be
developed and to adjust the constitutive
model required for the cover slab and underlying soil materials.
    This set-up resulted in the use of a 20 mm
thick hard polyethene slab overlying an in-
strumented sandbed with varying degree of
porosity and water content distribution.
The transient load was schematisized to a
peak pressure of $100kN/m^2$, (nominal) on a
circular area of radius 0.15 m with risetime
of 8-10 msec. and a duration of approx.
25 msec.

3 INSTRUMENTATION

The test-bin (dimensions $4 * 2.5 m^2$ in area,
1.5 m height) was constructed of assembled
concrete retaining-wall elements, reinforced
by means of steel girders. The inner walls
were clothed with a thin bituminous layer.
A drain system was installed at the bottom.
The sandbed was built up with a sand sprink-
ler system that enabled to vary relative
density of the sand used in dry state between 8 and 93% (porosity max.~ 46%, min.
~ 35%). To obtain a nearly saturated bed

the sand was sprinkled into a thin layer of de-aired water, in which case only porosities larger than 40% could be obtained (depending also on water layer thickness) To saturate a higher packing degree a method employing dry build up and replacement of pore air by easy soluble gases (carbondioxyde, ammonia) preceeding water saturation is proposed.

A falling weight system, that enables limited adjustment of impact rise time and magnitude, was chosen as load exitator (basic frame supplied by Ph ø nix A.S., Denmark). It includes a load-cell (Hottinger C2M/10M) and exchangeable loading area facilities. The impact chosen is characterised by a normative peak pressure of 100 kN/m^2 over a circular area of radius 150 mm, with risetime of order 8 msec. and extinction within 17 msec.

Transducer selection was based on a preceeding study on available information about soil-gauge interaction and "expected" order of magnitude of the different variables (local cover slab deflections, soil stresses, pore pressures and soil strains) within a pre-established measuring zone of interest. The pressure transducer should posses a resolution of at least 0.2kN/m^2, coupled to a high sensitivity and signal to noise ratio, good linearity within a high dynamic range (0.5-50 kN/m^2 at least) and transient response within 1 msec.
Strain transducers should have a resolution of 10^{-5} (10^{-3}%) and useful range of ~ 20%. Their dimensions should be restricted, in view of spatial limitations (mechanical or electrical interference), stress and strain gradient and placement technique, to about 30-50 mm max. gauge "diameter". Displacement transducers should possess high linearity and a resolution of at least 20 μm together with linear frequency response up to min. 1.2 kc.

Three types of (total) soil stress gauges were applied:
URS - free field soil-stress gauge: design and test evaluation are described in a final report by Walter etal (1971).
MRC - FFSST-200 soil stress gauge, of closely related design (see Mason Research Bulletin GSC1 and GCS1) except for the sensor suspension. Pressure sensitivity is of order 2 mV/Vbar.
Kulite 0234 - in contrast to the former one-sided sensitive, typical sensitivity 6 mV/Vbar.
All gauges employ a fluid-filled, foil covered void as stress transmitting system and a piezo-resistive sensor to monitor the fluid pressure. In fact they represent a modern adaptation of the approach of design, described by Plantema (1953),

whereby very small sensors of high volume stiffness and sensitivity allow for gauge-miniaturization.

Two pore-pressure gauges of comparable design were applied. The application of piezo-resistive sensors enable small measuring chamber volumes and limitation of dimensions. The first was a laboratory development employing a Kulite TQS-360-50 sensor and a filter cap of cemented glass ballotini (typical sensitivity ~ 3mV/Vbar). At a later stage the Druck Ltd. miniature transducer PDCR 42 appeared to be a welcome supply for reasons of smaller dimensions and better matched volumic weight. (typical sensitivity ~ 13mV/Vbar). Soil-strain measurements were performed conform to the system described by Truesdale and Schwab (1967) involving magnetic coupling of two coaxially spaced coils (man. Bison Instr. Ltd) of 1 inch diameter. A ten-channel system for synchronous driving and readout of max. ten gauges was developed by Royal Shell Labs.(Amsterdam). Vertical cover plate deflections were monitored by means of DCDT (displacement) transducers of type Sangano Western Control, type NDLR (∓ 1 inch. range and sensitivity 16mV/Vmm) attached to a reference bar, supported outside of the test bin. In one test accelerometers were used, mainly for qualitative purposes (Brüel and Kjaer type 4371). Data recording provided some difficulties when it realised that to monitor full information of impact-stress, 3-4 cover slab deflections and say 7 equivalent positions in the sandbed (not to mention desired duplicating gauges) about 70 recording channels are required with a band-width of about 1.2-1.5 kc to retain adequate transient response.
At the time (1973) such demands excluded the use of PCM (pulse code modulation) recording. As a compromise between several aspects, data recording was performed with use of two 14-channel 40 kC FM-recorders (Bell and Howell 4010) and an 8-channel galvanometric recorder (Honeywell 2208) implicating 33 useful channels and 3 for synchronisation purposes.
Late 1977 a multichannel PCM recording system (Bell and Howell PC-8/2) became available, which enabled to operate with 24 channels at frequency band-width 1.8 kc/channel.
Data processing means are still very limited.

During build-up of the sandbed and after the impact-tests several types of check tests were performed: volumetric and radiometric density measurements in dry sand and also an electrical soil conductivity meter as porosity-monitor in case of saturated sand.

A miniature cone penetrometer offered a very sensitive means for homogeneity check and densification monitoring. Random checks on degree of saturation were taken by means of a sample flask fitted with a gas catcher and calibrated capillary tube. The flask is closed after sampling and placed in a pressure cell, whereafter "isothermal" (small) pressure-volume change behaviour is observed to obtain information about the air content.

Sand surface settlement was recorded on film using known changes in waterlevel at the sandbed surface.

4 GAUGE-CALIBRATION AND PLACEMENT PROCEDURES

It follows from literature that soil-gauge interaction may severely affect the indicated magnitude of stress (f.e. Triandafilides (1974), Hvorslev (1976)). The influence of stress field configuration (Askegaard (1963), Baranov (1973)) and overall gauge geometry (Collins etal (1972), Forsyth, Jackura (1974)) add even more complications w.r.t. justified result interpretation, apart from operational problems like placement procedure (Hadala (1967), Hamilton (1960)) and mechanical gauge sensitivities to bending, acceleration, cross-stress etc. (f.e. Walter etal (1971), Gerrard, Morgan (1972)).

An attempt to summarize and evaluate some of these aspects, including the use of diaphragm point-load sensitivity as an additional gauge characteristic to approach its response, is described elsewhere (Ruygrok (1975)).

The practical conclusions to be dealt with are in fact a relevant calibration procedure, and sample preparation preferrably closely related to the test situation.

Some stress and strain gauges were tested in a large, so-called plantema-cell (Plantema (1953)) or a medium-large triaxial cell with low-friction end plattens at moderate loose porosity ($n \simeq 43\%$) and employing stress paths with different stress-ratio and axial or radial gauge orientation.

Because of high cost the other gauges were tested simultaneously together with the calibrated specimen in the large test bin at equivalent positions and loading to obtain a general performance picture.

The placement procedure was chosen so as to avoid unnecessary interference with the sand raining preparation method. Gauges were either simply set on surface and waterlevel-adjusted with fingertip-pressure only, or in case of vertical orientation (gauge placed on its edge) its position was maintained by means of thin glass, wooden or metal meedless.

When the sandbed had been raised at least over 80% of gauge height, the pins were removed by pulling out in line by means of cotton wire. To install the strain-coils horizontally a 2" parallel-faced perspex block was employed to approach coaxiality and nominal separation. In case of vertical measuring orientation a needle with 2 marks (spaced 2 inch) was carefully inserted through a centre hole in the coil first placed. After raising the sandbed to about the upper mark the second coil was placed (the needle is removed after at least 20 mm raise of sandbed level).

It should be realised that thin gauges also favour the use of sand-raining technique as in both horizontal and vertical gauge oriëntations irregular porosity around the gauge will be reduced.

The comparatively tested stress gauges showed response differences at first loading within a deviation band of ~ 16% (of mean response). For the strain gauges difference bands of about 25% at small strains (< 0.01) decreasing to ~ 12% at larger strain (>0.04) were observed. (This strain "levelling" effect may be related to locally poor stress gauge performance and emphasize the necessity to use duplicating gauges on equivalent positions).

For the few URS/MRC gauges, tested in the calibration cell, overall figures like ~ 4% non-linearity, ~ 3% hysteresis, and response factors between 1.04-1.24 (including unexpected low lateral stress sensitivity within 12% for stress ratios 1 to 3) resulted at first static loading. Significant decrease in non-linearity, hysteresis and zero offset (caused by improper placement) was observed at repeated loading. Other (detailed) results will not be reviewed here; further evaluation is awaiting new equipment.

5 CHARACTERISATION OF TEST-MATERIAL

The test-sand was chosen to serve as some "mean representative" for the different types used in hydraulic fill dike-embankments. Grain size distribution is presented in fig.(14). Further characterisations, like grain-shape and angularity, were studied for each sieve fraction separately (not presented here).

Results of standard triaxial and permeability tests are summarized in table 1. Further research will be continued in connection with an appropriate constitutive soil model. The cover material employed was carbon-black hard-polyethylene, with approximate characteristics $E \sim 1.8 * 10^6$ kN/m^2, $\nu \sim 0.44$ and $tg\ \delta \sim 0.03$ (material damping) in the relevant frequency range. Interface friction coëfficient w.r.t. sand is of order 0,8 (sand-paper roughened cover plate bottom).

TABLE 1.

n(%)	Φ^0	k(10^4 m/s)	conditions
35	44.3	1.1	temp. 19^0 C
37.5	40.4	1.5	initial stress
			30 kN/m^2
41.4	35.5	2.5	test type C.D.
			strain rate 2%/hr
43.5		3.4	

6 DISCUSSION OF TEST RESULTS

In most tests nearly the same porosity (n=42.5-43%) was used and the position of (nearly) saturated level below cover slab and water content distribution due to a specific sequence of simulated tidal movements on the dynamic response were chosen as initial parameter to be studied.
In addition a test with near-saturated very loose packing (n~45%) was performed. Only a very limited number of illustrative results will be discussed below.

The "zero" line of the recorded signals represents the state before each impact loading (initial state before the first impact or preceeding final state to any subsequent loading). Consider a simple liquefied state only (without complicating phenomena) to allow for simple discussion concerning changes in the post-impact gauge response with respect to the pre-impact value (artificial "zero"). To simplify and shorten the point, the pre-impact state at a measuring position is described by σ'_v (vertical eff. stress), u (pore pressure), horizontal effective stress $\sigma'_h = K\sigma'_v$ and C_r (stress gauge calibration factor for effective stress).

It may be shown that within mentioned conditions the following changes influence the readings at liquefaction:pore pressure gauge $\Delta U \sim + \sigma'_v$,vert. total stress gauge: $\Delta\sigma_k \sim - (C_r - 1)\sigma'_v$,hor. total stress gauge: $\Delta\sigma_h \sim (1-KC_r)\sigma'_v$.
Hereby it is assumed that in the liquefied state the fluid calibration of stress gauges ($C_r = 1$) applies (relieve of stress concentration).
If a measure for effective stress is obtained by synchronous substraction of the (offset zero) records it follows for the post-to pre-impact reading change:
vert. "effective" reading $\Delta\sigma'_v \simeq -C_r\sigma'_v$
hor. "effective" reading $\Delta\sigma'_v \simeq -KC_r\sigma'_v$.
Obviously this hypothetical response will rarely be met exactly in practice; the order of magnitude may however be of some help in

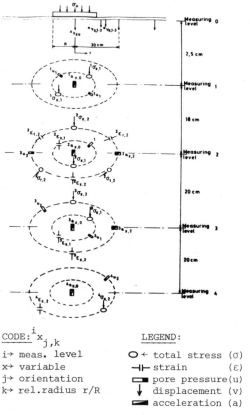

CODE: $^i x_{j,k}$

i→ meas. level
x→ variable
j→ orientation
k→ rel.radius r/R

LEGEND:

O← total stress (σ)
⊣⊢ strain (ϵ)
▭ pore pressure(u)
↓ displacement (v)
▬ acceleration (a)

Fig. 1 Gauge placement configuration (example).

the interpretation procedure (it may also call for gauges with moreover excellent static performance to enable "back calculations" from initial readings).

In fig. 2a) selected long time-base survey from the test with very loose packed, 98% saturated sand is presented (see fig. 1 for symbols). The impact sequence is denoted in Roman numerals.

Fig. 2b) shows fluid pressure rise clearly above minimum liquefaction level at the more shallow depths. The liquefaction boundary however is situated very near the fourth measuring depth, as was inferred from short time-base registrations. As the duration of the liquefied state at the third depth level is only of order of tenths of seconds it didn't show up in $^3u_2$ at this compressed time scale. The effect of densification is demonstrated by the decrease in duration of liquefaction following subsequent equally intense impacts and reduction in subsequent cover slab deflection (and its lateral gradient).
In evaluating the quick deflection of the cover slab in the early stage it should be

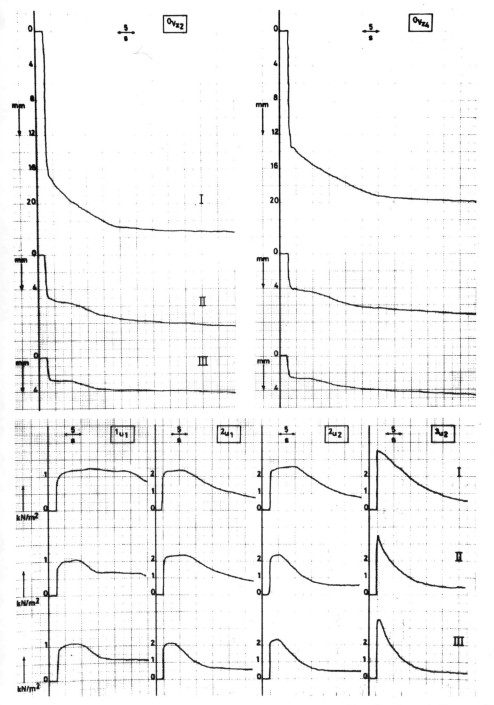

Fig. 2. Cover slab deflections and pore pressures during liquefaction at three subsequent impacts (long time-base) loose initial porosity (n = 0.45 S$_r$ ≈ 98%) - for symbols refer to fig. 1.

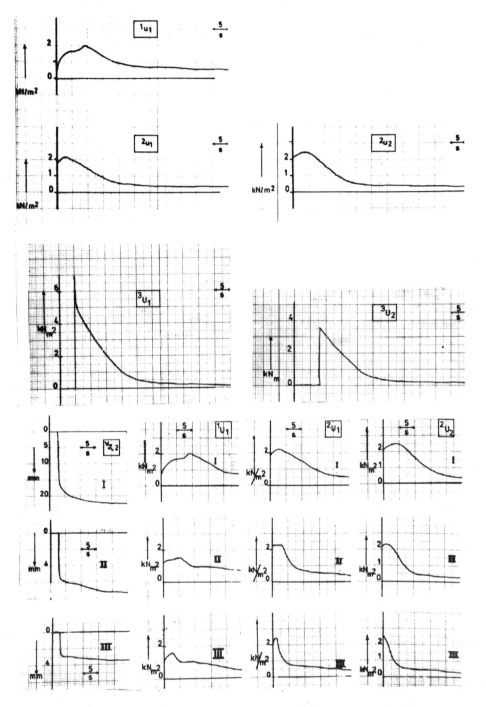

Fig.3. Liquefaction phenomena at moderate loose porosity (n = 0.43 S_r = 97%)- three subsequent impacts. (see fig. 1 for reference).

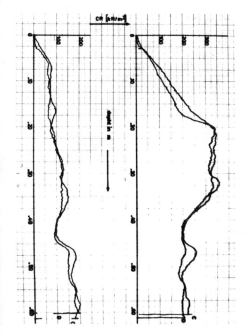

Fig. 4 Cone resistance before and after the test with initial porosity n ~ 0.43

mentioned that the dead weight of the falling weight system ex erts a static post-impact stress (~5 kN/m^2) on the loading area. Because floating drift phenomena of the strain gauge coils past onset of liquefaction it is not justified to correlate deflection and strain records beyond some early post-impact stage.

Fig. 3 shows much of the same overall features at a test with moderate loose porosity (n = 0.43). The duration of cover slab deformation is clearly shorter compared to the preceeding case. It may be useful to mark that the gauges (mainly in the upper layer) will experience significant settlement, leading to zero-offset of fluid pressure reading (complicating estimates of liquefaction duration (see $^1u_1$). Surprisingly $^3u_1$ clearly suggest liquefaction, in contrast to other readings at this depth. A possible explanation may be local inhomogeneity as can be seen in the left cone penetration soundings in fig. 4 (depth $\simeq$ 0.4 m) at two positions not very remote from the gauge (the local density decrease may be estimated to $\simeq$ 0.7 - 1.2% porosity). The increase in cone resistance, shown in the soundings on the right, reflect an impression of densification after 3 impacts of nominal 100kN/m^2) down to about the estimated liquefaction boundary. In spite of the significant increase in cone resistance, density measurements suggest densification of around 2% porosity only (a difficulty is the relatively low vertical resolution of most density-gauges). As cone resistance show high sensitivity to packing structure it seems to allow for refined qualitive monitoring, particularly suitable in small scale tests.

Fig. 5 presents part of some short time-base stress and pore pressure records of the same test.
The total stresses provide some idea of impact transmission (damping), while a part of the post-impact phenomena may be explained by the introductory remarks in this chapter.

An interesting phenomenon appears in the pore pressure records, suggesting underpressure peaks or trends during the total stress

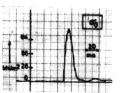

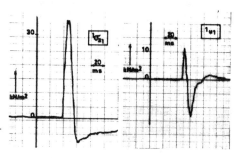

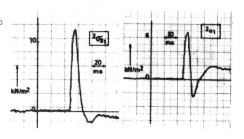

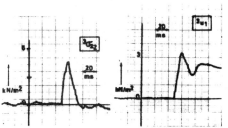

Fig. 5 Transient stresses and pore pressures at first impact (porosity n ~ 0.43, Sr ~ 97%)

217

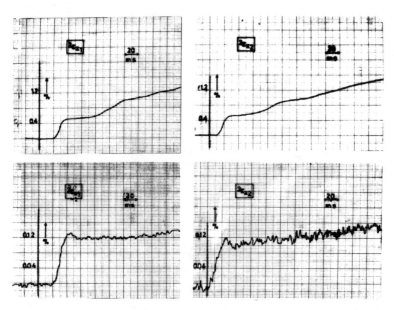

Fig. 6.Selection of vertical strain measurements (first impact, n = 0.43)

impulse (even during rise-time in $^1u_1$). As full discussion is not possible in the scope of this article, we confine ourselves to a few remarks.

1) Shock-tube tests with these type of gauges (in water and saturated cemented sand) did not show significant acceleration effects (suspected to be a probable cause of the phenomenon).

2) In the short time base records of the cover slab deflection, no specific vibration movement was observed that could induce something like cavitation tendency as source of such a pressure drop.

These remarks are of course not conclusive to reject any possible influence of soil-gauge or inertia interactions nor (undetected) cover plate dynamics as source of initiation, but it is at least allowed to consider also the possibility of some kind of temporary pore volume increase ("dilatancy", or, as suggested by a colleague, Mr. H.L. Koning, a parallel to the cryermandel effect).

Unfortunately soil-strain measurements in the upper layer were not feasible. At one equivalent point three orthogonal strain measurements are available (fig. 7). It presents many difficulties to obtain reliable volume-strain values, in view of their accuracy (certainly in case of data from subsequent impacts). On the other hand, when placed carefully, this type of strain gauge may provide surprising results as f.e. in fig. 6 (for configuration see fig. 1), records from the same test at first impact. Locally a nearly lateral homogenuous

distribution of vertical deformation is suggested, although it is not easy to point out at what time after impact (onset liquefaction) the recording ceases to indicate a relative displacement (instead of gauge drifting phenomena). Fig. 7 is fully illustrative for this kind of interpretation problem, certainly in case of the course of the radial strain record. It may be instructive to consider also $^2u_2$ in fig. 5 where the early stage of liquefaction is indicated after 40 msec., whereas the jump in $^2\varepsilon_{r.2}$ in fact suggest a sudden change to radial extension, starting during impact extinction, which may also be indicative to the underpressure phenomenon mentioned.

This effect is very clearly demonstrated in results (fig. 8) from an other test. The porosity (n ~ 41-42%) was somewhat lower than planned and a capillary pre-tension of 0.2 m (2 kN/m^2) was applied.

In short, the pre-tension was "destroyed" during the first impact and only limited liquefaction occurred. The development of the indicated pore pressure drop trends during the subsequent impacts show that some systematic effect is involved. It is f.e. seen in the response of $^4u_2$ during impact I compared to III that a barely visible post-impact fluctuation has underwent drastic change (no liquefaction occurs at this depth). It may be that the underpressure impulse partly propagates from above, however, comparison of $^2u_1$ and ^1u at impacts I and III seem to exclude pure^{1}cover vibration initiation as the negative peak in $^2u_1$ at impact III occurs earlier and deepens

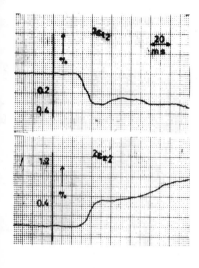

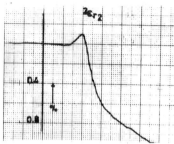

Fig. 7.Orthogonal strain-measurements at
the same equivalent position (n = 0.43
first impact)

whereas that of [1]u_1 seem to broaden (even
tends to decrease). At first impression the
phenomenon seems to be related to a downward increase in densification, thus increased "dilatancy" characteristics. Again,
irregular gauge-soil interactions are not
excluded, but it is then difficult to understand why equivalent-placed pore pressure
gauges show nearly identical behaviour (not
presented here) unless for complicated
change in boundary conditions below the
cover slab. Another remarkable point is the
propagation of the secondary pore pressure
generation (it far exeeds the primary impulse at the lower measuring levels)
may be due to a subsequently increased underpressure effect.

In fig. 9 the consequence of the indicated
transient negative pore pressure is markedly
reflected in the records of the difference
of total stress and pore pressure (σ^{*} presumed to furnish an idea of transient effective stress).
The validity of the record depends a.o. on:
1. synchronous response of both transducers.
(A similar requirement applies to any

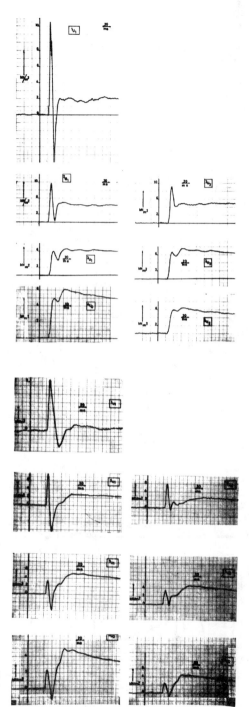

Fig.8. Pore pressure field at first and
third impact (n ~ 0.41 - 0.42) capillary
pre-tension 2kN/m^2 initial.

219

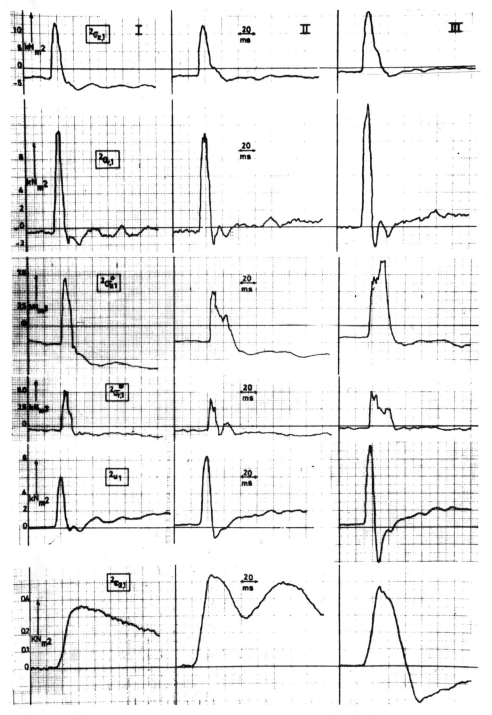

Fig. 9. Influence of transient negative pore pressure on stress difference records. (σ^{*} = total stress-pore pressure). Test on sand with loose porosity ($n = 0.45$, $S_r = 98\%$). Three subsequent impacts.

effective stress gauge w.r.t. the differential signal, see f.e. Studer and Prater
(1977)).
2. representative magnitude of stresses at
equivalent positions (if not closely spaced).
3. restrictions w.r.t. soil-gauge interaction (for both types of gauge).
The σ^{*}-records in fact suggest an extended
duration of transient "effective-stress"
during impacts II and III, compared to case
I, obviously due to the "underpressure"
effect. Some of the strain records at related positions show a behaviour that is not in
contradiction with a delay in initiation of
the liquefied state. However, the erratic
floating strain-coil phenomena, during liquefaction in fact obstruct a relevant observation, particularly when a volume-strain
record has to be composed (that could have
supported the existence of elongated duration of transient effective stress and consequently that of the negative transient
pore pressure).
    Finally it may be mentioned that the so-
called critical porosity (no volume strain
by deviatoric loading at constant mean
stress) established by means of a slow tri-
axial test is of order $n_{crit} \sim 0.42$ (at $\sigma_1 +$
$2\sigma_3 = 50$ kN/m^2.

    It was felt that a test with dry sand
could possibly add nearer information about
the associated mechanism of these phenomena.
Fig. 10 shows a selection of results in the
layer near the cover plate (see fig. 1).
    According to the pore (air)pressure
magnitude there will be insignificant difference between total and effective stress.
Also interesting is the acceleration record;
mark that it exeeds - 2 g. in the unloading
stage. (an idea of acceleration gradient
might be obtained by observing also fig. 12).
If we concentrate on pore pressure (now
being a nearly non-interacting variable
w.r.t. the situation with high saturation
degree) and cover slab movement, some
typical features are observed. Contrary to
the behaviour described in preceeding tests
the cover clearly vibrates with only
limited damping.
Here the post-impact pressure drop tendency
is associated with upward cover plate movement (suction effect overlying pressure
dissipation).
The following pore pressure generation with
a peak value significantly larger than the
primary cannot only be due to downward cover
slab movement. It is suggested that the
second, much smaller impact (after - 80msec)
caused additional pore compression (and a
further contribution in cover deflection.
The remaining cover plate vibrations exert
no further influence of importance. (At

impacts II and III however their amplitude
and influence on pore pressure increase.
Also the additional compression by the
secondary impact is clearly seen in the
strains).
Also remarkable are irregularities in the
soil stress decay stage (fig. 10, 11), that
may be connected with the irregular
phenomena in the air pressure records during
impact.
In fig. 11 the post-peak behaviour of radial
strain offers some indication, but it is not
clearly shown in volumetric strain as well
as only a small effect in $^2u_2$.
In this respect $^2u_1$ in fig. 12 is much more
illustrative: the first drop may be due to a
"dilatancy" or volumetric rebound effect.
The second coincides with cover plate
rebound but at this depth is rather to be
described as a consequence of ordinary pore

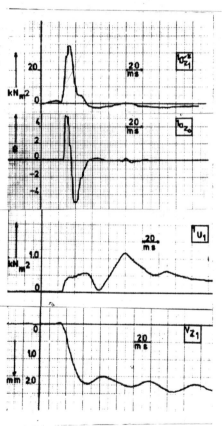

Fig. 10. Selection of results from test on
dry sand (n ~ 0.43, first impact (see
fig. 1 for symbols).

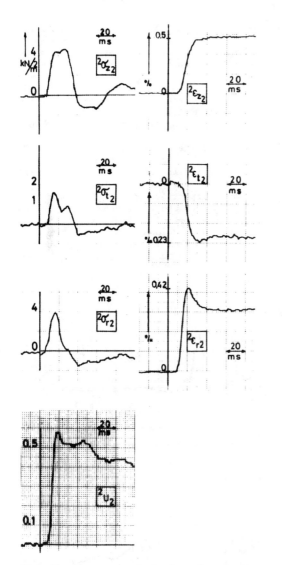

Fig. 11. Stresses and strains at an equivalent position. (dry sand, n = 0.43, first impact)

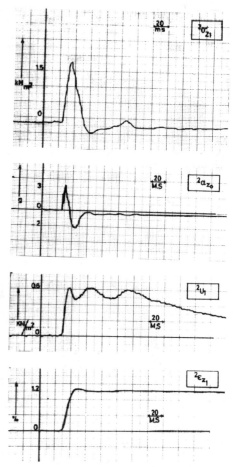

Fig. 12. Some results of the test on dry sand (first impact).

air pressure dissipation (decay) interrupted by secondary generation of air pressure (coincident with the small second impact mentioned above).
Unfortunately no other local strain information than $^2\varepsilon_{z,1}$ is available. It is tempting to put the assumption that the first pressure drop in $^2u_1$ (fig. 12) and the underpressure peak in the nearly saturated cases share a similar mechanism, occurring during the impact stress drop. Whether the secondary pressure generation in both conditions bear relations is an open question. In theory a secondary wave occur in the

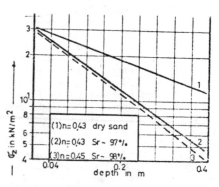

Fig. 13. Damping of vertical total stress within measured zone (first impact).

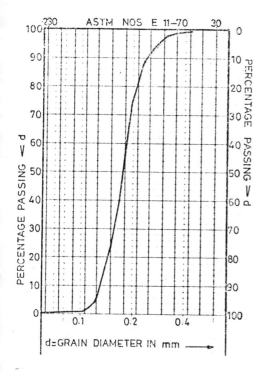

Fig. 14. Grain seize distribution.

saturated case, in the dry case it seems to be due to compression of a not too stable post-impact packing structure.
A mathematical model and numerical program in order to approach the observed phenomena (and eventually the dynamic dike design) are in development.

7 CONCLUSIONS

1) Soil stress and strain measurements employing gauges with piezo-resistive sensors are capable to fullfil high requirements concerning both dynamic and static response, together with excellent signal/noise ratio and very high dynamic range. The main difficulty in gauge design and operation is still the incompletely understood soil-gauge interaction problem, also in connection with placement techniques. The development of effective pressure gauges is in fact essential to improve on further experimental research. The mutual inductance strain gauges provided some interesting results.
Unfortunately the overall accuracy (roughly 14%) did not permit decisive conclusions w.r.t. volume strain phenomena. Their use is limited in case of liquefaction phenomena, because orientation-drift may occur.

2) Liquefaction was observed in all tests with high degree of saturation, even following several subsequent loadings of the same intensity. The influence of densification structure on duration of liquefaction (and thus eventually on extensiveness) were clearly shown.
3) During the impact stress-pulse an unexpected phenomenon, transient negative pore pressure fluctuation, was observed. Its occurrence provide problems in valuating effective stress records obtained from total stress-pore pressure readings, as it cannot be shown rigourously to be due to either specific boundary conditions or a soil "dilatancy" effect. (Some strain measurements indicate correlation with the radial strain behaviour). Even some unknown soil-gauge interaction effect is not completely excluded in spite of some efforts to eliminate this possibility.
4) In tests with dry sand similar phenomena were observed, together with irregularities during the fall of the soil stress-impulse. Unfortunately the number and position of strain measurements were again not enough adequate to provide decisive conclusions. Surprisingly also a secondary generation of pore air pressure, exceeding the magnitude during impact stress, was observed.
A numerical evaluation of the test results is in preparation.

ACKNOWLEDGEMENTS

The research is sponsored and supervised by the Centrum voor Onderzoek Waterkeringen (Water Retaining Structures Research Centre) as executive organ of Technische Adviescommissie Waterkeringen.
The authors like to express their gratitude for the permission to publish this paper and to the members of the adjoined projectgroup for continuing interest.

REFERENCES

Askegaard V. (1963), Measurement of pressure in solids by means of pressure cells (Copenhagen, 1963).
    Struct. Res. Lab. (Techn. Univ. of Denmark) Bulletin 17 (see also bulletin 14).
Baranov D.S.,
    Proc. 8th Conf. Soil Mech. and Fo. Eng. 4.3. (Moscow 1973), 11-20.
Collins R., K.J. Lee (1972), G.P. Lilly, R.A. Westman,
    Mechanics of pressure cells. Experimental Mechanics, 12 (1972), 514.
Forsyth, R.A. (1974), K. Jackura.
    Recent development in earthwork instrumentation.
    Proc. Specialty Conf. ASCE (1974), Subsurface Exploration pg. 254-268.

223

Gerrard, C.M., J.R. Morgan (1972), Initial loading of a sand layer under a circular pressure membrane. Geotechnique 22 no. 4, 635.

Hadala, P.F. (1967), The effect of placement method on the response of soil stress gauges, Proc. Symp. Wave Prop. Earth Mat. (Un. New Mex.) 1967 pg. 255.

Hamilton, J.J. (1960), Earth pressure cells-design, calibration and performance, Div. Build. of Nat. Res. Council Canada, Techn. pap. no. 109.

Hvorslev MJuul, The Changeable interaction between soils and pressure cells, W.E.S. report AD/A-029161 (Vicksburg).

Plantema, G. (1953), A soil pressure cell and calibration equipment. Proc. 3d. Int. Conf. Soil Mech. Found. Eng. Vol I, 283.

Ruygrok, P. (1975), Characterisation of soil stress gauge response (in Dutch) (1975, report to Stichting Studie Centrum Wegenbouw, Arnhem-Netherlands).

Studer, J., E.G. Prater (1977). An experimental and analytical study of the liquifaction of saturated sands under blast loading: Proc. of DMSR 77 - Karlsruhe, Vol 2, pg. 217.

Triandafilides, G.E. (1974), Soil stress gage design and evaluation Jnl. of Testing and Evaluation 2, (no. 3), 146.

Truesdale, W.B., R.B. Schwab, (1967) Soil strain gauge instrumentation. Proc. Int. Symp. Wave Prop. and dyn. prop. earth mat. (Un. New Mexico, 1967) pg. 931.

Walter, D., A.R. Krebel, K. Kaplan (1971) URS free-field soil-stress gage. Design and evaluation.
URS 758-6 Final Rep. to US Dept. of Transp. (Wash. D.C.).

Cyclic 3-D stress paths and superposition
of hysteresis loops

P.I. LEWIN
Building Research Station, Garston, Watford, Herts, UK

1 SUMMARY

The paper looks at the behaviour of soils under cyclic loading which exhibit hysteretic stress–strain loops. It considers what happens when the stress–strain loops for the three principal stresses are superimposed to give the strain behaviour for three–dimensional stress paths. It is shown that this synthetic approach successfully reproduces some of the characteristics of a test in which a clay sample in a true triaxial apparatus was subjected to a circular stress path in stress space.

2 INTRODUCTION

The assumption of elastic behaviour for soils, although convenient for calculation purposes, gives rise to difficulties when the stress paths being investigated differ from those used in the experiments to establish the values of the elastic parameters. The problem, of course, is that soils are neither linear nor elastic. Further complications will also arise if the stress paths being investigated are three–dimensional rather than the axisymmetric condition of the conventional triaxial apparatus. The purpose of this paper, therefore, is to see if one can get an improved understanding of a soil's behaviour under complex stress paths in order to make better use of data obtained from conventional laboratory tests.

The approach made here is to take from elastic theory the use of superposition of 'one–dimensional' stress–strain behaviour for each separate principal stress and to apply it to cases which involve hysteresis. The investigation is concerned only with variation in the magnitude of the principal stresses and not with any rotation in the direction of their axes. The particular stress paths to be examined are: (i) a conventional cyclic drained (ie constant σ'_3) triaxial test (marked AB in Figs 1 and 2) and (ii) a circular stress path in the octahedral plane (marked C in Figs 1 and 2), this being a repeat of a test carried out by Wood (1973) in the Cambridge True Triaxial Apparatus.

The tests described in this paper were part of a wider investigation, using both conventional triaxial apparatus and the BRS True Triaxial Apparatus. The results are reported more fully in Lewin (1978). The material used was a reconstituted clay made from Llyn Brianne slate dust (LL 29%, PL 18%, clay fraction 31%) which has the advantage of deforming with very little creep. The samples were prepared initially from a slurry with a water content of 70%.

3 SUPERPOSITION WITH ELASTICITY

If a material is linear elastic then it is permissible to make use of superposition to predict strains for complex stress paths such that the strains produced by each principal stress taken separately may be simply added together.

Consider a hypothetical drained (constant σ'_3) triaxial test marked AB in Fig 2 such that the deviator stress is cycled equally into compression and extension. The material may be assumed to be elastic and isotropic. Let the stress strain behaviour be as shown in Fig 3(a) and let the associated lateral strains ε_2 and ε_3 be governed by a Poisson's ratio of 0.3. If the cyclic loading is applied sinusoidally then the

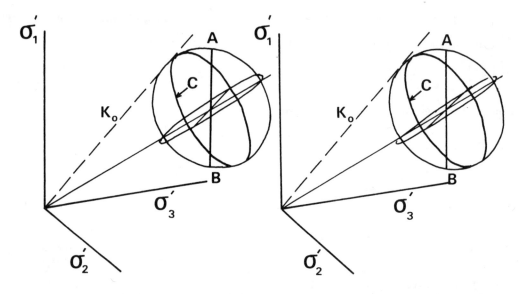

Fig 1   Stereoplot of stress paths

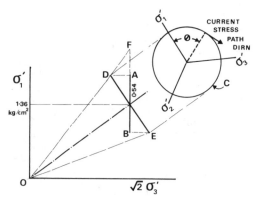

Fig 2   Circular stress path C

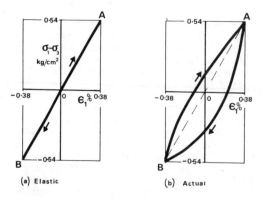

(a) Elastic   (b) Actual

Fig 3   Stress-strain loop for drained
cyclic test

strain against time will appear as a sin wave as shown by the dotted curve in Fig 4.

The case so far has considered just the strains produced by σ'_1 with the other two principal stresses σ'_2 and σ'_3 held constant. Suppose now that these two were also varying sinusoidally with the same amplitude but phased $120°$ and $240°$ with respect to σ'_1. The resulting stress path in stress space would be the sum of these three sin waves and this produces a circle in the octahedral plane as shown by C in Figs 1 and 2. (It should be noted here that this process brings the circular stress path closer to the failure condition than the component changes in each principal stress taken separately, ie D is closer to failure than A).

In the same way, the sinusoidal strain paths for each principal stress, including the Poisson's ratio effects, would sum to a similar circle in strain space (as shown in the octahedral view in Fig 5). The direction of the stress increment vector at any instant is necessarily coincident with the direction of the strain increment vector and this is shown by arrows which are tangential to the strain circle.

4   SUPERPOSITION WITH HYSTERESIS?

In practice, however, the actual stress-strain behaviour for a cyclic drained (constant σ'_3) triaxial test is not elastic but produces a hysteresis loop as shown in

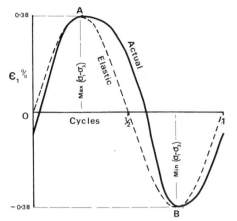

Fig 4 Cyclic strains

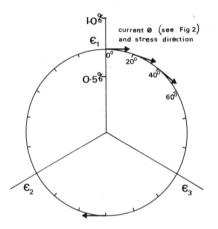

Fig 5 Strains for stress path C
assuming elastic behaviour

Fig 3(b). (This curve is interpolated from the results of tests carried out with different deviator stress ranges and has been drawn as for $\sigma'_1 = 1.36 \pm 0.54$ kg/cm^2). The value of the Poisson's ratio from these tests was found to be 0.3. The strain variation during the cycle has been plotted as the full line in Fig 4.

If this behaviour is now assumed to hold equally for the independent variation of σ'_2 and σ'_3 then we can again look at the effect of superposition of the strain curve for the simultaneous sinusoidal variation of σ'_1, σ'_2 and σ'_3 phased $120°$ apart to give the same circular stress path as before. The resulting strain path is shown in Fig 6(a). The variation in volumetric strain $(\varepsilon_1 + \varepsilon_2 + \varepsilon_3)$ is only 0.015% and can be neglected.

The strain path is seen to be not circular but to have a 'triangular' appearance. (If the basic hysteresis loop in Fig 3(b) had been such that loading and unloading were symmetrical then the strain path would have been 'hexagonal' rather than 'triangular'). It is also seen from Fig 6(a) that the direction of the strain increment lags behind the direction of the stress increment. This has the effect, by comparing the positions of the $\Theta = 0°$ point in Figs 5 and 6(a), of rotating the shape of the strain path in an anti-clockwise direction (where Θ is the direction of the stress path as defined in Fig 2). It may be noted that the lag does not have a constant value as the stress state rotates round its circular path nor are the strain increments of

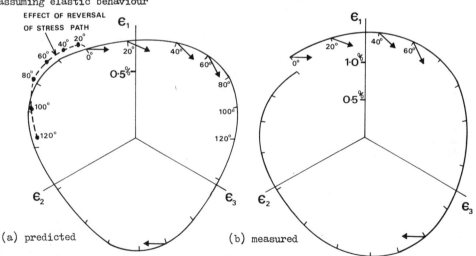

(a) predicted   (b) measured

Fig 6 Predicted and experimental octahedral strain paths for circular stress path C

227

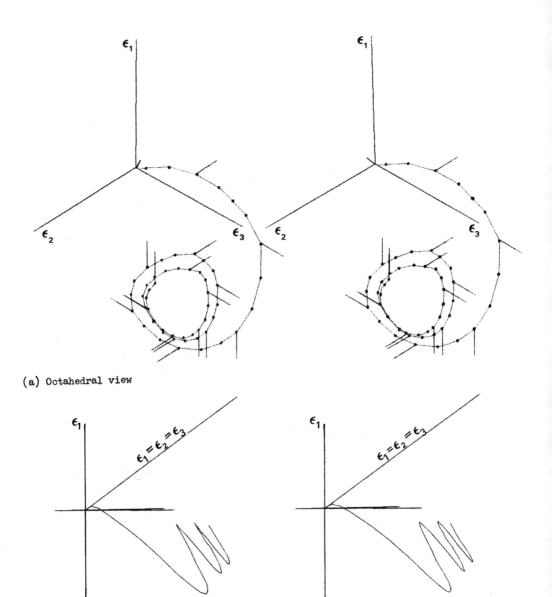

(a) Octahedral view

(b) View on triaxial plane

Fig 7   Stereoplots of strain path for circular stress path C

equal magnitude for equal increments of stress path rotation.

5 3D EXPERIMENT

With the Cambridge True Triaxial Apparatus, the circular stress path tested by Wood (1973) was achieved by computer control.

In the BRS True Triaxial Apparatus (Lewin 1971) the sample is a 6 cm cube with flexible membranes on each of the six faces. The initial stress history to produce the cubic sample is one-dimensional consolidation from 10 cm height of slurry. The membranes for each pair of opposite faces can be pressurised by a pair of mercury pot pressure developers arranged so that the upper pot of each pair is suspended at the end of a rotating arm. The desired circular stress path was then obtained by having the three arms of equal length but phased 120° apart. The test was carried out under drained conditions, against a back pressure of 2 kg/cm², with one circuit completed every four days.

The strain path for three circuits is presented 'three-dimensionally' in the stereo-pairs shown in Fig 7. Most of the volumetric strain, which is measured by the progress along the strain space diagonal, takes place in the first loop, although there is still some volume strain occurring in the third loop, which is shown in more detail in Fig 6(b).

Figure 7(a) clearly demonstrates a triangular feature. The fact that the triangle is substantially equilateral suggests that the sample retains very little memory of its stress history of one-dimensional consolidation and that the sample can no longer distinguish any difference between the vertical and horizontal stresses.

Comparing Figs 6(a) and 6(b) there is a marked similarity between the predicted and experimental results. However, the biggest difference is that the experimentally measured strains were nearly twice the predicted ones. This is almost certainly attributable to the fact that the experimental stress circle, as observed earlier in this paper, goes much closer to the failure condition than the component stress path used for the prediction. Tentatively, there might be some justification for suggesting that the predicted strains should be multiplied by a factor $\varepsilon_F/\varepsilon_A$, where (i) ε_F is the strain which would have been reached in a drained test (constant σ'_3) had it gone to the same stress ratio in the final circular stress path and (ii) ε_A is the strain reached in the component stress path used in the prediction. The subscripts F and A refer to the appropriate points marked in Fig 2.

6 REVERSAL OF CIRCULAR STRESS PATH

An interesting check on the stress path dependency of this superposition approach is to speculate on what would be expected to happen if the circular stress path were reversed, for instance taking the case at top dead centre when $\Theta = 0$. Since σ'_1 is then at its maximum value it will be about to reduce anyway so the strain predictions associated with σ'_1 will not be affected. However, σ'_2 will be reversed just after it has passed through its minimum value (ie it will just have started on the upward part of its $\sigma'_2 - \varepsilon_2$ hysteresis loop) and σ'_3 will be reversed just before reaching

its minimum value (ie it will be well into the downward part of its $\sigma'_3-\varepsilon_3$ loop) — and of course these will also affect the associated Poisson's ratio contributions. The net result is shown as a dotted line in Fig 6(a). It will be seen that the first four intervals of the strain path are very short (each interval representing a 20° increment of the rotation Θ of the stress path) and this is a result that Wood found when he performed such a reversal in his test J8. After 120° the reversal effect disappears and the strain path then continues as the mirror image of the right hand side.

7 CONCLUSIONS

It is clear there is a good qualitative relationship between the experimental results and the results predicted on the basis of superposition of hysteretic curves and that further work along these lines would be profitable. The particular test selected, a circular stress path, transverse to the isotropic axis, is quite a severe check on the method and yet it showed very successfully the main characteristic features of the strain behaviour. These features were not apparent when using the simple elastic solution. However, the quantitative comparison still needs further investigation.

8 ACKNOWLEDGEMENT

The work described has been carried out as part of the research programme of the Building Research Establishment of the Department of the Environment and this paper is published by permission of the Director.

9 REFERENCES

Lewin, P I (1971): A new apparatus for testing a one-dimensionally consolidated clay cube with independent stress control. Roscoe Memorial Symp, Foulis, Henley-on-Thames, pp 324-329.

Lewin, P I (1978): The deformation of soft clay under generalised stress conditions. PhD Thesis, University of London.

Wood, D M (1973): Truly triaxial stress-strain behaviour of kaolin. Proc Symp The Role of Plasticity in Soil Mechanics. Cambridge University, pp 67-93.

Vibrocreep and loose soil strength
under cyclic loading actions

P. L. IVANOV
Leningrad Polytechnic Institute, Leningrad, USSR

SYNOPSIS

Under the conditions of vibro-compression tests loose soils show strongly pronounced volume vibro-strains of a soil skeleton. In the paper are given test data of the investigation of the effects of the initial value of a soil density, compression stresses and oscillation acceleration on vibro-compression consolidation and on the development of a vibrocreep process. Summarizing these data and the results obtained by D.D. Barkan, O.A.Savinov, O.J.Shechter and N.N.Maslov, the author shows a feasible applicability of the linear theory of the inherited creep to describe vibrocreep processes, a complete irreversibility of vibrocreep volume strains being taken into account.

A sand soil strength under vibration effects has been studied with the application of vibrated plates moving along a sand foundation under the conditions of sand soil tests in triaxial apparatus with pulsating loads. A substantial development of vibrocreep strains in a sublimiting state is also shown.

The account of stressed state changes being taken under dynamic actions has shown that in a wide range of oscillation accelerations (up to 1 g) there is every reason to consider, that strength characteristics of loose soils, that is an internal friction angle do not change.

A characteristic feature of loose soil consolidation under vibration effects or the effects of repeated impulses is a pronounced accumulation (increment) of volume strains in time. With the increase of the frequency of impulse applications or vibrations, interdisplacements of particles superimpose and the process of continuous particle displacement takes place. Strain accumulations in time at constant oscillation accelerations and stresses were called by the author (1960) "vibrocreep" of loose soils. Barkan D.D. was the first to study a dry sand soil consolidation in time under vibrations effects.

Fig.1 shows the character of vibrocreep curves, that is a void ratio ($\Delta \varepsilon$) variations in time (t). The curves were obtained as a result of consolidation of fine-grained sands placed into compression apparatus (without any feasible lateral expansion) set on a vibro-table, having vertical amplitude direction. The load at rest was made practically inertialess by means of a hydraulic jack. They are satisfactorily described by an exponential function (see the dotted line in fig.1) of the form:

$$1 - e^{-\gamma_i t}.$$

The other not less substantial feature of a vibrocreep consolidation of loose soils is the existence of a soil vibrocreep consolidation limit ε_K dependent of the load at rest intensity. ε_K increases with the increase of the load at rest.

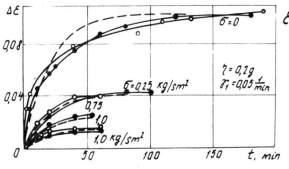

Fig.1. Vibrocreep curves at different compression stresses.

As a result of vibro-compression
tests of sand soils the relation
between the final attainable value
of a void ratio ε_K and the load
at rest intensity σ as well as
the oscillation accelerations (see
fig.2) has been obtained. In a wide
range of amplitude accelerations
(up to 0.8 g, where g is gravity
accelerations) the dependence of
the void ratio on compressive
stresses in a soil skeleton at
constant oscillation accelerations
can be taken as a linear one (according to the lettering of fig.2)
in the form of:

$$\varepsilon_K = \varepsilon_o - a\left[\sigma(\varepsilon_o) - \sigma\right],$$

where a - a coefficient of con-
                 solidation;
        $\sigma(\varepsilon_o)$- stresses that at the
                 initial void ratio
                 ε_o resist the soil
                 structure failure due
                 to dynamic effects
                 at the action of σ
                 stresses;
        η - oscillation accele-
                 rations.

Thus, the load at rest not only
reduces the feasibility of a soil
structure failure but also reduces
(and what is not less substantial)
a loose soil compactibility under
dynamic effects.
Taking into account the similarity of vibrocreep curves, experimentally confirmed at different
values of σ , it is possible to
draw a conclusion about the applicability of the model of a linear-
creeped body to vibrating loose
soils and also to represent a vibro-compression equation at constant

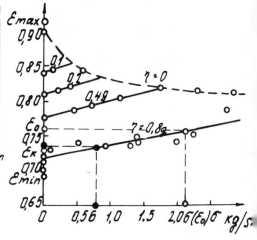

Fig.2. Void ratio (ε) - load at
rest (σ) curve at diffe-
rent oscillation accelerations . The dotted line of
a sand compression is only
for the load at rest.

oscillation accelerations and compressive stress in the form of:

$$\varepsilon(t) = \varepsilon_o - a\left[\sigma(\varepsilon_o) - \sigma\right]\left(1 - e^{-\eta_1 t}\right).$$

A loose soil moisture is known
to affect substantially a soil compactibility under the action of dynamic and particularly vibration
loads.
Laboratory investigations and
test experience testify to a much
better consolidation of dry and
particularly water-saturated sands.
The initial soil moisture substantially affects the density of
a soil placement. With the increase of the soil moisture the soil
void ratio increase can first be
seen (fig.3b), but as the moisture
further increases, the initial void
ratio decreases and at a considerable, almost complete water saturation, it attains a minimum value
close to the value of void ratio
during the dry soil placement.
Thus, for each loose soil the characteristic moisture is marked, at
which the least density of the soil,
placed by any other means, can be
noticed.
During vibrations a similar picture of the initial moisture effects
can be seen. Thus, for the case of
small moisture, less than W_K (fig.
3a), an initial horizontal part of

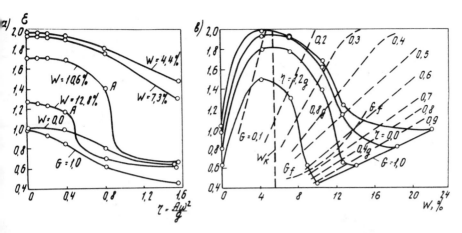

Fig.3. The effects of moisture (W) on vibro-compression curves (a)
and the relation of soil density to initial moisture at different amplitude accelerations (b).

vibro-compression curves can be
seen, corresponding to the absence
of consolidation up to the critical
value of oscillation accelerations,
which can be explained by the avail-
ability of cohesion, that attains
its maximum value at   .
    As the process of consolidation
of moisted soils proceeds, the increase of water saturation extent
of the soil G , occurs. At a lower initial moisture the increase
of water saturation extent results
in the increase of cohesion and
soil strength. In the case of
moisture, larger than w_K , the
soil consolidation gives rise to
reduction of cohesion and as a
certain value of water saturation
extent G_{γ} is attained (which corresponds to the beginning of water
seepage) the soil consolidation is
accompanied by pore water wringing
out of pores and by a sharp reduction of the soil structure strength.
    Therefore, with the increase of
the soil density and cohesion vibro-compression curves ($w < w_K$ in
fig.3a) are of smooth and decay
character, corresponding to the
whole soil consolidation. If the
increase of a structure strength
on the account of the density increase is not compensated by the
decrease of the structure strength
due to cohesion reduction, then
the process of an intensive increment of the soil compactibility
occurs and vibro-compression cur-

ves have a characteristic break
(point A in fig.5a). For all that,
during the soil consolidation the
so-to say softening of the soil
occurs and a sharp "slumping" exhibits. When the water saturation
extent attains the value of G_{γ} ,
corresponding to the beginning of
water seepage and cohesion disapp-
earance, the soil compactibility
greatly increases. The value of G_{γ}
for fine grained and flour sands
is of the order of 0.6-0.7.
    As all the tests and studies
show, the resistance to shear under vibration effects decreases
and a construction stability reduces. For many years it has been
thought that the decrease of resistance to shear is caused by the
reduction of the angle of internal
friction. Many investigators who
accepted this point of view, did
not account in their experiments
for the stress variations in the
period of vibration load actions.
    At the laboratory of soil mecha-
nics of the Leningrad polytechnic
institute a great number of tests
were carried out in order to study
the sand soil strength under the
action of vibrating and pulsating
loads in different tests, the characteristic feature of which is
taking into account and measuring
soil stressed state variations
under dynamic effects.
    In the shear tests of vibrating
plates (fig.4) variations of normal

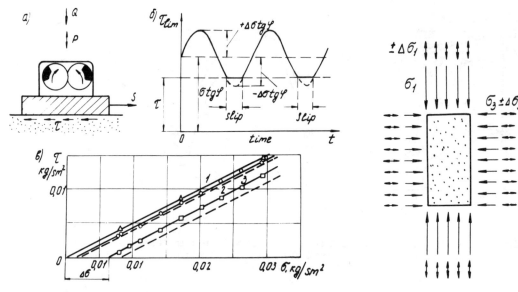

Fig.4. The model of shear tests with a settlement plate (a) and charts (b,c) of limit shearing resistance variations during settlement plate vibrations (1 - statics, 2,3 - during vibrations, the dotted line - according to calculations).

Fig.5. A stressed state of a sample in the dynamic triaxial apparatus.

stresses along the toe of the plate were taken into consideration. A mechanical vibrator of the vertical action changed stresses in the limits of $\sigma \pm \Delta\sigma$. Taking shear resistance reduction into account according to Coulumb relationship $\tau = (\sigma - \Delta\sigma) t_g \varphi$ resulted in a periodic shear, and practically speaking, in an exact agreement of the test and calculated data (fig.4b), φ obtained from static shear tests of the plates, was kept unchanged.

In the tests using triaxial apparatus some extra dynamic vertical $\Delta\sigma_1$ and lateral $\Delta\sigma_3$ pressures (fig.5), which were created by means of hydraulic pulsators and changed according to the sinusoidal relation, were added to the acting loads at rest, σ_1 and σ_3. The periodic movement of the pulsator plungers was determined by moving eccentrics with different frequency.

The condition of strength through the principal stresses with the account of extra effects of dyna-mic loads in the triaxial apparatus being taken, acquires the form of:

$$\sin\varphi = \frac{\sigma_1 + \Delta\sigma_1 - \sigma_3 - \Delta\sigma_3}{\sigma_1 + \Delta\sigma_1 + \sigma_3 + \Delta\sigma_3}.$$

In all the cases of dynamic tests, failure vertical pressure σ_1 was less than the static pressure ($\Delta\sigma \approx 0$). However, all the tests showed that the angle of internal friction of sands in a wide range of intensities of pulsating loads (up to 30-40% from the static loads, fig.7a) and frequencies (up to 50 cycles, fig.7b) remained unchanged. If a dynamic cimponent is not taken into account, a conventional (fictitious) angle of friction will reduce (fig.7, a dotted line). This circumstance, in the main, accounts for the reason of a wrongly marked reduction of the angle of friction, obtained in a number of works.

The treatment of the results of other aothors, who consider the

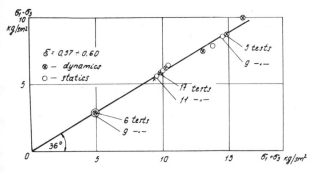

Fig.6. Test results on median grained sands
       at 1) static and 2) dynamic loads

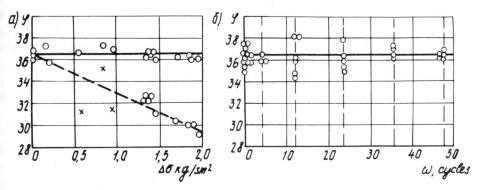

Fig.7. The relationship between the angle
       of internal friction φ and the in-
       tensity of the dynamic effects (a)
       and the vibration frequency ω (b).

friction coefficient of sands to
vary during vibration, also confirms their erroneous point of
view. For example, taking an account of the dynamic component made
by us on the basis of the test results in a triaxial apparatus (Bogolubchik 1975), shows (fig.8)
that the angle of internal friction both for dry and water-saturated sands didn't change (solid lines), though the authors came to
an opposite conclusion (fig.8,
dotted lines).

In the tests (Okomoto 1961, Zubanov 1958) a plane horizontal
shear of unloaded sands at vertical vibration effects was made.
That is why taking an account of
the dynamic stress component is of
no difficulty. The stability of a
real angle of friction during the

vibration of sands is clearly reflected in the tests in a linearity of shear strength variations
as well as of a fictitious friction coefficient in the range from
0 to 1g (fig.9,10).

Thus, in a wide range of oscillation accelerations (up to 1g)
there is every reason to consider,
that loose sand strength characteristics, that is the angle of internal friction, remain unchanged.
Under vibration and seismic effects
the resistence of structures to
shear or to plastic strain development should be checked up with
dynamic stress components taken
into consideration at constant values of the friction angle, obtained from common static tests.
In a general form, the condition
of the shear absence in a loose

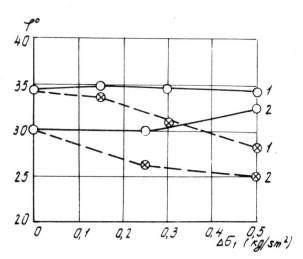

Fig.8. The relation of the angle of internal friction to the intensity of vertical pulsating load ($\Delta \sigma_1$); the dotted line – without dynamic stress component being taken into account (Bogolubcjik 1975).

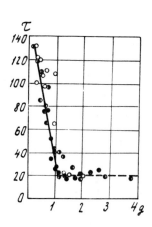

Fig.9. Sand shear strength (τ) at different oscillati accelerations (Okomoto 1961

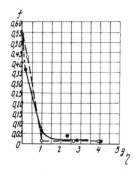

Fig.10. A fictitious coefficient of friction (f) at different oscillation accelerations (Zubanov 1958)

sand across any plane can be represented in the form of:

$$\tau \pm \Delta\tau(t) < [\sigma \pm p(t) \pm \Delta\sigma(t)] \, tg \, \varphi$$

where,

$\Delta\tau(t)$ and $\Delta\sigma(t)$ – additional stresses in a period of dynamic load effects;

$p(t)$ – an overpressure in a pore water, developing as a result of dynamic load actions.

It is to be noted that all the tests described above, were carried out on preconsolidated soils at a given level of dynamic and static stresses. As the coefficient of friction at rest is greater than of in a movement (during loose sand consolidations particle interdisplacements occur) it is feasible that during soil consolidations the angle of internal friction can reduce. On the other hand, after consolidations, the internal friction angle increase can be seen. In this connection, it is quite possible, that the reduction of the angle of internal friction during vibrations of moderate intensities (up to 1 g) is so small, that it can be neglected.

In the conclusions it is to be noted, that some works periodically published which deal with sand soil friction angle variations at moderate vibrations, require a thorough verification.

REFERENCES

Barkan, D.D. 1959, A vibration technique used in construction engineering. Moscow, Stroiizdat.
Ivanov, P.L. 1962, Liquefaction of sandy soils. Moscow, Gosenergo-izdat.
Bogolubchik, V.S. & V.J.Hein 1975, Subjects on geoengineering. Proc. 24, Dnepropetrovsk.
Okomoto, Sh. 1961, Sand soil bearing capacities during earth quakes. Intern. Conf. on seismo-stable construction engineering in San-Francisco. Moscow, Stroiizdat.
Zubanov, M.P. 1958, On coefficient of friction under vibration effects. Nauchno-Technicheskii Bulletin of the Leningrad Polytechnic institute, N 6, Mashinostrojenije.
Ivanov, P.L. & A.P.Sinitsyn 1977, Soil liquefaction and stability of foundations. Proc. of the IX Intern. Congress on soil mechanics and foundations, vol.II, Tokyo.

Experimental study of sand soil vibrocreeping

V. A. ILYICHEV
Scientific Research Institute of Bases and Underground Structures, Moscow, USSR

V. I. KERCHMAN & B. I. RUBIN
Seismic Test Field Station, Scientific Research Institute of Bases and Underground Structures, Kishinev, USSR

V. M. PIATETSKY
Leningradsky Promstroiproekt, Leningrad, USSR

1. Introduction

The cases of heavy settlements of the foundations for machines and the nearby building structures resting on sandy soils after putting into operation thereof, make to examine the traditional view of the sandy soils as beddings where settlements mainly occur during the construction period.

The survey of some known cases of the progressing settlements of foundations under dynamic effects is given in paper of Kerchman & Filippov (1977). The similar settlements were observed when making both the laboratory (Ivanov 1962; Goldstein, Hain & Bogolubchik 1974) and large-scale field (Kerchman 1977a, Bogolubchik, Goldstein & Hain 1977) experiments. The settlements occur due to vibrocreeping of sandy soil under the joint effect of static and dynamic loads. The term was introduced by O.A. Savinov (1964).

If static pressure under foundation footing is larger than threshold pressure σ_* for the given level of dynamic effect, viscoplastic shear deformations occur in soil that result in progressing settlement of foundation. It is worth to note that in the known experiments (Kerchman 1977ab, Savinov 1964) this threshold value or the long-term strength σ_*, and the reological features of the settlement process appeared dependent on the dynamic pressure amplitude but not on the vibration acceleration. This fact is in agreement with the results of three-axle tests and the current views on the long-term strength of soils.

The present paper deals with the results of the studies that were carried out due to the analysis of an instructive case – the development of settlements of the foundations of an industrial building caused by the dynamic effect of the mills work.

2. Description of the real case

The industrial building is a five-span structure, 276x128m in size. The large ball mills are installed in spans D-I and I-F. The ball mill foundations are of wall-type made of reinforced concrete, rest on the common lower cast in-situ slab and join the columns of the building frame at the top. The foundations for the frame columns are independent, cast in-situ, the footing sizes being 5.5x6m.

The foundations rest on the aluvial deposits up to 20m thick of fine and medium sands, ground water being above the foundation footings. The sands deformation modulus $E=230$ kg/cm^2, volume weight $\gamma =1.95$ tf/m^3, angle of internal friction $\varphi =28-30°$.

Non-uniform settlements of the columns along rows D, I.F were observed after the buildings had been put into operation and are being in progress now (Fig.1). The maximum settlement for seven years has reached 280mm. Approximately the same are the absolute settlements of the mills foundations, the maximum difference in settlements between the neighbou-

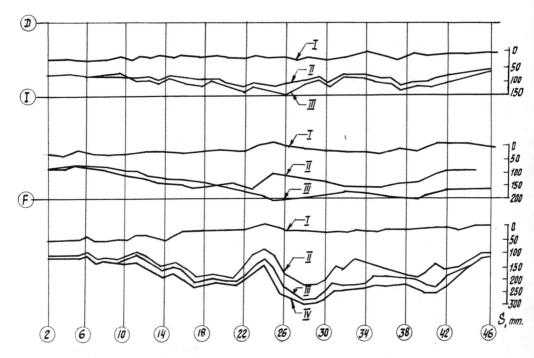

Fig. 1. Settlements of building column foundations along axes

    I - at the early stage of service
   II - in 26 months of service
  III - in 56 months
   IV - in 80 months

ring mills has reached 75mm during the same period, the largest deflection of the lower slab in this place is 33mm. In some of the foundation elements cracks have been observed. The maximum static pressure under the column foundations in the normal conditions reaches 2.5 to 3.0 kg/cm^2. The amplitude of the foundation vibration is within the range of 20 to 25 microns with prevailing frequences of 6 to 10 Hz.

The results of the permanent observations of the settlements show a direct relationship between the additional settlement and the static load. So, the columns along row D being under less loads than those in rows I and F have considerably less settlements. In conducting repair works in one of the mills when the static pressure under the foundation of the associated column of row F increased from 2.5 kg/cm^2 to 3.5 kg/cm2 due to the loaded crane, the mill irreversible settlement caused, evi-

dently, by the neighbouring mills vibration, reached 8mm for 24 hours.

3. Experimental studies

There was an assumption made, that non-uniform and continuous settlements in foundations result from the process of vibrocreeping of water-saturated sands under small dynamic but considerable static loads.

To verify this assumption, some field tests have been conducted in the area of the operating plant, the descriptions of which are represented below.

The experimental unit consists of a foundation with vibrator (sourse) and six foundations (receivers), see fig.2. For arrangement of the foundations and their sizes, refer to fig.3. Foundation F3 represents a square reinforced concrete block with a vibrator mounted on its top. The additional static load is created by reinfor-

Fig.2. General view of test set

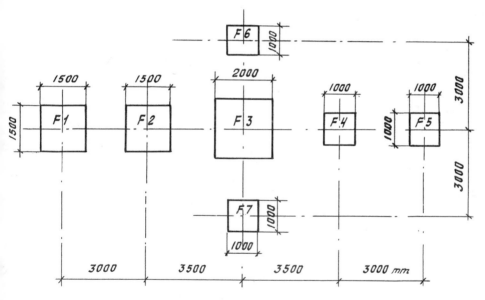

Fig.3. Arrangement of test foundations

ced concrete blocks mounted on springs. The static pressures in the footing of the test foundations are following: for F3-1.85 kg/cm², F1-2.0 kg/cm², F2-2.5 kg/cm², F4- and F5 -2.3 kg/cm², F6-1.3 kg/cm², F7-3.3 kg/cm². The system with the additional loads on springs made the test conditions close the natural ones, and allowed for changing the dynamic pressure on the soil irrespective of the static pressure.

There were four tests conducted on water saturated soil subjected to vibration during 40 to 100 hours at different levels of dynamic pressure. Two similar tests were carried out on soil of natural moisture (here the ground water table was 1.6m lower than the foundation footing). Observations were also carried out on the settlemenents of test foundations when soil vibrations were caused by the ball mills, located in the building at a distance of 100 meters.

The settlements of the foundations were measured to the accuracy of 0.01mm. The amplitudes of the foundation vibrations were also regularly measured during the test. The rigidity properties of the sandy soil were determined from the amplitude-frequency curve of F3 foundation vibration.

The coefficient of elastic uniform compression C_z became equal to 3.5 kg/cm³ for dry soils and 2.8 kg/cm3 for watersaturated soil.

The main tests were carried out on watersaturated soil. Here is the description of two tests— A and B. In this case the ground water level reached as far as the foundation footing by means of flooding of the test pit.

tudes make 60 to 80% of those given in the table.

The graphs of the development of the foundation settlements with time of test A are shown in fig.4

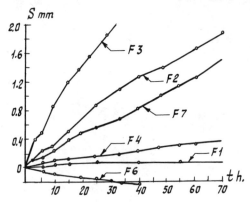

Fig.4. Graph of development of foundation settlements in test A.

Foundations F2, F3 and F4 are running into progressing settlement, its speed slowly decaying (to the end of the test making 0.025 mm/hr, 0.05 mm/hr and 0.02 mm/hr, respectively.

The settlement of foundation F3 reached 43 mm to the test end. The process of the settlement of foundation F1 stopped soon, but the settlement of foundations F5 and F6 was not observed. As it was observed during the test, foundation F6 raised due to the great settlements of foundations F2 and F3 located nearby. The zones of building of F2 and F3 caused the uplift of the soil together with the slightly loaded foundation F6.

It should be noted that the given values of settlements refer only to the additional settlements that

Table

Tests	Vibration frequency, Hz	Eccentric moment, kg.cm	Vertical vibration amplitures, mc						
			F1	F2	F3	F4	F5	F6	F7
A	12	240	30	60	125	55	28	58	62
B	9	440	25	43	80	42	18	38	46

It should be noted that the values of the horizontal vibration ampli-

occur after the settlement stabilization, caused only by the static loads.

The same features in the development of settlements have been found during test B.

The duration of the test was 48 hours. In this case the settlements of foundations F2,F3,F4 and F7 were observed, the speed of which to the time of the test completion made 0.02, 0.03, 0.01 and 0.02 mm/hr. The settlement of foundations F1 and F6 had a decaying character.

4. Analysis of the observed data

The curves of the additional settlement with time in tests A and B are well approximated by the power function

$$S(t)=At^{0.4} \qquad (1)$$

The same relationship, but with the index of power 0.45 approximates as well the curves of the settlements of the building column foundations (Fig.5).

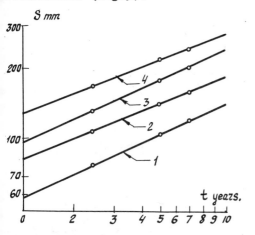

Fig.5. Graph of power function approximating the development of existing foundation settlement
1 - foundation of building frame in axes "D-22"
2 - ditto, in axes "I-34".
3 - ditto, in axes "I-26".
4 - ditto, in axes "F-28".

Expression (1) may as well be used for prediction settlements.

Two more continuous tests and some short-time tests (from two to four hours) were carried out using different parameters of dynamic load.

The results of all the tests show that the acceleration of vibrations does not influence decisively on the settlements of foundations. For example, the settlements of all the foundations in testa A and B developed identically, having different frequences and vibration accelerations but similar dynamic loads.

The settlements of the test foundations placed on watersaturated and dry soils were also observed during 10 days (on each soil) under the natural vibration condition produced by the ball mills located in the building (foundation vibration amplitudes 15 to 20 microns).

The results of these observations show that foundations placed on watersaturated soil practically gave no settlements, static pressure under the foundations making 1.3-2.3 kg/cm2, while foundation F7 developed the settlement with the speed of 0.2 to 0.3 mm/day and the speed of the settlement of foundation F2 was 0.05 mm/day.

The settlements of this kind have not been observed on the soil of natural moisture.

It might be well to point out that the tests with continuous vibration similar to those described above were carried out with the natural ground water level. Foundations on these types of soil have the same settlements as on watersaturated soil, however it occurs at rather high level of dynamic forces.

The tests described above as well as the other information on the settlements of foundations placed on sand soil subjected to the dynamic effects bring us to the conclusion about the necessity to reduce permissible static load on soil even at minor dynamic effect.

However at present there might be no any practical recommendations to take into consideration on the stage of designing.

The authors have analized the results of the tests described above, to find the practical recommendation for assessment of vibration effect on the decrease long-term bearing capacity.

In Fig.6, sign "x" shows the foundations with continuous settlements in the test conditions (the abciss and the ordinate in the graph show the static and dynamic stresses in soil). Sign "O" in

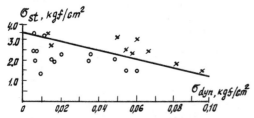

Fig.6. Results of test on water-saturated soil

    x-foundations with conti-
      nuous settlements;
    O-foundations practically
      without settlements.

the same graph indicates the cases when the additional settlements of the test foundations did not practically occur.

It can be seen that a stright line may be drawn, that disconnects the points corresponding to different behaviour of the foundations.

This results in the equation:

$$\sigma_{st} + \alpha \sigma_{dyn} = \sigma_* \qquad (2)$$

The value of the parameter σ_* appeared to be very close to the values of the permissible pressures R acting upon the soil under the test foundations.

Therefore, the conditions of long-term strength of sand in presence of dynamic loads may be assumed as follows:

$$\sigma_* < R \qquad (3)$$

where
    R – permissible pressure,
      determined by the stan-
      dards for design of foun-
      dations placed on natu-
      ral soil.

For the given type of soils it is recommended to take the value α equal to 20. It should be noted, that equation (3) may be used not only in designing of foundations for machines (vibratition producers) but also in designing of adjacent foundations acted upon by vibrations through the soil.

5. Conclusions

1. As described in this paper, the reason for continuous settlements of structures placed on sand soil is the effect of vibration produced by the working equipment (ball mills).

2. The assumption of the effect of dynamic forces on the vibro-resistance and continuous deformations of sandy soils, proposed earlier, is supported by experimental results.

3. The conclusion is as follows: it is necessary to reduce permissible pressure on sandy soil in presence of vibration effect. Proposed is the empiric equation that calls for the described effect to be taken into consideration at the stage of designing.

6. References

Kerchman,V.I. & O.R.Filippov 1977, Continuous settlements of structure foundations under dynamic loads, in Proceedings of All-Union Fourth Conference on Dynamics of soils, foundations and underground structures, p.206-211. Tashkent, Fan. (in Russian).

Ivanov,P.L. 1962, Liquefaction of sandy soils. Leningrad, Gosenergoizdat.(in Russian).

Goldstein M.N., V.Y.Hain & V.S.Bogolubchik 1974, Experimental laboratory study of sandy soil vibrocreeping, Bases, foundations and soil mechanics, 1:33-35.

Kerchman,V.I 1977, Experimental stamp studies of sandy soil vibrocreeping, in Proceedings of co-ordinating conferences on hydrotechnique 110: p.46-48. Leningrad, Energy. (in Russian).

Bogolubchik,V.S.,M.N.Goldstein & V.Y.Hain 1977, Experimental field studies of sands and sandy loams, in Proceedings of All-Union Fourth Conference on Dynamics of soils, foundations and underground structures, p.192-195. Tashkent, Fan. (in Russian).

Savinov,O.A 1964 Up-to-date construction and design of machine foundations. Leningrad, Stroyizdat. (in Russian).

Kerchman,V.I. 1977 Consideration
for vibrocreeping of sandy soils
in designing the foundations
under dynamic forces, in Proceedings of All-Union Fourth Conference on Dynamics of soils, Foundations and underground structures, p.110-115. Tashkent, Fan.
(in Russian).

2. Constitutive relations for soils

The rate process approach to the prediction
of in situ soil behaviour

M. J. KEEDWELL
Lanchester Polytechnic, Coventry, UK

1 INTRODUCTION

The theory of absolute reaction rates has been applied to various natural phenomena described as rate processes. Chemical reactions, diffusion and viscous movement are examples (see Glasstone, Laidler and Eyring, 1941) and a common feature is that relative movement of atoms occurs in a time-dependent manner. Since soils exhibit creep it seems likely that soil deformation is a rate process. Generally soil deformation results mainly from the relative movement of particles and is very little due to the deformation of the particles themselves. Consequently small deformations may be visualised as being the result of relative viscous movement of atoms or groups of atoms, called flow units, in the zones where particles are in contact.

Andersland and Akili, 1967; Andersland and Douglas, 1970; Christensen and Wu, 1964; Mitchell, 1964; Mitchell, Campanella and Singh, 1968; and Murayama and Shibata, 1964; have assumed that the theory of absolute reaction rates may therefore be applied to soil deformation.

In every case it was assumed by these workers that the activation energy ΔF for a given soil is a constant. The purpose of the present paper is to show that by making the assumption that for many soils ΔF is to some extent dependent on the rate of stress change it is possible to derive equations which model the behaviour of soils as determined in triaxial tests. A further objective is to illustrate how in situ soil behaviour may be analysed using the prediction of the effect of a cyclic loading on a soil mass as an example.

2 DERIVATION OF RHEOLOGICAL EQUATIONS

Application of the theory of rate processes to a single flow unit in any interparticle contact zone leads to the following equation for the rate of relative movement $\dot{\delta}$ of any typical pair of contacting particles (see any of the papers listed in the Introduction for further details of the steps leading to this equation):

$$\dot{\delta} = \lambda.(2kT/h).\exp(-\Delta F/RT).\sinh(\tau V_f/2\,kT) \quad \dots 1$$

in which λ is the thickness of a typical contact zone $(10^{-3}$ to 10^{-7} cm)

τ is the contact zone shear stress

T is the absolute temperature
V_f is the volume of a flow unit $(10^{-23}$ cm^3 approx.)
k is Boltzmann's Constant $(1.38 \times 10^{-16}$ erg/deg K/mol)
R is the Gas Constant $(1.987$ cal/deg K mole)
h is Planck's Constant $(6.624 \times 10^{-27}$ erg .sec)

Fig. 1. shows an assembly of spherical particles within a soil element. Following Rowe (1962), we may write:

$$\dot{\varepsilon}_1 = 2.\dot{\delta}(\cos\beta).S. \quad \dots 2$$

in which $\dot{\varepsilon}_1$ is the axial strain rate corresponding to $\dot{\delta}$.
β is as shown in the Figure and S is the number of contacts per unit length along the axis of symmetry of the stress system $(\sigma_2' = \sigma_3')$. Rowe explains that in general terms this equation is applicable to any typical soil composed of particles not necessarily of spherical shape.
Eliminating $\dot{\delta}$ from equations 1 and 2 yields:

$$\dot{\varepsilon}_1 = 2(\cos\beta).S.\lambda.(2kT/h).\exp(-\Delta F/RT).\sinh(\tau V_f/2kT) \quad \dots 3$$

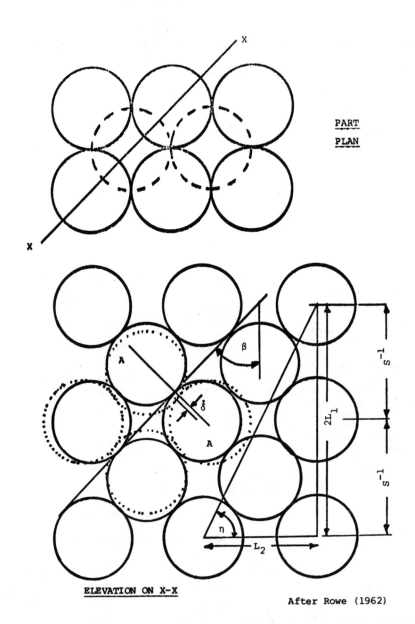

PART
PLAN

ELEVATION ON X-X

After Rowe (1962)

Fig. 1. Effect of an increment of axial compressive stress on an assembly of spherical particles

For a given soil at a particular temperature T, k, h, R and T are constants and β, S, λ and V_f may be assumed to be approximately constant over a wide range of axial strain.

The writer (1978) made the following assumptions regarding the values of τ and ΔF:

i) $\tau = \text{const.} (s' - s_o')$ 4

   in which the stress ratio $s' = s_1/p$

$s_1 = (\sigma_1 - p) = $ stress deviation

$p = $ mean effective pressure

   $s_o' = $ value of s' when $\dot{\varepsilon}_1 = 0$

const $= 10^4$ to 10^6 depending on soil type.

ii) for mineral-to-mineral (Zone 1) contacts: $-\Delta F = $ const
(e.g. 21 k cal/mole for Leighton Buzzard sand),

iii) for adsorbed water layer (Zone 2) contacts (e.g. see Fig. 2.):

   $-\Delta F \propto \ln \dot{s}'$ $(\dot{s}' = ds'/dt)$ 5

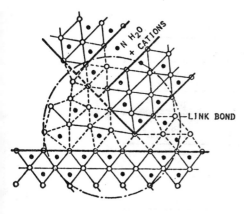

OXYGEN OR HYDROXYL
ION

After Arnold (1967)

Fig. 2. Two dimentional Analogue of the contact between particles of Montmorillonite.

Hence for a Zone 1 type soil:

$$\dot{\epsilon}_1 = const(1) . \sinh (b(s' - s_o')) \quad \ldots \ 6a$$

and for a Zone 2 type soil:

$$\dot{\epsilon}_1 = const(2) . |\dot{s}'| \ \sinh (b(s' - s_o')) \quad \ldots \ 6b$$

and the writer has further shown that for the more general Zone (1/2) type soil containing some mineral-to-mineral and some adsorbed water layer type contacts:

$$\dot{\epsilon}_1 = B_1 . |\dot{s}'|^m . \sinh (a (s' - s_o'))\quad \ldots \ 6c$$

in which B_1, m and a are soil parameters and $0 < m < 1$ 6d

Although the above derivation has related to an axially symmetrical system (i.e. $\sigma_2' = \sigma_3'$) it is not unreasonable to suppose that with suitable adjustments to the values of the soil parameters an equation of similar form may be applicable to any stress system, i.e.:

$$\dot{\epsilon}_i = B_i . |\dot{s}_i'|^m . \sinh (a_i (s_i' - s_o'))\quad \ldots \ 6e$$

where i = 1, 2, or 3.

3 PHYSICAL SIGNIFICANCE OF RHEOLOGICAL PARAMETERS

By comparing equations 3 and 6e in the derivation given above it may be readily seen that B_i is related to the energy ΔF required to activate a bond rupture in the interparticle contact zone, the absolute

temperature T, the average direction of contact zone shear strain as represented by the angle β in Fig. 1., the contact zone thickness λ and the number of contacts per unit length S. The nature of the interparticle contacts; Zone 1. (mineral to mineral) or Zone 2 (adsorbed water layer) or combinations of 1 and 2; is indicated by the value of m; and a is related to the number of contacts per unit length S.

Table 1 indicates the orders of magnitude of some of the contact zone parameters for two different extreme values of the product

$$a (s' - s_o')$$

The predicted effect of temperature variation on contact zone shear strain rate is shown in Fig. 3.

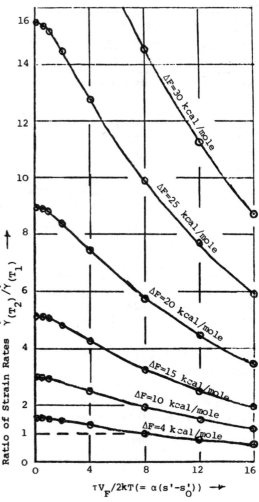

Fig. 3. The predicted ratio of contact zone strain rates at $40^{\circ}C$ ($\dot{\gamma}_{(T_2)}$) and 20° ($\dot{\gamma}_{(T_1)}$)

251

Table 1: Values of Contact Zone Parameters

Quantity	Value	
$\tau V_f / 2kT$ or $\alpha(s'-s_0')$	0.1	10
α corresponding to $s' = 1$	0.1	10
Contact zone shear stress corresponding to average shear stress of 100 kN/m^2	$10^6 \, kN/m^2$	$10^8 \, kN/m^2$
(Total area of shear surface/area occupied by contact zones) A_t/A	10^4	10^6
Product of contact zone thickness and number of contacts per unit length λS	10^{-2}	10^{-3}
(Contact zone shear strain rate/axial strain rate) $\dot{\gamma}/\dot{\varepsilon}$	10^2	10^3
Value of S for clay soil $\lambda = 10^{-7}$ cm	10^5	10^4
Value of S for uniform sand $\lambda = 10^{-3}$ cm	10	1
Volume of a flow unit V_f	$10^{-23} \, cm^3$	$10^{-23} \, cm^3$
Number of flow units* per contact for clay soil $\lambda = 10^{-7}$ cm	100	100
Number of flow units* per contact for uniform sand $\lambda = 10^{-3}$ cm	10^{14}	10^{14}

* Note the very great difference in the values for the two soils

4 EVALUATION OF RHEOLOGICAL PARAMETERS USING THE RESULTS OF TRIAXIAL TESTS

Equation 6e may be written in the form:

$$(\dot{\varepsilon}_i / \ |s_i'|^m) = B_i \sinh (\alpha_i (s_i' - s_0'))$$

Using the data from any triaxial test in which readings of $\dot{\varepsilon}_i, \sigma_1', \sigma_2'$, and σ_3' have been recorded at time intervals Δt it is possible to calculate corresponding values of $\dot{\varepsilon}_i$ ($= \Delta\varepsilon_i / \Delta t$), $\dot{s}_i'$ ($= \Delta s_i' / \Delta t$) and s_i'.

For values of $\alpha_i (s_i' - s_0') > 2$ or thereabouts

$$|\sinh (\alpha_i (s_i' - s_0'))| = 0.5 \exp |(\alpha_i (s_i' - s_0'))| \text{ approx.}$$

ie $\ln (\ |\dot{\varepsilon}_i| \ / \ |\dot{s}_i'|^m) = \ln (0.5 B_i) + \alpha_i (s_i' - s_0')$ approx.

Hence a plot of $\ln (|\dot{\varepsilon}_i| \ / \ |\dot{s}_i'|^m)$ versus $s_i' - s_0'$ should yield a straight line provided $(s_i' - s_0') > 2$

and the correct value of m is used.

By trying a number of different values of m the best straight line plot can be found by trial and error, and hence the optimum value of m is found. The slope of the straight line is α and the intercept on the $\ln (\ |\dot{\varepsilon}_i| \ / \ |\dot{s}_i'|^m)$ axis is the logarithm of $0.5 B_i$.

The values of the parameters can be further refined by comparing predicted stress-strain curves (examples are shown in Figs. 4 to 7) with the experimental data.

It will be readily appreciated that B_i mainly affects the initial slope of the predicted curves, and α_i the curvature.

From Figs. 4 to 7 it will be seen that typical values of the parameters are:

Soil type	B_i	α_i	m
Well-graded sand	7.0×10^{-5}	3.0	0.5
Uniform sand	1.3×10^{-6}	1.8	0.0
Slate dust	6.9×10^{-4}	4.3	0.6
Remoulded kaolin	6.0×10^{-2}	5.85	1.0

WELL-GRADED SAND - UNDRAINED TEST

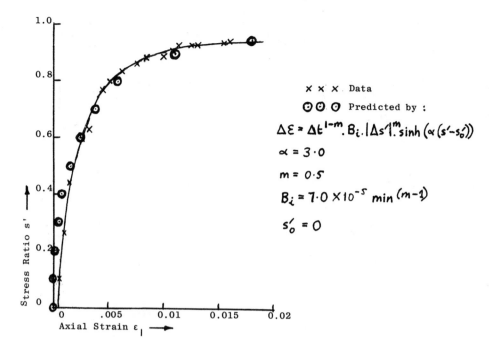

Calculations

$s'-s'_o$	$\sinh(\alpha(s'-s'_o))$	$B_i.|\Delta s'|^m$	Δt^{1-m}	$\Delta \varepsilon$	ε
0.1	0.30	2.21×10^{-5}	12.25	8.12×10^{-5}	8.12×10^{-5}
0.2	0.63	2.21×10^{-5}	12.25	1.70×10^{-4}	2.52×10^{-4}
0.3	1.03	2.21×10^{-5}	12.25	2.78×10^{-4}	5.30×10^{-4}
0.4	1.51	2.21×10^{-5}	12.25	4.09×10^{-4}	9.39×10^{-4}
0.5	2.13	2.21×10^{-5}	14.14	6.66×10^{-4}	1.60×10^{-3}
0.6	2.94	2.21×10^{-5}	14.14	9.19×10^{-4}	2.52×10^{-3}
0.7	4.02	2.21×10^{-5}	15.81	1.40×10^{-3}	3.93×10^{-3}
0.8	5.47	2.21×10^{-5}	30.82	5.05×10^{-3}	1.12×10^{-2}

Fig. 4   Well-graded sand - predicted and experimentally determined stress-strain curves

SLATE DUST - 1st UNDRAINED CYCLE

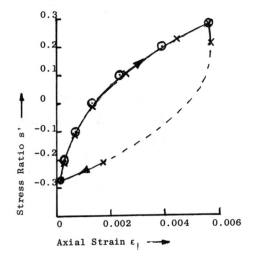

$x \; x \; x \; x$ Data

$\odot \; \odot \; \odot$ Predicted by :

$$\Delta \varepsilon = \Delta t^{1-m}. B_i .|\Delta s'|^m. \sinh\left(\alpha(s'-s_o')\right)$$

$$\alpha = 4.3$$

$$B_i = 6.9 \times 10^{-4} \; min \; (m-1)$$

$$m = 0.6$$

$$s_o' = -0.27$$

Experimental data kindly supplied by
P.I. Lewin,
Building Research Establishment.

Calculations

$s'-s_o'$	$\sinh\left(\alpha(s'-s_o')\right)$	$B_i.|\Delta s'|^m$	Δt^{1-m}	$\Delta \varepsilon$	ε
0	0				1.7×10^{-4}
0.07	0.30	1.40×10^{-4}	2.51	1.05×10^{-4}	2.75×10^{-4}
0.17	0.80	1.72×10^{-4}	2.51	3.47×10^{-4}	6.22×10^{-4}
0.27	1.42	1.72×10^{-4}	2.51	6.16×10^{-4}	1.24×10^{-3}
0.37	2.32	1.72×10^{-4}	2.51	1.01×10^{-3}	2.25×10^{-3}
0.47	3.65	1.72×10^{-4}	2.51	1.58×10^{-3}	3.83×10^{-3}
0.54	5.00	1.40×10^{-4}	2.51	1.74×10^{-3}	5.51×10^{-3}

Fig. 5   Slate dust - predicted and experimentally determined stress-strain curves

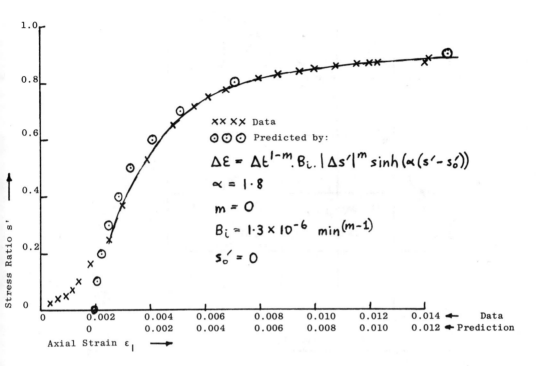

Calculations

$s'-s_o'$	$\sinh(\alpha(s'-s_o'))$	$B_i \cdot \lvert\Delta s'\rvert^m$	Δt^{1-m}	$\Delta\varepsilon$	ε
0.1	0.18	1.3×10^{-6}	350	8.19×10^{-5}	8.19×10^{-5}
0.2	0.36	1.3×10^{-6}	350	1.64×10^{-4}	2.46×10^{-4}
0.3	0.58	1.3×10^{-6}	350	2.64×10^{-4}	5.10×10^{-4}
0.4	0.78	1.3×10^{-6}	350	3.55×10^{-4}	8.64×10^{-4}
0.5	1.00	1.3×10^{-6}	400	5.20×10^{-4}	1.38×10^{-3}
0.6	1.30	1.3×10^{-6}	450	7.60×10^{-4}	2.14×10^{-3}
0.7	1.60	1.3×10^{-6}	500	1.04×10^{-3}	3.18×10^{-3}
0.8	1.95	1.3×10^{-6}	800	2.03×10^{-3}	5.21×10^{-3}
0.9	2.40	1.3×10^{-6}	2500	7.80×10^{-3}	1.30×10^{-2}

Fig. 6 Uniform sand - predicted and experimentally determined stress-strain curves

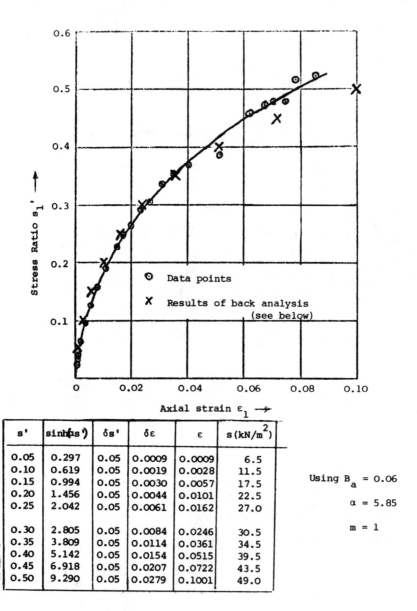

s'	sinh(αs')	δs'	$\delta\epsilon$	ϵ	s(kN/m^2)
0.05	0.297	0.05	0.0009	0.0009	6.5
0.10	0.619	0.05	0.0019	0.0028	11.5
0.15	0.994	0.05	0.0030	0.0057	17.5
0.20	1.456	0.05	0.0044	0.0101	22.5
0.25	2.042	0.05	0.0061	0.0162	27.0
0.30	2.805	0.05	0.0084	0.0246	30.5
0.35	3.809	0.05	0.0114	0.0361	34.5
0.40	5.142	0.05	0.0154	0.0515	39.5
0.45	6.918	0.05	0.0207	0.0722	43.5
0.50	9.290	0.05	0.0279	0.1001	49.0

Using B_a = 0.06

α = 5.85

m = 1

Fig. 7   Remoulded kaolin - predicted experimentally determined stress-strain curves

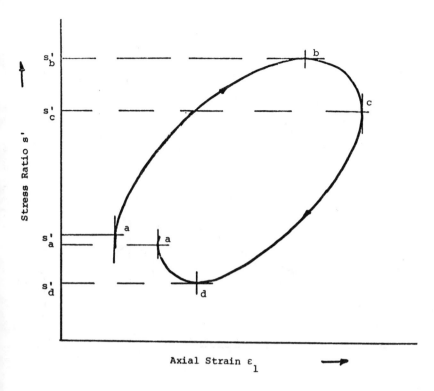

Fig. 8 Predicted behaviour of zone 1/2 type soil subjected to cyclic loading

5 APPLICATION OF RHEOLOGICAL EQUATIONS TO THE CASE OF CYCLIC LOADING

Fig. 8 shows the predicted behaviour of a soil element subjected to a cycle of stress. In this illustration s_a' and s_c' are respectively the values of s_o' for loading and unloading.

For comparison Fig. 9 shows experimental data from cyclic triaxial tests carried out by Lewin (1978). Analysis of this data has yielded the following parameters for the portions, $b \rightarrow c$, $c \rightarrow d$, $d \rightarrow a$ and $a \rightarrow b$ of the second loop:

Portion of loop	B_a or B_c $\frac{min}{}$	$\alpha_{o \rightarrow o}$ $(m - 1)$	m
$b \rightarrow c$)	10^{-3}	(12	0.6
$c \rightarrow d$)		3	0.6
$d \rightarrow a$)	10^{-3}	(12	0.6
$a \rightarrow b$)		2	0.6

(B_a and B_c are respectively the values of B_i for loading and unloading)

It will be seen that a change in the value of α from 3 (or 2) to 12 occurs

at the extremes of the cycles b and d, suggesting that a reversal of sign of the rate of change of stress ratio causes a temporary reduction in the number of contact zones per unit length; the loss being made up at the points of zero strain rate a and c.

6 EXAMPLE OF THE PREDICTION OF SETTLEMENT DUE TO CYCLIC LOADING

For this example the reader is asked to imagine that the footing shown in Fig. 10 is one leg of an oil-rig platform which has sunk under its own weight through soft mud of negligible strength to rest on a layer of slate dust underlain by a permeable rock. Due to wave action on the rig the footing applies a cyclic loading to the founding soil as shown in Fig. 11.

This figure also shows the predicted settlements and the corresponding calculations are in Table 2.

The calculations are based on the following assumptions:-

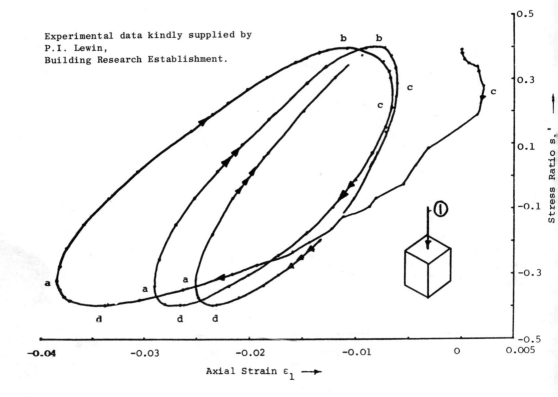

Experimental data kindly supplied by
P.I. Lewin,
Building Research Establishment.

Fig. 9   Slate dust - data from drained cyclic test

i)   The relevant rheological parameters
are as listed in the previous section
(except that a is taken as 3 for both
$a \rightarrow b$ and $c \rightarrow d$).

ii) The average vertical effective stress
change $\Delta \sigma_z'$ in the slate dust layer is at
all times one half of the contact pressure
p.

iii) The horizontal effective stress change
is at all times $0.3 \, \Delta\sigma_z'$.

iv) The average effective overburden
pressure for the slate dust is 80 kN/m^2
and K_o = 0.6.

Assumption (iv) means that s_a' = 0.363

and by assuming that at the peak b the same
strain rate is common to both $a \rightarrow b$ and
$b \rightarrow c$ we have

$$(s_b' - s_c')/(s_b' - s_a') = a_{b \rightarrow a}/ a_{b \rightarrow c} = 3/12$$

from which s_c' has been calculated.

Assumptions (ii) and (iii) above are

unlikely to accurately represent in situ
stress conditions, but the over-simplifi-
cation is considered to be justified in this
context since the objective here is merely
to illustrate one possible use of the
rheological equations.

If a more exact calculation were required
then the rheological equations could be used
in a finite element analysis using the variable stiffness method. Assuming isotopic
soil properties (not a necessary assumption)
then the relevant pseudo-elastic drained
parameters would be:

$$E' = (\dot{\sigma}_z' - \nu (\dot{\sigma}_x'+ \dot{\sigma}_y')) . \left[B_z . | \dot{s}_z' | ^m . \right.$$
$$\left. \sinh (a \, (s_z' - s_o')) \right]^{-1}$$

$$K' = \dot{p} . \left[B_v . | \dot{s}_z' | ^m . \sinh (a_z(s_z' - s_o')) \right]^{-1}$$

$$\nu' = 0.5 . (1 - (E'/3K'))$$

(in this context $\dot{\sigma}_z'$ (for example) may be

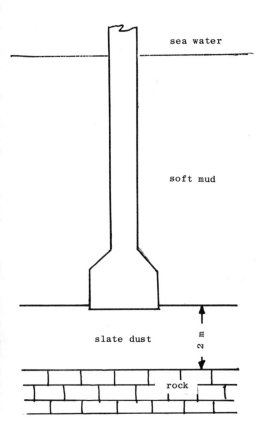

sea water

soft mud

slate dust

$\frac{m}{2}$

rock

Fig. 10 One footing of imaginary oil-rig
platform

8 REFERENCES

Andersland, O.B. and Akili, W. (1967),
"Stress Effect on Creep Rates of Frozen
Clay Soil," Geotechnique 17, No. 1,
pp. 27-39.

Andersland, O.B. and Douglas, A.G. (1970),
"Soil Deformation Rates and Activation
Energies," Geotechnique 20, No. 1,
pp. 1.16.

Keedwell, M.J. (1978), "Rheology and the
Analysis of In Situ Soil Deformations,"
PhD Thesis, C.N.A.A.

Mitchell, J.K. (1964), "Shearing Resist-
ance of Soils as a Rate Process,"
J. Soil Mech. Fdns. Div. Am. Soc. Civ.
Engrs. 90, SM1 pp. 29-61.

Mitchell, J.K. Campanella, R.G. et al
(1968), "Soil Creep as a Rate Process,"
Proc. Am. Soc. Civil Engrs. 94,
No. SM1, pp. 231-254.

Marayama, S. and Shibata, T. (1964),
"Flow and Stress Relaxation of Clays,"
Proc. Rheol. and Soil Mech. Symp. Int.
Unions of Theoretical and Appl. Mech.,
Grenoble, France, pp. 99-129.

Rowe, P.W. (1962), "The Stress-Dilatancy
Relation for Static Equilibrium of an
Assembly of Particles in Contact,"
Proc. Roy. Soc., A269, pp. 200-527.

interpreted as $\Delta\sigma'_z / \Delta t$ where Δt is a
time step relating to a change in boundary
conditions.)

7 CONCLUDING REMARKS
A rheological theory has been described
which is based on the proposition that the
mechanical behaviour of an element of soil
is partly governed by the manner
in which relative movements
of atoms occur in the interparticle contact zones. Evidence has been
presented which illustrates the potential
of this approach to soil mechanics both in
interpreting the results of tests and in
predicting in situ behaviour.

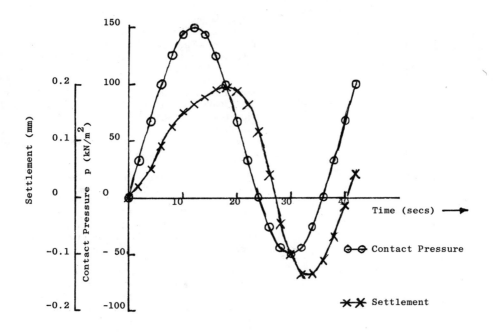

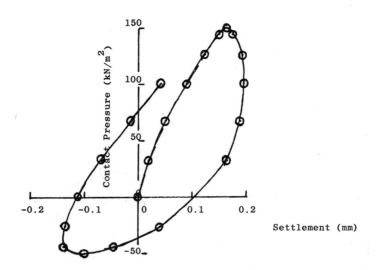

Fig. 11 Predicted Effect of Cyclic Loading on the Settlement of the footing illustrated in Fig. 10

Table 2 Calculations of settlement for one loading cycle

The equation used is:

$$\Delta\varepsilon_{z} = \Delta t \cdot B_i \, |\dot{s}'_z|^{m} \sinh\left(a\,(s'_z - s'_o)\right)$$

hence $\Delta\rho = 2000\ \Delta t \cdot B_i \, |\dot{s}'_z|^{m} \cdot \sinh\left(a\,(s'_z - s'_o)\right)$ mm

Time t(secs)	Contact Pressure $p\,(kN/m^2)$	Stress Ratio s'_z	s'_a	s'_c	$s' - s'_a$ or $s'-s'_c$	$|\dot{s}'_z|^m$ min^{-m}	a	$\Delta\rho$ mm	Settlement ρ mm
0	0	0.363	0.363		0		3		0
2	32.6	0.429	0.363		0.066	1.505	3	0.019	0.019
4	67.4	0.483	0.363		0.120	1.334	3	0.032	0.052
6	100.0	0.523	0.363		0.160	1.115	3	0.037	0.089
8	126.6	0.550	0.363		0.187	0.880	3	0.034	0.123
10	144.0	0.565	0.363		0.202	0.617	3	0.026	0.149
12	150.0	0.570	0.363		0.207	0.318	3	0.014	0.163
				0.518	0.052		12		
14	144.0	0.565		0.518	0.047	0.318	12	0.012	0.176
16	126.6	0.550		0.518	0.032	0.617	12	0.016	0.192
18	100.0	0.523		0.518	0.005	0.880	12	0.004	0.195
20	67.4	0.483		0.518	-0.035	1.115	3	-0.008	0.188
22	32.6	0.429		0.518	-0.089	1.334	3	-0.024	0.164
24	0.0	0.363		0.518	-0.155	1.505	3	-0.048	0.116
26	-26.6	0.293		0.518	-0.225	1.560	3	-0.075	0.041
28	-44.4	0.236		0.518	-0.282	1.377	3	-0.086	-0.045
30	-50.0	0.213		0.518	-0.305	0.800	3	-0.055	-0.101
			0.289		-0.076		12		
32	-44.4	0.236	0.289		-0.053	0.800	12	-0.036	-0.137
34	-26.6	0.293	0.289		0.004	1.377	3	0.001	-0.135
36	0.0	0.363	0.289		0.074	1.560	3	0.023	-0.112
38	32.6	0.429	0.289		0.140	1.505	3	0.043	-0.069
40	67.4	0.483	0.289		0.194	1.334	3	0.054	-0.015
42	100.0	0.523	0.289		0.234	1.115	3	0.056	0.041

Finite element solution of boundary value problems
in soil mechanics

JEAN H. PREVOST
Princeton University, Princeton, N.J., USA

THOMAS J.R. HUGHES
California Institute of Technology, Pasadena, Calif., USA

INTRODUCTION

Soil consists of an assemblage of particles with different sizes and shapes which form a skeleton whose voids are filled with various fluids. The stresses carried by the soil skeleton are conventionally termed "effective stresses" [1] in the soil mechanics literature, and those in the fluids are called the "pore-fluid pressures". In cases in which some flow of the pore fluids takes place, there is an interaction between the skeleton strains and the pore-fluid flow. The solution of these problems therefore requires that soil behavior be analyzed by incorporating the effects of the flow (transient or steady) of the pore-fluids through the voids, and thus requires that a multiphase continuum formulation be available for soils. Such a theory was first developed by Biot [2] for an elastic porous skeleton. However, it is observed experimentally that the stress-strain behavior of the soil skeleton is strongly non-linear, anisotropic, elasto-plastic and path-dependent. An extension of Biot's theory into the nonlinear anelastic range is therefore necessary in order to analyze the transient response of soil deposits. Such an extension of Biot's formulation [3] is adopted herein. The resulting coupled field equations [3] obtained by viewing soil as a multiphase medium consisting of an anelastic porous skeleton and viscous fluids, and by using the modern theories of mixtures developed by Green and Naghdi [4] and Eringen and Ingram [5], are presented in the following. In order to relate the changes in effective stresses carried by the soil skeleton to the skeleton rate of deformations, a general analytical model [5] which describes the nonlinear, anisotropic, elastoplastic, stress and strain dependent, stress-strain-strength properties

of the soil skeleton when subjected to complicated three-dimensional, and in particular to cyclic loading paths [7], is used. A brief summary of the model's basic principle is included in the following and the constitutive equations are provided. The model parameters required to characterize the behavior of any given soil can be derived entirely from the results of conventional soil tests. The model's extreme versatility and accuracy are demonstrated by applying it to represent the behavior of both cohesive and cohesionless soils under both drained and undrained, monotonic and cyclic loading conditions. The use of the proposed formulation for solving boundary value problems of interest in soil mechanics is thereafter illustrated by applying it to analyze (1) the interaction of an offshore gravity structure with its soil foundation when subjected to cyclic wave forces, (2) the penetration of a marine pipeline into its soil foundation, and (3) the localization of deformations into shear bands in soil media.

FIELD EQUATIONS

For a saturated soil consisting of a perfect fluid and a piecewise-linear time-independent porous skeleton wherein both the pore-fluid and the solid grains are incompressible, the coupled field equations take the following forms [3],

$$\text{div}[\overset{\nabla'}{\underset{\sim}{\sigma}}{}^{s} + \overset{'}{\underset{\sim}{\sigma}}{}^{s}\, \text{div}\,\underset{\sim}{v}{}^{s}] - \text{div}[(\overset{\cdot}{p}_{w} + p_{w}\,\text{div}\,\underset{\sim}{v}{}^{s})\underset{\sim}{1}] +$$
$$\text{div}[\underset{\sim}{D}:\underset{\sim}{L}{}^{s}] + \rho_{w}\,\text{div}\,\underset{\sim}{v}{}^{s}(\underset{\sim}{b} - \underset{\sim}{a}{}^{w}) + \rho\underset{\sim}{b} = \rho^{s}\underset{\sim}{a}{}^{s} + \rho^{w}\underset{\sim}{a}{}^{w}$$
$$(1)$$

$$-\,\text{div}[\frac{n^{w}}{\rho_{w}}\underset{\sim}{k}{}^{ws}\cdot(\text{grad}\,p_{w} - \rho_{w}\underset{\sim}{b} + \rho_{w}\underset{\sim}{a}{}^{w})] + \text{div}\,\underset{\sim}{v}{}^{s} = 0$$
$$(2)$$

263

in which

$$D_{abcd} = \frac{1}{2}[\sigma_{bd}\delta_{ac} - \sigma_{ad}\delta_{bc} - \sigma_{ac}\delta_{bd} - \sigma_{bc}\delta_{ad}] \quad (3)$$

is a tensor arising from geometric changes,

$$\underset{\sim}{\sigma} = \underset{\sim}{\sigma}'^s - p_w \underset{\sim}{1} \quad (4)$$

is the total stress tensor [1],

$$\dot{p}_w = \frac{\partial p_w}{\partial t} + \underset{\sim}{v}^s \cdot \text{grad } p_w \quad (5)$$

$$\dot{n}^w = (1-n^w)\text{ div } \underset{\sim}{v}^s \quad (6)$$

$$\rho^s = (1-n^w)\rho_s \quad (7)$$

$$\rho = \rho^s + n^w\rho_w \quad (8)$$

and the subscript s and w refer to the solid and fluid phases, respectively. In Eqs. 1 and 2, σ'^s = effective cauchy stress tensor, p_w = pore-fluid pressure, v^α = velocity of α-phase, $\underset{\sim}{a}^\alpha$ = acceleration of α-phase, $\underset{\sim}{d}^\alpha$ and $\underset{\sim}{w}^\alpha$ = symmetric and skew-symmetric parts of the velocity gradient $\underset{\sim}{L}^\alpha$, respectively, $\underset{\sim}{k}^{ws}$ = permeability tensor, ρ_α = microscopic mass density of α-phase, n^w = porosity, $\underset{\sim}{b}$ = body force density per unit mass, and a superimposed dot indicates the material derivative following the motion of the solid skeleton. In Eq. 1, $\overset{\triangledown}{\underset{\sim}{\sigma}}'^s$ denotes the Jaumann derivative viz.

$$\overset{\triangledown}{\underset{\sim}{\sigma}}'^s = \underset{\sim}{\dot{\sigma}}'^s + \underset{\sim}{\sigma}'^s \cdot \underset{\sim}{w}^s - \underset{\sim}{w}^s \cdot \underset{\sim}{\sigma}'^s \quad (9)$$

When neglecting inertia effects in both the solid and fluid phases, Eqs. 1 and 2 simplify to

$$\text{div}[\overset{\triangledown}{\underset{\sim}{\sigma}}'^s + \underset{\sim}{\sigma}'^s \text{ div } \underset{\sim}{v}^s] - \text{div}[(\dot{p}_w + p_w \text{ div } \underset{\sim}{v}^s)\underset{\sim}{1}] +$$

$$\text{div } [\underset{\sim}{D}:\underset{\sim}{L}^s] + \rho_w \text{ div } \underset{\sim}{v}^s \underset{\sim}{b} + \rho \underset{\sim}{\dot{b}} = 0 \quad (10)$$

$$- \text{div}[\frac{n^w}{p_w} \underset{\sim}{k}^{ws} \cdot (\text{grad } p_w - \rho_w \underset{\sim}{b})] + \text{div } \underset{\sim}{v}^s = 0 \quad (11)$$

In its linearized form, Eq. 10 simplifies to

$$\text{div}[\underset{\sim}{\dot{\sigma}}'^s - p_w \underset{\sim}{1}] + \rho_w \text{ div } \underset{\sim}{v}^s \underset{\sim}{b} + \rho \underset{\sim}{\dot{b}} = 0 \quad (12)$$

Of particular importance in soil mechanics are fully drained and undrained loading conditions. Clearly, both of these conditions are special cases of the general equations presented above, and the corresponding field equations can be derived from Eqs. 10 and 11 as shown in the following:

(i) Undrained (instantaneous) Loading
     Condition:

In that case, $\underset{\sim}{k}^{ws} = 0$ relative to the rate of loading, no drainage of the fluid phase can take place, and the pore fluid follows the motion of the solid skeleton. In that case, $\underset{\sim}{v}^s = \underset{\sim}{v}^w$, $\dot{n}^w = 0$ and div $\underset{\sim}{v}^s =$ div $\underset{\sim}{v}^w = 0$, and Eq. 10 simplifies to

$$\text{div}[\overset{\triangledown}{\underset{\sim}{\sigma}}'^s - \dot{p}_w \underset{\sim}{1}] + \text{div}[\underset{\sim}{D}:\underset{\sim}{L}^s] + \rho \underset{\sim}{\dot{b}} = 0 \quad (13)$$

(ii) Fully Drained (slow) Loading Conditions

In that case, $\underset{\sim}{k}^{ws} = \infty$ relative to the rate of loading, no excess pore-fluid pressure builds up (i.e. $\dot{p}_w = 0$), and Eqs. 10 and 11 then simplify to

$$\text{div}[\overset{\triangledown}{\underset{\sim}{\sigma}}'^s + (\underset{\sim}{\sigma}'^s - p_w \underset{\sim}{1}) \text{ div } \underset{\sim}{v}^s] + \text{div}[\underset{\sim}{D}:\underset{\sim}{L}^s]$$

$$+ \rho_w \text{ div } \underset{\sim}{v}^s \underset{\sim}{b} + \rho \underset{\sim}{\dot{b}} = 0 \quad (14)$$

$$- \text{grad } p_w + \rho_w \underset{\sim}{b} = 0 \quad (15)$$

(iii) Fully Drained Steady State Conditions

When the skeleton stresses and strains have reached constant values, $v^s = 0$, and the pore-fluid flow has reached steady state, $\dot{p}_w = 0$ and $\frac{\partial \underset{\sim}{v}^w}{\partial t} = 0$. Eq. 11 then simplifies to

$$\text{div}[\frac{n^w}{\rho_w} \underset{\sim}{k}^{ws} \cdot (\text{grad } p_w - \rho_w \underset{\sim}{b})] = 0 \quad (16)$$

CONSTITUTIVE EQUATIONS

The constitutive equations for the solid skeleton are written in one of the following forms:

$$C_{abcd} d^s_{cd} = \begin{cases} \dot{\sigma}'^s_{ab} & \text{small deformations} \\ \overset{\triangledown}{\sigma}'^s_{ab} + \sigma'^s_{ab} v^s_{c,c} & \text{finite deformations} \end{cases} \quad (17)$$

C_{abcd} is an (objective) tensor valued function of, possibly, σ'^s_{ab} and the deformation gradients. Many nonlinear material models of interest can be put in the above form (e.g. all nonlinear elastic materials, and many elastoplastic materials). The finite deformation form of the constitutive equation above was first proposed by Hill [8] in the context of plasticity theory and has been advocated by McMeeking and Rice [9].

For soil media, the form of the $\underset{\sim}{C}$ tensor is given as follows [6],

$$C = E - \frac{1}{H' + Q:E:P} (E:P)(Q:E)$$

in which H' is the plastic modulus; P and Q are dimensionless symmetric second-order tensors, normalized in such a way that $P:P = Q:Q = 1$ and such that P gives the direction of plastic deformations and Q the outer normal to the active yield surface; and E is the fourth-order tensor of elastic moduli, assumed isotropic for the particular class of material models implemented. The plastic potential is selected such that, in agreement with experimental observations, the plastic deviatoric rate of deformation vector remains normal to the projection of the yield surface onto the deviatoric stress subspace, i.e.

$$P - \frac{1}{3}(\text{trace } P)1 = Q - \frac{1}{3}(\text{trace } Q)1 = Q' \quad (18a)$$

$$\text{trace } P = \text{trace } Q + A(Q':Q')^{\frac{1}{2}} \quad (18b)$$

in which A is a material parameter which measures the departure from an associative flow rule. When $A = 0$, $P = Q$ and consequently the C tensor possesses the major symmetry. The yield function is selected of the following form [6]

$$f = \frac{3}{2}(s^s - \alpha):(s^s - \alpha) + c^2(p'^s - \beta)^2 - k^2 = 0 \quad (19)$$

where s^s is the deviatoric stress tensor (i.e. $s^s = \sigma'^s - p'^s 1$, $p'^s = \frac{1}{3}\text{trace } \sigma'^s$) α and β are the coordinates of the center of the yield surface in the deviatoric stress space and along the hydrostatic stress axis, respectively; k is the size of the yield surface; and C is a material parameter called the yield surface axis ratio. In order to allow for the adjustment of the plastic hardening rule to any kind of experimental data, for example data obtained from axial or simple shear soil tests, a collection of nested yield surfaces is used. A plastic modulus is associated with each of the yield surfaces, and

$$H' = h' + \sqrt{3}(\text{trace } Q)B' \quad (20)$$

where h' is the plastic shear modulus and $(h'+B')$ and $(h'-B')$ are the plastic bulk muduli associated with f which are mobilized in consolidation tests upon loading and unloading, respectively. The projections of the yield surfaces onto the deviatoric stress subspace thus define regions of constant plastic shear moduli.

The yield surfaces' initial positions and sizes reflect the past stress-strain history of the soil skeleton, and in particular their initial positions are a direct expression of the material "memory" of its past loading history. Because the α's are not necessarily all equal to zero, the yielding of the material is anisotropic. Direction is therefore of great importance and the physical reference axes (x,y,z) are fixed with respect to the material element and specified to coincide with the reference axes of consolidation. For a soil element whose anisotropy initially exhibits rotational symmetry about the y-axis, $\alpha_x = \alpha_z = -\alpha_y/2$ and Eq. 19 simplifies to

$$[(\sigma_y'^s - \sigma_x'^s) - \alpha_1]^2 + c^2(p'^s - \beta)^2 - k^2 = 0$$

in which $\alpha_1 = 3\alpha_y/2$. The yield surfaces then plot as ellipses in the axisymmetric stress plane $(\sigma_x'^s = \sigma_z'^s)$ as shown in Figure 1. Points C and E on the outermost yield surface define the critical state conditions (i.e. $H' = 0$) for axial compression and extension loading conditions, respectively [10]. It is assumed that the slopes of the critical state lines OC and OE remain constant during yielding.

The yield surfaces are allowed to change in size as well as to be translated by the stress point. Their associated plastic moduli are also allowed to vary and in general both k and H' are functions of the plastic strain history. They are conveniently taken as functions of invariant measures of the amount of plastic volumetric strains and/or plastic shear distortions, respectively [7].

Complete specification of the model parameters requires the determination of (i) the initial positions and sizes of the yield surfaces together with their associated plastic moduli; (ii) their size and/or plastic modulus changes as loading proceeds, and finally (iii) the elastic shear G and bulk B moduli. The soil's anisotropy originally develops during its deposition and subsequent consolidation which, in most practical cases, occurs under no lateral deformations. In the following, the y-axis is vertical and coincides with the direction of consolidation, the horizontal xz-plane is thus a plane of material's isotropy and the material's anisotropy initially exhibits rotational symmetry about the vertical y-axis. The model parameters required to characterize the behavior of any given soil can then be derived *entirely* from the results of conventional monotonic axial and cyclic strain-controlled simple shear soil tests. This is explained and further discussed in Refs. [6,7,11,12]. As an illustration, Tables 1 and 2 give the parameter values for the Cook's Bayou Sand [13] and the Drammen clay [14].

Figs. 2 and 3 show the model predictions for both consolidation/swelling and axial compression tests, respectively, performed on the Cook's Bayou sand. They agree very well with the experimental test results for all cases. Figs. 4 and 5 show the model predictions for both undrained axial and simple shear tests, respectively, performed on the Drammen clay. They agree very well with the experimental test results for all cases (in Fig. 6 τ_h denotes the *average* horizontal shear stress measured experimentally). Fig. 6 shows the model predictions for cyclic stress- and strain-controlled simple shear and axial tests performed on the Drammen clay. They agree very well with the experimental test results [14] for all cases.

APPLICATION TO SOLUTION OF BOUNDARY VALUE PROBLEMS

Attention is restricted in the following to quasi-static fully drained and undrained loading conditions. The class of constitutive equations assumed leads directly to the definition of tangent stiffness matrix (see e.g. [9]), and an incremental predictor-corrector type algorithm has been adopted [15]. The element and material model libraries are modularized and may be easily expanded without alteration of the main code. The present element library contains a two-dimensional element with plane stress/plain strain options, and full finite deformation effects may be accounted for. A three-dimensional element is also included. A contact element is available for two- and three-dimensional analysis. The present material library contains a linear elastic model and various elasto-plastic and soil models. Some features which are available in the program are

• Both symmetric and non-symmetric matrix equation solvers

• Reduced/selective integration procedures, for effective treatment of incompressibility constraints [16] which arise in undrained loading cases.

The formulation's use in solving boundary value problems is now illustrated by applying it to analyze the interaction of an offshore gravity structure with its soil foundation when subjected to cyclic wave forces. In order to make the present study quite specific, attention herein is concentrated on the behavior of a fully saturated clay foundation for which the repeated

loading resulting from wave forces during one individual storm is assumed to occur with no volume change. Furthermore, the foundation material is assumed to consist of a homogeneous deposit of Drammen clay.

Due to the necessity of working within limited computation budgets, the problem geometry is transposed into 2-dimensions by assuming that it is of plane strain. Its 2-dimensional finite representation is shown in Fig. 7. The structure foundation is represented by a strip footing which consists of a thin layer of elements 1000 times stiffer than the supporting soil. Fig. 8a shows the load system and notation. In addition to the static load W due to the dead weight of the structure, the foundation is also subjected to the cyclic inclined eccentric load F due to the wave forces, and Fig. 8b shows the time histories for its components V, H and M over a complete cycle of loading. For the purpose of illustration, $W/ACu = 3.42$ in the following, where $Cu = 0.584 \, \sigma'_{vc}$ denotes the static (unsoftened) undrained simple shear strength of the clay, and A the footing area. The computed load-displacement curves for the first 5 cycles of loading are shown in Fig. 9 from which the progress of the various deformations can easily be traced. In Fig. 9 δ, d and θ denote the vertical and horizontal displacements of the center of the foundation and θ its tilt, respectively. Fig. 10 shows the distorted configuration of the soil foundation and the direction of its flow at various instants of time during the 5th cycle of loading.

The use of contact elements is illustrated in Fig. 11 which shows the penetration of a marine pipeline into its soil foundation. The foundation material consists of a soft clay deposit whose strength increases linearly with depth.

It is observed experimentally that when materials are deformed sufficiently, a smooth and continuously varying deformation patern may give rise to highly localized deformation regions in the form of shear bands. These shear bands are commonly observed in mild steels testing where they are referred to as Lüders bands [17], but they are also observed to develop for a wider class of materials including geological materials like rocks [18] and soils [19]. Upon their formations, the subsequent deformations either proceed further in a markedly non-uniform manner or lead directly to ductile fracture. In the latter case, the onset of localization is synonymous with the inception of rupture. The basic theoretical principles for understanding the phenomenon are contained in [20-24] where it is shown that its existence in elastic-plastic solids is contingent upon a loss of ellipticity of the velocity equations of equilibrium. Numerical results which illustrate the phenomenon of localization of deformation into shear bands for a rectangular block constrained for plane deformations and subjected to tension in one direction are presented hereafter.

In order to make the present study quite specific, the material is modeled here as an incompressible isotropic elastic-plastic Prandtl-Reuss material. Fig.12a shows the two-dimensional finite element representation of the tensile specimen. The grid consists of 171 bilinear isoparametric rectangular elements. The specimen length to width ratio is equal to two, and 9 elements are placed across the width. Uniform longitudinal ends displacements are prescribed, and no shearing tractions are applied. The lower left corner of the specimen is fixed, and the loading is accomplished by imposing increments of end displacements at the upper end of the specimen.

For the particular case of the Prandtl-Reuss material, loss of ellipticity of the velocity equations of equilibrium in the small deformation regime is achieved simply by selecting a plastic modulus less or equal to zero, and in the following $H'/2G=-0.048$. The corresponding angle for the plane of localization is then 38.7°. The assumed stress-strain curve is shown in Fig. 12b.

In a first attempt to obtain localization, both the material properties and the ends' displacements were taken as uniform. These conditions resulted in smooth and continuously varying deformation patterns well into the softening range, but no localization occurred. This result may be interpreted by recalling that the loss of illipticity is a necessary but not a sufficient condition for localization. Some type of non-uniformity is therefore necessary in order to trigger the phenomenon. In the following, localization is achieved by introducing a weak element which plays the role of a local "imperfection" in the material properties. This element is located either at the center (series C) or at the corner (series D) of the specimen as shown in Fig. 11, and its plastic modulus is such that $H'/2G=-1/3$. Typical results are shown in Figs. 13-15. The spreading of the plastic zone is indicated by a shaded area. In both Figs. 13 and 14, the imperfection is located at the center of the specimen. In Fig. 13 it has a yield strength 5% smaller than the surrounding material. In Fig. 13a the axial strain = .099%, and only the weak element has yielded. Upon further loading, localization occurs and as shown in Fig. 13b (axial strain = 0.101%) results in a very symmetrical pattern. This pattern was found to remain stable upon further

loading and the specimen failed by necking. In Fig. 13c (axial strain = .150%) a slight non-uniformity in the end displacements was introduced to break this symmetry. This was achieved by making the upper right hand corner longitudinal displacement 1% larger than the remaining. By comparing Figs. 13b and 13c, it is apparent that as a result, some elements unloaded and one shear band emerged. Note that the angle of the shock line is very close to the predicted value (38.7°). In Fig. 14, the imperfection has the same strength as its surrounding. In Fig. 14a, the axial strain = .100%, and all the elements have yielded. In Figs. 14b and 14c, the axial strain = .103% and .120%, respectively. Note that the localization pattern is very different from the one found in Fig. 13. Again this pattern is stable, and the subsequent failure of the specimen is illustrated by Fig. 14c. In both Figs. 15 and 16, the imperfection is located at the corner of the specimen. In Fig. 15 it has a yield strength 5% smaller than the surrounding material. In Fig. 15a, the axial strain = .099%. Upon further loading, localization takes place and leads to the formation of two symmetrical shock lines as shown in Fig. 15b (axial strain = .101%). However, this configuration is not stable, and upon further loading, only one shear band remains as shown in Fig. 15c (axial strain = 0.140%). In Fig. 16 the imperfection has the same strength as its surrounding and this leads directly to one shear band as shown in Fig. 16b (axial strain = .102%). Again, note that for all cases, the angle of the shock line is very close to the predicted value (38.7°).

ACKNOWLEDGEMENTS

The authors are most grateful to M. Cohen who helped to perform some of the computations. The authors would also like to acknowledge with appreciation the attention and interest paid to this research by the Civil Engineering Laboratory, Port Hueneme, California. Computer time was provided by the California Institute of Technology Computer Center and Princeton University Computer Center.

REFERENCES

[1] Terzaghi, K., Theoretical Soil Mechanics, Wiley, New York, 1943.
[2] Biot, M.A., "Theory of Elasticity and Consolidation for a Porous Anisotropic Solid," J. Applied Physics, Vol. 26, 1955, pp. 182-185.

[3] Prevost, J.H., "Mechanics of Continuous Porous Media," Int. J. of Engineering Science, 1979 (to appear).

[4] Green, A.C. and P.M. Naghdi, "A Dynamical Theory of Interacting Continua," Int. J. of Eng. Science, Vol. 3, 1965, pp. 231-241.

[5] Eringen, A.C. and J.D. Ingram, "A Continuum Theory of Chemically Reaching Media, I and II," Int. J. Eng. Science, Vol. 3, 1965, pp. 197-212, and Vol. 5, 1967, pp. 289-322.

[6] Prevost, J.H., "Plasticity Theory for Soil Stress-Strain Behavior," J. Eng. Mech. Div., ASCE, Vol. 104, No. EM5, 1978, pp. 1177-1194.

[7] Prevost, J.H., "Mathematical Modeling of Monotonic and Cyclic Unchained Clay Behavior," Int. J. Num. Analyt. Methods in Geomechanics, Vol. 1, No. 2, 1977, pp. 195-216.

[8] Hill, R., "A General Theory of Uniqueness and Stability in Elastic-Plastic Solids," J. Mech. Phys. Solids, Vol. 6, 1958, pp. 236-249.

[9] McMeeking, R.M. and J.R. Rice, "Finite Element Formulation for Problems of Large Elastic-Plastic Deformation," Int. J. Solids Structures, Vol. 11, 1975, pp. 601-616.

[10] Hvorslev, M.J., "Uber die Festigkeitseigenschaften gestorler bindiger boden," Ingeniozvidenskabelige Skrifter, A. no. 45, Copenhagen, 1937.

[11] Prevost, J.H., "Anisotropic Undrained Stress-Strain Behavior of Clays," J. Geotech. Eng. Div., ASCE, Vol. 104, No. GT8, 1978, pp. 1075-1090.

[12] Prevost, J.H., "Mathematical Modeling of Soil Stress-Strain-Strength Behavior," Proceedings, 3rd Int. Conf. Num. Methods in Geomechanics, Aachen, Germany, Vol. 1, 1979, pp. 347-361.

[13] Forrest, J.H., et al., "Experimental Relationships Between Moduli for Soil Layers Beneath Concrete Pavements," Report No. FAA-RD-76-206, 1976.

[14] Anderson, K.H., "Behavior of Clay Subjected to Undrained Cyclic Loading," Proceedings, BOSS 76 Conf., Tronheim, Norway, Vol. 1, 1976, pp. 392-403.

[15] Hughes, T.J.R. and J.H. Prevost, "DIRT II - A Nonlinear Quasi-static Finite Element Analysis Program," California Institute of Technology, Pasadena, California, August 1979.

[16] Malkus, D.S. and T.J.R. Hughes, "Mixed Finite Element Methods - Reduced and Selective Integration Techniques: A Unification of Concepts," Computer Meths. Appl. Mech. Eng., Vol. 1979.

[17] Nadai, A., Theory of Flow and Fracture of Solids, Vol. 1, Second Edition, McGraw-Hill, New York, 1950.

[18] Brace, W.F., State of Stress in the Earth Crust, ed. by W.R. Judd. Elsevier, New York, 1964, pp. 111.

[19] Roscoe, K.H., "The Influence of Strains in Soil Mechanics," Tenth Rankine Lecture, Geotechnique, Vol. 20, No. 2, 1970, pp. 129-170.

[20] Hadamard, J., Lecon our la Propagation des Ondes et les Equations de l'Hydrodynamique, Paris, Chap. 6, 1903.

[21] Thomas, T.Y., Plastic Flow and Fracture in Solids, Academic Press, Inc., 1961.

[22] Hill, R., "Acceleration Waves in Solids," Journal of the Mechanics and Physics of Solids, Vol. 10, 1962, pp. 1-16.

[23] Mandel, J., "Conditions de Stabilite' et Postulat de Drucker," in Rheology and Soil Mechanics, Eds. J. Kravtchenko and P.M. Sirieys, Springer-Verlag, 1966, pp. 58-68.

[24] Rice, J.R., "The Localization of Plastic Deformations," in Theoretical and Applied Mechanics, Proceedings of the 14th IUTAM Congress, Delft, The Netherlands, North Holland Publishing Co., 1976, pp. 207-220.

TABLE 1 - COOK'S BAYOU SAND σ'_{vc} =50 psi. MODEL PARAMETERS

$(G=400\sigma'_{vc}, \ B_1=470.6\sigma'_{vc}, \ n=0.5)$

m	$\beta^{(m)}/\sigma'_{vc}$	$k^{(m)}/\sigma'_{vc}$	h_m/σ'_{vc}	B'_m/σ'_{vc}	A_m/σ'_{vc}
2	0.800	0.424	800.000	-12325.089	0.080
3	0.700	0.636	800.000	-8333.974	0.080
4	0.600	0.849	800.000	-2000.000	0.080
5	0.550	0.955	800.000	-2000.000	0.080
6	0.500	1.061	800.000	-2000.000	0.080
7	0.633	1.344	371.429	-2000.000	0.059
8	0.750	1.591	270.011	-2000.000	0.040
9	0.882	1.871	218.370	-2000.000	0.029
10	1.025	2.174	204.684	-2000.00	0.024
11	1.138	2.415	177.342	-2000.000	0.022
12	1.256	2.663	162.053	-2000.000	0.020
13	1.316	2.791	122.269	-2000.000	0.019
14	1.376	2.920	102.628	-2000.000	0.018
15	1.500	3.182	72.687	-2000.000	0.017
16	1.626	3.450	49.335	-2000.000	0.014
17	1.755	3.722	47.909	-2000.000	0.014
18	1.819	3.860	27.486	-2000.000	0.014
19	1.885	3.998	16.148	-2000.000	0.014
20	4.000	20.000	0.000	-2000.000	0.014

TABLE 2 - DRAMMEN CLAY OCR=4. MODEL PARAMETERS $(G=200\sigma'_{vc})$

m	$\alpha_1^{(m)}/\sigma'_{vc}$	$k^{(m)}/\sigma'_{vc}$	h_m/σ'_{vc}
2	0.100	0.300	266.667
3	0.150	0.350	133.333
4	0.300	0.600	100.000
5	0.400	0.70	73.333
6	0.475	0.775	54.667
7	0.525	0.875	40.000
8	0.550	0.950	31.000
9	0.575	1.025	24.333
10	0.600	1.050	17.333
11	0.575	1.125	13.333
12	0.550	1.200	10.000
13	0.550	1.250	6.667
14	0.525	1.275	3.333
15	0.467	1.373	0.000

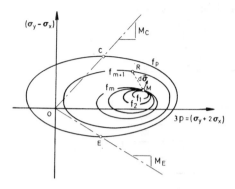

Fig. 1: Field of Yield Surfaces

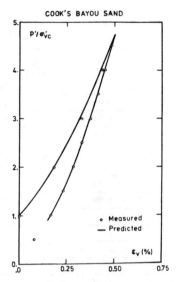

Fig. 2: Cook's Bayou Sand. Calculated and Measured Stress-Strain Behavior for Consolidation/Swelling Tests

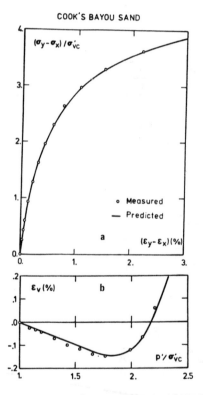

Fig. 3: Cook's Bayou Sand. Calculated and Measured Stress-Strain Behavior for Triaxial Soil Test

   (a) shear stress-strain behavior
   (b) volumetric stress-strain behavior

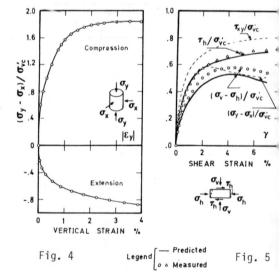

Fig. 4: Drammen Clay OCR=4. Calculated and Measured Stress-Strain Behavior for Triaxial Soil Test

Fig. 5: Drammen Clay OCR=4. Calculated and Measures Stress-Strain Behavior for Simple Shear Test

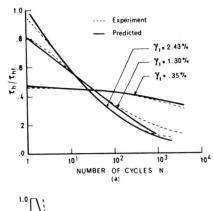

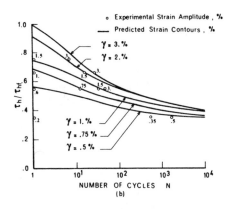

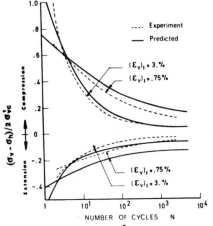

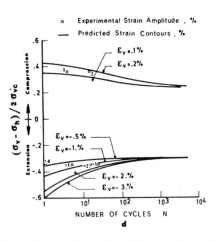

Fig. 6: Drammen Clay OCR=4. Calculated and Measured Stress-Strain Behavior for Cyclic

(a) Strain-Controlled Simple Shear Tests
(b) Stress-Controlled Simple Shear Tests
(c) Strain-Controlled Triaxial Soil Tests
(d) Stress-Controlled Triaxial Soil Tests

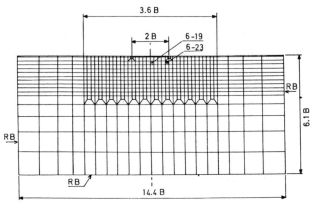

RB = REFLECTED BOUNDARY

Fig. 7: Problem Idealization-Finite Element Mesh

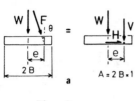

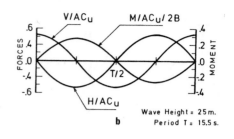

Fig. 8a

Wave Height = 25 m.
Period T = 15.5 s.

b

Fig. 8b

Fig. 8: Foundation Loading Conditions
(a) Notation
(b) Wave Load-Time Functions

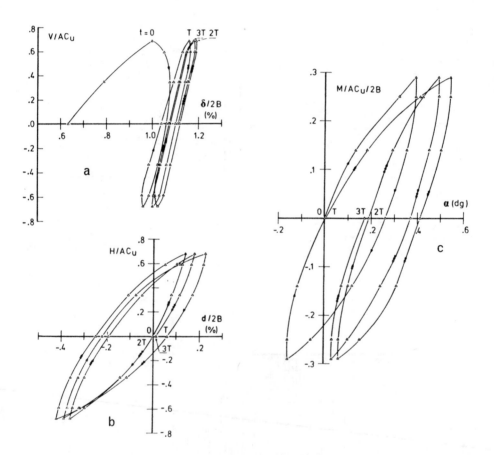

Fig. 9: Cyclic Wave Loading Conditions.
(Computed Displacement-Time Histories)

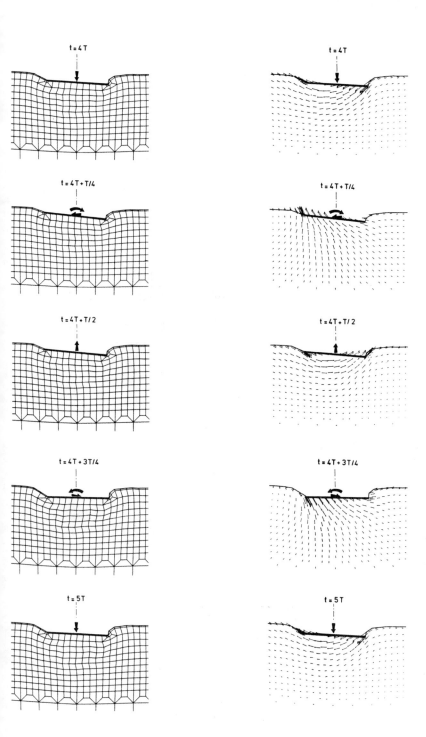

Fig. 10: Cyclic Wave Loading Conditions.

Computed Distorted Configuration and Velocity
Vields at Various Instants of Time During the
5th Cycle of Wave Loading

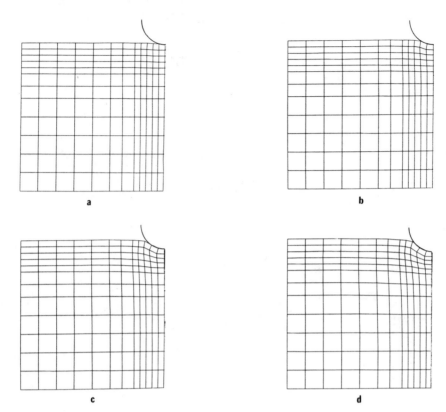

Fig. 11: Penetration of Marine Pipe Line in Soft Clay Deposit

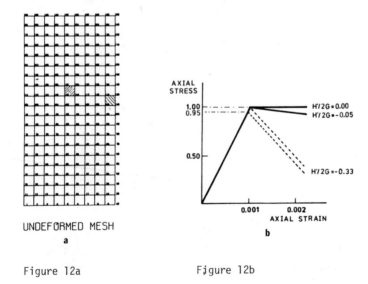

UNDEFORMED MESH
a

Figure 12a

Figure 12b

274

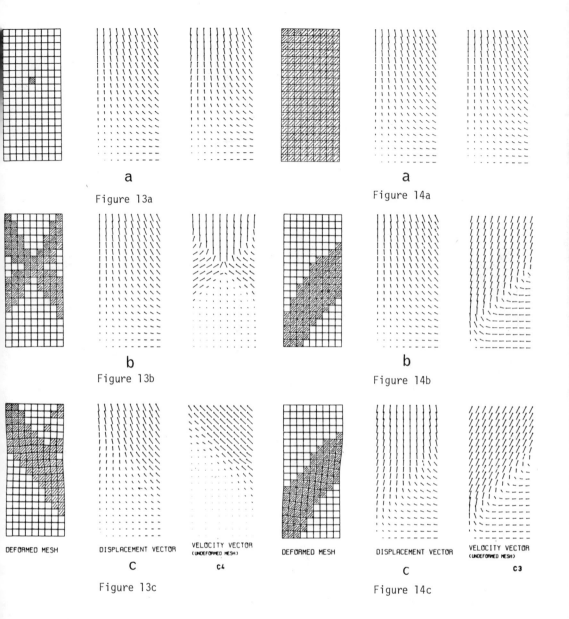

a

Figure 13a

a

Figure 14a

b

Figure 13b

b

Figure 14b

DEFORMED MESH   DISPLACEMENT VECTOR   VELOCITY VECTOR (UNDEFORMED MESH)

c

c4

Figure 13c

DEFORMED MESH   DISPLACEMENT VECTOR   VELOCITY VECTOR (UNDEFORMED MESH)

c

c3

Figure 14c

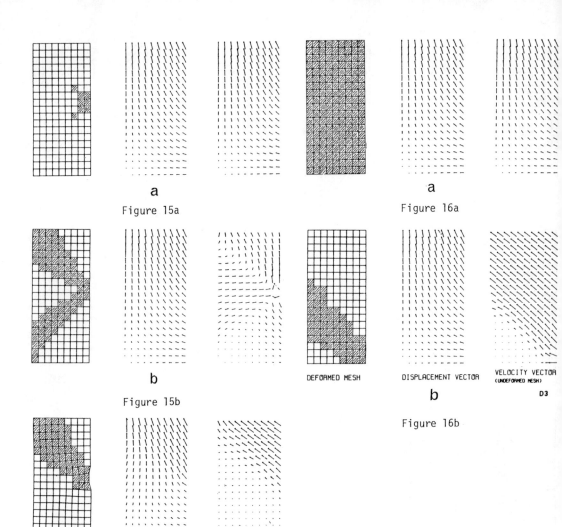

a

Figure 15a

a

Figure 16a

b

Figure 15b

DEFORMED MESH   DISPLACEMENT VECTOR   VELOCITY VECTOR
(UNDEFORMED MESH)

b   D3

Figure 16b

DEFORMED MESH   DISPLACEMENT VECTOR   VELOCITY VECTOR
(UNDEFORMED MESH)

c   D4

Figure 15c

International Symposium on Soils under Cyclic and Transient Loading / Swansea / 7-11 January 1980

Effect of work hardening rules on the elasto-plastic matrix

WEN-XI HUANG
Tsinghua University, Peking, China

1 SUMMARY

Different work hardening rules have been assumed in formulating the constitutional relationship of soils used in elasto-plastic analyses. How these different assumptions may affect the elasto-plastic matrix $[D]_{ep}$ are investigated. The general expressions of $[D]_{ep}$ in 3 dimensional, plane strain and axi-symmetrical cases are developed and made ready for use in F.E.M. analyses.

2 INTRODUCTION

In the incremental plasticity theory used in the elasto-plastic analyses of soil masses, it is generally assumed that the total strain increment is divided into an elastic and a plastic component such that

$$\{de\} = \{de^e\} + \{de^p\} \qquad (1)$$

These strains are then calculated seperately, the elastic strain by the generalized Hooke's law:

$$\{de^e\} = [D]^{-1}\{dS\} \qquad (2)$$

wherein $[D]$ is the elastic matrix, and S denotes stress.

The development of the plastic stress-strain relations for soils is based on the concepts of plasticity theory, e.g. as outlined by Hill, 1950. The three basic concepts used may be referred to as i) Theory of yield surface or yield criterion, ii) Theory of flow rule, iii) Theory of work hardening rule.

3 THEORY OF YIELD SURFACE OR YIELD CRITERION

In the theory of yield surface it is generally assumed that there exists in the stress space a yield surface, such that it may change in size, position or shape as the soil is loaded to successively higher stress levels. We term these kinds of changes as work hardening.

In soil mechanics the most commonly used mathematical expression for the yield surfaces is

$$f(S_{ij}) = F(H) \qquad (3)$$

wherein H is the hardening factor, and it may be taken as equal to the plastic energy W_p, or it may be otherwise defined as will be further discussed in Sect.5. Since soil tests are usually done on the conventional triaxial compression device, wherein $S_2 = S_3$, the yield surface function f is very often given as:

$$f(p,q) = F(H) \qquad (4)$$

wherein
$$p = 1/3(S_1 + 2S_3)$$

$$q = S_1 - S_3 \qquad (5)$$

Sometimes it may happen that the above simplified yield functions fail to reflect the empirical findings correctly, then it may be assumed that there are two yield surfaces passing through any point M in the stress space, and they may be expressed as:

$$f_1(S_{ij}) = F_1(H)$$

$$f_2(S_{ij}) = F_2(H) \qquad (6)$$

or $\quad f_1(p,q) = F_1(H)$

$$f_2(p,q) = F_2(H) \qquad (7)$$

The intersection of these two surfaces will form a ridge line passing through the point M in the stress space. In a similar manner in the p-q stress plane we shall have two yield loci, namely curve 1 and curve 2, as shown in Fig.1, passing through any given point M(p,q). M is thus known as a corner point. Roscoe & Burland, 1968; Prevost & Höeg, 1975; and Lade, 1977 have respectively suggested these kinds of yield functions or cap models.

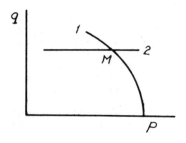

Fig.1. Yield loci through a corner point

In order to determine the yield locus from the experimental data, the following proceedure is suggested . On the system of S_3 = const.(for example) empirical curves, we may select a number of points, and calculate their corresponding H values (e.g. W_p values). Plot these points on the p-q diagram as shown in Fig.2, and label the corresponding value of H beside each point. Draw the H contours through these points, then these contours will represent the yield loci.
From the above discussion it can be seen that if H is any function of e^p other than W_p, different yield loci will be found. Thus the yield locus thus determined is not unique, and it depends upon how H is selected . That is to say the yield locus will be different if different work hardening rule is assumed.

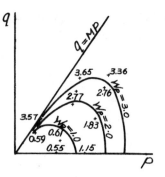

Fig.2. Determination of yield loci

4 THEORY OF FLOW RULE

Let g be the function representing the plastic potential surface, thus

$$g = g(S_{ij}) \qquad (8)$$

or $\quad g = g(p,q) \qquad (9)$

The flow rule or normality law may then be expressed as

$$de_{ij}^p = d\lambda \cdot \delta g / \delta S_{ij} \qquad (10)$$

or

$$de_v^p = d\lambda \cdot \delta g / \delta p$$

$$\overline{de}^p = d\lambda \cdot \delta g / \delta q \qquad (11)$$

wherein $d\lambda$ is a factor used to determine the magnitude of the plastic strain increment, and it will be discussed more fully in next section. The plastic volume strain increment

$$de_v^p = de_{ii}^p \qquad (12)$$

and the plastic shear strain increment
$$\overline{de}^p = (2/3 \, de_{ij}'^p \, de_{ij}'^p)^{1/2} \qquad (13)$$

wherein

$$de_{ij}'^p = de_{ij}^p - \delta_{ij}(de_v^p/3)$$

$$\delta_{ij} = 1 \quad i=j$$
$$= 0 \quad i \neq j$$

The plastic potential at any point M in the stress space may also be assumed as being composed of two

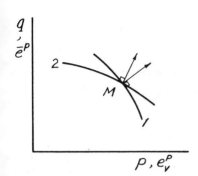

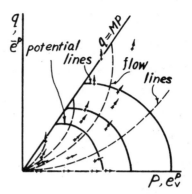

Fig.3. Directions of the plastic
strain increments at a corner point

Fig.4. Determination of the plastic
potential lines

parts, namely

$$g_1 = g_1(S_{ij})$$

$$g_2 = g_2(S_{ij})$$

or

$$g_1 = g_1(p,q)$$

$$g_2 = g_2(p,q) \qquad (14)$$

M will then be a corner point, and
the plastic strain increment at that
point will be given by

$$de^p_{ij} = d\lambda_1 \cdot \delta g_1 / \delta S_{ij} + d\lambda_2 \cdot \delta g_2 / \delta S_{ij}$$

or

$$de^p_v = d\lambda_1 \cdot \delta g_1 / \delta p + d\lambda_2 \cdot \delta g_2 / \delta p$$

$$de^{\bar{p}} = d\lambda_1 \cdot \delta g_1 / \delta q + d\lambda_2 \cdot \delta g_2 / \delta q$$
$$(15)$$

The direction of the resultant
plastic strain increment at the
corner point M will lie between the
normal of g_1 and that of g_2 as shown
in Fig.3.

The plastic potential surface g
may also be determined from conventional triaxial compression test
data. Thus corresponding to any
point M(p,q) in the p-q diagram,
we can calculate and determine the
direction of the plastic strain
increment from the test data.
Assuming that the e^p_v and $\bar{e}^p$ axes
coincide respectively with the p
and q axes, we can then use an arrow
head to represent this direction at
any point M on the p-q diagram as

shown in Fig.4. Through these arrow
heads we may trace the flow lines
as shown by the dotted lines in the
same figure. Draw lines orthogonal
to this set of flow lines, we can
find the system of plastic potential
lines as shown.

If this set of plastic potential
lines coincides with the yield loci
determined in Sect.3, i.e. if g = f ,
then we term the flow rule as an
associated flow rule. If f ≠ g , the
flow rule will be termed unassociated
flow rule.

Since f varies with H, therefore
it may happen that by properly chos-
ing H we can make f = g. In such
case all the calculations that follow
will be greatly simplified.

5 THEORY OF WORK HARDENING RULE

A work hardening rule is needed to
determine the magnitudes of the
plastic strain increments caused by
a given stress increment. In Equ.
(10) and (11) the giverning factor
dλ may be most generally expressed
as

$$d\lambda = df/A \qquad (16)$$

wherein the hardening parameter **A**
is a function of the hardening factor H. Different theories have been
proposed to evaluate **A**, thus we have
the following different hardening
rules:

1) W_p Work Hardening Rule assumes
(e.g. Hill,1050; Lade & Duncan,1975)

$$F(H) = F(W_p)$$

wherein the plastic energy

$$W_P = \int S_{ij} de_{ij}^p$$

Since $df = \delta F/\delta W_P \cdot dW_P = F' dW_P$

and by flow rule

$$de_{ij}^p = F'/A \cdot dW_P \cdot \delta g/\delta S_{ij}$$

which when multiplied both sides by S_{ij} will yield

$$S_{ij} de_{ij}^p = F'/A \cdot dW_P \cdot S_{ij} \cdot \delta g/\delta S_{ij}$$

Since $dW_P = S_{ij} de_{ij}^p$

therefore $\quad A = S_{ij} \cdot \delta g/\delta S_{ij} \cdot F' \quad (17)$

If g is a homogeneous function of nth order, then by Euler's theorem of homogeneous function, the above equation can be reduced to

$$A = ngF' \quad (18)$$

F' is the slope of the f-W_P curve, which can be derived from the experimental data.

ii) e^p Work Hardening Rule assumes (e.g. Mroz, 1967)

$$F(H) = F(e^p)$$

wherein $e^p = \int de^p = \int \sqrt{de_{ij}^p de_{ij}^p} \quad (19)$

Similar to case i) we can prove that

$$A = F' \sqrt{\delta g/\delta S_{ij} \cdot \delta g/\delta S_{ij}} \quad (20)$$

wherein F' is the slope of the f-e^p curve.

iii) e_v^p Work Hardening Rule assumes (e.g. Roscoe & Burland, 1968)

$$F(H) = F(e_v^p)$$

wherein $e_v^p = \int de_v^p$

Similar to case i) we can prove that

$$A = F' \cdot \delta g/\delta p \quad (21)$$

iv) $\overline{e}^p$ Work Hardening Rule assumes (e.g. Prevost & Höeg, 1975b)

$$F(H) = F(\overline{e}^p)$$

wherein

$$\overline{e}^p = \int d\overline{e}^p = \int \sqrt{2/3 \cdot de_{ij}^p de_{ij}^p}$$

Similar to case i) we can prove that

$$A = F' \cdot \delta g/\delta q \quad (22)$$

wherein F' is the slope of the f-$\overline{e}^p$ curve.

v) $(e_v^p, \overline{e}^p)$ Work Hardening Rule assumes (e.g. Prevost & Höeg, 1975a)

$$F(H) = F(e_v^p, \overline{e}^p)$$

Similar to case i) we can prove that

$$A = \delta F/\delta e_v^p \cdot \delta F/\delta p + \delta F/\delta \overline{e}^p \cdot \delta F/\delta q \quad (23)$$

4 ELASTO-PLASTIC MATRIX $[D]_{ep}$

The elasto-plastic stress-strain relationship for soils may be written as

$$\{dS_{ij}\} = [D]_{ep} \{de_{ij}\} \quad (24)$$

wherein $[D]_{ep}$ is termed elasto-plastic matrix.

It can be shown (e.g. Zienkiewicz, 1977) that

$$[D]_{ep} = [D] - \frac{[D]\{\delta g/\delta S\}\{\delta f/\delta S\}^T[D]}{A + \{\delta f/\delta S\}^T[D]\{\delta g/\delta S\}} \quad (25)$$

This equation after evaluation can be rewritten as

$$[D]_{ep} = [D] - \frac{G[X]}{A/G + \emptyset} \quad (26)$$

Let $\quad$ G = elastic shear modulus
$\qquad$ K = elastic bulk modulus

$$m = K/G + 4/3$$
$$n = K/G - 2/3 \quad (27)$$

$$a_1 = m \cdot \delta g/\delta S_{xx} + n \cdot \delta g/\delta S_{yy} + n \cdot \delta g/\delta S_{zz}$$

$$a_2 = n \cdot \delta g/\delta S_{xx} + m \cdot \delta g/\delta S_{yy} + n \cdot \delta g/\delta S_{zz}$$

$$a_3 = n \cdot \delta g/\delta S_{xx} + n \cdot \delta g/\delta S_{yy} + m \cdot \delta g/\delta S_{zz}$$

$$a_4 = \delta g/\delta S_{xy}, \quad a_5 = \delta g/\delta S_{yz}, \quad a_6 = \delta g/\delta S_{zx} \quad (28)$$

$$b_1 = m \cdot \delta f / \delta S_{xx} + n \cdot \delta f / \delta S_{yy} + n \cdot \delta f / \delta S_{zz}$$

$$b_2 = n \cdot \delta f / \delta S_{xx} + m \cdot \delta f / \delta S_{yy} + n \cdot \delta f / \delta S_{zz}$$

$$b_3 = n \cdot \delta f / \delta S_{xx} + n \cdot \delta f / \delta S_{yy} + m \cdot \delta f / \delta S_{zz}$$

$$b_4 = \delta f / \delta S_{xy}, \quad b_5 = \delta f / \delta S_{yz}, \quad b_6 = \delta f / \delta S_{zx}$$

$$(29)$$

then we have for:

1) 3 dimensional case

$$[D] = G \begin{bmatrix} m & & & & & \\ n & m & & & \text{sym.} & \\ n & n & m & & & \\ 0 & 0 & 0 & 2 & & \\ 0 & 0 & 0 & 0 & 2 & \\ 0 & 0 & 0 & 0 & 0 & 2 \end{bmatrix} \quad (30)$$

$$\emptyset = a_1 \cdot \delta f / \delta S_{xx} + a_2 \cdot \delta f / \delta S_{yy}$$

$$+ a_3 \cdot \delta f / \delta S_{zz} + a_4 \cdot \delta f / \delta S_{xy}$$

$$+ a_5 \cdot \delta f / \delta S_{yz} + a_6 \cdot \delta f / \delta S_{zx}$$

$$= b_1 \cdot \delta g / \delta S_{xx} + b_2 \cdot \delta g / \delta S_{yy}$$

$$+ b_3 \cdot \delta g / \delta S_{zz} + b_4 \cdot \delta g / \delta S_{xy}$$

$$+ b_5 \cdot \delta g / \delta S_{yz} + b_6 \cdot \delta g / \delta S_{zx} \quad (31)$$

$$[X] = \begin{bmatrix} a_1b_1 & a_1b_2 & a_1b_3 & a_1b_4 & a_1b_5 & a_1b_6 \\ a_2b_1 & a_2b_2 & a_2b_3 & a_2b_4 & a_2b_5 & a_2b_6 \\ a_3b_1 & a_3b_2 & a_3b_3 & a_3b_4 & a_3b_5 & a_3b_6 \\ a_4b_1 & a_4b_2 & a_4b_3 & a_4b_4 & a_4b_5 & a_4b_6 \\ a_5b_1 & a_5b_2 & a_5b_3 & a_5b_4 & a_5b_5 & a_5b_6 \\ a_6b_1 & a_6b_2 & a_6b_3 & a_6b_4 & a_6b_5 & a_6b_6 \end{bmatrix} \quad (32)$$

11) Plane strain case

$$[D] = G \begin{bmatrix} m & n & 0 \\ n & m & 0 \\ n & n & 0 \\ 0 & 0 & 2 \end{bmatrix} \quad (33)$$

$$\emptyset = a_1 \cdot \delta f / \delta S_{xx} + a_2 \cdot \delta f / \delta S_{yy}$$

$$+ a_3 \cdot \delta f / \delta S_{zz} + a_4 \cdot \delta f / \delta S_{xy}$$

$$= b_1 \cdot \delta g / \delta S_{xx} + b_2 \cdot \delta g / \delta S_{yy}$$

$$+ b_3 \cdot \delta g / \delta S_{zz} + b_4 \cdot \delta g / \delta S_{xy} \quad (34)$$

$$[X] = \begin{bmatrix} a_1b_1 & a_1b_2 & a_1b_4 \\ a_2b_1 & a_2b_2 & a_2b_4 \\ a_3b_1 & a_3b_2 & a_3b_4 \\ a_4b_1 & a_4b_2 & a_4b_4 \end{bmatrix} \quad (35)$$

iii) Axi-symmetrical case

$$[D] = G \begin{bmatrix} m & & & \\ n & m & \text{sym.} & \\ n & n & m & \\ 0 & 0 & 0 & 2 \end{bmatrix} \quad (36)$$

$$\emptyset = a_1 \cdot \delta f / \delta S_{rr} + a_2 \cdot \delta f / \delta S_{oo}$$

$$+ a_3 \cdot \delta f / \delta S_{zz} + a_4 \cdot \delta f / \delta S_{rz}$$

$$= b_1 \cdot \delta g / \delta S_{rr} + b_2 \cdot \delta g / \delta S_{oo}$$

$$+ b_3 \cdot \delta g / \delta S_{zz} + b_4 \cdot \delta g / \delta S_{rz}$$

$$(37)$$

$$[X] = \begin{bmatrix} a_1b_1 & a_1b_2 & a_1b_3 & a_1b_4 \\ a_2b_1 & a_2b_2 & a_2b_3 & a_2b_4 \\ a_3b_1 & a_3b_2 & a_3b_3 & a_3b_4 \\ a_4b_1 & a_4b_2 & a_4b_3 & a_4b_4 \end{bmatrix}$$

$$(38)$$

5 ELASTO-PLASTIC MATRIX AT A CORNER POINT

The plastic strain increment at a corner point consists of two parts, namely

$$\{de^p\} = d\lambda_1 \cdot \{\delta g_1 / \delta S\} + d\lambda_2 \cdot \{\delta g_2 / \delta S\}$$

therefore the total strain increment will be

$$\{de\} = [D]^{-1} \{dS\} + d\lambda_1 \{\delta g_1 / \delta S\}$$

$$+ d\lambda_2 \{\delta g_2 / \delta S\}$$

Since

$$df_1 = \{\delta f_1 / \delta S\}^T \{dS\}$$

$$df_2 = \{\delta f_2 / \delta S\}^T \{dS\}$$

we get from the above three equations

$$\{\delta S\} = [D] \{de\} - d\lambda_1 [D] \{\delta g_1 / \delta S\}$$

$$- d\lambda_2 [D] \{\delta g_2 / \delta S\}$$

281

$$d\lambda_1 = df_1/A_1 = 1/A_1 \{\delta f_1/\delta S\}^T \{dS\}$$

$$d\lambda_2 = df_2/A_2 = 1/A_2 \{\delta f_2/\delta S\}^T \{dS\}$$

Solving these three equations simultaneously for $d\lambda_1$, $d\lambda_2$ and $\{dS\}$, we get finally

$$\{dS\} = [D]_{ep} \{de\}$$

wherein

$$[D]_{ep} = [D] - \frac{G[X]}{B_1 B_2 - \emptyset_{f_1 g_2} \emptyset_{f_2 g_1}} \tag{39}$$

$$[X] = B_2 \left[X_{f_1 g_1}\right] + B_1 \left[X_{f_2 g_2}\right]$$
$$- \emptyset_{f_1 g_2} \left[X_{f_2 g_1}\right] - \emptyset_{f_2 g_1} \left[X_{f_1 g_2}\right] \tag{40}$$

$$B_1 = A_1/G + \emptyset_{f_1 g_1}$$

$$B_2 = A_2/G + \emptyset_{f_2 g_2} \tag{41}$$

Let

$$a_{j1} = m \cdot \delta g_j/\delta S_{xx} + n \cdot \delta g_j/\delta S_{yy} + n \cdot \delta g_j/\delta S_{zz}$$

$$a_{j2} = n \cdot \delta g_j/\delta S_{xx} + m \cdot \delta g_j/\delta S_{yy} + n \cdot \delta g_j/\delta S_{zz}$$

$$a_{j3} = n \cdot \delta g_j/\delta S_{xx} + n \cdot \delta g_j/\delta S_{yy} + m \cdot \delta g_j/\delta S_{zz}$$

$$a_{j4} = \delta g_j/\delta S_{xy}, \quad a_{j5} = \delta g_j/\delta S_{yz},$$

$$a_{j6} = \delta g_j/\delta S_{zx} \tag{42}$$

$$b_{11} = m \cdot \delta f_1/\delta S_{xx} + n \cdot \delta f_1/\delta S_{yy} + n \cdot \delta f_1/\delta S_{zz}$$

$$b_{12} = n \cdot \delta f_1/\delta S_{xx} + m \cdot \delta f_1/\delta S_{yy} + n \cdot \delta f_1/\delta S_{zz}$$

$$b_{13} = n \cdot \delta f_1/\delta S_{xx} + n \cdot \delta f_1/\delta S_{yy} + m \cdot \delta f_1/\delta S_{zz}$$

$$b_{14} = \delta f_1/\delta S_{xy}, \quad b_{15} = \delta f_1/\delta S_{yz},$$

$$b_{16} = \delta f_1/\delta S_{zx} \tag{43}$$

then we have for:

1) 3 dimensional case

$$\emptyset_{f_1 g_j} = a_{j1} \cdot \delta f_1/\delta S_{xx} + a_{j2} \cdot \delta f_1/\delta S_{yy}$$
$$+ a_{j3} \cdot \delta f_1/\delta S_{zz} + a_{j4} \cdot \delta f/\delta S_{xy}$$
$$+ a_{j5} \cdot \delta f_1/\delta S_{yz} + a_{j6} \cdot \delta f_1/\delta S_{zx}$$

$$= b_{11} \cdot \delta g_j/\delta S_{xx} + b_{12} \cdot \delta g_j/\delta S_{yy}$$
$$+ b_{13} \cdot \delta g_j/\delta S_{zz} + b_{14} \cdot \delta g_j/\delta S_{xy}$$
$$+ b_{15} \cdot \delta g_j/\delta S_{yz} + b_{16} \cdot \delta g_j/\delta S_{zx}$$

$$= \emptyset_{g_j f_1} \tag{44}$$

$$\left[X_{f_1 g_j}\right] = \begin{bmatrix} a_{j1} b_{11} & a_{j1} b_{12} & a_{j1} b_{13} \\ a_{j2} b_{11} & a_{j2} b_{12} & a_{j2} b_{13} \\ a_{j3} b_{11} & a_{j3} b_{12} & a_{j3} b_{13} \\ a_{j4} b_{11} & a_{j4} b_{12} & a_{j4} b_{13} \\ a_{j5} b_{11} & a_{j5} b_{12} & a_{j5} b_{13} \\ a_{j6} b_{11} & a_{j6} b_{12} & a_{j6} b_{13} \end{bmatrix}$$

$$\begin{bmatrix} a_{j1} b_{14} & a_{j1} b_{15} & a_{j1} b_{16} \\ a_{j2} b_{14} & a_{j2} b_{15} & a_{j2} b_{16} \\ a_{j3} b_{14} & a_{j3} b_{15} & a_{j3} b_{16} \\ a_{j4} b_{14} & a_{j4} b_{15} & a_{j4} b_{16} \\ a_{j5} b_{14} & a_{j5} b_{15} & a_{j5} b_{16} \\ a_{j6} b_{14} & a_{j6} b_{15} & a_{j6} b_{16} \end{bmatrix} \tag{45}$$

11) Plane strain case

$$\emptyset_{f_1 g_j} = a_{j1} \cdot \delta f_1/\delta S_{xx} + a_{j2} \cdot \delta f_1/\delta S_{yy}$$
$$+ a_{j3} \cdot \delta f_1/\delta S_{zz} + a_{j4} \cdot \delta f_1/\delta S_{yz}$$

$$= b_{11} \cdot \delta g_j/\delta S_{xx} + b_{12} \cdot \delta g_j/\delta S_{yy}$$
$$+ b_{13} \cdot \delta g_j/\delta S_{zz} + b_{14} \cdot \delta g_j/\delta S_{xy}$$

$$= \emptyset_{g_j f_1} \tag{46}$$

$$\left[X_{f_1 g_j}\right] = \begin{bmatrix} a_{j1} b_{11} & a_{j1} b_{12} & a_{j1} b_{14} \\ a_{j2} b_{11} & a_{j2} b_{12} & a_{j2} b_{14} \\ a_{j3} b_{11} & a_{j3} b_{12} & a_{j3} b_{14} \\ a_{j4} b_{11} & a_{j4} b_{12} & a_{j4} b_{14} \end{bmatrix} \tag{47}$$

111) Axi-symmetrical case

$$\phi_{f_i g_j} = a_{j1} \cdot \delta f_i / \delta S_{rr} + a_{j2} \cdot \delta f_i / \delta S_{oo}$$
$$+ a_{j3} \cdot \delta f_i / \delta S_{zz} + a_{j4} \cdot \delta f_i / \delta S_{rz}$$
$$= b_{11} \cdot \delta g_j / \delta S_{rr} + b_{12} \cdot \delta g_j / \delta S_{oo}$$
$$+ b_{13} \cdot \delta g_j / \delta S_{zz} + b_{14} \cdot \delta g_j / \delta S_{rz}$$
$$= \phi_{g_j f_i} \tag{48}$$

$$\left[X_{f_i g_j} \right] = \begin{bmatrix} a_{j1}b_{11} & a_{j1}b_{12} & a_{j1}b_{13} & a_{j1}b_{14} \\ a_{j2}b_{11} & a_{j2}b_{12} & a_{j2}b_{13} & a_{j2}b_{14} \\ a_{j3}b_{11} & a_{j3}b_{12} & a_{j3}b_{13} & a_{j3}b_{14} \\ a_{j4}b_{11} & a_{j4}b_{12} & a_{j4}b_{14} & a_{j4}b_{14} \end{bmatrix} \tag{49}$$

6 CONCLUSION

This paper investigates how the selection of the work hardening rule will affect the elasto-plastic matrix $[D]_{ep}$.

The general expressions of the ordinary $[D]_{ep}$ and the $[D]_{ep}$ at a corner point for 3 dimensional, plane strain and axi-symmetrical cases are developed and made ready for use in F.E.M/ analyses.

Through the discussion it has been shown that both the hardening parameter A and the yield surface function f, and thus also the elasto-plastic matrix $[D]_{ep}$, determined from a given set of experimental data, will be different if different work hardening rule is assumed.

It has also been shown that the plastic potential function g can be uniquely determined from the experimental data irrespective of the selection of the work hardening rule. Thus it may be argued that $[D]_{ep}$ can only be uniquely determined provided the selection of the work hardening rule is such, that it would lead to a yield surface function f, which is identical with the plastic potential function g. This finding agrees with the well-known theoretical conclusion that the solution of an elasto-plastic analysis will be unique when and only when an associated flow rule is assumed.

7 REFERENCES

Hill, R. 1950, Mathematical theory of plasticity. London, Oxford University Press.

Lade, P.V. 1977, Elasto-plastic stress-strain theory for cohesionless soil with curved yield surfaces. Intern. Journal of Solids & Structures, v13, No.11.

Lade, P.V. & Duncan, J.M. 1975, Elasto-plastic stress-strain theory for cohesionless soil. Journal of Geotechn. Engrg. Div., A.S.C.E., v101, GT10.

Mroz, Z. 1967, On the description of anisotropic workhardening. Journal of Mech. & Phys. of Solids, v15, p164.

Prevost, J.H. & Höeg, K. 1975a, Effective stress-strain-strength model for soils. Journal of Geotechn. Engrg. Div., A.S.C.E., v101, GT3.

Prevost, J.H. & Höeg, K. 1975b, Analysis of pressuremeter in strain softening soil. Journal of Geotechn. Engrg. Div., A.S.C.E., v101, GT8.

Roscoe, K.H. & Burland, J.B. 1968, On the generalized stress-strain behaviour of 'wet' clay. In Heyman, J. & Leckie, F.A. (ed.), Engineering plasticity, p.535-609. Cambridge, Cambridge University Press.

Zienkiewicz, O.C. 1977, The finite element method in engineering science, p.463. McGraw Hill, London.

8 APPENDIX

Fig.2 and Fig.4 of this paper are prepared by my colleague Jia-Liu Pu. Values of $H = W_p$ are calculated or determined from the test data of "Chengde" sand with e = 0.542 and $D_r = 64\%$. Description of this medium fine sand is: $D_{50} = 0.18$mm, $D_{60}/D_{10} = 2.8$, $e_{max} = 0.802$, $e_{min} = 0.396$. The H contours for H other than W_p are also determined, and they are found to be radically different from those given in Fig.2 for $H = W_p$. Detail study along this line is now still in progress.

283

A non-linear model for the elastic behaviour
of granular materials under repeated loading

H.R. BOYCE
Loughborough University of Technology, Loughborough, UK

1 SUMMARY

After a number of load repetitions along
the same stress path under drained
conditions the behaviour of cohesionless
soils and other granular materials becomes
essentially elastic. However, this elastic
behaviour differs from that commonly
encountered in being non-linear. The
stiffness increases with increasing
effective spherical pressure. A model of
this type of behaviour is developed and
the consequences of the model are explored.
The model material is entirely elastic;
that is all strain increments are
recoverable. This type of non-linear
elastic behaviour is distinct from the non-
linear relationships which apply to many
materials, including soils, in the transition from elastic to plastic behaviour.

The model is framed in terms of secant
bulk and shear moduli which are functions
of the first and second stress invariants.
The origin of the non-linearity is traced
to the particulate nature of the materials
from the theory of elastic spheres in
contact. Equations are developed for the
incremental stress-strain behaviour and
stiffness coefficients of the material,
and these equations are used to give predictions of material behaviour under
different applied stress paths in the
triaxial test. The predictions are in
reasonable agreement with the behaviour
of granular materials under repeated
loading observed by a number of workers.
In particular, values of resilient Poisson's
ratio greater than 0.5 which are often
observed are predicted by the model and are
a consequence of recoverable dilation which
occurs as the shear stress is applied. In
some tests the confining stress has been
cycled at the same time as the axial load
and the effect of this is correctly predicted.

The possibility of the stiffness of
granular materials being influenced by the
third stress invariant and the application
of the model to finite element analysis
are briefly discussed.

2 INTRODUCTION

Granular soils exhibit significant changes
in volume when subjected to shear stress.
Under undrained conditions, the associated
changes in pore pressure have a considerable effect on the stiffness and shear
strength of the soil. Rowe (1962) has
shown that this volume change, which he
called stress dilatancy, can be accounted
for by considering the energy dissipated
during frictional sliding between groups
of soil particles. Alternatively; volume
change can be deduced from the assumption
that soil is a continuum which deforms by
plastic yield (Schofield and Wroth 1968).
Although these two theories differ
considerably in their underlying concepts,
they lead to similar predictions of
irrecoverable strain in the triaxial test
and in many practical situations. Neither
theory gives any indication that soils
might exhibit dilation (that is changes of
volume caused by shear stress) in their
elastic behaviour.

In recent years the elastic behaviour of
soils has been the subject of increasing
interest in order to predict stresses
and deformations prior to failure and
under repeated loading. This paper
describes a model which represents the
elastic behaviour of granular materials
under repeated loading, and may also be
useful under other types of loading. In
particular, the conclusion that granular

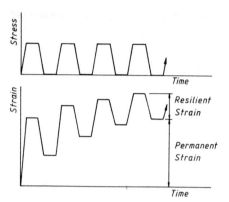

Fig. 1 Permanent and Resilient Strain due to repeated loading

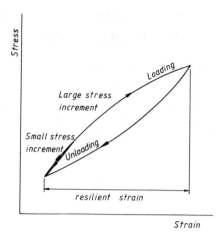

Fig. 2 Stress-strain hysteresis loop

materials exhibit dilation in their elastic behaviour is of general interest. Elastic dilation is a separate phenomenon from the irrecoverable dilation described by Rowe's stress dilatancy theory; however, the origin of both phenomena can be traced to the particulate nature of the material.

3 BEHAVIOUR OF GRANULAR MATERIALS UNDER REPEATED LOADING

The behaviour of soils and granular materials under repeated loading is complex. Each load application causes some deformation of the material which is only partially recovered on removal of the load. As loading continues, the strain recovered after each load application becomes more or less constant, and is termed resilient strain. The irrecoverable strain from each application accumulates as shown in Figure 1, and this accumulation is termed permanent strain.

When granular materials are subject to stress conditions which do no approach failure, the irrecoverable deformation diminishes with each load repetition, and eventually, under drained conditions, the behaviour becomes almost entirely resilient. For relatively small stress increments the resilient strain is more or less elastic, that is to say, the energy imparted by each load application is completely recovered on unloading. However, for the large stress increments commonly used in repeated load tests and, especially under conditions approaching failure, there is considerable hysteresis as shown in Figure 2. Hysteresis indicates that the resilient strain is not entirely elastic and that some energy is absorbed during each cycle of loading.

Also, resilient strain in granular materials is not a linear function of stress and the non-linearity is separate from the departures from elasticity described above. Under undrained conditions, both the resilient and permanent strain can be modified by changes in pore pressure.

3.1 Measurements of resilient strain

Many experimental studies of the resilient behaviour of granular material have been made in the past twenty years. These studies are reviewed in detail by Hicks (1970), Rowe (1971) and Boyce (1976). The major finding is that stiffness increases markedly with stress level.

Many of these experiments (Mitry 1964, Kallas and Riley 1967, Hicks and Monismith 1971, Allen and Thompson 1974) were tri-axial compression tests in which the confining stress, σ_3 was held constant while the deviator stress, $\sigma_1 - \sigma_3$, was repeatedly applied from zero. The results were generally expressed in the form

$$E^r = k_1 \sigma_3^{k_2} \qquad (1)$$

or

$$E^r = k_1' \theta^{k_2'} \qquad (2)$$

where θ is the sum of the principal stresses at maximum deviator stress, $3\sigma_3 + (\sigma_1 - \sigma_3)_{max}$. The resilient modulus, E^r, is defined as the repeated axial stress, σ_1^r, divided by the resilient axial strain, ε_1^r; k_1 and k_2 (or k_1' and k_2') are material constants.

In some tests (Morgan 1966, Hicks and

Monismith 1971) resilient radial strain, ε_3^r, was measured and it was found that the resilient Poisson's ratio, $\nu^r = -\varepsilon_3^r/\varepsilon_1^r$, varied with the stress ratio. Values greater than 0.5 were sometimes observed and it was suggested that this was due to anisotropy in the test specimens.

More recently, triaxial tests have been reported in which the confining stress was cycled in phase with the deviator stress (Allen and Thompson 1974, Brown and Hyde 1975). The results were analysed by using the generalised form of Hooke's Law

$$\varepsilon_1^r = \frac{1}{E^r} \sigma_1^r - 2\nu^r \sigma_3^r$$

$$\varepsilon_3^r = \frac{1}{E^r}\left[(1-\nu^r)\sigma_3^r - \nu_r\sigma_1^r\right]$$

(3)

From this form of analysis it appears that Equations 1 and 2 are still applicable provided that E^r is calculated from the mean value of σ_3. However, tests with cyclic confining stress indicate much lower values of resilient Poissons's ratio than similar tests with constant confining stress.

Boyce et al (1976) reported the results of triaxial tests to determine the resilient behaviour of a granular material under a much wider range of stress conditions than the tests above. An empirical formula for resilient strain was suggested which avoids the use of Hooke's law.

3.2 Non-linear models of soil behaviour

The main reason for the work summarised above was to provide data about the stiffness of granular materials for use in the analysis of highway pavements which are subject to repeated loading by traffic. In order to determine stress and strain in the structure, it is necessary to adopt some model of the material behaviour under general (three-dimensional) stress conditions. The most convenient model to use in a finite element analysis for granular materials is one based on Equation 2 (Duncan et al 1968). E^r is treated as a tangent modulus relating increments of stress and strain, and Poisson's ratio is assumed to be constant. A failure criterion can also be incorporated into the analysis which reduces the stiffness of elements subject to a high stress ratio (Barker 1976, Stock 1979).

Other models of soil behaviour (eg Nelson and Baron 1971) have used an incremental bulk modulus which increases with normal stress and an incremental shear modulus which decreases with increasing shear stress.

Such models are useful in analysing the stress distribution in a soil during the transition from elastic to plastic behaviour; however, the experimental evidence above does not support the use of such a model for repeated loading of granular materials.

4 THE G-K MODEL

This paper investigates a non-linear model applicable to granular materials under repeated load conditions. The model is based upon the specification of the secant bulk modulus, K and the secant shear modulus, G, as functions of stress. Strain computed from the model, which is elastic and isotropic, is in reasonable agreement with the resilient strain observed in repeated load triaxial tests, including tests with cyclic confining stress.

4.1 Fundamental equations

The model material is elastic, meaning that strain is entirely recoverable; but it is non-linear, the elastic moduli being functions of the state of stress in the material. The model is isotropic, that is the material properties are non-directional; and therefore, the elastic moduli can be specified as functions of the stress invariants. These moduli apply to any stress path, loading or unloading, and the state of strain is a unique function of the state of stress. This type of non-linear elastic behaviour should not be confused with the non-linear stress-strain relationships which occur in many materials during the transition from elastic to plastic behaviour. Hysteresis or plastic yield is excluded from the model, and if they play a significant role in the behaviour of a real material, the resulting strain must be calculated separately.

The expressions set out below (Equations 4 to 14) can be applied to any isotropic elastic material if suitable functions can be determined for the moduli. In Section 4.2 functions are chosen to represent a granular material under repeated loading (Equations 15 to 18) and the consequences of this choice are explored.

In an isotropic elastic material, states of stress and strain can be related by a secant bulk modulus, K, and a secant shear modulus, G.

$$\varepsilon_{ij} = \frac{1}{3K} \delta_{ij} p + \frac{1}{2G} S_{ij}$$

(4)

where ε_{ij} is a component of the strain tensor, δ_{ij} is the Kronecker delta, p is the normal stress (mean spherical pressure) and S_{ij} is a component of the deviator stress tensor $(\sigma_{ij} - \delta_{ij}p)$.

For a linear elastic material G and K are constants and, therefore, increments of stress and strain (denoted by a dot over the symbol) can be related by a similar expression.

$$\dot{\varepsilon}_{ij} = \frac{1}{3K_t}\delta_{ij}\dot{p} + \frac{1}{2G_t}\dot{S}_{ij} \qquad (5)$$

where the tangent bulk modulus, K_t, and the tangent shear modulus, G_t, are equal to the respective secant moduli.

For a non-linear elastic material the expression relating increments of stress and strain is more complex.

$$\dot{\varepsilon}_{ij} = \frac{1}{3K}\delta_{ij}\dot{p} - \frac{\dot{K}}{3K^2}\delta_{ij}p + \frac{1}{2G}\dot{S}_{ij} - \frac{\dot{G}}{2G^2}S_{ij} \qquad (6)$$

where $\dot{K}$ and $\dot{G}$ are changes in the moduli caused by the increment of stress. For an isotropic material

$$\dot{K} = (\frac{\partial K}{\partial I_1})\dot{I}_1 + (\frac{\partial K}{\partial I_2})\dot{I}_2 + (\frac{\partial K}{\partial I_3})\dot{I}_3$$

$$\dot{G} = (\frac{\partial G}{\partial I_1})\dot{I}_1 + (\frac{\partial G}{\partial I_2})\dot{I}_2 + (\frac{\partial G}{\partial I_3})\dot{I}_3 \qquad (7)$$

where I_1, I_2 and I_3 are the stress invariants.

For granular material under repeated loading, there is at present no evidence of any influence of I_3 on the stiffness of the material. Therefore, the remainder of the paper has been simplified by the assumption that G and K are functions of the first two stress invariants only. The invariants used are the normal stress, p, and deviator stress, q, which are defined as
$p = (\sigma_1 + \sigma_2 + \sigma_3)$

$$q = \frac{1}{\sqrt{2}}\left[(\sigma_1-\sigma_2)^2+(\sigma_2-\sigma_3)^2+(\sigma_3-\sigma_1)^2\right]^{0.5} \qquad (8)$$

where σ_1, σ_2 and σ_3 are the principal stresses. The corresponding strain invariants are the volumetric strain, ε_v, and the shear strain, ε_s, which are defined as

$\varepsilon_v = \varepsilon_1 + \varepsilon_2 + \varepsilon_3$

$$\varepsilon_s = \frac{\sqrt{2}}{3}\left[(\varepsilon_1-\varepsilon_2)^2+(\varepsilon_2-\varepsilon_3)^2+(\varepsilon_3-\varepsilon_1)^2\right]^{0.5} \qquad (9)$$

where $\varepsilon_1, \varepsilon_2$ and ε_3 are the principal strains.

In terms of these invariants, an elastic isotropic material is characterised by the equations

$$\varepsilon_v = \frac{1}{K}p$$

$$\varepsilon_s = \frac{1}{3G}q \qquad (10)$$

Increments of stress and strain are related by the equations

$$\dot{\varepsilon}_v = \frac{1}{K}\dot{p} - \frac{\dot{K}}{K^2}p$$

$$\dot{\varepsilon}_s = \frac{1}{3G}\dot{q} - \frac{\dot{G}}{3G^2}q \qquad (11)$$

By using Equation 7, this expression can be rewritten as

$$\dot{\varepsilon}_v = \left[\frac{1}{K} - \frac{p}{K^2}\frac{\partial K}{\partial p}\right]\dot{p} - \frac{p}{K^2}\frac{\partial K}{\partial q}\dot{q}$$

$$\dot{\varepsilon}_s = \left[\frac{1}{3G} - \frac{q}{3G^2}\frac{\partial G}{\partial q}\right]\dot{q} - \frac{q}{3G^2}\frac{\partial G}{\partial p}\dot{p} \qquad (12)$$

It can be seen that the expressions for the strain increments, ε_v and ε_s, each consist of two parts indicating a separate contribution from $\dot{p}$ and from $\dot{q}$ in both cases. Generally, an increment of normal stress, $\dot{p}$, gives rise to shear strain, ε_s, as well as volumetric strain, $\dot{\varepsilon}_v$; and an increment of deviator stress, $\dot{q}$, gives rise to volumetric strain, $\dot{\varepsilon}_v$ as well as shear strain $\dot{\varepsilon}_s$.

In these circumstances, there is a theorem of reciprocity which can be applied to any elastic, linear or non-linear. This theorem, proof of which is given in the Appendix, states that the following relationship must be satisfied at all states of stress:

$$\frac{\partial \varepsilon_v}{\partial q} = \frac{\partial \varepsilon_s}{\partial p} \qquad (13)$$

Therefore at all states of stress,

$$\frac{p}{K^2}\frac{\partial K}{\partial q} = \frac{q}{3G^2}\frac{\partial G}{\partial p} \qquad (14)$$

In effect the theorem of reciprocity limits the choice of functions for G and K which can be used to represent a non-linear elastic material. If K is a function of p alone and G is a function of q alone there is no difficulty, but otherwise care must be taken to ensure that Equation 14 is satisfied.

4.2 A G-K model for granular material

From the experimental work summarised in Section 3.1 it is clear that the stiffness of a granular material is proportional to the normal stress raised to a power less than one. There are also theoretical reasons for choosing this type relationship (see Section 5). Hence, in a G-K model, the stiffness might be expressed by the functions:

$$K = K_1 \, p^{(1-n)}$$
$$G = G_1 \, p^{(1-n)} \qquad (15)$$

It will be observed that these expressions do not conform to the theorem of reciprocity (Equation 14). Experimental evidence, in particular Brown and Hyde (1975) and Boyce (1976), indicates that the shear modulus does increase with the normal stress as indicated by Equation 15. It therefore follows that if the material is elastic, the bulk modulus is a function of the deviator stress.

Amending Equation 15 to conform with the theorem of reciprocity gives

$$K = K_1 p^{(1-n)} / \, (1 - \beta \frac{q^2}{p^2})$$
$$G = G_1 p^{(1-n)} \qquad (16)$$

where, $\beta = (1-n) K_1 / 6 G_1$

This indicates an increase in the bulk modulus at higher values of stress ratio. There are no observations of such an increase reported in the literature, but this is not altogether surprising when one considers the type of tests which have been performed. In the majority of repeated load tests the deviator stress was cycled from zero, and therefore the hysteresis occuring when high values of repeated deviator stress were applied would mask any increase in elastic modulus. However, some of the tests performed by Boyce (1976) did include measurements of resilient strain for relatively small changes of stress at different values of q/p. A detailed analysis (Pappin and Brown, 1980) indicates that there is indeed an increase in bulk modulus at higher values of q/p in the manner indicated by Equation 16.

For the G-K model, stress and strain invariants are related by the expressions:

$$\varepsilon_v = \frac{1}{K_1} p^n \, (1 - \beta \frac{q^2}{p^2})$$
$$\varepsilon_s = \frac{1}{3G_1} p^n \frac{q}{p} \qquad (17)$$

and increments of the stress and strain invariants by the expressions:

$$\dot{\varepsilon}_v = \frac{1}{K_t} \dot{p} - D_t \, \dot{q}$$
$$\dot{\varepsilon}_s = \frac{1}{3G_t} \dot{q} - D_t \, \dot{p} \qquad (18)$$

where,
$$K_t = \frac{K_1}{n} p^{(1-n)} / \, (1 + \beta \frac{2-n}{n} \frac{q^2}{p^2})$$
$$G_t = G_1 \, p^{(1-n)}$$

$$D_t = 2\beta \frac{q}{p} / K_1 p^{(1-n)}$$

It can be seen that an increase in normal stress causes a decrease in shear strain, and that an increase in deviator stress causes a decrease in volumetric strain. That is to say, the model material, which is isotropic and elastic, exhibits dilation. The size of the dilation, indicated by the coefficient, D_t, varies with the stress ratio and this explains the high values of resilient Poisson's ratio observed in repeated load triaxial tests (see Section 6.2).

5   ORIGIN OF NON-LINEARITY IN GRANULAR MATERIALS

There are three processes which may contribute to strain in a granular material:

1. Particle deformation around points of contact
2. Slip between particles at points of contact
3. Crushing of particles at points of contact

When considering elastic strain, crushing may be eliminated as an irreversible process. Slip between particles may also be eliminated, because even if it is reversible, some energy will be dissipated by friction. This leaves particle deformation as the process which determines elastic strain. During cyclic or repeated loading slip between particles may also occur, and some of this slip may be reversed during the return part of each load cycle. Hence the resilient strain observed during repeated loading may be greater than the strain calculated from a purely elastic model.

If elastic strain is assumed to be due to particle deformation, this strain can be calculated in the case of a regular array of spheres in contact. According to the classical theory of Hertz (Timoshenko and Goodier 1970) the relative displacement, Y of two spherical elastic particles in contact under a normal force, P, is given by:

$$Y = Y_1 \, P^{2/3} \qquad (19)$$

The stiffness of the contact is given by:

$$\frac{dP}{dY} = 1.5 \, P^{1/3} / Y_1 \qquad (20)$$

where Y_1 is a function of the radius and elastic properties of the spheres. Duffy and Mindlin (1957) used this expression to derive the stiffness of a face-centred

289

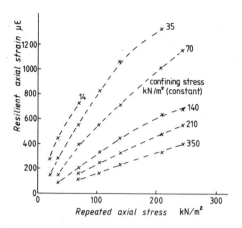

Fig. 3 Resilient axial strain measured
under repeated loading (after
Hicks and Monismith, 1971)

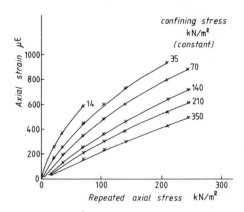

Fig.4 Axial strain computed from the
G-K Model

cubic array of spheres and found that each
component, C_{ij}, of the stiffness matrix
relating increments of stress and strain
is given by an expression of the form:

$$C_{ij} = G_{ij} p^{(1-n)} \qquad (21)$$

where p is the normal stress, the coefficient, n, is equal to $\frac{2}{3}$ and the constants,
G_{ij}, are dependent on Y_1 and the orientation of spheres in the array.

They attempted to confirm this expression
by measuring the stiffness of an assembly
of steel ball bearings using a resonance
method. They found that the theoretical
expression was valid for high tolerance
steel balls ($\bar{+}$ 0.25 µm on a diameter of
3.2 mm) in a 1,0,0 orientation except that
the stiffness was about 15% less than
predicted. However, for low tolerance
balls ($\bar{+}$ 1.25 µm) in this orientation and
for high tolerance balls in a 1,1,0
orientation, the coefficient, n, was about
0.5. This was explained by the fact that
the relative approach of the balls under
the highest pressures used (100 kN/m²) was
much less than the tolerance of the low
tolerance balls. Hence, as the pressure
was increased the number of contacts
increased, and the stiffness of the array
increased more rapidly than if the number
of contacts had remained constant.

For granular materials, which are irregular assemblies of particles, similar
relationships have been found between
stiffness and stress level in resonant
vibration tests and the coefficient, n,
is observed to be in the range 0.2 to 0.6
(Hardin and Black 1966, Robinson 1974).
It is of interest to note that Equation 19

is very similar to the initial expressions
used in developing the G-K model (Equation
15). After modifying the expressions to
satisfy the theorem of reciprocity it was
shown that the model material exhibits
dilation. It appears that this elastic
dilation is a consequence of a Hertzian
type of stiffness relationship at particle
contacts, whereas irrecoverable dilation
is a consequence of interparticle friction
(Rowe 1962).

6 COMPARISON OF THE MODEL WITH EXPERIMENTAL
OBSERVATIONS OF RESILIENT STRAIN

6.1 Resilient modulus

The main feature of the resilient behaviour
of granular material under repeated loading
is the increase in resilient modulus with
stress level (see Equations 1 and 2). In
the model, expressions for the secant
moduli (Equation 16) and the tangent
moduli (Equation 18) indicate a similar
increase in stiffness with stress level.
The extent to which the stiffness varies
with normal stress is determined by the
coefficient, 1-n.

As an example, Figure 3 shows some
published data (Hicks and Monismith 1971)
of the resilient axial strain measured in
a repeated load triaxial test on a
specimen of partially crushed aggregate.
The value of K_2 (see Equation 1)
determined from this test was 0.60 indicating a value of n in the model of 0.40.
The value of resilient Poisson's ratio
determined indicates a ratio of K_1 to G_1
of 0.45. Figure 4 shows the axial strain
under the same stress conditions computed
from the G-K model with these parameters.

290

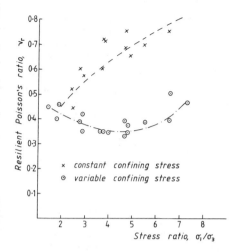

Fig. 5 Resilient Poisson's ratio
        calculated from repeated loading
        tests (after Allen and Thompson,
        1974)

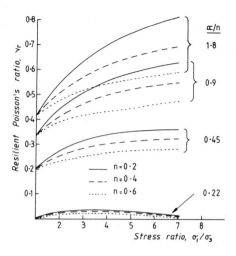

Fig. 6 Resilient Poisson's ratio
        computed from the G-K model for
        constant confining stress

Agreement is within 20% except for those
readings in which a very large resilient
strain was recorded (greater than 800 $\mu\varepsilon$).
For these readings, one would expect the
measured strain to exceed that calculated
from an elastic model because of the
hysteresis caused by reversible slip
between particles.

6.2 Resilient Poisson's Ratio

It is not strictly valid to use the term
'Poisson's ratio' to describe the
behaviour of a non-linear material,
especially a granular material which
exhibits dilation. However, the term has
been used in the analysis of repeated load
triaxial tests to access the amount of
radial strain measured. With a repeated
axial stress, σ_1^r, accompanied by a variable
confining stress, $\sigma_3^r = \sigma_3^{max} - \sigma_3^{min}$, values
of resilient Poisson's ratio have been
calculated using the expression

$$\nu_r = \frac{\sigma_1^r \varepsilon_3^r - \sigma_3^r \varepsilon_1^r}{2\sigma_3^r \varepsilon_3^r - \varepsilon_1^r (\sigma_1^r + \sigma_3^r)} \qquad (22)$$

Figure 5 shows the values calculated by
Allen and Thompson (1974) from repeated
load triaxial tests on a well graded,
partially saturated gravel (using total
stresses).

The value of Poisson's ratio in a test
with constant confining stress, σ_3,
computed from the G-K model is dependent
on the stress ratio, σ_1^r/σ_3. Figure 6

shows how this computed Poisson's ratio
varies with the stress ratio for different
values of n and of α/n, where $\alpha = K_1/3G_1$.
It can be seen that α/n determines the
value of ν^r under pure normal stress, and
that n determines how much ν^r increases
above this value as the stress ratio is
increased. Note that the resilient
Poisson's ratio can be greater than 0.5
even though the model specification is
isotropic. This is because of the
recoverable dilation which occurs as the
deviator stress is increased.

The computed values of ν^r for n = 0.2
and α/n = 1.8 are within the experimental
scatter of those reported by Allen and
Thompson. It is apparent that the high
values of resilient radial strain observed
in repeated load triaxial tests can be
explained by the non-linear character of
the material which also leads to elastic
dilation. However, some anisotropy as
a result of specimen preparation or
previous loading cannot be ruled out.

Consider now a repeated load triaxial
test in which the confining stress is
cycled in phase with the axial load. The
value of ν^r computed from the model is
lower than for a test with constant
confining stress. Figure 7 shows the
resilient Poisson's ratio computed for
n=0.2 and α=0.36 with different amounts
of cyclic confining stress. The highest
curve is for constant confining stress
$(\sigma_3^{min}/\sigma_3^{max} = 1.0)$ and the lowest curve,
which cannot be achieved in practice, is
for a minimum confining stress of zero
$(\sigma_3^{min}/\sigma_3^{max} = 0.0)$. The minimum deviator

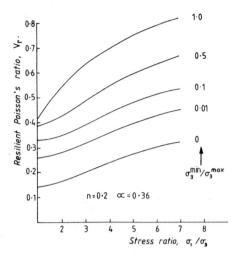

Fig. 7 Resilient Poisson's ratio
computed from the G-K model for
tests with variable confining
stress

stress was taken as zero throughout. These
computed values of v^r are in reasonable
agreement with the values derived by Allen
and Thompson from their experimental data
(see Figure 5). Detailed comparison is
not possible, because the minimum confining
stress applied in their tests is not given.

For all values of n and α, cyclic
confining stress gave a reduction in the
computed values of v^r, and this is in
agreement with the experimental data of
Allen and Thompson (1974) and Brown and
Hyde (1975). In some cases for values of
α/n less than 0.5 negative values of
resilient Poisson's ratio were computed
with cyclic confining stress. This
emphasises again that the concept of
Poisson's ratio has little meaning for a
non-linear material.

6.3 Triaxial extension tests

All the data given above is for triaxial
tests in which a repeated deviator stress
was applied in compression. Some tests
have also been carried out on a granular
material (a well graded, crushed limestone
aggregate) with additional stress paths in
triaxial extension. (Boyce et al 1976)
The results of these tests are also in
reasonable agreement with the model (Boyce
1978) but space does not permit a detailed
comparison to be given here.

The primary features of the G-K model are
the stress dependent stiffness of the
material and the consequent phenomenon of
elastic dilation. The possibilities of a

model which is anisotropic or one in which
the stiffness is also dependent on the third
stress invariant have not been investigated.
In theory these effects could be studied
by comparison of triaxial tests in extension
and compression (Thrower 1978). However,
in the triaxial compression tests described
above, anisotropy on any effect of I_3
could not be distinguished from the predominant effect of normal stress on stiffness.

7   USE OF THE G-K MODEL IN FINITE ELEMENT ANALYSIS

The principal application envisaged for the
model is in the finite element analysis of
pavement structures incorporating a layer
of granular material, or in other situations involving repeated loading.

In any finite element analysis involving
non-linear material characteristics, it
is necessary to use an iterative or incremental procedure to ensure that the stiffness
of every element is consistent with the
stresses within it. In an iterative
analysis, there is no undue difficulty
and the stiffness coefficients can be
calculated from the secant moduli given
by Equation 16. However, an incremental
procedure is often preferred if there is
an initial stress in the structure (due to
the overburden, perhaps) or it is desired
to incorporate a failure criterion to modify
elements which become overstressed (Stock
1979). In this case the tangent moduli
given by Equation 18 are required. In
this expression relating increments of
stress and strain, there is a dilation
coefficient, D_t, in addition to the usual
moduli, K_t and G_t. Facilities need to be
available within the finite element
program to incorporate the effect of
elastic dilation on the element stiffness
coefficients.

8   CONCLUSIONS

1. The resilient behaviour of granular
materials under repeated loading is in
reasonable agreement with a non-linear
elastic model in which the stiffness is
proportional to the normal stress raised
to a power less than one.
2. The non-linearity gives rise to
recoverable dilation under repeated
loading ie recoverable volumetric strain
occurs under repeated deviator stress.
3. This type of non-linear elastic
behaviour arises from particle deformation
around points of contact.

ACKNOWLEDGEMENTS

The author is grateful for many useful
discussions with colleagues at Loughborough
University of Technology and with former
colleagues at the University of Nottingham,
and especially for the encouragement of
Professor P S Pell.

REFERENCES

Allen, J. J. and M. R. Thompson 1974,
    Resilient response of granular materials
    subjected to time-dependent lateral
    stresses, Transportation Research Record
    510: 1-13.
Barker, W. R. 1976, Elasto-plastic analysis
    of a typical flexible airport pavement,
    unpublished report, U.S. Army Engineer
    Waterways Experiment Station, PO Box 631,
    Vicksburg, Miss. 39180, U.S.A.
Boyce, J. R. 1976, The behaviour of a
    granular material under repeated loading.
    Ph.D. Thesis, University of Nottingham.
Boyce, J. R. 1978, A non-linear elastic
    model for granular materials, unpublished
    report, Loughborough University of
    Technology.
Boyce, J. R., S. F. Brown and P. S. Pell
    1976, The resilient behaviour of a
    granular material under repeated
    loading, Proc. Australian Road Research
    Board 8: 8-19.
Brown, S. F. and A. F. L. Hyde 1975, The
    significance of cyclic confining stress
    in repeated load triaxial testing of
    granular material, Transportation
    Research Record 537: 49-58.
Duffy, J. and R. D. Mindlin 1957, Stress-
    strain relationships of a granular medium,
    Journal of Applied Mechanics 24: 585-593.
Duncan, J. M., C. L. Monismith and E. L.
    Wilson 1968, Finite element analysis of
    pavements, Highway Research Record 228:
    18-32.
Hardin, B. O. and W. L. Black 1966, Sand
    stiffness under various triaxial stresses,
    Jour. Soil Mech. and Found. Div., Proc.
    ASCE.
Hicks, R. G. 1970, Factors influencing the
    resilient properties of granular materials,
    Ph.D. Thesis, University of California,
    U.S.A.
Hicks, R.G. and C. L. Monismith 1971,
    Factors influencing the resilient response
    of granular materials, Highway Research
    Record 345: 15-31.
Kallas, B. F. and J. C. Riley 1967,
    Mechanical properties of asphalt paving
    materials, Proc. 2nd Int. Conf. Struct.
    Des. of Asphalt Pavements, University of
    Michigan, Ann Arbor, U.S.A.
Mitry, B. F. G. 1964, Determination of the
    modulus of resilient deformation of
    untreated base course materials, Ph.D.
    Thesis, University of California, U.S.A.
Morgan, J. R. 1966, The response of
    granular materials to repeated loading,
    Proc. Australian Road Research Board:
    1178-1191.
Nelson, I and M.L.Baron 1971, Application
    of variable moduli models to soil
    behaviour, Int.J.Solids Structures,
    7 : 399-417.
Pappin, J. W. and S. F. Brown 1980,
    Resilient stress-strain behaviour of a
    crushed rock, Proc. Int. Sym. Soils
    under Cyclic and Transient Loading, Swansea.
Robinson, R. G. 1974, Measurement of the
    elastic properties of granular materials
    using a resonance method, TRRL Supplemen-
    tary Report 111UC.
Rowe, P. W. 1962, The stress-dilatancy
    relation for static equilibrium of an
    assembly of particles in contact, Proc.
    Royal Society A269: 500-527.
Rowe, P. W. 1971. Theoretical meaning and
    observed values of deformation parameters
    for soil, Roscoe Memorial Symposium,
    Cambridge University.
Schofield, A. N. and Wroth, C. P. 1968,
    Critical state soil mechanics, London,
    McGraw-Hill.
Stock, A. F. 1979, Flexible Pavement Design,
    Ph.D. Thesis, University of Nottingham.
Thrower, E. N. 1978, Stress invariants and
    mechanical testing of pavement materials,
    TRRL Report 810.
Timoshenko, S.P. and J. N. Goodier 1970,
    Theory of Elasticity, McGraw-Hill, 3rd Ed.:
    409-414.

APPENDIX

Theorem of reciprocity applied to non-
    linear elastic materials

Consider unit volume of elastic material
subjected to the stress path shown in
Fig. 8, forming a closed loop beginning and
ending at the origin. During this process
the material is strained, and since the
material is elastic the strain at any point
on the path is uniquely determined by the
stress at that point. The material
arrives back at zero stress in the same
state as it left with zero strain and the
net energy absorbed during this cycle of
stress is zero. (If there was any net
absorption of energy, energy could be
continually released by cycling the
material in the opposite direction).
    Consider now an increment of this stress
path $\dot{p}$, $\dot{q}$ which is accompanied by an
increment of strain $\dot{\varepsilon}_v$, $\dot{\varepsilon}_s$. The energy
absorbed by the material during this
increment, $\dot{W}$ is given by:

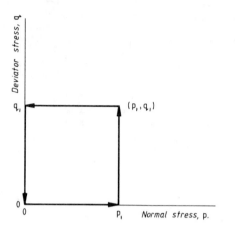

Fig. 8   Stress path forming a closed
         loop

$$\dot{W} = p\dot{\varepsilon}_v + q\dot{\varepsilon}_s \qquad (23)$$

It follows from the above that for an elastic material, linear or non-linear,

$$\oint \dot{W} = 0 \qquad (24)$$

Expanding equation (23) for a non-linear material gives:

$$\dot{W} = (p\cdot\frac{\partial\varepsilon_v}{\partial p} + \frac{\partial\varepsilon_s}{\partial p})\,\dot{p} + (p\cdot\frac{\partial\varepsilon_v}{\partial q} + q\frac{\partial\varepsilon_s}{\partial q})\dot{q} \qquad (25)$$

and using equation (24) for the stress path shown in Figure (8) gives

$$\int_0^{p_1} (p\frac{\partial\varepsilon_v}{\partial p}+q\frac{\partial\varepsilon_s}{\partial p})_{q=0}\,dp + \int_0^{q_1} (p\frac{\partial\varepsilon_v}{\partial q}+\frac{\partial\varepsilon_s}{\partial q})_{p=p_1}\,dq +$$

$$\int_{p_1}^{0} (p\frac{\partial\varepsilon_v}{\partial p}+q\frac{\partial\varepsilon_s}{\partial p})_{q=q_1}\,dp \quad \int_{q_1}^{0} (p\frac{\partial\varepsilon_v}{\partial q}+q\frac{\partial\varepsilon_s}{\partial q})_{p=0}\,dq$$

$$\qquad (26)$$

$$= 0$$

Reorganising the equation leads to,

$$\int_0^{p_1} \int_0^{q_1} \frac{\partial}{\partial q}(p\frac{\partial\varepsilon_v}{\partial p}+\frac{\partial\varepsilon_s}{\partial p})\,dq\,dp\;-$$

$$\qquad (27)$$

$$\int_0^{q_1} \int_0^{p_1} \frac{\partial}{\partial p}(p\frac{\partial\varepsilon_v}{\partial q}+q\frac{\partial\varepsilon_s}{\partial q})\,dp\,dq = 0$$

and since the equation applies for any value of p_1 and q_1,

$$\frac{\partial}{\partial q}(p\frac{\partial\varepsilon_v}{\partial p}+q\frac{\partial\varepsilon_s}{\partial p}) = \frac{\partial}{\partial p}(p\frac{\partial\varepsilon_v}{\partial q}+q\frac{\partial\varepsilon_s}{\partial q}) \qquad (28)$$

Evaluating the partial differentials gives,

$$p\frac{\partial^2\varepsilon_v}{\partial p\partial q} + \frac{\partial^2\varepsilon_s}{\partial p\partial q} + (\frac{\partial\varepsilon_s}{\partial p})_{q\;constant} =$$

$$p\frac{\partial\varepsilon_v}{\partial q\partial p} + \frac{\partial^2\varepsilon_s}{\partial q\partial p} + (\frac{\partial\varepsilon_v}{\partial q})_{p\;constant} \qquad (29)$$

which simplifies to

$$\frac{\partial\varepsilon_v}{\partial q} = \frac{\partial\varepsilon_s}{\partial p} \qquad (13\ bis)$$

For many non-linear elastic materials, the shear strain (and shear modulus) is independent of the normal stress ($\partial\varepsilon_s/\partial p = 0$) and the volumetric strain (and bulk modulus) is independent of the deviator stress ($\partial\varepsilon_v/\partial q = 0$); and therefore equation (13) becomes trivial. However, for non-linear materials which exhibit recoverable dilation, equation (13) requires that the functions of p and q which determine the moduli are partially related. If equation (13) is not satisfied, the energy stored by the material in some state of stress is dependent on the stress path taken to reach that stress. Such stress path dependence indicates some form of inelastic behaviour.

294

A dynamic pore pressure parameter A_n

S.K. SARMA & D.N. JENNINGS
Imperial College of Science and Technology, London, UK

ABSTRACT

A dynamic pore pressure parameter, A_n is proposed. The parameter A_n is investigated to determine whether or not there is some relatively simple mathematical expression which will define its numerical value. A study of the limited data available in the literature has established that the square root of A_n is linearly related to the logarithm of the number of the loading cycle.

The relatively simple expression will greatly reduce the problems associated with incremental pore pressure problems. An example is provided to show the use of the expression to determine the excess pore pressure at the end of a non uniform stress cycle history.

Although the fit with published data is excellent, this representation will benefit from a more rigorous investigation.

INTRODUCTION

The progressive development of pore pressure in cohesionless soils during cyclic loading has been of considerable interest to investigators for many years. Progressive development of pore pressure is the basis of the phenomenon more generally known as liquefaction which has been responsible for many catastrophic ground failures in past earthquakes. The problem of predicting the pore pressures developed during a regular or irregular cyclic loading sequence has proved difficult. Many complex expressions involving several parameters have been proposed in the past which have enjoyed some success in modelling the pore pressure generated in cyclic loading laboratory tests, Martin Finn and Seed (1975).

CONCEPT OF A DYNAMIC PORE PRESSURE PARAMETER

The concept of a dynamic pore pressure parameter, A_n, was proposed by Sarma (1976). He suggested that it was reasonable to define the parameter as follows
    a) for simple shear conditions
$$u = A_n \Delta \zeta$$
or
    b) for the triaxial test conditions
$$u = A_n \Delta \sigma_1 / 2$$

where u is the pore pressure after N cycles

   $\Delta \zeta$ is the applied shear stress

   $\Delta \sigma_1$ is the applied deviator stress

Sarma plotted the date from Martin et al (1975) in the form of u/σ_{vo} against $\Delta \zeta / \sigma'_{vo}$ for various numbers of cycles as shown in figure 1. From this plot, the parameter A_n can readily be determined for any given number of cycles.

Data from several other sources, Seed et al (1973), De Alba et al (1975), Lee & Albasia (1974), Lee & Focht (1975), Silver & Park (1976) was also studied and the plots all produced a similar pattern to that of figure 1. Two of these are shown in figures 2 and 3.

On the basis of these data, the corresponding A_n parameters were determined. These were then investigated in an attempt to establish whether or not the A_n parameters could be conveniently represented by some relatively simple mathematical expression. In all the cases studied, a straight line relationship was obtained when the square root of A_n was plotted against the logarithm of the number of cycles. Thus the parameter A_n obeys the expression

$$\sqrt{A_n} = \sqrt{A_1} + \beta \log N$$

where A_1 is the parameter corresponding to

the first cycle of loading and β is a parameter which is constant for a given material and state of testing, and N is the number of cycles.

It appears that both the parameters A_1 and β are independent of the magnitude of the applied shear stress or the deviatoric stress so long as the pore pressures are such that the Mohr's circle in terms of the effective stresses does not touch the Mohr envelope. Beyond this stage in the laboratory tests, the pore pressures rise very rapidly. However, it is doubtful whether this part of the test does represent the field behaviour of pore pressures.

The A_1 parameters appear to have little in common with the more familiar A parameter established by Skempton (1954) for monotonic loading. However, this is to be expected since A parameters generally represent those applicable at failure whereas the A_1 parameters do not necessarily represent the failure state. Note also that negative values of the A_1 parameter cannot be accommodated. The A_1 parameters measured in the triaxial tests and in simple shear tests seem to be very different.

The expression developed above can be used to determine the pore pressure response through a varied loading pattern. The technique is based on the assumption that the pore pressure response to a cycle of loading depends on the amount of excess pore pressure already present and also on the level of shear stress applied. The procedure is as follows:

1) For the first increment of loading $\Delta\zeta_1$ the pore pressure is $u = A_1 \Delta\zeta_1$

2) for the second increment of loading $\Delta\zeta_2$ determine the equivalent number of cycles of uniform stress history of $\Delta\zeta_2$ which would have been responsible for developing u. Thus we obtain

$$A_{ne} = u/\Delta\zeta_2$$

and therefore $N_e = \log^{-1}\left|(\sqrt{A_{ne}} - \sqrt{A_1})/\beta\right|$

3) The present cycle of $\Delta\zeta_2$ is therefore equivalent to $(N_e + 1)$ cycle of uniform stress history of $\Delta\zeta_2$. Therefore

$$A_{(N_e + 1)} = \left[\sqrt{A_1} + \beta\log(N_e + 1)\right]^2$$

and therefore

$$u = A_{(ne + 1)} \cdot \Delta\zeta_2$$

4) The steps 2 and 3 can be repeated for subsequent increments of shear stresses. IN the present study the increment of shear stress used is the peak value of the cycle.

Unfortunately, there are few instances in the literature where the data is suffi-

ciently detailed to enable the method to be checked. However, two cases are presented in figures 4 and 5 where the present method of prediction is compared with those predicted by Martin et al (1975). The parameters A_1 and β are those determined from their data. It appears that using a constant value of A_1 and β for all stress levels, the two predictions compare reasonably well up to a pore pressure ratio u/σ_{vo} of around 60%. The fit of data can be improved by using a variable A_1 parameter varying with the applied shear stress. However, it can be shown that with a pore pressure ratio of around 60% the sample will fail in the sense that the Mohr's circle of stress will touch the failure envelope in terms of effective stresses.

CONCLUSION

The authors believe that the concept of the dynamic pore pressure parameter is reasonable and the parameter can be generally defined by a simple expression. The relatively crude data used to establish this relationship is acknowledged and the need for a rigorous investigation with good quality data recommended.

The use of the dynamic pore pressure parameter, in the form presented, is recognised to have significant potential for use in the field of dynamic analysis because of its inherent simplicity. In attempting to apply this concept, one should not lose sight of the problems inherent in attempting to relate laboratory behaviour to field performance.

REFERENCES

De Alba, P., C.K. Chan, H.B. Seed (1975), "Determination of Soil liquefaction characteristics by large scale laboratory tests". E.E.R.C. Report No. 75-14, College of Engineering, University of California, Berkeley.

Lee, K.L., A.Albasia (1974) "Earthquake induced settlements in saturated sands" Proc. A.S.C.E. vol. 100, GT 4, pp 387-406

Lee, K.L. and Focht (1975) "Liquefaction potential at Ekifisk Tank in North Sea" Proc. A.S.C.E. vol. 101, No. GT1, pp 1-18

Martin, G.R., W.D.L. Finn, H.B. Seed (1975) "Fundamentals of liquefaction under cyclic loading". Proc. A.S.C.E. vol. 101, GT 5, pp 423 - 438.

Sarma, S.K. (1976) "Earth dam studies - Report on response and stability of earth dams during strong earthquakes". Report prepared for the U.S. Army Corps of

Engineers, Vicksburg, Mississippi, U.S.A.
Seed, H.B., K.L. Lee, I.M. Idriss, F.
Makdisi, (1973) "Analysis of the slides
in the San Fernando Dams during the
earthquake of February 9, 1971". E.E.R.C.
Report No. 73-2, College of Engineering,

University of California, Berkeley, USA
Silver, M.L., T.K.Park (1976) Liquefaction
potential evaluated from cyclic strain-
controlled properties tests on sands".
Soils and Foundations, vol. 16, No. 3,
pp 51 - 65.

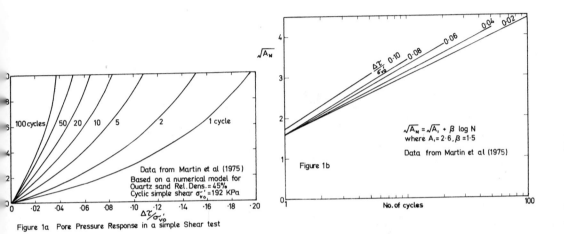

Figure 1a   Pore Pressure Response in a simple Shear test

Figure 1b

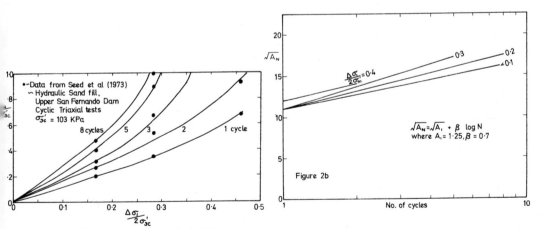

Figure 2a   Pore Pressure response in a cyclic triaxial test

Figure 2b

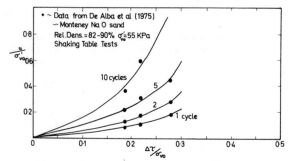

Figure 3a Pore Pressure Response in a shaking table test

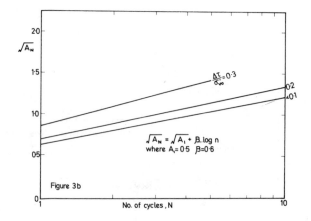

$$\sqrt{A_N} = \sqrt{A_1} + \beta \log n$$
where $A_1 = 0.5$ $\beta = 0.6$

Figure 3b

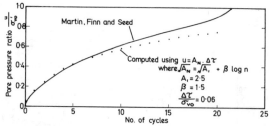

Martin, Finn and Seed

Computed using $u = A_N \cdot \Delta \tau$
where $\sqrt{A_N} = \sqrt{A_1} + \beta \log n$
$A_1 = 2.5$
$\beta = 1.5$
$\frac{\Delta \tau}{\sigma_{vo}} = 0.06$

Figure 4 - Comparison of pore pressure behaviour

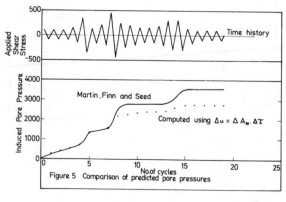

Time history

Martin, Finn and Seed

Computed using $\Delta u = \Delta A_N \cdot \Delta \tau$

Figure 5 Comparison of predicted pore pressures

Modeling and analysis of cyclic behavior of sands

JAMSHID GHABOUSSI & HASSAN MOMEN
University of Illinois at Urbana-Champaign, Urbana, Ill., USA

ABSTRACT

The intent of this study is to develop a
material model which can be used in finite
element analysis of transient behavior of
soil structures under dynamic loading.
For such an analysis to be successful, the
material model must accurately represent
the following aspects of the soil behavior.
  (a) Hysteretic energy dissipation
  (b) Accumulated irreversible deforma-
      tions
  (c) Volumetric deformations
The accurate representation of these three
aspects of the material behavior is possibly the key to successful dynamic analysis of the geotechnical systems. Material
hysteresis determines the amount of the
effective damping in the system. Energy
dissipation through hysteresis is probably
the major portion of the actual damping in
soils. The accumulated irreversible
deformations provide a measure of damage.
Under cyclic stresses the accumulated
deformations may reach magnitudes comparable to those resulting from failure under
static loading condition. Even in absence
of outright failure under cyclic stress
conditions, the accumulated irreversible
deformation may amount to unacceptable
levels of damage. Importance of accurate
modeling of volumetric deformations is in
relation to pore pressure generation and
possibility of liquefaction in masses of
saturated cohesionless soils. The material model proposed in this paper will be
evaluated and compared to experimental
results with an emphasis on the above
mentioned three aspect of soil behavior
under cyclic stress condition.
  The material model presented here is for
completely drained behavior of sand and is
formulated in terms of effective stress.
Thus, the proposed model will not directly

provide any information regarding the pore
pressures. Such an effective stress material model is intended for use in a two-
phase model of saturated sands in which the
volumetric strains of the sand structure
and the pore fluid are coupled. In such
an analysis the migration of pore fluid
and redistribution of pore pressures during
the analysis are taken into account. However, in completely undrained condition,
when uniform pore pressure throughout the
soil can be assumed, as in cyclic triaxial
tests, it is possible to compute the pore
pressures directly from the proposed model.
Several undrained triaxial tests have been
simulated in this paper, assuming that the
compressibility of the pore fluid can be
neglected in comparison with the compressibility of the soil structure. Thus, the
total volumetric strain is zero and any
potential volumetric strain will result in
changes in the effective stress. The proposed material model consists of a failure
surface and a yield surface and the laws
governing their behavior. Failure is
assumed to be an asymptotic state corresponding to very large strains. The failure surface has a conical shape in the
principal stress space, whose cross-section
on the octahedral plane is a rounded corner triangle. The yield surface has a
shape similar to the failure surface and
undergoes a combination of isotropic hardening and kinematic hardening. The general
form of the yield surface is given in the
following equation.

$$f\left(\sigma, \alpha, \kappa\right) = 0 \qquad (1)$$

in which $\alpha = \{\alpha_{ij}\}$ is a unit vector along
the axis of the yield surface and κ is a
scalar. In Equation 1 α and κ are the
kinematic and isotropic hardening parameters, respectively. The expression for

yield surface is as follows.

$$f = \bar{S}_{ij} \bar{S}_{ij} - R^2 \kappa^2 \bar{I}^2 = 0 \qquad (2)$$

in which

$$\bar{S}_{ij} = \sigma_{ij} - \alpha_{ij} \bar{I} , \qquad (3)$$

$$\bar{I} = \alpha_{ij} \alpha_{ij} , \qquad (4)$$

and R is function of stresses which defines the shape of cross-section of the yield surface on a plane normal to the α vector.

The isotropic and the kinematic hardening are defined through following two expressions for the rate of change of the hardening parameters κ and α.

$$d\kappa = g(\xi) \, d\xi \qquad (5)$$

$$d\alpha_{ij} = c \, (\sigma, \xi) \, \bar{S}_{ij} / (\bar{S}_{kl} \bar{S}_{kl})^{1/2} \qquad (6)$$

in which

$$d\xi = (de_{ij}^p \, de_{ij}^p)^{1/2} \; ; \; \xi = \int d\xi, \qquad (7)$$

and de_{ij}^p is the increment of the plastic deviatoric strains. The isotropic hardening function, g, is a monotonically decreasing function and is independent of stresses. This implies that the isotropic hardening is the same for all the points on the yield surface. However, the kinematic hardening function, c, is not uniform over the yield surface and in general depends on the stresses. The main parameter in the kinematic hardening function is the "degree of stress reversal", which makes the current value of the function c dependent on the stresses at the immediate past point of stress reversal. The most general form of the function c is given in the paper and its specific shape for a typical cyclic triaxial test is presented.

A non-associated flow rule is used to determine the plastic strains. The deviatoric plastic strains are determined from a flow rule using a potential function which has a cylindrical shape but its cross-section on the octahedral plane is the same as the yield surface. The volumetric plastic strains are determined from a semi-empirical relation for the plastic work.

In the development of this material model special care has been taken in order to minimize the number of the material parameters. The physical significance of each material parameter has been studied. Most of the material parameters used can

directly be measured from routine tests. It is recognized that by introducting more material parameters it is possible to reproduce test results more closely. However, the measurement of additional material parameters for fine tuning the response of the material model becomes exceedingly difficult.

The proposed material model is used to simulate several drained and undrained cyclic triaxial tests. Overall, the simulated results compare reasonably well with the test results. The accuracy in various aspects of the material model is discussed in the paper. In general, the proposed material model is capable of representing the hysteretic energy dissipation and the volumetric strains better than the accumulated plastic strains. However, it is recognized that the accumulated plastic strains are probably one of the most difficult aspects of the material behavior to simulate.

A "geotechnical" stress variables approach
to cyclic behaviour of soils

ROBERTO NOVA
Technical University (Politecnico), Milan, Italy

TOMASZ HUECKEL
Institute of Fundamental Technological Research, IPPT-PAN, Warsaw, Poland

1 INTRODUCTION

Till now two main ways have been developed to tackle with the mathematical description of the behaviour of solids, and
soils in particular, under cyclic loading.
The first one, to which belong the models
by Mròz (1967), Phillips (1972), Dafalias
and Popov (1975, 1976), Prévost (1977),
Mròz, Norris and Zienkiewicz (1978, 1979)
is essentially based on the concept of kinematic hardening plasticity and will be
called hereinafter kinematic hardening
approach. The second, to which belong the
approaches by Ramberg-Osgood (1943), Dobry (1970), Hardin and Drnevich (1972a,
1972b), Greenstreet and Phillips (1973),
Mròz and Lind (1975), Boisserie and Guelin
(1977), Idriss, Dobry and Singh (1978),
Hueckel and Nova (1979a, 1979b) is based
on a piecewise path independent description.
In the former any hysteretic effect due to
unloading and reloading is considered to
be consequence of the way in which plastic
deformations develop. Usually, this has
been accomplished by introducing within
the traditional yield surface for monotonic (virgin) loads (sometimes named bounding surface or consolidation surface) a
set of internal yield surfaces of varying
dimensions and/or position following numerous different hardening rules. As a
consequence the constitutive relations
are formulated in terms of rates as in the
plasticity theory. In this way the implied
path dependence of the behaviour is taken
into account.
The latter approach is that employed in
the archetypes of modelling of hysteresis in
which only uniaxial cyclic loading was considered. The central idea was that of repro-
ducing the unloading curve from the loading
one (the skeleton curve) by means of a
simple scaling procedure. In that context

total strains, not strain rates, were considered without any distinction between
their reversible and irreversible parts.
Path independence was therefore implicitly
assumed between subsequent stress reversals.
Different generalizations to multiaxial
cases have been more recently suggested
either by applying to "radial" paths the
scaling procedure with suitable modifications, or by employing such a procedure in
terms of stress and strain intensity invariants for more complex loading.
In this line a different approach may be
conceived by separating the elastoplastic
behaviour under monotonic loading and the
behaviour under unloading-reloading (within
the yield surface). In fact, a unique skeleton curve is hardly conceivable for soils
subjected to complex multidimensional loading and moreover unloading-reloading behaviour bears often little similarity to
virgin loading. It appears then reasonable
to treat the virgin loading by means of a
traditional elastic plastic hardening-softening model and unloading-reloading within
the yield surface by means of a piecewise
path independent model as in Hueckel and
Nova (1979b). A model that succesfully
copes with monotonic-not necessarily ra-
dial-loading of soils is that of Cam Clay
(Schofield and Wroth (1968)). It has been
shown by the authors (Nova and Hueckel
(1979)) that a combination of the aforementioned models for virgin loading and
unloading-reloading is possible and that
fairly good agreement between predicted
and experimental results may be achieved
in several tests involving cyclic loading
both within and at the yield surface.
In this paper a mathematical description
of the memory of the material will be given.
In this framework the stress-strain constitutive law will be written in terms of
"geotechnical" stress variables, using which

it is possible to take account of the logarithmic dependence of volumetric strain
on the effective isotropic pressure,
the pressure sensitivity of the material
behaviour in shear and finally of shear
compaction in cyclic loading. These variables are a generalization to multiaxial
conditions of those employed in the Cam
Clay type models, i.e. the logarithm of
the isotropic effective stress and the
stress ratio.

In the following the term stress will
mean effective stress in the usual sense
of Soil Mechanics if not otherwise specified. The theory presented is restricted
to small strains and no viscous effect is
taken into account. Stresses and strains
are taken positive in compression.

2 "GEOTECHNICAL" STRESS VARIABLES

In order to deal with one of the main pe-
culiarities of soil behaviour under shear
stress, i.e. its dependence on the mean
normal effective stress (since in this
note only effective stresses will be considered the usual dash denoting effective
stresses will be omitted for the sake of
simplicity of notation)
$p = \frac{1}{3} \sigma_{ii}$, a set of generalized stress

variables will be introduced.

The dependence on p is manifested in
several aspects. The shear stress at failure is a function of p as well as the
current yield locus. Moreover, at variance with the behaviour of non-frictional
materials, the shear strain rates do depend
on the mean effective stress rate. In addition, experimental evidence shows, e.g.
Wroth (1965), that for a given stress path
in a standard triaxial test, a unique relation can be established between the deviatoric strain and the ratio of the deviatoric stress and p , rather than with the
deviatoric stress itself. Finally, there
exists a non-linear relation between p
and volumetric strain in isotropic tests.
It is commonly accepted that the loading
stress-strain relationship can be well
approximated by a logarithmic function.

Let us define as generalized stress
variables the logarithmic measure of the
isotropic effective pressure

$$\ln p/p_a \equiv \ln \frac{\sigma_{ii}}{3 p_a} \qquad (2.1)$$

where p_a is the atmospheric pressure,
and the tensor of stress ratios

$$\eta_{ij} = \frac{s_{ij}}{p} \qquad (2.2)$$

where s_{ij} is deviator of the stress tensor
defined as usual as

$$s_{ij} = \sigma_{ij} - p \delta_{ij} \qquad (2.3)$$

Note that η_{ij} is a deviatoric stress
tensor normalized with respect to the
current isotropic pressure and thus enjoys
all properties of a deviator.

It is evident from their very definition
that the defined stress variables are a
generalization to the multidimensional case
of the two dimensional stress representation in the triaxial test plane, i.e.
$\ln p/p_a$ and the stress ratio $\eta \equiv q/p$,
where $q \equiv \sigma_1 - \sigma_3$. These latter variables
have been introduced for the Cam Clay
model and are now commonly used in Soil
Mechanics. Thus the quantities defined in
(2.1), (2.2) will be called hereinafter
"geotechnical" stress variables.

It will be shown in what follows that the
highly non linear behaviour of soils both
in the elastoplastic range and within the
yield surface may be described with reasonable accuracy by constitutive laws formulated still as linear relations between
strain rates and rates of "geotechnical"
stress variables.

For further purposes let us denote the
square root of the second invariant of the
η_{ij} tensor as

$$I_{2\eta} = (\eta_{kl} \eta_{kl})^{1/2} \qquad (2.4)$$

It is possible to note that in case of traditional triaxial compression

$$\sqrt{\frac{3}{2}} I_{2\eta} \equiv \eta \qquad (2.5)$$

3 THE MODELLING OF ELASTOPLASTIC BEHAVIOUR

The soil is considered as virgin at the
first application of a load, which means
that it exhibits some irrecoverable deformation from the very beginning of the loading process. Therefore no elastic domain
is supposed to exist for a virgin soil.
However, the yield locus expands homotheti-
cally from the origin of axes in the stress
space (isotropic hardening) as long as
plastic deformations occur. The equation
of the yield locus will be defined as
follows:

$$f = f_1 \equiv \frac{3}{2} a \, I_{2\eta}^2 + 1 - (p_c/p)^2 = 0 \qquad (3.1)$$

$$\text{if } \sqrt{\frac{3}{2}} \, I_{2\eta} \leq \frac{M}{2}$$

$$f = f_2 \equiv \sqrt{\frac{3}{2}} \, I_{2\eta} - M + m \ln p/p_u = 0 \qquad (3.2)$$

$$\text{if } \sqrt{\frac{3}{2}} \, I_{2\eta} \geq \frac{M}{2}$$

where M is determined by the value of the stress ratio at the critical state in triaxial compression, p_c is the isotropic consolidation pressure, p_u is given by

$$p_u = \frac{p_c}{\sqrt{1 + \mu}} \, \exp\left(- \frac{M}{2m} \right) \qquad (3.3)$$

where μ and m are material constants and a is given by

$$a = \frac{4 \mu}{M^2} \qquad (3.4)$$

Eqs.(3.1), (3.2) are supposed to correspond to two different mechanisms of shear. A representation of the yield locus in the triaxial test plane is given in Fig.1 .

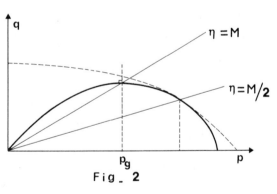

Fig 1

The surfaces defined by the equations (3.1), (3.2) are rotational surfaces generated by the curve of Fig.1 in the spirit of the Drucker-Prager (1952) extension of the Coulomb Criterion. Abundant experimental justification for this choice of the yield locus form in triaxial compression is given in the paper by Nova and Wood (1979).

In the same paper it is demostrated that

the flow rule for $\sqrt{\frac{3}{2}} \, I_{2\eta} \leq \frac{M}{2}$ may be treated as associated whilst for larger values of $I_{2\eta}$ the plastic potential is different from f_2 , i.e. the flow rule is non-associated. Thus in general

$$\dot{\varepsilon}_{ij}^p = \Lambda^+ \frac{\partial g}{\partial \sigma_{ij}} \qquad (3.5)$$

where Λ^+ is called plastic multiplier and

$$g \equiv f_1 \quad \text{if } \sqrt{\frac{3}{2}} \, I_{2\eta} \leq M/2 \qquad (3.6)$$

and

$$g = g_2 \equiv \sqrt{\frac{3}{2}} \, I_{2\eta} - \frac{M}{1 - \mu} \left\{1 - \mu\left(\frac{p}{p_g}\right)^{\frac{1-\mu}{\mu}}\right\} =$$

$$= 0 \quad \text{if } \sqrt{\frac{3}{2}} \, I_{2\eta} \geq \frac{M}{2} \qquad (3.7)$$

where p_g is linked to p_c in the following way:

$$p_g = \frac{p_c}{\sqrt{1 + \mu}} \left(\frac{2 \mu}{1 + \mu}\right)^{\frac{\mu}{1 - \mu}} \qquad (3.8)$$

A representation of the plastic potential in the triaxial test plane is given in Fig.2. Again the surfaces given by equations (3.6), (3.7) are rotational surfaces whose generatrix is the curve shown in Fig.2 .

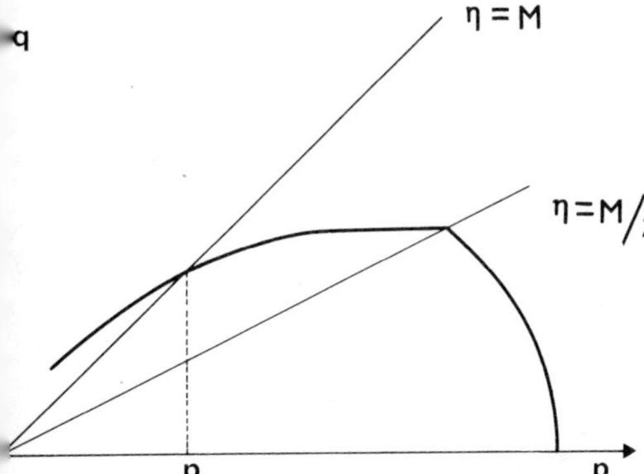

Fig. 2

The preconsolidation pressure p_c is assumed to be a function of the first invariant of plastic strains $I_1^p \equiv \varepsilon_{ii}^p$ and of the square root of the second invariant of the deviator of plastic strains

$$I_2^p = (\frac{2}{3} e_{ij}^p e_{ij}^p)^{1/2}$$

where

$$e_{ij}^p = \varepsilon_{ij}^p - 1/3 \ I_1^p \ \delta_{ij}$$

in the following way (see Nova (1977))

$$p_c = \exp (\frac{I_1^p + D \ I_2^p}{\lambda - B_o}) \qquad (3.9)$$

The constant appearing in (3.9) may be determined in an anisotropic consolidation test and in a $\eta = M$ constant test. The hardening moduli in these tests are identified with $\lambda - B_o$ and $(\lambda - B_o)/D$ respectively in similar plots. B_o is the virgin elastic bulk compliance modulus as specified at the end of this Section.

The plastic multiplier Λ^+ can be determined through the consistency equation so that

$$\Lambda^+ = \frac{1}{H} \frac{\partial f}{\partial \sigma_{ij}} \ \dot{\sigma}_{ij} \qquad (3.10)$$

where H is the hardening modulus given by

$$H = - \frac{\partial f}{\partial p_c} \{ \frac{\partial p_c}{\partial I_1^p} \frac{\partial I_1^p}{\partial \varepsilon_{ij}^p} +$$

$$+ \frac{\partial p_c}{\partial I_2^p} \frac{\partial I_2^p}{\partial \varepsilon_{ij}^p} \} \frac{\partial g}{\partial \sigma_{ij}} \qquad (3.11)$$

It is possible to note from Eq.(3.11) that the sign of the hardening modulus does not depend only on the first term within parenthesis which is connected with the dilatancy of the material, as in Cam Clay. As a consequence the hardening modulus may be positive even if this term is negative, i.e. if the material is dilating, what has been widely confirmed by experimental evidence on sand, see also Nova (1977), Nova and Wood (1979).

Taking account of Eqs.(3.1)-(3.11) the explicit expressions of Λ^+ read

$$\Lambda^+ = \frac{(\lambda - B_o)p \{ \frac{3}{2} a \ n_{ij} \ \dot{n}_{ij} +}{2 (\frac{3}{2} a \ I_{2\eta}^2 + 1)(1 + a \ D \frac{e_{ij}}{I_2^p} n_{ij})}$$

$$\frac{+ (\frac{3}{2} I_{2\eta}^2 a + 1) \ \dot{p} /p \}}{} \qquad (3.12)$$

if $\sqrt{\frac{3}{2}} I_{2\eta} \leq M/2$

$$\Lambda^+ = \frac{(\lambda - B_o)p \{ \sqrt{\frac{3}{2}} \frac{n_{ij}}{I_2 \eta} \dot{n}_{ij} +}{m \{ \sqrt{\frac{2}{3}} \ D \frac{e_{ij}}{I_2^p} \frac{n_{ij}}{I_{2\eta}} +}$$

$$\frac{+ m \ \dot{p}/p \}}{+ \frac{M}{\mu} - \frac{1}{\mu} \sqrt{3/2} \ I_{2\eta} \}} \qquad (3.13)$$

if $\sqrt{3/2} \ I_{2\eta} \geq \frac{M}{2}$

and from Eqs.(3.5), (3.6), (3.7)

$$\dot{\varepsilon}_{ij}^p = \frac{(\lambda - B_o) \{ \frac{9}{2} a \ n_{ij} \ n_{kl} \ \dot{n}_{kl} +}{2 (3/2 \ a \ I_{2\eta}^2 + 1)(1 + a \ D \frac{e_{ij}}{I_2^p} n_{ij})}$$

$$\frac{+ (1 + \frac{3}{2} a \ I_{2\eta}^2)(3a \ n_{ij} + \delta_{ij}) \dot{p}/p \}}{}$$

$$\qquad (3.14)$$

if $\sqrt{\frac{3}{2}} \ I_{2\eta} \leq \frac{M}{2}$

$$\dot{\varepsilon}_{ij}^p = \frac{(\lambda - B_o) \{ \frac{3}{2 \ I_{2\eta}^2} n_{ij} \ n_{kl} \ \dot{n}_{kl} +}{m \{ \sqrt{\frac{2}{3}} \ D \frac{e_{ij}^p}{I_2^p} \frac{n_{ij}}{I_{2\eta}} +}$$

$$+ m[\sqrt{3/2} \frac{n_{ij}}{I_{2\eta}} + \frac{1}{3\mu}$$

$$+ \frac{1}{\mu} (M - \sqrt{\frac{3}{2}} I_{2\eta}) \}$$

$$\frac{(M - \sqrt{\frac{3}{2}} I_{2\eta}) \ \delta_{ij} \ \dot{p}/p \}}{} \qquad (3.15)$$

if $\sqrt{3/2} \ I_{2\eta} \geq \frac{M}{2}$

It is possible to see from the above equations that the plastic strain rates as formulated within the framework of the usual linear plasticity theory are linearly dependent on the rates of the "geotechnical" stress variables.

Eqs.(3.14), (3.15) hold both for clay and sand, provided that the constant D is set to zero in the former case. For sand D depends on its initial density and increases with it. Eq.(3.15) can model either hardening or softening behaviour for sufficiently high values of $I_{2\eta}$. The softening process, modelled by setting D to zero for further strains, begins when $\sqrt{3/2}\, I_{2\eta}$ reaches the value

$$\sqrt{\frac{3}{2}}\, I_{2\eta} = M + \frac{B_o}{\lambda}\, \mu\, D \qquad (3.16)$$

as can be inferred from the paper by Nova and Wood (1979). Overconsolidated clays can be treated in a way similar to Cam Clay as shown by Nova (1979).

Along the processes at the yield surface total strain rates are given by

$$\dot{\varepsilon}_{ij} = \dot{\varepsilon}^e_{ij} + \dot{\varepsilon}^p_{ij} \qquad (3.17)$$

The elastic component is assumed to be given by a linear relation in terms of the rates of "geotechnical" stress variables

$$\dot{\varepsilon}^e_{ij} = L_o\, \dot{\eta}_{ij} + \frac{1}{3}\, B_o\, \dot{p}/p\, \delta_{ij} \qquad (3.18)$$

L_o will be called virgin elastic shear modulus
Eqs. (3.12), (3.13) hold at the yield surface, for

$$f = 0 \quad , \quad \dot{f} = 0 \qquad (3.19)$$

Eqs. (3.17), (3.18) are valid with $\Lambda^+ = 0$, at the yield surface, for

$$f = 0 \quad , \quad \dot{f} < 0 \qquad (3.20)$$

For processes internal to the yield surface, i.e. $f < 0$, the behaviour is described by the relations presented in Sec.4.

4 THE PARAELASTIC STRESS-STRAIN RELATION

Within the yield locus the behaviour is not considered as elastic but instead "hysteretic" or "paraelastic" as defined by the authors (Hueckel and Nova (1979a), (1979b)). The term paraelastic is justified by the fact that the material behaviour is treated as piecewise path independent between subsequent suitably defined

stress reversal points. This implies that the stress strain law is formulated in terms of differences of variables relative to the current state of the material and those relative to the last stress reversal point. The relation between paraelastic strain differences and differences of "geotechnical" stress variables, assumed to be tensorally linear is postulated as follows

$$\Delta^M \varepsilon_{ij} = \frac{1}{3}\, B\, (\ln \frac{p}{p^M} + \phi\, I_{2\eta}\,)\, \delta_{ij} +$$
$$+ L\, \Delta^M \eta_{ij} \qquad (4.1)$$

where M denotes the last, Mth stress reversal point and the symbol $\Delta^M x$ means

$$\Delta^M x \equiv x - x^M$$

ϕ is a positive material constant referred to as shear compaction parameters. B and L, the current bulk and shear compliance moduli, respectively, are assumed to be linear functions of the "strain amplitude parameter" χ^M defined as

$$\chi^M = (\Delta^M \varepsilon_{ij}\, \Delta^M \varepsilon_{ij})^{1/2} \qquad (4.2)$$

in the following way

$$\begin{cases} B = B^M\, (1 + \omega\, \chi^M) \\ L = L^M\, (1 + \omega_\varepsilon\, \chi^M) \end{cases} \qquad (4.3)$$

The terms ω_v and ω_ε are material constants whilst B^M and L^M are the values of B and L at the onset of the Mth stress path portion after the Mth stress reversal point. It is worth noting that the intrinsic non-linearity of any individual loop branch is taken into account by the dependence of B and L on the scalar parameter χ^M. The isotropic effects are coupled with the deviatoric ones by the introduction of the term $I_{2\eta}$ in Eq.(4.1). Thus volumetric strains will occur even in a p constant test and conversely the material will undergo an effective pressure change in constant volume test (undrained test if the material is saturated).

One of the fundamental features of the constitutive equation (4.1) is that on closed radial cycles ($\eta_{ij} = $ const) the deviatoric strains are recoverable, provided that B^{M+1}, L^{M+1} are equal to B^M, L^M. Moreover in these type of cycles the

305

stress-strain loop, in the plane of stress ratio and deviatoric strain invariants, exhibits a polar symmetry with respect to its center. The volumetric strain, however, is not restored after the cycle completion and, since the coupling term in (4.1) is insensitive to the load direction, it may be accumulated under repeated loading. The volumetric strain is completely recoverable only in a purely isotropic cycle, provided that again $B^{M+1} = B^M$ and $L^{M+1} = L^M$. The stress-strain loop in terms of ε_{ij} and of the introduced "geotechnical" variable $\ln p/p_a$ enjoys in this case the polar symmetry property.

In fact, experimental evidence shows, as discussed by the authors (Nova and Hueckel (1979)), that the polar symmetry does not hold for stress-strain loops in terms of the usual stress and strain invariants, but that an acceptable idealization of this type is possible when experimental results are plotted in terms of strain invariants and invariants of "geotechnical" stress variables.

The notion of stress reversal employed until now is selfevident only in case of radial cycles. In generic multiaxial conditions, however, it is not a priori clear what is the meaning of the terms stress reversal, unloading and reloading and thus a criterion for stress reversal must be postulated. The criterion that will be adopted in this paper is based on the concept of loading function. Define first the "geotechnical" stress tensor z_{ij} and the tensor of its differences $\Delta^M z_{ij}$ in the above specified sense

$$z_{ij} = \frac{1}{3} \ln \frac{p}{p_a} \delta_{ij} + \eta_{ij} \qquad (4.4)$$

$$\Delta^M z_{ij} = \frac{1}{3} \ln \frac{p}{p^M} \delta_{ij} + \Delta^M \eta_{ij} \qquad (4.5)$$

In the "geotechnical" stress space a locus $\overline{W}^M (\Delta^M z_{ij}) = 0$ passing through the current stress point is conceived to delimit between stress rates which give rise to the continuation of the current law or to the stress reversal. The loading function is defined as the scalar product of the gradient vector of the hypersurface $\overline{W}^M = 0$ and the stress rate vector:

$$\mathcal{L} (\Delta^M z_{ij} , \dot{z}_{ij}) = \frac{\partial \overline{W}^M}{\partial \Delta^M z_{ij}} \dot{z}_{ij} \qquad (4.6)$$

For $\mathcal{L} \geq 0$, i.e. for the stress rate vectors directed outward the locus, the process continues to follow the law

relative to the Mth stress reversal point, while for $\mathcal{L} < 0$, i.e. for inward oriented stress rate vectors, the M + 1th stress reversal point occurs. The continuation condition will be associated with the condition that the strain amplitude parameter can not decrease $\dot{\chi} \geq 0$, that means that the local compliance moduli B and L are always increasing for an individual branch of the stress path. The stress rate that would violate this condition is inadmissible within the current law and is then supposed to give rise to a stress reversal. The equation of the current stress reversal locus is consequently

$$\overline{W}^M = B^2 \ln^2 p/p^M + (\phi^2 B^2 + L^2)I_{2\eta}^2 +$$

$$+ 2 \phi B^2 \ln p/p^M I_{2\eta} - \chi^{M^2} = 0 \qquad (4.7)$$

A dual formulation of the loading function in the strain space may be conceived by introducing a corresponding locus

$$\overline{V}^M = \Delta^M \varepsilon_{ij} \Delta^M \varepsilon_{ij} - \chi^{M^2} = 0 \qquad (4.8)$$

which is a sphere.

The locus for which a stress reversal occurs is called dead locus and is stored in the memory of the material. Investigations of simple uniaxial cycles allow to infer that the material enjoys a discrete memory organised in two levels in a hierarchic way. The first level, active, governs the current constitutive law, the second remembers all the dead loci in stack. If a stress path, say z_{ij}^M , z_{ij}^{M+1} , z_{ij}^k (see Fig.3) touches and crosses one of these loci, say $\overline{W}^M (z_{ij}^{M+1} - z_{ij}^M) \equiv \overline{W}^{M,M+1} = 0$ the memory is updated. All the younger dead loci and the current locus, say $\overline{W}^{M+1}$, are forgotten, while the current origin is shifted back to the point which was the origin for the crossed dead locus, i.e. z_{ij}^M . This locus is no more a dead one but it is reactivated together with the corresponding stress-strain law. However, for the sake of continuity, at the reactivation point the strain difference $\Delta^M \varepsilon_{ij}$ must be complemented by an amount of residual strains which may arise along the stress path within the locus. The above rules may be formalized in the following way.

Let $\hat{z}_{ij}$ denote the stress at the last, Mth stress reversal $\hat{z}_{ij} = z_{ij}^M$, i.e. the origin to which refer the current law, and z_{ij} denote the current stress point. For a given $\dot{z}_{ij}$ the origin for further loading may remain unaltered or may shift to z_{ij} giving rise to a stress reversal,

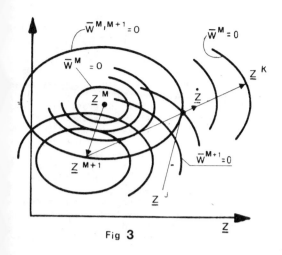

Fig 3

$$\text{i.e.} \quad \hat{z}_{ij} = z^{M+1}_{ij} = z_{ij} \quad . \text{ Thus}$$

$$\hat{z}_{ij} = \sum_{k}^{M} \left[z^{k}_{ij} + (z_{ij} - z^{k}_{ij}) \right.$$

$$\left. \psi\{\mathcal{L}^{k}\} \right] \xi^{k} Z^{k} \tag{4.9}$$

where

$$\psi\{ y \} = 1 \quad \text{if} \quad y < 0;$$
$$\psi\{ y \} = 0 \quad \text{if} \quad y \geq 0 \tag{4.10}$$

while $\mathcal{L}^{k}$ is the joint loading function

$$\mathcal{L}^{k} = \left(\frac{\partial \ \bar{w}^{k,k+1}}{\partial \ \Delta k_{z_{ij}}} \right) \dot{z}_{ij} \ , \ \text{if} \ 1 \leq k \leq M-1 \ ;$$

$$\mathcal{L}^{k} = \frac{\partial \ \bar{w}^{k}}{\partial \ \Delta k_{z_{ij}}} \ \dot{z}_{ij} \ , \quad \text{if} \ k = M \tag{4.11}$$

The function ξ_{k} is non-zero when the k-th locus is reached, in such a way

$$\xi^{k} = 1 - \psi\{ \bar{w}^{k,k+1} \} \quad \text{if} \quad 1 \leq k \leq M-1 \ ;$$
$$\xi^{k} = 1 - \psi\{ \bar{w}^{k} \} \quad \text{if} \quad k = M \tag{4.12}$$

The function Z^{k} represents the effect of forgetting stress reversal points

$$Z^{k} = \prod_{n=1}^{k-1} (1 - \xi_{n}) \tag{4.13}$$

where Π is the multiplication symbol over the values $n = 1, ..., k - 1$.

A detailed discussion of these formulae is given in Hueckel and Nova (1979b).

The number of the dead stress reversal loci stored in the material memory is theoretically not limited. In numerical applications, however, this number is limited both by the successive forgetting of some loci and by the restrictive form of most practical stress paths.

The plastic yield locus acts as the oldest dead locus, i.e. the hierarchically most important. If the stress path reaches the plastic yield locus all the hysteresis loci are then cancelled from the memory of the soil.

Finally, let us discuss the evolution rule for moduli B^{M+1} and L^{M+1} which, as already mentioned, are the initial compliance moduli after the (M+1)th stress reversal. It is assumed that each stress reversal provokes an updating of these quantities depending on a measure of the direction change of the stress path at the reversal. In fact, as indicated by numerous experimental data in radial paths the compliance moduli are fully restored to the virgin elastic value B_o , L_o . On the other hand, if the stress rate vector at a generic point is directed along the surface of the current stress reversal locus the moduli B and L are given by Eqs. (4.3). However, any stress rate $\dot{z}^{k}_{ij}$ along the surface which delineates the Mth and the (M+1)th branches of the stress path, i.e. for $\mathcal{L}^{M}(^{\Delta M}z_{ij}, \dot{z}_{ij}) = 0$, should give rise to a unique strain rate response regardless which law is applied, i.e.

$$\dot{\varepsilon}_{ij} = C_{ijkl} \ \dot{z}_{kl} = 1/3B (\chi) \ \delta_{ij} \ \dot{p}/p +$$

$$+ \left[\frac{1/3\phi \ B(\chi) \ n_{kl}}{I_{2\eta}} \ \delta_{ij} + L \ \delta_{ik} \ \delta_{lj} \right] \dot{n}_{kl} \tag{4.14}$$

An analogous condition is known in the theory of plasticity as the condition of continuity along neutral paths, see also Mróz (1979). In this paper, an updating rule is adopted such that the full restoration of compliance moduli on radial paths and the continuity on neutral paths is ensured, see Hueckel (1979).

Let z'_{ij} denote the normalized stress tensor defined as

$$z'_{ij} = \frac{^{\Delta M}z_{ij}}{\sqrt{^{\Delta M}z_{ij} \ ^{\Delta M}z_{ij}}} \tag{4.15}$$

307

It can be conceived as a unit vector in the stress space directed in the opposite sense of vector $^{\Delta M}z_{ij}$. Define the tensor z''_{ij} as

$$z''_{ij} = (\dot{z}^R_{ij} + b\, z'_{ij}) \left[(\dot{z}^R_{kl} + b\, z'_{kl}) \cdot (\dot{z}^R_{kl} + b\, z'_{kl}) \right]^{-1/2} \quad (4.16)$$

where

$$b = - \int (\dot{z}^R_{kl}) \left[\frac{\partial\, \overline{W}}{\partial\, {}^{\Delta M}z'_{ij}}\; z'_{ij} \right]^{-1} \quad (4.17)$$

The tensor z''_{ij} may be represented in the stress space as a vector complanar with z'_{ij} and $\dot{z}^R_{ij}$ and belonging to the plane tangential to the locus $\overline{W}$ at $^{\Delta M}z'_{ij}$.

Let c^M_{ijkl} denote the tensor of compliance moduli at the onset of the Mth stress path portion. A simple, linear scalar interpolation rule is adopted between the tensors c^M_{ijkl} and c_{ijkl} in terms of angles ϕ' and ϕ'' between vectors $\dot{z}^R_{ij}$ and z'_{ij} and between z''_{ij} and z^R_{ij} , respectively. Fig.4 .

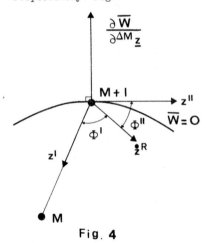

Fig. 4

Thus

$$c^{M+1}_{ijkl} = c^M_{ijkl} \left(1 - \frac{\phi'}{\phi' + \phi''}\right) + $$

$$ + c_{ijkl} \left(1 - \frac{\phi''}{\phi' + \phi''}\right) \quad (4.18)$$

where

$$\phi^{(i)} = \cos^{-1}\left[\dot{z}^R_{ij}\, z^{(i)}_{ij} (\dot{z}^R_{kl}\, \dot{z}^R_{kl})^{-1/2}\right] (4.19)$$

The compliance tensor just after the stress reversal is thus stress rate sensitive. Note however, that for the whole further portion of the stress path prior to the subsequent stress reversal the stress rate tensor $\dot{z}^R_{ij}$ remains constant. Therefore the rate dependence is limited to the stress reversal point only. This fact may be interpreted as a degenerate stress difference sensitivity at corners of the stress path where $\dot{z}^R_{ij}$ may be considered as $\dot{z}^R_{ij} = \lim (z_{ij} - z^M_{ij})$.

In the case of a loading path controlled by kinematic conditions, e.g. oedometric and undrained tests, because of the nonlinearity of aforementioned rule, an iterative procedure should be applied to determine c^{M+1}_{ijkl} .

Note finally that an analogous interpolation rule has to be applied at the reactivation point of a dead locus. In this case the vectors z'_{ij} and z_{ij} are directed towards the oldest stress reversal point and along the relative locus respectively.

The model presented enjoys some similarities and several differences with the kinematic hardening approach. The most striking difference is perhaps the use in the former approach of the "geotechnical" stress variables what leads to qualitative and quantitative differences in predictions obtained with the two kind of models. Their intrinsic structure remains anyway comparable.

Note first that plasticity, as traditionally understood, is treated in both the approaches in the same way. In the kinematic hardening approach the rule of the plastic yield surface is attributed to the "consolidation" or "bounding" surface, that usually is assumed to undergo isotropic hardening (or softening) with an associated flow rule. The peculiarity of the presented approach is the possibility to take account of the dilatancy phenomenon both in the hardening and the softening range, as experimentally observed to occur for sands. This is a consequence of the assumed dependence of the hardening function on two parameters. The critical state stress ratio does no more separate hardening and softening ranges and the critical state can be reached from above after a hardening-softening transition.

Moreover, the behaviour within the yield surface is understood in a similar way in the two approaches. Namely, in both cases a "microplastic", i.e. irreversible strain, is allowed for together with an elastic one. In the presented theory no domain of perfectly elastic behaviour exists, although, through the assumption of polar symmetry of the strain-geotechnical stress curve, a reversibility of deviatoric strains is ensured in radial paths. In the present approach the "microplastic" and "plastic" behaviour are uncoupled and the same happens in the "field of hardening moduli" approach. In contrast, in the "two surfaces" model the translation of the internal yield surface provokes an appropriate translation of the bounding surface. This implies the sensitivity of the current yield limit to microplastic processes within it.

The essential difference between the two approaches consists in the constitutive law within the plastic yield surface. First, in the present approach it is assumed that the material behaviour is piece-wise path independent and thus described in terms of strains and stress rather than in terms of their rates as in the kinematic hardening approach.

Another substantial difference stands in the hardening rules. In fact, in the presented model the non linearity of the material behaviour, i.e. the local compliance moduli, depends on the "distance" between the current strain and the strain at the last stress reversal, then entirely on the past history of the material. In the "two surfaces" model the current hardening modulus depends on the distance between the current stress point and a suitably chosen point on a bounding surface which the stress path is supposed to reach—not necessarily in that point. This surface has a a priori assumed shape and has been created in a previous, generally different, stress path.

In both approaches loading functions are employed together with the corresponding loci. In the present case the locus undergoes an expansion from the stress reversal point, while in the other it undergoes a shift according to the rules of kinematic plasticity or an isotropic expansion and shift, the reversal point remaining always at the locus border.

In the present approach the set of dead loci forms a domain of validity of the current law. The entire sequence of dead loci is remembered by the material rather than only the external locus, i.e. the consolidation surface, as in the other approach.

Finally, it is worth noting that to ensure the continuity of the rate response across the surface delimiting the validity of two subsequent constitutive laws, in the theory of plasticity it is assumed that the plastic strain rate is insensitive to the stress rate component tangential

to the yield surface. In the present approach an isolated non-linearity is introduced in the constitutive law by taking the material compliance at the stress reversal as dependent on the stress rate vector direction.

There is a general lack of experimental evidence which may give a reliable confirmation of most of the elements of both types of models. Authors believe that numerical computations and comparison with phenomenological data may give some indications on the adequateness of these elements.

5 COMPARISONS WITH EXPERIMENTAL RESULTS

By using the equations derived in Sections 3 and 4 it is possible to compare theoretical predictions with available experimental results. To make the comparison simpler only standard triaxial tests will be considered. It is necessary then to specialize the constitutive relations to conditions of axial symmetry:

$$\sigma_{22} = \sigma_{33} \quad ; \quad \sigma_{12} = \sigma_{13} = \sigma_{23} = 0 \quad ;$$

$$\varepsilon_{12} = \varepsilon_{13} = \varepsilon_{23} = 0 .$$

Define, as usual, $p \equiv 1/3 (\sigma_{11} + 2 \sigma_{33})$;

$q \equiv \sigma_{11} - \sigma_{33}$; $\eta \equiv q/p$; $v = \varepsilon_{11} + 2\varepsilon_{33}$;

$$\varepsilon = 2/3 (\varepsilon_{11} - \varepsilon_{33}).$$

The constitutive equations then become

a) in the elastoplastic range

$$\dot{v} = B_o \, \dot{p}/p + \Lambda \, d \qquad (5.1)$$

$$\dot{\varepsilon} = 2/3 \, L_o \dot{\eta} + \Lambda \qquad (5.2)$$

where $\Lambda = \Lambda^+ \, \partial g / \partial q$ and $d = \dfrac{\partial g / \partial p}{\partial g / \partial q}$

so that

$$\Lambda = (\lambda - B_o) \frac{(\eta + d)\dot{p}/p + \dot{\eta}}{(\eta + d)(d + D)} \quad ;$$

$$\qquad (5.3)$$

$$d = \frac{1}{a \, \eta} \qquad \text{if } \eta \leq M/2$$

$$\Lambda = (\lambda - B_o) \frac{m \, \dot{p}/p + \dot{\eta}}{m \, (D + d)} \quad ;$$

$$\qquad (5.4)$$

$$d = \frac{M - \eta}{\mu} \qquad \text{if } \eta \geq M/2$$

in the softening range $D = 0$.

b) in the paraelastic range

$$\Delta M_v = B (\ln p/p^M + \sqrt{2/3} \phi |^{\Delta M}\eta|) \qquad (5.5)$$

$$\Delta M_\varepsilon = 2/3 L^{\Delta M}_\eta \qquad (5.6)$$

where

$$B = B_0 (1 + \omega_v \chi^M)$$
$$L = L_0 (1 + \omega_\varepsilon \chi^M) \qquad (5.7)$$

where χ becomes

$$\chi^M = (1/3 \,^{\Delta M}_v{}^2 + 3/2 \,^{\Delta M}_\varepsilon{}^2)^{1/2} \qquad (5.8)$$

For a generic path portion that does not undergo any stress reversal it is possible to derive after some algebra from Eqs. (5.5), (5.6) that

$$\Delta M_v = \{ \frac{\omega_v}{\omega_\varepsilon} (\frac{\Delta M_\varepsilon}{2/3 L^M \Delta M_\eta} - 1) +$$

$$+ 1\} \,^{\Delta M}_\Sigma \qquad (5.9)$$

$$\Delta M_\varepsilon = \{ \frac{1 + 1/3 \omega_v (\omega_\varepsilon - \omega_v)^{\Delta M}\Sigma^2 + \omega_\varepsilon \cdot}{1 - 1/3 \omega_v^2 \,^{\Delta M}\Sigma^2 -}$$

$$\cdot \frac{[1/3 \,^{\Delta M}\Sigma^2 + 2/3 L^{M2} \Delta M_\eta{}^2}{- 2/3 L^{M2} \omega_\varepsilon^2 \Delta M_\eta{}^2}$$

$$\frac{\cdot (1 - 1/3 (\omega_\varepsilon - \omega_v) \,^{\Delta M}\Sigma^2)]^{1/2}}{\} \cdot$$

$$\cdot 2/3 L^M \Delta M_\eta \qquad (5.10)$$

where

$$\Delta M_\Sigma \equiv B^M(\ln p/p^M + \sqrt{2/3} \phi |^{\Delta M}\eta|) \qquad (5.11)$$

Although Eqs. (5.9), are apparently involved, it must be emphasized that they are closed form solutions in terms of differences of strains and of "geotechnical" stress variables for any path portion within the yield locus.

Consider then an undrained test on a normally consolidated clay $(D = 0)$ under cyclic loading with variable amplitude. Assume the programme of loading to be that imposed by Wroth and Loudon (1967)

on a kaolin. The first loading yields elastoplastic strains and the stress path can be found by integrating Eq. (5.1), (5.3) under the condition of no volume change:

$$\ln p/p_c = - 1/2 (1 - B_0/\lambda) \ln (1 + a \eta^2) \qquad (5.12)$$

The unloading reloading paths within the yield locus are given by

$$p = p^M \exp \{- \sqrt{2/3} \phi |^{\Delta M}\eta| \} \qquad (5.13)$$

Since, for the particular loading programme, the yield locus is reached every time at the end of an unloading reloading cycle, the hysteretic memory of the material is cleared after every cycle. If yielding occurs for $\eta \geq M/2$ the stress path for further loading becomes

$$\ln p/p_1 = - \frac{1 - B_0/\lambda}{m} (\eta - \eta_i) \qquad (5.14)$$

where p_i and η_i are the stress coordinates of onset of yielding.

Fig. 5a shows the comparison between calculated and experimental stress paths. The constitutive parameters employed are $M = .96, \lambda = .113, B_0 = .022, \mu = .67, m = .7, \omega_v = 23, \omega_\varepsilon = 150, L_0 = .00397, \phi = .1$.

It is seen that the travelling of the effective stress point towards the critical state is matched without substantial departures. In Fig. 5b the calculated pore pressure change with deviatoric strains is presented. Although it is not possible to judge the quantitative agreement with actual experimental results, which were not available to the authors, the qualitative agreement with results obtained in similar tests is rather satisfactory.

Consider now another specimen of a different kaolin subjected to a cyclic p constant test at constant axial stress amplitude. The "predicted" results can be obtained by integrating Eqs. (5.1), (5.2) under the condition $p = 0$ and by putting $\ln p/p^M = 0$ in Eq. (5.11). It is possible to note from Figures 6a and 6b that a reasonable qualitative modelling of ratcheting and shear compaction phenomena is possible.

Finally, consider a confined compression test (oedometric) with cycles of loading and reloading on a clay. By integrating Eqs. (5.1), (5.2) with the condition of no lateral strain $(\varepsilon_3 = 0)$ it is possible to find the equation of the stress path

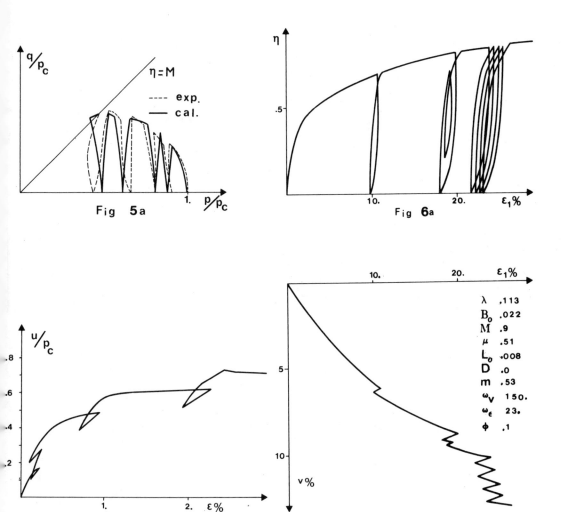

Fig 5a

Fig 6a

Fig 5b

Fig 6b

λ	.113
B_o	.022
M	.9
μ	.51
L_o	.008
D	.0
m	.53
ω_v	150.
ω_ϵ	23.
ϕ	.1

followed by the specimen. When virgin loading is considered   one has

$$m \ln p/p_o = \eta_o - \eta + (1 + \frac{L_o}{B_o} m) \ .$$

$$.\{\frac{B_o}{\lambda} (\eta - \eta_o) + 3/2 (\lambda /B_o - 1) \frac{B_o^2}{\lambda 2} \ .$$

$$.\mu \ln \left[\frac{\lambda/B_o (\eta - \eta_o)-3/2\mu (\lambda/B_o -1)}{\lambda/B_o (M - \eta) - 3/2\mu (\lambda/B_o - 1)} \right] \}$$

(5.15)

It is possible to show (Nova (1979)) that
starting from an isotropic pressure p_o
the stress path rapidly reaches an
asympthote given by

$$\eta_o = M - 3/2 \mu (1 - B_o/\lambda) \qquad (5.16)$$

from which is straightforward to get the
coefficient of earth pressure at rest

$$k_o = \frac{3 - M + 3/2 \mu (1 - B_o/\lambda)}{3 + 2M - 3\mu (1 - B_o/\lambda)} \qquad (5.17)$$

311

It is possible to show, see Nova and Hueckel (1979), that the unloading path is given by

$$p = p_M \exp \left\{ 1/B_o \ \frac{L_o \ \Delta M_\eta}{1 + (\omega_v - \omega_\epsilon) L_o| \ \Delta M_\eta|} - \right.$$

$$\left. - \sqrt{2/3} \ B_o \ \phi | \Delta M_\eta| \right\} \qquad (5.18)$$

The reloading path is given by a similar equation in which B_o and L_o are substituted by $\hat{B}$ and $\hat{L}$ obtained iteratively as discussed in the previous Section. The calculated and experimental paths for a silty clay at Gioia Tauro - courtesy of Studio Geotecnico Italiano - are shown in Fig. 7a. The corresponding strains are plotted in Fig. 7b. Very good agreement is found for both plots, although some discrepancies occur, mainly for the reloading stress path. It may be due to that in the calculation for the sake of simplicity, the constant B_o, L_o have been employed also for this part of the path.

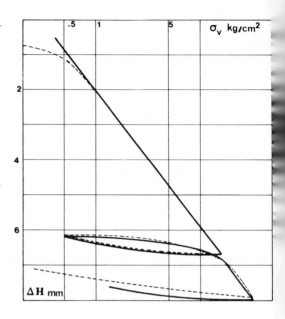

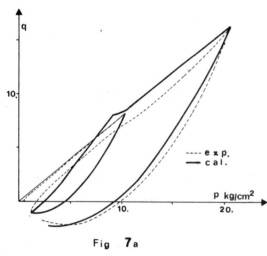

Fig **7**a

6 CONCLUSIONS

The presented mathematical model of soil behaviour under cyclic loading combines theories pertinent to the elastoplastic and paraelastic range. The model is formulated in terms of "geotechnical" variables, i.e. the tensor of stress ratios η_{kl} and the logarithmic measure of the isotropic pressure.

The "paraelastic" behaviour is treated by means of a piecewise path independent description between two subsequent, suitably defined, stress reversal loci. A discrete hierarchic structure of the material memory

is introduced which collects informations pertaining to the stress reversal points. The stress reversals induce also a change of the compliance depending on a "measure" of the reversal direction in order to fulfill the continuity requirement on neutral paths. In this way, the history dependence of the material behaviour within the yield locus is lumped in the stress reversal points, while the process within them is considered to have no consequences on further behaviour. Such assumption renders the tackling of the complex behaviour of the soil under cyclic loading much simpler. Above all it enables to obtain closed form solutions for strains even for complex stress paths, and viceversa. The theory requires only three material constants in addition to those necessary to define the elastoplastic model. Their experimental determination is straightforward as shown in Nova and Hueckel (1979). Closed form solution are available even for the elastoplastic range when common triaxial tests are considered. The material behaviour is completely specified by seven parameters only that can be determined experimentally without much effort, see Nova (1977b). Nevertheless, it must be emphasized that the Drucker-Prager type extension to multiaxial conditions of a model derived for triaxial compression leads inevitably to incorrect predictions if extension or plane strain tests are to be considered. A further sophistication of the model is then necessary if general complex threedimensional paths should be considered.

The presented model enjoys a chinese box structure. By setting to zero some constants the effect of strain accumulation in hysteresis, every hysteresis effect, the hardening dependence on the second invariant of plastic strain deviator and so on may be dropped in successive simplifications of the theory to obtain at the very end the simple Granta Gravel model. This allows to treat problems with various accuracy with a unified numerical algorithm.

Further effects may be taken into account with more complex constitutive laws. At this stage the model does not predict the stabilization of stress-strain loops so that the theory is limited to cases of low number of cycles. Acceptable qualitative results are obtained without involved computation. The phenomenon of liquefaction can in principle be modelled by the theory but at this stage predictions would be unrealistic for the assumed yield function symmetry in compression and extension. In addition the analysis of liquefaction requires a more detailed study of the form of the stress reversal locus and of its properties (e.g. convexity) in order to tackle with the complex effective stress path which involves reversals, reactivations and plasticity. This study will be pursued in a forthcoming paper.

7 REFERENCES

Boisserie, J.M., Guelin, P., 1977, Remarks on the tensorial formulation of constitutive law describing mechanical hysteresis, Proc. SMiRT IV, S. Francisco, L 1/9 .

Dafalias, Y.F., Popov, E.P., 1975, A model of non-linearly hardening materials for complex loadings, Acta Mech. 21:173-192.

Dafalias, Y.F., Popov, E.P., 1976, Plastic internal variables formalism of cyclic plasticity, J. Appl. Mech. 98:645-651.

Dobry, R., 1970, Damping in soils: its hysteretic nature and the linear approximation, Report R70-14, Dept. of Civil Engineering, MIT, Cambridge, Mass.

Drucker, D.C., Prager, W., 1952, Soil mechanics and plastic analysis or limit design, Quart. Appl. Math. 10:157-165.

Greenstreet, W.L., Phillips, A., 1973, A theory of an elastic-plastic continuum with special emphasis to artificial graphite, Acta Mech. 16:143-156.

Hardin, B.O., Drnevich, V.P., 1972a, Shear modulus and damping in soils: measurement and parameters effect, Proc. ASCE, 98:603-624 (SM6).

Hardin, B.O., Drnevich, V.P., 1972b, Shear modulus and damping in soils: design

equation and curves, Proc. ASCE 98:667-692 (SM7).

Hueckel, T., 1979, On a neutral path uniqueness in a piece-wise description of cyclic behaviour of inelastic solids, pending publication.

Hueckel, T., Nova, R., 1979a, On paraelastic hysteresis of soils and rocks, Bull. Acad. Polon. des Sciences. Sec. Sc. Tech. 27:1, 49-55.

Hueckel, T., Nova, R., 1979b, Some hysteresis effects of the behaviour of geologic media, Int. J. of Solids and Structures, 15:

Idriss, I.M., Dobry, R., Singh, R.D., 1978, Non linear behaviour of soft clays during cyclic loading, ASCE GT12:1427-1447.

Maier, G., Hueckel, T., 1979, Non associated and coupled flow rules of elastoplasticity for geotechnical media, Int. J. Rock Mech. Min. Sci.

Mróz, Z., 1967, On the description of anisotropic workhardening, Int. J. of Mech. and Phys. of Solids 15:163-175.

Mróz, Z., 1979, On hypoelasticity and plasticity approaches to the constitutive modelling of inelastic behaviour of soils, Int. J. Num. Anal. Meth. Geomech., to appear.

Mróz, Z., Lind, N.L., 1975, Simplified theories of cyclic plasticity, Acta Mech. 22:131-152.

Mróz, Z., Norris, V.A., Zienkiewicz, O.C., 1978, An anisotropic hardening model for soils and its application to cyclic loading, Int. J. Num. Anal. Meth. Geomechan. 2:203-221.

Mróz, Z., Norris, V.A., Zienkiewicz, O.C., 1979, Application of an anisotropic hardening model in the analysis of elastoplastic deformation of soils, Geotechnique 29 .

Nova, R., 1977a, On the hardening of soils, Arch. Mech. Stosow., 3, 29:445-458.

Nova, R., 1977b, Theoretical studies of constitutive relations for sand, MSc. Thesis Univ. of Cambridge.

Nova, R., 1979, Un modello costitutivo per l'argilla, Rivista Italiana di Geotecnica.

Nova, R., Hueckel, T., 1979, An engineering theory of soil behaviour in unloading and reloading, pending publication.

Nova, R., Wood, D.M., 1979, A constitutive model for sand in triaxial compression, Int. J. Num. Anal. Meth. in Geomechanics 3:3 .

Phillips, A., 1972, On rate independent continuum theories of graphite and their experimental verification, Nucl. Eng. and Design 18:143-156.

Prevost, J.H., 1977, Mathematical modelling of monotonic and cyclic undrained clay

behaviour, Int. J. Num. Anal. Meth. Geomech., 1:195-201.
Ramberg, W., Osgood, W.R., 1943, Description of stress strain curves by three
parameters, Techn. Note 902 Nat. Advisory
Committee for Aeronautics, Washington
D.C. .
Schofield, A.N., Wroth, C.P., 1968, Critical state soil mechanics, Wiley.
Wroth, C.P., 1965, The prediction of shear
strains in triaxial tests on normally
consolidated clays, 6th ICSMFE, Montreal,
2:417-420.
Wroth, C.P., Loudon, P.A., 1967, The correlations of strains within a family of
triaxial tests on overconsolidated samples
of kaolin, Proc. Geot. Conf., Oslo, 1:
159-163.

ACKNOWLEDGMENT

Authors gratefully acknowledge the assistance of the exchange agreement between
the National Research Council of Italy
(CNR) and the Polish Academy of Sciences.
The former Institute and the University
of Cambridge are also acknowledged by one
of the authors (R.N.) for financial support.

Stress-strain aspects of cohesionless soils under cyclic and transient loading

M. P. LUONG
Ecole Polytechnique, Palaiseau, France

1. INTRODUCTION

Many problems in geotechnology involve cyclic factors concerned with the response of earth materials to cyclic loads, either generated by the forces of nature (sea waves, currents, winds, earthquakes,...) or in the course of engineering operations such as blasting, pile driving, vibroflotation, rotating machines,... The rapid development of civil engineering design in these fields has fostered an increased interest for the stress-strain aspects of cohesionless soils under cyclic and transient loading.

The purpose of this study is to :
(1) identify the real soil stress-strain behaviour ;
(2) present test results on the characteristic properties of cohesionless soils depending on stress paths ;
(3) consolidate the experimental data in order to define a characteristic stress domain where the resultant effect of load cycling is contractancy ;
(4) establish a simple criterion for liquefaction of loose fine sand ; and
(5) interpret the different features of cyclic behaviour of cohesionless soils.

2. TEST PROCEDURE

The paper describes extensive laboratory tests undertaken to assess various rheological aspects of cohesionless soils under cyclic and transient loading. Notwithstanding the restrictions of the chosen experimental arrangement, the exploration of the mechanical soil behaviour during quite varied stress paths was possible, thus bringing to light interesting characteristic properties.

The loading parameters of axisymmetric triaxial tests are :

- mean stress $\qquad p = \dfrac{\sigma_1 + 2\sigma_3}{3}$

- deviator stress $\qquad q = \sigma_1 - \sigma_3$

- deviatoric level $\qquad \eta = q/p$

The corresponding deformation parameters may be defined by :

- volumetric strain $\qquad \varepsilon_v = \varepsilon_1 + 2\varepsilon_3$

- distortional strain $\qquad \varepsilon_q = \dfrac{2}{3}(\varepsilon_1 - \varepsilon_3)$

- distortional level $\qquad \beta = \dfrac{\varepsilon_q}{\varepsilon_v}$

Triaxial compression and extension tests are carried out either with a constant stress rate $\dot{\sigma}$ or a constant strain rate $\dot{\varepsilon}$ on saturated soils under drained or undrained conditions.

3. DEFORMATION MECHANISM

In order to analyse and predict response soil behaviour, it is necessary to understand how the individual microscopic constitute elements interact.

Deformations of soil under loading are mainly the result of three mechanisms :
(1) compressibility and shape change of the particle assemblies ; (2) bending of platy particles, sliding and rolling of rounded grains ; and (3) particle crushing and breakage modifying soil fabric.

- Volume changes depend on particle associations and arrangements. They are caused chiefly by two mechanisms : (1) generalized contraction or expansion of solid skeleton without modification of soil structure ; (2) variations of grain arrangements, grain orientations, particularly sensitive in shear tests or during the first hydrostatic loading of a loose sand.

- Distortional strains are governed by (1) sliding friction which consists of microscopic interlocking due to surface roughness of contacting particle surfaces, there is no significant volume expansion associated with this type of action ; (2) interlocking friction or physical restraint to relative particle translation afforded by adjacent particles.

4. CONVENTIONAL TRIAXIAL TEST RESULTS

4.1. Monotonous triaxial compression

Starting with an isotropic confinement, $\sigma_2 = \sigma_3$ is maintained constant while the deviator stress $q = \sigma_1 - \sigma_3$ increases. Initially, the volume decreases, then the volume decrease rate slows down and becomes zero for a loose sand. The tendency is reversed for a dense sand and the volume increases in order to disengage the particle interlocking and allow large relative movements. The ratio $-\dfrac{d\varepsilon_3}{d\varepsilon_1}$ between lateral and axial strains small, at first, increases monotonously from zero towards values largely greater than 0.5 (Fig.1). This volume increase called dilatancy is more pronounced the more dense is the initial arrangement and the smaller is confining pressure.

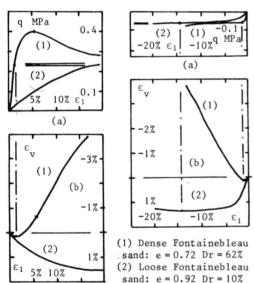

Fig.1.Conventional triaxial compression.

Fig.2.Conventional triaxial extension.

(1) Dense Fontainebleau sand: e = 0.72 Dr = 62%
(2) Loose Fontainebleau sand: e = 0.92 Dr = 10%

4.2. Monotonous triaxial extension

Triaxial extension is easily carried out by an algebraic reduction of the deviator stress q . The intermediary principal stress σ_2 instead of being equal to the minimum one is now equal to the

maximum. The volume at first decreases, however the mean stress p is decreasing.

As in the former case, the volume change rate either becomes zero or the tendency is inversed and dilatancy is observed. This initial volume change with opposite sense to p means that the ratio $-\dfrac{d\varepsilon_3}{d\varepsilon_1}$ is initially greater than 0.5. Reversible approximation is not applicable for this first loading path. (Fig.2).

4.3. Comparison between these two tests

The volume change similarity of the two cases hides big differences between rheological behaviours :
i) Soil deformability is greater in triaxial compression test than in extension test where the material is apparently more rigid.
ii) Initially, Poisson ratio $-\dfrac{d\varepsilon_3}{d\varepsilon_1}$ is smaller in triaxial compression (<0.5) than in triaxial extension (>0.5).
iii) Shear strength at the grain interlocking breakdown limit according to Mohr's representation is higher in triaxial compression than in triaxial extension ($\varphi_{TC} > \varphi_{TE}$) for the tested sands : Fontainebleau sand, Loire sand, carbonated Channel sand. (Fig.3).

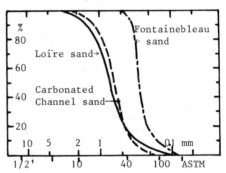

Fig.3.Grain size distribution curves.

iv) Stress path in the p,q diagram during a coupled compression-extension triaxial test is a straight line with a slope 3 : the triaxial compression corresponds to q> 0 and the triaxial extension to q < 0. In fact during the passage through q = 0, the bisecting plane changes relative to the hydrostatic stress axis.
v) The grain interlocking breakdown at the characteristic threshold in triaxial compression test gives an important dilatancy. In triaxial extension test, predominant distortion occurs with a relatively limited dilatancy.

5. THE CONCEPT OF A SOIL CHARACTERISTIC STATE

Having followed the deformation from the beginning of loading, three main points readily become evident : (i) strain hardening reflecting the difference of stress strain behaviour between first loading on virgin soil and following cycles of unloading-reloading ; (ii) hysteresis of soil under cyclic loading due to an irreversible behaviour during each load cycle ; (iii) stress path enhancing mechanisms of deformation associated either with great volumetric strain or with predominant distortional strain.

Thus, stress-strain study of soils under non-monotonous loading necessitates some new concepts in order that its rheological behaviour might be followed during the irreversible deformation process.

5.1. The existing concept : "Soil critical state"

The analogy between soils and metals at failure suggests the existence of an idealized state in the soil, the so-called critical state and accompanied by a constant specific volume v , defined as follows :

$$q = Mp \quad \text{and} \quad v = \Gamma - \lambda Lnp$$

M, Γ, λ are constants characterizing the material (Schofield and Wroth 1968).

This very fruitful concept has brought forward significant progress in the understanding and the construction of numerical algorithms for the mechanical behaviour of soils during static and monotonous loading.

The critical state represents an asymptotic soil behaviour at the yield point, where it flows at constant volume. However it concerns only cases where the void ratio e = v-1 has the critical value e_{cr}. Hence the concept of a critical state is not suitable for following the irreversible deformation from the start of loading.

5.2. A new concept : "Soil Characteristic state"

The conventional triaxial tests under drained condition either in compression ($\sigma_1 > \sigma_2 = \sigma_3$) or in extension ($\sigma_1 < \sigma_2 = \sigma_3$) allows quite simply the utilization of the deviatoric stress level at the moment of passage from compressive volume changes to dilatation in order to define a characteristic state in the soil, compatible with the critical state and related to the following facts :

i) a zero volumetric strain rate $\dot{\varepsilon}_v = 0$.

ii) a stress level where interlocking ceases and disruption of interlocking starts. This level is given by the ratio $\dfrac{\sigma_1}{\sigma_3} = tg^2\left(\dfrac{\Pi}{4} + \dfrac{\varphi_c}{2}\right)$ defining thus an angle φ_c which measures the interlocking capacity of the material (Kirkpatrick 1961). φ_c is an intrinsic factor of the material ;

iii) relatively small deformations in the soil mass ;

iv) independance of the initial porosity ;

v) non-influence of anisotropy of contact forces and chains of grains ;

vi) non-sensitivity to grain size.

For this kind of loading history under drained and constant confinement conditions, the characteristic state separates two types of soil rheological behaviour : contracting in the subcharacteristic region which is bounded in the p,q plane by the characteristic lines CL and dilating in the surcharacteristic region between those lines and the failure lines FL whether well defined or not as in the case of very loose sands. In this case the characteristic state merges with the critical state (Fig.4).

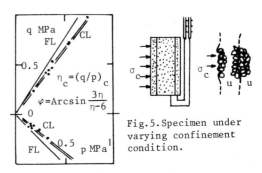

Fig.4.Characteristic criterion.

Fig.5.Specimen under varying confinement condition.

The following question is raised : having shown in a simple way the existence of a characteristic state in the case of conventional triaxial test at constant confining pressure, does it follow that this state fully characterizes the threshold of interlocking disrupture in the grain structure in any other loading path in the p,q plane ?

The following tests are performed at varying confining pressures. As the penetration of the protective membrane may cause significant errors in the measurements of volume changes, it was necessary to install the arrangement of figure 5, the purpose of which is the minimisation of the error and to check the calibration and the precision of the volumetric strains.

317

6. MONOTONOUS TEST RESULTS

Generally the different following experimental results show that the characteristic state is defined by the tendency to zero of the ratio $\frac{d\varepsilon_v}{d\varepsilon_q}$ for any chosen loading path. This means that the distortional level β increases considerably causing both interlocking disrupture (individual particle must be plucked from their interlocking seats and made to slide over the adjacent particles) and great distortions of the grain arrangement.

6.1. Drained shearing under constant p condition

Drained tests under constant p condition using cylindrical triaxial apparatus give similar results to tests for a loading path of slope 3 in the p,q plane. Fig. 6a-b and c-d show respectively a triaxial compression test and a triaxial extension test under constant p condition. The characteristic thresholds on the q, ε_v diagrams under constant p condition are quite the same as those given from the curves of loading path under constant confining pressure.

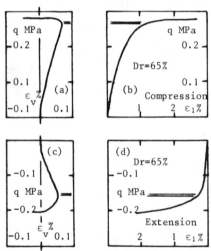

Fig.6. Drained triaxial tests under constant p condition (p=0.2MPa).

6.2. Drained shearing under constant q condition

Starting with a constant confining pressure $\sigma_2 = \sigma_3 = 0,2$ MPa the deviator stress $q = \sigma_1 - \sigma_3$ is increased progressively until 0,2 MPa. Under this deviator stress maintained constant, let us decrease the radial stress so that to perform a drained test under constant deviator load.

Fig. 7a shows with an arrowed line the drained shear load path under constant de-

viator stress q condition.

Fig. 7b gives the variation of volumetric strain ε_v against axial strain ε_1. Inside the characteristic boundary CL, ε_1 varies very little with ε_v. As soon as the characteristic threshold is reached, distortion increases rapidly with volumetric strain.

Fig. 7c describes the variation of ε_v with deviatoric level $\eta = q/p$. As soon as η reaches the characteristic level η_c, volumetric strain increases and gives dilatancy.

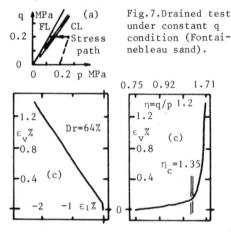

Fig.7.Drained test under constant q condition (Fontainebleau sand).

6.3. Stress-strain properties under drained radial loading

Stress paths at $\eta = q/p = ct$ are radial paths in stress space : the applied average obliquity at the points of contact is kept unchanged and consequently the change of geometry of the internal structure during the test would be very small (El Sohby, 1969).

i) Under hydrostatic loading or $\eta = 0$, stress-strain characteristics depend on initial density and confining pressure (or mean stress during loading). The looser is the material, the greater is volume change, and a greater part of irreversible deformation at first loading. This is due to tightening mechanism (Fig.8).

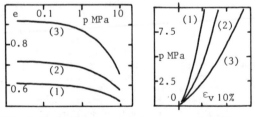

Fig.8. Fontainebleau sand compressibility under hydrostatic loading:(1)γ_d=16.8kN/m^3, Dr=90% ; (2)γ_d=15.7kN/m^3,Dr=62%; (3)Dr=10%, γ_d=14.1kN/m^3.

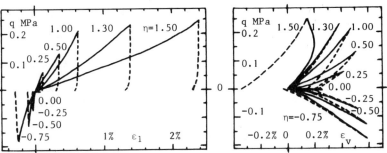

Fig.9.Drained tests under radial loading (triaxial compression and extension) on Fontainebleau sand γ_d=15.7kN/m³,Dr=62%, e=0.72

ii) If $\eta \leqslant \eta_c$ corresponding to characteristic state, the geometry change of packing of the particles comprising the assembly is relatively small. The soil contacts by interlocking under loading and dilates by slackening of contact forces during unloading.

iii) If the ratio η exceeds η_c of the characteristic threshold, a great number of particles move relative to eachother because of the great mean obliquity of contact forces. The physical processes contributing to motion include both sliding of particles and removal or displacement of particles from interlocking action between adjacent particles. Dilatancy occurs under loading and even under unloading. The volumetric dilatancy is predominant in triaxial compression whereas irreversible distortional strains are important in triaxial extension (Fig.9).

These tests under radial stress paths illustrate clearly the importance of characteristic state concept which defines a subcharacteristic domain of stable intergranular contacts. Beyond, in the surcharacteristic domain, the great obliquity of intergranular contact forces facilitates interlocking breakdown and induces large distortions of the grain assembly.

6.4. Drained shearing at varying stress obliquity

Radial loading maintains a constant mean stress obliquity, whereas spherical path in stress space or circular path in (p,q) diagram allows study of the influence of stress obliquity on the rheological behaviour of the material against constant stress modulus.

i) Spherical stress path
Fig. 10a and 10b show respectively the variation of ε_1 and ε_v against deviator stress q or stress level $\eta = q/p$. On the diagram (q,ε_v) the hysteresis loop does not occur under unloading if the characteristic threshold is not reached. If it is exceeded a dilatancy loop appears distinctly after unloading.

ii) Circular stress path
Fig. 11a-b illustrate respectively typical variation of axial strain ε_1 and volumetric strain ε_v with q in triaxial compression in a drained shear test under loading constant (p,q) stress modulus.

Like the spherical case, no hysteresis loop appears for the volume change in the characteristic domain. The dilatancy loop is observed only when the characteristic threshold is exceeded.

In these two cases, spherical or circular path, it is evident that when the characteristic threshold η_c is reached, the distortional level β increases suddenly

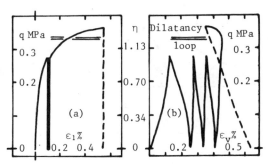

Fig.10.Drained shearing on spherical stress path (Fontainebleau sand Dr=64% ; σ=0.52MPa)

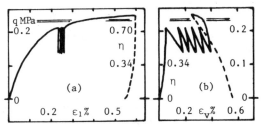

Fig.11.Drained shearing on circular stress path (Fontainebleau sand Dr=64% ; mod=0.3MPa)

319

showing the beginning of the interlocking breakdown of the granular structure. Unloading gives a significant volumetric contraction ε_v whereas axial strain ε_1 remains more limited.

The characteristic state is indicated in these two stress paths by the existence or not of a dilatancy loop under unloading.

6.5. Drained shearing at constant volume

This type of test recommended by Taylor (1948) presents more correctly the behaviour of large masses of saturated sand sheared quite rapidly so that no drainage of pore water is possible. Fig. 12 shows the stress paths in tests at constant volume in triaxial compression and extension.

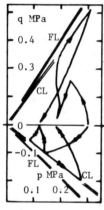 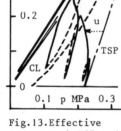

Fig.12.Drained tests at constant volume.

Fig.13.Effective stress path ESP under undrained condition.

As soon as the loading trajectory reaches the characteristic threshold either in compression or in extension, its slope changes following the characteristic line, which tends to prove that the characteristic state represents indeed the limit of interlocking beyond which large deformations are possible owing to dilatancy, i. e. to disrupture of grain interlocking structure.

Experimentally this test either in triaxial compression or extension at constant volume shows those increments of the load in the p,q plane which correspond either to a small expansion and weakening of the grain assembly or to compression and further increase in interlocking. Here again the residual deformation is larger the more the loading path followed the characteristic line. A few cycles of loading reaching the characteristic line followed by unloading may bring the operation loading point to the origin O, which means a total

loss of resistance.

6.6. Effective stress path under undrained condition

During a conventional undrained triaxial test, it is found that the pore pressure u first increases. Its value determines the effective stress path ESP in the p,q plane. Load-unload cycles show clearly workhardening effects in the soil.

As soon as the characteristic threshold is reached, test results show that the effective stress path follows the characteristic line CL giving thus a pore pressure generation law. On unloading the residual deformation observed is greater, the more extended was the movement of the effective loading point on the characteristic line. On reloading the behaviour is of a virgin sand having a renewed initial grain arrangement. This arises because movements of the effective loading point on the characteristic line disrupts the grain structure by a microscopic interlocking breakdown mechanism (Fig.13).

7. CYCLIC LOADING EFFECT ON COHESIONLESS SOILS

The characteristic state is shown to be a coherent concept for the study of different types of cyclic behaviour of sandy soils.

7.1. Deformations under radial load cycling

Figure 14 shows the total volumetric strain ε_v and the irreversible part ε_v^p, under hydrostatic loading ($\eta = 0$) under drained condition.

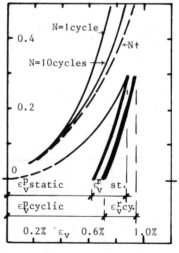

Fig.14.Cyclic hydrostatic compression on Fontainebleau sand ($\gamma_d = 15.7 \text{kN/m}^3$, e=0.72, Dr=62%).

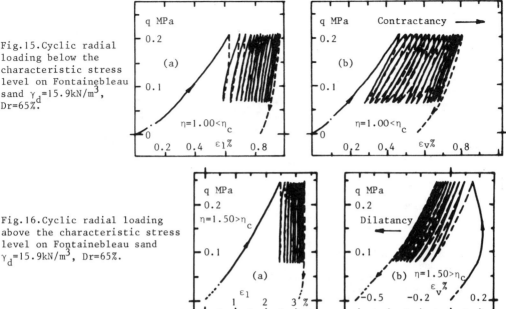

Fig.15.Cyclic radial loading below the characteristic stress level on Fontainebleau sand γ_d=15.9kN/m^3, Dr=65%.

Fig.16.Cyclic radial loading above the characteristic stress level on Fontainebleau sand γ_d=15.9kN/m^3, Dr=65%.

These are shown at the end of one cycle, ten cycles and finally at the end of a great number of cycles, where an asymptotic value is approached. Soil behaviour is then non linear quasi-elastic (Ko & Scott 1967). Adaptation is considered as obtained after a finite number of cycles of isotropic loading.

Under radial loading at $\eta = \dfrac{q}{p}$ other than zero, but smaller than the characteristic value, either in triaxial compression or extension, the soil shows little hysteresis susceptibility becoming negligible when the number of cycles increase. (Fig.15). For $\eta > \eta_c$, the hysteresis disappears. Cyclic loading causes ratchet behaviour, the volume increases reflecting the phenomenon of dilatancy of the granular structure, (Fig.16).

7.2. Stress-strain behaviour under deviator stress cycling in conventional triaxial test

Utilizing Fontainebleau sand at constant confinement (0,2 MPa), 20 cycles of deviatoric loading ($\Delta q = 0,2$ MPa) were carried out at each deviator stress level. Typical load cycling sequences are shown in Fig. 17a, from which it can be seen that :
  i. the residual axial strain ε_1^p increases with the mean deviator stress level ;
  ii. for triaxial extension (q < 0), the considerable stiffness of the sand is reflected in the q,ε_1 curve ;

iii. the hysteresis loop in the q,ε_1 diagram becomes stable when the stress level was higher at the preceeding stage.

Fig.17b is particularly clear and shows that :
  i. the contracting behaviour of the soil is obtained each time the mean deviator stress level is lower than the characteristic level ;
  ii. the dilating behaviour of the soil during load cycling is evident only when the mean deviatoric stress level becomes higher than the characteristic threshold.

Whenever the maximum stress level is below the characteristic value the soil rea-

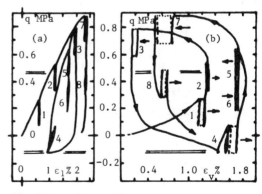

Fig.17.Cyclic loading under constant confinement condition (σ_3=0.2MPa) on Fontainebleau sand γ_d=15.8kN/m^3, Dr=64%.

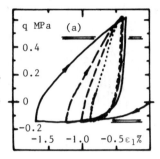
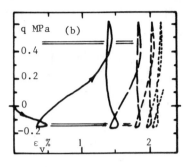

Fig.18.Densification of Fontainebleau sand $\gamma_d=15.7kN/m^3$, Dr=62% under constant confinement condition ($\sigma_3=0.2MPa$).

ches an accommodation as the number of cycles increases. If the characteristic threshold is exceeded, ratcheting takes place ;

iii. the contraction decreases when the characteristic threshold is approached ; and dilatation increases when deviator stresses increase further ;

iv. the contracting effect is more pronounced for extension loading having q<0 ($\sigma_1 < \sigma_2 = \sigma_3$) ;

v. dilatancy effects are more pronounced as the medium becomes denser.

7.3. Contracting behaviour of dense sands

Densification of dense sands may be obtained easily by cyclic loading exceeding both triaxial compression and extension characteristic thresholds. This high amplitude loading benefits in a partial loss of strain hardening during the dilating phase which breakdowns the granular interlocking assembly. On each reload, the tightening mechanism induces new irreversible volumetric strains and recurs with a renewed material becoming each time denser.

i) Conventional triaxial path under constant confining pressure.
Fig.18a shows a rapid stiffening of sand under cycles of alternating deviatoric stresses. It may be seen (Fig.18b) that following a few cycles, the void ratio passes from 0.720 to 0.682. This behaviour shows the predominance of the triaxial extension region because the characteristic threshold is rapidly reached. This experimental result is in agreement with direct shear tests carried out by Youd (1972) on Ottawa sand : each shear cycle formed a similar sequence of contractancy-dilatancy while an irreversible volumetric strain accumulated during cycles, reaching the relative density level of $D_r \simeq 128\%$ (ASTM norm D 3049-69) at the end of 10.000 shear cycles having an amplitude of ± 0.51mm.

ii) Triaxial path under constant p condition
Fig.19a presents typical (q,ε_1) hysteresis loop tightening increasingly during cycling Under this cyclic loading, the material seems to tend toward an accommodated state described by a stabilized (q,ε_1)hysteresis loop.
Fig.19b expresses clearly the densification process with its dilatancy loop at compression and extension characteristic levels. The intermediate part corresponds to an irreversible tightening between two sequences of granular assembly reinterlocking which fills up progressively the existing gaped voids.

iii) Circular path under constant (p,q) modulus condition
Fig.20a and b confirm the densification phenomenon under cyclic high amplitude loading exceeding triaxial compression and extension characteristic thresholds.
Wherever the stress path is, test curves are relatively weakly modified : "dilatancy loop followed by an irreversible tightening" mechanism remains perfectly preserved during load cycling.

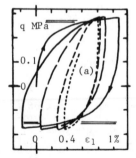
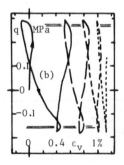

Fig.19.Densification of Fontainebleau sand $\gamma_d=15.9kN/m^3$, Dr=65% under constant mean stress p condition (p=0.2MPa).

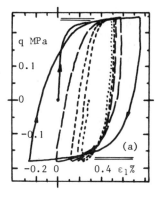
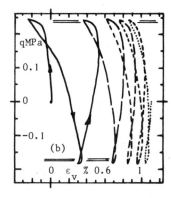

Fig.20.Densification of Fontainebleau sand $\gamma_d=15.9$kN/m³, Dr=65% on circular stress path (module {p,q} = 0.3MPa).

7.4. Drained cyclic shearing under constant deviator stress

Fig.21 points out clearly a progressive tightening process when the granular material is cyclically sheared under drained and constant deviator stress condition in the subcharacteristic domain $\eta < \eta_c$ whatever in triaxial extension (a-b) or compression (cd and ef). The characteristic state concept is one more verified and confirmed in the proposition of a contracting subcharacteristic domain for any followed stress path of the cyclic solicitation.

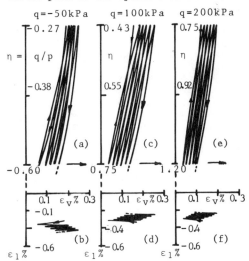

Fig.21.Densification of Fontainebleau sand $\gamma_d=15.9$kN/m³, Dr=65% under constant deviator stress q condition.

7.5. Liquefaction of saturated fine sand under undrained condition

Fig.22a.b.c show the behaviour of a saturated Fontainebleau sand specimen loaded cyclically by a deviator stress -0.01 MPa $\leqslant q \leqslant 0.2$ MPa under undrained condition and a constant strain rate $\dot{\varepsilon}_1 = 6.83.10^{-4}s^{-1}$. The liquefaction of the

sand occurs after N = 34 cycles shown by a sudden increase in the axial strain ε_1 accompanied by an excessive pore pressure being nearly equal to the confining stress, the deviator stress being near zero.

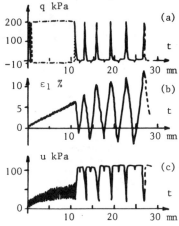

Fig.22.Liquefaction of saturated Fontainebleau sand $\gamma_d=15.7$kN/m³, Dr=62% e=0.72 under cyclic loading on conventional triaxial stress path.

The instability of the liquefaction phenomenon can be correctly interpreted using the concept of characteristic state. The effective loading point (q,p') must indeed remain inside the subcharacteristic region bounded by the characteristic lines in compression and extension. As the effective loading point reaches either the compression or extension characteristic lines it follows these lines. On unloading the deformability becomes greater considerably increasing the pore pressure which accumulates from cycle to cycle until the specimen collapses. Liquefaction may occur only if the alternating loading surpasses the point q = O on both sides.

7.6. Cyclic mobility and stabilization

323

During one sided conventional deviatoric loading cycles (q<0 or q>0) under undrained condition, either a partial softening of the soil owing to a cyclic increase in the pore pressure of constant amplitude (cyclic mobility) or a partial stiffening owing to a cyclic decrease in the pore pressure of constant amplitude (stabilization) is shown.

Figure 23 summarises the different rheological phenomena exhibited during conventional triaxial tests on sands under either drained or undrained condition

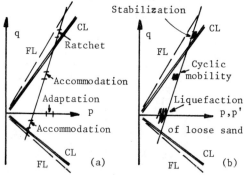

Fig.23.Cyclic effects observed on conventional triaxial path.

8. CONCLUSIONS

8.1. The essential parameter for studying the rheological behaviour of soils is the development of the volumetric strain during loading stages. The angle φ_c is an intrinsic factor characterizing the interlocking capacity of the soil.

8.2. The characteristic state concept, explained and simply formulated on the basis of ordinary laboratory loading path is further verified and enhanced by other different loading paths in the p,q plane. It can be defined by a distortional level β suddenly increasing to a very high value.

8.3. Under either undrained or constant volume conditions, the subcharacteristic region includes all possible effective loading points. As soon as this point reaches a characteristic line it follows it. The length of the section followed determines the degree of memory-loss of preceeding loading history, reloading being related to a new initial state.

8.4. This concept becomes all important in the domain of cyclic loading, facilitating the definition of a region of contraction behaviour for the soil. A criterion of liquefactibility follows : the

effective loading point reaches the origin only for cases of alternating total deviatoric loading on both sides of q = 0. Beyond that region, in the surcharacteristic state and up to the failure limit, the behaviour during cycles is dilating.

8.5. The different cases of granular soil behaviour during cyclic loading studied utilizing the conventional triaxial apparatus are easily interpreted within the framework of the characteristic state.

REFERENCES

Boutwell G.P. Jr, 1968, On the yield behaviour of cohesionless materials. Soil Mechanics, n°7, Duke University Thesis.
El Sohby M.A., 1969, Deformation of sands under constant stress ratios. 7th ICSMFE pp.111-119, Mexico.
Flavigny E. & Darve F., 1979, Membrane penetration and its effect on pore pressure. Discussion ASCE GT2, vol.105, january 1979, pp. 115-117.
Habib P. & Luong M.P., 1978, Sols pulvérulents sous chargements cycliques. Séminaire Matériaux et Structures sous Chargement Cyclique, 28-29 sept. 1978, Ecole Polytechnique, Palaiseau, France.
Kirkpatrick W.M., 1961, Discussion on soil properties and their measurement. Proc. 5th Int. Conf. Soil Mech. & Found. Eng. III, pp. 131-133, Paris.
Ko H.Y. & Scott R.F., 1967, Deformation of sand in hydrostatic compression. J. S.M. F. ASCE, vol.93, SM3, pp.137-156.
Luong M.P., 1978, Etat caractéristique du sol. C.R.A.S. Paris, t.287, série B,305.
Luong M.P., 1978, Comportements cycliques des sols pulvérulents. C.R.A.S. Paris, t.287, série B.313.
Luong M.P., 1979, Les phénomènes cycliques dans les sables. Journée de Rhéologie 1979, E.N.T.P.E. Vaulx en Velin, 25 avril 1979.
Rowe P.W., 1971, Theoretical meaning and observed values of deformation for soils. Stress-strain behaviour of soils, Cambridge, march 1971, pp.143-194, (ed. RGH Parry).
Schofield A.N. & Wroth C.P., 1968, Critical state soil mechanics. Mc Graw Hill, London G.B.
Taylor D.W., 1948, Fundamentals of soil mechanics. John Wiley, New-York.
Youd T.L., 1972, Compaction of sands by repeated shear straining . ASCE 98, SM7, pp. 709-725, july 1972.

Cyclic mobility
A critical state model

M. J. PENDER
University of Auckland, Auckland, New Zealand

1 INTRODUCTION

Two phenomena have been identified for the cyclic loading behaviour of saturated soils. Liquefaction refers to the substantial and irrecoverable loss of strength that occurs when a loose material is subjected to cyclic loading. On the other hand some soils, when in a dense state, also exhibit a positive pore pressure response at low levels of cyclic shear stress. This is well established for cohesionless soils, but also occurs for overconsolidated cohesive soils (Taylor and Bacchus 1969 & Ogawa et al 1977). For shear stresses approaching failure dense materials exhibit a dilatant pore pressure response. Thus even though the pore pressure build up under cyclic loading may cause the effective confining pressure to approach zero, with a consequent loss of stiffness, the material will not fail in the manner that occurs during liquefaction. This phenomenon is termed cyclic mobility. Further discussion of the distinction between liquefaction and cyclic mobility is given by Castro (1975), Castro & Poulos (1977) and Seed (1979). This paper reports preliminary work on the development of a critical state mathematical model for the phenomenon of cyclic mobility.

1.1 Previous work

Previous papers (Pender 1977a,b,c, 1978) set out a mathematical model, covering both static and cyclic loading, for the stress-strain behaviour of overconsolidated soil. The aim of the papers was to develop a model which provided good qualitative modelling of many aspects of soil stress-strain behaviour. An additional requirement was that the input data should be few and easily determined. Although the model describes well many aspects of soil

behaviour it does not predict cyclic mobility. The model is founded on two basic groups of concepts. These are:
1. the framework of critical state soil mechanics, and
2. the use of workhardening plasticity to describe the nonlinear and irrecoverable strains.

The constitutive equation for a workhardening plastic material requires that three mathematical functions be specified. These were provided by making three hypotheses about the behaviour of overconsolidated soil. Two of these are:
1. plastic strain occurs whenever there is a change in stress ratio, and
2. undrained stress paths are parabolic in the q,p space.

The form of the undrained stress path plus the requirement that, for such a path, the sum of the recoverable and plastic volumetric strain increments must be zero, makes it possible to determine the hardening function. Thus the plastic strains are related to the recoverable volumetric strains by the slope of the swelling line, κ. The third mathematical function specifies the ratio between the increments of plastic distortion and volume change. i.e. it provides information about the plastic potential. The mathematical model for cyclic mobility suggested in this paper follows the structure of this earlier model.

1.2 Outline of model

The writer's earlier model uses the idea that at all stages of loading, involving a change in stress ratio, the state of the soil is moving towards a critical state. This results in the cyclic undrained effective stress path shown in Figure 1. The assumptions of the model constrain the state of the soil always to move towards p_{cs}, so

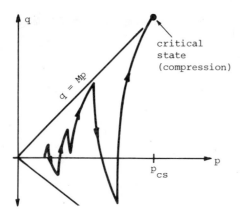

Figure 1. Cyclic undrained stress path
from Equation (1).

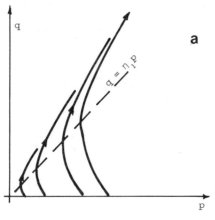

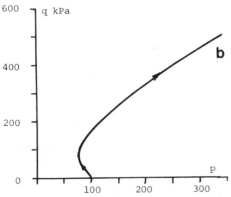

Figure 2. Undrained stress paths.
(a) For a compacted silty clay.
(b) For a saturated sand (Lade & Hernandez).

that, for a soil initially dense or overconsolidated, the pore pressure will tend to
decrease rather than increase under cyclic
loading. In Figure 2a the behaviour of a
compacted saturated silty clay (Pender 1971)
is given. For increasing shear stress there
is an initial decrease in p after which the
tendency for the state of the soil to move
towards the critical state gives negative
pore water pressures. Furthermore Figure 3
shows that when the direction of the shearing is reversed the 'unloading' part of the
stress path also moves away from the critical
state. The results of a test on sand
(Lade & Hernandez 1977) are replotted to
give a q,p stress path in Figure 2b. Once
again it is evident for increasing shear
stress that p decreases before moving towards
p_{cs}. These two diagrams confirm that the
undrained stress paths for dense sand and
overconsolidated cohesive soil, at shear
stress levels low in relation to the undrained strength of the material, give a positive
pore pressure response and, at shear stress
levels somewhat higher, a negative pore
pressure response. Figure 3 shows the
effective stress path, for the same material
as in Figure 2a, during cyclic loading. The
possibility of a progressive build up in pore
water pressure is evident. These two diagrams
suggest that the earlier undrained stress path
expression for overconsolidated soil (Equation
(1) below), although providing a good over-
view of the behaviour, is too simple to
represent the phenomenon of cyclic mobility.
Modification of the undrained stress path
enables the cyclic build up in pore water
pressure to be described. An additional
aspect of the undrained stress path that
emerges is the position of the critical state
line for extension failure. It is found that
better modelling results if it is not assumed
that the critical state line in extension is
a mirror image of that in compression.

The work reported in this paper models
cyclic mobility as a phenomenon which occurs
for initial states dry of critical. It is
suggested that liquefaction might be associated with initial states wet of critical.

Cyclic mobility is associated with an
apparent loss in shear stiffness as well
as a build-up in pore water pressure.
One criterion of cyclic mobility is a limiting value on the peak to peak strain which
occurs during a loading cycle, commonly 5%
(Castro 1975, Seed 1979). Other models for
soil behaviour under cyclic loading handle
the loss of shear stiffness by relating the
shear modulus to the current value of the
minor principal effective stress, e.g. $\sqrt{\sigma}'_3$
(Finn et al 1977, Bazant & Krizek 1976).
A similar idea is introduced herein. The
slope of the swelling line for the soil, κ,
is made a function of $\sqrt{(p_{cs}/p)}$. The

326

idealisation that κ is constant served as a reasonable approximation in the earlier work but something more sophisticated is required herein. There is ample evidence showing that, for cohesive soils, κ increases with overconsolidation ratio (Amerasinghe & Parry 1975, Mesri et al 1978).

1.3 Summary

The earlier model for the stress-strain behaviour of overconsolidated soil is modified to include the phenomenon of cyclic mobility. This is done by keeping to the earlier framework but making the following changes:
1. the shape of the undrained stress paths,
2. the slope of the swelling line, κ, a function of the mean principal effective stress p, and
3. adjusting the position of the critical state line so that (p_{cs}) extension is not equal to (p_{cs}) compression.
It is found that, although there are some difficulties, many of the features associated with cyclic mobility are modelled. The original model required four input parameters for a given soil. The model for cyclic mobility requires seven. Some of the difficulties discussed are expected to be resolved by inclusion of further developments to the earlier model that are in hand and which do not require additional parameters.

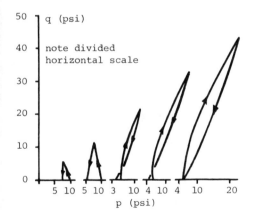

50 q (psi)

40 note divided horizontal scale

Figure 3. Cyclic undrained stress path for a compacted silty clay.

2 UNDRAINED STRESS PATHS

2.1 Mathematical expressions

The mathematical expression used in the earlier model for the undrained stress path is:

$$\left(\frac{\eta - \eta_o}{AM - \eta_o}\right)^2 = \frac{p_{cs}}{p}\left\{\frac{1 - p_o/p}{1 - p_o/p_{cs}}\right\} \qquad (1)$$

where:
η the stress ratio q/p
η_o the stress ratio at the start of the current undrained path
p mean principal effective stress, $1/3(\sigma'_1 + \sigma'_2 + \sigma'_3)$
p_o value of p at the start of the current undrained path
p_{cs} value of p at the critical state for the current void ratio
q the stress $(\sigma'_1 - \sigma'_3)$
A a parameter which adjusts the value of M for the critical state in extension or compression. For compression conditions it is +1. For extension, assuming the Mohr-Coulomb criterion, it is $-3/(3+M)$
M the value of the stress ratio q/p at the critical state in compression.

The stress path plotted in Figure 1 is calculated with Equation (1). Each time the path changes direction the values for A, η_o, p_o and p_{cs} are reset. As explained in the introduction the equation gives the general trend of the stress paths shown in Figure 2 but not a description that is accurate in detail. From Figure 2a it is suggested that there is a certain stress ratio at which p reaches a minimum value and above which p tends to move towards p_{cs}. This stress ratio is denoted herein as η_1. For states beneath the appropriate η_1 line the undrained stress path moves away from the critical state and only when the path crosses the η_1 line does the path head towards the critical state. This concept provides a simple way in which the effects illustrated in Figures 2 and 3 can be modelled.
As in the earlier work the undrained stress paths are assumed to be parabolic in the q,p space. The initial point on the path q_o,p_o is known, when $\eta = \eta_1$ $dq/dp \to \infty$, and at the critical state $\eta = M$ and $p = p_{cs}$. The equation for the stress paths is then:

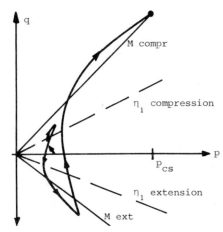

Figure 4. Cyclic undrained stress path from Equation (2).

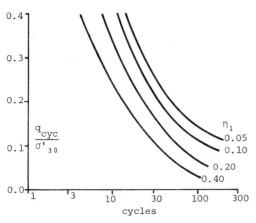

Figure 5. Effect of η_1 on the number of cycles to give $\sigma'_3 = 0$.

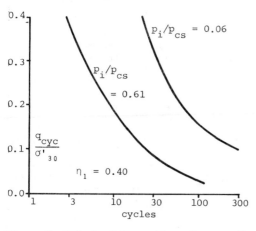

Figure 6. Effect of the ratio p_i/p_{cs} on the number of cycles to give $\sigma'_3 = 0$.

$$\left(\frac{p-p_o}{p_{cs}-p_o}\right) = \frac{(q-q_o)\{q+q_o-2A\eta_1 p_1\}}{(AMp_{cs}-q_o)\{AMp_{cs}+q_o-2A\eta_1 p_1\}} \qquad (2)$$

where:

q_o, p_o is the starting point for the current undrained path

η_1 is the value of the stress ratio at which the direction of the stress path changes

p_1 is the value of p when $\eta = \eta_1$.

Calculations for a given path require that the parameter η_1 be determined, presumably a constant for a given soil. Also the value of p_1 must be known before the stress path can be calculated. This is done, for each phase of the stress path starting from a known q_o, p_o and heading towards a known critical state M, p_{cs}, by putting $p = p_1$ and $\eta = \eta_1$ in Equation (2). The solution of the resulting quadratic equation then gives the required value of p_1. A cyclic stress path, like that shown in Figure 4, is calculated by resetting q_o, p_o, p_1 and A at each turning point. Investigation of Equation (2) shows that when $p_o = 0$ the stress path reduces to the same form as Equation (1). Thus both forms of the stress path have the same limiting curve.

2.2 Effects of various parameters on the number of cycles to $\sigma'_3 = 0$

Figure 5 shows the effect of η_1 on the number of cycles required to produce the condition $\sigma'_3 = 0$. In these calculations $M = 1$, $p_i = 0.5p_{cs}$ and η_1 covers the range $0.40 \to 0.05$. The results were calculated in the basis that $M_{extension}$ is given by the Mohr-Coulomb criterion and that $(p_{cs})_{compression} = (p_{cs})_{extension}$. As would be expected larger values of η_1 result in a smaller number of cycles to $\sigma'_3 = 0$. The ordinate in these diagrams is the ratio of the cyclic deviator stress to the effective minor principal stress at the start of the cyclic loading, q_{cyc}/σ'_{30}. This is in keeping with the literature on liquefaction and cyclic mobility.

One well documented effect on the number of cycles to reduce σ'_3 to zero is the relative density of the soil. A dense sand requires a larger number of cycles between fixed shear stress limits to reach cyclic mobility than a loose sand. Now critical state ideas do not contain explicit reference to the concept of relative density. However, the essential idea is covered by the ratio of the initial value of p for a particular path

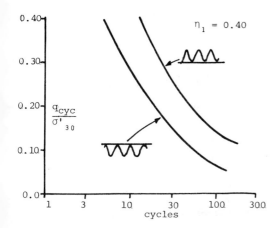

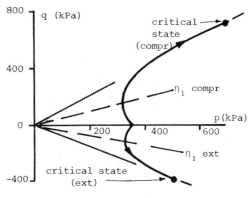

Figure 7. Effect of mean shear stress level on the number of cycles to cyclic mobility.

Figure 8. Undrained stress paths with differing positions of the critical state in compression and extension.

to the critical state p for that void ratio. When the starting value of p, p_i, is near the critical state value the state of the soil is towards the loose end of the range. On the other hand a value of p_i much smaller than p_{cs} means the material will behave in a dense manner. Figure 6 shows the number of cycles required to reduce σ'_3 to zero for two values of p_i/p_{cs}, it is clear that the effect is modelled.

Test results have shown (Annaki & Lee 1977, and Castro & Poulos 1977) that the number of cycles required to reduce the effective confining pressure to zero depends on the average shear stress level. As the average shear stress level decreases and moves from compression into extension the number of cycles to produce cyclic mobility decreases. The results of applying Equation (2) to this situation are shown in Figure 7. For the cycles with compressive stresses the curve in Figure 7 is for the condition $\sigma'_3 = 0$. For the extension case the strain amplitude reaches 5% before $\sigma'_3 = 0$. The earlier work was based on the assumption the p_{cs} for failure in extension is the same as that for failure in compression. This did not give satisfactory results so for the calculations in Figure 7 the value of p_{cs} has been determined from:

$$(p_{cs})_{extension} = |A| (p_{cs})_{compression}. \qquad (3)$$

This requirement in effect introduces another parameter into the model, although using Equation (3) means that no special test is required to determine it. In Figure 8 stress paths calculated with this

assumption are given.

3 SHEAR STRAINS

3.1 Stress-strain equations

As in the earlier work a constitutive equation for a work hardening plastic material is used:

$$d\varepsilon^p_{ij} = h \frac{\partial g}{\partial \sigma_{ij}} df \qquad (4)$$

where:
f   is the yield function
g   is the plastic potential
h   is the hardening function
$d\varepsilon^p_{ij}$ is the plastic strain increment tensor
σ_{ij} is the stress tensor.

In the earlier work the three functions f, g and h were determined by making some convenient hypothesis about the soil behaviour. A similar procedure is followed here.

The earlier hypothesis is maintained for f:

$$f_i = q - \eta_i p = 0 \qquad (5)$$

so that:

$$df = dq - \eta_i dp = pd\eta \qquad (6)$$

329

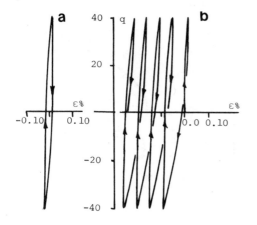

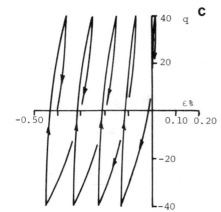

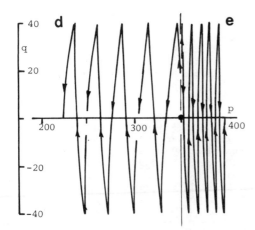

Figure 9. Comparison of stress-strain and stress path behaviour calculated with the earlier model and the cyclic mobility model.

The hardening function is determined from the concept that for an undrained test:

$$dv = 0 = dv^p + dv^r \qquad (7)$$

where:
dv^p is the increment of plastic volumetric strain
dv^r is the increment of recoverable volumetric strain.

The recoverable volumetric strain is given by:

$$dv^r = \frac{\kappa(p)\,dp}{p(1+e)} \qquad (8)$$

where:
e is the current void ratio of the soil
$\kappa(p)$ is minus the slope of the swelling line in the e, $\ln(p)$ plane.

This has the same form as the relationship used earlier but, as explained in the introduction, because of the large range of p covered by the soil approaching cyclic mobility, κ is no longer treated as a soil 'constant'. The plastic volumetric strain is:

$$dv^p = h\,\frac{\partial g}{\partial p}\,pd\eta$$

Combining this with Equations (7) and (8) gives:

$$h\,\frac{\partial g}{\partial p}\,pd\eta + \frac{\kappa(p)\,dp}{p(1+e)} = 0$$

so that:

$$h = -\,\frac{\kappa(p)\,(dp/d\eta)}{p^2(1+e)\,(\partial g/\partial p)} \qquad (9)$$

From Equation (2):

$$\frac{dp}{d\eta} = \frac{2(1-p_o/p_{cs})\,(\eta - A\eta_1 p_1/p)\cdot(p^2/p_{cs})}{(D-E)} \qquad (10)$$

where:

$$D = 2(1-p_o/p_{cs})\,(\eta - A\eta_1 p_1/p)\,(p/p_{cs})\eta$$

$$E = (AM - q_o/p_{cs})\,(AM + q_o/p_{cs} - 2A\eta_1 p_1/p_{cs})$$

330

Examination of the expressions for h and $dv^p/d\varepsilon^p$ in the earlier model, suggests that $\partial g/\partial p$ for the cyclic mobility model be defined so that:

$$d\varepsilon^p = \frac{2\kappa(p)(\eta-\eta_o)pd\eta}{p_{cs}(1+e)(D-E)[AM-\eta_o-(\eta-\eta_o)p/p_{cs}]} \quad (11)$$

The relationship used herein for $\kappa(p)$ is:

$$\kappa(p) = \kappa_{nc}\sqrt{(p_{cs}/p)} \quad (12)$$

where:

κ_{nc} is the value of κ immediately adjacent to the virgin compression line, i.e. when the overconsolidation ratio is just greater than unity.

The equation for the ratio $dv^p/d\varepsilon^p$, and hence the plastic volumetric strain, is not developed here because this model is concerned with the strains observed in the undrained situation only.

3.2 Applications

Figures 9 and 10 show results of calculations made using Equation (11).

Figure 9 gives a comparison of the stress-strain and stress path behaviour during a few cycles of loading with q ranging between ± 40 kPa. Differing assumptions about the stress path and the location of the critical state in extension are illustrated in the diagram. Figure 9a shows the closed stress-strain loop that is calculated with the previous model for the case where M has the same magnitude in extension and compression. Figure 9b shows the stress-strain behaviour calculated with the previous model for the case where the value of M in extension is controlled by the Mohr-Coulomb criterion:

$$M_{ext} = \frac{-3M_{comp}}{(3+M_{comp})} \quad (13)$$

This has the effect, when cycling between fixed limits, of giving greater strains for the extension part of the cycle. Thus the strains accumulate in the extension direction. These calculations were made with a fixed value of κ and the form of the stress path given in Equation (1). The stress path is plotted in Figure 9e. Figure 9c plots the strains calculated with Equation (11) using the form of stress path in Equation (2). Once again accumulation of extension strain is evident, primarily because the Mohr-Coulomb criterion controls the value of q_{cs} in extension. However, the

accumulation of extension behaviour, in this case, is augmented by the stress path, Figure 9d, which moves away from p_{cs} causing $\kappa(p)$ to increase.

Figure 10 gives the build up in pore water pressure and strain for loading which cycles between the q limits of 35.0 to 0.0 kPa. The strains build up considerably when $\sigma'_3 \to 0$. It is evident in Figure 10a that there is an accumulation of extension strain similar to that in Figure 9b and c. However the accumulation of extension strain is not, in principle, a serious defect as it has been noted in the literature as a real phenomenon (Annaki & Lee 1977). Also careful examination of many published diagrams giving the strain build-up during cyclic loading reveals a tendency for the strain to drift into the extension region.

4 PROBLEMS

The work reported in this paper is an initial attempt to extend the writer's earlier stress-strain model to the phenomenon of cyclic mobility. At this point it is appropriate to mention some of the difficulties that still exist.

The first defect is evident in Figures 6 and 7. The overall slope of the curves is too steep. This is because the undrained stress path, Equation (2), predicts that σ'_3 will always reach zero after a finite number of cycles. Thus for low values of

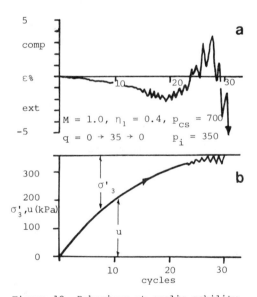

Figure 10. Behaviour at cyclic mobility.
(a) Strain build up during cyclic loading.
(b) Build up in pore water pressure.

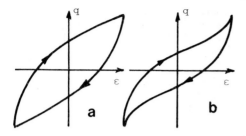

Figure 11. Shape of hysteresis loops.
(a) From the stress-strain model.
(b) Observed experimentally near cyclic mobility.

the cyclic stress ratio the model predicts that $\sigma'_3 = 0$ after too few cycles. Indeed in some cases experimental work has shown that if the cyclic stress ratio is sufficiently small then the state of the material reaches an equilibrium situation (Sangrey et al 1969). This defect is related to a similar problem in the original model where it was found that the state of the soil under cyclic loading approaches the critical state too rapidly. A further modification to the form of the undrained stress path has been found to remedy this problem. Essentially this supposes that the paths are parabolic only for the first loading. For subsequent cycles the degree of the curve is increased.

The second defect is the rate at which the strains accumulate under cyclic loading coupled with the drift in the extension direction. Once again this problem is related to a problem in the original model. There the calculated strains under cyclic loading were found to accumulate rather more rapidly than those observed in experiment. A modification to the original model increases the stiffness of the soil each time the shear stress reverses. This is also expected to remedy the problem with the cyclic mobility model.

The third difficulty relates to the shape of the hysteresis loops for the cyclic loading. It is illustrated in Figure 11. The convex hysteresis loops shown in Figure 11a are found for all conditions with the writer's earlier model and the present cyclic mobility model. At low values of p it has been found (Ogawa et al 1977, Taylor 1971) that the hysteresis loops have the pinched shape shown in Figure 11b. Two effects contribute to the stiffness of soil. Increasing q, so that the stress state moves nearer the critical state, produces a softening. Decreasing p produces softening

and increasing p produces hardening (Pender 1977b). For stress ratios with η less than η_1 the stress paths generated with Equation (2) have both of these effects leading to softening. When $\eta > \eta_1$ the increase in p causes hardening but the increase in η causes softening. Thus a mechanism is available for explaining the effect in Figure 11b, namely that the stiffening associated with increasing p is a more significant effect from the softening associated with increasing η. However it appears that for the present model the η effect dominates over the p effect, thus the calculated hysteresis loops retain their convex shape all the way to cyclic mobility.

5 CONCLUSIONS

An existing stress-strain model, based on the critical state theory, for the behaviour of overconsolidated soil has been extended to cover, at least qualitatively, the phenomenon of cyclic mobility. This is done by modification of the shape of the undrained stress path for initial states dry of critical.

It is found that the model does predict the cyclic build up in pore water pressure which is observed to occur for dense materials. The general form of the predictions is in reasonable agreement with the broad features of the phenomenon observed in laboratory tests.

Three additional items of information must be provided to specify a given soil. These are the stress ratio at which the undrained stress path changes direction, Figure 2a, the position of the critical state line for extension failure and the variation of κ with p. Thus the total number of parameters, in addition to initial conditions, which must be specified for a given soil is seven. Equations (3) and (12) provide reasonable assumptions about two of these additional items of soil data. The remaining five are determined by tests on soil samples.

Although the general phenomenon of cyclic mobility is described quite well by the model developed in the paper, there are some aspects of the phenomenon which are not modelled adequately. Further work is needed to investigate these defects, but it is hoped that refinements to the cyclic loading behaviour of the original model, at present under development, will be of assistance.

6 REFERENCES

Amerasinghe, S.F. & Parry, R.H.G. 1975, Anisotrophy of heavily overconsolidated kaolin, Proc. ASCE. Jnl. Geotech. Eng. Div., 101 GT12:1277-1293.

Annaki, M. & Lee, K.L. 1977, Equivalent uniform cycle concept for soil dynamics, Proc. ASCE. Jnl. Geotech. Eng. Div., 103 GT6:549-564.

Bazant, Z.P. & Krizek, R.J. 1976, Endochronic constitutive law for liquefaction of sand, Proc. ASCE. Jnl. Eng. Mech. Div., 102 EM2:225-238.

Castro, G. 1975, Liquefaction and cyclic mobility of saturated sands, Proc. ASCE. Jnl. Geotech. Eng. Div., 101 GT2:551-569.

Castro, G. & Poulos, S.J. 1977, Factors affecting liquefaction and cyclic mobility, Proc. ASCE. Jnl. Geotech. Eng. Div., 103 GT6:501-516.

Finn, W.D.L., Lee, K.L. & Martin, G.R. 1977, Constitutive equations for sand in simple shear, Proc. 9th ICSMFE, Tokyo, Specialty Session No. 9: Constitutive equations of soils:51-55.

Lade, P.V. & Hernandez, S.B. 1977, Membrane penetration effects in undrained tests, Proc. ASCE. Jnl. Geotech. Eng. Div., 103 GT2:109-125.

Mesri, G., Ullrich, C.R. & Choi, Y.K. 1978, The rate of swelling of overconsolidated clays subjected to unloading, Geotechnique, 28:281-307.

Ogawa, S., Shibayama, T. & Yamaguchi, H. 1977, Dynamic strength of saturated cohesive soil, Proc. 9th ICSMFE, Tokyo, Vol. 2:317-320.

Pender, M.J. 1971, The stress deformation behaviour of a compacted silty clay, Ph.D thesis, Civil Engineering Department, University of Canterbury, New Zealand.

Pender, M.J. 1977a, Modelling soil behaviour under cyclic loading, Proc. 9th ICSMFE, Tokyo, Vol II:325-331.

Pender, M.J. 1977b, The response of an oscillator with soil stress-strain behaviour, Proc. 6th Australasian Conference on the Mechanics of Structures and Materials, Christchurch, New Zealand, Vol I:201-207.

Pender, M.J. 1977c, Discussion of: Generalised cap model for geological materials, by I.S. Sandler, F.L. Dimaggio & G.Y. Baladi, Proc. ASCE. Jnl. Geotech. Eng. Div., 103 GT7:821-822.

Pender, M.J. 1978, A model for the behaviour of overconsolidated soil, Geotechnique, 28:1-25.

Sangrey, D.A., Henkel, D.J. & Esrig, M.I. 1969, The effective stress response of a saturated clay soil to repeated loading, Canadian Geotechnical Journal, 6:241-252.

Seed, H.B. 1979, Soil liquefaction and cyclic mobility evaluation for level ground during earthquakes, Proc. ASCE. Jnl. Geotech. Eng. Div., 105 GT2:201-255.

Taylor, P.W. & Bacchus, D.R. 1969, Dynamic cyclic strain tests on clay, Proc. 7th ICSMFE, Mexico, Vol I:401-409.

Taylor, P.W. 1971, The properties of soils under dynamic stress conditions, with applications to the design of foundations in seismic areas, Report No 79, School of Engineering, University of Auckland.

A bounding surface soil plasticity model

YANNIS F. DAFALIAS & LEONARD R. HERRMANN
University of California, Davis, USA

1 SUMMARY

The concept of the "Bounding Surface" in stress space is introduced within the framework of critical state soil plasticity. This yields a general plasticity model capable of describing the soil behavior under monotonic and cyclic loading conditions. The salient feature of this concept is that plastic deformation may occur when the stress state lies on or within the bounding surface, by allowing the plastic modulus to be a decreasing function of the distance of the stress state from a corresponding point on the bounding surface.

The model allows, among others, the description of the following phenomena:
1. The densification of normally consolidated or lightly overconsolidated soils with increasing deviatoric stress up to and including failure at the critical state.
2. The initial small densification with subsequent dilatation and unstable behavior of heavily overconsolidated soils with increasing deviatoric stress up to critical failure.
3. The decrease of the mean normal effective stress and increase of pore water pressure under undrained cyclic deviatoric loading of consolidated or overconsolidated soils.

In addition to the usual parameters of critical state soil mechanics, the present theory requires only one additional soil hardening function related to soil response within the bounding surface. The concept of the elastic nucleus is also introduced to account for cyclic soil stability. Comparison with experiments yields very good agreement.

2 HISTORICAL PERSPECTIVE

The concept of the bounding surface has been introduced earlier (Dafalias 1975, Dafalias and Popov 1974, 1975a, 1976, Krieg 1975) in conjunction with an enclosed yield surface for the description of the monotonic and cyclic behavior of metals. Subsequently, the concept of the yield surface was completely abandoned (Dafalias and Popov 1975b, 1977) in order to describe the behavior of materials with vanishing elastic range, in particular of artificial graphite. The rate effect was also described by a plasticity model (Dafalias, Ramey, and Sheikh 1977) where the size of the bounding surface was dependent on a properly invariant strain rate measure. Further work has improved the bounding/yield surface models (Petersson and Popov 1977), and recently the bounding surface concept found a microscopic interpretation for the uniaxial case within the framework of dislocation theory (Popov and Ortiz 1979).

A form of a bounding/yield surface model was briefly mentioned (Mroz, Norris, and Zienkiewicz 1978) and subsequently fully developed (Mroz, Norris, and Zienkiewicz 1979) within the framework of critical state soil mechanics, where also the case of vanishing elastic region has been discussed. Anisotropic properties have been also considered (Pietruszczak and Mroz 1979).

A direct bounding surface formulation in soil plasticity was qualitatively presented for the case of zero elastic range (Dafalias 1979a) and in conjunction with implied loading surfaces (Dafalias 1979b) and a quasi-elastic range. This second version is fully developed here and quantitatively applied.

3 GENERAL FORMULATION

In the following, compressive stress and strain are considered positive and all stresses are effective unless stated otherwise. A stress state σ_{ij} lies always within or on the bounding surface, Fig. 1, defined analytically by

$$F(\bar{\sigma}_{ij}, e^P) = 0 \qquad (1)$$

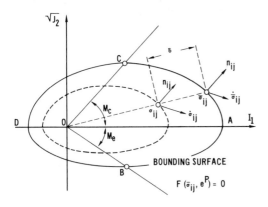

Fig. 1. Schematic Illustration of the Bounding Surface

where a bar over stress quantities indicates points on $F = 0$ and e^P, the plastic void ratio, is the only plastic internal variable considered here defining the hardening behavior of $F = 0$. The bounding surface is shown in the space of I_1 (first stress invariant) and $\sqrt{J_2}$ (root of second deviatoric stress invariant) as two different semi-ellipses intersecting the critical state lines OC (slope M_c) and OB (slope M_e) at C,B where the normal is parallel to the $\sqrt{J_2}$ axis. If ε_{ij}, ε'_{ij}, ε''_{ij} are the total, elastic and plastic strains respectively, e_i is the initial total void ratio and a dot indicates the rate, then

$$\dot{\varepsilon}_{ij} = \dot{\varepsilon}'_{ij} + \dot{\varepsilon}''_{ij} \tag{2}$$

$$\dot{e}^P = -(1 + e_i)\,\dot{\varepsilon}''_{kk} \tag{3}$$

The elastic constitutive relations are given by

$$\dot{\varepsilon}'_{ij} = C_{ijk\ell}\,\dot{\sigma}_{k\ell} \tag{4}$$

where $C_{ijk\ell}$ is the tensor of elastic compliance. To each σ_{ij} a corresponding $\bar{\sigma}_{ij}$ is defined on $F = 0$ according to a specific rule. Here $\bar{\sigma}_{ij}$ is the intersection of $F = 0$ with the line connecting the origin O and σ_{ij} (it is assumed that O lies always within or on the convex $F = 0$). If n_{ij} is the unit normal at $\bar{\sigma}_{ij}$, the plastic constitutive relations are given by

$$\dot{\varepsilon}''_{ij} = \langle L \rangle\, n_{ij} \tag{5a}$$

$$L = \frac{1}{K}\,\dot{\sigma}_{k\ell}\,n_{k\ell} = \frac{1}{K_b}\,\dot{\bar{\sigma}}_{k\ell}\,n_{k\ell} \tag{5b}$$

where K is the actual plastic modulus associated with $\dot{\sigma}_{kj}$, K_b is a plastic modulus

on the bounding surface associated with $\dot{\bar{\sigma}}_{ij}$ and the brackets $\langle \rangle$ define the operation $\langle z \rangle = z h_*(z)$, h_* being the heavyside step function. The loading function L includes the K to account for loading during the unstable material behavior when $\dot{\sigma}_{k\ell}\, n_{k\ell} < 0$ but also simultaneously $K < 0$. The K and K_b are related by

$$K = K_b + H(\sigma_{ij}, e^P)\,\frac{\delta}{\delta_o(\sigma_{ij}, e^P) - \delta} \tag{6}$$

where H is the hardening function, $\delta = \left(\mathrm{tr}\left[(\bar{\sigma}_{ij} - \sigma_{ij})^2\right]\right)^{1/2}$ is the distance between σ_{ij} and $\bar{\sigma}_{ij}$ and δ_o is a reference stress. For example, δ_o can be defined as the maximum isotropic stress OA or the distance $\bar{r} = (\bar{\sigma}_{ij}\,\bar{\sigma}_{ij})^{1/2}$. From Eqs. (3) and (5) and the consistency condition $\dot{F} = 0$, follows

$$K_b = \frac{1 + e_i}{\left[\dfrac{\partial F}{\partial \bar{\sigma}_{ij}}\dfrac{\partial F}{\partial \bar{\sigma}_{ij}}\right]^{1/2}}\,\frac{\partial F}{\partial e^P}\,\frac{\partial F}{\partial \bar{\sigma}_{kk}} \tag{7}$$

Observe that all quantities depend on σ_{ij} and the plastic loading history embodied in e^P. When $\delta = 0$, the σ_{ij} lies on $F = 0$ and $K = K_b$. With $(\partial F/\partial e^P) > 0$, Eq. (7) yields $K_b > 0$ for the domain OCAB (consolidation, n_{ij} has component along $+ I_1$), $K_b < 0$ for the domain OCDB (dilatation, n_{ij} has component along $- I_1$) and $K_b = 0$ at points C, B (no plastic volume change, n_{ij} along $J_2^{1/2}$). It follows from Eq. (5b) that during loading (L > 0), the bounding surface expands for $K_b > 0$, contracts for $K_b < 0$ and does not harden for $K_b = 0$. These values of K_b reflect into corresponding values of K through Eq. (6). It is conceivable to have $K_b < 0$ and $K > 0$ if δ is large enough which allows the description of a rising stress-strain curve and the subsequent unstable falling curve behavior when eventually both K_b, $K < 0$ for smaller δ in the case of heavily overconsolidated soils.

From the definition of n_{ij} and L, Eq. (5b), it follows that at each point σ_{ij} a surface homeothetic to the bounding surface with respect to point O is indirectly defined (shown by a dashed line in Fig. 1), which determines the paths of neutral loading from σ_{ij}. This surface defines a quasi-elastic domain but is not a yield surface since an inward motion of σ_{ij} will eventually induce loading after an initial path of unloading before σ_{ij} reaches the surface again. It is rather closer to the concept of a loading surface (Eisenberg and Phillips 1969) without exactly being one since no associated consistency condition is required. As a matter of fact it never enters the present formulation explicitly.

Inverting Eqs. (4) and (5) it follows

$$\dot{\sigma}_{k\ell} = E_{k\ell ij}(\dot{\epsilon}_{ij} - <L>n_{ij}) \tag{8a}$$

$$L = \frac{E_{rspq} \, n_{rs} \dot{\epsilon}_{pq}}{K + E_{abcd} \, n_{ab} \, n_{cd}} \tag{8b}$$

where $E_{k\ell ij}$ is the tensor of elastic moduli, inverse of $C_{ijk\ell}$. Observe from (8b) that loading-unloading events are not defined only from the sign of $E_{rspq} \, n_{rs} \dot{\epsilon}_{pq}$ but from the sign of the whole fraction since K may become negative. For $K = 0$ (beginning of instability), $\dot{\sigma}_{k\ell} \, n_{k\ell}$ must equal zero in Eq. (5b) and (8b) yields conveniently the limiting value of L. The above equations fully describe the model once the H and δ_o are defined.

4 TRIAXIAL SPACE

Further discussion will be restricted to triaxial compression. If σ_c is the total confining pressure, σ_a is the additional total axial stress and u is the pore water pressure, the total and effective mean pressures σ_m and p, and the effective deviatoric stress q are given respectively by

$$\sigma_m = (1/3)\sigma_a + \sigma_c \tag{9a}$$

$$p = \sigma_m - u \tag{9b}$$

$$q = \sigma_a \tag{9c}$$

It is also customary to set $\sigma_1 = \sigma_a + \sigma_c$ and $\sigma_2 = \sigma_3 = \sigma_c$. The corresponding strain measures are $\epsilon_p = \epsilon_1 + 2\epsilon_3$ and $\epsilon_q = (2/3)(\epsilon_1 - \epsilon_3)$ (primes will indicate elastic and plastic components). The elastic constitutive relations are given by

$$\dot{\epsilon}'_p = \dot{p}/B, \qquad \dot{\epsilon}'_q = \dot{q}/3G \tag{10}$$

where G is the shear modulus and the bulk modulus B is obtained from

$$B = (1 + e_i)p/\kappa \tag{11}$$

with κ the slope of the rebound curve in the e-lnp plot, for $p \geq 1$. For $p \leq 1$ a linear e -p relation can be assumed.

If p_o is the value of p for isotropic consolidation and λ is the slope of the consolidation line, from $\dot{e}^p = -(1 + e_i)\dot{\epsilon}''_p$ follows

$$\dot{p}_o = \frac{(1 + e_i)p_o}{\lambda - \kappa} \dot{\epsilon}''_p \tag{12}$$

4.1 The Bounding Surface Equation

Restricting to $q \geq 0$ and with $\eta = q/p = \bar{q}/\bar{p}$, a combination of 2 ellipses and a hyperbola gives the shape of the bounding surface for all η as eloquently is shown in Fig. 2, where also the other features of the model are shown in p - q space in direct correspondence with Fig. 1 (unit normal components, δ etc.). More specifically:

(i) For $0 \leq \eta < M$

If p_1, q_1 are the coordinates of the critical point C, the relation

$$p_1 = p_o/R \tag{13}$$

is a fundamental property of the material. The parameter R has been taken equal to e* = 2.72 (Schofield and Wroth 1968) or R = 2 (Roscoe and Burland 1968), but it is the opinion of the authors that R may assume other values according to the particular clay soil considered. The bounding surface is an ellipse (Ellipse 1 in Fig. 2) whose equation is

$$F = \left(\frac{\bar{p}}{p_o}\right)^2 + (R - 1)^2 \left(\frac{\bar{q}/p_o}{M}\right)^2$$

$$- \frac{2}{R}\frac{\bar{p}}{p_o} + \frac{2 - R}{R} = 0 \tag{14}$$

(ii) For $M \leq \eta < + \infty$

If the previous ellipse is extended to this range, it was found that the predictions of the model were totally unrealistic. A shape more parallel to the critical line OC was necessary, and as such a hyperbola is proposed whose apex C is at a distance a from its center G, and its asymptote is parallel to OC, Fig. 2. Assuming $a = Ap_o$, the equation of the hyperbola is defined in terms of one additional parameter A and given by:

$$F = \left(\frac{\bar{p}}{p_o}\right)^2 - \frac{2}{R}\frac{\bar{p}}{p_o} - \left(\frac{\bar{q}/p_o}{M}\right)^2$$

$$+ 2\left(\frac{1}{R} + \frac{A}{M}\right)\frac{\bar{q}/p_o}{M} - \frac{2}{R}\frac{A}{M} = 0 \tag{15}$$

(iii) For $-\infty < \eta \leq 0$
In an effort to describe in the future the

337

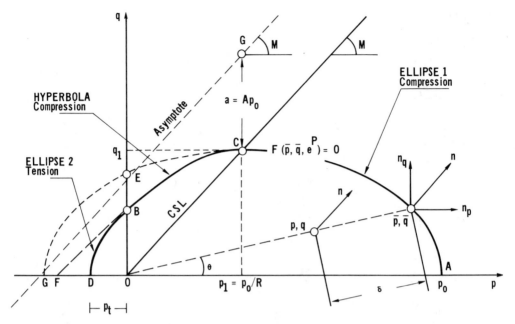

Fig. 2. The Bounding Surface in p-q Space

material behavior in tension (Al-Hussaini and Townsend 1974), the bounding surface has been extended into the $p < 0$ range as a second ellipse (Ellipse 2 in Fig. 2) with continuous tangent to the hyperbola at point B, which intersects the p axis at p_t (tension strength). Assuming $p_t = T p_0$ and defining the quantity

$$y = RA/M \qquad (16)$$

its analytical expression is given by Eq. (14) where now the M, R and p_0 must be changed respectively to M_t, R_t, P_{ot} where

$$M_t = \frac{Y - RT}{R^2 T^2} \left(\frac{Y(Y - 2RT)}{1 + y^2}\right)^{\frac{1}{2}} M \qquad (17a)$$

$$R_t = Y/RT \qquad (17b)$$

$$P_{ot} = \frac{TY}{Y - 2RT} P_0 \qquad \text{with} \qquad (17c)$$

$$Y = (1 + y)(1 + y^2)^{\frac{1}{2}} - (1 + y^2) \qquad (17d)$$

No further consideration will be given here to tension.

4.2 The Plastic Constitutive Relations

It is now possible to obtain explicitly the plastic constitutive relations as follows:

$$\dot{\varepsilon}_p'' = \langle L \rangle\, n_p \qquad (18a)$$

$$\dot{\varepsilon}_q'' = \langle L \rangle\, n_q \qquad (18b)$$

$$L = \frac{1}{K}(\dot{p}n_p + \dot{q}n_q) = \frac{1}{K_b}(\dot{\bar{p}}n_p + \dot{\bar{q}}n_q) \qquad (18c)$$

where if $f(\eta)$ is the function which yields $\bar{p} = fp_0$, $\bar{q} = \eta f p_0$ (note: $f(0) = 1$, $f(M) = 1/R$, $f(\pm\infty) = 0$) and $x = \eta/M$, then the following three cases can be distinguished:

(i) For $0 \le \eta \le M$ (Ellipse 1)

$$f(\eta) = \frac{1 \pm (R - 1)(1 + R(R - 2)x^2)^{\frac{1}{2}}}{R(1 + x^2 + R(R - 2)x^2)} \qquad (19)$$

taken with $+$ sign and

$$n_p = \frac{1}{g}(f - \frac{1}{R}) \qquad (20a)$$

$$n_q = \frac{1}{g}\, \eta f\, (\frac{R - 1}{M})^2 \qquad (20b)$$

$$g = \left[(f - \frac{1}{R})^2 + \eta^2 f^2 (\frac{R - 1}{M})^4\right]^{\frac{1}{2}} \qquad (20c)$$

338

and in correspondence to Eq. (7) from $\dot{F} = 0$ and using Eqs. (12), (14), (18a), and (20a) follows

$$K_b = \frac{1 + e_i}{\lambda - \kappa} \frac{P_o}{R} \frac{1}{g^2} (f - \frac{1}{R})(f + R - 2) \quad (21)$$

(ii) For $M \leq \eta < +\infty$ (Hyperbola)

The corresponding equations are (recall $x = \eta/M$):

$$f(\eta) =$$

$$\frac{x - 1 + xy - \left[(x - 1)^2 - 2(x - 1)y + x^2 y^2\right]^{\frac{1}{2}}}{R(x^2 - 1)}$$

$$(22)$$

$$n_p = \frac{1}{g}(f - \frac{1}{R}) \quad (23a)$$

$$n_q = \frac{1}{g} \frac{A + (M/R) - \eta f}{M^2} \quad (23b)$$

$$g = \left[(f - \frac{1}{R})^2 + \left(\frac{A + (M/R) - \eta f}{M^2}\right)^2\right]^{\frac{1}{2}} \quad (23c)$$

Again from $\dot{F} = 0$ and Eqs. (12), (15), (18a) and (23a) follows

$$K_b = \frac{1 + e_i}{\lambda - \kappa} \frac{P_o}{R} \frac{1}{g^2} (f - \frac{1}{R})$$

$$\left[\left[1 - x(1 + y)\right] f + \frac{2A}{M}\right] \quad (24)$$

(iii) For $-\infty < \eta \leq 0$ (Ellipse 2)

The same equations as for the ellipse 1 apply but with M_t, R_t substituting for M, R (Eqs. (17)) and the square root taken with the $-$ sign in Eq. (19).

For all the above cases, K is given from Eq. (6) which using $\delta_o = p_o$ and a particular form for $H(p, q, e^p)$ becomes

$$K = K_b + \frac{1 + e_i}{\lambda - \kappa} (1 + \left|\frac{M}{\eta}\right|^m) h \frac{\delta}{P_o - \delta} \quad (25)$$

and δ can be expressed as

$$\delta = (fp_o - p)(1 + \eta^2)^{\frac{1}{2}} \quad (26)$$

The m and the shape hardening factor h (Dafalias and Popov 1976) are material constants. The inclusion of $\left|M/\eta\right|^m$ in Eq. (25) guarantees that no plastic deformation occurs within $F = 0$ for $\eta = 0$ ($K = \infty$), unless $\delta = 0$. The variation of the normalized K_b with θ (Fig. 1), for $R = 2.72$, $M = 1.05$ and $A = 0.3$ is obtained from Eqs. (21), (24) and shown in Fig. 3. The inverted constitutive relations in direct correspondence to Eq. (8) are

$$\dot{p} = B(\dot{\epsilon}_p - <L>n_p) \quad (27a)$$

$$\dot{q} = 3G(\dot{\epsilon}_q - <L>n_q) \quad (27b)$$

$$L = \frac{Bn_p \dot{\epsilon}_p + 3Gn_q \dot{\epsilon}_q}{K + Bn_p^2 + 3Gn_q^2} \quad (27c)$$

4.3 Undrained Conditions

The undrained conditions impose the restriction $\dot{\epsilon}_p = \dot{\epsilon}_p' + \dot{\epsilon}_p'' = 0$, hence recalling Eqs. (9b) and (18)

$$\dot{p} = -B\dot{\epsilon}_p'' = \dot{\sigma}_m - \dot{u} \quad (28)$$

Inserting this value of $\dot{p}$ in Eq. (18a) we can solve for $\dot{\epsilon}_p''$ and using again Eqs. (9), (28) and (18) the undrained plastic constitutive relations can be obtained in terms of the total stress rates $\dot{\sigma}_a$, $\dot{\sigma}_c$ as follows

$$\dot{\epsilon}_p'' = <L>n_p, \quad \dot{\epsilon}_q'' = <L>n_q \quad (29a)$$

$$L = \frac{\dot{\sigma}_a n_q}{K + Bn_p^2} \quad (29b)$$

$$\dot{u} = (\dot{\sigma}_a/3) + \dot{\sigma}_c + <L>Bn_p \quad (30)$$

Observe now that the loading function L depends only on the $\dot{\sigma}_a = \dot{q}$ since $\dot{p}$ is not controlled independently. From Eqs. (28), (29a), (29b) the differential equation of the undrained stress path is (recall $dq = \dot{q}dt$, $dp = \dot{p}dt$)

$$\frac{dq}{dp} = -\frac{K}{Bn_p n_q} - \frac{n_p}{n_q} \quad (31)$$

339

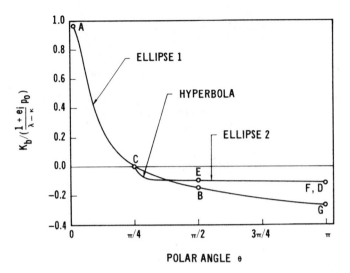

POLAR ANGLE θ

Fig. 3. Change of the Plastic Modulus K_b on the Bounding Surface with a Polar Angle θ.

It is interesting to study further Eq. (31) for different stress states and values of K. For $\eta = 0$ and $K > 0 \to n_q = 0$ and $n_p = 1$, hence $dq/dp = -\infty$, i.e., the undrained stress path is vertical to the p axis for any OCR. This is due to the elliptical shape of the bounding surface. For $\eta = M$ and $K > K_b = 0 \to n_p = 0$ and $n_q = 1$, hence $dq/dp = -\infty$, i.e., the stress path crosses the critical line parallel to the q axis. This is a property for any shape of bounding surface which has $n_p = 0$ at $\eta = M$. But when the stress state is on the bounding surface and $\eta = M$, $K = K_b = 0$, then $dq/dp = -(1/B) \lim (K_b/n_p)$, and using Eqs. (20a), (21) we readily find $dq/dp = -\kappa M/(\lambda -\kappa)$. This is independent of R and property of the intrinsic local differential geometry of the bounding surface on $\eta = M$.

Recall now from Eqs. (6) and (25) that when $\eta \ge M$ it is possible to have $K \to 0$ as $K_b < 0$ and δ is small enough. In that case $n_p < 0$, $n_q > 0$ and from Eq. (31) follows $dq/dp = -n_p/n_q$, i.e., the stress path moves towards increasing p along a near neutral loading direction while negative pore water pressure developes. This indicates the well known stabilizing effect of negative pore water pressure increase. As a result the unstable behavior of the model (falling $q - \epsilon_1$ curve) is weakly or not at all pronounced. This does not happen for drained tests.

Finally, when the stress state is on the bounding surface $K = K_b$ and recalling Eq. (21) the integration of Eq. (31) is possible in closed form yielding for the undrained stress path

$$\frac{q}{P_0} = \frac{M}{R-1}\left[\frac{2}{R}\left(\frac{p}{P_0}\right)^{\frac{\lambda - 2\kappa}{\lambda - \kappa}}\right.$$
$$\left. + \left(1 - \frac{2}{R}\right)\frac{p}{P_0}^{\frac{-2\kappa}{\lambda - \kappa}} - \left(\frac{p}{P_0}\right)^2\right]^{\frac{1}{2}} \tag{32}$$

5 GENERAL BEHAVIOR OF THE MODEL

In addition to the usual critical state soil mechanics parameters, only two additional ones h and m (Eq. (25)) must be introduced for the bounding surface formulation. The shape of the surface can be chosen in many different ways and a very simple combination of ellipses and hyperbola was proposed introducing the parameters R (usually taken as 2.72 or 2), A and T. For the elastic behavior B was related to κ through Eq. (11) and G will be computed from B and a constant Poisson's ratio ν through the usual interrelation of elastic moduli for isotropy. The objections, raised on the basis of energy dissipation, against such a change of G with p (Zytynski, Randolph, Nova and Wroth 1978) do not apply straight forward here since no purely elastic range exists in the usual sense. However, there may be still some cyclic stress paths causing problems, especially in relation to the elastic nucleus concept presented subsequently, and further investigation will be necessary.

The following two sets of material parametes will be used subsequently:

340

a. Set No. 1

M = 1.05	R = 2.72
κ = 0.05	A = 0.30
λ = 0.14	h = 30
ν = 0.15	m = 0.20

b. Set No. 2

M = 0.95	R = 2
κ = 0.05	A = 0.30
λ = 0.26	h = 90(or 30)
ν = 0.15	m = 0.20

5.1 Monotonic Loading

Using the set No. 1 the drained and undrained behavior of the model with increasing σ_a = q up to critical failure at OCR = 1, 1.2, 2, 5, 10 and 20 for initial void ratios e_i = 0.94, 0.95, 0.97, 1.02, 1.06 and 1.09 correspondingly, is shown in Figures 4 and 5 respectively. Stress or strain increments are imposed, and the model incremental response is obtained from the developed constitutive relations.

Consider first the p - q plot. Observe the characteristic "hook" of the undrained stress path for OCR = 2, 5, 10, 20. For OCR = 1, 1.2 the stress state reaches the bounding surface and moves with it until critical failure. For heavy overconsolidation and both drained and undrained tests the failure envelope lies between the critical state line projection and the initial position of the bounding surface.

Considering now the q - ε_1 plot, a distinct difference between drained and undrained behavior is observed for heavy overconsolidation. For the drained case when K = 0, because K_b < 0 and δ is small enough, the q reaches its maximum value when the stress state is between the critical state line and the contracting current bounding surface. Subsequently, K becomes negative yielding the unstable falling q - ε_1 curve in Fig. 4 for OCR = 10, 20, and q falls back progressively on the critical state line. For the undrained case, as K $\rightarrow$ 0 the stress path follows a near neutral loading direction as proved theoretically in section 4.3, and as the bounding surface contracts further the path hits the critical state line almost horizontally indicating very little or no unstable behavior of the q - ε_1 plot even for heavy overconsolidation.

Finally, the ε_p - ε_1 plot for the drained case, Fig. 4, and the u - ε_1 plot for the undrained case, Fig. 5, clearly shows that for heavy overconsolidation there is always an initial small positive increase of ε_p and u (consolidation) due to both elastic and plastic deformation, followed by a drastic plastic dilation as the stress state crosses the critical state line, leading to negative values of ε_p and u until critical failure.

5.2 Cyclic Loading

Using the set No. 2 of soil parameters with h = 30, the undrained cyclic deviatoric loading for 8 cycles and for amplitudes q/p_o = .25 and q/p_o = .42 yields the soil response as shown in Fig. 6. Observe the progressive motion of the effective stress towards the critical state line and the simultaneous expansion of the bounding surface shown by discontinuous lines. At the end of the 8th cycle and for q/p_o = .25 the stress state has not reached the critical state line. On the other hand for q/p_o = .42 four cycles suffice to bring the soil to a state where the effective stress loop does not progress anymore while pore water pressure increases and decreases cyclically with zero mean net increase, while axial strain ε_1 accumulates continuously until failure.

A deficiency of the present formulation is that such a behavior (as for q/p_o = .42) will be exhibited for any amplitude with a proper number of cycles, while it has been shown experimentally (Sangrey, Henkel and Espig 1969) that depending on the q amplitude, the soil may fail or be brought to a non-failure equilibrium. This can be very easily remedied, however, introducing the concept of the elastic nucleus in the present formulation, Fig. 6. The elastic nucleus defines a domain of purely elastic response (recall that irreversible plastic deformation is responsible for the cyclic creep phenomenon) within the bounding surface, in the sense that K = ∞ inside the nucleus. Its growth can depend on the magnitude of the plastic void ratio e^p. As the nucleus size increases, the lower amplitude loops will eventually enter its domain with full stabilization (no accumulation of ε_1) while the higher amplitude loops will still be capable of reaching the critical state line and yield the ε_1 accumulation which leads to failure. Observe that the elastic nucleus is not the same concept as that of a yield surface (no consistency condition, no loading-unloading criterion etc.).

As a matter of fact, it is not necessary to explicitly write an equation for the growth of the elastic nucleus. Its effect can be conveniently built into the form of Eqs. (6) or (25) as follows. With $\bar{r}$ = O$\bar{A}$, Fig. 6, purely elastic response, i.e., K = ∞, can be assumed whenever δ = A$\bar{A}$ > $\bar{r}$/s, where s > 1 is a stabilization factor, possibly a function of the state. This can be achieved by substituting

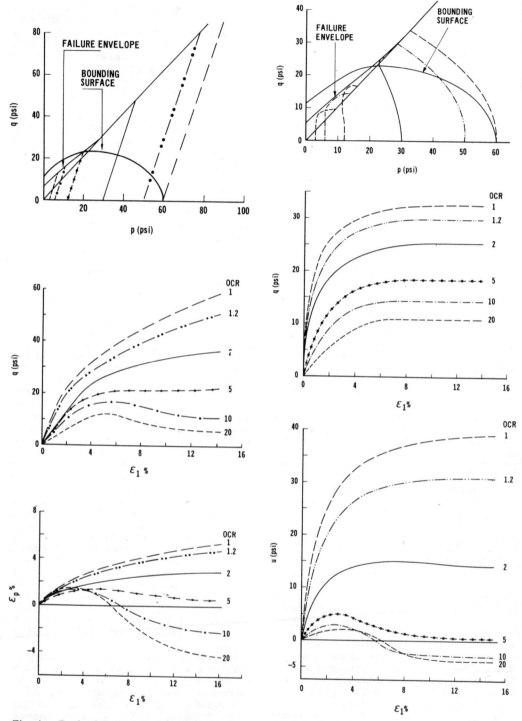

Fig. 4. Drained Behavior of the Model at Different OCR.

Fig. 5. Undrained Behavior of the Model at Different OCR.

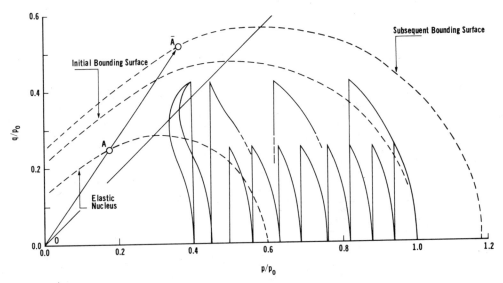

Fig. 6. Undrained Cyclic Behavior of the Model for Different Cyclic Stress Amplitudes.

for $\delta_o - \delta$ in Eq. (6) the quantity $<\bar{r} - s\delta>$, if $\delta_o = \bar{r}$, or the quantity $<p_o - (sp_o/\bar{r})\delta>$ if $\delta_o = p_o$. Then, whenever $\delta \geq \bar{r}/s$ the brackets yield a zero value and $K = \infty$. Observe that $K \rightarrow \infty$ in a continuous way as $\delta \rightarrow \bar{r}/s$. Further investigation of this modification is necessary.

6 COMPARISON WITH EXPERIMENTS

The soil parameters set No. 1 is used to predict the undrained soft clay response under monotonic deviatoric loading in compression as reported experimentally by Banerjee and Stipho (1978 and 1979). The classical parameters M, R, κ, λ are taken from the above works, ν was chosen equal to 0.15 instead of the suggested value 0.30, and only A, h and m were specified by a curve fitting for two loadings at different OCR. The comparison between measured and predicted behavior for OCR = 1, 1.2, 2, 5, 8 and 12 with initial void ratios e_i = 0.93, 0.95, 0.97, 0.95, 0.95 and 0.95 respectively is shown in Figures 7 and 8 where also the p_o is specified. For OCR = 2 no experiemental points were available for the $q - \varepsilon_1$ and $u - \varepsilon_1$ plots. The prediction can be considered very satisfactory, for the large spectrum of OCR considered.

The soil parameters set No. 2 with h = 90 is used to predict the undrained kaolin response under cyclic deviatoric loading as reported experimentally by Wroth and Loudon (1967). Again the classical critical state parameters are taken from the above work. The comparison of calculated versus experimental behavior is shown in Fig. 9. It is believed that the prediction is successful both qualitatively and quantitatively.

7 CONCLUSION

The concept of the bounding surface has been used to construct a simple plasticity model within the framework of critical state soil mechanics, capable of describing realistically the soil response under different monotonic and cyclic loading conditions, including unstable behavior and cyclic creep with further potential to include tension response. The present formulation introduces only two new parameters h and m associated wtih the general functioning of the model. A third parameter A aims at improving the shape of the used bounding surface but it is not essential to the general concept (other shapes can be used). The constitutive equations are given in closed incremental form and it is worthwhile to notice that only the current state p, q and e^p suffices to determine the next incremental response. This can be very important from a large numerical analysis point of view.

Finally, perhaps the value of this formulation is not so much its capability for realistic predictions, as is the simple idea that any sound classical yield surface soil plasticity model can be easily transformed into a corresponding and more flexible bounding surface model.

ACKNOWLEDGEMENTS

The research was in part carried out in conjunction with a project funded by the U.S. Army Engineer Waterways Experiment Station, Vicksburg, Mississippi.

343

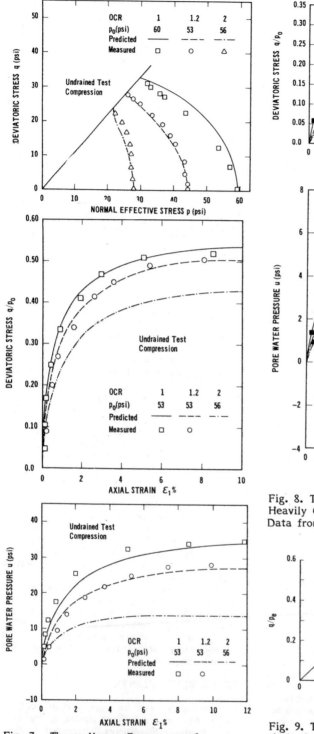

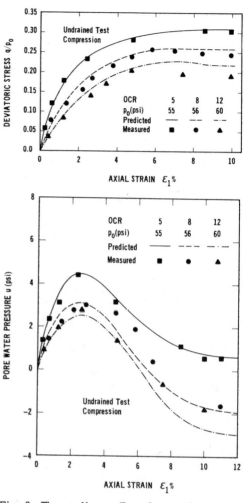

Fig. 8. Theory Versus Experiments for Heavily Overconsolidated Clay (Experimental Data from Banerjee and Stipho 1979).

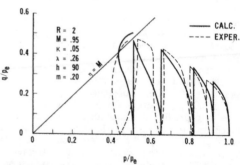

Fig. 7. Theory Versus Experiments for Lightly Overconsolidated Clay (Experimental Data from Banerjee and Stipho 1978).

Fig. 9. Theory Versus Experiments for Undrained Cyclic Loading (Experimental Data from Wroth and Loudon 1967).

REFERENCES

Al-Hussaini, M.M. and F.C. Townsend 1974, Investigation of tensile testing of compacted soils, Miscellaneous paper S-74-10, U.S. Army Engineer Waterways Experiment Station, Vicksburg, Mississippi.

Banerjee, P.K. and A.S. Stipho 1978, Associated and non-associated constitutive relations for undrained behavior of isotropic soft clays, Int. J. Numerical and Analytical Methods in Geomechanics, 2:35-56.

Banerjee, P.K. and A.S. Stipho 1979, An elasto-plastic model for undrained behaviour of heavily overconsolidated clays, Int. J. Numerical and Analytical Methods in Geo-mechanics (Short Communication), 3:97-103.

Dafalias, Y.F. 1975, On cyclic and anisotropic plasticity: i) A general model including material behavior under stress reversals, ii) anisotropic hardening for initially orthotropic materials, Ph.D. Thesis, University of California, Berkeley.

Dafalias, Y.F. 1979a, A bounding surface plasticity model, Proc. 7th Canadian Congress of Applied Mechanics, Sherbrooke, Canada.

Dafalias, Y.F. 1979b, A model for soil behavior under monotonic and cyclic loading conditions, Trans., 5th SMiRT, K 1/8, Berlin, Germany.

Dafalias, Y.F. and E.P. Popov 1974, A model of nonlinearly hardening materials for complex loadings, Proc., 7th U.S. National Congress of Applied Mechanics, p. 149 (Abstract), Boulder, USA.

Dafalias, Y.F. and E.P. Popov 1975a, A model of nonlinearly hardening materials for complex loadings, Acta Mech., 21:173-192.

Dafalias, Y.F. and E.P. Popov 1975b, A simple constitutive law for artificial graphite-like materials, Trans., 3rd SMiRT, C 1/5, London, U.K.

Dafalias, Y.F. and E.P. Popov 1976, Plastic internal variables formalism of cyclic plasticity, J. Applied Mechanics, 98(4):645-650.

Dafalias, Y.F. and E.P. Popov 1977, Cyclic loading for materials with a vanishing elastic region, Nuclear Engineering and Design, 41(2):293-302.

Dafalias, Y.F., M.R. Ramey and I. Sheikh 1977, A model for rate-dependent but time-independent material behavior in cyclic plasticity, 4th SMiRT, L 1/8, San Francisco, USA.

Eisenberg, M.A. and A. Phillips 1969, A theory of plasticity with non-coincident yield and loading surfaces, Acta Mech., 11:247-260.

Krieg, R.D. 1975, A practical two-surface plasticity theory, J. Applied Mechanics, 42:641-646.

Mroz, Z., V.A. Norris and O.C. Zienkiewicz 1978, An anisotropic hardening model for soils and its application to cyclic loading, Int. J. Numerical and Analytical Methods in Geomechanics, 2:203-221.

Mroz, Z., V.A. Norris and O.C. Zienkiewicz 1979, Application of an anisotropic hardening model in the analysis of elasto-plastic deformation of soils, Geotechnique, 29(1):1-34.

Petersson, H. and E.P. Popov 1977, Constitutive relations for generalized loadings, J. Eng. Mech. Division, ASCE, EM4:611-627.

Pietruszczak, St. and Z. Mroz 1979, Description of anisotropy of naturally K_o - - consolidated clays, Proc., Euromech Colloquium 115, Villard-de-Lans, France.

Popov, E.P. and M. Ortiz 1979, Macroscopic and microscopic cyclic metal plasticity, Proc., 3rd ASCE/EMD Specialty Conference, Austin, USA.

Roscoe, K.H. and J.B. Burland 1968, On the generalized stress-strain behaviour of "wet" clay, Engineering Plasticity, ed. J. Heyman and F.A. Leckie, Cambridge University Press, p. 535-609.

Sangrey, D.A., D.J. Henkel and M.I. Espig 1969, The effective stress response of a saturated clay soil to repeated loading, Canadian Geotechnical Journal, 6(3):241-252.

Schofield, A.N. and C.P. Wroth 1968, Critical State Soil Mechanics, McGraw-Hill, London.

Wroth, C.P. and P.A. Loudon 1967, The correlation of strains with a family of triaxial tests on overconsolidated samples of kaolin, Proc., Geotechnical Conference, Oslo, 1:159-163.

Zytynski, M., M.R. Randolph, R. Nova and C.P. Wroth 1978, On modelling the unloading-reloading behaviour of soils, Int. J. Numerical and Analytical Methods in Geomechanics (Short Comm.), 2:87-94.

Endochronic constitutive equation for soil

A.SZAVITS-NOSSAN
Faculty of Civil Engineering, Zagreb, Yugoslavia

1 INTRODUCTION

Endochronic constitutive equations were
first introduced by Valanis (Valanis, 1971)
who has shown that these equations describe
in an elegant and simple way some complex
aspects of mechanical behaviour of materials
as elastoplasticity, Bauschinger effect,
gradual transition from elastic to plastic
behaviour etc. In the original work of Valanis
endochronic equations were formulated via
linear functionals. Such formulations possess
certian shortcomings, particularly their nu ·
merical aspects as well as their adaptability
to particular material modeling. On the other
hand, by use of the differential forms of
these equations these shortcomings are greatly reduced. So far there is no established
general form of endocronic equations, but
most of the proposed (strain rate independent), (Cuellar et.al. 1977, Bažant 1976,
1977), are of the following form:

$$\overset{\circ}{T} = f\,(T,D,x_i),$$

$$x_i = d_i, \qquad i = 1,\ldots,n \qquad (1)$$

where $\overset{\circ}{T}$ is the corotational stress rate
$(\overset{\circ}{T} = \dot{T} - T\,W + W\,T\,)$, T is the effective stress
tensor, D is the strain rate tensor, (Trues-
dell, Noll 1965), d_i are some functions of
strain rate invariants such that

$$x_i = x_{io} + \int_{t=0}^{t=t} |d_i| \cdot dt \geqslant 0 \qquad (2)$$

where t denotes time. (In this paper the
time scale will be irrelevant as the strain
rate independent soil will be modeled). In
eq. 1. x_i are some directly unmeasurable
internal parameters. Function f must satisfy
basic principles of nonlinear continuum
mechanics e.g. principle of material indifference etc., (Truesdell, Noll 1965). In
this paper an endochronic constitutive equation will be constructed on the basis of real
sand stress – strain behaviour. As the first
simplification it will be assumed that the
function of eq. 1 is isotropic and that it
is linear in T and D.

2 COMPONENT OF THE EQUATION FOR ISOCHORIC (CONSTANT VOLUME) DEFORMATIONS

Typical effective stress paths in triaxial
undrained compression tests are shown on
fig. 1 in the p/q diagram:

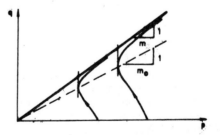

fig. 1 Typical effective stress paths for
consolidated undrained triaxial
compression tests

where the mean pressure p = - 1/3 tr T,
deviator invariant q = $1/\sqrt{3}$ $|Q|$,

$Q = \sqrt{\text{tr } Q^2}$, stress deviator $Q = T - 1/3 \cdot$
$\cdot (\text{tr } T)$ I, I is the unit tensor $(\text{tr } I = 3)$, tr
meaning the trace. Endochronic formulation
of such behaviour can approximately be modeled by

$$\overset{\circ}{Q}_1 = G p \, (E - m \, |E| \, s_1 \, Q/p) \qquad (3)$$

$$\overset{\circ}{T}_1 = - AGT \, (|E| - m_o \, \text{tr } (EQ)/p \qquad (4)$$

where $\overset{\circ}{Q}_1$ is a deviator component of $\overset{\circ}{T}$, E
being deviator of D, $(E = D - 1/3 \, (\text{tr } D) \, I)$,
$\overset{\circ}{T}_1$ component of $\overset{\circ}{T}$ in direction of T; G, m,
m_o are material constants, $m = (1+2 \, K_A)/$
$(1 - K_A) \sqrt{2}$, $K_A = \text{tg}^2 \, (45^\circ - \varphi/2)$, and A
and s_1 are quantities which will be later
defined (for the moment A may be considered as a constant and $s_1 = 1$).

Properties of eq. 3 can be analysed via the
stiffness locus. Here the stiffness locus is
defined as the surface in the stress space
traced by the stress rate vector positioned
on the top of a particular stress vector and
being the response of the material to unit
strain rate vectors. Two stiffness loci for
eq. 3 are shown fig. 2.

Stiffness locus in the deviatoric plane of
the stress space is a circle with radius Gp
shifted in the oposite direction of stress
deviator vector for the magnitude $G \, m \, q \, s_1$.
The envelope of all stiffness loci at failure
for a certain value p is a circle with radius
$r_f = p/m$.

Eq. 4. gives the shape of the stress path on
fig. 1. Parameter A has a function closely
related to the known pore pressure parameter A, (Skempton 1954).

Eqs. 3 and 4 are for the conditions of the
standard triaxial test equivalent to

$$\dot{\eta} = G \, (\dot{\varepsilon} - m \, \eta \, |\dot{\varepsilon}| \, s_1), \quad \eta = q/p \qquad (5)$$

where $\dot{\varepsilon} = \sqrt{2/3} \cdot (\dot{\varepsilon}_1 - \dot{\varepsilon}_3)$, $\dot{\varepsilon}_1$ and $\dot{\varepsilon}_3$
being vertical and lateral strain rates of the
triaxial sample. The role of the absolute
value $|\dot{\varepsilon}|$ in eq. 5 is shown on fig. 3. Parame-
ter s_1 has the role of the material deforma-
tion memory. Defining a parameter η_n as
equal to η for a monotoneous compression
history begining from $\eta = 0$, for which it
will be taken $\dot{x}_1 = |E| /\sqrt{3} = |\dot{\varepsilon}|$, it is conve-
nient to define:

$$s_1 = (1 - m \, (\eta_n - \eta) \,)^{n_1} \qquad (6)$$

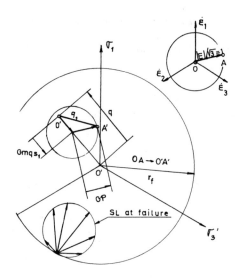

fig. 2 Stiffness locus for eq. 3 when T
and E are coaxial

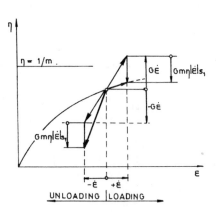

fig. 3 The mechanism of the endochronic
constitutive equation

where n_1 is a material constant. The form
of eq. 6. was chosen so that always for
$\eta = \eta_n$, $s_1 = 1$ i.e. for $\eta = \eta_n$ the material
has no memory of previous deformation
history. In the same way, for the maximal
difference $\eta_n - \eta = 1/ m$ (at $\eta_n = 1/m, \eta = 0$)
the material behaviour is totally elastic
since $s_1 = 0$. The equation for s_1 is obtained
by integrating eq. 5. and taking $s_1 = 1$:

$$\eta_n = \int_0^{\eta_n} d\eta = (1 - \exp(-G m x_1))/m \qquad (7)$$

where

$$x_1 = \int_o^{x_1} |E| \, dt / \sqrt{3}, \quad \dot{x}_1 = |E| / \sqrt{3} \qquad (8)$$

By inspection of effective stress paths in undrained cyclic triaxial tests (Castro, Poulos, 1976), it may be assumed that by successive cyclic straining parameter A from eq. 4. decreases to a certain minimum value and then begins to increase. It should be recalled that for $A = 0$ stress paths are in direction of the stress deviator Q and for $A \rightarrow +\infty$ stress paths are directed towards the origin of the stress space. The process of cyclic mobility starts when the parameter A begins to increase. The effective stress paths then start to approach the origin of the p/q space thus greatly decreasing the stiffness of the material. For materials with $m < m_o$ this process is valid only for small stress deviators since for monotoneous compression, stress path follows the failure envelope with increasing p value and the material stiffens again. On the other hand, for materials with $m = m_o$ successive stiffening under constant volume is not possible and the material liquefies. For these reasons it will be assumed that A takes the form:

$$A = A_o \, s_A \qquad (9)$$

where A_o is a material constant and

$$s_A = (1 - s_o) \, (x_1 - x_{1o})^2 / x_{1c}^2 + s_o \qquad (10)$$

where x_{1o} is the value for x_1 when $s_A = s_o \lessgtr 1$, s_o and x_{1o} being material constants.

The isochoric material behaviour is thus defined with constants G, m, m_o, A_o, s_o, x_{1o} and n_1. It should be pointed out that the proposed constitutive equation´s failure envelope is a circular cone with the vertex at $p = 0$ i.e. it is a Drucker - Prager (Drucker 1952) failure envelope. Some stress - strain curves and stress paths for a chosen set of material constants are shown on fig. 4.

3 COMPONENT OF THE CONSTITUTIVE EQUATION FOR NONISOCHORIC DEFORMATIOS

El - Sohby (El - Sohby, 1969) has shown that triaxial tests that best isolate elastic behavior of sand are the so called $\eta = const.$ unloading tests ($\dot{p} < 0$). At the same time it is known that Poisson ratios' for loading and unloading in $\eta = const.$ triaxial tests are not equal. To incorporate this behaviour in the endochronic constitutive equation it will be assumed that the strain rate tensor D has one component tensor (E´) that makes, and one (D - E´) that does not make active the equations 3 and 4. Equations 3 and 4 will be said to describe the shear stress - strain behaviour. It vill be assumed that the equation for E´ reads:

$$E' = E + (tr \, D) \, Q \, K_1 \, / \, 3 \, p \qquad (11)$$

For the component (D - E´) it will be assumed that it causes following stress rate:

$$\overset{\circ}{T}_2 = (K_2 \, p \, I + K_3 \, Q) \, tr \, D +$$
$$(K_4 \, pI + K_5 Q) \, |tr \, D| \, s_2 \qquad (12)$$

or in the component form

$$\dot{p}_2 = -K_2 \, p \, tr \, D - K_4 \, p |tr \, D| s_2 \qquad (13)$$

$$\dot{q}_2 = -K_3 \, q \, tr \, D - K_5 \, q |tr \, D| \, s_2 \qquad (14)$$

where $\dot{p}_2 = -(1/3) \, tr \, \overset{\circ}{T}_2$, $\dot{q}_2 = (1/\sqrt{3}) \cdot |\overset{\circ}{T}_2 - (1/3) \, (tr \, \overset{\circ}{T}_2) \, I|$, $tr \, (D-E') = tr \, D$; K_2, K_3, K_4 and K_5 are material constants that will be defined later, and s_2 is a memory parameter for volumetric strain history. For a normally consolidated material its value will be $s_2 = 1$. For materials in that state there will be for $- tr \, D < 0$ (unloading)

$$\dot{p}_2 = -(K_2 + K_4) \, p \, |tr \, D| \qquad (15)$$

where $K_2 + K_4 = 1/\varkappa$, $\varkappa$ being the constant known from Cam - Clay theory, (Schofield, Wroth, 1968). For $- tr \, D > 0$ (loading):

$$\dot{p}_2 = (K_2 - K_4) \, p \, |tr \, D| \qquad (16)$$

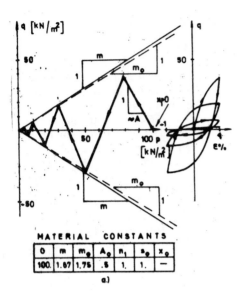

MATERIAL CONSTANTS

o	m	m_0	A_0	n_1	s_0	x_0
100.	1.07	1.75	.5	1.	1.	—

a)

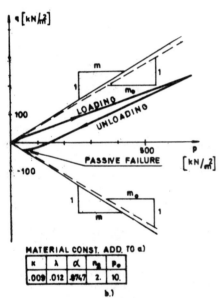

LOADING

UNLOADING

PASSIVE FAILURE $\left[kN/m^2\right]$

MATERIAL CONST. ADD. TO a)

κ	λ	α	n_2	p_0
.009	.012	.8747	2.	10.

b.)

fig. 4. Cyclic triaxial effective stress path and stress – strain curve for undrained conditions (a) and oedometer stress path (b)

where $K_2 - K_4 = 1/\lambda$, λ being also the constant from Cam – Clay theory.

From condition that the unloading η = const. triaxial test isolates best elastic strains i.e. when there is no activation of the shear mechanism via eqs. 3 and 4, it can be stated that:

$$\dot{q}_2 / \dot{p}_2 = \eta \tag{17}$$

and therefore from eqs. 13 and 14 that:

$$K_3 + K_5 = K_2 + K_4 = 1/\varkappa \tag{18}$$

Insted of constant $(K_3 - K_5)$, constant α will be introduced:

$$\alpha = \lambda (K_3 - K_5) \tag{19}$$

Constant α can be measured in η = const. loading triaxial test e.g. K_0 - test.

The value of constant K_1 will be determined by assuming that the Poisson ratio ν_e for the reversible part of the deformation process (for elastic deformations) is constant. As was earlier proposed, elastic properties of the material behaviour are best isolated by η = const. unloading triaxial tests. On the other hand, elastic shear properties are described by components of eqs. 3 and 4 not containing $|E'|$ at $\eta = 0$. Thus from eqs. 3 and 13 with the above propositions:

$$\dot{q}_1 = Gp\dot{\varepsilon} , \quad \dot{p}_2 = -p \, tr \, D /\varkappa \tag{20}$$

For the reversible part of the deformation process:

$$\dot{q}_1 / \dot{p}_2 = \eta = -\varkappa G\dot{\varepsilon}/ tr \, D , \tag{21}$$

and from eq. 11 for $E' = 0$ (irreversible shearing being isolated) and eq. 21:

$$K_1 = 3 / G\varkappa \tag{22}$$

being related to the classicaly defined Poisson's ratio:

$$K_1 = (1 + \nu_e) / (1 - 2 \nu_e) \tag{23}$$

where index "e" stays for "elastic".

Now, the constant α from eq. 19 can be related to the known coeficient of earth pressure at rest K_0 for normally consolidated sand. From eqs. 13 and 14:

$$\dot{p}_2 / p = - tr \, D /\lambda \tag{24}$$

$$\dot{q}_2 / p = - \alpha \, tr \, D /\lambda \tag{25}$$

As

$$\dot{\eta} = \dot{q} / p - \eta\dot{p} / p \tag{26}$$

350

so from eqs. 24 and 25:

$$\dot{\eta}_2 = - (\alpha - 1) \, \text{tr} \, D \, / \lambda \tag{27}$$

As for the triaxial loading K_o - test $\dot{\eta} = 0$, $\dot{\eta}_2$ must be compensated with shearing via eq. 5.

In a fictious triaxial test on a normally consolidated sand for $E' = 0$ and $- \text{tr} \, D < 0$, $\dot{\eta} = 0$ (eq. 21). To account for the known behaviour of sand that the coefficient of earth pressure at rest (K_o) increases with unloading, it must be $\dot{\eta}_2 \leqslant 0$ for $E' = 0$ and $- \text{tr} \, D > 0$. For this reason, shearing (eq. 5) that compensates $\dot{\eta}_2$ in loading K_o - test must act with $\dot{\varepsilon}' > 0$. Thus, from eqs. 5 and 27:

$$-(\alpha - 1) \eta \, \text{tr} \, D \, / \lambda + G \, (1 - m\eta) \, \dot{\varepsilon}' = 0 \tag{28}$$

Inserting the component form of eq. 11 into eq. 28, taking account of eq. 22 and observing that for K_o conditions:

$$- 3 \dot{\varepsilon} \, / \, \text{tr} \, D = \sqrt{2},$$

$$\eta = \sqrt{2} \, (1 - K_o) \, / \, (1 + 2 K_o) \tag{29}$$

the following formula for α is obtained:

$$\alpha = 1 - \lambda \, (1 - m \sqrt{2} (1 - K_o) \, / \, (1 + 2 K_o) \,) \cdot$$
$$\cdot (G \, (1 + 2 K_o) \, / \, (1 - K_o) \, - 3 / \varkappa) / 3 \tag{30}$$

As discussed previously, real values of parameter α are

$$\alpha \leqslant 1 \tag{31}$$

the equal sign being for materials with K_o independent of overconsolidation.

Yet the parameter s_2 has to be defined. Similar to parametar s_1, its values will be $s_2 \leqslant 1$, where equality holds for normally consolidated sand. This may be achieved by introducing parameter p_n:

$$s_2 = (p \, / \, p_n)^{n_2} \tag{32}$$

where n_2 is a material constant, and for normally consolidated sand $p_n = p$. For monotonous loading from eq. 13 follows

$$\dot{p}_n = -p_n \, \text{tr} \, D \, / \lambda \tag{33}$$

from which

$$p_n = p_o \, \exp \, (x_2 / \lambda) \tag{34}$$

where x_2 denotes

$$x_2 = \int_o^{x_2} | \text{tr} \, D | \, dt, \quad \dot{x}_2 = | \text{tr} \, D | \, , \tag{35}$$

p_o being a small pressure $(p_o > 0)$ for $x_2 = 0$.

It may happen that s_2 is slightly greater than 1 for $\eta > 1 / m_o$ since eq. 4 changes then its sign.

The total constitutive equation is now obtained as the sum of its components:

$$\overset{o}{T} = \overset{o}{Q}_1 + \overset{o}{T}_1 + \overset{o}{T}_2 \, ,$$
$$\dot{x}_1 = | E' | / \sqrt{3}$$
$$\dot{x}_2 = | \text{tr} \, D | \tag{36}$$

or explicitly:

$$\overset{o}{T} = C_o I + C_1 T + C_2 Q + C_3 E \, ;$$
$$\dot{x}_1 = | E' | / \sqrt{3}$$
$$\dot{x}_2 = | \text{tr} \, D |$$

where

$$C_o = -((1/\varkappa + 1/\lambda) \, \text{tr} \, D + (1/\varkappa - 1/\lambda) \cdot$$
$$\cdot | \text{tr} \, T | \, s_2) \, \text{tr} \, T$$

$$C_1 = - AG \, (| E' | / \sqrt{3} + m_o \, \text{tr} \, (E'Q) / \text{tr} \, T)$$

$$C_2 = -((1/\varkappa + \alpha / \lambda) \, \text{tr} \, D + (1/\varkappa - \alpha / \lambda) \cdot$$
$$\cdot | \text{tr} \, D | \, s_2) \, / \, 2 - G \, m | E' | s_1 / \sqrt{3}$$

$$C_3 = - G \, \text{tr} \, T \, / \, 3$$

$$A = A_o \, (\, (1 - s_o) \, (x_1 - x_{1o})^2 / x_{1o}^2 + s_o)$$

$$s_1 = (\exp \, (- G \, m \, x_1) \, - \sqrt{3} | Q | m \, / \, \text{tr} \, T)^{n_1}$$

$$s_2 = (- \, (\text{tr} \, T \, / 3 \, p_o) \exp \, (-x_2 / \lambda) \,)^{n_2}$$

$$\overset{o}{T} = \dot{T} + W \, T - T \, W$$
$$Q = T - (\text{tr} \, T) \, I \, / \, 3$$
$$E' = E - (\text{tr} \, D \, / \, G \varkappa \text{tr} \, T) \, Q$$
$$E = D - (\text{tr} \, D) \, I \, / \, 3 \tag{37}$$

To model a material behaviour, appart from eqs. 37, the inital values of T, x_1 and x_2 are to be known.

4 CONCLUSIONS

The proposed constitutive equation has following material constants: $\varkappa, \lambda$, A_o, s_o, x_{1o}, G, m, m_o, α, n_1, n_2 and p_o. Constants $\varkappa, \lambda$, A_o, G, m, m_o can be determined by conventional static triaxial tests, α is related to the coefficient of earth pressure at rest for normally consolidated material, while parameters n_1, n_2, s_o and x_{1o} are obtained from dynamic triaxial tests and have no great influence on the static behaviour. Constant p_o is an arbitrary small pressure e.g. the smallest mean pressure expected in a certain problem. Most of the material constants are equal or closely related to standard soil parameters, what is considered to be an advantage for the proposed equation.

On the other hand, the equation is able to describe rather complex soil behaviour as loading and unloading, failure, stress dilatancy, Bauschinger effect, cyclic mobility, liquefaction. At the same time there was no need for yield surfaces thus avoiding abrupt changes in material behaviour. The equation is considered not being too complex and being usable in finite element programs. At the same time, the proposed equation has overcome some shortcomings and over simplifications of earlier endochronic theories.

Practical application of the proposed constitutive equation for the solution of certain boundary value and initial value problems is demonstrated in another paper on the same conference (Szavits-Nossan, Kovačić, 1980).

REFERENCES

Bažant, Z.P., R.J.Krizek 1976, Endochronic constitutive law for liquefaction of sand, Journal of Engineering Mechanics Division ASCE, vol.102, No.EM 2,: 225-238.

Bažant, Z.P. 1976, On endochronic inelasticity and incremental plasticity, Structural Eng.Report, No.76-12/640/643 to Nat. Science Found., Northwestren University, Evanston, Illinois, USA, 44 p.

Bažant, Z.P. 1977, Endochronic and classical theories of plasticity in finite element analysis, Int. Conf. Finite Elements in Nonlinear Solid and Structural Mechanics, Geilo, Norway, Vol. 2

Castro, G.,S.J. Poulos 1976, Factors affecting liquefaction and cyclic mobility, in Liquefaction problems in geotechnical engineering, ASCE Anual Convention and Exposition, Philadelphia, USA: 105-137.

Cuellar, V., Z.P. Bažant, R.J. Krizek, M.L. Silver 1977, Densification and hysteresis of sand under cyclic shear, Journal of the Geotechnical Eng. Div., ASCE, Vol. 103, GT 5, Proc.Paper 12908: 399-416.

Drucker, D.C. 1952, A more fundamental approach to plastic stress - strain relations, Proceedings First U.S. Nat. Cong. Appl. Mech., ASME, New York: 487-491.

El-Sohby, M.A. 1969, Deformation of sands under constant stress ratios, Proc 7 th Int. Conf. SMFE, Mexico, 1,: 111-119.

Schofield, A., P. Wroth 1968, Critical state soil mechanics, London, Mc Graw - Hill

Skempton, A.W. 1954, The pore pressure coefficients A and B, Geotechnique, Vol. 4.: 143-147.

Szavits-Nossan,A., D. Kovačić 1980, Liquefaction analysis of a flexible strip foundation, Int.Symp. Soils under Cyclic and Transient Loading, Swansea, UK, in print

Truesdell, C., W. Noll, 1965, The nonlinear field theories of mechanics. in: Handbuch der Physik, S. Flugge ed., Vol.III/3, Springer Verlag

Valanis, K.C. 1971, A theory of viscoplasticity without a yield surface, I, II, Arch Mech. Stosowanej, 23, 4 Warszawa,: 517-551.

The constant stress ratio elastic model for sand

RICHARD J.W. McDERMOTT
County Planning Department, Gwynedd, North Wales, UK

1 SUMMARY

The basis and mathematical derivation of
the constant stress ratio (CR) elastic
model (Rowe 1971) to describe the non-
linear anisotropic elastic component of
deformation of sand under any axi-symmetric
loading are reviewed. The model is simplified by reference to CR stress path data
so that only three mass isotropic parameters m, S and μ together with Young's
modulus, E, of the parent material are
required for a complete elastic model for
a given sand and state of packing. The
anisotropic elastic behaviour of dense
sand can then be simply accounted for by
the non-linear variation of the Young's
moduli in the axial and radial directions
with the corresponding principal effective
stress. The variation of the elastic
compliances for use in the generalized
Hooke's law are given and it is shown that
when $\sigma_1' > \sigma_3'$ the symmetry condition
$A_{ij} = A_{ji}$ is not obeyed although a
positive strain energy function still
results. The elastic strains in the varying stress ratio (VR) stress paths
predicted from the CR model are seen to
give good agreement with the recoverable
strains obtained from the corresponding
cyclic data; exact agreement was not expected and is discussed. Results from loose
and dense sand (during the initial contract-
ive stage $R \leqslant K_{\phi\mu}$) indicate that the use
of a unique p-V^e relationship for sand is
acceptable and therefore lend support to
the Cam-clay models for 'wet'-clays. In
dense sand the increment in the elastic
shear strain during VR stress paths is
shown to be as significant as the corresponding elastic volumetric strain but being
path dependent does not permit a general
elastic shear strain energy term to be
included in an energy balance for sand.
Likewise the increment in elastic shear
strain for loose sand can be seen to be
significant.

It is concluded that the simplified model
gives sufficient accuracy for elastic
predictions in any axi-symmetric stress
path particularly during the contractive
stage in dense sand when the elastic
component is important.

2 INTRODUCTION

In both the plasticity and particulate
approaches to the study of the stress-
strain behaviour of soil a separation of
the two main components of total strain is
required namely,
1. the recoverable component, which is
stored as elastic strain energy within the
mass,
2. the non-recoverable component, which
is caused by the dissipation of energy in
overcoming friction within the mass.
These components being distinct mechanisms
of deformation follow different physical
laws and therefore may be expected to be
represented by different mathematical
formulae. Since both these mechanisms
depend on the magnitudes of the normal and
shearing stresses different proportions of
recoverable and non-recoverable deformation
may be expected to accompany different
stress paths (defined in p-q stress space
in fig. 1). In the case of sand and use of
total strain data obtained in axi-symmetric
compression and extension, plane strain and
hollow cylinder tests in the stress
dilatancy relationship of Rowe (1962),
which is strictly applicable to non-
recoverable strains due to rigid body sliding, has shown, that on first loading in
stress paths with the ratio q/p increasing
and p constant or increasing, the dominant
mechanism is the non-recoverable component
(Barden & Khayatt 1966, 1968, Procter 1967,

Tong 1970). However, these studies have also indicated that the recoverable component is important, at low shear stress levels in all stress paths, in pre-peak deformations of stress paths with p decreasing and is dominant in isotropic and anisotropic consolidation (CR) tests (El-Sohby 1964, Andrawes 1964). Similar results have been reported for the Cam-clay models (Roscoe et al 1958, 1963, Roscoe & Burland 1968, Wood 1973, 1975). On repeated (cyclic) loading to the previous maximum shear stress level, R_{max}, ($R_{max} < R_{peak}$) the recoverable component in any stress path becomes increasingly the dominant mechanism.

These are some of the important aspects which would need to be embodied in a comprehensive model accurately predicting the complex stress-strain behaviour of soil. As in the case of Cam-clay (Roscoe & Schofield 1963, Schofield & Wroth 1968) a complete and self-consistent model based on a few parameters obtained from simple tests which gives accurate qualitative and at best quantitative predictions (Wood 1973, 1975) forms a useful basic framework. Modifications in the treatment of certain aspects, for example the non-recoverable component (Roscoe & Burland 1968), stress history and induced anisotropy (Calladine 1971) or extensions to sand (Wroth & Bassett 1965) and over-consolidated clay (Calladine 1973, Pender 1978) can then be made to the basic framework to treat the case under consideration.

With the above in mind and in the context of the theme of this Symposium the following easily derived model for predicting the component of recoverable deformation for sands in all stress paths under axi-symmetric compression is presented. The model was reviewed by Rowe (1971) and further information (hitherto unreported) is now given based on research work carried out by McDermott (1972).

3 EXPERIMENTAL PROCEDURE

3.1 Modified apparatus

Previous modifications to the basic 'triaxial' cell (Bishop & Henkel 1962) for use with 100 mm high by 100 mm diameter samples with lubricated end platens (Rowe & Barden 1964, Barden & McDermott 1965) have been given by Barden & Khayatt (1966). In this study the important modification of two diametrically opposite internal dial gauges was maintained but these were sur-rounded by a waterproof brass housing to restore the use of hydraulic cell pressure.

A standard rotating bushing cell was used with external proving ring and was shown to give virtually indistinguishable readings to an internal mounted ring provided the bushing was constantly lubricated by a layer of oil introduced into the top of the cell before applying the cell pressure. A pressure transducer was used to measure cell pressure in order to obtain better control in tests with varying σ_3'. The effectiveness of the lubricated end platens (with a high vacuum silicone grease) to encourage greater uniformity of stress and strain was gauged during the tests by reference to the two vertical dial gauge supports adjacent to the sample. The experimental data for the elastic analysis were obtained at axial strains well below R_{peak} ($< 5\%$ axial strain). The layout and full details of the apparatus and photo-graphs of deformed samples together with error analyses are given in McDermott (1972).

3.2 Errors in strain measurements

It was important in this study to assess the extent of errors in axial and volumetric strains caused by axial bedding of sand grains and membrane penetration respectively and full details are given in McDermott (1972). The use of fine sand minimised these errors (El-Sohby 1964). In brief, it was found that using dummy samples and different numbers of disc coated with lubri-cated ends contained in the samples and by performing, a) isotropic consolidation tests in which the samples at first were cycled to obtain an isotropic state and, b) anisotropic consolidation tests, that the errors in axial and radial strains were dependent on the value of the corresponding principal effective stress and could there-fore be given by the same relationship.

3.3 Description of sand

The sand used principally in this investi-gation was a uniform fraction of River Welland (RW) sand passing a BSS 100 sieve ($150 \mu m$) and retained on a BSS 150 sieve ($100 \mu m$). The material was wet sieved after each test to maintain this grading and to assess the degree of particle crushing. The particle shape was essentially sub-ang-ular with maximum void ratio, $e_{max} = 0.92$ and minimum void ratio, $e_{min} = 0.62$; specific gravity 2.66. The value of $\phi_\mu = 28°$ was determined by Tong (1970) using the sliding friction test with a mass of free particles on a parent block of the material (Rowe 1962, Procter & Barton 1974).

A few comparitive tests were conducted on glass ballotini (spherical) of similar grading with $e_{max} = 0.79$, $e_{min} = 0.54$ specific gravity 3.03 and $\phi_\mu = 16°$. (ϕ_μ = angle of interparticle friction)

3.4 Testing programme

The stress paths reported herein were in axi-symmetric compression (see fig. 1) and were performed on samples prepared mainly in the dense state by vibration (Alyanak 1961) and using a constant height of deposition of particles.

4 THE CR ELASTIC MODEL

4.1 Background

A review of previous attempts to determine representative elastic parameters for clays and sands can be found in Rowe (1971) and McDermott (1972).

The basis of the model was derived by combining two hitherto separate approaches to determine the elastic parameters:

1. The anisotropic consolidation (CR>1) stress path. Results given by Thurairajah (1961) and reported by Roscoe et al (1963) for normally consolidated kaolin tested at constant $q/p = 0 \rightarrow 0.9$ (CR = 1→2.3) gave a family of parallel lines to the normal consolidation and critical state lines when plotted with $e - \log p$ axes thus indicating the essentially isotropic nature of kaolin throughout shear.

Results given by El-Sohby (1964) and Andrawes (1964) for dense sands gave a linear variation of the volumetric to axial strain on unload which varied with the value of R thus indicating the effects of the induced anisotropy (see fig. 3). Loose sands were found to remain essentially isotropic during shear.

For both clays and sands non-recoverable and recoverable components of deformation took place on first load and the unload curve was considered to be elastic.

2. The application of the generalised Hooke's law. Holubec (1966) and Barden & Khayatt (1968) used this relationship to describe the non-linear anisotropic elastic deformation of dense sand. Both attempts used cyclic data at common shear stress levels from CR and VR stress paths to determine values for the elastic compliances. The results were found to be stress path dependent (Coon & Evans 1969) and involve local slip. Holubec used the mean slope of the unload and reload curves and Khayatt (1967) used the reload curve.

4.2 Ideal assemblies

It is noteworthy that the results from the two approaches above gave broadly similar qualitative results to those predicted by Mindlin & Deresiewicz (1953) and Deresiewicz (1958) for regular packings of equal spherical particles subjected to different combinations of tangential, F, and normal, N, forces in the range F<N tanϕ where ϕ is the angle of static friction at point contact. In these studies it was predicted that local slip over the outer annulus of particle contact would invariably accompany elastic deformation in tests with a varying tangential to normal force loading path (varying obliquity paths) causing dissipation of energy and resulting in a hysteresis loop on repetitive loading. Only when the ratio of the tangential to normal force remained constant (constant obliquity paths) did the unload and reload curves become coincident after the initial portion of the cycle. The stress strain relationships were therefore not only non-linear but could involve non-recoverable deformations which being affected by the history of loading (stress path dependent) required to be treated in incremental form. This meant that the total stress strain relationship for use with incremental stresses of the same order of magnitude as the initial stress could only be obtained when the loading path was specified. The shape of the stress strain curve for a face-centred cubic array of like spheres under constant obliquity loading is shown in figure 2 and will be compared with those obtained for dense sand under similar stress paths in paragraph 4.3.

4.3 The CR stress path data

A typical family of volumetric-axial strain relationships for CR stress paths during first load and unload obtained by the author for dense River Welland sand is presented in figure 3. An essentially linear relationship is obtained between volumetric and axial strain on first load for $R \leqslant K_{\phi\mu}$. (i.e. during the contractive stage found by putting D = 1 in stress dilatancy relationship, $R = DK_{\phi\mu}$). The unload volumetric-axial strain relationship is seen to be linear for all values of R (El-Sohby 1964). The effect of cyclic loading along a CR stress path (CR = 2.97) on the stress strain behaviour can be seen in figures 4 and 5. First loading is seen to produce a much greater axial strain than subsequent reloads over the same stress range. The radial strain on first loading is seen to be opposite in direction (expansive) to subsequent reloads and unloads (compressive). This strain reversal on reloading was obtained in all CR>1 stress paths and was previously noted by El-Sohby (1964, 1969) who concluded that the slip deformation (mainly rigid body sliding) was in the reverse direction to the elastic component of deformation. It can be seen that both axial and radial strains become progressively coincident on successive cycling with the reload data approaching the almost unique unload curves. Complete coincidence was not obtained by the author even after seven cycles in a CR = 1.8 stress path but instead a residual cycle was obtained indicating that dissipation of energy was still taking place. The ratio of energy output on unload to energy input on reload, E_R, was found to rise from about 78% on the first cycle to a constant 86% (see fig. 6). At higher values of R ($> K\phi\mu$) more cycles would be required to achieve a residual cycle. It was decided to use the almost unique unload axial stress-strain curve with linear unload volumetric-axial strain relationship as the data for the CR elastic model.

In the light of the small magnitude of the elastic strains, corrections for axial bedding and membrane penetration are important and have been applied to the CR stress path data. As an example the corrected individual plots of axial stress against axial strain and axial stress against radial strain together with the corresponding uncorrected plots for CR = 2.97 are given in figure 7. It is apparent that the corrections are significant particularly at low stress levels. Similarly, corrected plots for the other CR stress paths $1 < CR \leqslant 4.67 (R_{peak} \simeq 2K_{\phi\mu} \simeq 5.5)$ were obtained.

The effect of these corrections on the volumetric-axial strain plots for CR = 2.97 is shown in figure 8. It can be seen that the original linear relationship obtained with uncorrected data is sensibly maintained. However the effect of the corrections on the ratio of the unload volumetric to axial strain ratio can be seen in figure 9 to reduce the ratio significantly and it is therefore important to use the corrected data in this analysis.

In order to compare the effect of increasing the value of the stress ratio on the unload curves obtained from CR stress paths a suitable common origin has to be determined. The corrected experimental data have been combined in figures 10 and 11 for an arbitrary axial stress datum of 35 kN/m^2 since this is the lowest value of axial stress common to all CR tests. On inspection it can be seen that for a given

increment in axial stress the 'elastic' axial strain increment increases whereas the radial strain increment decreases in magnitude with increase in R until R = 2.97. Thereafter the first unload radial strain data appear to be coincident (CR = 3.8) or show a reverse trend (CR = 4.67). As seen in figure 5 this is probably the effect of slip in the reverse direction and indicates that for $R > K\phi\mu$ first unload data may be suspect. By increasing the value of the arbitrary datum the curves in figures 10 and 11 become progressively flatter and may even appear to be unique. Therefore it is expected that the curves will further diverge by reducing the value of the common axial stress datum. For this reason it is essential to be able to extrapolate the experimental curves to a hypothetical 'zero' stress in order to give the absolute stress-strain curve as seen in paragraph 4.6.2.

It is informative at this stage to compare the similarity in shape of the axial stress-strain plots (fig. 10) obtained with sub-angular dense RW sand with the result obtained from a face-centred cubic array of equal spheres (Thurston & Deresiewicz, 1959) in figure 2.

4.4 The reduced form of the generalised Hooke's Law

It has been shown (Barden 1963, Pickering 1970) that for the axi-symmetric case with equal radial stresses $\sigma_2' = \sigma_3'$ the generalised Hooke's Law reduces to five independent constants as given in equations 1 and 2 to describe the axial and radial elastic deformations respectively

$$d\varepsilon_{r_1}^e = A_{11} d\sigma_1' + 2A_{13} d\sigma_3' \qquad 1$$

(subscript r = resultant strain)

$$d\varepsilon_{r_3}^e = A_{31} d\sigma_1' + A_{32} d\sigma_2' + A_{33} d\sigma_3' \qquad 2$$

Since A_{32} and A_{33} cannot be determined separately in the 'triaxial' apparatus and are dependent on σ_3' they can be designated by a summation term $\alpha = A_{32} + A_{33}$ thus reducing the constants to four. However, it must be emphasised that equations 1 and 2 are only valid for small deformations and in order to apply them to the non-linear elastic behaviour of soil the stress-strain curve is considered to be composed of a large number of infinitesimal linear increments.

Equations 1 and 2 are rewritten to apply to the four stress paths analysed.

356

CR stress path

This test is performed so that $d\sigma_1' = R d\sigma_3'$

thus $d\varepsilon_{r_1}^e = \left(A_{11} + {}^2/R\, A_{13}\right) d\sigma_1'$ 3

and $d\varepsilon_{r_3}^e = \left(A_{31} + {}^\alpha/R\right) d\sigma_1'$ 4

Constant mean principal stress test, CP

This test is performed so that $d\sigma_1' = 2 d\sigma_3'$

thus $d\varepsilon_{r_1}^e = \left(A_{11} - A_{13}\right) d\sigma_1'$ 5

and $d\varepsilon_{r_3}^e = \left(A_{31} - {}^\alpha/2\right) d\sigma_1'$ 6

Constant cell pressure test, $C\sigma_3'$

This test is performed so that $d\sigma_3' = 0$

thus $d\varepsilon_{r_1}^e = A_{11}\, d\sigma_1'$ 7

and $d\varepsilon_{r_3}^e = A_{31}\, d\sigma_1'$ 8

Constant axial stress test, $C\sigma_1'$

This test is performed so that $d\sigma_1' = 0$

thus $d\varepsilon_{r_1}^e = 2 A_{13}\, d\sigma_3'$ 9

and $d\varepsilon_{r_3}^e = \alpha\, d\sigma_3'$ 10

Equations 3 to 10 are seen to give the instantaneous slopes of the elastic stress-strain curves for the respective stress paths. In particular equations 3 and 4 for the CR stress path represent the instantaneous slope of the elastic stress-strain curve which as discussed in paragraph 4.3 can be obtained by unloading along the same CR stress path. By definition the elastic constants at a given stress level should be stress path independent and therefore the same values of the compliances should obtain in other stress paths at the same stress level. As a direct result it should be possible therefore to trace continuous functions for other stress paths with varying R using the unload CR stress path data

(McDermott 1970). These relationships would be the integrated form of equations 5 to 10 and represent the true elastic component of deformation to within the experimental limits defined in paragraph 4.3.

4.5 Empirical form of Hertz contact equation

It is necessary to extrapolate the experimental unload CR stress path curves to a hypothetical 'zero' stress in order to give the absolute stress-strain framework. In order to perform the necessary extrapolation and at the same time derive the elastic stress strain relationship for the various stress paths Rowe (1971) suggested the following semi-empirical form of the Hertz contact equation be applied to equations 3 to 10

$$\varepsilon_{1,3}^e = S_{1,3} \left(\frac{\sigma_{1,3}}{E}\right)^m \quad\quad 11$$

cf. $$\varepsilon = 0.615 \left(\frac{4\sigma}{E}\right)^{2/3} \quad\quad 12$$

for the axial compressive strain ε due to axial stress σ applied to equal spheres in contact after Hertz (Wilson & Sutton 1948) where superscript e refers to the 'elastic' strain components in the axial and radial perpendicular directions 1 and 3.

$S_{1,3}$ and m are 'elastic' coefficients which are expected to be functions of the structure of all the particle contacts and not just those at sliding equilibrium. Therefore $S_{1,3}$ and m are expected to be independent of R.

E is the Young's modulus of the parent mineral of the sand particles.

σ_1', σ_3' are the principal effective stresses in the two mutually perpendicular directions 1 and 3.

The value of m is assumed equal in the axial and radial directions since it has been shown in paragraph 4.3 that the ratio of the volumetric strain to axial strain is constant during a CR test. Equation 11 yields $\varepsilon_{1,3}^e = 0$ when $\sigma_{1,3}' = 0$ and is seen to reduce to the isotropic linear elastic form by putting $m = S_{1,3} = 1$.

4.6 The non-linear stress-strain relationships

These relationships are given in more detail by Rowe (1971) and McDermott (1972) and their derivation is briefly indicated below.

357

By definition

$$A_{11} = \frac{1}{E_1} \quad ; \quad A_{13} = -\frac{\mu_{13}}{E_3}$$

$$\alpha = A_{32} + A_{33} = \frac{1 - \mu_{32}}{E_3} \quad ; \quad A_{31} = -\frac{\mu_{31}}{E_1}$$

Substituting for A_{11} and A_{13} in equation 3 and for A_{31} and α in equation 4 gives the strain relationships 13 and 14

$$d\varepsilon^e_{r1} = d\varepsilon^e_1 - 2\mu_{13} d\varepsilon^e_3 \qquad 13$$

$$d\varepsilon^e_{r3} = d\varepsilon^e_3 - \mu_{31} d\varepsilon^e_1 - \mu_{32} d\varepsilon^e_2 \qquad 14$$

Differentiating equations 11

gives $\quad d\varepsilon^e_1 = S_1 . \frac{m}{G_1} \left(\frac{G_1'}{E}\right)^m dG_1' \qquad 15$

and $\quad d\varepsilon^e_3 = S_3 . \frac{m}{G_3'} \left(\frac{G_3'}{E}\right)^m dG_3' \qquad 16$

Substituting equations 15 and 16 in equation 13 gives

$$d\varepsilon^e_{r1} = S_1\left(\frac{m}{G_1}\right)\left(\frac{G_1'}{E}\right)^m dG_1' - 2\mu_{13} S_3\left(\frac{m}{G_3'}\right)\left(\frac{G_3'}{E}\right)^m dG_3' \quad 17$$

which can be simplified to give

$$d\varepsilon^e_{r1} = \frac{Q_1 m S_1}{G_1'} \left(\frac{G_1'}{E}\right)^m dG_1' \qquad 18$$

where

$$Q_1 = 1 - 2\mu_{13}. \frac{S_3}{S_1} . \frac{1}{R^{m-1}} . \frac{dG_3'}{dG_1'} \qquad 19$$

for a constant stress ratio test

$$\frac{dG_3'}{dG_1'} = \frac{1}{R}$$

therefore

$$(Q_1)_{CR} = 1 - 2\mu_{13}. \frac{S_3}{S_1} \frac{1}{R^m} \qquad 20$$

Similarly substituting equations 15 and 16 in equation 14 gives

$$d\varepsilon^e_{r3} = S_3 \frac{m}{G_3'}\left(\frac{G_3'}{E}\right)^m \frac{1}{R} dG_1' - $$
$$\mu_{31} S_1\left(\frac{m}{G_1'}\right)\left(\frac{G_1'}{E}\right)^m dG_1' - \mu_{32} S_3\left(\frac{m}{G_3'}\right)\left(\frac{G_3'}{m}\right) \frac{1}{R}. dG_1' \qquad 21$$

$$d\varepsilon^e_{r3} = \frac{Q_3.m.S_1}{G_1'}.\left(\frac{G_1'}{E}\right)^m. dG_1' \qquad 22$$

where

$$(Q_3)_{CR} = \frac{S_3}{S_1}. \frac{1}{R^m}.(1 - \mu_{32}) - \mu_{31}$$

$$\qquad 23$$

These equations are now applied to the CR stress path.

4.6.1 CR stress path

In these tests $\frac{dG_1'}{dG_3'} = R = $ constant.

By inspection of equation 20 the value of Q_1 will be constant for a given value of R provided S_1, S_3, m and μ_{13} are constant. By plotting the test data on reference axes ε^e_{r1} and G_1' the slopes of the curve will give values of $d\varepsilon^e_{r1}/dG_1'$ which can be related to equation 18. In particular by taking the logarithm of equation 18 whereby

$$\log\left(\frac{d\varepsilon^e_{r1}}{dG_1'}\right) = \log\left(\frac{Q_1 S_1 m}{E^m}\right) - (1-m)\log G_1' \quad 24$$

the plot of $\log G_1'$ against $\log\left(d\varepsilon^e_{r1}/dG_1'\right)$ should be linear of slope $-(1-m)$ provided $\frac{Q_1.S_1.m}{E^m}$ remains constant during the test. The results of different CR stress paths obtained by drawing tangents to points on the curve at different values of G_1' are presented in figure 12 and indicate that an essentially straight line relationship exists and this therefore confirms the use of equation 11. In addition an important point to notice is that the value of m as calculated from the slope of the graphs $(-(1-m))$ is sensibly constant for all R. This indicates the assumption that the value of m is sensibly independent of the value of R is valid and that the non-linear 'elastic' deformation is a function of the structure of all the point contacts. Also by reference to equation 24 the product $Q_1 S_1$ must be sensibly constant throughout a given CR test.

Similarly, the logarithm of equation 22 given by equation 25 can be applied to the radial strain-axial stress data

$$\log\left(\frac{d\varepsilon^e_{r3}}{dG_1'}\right) = \log\left(\frac{Q_3.S_1.m}{E^m}\right) - (1-m)\log G_1' \quad 25$$

and it can be shown likewise that $Q_3 S_1$ is sensibly constant throughout a given CR stress path. However a simpler method for predicting the radial strain-axial stress relationship is obtained by using the constant ratio of $V_r^e / \varepsilon_{r_1}^e$ for a given value of R. Now $dV_r^e = d\varepsilon_{r_1}^e + 2d\varepsilon_{r_3}^e$

for infinitesimal strains in axi-symmetric test conditions.
Substituting equations 18 and 22 gives

$$dV_r^e = \frac{mS_1}{G_1'}\left(\frac{G_1'}{E}\right)^m dG_1' \left(Q_1 + 2Q_3\right)$$

and therefore

$$\frac{dV_r^e}{d\varepsilon_{r_1}^e} \quad \frac{Q_1 + 2Q_3}{Q_1} = 1 + \frac{2Q_3}{Q_1} \quad 26(a)$$

Alternatively $\dfrac{d\varepsilon_{r_3}^e}{d\varepsilon_{r_1}^e} = \dfrac{Q_3}{Q_1}$ 26(b)

Since $dV_r^e/d\varepsilon_{r_1}^e$ (and hence $d\varepsilon_{r_3}^e/d\varepsilon_{r_1}^e$) is known to be constant for a given CR stress path then Q_3/Q_1 is constant throughout a given CR stress path.

An important initial use of equations 18 and 22 (together with the knowledge that the value of m remains constant throughout a CR stress path) lies in the extrapolation of the experimental stress-strain curves to zero datum stress.

4.6.2 The extrapolation of CR stress path data

Denoting the unknown initial strain corresponding to the arbitrary experimental datum stress in a constant stress ratio test by $(\varepsilon_{r_1}^e)_o$ and the experimentally determined increment of strain by $\Delta\varepsilon_{r_1}^e$ due to an increase in axial stress then the total strain $\varepsilon_{r_1}^e$ from zero datum will be given by

$$\varepsilon_{r_1}^e = (\varepsilon_{r_1}^e)_o + \Delta\varepsilon_{r_1}^e \quad 27$$

since both $\varepsilon_{r_1}^e$ and $(\varepsilon_{r_1}^e)_o$ represent finite strains equation 18 requires to be integrated before substitution into equation 27. Integration of 18 gives

$$\varepsilon_{r_1}^e = Q_1 S_1 \left(\frac{G_1'}{E}\right)^m + C_1 \quad 28$$

Since $\varepsilon_{r_1}^e = 0$ when $G_1' = 0$ then $C_1 = 0$ Substitution of equation 28 in equation 27 gives

$$Q_1 S_1 \left(\frac{G_1'}{E}\right)^m = Q_1 S_1 \left(\frac{G_{1o}'}{E}\right)^m + \Delta\varepsilon_{r_1}^e$$

on rearranging

$$\Delta\varepsilon_{r_1}^e = \frac{Q_1 S_1}{E^m} \cdot G_{1o}' \left(\left(\frac{G_1'}{G_{1o}'}\right)^m - 1\right)$$
$$= (\varepsilon_{r_1}^e)_o \left(\left(\frac{G_1'}{G_{1o}'}\right)^m - 1\right) \quad 29$$

Since the value of m is sensibly constant then by plotting $(G_1'/G_{1o}')^m - 1$ against $\Delta\varepsilon_{r_1}^e$ the value of the slope obtained will give the unknown value of $(\varepsilon_{r_1}^e)_o$. This value can be added to the measured experimental strains to give the absolute stress-strain plot.
Similarly it can be shown

$$\varepsilon_{r_3}^e = Q_3 S_1 \left(\frac{G_1'}{E}\right)^m \quad 30$$

$$\Delta\varepsilon_{r_3}^e = (\varepsilon_{r_3}^e)_o \left(\left(\frac{G_1'}{G_{1o}'}\right)^m - 1\right) \quad 31$$

The above procedure has been carried out for all the CR stress paths performed on River Welland sand using a datum value of 35 kN/m^2 and presented in figure 13. The value of m was found to be sensitive to the plot and by decreasing or increasing the value of m by 0.01 a non-linearity to the left or to the right of the straight line was obtained respectively. The approximate value of m determined previously by fitting tangents to the curves (fig. 12) could be therefore made more exact. It is seen from figure 13 that a value of $m = 0.35$ for River Welland sand can be taken as constant for all the CR stress paths. This is a significant result in this analysis and will enable further simplifications to be attempted later in paragraph 4.8. An additional check that the correct value of m and $(\varepsilon_{r_1}^e)_o$ has been chosen can be made by using other arbitrary datums of say 52 and 69 kN/m^2. The absolute axial stress axial strain plots for the different CR stress paths have been plotted together in figure 14.
In a similar manner the absolute plot for the axial stress radial strain data can be determined. However simplification is effected by using the $\varepsilon_{r_1}^e - G_1'$ absolute plot and the observed constancy between the ratio of the elastic radial and elastic axial strain for a given constant stress ratio as previously expressed in equation 26(b) and in a more direct form in the following equation

359

Since $\left(\dfrac{dV_r^e}{d\varepsilon_{r_1}^e}\right)_{CR} = \left(\dfrac{\Delta V_r^e}{\Delta\varepsilon_{r_1}^e}\right)_{CR}$ constant

and $2d\varepsilon_{r_3}^e = dV_r^e - d\varepsilon_{r_1}^e$

then $d\varepsilon_{r_3}^e = \dfrac{1}{2}\left(\dfrac{dV_r^e}{d\varepsilon_{r_1}^e} - 1\right)d\varepsilon_{r_1}^e$ 32

The relationship between $dV_r^e / d\varepsilon_{r_1}^e$ and R is given in figure 9. The resulting absolute axial stress-radial strain plot is given in figure 15. The absolute volumetric strain-axial stress plot given in figure 16 is obtained using the relationship between $dV_r^e / d\varepsilon_{r_1}^e$ and R and the absolute axial strain-axial stress plot given in figure 14.

Figures 14, 15 and 16 form the basic elastic stress strain model and in paragraph 4.7 it will be shown how the elastic strains during variable R stress paths can be predicted.

4.7 The prediction of elastic strains in VR stress paths

4.7.1 The CG_3' stress path

The elastic strains occurring during a CG_3' stress path are obtained from the CR model by joining points of constant cell pressure on each of the CR curves as shown in figures 17 and 18. The resulting axial stress-axial strain relationships are found to be continuous and non-linear with a gradual decreasing positive slope. The resulting axial stress-radial strain graphs are also non-linear but with a decreasing negative slope.

The form of these relationships can be derived as follows from the basic equations 18 to 23.

For the CG_3' stress path $dG_3' = 0$ therefore in equation 19 $Q_1 = 1$ and in equation 23 $Q_3 = -\mu_{31}$
Substitution in equation 18 gives

$\left(\dfrac{d\varepsilon_{r_1}^e}{dG_1'}\right)_{CG_3'} = \dfrac{mS_1}{G_1'}\left(\dfrac{G_1'}{E}\right)^m$ 33

which upon integration gives

$\left(\varepsilon_{r_1}^e\right)_{CG_3'} = S_1\left(\dfrac{G_1'}{E}\right)^m + \left(C_1\right)_{CG_3}$

The value of $\left(C_1\right)_{CG_3}$ is obtained from the isotropic consolidation test plotted in figure 14 since the elastic shear strains in all stress paths are measured from the value of the consolidation pressure. Then

denoting the cell pressure by
$\left(G_1'\right)_o = G_3' = $ constant and the initial strain on the isotropic consolidation line corresponding to this initial stress by $\left(\varepsilon_1^e\right)_o$ then

$\left(C_1\right)_{CG_3} = \left(\varepsilon_1^e\right)_o - S_1\left(\dfrac{G_{1o}'}{E}\right)^m$

The equation for the elastic stress-strain relationship for the CG_3' stress path becomes

$\left(\varepsilon_{r_1}^e\right)_{CG_3} = \left(\varepsilon_1^e\right)_o + S_1\left(\left(\dfrac{G_1'}{E}\right)^m - \left(\dfrac{G_{1o}'}{E}\right)^m\right)$ 34

Since $\left(\varepsilon_1^e\right)_o$ is the elastic strain undergone during the initial isotropic consolidation to the required cell pressure it can be evaluated using equation 28 with R = 1. Therefore equation 34 gives the combined isotropic and shear elastic axial strain at any stage during the stress path.

Similarly it can be shown that

$\left(\dfrac{d\varepsilon_{r_3}^e}{dG_1'}\right)_{CG_3} = \dfrac{-\mu_{31} mS_1}{G_1'}\left(\dfrac{G_1'}{E}\right)^m$

and

$\left(\varepsilon_{r_3}^e\right)_{CG_3} = -\mu_{31}S_1\left(\dfrac{G_1'}{E}\right)^m + \left(C_2\right)_{CG_3}$

and evaluating $\left(C_2\right)_{CG_3}$

gives

$\left(\varepsilon_{r_3}^e\right)_{CG_3} = \left(\varepsilon_3^e\right)_o - \mu_{31}S_1\left(\left(\dfrac{G_1'}{E}\right)^m - \left(\dfrac{G_{1o}'}{E}\right)^m\right)$ 35

where $\left(\varepsilon_3^e\right)_o$ is given by equation 30 with evaluated for R = 1 (equation 23). Likewise equation 35 allows the combined isotropic and shear elastic radial strain to be calculated at any stage during a stress path. The form of equations 34 and 35 have been adopted since they allow an analogy to be made with the principal of superposition in linear isotropic elasticity in which the hydrostatic and deviatoric components are superposed. In addition this form allows calculations to be made more easily.

4.7.2 The CP stress path

The elastic strains occurring during a CP stress path are obtained from the CR model by joining points of constant mean principal stress on each of the CR curves shown in figures 17 and 18. It can be seen that the resulting relationship of axial stress-axial strain is continuous and sensibly linear and that the relationship of axial stress-radial strain is slightly non-linear with an increasing negative gradient. The form of these equations are derived

360

using $d\sigma_1' = -2 d\sigma_3'$. Therefore the elastic stress strain equations for the CP stress path become

$$\left(\varepsilon^e_{r1}\right)_{cp} = Q_1 S_1 \left(\frac{\sigma_1'}{E}\right)^m \qquad 36$$

and

$$\left(\varepsilon^e_{r3}\right)_{cp} = Q_3 S_1 \left(\frac{\sigma_1'}{E}\right)^m \qquad 37$$

in which both Q_1 and Q_3 are variable. The initial strain due to isotropic consolidation is given by evaluating Q_1 and Q_3 when $R = 1$. The elastic volumetric strain-axial stress relationship is discussed further in paragraph

4.7.3 $c\sigma_1'$ stress path

The elastic axial strains and elastic radial strains occurring in a stress path can be found by simply constructing perpendicular lines in figures 14 and 15 respectively for a given value of axial stress. The form of the equations giving the variation of these strains with the variable, σ_3' can be shown to be

$$\left(\varepsilon^e_{r1}\right)_{c\sigma_1} = \left(\varepsilon^e_1\right)_o - 2\mu_{13} S_3 \left(\left(\frac{\sigma_3'}{E}\right)^m - \left(\frac{\sigma_{3o}'}{E}\right)^m\right) 38$$

where $\left(\varepsilon^e_1\right)_o$ is given by equation 28 with evaluated with $R = 1$

$$\left(\varepsilon^e_{r3}\right)_{c\sigma_1} = \left(\varepsilon^e_3\right)_o + S_3 \left(1-\mu_{32}\right)\left(\left(\frac{\sigma_3'}{E}\right)^m - \left(\frac{\sigma_{3o}'}{E}\right)^m\right) 39$$

where $\left(\varepsilon^e_3\right)_o$ is given by equation 30 with Q_3 evaluated with $R = 1$. Equations 38 and 39 can be compared with equations 35 and 34 respectively for the $c\sigma_3'$ stress path. Likewise it is seen that an analogy can be made with the principle of superposition of the hydrostatic and deviatoric components in isotropic linear elasticity.

4.8 Simplifying the CR elastic model

An inspection of the foregoing equations in paragraphs 4.6 and 4.7 for the four stress paths shows that for a complete specification of the elastic CR model the elastic parameters m, S_1, S_3, μ_{13}, μ_{31} and μ_{32} need to be evaluated for all states of stress. As shown the value of m can be obtained from the CR stress paths whilst the $c\sigma_3'$ stress paths yield the values of S_1 and μ_{31} directly. The parameters S_3, μ_{13} and μ_{32} can not however be determined directly and would require to be determined

in a true triaxial apparatus in which all three principal stresses can be independently controlled. It would however be possible to solve the elastic model for all stress paths using the parameters Q_1 and Q_3 whereby reducing the number of variables to four. Knowing the value of S_1 both Q_1 and Q_3 could be obtained from equations 24 and 25 respectively since E^m can be calculated and the variation of Q_1 and Q_3 with R could be determined. In paragraph 4.6.2 the value of m was found to be constant for all CR stress paths and the CR model will now be examined to see if further simplifications can be made.

Firstly, the variation of the ratio $V^e_r / \varepsilon^e_{1r}$ with R given in figure 9 will be examined. Equation 26(a) indicates that in order for the elastic strain ratio to be constant for a given value of stress ratio then the ratio of Q_3 / Q_1 must remain constant during a CR stress path. Hence

$$\frac{dv^e_r}{d\varepsilon^e_{r1}} = \frac{Q_1 + 2Q_3}{Q_1} = \left(con\right)_{CR} \qquad 26(a)$$

substituting for Q_1 and Q_3 using equations 19 and 23 respectively and rearranging gives

$$\frac{dv^e_r}{d\varepsilon^e_{r1}} = \frac{S_1/S_3 R^m\left(1-2\mu_{31}\right) + 2\left(1-\mu_{13}-\mu_{32}\right)}{S_1/S_3 \cdot R^m - 2\mu_{13}} \qquad 30$$

In the case of the ambient consolidation test, CR = 1, it is justifiable to assume that upon repeated cycling the sample becomes truly isotropic and therefore $S_1 = S_3$, $\mu_{13} = \mu_{31} = \mu_{32} = \mu$ and $R = 1$. Whereupon $dv^e_r/d\varepsilon^e_{r1} = 3$. The modified form of equation 30 for the isotropic state is therefore

$$\frac{dv^e_r}{d\varepsilon^e_{r1}} = \frac{\left(R^m + 2\right)\left(1-2\mu\right)}{R^m - 2\mu} \qquad 31$$

An identical relationship was given by El-Sohby (1964) and El-Sohby & Andrawes (1973) starting from a slightly different form of equation 11. It will now be interesting to see whether equation 31 can be fitted to the other experimentally observed points in figure 9 by an appropriate choice of μ , since $m = 0.35$ has already been found constant. By trial and error it was found that $\mu = 0.29$ gave a very good fit to the experimental curve for $R \leq 3$ as seen in figure 9. For $R > 3$ an increasing divergence of the theoretical and experimental values is noticed. This is attributed to the 'elastic' radial strain being underestimated owing to the presence of slip deformation in the opposite direction (see paragraph 4.3) and to the effect of the increasing anisotropic nature of the material. However the CR 'elastic' model will be developed at this stage on the basis of a constant μ for all values of the stress ratio since the departure from the experimental data is small beyond $R > 3$ and as mentioned in paragraph 2 the elastic strains are less dominant at higher shear stress levels $(R > K_{\phi\mu})$.

For the case of the isotropic stress state equations 19 and 23 can be modified as follows

$$Q_1 = 1 - \frac{2\mu}{R^m} \qquad 32$$

and

$$Q_3 = \frac{1}{R^m}(1-\mu) - \mu \qquad 33$$

These equations will be applied to all the CR tests. Since m and μ have been determined the variation of Q_1 and Q_3 with stress ratio can be evaluated and these relationships are given in figure 20.

In order to determine the value of $S = S_1 = S_3$ to fit the simplified CR 'elastic' model to the experimental values use is made of equations 28 and 30 and the absolute CR stress-strain plots in figures 14 and 15.

Hence

$$S = \frac{(\varepsilon_{r1}^e)_{exp}}{Q_1\left(\frac{\sigma_1'}{\sigma}\right)^m} = \frac{(\varepsilon_{r3}^e)_{exp}}{Q_3\left(\frac{\sigma_1'}{\sigma}\right)^m}$$

A few results are given in table 1 and indicate that $S = 0.28$ can be assumed constant for all values of CR.

In summary the equations of the CR and VR stress paths of the simplified model are given as in paragraphs 4.6.1 (CR), 4.7.1 $(c\sigma_3')$, 4.7.2 (cp) and 4.7.3 $(c\sigma_1')$ with Q_1 and Q_3 given by equations 32 and 33 respectively, m = constant, $S_1 = S_3 = S$ and $\mu_{13} = \mu_{31} = \mu_{32} = \mu$. The simplified CR elastic model with the specification $m = 0.35$, $S = 0.28$ and $\mu = 0.29$ is given in figures 21, 22 and 23.

4.9 The relationships for the elastic compliances, A_{ij}

The present treatment of the non-linear elastic behaviour of sand has been based on the use of the generalised Hooke's law for small linear strain-stress increments and the symmetry condition $A_{ij} = A_{ji}$ which has been shown to apply for linear elasticity (Love, 1892) is assumed to apply over these small increments in stress. Whether this concept can be readily applied to the non-linear elastic behaviour of sand can be shown by evaluating the constants A_{13} and

A_{31}.

By comparing equations 1, 13 and 17 it can be seen that

$$A_{13}\left(=-\frac{\mu_{13}}{E_3}\right) = -\mu_{13}.m.\frac{S_3}{E^m}.\ \sigma_3'^{\,m-1} \qquad 34$$

and $A_{11}\left(=\frac{1}{E_1}\right) = \frac{mS_1}{E^m}.\ \sigma_1'^{\,m-1} \qquad 35$

By comparing equations 2, 14 and 21 rearranged it can be seen that

$$A_{31}\left(=-\frac{\mu_{31}}{E_1}\right) = -\mu_{31}.\frac{mS_1}{E^m}.\ \sigma_1'^{\,m-1} \qquad 36$$

$$A_{32}\left(=-\frac{\mu_{32}}{E_3}\right) = -\mu_{32}.\frac{mS_3}{E^m}.\ \sigma_3'^{\,m-1} \qquad 37$$

$$A_{33}\left(=\frac{1}{E_3}\right) = \frac{mS_3}{E^m}.\ \sigma_3'^{\,m-1} \qquad 38$$

It has been shown in paragraph 4.8 that the simplified CR elastic model gives very good agreement with the experimental data for $R \leq 3$. By adopting the simplified model in which $S_1 = S_3 = S$ and $\mu_{13} = \mu_{31} = \mu_{32} = \mu$ then

$$\frac{-\mu_{13}.m\,S_3}{E^m} = \frac{-\mu_{31}\,m\,S_1}{E^m} = \text{constant}$$

since m is constant for all R
From equation 34 and 35
then $A_{13} \propto \sigma_3'^{\,m-1}$ and $A_{31} \propto \sigma_1'^{\,m-1}$
For the case of isotropic consolidation (CR = 1) then $\sigma_1' = \sigma_3'$ and $A_{13} = A_{31}$. Also from equations 35 and 38, $E_1 = E_3$.
For the general case of $R > 1$ then $A_{31} < A_{13}$ and $E_1 > E_3$.
This has the following implications:

1. For all anisotropic stress states in which $R > 1$ the coefficients A_{ij} are not symmetric in the simplified CR elastic model and need to be evaluated independently when using this form in the generalized Hooke's law. It would be interesting to determine if this applied when the compliances 34 and 36 are evaluated in a true triaxial apparatus.

2. The simplified CR elastic model is described by mass isotropic parameters m, S and μ with the anisotropic elastic behaviour accounted for when $\sigma_1' > \sigma_3'$ by the dependence of the elastic compliances on the non-linear variation in the corresponding principal effective stress. In particular for $\sigma_1' > \sigma_3'$ then $E_1 > E_3$ as given below.
From equation 35

$$E_1 = k_1.\sigma_1'^{\,1-m} \qquad \text{where } k_1 = \frac{E^m}{m\,S_1} \qquad 39$$

and from equation 38

$$E_3 = k_3 \, \sigma_3^{\,1-m} \quad \text{where} \quad k_3 = \frac{E^m}{m \, S_3} \qquad 40$$

In the simplified model $k_1 = k_3 = E^m / mS$.

Equations 39 and 40 indicate that the Young's modulus of the mass of sand particles is directly proportional to the power m of the Young's modulus of the parent material and the non-linear variation of the corresponding principal stress. This variation is plotted in figure 24 and is of the order of 10^3 less than that for the parent mineral in agreement with Murayama (1964) and Holubec (1968). Holubec (1968) found that the axial Young's modulus E_1 for a given void ratio varied with both the mean normal stress and deviator stress. The variation of E_1 given by equation 39 (using the simplified CR elastic model) has been compared in figure 25 with the result obtained by Holubec.

Since the symmetry condition does not apply by using the simplified model it is necessary to check that the model satisfies thermodynamic considerations in that the strain energy function is always positive. Pickering (1970) has given the following three necessary conditions for positive strain energy in a cross-anisotropic material:

1. E_1, E_3 and $G_{13} > 0$
2. $\mu_{33} > -1$
3. $E_3/E_1 (1-\mu_{33}) - 2\mu_{31}^2 > 0$

Since G_{13} can not be determined in the 'triaxial' apparatus reference to the approximate solution (Maini 1976)
$G_{13} = E_1 E_3 / E_1 + E_3(1+2\mu)$ indicates that the above conditions are satisfied with $\mu_{33} = \mu_{31} = \mu = 0.29$ in the simplified CR model.

The correlation of the predicted elastic stress-strain relationship in the varying R stress paths CG_3 and CP with the experimentally obtained values from cyclic loading will now be examined.

4.10 Comparison with VR cyclic stress paths

In analysing whether an elastic state can be approached in a material an important indication is the ratio of the energy recovered on unload to the work applied on loading, E_R, since in an ideal elastic material $E_R = 100\%$ and the unload and reload curves are coincident.

The net work done per unit volume in a sample during a small increment of axial stress $(\sigma_1')_1$ to $(\sigma_1')_2$ in the CG_3 stress path is

$$(\Delta W_{1 \to 2})_{CG_3} = \int_1^2 \frac{(\sigma_1')_1 + (\sigma_1')_2}{2} d\varepsilon_1 + 2\sigma_3' d\varepsilon_3$$

Cyclic behaviour of sands and clay has been studied under varying R stress paths by Lo

(1961), Biarez (1961), Murayama (1964), Makhlouf and Stewart (1965), Holubec (1966), Khayatt (1967), Cole (1967), McDermott (1972) and others and their results indicate the following:

1. The reload stress-strain curve is steeper than virgin loading curve indicating that much less deformation is taking place over a given shear stress range.
2. The width of the hysteresis loop for the first reload cycle increases with increase in cyclic shear stress level, R_{max}.
3. Repeated cycling at a given R_{max} level causes progressing reduction in the width of the hysteresis loop indicating that less energy is being dissipated and tends to an ultimate cycle with no further reduction in hysteresis.
4. Cole (1967) in the SSA MK6 and Khayatt (1967) independently noted that the reload curves after repetitive cycling are sensibly unique over their common shear stress ranges.

From the above it would be informative to compare a plot of the cyclic reload axial strain and the ratio E_R obtained from different CG_3 stress paths at different R_{max} with the prediction from the CR elastic model. This is presented in figure 26.

The predicted 'elastic' deformation over the same range of cycle from the CR model has been indicated in the figure at the hypothetical point where $E_R = 100\%$. From the work of Mindlin and Deresiewicz (1953) on regular packings it is expected that in varying obliquity tests it is impossible to reach a value of $E_R = 100\%$ by cycling and for this reason the curves have not been extrapolated. It was expected that the reload strain from the CG_3 stress paths would be greater than the corresponding predicted 'elastic' strain from the CR model over the same cycle due to the occurrence of local slip at point contacts on reload at varying R. However, it is seen that the predicted 'elastic' strain from the CR model is slightly greater than the reload strain at low values of R_{max}. As the value of R_{max} increases the predicted 'elastic' strain corresponds to the strains occurring after a number of cycles. At high R_{max} the CR model gives a lower bound to the value obtained experimentally in stress paths. A comparison of the corresponding radial reload strain has not been given in graphical form because of the small values of strain involved. These values are given in table 2 and indicate a similar trend.

A comparison of the CR elastic model prediction for axial and radial strains over a full cycle can be made in figures 27 and 28 for cyclic tests conducted with CG_3' and CP. It is seen that the CR model tends to predict a curve in the middle of the

363

hysteresis cycle. It is interesting to note in figure 28 that the energy recovered region in the CP test increased with number of cycles but in all cases was affected by slip deformation. Therefore slip deformation may give a false value for the tangent modulus of elasticity either too low at the start of unloading/reloading or too great at the latter part of the cycle depending whether the direction is reverse or the same as the elastic deformation.

4.11 Factors affecting the CR elastic model

4.11.1 Effect of void ratio

A preliminary investigation of the effect of increasing the void ratio on the mass 'elastic' parameters m, S and μ was made and resulting absolute axial and radial strain-axial stress plots are given in figure 29. The values of the mass parameters for loose River Welland sand are given in table 3.

Variation in the value of μ may be attributed to the difficulties of setting up representative loose samples. It is interesting to note that if a value of $\mu = 0.29$ is imposed then $S = 1.34$ for the three CR stress paths studied.

4.11.2 Effect of particle shape

A preliminary investigation of the effect of particle shape on the parameters m, S and μ was made by comparing results obtained from dense samples of glass ballotini with the dense samples of River Welland sand. The results are given in figure 30 and the values of the mass parameters are given in table 3. Some difficulty was experienced in assuming a constant value of m and the average value of $m = 0.52$ was used to calculate the average S and μ values.

In summary, table 3 indicates the value of m increases with increase in void ratio and particle roundness. The value of μ is less affected by increase in void ratio than increase in particle roundness. The value of S is more sensitive to an increase in void ratio than an increase in particle roundness.

4.12 Component of elastic volumetric strain

Using the absolute $V_r^e - \sigma_i'$ relationship in figure 16 with experimental data for $\sigma_i' > 35 \text{kN/m}^2$ and the corresponding simplified CR elastic prediction in figure 23 a plot of V_r^e with p for dense River Welland sand is given in figure 31.

It can be seen that for CR = 1, 1.8 and 2.97 the relationship is sensibly unique. Using the experimental data for CR = 3.8 and 4.67 there is a tendency for ΔV_r^e to increase with R whereas using the simplified CR model the reverse is obtained. It was pointed out in paragraph 4.8 that the effect of adopting the simplification of a constant μ value in the simplified CR model resulted in a discrepancy in the $V_r^e/\mathcal{E}_{rl}$- R relationship for CR > 3. Therefore in figure 31 the theoretical and experimental curves for CR = 3.8 and 4.67 show different departures from the CR = 1, 1.8 and 2.97 'unique' curve. It is proposed that the $V_r^e - p$ relationship be treated as unique for all values of R for dense River Welland sand. In figure 16 it can be seen that the ratio $\left(dV_r^e / d\sigma_i \right)_{CP} \neq 0$ and therefore strict uniqueness is not expected. However for contractive states, 'wet', of minimum void ratio $(R \leqslant K_{\phi_\mu})$ figures 23 and 31 indicate that uniqueness is approached. For $R > K_{\phi_\mu}$ the error in the value of V_r^e by assuming uniqueness will be negligible in comparison with the increasing non-recoverable component of deformation in VR stress paths. The 'uniqueness' 'wet' of minimum voids ratio is seen to be in agreement with the Cam - clay model for 'wet' clays. Thus different anisotropic consolidation tests $(R \leqslant K_{\phi_\mu})$ for dense sand give swelling lines essentially similar to the isotropic swelling lines (at a similar void ratio) when plotted in $V_r^e - p$ space. This result is in agreement with the result of Biarez (1961) and Ko and Scott (1967) who pointed out that the compressibility properties of sand specimens during shear remained essentially 'isotropic'. However as seen from paragraph 4.9 since the elastic compliances are related to the corresponding principal effective stress the elastic behaviour is not isotropic for R > 1 in the sense implied by the use of equal E values. El-Sohby (1964) using an arbitrary Pmean datum showed a more pronounced increase in the compressibility characteristics of dense sands with increase in R for CR stress paths. However a 'unique' relationship was obtained for loose sands. Holubec (1968) using an arbitrary datum stress related the change in compressibility to the applied cell pressure.

The 'unique' $V_r^e - p$ relationship implies that the changes in the volumetric strain are governed sensibly by the stress parameter p. This can be inferred from figure 23 for $R \leqslant 2.97$ where by joining points at CP the variation in ΔV_r^e is seen to be negligible. By assuming 'uniqueness' an important step can be taken in the formulation of a recovery energy term for all stress paths for dense sand equivalent to the Cam⁻-clay model.

An important point to make in relation to the CP stress path is that although the elastic volumetric strain remains essentially constant throughout shear an elastic correction is still required to the axial and radial elastic strains which vary with R as seen in figures 17 and 18. This implies that during CR stress paths of unity and 3, approximately the same increase in stored volumetric elastic energy occurs for the same increase in mean principal stress. In the case of the CR = 1 stress path equal components of the stored energy are introduced in the axial and radial directions. In the case of the CR = 3 stress path since $\sigma_1' = 3\sigma_3'$ a greater amount of elastic energy is put into the sample in the axial direction than the radial direction. If the elastic component remained truly isotropic throughout shear and there was no elastic shear term (i.e. $\mu = -1$) then these components would be equal. It is shown in paragraph 4.13 that the elastic shear strain term increases with increase in R.

The sensibly unique $V_r^e - P$ relationship for 'dry' sand allows a recoverable energy term to be derived and applied to all drained and undrained stress paths as in paragraph 5.1.

4.13 Component of elastic shear strain

From figures 14 and 15 (experimental) and figures 21 and 23 (theoretical) a variation of the elastic shear strain with axial stress as given in figure 33a can be obtained. Since the 'elastic' shear strain energy is given by the product $q\,d\epsilon$ a more useful plot is given by the variation of the elastic shear strain with the deviator stress as shown in figure 33b. As expected it can be seen that the elastic shear strain increases with the value of R and emphasises the anisotropic nature of the recoverable component of deformation in dense sands. By joining points of CP and $C\sigma_3'$ the corresponding elastic shear strain relationships are obtained for these stress paths. It is seen that the $\mathcal{E}_r^e - q$ relationship is stress path dependent and therefore a general term can not be included in an energy balance for sand. A comparison of the incremental changes in elastic volumetric and shear strains for $C\sigma_3'$, CP and $C\sigma_1'$ stress paths for dense RW sand initially consolidated to 275 kN/m² is given in figure 34. The elastic shear strain term is seen to be significant and even greater than the corresponding elastic volumetric

term which for the $C\sigma_1'$ and CP stress paths is negative. Similarly the elastic shear strain term is significant for loose sand as seen from inspection of figure 29.

5 DISCUSSION OF CR ELASTIC MODEL

5.1 Comparison with 'Cam'-clay

The form of the compressibility relationship appertaining to dense ('dry') sand is examined in figure 32. Using the $V_r^e - \sigma_1'$ relationship for CR = 1 it has been shown in paragraph 4.6.1 that a non-linear relationship is closely followed. Consequently for CR = 1 since $\sigma_1' = p$ a linear plot of V_r^e with p^m should be obtained. For comparison the corresponding $V_r^e - \log_e P$ (Cam-clay) relationship has been plotted in figure 32 using an equivalent scale. It can be seen that a constant slope for the $V_r^e - \log_e P$ plot is only obtained over restricted stress ranges.

The recoverable elastic volumetric energy term per unit volume, U_1 is given by

$$U_1 = \int_0^v p\,dv_r^e \quad . \quad \text{Since } V_r^e = k_4 . p^m$$

then $dv_r^e = k_4\, m\, p^{m-1} dp$

cf equation $dv_r^e = \dfrac{mS_1}{E^m} \sigma^{'m-1} d\sigma_1' (Q_1 + 2Q_3)$ (paragraph 4.6.1)

then $k_4 = S_1 E^m . (Q_1 + 2Q_3)$ since $\sigma_1' = p$

$$U_1 = \int_0^p k_4 . m. p. p^{m-1} dp = \frac{k_4 m}{m+1} p^{m+1}$$

The increase in recoverable energy due to an increase in the mean normal stress is given by $\delta U_1 = k_5\, p^m \delta p$

where $k_5 = \dfrac{mS_1}{E^m}(Q_1 + 2Q_3)$ in simplified model

Therefore the elastic volumetric component is linked to the fundamental Young's Modulus of the parent material and the elastic constants m and S. This can be compared with the equivalent recoverable energy term for the 'Cam'-clay models.

$$\delta U = k\,\delta p / 1 + e$$

the constant k already incorporates the term $1 + e$, since it is measured from the volumetric strain plot.

5.2 Considerations on elastic deformation

Zienkiewicz and Naylor (1971) mentioned the assumption in the Cam-clay model of neglecting the elastic shear strain term pointing out that the inherent value of $\mu = -1$ proved inconvenient in numerical analysis. However, Naylor (1970) showed that by varying the value of Poisson's ratio in the range $-1 \leqslant \mu \leqslant 0.3$ the effect on the total calculated strains was small because the elastic component was small in comparison with the non-recoverable component. Likewise, the microstructural model of normally consolidated clay of Calladine (1971) and the extension of Cam-clay to overconsolidated clays by Pender (1978) neglect the elastic shear strain term. Parry and Amerasinghe (1973), however, noted in cyclic axi-symmetric tests that for both normally consolidated and heavily over-consolidated kaolin the recovered shear strain on unload was a constant proportion of the total shear strain and independent of strain, stress history and stress path. In the light of paragraph 4.10 the first unload strain would include slip (dissipative) deformation. In the microstructural model of overconsolidated clay (Calladine 1973) a consideration of the tangential and normal compliances of random isotropic cross-cut planes led to the important conclusion that the Poisson's ratio μ_{13} and μ_{31} (defining anisotropy) were to a first approximation independent of the value of R (degree of induced anisotropy) and could be taken as equal, i.e. isotropic. An isotropic μ value is in keeping with the simplified CR elastic model. Values of μ were shown to vary with plasticity index P.I. from 0.2 to 0.4 and two values for sands (P.I.=0) were given of about $\mu = 0.18$. This value is seen to be closer to that obtained for spherical particles (dense glass ballotini) than sub-angular dense R. W. sand ($\mu = 0.29$). Ohta (1973) extended the Cam-clay model to the case of anisotropically preconsolidated clay and included both elastic volumetric and shear strain terms. The former term was determined from an unload CR stress path and the latter term from unload CP stress path. It is interesting to compare this approach with the CR elastic model for sand. As seen from figure 34 the predominant elastic deformation in the CP stress path would be the shear strain term and would only be slightly affected by the elastic volumetric component. However, the data from the CR stress path would by reference to figure 33(b) include elastic volumetric and shear strain terms which would need to be separated. The volumetric component could however be simply obtained by choosing the isotropic consolidation stress path since as seen in figure 31 the $V_r^e - p$ relationship is sensibly unique. Cairncross and James (1977) in discussing the anisotropic behaviour of overconsolidated clay suggest that the anisotropy may be a consequence of the variation of the stiffness moduli, $1/E$, with the direction of incremental shear and not require an elaborate treatment given by the use of the generalized Hooke's law. This is seen to

have a direct link with the simplified CR elastic model in which the Young's moduli vary with the magnitude of the corresponding principal effective stresses and the Poisson's ratio is isotropic.

5.3 Formation of the simplified CR elastic model

From paragraphs 4.6.1 and 4.8 the mass parameters m, S and μ can be determined from the unload curve of one CR stress path. In order to select a suitable CR stress path in the contractive stage for dense sand use is made of the stress-dilatancy relationship $R = D K_{\phi\nu}$ in which $D = 1$ and $D = 2$ give the values of R_{min} and R_{peak} respectively once ϕ_ν has been determined. A CR stress path is chosen so that $R < K_{\phi\nu}$ in order to obtain representative elastic parameters on first unload. In addition Young's Modulus of the parent material is required. The graphical simplified CR elastic model can then be prepared for a range of $R < R_{peak}$ using the equations 18, 20, 22 and 23. Alternatively, the equations of the required stress path can be used as given in paragraphs 4.7.1, 4.7.2 and 4.7.3 or the elastic compliances evaluated as given in paragraph 4.9 for use in the generalized Hooke's law.

5.4 Use of the CR elastic model

The graphical form of the model is most useful when elastic deformations are required in variable stress paths e.g. the undrained stress path. In these cases the effective stress path is plotted directly onto figures 14, 15 and 16 and the corresponding elastic strain increments read off directly. Alternatively, in constant stress paths the equations given in paragraphs 4.6 and 4.7 can be applied directly to known stress increments. Knowledge of the elastic strain component allows the total strain data to be corrected so as to yield the non-recoverable component of deformation. Correct isolation of non-recoverable (plastic) deformation is important in stress probe experiments (Lewin & Burland 1970, Lewin 1973) since decisions need to be made on the choice of a flow rule i.e. whether an associated or non-associated flow rule applies. The effect that the elastic component of deformation can have on the non-recoverable strain rate ratio particularly during the contractive stage during shear of dense sand can be gauged from the three stress paths given in figure 35.

366

As seen from figures 27 and 28 the CR elastic model can give an accurate measure of the elastic strain experienced in sands under cyclic loading. Although the direct application of the model to field situations is not yet proven an indication of the behaviour and magnitude of the parameters in a simple test apparatus may give useful leads in the treatment of field problems. It is interesting to note that the work of Gerrard (1972), Gerrard & Morgan (1972), Milovic (1972) and Hooper (1975) in seeking a more representative theoretical treatment in settlement studies on sands and over-consolidated clays displaying anisotropy have emphasised the importance of the ratio E_1/E_3 and a suitable choice of Poisson's ratio. The simplified CR elastic model enables these parameters to be calculated for a given sand, state of packing and state of stress (for example $E_1/E_3 = R^{1-m}$).

Finally, use of the semi-empirical form of the Hertz contact equation (equation 11) permits the elastic constants to be linked to Young's Modulus, the fundamental elastic constant of the parent material. In addition it has been shown that for dense sub-angular RW sand the non-linear power law, m, = 0.35 and this can be compared to the two-thirds power law of Hertz for like spheres in contact. The variation of m with particle shape and state of packing has been indicated.

It is for this reason that Rowe has concentrated on the fundamental interpretation of shear strength behaviour at the particulate level in order to derive model(s) of soil behaviour based on fundamental physical parameters.

ADDITIONAL NOTATION

$$K_{\phi\mu} = \tan^2\left(45 + \frac{\phi\mu}{2}\right)$$

ACKNOWLEDGEMENTS

The author is indebted to Professor P. W. Rowe for facilities provided during the period 1967-1970 when this research was undertaken. Discussions with Professor P. W. Rowe and Dr. I. M. Smith are remembered with gratitude.

REFERENCES

Alyanak, I. 1961, Vibration of sands with special reference to the minimum porosity test for sands, Proc.Midland Soil Found. Eng.Soc., Vol.4.

Andrawes, K.Z. 1964, The behaviour of particulate materials in the 'at rest' state, M.Sc. Thesis, Univ. of Manchester.

Barden, L. 1963, Stresses and displacements in a cross-anisotropic soil, Geotechnique 13, No.3, 198-210.

Barden, L. & Khayatt, A.J. 1966, Incremental strain rate ratios and strength of sand in the triaxial test, Geotechnique, 16, 4, 338-357.

Barden, L. & Khayatt, A.J. 1968, Incremental stress-strain relations for sand, Parts 1 & 11, Research Report No.5, Dept. of Civil Eng. Univ. of Manchester.

Barden, L. & McDermott, R.J.W., 1965, Use of free ends in the triaxial testing of clays, J.Soil Mech.Found.Eng. ASCE, Vol. 91, SM6, 1-23.

Bishop, A.W. & Henkel, D.J. 1962, The measurement of soil properties in the triaxial test. Ed. Arnold.

Cairncross A.M. & James, R.G. 1977, Anisotropy in overconsolidated clays, Geotechnique 27, No.1, 31-36.

Calladine, C.R., 1971, A microstructural view of the mechanical properties of saturated clay, Geotechnique 21, No.4, 391-415.

Calladine, C.R., 1973, Overconsolidated clay: A microstructural view, Proc.Symp. Role Plasticity in Soil Mech., Cambridge (ed. A.C. Palmer), 144-158.

Cole, E.R.L., 1967, The behaviour of soils in the simple shear apparatus, Ph.D. Thesis, Univ. of Cambridge.

Coon, H.D. & Evans, R.J. 1969, Disc. on elastic behaviour of cohesionless soil, J.Soil.Mech.Found.Div.ASCE.95 SM5, 1281-1283.

Deresiewicz, H. 1958, Mechanics of granular matter, Adv.App.Mech., Vol. 5, 233-306.

El-Sohby, M.A. 1964, The behaviour of particulate materials under stress. Ph.D. Thesis, Univ. of Manchester.

El-Sohby, M.A. 1969, Deformation of sands under constant stress ratios. Proc.7th Int.Conf. Soil Mech.Found.Eng. 1, 111-119.

El-Sohby, M.A. & Andrawes, K.Z., 1973, Experimental examination of sand anisotropy. Proc.8th Int.Conf.Soil Mech. Found.Eng. 1, 103-109.

Gerrard, C.M., 1972, Disc. Barden L. 1963 Geotechnique 13, 3, 198-210. Geotechnique 22, 2, 372-376.

Gerrard, C.M. & Morgan, J.R. 1972, Initial loading of a sand layer under a circular pressure membrane. Geotechnique 22, 4, 635-661.

Holubec, I. 1966, The yielding of cohesionless soils. Ph.D. Thesis, Univ. of Waterloo, Ontario.

Holubec, I. 1968, Elastic behaviour of cohesionless soil. J.Soil Mech.Found. Div., ASCE 94 SM6, 1215-1231.

Hooper, J.A. 1975, Elastic settlement of a

circular raft in adhesive contact with a transverse isotropic medium. Geotechnique 25, 4, 691-711.

Khayatt, A.J. 1967, Some incremental stress-strain relations for sand. Ph.D. Thesis, Univ. of Manchester.

Ko, H.Y. & Scott, R.F. 1967, Deformation of sand in hydrostatic compression. J.Soil Mech.Found.Div.ASCE 93, SM3, 137-156.

Lewin, P.I. 1973, The influence of stress history on the plastic potential. Proc. Symp.Role plasticity in soil mech. Cambridge (ed. A.C. Palmer) 96-106.

Lewin, P.I. & Burland, J.B. 1970, Stress probe experiments on saturated normally consolidated clay. Geotechnique, 20, 1, 38-56.

Lo, K.Y. 1961, Stress-strain relationship and pore water pressure characteristics of a normally consolidated clay. Proc. 5th Int.Conf.Soil Mech.Found.Eng. 1, 219.

Love, A.E.H. 1892, A treatise on the mathematical theory of elasticity. Cambridge University Press.

Makhlouf, H.M. & Stewart, J.J. 1965, Factors influencing the modulus of elasticity of dry sand. Proc.6th Int. Conf.Soil Mech.Found.Eng., 1, 298-302.

Maini, K.S., 1976, Disc. Hooper, J.A. 1975 Geotechnique 25, 4, 691, Geotechnique 26, 3, 540-541.

McDermott, R.J.W. 1970, Assessment of elastic deformation in 'Triaxial' compression tests. Internal note (unpublished), Univ. of Manchester.

McDermott, R.J.W. 1972, Deformation characteristics of sand under axi-symmetric loading. Ph.D. Thesis, Univ. of Manchester.

Milovic, D.M. 1972, Stresses and displacements in an anisotropic layer due to a rigid circular foundation. Technical note Geotechnique 22 No. 1, 169-174.

Mindlin, R.D. & Deresiewicz, H. 1953, Elastic spheres in contact under oblique forces. J.Appl.Mech. 20, 327-344.

Murayama, S. 1964, A theoretical consideration on a behaviour of sand. I.V.T.A.M. Symposium, Grenoble. 146-159.

Naylor, D.J. 1970, Disc. Lewin, P.I. & Burland, J.B. 1970 (Geotechnique 20, No. 1, 38-56) Geotechnique 20, No. 3, 336-339.

Ohta, H. 1973, An extended Cam-clay model. Proc.Symp.Role Plasticity in Soil Mech. Cambridge (ed. A.C. Palmer) 211-214.

Parry, R.H.G. & Amerasinghe, S.F. 1973, Components of deformation in clays. Proc. Symp.Role Plasticity in Soil Mech. Cambridge (ed. A.C. Palmer) 108-126.

Pender, M.J. 1978, A model for the behaviour of overconsolidated soil. Geotechnique 28, No. 1, 1-25.

Pickering, D.J. 1970, Anisotropic elastic parameters for soil. Geotechnique 20,

No. 3, 271-276.

Proctor, D.C. 1967, The stress dilatancy behaviour of dense sand in the hollow cylinder test. M.Sc. Thesis, Univ. of Manchester.

Proctor, D.C. & Barton, R.R. 1974, Measurements of the angle of interparticle friction. Geotechnique 24, No.4, 581-604.

Roscoe, K.H. & Burland J.B. 1968, On the generalised stress strain behaviour of 'wet' clay. Engineering Plasticity, Cambridge University Press. 535-609.

Roscoe, K.H. & Schofield, A.N. 1963, Mechanical behaviour of an idealized 'wet' clay. Proc.2nd European Conf.Soil Mech., Wiesbaden. 1, 47-54.

Roscoe, K.H., Schofield, A.N. & Thurairajah, A. 1963, Yielding of clays in states wetter than critical. Geotechnique 13. 211-240.

Roscoe, K.H., Schofield, A.N. & Wroth, C.P. 1958, On the yielding of soils. Geotechnique 8. 1, 22-53.

Rowe, P.W. 1962, The stress dilatancy relation for static equilibrium of an assembly of particles in contact. Proc. Roy.Soc.London A, 269, 500-527.

Rowe, P.W. 1971, Theoretical meaning and observed values of deformation parameters for soil. Proc.Roscoe Memorial Symposium, Univ.of Cambridge, Foulis, 1972.

Rowe, P.W. & Barden, L. 1964, Importance of free ends in triaxial testing. J.Soil Mech.Found.Div. ASCE 90, SMI, 1-27.

Schofield, A.N. & Wroth, C.P. 1968, Critical state soil mechanics. McGraw-Hill.

Thurairajah, A. 1961, Some shear properties of kaolin and of sand. Ph.D. Thesis, Univ. of Cambridge.

Thurston, C.W. & Deresiewicz, H. 1959, Analysis of a compression test of a model of a granular medium. J.Appl.Mech. 26, 251-258.

Tong, P.Y.L. 1970, Plane strain deformation of sands. Ph.D. Thesis, Univ. of Manchester.

Wilson, G. & Sutton, J.L.E. 1948, A contribution to the study of the elastic properties of sand. Proc.2nd Int.Conf. Soil Mech. 1, 197-202.

Wood, D.M. 1973, Truly triaxial stress strain behaviour of kaolin. Proc.Symp. Role Plasticity in Soil Mech. Cambridge (ed. A.C. Palmer) 67-93.

Wood, D.M. 1975, Exploration of principal stress space with kaolin in a true triaxial apparatus. Geotechnique 25, No. 4, 783-797.

Wroth, C.P. & Bassett, R.H. 1965, A stress strain relationship for the shearing behaviour of a sand. Geotechnique 15, 2, 32-56.

Zienkiewicz, O.C. & Naylor, D.J. 1971, The adaptation of critical state soil mech-

anics theory for use in finite elements.
Roscoe Mem.Symp.Stress-strain behaviour
of soils. Foulis. 537-547.

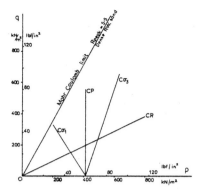

FIG. I Typical Stress Paths Studied

Table 1. Determination of S for dense R
River Welland sand.

$$\varepsilon^e_{r_1} = Q_1 . S \left(\frac{G'_1}{E}\right)^m \qquad \text{eqn. 28}$$

Values of Q_1 from figure 20.

1	2	3	4	5
R	G'_1 lbf/in^2	$Q_1\left(\frac{G'_1}{E}\right)^M$ (x10^{-4})	$\varepsilon^e_{r_1}$ (x10^{-4})	$S = \frac{4}{3}$(2dec. 3plac-es)
1	10	33.4	9.3	0.28
	20	42.6	12	0.28
	40	54.2	15.3	0.28
	100	74.8	20.85	0.28
2.97	10	47.3	13.3	0.28
	20	60.4	17	0.28
	40	76.9	21.6	0.28
	100	106.1	29.5	0.28
4.67	10	51.7	14.5	0.28
	20	66	18.7	0.28
	40	84	23.8	0.28
	100	116	32.5	0.28

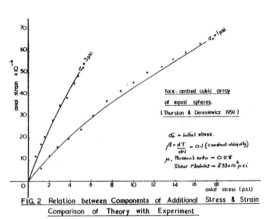

FIG. 2 Relation between Components of Additional Stress & Strain
Comparison of Theory with Experiment

Table 2. Comparison of ε_3 reload with
predicted $\varepsilon^e_{r_3}$.

From figure 26. 6th cycle test 1/1/32.

R_{limits}	reload ε_3 x10^{-4}	$\varepsilon^e_{r_3}$(CRmodel) x10^{-4} fig.22
1.84 - 1.10	0.2	1
2.90 - 1.10	1.3	2.2
4.00 - 1.10	3.1	3.5
4.70 - 1.1	5.7	4.75
5.35 - 1.1	21.8(slip)	6.0

Table 3.

		e	m	S	μ
RW sand	Dense	0.62	0.35	0.28	0.29
	Loose	0.86	0.45	1.26	0.26
					R=1.98
			0.45	1.14	0.21
					R=2.5
			R=1→2.5	1.34	assume =0.29
Glass B.	Dense	0.54	0.46R=1		
			0.58R=1.8		
			0.52aver~0.80 age	0.13	

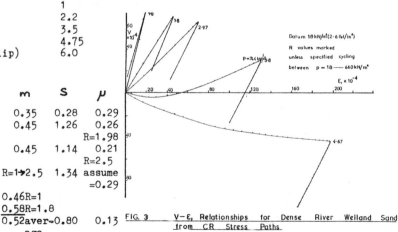

FIG. 3 V – ε_1 Relationships for Dense River Welland Sand
from CR Stress Paths

369

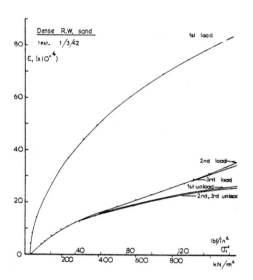

FIG. 4 $\varepsilon_1 - \sigma_1'$ Plot for   CR = 2·97

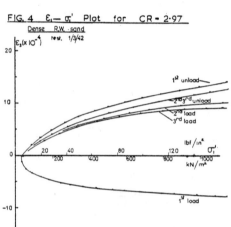

FIG. 5 $\varepsilon_r \sigma_1'$ Plot for   CR = 2·97

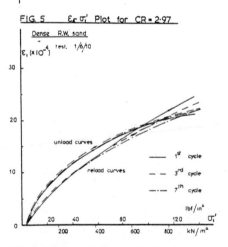

FIG. 6   Effect of Cycling on the $\varepsilon_1 - \sigma_1'$
Relationship   for   CR = 1·8

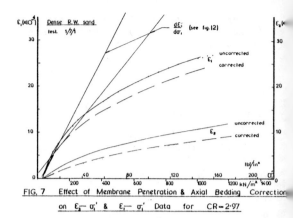

FIG. 7   Effect of Membrane Penetration & Axial Bedding Correction
on $\varepsilon_3 - \sigma_1'$ & $\varepsilon_1 - \sigma_1'$ Data   for   CR = 2·97

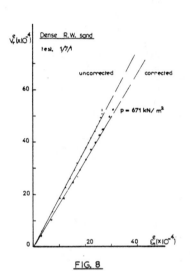

FIG. 8

Effect of Membrane Penetration & Axial
Bedding Corrections on the $V_r^e - \varepsilon_{r1}^e$ Relationship
for   CR = 2·97

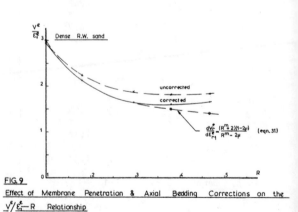

FIG. 9
Effect of Membrane Penetration & Axial Bedding Corrections on the
$V^e/\varepsilon_1^e - R$ Relationship

370

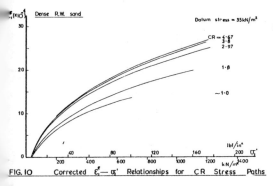

FIG. 10   Corrected $\varepsilon_1^e - \sigma_1'$ Relationships for CR Stress Paths

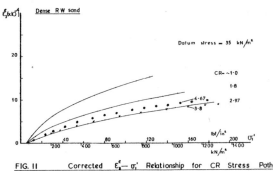

FIG. 11   Corrected $\varepsilon_3^e - \sigma_1'$ Relationship for CR Stress Paths

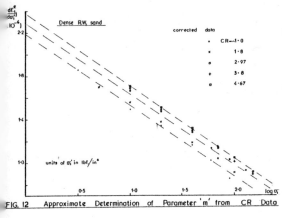

FIG. 12   Approximate Determination of Parameter 'm' from CR Data

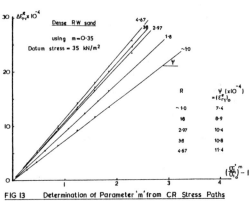

FIG 13   Determination of Parameter 'm' from CR Stress Paths

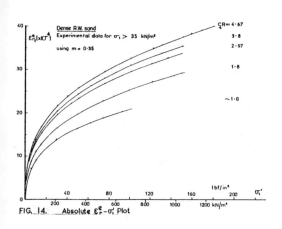

FIG. 14.   Absolute $\varepsilon_r^e - \sigma_1'$ Plot

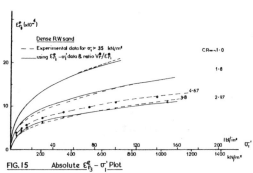

FIG. 15   Absolute $\varepsilon_{r_3}^e - \sigma_1'$ Plot

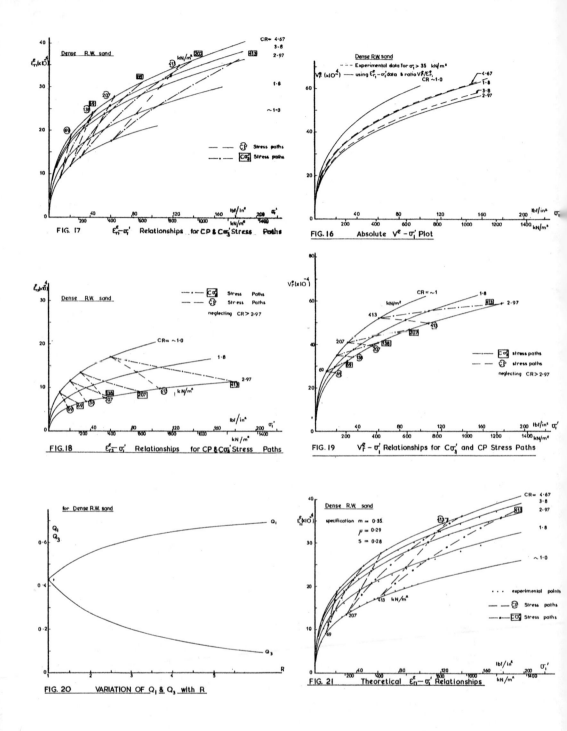

FIG. 17 $\varepsilon_{r_1}^e - \sigma_1'$ Relationships   for CP & Cσ_3' Stress   Paths

FIG. 16   Absolute $V^e - \sigma_1'$ Plot

FIG. 18 $\varepsilon_{r_3}^e - \sigma_1'$ Relationships   for CP & Cσ_3' Stress   Paths

FIG. 19 $V_f^2 - \sigma_1'$ Relationships for Cσ_3' and CP Stress Paths

FIG. 20   VARIATION OF Q_1 & Q_3 with R

FIG. 21   Theoretical $\varepsilon_{r_1}^e - \sigma_1'$ Relationships

372

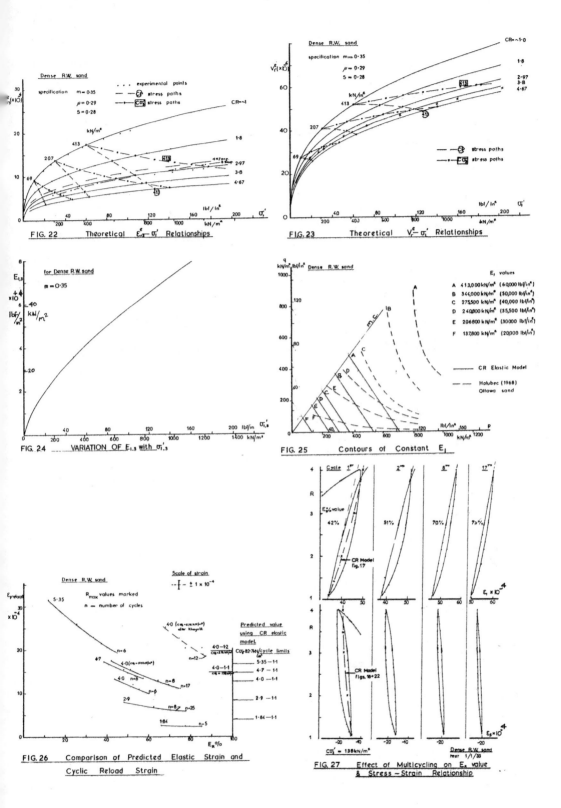

FIG. 22   Theoretical $\varepsilon_{1x}^e - \sigma_1'$ Relationships

FIG. 23   Theoretical $V_r^e - \sigma_1'$ Relationships

FIG. 24   VARIATION OF $E_{1.3}$ with $\sigma_{1.3}'$

FIG. 25   Contours of Constant E_1

FIG. 26   Comparison of Predicted Elastic Strain and Cyclic Reload Strain

FIG. 27   Effect of Multicycling on E_R value & Stress – Strain Relationship

373

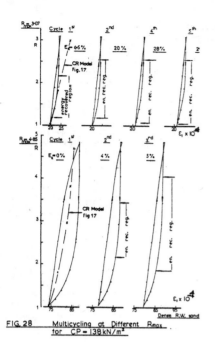

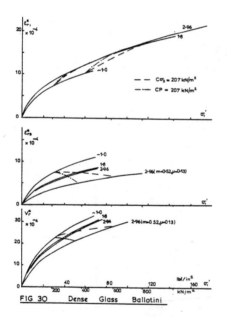

FIG. 28   Multicycling at Different R_{max}
for CP = 138 kN/m²

FIG. 30   Dense Glass Ballotini

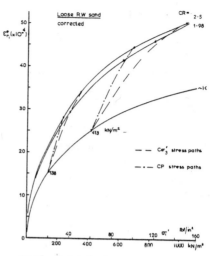

FIG. 29 a Absolute $\varepsilon_{r_1}^a - \sigma_1'$ Plot for Loose R.W. sand

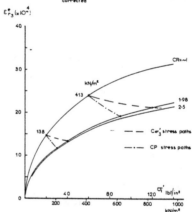

FIG. 29 b   Absolute $\varepsilon_{r_3}^a - \sigma_1'$ Plot for Loose R.W. sand

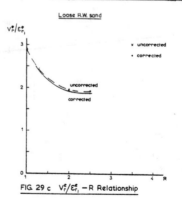

FIG. 29 c $V_r^a / \varepsilon_{r_1}^a$ — R Relationship

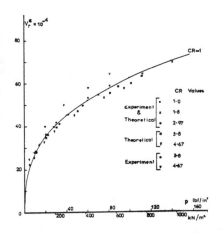

FIG. 31　V_r^e—P Relationships for CR Stress
　　　Paths for Dense River Welland
　　　Sand

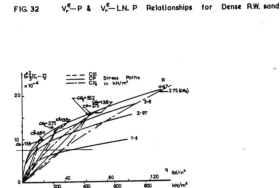

FIG. 32　V_r^e—P & V_r^e—LN. P Relationships for Dense R.W. sand

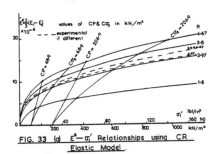

FIG. 33 (a) ε^e—σ_1' Relationships using CR
　　　Elastic Model

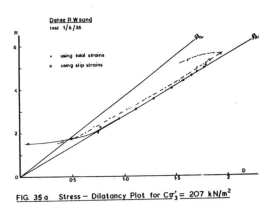

FIG. 33 (b) ε^e—q Relationships using
　　　CR Elastic Model

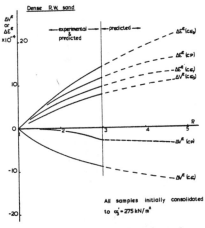

Dense R.W. sand

All samples initially consolidated
to $\sigma_3' = 275$ kN/m²

FIG. 34 Comparison of Elastic Volumetric &
　　　Shear Strains During Shear

Dense R.W.sand
test 1/6/36

• using total strains
× using slip strains

FIG. 35 a　Stress — Dilatancy Plot for Cσ_3' = 207 kN/m²

375

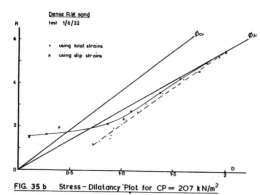

FIG. 35 b   Stress – Dilatancy ˙Plot for CP = 207 kN/m²

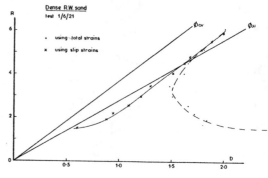

FIG. 35 c   Stress – Dilatancy Plot for Cσ′₁ = 276 kN/m²

376

Constitutive behavior of cemented alluvium

Part I. Theory and experiments

A.S. ABOU-SAYED
Terra Tek, Inc., Salt Lake City, Utah, USA

1 SUMMARY

An analytical and experimental investigation of the elastic-plastic response of loosely cemented alluvium is presented. The study is aimed at the characterization of the mechanical properties of near surface material for use in the prediction of ground motion and soil/structure interaction. The presentation consists of two parts. Part I concerns Experiment and Theory while Part II consists of comparison of Theory with Experiments.

The experimental program involves the determination of the effects of saturation levels and load path on the drained and undrained response of core samples. The elastic-plastic response as well as the failure surface, including the cap parameters, of the material were defined. Hydrostatic compression, triaxial compression and uniaxial strain tests were performed. The samples were also subjected to cyclic loading. Sample volume strain was measured directly using a dilatometer. During the undrained tests, the pore-pressure build up was monitored and correlated to the applied mean stress for various saturation levels. The strain paths corresponding to prescribed stress paths were determined. Part I reports on the experimental results and theoritical background.

In Part II, the results of the experimental program have been used to obtain a description of the constitutive behavior of this material. An elastic-plastic equation of saturated media based on the effective stress law is developed which may be extended to the onset of liquification. The Cap model parameters were determined based on a pseudo-complimentary

energy description of the Cap. The equations are valid for friable rocks as well as cemented alluvial materials. The implication of the work on ground motion predictions for wet sites are delineated.

2 INTRODUCTION

A comprehensive evaluation of the mechanical properties of low-cohesion soils and cemented alluvium at low levels of confining pressure is essential to the prediction of ground motion due to near-surface explosions. Elastic and inelastic behavior of the material needs to be determined. Furthermore, since moisture variation may drastically affect ground motion and stress wave attenuation, this determination must include the pore pressure response of the material at various saturation levels. The present paper summarizes recent work performed at Terra Tek concerning the characterization of soft geologic materials response when subjected to *in situ* conditions of stress and pore pressure.

It has been recognized that the response of weakly cemented alluvium and soft rocks is affected by the stress state within the material, including the pore pressure. Constitutive laws depicting mean-stress dependent failure surfaces, cap models and effective stress laws are reflective of this phenomenon. Most recently, the undrained elastic-plastic response of saturated porous material was modeled (Yamada, *et al.* 1979a) using an incremental plasticity theory that accounts for pore-pressure build up due to pore collapse. The model allows for prediction of the undrained material behavior from the know-ledge of dry material response. It is general enough to use any specified yield

function, however, it used the Terzaghi
effective stress(c.f. Biot 1956) for
determining the plastic deformation while
the elastic strains are calculated using
Nur's effective stress (Gorg and Nur,
1973). The formulation presented by
Yamada *et al.*(1979a) provides some conditions for unstable deformation response
of saturated material under undrained
conditions.

Dropek *et al.* (1978) have provided some
experimental evaluation of the effective
stress concept where they measured the
pore-pressure change in undrained test
Their work presented measured coefficients
for the pore-pressure-mean-stress relationship (c.f. Rice and Cleary 1978) for
Kayenta Sandstone samples. In the present
work, their experimental techniques have
been used to measure the change in pore
pressure within alluvium samples due to
externally applied triaxial stress conditions. or uniaxial strain deformation.

While triaxial-stress states adquately
represents the *in-situ* conditions for
geologic material, the laboratory observation of material response is often affected
by the experimental configuration. End-cap
effects, sample size effects, specimen
boundary conditions all tend to create non-
homogeneous deformation field, specially
near sample failure. Rudnicki and Rice,
(1978) have pointed out the condition for
shear band formation and deformation localization in triaxially loaded samples. To
avoid such complications in modeling
geologic material response from laboratory
investigation, Yamada *et al.* 1979b) have
suggested the use of the uniaxial-strain
test configuration to formulate the failure
surface. They identified the flow surface
as a two-parameter function. Both parameters can be determined from a single uniaxial strain test. The present paper will
summarize this formulation through its
application to the characterization of
the tested alluvium material.

In summary, the reported work consists
of some experimental and analytical evaluation of the mechanical response of
loosely cemented soil-type material to
applied *in-situ* conditions. The experiments involve the measurements of stress,
strain, pre-pressure as well as the
volume-strain of the sample. In the presentation, some recent work at Terra Tek
concerning the development of constitutive
modeling of geologic materials, including
cemented alluvium, in the presence of pore
fluid is summarized.

3 EXPERIMENTAL SET-UP

The test apparatus used in the present
investigation is the Terra Tek soil-
testing machine. The equipment (shown in
Figure 1) was reported in a previous publication (Yamada and Abou-Sayed 1978).
Briefly, it is capable of providing a
maximum confining pressure of 27 MPa and
a total axial force of 14,000 kg. The
vessel can accomodate samples up to 7.5 cm
in diameter and 15 cm long. Axial deformation rates range from 10^{-5} cm/sec to 10^{-2}
cm/sec using a closed loop-servo controlled
screw actuator.

Figure 1. Low pressure triaxial test
machine.

During the experiment, applied axial
load, confining and pore pressures, as well
as, axial and transverse strains were
monitored inside the vessel using a load
cell, pressure diaphrams, linear variable
displacement transducer (LVDT) and strain
gaged cantilevers respectively. Accuracy
of these respective measurements was ± 6
Kg, ± 0.003 MPa, ± 0.0006 cm/cm and ± 0.03
cm/cm. The details of the instrumentations
can be found elsewhere (Yamada and Abou-
Sayed 1979; and Dropek *et al.* 1978) and
will not be reported here. Instead, there
is one new feature in the reported experiments that deserves further discussion.
Since the sample volume strain is of
particular interest to this study, the
experimental set-up was designed so that
the volume strain, ε_V, was measured during
triaxial tests by two independent methods.
The first method used the volume strain
relationship $\varepsilon_V = \varepsilon_a + 2\varepsilon_t$ to determine ε_V
from measurement of the axial strain, ε_a,

378

and the transverse strain, ε_t. The second technique consists of using a dilatometer as a direct volume strain measuring device. The dilatometer is a pressure vessel with internal dimensions that are only slightly larger than the external dimensions of the test sample to accomodate the cantilever used for transverse strain measurements. It determines the actual sample volume change by measuring the displaced fluid surrounding the sample in or out of the vessel while the fluid pressure is maintained constant. A small confining volume is necessary to minimize any volume displacement due to the compressibility of the fluid. It should be noted, however, that the material tested in this program exhibited large volume strains, eliminating resolution problems. The other major components of the dilotometer are the volume-measuring pressure pump and a pressure measuring transducer. Figure 2 is a photograph of the various components constituting the dilatometer.

Figure 2. Dilatometer apparatus.

The developed apparatuses were used to perform compression, triaxial compression and uniaxial strain tests on cemented alluvium samples. Tests at various saturation levels from "as-cored" to 100 percent were completed and pore pressure data monitored. The effective stress law was also checked for this material. Finally, strain-path tests were completed to investigate the effect of path on material behavior.

4 THEORETICAL CONSIDERATIONS

The theoretical bases of the present investigation have been worked out in two other separate publications (see Yamada *et al.* 1979a and 1979b). Both the detailed elastic-plastic deformation response of saturated porous media under triaxial loading conditions as well as the formulation of the failure surface from uniaxial strain test data have been discussed. Here only a summary will be presented and the reader is referred to the original papers for further details.

4.1 Incremental plasticity theory for saturated porous media

Consider an elastic-plastic time-independent isotropic porous media subjected to the stress field given by σ_{ij}. The pore pressure within the material is given by p. Let $d\varepsilon_{ij}$ depicts the strain increment corresponding to a stress change $d\sigma_{ij}$. The accompanying pore pressure change is given by dp. Assume that the total strain tensor can be decomposed into recoverable, drainage and plastic strains denoted by $d\varepsilon_{ij}'$, $d\varepsilon_{ij}''$ and $d\varepsilon_{ij}{}^p$ respectively, *i.e.*
$$d\varepsilon_{ij} = d\varepsilon_{ij}' + d\varepsilon_{ij}'' + d\varepsilon_{ij}{}^p. \qquad (1)$$

Yamada *et al.* (1979a) have introduced the drainage strain increment into their analysis. Furthermore, they showed that each strain increment has a physical meaning and is related to either known or measurable quantity. The elastic portion of the strain increment, $d\varepsilon^e$, consists of the sun,
$$d\varepsilon_{ij}{}^e = d\varepsilon_{ij}' + d\varepsilon_{ij}''$$
However, during unloading the undrained elastic response is only given by $d\varepsilon_{ij}'$, hence the nomenclature, recoverable. The drainage strain increment $d\varepsilon_{ij}''$, is recoverable only upon removal of the pore pressure build up, dp', that results due to the irreversible deformation, $d\varepsilon_{ij}{}^p$. Therefore, the constitutive relation for this material can be summarized by the following equations,

$$d\varepsilon_{ij} = S_{ijRe}\left[d\sigma_{Re} - B\left(1-\frac{K}{K_s}\right)S_{ij}d\sigma_{mm}\right] \quad (2)$$

$$d\varepsilon_{ij}^{P} = h\,\frac{\partial f}{\partial \sigma_{ij}}\,df \quad (3)$$

$$d\varepsilon_{ij}'' = -\frac{3(1-\nu)n}{2(1-n)E}\,H'\delta_{ij}\left[d\varepsilon_{mm}^{P}\right] \quad (4)$$

in which,

$$B = \frac{\frac{1}{3}\left(\frac{1}{K}-\frac{1}{K_s}\right)}{\frac{1}{K}-\frac{1}{K_s}+n\left(\frac{1}{K_w}-\frac{1}{K_s}\right)} \quad (5)$$

$$H = \frac{K_w}{(n+m_2\,K_w)}\;;$$
$$m_2 = \frac{3n}{E}\left[\frac{(1-\nu)(1+2n)}{2(1-n)}+\nu\right]. \quad (6)$$

and where,

$S_{ijk\ell}$ is an elastic compliance tensor, S_{ij} is the Kroneker delta, K, K and K_w are bulk moduli for the matrix, the grain and the pore fluid respectively. E and ν are Young's modulus and Poisson's ratio; n represents the porosity of the material and h is a positive scalar. The yield surface f in the stress space is assumed to be a function of the effective stress $T_{ij} = \sigma_{ij} - \delta_{ij}P$. Assuming normality of the plastic strain vector d $_{ij}$ to the surface $f(T_{ij})$ as implied by equation (3) and neglecting any plastic deformation of the grain (an unlikely phenomenon due to the low stress involved) one can relate the volumetric plastic deformation, $d\varepsilon_v^P$, to the variables P, T_{ij}, σ_{ij} and the yield surface f via the equation*

$$d\varepsilon_v^P = 3h\,\frac{\partial f}{\partial T_{nn}}\left(1-\frac{\partial P}{\partial \sigma_{nn}}\right)\left(\frac{\partial f}{\partial T_{ij}}-3\frac{\partial f}{\partial T_{nn}}\right)d\sigma_{ij} \quad (7)$$

Equations 2 through 7 describe the elastic-plastic response of a porous material subjected to undrained conditions. Only a relationship between the porosity n and the applied stress, σ_{ij} (c.f. the spherical pore model by Schatz 1976) would be required to complete the formulation.

Finally, this formulation as discussed by Yamada $et\ al.$ (1979a) leads to the necessary condition for stable material behavior expressed by

$$\frac{\partial P}{\partial \sigma_{ii}} \leq 1 \quad (8)$$

Equation (8) is useful in depicting unstability for undrained porous media.

---

* repeated indices imply summation

During triaxial and uniaxial-strain experiments, the quantities $S_{ijk\ell}$, E, ν, $\frac{\partial P}{\partial \sigma_{ij}}$ and n can be measured or determined. Therefore, the complete material description can be obtained if, besides the usual mechanical moduli, one can obtain the relationship between the increase in pore pressure within the material, due to the applied loads, and pore compaction. In the present experimental work pore pressure-mean stress and pore pressure-deviatoric stress relationships were both monitored.

4.2 Porous material modeling by uniaxial-strain tests under drained conditions

Cap models have been developed to represent the irreversible volume compaction phase of geologic materials (Nelson $et\ al.$ 1971). Recently, Yamada and Abou-Sayed (1979) presented a pseudo-complimentary energy surface, f, in the stress space σ_{ij} to delineate the cap as follows in equation (9).

$$f^2(\alpha) = \tfrac{1}{2}\left(\sigma_{ij}\sigma_{ij} - \tfrac{2}{3}\alpha\,\sigma_{KK} + \tfrac{1}{3}\alpha\right) + \frac{(1-2\nu)}{6(1+\nu)}(\sigma_{KK}-\alpha)^2$$

where α, represents a hydrostatic state of stress, is a hardening indicator and is a function of the platic work, W^P. Equivalently, the cap is represented by an ellipse (see Figure 3) that is centered at the stress state K and has an aspect ratio, R, given by the equation,

$$R^2 = \frac{1-\nu}{6(1+\nu)} \quad (10)$$

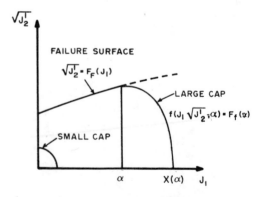

Figure 3. Yield surface and cap

in which ν is Poisson's ratio of the material during unloading in a uniaxial-strain test, E denotes Young's modulus.

During the loading portion of a uniaxial strain test on dry porous media, the axial strain increment, $d\varepsilon_1$, is given by

$$d\varepsilon_1 = \frac{1+\nu}{E} d\sigma_1 - \frac{\nu}{E}(d\sigma_1 + 2d\sigma_3) + h \frac{\partial f}{\partial \sigma_1} df \quad (9)$$

where ν, E, defined earlier, can be obtained from the unloading portion. h is a positive proportionality scalar. If the plastic work increment due to a small plastic strain $d\varepsilon_{ij}{}^P$ is constant for any stress state $(\sigma_{ij} - \alpha\delta_{ij})$ along the cap, then h becomes a function of the parameter α only. Hence h is constant along any particular cap surface. Finally, df is expressed by the following relationship,

$$df = \frac{\partial f}{\partial \sigma_1} d\sigma_1 + 2 \frac{\partial f}{\partial \sigma_3} d\sigma_3 \quad (10)$$

in which σ_1 and σ_3 are the axial and confining stress respectively.

To determine α during the loading portion of the test the following equation is used,

$$\alpha \left[\frac{(1-2\nu)}{3}\left\{(1+\nu)\left(\frac{d\sigma_3}{d\sigma_1}-1\right)+E\frac{d\varepsilon_1}{d\sigma_1}\right\}\right]$$

$$= \left[\left\{E\frac{d\varepsilon_1}{d\sigma_1}-1+2\nu\frac{d\sigma_3}{d\sigma_1}\right\} + \left\{\sigma_3-\nu(\sigma_1+\sigma_3)\right\}\right.$$

$$\left. + \left\{\nu-(1-\nu)\frac{d\sigma_3}{d\sigma_1}\right\}\left\{\sigma_1-2\nu\sigma_3\right\}\right] \quad (11)$$

where $\frac{d\sigma_1}{d\varepsilon_1}$ is the slope of axial stress-strain curve, $\frac{d\sigma_3}{d\sigma_1}$ is the slope of confining pressure axial-stress curve. The function f is given by equation (9), therefore,

$$\frac{\partial f}{\partial \sigma_1} = \frac{1}{(1+\nu)f}\left\{S_1'-2\nu S_3'+2(S_3'-\nu(S_1'+S_3'))\frac{d\sigma_3}{d\sigma_1}\right\} \quad (12)$$

where $S_i' = \sigma_i' - \alpha$ is a deviatoric portion of the principal stress σ_i.

Finally, h is given by

$$h = \left\{\frac{d\varepsilon_1}{d\sigma_1} - \left(\frac{1}{E} 1-2\nu\frac{d\sigma_3}{d\sigma_1}\right)\right\}\left(\frac{\partial f}{\partial \sigma_1}\frac{df}{d\sigma_1}\right) \quad (13)$$

Equations 9 through 13 complete the constitutive behavior representation for the dry porous material. A numerical scheme has been developed to use the experimental results for the determination of the various parameter in these constitutive equations. For more detailed derivations, the reader should refer to the paper by Yamada *et al.* (1979b).

5 EXPERIMENTS

Triaxial and uniaxial strain experiments were performed on cores recovered from an ulluvial valley at a depth of 0.5m, 1.0m and 2.0m. The material can be described as loosely cemented soil. During the laboratory invesitgation, the pore pressure change was monitored in several experiments as the sample compacted. Average physical properties of the material are given in Table I.

TABLE I: Average Physical Properties of Cores

Depth m	Wet Density gm/cc	Dry Density gm/cc	Grain Density gm/cc	Porosity %	Saturation %
0.5	2.56	2.335	3.09	24.4	92.1
1.0	2.19	1.816	3.17	42.7	87.6

5.1 Sample preparation

Sample preparation without soil disturbance was extremely difficult. The material arrived in metal tubes. Before the test, the sample was pushed out of the tube into a rubber jacket. End caps were installed after sample was resaturated to prescribed level when needed.

For the samples tested in the "as-cored" condition, the ends of the tube were wired to seal the sample from the confining fluid. The sample was then ready to be placed into the pressure vessel for testing. Figure 4 shows a typical sample before and after testing. Several samples were tested at various levels of saturation with water. The method of saturation was as follows. The samples were allowed to dry completely. The weights of the samples were taken in "as-cored" and dried. Vacuum was drawn on the sample, whereupon, the samples were saturated to 100 percent condition with water. The samples were again weighed. The desired percentage sample saturation was achieved by allowing the samples to dry out by evaporation to the proper weight. When this was obtained, the final end cab was wired in place.

5.2 Volume-strain measurements

Volume strain determination by summing the axial strain with the two transverse strains as determined by strain-gaged cantilevers is questionable during large or non-uniform deformation. This is mainly due to barreling of the sample. Therefore, sample dilatancy may be less than that

Figure 4. Sample before and after testing.

measured with the cantilevers. In the reported work, a new apparatus of measuring volume strain was designed and implemented. The apparatus, a dilatometer, (Fig. 2) consists of a small pressure vessel, a pressure transducer and a volume measuring transducer. As the sample is loaded, its volume compaction causes the pressure in the vessel to change. The pressure is returned to its initial value by a volume pump. This change in fluid volume (which is equal to the sample change in volume) is monitored by the volume transducer. Several samples were tested with volume strains measured using both cantilevers and the dilatometer. Figure 5 shows comparison of the two measurements. It was

observed that as long as the transverse strain were relatively small (under 3%) the volume strain measurements compare quite well. As the transverse strain increased the dilatancy or volume increase measured by the cantilever was, as expected, more than that shown by the dilatometer. Thus, in tests where transverse strains are small (such as uniaxial strain, hydrostatic compression and triaxial compression within limits) the volume strain can be measured using either system with accuracy. In the remainder of this paper the dilatometer-measured values are used unless otherwise stated.

6 EXPERIMENTAL RESULTS

This section describes in detail the experimental results from hydrostatic compression, triaxial compression and uniaxial strain tests performed on core samples from depths of up to 2.0 m. The majority of the tests were completed on samples in the as-cored condition. Some tests were run on samples saturated to varying amounts. Special tests to determine effect of load paths were also completed. Tests to check the validity of the effective stress law for this material were also completed. Variables measured were confining pressure, pore pressure, axial stress, axial strain, transverse strains and volume strain. The materials tested underwent extremely high strains (greater than 20 percent).

6.1 "As-Cored" samples

Samples were tested in the conditions at which they were removed from the coring tubes ("as-cored"). Hydrostatic compression tests of samples from 0.5m (1.7 ft.) and 1.0m (3.3 ft.) depth are shown in Figure 6.

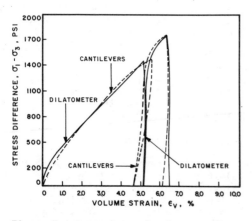

Figure 5. Comparison of volume strain measured by cantilevers on a dilatometer during a triaxial compression test.

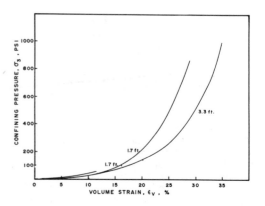

Figure 6. Hydrostats for "as-cored" samples.

Triaxial compression tests of the above material were completed at severl confining pressures. The stress difference as a function axial, transverse, and volume strains is plotted in Figures 7 and 8. The unaxial strain test data is presented in Figures 9 and 10. Figure 9 shows a plot of stress difference as a function of confining pressure. Data from two core depths are compared. Figure 10 shows stress difference as a function of volume strain from the same samples.

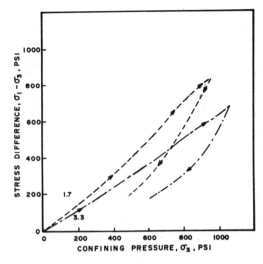

Figure 9. Uniaxial-strain test data for "as-cored" samples.

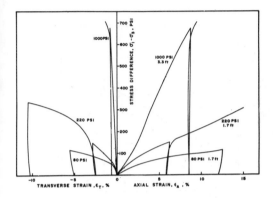

Figure 7. Triaxial compression test data for "as-cored" samples.

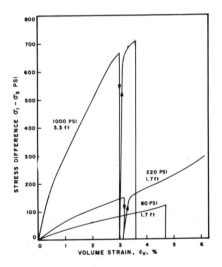

Figure 8. Triaxial compression test data for "as-cored" samples.

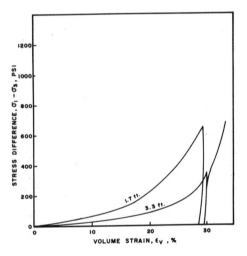

Figure 10. Uniaxial-strain test data for "as-cored" samples.

In both the triaxial compression and uniaxial strain tests, it was noted that cycling had very little effect. The sample would unload and reload at a very high modulus and then the stress-strain path follows the original curve almost immediately after reaching the start of the unloading point (c.f. Figure 7).

6.2 Effect of saturation tests

The effect of saturation level on the stress-strain behavior is report in this section. Tests were performed on samples from a depth of 2.0m (6.8 ft.) Samples were saturated to various levels as explained in the smaple preparation section.

Figure 11 shows triaxial data for various saturation levels. Results for 100 percent, 80 percent and "as-cored" saturation are shown. The stress difference is plotted against axial and transverse strains in the figure. The pore pressure as a function of mean stress for confining pressure of 69 MPa (1,000 psi) is depicted in Figure 12.

The effect of saturation level on sample behavior under uniaxial strain conditions is shown in Figures 13 through 15. Cores with water-saturation levels of 100 percent, 95 percent, 90 percent and 60 percent and as-cored were tested. Plots of stress difference verses confining pressure (Figure 13) and pore pressure verses confining pressure (Figure 14) are shown. Figure 15 is a comparison of the unloading behavior of 60 percent - saturated and "as-cored" samples.

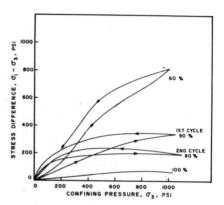

Figure 13. Uniaxial-test data for cores from 2.0m (6.8 ft) depth at various saturation levels.

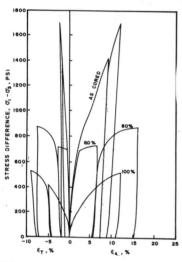

Figure 11. Triaxial compression test data (1000 psi) for samples from 2.0m (6.8 ft) depth at various saturation levels.

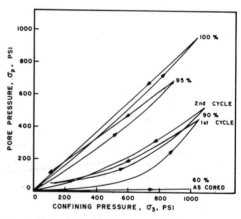

Figure 14. Uniaxial-test data for cores from 2.0m (6.8 ft) depth at various saturation levels.

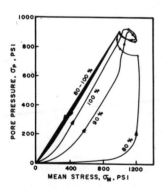

Figure 12. Triaxial compression test data (1000 psi) for samples from 2.0m (6.8 ft) depth at various saturation levels.

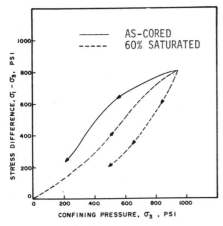

Figure 15. Uniaxial-strain test data for cores from 2.0m (6.8 foot) depths at 60 percent saturation and "as-cored" conditions.

6.3 Effective-stress law

Tests to determine if the material at 2.0m (6.8 foot) depth could be modeled using the effective stress law was performed. The sample (40 percent saturated) was tested triaxially at a confining pressure 69 MPa (1,000 psi). A pore pressure of 8 MPa (550 psi) was superimposed on the sample using nitrogen gas to pressurize the pores. The results of the tests are shown in Figure 16. The figure shows the stress difference as a function of axial and transverse strains along with a comparison with the results of triaxial tests where the confining pressure was set at 31 MPa (450 psi) while the pore pressure was maintained at zero. The data from both tests compared quite well, as seen in Figure 16.

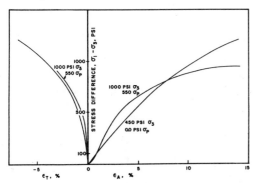

Figure 16. Triaxial test data to check effective stress law for cores from 2.0m (6.8 foot) depth.

6.4 Controlled-path tests

Tests following a prescribed loading path were performed on core samples from 0.5m (1.7 ft.) depth. The loading path followed a uniaxial strain condition to 3.5 MPA (500 psi) confining pressure. Then the sample was unloaded uniaxially to 1.4 MPA. At that stress state, the confining pressure was kept constant and triaxial test was carried out. Atypical result from these tests is shown in Figure 17. Figure 18 depicts the strain path followed by the sample.

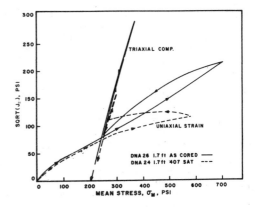

Figure 17. Samples from 0.5m (1.7 ft) depth tested in uniaxial strain to a confining pressure of 3.5 MPa (500 psi) unloaded to 1.4 MPa (200 psi) and then tested in triaxial compression (stress path).

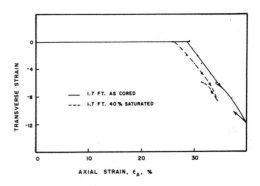

Figure 18. Sample from 0.5m (1.7 ft) depth tested in uniaxial strain to a confining pressure of 3.5 MPa (500 psi) unloaded to 1.4 MPa (200 psi) and then tested in triaxial compression (strain path).

385

7 DATA ANALYSIS

The experimental results in Section 6 have been analyzed in view of the theoritical considerations presented earlier. Value of elastic moduli, E and ν, as well as the cap parameter R would be determined from the results of uniaxial strain test during unloading. Equations (11) and (13) can be used to determine the parameters α and h as functions of the axial stress σ_1 from the uniaxial strain data.

Data from Figure (10) and (13) may be used to determine the functional relationship $\frac{d\varepsilon_1}{d\sigma_1}(\sigma_1)$, $\frac{d\sigma_2}{d\sigma_1}(\sigma_1)$ as well as $F(\alpha)$ and $h(\alpha)$. The procedure requires numerical analysis as indicated in the work by Yamada *et al.* (1979b). At this stage, this step has not been implemented yet and hence is deferred to Part II of the paper.

8 CONCLUDING REMARKS

Test data for hydrostatic, triaxial compression and uniaxial strain tests have been presented. Samples at several core depths from 0.5m to 2.0m and various levels of saturation from "as-cored" condition to 100 percent saturation were tested. The experimental results obtained indicate the following.
1. The materials showed the usual increased strength with pressure in the "as-cored" condition.
2. The effect of level of saturation was investigated. The strength reduced with increased saturation levels. The deep material exhibited some shear capability even at 100 percent saturation while the shallower material showed essentially no shear strength even at 80 percent saturation. The compressibility also was reduced as the level of saturation increased. Unloading behavior of dry and saturated samples showed the effect of pore pressure build up on strength.
3. The pore pressure was measured for all tests. The pore pressure was observed to increase in both the hydrostat and compression portions of a triaxial compression test. The effective stress law was also shown to hold reasonably well.
4. Several samples were cycled. The cycling had essentially no effect on the stress strain behavior of both the dry and saturated material.
5. Strain path tests were completed. Samples were tested in uniaxial-strain

to 3.5 MPa (500 psi) confining pressure unloaded to 1.4 MPa (200 psi) and tested triaxially. The correlation of the stress and strain paths require further investigation.
6. Measurements of volume strain using the new apparatus (a dilatometer) to determine actual volume strain compared favorably with cantilever-measured strain. Both systems gave similar results for small transverse strains.

ACKNOWLEDGEMENT

The author would like to express his thanks to Mr. S. Green and Dr. A. Jones of Terra Tek for their guidance and support and to his co-worker Dr. S. Yamada for his contribution to the theoretical aspect of this work. Dr. R. Christensen and Mr. K. Bergstrom have provided support during the experimental phase of the study. Appreciation is hereby extended to Ms. Colleen Hollingshead for her tireless scretarial and typing support.

This work was supported by the Defense Nuclear Agency (Dr. George Ullrich, Technical Officer) through various contracts with Terra Tek, Incorporated.

REFERENCES

Dropek, R. K. and J. N. Johnson, 1978, The Influence of Pore Pressure on the Mechanical Properties of Kayenta Sandstone, J. of Geophys. Res., Vol. 83.

Garg, S. K. and A. Nur, 1973, Effective Stress Laws for Fluid-Saturated Porous Rocks, J. of Geophys. Res., Vol. 78.

Rice, J. R. and Cleary, M. P., 1976, Some Basic Stress Diffusion Solutions for Fluid-Saturated Elastic Porous Media with Compressible Constituents, Rev. Geophys. and Space Phys., 14(2), pp. 227-241.

Rudnicki, J. W. and Rice, J. R., 1975, Conditions for the Localization of Deformation in Pressure-Sensitive Dilatent Material, J. of Mech. Phys. Solids, Vol. 25, pp. 371-394.

Schatz J. F., 1976, Models of Inelastic Volume Deformation for Porous Geologic Materials, The Effects of Voids on Material Deformation, ASME AMD-Vol. 16, 1976, Applied Mechanics Division Meeting, Salt Lake City, Utah, pp. 141-170.

Yamada, S. E. and Abou-Sayed, A. S., 1979,
Cap Model Guided by Energy Concept, J.
of Geotech. Eng., ASCE, Vol. 105, No.
GT2, pp. 183-200.

Yamada, E. E. Abou-Sayed, A. S. and Jones,
A. H., 1979a, Undrained Elastic-Plastic
Behavior of Porous Material, submitted
for publication to J. Appl. Mech., ASME
Trans.

Yamada, S. E., Abou-Sayed, A. S. and Jones,
A. H., 1979b, Elasto-Plastic Modeling of
Geologic Materials by Uniaxial Strain
Test, submitted for publication to J.
Geotech. Eng., ASCE.

Formulation of constitutive equations for sand

K.B. AGARWAL
University of Roorkee, Roorkee, India

B. SIVA RAM
Central Building Research Institute, Roorkee, India

1. INTRODUCTION

When the dynamic stress in the form of blast pressure is applied on the surface of sand, the stress pulse is transmitted into the medium according to equation of motion. The governing constitutive equations, for solution of equation of motion have been formulated on the basis of stress-strain behaviour of sand.

Fig. 1(a) depicts the behaviour of the soil in true shear when it is loaded, unloaded partly, reloaded and then unloaded completely. A full cycle of loading and equal reverse loading cycle is depicted in Fig. 1(b). Fig. 1(c) shows a typical curve on number of cycles of loading and unloading up to a certain level. Pure compressive stresses in loading, unloading and reloading result in stress-strain curves as shown in Fig.1(d). A reverse i.e., tensile loading produces large strain in cohesionless soil.

Some important parameters influencing the stress strain behaviour of cohesionless soil and their relative importance as given by Hardin (1972) are the strain amplitude, effective mean principal stress and void ratio. In view of less importance of frequency of loading, it can be assumed that the constitutive equation in slow loading cycle is not much different from that for fast loading. For sands Cassagrande and Shannon (1948,

1949) showed negligible increase in the value of modulus of deformation, whereas, increase in strength was of the order of 10 per cent in transient loading. Whitman (1969) and Whitman and Healy (1962) also showed under transient loading condition for sand an increase of 10-15 per cent in strength over static case. Therefore in this study laboratory triaxial tests were conducted to obtain the cyclic loading behaviour of sand.

2 BRIEF REVIEW OF CONSTITUTIVE LAWS

On reviewing the literature, it can be found that the constitutive relationships for soils have been arrived at, on the basis of rheological and mathematical models. Mathematical models have been more commonly used. Kondner(1963) and Kondner and Zelasko(1963) have considered the non-linear stress-strain curves for both clay and sand to be approximated by hyperbolae. Duncan and Chang(1970) have expressed the Kondner's expression in terms of the shear strength and initial tangent modulus. Janbu (1963) developed a relationship between intial tangent modulus and the confining pressure. Hansen (1963) proposed the use of parabolic functions to represent the stress-strain characteristics. Desai (1971) proposed the use of spline functions which approximate the given non-linear stress-strain

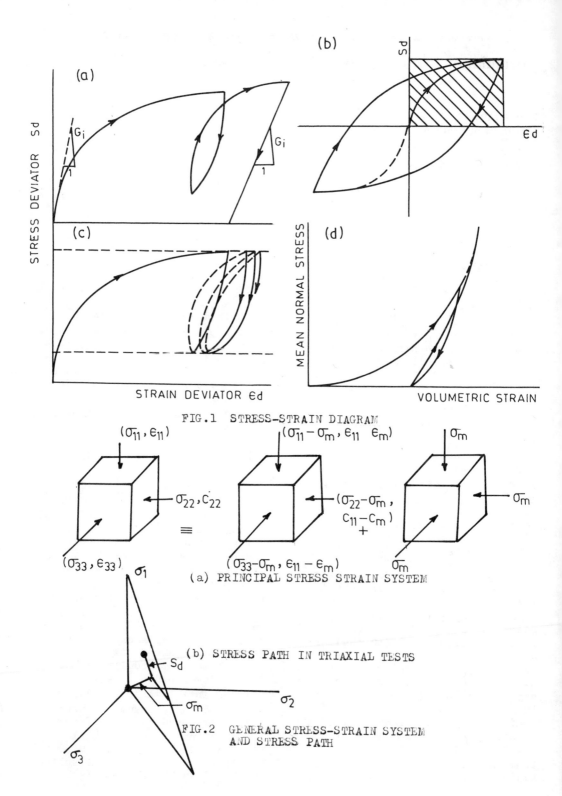

FIG.1   STRESS-STRAIN DIAGRAM

(a) PRINCIPAL STRESS STRAIN SYSTEM

(b) STRESS PATH IN TRIAXIAL TESTS

FIG.2   GENERAL STRESS-STRAIN SYSTEM
        AND STRESS PATH

behaviour by a number of polynomials of a given degree spanning a number of data points.

3 STRESS-STRAIN AND MODULI

Any normal effective stress σ_{11} in arbitrary direction 1 can be obtained in terms of volumetric stresses and deviatoric shear stresses.

$$\sigma_{11} = \frac{(\sigma_{11} + \sigma_{22} + \sigma_{33})}{3} + \sigma_{11} - \frac{(\sigma_{11} + \sigma_{22} + \sigma_{33})}{3} \qquad (1)$$

Volumetric   Deviatoric

where σ_{11}, σ_{22} and σ_{33} are the effective normal stresses in mutually perpendicular directions. An equation can also be written for strains in a general strain system.

$$\epsilon_{11} = \frac{\epsilon_{11} + \epsilon_{22} + \epsilon_{33}}{3} + \epsilon_{11} - \left(\frac{\epsilon_{11} + \epsilon_{22} + \epsilon_{33}}{3}\right) \qquad (2)$$

Volumetric   Deviatoric

where, $\epsilon_{11}, \epsilon_{22}, \epsilon_{33}$ are strains in the directions parallel to σ_{11}, σ_{22} and σ_{33}. The stress-strain relationship for homogeneous, isotropic, elastic solid, when subjected to general stress system can be represented by tensor equation in terms of elastic constants the tangent bulk modulus K_t and the tangent shear modulus G_t for mean normal stress and deviatoric stress respectively.

$$\sigma_{ij} = K_t \epsilon_{kk} \delta_{ij} + 2G_t (\epsilon_{ij} - \tfrac{1}{3} \epsilon_{kk} \delta_{ij}) \qquad (3)$$

where σ_{ij} and ϵ_{ij} are the stress and strain components respectively; δ_{ij} is the kronecker delta, i and j varying from 1 to 3. K_t is tangent bulk modulus which is defined by

$$K_t = \frac{\sigma_m}{\epsilon_v}, \text{ for elastic case;} \qquad (4)$$

$$K_t = \frac{\Delta \sigma_m}{\Delta \epsilon_v}, \ \Delta \epsilon_v \to 0 = \frac{d \sigma_m}{d \epsilon_v} \qquad (5)$$

Secant modulus for inelastic case is defined by:

$$K = \frac{\sigma_m}{\epsilon_v} \qquad (6)$$

where, σ_m is mean normal stress and ϵ_v is the volumetric strain at a point on stress strain curve. Deviatoric shear stress S_{ij} for condition of pure shear and deviatoric shear strain D_{ij} are related by shear modulus as per following relation:

$$S_{ij} = 2 G_t D_{ij} \qquad (7)$$

where,

$$S_{ij} = (\sigma_{ij} - \tfrac{1}{3} \sigma_{kk} \delta_{ij}) \qquad (8)$$

$$D_{ij} = (\epsilon_{ij} - \tfrac{1}{3} \epsilon_{kk} \delta_{ij}) \qquad (9)$$

Let a system of stresses, represented by Fig. 2(a) consisting of principal stresses σ_1, σ_2, σ_3 be considered. Resulting S_d of S_{ij} (i.e., S_{ij} in principal direction, with i = j = 1,2,3 is defined by Eq. 8 and resultant of deviatoric strains ϵ_d of D_{ij} (i.e., D_{ij} in principal directions i = j = 1,2,3 be defined by Eq. 9) are related by Equation 10.

$$G_t = \frac{\Delta S_d}{\Delta \epsilon_d} \ ; \ \Delta \epsilon_d \to 0 \qquad (10)$$

where,

$$\epsilon_d = 2\sqrt{(\epsilon_1 - \epsilon_m)^2 + (\epsilon_2 - \epsilon_m)^2 + (\epsilon_3 - \epsilon_m)^2} \qquad (11)$$

$$S_d = \sqrt{(\sigma_1 - \sigma_m)^2 + (\sigma_2 - \sigma_m)^2 + (\sigma_3 - \sigma_m)^2} \qquad (12)$$

and ϵ_m = mean strain, viz., $\epsilon_m = \epsilon_v / 3$

For triaxial test $\sigma_2 = \sigma_3$, therefore

$$S_d = \sqrt{\tfrac{2}{3}} (\sigma_1 - \sigma_3), \qquad (13)$$

$$\epsilon_d = 2\sqrt{\tfrac{2}{3}} (\epsilon_1 - \epsilon_3), \qquad (14)$$

and tangent shear modulus for inelastic case

$$G_t = \frac{1}{2} \frac{\Delta (\sigma_1 - \sigma_3)}{\Delta (\epsilon_1 - \epsilon_3)}; \Delta(\epsilon_1 - \epsilon_3) \to 0 \qquad (15)$$

4 PROPERTIES OF SAND STUDIED

The sand used in this study was obtained from river known as Ranipur and had the following properties:

Maximum dry density= 1.68 g/cm³.
Minimum dry density= 1.38 g/cm³.
Specific gravity of sand = 2.628

The sand was poorly graded, medium to fine grained sand with coeff. of uniformity of 1.86 and coeff. of curvature of 0.75.

5 DEVELOPMENT OF EMPIRICAL CONSTITUTIVE RELATIONSHIP FOR RANIPUR SAND

5.1 Test procedure

Triaxial tests were conducted with an aim to obtain separate constitutive equations for pure hydrostatic compression and for pure shear. The stress path, therefore, was followed in triaxial tests as shown in Fig. 2(b). The shearing of sample was done by maintaining the mean normal stress equal to the initial isotropic consolidation pressure, (i.e., $\sigma_{11} = \sigma_{22} = \sigma_{33}$). The shearing was possible by following the stress path so as to increase σ_1 and at the same time decrease σ_3 in steps.

The samples for triaxial test were of size 7.5 cm diameter X 15 cm height and were prepared by depositing dry Ranipur sand at 40 per cent, 60 per cent and 80 per cent relative densities. Four samples at each relative density were prepared and were subjected to cell pressure of 1.5, 2,3 and 4 kg/cm² respectively for isotropic consolidation. After allowing sufficient time for samples to be fully consolidated, they were sheared along the desired stress path under drained conditions. The volume changes were measured throughout. When the cell pressure was reduced by $\Delta\sigma_3$ decreasing vertical stress also by $\Delta\sigma_3$, the constant mean normal stress, was restored by increasing the vertical stress through Proving-ring to $3\Delta\sigma_3$. An additional margin in the vertical load recorded by proving-ring was allowed for the area correction, as for drained condition at each step. For finding the resultant deviatoric strain ϵ_d by Eq. 14, the value of ϵ_3 was required. The lateral strain, ϵ_3 was found out indirectly from the volumetric strain ϵ_v and the axial strain ϵ_1 readings by the following equations;

$$\epsilon_3 = \frac{1}{2}(\epsilon_v - \epsilon_1) \quad \therefore \epsilon_v = (\epsilon_1 + 2\epsilon_3) \qquad (16)$$
$$\text{for } \sigma_2 = \sigma_3$$

5.2 Equation for pure compression

From the observed volume change, the calculated volumetric strains, were plotted against mean stress in Fig. 3(a). A linear variation was found between the tangent modulus calculated from Eq. 5 and mean normal stress, Fig. 3(b), expressed by the form:

$$K_t = K_i + \alpha\sigma_m \qquad (17)$$

where, K_i is initial tangent modulus (i.e., at $\sigma_m = 0$) σ_m is the mean normal stress and α the slope of linear plot. The initial tangent modulus K_i in Kg/cm² was found to be proportional to initial relative density Dr in percentage, Fig. 3(c), expressed by

$$K_i = 0.375 \quad Dr \qquad (18)$$

The slope α has been found to vary linearly with Dr as shown in Fig. 3(d), expressed by

$$\alpha = -475 + 337.5 \text{ Log Dr} \qquad (19)$$

Eq. 5 may be rewritten for the Ranipur sand as

$$K_t = \frac{d\sigma_m}{d\epsilon_v} = K_i + \alpha\sigma_m \qquad (20)$$

By integrating Eq. 20 we get the constitutive equation in compression:

$$\sigma_m = \frac{K_i}{\alpha}\left[\text{Exp}(\alpha\epsilon_v) - 1\right] \qquad (21)$$

The stress strain plot during virgin loading has been shown in Fig.4(a).

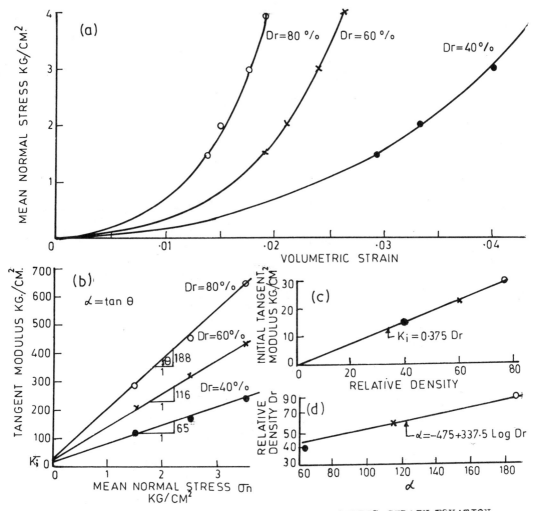

FIG.3 EMPIRICAL EVALUATION OF COMPRESSIVE STRESS STRAIN EQUATION

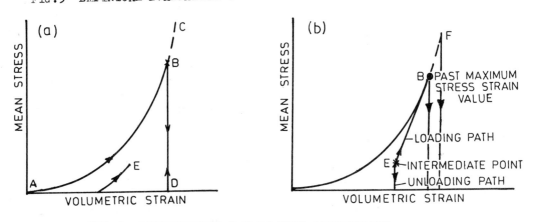

FIG.4 CONSTITUTIVE LAW IN PURE COMPRESSION

It follows path AC. The Eq. 21 is the constitutive equation for first cycle of loading. Though there was small proportion of recovery of the order of 1/25 th of compressive strain at which unloading takes place, the same has been considered negligible. Therefore the stress strain curve has been idealized and assumed to follow the unloading path with no recovery of strain as shown in Fig. 4(a) by the path BD from the unloading point B. On reloading vertical path is retraced up to the initial unloading point B and beyond B the stress strain curve again follows the Eq. 21. The other simplifying assumptions made for defining the constitutive equation are as follows:

(a) Full recovery of the residual strain may be assumed to take place at a very small tensile stress and this may follow the volumetric strain axis. The recovery of strain may continue even after zero strain leading to negative strain.

(b) As a result of unloading under negligible tensile stresses and subsequent reloading, an intermediate point, E is reached upon reloading. Reloading from this intermediate point is assumed to occur along straight line joining this intermediate point E to the point B from where unloading started, Fig. 4(b), the stress strain function follows Eq. 21.

5.3 Equation for pure shear

The shear stress strain relationship has been characterized by extending hyperbolic curve using the parameters as used by Kondner and Zelaskó. These parameters are depicted in Fig. 5 wherein, G_i is the initial slope of the stress strain curve and b is the inverse of ultimate strength at failure keeping the mean normal stress constant. Using the Eqs. 13 and 14 for S_d and ϵ_d respectively the constitutive equation is given by

$$S_d = \frac{\epsilon_d \, G_i}{1 + b \epsilon_d G_i} \qquad (22)$$

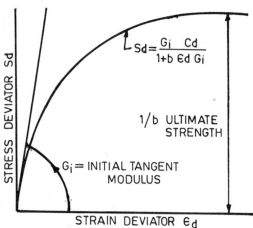

FIG. 5 CONSTITUTIVE LAW IN PURE SHEAR

or alternatively

$$\frac{1}{S_d} = \frac{1}{G_i} \frac{1}{\epsilon_d} + b \qquad (23)$$

The tangent modulus

$$G_t = \frac{\Delta S_d}{\Delta \epsilon_d}, \Delta \epsilon_d \to 0 \qquad (24)$$

By differentiating with respect to ϵ_d the Eq. 22 and using Eq. 24, G_t can be obtained as

$$G_t = G_i \left(1 - b \, S_d \right)^2 \qquad (25)$$

The experimental stress-strain curves were plotted for the triaxial test data, Eqs. 13 and 14 were used to evaluate S_d and ϵ_d from known values of σ_1, σ_3, ϵ_1 and ϵ_3. The values of G_i and b are graphically obtained by plotting $1/S_d$ on ordinate vs. $1/\epsilon_d$ on abscissa. It can be seen that for each normal stress the best fit is straight line; Fig. 6(a) shows typical plots for Dr = 80 per cent. Using Eq. 23, the values of b can be found as the intercept on the ordinate and Gi as the inverse of the slope of the best fit straight line for each normal stress. Fig. 6 (b) shows approximately linear variation of G_i with respect to σ_m on log-log plot. The value of G_i can be obtained in terms of σ_m in kg/cm2 as

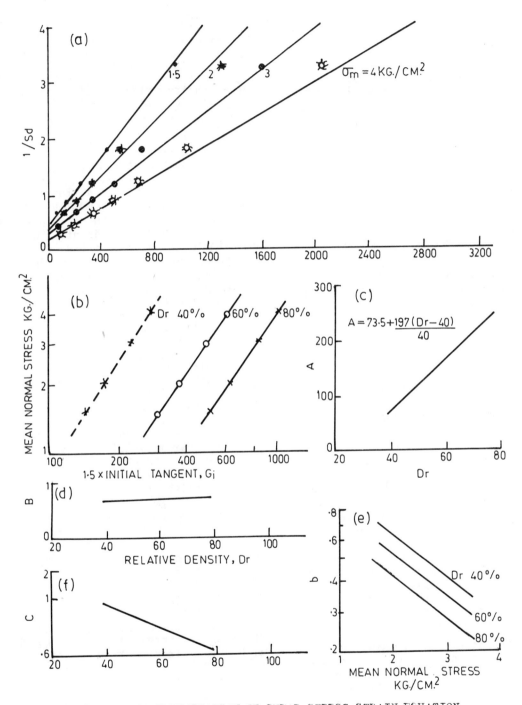

FIG.6 EMPIRICAL EVALUATION OF SHEAR STRESS STRAIN EQUATION

$$G_i = A (\sigma_m)^B \qquad (26)$$

where, A, corresponds to the intercept at $\sigma_m = 1$ kg/cm^2 and B corresponds to the slope of the straight line fitted.

The values of A and B were found to be related to Dr through a linear fit, Fig. 6(c) and (d). The following correlations are found

$$A = 73.5+177 \ (Dr-40)/40 \qquad (27)$$
$$B = 0.664+0.078(Dr-40)/40 \qquad (28)$$

A linear correlation is found to exist between b and σ_m on log-log plot, Fig. 6(e):

$$b = C(\sigma_m)^D \qquad (29)$$

where C and D are constants and can be found from:

$$C = 0.96 -0.34(Dr-40)/40 \qquad (30)$$
$$D = - 0.75 \qquad (31)$$

For defining constitutive equation completely, certain idealizations have been made to achieve simple model amenable to their use in solving the equation of motion. The Eq. 22 with parameters G_i and b defined by Eqs. 26 to 31, represents the constitutive equation for virgin loading irrespective of direction of shear. This stress-strain relationship holds good provided the soil element experiences such stress which has never before been exceeded before.

The unloading in shear is considered to have taken place when the soil element experiences less absolute value of shear stress compared to what it has been previously subjected to. The reloading is considered to take place when the absolute value of shear stress increases compared to immediately previous stress.

The direction of shear would change with rotation of the major principal stress axis. The question of positive and negative deviatoric shear is resolved in plane strain case using the following convention. The sign of deviatoric stress whe-

ther positive or negative, seems to depend upon the direction of the major principal stress axis with respect to the vertical direction. The direction of the major principal stress axis for a given set of stress may be found out conventionally, Harr (1966). As far as the corresponding major principal stress or strain axis is within 45° either side of horizontal, it will bear a negative sign.

The magnitudes and directions of principal stresses can be determined from general stress σ_x, σ_z and τ_{xz} in two dimensional plane strain problem by routine equations. Based on the above sign convention, the stress strain values can lie in any one of the four quadrants. The coordinates of stress-strain event can be uniquely located in stress-strain domain as shown in Fig. 7. The stress strain event in two successive time intervals, can change its position from one quadrant to another. Let these quadrant be numbered according to Fig. 7 designating the former with I_o and the latter with I_n.

The virgin stress strain curve is shown in Fig. 7(a), when the soil experiences stress which has not been exceeded before. There is a possibility of unloading from a point P as after travelling along virgin curve AP. The unloading from P takes place along a line PQ parallel to the initial tangent of stress-strain curve. Any reloading during the course of unloading along PQ retraces itself. The stress strain may be located at intermediate point O through a stress strain history as depicted in Fig. 7(a) or by any other path. The loading from this point follows OP, while, unloading follows OM. When the unloading has occured a number of times from virgin curve, the point P corresponds to the point of highest stress from which the unloading started. Similar conditions as of $I_o = I_n = 1$ would exist for I_o and $I_n = 3$ due to symmetry.

Now let the stress-strain change from quadrant $I_o = 1$ to $I_n = 4$ due to unloading process. A point R is

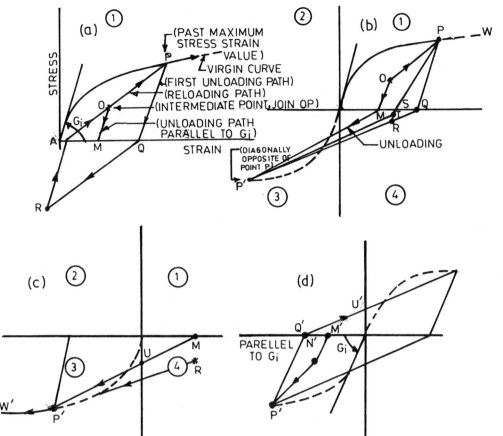

FIG. 7 DEVIATORIC STRESS-STRAIN DIAGRAM FROM INTERMEDIATE STRESS-STRAIN

reached through unloading of stress along P-Q(I_o = 1) and then along Q-R. Further reloading takes path along RS parallel to PQ followed by path joining point SP of quadrant 1. Similarly any intermediate point T in quadrant 4 is unloaded along TP' and reloaded along TS parallel PQ. Due to symmetry case involving I_o= 3 and I_n = 2 will be treated similar to the case I_o = 1 and I_n = 4. As most of these stress strain paths are straight lines, it is quite easy to formulate the corresponding equations. The coordinates of point P, Q, M and U(P',Q',M', and U') are defined in above discussion. One cycle of loading, unloading and then reloading is depicted in Fig. 7(d).

6 CONCLUSION

The constitutive equations for pure compression and shear have been developed for Ranipur sand at 40 to 80 per cent initial relative density, on the basis of laboratory triaxial shear tests. Influence of the parameters like the initial density and mean stress level have been suitably incorporated.

7 REFERENCES

Casagrande,A and Shannon,W.L.1948, Stress Deformation and Strength Characteristics of Soil under dynamic loads, Proc. Second Int. Conf. on Soil Mech. and Found. Engg.,Rotterdam Vol.I,p 29-34.
Ibid 1949, Strength of Soil Under dynamic loads, Trans.ASCE,Vol.114, p. 775.

Desai, C.S. 1971, Non linear analysis using spline functions, Proc.
ASCE, SM-10, p.1461.

Duncan, J.M. and Chang,C.Y. 1970,
Non linear analysis of stress strain in soils, Proc. ASCE SM-5,
p. 1629.

Hansen,J.B. 1963, Discussion of hyperbolic stress-strain response
of cohesive soils, Proc. ASCE,
SM-4, p. 241-242.

Hardin, B.O. Vincent P.Drenvich
1972, Shear modulus and damping
of soils measurement and paramet-
eric effect, Jnl. of Soils and
Found. Engg., Proc. ASCE, SM 6,
p. 603.

Harr, M.E. 1966, Foundations of
Theoretical Soil Mechanics, Mc,
Graw-Hill Book Company, Newyork.

Kondner, R.L. 1963, Hyperbolic
stress-strain response of cohesive soils, Jnl. of Soils and Found
Engg., Proc. of ASCE, Vol. 89,
SM 1, p. 115-143.

Kondner, R.L. and Zelasko, S.1963,
A hyperbolic stress-strain formu-
lation for sands, Proc. 2nd Pan
American Conf. on SM and FE,
Brazil, Vol. 1, p.289-324.

Whitman, R.V. 1969, Preliminary
high speed triaxial tests on compacted Buckshot clay, MIT, Soil
lab. Report.

Whitman, R.V. and Healy K.A. 1962,
Shear strength of sand during
rapid loading, Jnl. of Soil and
Found., Proc. ASCE, Vol. 88,
SM 2, p. 99-132.

Incremental constitutive equations for soils and application to cyclic loadings

R. CHAMBON & F. DARVE
Institut de Mécanique de Grenoble, Grenoble, France

1 INTRODUCTION

Different constitutive laws were recently proposed to predict the cyclic behaviour of soils. In order to describe the irreversible part of the deformations, we will see later that it is necessary to solve the problem of the "incremental non-linearity" of the constitutive law. Several authors have generalized the elasto-plastic models by adjunction of an anisotropic hardening (Mroz et al. (1978), Prevost (1977)) or with two yield functions (Maier and Hueckel (77), Lade (1976)). These laws are bi-linear or quadri-linear. Others authors have proposed non-linear laws : Bazant et al. (1976) with an endochronic model, Gudehus and Kolymbas (1979) with his non-linear rate-type constitutive law.

The notion of "tensorial zone" (zone of the incremental loadings space in which the incremental law is a linear transformation, Darve, Labanieh, Chambon (1976) and Darve, Boulon, Chambon (1978), permits to study the multi-linear laws and to verify the essential condition of continuity between two adjacent zones. Some constitutive models do not verify this condition (Gudehus (1979) gives examples). We showed (Darve et al. (1978)) that the multi-linear incremental law with eight tensorial zones respect this continuity condition only if the eight matrices of the eight associated linear transformations are chosen non-symmetrical. Recently we generalized (Chambon, Renoud-Lias (1979)) this law by considering a directional interpolation : in this way the law appears to be completely incrementally non-linear. In this article we present the principal assumptions and features of this law.

For an explicit formulation of the law it is necessary to know the behaviour of the soils for conventional triaxial tests in compression and in extension. Therefore, in order to describe the cyclic soil behaviour, analytical expressions for cyclic conventional triaxial tests must be given. Previously we gave (Darve (1978) and Darve, Flavigny (1979)) the first choiced rudimentary assumptions and examples of liquefaction. A boundary problem with loading, unloading, reloading in the case of the expansion of a cylindrical cavity is presented (Chambon et al. (1979)).

Hereafter we show the described cyclic behaviour with new analytical formulations. Some examples of cyclic drained and undrained conventional triaxial loadings are given. A cyclic liquefaction is also presented.

2 CONSTITUTIVE EQUATIONS

2.1 Incremental constitutive law

If we assume that soils are simple materials, the history of the strains of an homogeneous sample determines its state of stress. But is is very difficult to explicit this relation called functional equation. An incremental (or step by step) formulation is prefered. For a given loading history, and for a given loading increment the determinism principle implies the knowledge of the response. (For instance to each strain rate tensor $\dot{\varepsilon}$ corresponds one stress rate tensor $\dot{\sigma}$). In this paper we shall consider only non-viscous materials such as sands. Then there exists a function F depending on the strain history of the material such that

$$F(\delta\sigma, \delta\varepsilon) = 0$$

$\delta\sigma$ is the stress increment tensor and $\delta\varepsilon$ the strain increment tensor. The previously quantities have to be objective, for instance the stress increment tensor can be the one with respect to a corotational frame.

However in this paper, only cases without rotation and for which the strains are infinitesimal are studied. The first obvious property of the tensorial function F is to be homogeneous only in the following restricted sense. Since $\delta\sigma$ and $\delta\varepsilon$ are small for any $\lambda \geqslant 0$

$$F(\delta\sigma, \delta\varepsilon) = 0 \implies F(\lambda\delta\sigma, \lambda\delta\varepsilon) = 0$$

Moreover we will assume that there is a bijection between $\delta\sigma$ and $\delta\varepsilon$ so that

$$F(\delta\sigma, \delta\varepsilon) = 0 \iff \begin{cases} \delta\varepsilon = G(\delta\sigma) \\ \delta\sigma = G^{-1}(\delta\varepsilon) \end{cases}$$

except if the stress tensor σ is on the failure surface. This restriction is not very important because in applications of our law states of stress belonging to the failure surface are reached assymptotically. Since the function F is homogeneous, the tensorial functions G and G^{-1} are also homogeneous like F. For any $\lambda \geqslant 0$

$$\delta\varepsilon = G(\delta\sigma) \iff \lambda\delta\varepsilon = G(\lambda\delta\sigma)$$

$\delta\varepsilon = G(\delta\sigma)$ for which G depends on the loading history is the mathematical formulation of the incremental constitutive law. Such a constitutive law needs to be integrated along a loading path, but on the other hand it can take full account for the stress path dependancy. This implies the definition of a state of reference from which the integration is made.

At last it must be outlined that an incremental way does not solve completely the problem of the great non-linearity of the soil-behaviour an incremental constitutive law is only homogeneous, but it is non linear. This is necessary to represent an irreversibility of the material because it is well known that the incremental response to a given loading $\delta\sigma$ is very different from that corresponding to the opposite loading $-\delta\sigma$. This fact explains that strict hypoelasticity cannot be used to describe soil behaviour. An incremental constitutive law is not the linearization of a constitutive law. Assumptions and determination of the law are studied hereafter.

2.2 General assumptions

Firstly an orthotropy of the incremental constitutive equations is assumed. The orthotropy axes are supposed to be in coincidence with the stress principal axes. In this paper, we will only consider loadings for which the principal axes of the stress

increment tensor $\delta\sigma$ coincide with orthotropy axes. Hence it is the same for the principal axes of the strain increment tensor $\delta\varepsilon$ and also for the principal axes of the strain tensor ε. Thus all the tensors can be represented in a three-dimensional space. A more general law can however be developped, a restriction of which is the law studied here.

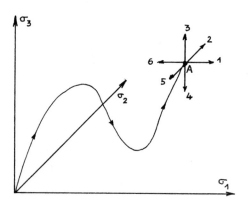

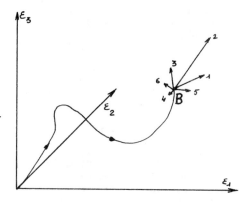

Fig. 1 Illustration of the constitutive law as a transformation from stress-space into strain-space

In order to explain others assumptions used to build up the law, let us consider figure 1 where the principal stresses space and the principal strains space are represented. In the first space the stress history is drawn. In the second space it is the strain history. Point A of the first space corresponds to the current state of stress and point B of the second space is the current state of strain. Because of its property of homogeneity the incremental constitutive law can be understood as a bijection between a unit increment (i.e. whose norm is 1) of stress, the origin of which is A and the associated increment in the strain space the origin of which is B. Thus it appears that the law is also a correspondance between a direction in the stress increment space and a direction in the strain increment space. We have drawn six particular increments in stress space. For instance the number 1 is the beginning of a loading for which the components of the stress tensor σ_2 and σ_3 stay each at the values that they have in A and the component σ_1 increases. For the number 2 it is the values of σ_3 and σ_1 which are constant whereas σ_2 is increasing. For the number 6 the values of σ_2 and σ_3 are constant but σ_1 is decreasing... and so on. Corresponding responses are drawn in the strain space. The six particular loadings, the beginning of which are A which extend these six increments are called "generalized triaxial tests". There are two steps to describe the main features of the law used in the present paper.

1) Based on experimental results assumptions are made in order to simulate generalized triaxial tests with constitutive parameters calculated from conventional compression and extension triaxial tests. These assumptions may take into account memory and cyclic effects. They are explained latter in 2.4.

2) The first step brings the knowledge in strain space of the responses of the six stress increments described previously. In order to know the response of any stress increment the origin of which is A an assumption has to be made for a directional interpolation between the six previous stress increments. This is described in section 2.3.

These two steps are almost independant then as soon as experimental results are firmly established they can be included in the law which have so a great flexibility.

2.3 Directional interpolation

This problem is very important. In order to perform good simulation, it is not sufficient to have good formulation of responses corresponding to particular loadings increments. For instance laws built from infinitesimal elasticity with a loading modulus and a different unloading modulus associated with a loading criterion have a great default. There are two responses for a loading for which the loading criterion has a null value. There is a discontinuity of the response as the loading increment cross over the boundary which separates loading zone from unloading zone as it is illustrated by Gudehus (1979). Consequently there is no uniquiness of the solution for a boundary

value problem as it is pointed out by Nelson (1978). Such a law violates the assumption of bijection between $\delta\sigma$ and $\delta\varepsilon$ made in 2.1. Then our law avoids the previous default.

A first attempt to realize the interpolation is described in (Darve, Boulon & Chambon (1978) or in Darve (1978) but through an other presentation. The stress increment space the origine of which is A is devided into 8 zones. The first is the set of points the coordinates of which $\delta\sigma$, $\delta\sigma_2$, $\delta\sigma_3$ are positive or null. The second is the set of points for which $\delta\sigma_1 \geqslant 0$ $\delta\sigma_2 \geqslant 0$ $\delta\sigma_3 \leqslant 0$ ... and so on. Three of our six particular increments belong to each zone. Then a linear interpolation can be performed for each zone which is so called "tensorial zone". We denote $\{\delta\sigma\}$ the column matrix of the three components of tensor $\delta\sigma$ and $\{\delta\varepsilon\}$ the column matrix of the three components of tensor $\delta\varepsilon$

$$\{\delta\sigma\} = \begin{Bmatrix} \delta\sigma_1 \\ \delta\sigma_2 \\ \delta\sigma_3 \end{Bmatrix} \qquad \{\delta\varepsilon\} = \begin{Bmatrix} \delta\varepsilon_1 \\ \delta\varepsilon_2 \\ \delta\varepsilon_3 \end{Bmatrix}$$

The linear interpolation into a zone may be written $\{\delta\varepsilon\} = [M]\{\delta\sigma\}$. $[M]$ is a 3 x 3 Matrix. A matrix $[M]$ corresponds to each zone. So the incremental constitutive law is written

$$\{\delta\varepsilon\} = [M_n]\{\delta\sigma\}$$

where the non linearity results of the discret variation of $[M_n]$ according to the tensorial zone n to which the direction of stress increment belongs. This law satisfies our assumption concerning bijection between $\delta\sigma$ and $\delta\varepsilon$ as it can be seen in Gudehus (1979). Such an interpolation gave good results (Darve, Labanieh, Chambon 1976, Darve 1978) for numerous simulations of homogeneous samples. But some loadings the stress increment of which changes gradually from a zone to the adjacent one implies non regular response. Integrated in Finite Element Method computationnal routines the law involving zones do not cause any problem if for each finite element the stress increments stays in the same zone during the computation (Darve, Labanieh, Chambon 1976). But other cases are almost impossible to calculate with this law (Chambon, Renoud-Lias, 1979). A more reguliar interpolation is necessary. A new interpolation technique is used in this work it is detailed in Chambon, Renoud-Lias (1979). Let us recall its main features. The vectorial relationship $\{\delta\varepsilon\} = \mathcal{G}(\{\delta\sigma\})$ is searched for which vectorial function $\mathcal{G}$ is

401

homogeneous. The values of $\mathcal{G}$ are known at the extremity of the six increments drawn in Fig. 1. The interpolation is performed inside each zone as for the linear interpolation. The function $\mathcal{G}$ is wanted to be differenciable for every point except the origin (A). As the function $\mathcal{G}$ is homogeneous, it is completely determined in a zone by the knowledge of its values for the points of the equilateral triangle the apexes of which are the extremity of three our six particular increments. Then instead of performing interpolation in the three dimensional space it is possible to do it in the triangle which is obviously two-dimensioned. The technique used is the Razzaque's one (Razzaque (1973)) elaborated to generate conformal bending finite elements. We do not recall the assumptions needed by this technique. They can be found in Chambon, Renoud-Lias (1979). The result can be written

$$\{\delta\varepsilon\} = \left[M(\delta\sigma)\right]\{\delta\sigma\}$$

$\left[M(\delta\sigma)\right]$ is a 3 x 3 Matrix which depends on the direction of $\delta\sigma$. But because all incremental laws can be written in this manner we must point out that
 1) $\left[M(\delta\sigma)\right]$ varies continuously with the direction of $\delta\sigma$
 2) Moreover $[M]$ corresponds to the derivative of the incremental constitutive law along the direction associated with $\delta\sigma$
So we have performed a linearization of the constitutive law, but not only with an incremental formulation. $\{\delta\varepsilon\} = [M(\delta\sigma)]\{\delta\sigma\}$ is for our law a linearization for a given history and for the current direction of the stress increment. Finally it appears
 1) That this incremental constitutive law is completely non linear, which means that the superposition is not valid anymore (Gudehus, Goldscheider & Winter 1977) even in a restricted sense as it was in every zone for the law involving zones (linear interpolation).
 2) That this law can be linearized in the previous sense which is very important to integrate this law in a finite element method routine.
Since $[M(\delta\sigma)]$ represents the derivation along a direction the approximation $\{\delta\varepsilon\} = [M(\overline{\delta\sigma})]\{\delta\sigma\}$ induces the best accuracy if the direction associated with $\{\delta\sigma\}$ is in the neighbourhood of the one associated with $\{\overline{\delta\sigma}\}$. This property is very useful for finite-element method With an incremental constitutive law a computation are obviously a step by step one. Boundary conditions are devided into increments. This can simulate complex

loadings (construction, excavation, cyclic loadings...). When the history of the boundary conditions is linear, which means that all boundary conditions increment is identical or is positively proportional to the previous one, it can be expected that for almost all finite element stress increment directions vary continuously. Thus the previous stress increment direction is very close to the current one and can be a good approximation of it. This property used in the law to explicit $[M]$ it leads to throught the F.E.M. a linear problem very close to the real non linear one. Besides classical resolution methods of weak non-linearity (Zienkiewicz 1973) lead to the correct solution for such increments of boundary conditions. When it can be expected that stress increment direction undergoes discontinuity (for instance in cyclic problem when the sign of boundary conditions changes) an iterative technique can be used. It is described in Chambon, Renoud-Lias (1979). Thus we solve the problem of cyclic pressure-meter with our incremental non linear constitutive law and associated algorithm. This is shown in Chambon, Renoud-Lias (1978). Let us notice that the law used in this work differs from that which is presented in the previous paper. Assumptions made to describe generalized triaxial tests are rather different as it can be seen in 2.4.

2.4 Triaxial conventional tests

In order to know the response of the six stress increment described previously, the behaviour of the material for generalized triaxial tests has to be exhibited. This behaviour can be described by the three following families of functions :

$$\sigma_1 = f_{\sigma_2,\sigma_3}(\varepsilon_1) \qquad \varepsilon_2 = g_{\sigma_2,\sigma_3}(\varepsilon_1)$$
$$\varepsilon_3 = h_{\sigma_2,\sigma_3}(\varepsilon_1)$$

in which σ_2 and σ_3 are parameters remaining constant for a given test.
We gave (Darve et al. (1978)) the supplementary assumptions which permit to express the three families of functions f, g, h from the knowledge of the stress-strain curves : $\sigma_1 = f^*(\varepsilon_1,\sigma_3)$ and the void ratio variations : $e = g^*(\varepsilon_1,\sigma_3)$ for triaxial conventional tests in both compression case $\dot{\varepsilon}_1 > 0$ and the extension case $\dot{\varepsilon}_1 < 0$. Previously (Darve et al. (1978)) the analytical expressions of the stress-stain curves permitted to describe the stress maxima for the compression case and the stress minima for the extension case. The void

402

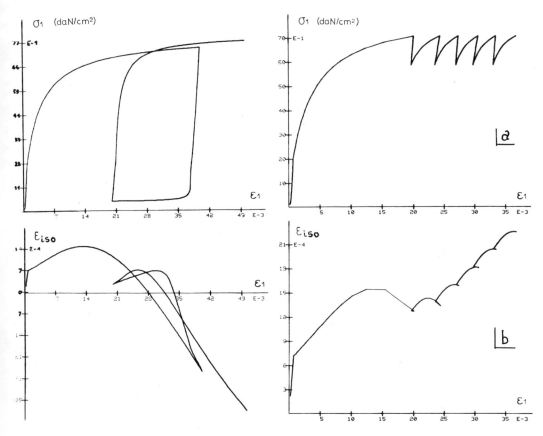

Figure 2. Theoretical conventional triaxial
loading for a compression-extension
cycle with the Hostun sand (e_i = 0.6 ;
$\sigma_2 = \sigma_3$ = 2 daN/cm^2).

Figure 3. Theoretical conventional triaxial
loading for small compression cycles
with the Hostun sand (e_i= 0.6 ;
$\sigma_2 = \sigma_3$ = 2 daN/cm^2)

ratio had an asymptotic value that has been
called : "critical void ratio". On the basis of recent studies which analyze the
sliding surfaces as a bifurcation problem
(Darve (1978) and Darve, Desrues, Jacquet
(1980)), it seems to us preferable to consider the behaviour after peak for initial
homogeneous samples of dense soils as a
consequence of heterogeneities which have
appeared at the failure (Vardoulakis et al.
(1978)).
New analytical formulations of f* and g*
are proposed here. In the case of compression triaxial tests the maximum of stress
is obtained in an asymptotic manner and
after a small preliminary contraction the
materials expands with a given dilatancy
angle. For extension triaxial tests the
minimum of stress is also reached asymptotically and the soil expands until a maximum value of the void ratio ; then it contracts asymptotically a little. In the case
of cyclic triaxial loadings, if an

extension follows a compression test, there
is a preliminary contraction before the
dilatancy phase (except for an extension,
which begins in the initial contraction
domain). But for compression loadings
following an extension, a contraction at
the load reversal always exists.
For the stress-strain curves of conventional triaxial tests, a hyperbolic formulation was chosen :

$$\sigma_1 - \sigma_i = \sigma_3 C_N \frac{\varepsilon_1 - \varepsilon_i}{A_s + \varepsilon_1 - \varepsilon_i}$$

where ε_i is the isotropic strain under
the confining pressure $\sigma_i = \sigma_3$ in the case of
a monotonic loading, and for cyclic loadings
σ_i and ε_i represent the values of the
axial stress σ_1 and the axial strain ε_1
at each load reversal.
C_N is determined by the knowledge of the

403

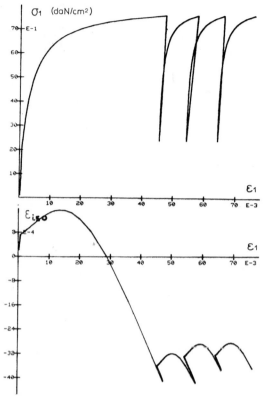

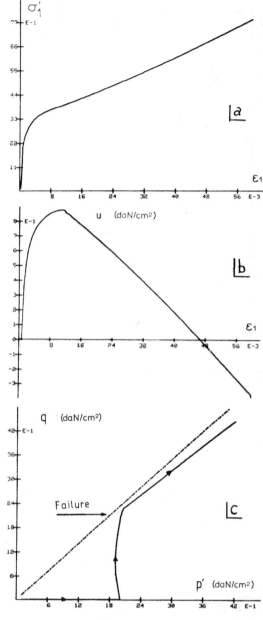

Figure 4. Theoretical conventional triaxial loading for large compression cycles with the Hostun sand (e_i = 0.6 ; $\sigma_2 = \sigma_3$ = 2 daN/cm^2).

friction angle :

$$tg\ \phi\ =\ \frac{A_4}{A_5 + \sigma_3} + A_6$$

The influence of the intermediate stress permits to calculate the friction angle in extension (Darve et al. (1978)).
A$_5$ depends on the slope at the origin, equal to :

$$U_i(\sigma_3) = K\sqrt{\sigma_3}$$

For cyclic loadings, K varies with ε_i in a continuous manner and with the type of load reversal (reloading or unloading) in a discret manner.
The void ratio variations for conventional triaxial tests are described by :

$$e = e_i - Y_M(1 - \exp(-A_y(\varepsilon_1 - \varepsilon_i))) + B_y(\varepsilon_1 - \varepsilon_i) - C_y(\varepsilon_1 - \varepsilon_i)^2 \exp(-D_y(\varepsilon_1 - \varepsilon_i))$$

Figure 5. Theoretical simulation of undrained conventional triaxial test for the Hostun sand (e_i = 0.6 ; $\sigma_2 = \sigma_3$ = 2 daN/cm^2)

404

The coefficient A_y is determined by the value of the initial Poisson's ratio, which is given by : $\nu_i^c = B_4 / (B_5 + \sigma_3)$ in the case of a monotonic compression and by :

$$\nu_i^e = 0.5 - B_3 \frac{0.5 - \nu_i^c}{B_3 + \sigma_3}$$

in the case of a monotonic extension. For the cycles the value of Poisson's ratio at the load reversal depends also on the last calculated slope of the $e - \varepsilon_1$ curve. C_y and D_y are functions of the values of the strain and the void ratio at his minimum (for compression tests) or at his maximum (for monotonic extension tests). B_y depends on an asymptotic Poisson's ratio :

$$\nu_M = 0.5 + \frac{B_1}{\sigma_m + 1} \qquad \text{for compression}$$

and $\nu_M = 0.5 + B_2 \sigma_m$ for monotonic extension where σ_m is the mean pressure for the plastic failure criterion.

The figures 2, 3 and 4 present the results obtained for three cases of cyclic loading. They illustrate some of the previous formulae in the case of the Hostun sand. ε_{iso} is the isotropic strain : $\varepsilon_{iso} = 1/3 (\varepsilon_1 + \varepsilon_2 + \varepsilon_3)$ The results of the virgin isotropic compression till $\sigma_2 = \sigma_3 = 2$ daN/cm^2 is plotted. The principal hysteretic phenomena seem to be described but many points are left to be discussed. Unfortunately, there presently is a lack of experimental results. However we think that the axial strain for each cycle is too large in figure 3(a) and the total contractancy too high in figure 3(b).

3 APPLICATIONS OF THE CONSTITUTIVE LAW

We have seen in section 2.4 some aspects of the determination of the Hostun sand constitutive parameters. For each current loading increment, from the knowledge of the load history and the current stress and strain states, we can express the associated functions f^* and g^* , then the functions f, g, h for the six directions in the $(\delta\sigma_1, \delta\sigma_2, \delta\sigma_3)$ space (figure 1). The six transformed vectors of these six unit directions in the $(\delta\varepsilon_1, \delta\varepsilon_2, \delta\varepsilon_3)$ space (figure 1) are therefore determined. Then the chosen directional interpolation permits to calculate the incremental response for a given incremental loading. The incremental constitutive law can then be integrated for a given loading history.

3.1 Simulations of undrained conventional triaxial tests

Figure 5 shows the results with the Hostun sand. In this paragraph 3, σ represents total

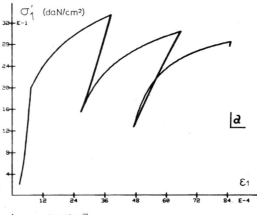

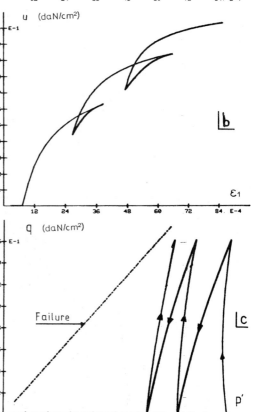

Figure 6. Theoretical simulation of cyclic undrained conventional triaxial test for a small value of the deviator q (Hostun sand ; $e_i = 0.6$; $\sigma_2 = \sigma_3 = 2$ daN/cm^2)

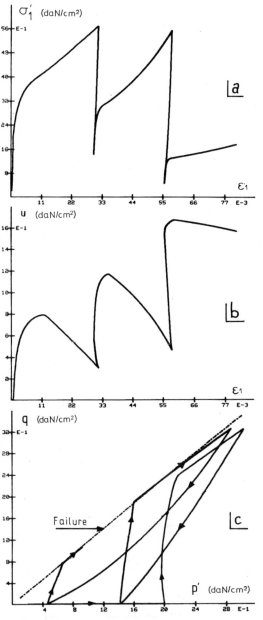

stresses and $\boldsymbol{\sigma'}$ effective stresses. u is the pore water pressure. Traces of the failure surfaces for compression and extension axi-symmetrical tests are plotted with dashes in the figure 5 (c) $\boldsymbol{p'}$ is the effective mean pressure :

$$p' = \frac{\sigma'_1 + 2\sigma'_2}{3}$$

and $\boldsymbol{q}$ the stress deviator :

$$q = \sqrt{\frac{2}{3}}\,(\sigma_1 - \sigma_2)$$

This dense sand presents firstly a contraction then it expands for monotonic conventional triaxial compression. The variations of the pore-water pressure for an undrained test (figure 5(b)) seem so to be correct. Figure 5(c) shows that this sand will obviously not liquefy.

The figure 6 presents undrained cycles for the Hostun sand. The maximum value of the stress deviator $\boldsymbol{q}$ is not too high : liquefaction seems to be difficult (figure 6(c)), but pore-water pressure increases more for the reloadings than it decreases for the unloadings (figure 6(b)).

The same theoretical simulations are plotted in figure 7 but the maximum of the stress deviator is more important : liquefaction will occur perhaps during the following cycle, as suggested by figure 7(c).

3.2 An example of liquefaction

We have also determined the constitutive parameters of the Monterey sand from the experimental results of Lade and Duncan (1973) and we present hereafter the theoretical results for cyclic undrained tests with the loose Monterey sand.

Liquefaction is theoretically obtained for a null effective stress state. Figure 8(a) and 8(d) show such a liquefaction. The pore-water pressure in figure 8(b) increases until the lateral effective stress is null that is until the pore-water pressure reaches a value equal to 1 daN/cm². The figure 9 presents a theoretical loading-unloading for this Monterey sand. It seems that liquefaction will occur at the end of the unloading in the extension domain. In conclusion, these results show that the volume variations for cyclic drained triaxial tests have a major importance in the appearance of the liquefaction.

Figure 7. Theoretical simulation of cyclic undrained conventional triaxial test for a big value of the deviator q (Hostun sand ; $e_i = 0.6$; $\sigma_2 = \sigma_3 = 2$ daN/cm²)

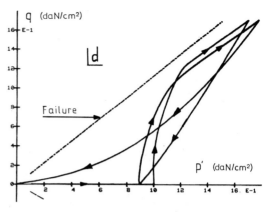

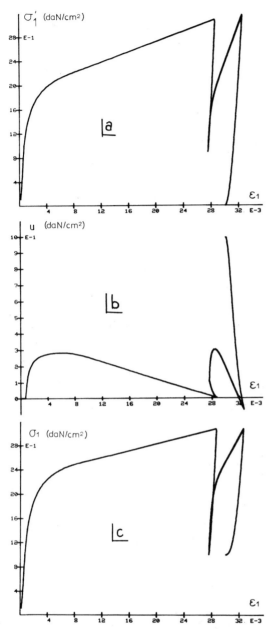

Figure 8. Theoretical cyclic liquefaction for undrained conventional triaxial test with the loose Monterey sand ($e_i = 0.783$; $\sigma_2 = \sigma_3 = 1$ daN/cm^2)

4 CONCLUSION

The great flexibility of the non linear interpolation has been illustrated once again with new formulations for simulation of generalized triaxial tests. Then introductions of new experimental results can be hoped as soon as they are formulated. This paper may be considered as reflecting the present state of a work in progress in Grenoble. Many laboratory tests must be performed and then their results have to be compared with theoretical computation. Perhaps good qualitative agreements between experiment and theory that we have noticed in this paper shall become good quantitative agreements. Even in this case it is presently impossible to solve a geotechnical problem by the F.E.M. and an incremental constitutive law with 1000 or 100 000 cycles. But an extrapolation of good results for the first cycles can be available. Others procedures can be used, for instance as in Boulon (1980).

5 REFERENCES

Bazant, Z.P. & Krizek , R.J. 1976, Endochronic constitutive law for liquefaction of sand, Jour. of the Eng. Mech. Div., 102 : 225-238.

Boulon, M. Chambon, R. & Darve, F. 1977, Loi rhéologique incrémentale pour les sols et applications par la méthode des éléments finis, Revue Française de Géotechnique, 2 : 5-22.

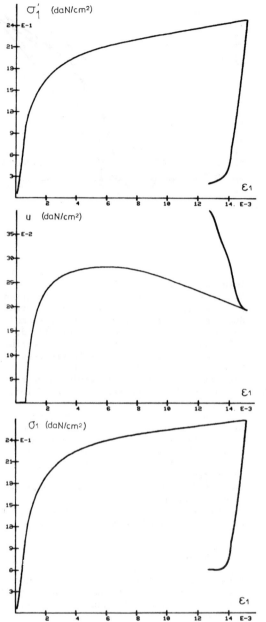

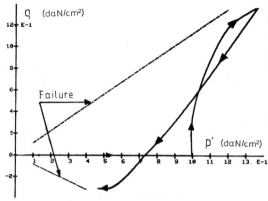

Figure 9. Theoretical simulation of a undrained compression-extension conventional triaxial test with the loose
  Monterey sand ($e_i = 0.783$;
  $\sigma_2 = \sigma_3 = 1$ daN/cm^2)

Boulon, M. & al 1980, Numerical model for
  foundations under cyclic loading ; on
  application to piles. International
  Symposium on soils under cyclic and
  transient loading , Swansea, Rotterdam
  A. A. Balkema.
Chambon, R. & Renoud-Lias, B. 1979, Incremental non-linear stress-strain relationship for soil and integration by
  F.E.M., in W. Wittke (ed.), 3rd Int. Conf.
  on Num. Meth. in Geomechanics, 1 : 405-
  413, Rotterdam, A. A. Balkema.
Darve, F. 1978, Une formulation incrémentale des lois rhéologiques Applications aux
  sols, Thèse d'Etat, Grenoble.
Darve, F., Labanieh, S. & Chambon, R. 1976,
  Incremental stress-strain relationship
  for cohesionless soil. In C.S. Desai
  (ed.), Numerical methods in geomechanics
  1 : 264-269 New York A.S.C.E.
Darve, F., Boulon, M & Chambon, R. 1978,
  Loi rhéologique incrémentale des sols,
  Journal de Mécanique, 17 n° 5 : 679-716.
Darve, F., Desrues, J. & Jacquet, M. 1980,
  Les surfaces de rupture en mécanique des
  sols en tant qu'instabilité de déforma-
  tion, Cahiers du Groupe Français de Rhéologie.
Darve, F., Flavigny, E. 1979, Simulation
  de la liquéfaction des sols avec une loi
  incrémentale, Proceedings of the symposium, Matériaux et structures sous chargement cyclique : 101-105 Paris, Ecole
  Nationale des Ponts et Chaussées.
Gudehus, G. 1979, A comparison of some
  constitutive laws for soils under radial-
  ly symmetric loading and unloading,

3rd Int. Conf. on Num. Methods in Geomechanics, 4, Rotterdam A. A. Balkema.

Gudehus, G., Goldscheider, M. & Winter, H.
1977, Mechanical properties of sand and
numerical integration methods : some
sources of errors and bounds of accuracy.
In G. Gudehus (ed.), Finite elements in
Geomechanics 121-151 J. Wiley & sons.

Gudehus, G. and Kolymbas, D. 1979, A constitutive law of the rate-type for soils,
in W. Wittke (ed.), 3rd Int. Conf. on
Num. Meth. in Geomechanics, 1, Rotterdam
A. A. Balkema.

Lade, P.V. 1976, Stress-path dependent
behaviour of cohesionless soil, Jour. of
the Geo. Eng. Div., 1, 51-68.

Lade, P., Duncan, J. M. 1973, Cubical triaxial tests on cohesionless soil, Journal
of the soil foundation division 92.

Maier, G. and Hueckel, T. 1977, Non-associated and coupled flow rules of elastoplasticity for geotechnical media, spec.
session, Int. Cong. of Soil Mechanics and
Eng. Foundations, Tokyo.

Mroz, Z., Norris, V.A., Zienkiewicz, O.C.
1978, An anisotropic hardening model for
soils and its application to cyclic loading, Int. Jour. for Num. and Anal. Methods in Geomechanics, 2 : 203-221.

Prevost, J.H. 1977, Mathematical modelling
of monotonic and cyclic undrained clay
behaviour, Int. Jour. for Num. and Anal.
Meth. in Geomechanics, 1 : 195-216.

Nelson, I. 1978, Constitutive models for
use in numerical computations. In G. Gudehus (ed.), Plastic and long-term
effects in soils 2 : 45-101, Rotterdam,
A. A. Balkema.

Razzaque, A. 1973, Program for triangular
bending elements with derivative smoo-
thing, International Journal for numerical methods in engineering 6 : 333-343.

Vardoulakis, I., Goldscheider, M., Gudehus,
G. 1978, Formation of shear hands in sand
bodies as a bifurcation problem, Int. J.
Num. Anal. Meth. Geom., 2 : 99-128, J.
Wiley & sons.

Zienkiewicz, O.C. 1973, La méthode des
éléments finis, Paris, Edisciences.

Some aspects of constitutive equations for sand

F. MOLENKAMP
Delft Soil Mechanics Laboratory, Delft, Netherlands

1. INTRODUCTION

With computers becoming both bigger and less expensive the possibilities of numerical predictions increase. In soil mechanics non-linear problems with alternating loading as coastal structures, earthquake engineering and pile-driving have become solvable for a limited number of cycles. After having overcome the numerical problems the validity of the results will remain dependent on the quality of the constitutive equations. From a practical point of view the constitutive model need just have sufficient parameters to be able to simulate the main features of the actual behaviour. With this the actual behaviour can be approximated by adapting the magnitudes of the parameters in some characteristic regions using stress path tests. The total number of adaptions and tests needed will be reduced by using a constitutive model of increased quality. Several models are being used (eg. Prevost 1978, Sandler, Baron 1979); the others are being developed further (eg. Ghaboussi, Momen 1979). To get more insight in the range of validity of the models and so in possible improvements some theoretical aspects of the constitutive equations of sand were considered.

First the type of parameters occurring in the macroscopic properties of a uniform packing of frictional rigid discs were determined. Obviously these parameters do not resemble those of sand exactly because the effects of irregularity of the packing are not included. For this special case it is found that the angle of non-coaxiality between the principal directions of stress and strain rate can be varied independently. For a random packing of discs it was already shown (Drescher, De Josselin de Jong 1972) that this angle could vary between $\pm \frac{1}{2} \phi$, ϕ being the friction angle, in the limit state.

It is also found that a yield surface exists having kinematic properties; during irreversible deformation due to rotation of the principal stresses and/or due to a change of the shear stress level the nonplastic area is shifted in the state space. Because in the case of the uniform packing the angle of non-coaxiality is an independent variable the rates of principal stresses and strains cannot be related directly. In such a case dilatancy ratios could be used.

The influence of the load history is completely described by the instantaneous geometry. The change of the load-history parameter can also be expressed in macroscopic quantities; it can increase and decrease.

Also the limitations of the elasto-plastic model are considered. In this model the total strains are separated into reversible and irreversible parts. For this separation to be possible the existence of an elastic region must be assumed. This assumption cannot be verified by experiment not even by including the thermal effects. To satisfy the entropy production law the plastic work has to be positive if it is assumed that the internal energy e and the entropy s are only dependent on the elastic strains and the temperature. In such case the plastic potential has to enclose the origin. Whether for sand the aforesaid assumptions are justified is still disputable.

Finally it is shown for coaxiality to exist that it is sufficient to describe the plastic potential in principal stress space.

2. UNIFORM STACK OF RIGID DISCS WITH FRICTIONAL INTERACTIONS

2.1. Geometry and interactions

The geometry of a uniform stack of discs is described by the radius r, the smallest angle $\psi (> 0)$ between the x-axis and the nearest centre line of the joining discs and the angle α between both centre lines (see fig. 1).

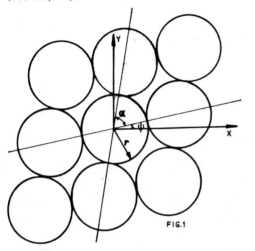

FIG.1

The interactions between the discs are the forces T_ξ and T_γ acting at the contact points on the centre line with angles ψ,

$$\sigma_{ij} = \frac{-1}{2r \sin \alpha} \begin{vmatrix} T_\xi \cos \psi \cos(\psi + \xi) + T_\gamma \cos(\psi + \alpha) \cos(\psi + \alpha + \gamma) \\ T_\xi \cos \psi \sin(\psi + \xi) + T_\gamma \cos(\psi + \alpha) \sin(\psi + \alpha + \gamma) \\ T_\xi \sin \psi \cos(\psi + \xi) + T_\gamma \sin(\psi + \alpha) \cos(\psi + \alpha + \gamma) \\ T_\xi \sin \psi \sin(\psi + \xi) + T_\gamma \sin(\psi + \alpha) \sin(\psi + \alpha + \gamma) \end{vmatrix} \quad (3)$$

and $\psi + \alpha$ respectively. Forces T_ξ and T_γ intersect at angles ξ respectively γ with the centre lines (see fig. 2). Due to the frictional character of the interactions the following holds:

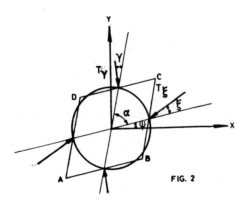

FIG. 2

$$|\xi| \leqslant \phi, |\gamma| \leqslant \phi \quad (1)$$

where ϕ is the maximum friction-angle.

2.2. Stress tensor

From the interactions T_ξ and T_γ and the geometry the stress tensor σ_{ij} can be calculated. For a homogeneous state of stress it holds (Drescher, De Josselin de Jong 1972):

$$\sigma_{ij} = \frac{1}{V} \int_V \sigma_{ij} \, dV = \frac{1}{V} \int_V \sigma_{kj} x_{i,k} \, dV =$$

$$\frac{1}{V} \int_S \sigma_{kj} x_i n_k ds = \frac{1}{V} \int_S t_j x_i ds$$

where the divergence theorem and the equilibrium condition are used. Here: V = volume, S = boundary of V, n_k = external normal on S, t_j = traction on S. Replacing the traction by discrete external forces with $T_j = t_j dS$ gives:

$$\sigma_{ij} = \frac{1}{V} \sum_m x_i^m T_j^m \quad (2)$$

In this case for volume V the diamond ABCD (fig. 2) can be chosen.

The stress tensor becomes, using eq. 2:

From the balance of the moment of forces on a disc it follows:

$$T_\xi \sin \xi + T_\gamma \sin \gamma = 0 \quad (4)$$

This causes symmetry of the stress tensor so no couple stress occurs. Rotation of the axes of reference over an angle β transforms the stress matrix into:

$$\begin{vmatrix} \sigma_{11}^* & \sigma_{12}^* \\ \sigma_{21}^* & \sigma_{22}^* \end{vmatrix} = \begin{vmatrix} \cos\beta & \sin\beta \\ -\sin\beta & \cos\beta \end{vmatrix} \begin{vmatrix} \sigma_{11} & \sigma_{12} \\ \sigma_{21} & \sigma_{22} \end{vmatrix} \begin{vmatrix} \cos\beta & -\sin\beta \\ \sin\beta & \cos\beta \end{vmatrix}$$

$$(5)$$

The principal directions are obtained if:

$$\sigma_{12}^* = \sigma_{21}^* = 0$$

Then $\tan 2\beta = \dfrac{2\sigma_{12}}{\sigma_{11} - \sigma_{22}} \quad (6)$

412

and the principal stresses become:

$$\sigma_1 = \sigma_{11} \cos^2\beta + \sigma_{22} \sin^2\beta + 2\sigma_{12} \sin\beta \cos\beta$$

$$\sigma_2 = \sigma_{11} \sin^2\beta + \sigma_{22} \cos^2\beta - 2\sigma_{12} \sin\beta \cos\beta$$

2.3 Limit equilibrium

The stress states at limit equilibrium enclose the region of admissible stress states. They are calculated as a function of α and β with $\frac{\pi}{3} < \alpha < \frac{2\pi}{3}$ and $0 < \beta < \frac{\pi}{2}$.
Substituting eq. 3 in eq. 6 gives a relation between T_ξ/T_γ, ξ and γ for fixed values of ψ, α and β. The following relation between ξ and γ is obtained after T_ξ/T_γ is eliminated using eq. 4:

$$\{-\sin 2\beta \cos(2\psi+\xi) + 2 \cos 2\beta \cos\psi \sin(\psi+\xi)\}$$
$$\sin \gamma + \{(\sin 2\beta \cos(2\psi+2\alpha+\gamma) +$$
$$-2 \cos 2\beta \cos(\psi+\alpha)\sin(\psi+\alpha+\gamma)\} \sin \xi = 0$$
(8)

From this equation for any ψ, α and β the limit states can be calculated by assigning the limit values ϕ and $-\phi$ to either ξ or γ and then calculating γ or ξ. From the four limit states obtained in this way only 2 are left after applying eq. 1. For these cases T_ξ/T_γ is calculated with eq. 4. Substitution of T_ξ/T_γ, ξ and γ in eq. 3 gives the relevant stress tensor. Substituting this stress tensor in eq. 7 gives

the principal stresses. They turn out to be proportional to T_γ. This makes it possible to define the shear stress level S1 as a relevant quantity:

$$S1 = \frac{\dfrac{\sigma_1}{\sigma_2} - 1}{\dfrac{\sigma_1}{\sigma_2} + 1}$$
(9)

Thus the state of limit equilibrium can be described with the shear stress level S1 as a function of α and $\beta - \psi$. An example of the results for $\phi = 30^0$ and $\alpha = 70^0$ is shown in fig. 3. For ψ is taken: $\psi = -\frac{1}{2}\alpha$; then the x-axis is placed along one of the diagonal directions of the stack. In fig. 3 the value of the angle ξ or γ associated with limit equilibrium are given. Besides the states of neutral stress ($\xi = \gamma = 0$) are shown. They can be calculated in a similar way as limit equilibrium.

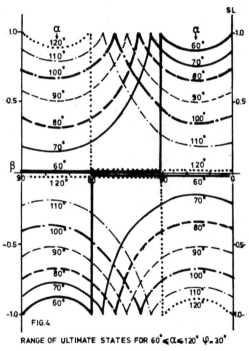

RANGE OF ULTIMATE STATES FOR $60^\circ \leqslant \alpha \leqslant 120^\circ$ $\psi = 30^\circ$

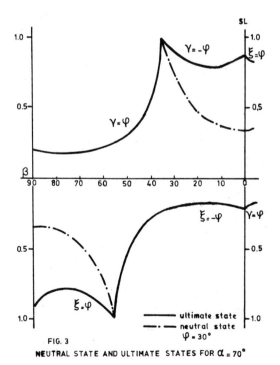

FIG. 3
NEUTRAL STATE AND ULTIMATE STATES FOR $\alpha = 70^\circ$

In fig. 4 the limit states are shown for $\phi = 30^0$ and the whole possible range of α. For the cases of $\alpha = 60^0$ and $\alpha = 120^0$ no contact is assumed between the nearby diagonally facing discs. Apparently the area of admissible stress states shifts in state space (β, S1) when α is changed.

413

2.4 Strain rate tensor

To be able to relate any deformation to changes of stress the macroscopic strain rate tensor is derived starting from the relative displacements of the discs. First the velocity gradient is calculated with:

$$\dot{u}_{i,j} = \frac{1}{V} \int_V \dot{u}_{i,j} \, dv = \frac{1}{V} \int_V \dot{u}_i \, n_j \, ds \quad (10)$$

This tensor can be decomposed in the symmetric strain rate tensor $\dot{\varepsilon}_{ij}$ and the anti-symmetric rotation rate tensor $\dot{\omega}_{ij}$. Due to the uniformity of the stack deformation occurs only because the discs translate (no rotation) relative to each other at the points of contact E or F (fig. 5) at the centre lines under angle ψ or $\psi + \alpha$ respectively. In fig. 5 the relative velocities alongside the boundary S of the diamond ABCD are shown for the case of slip at points of contact E. The magnitude of the velocity is $-2r\dot{\alpha}$, the direction is parallel to the tangent of the discs at points E, thus:

$$\dot{u}_1 = 2r\dot{\alpha} \sin \psi, \quad \dot{u}_2 = -2r\dot{\alpha} \cos \psi, \quad V = 4r^2 \sin \alpha \quad (11)$$

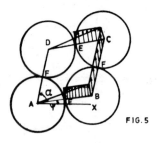

FIG.5

Elaboration of eq. 10 gives:

$$\dot{u}_{i,j} = \frac{\dot{\alpha}}{\sin \alpha} \begin{vmatrix} \sin\psi \sin(\alpha+\psi) & -\sin\psi \cos(\alpha+\psi) \\ -\cos\psi \sin(\alpha+\psi) & \cos\psi \cos(\alpha+\psi) \end{vmatrix} \quad (12)$$

For the case of slip at points E the symmetric strain rate tensor

$$\dot{\varepsilon}_{ij} = (\dot{u}_{i,j} + \dot{u}_{j,i})/2 \text{ becomes:}$$

$$\dot{\varepsilon}_{ij} = \frac{\dot{\alpha}}{\sin \alpha} \begin{vmatrix} \sin\psi \sin(\alpha+\psi) & -\tfrac{1}{2} \sin(\alpha+2\psi) \\ -\tfrac{1}{2} \sin(\alpha+2\psi) & \cos\psi \cos(\alpha+\psi) \end{vmatrix} \quad (13)$$

and the anti-symmetric rotation rate tensor $\dot{\omega}_{ij} = (\dot{u}_{i,j} - \dot{u}_{j,i})/2$:

$$\dot{\omega}_{ij} = \frac{\dot{\alpha}}{2} \begin{vmatrix} 0 & 1 \\ -1 & 0 \end{vmatrix} \quad (14)$$

The corresponding rotation rate vector has a clockwise rotation for positive $\dot{\alpha}$. For the case of slip at points F the same strain rate tensor is obtained but the sign of the rotation rate is opposite corresponding to counter clockwise rotation.

Rotation of the axes of reference over an angle ζ puts the strainrate tensor at its principal axes if (see eqs. 5 and 6).

$$\tan 2\zeta = \frac{2 \dot{\varepsilon}_{12}}{\dot{\varepsilon}_{11} - \dot{\varepsilon}_{22}} = \tan(\alpha + 2\psi)$$

$$\text{thus: } \zeta = \frac{\alpha}{2} + \psi. \quad (15)$$

The principal directions of the strainrate coincide with the diagonal directions of the stack. For the principal strainrates it is found (see eq. 7):

$$\dot{\varepsilon}_1 = \dot{\varepsilon}_{11} \cos^2\zeta + \dot{\varepsilon}_{22} \sin^2\zeta + 2\dot{\varepsilon}_{12} \sin\zeta \cos\zeta = -\frac{\dot{\alpha}}{2} \tan \frac{\alpha}{2} .$$

$$\dot{\varepsilon}_2 = \dot{\varepsilon}_{11} \sin^2\zeta + \dot{\varepsilon}_{22} \cos^2\zeta - 2\dot{\varepsilon}_{12} \sin\zeta \cos\zeta = \frac{\dot{\alpha}}{2} \frac{1}{\tan \frac{\alpha}{2}} \quad (16)$$

2.5 Equivalent continuum model

A continuum model is supposed to relate the rates of the strains to the rates of stresses.

The model satisfies the condition of objectivity of the observer if both eqs. (16) and (17) are satisfied. Equation (17) relates the general strain rate tensor $\dot{\varepsilon}_{ij}$ to the principal strain rate tensor $\dot{\varepsilon}_{kl}^*$:

$$\dot{\varepsilon}_{ij} = R_{ki} \, \dot{\varepsilon}_{kl}^* \, R_{lj} \quad (17)$$

$$\text{in which: } R_{ij} = \begin{vmatrix} \cos \zeta & \sin \zeta \\ -\sin \zeta & \cos \zeta \end{vmatrix}$$

rotation tensor (see figure 6)
ζ: angle between x axis and one of the directions of principal strain rates. These directions of principal strain rates are properties of the stack, actually they are the diagonal directions of the stack. They could also be called principal directions of anisotropy. Equation (16) relates the principal strain rates to α and $\dot{\alpha}$. How-

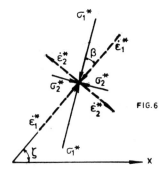

FIG.6

ever, in the continuum model they have to be related to the stress rates. To this end the concept of the yield surface as used in plasticity theory can be applied.

The curves of limit equilibrium in fig. 4 can be interpreted as yield surfaces.

$$F(\alpha, \beta, S1) = F'(\beta, S1) - \alpha = 0 \quad (18)$$

in which: β = angle between principal directions of stresses and strain rates. For the uniform stack β is an independent variable. However, it should be noted that β not only changes because the stress tensor is rotated but also because the principal directions of anisotropy rotate (see eq. 14).
Then:

$$\dot{\alpha} = \frac{\partial F}{\partial S1} \dot{S}1 + \frac{\partial F}{\partial \beta} \dot{\beta} \quad (19)$$

The partial derivatives can be calculated as a function of stress; $\dot{S}1$ and $\dot{\beta}$ depend on the rates of stress changes.
Trying to continue along the lines of the plastic model the terms:

$\tan \frac{\alpha}{2}$ and $1/\tan \frac{\alpha}{2}$ in eq. 16 can be tried

to be expressed as partial derivatives of a plastic potential $G(\beta, S1)$ with respect to the principal stresses. Using this concept the ratio of the rates of the principal strains become:

$$\frac{\dot{\varepsilon}_1^*}{\dot{\varepsilon}_2^*} = \frac{\dfrac{\partial G}{\partial \sigma 1} \dfrac{\partial G}{\partial S1} \dfrac{\partial S1}{\partial \sigma 1}}{\dfrac{\partial G}{\partial \sigma_2} \dfrac{\partial G}{\partial S1} \dfrac{\partial S1}{\partial \sigma_2}} = f(S1) \quad (20)$$

being a function of S1 only. According to eq. 16 the ratio of principal strains should be a function of both S1 and β. This means that the concept of the plastic potential cannot be used to relate rates of principal strains to rates of principal stresses only.
Instead for each $\tan \frac{\alpha}{2}$ and $1/\tan \frac{\alpha}{2}$ a separate function of S1 and β could be used,

for instance a dilatancy ratio D and a hardening H. The dilatancy ratio can be defined by:

$$D = \frac{\dot{V}}{\dot{\gamma}} = \frac{\dot{\varepsilon}_1 + \dot{\varepsilon}_2}{\dot{\varepsilon}_1 - \dot{\varepsilon}_2} \quad (21)$$

Then the principal strain rates can be expressed by:

$$\dot{\varepsilon}_1 = \frac{D+1}{H} (\frac{\partial F}{\partial S1} \dot{S} + \frac{\partial F}{\partial \beta} \dot{\beta})$$
$$\dot{\varepsilon}_2 = \frac{D-1}{H} (\frac{\partial F}{\partial S1} \dot{S}1 + \frac{\partial F}{\partial \beta} \dot{\beta}) \quad (22)$$

In this case of the uniform stack the effect of the load history is expressed by the angle α. The rate of change of this load history parameter can also be expressed in macroscopic quantities. From eq. 16 it follows:

$$\dot{\alpha} = \text{sign} (\dot{\varepsilon}_1) \cdot 2 \sqrt{|\dot{\varepsilon}_1 \dot{\varepsilon}_2|} \quad (23)$$

This load history parameter can both increase and decrease.

3. ELASTO PLASTICITY AND THERMODYNAMICS

In a constitutive model of the elasto-plastic type reversible and irreversible deformations are being distinguished. For metals this is realistic because unloading-reloading stress-strain paths coincide and so the behaviour along these paths is reversible.
However, in soils even at low amplitudes of cyclic stresses some irreversible deformation occurs (fig. 7).

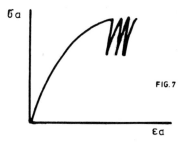

FIG.7

When assuming intuitively that the steepest parts of the stress-strain paths represent elastic behaviour it is found that the plastic work W in (eg. isotropic) unloading can be negative (see fig. 8), which seems to contradict the second law of thermodynamics (eg. Prager 1949).
In the following it is checked whether possibly a more fundamental basis for the separation of total strain could be obtained

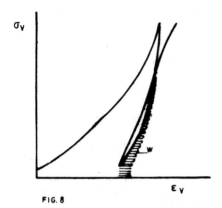

FIG. 8

using thermodynamics. In a continuum model including thermal effects three independent quantities occur namely internal energy e, entropy s and entropy production τ, all per unit of mass. These quantities have to satisfy the laws of thermodynamics. From the conservation of energy, the balance of momentum and the moment of momentum it follows that:

$$\rho \dot{e} = \sigma_{ij} \, \dot{\varepsilon}_{ij} - \frac{\partial q_i}{\partial x_i} \qquad (24)$$

in which $q_i = T\tilde{n}_i$ heat flux
T = temperature
$\tilde{n}_i$ = entropy flux
ρ = specific density

The entropy production law is:

$$\rho \tau = \rho \dot{s} + \frac{\partial h_i}{\partial x_i} \geqslant 0 \qquad (25)$$

For an elastoplastic material the internal energy e and the entropy s can be assumed to be functions of the elastic strains ε_{ij}^e and the temperature T only, just as for a purely elastic material. The effect of plastic deformation is accounted for by introducing the plastic work as an internal source of heat. Using classical thermodynamics (eg. Ziegler 1977) the temperature T and the stress tensor σ_{ij} can be calculated from the internal energy $e(\varepsilon_{ij}^e, s)$:

$$T(\varepsilon_{ij}^e, s) = \frac{\partial e(\varepsilon_{ij}^e, s)}{\partial s} \qquad (26)$$

$$\sigma_{ij}(\varepsilon_{ij}^e, s) = \rho \frac{\partial e(\varepsilon_{ij}^e, s)}{\partial \varepsilon_{ij}^e} \qquad (27)$$

So it is easiest to express the internal energy as a function of the elastic strains ε_{ij}^e and the entropy s. For instance for an isotropic linear elastic material the internal energy per unit of volume is des-

cribed by: (eg. Besseling 1967)

$$\rho e(\varepsilon_{ij}^e, s) = \mu \, \varepsilon_{ij}^e \, \varepsilon_{ij}^e + \{\frac{\lambda}{2} + \frac{(3\lambda+2\mu)^2 \alpha^2 T_o}{2\rho \, C_v}\} \varepsilon_{kk}^{e^2} +$$

$$- \frac{(3\lambda+2\mu)\alpha \, T_o \cdot s \cdot \varepsilon_{kk}^e}{C_v} + \rho \, T_o \, s +$$

$$+ \frac{\rho \, T_o \, s^2}{2 C_v} + \rho \, C_v \, T_o \qquad (28)$$

in which:
λ, μ = Lame's constants
α = coefficient of cubic thermal expansion
C_v = specific heat per unit mass at constant volume
T_o = reference temperature

Using eqs. (26) and (27) the temperature and the stress in an isotropic linear elastic material become:

$$T = \frac{-(3\lambda+2\mu)\alpha \, T_o \, \varepsilon_{kk}^e}{\rho \, C_v} + \frac{T_o \, s}{C_v} + T_o \qquad (29)$$

$$\sigma_{ij} = 2\mu \, \varepsilon_{ij}^e + \{\lambda + \frac{(3\lambda+2\mu)^2 \, \alpha^2 \, T_o}{\rho \, C_v}\} \, \varepsilon_{kk}^e \, \delta_{ij} +$$

$$- \frac{(3\lambda+2\mu)\alpha \, T_o}{C_v} \, s \, \delta_{ij} \qquad (30)$$

In an elastoplastic material the external work can be separated into elastic work $\sigma_{ij} \, \dot{\varepsilon}_{ij}^e$ and plastic work $\sigma_{ij} \, \dot{\varepsilon}_{ij}^p$. Substitution in eq. (24) gives for the rate of change of the internal energy per unit of volume:

$$\rho \dot{e} = \sigma_{ij} \, \dot{\varepsilon}_{ij}^e + \sigma_{ij} \, \dot{\varepsilon}_{ij}^p - \frac{\partial q_i}{\partial x_i} \qquad (31)$$

This rate of change of the internal energy equals:

$$\rho \dot{e} = \rho \frac{\partial e(\varepsilon_{ij}^e, s)}{\partial \varepsilon_{ij}^e} \, \dot{\varepsilon}_{ij}^e +$$

$$+ \rho \frac{\partial e(\varepsilon_{ij}^e, s)}{\partial s} \{\frac{\partial s(\varepsilon_{ij}^e, T)}{\partial \varepsilon_{ij}^e} \, \dot{\varepsilon}_{ij}^e + \frac{\partial s(\varepsilon_{ij}^e, T)}{\partial T} \, \dot{T}\} \qquad (32)$$

Substitution of eqs. (26) and (27) gives:

$$\rho \dot{e} = \sigma_{ij} \, \dot{\varepsilon}_{ij}^e + \rho \, T\dot{s} \qquad (33)$$

From eqs. (31) and (33) it follows:

$$\rho T\dot{s} = \sigma_{ij} \, \dot{\varepsilon}_{ij}^p - \frac{\partial q_i}{\partial x_i} \qquad (34)$$

By substituting this result in the entropy production law eq. (25) the following expression for the entropy production is obtained:

416

$$\rho T \tau = \sigma_{ij} \dot{\varepsilon}^P_{ij} - \frac{q_i}{T} \frac{\partial T}{\partial x_i} \geqslant 0 \qquad (35)$$

As $q_i = -k \frac{\partial T}{\partial x_i}$ (Fourier's law of heat condition) the entropy production law is satisfied if the plastic work is dissipative thus:

$$\sigma_{ij} \dot{\varepsilon}^P_{ij} \geqslant 0 \qquad (36)$$

Here it should be noted that this result is a consequence of the assumptions that both the internal energy e and the entropy s depend on the elastic strains ε^e_{ij} and the temperature T only. Recently also other assumptions have been considered (eg. Zuev 1978).
To satisfy eq. (36) the angle α between the vectors of stress and strain rate should not be larger than 90^0. Besides the plastic potential has to enclose the origin (Prager 1949) (see fig. 9).
When considering consecutive incremental loading and unloading at constant temperature the stress σ_{ij}, the stress rate $\dot{\sigma}_{ij}$, the strain rate $\dot{\varepsilon}_{ij}$ and the flux of thermal energy per unit of volume $\frac{\partial q_i}{\partial x_i} = \int_{\partial \beta} q_i n_i dA$ can be measured.

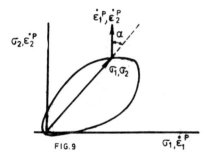

FIG.9

Applying eq. (34) for loading:

$$\frac{\partial q_i}{\partial x_i} = \sigma_{ij} \dot{\varepsilon}^P_{ij} - \rho T \frac{\partial s(\varepsilon^e_{ij}.T)}{\partial \varepsilon^e_{ij}} \dot{\varepsilon}^e_{ij} \qquad (37)$$

and also assuming plastic deformation during unloading eq. 34 gives:

$$\frac{\partial q_i^*}{\partial x_i} = \sigma_{ij} \dot{\varepsilon}^P_{ij}{}^* + \rho T \frac{\partial s(\varepsilon^e_{ij}.T)}{\partial \varepsilon^e_{ij}} \dot{\varepsilon}^e_{ij} \qquad (38)$$

After both incremental loading and unloading the total irreversible rate of strain $\dot{\varepsilon}^P_{ij} + \dot{\varepsilon}^P_{ij}{}^*$ can be determined. From eqs. (37) and (38) follows the condition:

$$\frac{\partial q_i}{\partial x_i} + \frac{\partial q_i^*}{\partial x_i} = \sigma_{ij}(\dot{\varepsilon}^P_{ij} + \dot{\varepsilon}^P_{ij}{}^*) \qquad (39)$$

By subtracting eqs. (37) and (38) is obtained:

$$\rho T \frac{\partial s}{\partial \varepsilon^e_{ij}} \dot{\varepsilon}^e_{ij} = \frac{1}{2}(\frac{\partial q_i^*}{\partial x_i} - \frac{\partial q_i}{\partial x_i}) + \frac{1}{2} \sigma_{ij}(\dot{\varepsilon}^P_{ij} - \dot{\varepsilon}^P_{ij}{}^*) \qquad (40)$$

From this equation $\dot{\varepsilon}^P_{ij} - \dot{\varepsilon}^P_{ij}{}^*$ and $\frac{\partial s}{\partial \varepsilon^e_{ij}}$ cannot be solved and thus $\dot{\varepsilon}^P_{ij}$ and $\dot{\varepsilon}^P_{ij}{}^*$ cannot be separated. So even by using classical thermodynamics it is not possible to separate reversible and irreversible strains other than by assuming the existence of an elastic region. Thus the apparent contradiction of local negative dissipative work remains. This contradiction is also found from the one-dimensional representation of an elastoplastic model (Iwan 1967) (see fig. 10).
The non-homogeneous model consists of two or more parallel sets each consisting of a linear spring and a frictional damper, all with different magnitude. A possible force displacement relation is shown in fig. 11. The released energy along the path DE is partly caused by the slip of damper P1. So also this model does not satisfy eq. (36). Apparently this condition is not necessarily true neither are the assumptions that the internal energy e and the entropy s only depend on the elastic strains ε^e_{ij} and the temperature T.

4. COAXIALITY AND PLASTICITY

4.1. Published experimental results and mathematical modeling

From experiments with rotating principal stresses it is known that even at low shear stress levels the principal directions of stress and strain rate do not necessarily coincide. At reloading after rotation Arthur, Chua, Dunstan (1977) found an average deviation (angle of non-coaxiality) of $7,5^0$ at low shear levels.
In experiments on a random packing of discs with frictional interaction Drescher, De Josselin de Jong (1972) found for the limit state a variable angle of non-coaxiality between $\frac{1}{2}\phi$ and $-\frac{1}{2}\phi$ in which ϕ is friction angle.
Having observed such material behaviour the application of coaxiality of principal directions of stress and plastic strain rates in an elastoplastic model is quite approximate. However, such an assumption simplifies the description of the plastic potential considerably. Besides at this stage other assumptions are hardly possible. If for instance it would be assumed that the angle of non-coaxiality could have any value between certain extremes by what criterion would the actual value be determined?

417

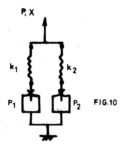

FIG.10

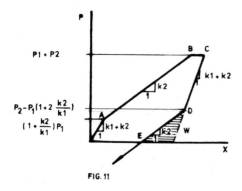

FIG. 11

Should the total external work be taken as a maximum? It is felt that before a better assumption can be made more research in the phenomenon of non-coaxiality is needed. Flügge (1958) has shown that for an isotropic elastoplastic material coaxiality always occurs. In the next section 4.2 this result is extended somewhat by showing that for coaxiality to occur the plastic potential has to be described in the principal stress space only.

4.2. Coaxiality and the plastic potential

In an elastoplastic model the usual relation between the rates of stress $\dot{\sigma}_{ij}$ and plastic strains $\dot{\varepsilon}^p_{ij}$ is given by (eg. Hill 1950):

$$\dot{\varepsilon}^p_{ij} = \frac{\dfrac{\partial G}{\partial \sigma_{ij}} \dfrac{\partial F}{\partial \sigma_{kl}} \dot{\sigma}_{kl}}{H} \qquad (41)$$

in which: F = 0: yield surface
G = 0: plastic potential
H   : hardening function

If the plastic potential is described in the principal stress space then it can be expressed as a function of the stress invariant $G(I_1, I_2, I_3)$ with

$$I_1 = \sigma_{ii}, \quad I_2 = \tfrac{1}{2} s_{ij} s_{ij} \quad \text{and} \quad I_3 = \tfrac{1}{3} s_{ij} s_{jk} s_{ki}$$

$$\text{while } s_{ij} = \sigma_{ij} - \tfrac{1}{3} I_1 \delta_{ij} \qquad (42)$$

From eq. (42) it follows that:

$$\frac{\partial I_1}{\partial \sigma_{ij}} = \delta_{ij}, \quad \frac{\partial I_2}{\partial \sigma_{ij}} = \frac{s_{ij}}{2I_2}$$

$$\text{and } \frac{\partial I_3}{\partial \sigma_{ij}} = s_{il} s_{lj} - 2/3 \, I_2^2 \delta_{ij} \qquad (43)$$

Using eqs. 42 the partial derivative of the plastic potential can be expressed as:

$$\frac{\partial G}{\partial \sigma_{ij}} = \frac{\partial G}{\partial I_1} \frac{\partial I_1}{\partial \sigma_{ij}} + \frac{G}{\partial I_2} \frac{\partial I_2}{\partial \sigma_{ij}} + \frac{\partial G}{\partial I_3} \frac{\partial I_3}{\partial \sigma_{ij}} =$$

$$= \left(\frac{\partial G}{\partial I_1} - \frac{\partial G}{\partial I_3} \frac{2I_2^2}{3} \right) \delta_{ij} - \frac{\partial G}{\partial I_2} \cdot \frac{1}{2I_2} s_{ij} +$$

$$\frac{\partial G}{\partial I_3} s_{il} s_{lj} \qquad (44)$$

In the following it is shown that in case of coaxiality the partial derivative can be expressed in the form of eq. 44, thus as:

$$\frac{\partial G}{\partial \sigma_{ij}} = a_1 \delta_{ij} + a_2 s_{ij} + a_3 \, s_{ik} s_{kj} \qquad (45)$$

The stress tensor σ_{ij} is transformed to a diagonal form σ^*_{ij} by the transformation:

$$\sigma^*_{ij} = Q_{ik} \sigma_{kl} Q_{jl} = \sigma^*_i \delta_{ij} \qquad (46)$$

in which Q_{ik} is the rotation tensor. The tensor of the plastic strain rate $\dot{\varepsilon}^p_{ij}$ is coaxial to the stress tensor σ_{ij} if the transformed tensor of plastic strain rate is diagonal too thus:

$$\dot{\varepsilon}^{p*}_{ij} = Q_{ik} \dot{\varepsilon}^p_{kl} Q_{jl} = \dot{\Sigma}^{p*}_i \delta_{ij} \qquad (47)$$

From eq. 41 the partial derivative of the plastic potential can be expressed as:

$$\frac{\partial G}{\partial \sigma_{ij}} = \frac{H}{\dfrac{\partial F}{\partial \sigma_{kl}} \dot{\sigma}_{kl}} \dot{\varepsilon}^p_{ij}$$

Substitution of $\dot{\varepsilon}^p_{kl}$ from eq. 47 gives:

$$\frac{\partial G}{\partial \sigma_{ij}} = \frac{H}{\dfrac{\partial F}{\partial \sigma_{kl}} \dot{\sigma}_{kl}} Q_{mi} \dot{\varepsilon}^{p*}_{mn} Q_{nj} = \frac{H}{\dfrac{\partial F}{\partial \sigma_{kl}} \dot{\sigma}_{kl}} \dot{\Sigma}^{p*}_m \cdot$$

$$Q_{mi} \delta_{mn} Q_{ni}$$

$$\frac{\partial G}{\partial \sigma_{ij}} = \frac{H}{\dfrac{\partial F}{\partial \sigma_{kl}} \dot{\sigma}_{kl}} \dot{\Sigma}^{p*}_m \cdot Q_{mi} Q_{mj} \qquad (49)$$

Now the right-hand side of eq. 47 can be put in the form of eq. 45 because:

418

$$\delta_{ij} = \alpha_m \, \varrho_{mi} \, \varrho_{mj} \tag{50}$$

$$s_{ij} = \beta_m \, \varrho_{mi} \, \varrho_{mj} \tag{51}$$

$$s_{ik} \, s_{kj} = \gamma_m \, \varrho_{mi} \, \varrho_{mj} \tag{52}$$

in which α_m, β_m, γ_m are functions of stress and $m = 1, 2, 3$. Namely multiplying eqs. (50), (51) and (52) respectively by ϱ_{ni} on the left-hand side and by ϱ_{pj} on the right-hand side the functions α, β and γ can be calculated.

$$\varrho_{ni} \, \delta_{ij} \, \varrho_{pj} = \delta_{np} = \alpha_n \, \delta_{np} \tag{53}$$

thus: $\alpha_n = 1$

and :

$$\varrho_{ni} \, s_{ij} \, \varrho_{pj} = \varrho_{ni} (\sigma_{ij} - 1/3 \, I_1 \delta_{ij}) \varrho_{pj} =$$

$$\sigma_n^* \, \delta_{np} - 1/3 \, I_1 \, \delta_{np} = \beta_n \, \delta_{np} \tag{54}$$

thus: $\beta_n = \sigma_n^* - 1/3 \, I_1$

and :

$$\varrho_{ni} \, s_{ik} \, s_{kj} \, \varrho_{pj} = \varrho_{ni} \, s_{ik} \, \delta_{ql} \, s_{lj} \, \varrho_{pj} =$$

$$= \varrho_{ni} \, s_{ik} \, \varrho_{qk} \, \varrho_{ql} \, s_{lj} \, \varrho_{pj} =$$

$$= (\sigma_n^* - \tfrac{1}{3} I_1) \delta_{nq} (\sigma_q^* - \tfrac{1}{3} I_1) \delta_{qp} =$$

$$= (\sigma_n^* - \tfrac{1}{3} I_1)^2 \delta_{np} = \gamma_n \, \delta_{np}$$

thus: $\gamma_n = (\sigma_n^* - \tfrac{1}{3} I_1)^2 \tag{55}$

Eq. 49 can be put in the form of eq. 45 if the constant a_1, a_2 and a_3 can be calculated. Substituting eqs. (50) until and included (55) in eq. 45 and putting it equal to eq. 49 the following equations are obtained:

$$\begin{vmatrix} (1) & (\sigma_1 - 1/3 I_1) & (\sigma_1 - 1/3 I_1)^2 \\ (1) & (\sigma_2 - 1/3 I_1) & (\sigma_2 - 1/3 I_1)^2 \\ (1) & (\sigma_3 - 1/3 I_1) & (\sigma_3 - 1/3 I_1)^2 \end{vmatrix} \begin{vmatrix} a_1 \\ a_2 \\ a_3 \end{vmatrix} =$$

$$= \frac{H}{\dfrac{\partial F}{\partial \sigma_{kl}} \dot{\sigma}_{kl}} \begin{vmatrix} \dot{\varepsilon}_1^{p*} \\ \dot{\varepsilon}_2^{p*} \\ \dot{\varepsilon}_3^{p*} \end{vmatrix} \tag{56}$$

From these a_1, a_2 and a_3 can be calculated if the equations are linearly independent and thus if the determinant of the matrix is not zero. The determinant equals:

$$(\sigma_2^* - \sigma_1^*)(\sigma_3^* - \sigma_1^*)(\sigma_3^* - \sigma_2^*)$$

In the general case of different principal stresses this determinant is not zero. Then eq. 48 can be expressed in the form of eq. 45 and so in case of coaxiality the plastic potential is described in principal stress space.

5. CONCLUSIONS

Several aspects of constitutive relations of frictional materials have been considered. For the special case of the uniform stack of frictional rigid discs it was shown that the angle of non-coaxiality between the principal directions of stress and strain-rate is an independent variable. In experiments on a random stack of discs (Drescher, De Josselin de Jong 1972) a variation between $\pm \tfrac{1}{2} \phi$ was observed. For the uniform stack of discs a yield surface with kinematic properties was found. The rates of principal stresses and strains could not be related directly because the angle of non-coaxiality was found to be an independent variable. Possibly a plastic potential could be defined in total stress space. Instead also a dilatancy ratio could be used.
The influence of the load history is completely described by the instantaneous geometry. For the case of the uniform stack it can be expressed in a macroscopic load history parameter which can increase and decrease. When describing, measured behaviour of sand under varying load with an elastoplastic model the total strains have to be separated into reversible and irreversible parts. To make this possible in state space an elastic region must be assumed. This assumption can not be verified by experiment, not even by including thermal effects. If it is assumed that the internal energy and the entropy are only dependent on the elastic strains and the temperature then the plastic work has to be positive in order to satisfy the entropy production law.
Whether for soil such assumptions are justified is still disputable. At present the phenomenon of non-coaxiality is still a matter of research. The assumption of coaxiality is known not to be justified by experiment but at present other assumptions can hardly be justified better. It has been shown that in case of coaxiality the plastic potential can always be described in principal stress space.

419

ACKNOWLEDGEMENTS

The author wishes to express his gratitude
to the Committee for Applied Scientific
Research, "Rijkswaterstaat", Department of
the Ministry of Transport and Public Works,
for its financial support.

6 REFERENCES

Arthur, J.R.F., K.S. Chua and T. Dunstan
1977, Induced anisotropy in a sand, Géotechnique 27, No. 1: 13-30.
Besseling, J.F. 1966, A thermodynamic approach to rheology, Prc. IUTAM-Symposium
on irreversible aspects of continuum mechanics, Vienna: 16-53.
Drescher, A. and G. de Josselin de Jong,
1972, Photo-elastic verification of a
mechanical model for the flow of a granular material. Journal Mech.Phys.Solids,
Vol. 20: 337-351.
Flügge, S., 1958, Handbuch der Physik.
Elastizität und Plastizität, Springer
Verlag: 256-259.
Ghaboussi, J. and H. Momen 1979, Plasticity
model for cyclic behaviour of sand, Third
Int.Conf. on Numerical Methods in Geomechanics, Aacken: 423-434.
Hill, R. 1950, The mathematical theory of
plasticity, Oxford University Press.
Iwan, W.D. 1967, On a class of models for
the yielding behaviour of continuous and
composite systems, Transactions of the
ASME, September: 612-617.
Prager, W. 1949, Recent developments in
the mathematical theory of plasticity,
Journal of Applied Physics, Vol. 20:
235-241.
Prevost, J.H. 1978, Plasticity theory for
soil stress-strain behaviour, Journal of
the Engineering Mechanics Division, ASCE,
Vol. 104, No. EM5, Oct.: 1177-1194.
Sandler, I.S. and M.L. Baron 1979, Recent
developments in the constitutive modeling of geological materials, Third Int.
Conf. on Numerical Methods in Geomechan-
ics, Aachen: 363-376.
Ziegler, H. 1977, An introduction to thermodynamics, North-Holland Publishing Cy.
Zuev, V.V. 1978, Governing relationships
of plasticity theory in strain and stress
spaces, Dokl. Akad. Nauk SSSR, Vol. 242,
No. 4-6: 792-795, Russian Original;
English translation: Sov.Phys.Dokl.,
Vol. 23, No. 10, October: 774-775.

Peak strength of clay soils
after a repeated loading history

DWIGHT A.SANGREY
Cornell University, Ithaca, USA

JOHN W.FRANCE
Geotechnical Engineers, Inc., Winchester, USA

1 INTRODUCTION

An important group of geotechnical engineering problems requires knowledge of the
peak strength of soil after a limited number of repeated loading cycles. These
problems include: loading of structural
foundations after earthquakes, stability
of slopes and foundations experiencing environmental loading by wind and waves,
loading of soils by traffic and changes in
the ultimate capacity of driven piles.
Since the time interval between a repeated
loading history and a subsequent loading
may vary, both the strength immediately
after repeated loading and the strength
after a period of drainage are of
interest.

The concern with liquefaction of sands
under earthquake loading has produced many
studies on the response of this material.
The large literature on the subject is
very comprehensive (Seed, 1978). There
also have been studies of the behavior of
silt and clay soils under repeated loading
(Sangrey et al, 1978) and a brief review
of the general conclusions of this work
will provide a basis for the arguments
presented herein.

In this paper, a qualitative model of
the expected undrained strength behavior
after repeated loading both with and without drainage will be developed from the
results of previous studies. A theoretical effective stress model will be proposed to predict this behavior and test
results from several clay soils will be
used to illustrate the applicability of
the theory. Finally, some examples of
field applications will be discussed.

2 PREVIOUS STUDIES

The most dramatic effect of repeated loading on most saturated soils is a loss of
strength or failure after some number of
loading cycles. The potential for strength
loss and failure increases as the level of
cycled stress increases and lower levels
of undrained repeated loading do not produce failure even under a large number of
stress cycles. This behavior has been
particularly noted for clay soils and the
term 'critical level of repeated loading'
has been applied to the stress level separating failure and nonfailure behavior
(Sangrey et al, 1978).

It is generally accepted that the failure of soils under repeated loading is a
consequence of accumulating excess pore
pressure during cyclic loading. These
pressures are often measured in sands but
few studies of clay soils have included
accurate measurement of pore pressure because of the low permeability of clays and
the corresponding long response time for
pore pressure measurement. In one early
effective stress study of clays (Sangrey
et al, 1969) it was shown that the critical level of repeated loading separates
the higher cycled stress levels, which develop sufficient excess pore pressure to
reach the effective stress failure conditions, from those cycled stress levels
which do not. Furthermore the behavior,
and the amount of excess pore pressure
accumulated during cyclic loading at lower
stress levels, is predictable.

When cycled at stress levels below the
critical level of repeated loading soil
samples reach a nonfailure equilibrium
condition (Fig. 1). At this state there
is no further accumulation of either strain
or excess pore pressure with additional
stress cycles. Sangrey et al (1969)

proposed that the excess pore pressures accumulated to equilibrium were related to the cycled stress level and that when presented in an effective stress space these equilibrium effective stress conditions defined a locus called the equilibrium line (Fig. 1c). The intersection of the equilibrium line and the failure envelope defined the critical level of repeated loading.

In the same study it was also reported that repeated loading of very dilative (heavily overconsolidated) clay specimens resulted in small accumulations of excess negative pore pressures. Consequently, the critical level of repeated loading was not very different from the conventional undrained shearing resistance. Similar behavior for dilative clay specimens was reported by Brown et al (1975).

Effective stress studies or repeated loading which included drainage intervals have also been reported (France and Sangrey, 1977). The results of these studies indicate that when normally consolidated and lightly overconoslidated clays are loaded in this way they do not fail under the repeated loading but, instead, the water content decreases and the critical level of repeated loading increases. For heavily overconsolidated clays, however, the repeated loading with drainage results in significant strength decrease.

Several previous studies have been concerned with the strength and shearing resistance of clay soils after a period of repeated loading. Most of these were concerned with the peak undrained shearing resistance after undrained repeated loading although Eide (1974) introduced a drainage interval into the loading sequence. Thiers and Seed (1968) presented a comprehensive review of strength after undrained repeated loading. They concluded that the peak strength after cyclic loading decreased as the level of cyclic loading increased. They also proposed that a useful design approach could be based on a predictable relationship between a nondimensionalized strength after repeated loading and strain ratio. Lee and Focht (1976) discussed this concept and considered some additional data. In a later part of this paper the proposal of Thiers and Seed will be critically reviewed with consideration of new experimental data.

Based on previous studies a qualitative model of the expected peak strength behavior after repeated loading can be proposed (Fig. 2). For the completely undrained case it has been shown that the peak shearing resistance decreases as the level of cycled stress increases. The form of this relationship is not clear and it has not

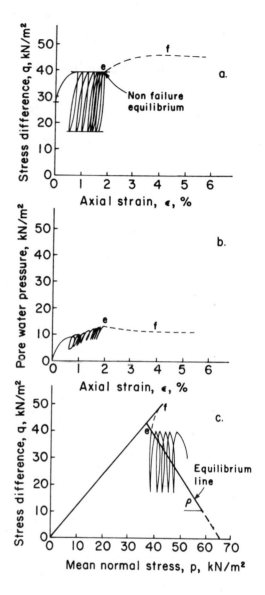

Fig. 1 Undrained cyclic loading at low stress levels produces non failure equilibrium (a) with an associated limit to pore pressure increase (b) defining an equilibrium line (c).

been shown how this peak shearing resistance relates to the accumulation of excess pore pressure and the effective stress for the soil. When drainage is introduced into the repeated loading sequence, water content decreases for normally consolidated and lightly overconsolidated soils. It is reasonable to expect that this should lead to an increase in shearing resistance as shown in Fig. 2. The form and the limits of this strength increase have not been described.

An analogous model could be developed for dilative clay with the major exception that the strength after repeated loading with drainage would decrease with increasing cycled stress. However, for long term problems in dilative clay, both before and after cyclic loading, it is the drained strength which is critical.

Similarly, although the writers believe the model presented in Fig. 2 to be valid for sands, it is somewhat academic since the high permeability of sands leads to drained behavior which governs almost immediately after the cessation of cyclic loading. The drained strength is a relatively simple effective stress problem requiring knowledge of the effective stress state and the effective stress strength parameters.

3   THEORETICAL MODEL OF BEHAVIOR

The qualitative behavior illustrated in Figure 2 can be described by extending presently used effective stress theories. Among the several similar methods used to describe effective stress behavior of soils, the concepts and terminology evolving from the research group at Cambridge University have been used in a wide range of applications (Schofield and Wroth, 1968), including repeated loading behavior of soils [Sangrey et al (1978)]. This approach will be the reference framework for developing a theoretical model of soil strength after repeated loading.

The theoretical model will be developed first for completely undrained behavior, undrained repeated loading followed by undrained loading to failure. That basic model can then be expanded to account for drainage intervals within the sequence such as total or partial reconsolidation. In the development of these arguments it will also be assumed that the soils are saturated and most illustrations will be appropriate for normally consolidated and lightly overconsolidated soils. The theory itself is not limited and could be applied to partially saturated soils and heavily

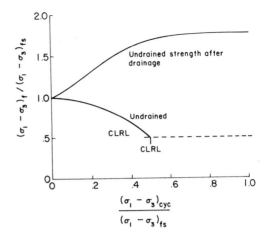

Fig. 2   Expected peak strength after cyclic loading with and without drainage for contractive soils.

overconsolidated clays as well; however, other, more difficult, problems may be involved with repeated loading of these soils (France and Sangrey, 1977).

3.1   Undrained Problem

The initial state of stress for an element of soil is illustrated by Point 0 in Fig. 3. This is a completely general state of stress which might represent a soil in a natural deposit or in the laboratory. The effective stress failure envelope illustrated is assumed to be known and a unique property of the soil.

The undrained response of this soil element when subjected to repeated loading or other types of loading such as undrained creep (Hyde and Brown, 1976) can be extremely complicated. Undrained repeated loading itself presents several alternatives as many studies have shown. In general for contractive soils, with increasing numbers of cycles and/or time of loading there will be an increase in both excess pore pressures and strain, Fig. 1. A complete description of these changes will be a function of the number of stress cycles, the time history of the loading, the form of the stress pulse and other variables which have been extensively studied for both sands and clays. A major factor in the behavior, however, is the level of cycled stress.

The point 1 in Fig. 3 represents the state of effective stress after some undrained repeated loading history. The change of stress to lower values of mean normal stress reflects the increase of pore water pressures. Point 1 could represent either an equilibrium position after a large number of cycles at stress levels below the critical level of repeated loading or it could represent the state of stress after a few cycles of stress at any level, even a stress level which would eventually cause failure under repeated loading. In the second case the state of effective stress, point 1, would change if there were additional cycles of loading. The change of stress between points 0 and 1 is entirely general and includes a change in deviator stress as illustrated in Fig. 3.

To complete the undrained problem, the stress path from point 1 to point r for monotonic, undrained loading to failure is illustrated in Fig. 3.

The stress difference at failure (point r) may be expressed in a general equation. Failure will occur at a particular ratio of effective stresses.

$$q/p = \frac{6 \sin\phi'}{3 \sin\phi'} = M \qquad (1)$$

where ϕ' is the conventional effective stress friction angle. Hence,

$$q_r = Mp_r \qquad (2)$$

where p_r can be expressed as

$$p_r = p_0 + \Delta p_{o-1} + \Delta p_{1-r} \qquad (3)$$

where Δp_{o-1} and Δp_{1-r} equal the changes in mean normal stress during undrained repeated loading and during undrained loading to failure, respectively.

From the definition of effective stress

$$\Delta p = \Delta p^{(T)} - \Delta u \qquad (4)$$

where $\Delta p^{(T)}$ is the change in total mean normal stress and Δu is the corresponding change in pore pressure. Combining equations 2 and 3 yields

$$q_r = M(p_0 + \Delta p_{o-1}^{(T)} - \Delta u_{o-1} + \Delta p_{1-r}^{(T)} - \Delta u_{1-r}) \qquad (5)$$

From a practical standpoint the parameter of most importance is how much additional deviator stress, Δq_r, can be added to the deviator stress at point 1 before failure occurs. This can be expressed as

$$\Delta q_r = q_r - q_1 = q_r - q_0 - \Delta q_{o-1} \qquad (6)$$

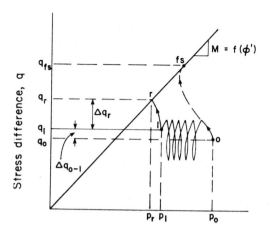

Mean normal stress, p

Fig. 3 Idealized model of undrained failure after undrained cyclic loading.

Combining equations 5 and 6 yields

$$\Delta q_r = M(p_0 + \Delta p_{o-1}^{(T)} - \Delta u_{o-1} + \Delta p_{1-r}^{(T)} - \Delta u_{1-r}) - q_0 - \Delta q_{o-1} \qquad (7)$$

which is an entirely general equation for the undrained shearing resistance after undrained repeated loading. Use of this equation requires knowledge or prediction of (1) the initial effective stress state, p_0 and q_0, (2) total stress changes during repeated loading, $\Delta p_{o-1}^{(T)}$ and $\Delta q_{o-1}^{(T)}$, and during loading to failure, $\Delta p_{1-r}^{(T)}$, and (3) pore pressure changes during repeated loading, Δu_{o-1}, and during loading to failure, Δu_{1-r}.

In current geotechnical engineering practice the initial effective stresses and the subsequent total stress changes can be predicted to varying degrees of accuracy depending on the soil deposit in question and on the specific design loadings.

Within our present state of knowledge, the pore pressure changes during undrained repeated loading may be predicted with greater or lesser degrees of confidence depending on the particular loading sequence. For soils achieving equilibrium at stress levels below the critical level of repeated loading, methods have been proposed and used to predict the equilibrium excess pore pressures (Sangrey, 1972; Sangrey, et al, 1969; Seed et al 1976). The excess pore pressures resulting from undrained creep (Hyde and Brown, 1976) have also been predicted and could be considered as well as those due to repeated loading.

For the transient condition of a limited number of stress cycles not yet reaching either equilibrium or failure, our state of knowledge is less precise. The general behavior has been shown in numerous examples, pore pressures accumulate progressively as the number of repeated loading cycles increases. Some studies have concluded that the pore pressure increase is related most easily to accumulated strain. In other studies the relationship between pore pressure increase and the number of loading cycles has been treated empirically [Seed et al, 1976]. In field applications the option to measure pore pressures during or after the repeated loading sequence may also be a practical alternative.

Pore pressure changes during loading to failure will be discussed in a later section of this paper.

The following example of how equation (7) can be simplified with a more precise knowledge of stress changes will be informative. First it will be assumed that the deviator stresses before and after cyclic loading are the same ($\Delta q_{o-1} = 0$) and triaxial compression conditions ($\sigma_2 = \sigma_3 =$ const.) apply. This later assumption implies that

$$\Delta p^{(T)} = \frac{\Delta q}{3} \tag{8}$$

and
$$\Delta u_{i-j} = A_{i-j} \Delta q_{i-j} \tag{9}$$

where A is Skempton's pore pressure coefficient.

Equation 7 then simplifies to

$$\Delta q_r = \frac{M(p_o - \Delta u_{o-1}) - q_o}{M(A_{1-r} - 1/3) + 1} \tag{10}$$

Further assuming the repeated loading achieves a condition of nonfailure equilibrium, the conclusions of Sangrey et al [1969] can be employed and the change in pore pressure during repeated loading can be expressed as a function of the repeated stress level:

$$\Delta u_{o-1} = \frac{1}{\rho} \Delta q_c \tag{11}$$

where: $\rho = -\dfrac{\Delta q}{\Delta p}$ = negative slope of the equilibrium line
Δq_c = the peak cycled deviator stress-initial deviator stress. Equation (10) then transforms to

$$\Delta q_r = \frac{M(p_o - \frac{1}{\rho} \Delta q_c) - q_o}{M[A_{1-r} - 1/3] + 1} \tag{12}$$

Prediction of undrained strength based on equation 12 requires knowledge of three empirical parameters:

M - which requires determination of the traditional effective stress friction angle

ρ - which requires a limited number of repeated loading test as described by Sangrey et al (1969) and others

and A_{1-r} - which will be discussed in a later section

3.2 Partly Drained Problem

The undrained peak shearing resistance following a sequence of undrained repeated loading is an important practical problem but more frequently actual problems include some amount of drainage in the total loading history. Dissipation of the excess pore pressure accumulated during undrained loading is essentially a reconsolidation. Following earthquakes or repeated environmental loading there will often be sufficient time for pore pressure dissipation before the application of loads which will require the peak soil strength. The very important problem of pile 'set up', the increase in capacity of driven piles with time, is a case where drainage follows repeated loading. A theoretical model can be developed for this problem using the same basic approach as that used for the totally undrained case.

The sequence of effective stress changes for the repeated loading problem with drainage is illustrated in Fig. 4. From some initial state of stress, 0, the undrained repeated loading experience results in a state of stress at point 1 which reflects the accumulation of excess pore pressure. The undrained repeated loading could have resulted in stresses beyond the strength of the soil, and failure, as in the case of driving a pile or the stresses could be within the soil strength limits. Reconsolidation and drainage of the excess pore pressures will result in a change of effective stress to point 2. Stress states along this reconsolidation path may be of interest but for most problems the limit state of full reconsolidation is most important. When subjected to monotonic undrained loading to failure the soil then follows a stress path to point d.

Emplying the same arguments used in developing equations 5 and 7 yields

425

$$q_d = M[p_o + \Delta p_{o-1}^{(T)} + \Delta p_{1-2}^{(T)} + \Delta p_{2-d}^{(T)} - \Delta u_{2-d}] \tag{13}$$

and

$$\Delta q_d = M[p_o + \Delta p_{o-2}^{(T)} + \Delta p_{2-d}^{(T)} - \Delta u_{2-d}] - q_o$$
$$- \Delta q_{o-2} \tag{14}$$

It should be noted the only pore pressure change in equations 13 and 14 is that during the loading to failure since full reconsolidation at point 2 is assumed.

Determining the state of stress at point 2 after reconsolidation is a key step in predicting the subsequent peak strength. In many cases, such as in earthquake loading of flat-lying deposits of soft soil, the state of stress after reconsolidation may be nearly the same as the initial state of stress, point 0. In other cases, such as driving piles through heavily overconsolidated clays, the state of stress after reconsolidation will be very different from the initial one.

Again as an example, equation (14) can be greatly simplified by assuming triaxial compression conditions. The result is

$$\Delta q_d = \frac{Mp_2 - q_2}{M(A_{2-d} - 1/3) + 1} \tag{15}$$

Prediction of undrained shearing resistance from equation 15 requires knowledge of the following:

M - which requires determination of ϕ'

$P_2 = p_o + \Delta p_{o-1}^{(T)} + \Delta p_{1-2}^{(T)}$ and $q_2 = q_o + \Delta q_{o-2}$ the reconsolidation stresses which were discussed above

and A_{2-d} which will be discussed later.

4 EXPERIMENTAL STUDIES OF SHEARING RESISTANCE AFTER REPEATED LOADING

The key to predicting the undrained shearing resistance of soil after a repeated loading stress history is the magnitude of excess pore pressure produced during subsequent loading. Experimental results from laboratory cyclic triaxial tests on a variety of fine grained soils have shown that these excess pore pressures and the consequent undrained behavior are very similar in form. As a result, general patterns of expected behavior can be

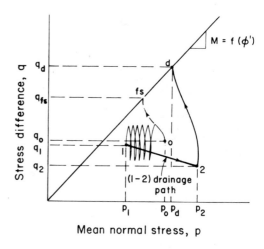

Fig. 4 Idealized model of undrained failure after cyclic loading and subsequent drainage.

defined along with a methodology for determining the empirical pore pressure parameters.

4.1 Undrained Behavior

Some tests in which the undrained shearing resistance was determined after a sequence of undrained repeated loadings can be considered first. A series of tests was done on samples of undisturbed Newfield clay (Sangrey et al, 1969) reconsolidated to an anisotropic state of stress approximately equal to the insitu state of stress for the normally consolidated deposit. Then under undrained conditions the specimens were subjected to slow cyclic loading at various stress levels. In some specimens the level of cyclic stress was sufficiently high to cause failure under the repeated loading but at lower levels of cycled stress nonfailure equilibrium was achieved. The complete test series defined both a critical level of repeated loading and the slope of the equilibrium line $-\rho$ (Fig. 1c). The critical level of repeated loading defined on the basis of stress increment, Δq, was slightly less than half of the peak shearing resistance when loaded to failure without any repeated loading.

The samples at equilibrium were then loaded to failure maintaining the initial undrained conditions (Fig. 1). A summary of the resulting peak shearing resistance

426

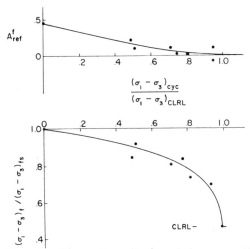

Fig. 5   Illustration of the pore pressure
response and strength decrease associated with undrained loading
after undrained cyclic loading.

after repeated loading is presented in
Fig. 5 in a non-dimensional plot of shear-
ing resistance, $(\sigma_1 - \sigma_3)_f$, divided by the
shearing resistance without repeated load-
ing, $(\sigma_1 - \sigma_3)_{fs}$, versus the ratio of
cycled stress to the critical level of repeated loading. Other reference stresses
for the nondimensional plot could be used
without changing the overall conclusions.

The limits of these data are a test to
failure without any repeated loading
$\dfrac{\Delta q_{cyc}}{\Delta q_{CLRL}} = 0$, and a specimen cycled at the
critical level of repeated loading for
which there would be no additional strength
beyond the stress being cycled. These
data confirm the anticipated pattern of the
decrease in peak strength as the level of
cycled stress increases.

In Fig. 5 the pore pressure coefficient,
A^f_{ref} measured in these tests is shown.
The general form of this relationship is
a decrease in A^f_{ref}, as the level of the
cycled stress ratio increases. For a
cycled stress ratio of 0 the value of A^f_{ref}
is the value for the soil loaded to failure without any repeated stress history.
The other limit is approximately $A^f_{ref} = 0$.
It should be noted that the data presented
in Fig. 5 come from a test series in which
some samples were subjected to one-way
stress cycling and others to two-way stress
cycling, the rate of loading was also variable. None of these variables appear to

have been significant in determining the
final behavior.

The pore pressure data (Fig. 5) define
an empirical relationship. A practical
question is whether this is a typical form
of this empirical relationship for soils
and, if so, whether it is possible to simplify the relationship. Tests on other
soils do confirm that this general form is
typical for normally consolidated and
lightly overconsolidated soils at least.
For heavily overconsolidated soils there are
other more significant problems resulting
from repeated loading (France and Sangrey,
1977). All of these data indicate that the
pore pressure coefficient, A^f_{ref} varies
from the value for the soil loaded continuously to failure without any repeated
loading, to a value of 0 at stress cycling
equal to the critical level of repeated
loading. For the data presented here it
might also be reasonable to describe the
variation as linear with cycled stress
ratio if the soils were loaded to failure
after achieving nonfailure equilibrium.

Many other loading histories could have
resulted in the reference stress state at
point 1 in Fig. 3. As described in the
early sections of this paper, the theory
and equation 7 are completely general. It
is the authors opinion that while the
limits of pore pressure response shown in
Fig. 5 is generally valid, the apparent
linearity of this relationship is not.

4.2   Strength After Repeated Loading with
       Drainage

The second limit to behavior for the shearing resistance of soils after repeated
loading is to allow complete dissipation
of excess pore pressures accumulated during
the repeated loading. The results from a
series of tests on a resedimented illite
clay (France and Sangrey, 1977) can be used
to illustrate this behavior (Fig. 6). In
these tests the soil specimens were subjected to a variety of repeated loading
sequences with complete drainage of excess
pore pressures permitted periodically.
The final state of stress prior to loading
to failure was a stable reconsolidation
equivalent to point 2 in Fig. 4. From this
point the samples were loaded to failure
under undrained conditions. The resulting
shearing resistance and pore pressure coefficients (Fig. 6) are presented.

As anticipated, drainage after repeated
loading leads to an increase in shearing
resistance. The amount of strength increase, expressed as a ratio, is related
to the level of repeated loading and

427

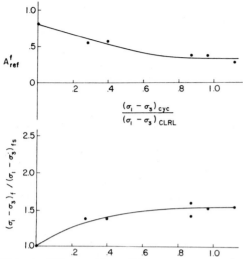

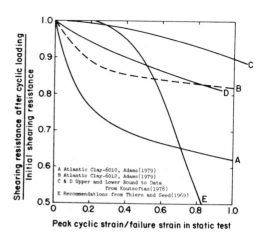

Fig. 6 Illustration of the pore pressure response and strength increase associated with undrained loading after cyclic loading including drainage.

Fig. 7 The relationship of strain ratio to stress ratio for soils loaded to failure after cyclic loading is similar but not unique.

increases from a value of 1 to a maximum after which there is little strength increase even though the level of repeated loading is higher. The corresponding pore pressure coefficients are consistent with these results. A_{ref}^f is bounded by the value for the undrained compression to failure without repeated loading and a value of 1/3.

Tests on other natural soils indicate the same general form. The specific values for each soil are different and there is no evidence that variation in either strength ratio or pore pressure coefficient is a simple function such as linear with cycled stress level. Consequently, the prediction of undrained strength after repeated loading using equations 13, 14 or 15 will require an empirical parameter measured for each soil. The limiting value of $A_{ref}^f = 1/3$ may be general, however, and this would have some practical applications for problems involving high levels of repeated loading. Sangrey (1977) has applied this approach successfully to predict the capacity of driven piles in clay.

5 THE SIGNIFICANCE OF STRAIN RELATIONSHIPS

Thiers and Seed (1968) in their comprehensive study of the response of clays to seismic loading proposed that the relationship of strain ratios to strength

ratios was useful in predicting the peak strength after cyclic loading under completely undrained conditions.

The data from the completely undrained tests reported in this paper plus other available data can also be presented in the form suggested by Thiers and Seed (Fig. 7). In general, the conclusion drawn by Thiers and Seed that strain ratio and strength ratio are related is confirmed by these data. At lower levels of strain ratio there is a relatively high level of strength ratio with a rather abrupt decrease in strength as the strain ratio is increased.

Thiers and Seed suggested that their relationship might be typical for many soils, however, the results presented in Fig. 7 show that this is clearly not the case. If the relationship of strain ratio to strength ratio is to be used in practice it is apparently an empirical property to be determined for each soil.

The strength ratio for soils having drainage as part of their repeated loading stress history clearly do not fit the pattern proposed by Thiers and Seed.

6 SUMMARY AND CONCLUSIONS

An effective stress theory is presented which can be used to predict the undrained shearing resistance of clay soils after some repeated loading. The theory is developed for two limiting conditions:

completely undrained repeated loading followed by undrained loading to failure and for repeated loading followed by drainage before being loaded to failure. The theoretical development is completely general with specific equations also given for the cyclic triaxial test condition of $\sigma_2 = \sigma_3$.

Data are presented in the paper illustrating the behavior from the end of the repeated loading sequence to failure. For the completely undrained case the final peak strength decreases as the level of cycled stress increases. The relationship is empirical but for some cases the form of the strength variation and particularly the pore pressure response during loading to failure is predictable. The pore pressure coefficient varies between predictable limits. The upper limit is a value equal to that for direct undrained loading to failure, the lower limit is zero.

When the excess pore pressures produced by repeated loading are allowed to dissipate by drainage, the subsequent peak strength of the soil will be larger than the original strength prior to repeated loading unless the soil was originally heavily overconsolidated. The specific increase in strength and the associated pore pressure response are empirically determined characteristics but predictable in form. The upper limit of pore pressure coefficient is equal to that for direct undrained loading to failure and the lower limit is equivalent to elastic pore pressure resonse, $A = 1/3$ for the boundary condition of $\sigma_2 = \sigma_3$.

The relationship between strain ratio and strength ratio originally proposed by Thiers and Seed has been confirmed in general form. Their specific curve is not generally valid, however, with each soil having a specific empirically determined relationship.

Practical applications of this theory have been demonstrated.

ACKNOWLEDGMENTS

Support for the research described in this paper was provided in part by the National Science Foundation under Grant No. AEN 75-03217. The writers appreciate the help of J. D. Adams and J. A. Egan who contributed some of the data reported and W. R. Sawbridge who assisted in the research and in the preparation of drawings.

REFERENCES

1. Brown, S.F., A.D.F. Lashine, and A.F.L. Hyde, (1975), Repeated Load Testing of a Silty Clay, Geo. 25, (1): 95-114.

2. Eide, O., (1974), Marine Soil Mechanics, Norwegian Geotechnical Institute Publication 103:1-20, Oslo.

3. France, J.W. and Sangrey, D.A., (1977), Effects of Drainage in Repeated Loading of Clays, Jour. Geot. Eng. Div., ASCE 103 (GT7):769-785.

4. Hyde, A.F.L. and S.F. Brown, (1976), The Plastic Deformation of a Silty Clay Under Creep and Repeated Loading, Geo. 26 (1):173-184.

5. Koutsoftas, D.C. (1978), Effect of Cyclic Loads on Undrained Strength of Two Marine Clays, Jour. Geo. Eng. Div., ASCE 104 (GT5):609-620.

6. Lee, K.L. and J.A. Focht, (1976), Strength of Clay Subjected to Cyclic Loading, Marine Geotechnology, 1 (3): 165-186.

7. Sangrey, D.A. (1972), Obtaining Strength Profiles with Depth for Marine Soil Deposits Using Disturbed Samples, ASTM STP 501:106-121.

8. Sangrey, D.A. (1977), Response of Offshore Piles to Cyclic Loading, 9th OTC 1:507-512.

9. Sangrey, D.A., Castro, G., Poulos, S.J. and France, J.W. (1978), Cyclic Loading of Sands, Silts and Clays, ASCE Spec. Conf. on Earthquake Engineering and Soil Dynamics 2:836-851.

10. Sangrey, D.A., D.J. Henkel, and M.I. Esrig, (1969), The Effective Stress Response of a Saturated Clay Soil to Repeated Loading, Can. Geo. Jour., 6 (3):241-252.

11. Schofield, A. and P. Wroth, (1968), Critical States Soil Mechanics, McGraw-Hill Book Co..

12. Seed, H.B. (1978), Soil Liquefaction and Cyclic Mobility Evaluation for Level Ground During Earthquakes, Jour. Geo. Eng. Div., ASCE 105 (GT2):201-255.

13. Seed H.B., P.P. Martin, and J. Lysmer, (1976), Pore-Water Pressure Changes During Soil Liquefaction, Jour. Geo. Eng. Div., ASCE, 102 (GT4):323-347.

14. Thiers, G.R. and H.B. Seed, (1968), Strength and Stress-Strain Characteristics of Clays Subjected to Seismic Loading Conditions, ASTM STP 450:3-56.

LIST OF SYMBOLS

A_{ref}^{f} = pore pressure coefficeint applicable from reference to f

p = mean normal effective stress,
$$\frac{\sigma_1' + \sigma_2' + \sigma_3'}{3}$$

Δp = change in p

$\Delta p^{(T)}$ = change in mean total stress

q = effective stress difference,
$\sigma_1 - \sigma_3$

Δq = change in q

u = pore pressure (change)

M = ratio of effective stresses at failure

ρ = $- \Delta q/\Delta p$

ϕ' = effective stress friction angle

$\sigma_1, \sigma_2, \sigma_3$ = principal stresses, effective stress if σ'

Elastoplastic and viscoplastic constitutive models
for soils with application to cyclic loading

Z. MRÓZ

Institute of Fundamental Technological Research, Warsaw, Poland

ABSTRACT

For soils under monotonic loading, the simple elastic plastic or non-linear elastic material models can be used to simulate deformational response of the material with sufficient accuracy. For a perfectly-plastic model both hardening and softening phenomena are neglected and the flow law is associated with the yield condition or the plastic potential by the gradiental rule. For an isotropic hardening model, the irreversible void ratio or density are assumed as state parameters. Using this model, one is able to predict hardening, softening and critical state response. However, for more complex loading programmes ivolving loading followed by unloading or for repetitive action of loads, more complex hardening ruleshould be proposed in order to describe more realistically the material response.

In order to for-mulate a complete set of constitutive equations one has to

i/ formulate the yield condition depending on the stress /or strain/ and some selected hardening parameters whose evolution defines the varying state of hardening

ii/ formulate the flow rule relating strain and stress increments

iii/formulate the evolution rule for hardening parameters

The last task is most difficult since such rule should incorporate properly the material memory of particular loading events. In particular, some loading events should be erased as inessential for subsequent material response whereas other events such as corresponding to maximal load amplitudes should be stored as they affect essentially the subsequent response.

In the first part of this paper a class of material models is discussed following the previous work /1-5/. The closed yield surface is assumed to translate and expand or contract in the stress space, thus

$$f_o / \underline{\sigma} - \underline{\alpha}^{(o)}, \eta^p / = 0 \qquad /1/$$

where $\underline{\alpha}^{(o)}$ and η^p are the tensor and scalar hardening parameters. Here $\underline{\alpha}^{(o)}$ can be visualized as the cetre of the yield surface and η^p is the plastic portion of relative density /or void ratio/.Using the flow rule associated with /1/

$$d\underline{\varepsilon}^p = \frac{1}{H} \frac{\partial f_o}{\partial \underline{\sigma}} df_o \qquad /2/$$

the evolution rule for   is specified by introducing the set of nesting surfaces

$$f_i (\underline{\sigma} - \underline{\alpha}^{(i)}, \eta^p) = 0 \qquad /3/$$

and requiring the consecutive translation of particular surfaces to occur so that successive engagement of surfaces occurs without intersection. Further, the hardening modulus occuring in the flow rule (2) is assumed to vary between the initial infinite or very high value on the yield surface and the final value on the outermost consolitation surface $f_k = F_c = 0$. The consolidation surface corresponds to maximal prestress in the past loading history and all stress paths within these surface correspond to overconsolidated behaviour. Two simplified versions of this multisurface model are next considered: 1/ a two-surface model in which only the yield surface and the exterior consolidation surface are retained. The evolution rule for $\underline{\alpha}^{(o)}$ and the variation of hardening modulus depend now on the relative configuration of these surfaces. 2/ a model with infinite number of surfaces contained between the yield and the consolidation surface. In this case the diameter of the active surface with respect to the diameter of consolidation surface define the value of hardening modulus H.

Using these models, first the monotonic

loading processes were considered such as isotropic and K_o consolidation, drained and undrained shear response after initial consolidation of clays and the study of anisotropy of K_o consolidated clays. The isotropic consolidation test was used for preliminary identification of material parameters whose values were next readjusted to fit more complex loading programmes.

The cyclic loading processes were grouped in two classes:
i/cyclic behaviour for small number of cycles
ii/processes occuring for large number of cycles.

Though this classification is not precise, it is convenient from the viewpoint of material models.In fact, for small number of cycles, the cyclic response could be predicted by using the models identified from monotonic loading programmes. On the other hand, progressive cyclic degradation and further consolidation accompanied by pore pressure growth can be described by introducing an additional scalar parameter $\mathfrak{æ}$ in the yield condition and the equation of consolidation surface, thus

$$f_o\left(\underset{\sim}{\sigma} - \underset{\sim}{\alpha}^{(o)}, \eta^p, \mathfrak{æ}\right) = 0 , \quad F_c\left(\underset{\sim}{\sigma} - \underset{\sim}{\alpha}, \eta^p, \mathfrak{æ}\right) = 0 \quad /4/$$

Growth of the parameter $\mathfrak{æ}$ depending on both deviatoric and volumetric strain changes during cyclic loading induces further softening effects and progressive build-up of pore pressure resulting in final failure and liquefaction.

Numerical tests carried out for both consolidated and overconsolidated clays subjected to stress controlled or strain-controlled loading indicate good qualitative agreement with observed deformational response. The progressive pore pressure build-up induces growth of amplitudes of cyclic deformation with subsequent failure , both in consolidated and overconsolidated clays.

The viscoplastic version of this model was also formulated and tested for undrained cyclic loading programmes. The total strain increment is composed of instantaeous elastic increment and viscous increment developing in time.The effect of frequency of cycles is now essential in generating pore pressure accumulation.

The model was applied to clays but its application to sandsis possible with proper identification of material parameters.

REFERENCES

1. Mroz, Z.,Norris V.A. and Zienkiewicz O.C. 1978, An anisotropic hardening model for soils and its application to cyclic loading, Int. J. Num. Anal. Meth. Geomech. 2, 203-221.
2. Mróz,Z.,Norris V. A. and O. C. Zienkievicz, 1979, Application of an anisotropic hardening model in the analysis of the elastoplastic deformation of soils, 1, 1-34
3. Mróz, Z., Norris, V.A., Zienkiewicz O. C. 1978, Simulation of soil behaviour unde cyclic loading by using a more general hardening rule, Univ. Swansea Rep. C/R/340/78
4. Pietruszczak, S. and Mróz, Z., 1979, Description of anisotropy of K_o consolidated clays, Proc. Euromech. Coll. 115 on "Mechanical Behaviour of Anisotropic Solids", Grenoble, 1979
5. Mróz, Z. and Sharma K. G. 1979, A viscoplastic anisotopic hardening model for soils, Univ. Swansea Rep. 1979.

432

A critical state soil model for cyclic loading

J. P. CARTER
University of Queensland, Brisbane, Australia

J. R. BOOKER
University of Sydney, Sydney, Australia

C. P. WROTH
University of Oxford, Oxford, UK

A problem of considerable importance in geotechnical engineering concerns the prediction of the behaviour of soils under repeated loading. Situations which are of interest include the response of soil under earthquake conditions, wave induced loading on offshore structures and repeated loads of the type caused by rolling vehicles.

There exists a considerable body of laboratory data on the behaviour of sands and clays under cyclic loading conditions (e.g. Seed and Lee 1966; Taylor and Bacchus 1969; Andersen 1975). Although the conclusions of these examinations differ, several facts emerge. The most important of these is that under undrained loading excess pore pressures are generated and, if cyclic loading is continued for a sufficiently long time, a failure or critical state condition may be reached.

A natural consequence of this interest in cyclic loading has been the attempt to develop constitutive models to predict this type of behaviour (Mróz et al 1979; Prévost 1977, 1978). Generally, these models are complex, involving nested yield surfaces and both kinematic and isotropic hardening, and depend on the specification of a number of parameters. There seems to be no straightforward way of determining values for these parameters directly and this places severe limitations on the use of these models in practical situa-tions. A less complicated model, which is potentially applicable to cyclic loading has been suggested by Pender (1977,1978).

In this paper the concepts of critical state soil mechanics have been used to develop a simple model which may be used to predict many aspects of the behaviour of clays under repeated loading. The model employs the parameters that are usually associated with the Cam-clay family of models and as such it possesses most of the characteristics of these models; but there

is one simple, yet important, modification. This involves a specified contraction of the yield surface as the soil sample is unloaded. With the introduction of this modification an additional parameter must also be defined. For reasons which are explained in the paper, this parameter is called the OCR degradation parameter, and it is shown how its value may be determined in a straightforward manner from a laboratory triaxial test involving repeated, undrained loading.

Calculations have been performed using this new model and the results have been presented in parametric form. The undrained behaviour of soils which are either initially normally consolidated or initially overconsolidated, is investigated for both stress controlled and strain controlled loadings in the triaxial test.

For stress controlled loading the model predicts that failure will eventually occur if the soil sample is subjected to a sufficient number of load repetitions. In soils which are initially normally consolidated failure occurs when the material comes into a critical state condition. Associated with this is a steady increase in excess pore pressure as the loading is repeated. In soils which are initially overconsolidated there is a possibility that peak failure will occur before the critical state condition is reached (i.e. in fewer cycles). Factors which affect the number of cycles to failure in any given soil include the amplitude of the cyclic deviator stress and the initial overconsolidation ratio. Initially overconsolidated soils tend to fail in fewer cycles than normally consolidated soils tested at the same cyclic stress level. The number of cycles to failure decreases as the amplitude of the cyclic deviator stress is increased and two-way stress-controlled cycling, with the deviator stress varying in the range $-q_c \leq q \leq q_c$, is more 'damaging' than one-way cycling in the range $0 \leq q \leq q_c$.

For triaxial tests in which the cell pressure is held constant and the axial strain is varied with a fixed amplitude, the following results have been found. If the soil is initially normally consolidated then it tends to liquefy as the load is repeated, i.e. the mean effective stress p' is steadily reduced and approaches zero, and positive excess pore pressures are built up as the cycling continues. Associated with this is a hysteresis of the stress-strain response, a gradual reduction in the value of the peak deviator stress which is achieved each cycle, and a decrease in the apparent shear modulus as the number of cycles is increased. A given amount of 'damage' to the soil occurs in fewer cycles as the cyclic strain amplitude is increased. The predictions of the soil response in a strain controlled test are, unlike the stress controlled test, very much dependent on the value assigned to the elastic shear modulus.

Most of the trends in behaviour predicted by the model are in agreement with those observed in laboratory experiments by many workers (Thiers and Seed 1968; Taylor and Bacchus 1969; Andersen 1975). In addition to the parametric study, predictions have been made of the behaviour of two particular clays and the results have been compared with actual test results. Reasonable agreement was found between the measured and predicted behaviour.

As a result of these comparisons with actual soil behaviour some suggestions for future research have been made. The most important of these is concerned with the need for an accurate determination of the yield surface and plastic potential, for any particular clay, under conditions of static loading. It has been suggested that the shape of this surface must be known in some detail before good quality predictions can be expected for the behaviour of the same soil under repeated loading.

REFERENCES

Andersen, K.H. 1975, Research project, repeated loading on clay - Summary and interpretation of test results. Norwegian Geotechnical Institute, Report No.74037-9, Oslo.

Mróz, Z., Norris, V.A. & Zienkiewicz, O.C. 1979, Application of an anistropic hardening model in the analysis of elastoplastic deformation. Géotechnique, 29:1-34.

Pender, M.J. 1977, Modelling soil behaviour under cyclic loading. Proc. 9th Int. Conf. Soil Mech. Found. Eng., Tokyo, 2:325-331.

Pender, M.J. 1978, A model for the behaviour of over-consolidated clay. Géotechnique, 28:1-25.

Prévost, J.H. 1977, Mathematical modelling of monotonic and cyclic undrained clay behaviour. Int. J. Numerical Analytical Methods in Geomechanics, 1:195-216.

Prévost, J.H. 1978, Plasticity theory for soil stress-strain behaviour. J. Eng. Mech. Divn., ASCE, 104:1177-1194.

Seed, H.B. & Lee, K.H. 1966, Liquefaction of saturated sands during cyclic loading. J. Soil. Mech. Found. Divn., ASCE, 92:105-134.

Taylor, P.W. & Bacchus, D.R. 1969, Dynamic cyclic strain tests on a clay. Proc. 7th Int. Conf. Soil Mech. Found. Eng., Mexico, 1:401-409.

Thiers, G.R. & Seed, H.B. 1968, Cyclic stress-strain characteristics of clay. J. Soil Mech. Found. Divn., ASCE, 94:555-569.

On dynamic and static behavior of granular materials

S. NEMAT-NASSER
Northwestern University, Evanston, Ill., USA

SUMMARY

The lecture summarizes some recent work by
the author and his associates on the dynamic and static behavior of granular
materials (saturated undrained, or drained
sands), consisting of three complementary
aspects: (1) densification and liquefaction of sand in cyclic shearing
(strain- or stress-controlled cases); (2)
application of finite plasticity theory
(with plastic volumetric changes and including internal friction) for the description of the sand behavior in monotone
loading regimes; and (3) development of
the basic rate constitutive relations by a
statistical averaging which is based on
the behavior of individual grains at
microscale.

The analysis of densification (dry or
saturated but drained) and liquefaction
(saturated and undrained sand) is based on
an energy consideration and some physical
observation pertaining to the behavior of
grains at microscale. It is observed
that, in order to reduce the void volume
by a given increment, a certain amount of
energy is consumed in the frictional
losses involved in the rearrangement of
the sand grains, as the sample undergoes
cyclic shearing. The required incremental
energy increases as the density of the
sample increases, becoming very large when
the sample tends to attain its maximum
density. If the sample is saturated and
undrained, the tendency toward densification results in an increase in the pore
water pressure, and therefore a decrease
in the contact forces between grains. The
reduction in contact forces leads to a
smaller frictional loss, as the grains
move relative to each other. Therefore,
in this case the required increment of
energy decreases with increasing pore
water pressure. Using the number of
cycles as the measure of time, the present author and his associates have developed differential equations which, in
a unified manner, quantify the above
physical observations. These differential
equations are then integrated and explicit
results are obtained for the change in
density as a function of the number of
cycles and strain amplitude in cyclic
shearing of the dry sand, and for pore
water pressure build-up as a function of
the number of cycles and stress amplitude
in cyclic shearing of saturated undrained
sand. In addition, explicit equations
for the stress amplitude in terms of the
strain amplitude and the number of cycles
(for strain-controlled test), and for the
strain amplitude in terms of the stress
amplitude and the number of cycles (for
the stress-controlled test), are obtained.
The theoretical results are compared with
some existing experiments, in an effort to
identify the material parameters and to
test the validity of the theory.

For monotone loading of granular materials under moderate (a few bars) or
large (a few kilobars) pressure, finite
deformation plasticity theory provides a
reasonable approach. Because of the large
plastic volume changes and the existence
of internal friction, the usual assumptions of incompressibility and normality
no longer hold. The author and his co-
workers have developed a finite plasticity
theory which is based on the assumption of
nonassociative flow rule, includes internal friction, and accounts for large
plastic distortions and volume changes.
The results are then applied to examine
the behavior of granular materials in
simple shear and in triaxial tests. Comparison with experimental results on
crushed Westerly granite and Ottawa sand

under high pressure, supports the theory. The attractive feature of this approach is that it is based on some minor modification of the usual J_2 plasticity theory.

If one assumes that individual grains are rigid, then one expects to be able to construct on a purely statistical basis the macroscopic rate constitutive behavior of granular materials, the corresponding constitutive relations involving only the coefficient of friction and some parameters which characterize the shape and size distribution of grains. Recently, the author and his associates have examined this rather challenging problem with some success. The lecture will be concluded with a presentation of some relevant results obtained in this manner.

ACKNOWLEDGMENT

This work has been supported, in part, by the U. S. Geological Survey under Contract No. 14-08-0001-17770.

REFERENCES

Nemat-Nasser, S. and A. Shokooh 1977, A unified approach to densification and liquefaction of cohesionless sand, Earthquake Research and Engineering Laboratory, Technical Report No. 77-10-3, Dept. of Civil Engineering, Northwestern University, Evanston, IL; a revised version to appear in Canadian Geotechnical Journal.

Nemat-Nasser, S. and A. Shokooh 1978, A new approach for the analysis of liquefaction of sand in cyclic shearing, Proc. of the Second International Conference on Microzonation, San Francisco, CA, 2: 957-969.

Nemat-Nasser, S. and A. Shokooh 1979, On finite plastic flows of compressible materials with internal friction, Earthquake Research and Engineering Laboratory, Technical Report No. 79-5-16, Dept. of Civil Engineering, Northwestern University, Evanston, IL; to appear in Int. J. Solids Structures.

Nemat-Nasser, S. 1979, On behavior of granular materials in simple shear, Earthquake Research and Engineering Laboratory, Technical Report No. 79-6-19, Dept. of Civil Engineering, Northwestern University, Evanston, IL.

Behavior of normally consolidated clay
under simulated earthquake and ocean wave loading conditions

I. M. IDRISS & Y. MORIWAKI
Woodward-Clyde Consultants, San Francisco, Calif., USA

S. G. WRIGHT
University of Texas, Austin, Texas, USA

E. H. DOYLE
Shell Development Company, Houston, Texas, USA

R. S. LADD
Woodward-Clyde Consultants, Clifton, N. J., USA

1 INTRODUCTION

Numerous facilities, such as drilling platforms and pipelines, are being designed and constructed offshore throughout the world. Many of these sites where these facilities are constructed are underlain by clay soils. The behavior of these soils under wave loading conditions, and, in some parts of the world, under earthquake loading conditions is of significance to the performance of the facility. Thus, the stress-strain-strength behavior of clays under earthquake and wave loading conditions has been receiving increasing attention in recent years.

The purpose of this paper is to present a summary of some of the results of a series of studies conducted in the period 1976 to 1979, in which the behavior of a marine clay was evaluated under simulated earthquake and wave loading conditions. These studies as a whole are referred to as the EQWS (Earthquake and Wave Loading Stability) study. The EQWS study consisted of: (1) static, cyclic and dynamic laboratory triaxial and direct simple shear tests under controlled-strain and controlled-stress conditions; (2) formulation of non-linear stress-strain models based on these tests that incorporate the modulus degradation of clays during cyclic loading for the analyses of horizontal and mildly-sloping sites; (3) incorporation of these non-linear, degrading stress-strain models into two computer programs, one program capable of non-linear seismic response analyses and the other program capable of non-linear wave-loading analyses.

Specimens used in all laboratory tests were prepared from undisturbed block samples obtained from Icy Bay in Alaska. Special emphasis was placed on the use of the normalized soil behavior (NSB) approach (Ladd and Foote, 1974) to minimize the effects of possible sample disturbance on the behavior of laboratory specimens. The results of laboratory tests covered the behavior of soft, normally consolidated clays to stiff clays by incorporating the effects of stress history in terms of the overconsolidation ratio (OCR).

Because of length limitations, only the results of the direct simple shear (DSS) tests on normally consolidated Icy Bay marine clay, mostly loaded cyclically under controlled-stress conditions, are presented in this paper. Some specimens were tested without any initial static shear stress but others were tested using an initial static shear stress equal to 10 percent of the initial static vertical effective stress applied to the specimen to assess the influence of a mildly sloping profile. Two frequencies of cyclic loading, 1 Hz and 0.05 Hz, were used to simulate the earthquake and wave loading conditions, respectively.

2 BASIC DATA, DEFINITIONS AND DATA REDUCTIONS

The basic information on Icy Bay marine clay and laboratory tests presented herein is summarized in Table 1. The pertinent definitions related to controlled-stress DSS tests are summarized in Fig. 1.

In a controlled-stress cyclic DSS test, the specimen is first consolidated to some vertical effective consolidation pressure, $\bar{\sigma}_{vc}$. A cyclic shear stress, τ_c, is then applied under constant volume conditions and the resulting shear strains, as well as the change in vertical stress (interpreted as the change in pore pressure), are measured. For those tests requiring an initial shear stress, an initial sustained static shear stress, τ_s, of known magnitude is imposed under drained conditions ($\tau_s/\bar{\sigma}_{vc}$ is defined as α_s) and then the cyclic load is applied.

The results of controlled-stress cyclic DSS tests can be summarized by the following quantities expressed as a function of the number of cycles, N: (1) cyclic shear stress, τ_c (which is controlled to be reasonably constant and expressed in a normalized form, $\beta = \tau_c/\bar{\sigma}_{vc}$); (2) peak-to-peak cyclic shear strain, γ_{pp} (see Fig. 1); (3) residual shear strain, γ_r (see Fig. 1); and (4) cyclically-induced excess pore water pressure, Δu_c (which is expressed in a normalized form, $\Delta u_c/\bar{\sigma}_{vc}$. The results of a typical test are shown in Fig. 2.

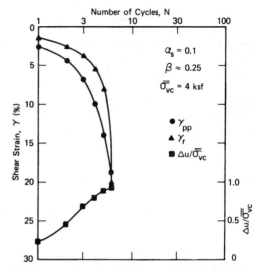

Fig. 2 - RESULTS OF A TYPICAL CONTROLLED-STRESS CYCLIC DSS TEST

Define for a Given Cycle:

$$\gamma_r = \gamma_f - (\gamma_{pp}/2)$$

$$\gamma_{pp} = \gamma_f - \gamma_b$$

$$\gamma_c = \gamma_{pp}/2$$

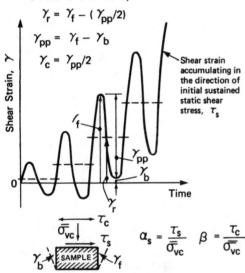

Fig. 1 - CONTROLLED-STRESS CYCLIC DSS TEST WITH INITIAL SUSTAINED STATIC STRESS

3 BASIC MODEL AND DETERMINATION OF MODEL PARAMETERS FROM CONTROLLED-STRESS TESTS

A non-linear, degrading stress-strain model based on controlled-strain cyclic tests was proposed previously (Idriss, Dobry, and Singh 1978; Moriwaki and Doyle 1978). The model is basically a single-parameter stress-strain model (Van Eckelen 1977) consisting of: (1) an initial backbone curve using a Ramberg-Osgood equation; (2) a modified Masing criterion to form arbitrary hysteretic loops; (3) a criterion for degradation of backbone

curve under constant cyclic strain conditions represented by a degradation index, δ, and t-parameter; and (4) a criterion for degradation under arbitrary cyclic loading conditions. In this section a brief summary of procedures to obtain model parameter values based on controlled-stress (rather than controlled-strain) cyclic tests is presented.

The data points for the initial backbone curve can be obtained by pairing τ_c with the single-amplitude cyclic shear strain, γ_c ($\gamma_c = \gamma_{pp}/2$). Both τ_c and γ_c are from the first-cycle data of each controlled-stress cyclic DSS test; typical data points of τ_c versus γ_c are schematically shown in Fig. 3 and designated by the symbols D_1 through D_6. A Ramberg-Osgood equation usually provides a reasonable fit to represent these data points mathematically.

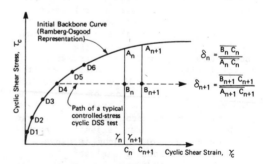

Fig. 3 - SCHEMATIC REPRESENTATION OF BACKBONE CURVE AND DEGRADATION INDEX

The nth and (n + 1)th cycle data points of τ_c versus γ_c from a typical controlled-stress cyclic DSS test are also shown schematically in Fig. 3 as B_n and B_{n+1}. A vertical line through the nth cycle data point B_n intersects the initial Ramberg-Osgood backbone curve at A_n and the cyclic shear strain axis at C_n; the nth cycle cyclic shear strain is designated γ_n. The value of the nth cycle degradation index, δ_n, is then equal to the ratio of the distance from B_n to C_n divided by the distance from A_n to C_n ($\delta_n = B_nC_n/A_nC_n$). δ_{n+1} is similarly defined as shown in Fig. 3. Since δ_n and δ_{n+1} are related by the following equation in the existing model (Idriss, et al 1978):

$$\delta_{n+1} = \delta_n \left[1 + \delta_n\right]^{\frac{1}{t_n} - t_n}$$

the value of the t-parameter, t_n, corresponding to γ_n can be determined from this equation.

The initial backbone curve and the t-parameter obtained using this procedure from a set of five controlled-stress cyclic DSS tests for the earthquake loading conditions (frequency = 1 Hz) with no initial static shear stress ($\alpha_s = 0$) are shown in Figs. 4 and 5, respectively. The values of Ramberg-Osgood parameters for the initial backbone curve and a modified hyperbolic equation for the t-parameter are also shown in Figs. 4 and 5, respectively. The data points are consistent and the mathematical representa-

tions of the data points appear to be reasonable.

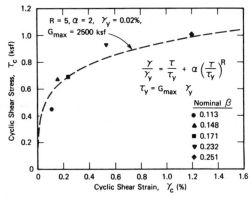

Fig. 4 - INITIAL BACKBONE CURVE WITH $\alpha_s = 0$ BASED ON HALF PEAK-TO-PEAK FIRST CYCLE STRESS-STRAIN DATA FROM CONTROLLED-STRESS CYCLIC DSS TESTS

4 ASSESSMENT OF THE MODEL

The Ramberg-Osgood representation of the initial backbone curve and the hyperbolic representation of the t-parameter (shown in Figs. 4 and 5, respectively) completely specify the model for the case without any initial sustained static shear stress ($\alpha_s = 0$). Using the specified model, the results of controlled-stress cyclic DSS tests which were used to derive the model parameter values were "predicted" to assess the quality of the fit.

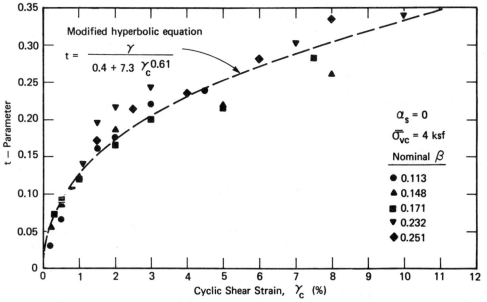

Fig. 5 - t—PARAMETER FROM CONTROLLED-STRESS CYCLIC DSS TESTS

439

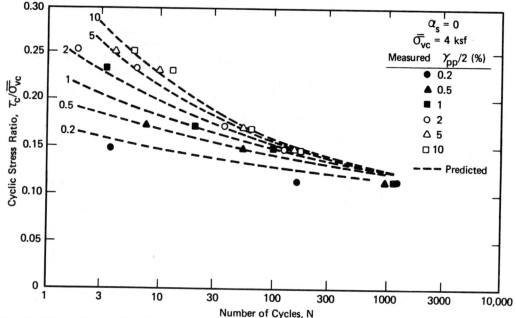

Fig. 6 - PREDICTED AND MEASURED CYCLIC STRESS RATIO REQUIRED TO CAUSE GIVEN VALUE OF $\gamma_{pp}/2$ VERSUS NUMBER OF CYCLES

For various values of cyclic stress ratio, $\tau_c/\bar{\sigma}_{vc}$, the values of γ_c (equal to $\gamma_{pp}/2$) were calculated for different number of cycles, N, using the model. The results are shown in Fig. 6 as a plot of cyclic stress ratio, $\tau_c/\bar{\sigma}_{vc}$, versus log N curves representing constant values of γ_c varying from 0.2 percent to 10 percent. Also shown in Fig. 6 are the data points representing the same constant values of γ_c from controlled-stress cyclic DSS tests. The agreement between the measured data and the "predicted" curves shown in Fig. 6 is reasonably good. Thus, the procedure described earlier to obtain the model parameter values from a set of controlled-stress rather than controlled-strain cyclic DSS tests appears to be acceptable for engineering purposes. This procedure makes it possible to extend the previously proposed model to incorporate the effects of the constant initial sustained static shear stress during cyclic loading. Note that the presence of initial static shear stress makes it difficult to conduct or interpret controlled-strain cyclic tests.

While a type of plot presented in Fig. 6 shows the consistency in the model formulation, a true test of the cyclic stress-strain model lies in its ability to predict the results of a cyclic DSS test with variable cyclic stress shown in Figs. 7, 8 and 9.

In Fig. 7, the values of cyclic shear strain, γ_c, measured during the variable cyclic DSS test is plotted versus number of cycles, N.

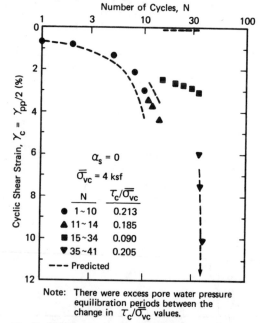

Note: There were excess pore water pressure equilibration periods between the change in $\tau_c/\bar{\sigma}_{vc}$ values.

Fig. 7 - PREDICTED AND MEASURED CYCLIC SHEAR STRAIN VERSUS NUMBER OF CYCLES FOR TRANSIENT CONTROLLED-STRESS TEST

440

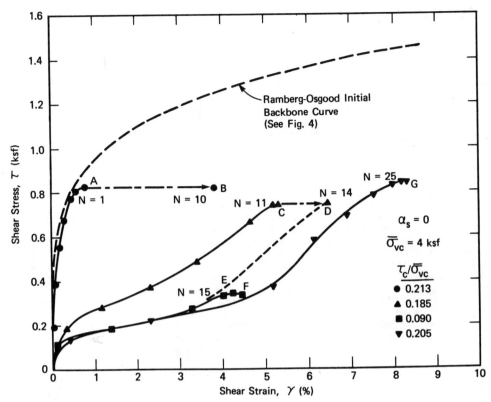

Fig. 8 - BACKBONE CURVES FOR TRANSIENT CONTROLLED-STRESS TEST

Also shown in Fig. 7 are the corresponding predicted curves using the model summarized in Figs. 4 and 5.

The comparison of the measured data and the predicted curves in Fig. 7 provides the following observations: (1) the predicted curve and the measured data points show a reasonable agreement up to N = 10; (2) between N = 11 to N = 14, the predicted curve appears to give somewhat lower values of γ_c compared to the measured data; (3) from N = 15 to N = 34, the model severely underpredicts the values of γ_c; and finally (4) beyond N = 35, the model predicts the measured data in a reasonable way.

Most of these observations can be explained in terms of the shape of the actual backbone curves as shown in Fig. 8 and the changes in excess pore water pressures as shown in Fig. 9. The first quarter cycle stress–strain curve from the variable stress cyclic DSS test (0 to A in Fig. 8) is quite consistent with the initial backbone curve shown in Fig. 4. After ten cycles of loading, the tip of the hysteresis loop moves from point A (N = 1) to point B (N = 10) as shown in Fig. 8.

The excess pore water pressure equalization period after the tenth cycle resulted in a significant increase of pore pressure as shown in Fig. 9. This increase in excess

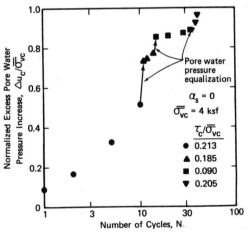

Fig. 9 - EXCESS PORE WATER PRESSURE VERSUS NUMBER OF CYCLES FOR TRANSIENT CONTROLLED-STRESS TEST

pore water pressure during the equalization period was accompanied by a slight increase in the volume of the specimen as measured by a pipette. The increase apparently had some effect on the specimen since for N = 11, the tip of the backbone curve was at point C (see Fig. 8), which represents a significant change from point B for N = 10. This resulted in a higher measured γ_c for N = 11 compared to the predicted value as shown in Fig. 7. Note also the double-curvatured backbone curve for N = 11 as shown in Fig. 8. From N = 11 to N = 14, the tip of the backbone curve degraded from point C to point D. A possible backbone curve for N = 14 is shown in Fig. 8 by a dashed line from point D to E and back to the origin. For N = 15, the measured backbone curve followed along the backbone curve for N = 14. Because of the double curvature of the backbone curve, the measured γ_c for N = 15 was significantly higher (Fig. 7) than that calculated from the model that assumes a single curvature Ramberg-Osgood backbone curve. When the applied cyclic stress ratio was increased significantly for N = 25, the backbone curve followed that of N = 15 to approximately, γ_c, = 4 percent and then increased (exhibiting dilatant behavior) at a faster rate preserving the general shape of the double-curvatured backbone curve of N = 11.

Based on the measured data and the predicted curves shown in Figs. 7, 8, and 9, a number of observations can be made:

1. The model appears to provide reasonably satisfactory peak values of γ_c for engineering purposes; note that the model predicts high peak strain values quite well.

2. A single parameter stress-strain model, such as presented herein, that assumes uniform degradation of the backbone curve may have a limited applicability in predicting the actual incremental stress-strain behavior at high strain levels because of the double curvature shape exhibited by the backbone curves at high strain levels.

3. Care must be exercised in interpreting the results of a DSS test when the conditions involving the stress-strain behavior, pore water pressure changes, and volume changes in the DSS test specimen possibly deviate from the assumed constant-volume testing conditions inherent in the Geonor-type DSS tests (Bjerrum and Landva 1966).

5 COMPARISON WITH WAVE LOADING CONDITIONS

As mentioned earlier, a number of controlled-stress cyclic DSS tests were conducted to simulate wave loading conditions; static DSS tests were also conducted. The difference among static stress-strain, cyclic stress-strain under wave loading conditions and cyclic stress-strain under earthquake loading conditions is basically that of strain rate. The strain rate of a static DSS test was 5 percent per hour and the loading frequencies of cyclic DSS tests were 0.05 Hz for simulated wave loading conditions. The first one-quarter cycle stress-strain data from cyclic DSS tests under simulated earthquake loading conditions and under simulated wave loading conditions together with stress-strain curve from a static test are presented in Fig. 10. All data shown in this figure correspond to tests with α_s = 0.1.

Fig. 10 shows clearly the effects of strain rate: for example, at 2 percent shear strain the cyclic shear stress under earthquake loading conditions is about 22 percent higher than that under wave loading conditions and about 68 percent higher than the static shear stress. These values are consistent with previously reported effects of strain rate on the shear resistance of clays (Bjerrum 1973).

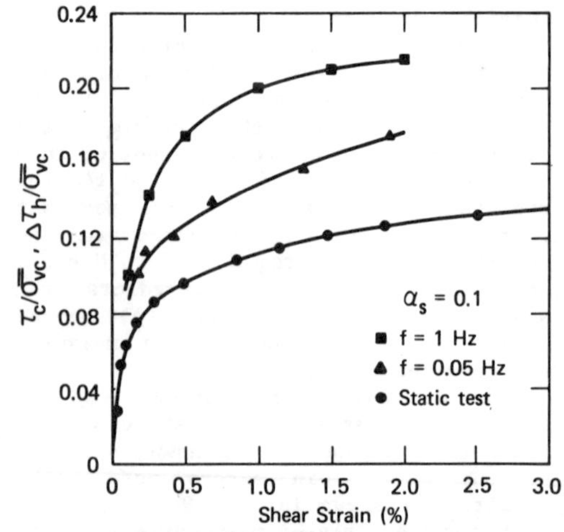

Note: $\Delta\tau_h$ is the applied static shear stress on the horizontal plane in a static DSS test.

Fig. 10 - COMPARISON OF CYCLIC STRESS-STRAIN CURVES BASED ON FIRST ONE-QUARTER CYCLE WITH STATIC STRESS-STRAIN CURVE

6 EFFECTS OF INITIAL SHEAR STRESS

When an initial sustained static shear stress, τ_s, is present, permanent residual shear strains, γ_r, accumulate as the specimens are cyclically loaded. These permanent residual shear strains can be used in assessing the amount of permanent deformations at mildly sloping sites under earthquake or wave loading conditions.

The calculated values of γ_r, (using the equation in Fig. 1) for 0.5 percent and 4 percent based on controlled-stress cyclic DSS tests under wave loading conditions and under earthquake loading conditions are compared in Fig. 11; all tests were conducted with $\alpha_s = 0.1$.

While a certain amount of scatter in data is apparent in this figure, it is clear that higher cyclic shear stresses are required to cause a given value of γ_r under earthquake loading conditions when compared to those under wave loading conditions. It is interesting to note, however, that the differences in $\tau_c/\bar{\sigma}_{vc}$ between γ_r data under wave loading conditions and under earthquake loading

conditions shown in Fig. 11 is similar to the difference in $\tau_c/\bar{\sigma}_{vc}$ values between the first one-quarter cycle shear stress data shown in Fig. 10.

7 CONCLUSIONS

Some results on normally consolidated Icy Bay marine clay tested under controlled-stress cyclic DSS conditions were presented. The main conclusions of this study are the following:

1. A non-linear, degrading stress-strain model, proposed previously, was based on the results of a series of controlled-strain cyclic tests. A procedure to derive the model parameters completely consistent with the previously proposed model but based on the results of a series of stress-controlled cyclic tests is developed. This procedure makes it possible to extend the model to incorporate the effects of constant, sustained static shear stress during cyclic loading to simulate the effects of mildly sloping sites.

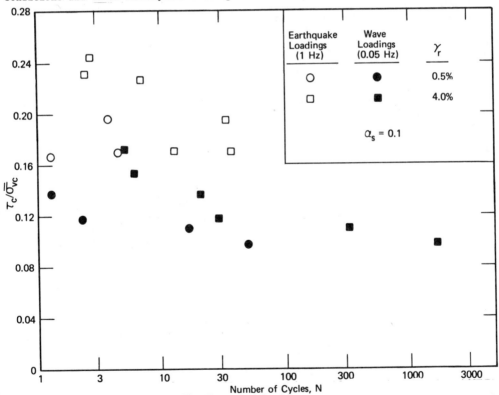

Fig. 11 - COMPARISON OF MEASURED VALUES OF CYCLIC STRESS RATIO AND NUMBER OF CYCLES REQUIRED TO PRODUCE 0.5 AND 4 PERCENT SHEAR STRAIN, γ_r, FROM EARTHQUAKE AND WAVE LOADING CYCLIC DSS TESTS

2. Reasonably good agreement between the measured data and the curves calculated from the model was obtained. In particular, (a) a consistent initial backbone curve and degradation parameters (t-parameter versus strain) representing the rate of backbone curve degradation were obtained, and (b) curves of constant cyclic strain calculated from the model showed a reasonably good agreement with the measured data points when compared in a plot of cyclic stress ratio, $\tau_c/\bar{\sigma}_{vc}$, versus log N.

3. The model was also used for predicting the results of a controlled-stress cyclic test with varying stress level and $\alpha_s = 0$. The agreement between the predicted and the measured values were reasonably satisfactory for engineering purposes. The observed disparities between the predicted and the measured values were partly explainable in terms of the non-uniform degradation of the backbone curve.

4. The results of tests under simulated wave loading conditions (cyclic loading at 0.05 Hz) when compared to those under simulated earthquake loading conditions (cyclic loading at 1 Hz) showed a somewhat softer backbone curve and a reduction in the cyclic stress ratio $(\tau_c/\bar{\sigma}_{vc})$ required to cause a given amount of permanent shear strain.

8 ACKNOWLEDGEMENTS

The EQWS study was supported by a group of participants with Shell as the administrator of the study. The participants included Atlantic-Richfield Company, Mobil Research and Development Company, Gulf Research and Development Company, Oxy Petroleum, Inc., Shell Development Company, and the US Geological Survey. The following individuals contributed to the study: Prof. R. Dobry of Rensselaer Polytechnic Institute, Prof. J. K. Mitchell of University of California at Berkeley, and Robert Green and Ram Singh of Woodward-Clyde Consultants.

TABLE 1

BASIC INFORMATION ON ICY BAY MARINE CLAY AND LABORATORY TESTS

Icy Bay Marine Clay

   85% to 90% passing #200 sieve

   CL - Unified Soil Classification System

Controlled-Stress Cyclic DSS Tests ($\bar{\sigma}_{vc}$ = 4 ksf for all)

	Liquid Limit (percent)	Plastic Limit (percent)	Water Content (percent)	Total Unit Weight (pcf)
Simulated Earthquake Loading (f = 1 Hz)				
α_s = 0 Total of Six Tests (one with variable $\tau_c/\bar{\sigma}_{vc}$)	30.2-31.5	16.5-18.2	17.9-21.0	133-137
α_s = 0.1 Total of Six Tests	29.1-30.7	16.1-17.0	18.7-20.2	134-138
Simulated Wave Loading (f = 0.05 Hz)				
α_s = 0.1 Total of Six Tests	28.9-30.4	15.9-16.5	19.0-20.1	134-137
Static DSS Test ($\bar{\sigma}_v$ = 4.0 ksf)				
α_s = 0.1	30.2	15.4	19.2	136

9 REFERENCES

Bjerrum, L. 1973, General Report, Proc. 8th Conf. SMFE, Vol. 3: 111-159.

Bjerrum, L. and Landva, A. 1966, Direct Simple Shear Tests on a Norwegian Quick Clay, Geotechnique, Vol. 26(1): 1-20.

Idriss, I. M., Dobry, R. & Singh, R. D. 1978, Nonlinear Behavior of Soft Clays During Cyclic Loading, Journal of the Geotechnical Engineering Div., ASCE, Vol. 104(GT12): 1427-1447, Proc. Paper 14265.

Ladd, C. C. and Foote, R. 1974, New Design Procedures for Stability of Soft Clays, Journal of the Geotechnical Engineering Div., ASCE, Vol. 99(GT7): 763-786, Proc. Paper 19664.

Moriwaki, Y. and Doyle, E.H. 1978, Site Effects on Microzonation in Offshore Areas, Proceedings of 2nd Int'l Conf. on Microzonation, Vol. 3: 1433-1445, San Francisco, California.

Van Eckelen, H.A.M. 1977, Single-Parameter Models for Progressive Weakening Of Soils by Cyclic Loading, Geotechnique, Vol. 27(3): 357-368.

Fatigue models for cyclic degradation of soils

H. A. M. VAN EEKELEN
Koninklijke/Shell Exploratie en Produktie Laboratorium, Rijswijk, Netherlands

SUMMARY

Theories for the behaviour of soils under persistent cyclic loading usually make implicit or explicit use of a memory parameter. In this paper, the structure of these theories is elucidated, alternative choices for the memory parameter are discussed, and a review of existing models and methods is presented.

1 INTRODUCTION

Modern theories of soil mechanics are usually based on yield surfaces, flow rules and hardening laws. These 'static' soil models are basically meant to describe the effects of monotonically increasing loading. Kinematic hardening may be included to describe unloading and damping, and, with some luck, shakedown or incremental collapse may be achieved for a soil structure under inhomogeneous repeated loading. The most advanced (and complicated) model uses sets of shifting yield surfaces within an isotropic bounding surface (Mroz et al. 1978). In this way some of the characteristic phenomena associated with persistent cyclic loading of clay may be reproduced, but agreement with experiment is less than perfect; for instance, the calculated pore pressure generated by undrained cyclic loading is too small for normally consolidated clays, and has the wrong sign for heavily overconsolidated clays (Mroz et al. 1979). Hence, the cumulative effects of a large number of loading cycles (soil weakening, liquefaction) are not described accurately by any of these static soil models.

In most models which have been constructed specifically for soils under cyclic loading, it is assumed (implicitly or explicitly) that a single state variable or memory parameter k (called 'fatigue') may be used to represent all effects of past cyclic loading. In such a model, one only needs to know the initial state of the sample (in terms of its response characteristics), and the present value of the state variable k, to predict the response of the soil to further static or cyclic loading. Details such as number, order, intensity or direction of past loading cycles then are irrelevant. It is also commonly assumed that the parameter k is monotonically increasing. This excludes materials with a fading memory, for which k could decrease during a period of relative quiet.

2 BASIC THEORY

Apart from a description of the initial state of the soil in terms of a static soil model, the basic elements of a theory for cyclic loading of soils are a relation which gives the increase of k per cycle (dependent on the nature of the loading cycle and on the current value of k), and a set of relations to indicate the changes in soil response characteristics with increasing value of k. In view of the above assumptions the first relation may be written in the form (van Eekelen 1977)

$$\frac{dk}{dN} = Q(k, \tau_c) , \quad \text{or} \quad \frac{dk}{dN} = \hat{Q}(k, \gamma_c) , \quad (1)$$

where the nature of a loading cycle is specified by either a generalised stress amplitude τ_c, or a generalised strain amplitude γ_c. The function Q or $\hat{Q}$ will depend on the initial state of the sample; for a given initial state it provides us with a yardstick to compare the effect of cycles with different intensities or loading geometries. If the function Q is separable, $Q(k, \tau_c) = g(k) \, h(\tau_c)$, then Miner's rule of superposition applies (Miner 1945): the cumulative effect of a number of stress cycles of different intensities may be obtained by superposition, independent of the order in which individual cycles occur. The same is true of strain cycles if the function Q is separable; it should be noted that the two versions of Miner's rule are not equivalent (van Eekelen 1977). In many empirical models for the behaviour of sands during earthquakes, Miner's rule is used to establish a simple conversion from the erratic actual stress history to an equivalent number of uniform stress cycles, from which the likelihood of soil liquefaction is estimated. These models have been reviewed by Seed (1979).

Once the function Q or $\hat{Q}$ has been specified, one only needs to indicate how the soil response to static or cyclic loading changes with increasing value of k. In the simplest models (Idriss et al. 1976, Martin et al. 1975, van Eekelen & Potts 1978) the change in cyclic response properties is given as a relation between stress and strain amplitudes

$$\gamma_c = F(k, \tau_c) , \quad \text{or} \quad \tau_c = \hat{F}(k, \gamma_c) \quad (2)$$

In addition, a relation may be given for the ultimate strength after cycling, in terms of the initial strength and the value of k. In more elaborate models, the state variable k is introduced as a parameter in a complete mechanical model for soil response, which may be a hardening elastoplastic model (Prévost 1977), or an 'endochronic' theory of soil behaviour (Cuellar et al. 1977).

3. REVIEW OF EXISTING MODELS FOR CYCLIC LOADING OF SOILS

Existing models differ in the manner in which the parameter k is related to observable quantities. The simplest procedure is to identify k with one of the observables of the system; the function Q or $\hat{Q}$ may then be obtained by monitoring this observable in a number of constant amplitude tests. The most obvious choice for this observable is the plastic volume strain e for drained cyclic loading, or the cumulative pore pressure u due to undrained cyclic loading; the two are related by the elastic rebound modulus. For drained simple shearing of sand, Martin et al. (1975) construct the following model:

$$k = e = c\, \gamma_c \ln\left(1 + \frac{N}{N_0}\right) \quad \text{for constant } \gamma_c$$

$$\tau_c = \sqrt{\sigma_v}\; \gamma_c / (a + b\gamma_c) \quad (3),(4)$$

where a and b are given functions of k. Formula (3) is an approximation (with less than two percent error) to the incremental expression given by Martin et al.; the parameters c and N_0 are independent of the vertical effective stress σ_v. By taking the derivative of (3) with respect to N and writing the result in terms of k and γ_c, one finds that $Q(k, \gamma_c)$ is not separable, so that Miner's rule of superposition (of strain cycles) does not apply for this model. By introducing an elastic rebound modulus $E_r = du/de$ the model may be used for undrained cyclic loading. With $E_r = \sigma_v/C$ one obtains, for instance,

$$u/\sigma_{vo} = 1 - e^{-k/C} , \quad \sigma_v = \sigma_{vo}\, e^{-k/C} \quad (5)$$

where σ_{vo} is the initial vertical effective stress. The pore pressure u is a monotonically increasing function of k; hence, for undrained loading the model is equivalent to one in which the pore pressure is used as memory parameter. It is very similar to a model proposed by Idriss et al. (1976) for soft clays under earthquake loading conditions, which may be written as (van Eekelen 1977)

$$k = t(\gamma_c)\, \ln(1 + N) \quad \text{for constant } \gamma_c$$

$$\tau_c = e^{-k}\, f(\gamma_c) \quad (6),(7)$$

$t(\gamma_c)$ in (6) is an empirical function of γ_c, which again seems to be independent of confining stress. On the basis of fourteen data points Idriss et al. (1978) select a nonlinear function for $t(\gamma_c)$, but the data do not deviate far from a straight line $t = c'\gamma_c$, consistent with eq. (3) which was based on observations by Silver & Seed (1971). The initial stress-strain relation $\tau = f(\gamma)$ in (7) is given by a Ramberg-Osgood curve $\gamma = a\tau + b\tau^m$, and damping is introduced by using Masing's rules (Masing 1926) to construct hysteresis loops (it should be noted that some difficulties inherent in the use of Masing's rules were recently pointed out by Pyke (1979)). Originally no attempt was made to identify k with an observable of the system, but in a later paper Doyle (1978) has related k to the pore pressure u, by a relation similar to eq. (5).

Some other models in which the fatigue parameter k in undrained cyclic loading is identified with the cumulative pore pressure are summarized elsewhere (van Eekelen 1977). More recently, van Eekelen & Potts (1978) analysed extensive cyclic loading data on Drammen clay (Andersen 1976) in this same way; they put

$$k = u^+, \quad Q = \frac{du^+}{dN} = A \exp \left(\frac{1}{B} \frac{J_c}{\langle J_f \rangle} \right) \quad ,$$

where u^+ is the part of the pore pressure due to cyclic loading (excluding 'static' pore pressure effects), J_c is the second invariant of the stress amplitude, and $\langle J_f \rangle$ is a suitably defined failure strength of the sample. It should be noted that in this model Q does not depend on $k = u^+$ ($u^+ \sim N$ in constant stress amplitude tests), so that Miner's rule of superposition applies for stress cycles of varying intensity. For eq. (2) a hyperbolic relation is adopted between the second invariant of the strain amplitude and a suitably normalised (k-dependent) stress amplitude. The model is meant to apply to arbitrary loading geometries; it includes an expression for the decrease in ultimate strength in terms of $k = u^+$, which also agrees with data by Doyle (1978). The relation may be interpreted as a shift in the 'virgin consolidation line' of the clay.

A different class of models are those in which the memory parameter k is linked to a cumulative deviatoric strain measure. Bonin et al. (1976) use the induced permanent strain for this purpose, but their procedure can only be used for cyclic triaxial compression. Zienkiewicz et al. (1978) seem to use the total length of the deviatoric strain path as a damage parameter; for variable stress amplitude, however, the relation between k and the total strain path length is no longer valid. A similar situation arises in a model proposed by Prévost (1977) : the memory parameter k appears to be the total length of the plastic deviatoric strain path, but a closer analysis of the formulae used reveals that k is basically a strain softening parameter, not directly related to any observable of the system (van Eekelen & Potts, 1978b). As a consequence, the function Q has to be obtained by a rather elaborate process of curve fitting.

The dilemma of identifying the fatigue parameter can be circumvented by formulating instead an hypothesis on the tangential compliance $L_t = d\gamma_c/d\tau_c$, which is the change in strain amplitude resulting from a sudden change in stress amplitude. In general, of course, L_t will still depend on k, but Andersen et al. (1976,1979) get around that difficulty by assuming that 'the immediate change in shear strain due to a change in cyclic shear stress is the same as it would have been in the first load cycle'. This means that, despite cyclic weakening, the material stiffness with respect to variations in stress amplitude remains unchanged, which seems somewhat improbable.

All models discussed above are based on the use of one single memory parameter k, although in some cases the material also 'remembers' the previous maximum deviatoric load. In one case (Prévost 1977) it also has a detailed memory for the direction of past loading cycles, as it manifests itself in the shifted pattern of a system of nested yield circles. In one recent model by Cuellar et al. (1977) two memory parameters are used, one for deviatoric strain and one for volumetric strain (van Eekelen & Potts 1978b). The model satisfies Miner's law for strain cycles; the deviatoric stress-strain relation is of the 'endochronic' type, and thus describes damping. Some objections against endochronic models (non-uniqueness and instability) have recently been raised by Sandler (1978).

CONCLUSION

To describe the cumulative effect of large numbers of loading cycles on the mechanical response properties of soil, one needs to introduce a memory parameter which takes account of past cyclic loading. If this memory parameter is identified with an observable of the system, the basic functions (1) and (2) may be determined relatively easily. The most obvious choice for the memory parameter is the plastic volume strain or (for undrained loading) the cumulative pore pressure generated by cyclic loading. An alternative is to use the total length of the deviatoric strain path. Unless experimental data clearly indicate otherwise it may be useful to formulate the model in such a way that Miner's rule of superposition applies, either for strain cycles or for stress cycles.

REFERENCES

Andersen, K.H. 1976, Behaviour of clay subjected to undrained cyclic loading. Proc. Conf Behaviour of Offshore Structures, Trondheim, I, 392-403.
Andersen, K.H., Selnes, P.B., Rowe, P.W. & Craig, W.H. 1979, Prediction and Observation of a Model Gravity Platform on Drammen clay. Proc. second Int. Conf. Behaviour of Offshore Structures, London, I, 427-446.
Bonin, J.P., Deleuil, G. & Zaleski-Zamenhof, L.C. 1976, Foundation Analysis of marine Gravity Structures submitted to Cyclic Loading. Offshore Techn. Conf., OTC 2475,

Dallas, Texas.

Cuellar, V., Bazant, Z.P., Krizek, R.J. & Silver, M.L. 1977, Densification and Hysteresis of sand under cyclic shear. Jnl. Geot. Eng. Div., ASCE, 103, 399-416.

Doyle, E.H. 1978, Free-field soil stability under cyclic loading. Offshore Engineering Conference, Houston, Texas.

Eekelen, H.A.M. van 1977, Single parameter models for progressive weakening of soils by cyclic loading. Géotechnique 27, 357-368.

Eekelen, H.A.M. van, & Potts, D.M. 1978, The behaviour of Drammen clay under cyclic loading. Géotechnique 28, 173-196.

Eekelen, H.A.M. van, & Potts, D.M. 1978b, Clay behaviour under cyclic loading. Dynamical methods in soil and rock mechanics, Karlsruhe.

Idriss, I.M., Dobry, R., Doyle, E.H. & Singh, R.D. 1976, Behaviour of soft clays under earthquake loading conditions. Offshore Techn. Conf., OTC 2671, Dallas, Texas.

Idriss, I.M., Dobry, R. & Singh, R.D. 1978, Nonlinear behaviour of soft clays during cyclic loading. Jnl. Geot. Eng. Div., ASCE, 104, 1427-1447.

Masing, G. 1926, Eigenspannungen und Verfestigung beim Messing. Proc. sec. Int. Congr. of Applied Mechanics, 332-335.

Martin, G.R., Finn, W.D.L. & Seed, H.B. 1975, Fundamentals of liquefaction under cyclic loading. Jnl. Geot. Eng. Div., ASCE, 101, 423-438.

Miner, M.A. 1945, Cumulative damage in fatigue. Trans. ASME, 67, A159-164.

Mroz, Z., Norris, V.A. & Zienkiewicz, O.C. 1978, An anisotropic hardening model for soils and its application to cyclic loading. Int. Jnl. for Num. Anal. Meth. in Geomech. 2, 203-221.

Mroz, Z. Norris, V.A. & Zienkiewicz, O.C. 1979, Application of an anisotropic hardening model in the analysis of elasto-plastic deformation of soils. Géotechnique 29, 1-34.

Prévost, J.H. 1977, Mathematical modelling of monotonic and cyclic undrained clay behaviour. Int. Jnl. for Num. Anal. Meth. in Geomech. 1, 195-216.

Pyke, R. 1979, Nonlinear soil models for irregular cyclic loading. Jnl. Geot. Eng. Div., ASCE, 105, 715-726.

Sandler, I.S. 1978, On the uniqueness and stability of endochronic theories of material behavior. Jnl. Appl. Mech. 45, 263-266.

Seed, H.B. 1979, Soil liquefaction and cyclic mobility evaluation for level ground during earthquakes. Jnl. Geot. Eng. Div., ASCE, 105, 201- 255.

Silver, M.L. & Seed, H.B. 1971, Volume changes in sand during cyclic loading. Jnl. Soil Mech. Fdn Div., ASCE, 97, 1171-1182.

Zienkiewics, O.C., Chang, C.T. & Hinton, E. 1978, Non-linear seismic response and liquefaction. Int. Jnl. Num. Anal. Meth. Geomech. 2, 381-404.

450

A micro-structural model for soils under cyclic loading

G.N. PANDE & K.G. SHARMA
University College of Swansea, Swansea, UK

INTRODUCTION

A soil sample when subjected to a number of cycles of stress fails at a stress level lower than its monotonic failure stress. It appears that five factors contribute to this behaviour in varying degrees depending upon the nature of the soil. These factors are (a) generation of pore pressures (b) cyclic degradation or fatigue (c) cyclic strain softening (d) ratchetting or progressive deformation (e) cyclic rotation of principal axes of stress. It may not be appreciated that the last two items relate to a structure and do not arise in an ideal conventional tri-axial compression experiment but may have an important influence as the stress state in the sample is hardly ever truly homogenous (Lee and Vernese 1978). While the aspects of generation of pore pressure and rotation of principal axes are peculiar to soils, theories already exist for cyclic fatigue and ratchetting/shakedown in other areas of engineering science. In recent years a number of constitutive models of soils have been proposed. The main line of approach has been to treat soil as an elasto/plastic material, define mathematical expressions for yield surface, flow and hardening rules and usually introduce one of the factors discussed above to best fit the experimental data.

Among others the notable contributions have been made by Schofield and Wroth (1968), Zienkiewicz, Humpheson and Lewis (1975), Lade and Duncan (1975), Mroz, Norris and Zienkiewicz (1978, 1979), Van Eekelen (1978), Prevost and Hoeg (1975), Prevost and Hughes (1978), Nova and Wood (1979) and a wide variety of models for sands and clays under cyclic loading are now

available. Some of these models have been developed to the extent that practical boundary value problems can be solved using numerical procedures like finite element while others are confined to tri-axial situations only either due to inadequacy of formulations or computational complexities. None of the models are entirely satisfactory in all respects and do not take into account the rotation of principal axes of stress tensor which invariably take place to some extent in most problems. They do not account for the strength anisotropy induced due to plastic flow and this induce an error of unknown magnitude making it impossible to assess the prediction even in simpler bench mark problems.

This paper presents a micro-structural model which in principle can incorporate all the five factors. However, though the mathematical formulation is complete, attention is devoted here to two aspects only viz. rotation of principal axes and cyclic strain softening.

The proposed model has conceptual links with the polycrystalline models of plasticity which were proposed a few decades ago by Taylor (1938), Batdorf and Budiansky (1949) and Sanders (1955) and others. It is also a numerical implementation and further extension of 'micro-structural' view of the clays as proposed by Calladine (1971). The first part of the paper lays the mathematical foundations of the model, the second part is devoted to the details of numerical implementation and solution of bench mark problems like a tri-axial test in monotonic loading and the third demonstrates the characteristics of the model in cases of cyclic loading.

The influence of the rotation of principal axes in the model in comparison to the conventional Critical State Model is demonstrated.

The software development involved in this model is extremely simple and many relatively complex models incorporating other factors like cyclic deformation, pore pressure generation as in transient loadings can be readily incorporated.

Part I

A SIMPLE CONCEPTUAL MODEL OF PARTICULATE MEDIA

Let us consider a solid block of arbitrary shape of homogenous isotropic, linear elastic material intersected by K number of randomly oriented planes. These planes render the solid block into an assemblage of perfectly fitting polyhedral blocks (Fig.1). Let us further assume that by some process the boundaries of the microscopic polyhedral blocks are roughened creating asperities and spot welded without inducing any stresses in the block. We would not go into details of justification for this rather simple conceptual model but it is clear that the key to the deformational behaviour of the block lies in an accurate description of the sliding phenomenon under the current effective normal and shear stresses (σ'_n, τ) along the boundaries of the polyhedral blocks and opening/closing of the inter-boundary gap (void ratio) in relation to the initial gap (initial void ratio) caused by the process of creation of asperities. Calladine (1971) has used this model to explain mechanical properties of saturated clays and has obtained remarkably good comparison with experimental data. A similar model has been used by Zienkiewicz and Pande (1977a) and Pande (1977) for solving practical problems of jointed rock masses although their main purpose was to model strong strength anisotropy of rocks developed due to the presence of 2 or 3 sets of families of planes of weakness. It is apparent that as the number of contact boundaries approaches infinity the isotropy of the material is restored.

The philisophy of this conceptual model is analogous to the finite element procedure. We look in detail the mathematical expressions involved in accurately describing the behaviour of one contact surface and assume that same relationships hold good for all contact surfaces, the global behaviour being obtained by integration of contributions made by each contact surface.

ASSUMPTIONS AND MATHEMATICAL FORMULATION

We restrict ourselves to the case of small strains and deformation only. We assume that the behaviour of soil grains is purely elastic while that of the contact boundaries is elasto/visco-plastic. We have developed this model using elasto/visco-plasticity assumptions. However, the elasto/plasticity assumptions are equally valid. The implication of this assumption is that when a soil element is subjected to any load, the stresses manifested are instantaneously elastic. We further assume that total strain (ε) can be split into (a) an elastic component (ε^e), (b) a visco-plastic component (ε^{vp}) and (c) an initial strain component (ε^o) which is nonstructural in origin

$$\underset{\sim}{\varepsilon} = \underset{\sim}{\varepsilon}^e + \underset{\sim}{\varepsilon}^{vp} + \underset{\sim}{\varepsilon}^o \qquad (1)$$

and in incremental form

$$d\underset{\sim}{\varepsilon} = d\underset{\sim}{\varepsilon}^e + d\underset{\sim}{\varepsilon}^{vp} + d\underset{\sim}{\varepsilon}^o \qquad (2)$$

We develop our mathematical framework suitable for effective stress analysis. Thus we assume that the total stress applied on a saturated soil element can be split into two components viz a component which is carried by the soil grains (effective stress (σ') and the other carried by the pore water (pore water pressure (p)). Thus,

$$\underset{\sim}{\sigma} = \underset{\sim}{\sigma}' + \underset{\sim}{m}p \qquad (3)$$

and in incremental form

$$d\underset{\sim}{\sigma} = d\underset{\sim}{\sigma}' + \underset{\sim}{m}dp \qquad (4)$$

where $\underset{\sim}{m}^T = (1,1,1,0,0,0)$

and $\underset{\sim}{\sigma}^T = (\sigma_x, \sigma_y, \sigma_z, \tau_{xy}, \tau_{yz}, \tau_{zx}) \qquad (5)$

The increments of effective stress are related to various strain components by

$$d\underset{\sim}{\sigma}' = \underset{\sim}{D}_T (d\underset{\sim}{\varepsilon} - d\underset{\sim}{\varepsilon}^{vp} - d\underset{\sim}{\varepsilon}^o) \qquad (6)$$

where $\underset{\sim}{D}_T$ is the tangential drained elasticity matrix. If the soil grains are assumed as incompressible then for the undrained condition any volumetric strain is related to the increment (or decrement) of pore pressure by

$$dp = \frac{K_f}{\eta} d\varepsilon_y \qquad (7)$$

where $d\varepsilon_v$ is the increment of volumetric strain

K_f is the bulk modulus of the fluid

η is the porosity of the soil

Equations (1) to (7) are fundamental postulates of soil plasticity and have been extensively used by many researchers (Naylor 1975 , Humpheson 1976 and Chang 1979). In the theory of elasto/visco-plasticity it is assumed that rate of visco-plastic strain is given by

$$\dot{\underset{\sim}{\varepsilon}}^{vp} = \gamma < \Phi\ (F)\ > \frac{\partial Q}{\partial \underset{\sim}{\sigma}} \qquad (8)$$

where γ is the fluidity parameter

F is the scalar yield function

Φ represents a monotonic function of F

Q is the scalar plastic potential function

<> indicates that $\Phi(F) = \Phi(F)$ if $F > 0$
$= 0$ if $F \leqslant 0$

Most plasticity based models assume that soil is isotropic and thus the scalar yield function (F) is a function of principal stresses $(\sigma_1, \sigma_2, \sigma_3)$ alone and not their directions. Yield functions are written in terms of invariant of stress (Nayak and Zienkiewicz 1972, Zienkiewicz and Pande 1977b) thus precluding the possibility of anisotropic behaviour. On the other hand it is well-known (Drucker 1966) that originally isotropic soils become strongly anisotropic after some plastic flow. To model this behaviour we will express the yield function in terms of certain stress components rather than invariants.

We formulate the yield function for K contact boundaries encompassing a point in our conceptual model. Considering i^{th} plane the effective normal stress (σ_n') and shear stress (τ) can be written as

$$\left| \begin{matrix} \sigma'_n \\ \tau \end{matrix} \right| = T_i \underset{\sim}{\sigma} \qquad (9)$$

Where T_i represents a transformation matrix which is a function of the direction cosines of the unit normal to the ith plane or contact boundary. The current yield function can now be expressed as

$$F_i = f(\sigma'_n, \tau, \underset{\sim}{\varepsilon}^{vp})_i = 0 \qquad (10)$$

Equation (10) represents the conditions of sliding on the ith contact plane. If an associated form of flow is assumed then from equation (8) the contribution made by the ith contact plane to the $\dot{\underset{\sim}{\varepsilon}}^{vp}$ vector (partial visco-plastic strain rate $\dot{\underset{\sim}{\varepsilon}}^{vp}_i$), is

$$\dot{\underset{\sim}{\varepsilon}}^{vp}_i = \gamma < \Phi\ (F_i)\ > \frac{\partial F_i}{\partial \underset{\sim}{\sigma}} \qquad (11)$$

To obtain the expression for the visco-plastic strain rate, we sum up the partial viscoplastic strain rates for all the K contact planes and let $K \to \infty$ Thus,

$$\dot{\underset{\sim}{\varepsilon}}^{vp} = \sum_{i=1}^{K} \gamma < \Phi(F_i)\ > \frac{\partial F_i}{\partial \underset{\sim}{\sigma}} \qquad (12)$$

Alternatively if we assume a continuous random distribution of contact planes, the discrete summation in equation (12) can be replaced by integration over the solid angle subtended by a sphere of unit radius at its centre. Thus,

$$\dot{\underset{\sim}{\varepsilon}}^{vp} = \int_{\Omega} \gamma < \Phi\ (F)\ > \frac{\partial F}{\partial \underset{\sim}{\sigma}}\ d\Omega \qquad (13)$$

where $d\Omega$ represents an infinitismal solid angle. Equations (12) and (13) show that a visco-plastic strain rate equation can be derived from the basic mechanism of sliding on the contact planes which is more fundamental in nature. Once we specify the equation (10) in explicit terms of effective normal stress, shear stress and visco-plastic strains, the flow equation of soil behaviour is automatically defined and can be computed.

If the form of the equation (10) is chosen according to ideal plasticity theory (no hardening/softening) then it reduces to

$$F_i = f(\sigma'_n, \tau)_i = 0 \qquad (14)$$

The yield function for the soil material is not required in this model as all computations can be done for randomly oriented contact boundaries. However, it can be expressed as follows

$$F = F_1 \cap F_2 \cap \dots F_i \cap \dots F_k = 0 \qquad (15)$$

This yield function is isotropic for $K \to \infty$ and for ideal plasticity (equation 14) since if the axes of principal stresses are rotated, values of F for each plane would change but the sum total would remain unaltered. However, if hardening/softening is considered, preferential hardening/softening would take place based on the direction of principal stresses and their subsequent rotation during the process of incremental loading and plastic flow.

SPECIFIC FORMS OF YIELD FUNCTIONS IN $\sigma'_n - \tau$ SPACE

A key to the constitutive modelling in the

proposed framework is the yield function in $\sigma_n' - \tau$ space. Much experimental as well as theoretical work (Bowden and Tabor 1950, 1956, Tabor 1951, 1959) has been done chiefly in tribology to explain the behaviour of one metal plate sliding over the other. Here we will keep our approach to that of simple model parameters which can be derived from conventional soil test data. In the simplest form the yield function in σ_n', τ space (Figure 2) can be written as:

$$F_i = |\tau| + \sigma_n' \tan\phi' - c' = 0 \qquad (16)$$

which is a representation of classical Mohr-Coulomb yield function. In equation (16) ϕ' is the apparent angle of friction and c' is the apparent cohesion. $\phi' = 0$ case is included in equation (15) when it becomes

$$F_i = |\tau| \qquad - c' = 0 \qquad (17)$$

Equation (16) when substituted in equation (12) leads to

$$\dot{\varepsilon}^{vp} = \sum_{i=1}^{K} \gamma < \Phi \ (|\tau| + \sigma_n' \tan\phi' - c')_i > \frac{\partial F_i}{\partial \sigma} \qquad (18)$$

and can be conveniently used in computation and as has been shown elsewhere (Pande and Sharma 1979) leads to exactly the same results as obtained by the use of Mohr-Coulomb yield function in invariant form.

A critical state model accounting for rotation of principal stress axes and load cycling: Critical state model which interprets isotropic strain hardening was originally developed for normally consolidated clays by Schofield and Wroth (1968) and there are now a number of variants of this model (Naylor 1975 , Roscoe and Burland 1968). Nova and Wood (1979) have extended the concepts of this model to sands. Gerogiannopoulos and Brown (1978) have attempted to demonstrate that this model is applicable to rocks as well. All authors have used an invariant formulation to the yield function. Here we shall specify the yield function for each randomly oriented plane in $\sigma_n' - \tau$ space. This would induce anisotropic hardening. Moreover to account for the cycling of load we shall introduce cyclic strain softening. In our model cyclic strain softening takes place as a function of cumulative value of absolute plastic strains on each of the planes. Thus the yield function and strain hardening/softening rules are given by

(i) On the wet side $(-\sigma_n' - p_c^j \geqslant 0)$

$$F_i = \left(\frac{\tau}{\mu p_c^j}\right)^2 + \left(\frac{-\sigma_n' - p_c^j}{p_c^j}\right)^2 - 1 = 0 \qquad (19)$$

where $\mu = \tan\phi'$

$$p_c^j = p_{co}^j \ e^{\chi \varepsilon_n^{vp}}$$

in which ε_n^{vp} = normal component of the visco-plastic strain on the ith plane

p_c^j = current value of pre-consolidation pressure for the jth cycle

p_{co}^j = initial isotropic pre-consolidation pressure for the jth cycle

χ = hardening parameter

$\qquad = \dfrac{1 + e_o}{\lambda - \kappa} \qquad (20a)$

where e_o = initial void ratio

λ, κ = compression and swelling indices respectively

To account for the influence of the cyclic softening the value of p_{co} for the jth cycle (p_{co}^j) is defined as

$$p_{co}^j = p_{co}^{j-1} (K_1 - K_2 e^{\frac{-\varepsilon_n^{vp}}{\gamma^{vp}}})^\beta \qquad (20b)$$

where K_1, K_2, β are constants and γ^{vp} is the absolute value of the cumulative visco-plastic shear strain on the ith plane.

(ii) On the dry side $(-\sigma_n' - p_c^j < 0)$

$$F_i = |\tau| + \sigma_n \tan\phi_o - \tau_o = 0 \qquad (21a)$$

where ϕ_o, τ_o are experimented constants defining peak strength. and a non-associative flow rule is used i.e.

$$Q_i = |\tau| p_c^j + (-\sigma_n - p_c^j)^2 \tan\phi' = 0 \qquad (21b)$$

where Q_i represents the plastic potential function. The form of this function is such that it ensures the continuity of strain rates at $-\sigma_n = p_c^j$ which is important

454

from computational point of view.

Figure (3) shows the yield function represented by equations (19) and (21).

It will be noted that there is an important difference between this model and the conventional critical state model. In the proposed model the evaluation of yield loci is traced individually for each plane which is a function of effective normal stress, shear stress and normal component of visco-plastic strain on that plane. The effective normal stress and shear stress on a plane are in turn a function of not only magnitude of principal stresses but of their directions as well. If loading is such that the direction of principal stresses remain fixed (such is the case with tri-axial tests), the hardening/softening progresses on some fixed planes and leads to isotropic hardening. On the other hand if the direction of principal stresses rotates during loading only selective hardening/softening will take place involving anistropy induced by plastic flow. The utmost simplicity is the main advantage of the proposed model which we shall term as 'critical state model accounting for rotation' (CSMAR) of principal stresses and cyclic softening.

PART II

NUMERICAL IMPLEMENTATION

The proposed numerical framework for investigating the constitutive relationship starting with a definition of yield surface on $\sigma_n' - \tau$ plane, flow rule and hardening rule is extremely simple in computer programming.

The summation implied by equation (12) is done numerically. A number of trial and error studies have led to a number of contact planes surrounding a point being chosen as 10 in one octant. Another 10 planes are taken as mirror image in the adjacent octant making K = 20. Due to symmetry only one quadrant of upper or lower hemisphere need be considered. The details of direction cosines of these planes and associated transformation matrices etc. are given in Appendix I. The computations involved are made once for all and stored. The equation (12) also involves the computations of the derivatives of σ_n' and τ with respect to the stress vector. The associated algebra is indicated in Appendix II. Once the $\dot{\varepsilon}^{vp}$ is computed for each Gauss point based on the contribution of 20 planes, the standard visco-plastic algorithm (Cormeau 1976, Zienkiewicz 1977) is adopted

involving repeated solution of   e equations of the type

$$ \underset{\sim}{K} \, \underset{\sim}{\delta} \quad - \int_{v} \underset{\sim}{B}^{T} \, \underset{\sim}{D} \, \underset{\sim}{\varepsilon}^{vp} \; dv = \underset{\sim}{R} \qquad (22) $$

where $\underset{\sim}{K}$ is the overall stiffness matrix of the system, $\underset{\sim}{D}$ is the drained elasticity matrix and the right hand side term $\underset{\sim}{R}$ includes the influence of initial effective stress, initial pore water pressure, body forces, imposed tractions etc. $\underset{\sim}{B}$ is the strain matrix in

$$ \underset{\sim}{\varepsilon} = \underset{\sim}{B} \, \underset{\sim}{\delta} \qquad (23) $$

δ being the nodal displacement vector of the system.

NUMERICAL EXAMPLES

The numerical examples in this section are designed with following aims in mind

(a) To show that with the fixed directions of principal stresses, and monotonic loading, the proposed model gives same results as the conventional critical state model but in truly tri-axial situation in contrast to critical state model the proposed model (CSMAR) gives different results when axes of principal stresses are rotated.

(b) To show that with proper choice of cyclic softening parameters experimentally observed behaviour of pore pressure generation can be simulated. At low intensities of stress it leads to purely elastic behaviour which at higher intensities, the failure takes after a number of cycles which can be controlled by the constant β.

EXAMPLE 1: TRI-AXIAL TEST UNDER MONOTONIC
            LOADING

A tri-axial compression test under isotropic consolidation is simulated adopting the following data:
   $E = 210000$ kN/m^2
   $\nu = 0.3$
   $p_{co} = 280$ kN/m^2
   $M = 0.772$
   $\chi = 700 (7000$ for CSMAR)
   $\phi' = 20°$
Computations are made using the conventional critical state model (CSM) and the proposed model (CSMAR). Figure 4 shows the (σ_1, σ_3) vrs. axial strain (ε_1) and volumetric strain vrs. ε_1 under drained conditions. Figure 5 shows the $(\sigma_1 - \sigma_3)$ vrs. ε_1 and excess pore pressure versus ε_1 for the undrained case. The comparison shows that results from the CSMAR are nearly the same as from CSM. However, if the direction

of principal stress is rotated through 180°- as discussed in example 2 at some point in the stress path (Point z in Figs.4 and 5) the stress-strain curve for CSMAR model would be affected while that of CSM would remain unchanged. Figure 6 shows the comparison of $(\sigma_1-\sigma_3)$ vrs.$\varepsilon_d(=\varepsilon_1-\frac{\varepsilon v}{3})$ curves for CSM and CSMAR with experimental results of Boulder clay (Naylor 1975, Jones 1975). The comparison of $(\sigma_1-\sigma_3)$ vrs.ε_d is very good while that ε_v vrs. ε_d is reasonable

EXAMPLE 2: A TRULY TRI-AXIAL TEST

In order to show the response of the model to rotation of principal axes, an imaginary truly tri-axial test is devised which could in principle be performed in laboratory. It is imagined that a clay specimen is isotropically or anistropically consolidated to a specified degree. One of the principal stresses is then incrementally raised keeping the other two principal stresses equal and unchanged. After a few increments, a system of stresses are successively applied in such a manner that magnitude of principal stresses are unchanged but their directions are progressively rotated in the plane passing through major and minor stress as shown in Figure 7.

Figure 8 shows the total and effective stress paths followed by the clay element in consolidated drained and undrained tests. The final stress situation at which the rotation of principal stress axes is induced is shown by a large dot. Computations for plastic strains were made using the same data as given in Example 1.

Two parameters r and s defined as below were studied

$$r = \frac{\varepsilon_{II}^{vp} \text{ at } \alpha = \theta}{\varepsilon_{II}^{vp} \text{ at } \alpha = 0} \qquad (24)$$

$$s = \frac{\varepsilon_{v}^{vp} \text{ at } \alpha = \theta}{\varepsilon_{v}^{vp} \text{ at } \alpha = 0} \qquad (25)$$

where ε_{II}^{vp} = second invariant of deviatoric visco-plastic strains

$$= \sqrt{2((\varepsilon_x^{vp})^2+(\varepsilon_y^{vp})^2+(\varepsilon_z^{vp})^2)+\gamma_{xy}^2}$$

ε_v^{vp} = volumetric visco-plastic strains

θ = the angle which the current major principal stress axis makes with the initial position

Figures (9) and (10) show variation of r and s with θ for consolidated drained and consolidated undrained tests for various K_o conditions.

These results clearly indicate that influence of rotation of principal axis could be considerable.

r and s which are obviously constant and equal to 1 for critical state model vary up to nearly 100% in some cases when CSMAR model is used.

EXAMPLE 3: 'CSMAR' IN CYCLIC TRIAXIAL COMPRESSION TEST

This example shows the behaviour of the model in cyclic triaxial compression test. The parameters chosen for the soil sample are as follows:

E = 3840 kN/m^2
ν = 0.3 $\qquad \lambda$ = 0.25
ϕ' = 30° $\qquad \kappa$ = 0.05
P_{co} = 40 kN/m^2 $\qquad e_o$ = 1.0
χ = 100

With these parameters, the monotonic failure stress under consolidated undrained conditions is given by q = σ_1', - σ_3' = 54·0 kN/m^2 and $p=\frac{\sigma_1'+2\sigma_3'}{3}$ =44·5 kN/m^2. Cyclic

loading test was performed with q = The value of other constants related to cyclic loading are taken as follows:

K_1 = 0.95 $\qquad \beta$ = 0.025
K_2 = 0.5

The effective stress path followed by the specimen is shown on a q-p plot in Fig. 11. Fig. 12 shows the pore pressures vrs. the number of cycles. Fig. 13 shows the q vrs. deviatoric strain curve, deviatoric strain being defined as $\frac{2}{3}$ $(\varepsilon_1-\varepsilon_3)$ where ε_1 and ε_3 are axial and radial strains.

It may be noted that in this model the generation of pore pressures is initially very gradual. The stress-strain and pore pressure generation curves closely follow the experimentally observed pattern and the cyclic strain softening is dependent on the level of cumulative cyclic plastic strains on individual planes.

CONCLUSIONS

A micro-structure model responding to the rotation of principal axes and accounting for the cyclic strain softening has been presented. It can be further extended to account for cyclic degradation. A non-structural autogenous strain component has been retained in the formulation to model transient loading conditions. The proposed model is conceptually very simple and can be incorporated in the finite element code very easily. Further work on the model is continuing and would be presented at the symposium.

REFERENCES

Batdorf, S.B. and Budiansky, B. 1949, "A Mathematical Theory of Plasticity based on the concept of slip" National Advisory Committee for Aeronautics, TN1871

Bowden, F.P. and Tabor, D. 1950, "The friction and lubrication of solids" Oxford University Press.

Bowden, F.P. and Tabor, D. 1956, "Friction and Lubrication" London - Methuen.

Calladine, C.R. 1971, "A microstructural view of the mechanical properties of saturated clay", Geotechnique, Vol.21, 391-415.

Chang, C.T. 1979, "Non-linear response of earth dams and foundations in earthquakes", Ph.D. Thesis, University of Wales, Swansea.

Cormeau, I.C. 1976, "Viscoplasticity and plasticity in the finite element methods" Ph.D. Thesis, University of Wales, Swansea.

Drucker, D.C. 1966, "Concepts of path independence and material stability for soils" Proc. IUTAM Symp. on Rheology and Soil Mech., Grenbole, 1964 23-46, Springer-Verlag, Berlin.

Cerogiannopoulos, N.G. and Brown, E.T. 1978, "The critical state concept applied to rock", Int. J. Rock Mech. Min. Sci. & Geomech., Vol.15, 1-10

Humpheson, C. 1976, "Finite element analysis of elasto-viscoplastic soils" Ph.D. Thesis, University of Wales, Swansea.

Jones, D.B. 1975, "A Non-linear elastic anisotropic analysis of Llyn Brianne Dam by finite element", Ph.D. Thesis, University of Wales, Swansea.

Lade, P.V. and Duncan, J.M. 1975, "Elasto-plastic stress-strain theory for cohesionless soil" J. Geotech. Engg. Div. ASCE, Vol.101, No. GT10, 1037-1053.

Lee, K.L. and Vernese, F.J. 1978, "End restraint effects on cyclic tri-axial strength of sand", J.Geotechn. Engg. Div., ASCE, Vol. 104, No.GT6, 705-719.

Mroz, Z., Norris, V.A. and Zienkiewicz,O.C. 1978, "An anisotropic hardening model for soils and its application to cyclic loading". Int. J. Num. Anal. Meth. Geomech., Vol. 2, 203-221.

Mroz, Z., Norris, V.A. and Zienkiewicz, O.C. 1979, "Application of an anisotropic hardening model in the analysis of elasto-plastic deformation of soils", Geotechnique, Vol. 29, 1-34.

Nayak, G.C. and Zienkiewicz, O.C. 1972, "Convenient form of stress invariants for plasticity", Proc. ASCE, Vol. 98, 949-954.

Naylor, D.J. 1975, "Non-linear finite element models for soils", Ph.D. Thesis, University of Wales, Swansea.

Nova, R. and Wood, D.M. 1979, "A Constitutive Model for Sand in Tri-axial Compression", Int. J. Num. Anal. Meth. Geomech. Vol. 3, 255-278.

Pande, G.N. 1977, "Non-linear finite element analysis of jointed rock masses" Ph.D. Thesis, University of Wales, Seansea.

Pande,G.N. and Sharma, K.G. 1979, "A numerical framework for the development of constitutive relations of soils" to be published.

Prevost, J.H. and Hoeg, K. 1975, "Effective stress-strain strength models for soils" J. Goetech. Engg. Div., ASCE, Vol. 101, No. GT3, 259-278.

Prevost, J.H. and Hughes, T.J.R. 1978, "Analysis of gravity offshore structure foundations subjected to cyclic wave loading", Proc. 10th Annual Offshore Technology Conference in Houston, Texas.

Roscoe, K.H. and Burland, J.B.1968, "On the generalized stress-strain behaviour of wet clay" in Engineering Plasticity (ed. J. Heyman and F.A. Leckie), 535-605 Cambridge University Press.

Sanders, J.L. Jr., 1955, "Plastic stress-strain relations based on linear loading functions" Proc. 2nd U.S. Nat. Congr. Appl. Mech., Am. Soc. Mech. Engrs., 455-460.

Schofield, A.N. and Wroth, C.P. 1968, "Critical State Soil Mechanics" McGraw-Hill, London

Tabor, D. 1951, "The hardness of metals" Oxford University Press.

Tabor, D. 1959, "Junction growth in metallic friction: the role of combined stresses and surface contamination", Proc. Roy. Soc., Series A251, 378-393.

TAYLOR, G.I. 1938, "Plastic strains in metals", J. Inst. Metals, Vol. 62, 307-324.

Van Eekelen, H.A.M. and Potts, D.M. 1978, "The Behaviour of Drammen clay under cyclic loading", Geotechnique, Vol. 28, 173-196

Wiebols, G.A. and Cook, N.G.W. 1968
"An energy criterion for the strength
of rock in polyaxial compression"
Int. J. Rock Mech. Min. Sci., Vol.5
529-549

Zienkiewicz, O.C. 1977, "The finite
element method", McGraw-Hill Book
Company (U.K.) Ltd., London.

Zienkiewicz, O.C., Humpheson, C. and
Lewis, R.W. 1975, "Associated and
Non-Associated Visco-plasticity and
Plasticity in Soil Mechanics"
Geotechnique, Vol.25, 671-689.

Zienkiewicz, O.C. and Pande, G.N. 1977a
"Time dependent multilaminate model
of rocks - a numerical study of deformation and failure of rock masses"

Int. J. Num. Anal. Meth. Geomech., Vol.1
219-247.

Zienkiewicz, O.C. and Pande, G.N. 1977b
"Some useful forms of isotropic yield
surfaces for soil and rock mechanics"
Finite Elements in Geomechanics (ed.
G. Gudehus), 179-190, John Wiley & Sons
London.

APPENDIX I

For the purpose of subdivision of unit octant, the summation points are considered over the surface of the unit sphere in a regular pattern. Let each point be specified by its longitudinal angle θ_o and polar angle ψ (Fig.I.1) as follows (Wiebols and Cook 1968)

$$\psi = \sqrt{\frac{\pi}{2K}} (a + \tfrac{1}{2}) \qquad a = 0,1,2\ldots \quad (I.1)$$

$$\theta_o = \frac{1}{\sin\psi} \sqrt{\frac{\pi}{2K}}(b + \tfrac{1}{2}) \quad b = 0,1,2\ldots$$

where a and b are assigned integer values between the limits

$$0 < a \leqslant \sqrt{\frac{\pi K}{2}} - \tfrac{1}{2}$$
$$0 < b \leqslant \sqrt{\frac{K \sin\psi}{2}} - \tfrac{1}{2} \qquad (I.2)$$

Then the direction cosines corresponding to each point can be written as

$$\begin{aligned}\ell &= \cos\psi \\ m &= \sin\psi \cos\theta_o \\ n &= \sin\psi \sin\theta_o\end{aligned} \qquad (I.3)$$

and $\sigma_n{}'$ and τ, on the plane whose normal has direction cosines given by equation (I.3), is given by

$$\sigma_n{}' = \ell p_x + m p_y + n p_z \qquad (I.4)$$
$$\tau = \sqrt{p_x^2 + p_y^2 + p_z^2 - \sigma_n{}'^2}$$

when $p_x = \ell\sigma_x{}' + m\tau_{xy}$, $p_y = \ell\tau_{xy} + m\sigma_y{}'$, $p_z = n\sigma_z{}'$ (I.5)

It is to be kept in mind that equation (I.5) applies only to plane strain and axisymmetric conditions.

APPENDIX II

Substituting equation (I.4) in equation (16) we get

$$F_i = \sqrt{p_x^2 + p_y^2 + p_z^2 - \sigma_n{}'^2} + \mu\sigma_n{}' - c' = 0 \quad (II.1)$$

Differentiating (II.1) with respect to $\underset{\sim}{\sigma}$ and using equation (I.5) we get

$$\frac{\partial F_i}{\partial \sigma_x} = \frac{\ell p_x - \ell^2(\sigma_n{}' - \mu\tau)}{\tau}$$

$$\frac{\partial F_i}{\partial \sigma_y} = \frac{m p_y - m^2(\sigma_n{}' - \mu\tau)}{\tau} \qquad (II.2)$$

$$\frac{\partial F_i}{\partial \sigma_z} = \frac{n p_z - n^2(\sigma_n - \mu\tau)}{\tau}$$

$$\frac{\partial F_i}{\partial \tau_{xy}} = \frac{\ell p_y + m p_x - 2\ell m(\sigma_n{}' - \mu\tau)}{\tau}$$

and if equation (I.4) is substituted in equation (19) we get

$$F_i = \sqrt{p_x^2 + p_y^2 + p_z^2 - \sigma_n{}'^2 + \mu^2(-\sigma_n{}' - p_c^j)^2} - \mu p_c^j = 0 \qquad (II.3)$$

Differentiating II.3 with respect to $\underset{\sim}{\sigma}$ and using equation (I.5) we get

$$\frac{\partial F_i}{\partial \sigma_x} = \frac{\ell p_x - \ell^2(\sigma_n{}' + \mu^2(-\sigma_n{}' - p_c^j))}{x}$$

$$\frac{\partial F_i}{\partial \sigma_y} = \frac{m p_y - m^2(\sigma_n{}' + \mu^2(-\sigma_n{}' - p_c^j))}{x}$$
$$\qquad (II.4)$$

$$\frac{\partial F_i}{\partial \sigma_z} = \frac{n p_z - n^2(\sigma_n{}' + \mu^2(-\sigma_n{}' - p_c^j))}{x}$$

$$\frac{\partial F_i}{\partial \tau_{xy}} = \frac{\ell p_y + m p_x - 2\ell m(\sigma_n{}' + \mu^2(-\sigma_n{}' - p_c^j))}{x}$$

where

$$x = \sqrt{p_x^2 + p_y^2 + p_z^2 - \sigma_n{}'^2 + \mu^2(-\sigma_n{}' - p_c^j)^2} \quad (II.5)$$

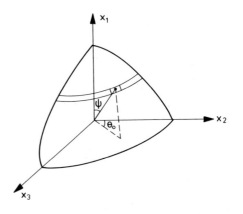

FIG I.1 : Spherical polar coordinate system.

459

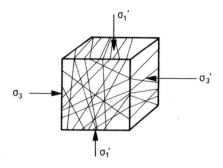

Fig.1 A block of isotropic elastic material intersected by a large number of randomly oriented planes

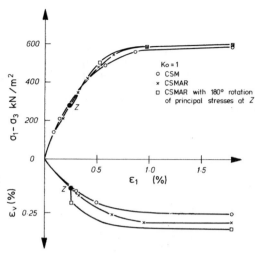

Fig.4 Isotropically consolidated drained triaxial test

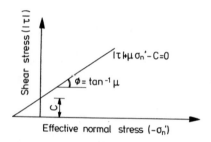

Fig.2 Mohr-Coulomb yield function in $\sigma_n' - \tau$ space

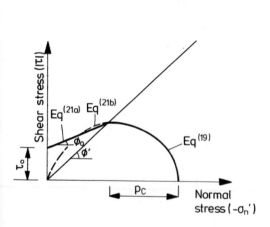

Fig.3 Critical state yield criterion in $\sigma_n' - \tau$ space (CSMAR)

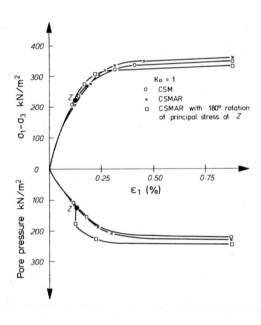

Fig. 5 Isotropically consolidated undrained triaxial test

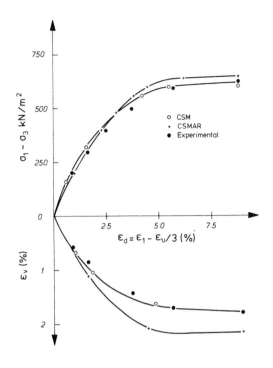

Fig.6 Consolidated drained triaxial test on Boulder Clay

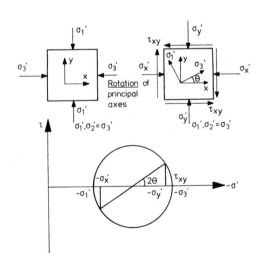

Fig. 7 Rotation of principal stress system in an imaginary test

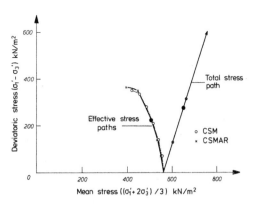

Fig. 8 Total and effective stress paths for isotropically consolidated drained and undrained triaxial tests

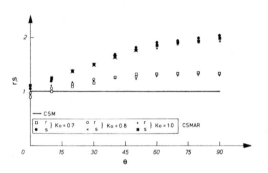

Fig. 9 Variation of r and s with θ for consolidated drained triaxial tests for various K_o conditions

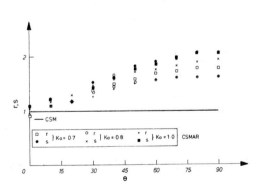

Fig.10 Variation of r and s with θ for consolidated undrained triaxial tests for various K_o conditions

461

DEVJAT. STRESS(KN/M^2)

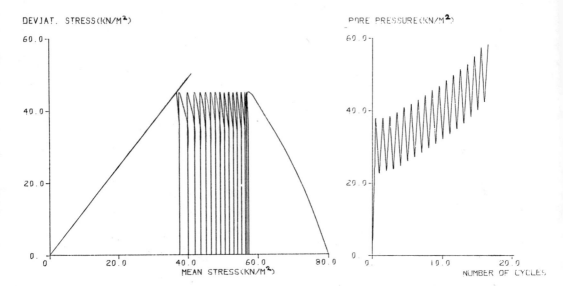

FIG 11 CYCLIC STRESS PATH

FIG 12 PORE PRESSURE V/S NO. OF CYCLES

DEVJAT. STRESS(KN/M^2)

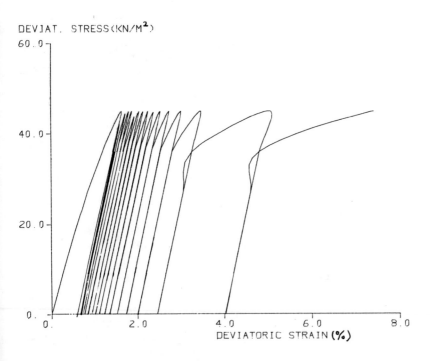

FIG 13 CYCLIC STRESS STRAIN CURVE

A simple stress-dependent constitutive law for sequential plane strain problems

P.SCHARLE
Hungarian Institute for Building Science, Budapest, Hungary

1 GENERAL CONSIDERATIONS

The different constitutive laws elaborated recently to describe the soil behaviour make possible to follow more and more difficult effects. At the same time, further lasts the demand on the simple computing methods working with modest outfit, because many people think the exactness attainable by complex analytical and numerical tools may be illusory due to the difficulties arising at the interpretation and determination of the physical parameters.

Indeed, at least two fundamental circumstances seem to prevent the establishing of general constitutive laws:

1. The behaviour of the soils is influenced by extremely many /mechanical and non-mechanical/ interactions. Their energy-level and that of the cross-effects may be very different. The general model taking into account all of the possible interactions would be unacceptably complicated; the range of validity of the models describing several particular interactions remains inevitably restricted.

2. The traditional variables of the governing equations /displacements, strains, stresses/ and their combinations do not describe sufficiently the behaviour of the multiphase materials. Introduction of further independent variables and expansion of the governing equations could result in better phenomenological models.

For these reasons mentioned the question of the advance is not restricted to the problems of improved constitutive laws and better methods for determining the physical parameters /Goldstejn,1978,Páti,Scharle,1979/. Many papers seem to confirm here the central role of the void ratio /Cowin, Goodman,1976,Davis,Muellenger,1979/.

Nevertheless, the demand of the less complex models continues to exist and even may be justified /Nova,Wood,1979/. The advantage of the simple engineering models lies in their lucidity and the low number of material parameters. Accepting a simple model the range of its validity could and should be checked by numerical and site experiments, and may prove wide enough.

Setting out from these considerations a reinterpretation of the Gibson-soil is given and a nonlinear physical constitutive law is established in the followings. The model takes into account the initial /geostatic/ stresses.Naturally, the possibilities provided by the finite element technique /consideration of the arbitrary geometry, inhomogeneity, etc/ are at hand and will not be discussed.

2 THE REINTERPRETATION OF THE GIBSON-SOIL

The cases of the soil moduli changing with depth were discussed by Harr/1966/, who presented several variations. The most important case /because of its physical background and mathematical simplicity/ is that of the linear one, investigated extensively by Gibson /1967/. For this type of soil the elastic inhomogeneous half space is considered with a shear modulus G varying linearly with depth /Fig.1/ as

$$G(z^2) = G(o) + \tilde{G}z^2 \qquad /1/$$

and the Poisson ratio may have any constant value. This assumption has been used by many authors even recently for the solution of different practical problems /Randolph,Wroth,1978, Brown,Gibson, 1972,1979, Jancsecz,1979/.

It would be straightforward to take into account the inhomogeneity using an available finite element model. However, it is worth noticing the change of the moduli with depth is not simply a geometric rela-

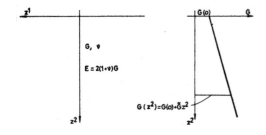

Fig.1. Half-plane with inhomogeneous soil

tion but a consequence of the initial soil state. The settling circumstances, the geostatic stresses result in an initial state, in which the assumption /1/ may prove to be valid. This fact is long well known in the soil physics /Kézdi,1955/. For this reason the assumption of

$$G(z^2) = G(o) + \tilde{G}\,\tilde{6}_o \qquad /2/$$

is obvious / $\tilde{6}_o$ stands for some combination of the initial 6_{ij}^o stresses/.Generally, the correlation between the depth and the initial stresses is very strong, the expressions of /1/ and /2/ are equivalent practically. In the most simple case of homogeneous soil the vertical normal stress 6_{22}^o can be expressed as

$$6^o = -\gamma\, z^2 = \tilde{6}_o \qquad /3/$$

by which we get from /1/

$$G(z^2) = G(o) - \frac{\tilde{G}}{\gamma}\,6_{22}^o = G(o) - G'\,6_{22}^o \qquad /4/$$

In this manner the values of the initial moduli can be connected with the initial state of stress.

When considering the subsequent changes in the soil state caused by loads or other effects new assumptions must be taken for the relationship between the stresses and

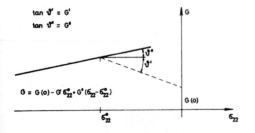

Fig.2. Changes of elastic moduli depending on the stress component 6_{22}

the moduli. The most obvious way is to modify the factor G'. Depending on the softening or hardening behaviour of the soil at a given stress level even the sign of G' may change. A simple choice is shown on Fig.2 - stress-softening soil behaviour is assumed, the modulus G'' decreases with increasing pressure stress 6_{22} /in this presentation the tensile stresses are considered as positive/.

The equation /4/ is not, in the usual sense, a constitutive law, since it does not express directly a stress-strain rela-

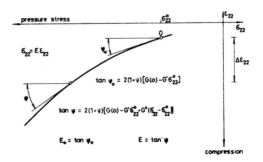

Fig.3. Stress-strain curve for the one-dimensional relationship of /4/

tionship. It is of worth to notice in the one dimensional stress state the conjugate strain-stress relationship can be obtained by integration as

$$\Delta\varepsilon_{22} = \frac{1}{2(1+v)G''}\,\ln\left[1 + \frac{G''}{G(o) - G'6_{22}^o}\,(6^{22} - 6_{22}^o)\right] \qquad /5/$$

Therefore it is of logarithmic type /see Fig.3/.However, the explicit strain-stress equations are not to be formulated in the model.

3 THE ALGORITHM

The equation /4/ is very advantageous from the computational point of view, since it determines a tangent-type coefficient with respect to the known initial state. Therefore, the formulation of a step by step algorithm is at hand.This strategy is in accordance with the nature of all practical problems being of sequential character.

Considering the governing equations of equilibrium and continuity the appropriate expressions of the algorithm could be formulated using any of the error principles /Scharle,1976/. However, this way of constructing an approximating procedure can be avoided now because the energetic formulation related to the case of initial stresses is well known /Washizu,1968/. In the general three-dimensional problem the stationarity conditions of the functional

$$\pi = \int\limits_{V}\left[W(e_{ij};\sigma_{ij}^{o}) + \tfrac{1}{2}\sigma_{ij}^{o}u_{k,i}u_{k,j}\right]dV -$$
$$- \int\limits_{V}\sigma_{ij}\left[e_{ij} - \tfrac{1}{2}(u_{i,j}+u_{j,i}+u_{k,i}u_{k,j})\right]dV - \int\limits_{S_{\sigma}}\bar{p}_{i}u_{i}ds - \int\limits_{S_{u}}p_{i}(u_{j}-\bar{u}_{j})ds \quad /6/$$

are the governing /field and boundary/ e-
quations of the initial stress problem
/the usual cartesian tensor symbolism is
used with the denotions given in the nomen-
clature/.The independent fields subject to
variation are e_{ij}, u_i, σ_{ij} and p_j with no
subsidiary conditions. To obtain a less ge-
neral principle for the plane-strain problem /Fig.l/ it is expedient to assume that
the dispalcements are of infinitesimal magnitude and the initial stresses are of finite magnitude. In this case the continuity and constitutive relations may be linearized. The work of the stress increments
is computed according to the linear theory,
the work of the strains is expressed in accordance with the finite strain theory.

Assuming the field and boundary equations of continuity are fulfilled a priori
by the functions /finite elements/ admitted into variation, the functional /6/ reduces to

$$\pi = \int\limits_{A_{o}}(\tfrac{1}{2}\sigma_{\lambda\mu}\varepsilon_{\lambda\mu} + \tfrac{1}{2}\sigma_{\lambda\mu}^{o}u_{\kappa,\lambda}u_{\kappa\mu})dA - \int\limits_{L_{o\sigma}}\bar{p}_{\lambda}u_{\lambda}dL \quad /7/$$

Here both $\varepsilon_{\lambda\mu}$ and $\sigma_{\lambda\mu}$ /$\mu=1,2$/ are expressed with the approximating displacement
functions, the stresses are computed using
the moduli determined by the preceding $\sigma_{\lambda\mu}^{o}$
stress level.

Obviously, taking into consideration the
effect of the anisotropy does not cause a-
ny dificulty in this order of ideas. For
example, in the case of the transversal a-

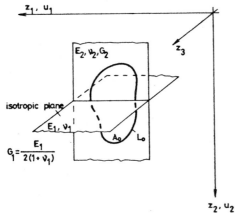

Fig.4. Plane strain problem with transversal anisotropy

nisotropy five independent parameters of E_1,
E_2, G_2 moduli and ν_1, ν_2 Poisson ratios are

to be considered /Fig.4/. In an elaborated
version instead of the modulus E_1 the K_o
lateral earth pressure coefficient at rest
was used, with the condition of

$$E_1 = \frac{K_o E_2}{\nu_2 + K_o \nu_1} \quad /8/$$

In the same version the expressions for
the elastic moduli E_2 and G_2 were given a-
nalogously to /2/ as

$$E_2 = E_2^o - E'\sigma_{22}^o + E''(\sigma_{22}-\sigma_{22}^o) \quad /9/$$

$$G_2 = G_2^o - G'\sigma_{22}^o + G''(\sigma_{22}-\sigma_{22}^o) \quad /10/$$

where E', E'', G', G'' denote the slopes of the
$E_2-\sigma_{22}$ and $G_2-\sigma_{22}$ lines, respectively /Fig.5/.

$$\tan \vartheta' = E'$$
$$\tan \vartheta'' = E''$$

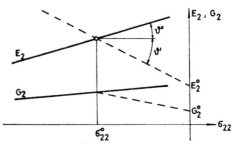

Fig.5. Interpretation of the material parameters

This assumption can be interpreted as if
the soil behaviour were approximated in
the neighbourhood of the inflexion in the
load-settlement diagram /Fig.6/. Obviously,

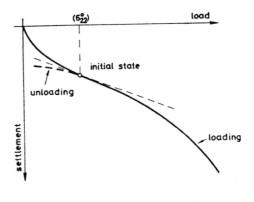

Fig.6. Usual load-settlement diagram of site experiments

there are many other alternatives when determining the slope parameters.

465

The tangential type of the approximation gives a simple possibility for taking into consideration the sequential building process, hole opening, changing the structural stiffness or auxiliary /even provisional/ effect.Their discussion is omitted here because it is independent of the idea of the constitutive law.

The numerical exactness can be improved by restricting the loading steps, the convergency may be checked similarly. On the Fig.7 the settlements of a medium-high, three sectional panel building are shown

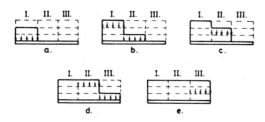

Three-sectional panel building constructed in five consecutive steps

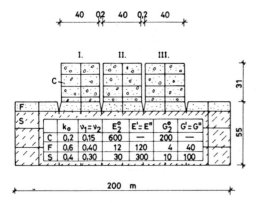

Mesh and material parameters (E_2^o, G_2^o in MPa)

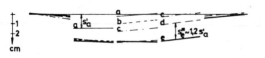

Consecutive surface settlements under the three-sectional building

Fig.7. Constructional sequence, computing scheme and settlements for a panel building

during the sequential construction.The effect of the changes in the structural stiffness and the nonlinear soil behaviour were investigated /Scharle,1979/.Their role being contrasted the simultaneous computation was necessary. In the given case the nonlinear soil behaviour proved to be dominant, the stiffening of the structure did not compensate the softening of the soil.

4 CONCLUSIONS

The material law presented is a simple modification to the Gibson-soil.Therefore it can be ranged among the piecewise linear models /Desai,1979/.Nevertheless, the physical parameters can be estimated easily, and the model gives a good possibility to take into account
- the initial stresses,
- the nonlinear soil behaviour,
- the sequential building process

simultaneously. In this manner it may give satisfactory approximations in many practical problems.

5 NOMENCLATURE

u_i displacement vector components
e_{ij} finite strain tensor components
ε_{ij} linear strain tensor components
6_{ij} stress tensor components
6_{ij}^o initial stress tensor components
P_i surface traction vector components
W   internal work function
V   volume of the body under investigation
S   surface consisting of the complementary parts with $\bar{u}_i$ prescribed displacements on S_u and $\bar{p}_i$ prescribed tractions on S_6 respectively
γ unit weight of soil
K_o lateral earth pressure coefficient at rest

6 REFERENCES

Brown,P.T. R.E.Gibson 1972, Surface settlements of a deep elastic stratum whose modulus increases linearly with depth, Canad. Geot. J., Vol.9:467-476.
Brown,P.T. R.E.Gibson 1979, Surface settlement of a finite elastic layer whose modulus increases linearly with depth,Int. J.for Num.and Anal.Meth.in Geomechanics, Vol.3:37-47.
Cowin,S.C. M.A.Goodmann 1976, A variational principle for granular materials,Zeitschrift für Angewandte Mathematik und Mechanik, Vol.56:281-286.
Davis,R.O. G.Mullenger 1979, A simple rate-

type constitutive representation for granular media. In W.Wittke /ed./,Numerical methods in geomechanics, Vol.I:415-421, Rotterdam, Balkema.

Desai,C.S. 1979, Some aspects of constitutive models for geologic media. In W. Wittke /ed./, Numerical methods in geomechanics, Vol.I:299-308, Rotterdam,Balkema.

Gibson,R.E. 1967, Some results concerning displacements and stresses in a non-homogeneous elastic half-space, Géotechnique, 17:58-67.

Goldstejn,M.M. 1978, On up-to-date tendencies in the development of soil mechanics, Osnovania, Fundamenty i Mehanika Gruntov, 2:24-27 /in Russian/.

Harr,M.E. 1966, Foundations of Theoretical Soil Mechanics, New York, McGraw Hill.

Jancsecz,S. 1979, Dependence of settlements on the soil inhomogeneity /Manuscript in Hungarian/.

Kézdi Á. 1955, Mechanics of plastic solids, the Poisson-number in the light of the rock mechanics, Budapest, Postgraduating Institute for Engineers /in Hungarian/.

Nova,R. D.M.Wood 1979, A constitutive model for sand in triaxial compression,Int. J.for Num. and Anal.Meth.in Geomechanics, Vol.3:255-278.

Páti,G. P.Scharle 1979, Interesting recommendations, Osnovania, Fundamenty i Mehanika Gruntov, 2:24-25 /in Russian/.

Randolph,M.F. C.P.Wroth 1978, Analysis of deformation of vertically loaded piles, Journal of the Geotechnical Div., ASCE, Vol.104:1465-1500.

Scharle,P. 1976, On the relationship between different approximating methods, Acta Technica Ac.Sci.Hun.82:53-59.

Scharle,P. 1979, The effect of the sequential construction process on the settlements, Hungarian Institute for Building Science /in Russian/

Washizu,K.1968, Variational methods in elasticity and plasticity, p.93-98.Oxford, Pergamon.

Anisotropic hardening model for granular model

K.HASHIGUCHI
Kyushu University, Fukuoka, Japan

1 INTRODUCTION

The author (1979) proposed previously the constitutive equations of elastoplastic materials with an elastic-plastic transition observed in the loading state after a first yield, introducing a new parameter denoting the ratio of the size of a loading surface to that of a yield surface in the classical idealization which ignores the transitional state. They describe reasonably not only the hardening but also the softening behavior which requires careful consideration about the elastic-plastic transition. And plastic constitutive equations of granular media were obtained from these equations.

In this paper the plastic constitutive equations of granular media will be extended to describe not only the isotropic but also the anisotropic hardening by incorporating further the relative translation of the loading and the yield surface.

2 BASIC CONSTITUTIVE EQUATIONS

First assume the yield condition

$$f(\sigma - \alpha) - F(\varepsilon_v^p) = 0 \qquad (1)$$

or

$$f(\hat{\sigma}) - F(\varepsilon_v^p) = 0, \qquad (1)'$$

where we set

$$\hat{\sigma} \equiv \sigma - \alpha \qquad (2)$$

and ε_v^p is a plastic volumetric strain, i.e.,

$$\dot{\varepsilon}_v^p \equiv tr(\dot{\varepsilon}^p), \qquad (3)$$

$\dot{\varepsilon}^p$ being a plastic strain increment. The second-order tensor σ is a stress, and the second-order tensor $\hat{\alpha}$ and the scalar K are parameters to describe the translation and

the size-change respectively of the yield surface. Here, let $\hat{\alpha}$ be given by

$$\hat{\alpha} \equiv -C1, \qquad (4)$$

where C (≥ 0) is a scalar function of ε_v^p.
Adopting the associated flow rule

$$\dot{\varepsilon}^p = \lambda \frac{\partial f}{\partial \sigma} \quad (\lambda > 0), \qquad (5)$$

where λ is a proportional factor, the plastic strain rate is given from (1)-(4) as follows:

$$\dot{\varepsilon}^p = \frac{tr(\frac{\partial f}{\partial \hat{\sigma}} \dot{\sigma})}{\{F' - C' tr(\frac{\partial f}{\partial \hat{\sigma}})\} tr(\frac{\partial f}{\partial \hat{\sigma}})} \cdot \frac{\partial f}{\partial \hat{\sigma}}. \qquad (6)$$

In what follows, introducing the concept of a loading surface in the subyield state (Hashguchi, 1979) and considering the relative translation of the yield and the loading surface, the above constitutive equation will be extended to describe the elastic-plastic transition and the development of anisotropy more accurately.

Consider the surface which passes through the current stress state and is similar to the yield surface (1). Such a surface is represented by

$$f(\sigma - \overline{\alpha}) - f = 0 \qquad (7)$$

or

$$f(\overline{\sigma}) - f = 0, \qquad (7)'$$

where we set

$$\overline{\sigma} \equiv \sigma - \overline{\alpha}. \qquad (8)$$

Hereinafter, let f be a homogeneous function of its argument. f is a scalar parameter which has a current value of $f(\overline{\sigma})$, while we call the surface (7) or (7)' a *loading surface*. Refering to Fig.1, $\overline{\alpha}$ which is a parameter to describe the translation of the loading surface is given as follows:

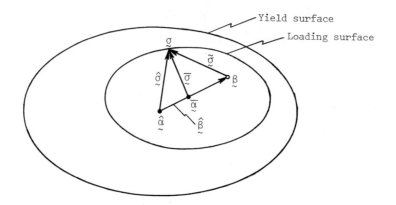

Fig.1 The yield and loading surface
and centers of them ($\hat{\alpha}$ and $\bar{\alpha}$) and similarity (β).

$$\bar{\alpha} = \beta + R(\hat{\alpha} - \beta), \qquad (9)$$

where R denotes a ratio of the size of a loading surface to that of a yield surface, i. e.,

$$R \equiv \frac{f(\bar{\sigma})}{F} \quad (0 \le R \le 1). \qquad (10)$$

β is a center of similarity of these surfaces and let its rate $\dot{\beta}$ be given by

$$\dot{\beta} = \dot{B}\tilde{\sigma} \qquad (11)$$

where we set

$$\tilde{\sigma} \equiv \sigma - \beta. \qquad (12)$$

$\dot{B}$ is a scalar function satiafying the condition

$$\dot{B} = 0 \text{ when } \dot{\varepsilon}^p = 0. \qquad (13)$$

An example of $\dot{B}$ is

$$\dot{B} = \exp[\zeta\{1 - \frac{f(\hat{\beta})}{F}\}]\frac{F' - C'\mathrm{tr}(\frac{\partial f}{\partial \hat{\beta}})}{\mathrm{tr}(\frac{\partial f}{\partial \hat{\beta}}\tilde{\sigma})}\dot{\varepsilon}^p_v, \qquad (14)$$

where

$$\hat{\beta} \equiv \beta - \hat{\alpha}, \qquad (15)$$

$$C' = \frac{dC}{d\varepsilon^p_v}, \quad F' = \frac{dF}{d\varepsilon^p_v}, \qquad (16)$$

and ζ is a material constant. The tensor $\dot{\beta}$ formulated by (11) and (14) has a tangential direction of the yield surface when $f(\hat{\beta}) = F$, and therefore β does not project from it.

Extending the constitutive equation (6) which holds in the yield state $R = 1$ to the general state including the yield and the subyield states, i. e., $R \le 1$, we as-

sume the following equation (cf. Appendix A).

$$\dot{\varepsilon}^p = U\frac{\mathrm{tr}(\frac{\partial f}{\partial \sigma}\dot{\sigma})}{\{F' - C'\mathrm{tr}(\frac{\partial f}{\partial \sigma})\}\mathrm{tr}(\frac{\partial f}{\partial \sigma})}\cdot\frac{\partial f}{\partial \sigma}, \qquad (17)$$

where U $(0 \le U \le 1)$ is a monotoneously increasing function of R satisfying the following condition so that (17) coincides with (6) at the yield state.

$$U = 1 \text{ when } R = 1. \qquad (18)$$

2 SPECIAL CONSTITUTIVE EQUATION

We first assume the special form of the function $f(\hat{\sigma})$ in the yield condition (1) as follws:

$$f(\hat{\sigma}) = \hat{P}^2 + (\frac{|\hat{\sigma}^*|}{m})^2, \qquad (19)$$

where

$$\hat{P} = \frac{1}{3}\mathrm{tr}(\hat{\sigma}), \qquad (20)$$

$$\hat{\sigma}^* = \hat{\sigma} - \hat{P}1.$$

While the yield condition (1)' with (19) is represented by an ellipsoidal surface whose center locates on the mean stress in a stress space as shown in Fig.2, m is a ratio of the minor axis $m\sqrt{F}$ to the major axis $\sqrt{F}$. In accordance with (19), the function (7)' is given as follows:

$$f(\bar{\sigma}) = \bar{P}^2 + (\frac{|\bar{\sigma}^*|}{m})^2, \qquad (21)$$

where

$$\bar{P} = \frac{1}{3}\mathrm{tr}(\bar{\sigma}), \qquad (22)$$

$$\bar{\sigma}^* = \bar{\sigma} - \bar{P}1.$$

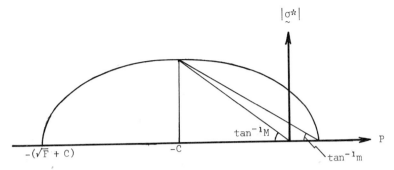

Fig.2 Assumed yield surface.

Now assume that the yield surface holds a similarity about an origin of stress space. Therefore the ratio of the minor axis $m\sqrt{F}$ to the distance C between the origin and the centor of the ellipsoid is constant. Let the ratio be denoted by M. Then it holds that

$$C = \frac{m}{M}\sqrt{F}. \tag{23}$$

In what follows we formulate the function $F(\varepsilon_v^P)$ according to the following relation between ε_v^P and P ($\equiv \frac{1}{3}\mathrm{tr}(\underset{\sim}{\sigma})$) in the isotropic normal consolidation (Hashiguchi, 1978).

$$\varepsilon_v^P = -\mu\ln(\frac{P}{P_o}) \tag{24}$$

where μ is a material constant and P_o is a value of P in the reference state $\varepsilon_v^P = 0$.
On the other hand, since it holds that

$$P = -(C + \sqrt{F}) \tag{25}$$

in the isotropic normal consolidation, F and C and their derived functions are given as follows:

$$F = F_o\exp(-\frac{2}{\alpha}\varepsilon_v^P), \tag{26}$$

$$C = \frac{m}{M}\sqrt{F_o}\exp(-\frac{1}{\alpha}\varepsilon_v^P), \tag{27}$$

$$F' = -\frac{2}{\alpha}F_o\exp(-\frac{2}{\alpha}\varepsilon_v^P), \tag{28}$$

$$C' = -\frac{1}{\alpha}\frac{m}{M}\sqrt{F_o}\exp(-\frac{1}{\alpha}\varepsilon_v^P), \tag{29}$$

where F_o is a value of F in the reference state $\varepsilon_v^P = 0$.
By substituting (21) into (17), the plastic strain increment is expressed as

$$\dot{\underset{\sim}{\varepsilon}}^P = U\frac{2\overrightarrow{PP} + \frac{1}{m^2}\mathrm{tr}(\underset{\sim}{\sigma}*\underset{\sim}{\sigma})}{(F' - 2C'\overline{P})\overline{P}}(\frac{1}{3}\overline{P}\underset{\sim}{1} + \frac{1}{m^2}\overline{\underset{\sim}{\sigma}}*). \tag{30}$$

And since, in accordance with (19), $f(\hat{\underset{\sim}{\beta}})$ is given by

$$f(\hat{\underset{\sim}{\beta}}) = \hat{B}^2 + (\frac{|\hat{\underset{\sim}{\beta}}*|}{m})^2, \tag{31}$$

where

$$\hat{B} = \frac{1}{3}\mathrm{tr}(\hat{\underset{\sim}{\beta}}), \tag{32}$$

$$\hat{\underset{\sim}{\beta}}* = \hat{\underset{\sim}{\beta}} - \hat{B}\underset{\sim}{1},$$

$\dot{\underset{\sim}{\beta}}$ is expressed from (11), (14) and (31) by

$$\dot{\underset{\sim}{\beta}} = \exp[\zeta\{1 - \frac{\hat{B}^2 + (\frac{|\hat{\underset{\sim}{\beta}}*|}{m})^2}{F}\}]$$
$$\cdot\frac{F' - 2C'\hat{B}}{2\{\frac{1}{3}\hat{B}\mathrm{tr}(\hat{\underset{\sim}{\sigma}}) + \frac{1}{m^2}\mathrm{tr}(\hat{\underset{\sim}{\beta}}*\hat{\underset{\sim}{\sigma}})\}}\dot{\varepsilon}_v^P\overline{\underset{\sim}{\sigma}}. \tag{33}$$

In the isotropic stress state the plastic volumetric strain increment and the increment of β are given as

$$\dot{\varepsilon}_v^P = U\frac{2\overrightarrow{PP}}{F' - 2C'\overline{P}}, \tag{34}$$

$$\frac{1}{3}\mathrm{tr}(\dot{\underset{\sim}{\beta}}) = \exp\{\zeta(1 - \frac{\hat{B}^2}{F})\}\{\frac{F'}{2\hat{B}} - C'\}\dot{\varepsilon}_v^P. \tag{35}$$

In the axisymmetric stress state it can be written that

$$\left.\begin{matrix}\dot{\varepsilon}_a^P\\\dot{\varepsilon}_1^P\end{matrix}\right\} = \frac{2}{9}U\frac{(1+\frac{1}{m^2}\frac{\overline{\sigma}_a-\overline{\sigma}_1}{\frac{\overline{\sigma}_a+2\overline{\sigma}_1}{3}})\dot{\sigma}_a + 2(1-\frac{1}{m^2}\frac{\overline{\sigma}_a-\overline{\sigma}_1}{\frac{\overline{\sigma}_a+2\overline{\sigma}_1}{3}})\dot{\sigma}_1}{\frac{F'}{\frac{\overline{\sigma}_a+2\overline{\sigma}_1}{3}} - 2C'}$$

$$\cdot[1 + \frac{1}{m^2}\{\frac{\overline{\sigma}_a-\overline{\sigma}_1}{\frac{\overline{\sigma}_a+2\overline{\sigma}_1}{3}}], \tag{35}$$

and

$$\left.\begin{aligned}\dot{\beta}_a \\ \dot{\beta}_1\end{aligned}\right\} = \exp[\zeta\{1 - \frac{(\frac{\hat{\beta}_a+2\hat{\beta}_1}{3})^2 + \frac{2}{3}m(\hat{\beta}_a-\hat{\beta}_1)^2}{F}\}]$$

$$\cdot\frac{F' - 2C'\frac{\hat{\beta}_a+2\hat{\beta}_1}{3}}{2\{\frac{\hat{\beta}_a+2\hat{\beta}_1}{3}\frac{\tilde{\sigma}_a+2\tilde{\sigma}_1}{3} + \frac{2}{3m^2}(\beta_a-\beta_1)(\tilde{\sigma}_a-\tilde{\sigma}_1)\}}\dot{\varepsilon}_v^p\left\{\begin{aligned}\tilde{\sigma}_a \\ \tilde{\sigma}_1\end{aligned}\right.$$

$$\tag{36}$$

where the quantities appended the subscript a or l denote the axial or the lateral component of them.

In this paper an anisotropic hardening model of granular media was formulated by incorporating the concept of a loading surface in the subyield state, which is similar to the yield surface in the yield state. On the other hand, Mróz, Norris and Zienkiewicz (1978) have presented the anisotropic hardening model of granular media in accordance with the model of a field of hardening moduli proposed by Mróz (1967), and further the same authors (1979) revised it in the form of the two-surface theory.
Which one of these models can describe the plastic behavior of granular media more suitably will be discussed in another paper comparing with experimental results.

ACKNOWLEDGEMENT

The author sincerely acknowledges his indebtedness to Professor Hakuju Yamaguchi of the Tokyo Institute of Technology for his continual guidance and encouragement given throughout this study. This study was supported in part by the Aid for Scientific Research from the Ministry of Education of Japan under Grant 355220 to the author.

REFERENCES

Hashiguchi, K. 1978, Constitutive equations of soils -Theories based on elastoplasticity-, Journ. Japan. Soci. Soil Mech. & Found. Engng, Vol.18, No.4, pp.131-142.
Hashiguchi, K. 1979, Constitutive equations of elastoplastic materials with elastic/-/ plastic transition, J. Appl. Mech., Trans. ASME, (in press).
Mróz, Z. 1967, On the description of Aniso- tropic workhardening, J. Mech. Phys. Solids, Vol.15, p.163-175.
Mróz, Z., Norris, V. A. and Zienkiewicz, O. C. 1978, An anisotropic hardening model for soils and its application to cyclic loading, Int. J. Numer. Anal. Method in Geomech., Vol.2, p.203-221.
Mróz, Z., Norris, V. A. and Zienkiewicz, O. C. 1979, Application of an anisotropic hardening model in the analysis of elasto-plastic deformation of soils, Geotechnique, Vol.29, No.1, p.1-34.

APPENDIX A

The past proofs for the associated flow rule are premised on the classical idealization to ignore the elastic-plastic transition and the alteration of an elastic response. On the other hand, the classical idealization is ignored in the extended plastic constitutive equation (17). This equation neverthless follows the associated flow rule.
In what follows, based on physically plausible assumptions which can be regarded as extensions of the Prager's conditions of uniqueness and continuity, the associated flow rule is derived for the generalized elastoplastic continuum with the elastic/-/ plastic transition and the hardening, the perfectly-plastic and the softening behaviors.
Preliminarily consider the uniaxial loading of a bar. Observing the stres-strain curves schematically illustrated in Fig.1, we could assume that in a loading state

$$\begin{aligned}\bar{\sigma}_a d\sigma_a > 0 &: \sigma_a d\varepsilon_a^p > 0, \\ \bar{\sigma}_a d\sigma_a = 0 &: \sigma_a d\varepsilon_a^p = 0, \\ \bar{\sigma}_a d\sigma_a < 0 &: \sigma_a d\varepsilon_a^p < 0\end{aligned}\tag{A1}$$

and

$$d\sigma_a \to 0 \text{ as } d\varepsilon_a^p \to 0,\tag{A2}$$

where

$$\bar{\sigma}_a \equiv \sigma_a - \bar{\alpha}_a.\tag{A3}$$

σ_a, $\bar{\alpha}_a$ and ε_a^p are the normal components of the stress, the translation of the loading surface and the strain along the axis respectively.
By extending the above notions to the general loading state, we introduce the following assumptions 1 and 2 from (A1) and (A1) respectively, in which the nine-dimensional stress space with superposed coordinate of stress components σ_{ij} and plastic strain increment components $d\varepsilon_{ij}^p$ is introduced.
Assumption 1 When, accompanying a plastic deformation (loading state), the state of stress moves to the exterior, the tangency and the interior of the existing loading surface in a stress space, the plastic work done by the stress increment is positive, zero and negative respectively.
Assumption 2 As the plastic strain increment vanishes, the normal component of

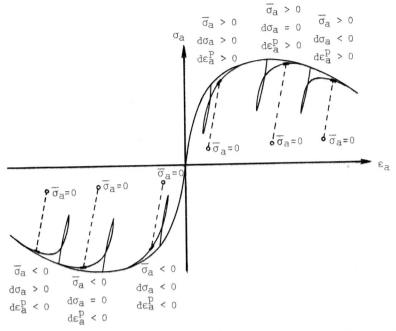

Fig.3 Signs of $\overline{\sigma}_a$, $d\sigma_a$ and $d\varepsilon_a^p$ in the uniaxial loading.

stress increment to the loading surface also vanishes.

Assumption 1 can be written in analytical form as follows:

$$\text{tr}(\frac{\partial f}{\partial \underset{\sim}{\sigma}}d\underset{\sim}{\sigma}) > 0 : \text{tr}(d\underset{\sim}{\sigma}d\underset{\sim}{\varepsilon}^p) > 0,$$

$$\text{tr}(\frac{\partial f}{\partial \underset{\sim}{\sigma}}d\underset{\sim}{\sigma}) = 0 : \text{tr}(d\underset{\sim}{\sigma}d\underset{\sim}{\varepsilon}^p) = 0, \qquad (A4)$$

$$\text{tr}(\frac{\partial f}{\partial \underset{\sim}{\sigma}}d\underset{\sim}{\sigma}) < 0 : \text{tr}(d\underset{\sim}{\sigma}d\underset{\sim}{\varepsilon}^p) < 0$$

in the loading state, from which we obtain

$$\text{tr}(d\underset{\sim}{\sigma}d\underset{\sim}{\varepsilon}^p) = S\text{tr}(\frac{\partial f}{\partial \underset{\sim}{\sigma}}d\underset{\sim}{\sigma}) \ (S > 0), \qquad (A5)$$

where S is a scalar function of stress and some internal variables. Assumption 1 can be regarded as the generalization of Prager's uniqueness condition (J. Appl. Physics, Vol.20, p.235-241) for hardening materials ($\text{tr}(d\underset{\sim}{\sigma}d\underset{\sim}{\varepsilon}^p) > 0$).

Assumption 2 can be expressed as

$$\text{tr}(\frac{\partial f}{\partial \underset{\sim}{\sigma}}d\underset{\sim}{\sigma}) \rightarrow 0 \text{ as } d\underset{\sim}{\varepsilon}^p \rightarrow \underset{\sim}{0}, \qquad (A6)$$

which is the inverse relation of Prager's continuity condition ($d\underset{\sim}{\varepsilon}^p \rightarrow \underset{\sim}{0}$ as $\text{tr}(\partial f/\partial \underset{\sim}{\sigma}$ $d\underset{\sim}{\sigma}) \rightarrow \underset{\sim}{0}$) that does not hold in the state satisfying $\text{tr}(\partial f/\partial \underset{\sim}{\sigma}d\underset{\sim}{\sigma}) = 0$. In addition to Assumption 2, we assume that there exists a linear relation between $d\underset{\sim}{\varepsilon}^p$ and $d\underset{\sim}{\sigma}$. Thus it can be written that

$$d\underset{\sim}{\varepsilon}^p = \text{tr}(\frac{\partial f}{\partial \underset{\sim}{\sigma}}d\underset{\sim}{\sigma})\underset{\sim}{Q} \ (\underset{\sim}{Q} \neq \underset{\sim}{0}), \qquad (A7)$$

where the second-order tensor $\underset{\sim}{Q}$ is a function of stress and some internal variables. The relation (A7) means that the direction of plastic strain increment is independent of the stress increment.

Substitution (A7) into (A5) leads to

$$\text{tr}(\underset{\sim}{Q}d\underset{\sim}{\sigma}) = S \ (> 0). \qquad (A8)$$

Then, noting that the stress increment can have any direction, it must hold that

$$\underset{\sim}{Q} = G\frac{\partial f}{\partial \underset{\sim}{\sigma}}, \qquad (A9)$$

with the condition

$$G\text{tr}(\frac{\partial f}{\partial \underset{\sim}{\sigma}}d\underset{\sim}{\sigma}) > 0, \qquad (A10)$$

where G is a scalar function of stress and some internal variables. Thus, the plastic strain increment is given from (A7)-(A10) as follows:

$$d\underset{\sim}{\varepsilon}^p = d\lambda\frac{\partial f}{\partial \underset{\sim}{\sigma}}, \qquad (A11)$$

where

$$d\lambda \equiv G\text{tr}(\frac{\partial f}{\partial \underset{\sim}{\sigma}}d\underset{\sim}{\sigma}) \ (> 0). \qquad (A12)$$

The equation (A11) is to be the associated flow rule, while the restriction for $d\lambda$ to take the form (A12) is imposed by as-

473

sumption 2 and the linearity of the constitutive equation. Here, note that the convexity of the loading surface is not imposed in this approach.

The above approach leads to the fact
that this rule is applicable to the elasto-
plastic continuum with the elastic-plastic
transition and the hardening, perfectly/-/
plastic and the softening behaviors independently of the elastic response. It refers to a more general elastoplastic continuum than those assumed in the past approaches. Here, remind that it is based on
two assumptions which are generalization
of distinct characteristics of the plastic
deformation in the uniaxial loading. While
these assumptions seem physically plausible,
still it would be desirable to verify ra-
tionalities of them by any other fundamental principles on the plastic deformation.

Endochronic models for soils

ATILLA M. ANSAL
Istanbul Technical University, Istanbul, Turkey

ZDENĚK P. BAŽANT & RAYMOND J. KRIZEK
Northwestern University, Evanston, Ill., USA

A comprehensive constitutive relation should ideally be based on microscopic material characteristics; however, the present state of knowledge precludes such a treatment for soils. Although endochronic theory is a macro-theory by nature, it is possible to develop interpretations of the material parameters in terms of the microscopic characteristics and physical properties of soils. The basic concept underlying the endochronic formulations for material behavior is the characterization of inelastic strains in terms of intrinsic time, which is a nondecreasing scalar variable that depends on increments of inelastic strain, as well as increments of time, and geometrically represents the length of the path traced by the states of the material in strain-time space of suitable metric. The term "endochronic" was coined by Valanis (1971) who developed the theory to model the cross-hardening behavior of metal alloys. Valanis' version of endochronic theory was later extended to describe the behavior of concrete (Bazant, 1974; Bazant and Bhat, 1976; Bazant and Shieh, 1977) and both cohesionless and cohesive soils (Bazant and Krizek, 1976; Cuellar et al, 1977; Krizek, Ansal and Bazant, 1978; Ansal Bazant and Krizek, 1979).

The notion of intrinsic time is introduced to account for both real time and the inelastic dissipative effects of accumulated strain. The intrinsic time parameter depends on time or strain rate only for rate-dependent materials, such as cohesive soils, and it must be a monotonically increasing function of strain and external time. Following Valanis, the increment of intrinsic time, z, is expressed as $dz = (d\zeta/z_1)^2 + (dt/\tau_1)^2$, in which z_1 and τ_1 are material parameters and dt is the increment of real time. The parameter ζ, called the rearrangement measure, may be visualized as a

parameter representing on a macroscopic scale the accumulation of microstructural changes that take place as deformation occurs; these changes may lead to either strain softening or strain hardening. The increment $d\zeta$ is assumed to be a function of only the current state of stress, $\underset{\sim}{\sigma}$, and strain, $\underset{\sim}{\varepsilon}$, and the cumulative value of the distortion measure, ξ; this function is formulated in differential form as $d\zeta = F(\underset{\sim}{\varepsilon}, \underset{\sim}{\sigma}, \zeta)d\xi$. The Function F, which must be determined semi-empirically from experimental data, represents hardening and softening behavior and ξ is defined as $d\xi = \sqrt{J_2(d\underset{\sim}{\varepsilon})}$, in which $J_2(d\underset{\sim}{\varepsilon})$ is the second invariant of incremental deviatoric strain. The inelastic volumetric strains caused by shear are represented by the variable λ, called the densification-dilatancy measure. If it is assumed that the inelastic volumetric change is due only to shear, which means that inelastic volumetric strains due to a change in hydrostatic stress are excluded from the formulation, $d\lambda$ can be expressed as $d\lambda = L(\underset{\sim}{\varepsilon}, \underset{\sim}{\sigma}, \lambda)d\xi$, in which the function L must be determined empirically from experimental data.

Although various characteristics of endochronic theory, such as the possible violation of Drucker's postulate, questions of stability and uniqueness, and the shape of the inelastic stiffness locus (Bazant, 1977, 1978) are currently being debated, it appears that these criticisms can be avoided by suitable refinements. In any event, the endochronic formulation provides a powerful approach with great flexibility for treating unloading, cyclic loading, strain softening, and cross-hardening. In the case of nonmetalic materials, such as soils, it is also important to note that there presently exists no microstructural mechanism that would favor some other theory over endochronic theory. This is in contrast to metals, where dislocation theory, the plastic slip mechanism, and the validity of Schmid's law point toward plasticity, while in soils and concrete the inapplicability of these concepts and the existence of friction, particle separation, and microcracking actually point against classical plasticity which is based on Druker's postulate. However, it appears (Bažant, 1978, 1980) that endochronic theory can also be associated with a loading function. A comparison between the expressions for plastic isotropic hardening and that in available endochronic formulations indicates that the endochronic loading function is the same as that for isotropic hardening in plasticity and a yield surface is always centered at the origin of stress space and dilates while retaining the same shape. A more realistic and accurate description would require non-isotropic (kinematic) hardening, where the yield surface not only dilates but also moves as a rigid body. A special type of the kinematic hardening, called jump-kinematic hardening, has recently been introduced by Sener, Krizek, and Bažant (1979) to obtain an improved model that agreed quite well with cyclic test data for sands. In this work, a further refinement was made by taking into account the inelastic volumetric strains due to the hydrostatic (mean) stress, in addition to those due to shear deformation.

The developed constitutive relations for both cohesive and cohesionless soils are reviewed, and the capabilities of the model are demonstrated qualitatively by predicting the stress-strain-pore pressure response of both cohesive and cohesionless soils for various stress and time histories. A two-dimensional plane-strain finite element program using endochronic theory has been developed and applied to describe the stress and displacement fields that arise in an earth dam subjected to an earthquake load.

REFERENCES

Ansal, A.M., Bazant, Z.P. & Krizek, R.J. 1979, Viscoplasticity of Normally Consolidated Clays, Proceedings of the American Society of Civil Engineers, Journal of the Geotechnical Engineering Div., Vol. 105, No. GT4, 519-537.

Bazant, Z.P. 1980, Work Inequalities for Plastic Hardening Materials, International Journal of Solids and Structures, in press

Bazant, Z.P. 1978, Endochronic Inelasticity and Incremental Plasticity, International Journal of Solids and Structures, Vol. 52, 1-24.

Bazant, Z.P. 1977, Endochronic and Classical Theories of Plasticity in Finite Element Analysis, International Conference on Finite Elements in Nonlinear Solid and Structural Mechanics, Geils, Norway, Vol. 1: 1-15.

Bazant, Z.P. 1974, A New Approach to Inelasticity and Failure of Concrete, Sand and Rock: Endochronic Theory, Proceedings of the 11th Annual Meeting, Society of Engineering Sciences, G.J. Dvorak, ed. Duke University, N.C. pp. 158-159.

Bazant, Z.P. & Shieh, C.-L. 1977, Endochronic Model for Nonlinear Triaxial Behavior of Concrete, Nuclear Engineering and Design, Vol. 47, 305-315.

Bazant, Z.P. & Krizek, R.J. 1976, Endochronic Constitutive Law for Liquefaction of Sand, Proceedings of the American Society of Civil Engineers, Journal of the Engineering Mechanics Div., Vol. 102, No. EM2, 225-238.

Bazant, Z.P. & Bhat, P.D. 1976, Endochronic Theory of Inelasticity and Failure of Concrete, Proceedings of the American Society of Civil Engineers, Journal of the Engineering Mechanics Division, Vol. 102, No. EM4, 701-722.

Cuellar, V., Bazant, Z.P., Krizek, R.J. & Silver, M.L. 1977, Densification and Hysteresis of Sand Under Cyclic Shear, Proceedings of the American Society of Civil Engineers, Journal of the Geotechnical Div. Vol, 103, No. GT5, 399-416.

Krizek, R.J., Ansal, A.M. & Bazant, Z.P. 1978, Constitutive Equation for Cyclic Behavior of Cohesive Soils, Proceedings of the American Society of Civil Engineers, Geotechnical Engineering Div. Specialty Conference, Earthquake Engineering and Soil Dynamics, Vol. II, 557-568.

Sener, C., Krizek, R.J. & Bazant, Z.P. 1980, An Endochronic Nonlinear Inelastic Constitutive Law for Cohesionless Soils Subjected to Dynamic Loading, Geotechnical Engineering Report, Northwestern University Evanston, Illinois Jan. 1980.

Valanis, K.C. 1975, On the Foundations of the Endochronic Theory of Viscoplasticity, Archivum Mechaniki Stossowanej (Archives of Mechanics) Vol. 27, Issue 5-6 857-868.

Equation for compression of noncoheisive soils
under transient loads with the stationary intensity

R. G. YURKIN
All-Union Scientific and Research Institute for Transport Construction, Nahodka, USSR

Basing on the numerous generalized experimental studies (Birulia 1964, Harhuta 1961, Shehter 1953, Rapoport 1978) and on the theoretical researches (Yurkin 1977,1979) it was proved that under compression with cyclic loads having constant parameters the incriment of soil compaction from these loads, as well as by vibrocompaction under the same conditions, could be expressed as such:

$$\Delta\gamma_d = \beta(1 - \beta_o \ln n)(\ln p/p_o) \ln t/t_o +$$
$$+ \beta_c(1 + \beta_n \ln n)(\ln p/p_c) \ln t/t_c, \qquad (1)$$

where $\Delta\gamma_d$ – complete change of the volumetric mass of soil skeleton for n cycles considering reversible and non-reversible compaction;

p – the maximum intensity of the compressing pressure;

p_o – the initial compressing pressure;

p_c – the point of the initial structural strength;

t – time from the moment when the load was applied until it was taken off;

t_o – delay in the beginning of the elastic strain;

t_c – delay in the beginning of the residual strain;

$\beta, \beta_o, \beta_c, \beta_n$ – the constants.

This eqation may be used at $t > t_o$ and $p > p_o$, but considering that $t \leqslant t_c$ or $p \leqslant p_c$, the second member in the right part of the equation should be missing.

The first member of the right part of Eq. (1) is the reversible part of the compaction deformation, and the second member is the residual part.

The experimental data, which are used to determine the constants of Eq. (1), should necesserialy contain information on reversible and residual parts of deformation.

The determination of the constants, depending on one factor, should be performed accordingly to the experimental data, in which only the values of constants, which characterize this factor, are changeble. The definition of two constant coefficients, depending on one factor, is possible if necessary experimental data are available, which are derived for two different digital values of this factor.

The constants of Eq. (1) could be determined with more precision using the experimental data with the help of the Least square method in the definite sequense. For this purpuse experimental data are necessary, the volume of which should surfice the precise determination of each characteristics.

Using the results of the tests performed on twin-samples for the determination of constant equation, characterizing the influense of load intensity (p) and the duration of load action in each cycle (t); it is possible to use the experimental data only for the 1-st cycle. In that case the formulas, obiained under the condition of minimization of deviation of squares sum of experimental data from the accepted regression equation, will be as such:

$$p = \exp \frac{\sum_1^k \ln p_i \sum_1^k (\Delta\gamma_{di})^2 - \sum_1^k \Delta\gamma_{di} \sum_1^k \Delta\gamma_{di} \ln p_i}{k \sum_1^k (\Delta\gamma_{di})^2 - \left(\sum_1^k \Delta\gamma_{di}\right)^2}; \qquad (2)$$

$$t_o = \exp \frac{\sum_1^k \ln t_i \sum_1^k (\Delta \gamma_{di})^2 - \sum_1^k \Delta \gamma_{di} \sum_1^k \Delta \gamma_{di} \ln t_i}{k \sum_1^k (\Delta \gamma_{di})^2 - \left(\sum_1^k \Delta \gamma_{di}\right)^2}; \quad (3)$$

$$\beta = \frac{k \sum_1^k (\Delta \gamma_{di})^2 - \left(\sum_1^k \Delta \gamma_{di}\right)^2}{\left[k \sum_1^k \Delta \gamma_{di} \ln p_i - \sum_1^k (\ln p_i) \sum_1^k \Delta \gamma_{di}\right](\ln t - \ln t_o)}; \quad (4)$$

$$\beta_o = \frac{\sum_1^k \Delta \gamma_{di} \sum_1^k \ln n_i - k \sum_1^k \Delta \gamma_{di} \ln n_i}{\beta \left[k \sum_1^k (\ln n_i)^2 - \left(\sum_1^k \ln n_i\right)^2\right](\ln p - \ln p_o)(\ln t - \ln t_o)}; \quad (5)$$

$$p_c = \exp \frac{\sum_1^k \ln p_i \sum_1^k (\Delta \gamma_{di})^2 - \sum_1^k \Delta \gamma_{di} \sum_1^k \Delta \gamma_{di} \ln p_i}{k \sum_1^k (\Delta \gamma_{di})^2 - \left(\sum_1^k \Delta \gamma_{di}\right)^2}; \quad (6)$$

$$t_c = \exp \frac{\sum_1^k \ln t_i \sum_1^k (\Delta \gamma_{di})^2 - \sum_1^k \Delta \gamma_{di} \sum_1^k \Delta \gamma_{di} \ln t_i}{k \sum_1^k (\Delta \gamma_{di})^2 - \left(\sum_1^k \Delta \gamma_{di}\right)^2}; \quad (7)$$

$$\beta_c = \frac{k \sum_1^k (\Delta \gamma_{di})^2 - \left(\sum_1^k \Delta \gamma_{di}\right)^2}{\left(k \sum_1^k \Delta \gamma_{di} \ln p_i - \sum_1^k \Delta \gamma_{di} \sum_1^k \ln p_i\right)(\ln t - \ln t_c)}; \quad (8)$$

$$\beta_n = \frac{k \sum_1^k \Delta \gamma_{di} \ln n_i - \sum_1^k \Delta \gamma_{di} \sum_1^k \ln n_i}{\beta_c \left[k \sum_1^k (\ln n_i)^2 - \left(\sum_1^k \ln n_i\right)^2\right](\ln p - \ln p_c)(\ln t - \ln t_c)}, \quad (9)$$

where i – index of experimental values;

k – number of experimental values under consideration.

In formulas (2),(4),(6),(8) – k is the number of $\Delta \gamma_{di}$ experimental values used, induced by p_i, with constant n and t. In formulas (3), (7) – k – is the number of $\Delta \gamma_{di}$ experimental values used, induced by t_i, with constant p and n. Accordingly in formulas (5), (9) – k is the number of $\Delta \gamma_{di}$ values, indused by n_i with constant p and t in the cycles considered.

With $k = 2$ the formulas are considerably simplified:

$$p_o = \exp \frac{\Delta \gamma_{d(i+1)} \ln p_i - \Delta \gamma_{d(i)} \ln p_{i+1}}{\Delta \gamma_{d(i+1)} - \Delta \gamma_{d(i)}}; \quad (10)$$

$$t_o = \exp \frac{\Delta \gamma_{d(i+1)} \ln t_i - \Delta \gamma_{d(i)} \ln t_{i+1}}{\Delta \gamma_{d(i+1)} - \Delta \gamma_{d(i)}}; \quad (11)$$

$$\beta = \frac{\Delta \gamma_{d(i+1)} - \Delta \gamma_{d(i)}}{(\ln p_{i+1} - \ln p_i)(\ln t - \ln t_o)}; \quad (12)$$

$$\beta_o = \frac{\Delta \gamma_{d(i)} - \Delta \gamma_{d(i+1)}}{\beta (\ln n_{i+1} - \ln n_i)(\ln p - \ln p_o)(\ln t - \ln t_o)}; \quad (13)$$

$$p_c = \exp \frac{\Delta \gamma_{d(i+1)} \ln p_i - \Delta \gamma_{d(i)} \ln p_{i+1}}{\Delta \gamma_{d(i+1)} - \Delta \gamma_{d(i)}}; \quad (14)$$

$$t_c = \exp \frac{\Delta \gamma_{d(i+1)} \ln t_i - \Delta \gamma_{d(i)} \ln t_{i+1}}{\Delta \gamma_{d(i+1)} - \Delta \gamma_{d(i)}}; \quad (15)$$

$$\beta_c = \frac{\Delta \gamma_{d(i+1)} - \Delta \gamma_{d(i)}}{(\ln p_{i+1} - \ln p_i)(\ln t - \ln t_c)}; \quad (16)$$

$$\beta_n = \frac{\Delta \gamma_{d(i+1)} - \Delta \gamma_{d(i)}}{\beta_o (\ln n_{i+1} - \ln n_i)(\ln p - \ln p_c)(\ln t - \ln t_c)}. \quad (17)$$

In the course of determination of p_o, p_c, β, β_c the time of loading in cycle (t) and the number of cycles (n) should be constants, and while determining t_o and t_c-the load intensity (p) and the number of cycles (n). Computation of β_o and β_n is performed with the steady pressure (p) and constant load in cycle (t).

To compute the characteristics of the reversible part of deformation, only the value of reversible deformation is used, and for the rest part – residual. Computation of the volumetric mass of soil skeleton is performed using the following formula:

$$\gamma_d = \frac{\gamma_{do}}{1 - \Delta h / h_o}, \quad (18)$$

where γ_{do} – the value of the volumetric mass in the soil skeleton before the experiment;

Δh – deformation of a sample using compression equipment;

h_o – the initial height of the sample.

In non-coheisive soil with $t > 1$ min

478

and if computation is done with approximity, time-dependance could be neglected and instead of Eq. (1) the following equation should be used;

$$\Delta\gamma_d = (\bar\beta - \bar\beta_o \ln n)\ln\rho/\rho_o + (\bar\beta_c + \bar\beta_n \ln n)\ln\rho/\rho_c, (19)$$

where $\bar\beta$, $\bar\beta_o$, $\bar\beta_c$, $\bar\beta_n$ – are constant coefficients of the equation, depending on t.

To define the coefficients $\bar\beta$, $\bar\beta_c$, ρ_o, ρ_c of Eq. (19) formulas (6),(8), (14), (16) could be used.

The value of $\bar\beta_o$ in many cases approximates 0 and could be neglected.

Considering the difficulty of obtaining the samples with the same psysicomechanical characteristics for the determination of the coefficients under consideration, it is advisable to use the method of a single sample (Rapoport 1978),which envizages the studying of different factors influence one and the same sample.

While performing combined tests to investigate the influence of load intensity (ρ) and the number of cycles (n), the determination of $\bar\beta_c$ and $\bar\beta_n$ should be performed with n equal to the full number of cycles, corresponding to each degree of intensity (ρ). In this case formulas (8),(9),(16),(17) could not be used. In the simplest case to determine $\bar\beta_c$ and $\bar\beta_n$ instead of those formulas the following formulas could be used.

$$\bar\beta_n = \frac{\Delta\gamma_{d(i+1)} - \Delta\gamma_{di}}{(\ln n_{i+1} - \ln n_i)(\ln\rho - \ln\rho_c)};\qquad(20)$$

$$\bar\beta_c = \frac{\Delta\gamma_{di}\ln n_{i+1} - \Delta\gamma_{d(i+1)}\ln n_i}{(\ln n_{i+1} - \ln n_i)(\ln\rho - \ln\rho_c)}.\qquad(21)$$

For n_{i+1} it is advisable to use the greatest values of n in the test under consideration, where n_i equals 2.

As an example to illustrate the values of other constant coefficients, below could be found the values of constants in Eq. (19), obtained as a result of data processing of sand,tested under repeated increasing action with the intensity from 0,5-0 MPa to 1-0 MPa (Rapoport 1978). The full cycle time was 4 min. (t = 2 min). The volumetric soil skeleton mass before the first load cycle was 1,5614 g/cm^2

$\rho_o = 0,0689$ MPa $\rho_c = 0,3707$ MPa
$\bar\beta = 0,00406$ g/cm^3 $\bar\beta_c = 0,001455$ g/cm^3
$\bar\beta_o = 0,0;$ $\bar\beta_n = 0,01505$ g/cm^3

The numerous experimental data processing performed demonstrated that Eq. (1) is quite adequate for qualitative and quantitative description of the reological process in non-coheisive soils volumetric deformation from cycling loads.

According to the results of researches,the compaction of non-coheisive soils indused by cyclic loads is a volumetric creep process (Harhuta 1961).

The given equations could be used to determine the incriment of non-coheisive soil compactness indused by cyclic loads independently as a result of reversible and non-reversible parts of the whole deformation considering at the same time the number of cycles, the load intensity and its duration.

REFERENCES

Birulia,A.K. 1964, Analisys and construction of nonrigid highway pavements. Moscow, Transport.
Harhuta, N.Y., V.M.Evlev 1961, Reological characteristics of soils. Moscow, Avtotransizdat.
Rapoport,S.G., V.E.Kurmes, R.O.Ziangirov 1978, Changebility of clay soils deformability under prolonged repeated actions in book Theoretical foundations of change predictibility of clay soils geological properties. Works of PNIIIS, issue 53, Moscow, Stroyizdat; 67-104.
Shehter, O.Y. 1953, Experimental study of vibrocompressible properties of sands. Works of Research Inst. of Footings and Foundations, No 22, Moscow, Gosstroyizdat:88-95.
Yurkin, R.G. 1977, The simplest characteristics of soil vibrocompaction and their determination in book The footings, foundations and underground structures dynamics (Materials of the 4 Interunion conference), Book 1, Tashkent, "Fan":190-193.
Yurkin, R.G., L.I.Vedenistova 1979, On some regularities of sands volumetric deformations in book Perspectives of development and inculcation experience of construction materials and structures in the Far East (the thesis of scientific and technical conference). Vladivostok, Dalpromstroyniiproect: 18-21.

WINTER OF THE WORLD

Also by Ken Follett

The Modigliani Scandal

Paper Money

Eye of the Needle

Triple

The Key to Rebecca

The Man from St Petersburg

On Wings of Eagles

Lie Down with Lions

The Pillars of the Earth

Night Over Water

A Dangerous Fortune

A Place Called Freedom

The Third Twin

The Hammer of Eden

Code to Zero

Jackdaws

Hornet Flight

Whiteout

World Without End

Fall of Giants

KEN FOLLETT

WINTER OF THE WORLD

MACMILLAN

First published 2012 by Macmillan
an imprint of Pan Macmillan, a division of Macmillan Publishers Limited
Pan Macmillan, 20 New Wharf Road, London N1 9RR
Basingstoke and Oxford
Associated companies throughout the world
www.panmacmillan.com

ISBN 978-0-230-71010-8

1 3 5 7 9 8 6 4 2

A CIP catalogue record for this book is available from
the British Library.

Pocket watch image on cover © Getty Images. The watch was found on the
body of a victim of the atomic bombing of Hiroshima on 6 August 1945.
The watch stopped at 8.15 a.m., when the bomb exploded.

Typeset by SetSystems Ltd, Saffron Walden, Essex
Printed and bound by CPI Group (UK) Ltd, Croydon, CR0 4YY

To the memory of my grandparents,

Tom and Minnie Follett

Arthur and Bessie Evans

Cast of characters

American

Dewar Family

Senator Gus Dewar
Rosa Dewar, *his wife*
Woody Dewar, *their elder son*
Chuck Dewar, *their younger son*
Ursula Dewar, *Gus's mother*

Peshkov Family

Lev Peshkov
Olga Peshkov, *his wife*
Daisy Peshkov, *their daughter*
Marga, *Lev's mistress*
Greg Peshkov, *son of Lev and Marga*
Gladys Angelus, *film star, also Lev's mistress*

Rouzrokh Family

Dave Rouzrokh
Joanne Rouzrokh, *his daughter*

Buffalo Socialites

Dot Renshaw
Charlie Farquharson

Cast of characters

Others

Joe Brekhunov, *a thug*

Brian Hall, *union organizer*

Jacky Jakes, *starlet*

Eddie Parry, *sailor, friend of Chuck's*

Captain Vandermeier, *Chuck's superior*

Margaret Cowdry, *beautiful heiress*

Real Historical Characters

President F. D. Roosevelt

Marguerite 'Missy' LeHand, *his assistant*

Vice-President Harry Truman

Cordell Hull, *Secretary of State*

Sumner Welles, *Undersecretary of State*

Colonel Leslie Groves, *Army Corps of Engineers*

English

Fitzherbert Family

Earl Fitzherbert, *called Fitz*

Princess Elizaveta, *called Bea, his wife*

'Boy' Fitzherbert, Viscount Aberowen, *their elder son*

Andy, *their younger son*

Leckwith-Williams Family

Ethel Leckwith (née Williams), *Member of Parliament for Aldgate*

Bernie Leckwith, *Ethel's husband*

Lloyd Williams, *Ethel's son, Bernie's stepson*

Millie Leckwith, *Ethel and Bernie's daughter*

Cast of characters

Others

Ruby Carter, *friend of Lloyd's*

Sir Bartholomew ('Bing') Westhampton, *friend of Fitz's*

Lindy and Lizzie Westhampton, *Bing's twin daughters*

Jimmy Murray, *son of General Murray*

May Murray, *his sister*

Marquis of Lowther, *called Lowthie*

Naomi Avery, *Millie's best friend*

Abe Avery, *Naomi's brother*

Real Historical Characters

Ernest Bevin, *MP, Foreign Secretary*

German & Austrian

Von Ulrich Family

Walter von Ulrich

Maud (née Lady Maud Fitzherbert), *his wife*

Erik, *their son*

Carla, *their daughter*

Ada Hempel, *their maid*

Kurt, *Ada's illegitimate son*

Robert von Ulrich, *Walter's second cousin*

Jörg Schleicher, *Robert's partner*

Rebecca Rosen, *an orphan*

Franck Family

Ludwig Franck

Monika (née Monika von der Helbard), *his wife*

Werner, *their elder son*

Frieda, *their daughter*

Axel, *their younger son*

Cast of characters

Ritter, *chauffeur*
Count Konrad von der Helbard, *Monika's father*

Rothmann Family

Dr Isaac Rothmann
Hannelore Rothmann, *his wife*
Eva, *their daughter*
Rudi, *their son*

Von Kessel Family

Gottfried von Kessel, *deputy for the Centre Party*
Heinrich von Kessel, *his son*

Gestapo

Commissar Thomas Macke
Inspector Kringelein, *Macke's boss*
Reinhold Wagner
Klaus Richter
Günther Schneider

Others

Hermann Braun, *Erik's best friend*
Sergeant Schwab, *gardener*
Wilhelm Frunze, *scientist*

Russian

Peshkov Family

Grigori Peshkov
Katerina, *his wife*
Vladimir, always called Volodya, *their son*
Anya, *their daughter*

Cast of characters

Others

Zoya Vorotsyntsev, *physicist*
Ilya Dvorkin, *officer of the secret police*
Colonel Lemitov, *Volodya's boss*
Colonel Bobrov, *Red Army officer in Spain*

Real Historical Characters

Lavrentiy Beria, *head of the secret police*
Vyacheslav Molotov, *Foreign Minister*

Spanish

Teresa, *literacy teacher*

Welsh

Williams Family

David Williams, *called Dai*, 'Granda'
Cara Williams, 'Grandmam'
Billy Williams, *MP for Aberowen*
Mildred, *Billy's wife*
Dave, *Billy's elder son*
Keir, *Billy's younger son*

Griffiths Family

Tommy Griffiths, *Billy Williams's political agent*
Lenny Griffiths, *Tommy's son*

Part One

THE OTHER CHEEK

1

1933

Carla knew her parents were about to have a row. The second she walked into the kitchen she felt the hostility, like the bone-deep cold of the wind that blew through the streets of Berlin before a February snowstorm. She almost turned and walked back out again.

It was unusual for them to fight. Mostly they were affectionate – too much so. Carla cringed when they kissed in front of other people. Her friends thought it was strange: their parents did not do that. She had said that to her mother, once. Mother had laughed in a pleased way and said: 'The day after our wedding, your father and I were separated by the Great War.' She had been born English, though you could hardly tell. 'I stayed in London while he came home to Germany and joined the army.' Carla had heard this story many times, but Mother never tired of telling it. 'We thought the war would last three months, but I didn't see him again for five years. All that time I longed to touch him. Now I never tire of it.'

Father was just as bad. 'Your mother is the cleverest woman I have ever met,' he had said here in the kitchen just a few days ago. 'That's why I married her. It had nothing to do with ...' He had tailed off, and Mother and he had giggled conspiratorially, as if Carla at the age of eleven knew nothing about sex. It was so embarrassing.

But once in a while they had a quarrel. Carla knew the signs. And a new one was about to erupt.

They were sitting at opposite ends of the kitchen table. Father was sombrely dressed in a dark-grey suit, starched white shirt and black satin tie. He looked dapper, as always, even though his hair was receding and his waistcoat bulged a little beneath the gold watch chain. His face was frozen in an expression of false calm. Carla knew that look. He wore it when one of the family had done something that angered him.

He held in his hand a copy of the weekly magazine for which Mother worked, *The Democrat*. She wrote a column of political and diplomatic gossip under the name of Lady Maud. Father began to read aloud. '"Our new chancellor, Herr Adolf Hitler, made his debut in diplomatic society at President Hindenburg's reception."'

The President was the head of state, Carla knew. He was elected, but he stood above the squabbles of day-to-day politics, acting as referee. The Chancellor was the premier. Although Hitler had been made chancellor, his Nazi party did not have an overall majority in the Reichstag – the German parliament – so, for the present, the other parties could restrain Nazi excesses.

Father spoke with distaste, as if forced to mention something repellent, like sewage. '"He looked uncomfortable in a formal tailcoat."'

Carla's mother sipped her coffee and looked out of the window to the street, as if interested in the people hurrying to work in scarves and gloves. She, too, was pretending to be calm, but Carla knew that she was just waiting for her moment.

The maid, Ada, was standing at the counter in an apron, slicing cheese. She put a plate in front of Father, but he ignored it. '"Herr Hitler was evidently charmed by Elisabeth Cerruti, the cultured wife of the Italian ambassador, in a rose-pink velvet gown trimmed with sable."'

Mother always wrote about what people were wearing. She said it helped the reader to picture them. She herself had fine clothes, but times were hard and she had not bought anything new for years. This morning, she looked slim and elegant in a navy-blue cashmere dress that was probably as old as Carla.

'"Signora Cerruti, who is Jewish, is a passionate Fascist, and they talked for many minutes. Did she beg Hitler to stop whipping up hatred of Jews?"' Father put the magazine down on the table with a slap.

Here it comes, Carla thought.

'You realize that will infuriate the Nazis,' he said.

'I hope so,' Mother said coolly. 'The day they're pleased with what I write, I shall give it up.'

'They're dangerous when riled.'

Mother's eyes flashed anger. 'Don't you dare condescend to me, Walter. I know they're dangerous – that's why I oppose them.'

'I just don't see the point of making them irate.'

'You attack them in the Reichstag.' Father was an elected parliamentary representative for the Social Democratic Party.

'I take part in a reasoned debate.'

This was typical, Carla thought. Father was logical, cautious, law-abiding. Mother had style and humour. He got his way by quiet persistence; she with charm and cheek. They would never agree.

Father added: 'I don't drive the Nazis mad with fury.'

'Perhaps that's because you don't do them much harm.'

Father was irritated by her quick wit. His voice became louder. 'And you think you damage them with jokes?'

'I mock them.'

'And that's your substitute for argument.'

'I believe we need both.'

Father became angrier. 'But Maud, don't you see how you're putting yourself and your family at risk?'

'On the contrary: the real danger is *not* to mock the Nazis. What would life be like for our children if Germany became a Fascist state?'

This kind of talk made Carla feel queasy. She could not bear to hear that the family was in danger. Life must go on as it always had. She wished she could sit in this kitchen for an eternity of mornings, with her parents at opposite ends of the pine table, Ada at the counter, and her brother, Erik, thumping around upstairs, late again. Why should anything change?

She had listened to political talk every breakfast-time of her life and she thought she understood what her parents did, and how they planned to make Germany a better place for everyone. But lately they had begun to talk in a different way. They seemed to think that a terrible danger loomed, but Carla could not quite imagine what it was.

Father said: 'God knows I'm doing everything I can to hold back Hitler and his mob.'

'And so am I. But when you do it, you believe you're following a sensible course.' Mother's face hardened in resentment. 'And when I do it, I'm accused of putting the family at risk.'

'And with good reason,' said Father. The row was only just getting started, but at that moment Erik came down, clattering like a horse on the stairs, and lurched into the kitchen with his school satchel swinging from his shoulder. He was thirteen, two years older than Carla, and there were unsightly black hairs sprouting from his upper lip. When they were small, Carla and Erik had played together all the time; but those days were over, and since he had grown so tall he had pretended to think that she was stupid and childish. In fact, she was smarter than he, and knew about a lot of things he did not understand, such as women's monthly cycles.

'What was that last tune you were playing?' he said to Mother.

The piano often woke them in the morning. It was a Steinway grand – inherited, like the house itself, from Father's parents. Mother played in the morning because, she said, she was too busy during the rest of the day and too tired in the evening. This morning, she had performed a Mozart sonata then a jazz tune. 'It's called "Tiger Rag",' she told Erik. 'Do you want some cheese?'

'Jazz is decadent,' Erik said.

'Don't be silly.'

Ada handed Erik a plate of cheese and sliced sausage, and he began to shovel it into his mouth. Carla thought his manners were dreadful.

Father looked severe. 'Who's been teaching you this nonsense, Erik?'

'Hermann Braun says that jazz isn't music, just Negroes making a noise.' Hermann was Erik's best friend; his father was a member of the Nazi Party.

'Hermann should try to play it.' Father looked at Mother, and his face softened. She smiled at him. He went on: 'Your mother tried to teach me ragtime, many years ago, but I couldn't master the rhythm.'

Mother laughed. 'It was like trying to get a giraffe to roller-skate.'

The fight was over, Carla saw with relief. She began to feel better. She took some black bread and dipped it in milk.

But now Erik wanted an argument. 'Negroes are an inferior race,' he said defiantly.

'I doubt that,' Father said patiently. 'If a Negro boy were brought up in a nice house full of books and paintings, and sent to an expensive school with good teachers, he might turn out to be smarter than you.'

'That's ridiculous!' Erik protested.

Mother put in: 'Don't call your father ridiculous, you foolish boy.' Her tone was mild: she had used up her anger on Father. Now she just sounded wearily disappointed. 'You don't know what you're talking about, and neither does Hermann Braun.'

Erik said: 'But the Aryan race must be superior – we rule the world!'

'Your Nazi friends don't know any history,' Father said. 'The Ancient Egyptians built the pyramids when Germans were living in caves. Arabs ruled the world in the Middle Ages – the Muslims were doing algebra when German princes could not write their own names. It's nothing to do with race.'

Carla frowned and said: 'What is it to do with, then?'

Father looked at her fondly. 'That's a very good question, and you're a bright girl to ask it.' She glowed with pleasure at his praise. 'Civilizations rise and fall – the Chinese, the Aztecs, the Romans – but no one really knows why.'

'Eat up, everyone, and put your coats on,' Mother said. 'It's getting late.'

Father pulled his watch out of his waistcoat pocket and looked at it with raised eyebrows. 'It's not late.'

'I've got to take Carla to the Francks' house,' Mother said. 'The girls' school is closed for a day – something about repairing the furnace – so Carla's going to spend today with Frieda.'

Frieda Franck was Carla's best friend. Their mothers were best friends, too. In fact, when they were young, Frieda's mother, Monika, had been in love with Father – a hilarious fact that Frieda's grandmother had revealed one day after drinking too much Sekt.

Father said: 'Why can't Ada look after Carla?'

'Ada has an appointment with the doctor.'

'Ah.'

Carla expected Father to ask what was wrong with Ada, but he nodded as if he already knew, and put his watch away. Carla wanted to ask, but something told her she should not. She made a mental note to ask Mother later. Then she immediately forgot about it.

Father left first, wearing a long black overcoat. Then Erik put on his cap – perching it as far back on his head as it would go without falling off, as was the fashion among his friends – and followed Father out of the door.

Carla and her mother helped Ada clear the table. Carla loved Ada almost as much as she loved her mother. When Carla was little, Ada had taken care of her full-time, until she was old enough to go to school, for Mother had always worked. Ada was not married yet. She was twenty-nine and homely looking, though she had a lovely, kind smile. Last summer, she had had a romance with a policeman, Paul Huber, but it had not lasted.

Carla and her mother stood in front of the mirror in the hall and put on their hats. Mother took her time. She chose a dark-blue felt, with a round crown and a narrow brim, the type all the women were wearing; but she tilted hers at a different angle, making it look chic. As Carla put on her knitted wool cap, she wondered whether she would ever have Mother's sense of style. Mother looked like a goddess of war, her long neck and chin and cheekbones carved out of white marble; beautiful, yes, but definitely not pretty. Carla had the same dark hair and green eyes, but looked more like a plump doll than a statue. Carla had once accidentally overheard her grandmother say to Mother: 'Your ugly duckling will grow into a swan, you'll see.' Carla was still waiting for it to happen.

When Mother was ready, they went out. Their home stood in a row of tall, gracious town houses in the Mitte district, the old centre of the city, built for high-ranking ministers and army officers such as Carla's grandfather, who had worked at the nearby government buildings.

Carla and her mother rode a tram along Unter den Linden, then took the S-train from Friedrich Strasse to the Zoo Station. The Francks lived in the south-western suburb of Schöneberg.

Carla was hoping to see Frieda's brother Werner, who was fourteen. She liked him. Sometimes Carla and Frieda imagined that they had each married the other's brother, and were next-door neighbours, and their children were best friends. It was just a game to Frieda, but secretly Carla was serious. Werner was handsome and grown-up and not a bit silly like Erik. In the doll's house in Carla's bedroom, the mother and father sleeping side by side in the miniature double bed were called Carla and Werner, but no one knew that, not even Frieda.

Frieda had another brother, Axel, who was seven; but he had been born with spina bifida, and had to have constant medical care. He lived in a special hospital on the outskirts of Berlin.

Mother was preoccupied on the journey. 'I hope this is going to be all right,' she muttered, half to herself, as they got off the train.

'Of course it will,' Carla said. 'I'll have a lovely time with Frieda.'

'I didn't mean that. I'm talking about my paragraph about Hitler.'

'Are we in danger? Was Father right?'

'Your father is usually right.'

'What will happen to us if we've annoyed the Nazis?'

Mother stared at her strangely for a long moment, then said: 'Dear God, what kind of a world did I bring you into?' Then she went quiet.

After a ten-minute walk they arrived at a grand villa in a big garden. The Francks were rich: Frieda's father, Ludwig, owned a factory making radio sets. Two cars stood in the drive. The large shiny black one belonged to Herr Franck. The engine rumbled, and a cloud of blue vapour rose from the tail pipe. The chauffeur, Ritter, with uniform trousers tucked into high boots, stood cap in hand ready to open the door. He bowed and said: 'Good morning, Frau von Ulrich.'

The second car was a little green two-seater. A short man with a grey beard came out of the house carrying a leather case, and touched his hat to Mother as he got into the small car. 'I wonder what Dr Rothmann is doing here so early in the morning,' Mother said anxiously.

They soon found out. Frieda's mother, Monika, came to the door; she was a tall woman with a mass of red hair. Anxiety showed on her pale face. Instead of welcoming them in, she stood squarely in the doorway as if to bar their entrance. 'Frieda has measles!' she said.

'I'm so sorry!' said Mother. 'How is she?'

'Miserable. She has a fever and a cough. But Rothmann says she'll be all right. However, she's quarantined.'

'Of course. Have you had it?'

'Yes – when I was a girl.'

'And Werner has, too – I remember he had a terrible rash all over. But what about your husband?'

'Ludi had it as a boy.'

Both women looked at Carla. She had never had measles. She realized this meant that she could not spend the day with Frieda.

Carla was disappointed, but Mother was quite shaken. 'This week's magazine is our election issue – I *can't* be absent.' She looked distraught. All the grown-ups were apprehensive about the general election to be held next Sunday. Mother and Father both feared the Nazis might do well enough to take full control of the government. 'Plus my oldest friend is visiting from London. I wonder whether Walter could be persuaded to take a day off to look after Carla?'

Monika said: 'Why don't you telephone to him?'

Not many people had phones in their homes, but the Francks did, and Carla and her mother stepped into the hall. The instrument stood on a spindly legged table near the door. Mother picked it up and gave the number of Father's office at the Reichstag, the parliament building. She got through to him and explained the situation. She listened for a minute, then looked angry. 'My magazine will urge a hundred thousand readers to campaign for the Social Democratic Party,' she said. 'Do you really have something more important than that to do today?'

Carla could guess how this argument would end. Father loved her dearly, she knew, but in all her eleven years he had never looked after her for a whole day. All her friends' fathers were the same. Men did not do that sort of thing. But Mother sometimes pretended not to know the rules women lived by.

'I'll just have to take her to the office with me, then,' Mother said into the phone. 'I dread to think what Jochmann will say.' Herr Jochmann was her boss. 'He's not much of a feminist at the best of times.' She replaced the handset without saying goodbye.

Carla hated it when they fought, and this was the second time in a day. It made the whole world seem unstable. She was much more scared of quarrels than of the Nazis.

'Come on, then,' Mother said to her, and she moved to the door.

I'm not even going to see Werner, Carla thought unhappily.

Just then Frieda's father appeared in the hall, a pink-faced man with a small black moustache, energetic and cheerful. He greeted Mother pleasantly, and she paused to speak politely to him while Monika helped him into a black topcoat with a fur collar.

He went to the foot of the stairs. 'Werner!' he shouted. 'I'm going without you!' He put on a grey felt hat and went out.

'I'm ready, I'm ready!' Werner ran down the stairs like a dancer. He was as tall as his father and more handsome, with red-blond hair worn too long. Under his arm he had a leather satchel that appeared to be full of books; in the other hand he held a pair of ice skates and a hockey stick. He paused in his rush to say: 'Good morning, Frau von Ulrich', very politely. Then in a more informal tone: 'Hello, Carla. My sister's got the measles.'

Carla felt herself blush, for no reason at all. 'I know,' she said. She tried to think of something charming and amusing to say, but came up with nothing. 'I've never had it, so I can't see her.'

'I had it when I was a kid,' he said, as if that was ever such a long time ago. 'I must hurry,' he added apologetically.

Carla did not want to lose sight of him so quickly. She followed him outside. Ritter was holding the rear door open. 'What kind of car is that?' Carla asked. Boys always knew the makes of cars.

'A Mercedes-Benz W10 limousine.'

'It looks very comfortable.' She caught a look from her mother, half surprised and half amused.

Werner said: 'Do you want a lift?'

'That would be nice.'

'I'll ask my father.' Werner put his head inside the car and said something.

Carla heard Herr Franck reply: 'Very well, but hurry up!'

She turned to her mother. 'We can go in the car!'

Mother hesitated for only a moment. She did not like Herr Franck's politics – he gave money to the Nazis – but she was not going to refuse a lift in a warm car on a cold morning. 'How very kind of you, Ludwig,' she said.

They got in. There was room for four in the back. Ritter pulled away smoothly. 'I assume you're going to Koch Strasse?' said Herr Franck. Many newspapers and book publishers had their offices in the same street in the Kreuzberg district.

'Please don't go out of your way. Leipziger Strasse would be fine.'

'I'd be happy to take you to the door – but I suppose you don't want your leftist colleagues to see you getting out of the car of a bloated plutocrat.' His tone was somewhere between humorous and hostile.

Mother gave him a charming smile. 'You're not bloated, Ludi – just a little plump.' She patted the front of his coat.

He laughed. 'I asked for that.' The tension eased. Herr Franck picked up the speaking tube and gave instructions to Ritter.

Carla was thrilled to be in a car with Werner, and she wanted to make the most of it by talking to him, but at first she could not think what to speak about. She really wanted to say: 'When you're older, do you think you might marry a girl with dark hair and green eyes, about three years younger than yourself, and clever?' Eventually she pointed to his skates and said: 'Do you have a match today?'

'No, just practice after school.'

'What position do you play in?' She knew nothing about ice hockey, but there were always positions in team games.

'Right wing.'

'Isn't it a rather dangerous sport?'

'Not if you're quick.'

'You must be ever such a good skater.'

'Not bad,' he said modestly.

Once again, Carla caught her mother watching her with an enigmatic little smile. Had she guessed how Carla felt about Werner? Carla felt another blush coming.

Then the car came to a stop outside a school building, and Werner got out. 'Goodbye, everyone!' he said, and ran through the gates into the yard.

Ritter drove on, following the south bank of the Landwehr Canal. Carla looked at the barges, their loads of coal topped with snow like mountains. She felt a sense of disappointment. She had contrived to spend longer with Werner, by hinting that she wanted a lift, then she had wasted the time talking about ice hockey.

What would she have liked to have talked to him about? She did not know.

Herr Franck said to Mother: 'I read your column in *The Democrat*.'

'I hope you enjoyed it.'

'I was sorry to see you writing disrespectfully about our chancellor.'

'Do you think journalists should write respectfully about politicians?' Mother replied cheerfully. 'That's radical. The Nazi press would have to be polite about my husband! They wouldn't like that.'

'Not all politicians, obviously,' Franck said irritably.

They crossed the teeming junction of Potsdamer Platz. Cars and trams vied with horse-drawn carts and pedestrians in a chaotic melee.

Mother said: 'Isn't it better for the press to be able to criticize everyone equally?'

'A wonderful idea,' he said. 'But you socialists live in a dream world. We practical men know that Germany cannot live on ideas. People must have bread and shoes and coal.'

'I quite agree,' Mother said. 'I could use more coal myself. But I want Carla and Erik to grow up as citizens of a free country.'

'You overrate freedom. It doesn't make people happy. They prefer leadership. I want Werner and Frieda and poor Axel to grow up in a country that is proud, and disciplined, and united.'

'And in order to be united, we need young thugs in brown shirts to beat up elderly Jewish shopkeepers?'

'Politics is rough. Nothing we can do about it.'

'On the contrary, you and I are leaders, Ludwig, in our different ways. It's our responsibility to make politics less rough – more honest, more rational, less violent. If we do not do that, we fail in our patriotic duty.'

Herr Franck bristled.

Carla did not know much about men, but she realized that they did not like to be lectured on their duty by women. Mother must have forgotten to press her charm switch this morning. But everyone was tense. The coming election had them all on edge.

The car reached Leipziger Platz. 'Where may I drop you?' Herr Franck said coldly.

'Just here will be fine,' said Mother.

Franck tapped on the glass partition. Ritter stopped the car and hurried to open the door.

Mother said: 'I do hope Frieda gets better soon.'

'Thank you.'

They got out and Ritter closed the door.

The office was several minutes' walk away, but Mother clearly had not wanted to stay any longer in the car. Carla hoped Mother was not going to quarrel permanently with Herr Franck. That might make it difficult for her to see Frieda and Werner. She would hate that.

They set off at a brisk pace. 'Try not to make a nuisance of yourself at the office,' Mother said. The note of genuine pleading in her voice touched Carla, making her feel ashamed of causing her mother worry. She resolved to behave perfectly.

Mother greeted several people on the way: she had been writing her column for as long as Carla could remember, and was well known in the press corps. They all called her 'Lady Maud' in English.

Near the building in which *The Democrat* had its office, they saw someone they knew: Sergeant Schwab. He had fought with Father in the Great War, and still wore his hair brutally short in the military style. After the war he had worked as a gardener, first for Carla's grandfather and later for her father; but he had stolen money from Mother's purse and Father had sacked him. Now he was wearing the ugly military uniform of the Storm troopers, the Brownshirts, who were not soldiers but Nazis who had been given the authority of auxiliary policemen.

Schwab said loudly: 'Good morning, Frau von Ulrich!' as if he felt no shame at all about being a thief. He did not even touch his cap.

Mother nodded coldly and walked past him. 'I wonder what he's doing here,' she muttered uneasily as they went inside.

The magazine had the first floor of a modern office building. Carla knew a child would not be welcome, and she hoped they could reach Mother's office without being seen. But they met Herr Jochmann on the stairs. He was a heavy man with thick spectacles. 'What's this?' he said brusquely, speaking around the cigarette in his mouth. 'Are we running a kindergarten now?'

Mother did not react to his rudeness. 'I was thinking over your comment the other day,' she said. 'About how young people imagine journalism is a glamorous profession, and don't understand how much hard work is necessary.'

He frowned. 'Did I say that? Well, it's certainly true.'

'So I brought my daughter here to see the reality. I think it will be good for her education, especially if she becomes a writer. She will make a report on the visit to her class. I felt sure you would approve.'

Mother was making this up as she went along, but it sounded convincing, Carla thought. She almost believed it herself. The charm switch had been turned to the On position at last.

Jochmann said: 'Don't you have an important visitor from London coming today?'

'Yes, Ethel Leckwith, but she's an old friend – she knew Carla as a baby.'

Jochmann was somewhat mollified. 'Hmm. Well, we have an editorial meeting in five minutes, as soon as I've bought some cigarettes.'

'Carla will get them for you.' Mother turned to her. 'There is a tobacconist three doors down. Herr Jochmann likes the Roth-Händle brand.'

'Oh, that will save me a trip.' Jochmann gave Carla a one-mark coin.

Mother said to her: 'When you come back, you'll find me at the top of the stairs, next to the fire alarm.' She turned away and took Jochmann's arm confidentially. 'I thought last week's issue was possibly our best ever,' she said as they went up.

Carla ran out into the street. Mother had got away with it, using her characteristic mixture of boldness and flirting. She sometimes said: 'We women have to deploy every weapon we have.' Thinking about it, Carla realized that she had used Mother's tactics to get a lift from Herr Franck. Perhaps she was like her mother after all. That might be why Mother had given her that curious little smile: she was seeing herself thirty years ago.

There was a queue in the shop. Half the journalists in Berlin seemed to be buying their supplies for the day. At last Carla got a pack of Roth-Händle and returned to the *Democrat* building. She found the fire alarm easily – it was a big lever fixed to the wall – but Mother was not in her office. No doubt she had gone to that editorial meeting.

Carla walked along the corridor. All the doors were open, and most of the rooms were empty but for a few women who might have been typists and secretaries. At the back of the building, around a corner, was a closed door marked 'Conference Room'. Carla could hear male voices raised in argument. She tapped on the door, but there was no response. She hesitated, then turned the handle and went in.

The room was full of tobacco smoke. Eight or ten people sat around a long table. Mother was the only woman. They fell silent, apparently surprised, when Carla went up to the head of the table and handed Jochmann the cigarettes and change. Their silence made her think she had done wrong to come in.

But Jochmann just said: 'Thank you.'

'You're welcome, sir,' she said, and for some reason she gave a little bow.

The men laughed. One said: 'New assistant, Jochmann?' Then she knew it was all right.

She left the room quickly and returned to Mother's office. She did not take off her coat – the place was cold. She looked around. On the desk were a phone, a typewriter, and stacks of paper and carbon paper.

Next to the phone was a photograph in a frame, showing Carla and Erik with Father. It had been taken a couple of years ago on a sunny day at the beach by the Wannsee lake, fifteen miles from the centre of Berlin. Father was wearing shorts. They were all laughing. That was before Erik had started to pretend to be a tough, serious man.

The only other picture, hanging on the wall, showed Mother with the social-democratic hero Friedrich Ebert, who had been the first President of Germany after the war. It had been taken about ten years ago. Carla smiled at Mother's shapeless, low-waisted dress and boyish haircut: they must have been fashionable at the time.

The bookshelf held social directories, phone books, dictionaries in several languages, and atlases, but nothing to read. In the desk drawer were pencils, several new pairs of formal gloves still wrapped in tissue paper, a packet of sanitary towels, and a notebook with names and phone numbers.

Carla reset the desk calendar to today's date, Monday 27 February 1933. Then she put a sheet of paper into the typewriter. She typed her full name, Heike Carla von Ulrich. At the age of five she had announced that she did not like the name Heike and she wanted everyone to use her second name, and somewhat to her surprise her family had complied.

Each key of the typewriter caused a metal rod to rise up and strike the paper through an inky ribbon, printing a letter. When by accident she pressed two keys, the rods got stuck. She tried to prise them apart but she could not. Pressing another key did not help: now there were three jammed rods. She groaned: she was in trouble already.

A noise from the street distracted her. She went to the window. A dozen Brownshirts were marching along the middle of the road, shouting slogans: 'Death to all Jews! Jews go to hell!' Carla could not understand why they got so angry about Jews, who seemed the same as everyone else apart from their religion. She was startled to see Sergeant Schwab at the head of the troop. She had felt sorry for him when he was sacked, for she knew he would find it hard to get another job. There were millions of men looking for jobs in Germany: Father said it was a depression. But Mother had said: 'How can we have a man in our house who steals?'

Their chant changed. 'Smash Jew papers!' they said in unison. One of them threw something, and a rotten vegetable splashed on the door of a national newspaper. Then, to Carla's horror, they turned towards the building she was in.

She drew back and peeped around the edge of the window frame, hoping that they could not see her. They stopped outside, still chanting. One threw a stone. It hit Carla's window without breaking it, but all the same she gave a little scream of fear. A moment later, one of the typists came in, a young woman in a red beret. 'What's the matter?' she said, then she looked out of the window. 'Oh, hell.'

The Brownshirts entered the building, and Carla heard boots on the stairs. She was scared: what were they going to do?

Sergeant Schwab came into Mother's office. He hesitated, seeing the two females; then seemed to screw up his nerve. He picked up the typewriter and threw it through the window, shattering the glass. Carla and the typist both screamed.

More Brownshirts passed the doorway, shouting their slogans.

Schwab grabbed the typist by the arm and said: 'Now, darling, where's the office safe?'

'In the file room!' she said in a terrified voice.

'Show me.'

'Yes, anything!'

He marched her out of the room.

Carla started to cry, then stopped herself.

She thought of hiding under the desk, but hesitated. She did not want to show them how scared she was. Something inside her wanted to defy them.

But what should she do? She decided to warn Mother.

She stepped to the doorway and looked along the corridor. The Brownshirts were going in and out of the offices but had not reached the far end. Carla did not know whether the people in the conference room could hear the commotion. She ran along the corridor as fast as she could, but a scream stopped her. She looked into a room and saw Schwab shaking the typist with the red beret, yelling: 'Where's the key?'

'I don't know, I swear I'm telling the truth!' the typist cried.

Carla was outraged. Schwab had no right to treat a woman that way. She shouted: 'Leave her alone, Schwab, you thief!'

Schwab looked at her with hatred in his eyes, and suddenly she was ten times more frightened. Then his gaze shifted to someone behind her, and he said: 'Get the kid out of the damn way.'

She was picked up from behind. 'Are you a little Jew?' said a man's voice. 'You look it, with all that dark hair.'

That terrified her. 'I'm not Jewish!' she screamed.

The Brownshirt carried her back along the corridor and put her down in Mother's office. She stumbled and fell to the floor. 'Stay in here,' he said, and he went away.

Carla got to her feet. She was not hurt. The corridor was full of Brownshirts now, and she could not get to her mother. But she had to summon help.

She looked out of the smashed window. A small crowd was gathering on the street. Two policemen stood among the onlookers, chatting. Carla shouted at them: 'Help! Help, police!'

They saw her and laughed.

That infuriated her, and anger made her less frightened. She looked outside the office again. Her gaze lit on the fire alarm on the wall. She reached up and grasped the handle.

She hesitated. You were not supposed to sound the alarm unless there was a fire, and a notice on the wall warned of dire penalties.

She pulled the handle anyway.

For a moment nothing happened. Perhaps the mechanism was not working.

Then there came a loud, harsh klaxon sound, rising and falling, which filled the building.

Almost immediately the people from the conference room appeared at the far end of the corridor. Jochmann was first. 'What the devil is going on?' he said angrily, shouting over the noise of the alarm.

One of the Brownshirts said: 'This Jew Communist rag has insulted our leader, and we're closing it down.'

'Get out of my office!'

The Brownshirt ignored him and went into a side room. A moment later there was a female scream and a crash that sounded like a steel desk being overturned.

Jochmann turned to one of his staff. 'Schneider – call the police immediately!'

Carla knew that would be no good. The police were here already, doing nothing.

Mother pushed through the knot of people and came running along the corridor. 'Are you all right?' she cried. She threw her arms around Carla.

Carla did not want to be comforted like a child. Pushing her mother away, she said: 'I'm fine, don't worry.'

Mother looked around. 'My typewriter!'

'They threw it through the window.' Carla realized that now she would not get into trouble for jamming the mechanism.

'We must get out of here.' Mother snatched up the desk photo then took Carla's hand, and they hurried out of the room.

No one tried to stop them running down the stairs. Ahead of them, a well-built young man who might have been one of the reporters had a Brownshirt in a headlock and was dragging him out of the building. Carla and her mother followed the pair out. Another Brownshirt came behind them.

The reporter approached the two policemen, still dragging the Brownshirt. 'Arrest this man,' he said. 'I found him robbing the office. You will find a stolen jar of coffee in his pocket.'

'Release him, please,' said the older of the two policemen.

Reluctantly, the reporter let the Brownshirt go.

The second Brownshirt stood beside his colleague.

'What is your name, sir?' the policeman asked the reporter.

'I am Rudolf Schmidt, chief parliamentary correspondent of *The Democrat.*'

'Rudolph Schmidt, I am arresting you on a charge of assaulting the police.'

'Don't be ridiculous. I caught this man stealing!'

The policeman nodded to the two Brownshirts. 'Take him to the station house.'

They grabbed Schmidt by the arms. He seemed about to struggle, then changed his mind. 'Every detail of this incident will appear in the next edition of *The Democrat!*' he said.

'There will never be another edition,' the policeman said. 'Take him away.'

A fire engine arrived and half a dozen firemen jumped out. Their leader spoke brusquely to the police. 'We need to clear the building,' he said.

'Go back to your fire station, there's no fire,' said the older policeman. 'It's just the Storm troopers closing down a Communist magazine.'

'That's no concern of mine,' the fireman said. 'The alarm has been sounded, and our first task is to get everyone out, Storm troopers and all. We'll manage without your help.' He led his men inside.

Carla heard her mother say: 'Oh, no!' She turned and saw that Mother was staring at her typewriter, which lay on the pavement where it had fallen. The metal casing had dropped away, exposing the links between keys and rods. The keyboard was twisted out of shape, one end of the roller had become detached, and the bell that sounded for the end of a line lay forlornly on the ground. A typewriter was not a precious object, but Mother looked as if she might cry.

The Brownshirts and the staff of the magazine came out of the building, herded by firemen. Sergeant Schwab was resisting, shouting angrily: 'There's no fire!' The firemen just shoved him on.

Jochmann came out and said to Mother: 'They didn't have time to do much damage – the firemen stopped them. Whoever sounded the alarm did us a great service!'

Carla had been worried that she would be reprimanded for causing a false alarm. Now she realized that she had done exactly the right thing.

She took her mother's hand. That seemed to jerk Mother out of her momentary fit of grief. She wiped her eyes with her sleeve, an unusual act that revealed how badly shaken she was: if Carla had done that, she would have been told to use her handkerchief. 'What do we do now?' Mother never said that – she always knew what to do next.

Carla became aware of two people standing nearby. She looked up. One was a woman about the same age as Mother, very pretty, with an air of authority. Carla knew her, but could not place her. Beside her was a man young enough to be her son. He was slim, and not very tall, but he looked like a movie star. He had a handsome face that would have been almost too pretty except that his nose was flattened and misshapen. Both newcomers looked shocked, and the young man was white with anger.

The woman spoke first, and she used the English language. 'Hello,

Maud,' she said, and the voice was distantly familiar to Carla. 'Don't you recognize me?' she went on. 'I'm Eth Leckwith, and this is Lloyd.'

(ii)

Lloyd Williams found a boxing club in Berlin where he could do an hour's training for a few pennies. It was in a working-class district called Wedding, north of the city centre. He exercised with the Indian clubs and the medicine ball, skipped rope, hit the punch bag, and then put on a helmet and did five rounds in the ring. The club coach found him a sparring partner, a German his own age and size – Lloyd was a welterweight. The German boy had a nice fast jab that came from nowhere and hurt Lloyd several times, until Lloyd hit him with a left hook and knocked him down.

Lloyd had been raised in a rough neighbourhood, the East End of London. At the age of twelve he had been bullied at school. 'Same thing happened to me,' his stepfather, Bernie Leckwith, had said. 'Cleverest boy in school, and you get picked on by the class *shlammer.*' Dad was Jewish – his mother had spoken only Yiddish. He had taken Lloyd to the Aldgate Boxing Club. Ethel had been against it, but Bernie had overruled her, something that did not happen often.

Lloyd had learned to move fast and punch hard, and the bullying had stopped. He had also got the broken nose that made him look less of a pretty boy. And he had discovered a talent. He had quick reflexes and a combative streak, and he had won prizes in the ring. The coach was disappointed that he wanted to go to Cambridge University instead of turning professional.

He showered and put his suit back on, then went to a workingmen's bar, bought a glass of draft beer, and sat down to write to his half-sister, Millie, about the incident with the Brownshirts. Millie was envious of him taking this trip with their mother, and he had promised to send her frequent bulletins.

Lloyd had been shaken by this morning's fracas. Politics was part of everyday life for him: his mother had been a Member of Parliament, his father was a local councillor in London, and he himself was London Chairman of the Labour League of Youth. But it had always been a matter of debating and voting – until today. He had never before seen an office trashed by uniformed thugs while the police looked on smiling. It was politics with the gloves off, and it had shocked him.

'Could this happen in London, Millie?' he wrote. His first instinct was to think that it could not. But Hitler had admirers among British industrialists and newspaper proprietors. Only a few months ago the rogue MP Sir Oswald Mosley had started the British Union of Fascists. Like the Nazis, they had to strut up and down in military-style uniforms. What next?

He finished his letter and folded it, then caught the S-train back into the city centre. He and his mother were going to meet Walter and Maud von Ulrich for dinner. Lloyd had been hearing about Maud all his life. She and his mother were unlikely friends: Ethel had started her working life as a maid in a grand house owned by Maud's family. Later they had been suffragettes together, campaigning for votes for women. During the war they had produced a feminist newspaper, *The Soldier's Wife*. Then they had quarrelled over political tactics and become estranged.

Lloyd could remember vividly the von Ulrich family's trip to London in 1925. He had been ten, old enough to feel embarrassed that he spoke no German while Erik and Carla, aged five and three, were bilingual. That was when Ethel and Maud had patched up their quarrel.

He made his way to the restaurant, Bistro Robert. The interior was art deco, with unforgivingly rectangular chairs and tables, and elaborate iron lampstands with coloured glass shades; but he liked the starched white napkins standing to attention beside the plates.

The other three were already there. The women were striking, he realized as he approached the table: both poised, well dressed, attractive and confident. They were getting admiring glances from other diners. He wondered how much of his mother's modish dress sense had been picked up from her aristocratic friend.

When they had ordered, Ethel explained her trip. 'I lost my parliamentary seat in 1931,' she said. 'I hope to win it back at the next election, but meanwhile I have to make a living. Fortunately, Maud, you taught me to be a journalist.'

'I didn't teach you much,' Maud said. 'You had a natural talent.'

'I'm writing a series of articles about the Nazis for the *News Chronicle*, and I have a contract to write a book for a publisher called Victor Gollancz. I brought Lloyd as my interpreter – he's studying French and German.'

Lloyd observed her proud smile and felt he did not deserve it. 'My translation skills have not been much tested,' he said. 'So far, we've mostly met people like you, who speak perfect English.'

Lloyd had ordered breaded veal, a dish he had never even seen in England. He found it delicious. While they were eating, Walter said to him: 'Shouldn't you be at school?'

'Mam thought I would learn more German this way, and the school agreed.'

'Why don't you come and work for me in the Reichstag for a while? Unpaid, I'm afraid, but you'd be speaking German all day.'

Lloyd was thrilled. 'I'd love to. What a marvellous opportunity!'

'If Ethel can spare you,' Walter added.

She smiled. 'Perhaps I can have him back now and again, when I really need him?'

'Of course.'

Ethel reached across the table and touched Walter's hand. It was an intimate gesture, and Lloyd realized that the bond between these three was very close. 'How kind you are, Walter,' she said.

'Not really. I can always use a bright young assistant who understands politics.'

Ethel said: 'I'm not sure I understand politics any more. What on earth is happening here in Germany?'

Maud said: 'We were doing all right in the mid-twenties. We had a democratic government and a growing economy. But everything was ruined by the Wall Street crash of 1929. Now we're in the depths of a depression.' Her voice shook with an emotion that seemed close to grief. 'You can see a hundred men standing in line for one advertised job. I look at their faces. They're desperate. They don't know how they're going to feed their children. Then the Nazis offer them hope, and they ask themselves: What have I got to lose?'

Walter seemed to think she might be overstating the case. In a more cheerful tone he said: 'The good news is that Hitler has failed to win over a majority of Germans. In the last election the Nazis got a third of the votes. Nevertheless, they were the largest party, but fortunately Hitler only leads a minority government.'

'That's why he demanded another election,' Maud put in. 'He needs an overall majority to turn Germany into the brutal dictatorship he wants.'

'Will he get it?' Ethel asked.

'No,' said Walter.

'Yes,' said Maud.

Walter said: 'I don't believe the German people will ever actually vote for a dictatorship.'

'But it won't be a fair election!' Maud said angrily: 'Look what happened to my magazine today. Anyone who criticizes the Nazis is in danger. Meanwhile, their propaganda is everywhere.'

Lloyd said: 'Nobody seems to fight back!' He wished that he had arrived a few minutes earlier at the *Democrat* office that morning, so that he could have punched a few Brownshirts. He realized he was making a fist, and forced himself to open his hand. But the indignation did not go away. 'Why don't left-wingers raid the offices of Nazi magazines? Give them a taste of their own medicine!'

'We must not meet violence with violence!' Maud said emphatically. 'Hitler is looking for an excuse to crack down – to declare a national emergency, sweep away civil rights, and put his opponents in jail.' Her voice took on a pleading note. 'We must avoid giving him that pretext – no matter how hard it is.'

They finished their meal. The restaurant began to empty out. As their coffee was served, they were joined by the owner, Walter's distant cousin Robert von Ulrich, and the chef, Jörg. Robert had been a diplomat at the Austrian Embassy in London before the Great War, while Walter was doing the same thing at the German Embassy there – and falling in love with Maud.

Robert resembled Walter, but was more fussily dressed, with a gold pin in his tie, seals on his watch chain, and heavily slicked hair. Jörg was younger, a blond man with delicate features and a cheerful smile. The two had been prisoners of war together in Russia. Now they lived in an apartment over the restaurant.

They reminisced about the wedding of Walter and Maud, held in great secrecy on the eve of the war. There had been no guests, but Robert and Ethel had been best man and bridesmaid. Ethel said: 'We had champagne at the hotel, then I tactfully said that Robert and I would leave, and Walter –' she suppressed a fit of giggles – 'Walter said: "Oh, I assumed we would all have dinner together"!'

Maud chuckled. 'You can imagine how pleased I was about that!'

Lloyd looked into his coffee, feeling embarrassed. He was eighteen and a virgin, and honeymoon jokes made him uncomfortable.

More sombrely, Ethel asked Maud: 'Do you ever hear from Fitz these days?'

Lloyd knew that the secret wedding had caused a terrible rift between Maud and her brother, Earl Fitzherbert. Fitz had disowned her because she had not gone to him, as head of the family, and asked his permission to marry.

Maud shook her head sadly. 'I wrote to him that time we went to London, but he refused even to see me. I hurt his pride by marrying Walter without telling him. My brother is an unforgiving man, I'm afraid.'

Ethel paid the bill. Everything in Germany was cheap if you had foreign currency. They were about to get up and leave when a stranger came to the table and, uninvited, pulled up a chair. He was a heavy man with a small moustache in the middle of a round face.

He wore a Brownshirt uniform.

Robert said coldly: 'What may I do for you, sir?'

'My name is Criminal Commissar Thomas Macke.' He grabbed a passing waiter by the arm and said: 'Bring me a coffee.'

The waiter looked enquiringly at Robert, who nodded.

'I work in the political department of the Prussian police,' Macke went on. 'I am in charge of the Berlin intelligence section.'

Lloyd translated for his mother in a low voice.

'However,' said Macke, 'I wish to speak to the proprietor of the restaurant about a personal matter.'

Robert said: 'Where did you work a month ago?'

The unexpected question startled Macke, and he replied immediately: 'At the police station in Kreuzberg.'

'And what was your job there?'

'I was in charge of records. Why do you ask?'

Robert nodded as if he had expected something like this. 'So you have gone from a job as a filing clerk to head of the Berlin intelligence section. Congratulations on your rapid promotion.' He turned to Ethel. 'When Hitler became Chancellor at the end of January, his henchman Hermann Göring took the role of Interior Minister of Prussia – in charge of the largest police force in the world. Since then, Göring has been firing policemen wholesale and replacing them with Nazis.' He turned back to Macke and said sarcastically: 'However, in the case of our surprise guest I'm sure the promotion was purely on merit.'

Macke flushed, but kept his temper. 'As I said, I wish to speak to the proprietor about something personal.'

'Please come and see me in the morning. Would ten o'clock suit you?'

Macke ignored this suggestion. 'My brother is in the restaurant business,' he ploughed on.

'Ah! Perhaps I know him. Macke is the name? What kind of establishment does he run?'

'A small place for working men in Friedrichshain.'

'Ah. Then it isn't likely that I have met him.'

Lloyd was not sure that it was wise for Robert to be so waspish. Macke was rude, and did not deserve kindness, but he could probably make serious trouble.

Macke went on: 'My brother would like to buy this restaurant.'

'Your brother wants to move up in the world, as you have.'

'We are prepared to offer you twenty thousand marks, payable over two years.'

Jörg burst out laughing.

Robert said: 'Permit me to explain something to you, Commissar. I am an Austrian count. Twenty years ago, I had a castle and a large country estate in Hungary where my mother and sister lived. In the war I lost my family, my castle, my lands, and even my country, which was ... miniaturized.' His tone of amused sarcasm had gone, and his voice became gruff with emotion. 'I came to Berlin with nothing but the address of Walter von Ulrich, my distant cousin. Nevertheless, I managed to open this restaurant.' He swallowed. 'It is all I have.' He paused, and drank some coffee. The others around the table were silent. He regained his poise, and something of his superior tone of voice. 'Even if you offered a generous price – which you have not – I would still refuse, because I would be selling my whole life. I have no wish to be rude to you, even though you have behaved unpleasantly. But my restaurant is not for sale at any price.' He stood up and held out his hand to shake. 'Goodnight, Commissar Macke.'

Macke automatically shook hands, then looked as if he regretted it. He stood up, clearly angry. His fat face was a purplish colour. 'We will talk again,' he said, and he walked out.

'What an oaf,' said Jörg.

Walter said to Ethel: 'You see what we have to put up with? Just because he wears that uniform, he can do anything he likes!'

What had bothered Lloyd was Macke's confidence. He had seemed to feel sure that he could buy the restaurant at the price he named. He reacted to Robert's refusal as if it was no more than a temporary setback. Were the Nazis already so powerful?

This was the kind of thing Oswald Mosley and his British Fascists wanted – a country in which the rule of law was replaced by bullying and beating. How could people be so damn stupid?

They put on their coats and hats and said goodnight to Robert and Jörg. As soon as they stepped outside, Lloyd smelled smoke – not tobacco, but something else. The four of them got into Walter's car, a BMW Dixi 3/15, which Lloyd knew was a German-manufactured Austin Seven.

As they drove through the Tiergarten park, two fire engines overtook them, bells clanging. 'I wonder where the fire is,' said Walter.

A moment later, they saw the glow of flames through the trees. Maud said: 'It seems to be near the Reichstag.'

Walter's tone changed. 'We'd better take a look,' he said worriedly, and he made a sudden turn.

The smell of smoke grew stronger. Over the tops of the trees Lloyd could see flames shooting skywards. 'It's a *big* fire,' he said.

They emerged from the park on to the Königs Platz, the broad plaza between the Reichstag building and the Kroll Opera House opposite. The Reichstag was ablaze. Red and yellow light danced behind the classical rows of windows. Flame and smoke jetted up through the central dome. 'Oh, no!' said Walter, and to Lloyd he sounded stricken with grief. 'Oh, God in heaven, no.'

He stopped the car and they all got out.

'This is a catastrophe,' said Walter.

Ethel said: 'Such a beautiful old building.'

'I don't care about the building,' Walter said surprisingly. 'It's our democracy that's on fire.'

A small crowd watched from a distance of about fifty yards. In front of the building, fire engines were lined up, their hoses already playing on the flames, water jetting in through broken windows. A handful of policemen stood around doing nothing. Walter spoke to one of them. 'I am a Reichstag deputy,' he said. 'When did this start?'

'An hour ago,' the policeman said. 'We've got one of them that did it – a man with nothing on but his trousers! He used his clothes to start the fire.'

'You should put up a rope cordon,' Walter said with authority. 'Keep people at a safe distance.'

'Yes, sir,' said the policeman, and went off.

Lloyd slipped away from the others and moved nearer to the building. The firemen were bringing the blaze under control: there was less flame and more smoke. He walked past the fire engines and approached a window. It did not seem very dangerous, and anyway his curiosity overcame his sense of self-preservation – as usual.

When he peered through a window he saw that the destruction was severe: walls and ceilings had collapsed into piles of rubble. As well as firemen he saw civilians in coats – presumably Reichstag officials – moving around in the debris, assessing the damage. Lloyd went to the entrance and climbed the steps.

Two black Mercedes cars roared up just as the police were erecting their cordon. Lloyd looked on with interest. Out of the second car jumped a man in a light-coloured trench coat and a floppy black hat. He had a narrow moustache under his nose. Lloyd realized that he was looking at the new Chancellor, Adolf Hitler.

Behind Hitler followed a taller man in the black uniform of the *Schutzstaffel*, the SS, his personal bodyguard. Limping after them came the Jew-hating propaganda chief Joseph Goebbels. Lloyd recognized them from newspaper photographs. He was so fascinated to see them close up that he forgot to be horrified.

Hitler ran up the steps two at a time, heading directly towards Lloyd. On impulse, Lloyd pushed open the big door and held it wide for the Chancellor. With a nod to him, Hitler walked in, and his entourage followed.

Lloyd joined them. No one spoke to him. Hitler's people seemed to assume he was one of the Reichstag staff, and vice versa.

There was a foul smell of wet ashes. Hitler and his party stepped over charred beams and hosepipes, treading in mucky puddles. In the entrance hall stood Hermann Göring, a camel-hair coat covering his huge belly, his hat turned up in front, Potsdam-fashion. This was the man who was packing the police force with Nazis, Lloyd thought, recalling the conversation in the restaurant.

As soon as Göring saw Hitler he shouted: 'This is the beginning of the Communist uprising! Now they'll strike out! There's not a minute to waste!'

Lloyd felt weirdly as if he were in the audience at the theatre, and these powerful men were being played by actors.

Hitler was even more histrionic than Göring. 'There will be no mercy now!' he shrieked. He sounded as if he was addressing a stadium. 'Anyone who stands in our way will be butchered.' He trembled as he worked himself up into a fury. 'Every Communist functionary will be shot where he is found. The Communist deputies to the Reichstag must be hanged this very night.' He looked as if he would burst.

But there was something artificial about it all. Hitler's hatred seemed real, but the outburst was also a performance, put on for the benefit of those around him, his own people and others. He was an actor, feeling a genuine emotion but amplifying it for the audience. And it was working, Lloyd saw: everyone within earshot was staring, mesmerized.

Göring said: 'My Führer, this is my chief of political police, Rudolf Diels.' He indicated a slim, dark-haired man at his side. 'He has already arrested one of the perpetrators.'

Diels was not hysterical. Calmly he said: 'Marinus van der Lubbe, a Dutch construction worker.'

'And a Communist!' Göring said triumphantly.

Diels said: 'Expelled from the Dutch Communist Party for starting fires.'

'I knew it!' said Hitler.

Lloyd saw that Hitler was determined to blame the Communists, regardless of the facts.

Diels said deferentially: 'From my first interrogation of the man, I have to say that it is clear that he is a lunatic, working alone.'

'Nonsense!' Hitler cried. 'This was planned long in advance. But they miscalculated! They don't understand that the people are on our side.'

Göring turned to Diels. 'The police are on an emergency footing from this moment,' he said. 'We have lists of Communists – Reichstag deputies, local government elected representatives, Communist Party organizers and activists. Arrest them all – tonight! Firearms should be used ruthlessly. Interrogate them without mercy.'

'Yes, Minister,' said Diels.

Lloyd realized that Walter had been right to worry. This was the pretext the Nazis had been looking for. They were not going to listen to anyone who said that the fire had been started by a lone madman. They wanted a Communist plot so that they could announce a crackdown.

Göring looked down with distaste at the muck on his shoes. 'My official residence is only a minute away, but it is fortunately unaffected by the fire, my Führer,' he said. 'Perhaps we should adjourn there?'

'Yes. We have much to discuss.'

Lloyd held the door and they all went out. As they drove away, he stepped over the police cordon and rejoined his mother and the von Ulrichs.

Ethel said: 'Lloyd! Where have you been? I was worried sick!'

'I went inside,' he said.

'What? How?'

'No one stopped me. It's all chaos and confusion.'

His mother threw her hands in the air. 'He has no sense of danger,' she said.

'I met Adolf Hitler.'

Walter said: 'Did he say anything?'

'He's blaming the Communists for the fire. There's going to be a purge.'

'God help us,' said Walter.

(iii)

Thomas Macke was still smarting from the sarcasm of Robert von Ulrich. 'Your brother wants to move up in the world, as you have,' von Ulrich had said.

Macke wished he had thought to reply: 'And why should we not? We are as good as you, you arrogant popinjay.' Now he yearned for revenge. But for a few days he was too busy to do anything about it.

The headquarters of the Prussian secret police were in a large, elegant building of classical architecture at No. 8 Prinz Albrecht Strasse in the government quarter. Macke felt proud every time he walked through the door.

It was a hectic time. Four thousand Communists had been arrested within twenty-four hours of the Reichstag fire, and more were being rounded up every hour. Germany was being cleansed of a plague, and to Macke the Berlin air already tasted purer.

But the police files were not up to date. People had moved house, elections had been lost and won, old men had died and young men had taken their places. Macke was in charge of a group updating the records, finding new names and addresses.

He was good at this. He liked registers, directories, street maps, news clippings, any kind of list. His talents had not been valued at the Kreuzberg police station, where criminal intelligence was simply beating up suspects until they named names. He was hoping to be better appreciated here.

Not that he had any problem with beating up suspects. In his office at the back of the building he could hear the screams of men and women being tortured in the basement, but it did not bother him. They were traitors, subversives and revolutionaries. They had ruined Germany with their strikes, and they would do worse if they got the chance. He had no sympathy for them. He only wished Robert von Ulrich was among them, groaning in agony and begging for mercy.

It was eight o'clock in the evening on Thursday 2 March before he got a chance to check on Robert.

He sent his team home, and took a sheaf of updated lists upstairs to his boss, Criminal Inspector Kringelein. Then he returned to the files.

He was in no hurry to go home. He lived alone. His wife, an undisciplined woman, had gone off with a waiter from his brother's restaurant, saying she wanted to be free. There were no children.

He began to comb the files.

He had already established that Robert von Ulrich had joined the Nazi Party in 1923 and had left two years later. That in itself did not mean much. Macke needed more.

The filing system was not as logical as he would have liked. All in all, he was disappointed in the Prussian police. The rumour was that Göring was equally unimpressed, and planned to detach the political and intelligence departments from the regular force and form them into a new, more efficient secret police force. Macke thought that was a good idea.

Meanwhile, he failed to find Robert von Ulrich in any of the regular files. Perhaps that was not merely a sign of inefficiency. The man might be blameless. As an Austrian count, he was unlikely to be a Communist or a Jew. It seemed the worst that could be said of him was that his cousin Walter was a Social Democrat. That was not a crime – not yet.

Macke now realized that he should have done this research before approaching the man. But he had gone ahead without full information. He might have known that was a mistake. In consequence, he had been forced to submit to condescension and sarcasm. He had felt humiliated. But he would get his own back.

He began to go through miscellaneous papers in a dusty cupboard at the back of the room.

The name of von Ulrich did not appear here either, but there was one document missing.

According to the list pinned to the inside of the cupboard door, there should have been a file of 117 pages entitled 'Vice Establishments'. It sounded like a survey of Berlin's nightclubs. Macke could guess why it was not here. It must have been in use recently: all the more decadent night spots had been closed down when Hitler became chancellor.

Macke went back upstairs. Kringelein was briefing uniformed police who were to raid the updated addresses Macke had provided of Communists and their allies.

Macke did not hesitate to interrupt his boss. Kringelein was not a Nazi, and would therefore be afraid to reprimand a Storm trooper. Macke said: 'I'm looking for the Vice Establishments file.'

Kringelein looked annoyed but made no protest. 'On the side table,' he said. 'Help yourself.'

Macke took the file and returned to his own room.

The survey was five years old. It detailed the clubs then in existence and stated what activities went on in them: gambling, indecent displays, prostitution, sale of drugs, homosexuality and other depravities. The file named owners and investors, club members and employees. Macke patiently read each entry: perhaps Robert von Ulrich was a drug addict or a user of whores.

Berlin was famous for its homosexual clubs. Macke ploughed through the dreary entry on the Pink Slipper, where men danced with men and the floor show featured transvestite singers. Sometimes, he thought, his work was disgusting.

He ran his finger down the list of members, and found Robert von Ulrich.

He gave a sigh of satisfaction.

Looking farther down, he saw the name of Jörg Schleicher.

'Well, well,' he said. 'Let's see how sarcastic you are now.'

(iv)

The next time Lloyd saw Walter and Maud he found them angrier – and more scared.

It was the following Saturday, 4 March, the day before the election. Lloyd and Ethel were planning to attend a Social Democratic Party rally organized by Walter, and they went to the von Ulrichs' home in Mitte for lunch beforehand.

It was a nineteenth-century house with spacious rooms and large windows, though much of the furniture was worn. The lunch was plain, pork chops with potatoes and cabbage, but there was good wine with it. Walter and Maud talked as if they were poor, and no doubt they were living more simply than their parents had but, all the same, they were not going hungry.

However they were frightened.

But had persuaded Germany's aging President, Paul von Hindenburg, to approve the Reichstag Fire Decree, which gave the Nazis authority for what they were already doing, beating and torturing their political opponents. 'Twenty thousand people have been arrested since Monday night!' Walter said, his voice shaking. 'Not just Communists, but people the Nazis call "Communist sympathizers".'

'Which means anyone they dislike,' said Maud.

Ethel said: 'How can there be a democratic election now?'

'We must do our best,' Walter said. 'If we don't campaign, it will only help the Nazis.'

Lloyd said impatiently: 'When will you stop accepting this and start to fight back? Do you still believe it would be wrong to meet violence with violence?'

'Absolutely,' said Maud. 'Peaceful resistance is our only hope.'

Walter said: 'The Social Democratic Party has a paramilitary wing, the Reichsbanner, but it's weak. A small group of Social Democrats proposed a violent response to the Nazis, but they were outvoted.'

Maud said: 'Remember, Lloyd, the Nazis have the police and the army on their side.'

Walter looked at his pocket watch. 'We must get going.'

Maud said suddenly: 'Walter, why don't you cancel?'

He stared at her in surprise. 'Seven hundred tickets have been sold.'

'Oh, to blazes with the tickets,' Maud said. 'I'm worried about *you*.'

'Don't worry. Seats have been carefully allocated, so there should be no troublemakers in the hall.'

Lloyd was not sure that Walter was as confident as he pretended.

Walter went on: 'Anyway, I cannot let down people who are still willing to come to a democratic political meeting. They are all the hope that remains to us.'

'You're right,' Maud said. She looked at Ethel. 'Perhaps you and Lloyd should stay home. It's dangerous, no matter what Walter says; and this isn't your country, after all.'

'Socialism is international,' Ethel said stoutly. 'Like your husband, I appreciate your concern, but I'm here to witness German politics first hand, and I'm not going to miss this.'

'Well, the children can't go,' Maud said.

Erik said: 'I don't even want to go.'

Carla looked disappointed but said nothing.

Walter, Maud, Ethel and Lloyd got into Walter's little car. Lloyd was nervous but excited too. He was getting a perspective on politics superior to anything his friends back home had. And if there was going to be a fight, he was not afraid.

They drove east, crossing Alexander Platz, into a neighbourhood of poor houses and small shops, some of which had signs in Hebrew letters. The Social Democratic Party was working-class but, like the British Labour Party, it had a few affluent supporters. Walter von Ulrich was in a small upper-class minority.

The car pulled up outside a marquee that said: 'People's Theatre'. A line had already formed outside. Walter crossed the pavement to the door, waving to the waiting crowd, who cheered. Lloyd and the others followed him inside.

Walter shook hands with a solemn young man of about eighteen. 'This is Wilhelm Frunze, secretary of the local branch of our party.' Frunze was one of those boys who looked as if they had been born middle-aged. He wore a blazer with buttoned pockets that had been fashionable ten years ago.

Frunze showed Walter how the theatre doors could be barred from the inside. 'When the audience is seated, we will lock up, so that no troublemakers can get in,' he said.

'Very good,' said Walter. 'Well done.'

Frunze ushered them into the auditorium. Walter went up on stage and greeted some other candidates who were already there. The public began to come in and take their seats. Frunze showed Maud, Ethel and Lloyd to reserved places in the front row.

Two boys approached. The younger, who looked about fourteen but was taller than Lloyd, greeted Maud with careful good manners and made a little bow. Maud turned to Ethel and said: 'This is Werner Franck, the son of my friend Monika.' Then she said to Werner: 'Does your father know you're here?'

'Yes – he said I should find out about social democracy myself.'

'He's broad-minded, for a Nazi.'

Lloyd thought this was a rather tough line to take with a fourteen-year-old, but Werner was a match for her. 'My father doesn't really believe in Nazism, but he thinks Hitler is good for German business.'

Wilhelm Frunze said indignantly: 'How can it be good for business to throw thousands of people into jail? Apart from the injustice, they can't work!'

Werner said: 'I agree with you. And yet Hitler's crackdown is popular.'

'People think they're being saved from a Bolshevik revolution,' Frunze said. 'The Nazi press has them convinced that the Communists were about to launch a campaign of murder, arson and poison in every town and village.'

The boy with Werner, who was shorter but older, said: 'And yet it is the Brownshirts, not the Communists, who drag people into basements and break their bones with clubs.' He spoke German fluently with a slight accent that Lloyd could not place.

Werner said: 'Forgive me, I forgot to introduce Vladimir Peshkov. He goes to the Berlin Boys' Academy, my school, and he's always called Volodya.'

Lloyd stood up to shake hands. Volodya was about Lloyd's age, a striking young man with a frank blue-eyed gaze.

Frunze said: 'I know Volodya Peshkov. I go to the Berlin Boys' Academy too.'

Volodya said: 'Wilhelm Frunze is the school genius – top marks in physics and chemistry and maths.'

'It's true,' said Werner.

Maud looked hard at Volodya and said: 'Peshkov? Is your father Grigori?'

'Yes, Frau von Ulrich. He is a military attaché at the Soviet Embassy.'

So Volodya was Russian. He spoke German effortlessly, Lloyd thought with a touch of envy. No doubt that came from living here.

'I know your parents well,' Maud said to Volodya. She knew all the diplomats in Berlin, Lloyd had already gathered. It was part of her job.

Frunze checked his watch and said: 'Time to begin.' He went up on stage and called for order.

The theatre went quiet.

Frunze announced that the candidates would make speeches and then take questions from the audience. Tickets had been issued only to Social Democratic Party members, he added, and the doors were now closed, so everyone could speak freely, knowing they were among friends.

It was like being a member of a secret society, Lloyd thought. This was not what he called democracy.

Walter spoke first. He was no demagogue, Lloyd observed. He had no rhetorical flourishes. But he flattered his audience, telling them that they were intelligent and well-informed men and women who understood the complexity of political issues.

He had been speaking for only a few minutes when a Brownshirt walked on stage.

Lloyd cursed. How had he got in? He had come from the wings: someone must have opened the stage door.

He was a huge brute with an army haircut. He stepped to the front of the stage and shouted: 'This is a seditious gathering. Communists and subversives are not wanted in today's Germany. The meeting is closed.'

The confident arrogance of the man outraged Lloyd. He wished he could get this great oaf in a boxing ring.

Wilhelm Frunze leaped to his feet, stood in front of the intruder, and yelled furiously: 'Get out of here, you thug!'

The man shoved him in the chest powerfully. Frunze staggered back, stumbled, and fell over backwards.

The audience were on their feet, some shouting in angry protest, some screaming in fear.

More Brownshirts appeared from the wings.

Lloyd realized with dismay that the bastards had planned this well.

The man who had shoved Frunze shouted: 'Out!' The other Brownshirts took up the cry: 'Out! Out! Out!' There were about twenty of them, now, and more appearing all the time. Some carried police nightsticks or improvised clubs. Lloyd saw a hockey stick, a wooden sledgehammer, even a chair leg. They strutted up and down the stage, grinning fiendishly and waving their weapons as they chanted, and Lloyd had no doubt that they were itching to start hitting people.

He was on his feet. Without thinking, he, Werner and Volodya had formed a protective line in front of Ethel and Maud.

Half the audience were trying to leave, the other half shouting and shaking their fists at the intruders. Those attempting to get out were shoving others, and minor scuffles had broken out. Many of the women were crying.

On stage, Walter grasped the lectern and shouted: 'Everyone try to keep calm, please! There is no need for disorder!' Most people could not hear and the rest ignored him.

The Brownshirts began to jump off the stage and to wade into the audience. Lloyd took his mother's arm, and Werner did the same with Maud. They moved towards the nearest exit in a group. But all the doors were already jammed with knots of panicking people trying to leave. That made no difference to the Brownshirts, who kept yelling at people to get out.

The attackers were mostly able-bodied, whereas the audience included women and old men. Lloyd wanted to fight back, but it was not a good idea.

A man in a Great War steel helmet shouldered Lloyd, and he lurched forward and bumped into his mother. He resisted the temptation to turn and confront the man. His priority was to protect Mam.

A spotty-faced boy carrying a truncheon put a hand on Werner's back and shoved energetically, yelling: 'Get out, get out!' Werner turned quickly and took a step towards him. 'Don't touch me, you Fascist pig,' he said. The Brownshirt suddenly stopped dead and looked scared, as if he had not been expecting resistance.

Werner turned away again, concentrating, like Lloyd, on getting the two women to safety. But the huge man had heard the exchange and yelled: 'Who are you calling a pig?' He lashed out at Werner, hitting the back of his head with his fist. His aim was poor and it was a glancing blow, but all the same Werner cried out and staggered forward.

Volodya stepped between them and hit the big man in the face, twice. Lloyd admired Volodya's rapid one-two, but turned his attention back to his task. Seconds later the four of them reached the doorway. Lloyd and Werner managed to help the women out into the theatre foyer. Here the crush eased and the violence stopped – there were no Brownshirts.

Seeing the women safe, Lloyd and Werner looked back into the auditorium.

Volodya was fighting the big man bravely, but he was in trouble. He kept punching the man's face and body, but his blows had little effect, and the man shook his head as if pestered by an insect. The Brownshirt was heavy-footed and slow-moving, but he hit Volodya in the chest and then the head, and Volodya staggered. The big man drew back his fist for a massive punch. Lloyd was afraid it could kill Volodya.

Then Walter took a flying leap off the stage and landed on the big man's back. Lloyd wanted to cheer. They fell to the floor in a blur of arms and legs, and Volodya was saved, for the moment.

The spotty youth who had shoved Werner was now harassing the people trying to leave, hitting their backs and heads with his truncheon. 'You fucking coward!' Lloyd yelled, stepping forward. But Werner was ahead of him. He shoved past Lloyd and grabbed the truncheon, trying to wrestle it away from the youth.

The older man in the steel helmet joined in and hit Werner with a pickaxe handle. Lloyd stepped forward and hit the older man with a straight right. The blow landed perfectly, next to the man's left eye.

But he was a war veteran, and not easily discouraged. He swung around and lashed out at Lloyd with his club. Lloyd dodged the blow easily and hit him twice more. He connected in the same area, around the man's eyes, breaking the skin. But the helmet protected the man's head, and Lloyd could not land a left hook, his knockout punch. He ducked a swing of the pickaxe handle and hit the man's face again, and the man backed away, blood pouring from cuts around his eyes.

Lloyd looked around. He saw that the Social Democrats were fighting back now, and he got a jolt of savage pleasure. Most of the audience had passed through the doors, leaving mainly young men in the auditorium, and they were coming forward, clambering over the theatre seats to get at the Brownshirts; and there were dozens of them.

Something hard struck his head from behind. It was so painful that he roared. He turned to see a boy of his own age holding a length of timber, raising it to strike again. Lloyd closed with him and hit him hard in the stomach twice, first with his right fist then with his left. The boy gasped for breath and dropped the wood. Lloyd hit him with an uppercut to the chin and the boy passed out.

Lloyd rubbed the back of his head. It hurt like hell but there was no blood.

The skin on his knuckles was raw and bleeding, he saw. He bent down and picked up the length of timber dropped by the boy.

When he looked around again, he was thrilled to see some of the Brownshirts retreating, clambering up on to the stage and disappearing into the wings, presumably aiming to leave through the stage door by which they had entered.

The big man who had started it all was on the floor, groaning and holding his knee as if he had dislocated something. Wilhelm Frunze stood over him, hitting him with a wooden shovel again and again, repeating at the top of his voice the words the man had used to start the riot: 'Not! Wanted! In! Today's! Germany!' Helpless, the big man tried to roll away from the blows, but Frunze went after him, until two more Brownshirts grabbed the man's arms and dragged him away.

Frunze let them go.

Did we beat them? Lloyd thought with growing exultation. Maybe we did!

Several of the younger men chased their opponents up on to the stage, but they stopped there and contented themselves with shouting insults as the Brownshirts disappeared.

Lloyd looked at the others. Volodya had a swollen face and one closed eye. Werner's jacket was ripped, a big square of cloth dangling. Walter was sitting on a front-row seat, breathing hard and rubbing his elbow, but he was smiling. Frunze threw his shovel away, sailing it across the rows of empty seats to the back.

Werner, who was only fourteen, was exultant. 'We gave them hell, didn't we?'

Lloyd grinned. 'Yes, we certainly did.'

Volodya put his arm around Frunze's shoulders. 'Not bad for a bunch of schoolboys, eh?'

Walter said: 'But they stopped our meeting.'

The youngsters stared resentfully at him for spoiling their triumph.

Walter looked angry. 'Be realistic, boys. Our audience has fled in terror. How long will it be before those people have the nerve to go to a political meeting again? The Nazis have made their point. It's dangerous even to listen to any party other than theirs. The big loser today is Germany.'

Werner said to Volodya: 'I hate those fucking Brownshirts. I think I might join you Communists.'

Volodya looked at him hard with those intense blue eyes and spoke in a low voice. 'If you're serious about fighting the Nazis, there might be something more effective you could do.'

Lloyd wondered what Volodya meant.

Then Maud and Ethel came running back into the auditorium, both speaking at the same time, crying and laughing with relief; and Lloyd forgot Volodya's words and never thought of them again.

(v)

Four days later, Erik von Ulrich came home in a Hitler Youth uniform.

He felt like a prince. He had a brown shirt just like the one worn by Storm troopers, with various patches and a swastika armband. He also had the regulation black tie and black shorts. He was a patriotic soldier dedicated to the service of his country. At last he was one of the gang.

This was even better than supporting Hertha, Berlin's favourite soccer team. Erik was taken to matches occasionally, on Saturdays when his father did not have a political meeting to attend. That gave him a similar sense of belonging to a great big crowd of people all feeling the same emotions. But Hertha sometimes lost, and he came home disconsolate.

The Nazis were winners.

He was terrified of what his father was going to say.

His parents infuriated him by insisting on marching out of step. All the boys were joining the Hitler Youth. They had sports and singing and adventures in the fields and forests outside the city. They were smart and fit and loyal and efficient.

Erik was deeply troubled by the thought that he might have to fight in battle some day – his father and grandfather had – and he wanted to be ready for that, trained and hardened, disciplined and aggressive.

The Nazis hated Communists, but so did Mother and Father. So what if the Nazis hated Jews as well? The von Ulrichs were not Jewish, why should they care? But Mother and Father stubbornly refused to join in. Well, Erik was fed up with being left out, and he had decided to defy them.

He was scared stiff.

As usual, neither Mother nor Father was at the house when Erik and Carla came home from school. Ada pursed her lips disapprovingly as she served their tea, but she said: 'You'll have to clear the table yourselves today – I've got a terrible backache, I'm going to lie down.'

Carla looked concerned. 'Is that what you had to see the doctor about?'

Ada hesitated before replying: 'Yes, that's right.'

She was obviously hiding something. The thought of Ada being ill

– and lying about it – made Erik uneasy. He would never go as far as Carla and say he loved Ada, but she had been a kindly presence all his life, and he was more fond of her than he liked to say.

Carla was just as concerned. 'I hope it gets better.'

Lately Carla had become more grown-up, somewhat to Erik's bewilderment. Although he was two years older, he still felt like a kid, but she acted like an adult half the time.

Ada said reassuringly: 'I'll be fine after a rest.'

Erik ate some bread. When Ada left the room, he swallowed and said: 'I'm only in the junior section, but as soon as I'm fourteen I can move up.'

Carla said: 'Father's going to hit the roof! Are you mad?'

'Herr Lippmann said that Father will be in trouble if he tries to make me leave.'

'Oh, brilliant,' said Carla. She had developed a streak of withering sarcasm that sometimes stung Erik. 'So you'll get Father into a row with the Nazis,' she said scornfully. 'What a great idea. So good for the whole family.'

Erik was taken aback. He had not thought of it that way. 'But all the boys in my class are members,' he said indignantly. 'Except for Frenchy Fontaine and Jewboy Rothmann.'

Carla spread fish paste on her bread. 'Why do you have to be the same as the others?' she said. 'Most of them are stupid. You told me Rudi Rothmann was the cleverest boy in the class.'

'I don't want to be with Frenchy and Rudi!' Erik cried, and to his mortification he felt tears come to his eyes. 'Why should I have to play with the boys no one likes?' This was what had given him the courage to defy his father: he could no longer bear to walk out of school with the Jews and the foreigners while all the German boys marched around the playing field in their uniforms.

They both heard a cry.

Erik looked at Carla and said: 'What was that?'

Carla frowned. 'It was Ada, I think.'

Then, more distinctly, they heard: 'Help!'

Erik got to his feet, but Carla was ahead of him. He went after her. Ada's room was in the basement. They ran down the stairs and into the small bedroom.

There was a narrow single bed up against the wall. Ada was lying there, her face screwed up in pain. Her skirt was wet and there was a puddle on the floor. Erik could hardly believe what he was seeing. Had she pissed herself? It was scary. There were no other grown-ups in the house. He did not know what to do.

Carla was scared, too – Erik could see it in her face – but she was not panicked. She said: 'Ada, what's wrong?' Her voice sounded strangely calm.

'My waters broke,' Ada said.

Erik had no idea what that meant.

Nor did Carla. 'I don't understand,' she said.

'It means my baby is coming.'

'You're pregnant?' Carla said in astonishment.

Erik said: 'But you're not married!'

Carla said furiously: 'Shut up, Erik – don't you know anything?'

He did know, of course, that women could have babies when they were not married – but surely not Ada!

'That's why you went to the doctor last week,' Carla said to Ada.

Ada nodded.

Erik was still trying to get used to the idea. 'Do you think Mother and Father know?'

'Of course they do. They just didn't tell us. Fetch a towel.'

'Where from?'

'The airing cupboard on the upstairs landing.'

'A clean one?'

'Of course a clean one!'

Erik ran up the stairs, took a small white towel from the cupboard, and ran down again.

'That's not much good,' Carla said, but she took it and dried Ada's legs.

Ada said: 'The baby's coming soon, I can feel it. But I don't know what to do.' She started to cry.

Erik was watching Carla. She was in charge now. It did not matter that he was the older one: he looked to her for leadership. She was being practical and staying calm, but he could tell that she was terrified, and her composure was fragile. She could crack at any minute, he thought.

Carla turned to Erik again. 'Go and fetch Dr Rothmann,' she said. 'You know where his office is.'

Erik was hugely relieved to have been given a task he could manage. Then he thought of a snag. 'What if he's out?'

'Then ask Frau Rothmann what you should do, you idiot!' Carl said. 'Get going – run!'

Erik was glad to get out of the room. What was happening there was mysterious and frightening. He went up the stairs three at a time and flew out of the front door. Running was one thing he did know how to do.

The doctor's surgery was half a mile away. He settled into a fast trot. As he ran he thought about Ada. Who was the father of her baby? He recalled that she had gone to the movies with Paul Huber a couple of times last summer. Had they had sexual intercourse? They must have! Erik and his friends talked about sex a lot, but they did not really know anything about it. Where had Ada and Paul done it? Not in a movie theatre, surely? Didn't people have to lie down? He was baffled.

Dr Rothmann's place was in a poorer street. He was a good doctor, Erik had heard Mother say, but he treated a lot of working-class people who could not pay high fees. The doctor's house had a consulting room and a waiting room on the ground floor, and the family lived upstairs.

Outside was parked a green Opel 4, an ugly little two-seater unofficially called the Tree Frog.

The front door of the house was unlatched. Erik walked in, breathing hard, and entered the waiting room. There was an old man coughing in a corner and a young woman with a baby. 'Hello!' Erik called, 'Dr Rothmann?'

The doctor's wife stepped out of the consulting room. Hannelore Rothmann was a tall, fair woman with strong features, and she gave Erik a look like thunder. 'How dare you come to this house in that uniform?' she said.

Erik was petrified. Frau Rothmann was not Jewish, but her husband was: Erik had forgotten that in his excitement. 'Our maid is having a baby!' he said.

'And so you want a Jewish doctor to help you?'

Erik was taken completely by surprise. It had never occurred to him that the Nazis' attacks might cause the Jews to retaliate. But suddenly he saw that Frau Rothmann made total sense. The Brownshirts went around shouting: 'Death to Jews!' Why should a Jewish doctor help such people?

Now he did not know what to do. There were other doctors, of course, plenty of them, but he did not know where, nor whether they would come out to see a total stranger. 'My sister sent me,' he said feebly.

'Carla's got a lot more sense than you.'

'Ada said the waters have broken.' Erik was not sure what that meant, but it sounded significant.

With a disgusted look, Frau Rothmann went back into the consulting room.

The old man in the corner cackled. 'We're all dirty Jews until you need our help!' he said. 'Then it's: "Please come, Dr Rothmann", and "What's your advice, Lawyer Koch?" and "Lend me a hundred marks, Herr Goldman", and—' He was overcome by a fit of coughing.

A girl of about sixteen came in from the hall. Erik thought she must be the Rothmanns' daughter, Eva. He had not seen her for years. She had breasts, now, but she was still plain and dumpy. She said: 'Did your father let you join the Hitler Youth?'

'He doesn't know,' said Erik.

'Oh, boy,' said Eva. 'You're in trouble.'

He looked from her to the consulting-room door. 'Do you think your father's going to come?' he said. 'Your mother was awfully cross with me.'

'Of course he'll come,' Eva said. 'If people are sick, he helps them.' Her voice became scornful. 'He doesn't check their race or politics first. We're not Nazis.' She went out again.

Erik felt bewildered. He had not expected this uniform to get him into so much trouble. At school everyone thought it was wonderful.

A moment later, Dr Rothmann appeared. Speaking to the two waiting patients, he said: 'I'll be back as soon as I can. I'm sorry, but a baby won't wait to be born.' He looked at Erik. 'Come on, young man, you'd better ride with me, despite that uniform.'

Erik followed him out and got into the passenger seat of the Tree Frog. He loved cars and was desperate to be old enough to drive, and normally he enjoyed riding in any vehicle, watching the dials and studying the driver's technique. But now he felt as if he were on display, sitting beside a Jewish doctor in his brown shirt. What if Herr Lippmann should see him? The trip was agony.

Fortunately, it was short: in a couple of minutes they were at the von Ulrich house.

'What's the young woman's name?' Rothmann asked.

'Ada Hempel.'

'Ah, yes, she came to see me last week. The baby's early. All right, take me to her.'

Erik led the way into the house. He heard a baby cry. It had come already! He hurried down to the basement, the doctor following.

Ada lay on her back. The bed was soaked with blood and something else. Carla stood holding a tiny baby in her arms. The baby was covered in slime. Something that looked like thick string ran from the baby up Ada's skirt. Carla was wide-eyed with terror. 'What must I do?' she cried.

'You're doing exactly the right thing,' Doctor Rothmann reassured her. 'Just hold that baby close a minute longer.' He sat beside Ada. He listened to her heart, took her pulse, and said: 'How do you feel, my dear?'

'I'm so tired,' she said.

Rothmann gave a satisfied nod. He stood up again and looked at the baby in Carla's arms. 'A little boy,' he said.

Erik watched with a mixture of fascination and revulsion as the doctor opened his bag, took out some thread and tied two knots in the cord. While he was doing so he spoke to Carla in a soft voice. 'Why are you crying? You've done a marvellous job. You've delivered a baby all on your own. You hardly needed me! You'd better be a doctor when you grow up.'

Carla became calmer. Then she whispered: 'Look at his head.' The doctor had to lean towards her to hear. 'I think there's something wrong with him.'

'I know.' The doctor took out a pair of sharp scissors and cut the cord between the two knots Then he took the naked baby from Carla and held him at arm's length, studying him. Erik could not see anything wrong, but the baby was so red and wrinkled and slimy that it was hard to tell. However, after a thoughtful moment, the doctor said: 'Oh, dear.'

Looking more carefully, Erik could see that there was something wrong. The baby's face was lopsided. One side was normal, but on the other the head seemed dented and there was something strange about the eye.

Rothmann handed the baby back to Carla.

Ada groaned again, and seemed to strain.

When she relaxed, Rothmann reached under her skirt and drew out a lump of something that looked disgustingly like meat. 'Erik,' he said. 'Fetch me a newspaper.'

Erik said: 'Which one?' His parents took all the main papers every day.

'Any one, lad,' said Rothmann gently. 'I don't want to read it.'

Erik ran upstairs and found yesterday's *Vossische Zeitung*. When he returned, the doctor wrapped the meaty thing in the paper and put it on the floor. 'It's what we call the afterbirth,' he said to Carla. 'Best to burn it, later.'

Then he sat on the edge of the bed again. 'Ada, my dear girl, you must be very brave,' he said. 'Your baby is alive, but there may be something wrong with him. We're going to wash him and wrap him up warmly, then we must take him to the hospital.'

Ada looked frightened. 'What's the matter?'

'I don't know. We need to have him checked.'

'Will he be all right?'

'The hospital doctors will do everything they can. The rest we must leave to God.'

Erik remembered that Jews worshipped the same God as Christians. It was easy to forget that.

Rothmann said: 'Do you think you could get up and come to the hospital with me, Ada? Baby needs you to feed him.'

'I'm so tired,' she said again.

'Take a minute or two to rest, then. But not much more, because Baby needs to be looked at soon. Carla will help you get dressed. I'll wait upstairs.' He addressed Erik with gentle irony. 'Come with me, little Nazi.'

Erik wanted to squirm. Dr Rothmann's forbearance was even worse than Frau Rothmann's scorn.

As they were leaving, Ada said: 'Doctor?'

'Yes, my dear.'

'His name is Kurt.'

'A very good name,' said Dr Rothmann. He went out, and Erik followed.

(vi)

Lloyd Williams's first day working as assistant to Walter von Ulrich was also the first day of the new parliament.

Walter and Maud were struggling frantically to save Germany's fragile democracy. Lloyd shared their desperation, partly because they were good people whom he had known on and off all his life, and partly because he feared that Britain could follow Germany down the road to hell.

The election had resolved nothing. The Nazis got 44 per cent, an increase but still short of the 51 per cent they craved.

Walter saw hope. Driving to the opening of the parliament, he said: 'Even with massive intimidation, they failed to win the votes of most Germans.' He banged his fist on the steering wheel. 'Despite everything they say, they are *not* popular. And the longer they stay in government, the better people will get to know their wickedness.'

Lloyd was not so sure. 'They've closed opposition newspapers, thrown Reichstag deputies in jail, and corrupted the police,' he said. 'And yet forty-four per cent of Germans approve? I don't find that reassuring.'

The Reichstag building was badly fire-damaged and quite unusable, so the parliament assembled in the Kroll Opera House, on the opposite side of the Königs Platz. It was a vast complex with three concert halls and fourteen smaller auditoria, plus restaurants and bars.

When they arrived, they had a shock. The place was surrounded by Brownshirts. Deputies and their aides crowded around the entrances, trying to get in. Walter said furiously: 'Is this how Hitler plans to get his way – by preventing us from entering the chamber?'

Lloyd saw that the doors were barred by Brownshirts. They admitted those in Nazi uniform without question, but everyone else had to produce credentials. A boy younger than Lloyd looked him up and down contemptuously before grudgingly letting him in. This was intimidation, pure and simple.

Lloyd felt his temper beginning to simmer. He hated to be bullied. He knew he could knock the Brownshirt boy down with one good left hook. He forced himself to remain calm, turn away, and walk through the door.

After the fight in the People's Theatre, his mother had examined the egg-shaped lump on his head and ordered him to go home to England. He had talked her round, but it had been a close thing.

She said he had no sense of danger, but that was not quite right. He did get scared sometimes, but it always made him feel combative. His instinct was to go on the attack, not to retreat. This scared his mother.

Ironically, she was just the same. She was not going home. She was frightened, but she was also thrilled to be here in Berlin at this turning point in German history, and outraged by the violence and repression she was witnessing; and she felt sure she could write a book that would forewarn democrats in other countries about Fascist tactics. 'You're worse than me,' Lloyd had said to her, and she had had no answer.

Inside, the opera house was swarming with Brownshirts and SS men, many of them armed. They guarded every door and showed, with looks and gestures, their hatred and contempt for anyone not supporting the Nazis.

Walter was late for a Social Democratic Party group meeting. Lloyd hurried around the building looking for the right room. Glancing into the debating chamber, he saw that a giant swastika hung from the ceiling, dominating the room.

The first matter to be discussed, when proceedings began that afternoon, was to be the Enabling Act, which would permit Hitler's cabinet to pass laws without the approval of the Reichstag.

The Act offered a dreadful prospect. It would make Hitler a dictator. The repression, intimidation, violence, torture and murder that Germany had seen in the past few weeks would become permanent. It was unthinkable.

But Lloyd could not imagine that any parliament in the world would pass such a law. They would be voting themselves out of power. It was political suicide.

He found the Social Democrats in a small auditorium. Their meeting had already begun. Lloyd hurried Walter to the room, then he was sent for coffee.

Waiting in the queue, he found himself behind a pale, intense-looking young man dressed in funereal black. Lloyd's German had become more fluent and colloquial, and he now had the confidence to strike up a conversation with a stranger. The man in black was Heinrich von Kessel, he learned. He was doing the same sort of job as Lloyd, working as an unpaid aide to his father, Gottfried von Kessel, a deputy for the Centre Party, which was Catholic.

'My father knows Walter von Ulrich very well,' Heinrich said. 'They were both attachés at the German embassy in London in 1914.'

The world of international politics and diplomacy was quite small, Lloyd reflected.

Heinrich told Lloyd that a return to the Christian faith was the answer to Germany's problems.

'I'm not much of a Christian,' Lloyd said candidly. 'I hope you don't mind my saying so. My grandparents are Welsh Bible-punchers, but my mother is indifferent and my stepfather's Jewish. Occasionally we go to the Calvary Gospel Hall in Aldgate, mainly because the pastor is a Labour Party member.'

Heinrich smiled and said: 'I'll pray for you.'

Catholics were not proselytizers, Lloyd remembered. What a contrast to his dogmatic grandparents in Aberowen, who thought that people who did not believe as they did were wilfully blinding themselves to the gospel, and would be condemned to eternal damnation.

When Lloyd re-entered the Social Democratic Party meeting, Walter was speaking. 'It can't happen!' he said. 'The Enabling Act is a constitutional amendment. Two thirds of the representatives must be present, which would be 432 out of a possible 647. And two thirds of those present must approve.'

Lloyd added up the numbers in his head as he put the tray down on the table. The Nazis had 288 seats, and the Nationalists who were their close allies had 52, making 340 – nearly 100 short. Walter was right. The Act could not be passed. Lloyd was comforted and sat down to listen to the discussion and to improve his German.

But his relief was short-lived. 'Don't be so sure,' said a man with a working-class Berlin accent. 'The Nazis are caucusing with the Centre Party.' That was Heinrich's lot, Lloyd recalled. 'That could give them another seventy-four,' the man finished.

Lloyd frowned. Why would the Centre Party support a measure that would take away all its power?

Walter voiced the same thought more bluntly. 'How could the Catholics be so stupid?'

Lloyd wished he had known about this before he went for coffee – then he could have discussed it with Heinrich. He might have learned something useful. Damn.

The man with the Berlin accent said: 'In Italy, the Catholics made a deal with Mussolini – a concordat to protect the Church. Why not here?'

Lloyd calculated that the Centre Party's support would bring the Nazis' votes up to 414. 'It's still not two thirds,' he said to Walter with relief.

Another young aide heard him and said: 'But that doesn't take into account the Reichstag president's latest announcement.' The Reichstag president was Hermann Göring, Hitler's closest associate. Lloyd had not heard about an announcement. Nor had anyone else, it seemed. The deputies went quiet. The aide went on: 'He has ruled that Communist deputies who are absent because they are in jail don't count.'

There was an outburst of indignant protest all around the room. Lloyd saw Walter go red in the face. 'He can't do that!' Walter said.

'It's completely illegal,' said the aide. 'But he has done it.'

Lloyd was dismayed. Surely the law could not be passed by a trick? He did some more arithmetic. The Communists had 81 seats. If they were discounted, the Nazis needed only two thirds of 566, which was 378. Even with the Nationalists they still did not have enough – but if they won the support of the Catholics they could swing it.

Someone said: 'This is all completely illegal. We should walk out in protest.'

'No, no!' said Walter emphatically. 'They would pass the Act in our absence. We've got to talk the Catholics out of it. Wels must speak to Kaas immediately.' Otto Wels was the leader of the Social Democratic Party; Prelate Ludwig Kaas the head of the Centre Party.

There was a murmur of agreement around the room.

Lloyd took a deep breath and spoke up. 'Herr von Ulrich, why don't you take Gottfried von Kessel to lunch? I believe you two worked together in London before the war.'

Walter laughed mirthlessly. 'That creep!' he said.

Maybe the lunch was not such a good idea. Lloyd said: 'I didn't realize you disliked the man.'

Walter looked thoughtful. 'I hate him – but I'll try anything, by God.'

Lloyd said: 'Shall I find him and extend the invitation?'

'All right, give it a try. If he accepts, tell him to meet me at the Herrenklub at one.'

'Very good.'

Lloyd hurried back to the room into which Heinrich had disappeared. He stepped inside. A meeting was going on similar to the one he had left. He scanned the room, spotted the black-clad Heinrich, met his gaze, and beckoned him urgently.

They both stepped outside, then Lloyd said: 'They're saying your party is going to support the Enabling Act!'

'It's not certain,' said Heinrich. 'They're divided.'

'Who's against the Nazis?'

'Brüning and some others.' Brüning was a former chancellor and a leading figure.

Lloyd felt more hopeful. 'Which others?'

'Did you call me out of the room to pump me for information?'

'Sorry, no, I didn't. Walter von Ulrich wants to have lunch with your father.'

Heinrich looked dubious. 'They don't like each other – you know that, don't you?'

'I gathered as much. But they'll put their differences aside today!'

Heinrich did not seem so sure. 'I'll ask him. Wait here.' He went back inside.

Lloyd wondered whether there was any chance this would work. It was a shame Walter and Gottfried were not bosom buddies. But he could hardly believe the Catholics would vote with the Nazis.

What bothered him most was the thought that if it could happen in Germany, it could happen in Britain. This grim prospect made him shiver with dread. He had his whole life in front of him, and he did not want to live it in a repressive dictatorship. He wanted to work in politics, like his parents, and make his country a better place for people such as the Aberowen coal miners. For that he needed political meetings where people could speak their minds, and newspapers that could attack the government, and pubs where men could have arguments without looking over their shoulders to see who was listening.

Fascism threatened all that. But perhaps Fascism would fail. Walter might be able to talk Gottfried around, and prevent the Centre Party supporting the Nazis.

Heinrich came out. 'He'll do it.'

'Great! Herr von Ulrich suggested the Herrenklub at one o'clock.'

'Really? Is he a member?'

'I assume so – why?'

'It's a conservative institution. I suppose he is Walter *von* Ulrich, so he must come from a noble family, even if he is a socialist.'

'I should probably book a table. Do you know where it is?'

'Just around the corner.' Heinrich gave Lloyd directions.

'Shall I book for four?'

Heinrich grinned. 'Why not? If they don't want you and me, they can just ask us to leave.' He went back into the room.

Lloyd left the building and walked quickly across the plaza, passing the burned-out Reichstag building, and made his way to the Herrenklub.

There were gentlemen's clubs in London, but Lloyd had never been inside one. This place was a cross between a restaurant and a funeral parlour, he thought. Waiters in full evening dress padded about, laying silent cutlery on tables shrouded in white. A head waiter took his reservation and wrote down the name 'von Ulrich' as solemnly as if he were making an entry in the Book of the Dead.

He returned to the opera house. The place was getting busier and noisier, and the tension seemed higher. Lloyd heard someone say excitedly that Hitler himself would open the proceedings this afternoon by proposing the Act.

A few minutes before one, Lloyd and Walter walked across the plaza. Lloyd said: 'Heinrich von Kessel was surprised to learn that you are a member of the Herrenklub.'

Walter nodded. 'I was one of the founders, a decade or more ago. In those days it was the Juniklub. We got together to campaign against the Versailles Treaty. It's become a right-wing bastion, and I'm probably the only Social Democrat, but I remain a member because it's a useful place to meet with the enemy.'

Inside the club Walter pointed to a sleek-looking man at the bar. 'That's Ludwig Franck, the father of young Werner, who fought alongside us at the People's Theatre,' Walter said. 'I'm sure he's not a member here – he isn't even German-born – but it seems he's having lunch with his father-in-law, Count von der Helbard, the elderly man beside him. Come with me.'

They went to the bar and Walter performed introductions. Franck said to Lloyd: 'You and my son got into quite a scrap a couple of weeks back.'

Lloyd touched the back of his head reflexively: the swelling had gone down, but the place was still painful to touch. 'We had women to protect, sir,' he said.

'Nothing wrong with a bit of a punch-up,' Franck said. 'Does you lads good.'

Walter cut in impatiently: 'Come on, Ludi. Busting up election meetings is bad enough, but your leader wants to completely destroy our democracy!'

'Perhaps democracy is not the right form of government for us,' said Franck. 'After all, we're not like the French or the Americans – thank God.'

'Don't you care about losing your freedom? Be serious!'

Franck suddenly dropped his facetious air. 'All right, Walter,' he said coldly. 'I will be serious, if you insist. My mother and I arrived here from Russia more than ten years ago. My father was not able to come with us. He had been found to be in possession of subversive literature, specifically a book called *Robinson Crusoe*, apparently a novel that promotes bourgeois individualism, whatever the hell that might be. He was sent to a prison camp somewhere in the Arctic. He may—' Franck's voice broke for a moment, and he paused, swallowed, and at last finished quietly: 'He may still be there.'

There was a moment of silence. Lloyd was shocked by the story. He knew that the Russian Communist government could be cruel, in general, but it was quite another thing to hear a personal account, told simply by a man who was clearly still grieving.

Walter said: 'Ludi, we all hate the Bolsheviks – but the Nazis could be worse!'

'I'm willing to take that risk,' said Franck.

Count von der Helbard said: 'We'd better go in for lunch. I've got an afternoon appointment. Excuse us.' The two men left.

'It's what they always say!' Walter raged. 'The Bolsheviks! As if they were the only alternative to the Nazis! I could weep.'

Heinrich walked in with an older man who was obviously his father: they had the same thick, dark hair combed with a parting, except that Gottfried's was shorter and tweeded with silver. Although their features were similar, Gottfried looked like a fussy bureaucrat in an old-fashioned collar, whereas Heinrich was more like a romantic poet than a political aide.

The four of them went into the dining room. Walter wasted no time. As soon as they had ordered, he said: 'I can't understand what your party hopes to gain by supporting this Enabling Act, Gottfried.'

Von Kessel was equally direct. 'We are a Catholic party, and our first duty is to protect the position of the Church in Germany. That's what people hope for when they vote for us.'

Lloyd frowned in disapproval. His mother had been a Member of Parliament, and she always said it was her duty to serve the people who did *not* vote for her, as well as those who did.

Walter employed a different argument. 'A democratic parliament is the best protection for all our churches – yet you're about to throw that away!'

'Wake up, Walter,' Gottfried said testily. 'Hitler won the election. He has come to power. Whatever we do, he's going to rule Germany for the foreseeable future. We have to protect ourselves.'

'His promises are worth nothing!'

'We have asked for specific assurances in writing: the Catholic Church to be independent of the state, Catholic schools to operate unmolested, no discrimination against Catholics in the civil service.' He looked enquiringly at his son.

Heinrich said: 'They promised the agreement would be with us first thing this afternoon.'

Walter said: 'Weigh the options! A scrap of paper signed by a tyrant, against a democratic parliament – which is better?'

'The greatest power of all is God.'

Walter rolled his eyes. 'Then God save Germany,' he said.

The Germans had not had time to develop faith in democracy, Lloyd reflected as the argument surged back and forth between Walter and Gottfried. The Reichstag had been sovereign for only fourteen years. They had lost a war, seen their currency devalued to nothing, and suffered mass unemployment: to them, the right to vote seemed inadequate protection.

Gottfried proved immovable. At the end of lunch his position was as firm as ever. His responsibility was to protect the Catholic Church. It made Lloyd want to scream.

They returned to the opera house and the deputies took their seats in the auditorium. Lloyd and Heinrich sat in a box looking down.

Lloyd could see the Social Democratic Party members in a group on the far left. As the hour approached, he noticed Brownshirts and SS men placing themselves at the exits and around the walls in a threatening arc behind the Social Democrats. It was almost as if they planned to prevent the deputies leaving the building until they had passed the Act. Lloyd found it powerfully sinister. He wondered, with a shiver of fear, whether he, too, might find himself imprisoned here.

There was a roar of cheering and applause, and Hitler walked in, wearing a Brownshirt uniform. The Nazi deputies, most of them similarly dressed, rose to their feet in ecstasy as he mounted the rostrum. Only the Social Democrats remained seated; but Lloyd noticed that one or two looked uneasily over their shoulders at the armed guards. How could they speak and vote freely if they were nervous even about not joining in the standing ovation for their opponent?

When at last they became quiet, Hitler began to speak. He stood straight, his left arm at his side, gesturing only with his right. His voice was harsh and grating but powerful, reminding Lloyd of both a machine gun and a barking dog. His tone thrilled with feeling as he spoke of the

'November traitors' of 1918 who had surrendered when Germany was about to win the war. He was not pretending: Lloyd felt he sincerely believed every stupid, ignorant word he spoke.

The November traitors were a well-worn topic for Hitler, but then he took a new tack. He spoke of the churches, and the important place of the Christian religion in the German state. This was an unusual theme for him, and his words were clearly aimed at the Centre Party, whose votes would determine today's result. He said that he saw the two main denominations, Protestant and Catholic, as the most important factors for upholding nationhood. Their rights would not be touched by the Nazi government.

Heinrich shot a triumphant look at Lloyd.

'I'd still get it in writing, if I were you,' Lloyd muttered.

It was two and a half hours before Hitler reached his peroration.

He ended with an unmistakable threat of violence. 'The government of the nationalist uprising is determined and ready to deal with the announcement that the Act has been rejected – and with it, that resistance has been declared.' He paused dramatically, letting the message sink in: voting against the Act would be a declaration of resistance. Then he reinforced it. 'May you, gentlemen, now take the decision yourselves as to whether it is to be peace or war!'

He sat down to roars of approval from the Nazi delegates, and the session was adjourned.

Heinrich was elated; Lloyd depressed. They went off in different directions: their parties would now hold desperate last-minute discussions.

The Social Democrats were gloomy. Their leader, Wels, had to speak in the chamber, but what could he say? Several deputies said that if he criticized Hitler, he might not leave the building alive. They feared for their own lives, too. If the deputies were killed, Lloyd thought in a moment of cold dread, what would happen to their aides?

Wels revealed that he had a cyanide capsule in his waistcoat pocket. If arrested, he would commit suicide to avoid torture. Lloyd was horrified. Wels was an elected representative, yet he was forced to behave like some kind of saboteur.

Lloyd had started the day with false expectations. He had thought the Enabling Act a crazy idea that had no chance of becoming reality. Now he saw that most people expected the Act to become a reality today. He had misjudged the situation badly.

Was he equally wrong to believe that something like this could not happen in his own country? Was he fooling himself?

Someone asked if the Catholics had made a final decision. Lloyd stood up. 'I'll find out,' he said. He left and ran to the Centre Party's meeting room. As before, he put his head around the door and beckoned Heinrich outside.

'Brüning and Ersing are wavering,' Heinrich said.

Lloyd's heart sank. Ersing was a Catholic union leader. 'How can a trade unionist even think about voting for this bill?' he said.

'Kaas says the Fatherland is in danger. They all think there will be bloody anarchy if we reject this Act.'

'There'll be bloody tyranny if you pass it.'

'What about your lot?'

'They think they will all be shot if they vote against. But they're going to do it anyway.'

Heinrich went back inside and Lloyd returned to the Social Democrats. 'The diehards are weakening,' Lloyd told Walter and his colleagues. 'They're afraid of a civil war if the Act is rejected.'

The gloom deepened.

They all returned to the debating chamber at six o'clock.

Wels spoke first. He was calm, reasonable and unemotional. He pointed out that life in a democratic republic had been good for Germans, overall, bringing freedom of opportunity and social welfare, and reinstating Germany as a normal member of the international community.

Lloyd noticed Hitler making notes.

At the end Wels bravely professed allegiance to humanity and justice, freedom and socialism. 'No Enabling Law gives you the power to annihilate ideas that are eternal and indestructible,' he said, gaining courage as the Nazis began to laugh and jeer.

The Social Democrats applauded, but they were drowned out.

'We greet the persecuted and oppressed!' Wels shouted. 'We greet our friends in the Reich. Their steadfastness and loyalty deserve admiration.'

Lloyd could just make out his words over the hooting and booing of the Nazis.

'The courage of their convictions and their unbroken optimism guarantee a brighter future!'

He sat down amid raucous heckling.

Would the speech make any difference? Lloyd could not tell.

After Wels, Hitler spoke again. This time his tone was quite different. Lloyd realized that in his earlier speech the Chancellor had only been warming up. His voice was louder now, his phrases more intemperate, his tone full of contempt. He used his right arm constantly to make aggressive gestures – pointing, hammering, clenching his fist, putting his hand on his heart, and sweeping the air in a gesture that seemed to brush all opposition aside. Every impassioned phrase was cheered uproariously by his supporters. Every sentence expressed the same emotion: a savage, all-consuming, murderous rage.

Hitler was also confident. He claimed he had not needed to propose the Enabling Act. 'We appeal in this hour to the German Reichstag to grant us something we would have taken anyway!' he jeered.

Heinrich looked worried, and left the box. A minute later Lloyd saw him on the floor of the auditorium, whispering in his father's ear.

When he returned to the box he looked stricken.

Lloyd said: 'Have you got your written assurances?'

Heinrich could not meet Lloyd's eye. 'The document is being typed up,' he replied.

Hitler finished by scorning the Social Democrats. He did not want their votes. 'Germany shall be free,' he screamed. 'But not through you!'

The leaders of the other parties spoke briefly. Every one appeared crushed. Prelate Kaas said the Centre Party would support the bill. The rest followed suit. Everyone but the Social Democrats was in favour.

The result of the vote was announced, and the Nazis cheered wildly.

Lloyd was awestruck. He had seen naked power brutally wielded, and it was an ugly sight.

He left the box without speaking to Heinrich.

He found Walter in the entrance lobby, weeping. He was using a large white handkerchief to wipe his face, but the tears kept coming. Lloyd had not seen men cry like that except at funerals.

Lloyd did not know what to say or do.

'My life has been a failure,' Walter said. 'This is the end of all hope. German democracy is dead.'

(vii)

Saturday 1 April was Boycott Jew Day. Lloyd and Ethel walked around Berlin, staring in incredulity, Ethel making notes for her book. The Star of David was crudely daubed on the windows of Jewish-owned shops. Brownshirts stood at the doors of Jewish-owned department stores, intimidating people who wanted to go in. Jewish lawyers and doctors were picketed. Lloyd happened to see a couple of Brownshirts stopping patients going in to see the von Ulrichs' family physician, Dr Rothmann, but then a hard-handed coal-heaver with a sprained ankle told the Brownshirts to fuck off out of it, and they went in search of easier prey. 'How can people be so mean to each other?' Ethel said.

Lloyd was thinking of the stepfather he loved. Bernie Leckwith was Jewish. If Fascism came to Britain, Bernie would be the target of this kind of hatred. The thought made Lloyd shudder.

A sort of wake was held at Bistro Robert that evening. Apparently no one had organized it, but by eight o'clock the place was full of Social Democrats, Maud's journalistic colleagues, and Robert's theatrical friends. The more optimistic among them said that liberty had merely gone into hibernation for the duration of the economic slump, and one day it would awaken. The rest just mourned.

Lloyd drank little. He did not enjoy the effect of alcohol on his brain. It blurred his thinking. He was asking himself what German left-wingers could have done to prevent this catastrophe, and he did not have an answer.

Maud told them about Ada's baby, Kurt. 'She's brought him home from the hospital, and he seems to be happy enough for now. But his brain is damaged and he will never be normal. When he's older he will have to live in an institution, poor mite.'

Lloyd had heard how the baby had been delivered by eleven-year-old Carla. That little girl had grit.

Commissar Thomas Macke arrived at half past nine, wearing his Brownshirt uniform.

Last time he was here, Robert had treated him as a figure of fun, but Lloyd had sensed the menace of the man. He looked foolish, with the little moustache in the middle of his fat face, but there was a glint of cruelty in his eyes that made Lloyd nervous.

Robert had refused to sell the restaurant. What did Macke want now?

Macke stood in the middle of the dining area and shouted: 'This restaurant is being used to promote degenerate behaviour!'

The patrons went quiet, wondering what this was about.

Macke raised a finger in a gesture that meant *You'd better listen!* Lloyd felt there was something horribly familiar about the action, and realized that Macke was mimicking Hitler.

Macke said: 'Homosexuality is incompatible with the masculine character of the German nation!'

Lloyd frowned. Was he saying that Robert was queer?

Jörg came into the restaurant from the kitchen, wearing his tall chef's hat. He stood by the door, glaring at Macke.

Lloyd was struck by a shocking thought. Maybe Robert *was* queer. After all, he and Jörg had been living together since the war.

Looking around at their theatrical friends, Lloyd noticed that they were all men in pairs, except for two women with short hair . . .

Lloyd felt bewildered. He knew that queers existed, and as a broad-minded person he believed that they should not be persecuted but helped. However, he thought of them as perverts and creeps. Robert and Jörg seemed like normal men, running a business and living quietly – almost like a married couple!

He turned to his mother and said quietly: 'Are Robert and Jörg really . . . ?'

'Yes, dear,' she said.

Maud, sitting next to her, said: 'Robert in his youth was a menace to footmen.'

Both women giggled.

Lloyd was doubly shocked: not only was Robert queer, but Ethel and Maud thought it a matter for light-hearted banter.

Macke said: 'This establishment is now closed!'

Robert said: 'You have no right!'

Macke could not close the place on his own, Lloyd thought; then he remembered how the Brownshirts had crowded on to the stage at the People's Theatre. He looked towards the entrance – and was aghast to see Brownshirts pushing through the door.

They went around the tables knocking over bottles and glasses. Some customers sat motionless and watched; others got to their feet. Several men shouted and a woman screamed.

Walter stood up and spoke loudly but calmly. 'We should all leave quietly,' he said. 'There's no need for any rough stuff. Everybody just get your coats and hats and go home.'

The customers began to leave, some trying to get their coats, others just fleeing. Walter and Lloyd ushered Maud and Ethel towards the door.

The till was near the exit, and Lloyd saw a Brownshirt open it and begin stuffing money into his pockets.

Until then Robert had been standing still, watching miserably as a night's business hurried out of the door; but this was too much. He gave a shout of protest and shoved the Brownshirt away from the till.

The Brownshirt punched him, knocking him to the floor, and began to kick him as he lay there. Another Brownshirt joined in.

Lloyd leaped to Robert's rescue. He heard his mother shout 'No!' as he shoved the Brownshirts aside. Jörg was almost as quick, and the two of them bent to help Robert up.

They were immediately attacked by several more Brownshirts. Lloyd was punched and kicked, and something heavy hit him over the head. As he cried out in pain he thought: *No, not again.*

He turned on his attackers, punching with his left and right, making every blow connect hard, trying to punch *through* the target as he had been taught. He knocked two men down, then he was grabbed from behind and thrown off balance. A moment later he was on the floor with two men holding him down while a third kicked him.

Then he was rolled over on to his front, his arms were pulled behind his back, and he felt metal on his wrists. He had been handcuffed for the first time in his life. He felt a new kind of fear. This was not just another rough-house. He had been beaten and kicked, but worse was in store.

'Get up,' someone told him in German.

He struggled to his feet. His head hurt. Robert and Jörg were also in handcuffs, he saw. Robert's mouth was bleeding and Jörg had one closed eye. Half a dozen Brownshirts were guarding them. The rest were drinking from the glasses and bottles left on the tables, or standing at the dessert cart stuffing their faces with pastries.

All the customers seemed to have gone. Lloyd felt relieved that his mother had got away.

The restaurant door opened and Walter came back in. 'Commissar Macke,' he said, displaying a typical politician's facility for remembering names. With as much authority as he could muster he said: 'What is the meaning of this outrage?'

Macke pointed to Robert and Jörg. 'These two men are homosexuals,' he said. 'And that boy attacked an auxiliary policeman who was arresting them.'

Walter pointed to the till, which was open, its drawer sticking out and empty except for a few small coins. 'Do police officers commit robbery nowadays?'

'A customer must have taken advantage of the confusion created by those resisting arrest.'

Some of the Brownshirts laughed knowingly.

Walter said: 'You used to be a law enforcement officer, didn' you, Macke? You might have been proud of yourself, once. But what are you now?'

Macke was stung. 'We enforce order, to protect the Fatherland.'

'Where are you planning to take your prisoners, I wonder?' Walter persisted. 'Will it be a properly constituted place of detention? Or some half-hidden unofficial basement?'

'They will be taken to the Friedrich Strasse Barracks,' Macke said indignantly.

Lloyd saw a look of satisfaction pass briefly across Walter's face, and realized that Walter had cleverly manipulated Macke, playing on whatever was left of his professional pride in order to get him to reveal his intentions. Now, at least, Walter knew where Lloyd and the others were being taken.

But what would happen at the barracks?

Lloyd had never been arrested. However, he lived in the East End of London, so he knew plenty of people who got into trouble with the police. Most of his life he had played street football with boys whose fathers were arrested frequently. He knew the reputation of Leman Street police station in Aldgate. Few men came out of that building uninjured. People said there was blood all over the walls. Was it likely that the Friedrich Strasse Barracks would be any better?

Walter said: 'This is an international incident, Commissar.' Lloyd guessed he was using the title in the hope of making Macke behave more like an officer and less like a thug. 'You have arrested three foreign citizens – two Austrians and one Englishman.' He held up a hand as if to fend off a protest. 'It is too late to back out now. Both embassies are being informed, and I have no doubt that their representatives will be knocking on the door of our Foreign Office in Wilhelm Strasse within the hour.'

Lloyd wondered whether that was true.

Macke grinned unpleasantly. 'The Foreign Office will not hasten to defend two queers and a young hooligan.'

'Our foreign minister, von Neurath, is not a member of your party,' Walter said. 'He may well put the interests of the Fatherland first.'

'I think you will find that he does what he's told. And now you are obstructing me in the course of my duty.'

'I warn you!' Walter said bravely. 'You had better follow procedure by the book – or there will be trouble.'

'Get out of my sight,' said Macke.

Walter left.

Lloyd, Robert and Jörg were marched outside and bundled into the back of some kind of truck. They were forced to lie on the floor while Brownshirts sat on benches guarding them. The vehicle moved off. It was painful being handcuffed, Lloyd discovered. He felt constantly that his shoulder was about to become dislocated.

The trip was mercifully short. They were shoved out of the truck and into a building. It was dark, and Lloyd saw little. At a desk, his name was written in a book and his passport was taken away. Robert lost his gold tie pin and watch chain. At last the handcuffs were removed and they were pushed into a room with dim lights and barred windows. There were about forty other prisoners there already.

Lloyd hurt all over. He had a pain in his chest that felt like a cracked rib. His face was bruised and he had a blinding headache. He wanted an aspirin, a cup of tea and a pillow. He had a feeling it might be some hours before he got any of those things.

The three of them sat on the floor near the door. Lloyd held his head in his hands while Robert and Jörg discussed how soon help would come. No doubt Walter would phone a lawyer. But all the usual rules had been suspended by the Reichstag Fire Decree, so they had no proper protection under the law. Walter would also contact the embassies: political influence was their main hope now. Lloyd thought his mother would probably try to place an international phone call to the British Foreign Office in London. If she could get through, the government would surely have something to say about the arrest of a British schoolboy. It would all take time – an hour at least, probably two or three.

But four hours passed, then five, and the door did not open.

Civilized countries had a law about how long the police could keep someone in custody without formalities: a charge, a lawyer, a court. Lloyd now realized that such a rule was no mere technicality. He could be here for ever.

The other prisoners in the room were all political, he discovered: Communists, Social Democrats, trade union organizers and one priest.

The night passed slowly. None of the three slept. To Lloyd, sleep seemed unthinkable. The grey light of morning was coming through the barred windows when at last the cell door opened. But no lawyers or diplomats came in, just two men in aprons pushing a trolley on which stood a large urn. They ladled out a thin oatmeal. Lloyd did not eat any, but he drank a tin mug of coffee that tasted of burnt barley.

He surmised that the staff on duty overnight at the British embassy were junior diplomats who carried little weight. This morning, as soon as the ambassador himself got up, action would be taken.

An hour after breakfast the door opened again, but this time only Brownshirts stood there. They marched all the prisoners out and loaded them on to a truck, forty or fifty men in one canvas-sided vehicle, packed so tightly that they had to remain standing. Lloyd managed to stay close to Robert and Jörg.

Perhaps they were going to court, even though it was Sunday. He hoped so. At least there would be lawyers, and some semblance of due process. He thought he was fluent enough to state his simple case in German, and he practised his speech in his head. He had been dining in a restaurant with his mother; he had seen someone robbing the till; he had intervened in the resulting fracas. He imagined his cross-examination. He would be asked if the man he attacked was a Brownshirt. He would answer: 'I didn't notice his clothing – I just saw a thief.' There would be laughter in court, and the prosecutor would look foolish.

They were driven out of town.

They could see through gaps in the canvas sides of the truck. It seemed to Lloyd that they had gone about twenty miles when Robert said: 'We're in Oranienburg', naming a small town north of Berlin.

The truck came to a halt outside a wooden gate between brick pillars. Two Brownshirts with rifles stood guard.

Lloyd's fear rose a notch. Where was the court? This looked more like a prison camp. How could they put people in prison without a judge?

After a short wait, the truck drove in and stopped at a group of derelict buildings.

Lloyd was becoming even more anxious. Last night at least he had the consolation that Walter knew where he was. Today it was possible no one would know. What if the police simply said he was not in custody and they had no record of his arrest? How could he be rescued?

They got out of the truck and shuffled into what looked like a factory of some sort. The place smelled like a pub. Perhaps it had been a brewery.

Once again all their names were taken. Lloyd was glad there was some record of his movements. They were not tied up or handcuffed, but they were constantly watched by Brownshirts with rifles, and Lloyd had a grim feeling that those young men were only too eager for an excuse to shoot.

They were each given a canvas mattress filled with straw and a thin blanket. They were herded into a tumbledown building that once might have been a warehouse. Then the waiting began.

No one came for Lloyd all that day.

In the evening there was another trolley and another urn, this one containing a stew of carrots and turnips. Each man got a bowlful and a piece of bread. Lloyd was now ravenous, not having eaten for twenty-four hours, and he wolfed down his meagre supper and wished for more.

Somewhere in the camp there were three or four dogs that howled all night.

Lloyd felt dirty. This was the second night he had spent in the same clothes. He needed a bath and a shave and a clean shirt. The toilet facilities, two barrels in the corner, were absolutely disgusting.

But tomorrow was Monday. Then there would be some action.

Lloyd fell asleep around four. At six they were awakened by a Brownshirt bawling: 'Schleicher! Jörg Schleicher! Which one is Schleicher?'

Maybe they were going to be released.

Jörg stood up and said: 'Me, I'm Schleicher.'

'Come with me,' said the Brownshirt.

Robert said in a frightened voice: 'Why? What do you want him for? Where is he going?'

'What are you, his mother?' said the Brownshirt. 'Lie down and shut your mouth.' He poked Jörg with his rifle. 'Outside, you.'

Watching them go, Lloyd asked himself why he had not punched the Brownshirt and snatched the rifle. He might have escaped. And if he had failed, what would they do to him – throw him in jail? But at the crucial moment the thought of escape had not even occurred to him. Was he already taking on the mentality of a prisoner?

He was even looking forward to the oatmeal.

Before breakfast, they were all taken outside.

They stood around a small wire-fenced area a quarter the size of a tennis court. It looked as if it might have been used to store something not very valuable, timber or tyres perhaps. Lloyd shivered in the cold morning air: his overcoat was still at Bistro Robert.

Then he saw Thomas Macke approaching.

The police detective wore a black coat over his Brownshirt uniform. He had a heavy, flat-footed stride, Lloyd noticed.

Behind Macke were two Brownshirts holding the arms of a naked man with a bucket over his head.

Lloyd stared in horror. The prisoner's hands were tied behind his back, and the bucket was tightly tied with string under his chin so that it would not fall off.

He was a slight, youngish man with blond pubic hair.

Robert groaned: 'Oh, sweet Jesus, it's Jörg.'

All the Brownshirts in the camp had gathered. Lloyd frowned. What was this, some kind of cruel game?

Jörg was led into the fenced compound and left there, shivering. His two escorts withdrew. They disappeared for a few minutes then returned, each of them leading two Alsatian dogs.

That explained the all-night barking.

The dogs were thin, with unhealthy bald patches in their tan fur. They looked starved. The Brownshirts led them to the fenced compound.

Lloyd had a vague but dreadful premonition of what was to come.

Robert screamed: 'No!' He ran forward. 'No, no, no!' He tried to open the gate of the compound. Three or four Brownshirts pulled him away roughly. He struggled, but they were strong young thugs, and Robert was approaching fifty years old: he could not resist them. They threw him contemptuously to the ground.

'No,' said Macke to his men. 'Make him watch.'

They lifted Robert to his feet and held him facing the wire fence.

The dogs were led into the compound. They were excited, barking and slavering. The two Brownshirts handled them expertly and without fear, clearly experienced. Lloyd wondered dismally how many times they had done this before.

The handlers released the dogs and hurried out of the compound.

The dogs dashed for Jörg. One bit his calf, another his arm, a third his thigh. From behind the metal bucket there was a muffled scream of agony and terror. The Brownshirts cheered and applauded. The prisoners looked on in mute horror.

After the first shock, Jörg tried to defend himself. His hands were tied and he was unable to see, but he could kick out randomly. However, his bare feet made little impact on the starving dogs. They dodged and came again, ripping his flesh with their sharp teeth.

He tried running. With the dogs at his heels he ran blindly in a straight line until he crashed into the wire fence. The Brownshirts cheered raucously. Jörg ran in a different direction with the same result. A dog took a chunk out of Jörg's behind, and they hooted with laughter.

A Brownshirt standing next to Lloyd was shouting: 'His tail! Bite his tail!' Lloyd guessed that 'tail' in German – *der Schwanz* – was slang for penis. The man was hysterical with excitement.

Jörg's white body was now running with blood from multiple wounds. He pressed himself up against the wire, face-first, protecting his genitals, kicking out backwards and sideways. But he was weakening. His kicks became feeble. He was having trouble staying upright. The dogs became bolder, tearing at him and swallowing bloody chunks.

At last Jörg slid to the ground.

The dogs settled down to feed.

The handlers re-entered the compound. With practised motions they reattached the dogs' leads, pulled them off Jörg, and led them away.

The show was over, and the Brownshirts began to move away, chattering excitedly.

Robert ran into the compound, and this time no one stopped him. He bent over Jörg, moaning.

Lloyd helped him to untie Jörg's hands and remove the bucket. Jörg was unconscious but breathing. Lloyd said: 'Let's get him indoors. You take his legs.' Lloyd grasped Jörg under the arms and the two of them carried him into the building where they had slept. They put him on a mattress. The other prisoners gathered around, frightened and subdued. Lloyd hoped one of them might announce that he was a doctor, but no one did.

Robert stripped off his jacket and waistcoat, then took off his shirt and used it to wipe the blood. 'We need clean water,' he said.

There was a standpipe in the yard. Lloyd went out, but he had no container. He returned to the compound. The bucket was still there on the ground. He washed it out then filled it with water.

When he returned, the mattress was soaked in blood.

Robert dipped his shirt in the bucket and continued to wash Jörg's wounds, kneeling beside the mattress. Soon the white shirt was red.

Jörg stirred.

Robert spoke to him in a low voice. 'Be calm, my beloved,' he said. 'It's over now, and I'm here.' But Jörg seemed not to hear.

Then Macke came in, with four or five Brownshirts following. He grabbed Robert's arm and pulled him. 'So!' he said. 'Now you know what we think of homosexual perverts.'

Lloyd pointed at Jörg and said angrily: 'The pervert is the one who caused this to happen.' Mustering all his rage and contempt, he said: 'Commissar Macke.'

Macke gave a slight nod to one of the Brownshirts. In a movement that was deceptively casual, the man reversed his rifle and hit Lloyd over the head with the butt.

Lloyd fell to the ground, holding his head in agony.

He heard Robert say: 'Please, just let me look after Jörg.'

'Perhaps,' said Macke. 'First come over here.'

Despite his pain, Lloyd opened his eyes to see what was happening.

Macke pulled Robert across the room to a rough wooden table. From his pocket he drew a document and a fountain pen. 'Your restaurant is now worth half of what I last offered you – ten thousand marks.'

'Anything,' said Robert, weeping. 'Leave me to be with Jörg.'

'Sign here,' said Macke. 'Then the three of you can go home.'

Robert signed.

'This gentleman can be a witness,' Macke said. He gave the pen to one of the Brownshirts. He looked across the room and met Lloyd's eye. 'And perhaps our foolhardy English guest can be the second witness.'

Robert said: 'Just do what he wants, Lloyd.'

Lloyd struggled to his feet, rubbed his sore head, took the pen, and signed.

Macke pocketed the contract triumphantly and went out.

Robert and Lloyd returned to Jörg.

But Jörg was dead.

(viii)

Walter and Maud came to the Lehrte Station, just north of the burned-out Reichstag, to see Ethel and Lloyd off. The station building was in the neo-Renaissance style and looked like a French palace. They were early, and they sat in a station café while they waited for the train.

Lloyd was glad to be leaving. In six weeks he had learned a lot, about the German language and about politics, but now he wanted to get home, tell people what he had seen, and warn them against the same thing happening to them.

All the same he felt strangely guilty about departing. He was going to a place where the law ruled, the press was free, and it was not a crime to be a social democrat. He was leaving the von Ulrich family to live on in a cruel dictatorship where an innocent man could be torn to pieces by dogs and no one would ever be brought to justice for the crime.

The von Ulrichs looked crushed; Walter even more than Maud. They were like people who have heard bad news, or suffered a death in the family. They seemed unable to think much about anything other than the catastrophe that had happened to them.

Lloyd had been released with profuse apologies from the German Foreign Ministry, and an explanatory statement that was abject yet at the same time mendacious, implying that he had got into a brawl through his own foolishness and then had been held prisoner by an administrative error for which the authorities were deeply sorry.

Walter said: 'I've had a telegram from Robert. He's arrived safely in London.'

As an Austrian citizen Robert had been able to leave Germany without much difficulty. Getting his money out had been more tricky. Walter had demanded that Macke pay the money to a bank in Switzerland. At first Macke had said that was impossible, but Walter had put pressure on him, threatening to challenge the sale in court, saying that Lloyd was prepared to testify that the contract had been signed under duress; and in the end Macke had pulled some strings.

'I'm glad Robert got out,' Lloyd said. He would be even happier when he himself was safe in London. His head was still tender and he got a pain in his ribs every time he turned over in bed.

Ethel said to Maud: 'Why don't you come to London? Both of you. The whole family, I mean.'

Walter looked at Maud. 'Perhaps we should,' he said. But Lloyd could tell that he did not really mean it.

'You've done your best,' Ethel said. 'You've fought bravely. But the other side won.'

Maud said: 'It's not over yet.'

'But you're in danger.'

'So is Germany.'

'If you came to live in London, Fitz might soften his attitude, and help you.'

Earl Fitzherbert was one of the wealthiest men in Britain, Lloyd knew, because of the coal mines beneath his land in South Wales.

'He wouldn't help me,' Maud said. 'Fitz doesn't relent. I know that, and so do you.'

'You're right,' Ethel said. Lloyd wondered how she could be so sure, but he did not get a chance to ask. Ethel went on: 'Well, you could easily get a job on a London newspaper, with your experience.'

Walter said: 'And what would I do in London?'

'I don't know,' Ethel said. 'What are you going to do here? There's not much point in being an elected representative in an impotent parliament.' She was being brutally frank, Lloyd felt, but characteristically she was saying what had to be said.

Lloyd sympathized, but felt that the von Ulrichs should stay. 'I know it will be hard,' he said. 'But if decent people flee from Fascism it will spread all the faster.'

'It's spreading anyway,' his mother rejoined.

Maud startled them all by saying vehemently: 'I will not go. I absolutely refuse to leave Germany.'

They all stared at her.

'I'm German, and have been for fourteen years,' she said. 'This is my country now.'

'But you were born English,' said Ethel.

'A country is mostly the people in it,' Maud said. 'I don't love England. My parents died a long time ago, and my brother has disowned me. I love Germany. For me, Germany is my wonderful husband, Walter; my misguided son, Erik; my alarmingly capable daughter, Carla; our maid, Ada, and her disabled son; my friend Monika and her family; my journalistic colleagues . . . I'm staying, to fight the Nazis.'

'You've already done more than your share,' Ethel said gently.

Maud's tone became emotional. 'My husband has dedicated himself, his life, his entire being to making this country free and prosperous. I will not be the cause of his giving up his life's work. If he loses that, he loses his soul.'

Ethel pushed the point in a way that only an old friend could. 'Still,' she said, 'there must be a temptation to take your children to safety.'

'A temptation? You mean a longing, a yearning, a desperate desire!'

She began to cry. 'Carla has nightmares about Brownshirts, and Erik puts on that shit-coloured uniform every chance he gets.' Lloyd was startled by her fervour. He had never heard a respectable woman say 'shit'. She went on: 'Of course I want to take them away.' Lloyd could see how torn she was. She rubbed her hands together as if washing them, turned her head from side to side in distraction, and spoke in a voice that shook violently with her inner conflict. 'But it's the wrong thing to do, for them as well as for us. I will not give in to it! Better to suffer evil than to stand by and do nothing.'

Ethel touched Maud's arm. 'I'm sorry I asked. Perhaps it was silly of me. I might have known you wouldn't run away.'

'I'm glad you asked,' Walter said. He reached out and took Maud's slim hands in his own. 'The question has been hanging in the air between Maud and me, unspoken. It was time we faced it.' Their joined hands rested on the café table. Lloyd rarely thought about the emotional lives of his mother's generation – they were middle-aged and married, and that seemed to say it all – but now he saw that between Walter and Maud there was a powerful connection that was much more than the familiar habit of a mature marriage. They were under no illusions: they knew that by staying here they were risking their lives and the lives of their children. But they had a shared commitment that defied death.

Lloyd wondered whether he would ever have such a love.

Ethel looked at the clock. 'Oh, my goodness!' she said. 'We're going to miss the train!'

Lloyd picked up their bags and they hurried across the platform. A whistle blew. They boarded the train just in time. They both leaned out of the window as it pulled out of the station.

Walter and Maud stood on the platform, waving, getting smaller and smaller in the distance, until finally they disappeared.

2

'Two things you need to know about girls in Buffalo,' said Daisy Peshkov. 'They drink like fish, and they're all snobs.'

Eva Rothmann giggled. 'I don't believe you,' she said. Her German accent had almost completely vanished.

'Oh, it's true,' said Daisy. They were in her pink-and-white bedroom, trying on clothes in front of a full-length three-way mirror. 'Navy and white might look good on you,' Daisy said. 'What do you think?' She held a blouse up to Eva's face and studied the effect. The contrasting colours seemed to suit her.

Daisy was looking through her closet for an outfit Eva could wear to the beach picnic. Eva was not a pretty girl, and the frills and bows that decorated many of Daisy's clothes only made Eva look frumpy. Stripes better suited her strong features.

Eva's hair was dark, and her eyes deep brown. 'You can wear bright colours,' Daisy told her.

Eva had few clothes of her own. Her father, a Jewish doctor in Berlin, had spent his life savings to send her to America, and she had arrived a year ago with nothing. A charity paid for her to go to Daisy's boarding school – they were the same age, nineteen. But Eva had nowhere to go in the summer vacation, so Daisy had impulsively invited her home.

At first Daisy's mother, Olga, had resisted. 'Oh, but you're away at school all year – I so look forward to having you to myself in the summer.'

'She's really great, Mother,' Daisy had said. 'She's charming and easygoing and a loyal friend.'

'I suppose you feel sorry for her because she's a refugee from the Nazis.'

'I don't care about the Nazis, I just like her.'

'That's fine, but does she have to live with us?'

'Mother, she has nowhere else to go!'

As usual, Olga let Daisy have her way in the end.

Now Eva said: 'Snobs? No one would be snobby to you!'

'Oh, yes, they would.'

'But you're so pretty and vivacious.'

Daisy did not bother to deny it. 'They hate that about me.'

'And you're rich.'

It was true. Daisy's father was wealthy, her mother had inherited a fortune, and Daisy herself would come into money when she was twenty-one. 'It doesn't mean a thing. In this town it's about how long you've been rich. You're nobody if you work. The superior people are those who live on the millions left by their great-grandparents.' She spoke in a tone of gay mockery to hide the resentment she felt.

Eva said: 'And your father is famous!'

'They think he's a gangster.'

Daisy's grandfather, Josef Vyalov, had owned bars and hotels. Her father, Lev Peshkov, had used the profits to buy ailing vaudeville theatres and convert them into cinemas. Now he owned a Hollywood studio, too.

Eva was indignant on Daisy's behalf. 'How can they say such a thing?'

'They believe he was a bootlegger. They're probably right. I can't see how else he made money out of bars during Prohibition. Anyway, that's why Mother will never be invited to join the Buffalo Ladies' Society.'

They both looked at Olga, sitting on Daisy's bed, reading the *Buffalo Sentinel*. In photographs taken when she was young, Olga was a willowy beauty. Now she was dumpy and drab. She had lost interest in her appearance, though she shopped energetically with Daisy, never caring how much she spent to make her daughter look fabulous.

Olga looked up from the newspaper to say: 'I'm not sure they mind your father being a bootlegger, dear. But he's a Russian immigrant, and on the rare occasions he decides to attend divine service, he goes to the Russian Orthodox church on Ideal Street. That's almost as bad as being Catholic.'

Eva said: 'It's so unfair.'

'I might as well warn you that they're not too fond of Jews, either,' Daisy said. Eva was, in fact, half Jewish. 'Sorry to be blunt.'

'Be as blunt as you like – after Germany, this country feels like the Promised Land.'

'Don't get too comfortable,' Olga warned. 'According to this paper, plenty of American business leaders hate President Roosevelt and admire Adolf Hitler. I know that's true, because Daisy's father is one of them.'

'Politics is boring,' said Daisy. 'Isn't there something interesting in the *Sentinel*?'

'Yes, there is. Muffie Dixon is to be presented at the British court.'

'Good for her,' Daisy said sourly, failing to conceal her envy.

Olga read: '"Miss Muriel Dixon, daughter of the late Charles 'Chuck' Dixon, who was killed in France during the war, will be presented at Buckingham Palace next Tuesday by the wife of the United States ambassador, Mrs Robert W. Bingham."'

Daisy had heard enough about Muffie Dixon. 'I've been to Paris, but never London,' she said to Eva. 'What about you?'

'Neither,' said Eva. 'The first time I left Germany was when I sailed to America.'

Olga suddenly said: 'Oh, dear!'

'What's happened?' Daisy asked.

Her mother crumpled the paper. 'Your father took Gladys Angelus to the White House.'

'Oh!' Daisy felt as if she had been slapped. 'But he said he would take me!'

President Roosevelt had invited a hundred businessmen to a reception in an attempt to win them over to his New Deal. Lev Peshkov thought Franklin D. Roosevelt was the next thing to a Communist, but he had been flattered to be asked to the White House. However, Olga had refused to accompany him, saying angrily: 'I'm not willing to pretend to the President that we have a normal marriage.'

Lev officially lived here, in the stylish pre-war prairie home built by Grandfather Vyalov, but he spent more nights at the swanky downtown apartment where he kept his mistress of many years, Marga. On top of that everyone assumed he was having an affair with his studio's biggest star, Gladys Angelus. Daisy understood why her mother felt spurned. Daisy, too, felt rejected when Lev drove off to spend his evenings with his other family.

She had been thrilled when he had asked her to accompany him to the White House instead of her mother. She had told everyone she was going. None of her friends had met the President, except the Dewar boys, whose father was a senator.

Lev had not told her the exact date, and she had assumed that he would let her know at the last minute, which was his usual style. But he had changed his mind, or perhaps just forgotten. Either way, he had rejected Daisy again.

'I'm sorry, honey,' said her mother. 'But promises never did mean much to your father.'

Eva was looking sympathetic. Her pity stung Daisy. Eva's father was thousands of miles away, and she might never see him again, but she felt sorry for Daisy, as if Daisy's plight was worse.

It made Daisy feel defiant. She would not let this ruin her day. 'Well, I'll be the only girl in Buffalo who has been stood up for Gladys Angelus,' she said. 'Now, what shall I wear?'

Skirts were dramatically short this year in Paris, but the conservative Buffalo set followed fashion at a distance. However, Daisy had a knee-length tennis dress in a shade of baby-blue the same as her eyes. Maybe today was the day to bring it out. She slipped off her dress and put on the new one. 'What do you think?' she said.

Eva said: 'Oh, Daisy, it's beautiful, but . . .'

Olga said: 'That'll make their eyes pop.' Olga liked it when Daisy dressed to kill. Perhaps it reminded her of her youth.

Eva said: 'Daisy, if they're all so snobbish, why do you want to go to the party?'

'Charlie Farquharson will be there, and I'm thinking of marrying him,' Daisy said.

'Are you serious?'

Olga said emphatically: 'He's a great catch.'

Eva said: 'What's he like?'

'Absolutely adorable,' Daisy said. 'Not the handsomest boy in Buffalo, but sweet and kind, and rather shy.'

'He sounds very different from you.'

'It's the attraction of opposites.'

Olga spoke again. 'The Farquharsons are among the oldest families in Buffalo.'

Eva raised her dark eyebrows. 'Snobby?'

'Very,' Daisy said. 'But Charlie's father lost all his money in the Wall Street crash, then died – killed himself, some say – so they need to restore the family fortunes.'

Eva looked shocked. 'You're hoping he'll marry you for your money?'

'No. He'll marry me because I will bewitch him. But his mother will accept me for my money.'

'You say you *will* bewitch him. Does he know about any of this?'

'Not yet. But I think I might make a start this afternoon. Yes, this is definitely the right dress.'

Daisy wore the baby-blue and Eva the navy-and-white stripes. By the time they had got ready they were late.

Daisy's mother would not have a chauffeur. 'I married my father's chauffeur, and it ruined my life,' she sometimes said. She was terrified Daisy might do something similar – that was why she was so keen on Charlie Farquharson. If she needed to go anywhere in her creaking 1925 Stutz she made Henry, the gardener, take off his rubber boots and put on a black suit. But Daisy had her own car, a red Chevrolet Sport Coupe.

Daisy liked driving, loved the power and speed of it. They headed south out of the city. She was almost sorry it was only five or six miles to the beach.

As she drove she thought about life as Charlie's wife. With her money and his status they would become the leading couple in Buffalo society. At their dinner parties the table settings would be so elegant that people would gasp in delight. They would have the biggest yacht in the harbour, and throw on-board parties for other wealthy, fun-loving couples. People would yearn for an invitation from Mrs Charles Farquharson. No charity function would be a success without Daisy and Charlie at the top table. In her head she watched a movie of herself, in a ravishing Paris gown, walking through a crowd of admiring men and women, smiling graciously at their compliments.

She was still daydreaming when they reached their destination.

The city of Buffalo was in upstate New York, near the Canadian border. Woodlawn Beach was a mile of sand on the shore of Lake Erie. Daisy parked and they walked across the dunes.

Fifty or sixty people were already there. These were the adolescent children of the Buffalo elite, a privileged group who spent their summers sailing and water-skiing in the daytime and going to parties and dances at night. Daisy greeted the people she knew, which was just about everyone, and introduced Eva around. They got glasses of punch. Daisy tasted it cautiously: some of the boys would think it hilarious to spike the drink with a couple of bottles of gin.

The party was for Dot Renshaw, a sharp-tongued girl whom no one wanted to marry. The Renshaws were an old Buffalo family, like the Farquharsons, but their fortune had survived the crash. Daisy made sure to approach the host, Dot's father, and thank him. 'I'm sorry we're late,' she said. 'I lost track of time!'

Philip Renshaw looked her up and down. 'That's a very short skirt.' Disapproval vied with lasciviousness in his expression.

'I'm so glad you like it,' Daisy replied, pretending he had paid her a straightforward compliment.

'Anyway, it's good that you're here at last,' he went on. 'A photographer from the *Sentinel* is coming and we must have some pretty girls in the picture.'

Daisy muttered to Eva: 'So that's why I was invited. How kind of him to let me know.'

Dot came up. She had a thin face with a pointed nose. Daisy always thought she looked as if she might peck you. 'I thought you were going with your father to meet the President,' she said.

Daisy felt mortified. She wished she had not boasted to everyone about this.

'I see he took his, ahem, leading lady,' Dot went on. 'Unusual, that sort of thing, in the White House.'

Daisy said: 'I guess the President likes to meet movie stars occasionally. He deserves a little glamour, don't you think?'

'I can't imagine that Eleanor Roosevelt approved. According to the *Sentinel*, all the other men took their wives.'

'How thoughtful of them.' Daisy turned away, desperate to escape.

She spotted Charlie Farquharson, trying to erect a net for beach tennis. He was too good-natured to mock her about Gladys Angelus. 'How are you, Charlie?' she said brightly.

'Fine, I guess.' He stood up, a tall man of about twenty-five, a little overweight, stooping slightly as if he feared his height might be intimidating.

Daisy introduced Eva. Charlie was sweetly awkward in company, especially with girls, but he made an effort and asked Eva how she liked America, and what she heard from her family back in Berlin.

Eva asked him if he was enjoying the picnic.

'Not much,' he said candidly. 'I'd rather be at home with my dogs.'

No doubt he found pets easier to deal with than girls, Daisy thought. But the mention of dogs was interesting. 'What kind of dogs do you have?' she asked.

'Jack Russell terriers.'

Daisy made a mental note.

An angular woman of about fifty approached. 'For goodness' sake, Charlie, haven't you got that net up yet?'

'Almost there, Mom,' he said.

Nora Farquharson was wearing a gold tennis bracelet, diamond ear studs, and a Tiffany necklace; more jewellery than she really needed for a picnic. The Farquharsons' poverty was relative, Daisy reflected. They said they had lost everything, but Mrs Farquharson still had a maid and a chauffeur and a couple of horses for riding in the park.

Daisy said: 'Good afternoon, Mrs Farquharson. This is my friend Eva Rothmann from Berlin.'

'How do you do,' said Nora Farquharson without offering her hand. She felt no need to be friendly towards arriviste Russians, much less their Jewish guests.

Then she seemed to be struck by a thought. 'Ah, Daisy, you could go round and find out who wants to play tennis.'

Daisy knew she was being treated somewhat as a servant, but she decided to be compliant. 'Of course,' she said. 'Mixed doubles, I suggest.'

'Good idea.' Mrs Farquharson held out a pencil stub and a scrap of paper. 'Write the names down.'

Daisy smiled sweetly and took a gold pen and a little beige leather notebook from her bag. 'I'm equipped.'

She knew who the tennis players were, good and bad. She belonged to the Racquet Club, which was not as exclusive as the Yacht Club. She paired Eva with Chuck Dewar, the fourteen-year-old son of Senator Dewar. She put Joanne Rouzrokh with the older Dewar boy, Woody, only fifteen but already as tall as his beanpole father. Naturally she herself would be Charlie's partner.

Daisy was startled to come across a somewhat familiar face and to recognize her half-brother, Greg, the son of Marga. They did not meet often, and she had not seen him for a year. In that time he seemed to have become a man. He was six inches taller, and although still only fifteen he had the dark shadow of a beard. As a child he had been dishevelled, and that had not changed. He wore his expensive clothes carelessly: the sleeves of the blazer rolled up, the striped tie loose at the neck, the linen pants sea-wet and sandy at the cuffs.

Daisy was always embarrassed to run into Greg. He was a living reminder of how their father had rejected Daisy and her mother in favour of Greg and Marga. Many married men had affairs, she knew; but *her* father's indiscretion showed up at parties for everyone to see. Father should have moved Marga and Greg to New York, where nobody knew anybody, or to California, where no one saw anything wrong with adultery. Here they were a permanent scandal, and Greg was part of the reason people looked down on Daisy.

He asked her politely how she was, and she answered: 'Angry as heck, if you want to know. Father's let me down – again.'

Greg said guardedly: 'What did he do?'

'Asked me to go to the White House with him – then took that tart Gladys Angelus. Now everyone's laughing at me.'

'It must have been good publicity for *Passion*, her new film.'

'You always take his side because he prefers you to me.'

Greg looked irritated. 'Maybe that's because I admire him instead of complaining about him all the time.'

'I don't—' Daisy was about to deny complaining all the time when she realized it was true. 'Well, maybe I do complain, but he should keep his promises, shouldn't he?'

'He has so much on his mind.'

'Maybe he shouldn't have two mistresses as well as a wife.'

Greg shrugged. 'It's a lot to handle.'

They both noticed the unintentional double entendre, and after a moment they giggled.

Daisy said: 'Well, I guess I shouldn't blame you. You didn't ask to be born.'

'And I should probably forgive you for taking my father away from me three nights a week – no matter how I cried and begged him to stay.'

Daisy had never thought of it that way. In her mind, Greg was the usurper, the illegitimate child who kept stealing her father. But now she realized that he felt as hurt as she did.

She stared at him. Some girls might find him attractive, she guessed. He was too young for Eva, though. And he would probably turn out as selfish and unreliable as their father.

'Anyway,' she said, 'do you play tennis?'

He shook his head. 'They don't let people like me into the Racquet Club.' He forced an insouciant grin, and Daisy realized that, like her, Greg felt rejected by Buffalo society. 'Ice hockey's my sport,' he said.

'Too bad.' She moved on.

When she had enough names, she returned to Charlie, who had finally got the net up. She sent Eva to round up the first foursome. Then she said to Charlie: 'Help me make a competition tree.'

They knelt side by side and drew a diagram in the sand with heats, semi-finals and a final. While they were entering the names, Charlie said: 'Do you like the movies?'

Daisy wondered if he was about to ask her for a date. 'Sure,' she said.

'Have you seen *Passion*, by any chance?'

'No, Charlie, I haven't seen it,' she said in a tone of exasperation. 'It stars my father's mistress.'

He was shocked. 'The papers say they're just good friends.'

'And why do you think Miss Angelus, who is barely twenty, is so *friendly* with my forty-year-old father?' Daisy asked sarcastically. 'Do you think she likes his receding hairline? Or his little paunch? Or his fifty million dollars?'

'Oh, I see,' said Charlie, looking abashed. 'Sorry.'

'You shouldn't be sorry. I'm being kind of bitchy. You're not like everyone else – you don't automatically think the worst of people.'

'I guess I'm just dumb.'

'No. You're just nice.'

Charlie looked embarrassed, but pleased.

'Let's get on with this,' Daisy said. 'We have to rig it so that the best players get through to the final.'

Nora Farquharson reappeared. She looked at Charlie and Daisy kneeling side by side in the sand, then studied their drawing.

Charlie said: 'Pretty good, Mom, don't you think?' He longed for approval from her, that was obvious.

'Very good.' She gave Daisy an appraising look, like a mother dog seeing a stranger approach her puppies.

'Charlie did most of it,' Daisy said.

'No, he didn't,' Mrs Farquharson said bluntly. Her gaze went to Charlie and back. 'You're a smart girl,' she said. She looked as if she were about to add something, but hesitated.

'What?' said Daisy.

'Nothing.' She turned away.

Daisy stood up. 'I know what she was thinking,' she murmured to Eva.

'What?'

'You're a smart girl – almost good enough for my son, if you came from a better family.'

Eva was sceptical. 'You can't know that.'

'I sure can. And I'll marry him if only to prove his mother wrong.'

'Oh, Daisy, why do you care so much what these people think?'

'Let's watch the tennis.'

Daisy sat on the sand beside Charlie. He might not be handsome, but he would worship his wife and do anything for her. The mother-in-law would be a problem, but Daisy thought she could handle her.

Tall Joanne Rouzrokh was serving, in a white skirt that flattered her long legs. Her partner, Woody Dewar, who was even taller, handed her a tennis ball. Something in the way he looked at Joanne made Daisy think he was attracted to her, maybe even in love with her. But he was fifteen and she eighteen, so there was no future in that.

She turned to Charlie. 'Maybe I should see *Passion* after all,' she said.

He did not take the hint. 'Maybe you should,' he said indifferently. The moment had passed.

Daisy turned to Eva. 'I wonder where I could buy a Jack Russell terrier?'

(ii)

Lev Peshkov was the best father a guy could have – or, at least, he would have been, if he had been around more. He was rich and generous, he was smarter than anybody, he was even well dressed. He had probably been handsome when he was younger, and even now women threw themselves at him. Greg Peshkov adored him, and his only complaint was that he did not see enough of him.

'I should have sold this fucking foundry when I had the chance,' Lev said as they walked around the silent, deserted factory. 'It was losing money even before the goddamned strike. I should stick to cinemas and bars.' He wagged a didactic finger. 'People always buy booze, in good times and bad. And they go to the movies even when they can't afford to. Never forget that.'

Greg was pretty sure his father did not often make mistakes in business. 'So why did you keep it?' he said.

'Sentiment,' Lev replied. 'When I was your age, I worked in a place like this, the Putilov Machine Factory in St Petersburg.' He looked around at the furnaces, moulds, hoists, lathes and workbenches. 'Actually, it was a lot worse.'

The Buffalo Metal Works made fans of all sizes, including huge propellers for ships. Greg was fascinated by the mathematics of the curved blades. He was top of his class in math. 'Were you an engineer?' he asked.

Lev grinned. 'I tell people that, if I need to impress them,' he said. 'But the truth is, I looked after the horses. I was a stable boy. I was never good with machines. That was my brother Grigori's talent. You take after him. All the same, never buy a foundry.'

'I won't.'

Greg was to spend the summer shadowing his father, learning the business. Lev had just got back from Los Angeles, and Greg's lessons had begun today. But he did not want to know about the foundry. He was good at math but he was interested in power. He wished his father would take him on one of his frequent trips to Washington to lobby for the movie industry. That was where the real decisions were made.

He was looking forward to lunch. He and his father were to meet Senator Gus Dewar. Greg wanted to ask a favour of Senator Dewar. However, he had not yet cleared this with his father. He was nervous about asking, and instead he said: 'Do you ever hear of your brother in Leningrad?'

Lev shook his head. 'Not since the war. I wouldn't be surprised if he's dead. A lot of old Bolsheviks have disappeared.'

'Speaking of family, I saw my half-sister on Saturday. She was at the beach picnic.'

'Did you have a good time?'

'She's mad at you, did you know that?'

'What have I done now?'

'You said you'd take her to the White House, then you took Gladys Angelus.'

'That's true. I forgot. But I wanted the publicity for *Passion*.'

They were approached by a tall man whose striped suit was loud even by current fashions. He touched the brim of his fedora and said: 'Morning, boss.'

Lev said to Greg: 'Joe Brekhunov is in charge of security here. Joe, this is my son Greg.'

'Pleased to meet ya,' said Brekhunov.

Greg shook his hand. Like most factories, the foundry had its own police force. But Brekhunov looked more like a hoodlum than a cop.

'All quiet?' Lev asked.

'A little incident in the night,' Brekhunov said. 'Two machinists tried to heist a length of fifteen-inch steel bar, aircraft quality. We caught them trying to manhandle it over the fence.'

Greg said: 'Did you call the police?'

'It wasn't necessary.' Brekhunov grinned. 'We gave them a little talk about the concept of private property, and sent them to the hospital to think about it.'

Greg was not surprised to learn that his father's security men beat thieves so badly that they had to go to hospital. Although Lev had never struck him or his mother, Greg felt that violence was never far below his father's charming surface. It was because of Lev's youth in the slums of Leningrad, he guessed.

A portly man wearing a blue suit with a workingman's cap appeared from behind a furnace. 'This is the union leader, Brian Hall,' said Lev. 'Morning, Hall.'

'Morning, Peshkov.'

Greg raised his eyebrows. People usually called his father Mr Peshkov.

Lev stood with his feet apart and his hands on his hips. 'Well, have you got an answer for me?'

Hall's face took on a stubborn expression. 'The men won't come back to work with a pay cut, if that's what you mean.'

'But I've improved my offer!'

'It's still a pay cut.'

Greg began to feel nervous. His father did not like opposition, and he might explode.

'The manager tells me we aren't getting any orders, because he can't tender a competitive price at these wage levels.'

'That's because you've got outdated machinery, Peshkov. Some of these lathes were here before the war! You need to re-equip.'

'In the middle of a depression? Are you out of your mind? I'm not going to throw away more money.'

'That's how your men feel,' said Hall, with the air of one who plays a trump card. 'They're not going to give money to you when they haven't got enough for themselves.'

Greg thought workers were stupid to strike during a depression, and he was angered by Hall's nerve. The man spoke as if he were Lev's equal, not an employee.

Lev said: 'Well, as things are, we're all losing money. Where's the sense in that?'

'It's out of my hands now,' said Hall. Greg thought he sounded smug. 'The union is sending a team from headquarters to take over.' He pulled a large steel watch out of his waistcoat pocket. 'Their train should be here in an hour.'

Lev's face darkened. 'We don't need outsiders stirring up trouble.'

'If you don't want trouble, you shouldn't provoke it.'

Lev clenched a fist, but Hall walked away.

Lev turned to Brekhunov. 'Did you know about these men from headquarters?' he said angrily.

Brekhunov looked nervous. 'I'll get on it right away, boss.'

'Find out who they are and where they're staying.'

'Won't be difficult.'

'Then send them back to New York in a fucking ambulance.'

'Leave it to me, boss.'

Lev turned away, and Greg followed him. Now that was power, Greg thought with a touch of awe. His father gave the word, and union officials would be beaten up.

They walked outside and got into Lev's car, a Cadillac five-passenger sedan in the new streamlined style. Its long curving fenders made Greg think of a girl's hips.

Lev drove along Porter Avenue to the waterfront and parked at the Buffalo Yacht Club. Sunlight played prettily on the boats in the marina. Greg was pretty sure that his father did not belong to this elite club. Gus Dewar must be a member.

They walked on to the pier. The clubhouse was built on pilings over the water. Lev and Greg went inside and checked their hats. Greg immediately felt uneasy, knowing he was a guest in a club that would not have him as a member. The people here probably thought he must feel privileged to be allowed in. He put his hands in his pockets and slouched, so they would know he was not impressed.

'I used to belong to this club,' Lev said. 'But in 1921 the chairman told me I had to resign because I was a bootlegger. Then he asked me to sell him a case of Scotch.'

'Why does Senator Dewar want to have lunch with you?' Greg asked.

'We're about to find out.'

'Would you mind if I asked him a favour?'

Lev frowned. 'I guess not. What are you after?'

But, before Greg could answer, Lev greeted a man of about sixty. 'This is Dave Rouzrokh,' he said to Greg. 'He's my main rival.'

'You flatter me,' the man said.

Roseroque Theatres was a chain of dilapidated movie houses in New York State. The owner was anything but decrepit. He had a patrician air: he was tall and white-haired, with a nose like a curved blade. He wore a blue cashmere blazer with the badge of the club on the breast pocket. Greg said: 'I had the pleasure of watching your daughter, Joanne, play tennis on Saturday.'

Dave was pleased. 'Pretty good, isn't she?'

'Very.'

Lev said: 'I'm glad I ran into you, Dave – I was planning to call you.'

'Why?'

'Your theatres need remodelling. They're very old-fashioned.'

Dave looked amused. 'You were planning to call me to give me this news?'

'Why don't you do something about it?'

He shrugged elegantly. 'Why bother? I'm making enough money. At my age, I don't want the strain.'

'You could double your profits.'

'By raising ticket prices. No, thanks.'

'You're crazy.'

'Not everyone is obsessed with money,' Dave said with a touch of disdain.

'Then sell to me,' Lev said.

Greg was surprised. He had not seen that coming.

'I'll give you a good price,' Lev added.

Dave shook his head. 'I like owning cinemas,' he said. 'They give people pleasure.'

'Eight million dollars,' Lev said.

Greg felt bemused. He thought: Did I just hear Father offer Dave eight million dollars?

'That is a fair price,' Dave admitted. 'But I'm not selling.'

'No one else will give you as much,' Lev said with exasperation.

'I know.' Dave looked as if he had taken enough browbeating. He swallowed the rest of his drink. 'Nice to see you both,' he said, and he strolled out of the bar into the dining room.

Lev looked disgusted. '"Not everyone is obsessed with money,"' he quoted. 'Dave's great-grandfather arrived here from Persia a hundred years ago with nothing but the clothes he wore and six rugs. He wouldn't have turned down eight million dollars.'

'I didn't know you had that much money,' Greg said.

'I don't, not in ready cash. That's what banks are for.'

'So you'd take out a loan to pay Dave?'

Lev raised his forefinger again. 'Never use your own money when you can spend someone else's.'

Gus Dewar walked in, a tall figure with a large head. He was in his mid-forties, and his light-brown hair was salted with silver. He greeted them with cool courtesy, shaking hands and offering them a drink. Greg saw immediately that Gus and Lev did not like one another. He feared that would mean Gus would not grant the favour Greg wanted to beg. Maybe he should give up the thought.

Gus was a big shot. His father had been a senator before him, a dynastic succession that Greg thought was un-American. Gus had helped Franklin Roosevelt become Governor of New York and then President. Now he was on the powerful Senate Foreign Relations Committee.

His sons, Woody and Chuck, went to the same school as Greg. Woody was brainy, Chuck was a sportsman.

Lev said: 'Has the President told you to settle my strike, Senator?'

Gus smiled. 'No – not yet, anyway.'

Lev turned to Greg. 'Last time the foundry was on strike, twenty years ago, President Wilson sent Gus to browbeat me into giving the men a raise.'

'I saved you money,' Gus said mildly. 'They were asking for a dollar – I made them take half that.'

'Which was exactly fifty cents more than I intended to give.'

Gus smiled and shrugged. 'Shall we have lunch?'

They went into the dining room. When they had ordered, Gus said: 'The President was glad you could make it to the reception at the White House.'

'I probably shouldn't have taken Gladys,' Lev said. 'Mrs Roosevelt was a bit frosty with her. I guess she doesn't approve of movie stars.'

She probably doesn't approve of movie stars who sleep with married men, Greg thought, but he kept his mouth shut.

Gus made small talk while they ate. Greg looked for an opportunity to ask his favour. He wanted to work in Washington one summer, to learn the ropes and make contacts. His father might have been able to get him an internship, but it would have been with a Republican, and they were out of power. Greg wanted to work in the office of the influential and respected Senator Dewar, personal friend and ally of the President.

He asked himself why he was nervous about asking. The worst that could happen was that Dewar would say no.

When the dessert was finished, Gus got down to business. 'The President has asked me to speak to you about the Liberty League,' he said.

Greg had heard of this organization, a right-wing group opposed to the New Deal.

Lev lit a cigarette and blew out smoke. 'We have to guard against creeping socialism.'

'The New Deal is all that is saving us from the kind of nightmare they're having in Germany.'

'The Liberty League aren't Nazis.'

'Aren't they? They have a plan for an armed insurrection to overthrow the President. It's not realistic, of course – not yet, anyway.'

'I believe I have a right to my opinions.'

'Then you're supporting the wrong people. The League is nothing to do with liberty, you know.'

'Don't talk to me about liberty,' Lev said with a touch of anger. 'When I was twelve years old I was flogged by the Leningrad police because my parents were on strike.'

Greg was not sure why his father had said that. The brutality of the Tsar's regime seemed like an argument for socialism, not against.

Gus said: 'Roosevelt knows you give money to the League, and he wants you to stop.'

'How does he know who I give money to?'

'The FBI told him. They investigate such people.'

'We're living in a police state! You're supposed to be a liberal.'

There was not much logic to Lev's arguments, Greg perceived. Lev was just trying everything he could think of to wrong-foot Gus, and he did not care if he contradicted himself in the process.

Gus remained cool. 'I'm trying to make sure this doesn't become a matter for the police,' he said.

Lev grinned. 'Does the President know I stole your fiancée?'

This was news to Greg – but it had to be true, for Lev had at last succeeded in throwing Gus off balance. Gus looked shocked, turned his gaze aside, and reddened. Score one for our team, Greg thought.

Lev explained to Greg: 'Gus was engaged to Olga, back in 1915,' he said. 'Then she changed her mind and married me.'

Gus recovered his composure. 'We were all terribly young.'

Lev said: 'You certainly got over Olga quickly enough.'

Gus gave Lev a cool look and said: 'So did you.'

Greg saw that his father was embarrassed now. Gus's shot had hit home.

There was a moment of awkward silence, then Gus said: 'You and I fought in a war, Lev. I was in a machine-gun battalion with my school friend Chuck Dixon. In a little French town called Château-Thierry he was blown to pieces in front of my eyes.' Gus was speaking in a conversational tone, but Greg found himself holding his breath. Gus went on: 'My ambition for my sons is that they should never have to go through what we went through. That's why groups such as the Liberty League have to be nipped in the bud.'

Greg saw his chance. 'I'm interested in politics, too, Senator, and I'd like to learn more. Might you be able to take me as an intern one summer?' He held his breath.

Gus looked surprised, but said: 'I can always use a bright young man who's willing to work in a team.'

That was neither a yes nor a no. 'I'm top in math, and captain of ice hockey,' Greg persisted, selling himself. 'Ask Woody about me.'

'I will.' Gus turned to Lev. 'And will you consider the President's request? It's really very important.'

It almost seemed as if Gus was suggesting an exchange of favours. But would Lev agree?

Lev hesitated a long moment, then stubbed out his cigarette and said: 'I guess we have a deal.'

Gus stood up. 'Good,' he said. 'The President will be pleased.'

Greg thought: I did it!

They walked out of the club to their cars.

As they drove out of the parking lot, Greg said: 'Thank you, Father. I really appreciate what you did.'

'You chose your moment well,' Lev said. 'I'm glad to see you're so smart.'

The compliment pleased Greg. In some ways he was smarter than his father – he certainly understood science and math better – but he feared he was not as shrewd and cunning as his old man.

'I want you to be a wise guy,' Lev went on. 'Not like some of these dummies.' Greg had no idea who the dummies were. 'You've got to stay ahead of the curve, all the time. That's the way to get on.'

Lev drove to his office, in a modern block downtown. As they walked through the marble lobby, Lev said: 'Now I'm going to teach a lesson to that fool Dave Rouzrokh.'

Going up in the elevator, Greg wondered how Lev would do that.

Peshkov Pictures occupied the top floor. Greg followed Lev along a broad corridor and through an outer office with two attractive young secretaries. 'Get Sol Starr on the phone, will you?' Lev said as they walked into the inner office.

Lev sat behind the desk. 'Solly owns one of the biggest studios in Hollywood,' he explained.

The phone on the desk rang and Lev picked it up. 'Sol!' he said. 'How are they hanging?' Greg listened to a minute or two of masculine joshing, then Lev got down to business. 'Little piece of advice,' he said. 'Here in New York State we have a crappy chain of fleapits called Roseroque Theatres ... yeah, that's the one ... take my tip, don't send them your top-of-the-line first-run pictures this summer – you may not get paid.' Greg realized that would hit Dave hard: without exciting new movies to show, his takings would tumble. 'A word to the wise, right? Solly, don't thank me, you'd do the same for me ... bye.'

Once again, Greg was awestruck by his father's power. He could have people beaten up. He could offer eight million dollars of other people's money. He could scare a president. He could seduce another man's fiancée. And he could ruin a business with a single phone call.

'You wait and see,' said his father. 'In a month's time, Dave Rouzrokh will be begging me to buy him out – at half the price I offered him today.'

(iii)

'I don't know what's wrong with this puppy,' Daisy said. 'He won't do anything I tell him. I'm going crazy.' There was a shake in her voice and a tear in her eye, and she was exaggerating only a little.

Charlie Farquharson studied the dog. 'There's nothing wrong with him,' he said. 'He's a lovely little fellow. What's his name?'

'Jack.'

'Hmm.'

They were sitting on lawn chairs in the well-kept two-acre garden of Daisy's home. Eva had greeted Charlie then tactfully retired to write a letter home. The gardener, Henry, was hoeing a bed of purple and yellow pansies in the distance. His wife, Ella, the maid, brought a pitcher of lemonade and some glasses, and set them on a folding table.

The puppy was a tiny Jack Russell terrier, small and strong, white with tan patches. He had an intelligent look, as if he understood every word, but he seemed to have no inclination to obey. Daisy held him on her lap and stroked his nose with dainty fingers in a way that she hoped Charlie would find strangely disturbing. 'Don't you like the name?'

'A bit obvious, perhaps?' Charlie stared at her white hand on the dog's nose and shifted uneasily in his chair.

Daisy did not want to overdo it. If she inflamed Charlie too much he would just go home. This was why he was still single at twenty-five: several Buffalo girls, including Dot Renshaw and Muffie Dixon, had found it impossible to nail his foot to the floor. But Daisy was different. 'Then you shall name him,' she said.

'It's good to have two syllables, as in Bonzo, to make it easier for him to recognize the name.'

Daisy had no idea how to name dogs. 'How about Rover?'

'Too common. Rusty might be better.'

'Perfect!' she said. 'Rusty he shall be.'

The dog wriggled effortlessly out of her grasp and jumped to the ground.

Charlie picked him up. Daisy noticed he had big hands. 'You must show Rusty you're the boss,' Charlie said. 'Hold him tight, and don't let him jump down until you say so.' He put the dog back on her lap.

'But he's so strong! And I'm afraid of hurting him.'

Charlie smiled condescendingly. 'You probably couldn't hurt him if you tried. Hold his collar tightly – twist it a bit if you need to – then put your other hand firmly on his back.'

Daisy followed Charlie's orders. The dog sensed the increased pressure in her touch and became still, as if waiting to see what would happen next.

'Tell him to sit, then press down on his rear end.'

'Sit,' she said.

'Say it louder, and pronounce the letter "t" very clearly. Then press down hard.'

'Sit, Rusty!' she said, and pushed him down. He sat.

'There you are,' said Charlie.

'You're so clever!' Daisy gushed.

Charlie looked pleased. 'It's just a matter of knowing what to do,' he said modestly. 'You must always be emphatic and decisive with dogs. You have to almost bark at them.' He sat back, looking content. He was quite heavy, and filled the chair. Talking about the subject in which he was expert had relaxed him, as Daisy had hoped.

She had called him that morning. 'I'm in despair!' she had said. 'I have a new puppy and I can't manage him at all. Can you give me any advice?'

'What breed of puppy?'

'It's a Jack Russell.'

'Why, that's the kind of dog I like best – I have three!'

'What a coincidence!'

As Daisy had hoped, Charlie volunteered to come over and help her train the dog.

Eva had said doubtfully: 'Do you really think Charlie is right for you?'

'Are you kidding?' Daisy had replied. 'He's one of the most eligible bachelors in Buffalo!'

Now Daisy said to Charlie: 'I bet you'd be really good with children, too.'

'Oh, I don't know about that.'

'You love dogs, but you're firm with them. I'm sure that works with children, too.'

'I have no idea.' He changed the subject. 'Are you intending to go to college in September?'

'I might go to Oakdale. It's a two-year finishing college for ladies. Unless . . .'

'Unless what?'

Unless I get married, she meant, but she said: 'I don't know. Unless something else happens.'

'Such as what?'

'I'd like to see England. My father went to London and met the Prince of Wales. What about you? Any plans?'

'It was always assumed that I would take over Father's bank, but now there is no bank. Mother has a little money from her family, and I manage that, but otherwise I'm kind of a loose wheel.'

'You should raise horses,' Daisy said. 'I know you'd be good at it.' She was a good rider and had won prizes when younger. She pictured herself and Charlie in the park on matching greys, with two children on ponies following behind. The vision gave her a warm glow.

'I love horses,' Charlie said.

'So do I! I want to breed racehorses.' Daisy did not have to feign this enthusiasm. It was her dream to raise a string of champions. She saw racehorse owners as the ultimate international elite.

'Thoroughbreds cost a lot of money,' Charlie said lugubriously.

Daisy had plenty. If Charlie married her, he would never have to worry about money again. She naturally did not say so, but she guessed that Charlie was thinking it, and she let the thought hang unspoken in the air for as long as possible.

Eventually Charlie said: 'Did your father really have those two union organizers beaten up?'

'What a strange idea!' Daisy did not know whether Lev Peshkov had done any such thing, but in truth it would not have surprised her.

'The men who came from New York to take over the strike,' Charlie persisted. 'They were hospitalized. The *Sentinel* says they quarrelled with local union leaders, but everyone thinks your father was responsible.'

'I never talk about politics,' Daisy said gaily. 'When did you get your first dog?'

Charlie began a long reminiscence. Daisy considered what to do next. I've got him here, she thought, and I've put him at ease; now I have to get him aroused. But stroking the dog suggestively had unnerved him. What they needed was some casual physical contact.

'What should I do next with Rusty?' she asked when Charlie had finished his story.

'Teach him to walk to heel,' Charlie said promptly.

'How do you do that?'

'Do you have some dog biscuits?'

'Sure.' The kitchen windows were open, and Daisy raised her voice so that the maid could hear her. 'Ella, would you kindly bring me that box of Milk-bones?'

Charlie broke up one of the biscuits, then took the dog on his lap. He held a piece of biscuit in his closed fist, letting Rusty sniff it, then opened his hand and allowed the dog to eat the morsel. He took another piece, making sure the dog knew he had it. Then he stood up and put the dog at his feet. Rusty kept an alert gaze on Charlie's closed fist. 'Walk to heel!' Charlie said, and walked a few steps.

The dog followed him.

'Good boy!' Charlie said, and gave Rusty the biscuit.

'That's amazing!' Daisy said.

'After a while you won't need the biscuit – he'll do it for a pat. Then eventually he'll do it automatically.'

'Charlie, you are a genius!'

Charlie looked pleased. He had nice brown eyes, just like the dog, she observed. 'Now you try,' he said to Daisy.

She copied what Charlie had done, and achieved the same result.

'See?' said Charlie. 'It's not so hard.'

Daisy laughed with delight. 'We should go into business,' she said. 'Farquharson and Peshkov, dog trainers.'

'What a nice idea,' he said, and he seemed to mean it.

This was going very well, Daisy thought.

She went to the table and poured two glasses of lemonade.

Standing beside her, he said: 'I'm usually a bit shy with girls.'

No kidding, she thought, but she kept her mouth firmly closed.

'But you're so easy to talk to,' he went on. He imagined that was a happy accident.

As she handed a glass to him she fumbled, spilling lemonade on him. 'Oh, how clumsy!' she cried.

'It's nothing,' he said, but the drink had wet his linen blazer and his white cotton trousers. He pulled out a handkerchief and began to mop it.

'Here, let me,' said Daisy, and she took the handkerchief from his large hand.

She moved intimately close to pat his lapel. He went still, and she knew he could smell her Jean Naté perfume – lavender notes on top, musk underneath. She brushed the handkerchief caressingly over the front of his jacket, though there was no spill there. 'Almost done,' she said as if she regretted having to stop soon.

Then she went down on one knee as if worshipping him. She began to blot the wet patches on his pants with butterfly lightness. As she stroked his thigh she put on a look of alluring innocence and looked up. He was staring down at her, breathing hard through his open mouth, mesmerized.

(iv)

Woody Dewar impatiently inspected the yacht *Sprinter*, checking that the kids had made everything shipshape. She was a forty-eight-foot racing ketch, long and slender like a knife. Dave Rouzrokh had loaned her to the Shipmates, a club Woody belonged to that took the sons of Buffalo's unemployed out on Lake Erie and taught them the rudiments of sailing. Woody was glad to see that the dock lines and fenders were set, the sails furled, the halyards tied off, and all the other lines neatly coiled.

His brother Chuck, a year younger at fourteen, was on the dock already, joshing with a couple of coloured kids. Chuck had an easygoing manner that enabled him to get on with everyone. Woody, who wanted to go into politics like their father, envied Chuck's effortless charm.

The boys wore nothing but shorts and sandals, and the three on the dock looked a picture of youthful strength and vitality. Woody would have liked to have taken a photograph, if he had had his camera with him. He was a keen photographer and had built a darkroom at home so that he could develop and print his own pictures.

Satisfied that the *Sprinter* was being left as they had found her this morning, Woody jumped on to the dock. A group of a dozen youngsters left the boatyard together, windswept and sunburned, aching pleasantly from their exertions, laughing as they relived the day's blunders and pratfalls and jokes.

The gap between the two rich brothers and the crowd of poor boys had vanished when they were out on the water, working together to control the yacht, but now it reappeared in the parking lot of the Buffalo Yacht Club. Two vehicles stood side by side: Senator Dewar's Chrysler Airflow, with a uniformed chauffeur at the wheel, for Woody and Chuck; and a Chevrolet Roadster pickup truck with two wooden benches in the back for the others. Woody felt embarrassed, saying goodbye as the chauffeur held the door for him, but the boys did not seem to care, thanking him and saying: 'See you next Saturday!'

As they drove up Delaware Avenue, Woody said: 'That was fun, though I'm not sure how much good it does.'

Chuck was surprised. 'Why?'

'Well, we're not helping their fathers find jobs, and that's the only thing that really counts.'

'It might help the sons get work in a few years' time.' Buffalo was a port city: in normal times there were thousands of jobs on merchant ships plying the Great Lakes and the Erie Canal, as well as on pleasure craft.

'Provided the President can get the economy moving again.'

Chuck shrugged. 'So go work for Roosevelt.'

'Why not? Papa worked for Woodrow Wilson.'

'I'll stick with the sailing.'

Woody checked his wristwatch. 'We've got time to change for the ball – just.' They were going to a dinner-dance at the Racquet Club. Anticipation made his heart beat faster. 'I want to be with humans that have soft skin, speak with high voices, and wear pink dresses.'

'Huh,' Chuck said derisively. 'Joanne Rouzrokh never wore pink in her life.'

Woody was taken aback. He had been dreaming about Joanne all day and half the night for a couple of weeks, but how did his brother know that? 'What makes you think—'

'Oh, come on,' Chuck said scornfully. 'When she arrived at the beach party in a tennis skirt you practically fainted. Everyone could see you were crazy about her. Fortunately *she* didn't seem to notice.'

'Why was that fortunate?'

'For God's sake – you're fifteen, and she's eighteen. It's embarrassing! She's looking for a husband, not a schoolboy.'

'Oh, gee, thanks, I forgot what an expert you are on women.'

Chuck flushed. He had never had a girlfriend. 'You don't have to be an expert to see what's under your goddamn nose.'

They talked like this all the time. There was no malice in it: they were just brutally frank with each other. They were brothers, so there was no need to be nice.

They reached home, a mock-Gothic mansion built by their late grandfather, Senator Cam Dewar. They ran inside to shower and change.

Woody was now the same height as his father, and he put on one of Papa's old dress suits. It was a bit worn, but that was all right. The younger boys would be wearing school suits or blazers, but the college men would have tuxedos, and Woody was keen to look older. Tonight he would dance with her, he thought as he slicked his hair with brilliantine. He would be allowed to hold her in his arms. The palms of his hands would feel the warmth of her skin. He would look into her eyes as she smiled. Her breasts would brush against his jacket as they danced.

When he came down, his parents were waiting in the drawing room, Papa drinking a cocktail, Mama smoking a cigarette. Papa was long and thin, and looked like a coat-hanger in his double-breasted tuxedo. Mama was beautiful, despite having only one eye, the other being permanently closed – she had been born that way. Tonight she looked stunning in a floor-length dress, black lace over red silk, and a short black velvet evening jacket.

Woody's grandmother was the last to arrive. At sixty-eight she was poised and elegant, as thin as her son but petite. She studied Mama's dress and said: 'Rosa, dear, you look wonderful.' She was always kind to her daughter-in-law. To everyone else she was waspish.

Gus made her a cocktail without being asked. Woody hid his impatience while she took her time drinking it. Grandmama could never be hurried. She assumed no social event would begin before she arrived: she was the grand old lady of Buffalo society, widow of a senator and mother of another, matriarch of one of the city's oldest and most distinguished families.

Woody asked himself when he had fallen for Joanne. He had known her most of his life, but he had always regarded girls as uninteresting spectators to the exciting adventures of boys – until two or three years ago, when girls had suddenly become even more fascinating than cars and speedboats. Even then he had been more interested in girls his own age or a little younger. Joanne, for her part, had always treated him as a kid – a bright kid, worth talking to now and again, but certainly not a possible boyfriend. But this summer, for no reason he could put a finger on, he had suddenly begun to see her as the most alluring girl in the world. Sadly, her feelings for him had not undergone a similar transformation.

Not yet.

Grandmama addressed a question to his brother. 'How is school, Chuck?'

'Terrible, Grandmama, as you know perfectly well. I'm the family cretin, a throwback to our chimpanzee forbears.'

'Cretins don't use phrases such as "our chimpanzee forbears" in my experience. Are you quite sure laziness plays no part?'

Rosa butted in. 'Chuck's teachers say he works pretty hard at school, Mama.'

Gus added: 'And he beats me at chess.'

'Then I ask what the problem is,' Grandmama persisted. 'If this goes on, he won't get into Harvard.'

Chuck said: 'I'm a slow reader, that's all.'

'Curious,' she said. 'My father-in-law, your paternal great-grandfather, was the most successful banker of his generation, yet he could barely read or write.'

Chuck said: 'I didn't know that.'

'It's true,' she said. 'But don't use it as an excuse. Work harder.'

Gus looked at his watch. 'If you're ready, Mama, we'd better go.'

At last they got into the car and drove to the club. Papa had taken a table for the dinner and had invited the Renshaws and their offspring, Dot and George. Woody looked around but, to his disappointment, he did not see Joanne. He checked the table plan, on an easel in the lobby, and was dismayed to see that there was no Rouzrokh table. Were they not coming? That would ruin his evening.

The talk over the lobster and steak was of events in Germany. Philip Renshaw thought Hitler was doing a good job. Woody's father said: 'According to today's *Sentinel*, they jailed a Catholic priest for criticizing the Nazis.'

'Are you Catholic?' asked Mr Renshaw in surprise.

'No, Episcopalian.'

'It's not about religion, Philip,' said Rosa crisply. 'It's about freedom.' Woody's mother had been an anarchist in her youth, and she was still a libertarian at heart.

Some people skipped the dinner and came later for the dancing, and more revellers appeared as the Dewars were served dessert. Woody kept his eyes peeled for Joanne. In the next room a band started to play 'The Continental', a hit from last year.

He could not say what it was about Joanne that had so captivated him. Most people would not call her a great beauty, though she was certainly striking. She looked like an Aztec queen, with high cheekbones and the same knife-blade nose as her father, Dave. Her hair was dark and thick and her skin an olive shade, no doubt because of her Persian ancestry. There was a brooding intensity about her that made Woody long to know her better, to make her relax and hear her murmur softly about nothing in particular. He felt that her formidable presence must signify a capacity for deep passion. Then he thought: Now who's pretending to be an expert on women?

'Are you looking out for someone, Woody?' said Grandmama, who did not miss much.

Chuck sniggered knowingly.

'Just wondering who's coming to the dance,' Woody replied casually, but he could not help blushing.

He still had not spotted her when his mother stood up and they all left the table. Disconsolate, he wandered into the ballroom to the strains of Benny Goodman's 'Moonglow' – and there Joanne was: she must have come in when he wasn't looking. His spirits lifted.

Tonight she wore a dramatically simple silver-grey silk dress with a deep V-neck that showed off her figure. She had looked sensational in a tennis skirt that revealed her long brown legs, but this was even more arousing. As she glided across the room, graceful and confident, she made Woody's throat go dry.

He moved towards her, but the ballroom had filled up, and suddenly he was irritatingly popular: everyone wanted to talk to him. During his progress through the crowd he was surprised to see dull old Charlie Farquharson dancing with the vivacious Daisy Peshkov. He could not recall seeing Charlie dance with anyone, let alone a tootsie like Daisy. What had she done to bring him out of his shell?

By the time he reached Joanne, she was at the end of the room farthest from the band, and to his chagrin she was deep in discussion with a group of boys four or five years older than he. Fortunately, he was taller than most of them, so the difference was not too obvious. They were all holding Coke glasses, but Woody could smell Scotch: one of them must have a bottle in his pocket.

As he joined them, he heard Victor Dixon say: 'No one's in favour of lynching, but you have to understand the problems they have in the South.'

Woody knew that Senator Wagner had proposed a law to punish sheriffs who permitted lynchings – but President Roosevelt had refused to back the bill.

Joanne was outraged. 'How can you say that, Victor? Lynching is murder! We don't have to understand their problems, we have to stop them killing people!'

Woody was pleased to learn how much Joanne shared his political values. But clearly this was not a good time to ask her to dance, which was unfortunate.

'You don't get it, Joanne, honey,' said Victor. 'Those Southern Negroes are not really civilized.'

I might be young and inexperienced, Woody thought, but I wouldn't have made the mistake of speaking so condescendingly to Joanne.

'It's the people who carry out lynchings who are uncivilized!' she said.

Woody decided this was the moment to make his contribution to the argument. 'Joanne is right,' he said. He made his voice lower in pitch, to sound older. 'There was a lynching in the home town of our help, Joe and Betty, who have looked after me and my brother since we were babies. Betty's cousin was stripped naked and burned with a blowtorch, while a crowd watched. Then he was hanged.' Victor glared at him, resentful of this kid who was taking Joanne's attention away; but the others in the group listened with horrified interest. 'I don't care what his crime was,' Woody said. 'The white people who did that to him are savages.'

Victor said: 'Your beloved President Roosevelt didn't support the anti-lynch bill, though, did he?'

'No, and that was very disappointing,' said Woody. 'I know why he made that decision: he was afraid that angry Southern congressmen would retaliate by sabotaging the New Deal. All the same, I would have liked him to tell them to go to hell.'

Victor said: 'What do you know? You're just a kid.' He took a silver flask from his jacket pocket and topped up his drink.

Joanne said: 'Woody's political ideas are more grown-up than yours, Victor.'

Woody glowed. 'Politics is kind of the family business,' he said. Then he was irritated by a tug at his elbow. Too polite to ignore it, he turned to see Charlie Farquharson, perspiring from his exertions on the dance floor.

'Can I talk to you for a minute?' said Charlie.

Woody resisted the temptation to tell him to buzz off. Charlie was a likeable guy who did no harm to anyone. You had to feel sorry for a man with a mother like that. 'What is it, Charlie?' he said with as much good grace as he could muster.

'It's about Daisy.'

'I saw you dancing with her.'

'Isn't she a great dancer?'

Woody had not noticed but, to be nice, he said: 'You bet she is!'

'She's great at everything.'

'Charlie,' said Woody, trying to suppress a tone of incredulity, 'are you and Daisy courting?'

Charlie looked bashful. 'We've been horse riding in the park a couple of times, and so on.'

'So you *are* courting.' Woody was surprised. They seemed an unlikely pair. Charlie was such a lump, and Daisy was a poppet.

Charlie added: 'She's not like other girls. She's so easy to talk to! And she loves dogs and horses. But people think her father is a gangster.'

'I guess he is a gangster, Charlie. Everyone bought their liquor from him during Prohibition.'

'That's what my mother says.'

'So your mother doesn't like Daisy.' Woody was not surprised.

'She likes Daisy fine. It's Daisy's family she objects to.'

An even more surprising thought occurred to Woody. 'Are you thinking of *marrying* Daisy?'

'Oh, God, yes,' said Charlie. 'And I think she might say yes, if I asked her.'

Well, Woody thought, Charlie had class but no money, and Daisy was the opposite, so maybe they would complement one another. 'Stranger things have happened,' he said. This was kind of fascinating, but he wanted to concentrate on his own romantic life. He looked around, checking that Joanne was still there. 'Why are you telling me this?' he asked Charlie. It was not as if they were great friends.

'My mother might change her mind if Mrs Peshkov were invited to join the Buffalo Ladies Society.'

Woody had not been expecting that. 'Why, it's the snobbiest club in town!'

'Exactly. If Olga Peshkov were a member, how could Mom object to Daisy?'

Woody did not know whether this scheme would work or not, but there was no doubting the earnest warmth of Charlie's feelings. 'Maybe you're right,' Woody said.

'Would you approach your grandmother for me?'

'Whoa! Wait a minute. Grandmama Dewar is a dragon. I wouldn't ask her for a favour for myself, let alone for you.'

'Woody, listen to me. You know she's really the boss of that little clique. If she wants someone, they're in – and if she doesn't, they're out.'

This was true. The Society had a chairwoman and a secretary and a treasurer, but Ursula Dewar ran the club as if it belonged to her. All the same, Woody was reluctant to petition her. She might bite his head off. 'I don't know,' he said apologetically.

'Oh, come on, Woody, please. You don't understand.' Charlie lowered his voice. 'You don't know what it's like to love someone this much.'

Yes, I do, Woody thought; and that changed his mind. If Charlie feels as bad as I do, how can I refuse him? I hope someone else would do the same for me, if it meant I had a better chance with Joanne. 'Okay, Charlie,' he said. 'I'll talk to her.'

'Thanks! Say – she's here, isn't she? Could you do it tonight?'

'Hell, no. I've got other things on my mind.'

'Okay, sure . . . but when?'

Woody shrugged. 'I'll do it tomorrow.'

'You're a pal!'

'Don't thank me yet. She'll probably say no.'

Woody turned back to speak to Joanne, but she had gone.

He began to look for her, then stopped himself. He must not appear desperate. A needy man was not sexy, he knew that much.

He danced dutifully with several girls: Dot Renshaw, Daisy Peshkov, and Daisy's German friend Eva. He got a Coke and went outside to where some of the boys were smoking cigarettes. George Renshaw poured some Scotch into Woody's Coke, which improved the taste, but he did not want to get drunk. He had done that before and he did not like it.

Joanne would want a man who shared her intellectual interests, Woody believed – and that would rule out Victor Dixon. Woody had heard Joanne mention Karl Marx and Sigmund Freud. In the public library he had read the *Communist Manifesto*, but it just seemed like a political rant. He had had more fun with Freud's *Studies in Hysteria*, which made a kind of detective story out of mental illness. He was looking forward to letting Joanne know, in a casual way, that he had read these books.

He was determined to dance with Joanne at least once tonight, and after a while he went in search of her. She was not in the ballroom or the bar. Had he missed his chance? In trying not to show his desperation, had he been too passive? It was unbearable to think that the ball could end without his even having touched her shoulder.

He stepped outside again. It was dark, but he saw her almost immediately. She was walking away from Greg Peshkov, looking a little flushed, as if she had been arguing with him. 'You might be the only person here who isn't a goddamned conservative,' she said to Woody. She sounded a little drunk.

Woody smiled. 'Thanks for the compliment – I think.'

'Do you know about the march tomorrow?' she asked abruptly.

He did. Strikers from the Buffalo Metal Works planned a demonstration to protest against the beating up of union men from New York. Woody guessed that was the subject of her argument with Greg: his father owned the factory. 'I was planning to go,' he said. 'I might take some photographs.'

'Bless you,' she said, and she kissed him.

He was so surprised that he almost failed to respond. For a second he stood there passively as she crushed her mouth to his, and he tasted whisky on her lips.

Then he recovered his composure. He put his arms around her and pressed her body to his, feeling her breasts and her thighs press delightfully against him. Part of him feared she would be offended, push him away, and angrily accuse him of treating her disrespectfully; but a deeper instinct told him he was on safe ground.

He had little experience of kissing girls – and none of kissing mature women of eighteen – but he liked the feel of her soft mouth so much that he moved his lips against hers in little nibbling motions that gave him exquisite pleasure, and he was rewarded by hearing her moan quietly.

He was vaguely aware that if one of the older generation should walk by, there might be an embarrassing scene, but he was too aroused to care.

Joanne's mouth opened and he felt her tongue. This was new to him: the few girls he had kissed had not done that. But he figured she must know what she was doing, and anyway he really liked it. He imitated the motions of her tongue with his own. It was shockingly intimate and highly exciting. It must have been the right thing to do, because she moaned again.

Summoning his nerve, he put his right hand on her left breast. It was wonderfully soft and heavy under the silk of her dress. As he caressed it he felt a small protuberance and thought, with a thrill of discovery, that it must be her nipple. He rubbed it with his thumb.

She pulled away from him abruptly. 'Good God,' she said. 'What am I doing?'

'You're kissing me,' Woody said happily. He rested his hands on her round hips. He could feel the heat of her skin through the silk dress. 'Let's do it some more.'

She pushed his hands away. 'I must be out of my mind. This is the Racquet Club, for Christ's sake.'

Woody could see that the spell had been broken, and sadly there would be no more kissing tonight. He looked around. 'Don't worry,' he said. 'No one saw.' He felt enjoyably conspiratorial.

'I'd better go home, before I do something even more stupid.'

He tried not to be offended. 'May I escort you to your car?'

'Are you crazy? If we walk in there together everyone will guess what we've been doing – especially with that dumb grin all over your face.'

Woody tried to stop grinning. 'Then why don't you go inside and I'll wait out here for a minute?'

'Good idea.' She walked away.

'See you tomorrow,' he called after her.

She did not look back.

(v)

Ursula Dewar had her own small suite of rooms in the old Victorian mansion on Delaware Avenue. There was a bedroom, a bathroom and a dressing room; and after her husband died she had converted his dressing room into a little parlour. Most of the time she had the whole house to herself: Gus and Rosa spent a lot of time in Washington, and Woody and Chuck went to a boarding school. But when they came home she spent a good deal of the day in her own quarters.

Woody went to talk to her on Sunday morning. He was still walking on air after Joanne's kiss, though he had spent half the night trying to figure out what it had meant. It could signify anything from true love to true drunkenness. All he knew was that he could hardly wait to see Joanne again.

He walked into his grandmother's room behind the maid, Betty, as she took in the breakfast tray. He liked it that Joanne got angry about the way Betty's Southern relations were treated. In politics, dispassionate argument was overrated, he felt. People *should* get angry about cruelty and injustice.

Grandmama was already sitting up in bed, wearing a lace shawl over a mushroom-coloured silk nightgown. 'Good morning, Woodrow!' she said, surprised.

'I'd like to have a cup of coffee with you, Grandmama, if I may.' He had already asked Betty to bring two cups.

'This is an honour,' Ursula said.

Betty was a grey-haired woman of about fifty with the kind of figure that was sometimes called comfortable. She set the tray in front of Ursula, and Woody poured coffee into Meissen cups.

He had given some thought to what he would say, and had marshalled his arguments. Prohibition was over, and Lev Peshkov was now a legitimate businessman, he would contend. Furthermore, it was not fair to punish Daisy because her father had been a criminal – especially since most of the respectable families in Buffalo had bought his illegal booze.

'Do you know Charlie Farquharson?' he began.

'Yes.' Of course she did. She knew every family in the Buffalo 'Blue Book'. She said: 'Would you like a piece of this toast?'

'No, thank you, I've had breakfast.'

'Boys of your age never have enough to eat.' She looked at him shrewdly. 'Unless they're in love.'

She was on good form this morning.

Woody said: 'Charlie is kind of under the thumb of his mother.'

'She kept her husband there, too,' Ursula said drily. 'Dying was the only way he could get free.' She drank some coffee and started to eat her grapefruit with a fork.

'Charlie came to me last night and asked me to ask you a favour.'

She raised an eyebrow, but said nothing.

Woody took a breath. 'He wants you to invite Mrs Peshkov to join the Buffalo Ladies Society.'

Ursula dropped her fork, and there was a chime of silver on fine porcelain. As if covering her discomposure, she said: 'Pour me some more coffee, please, Woody.'

He did her bidding, saying nothing for the moment. He could not recall ever seeing her discombobulated.

She sipped the coffee and said: 'Why in the name of heaven would Charles Farquharson, or anyone else for that matter, want Olga Peshkov in the Society?'

'He wants to marry Daisy.'

'Does he?'

'And he's afraid his mother will object.'

'He's got that part right.'

'But he thinks he might be able to talk her around . . .'

'. . . if I let Olga into the Society.'

'Then people might forget that her father was a gangster.'

'A gangster?'

'Well, a bootlegger at least.'

'Oh, that,' Ursula said dismissively. 'That's not it.'

'Really?' It was Woody's turn to be surprised. 'What is it, then?'

Ursula looked thoughtful. She was silent for such a long time that Woody wondered if she had forgotten he was there. Then she said: 'Your father was in love with Olga Peshkov.'

'Jesus!'

'Don't be vulgar.'

'Sorry, Grandmama, you surprised me.'

'They were engaged to be married.'

'Engaged?' he said, astonished. He thought for a minute, then said: 'I suppose I'm the only person in Buffalo who doesn't know about this.'

She smiled at him. 'There is a special mixture of wisdom and innocence that comes only to adolescents. I remember it so clearly in your father, and I see it in you. Yes, everyone in Buffalo knows, though your generation undoubtedly regard it as boring ancient history.'

'Well, what happened?' Woody said. 'I mean, who broke it off?'

'She did, when she got pregnant.'

Woody's mouth fell open. 'By Papa?'

'No, by her chauffeur – Lev Peshkov.'

'He was the chauffeur?' This was one shock after another. Woody was silent, trying to take it in. 'My goodness, Papa must have felt such a fool.'

'Your Papa was never a fool,' Ursula said sharply. 'The only foolish thing he did in his life was to propose to Olga.'

Woody remembered his mission. 'All the same, Grandmama, it was an awful long time ago.'

'Awfully. You require an adverb, not an adjective. But your judgement is better than your grammar. It *is* a long time.'

That sounded hopeful. 'So you'll do it?'

'How do you think your father would feel?'

Woody considered. He could not bullshit Ursula – she would see through it in a heartbeat. 'Would he care? I guess he might be embarrassed, if Olga were around as a constant reminder of a humiliating episode in his youth.'

'You guess right.'

'On the other hand, he's very committed to the ideal of behaving fairly to the people around him. He hates injustice. He wouldn't want to punish Daisy for something her mother did. Even less to punish Charlie. Papa has a pretty big heart.'

'Bigger than mine, you mean,' said Ursula.

'I didn't mean that, Grandmama. But I bet if you asked him he wouldn't object to Olga joining the Society.'

Ursula nodded. 'I agree. But I wonder whether you've worked out who is the real originator of this request?'

Woody saw what she was driving at. 'Oh, you're saying Daisy put Charlie up to it? I wouldn't be surprised. Does it make any difference to the rights and wrongs of the situation?'

'I guess not.'

'So, will you do it?'

'I'm glad to have a grandson with a kind heart – even if I do suspect he's being used by a clever and ambitious girl.'

Woody smiled. 'Is that a yes, Grandmama?'

'You know I can't guarantee anything. I'll suggest it to the committee.'

Ursula's suggestions were regarded by everyone else as royal commands, but Woody did not say so. 'Thank you. You're very kind.'

'Now give me a kiss and get ready for church.'

Woody made his escape.

He quickly forgot about Charlie and Daisy. Sitting in the Cathedral of St Paul in Shelton Square, he ignored the sermon – about Noah and the Flood – and thought about Joanne Rouzrokh. Her parents were in church, but she was not. Would she really show up at the demonstration? If she did, he was going to ask her for a date. But would she accept?

She was too smart to care about the age difference, he reckoned. She must know she had more in common with Woody than with boneheads such as Victor Dixon. And that kiss! He was still tingling from it. What she had done with her tongue – did other girls do that? He wanted to try it again, as soon as he could.

Thinking ahead, if she did agree to date him, what would happen in September? She was going to Vassar College, in the town of Poughkeepsie, he knew that. He would return to school and not see her until Christmas. Vassar was for girls only but there must be men in Poughkeepsie. Would she date other guys? He was jealous already.

Outside the church he told his parents he was not coming home for lunch, but was going on the protest march.

'Good for you,' his mother said. When young she had been the editor of the *Buffalo Anarchist*. She turned to her husband. 'You should go, too, Gus.'

'The union has brought charges,' Papa said. 'You know I can't prejudge the result of a court case.'

She turned back to Woody. 'Just don't get beaten up by Lev Peshkov's goons.'

Woody got his camera out of the trunk of his father's car. It was a Leica III, so small he could carry it on a strap around his neck, yet it had shutter speeds as fast as one five-hundredth of a second.

He walked a few blocks to Niagara Square, where the march was to begin. Lev Peshkov had tried to persuade the city to ban the demonstration on the grounds that it would lead to violence, but the union had insisted it would be peaceful. The union seemed to have won that argument, for several hundred people were milling around outside City Hall. Many carried lovingly embroidered banners, red flags, and placards reading: SAY NO TO BOSS THUGS. Woody looked around for Joanne but did not see her.

The weather was fine and the mood was sunny, and he took a few shots: workmen in their Sunday suits and hats; a car festooned with banners; a young cop biting his nails. There was still no sign of Joanne, and he began to think that she would not appear. She might have a headache this morning, he guessed.

The march was due to move off at noon. It finally got going a few minutes before one. There was a heavy police presence along the route, Woody noted. He found himself near the middle of the procession.

As they walked south on Washington Street, heading for the city's industrial heartland, he saw Joanne join the march a few yards ahead, and his heart leaped. She was wearing tailored pants that flattered her figure. He hurried to catch up with her. 'Good afternoon!' he said happily.

'Good grief, you're cheerful,' she said.

It was an understatement. He was delirious with happiness. 'Are you hungover?'

'Either that or I've contracted the Black Death. Which do you think it is?'

'If you have a rash, it's the Black Death. Are there any spots?' Woody hardly knew what he was saying. 'I'm not a doctor, but I'd be happy to check you over.'

'Stop being irrepressible. I know it's charming, but I'm not in the mood.'

Woody tried to calm down. 'We missed you in church,' he said. 'The sermon was about Noah.'

To his consternation she burst out laughing. 'Oh, Woody,' she said. 'I like you so much when you're funny, but please don't make me laugh today.'

He thought this remark was probably favourable, but he was far from certain.

He spotted an open grocery store on a side street. 'You need fluids,' he said. 'I'll be right back.' He ran into the store and bought two bottles of Coke, ice-cold from the refrigerator. He got the clerk to open them, then returned to the march. When he handed a bottle to Joanne, she said: 'Oh, boy, you're a life saver.' She put the bottle to her lips and drank a long draught.

Woody felt he was ahead, so far.

The march was good-humoured, despite the grim incident they were protesting about. A group of older men were singing political anthems and traditional songs. There were even a few families with children. And there was not a cloud in the sky.

'Have you read *Studies in Hysteria?*' Woody asked as they walked along.

'Never heard of it.'

'Oh! It's by Sigmund Freud. I thought you were a fan of his.'

'I'm interested in his ideas. I've never read one of his books.'

'You should. *Studies in Hysteria* is amazing.'

She looked curiously at him. 'What made you read a book such as that? I bet they don't teach psychology at your expensively old-fashioned school.'

'Oh, I don't know. I guess I heard you talking about psychoanalysis and thought it sounded really extraordinary. And it is.'

'In what way?'

Woody had the feeling she was testing him, to see whether he had really understood the book or was merely pretending. 'The idea that a crazy act, such as obsessively spilling ink on a tablecloth, can have a kind of hidden logic.'

She nodded. 'Yeah,' she said. 'That's it.'

Woody knew instinctively that she did not understand what he was talking about. He had already overtaken her in his knowledge of Freud, but she was embarrassed to admit it.

'What's your favourite thing to do?' he asked her. 'Theatre? Classical music? I guess going to a film is no big treat for someone whose father owns about a hundred movie houses.'

'Why do you ask?'

'Well . . .' He decided to be honest. 'I want to ask you out, and I'd like to tempt you with something you really love to do. So name it, and we'll do it.'

She smiled at him, but it was not the smile he was hoping for. It was friendly but sympathetic, and it told him that bad news was coming. 'Woody, I'd like to, but you're fifteen.'

'As you said last night, I'm more mature than Victor Dixon.'

'I wouldn't go out with him, either.'

Woody's throat seemed to constrict, and his voice came out hoarse. 'Are you turning me down?'

'Yes, very firmly. I don't want to date a boy three years younger.'

'Can I ask you again in three years? We'll be the same age then.'

She laughed, then said: 'Stop being witty, it hurts my head.'

Woody decided not to hide his pain. What did he have to lose? Feeling anguished, he said: 'So what was that kiss about?'

'It was nothing.'

He shook his head miserably. 'It was something to me. It was the best kiss I've ever had.'

'Oh, God, I knew it was a mistake. Look, it was just a bit of fun. Yes, I enjoyed it – be flattered, you're entitled. You're a cute kid, and smart as a whip, but a kiss is not a declaration of love, Woody, no matter how much you enjoy it.'

They were near the front of the march, and Woody saw their destination up ahead: the high wall around the Buffalo Metal Works. The gate was closed and guarded by a dozen or more factory police, thuggish men in light-blue shirts that mimicked police uniform.

'And I was drunk,' Joanne added.

'Yeah, I was drunk, too,' Woody said.

It was a pathetic attempt to salvage his dignity, but Joanne had the grace to pretend to believe him. 'Then we both did something a little foolish, and we should just forget it,' she said.

'Yeah,' said Woody, looking away.

They were outside the factory now. Those at the head of the march stopped at the gates, and someone began to make a speech through a bullhorn. Looking more closely, Woody saw that the speaker was a local union organizer, Brian Hall. Woody's father knew and liked the man: at some time in the dim past they had worked together to resolve a strike.

The rear of the procession kept coming forward, and a crush developed across the width of the street. The factory police were keeping the entrance clear, though the gates were shut. Woody now saw that they were armed with police-type nightsticks. One of them was shouting: 'Stay away from the gate! This is private property!' Woody lifted his camera and took a picture.

But the people at the front were being pushed forward by those behind. Woody took Joanne's arm and tried to steer her away from the focus of tension. However, it was difficult: the crowd was dense, now, and no one wanted to move out of the way. Against his will, Woody found himself edging closer to the factory gate and the guards with nightsticks. 'This is not a good situation,' he said to Joanne.

But she was flushed with excitement. 'Those bastards can't keep us back!' she cried.

A man next to her shouted: 'Right! Damn right!'

The crowd was still ten yards or more from the gate but, just the same, the guards unnecessarily began to push demonstrators away. Woody took a photograph.

Brian Hall had been yelling into his bullhorn about boss thugs and pointing an accusing finger at the factory police. Now he changed his tune and began to call for calm. 'Move away from the gates, please, brothers,' he said. 'Move back, no rough stuff.'

Woody saw a woman pushed by a guard hard enough to make her stumble. She did not fall over, but she cried out, and the man with her said to the guard: 'Hey, buddy, take it easy, will you?'

'Are you trying to start something?' the guard said challengingly.

The woman yelled: 'Just stop shoving!'

'Move back, move back!' the guard shouted. He raised his nightstick. The woman screamed.

As the nightstick came down, Woody took a picture.

Joanne said: 'The son of a bitch hit that woman!' She stepped forward.

But most of the crowd began to move in the opposite direction, away from the factory. As they turned, the guards came after them, shoving, kicking, and lashing out with their truncheons.

Brian Hall said: 'There is no need for violence! Factory police, step back! Do not use your clubs!' Then his bullhorn was knocked out of his hands by a guard.

Some of the younger men fought back. Half a dozen real policemen moved into the crowd. They did nothing to restrain the factory police, but began to arrest anyone fighting back.

The guard who had started the fracas fell to the ground, and two demonstrators started kicking him.

Woody took a picture.

Joanne was screaming with fury. She threw herself at a guard and scratched his face. He put out a hand to shove her away. Accidentally or otherwise, the heel of his hand connected sharply with her nose. She fell back with blood coming from her nostrils. The guard raised his nightstick. Woody grabbed her by the waist and jerked her back. The stick missed her. 'Come on!' Woody yelled at her. 'We have to get out of here!'

The blow to her face had deflated her fury, and she offered no resistance as he half pulled, half carried her away from the gates as fast as he could, his camera swinging on the strap around his neck. The crowd was panicking now, people falling over and others trampling them as everyone tried to flee.

Woody was taller than most and he managed to keep himself and

Joanne upright. They fought their way through the crush, staying just ahead of the nightsticks. At last the crowd thinned out. Joanne detached herself from his grasp and they both began to run.

The noise of the fight receded behind them. They turned a couple of corners and, a minute later, found themselves on a deserted street of factories and warehouses, all closed on Sunday. They slowed to a walk, catching their breath. Joanne began to laugh. 'That was so exciting!' she said.

Woody could not share her enthusiasm. 'It was nasty,' he said. 'And it could have got worse.' He had rescued her, and he half hoped that might cause her to change her mind about dating him.

But she did not feel she owed him much. 'Oh, come on,' she said in a tone of disparagement. 'Nobody died.'

'Those guards deliberately provoked a riot!'

'Of course they did! Peshkov wants to make union members look bad.'

'Well, we know the truth.' Woody tapped his camera. 'And I can prove it.'

They walked half a mile, then Woody saw a cruising cab and hailed it. He gave the driver the address of the Rouzrokh family home.

Sitting in the back of the taxi, he took a handkerchief from his pocket. 'I don't want to bring you home to your father looking like this,' he said. He unfolded the white cotton square and gently dabbed at the blood on her upper lip.

It was an intimate act, and he found it sexy, but she did not indulge him for long. After a second she said: 'I've got it.' She took the handkerchief from his grasp and cleaned herself up. 'How's that?'

'You've missed a bit,' he lied. He took the handkerchief back. Her mouth was wide, she had even, white teeth, and her lips were enchantingly full. He pretended there was something under her lower lip. He wiped it gently, then said: 'Better.'

'Thanks.' She looked at him with an odd expression, half fond, half annoyed. She knew he had been lying about the blood on her chin, he guessed, and she was not sure whether to be cross with him or not.

The cab halted outside her house. 'Don't come in,' she said. 'I'm going to lie to my parents about where I've been, and I don't want you blabbing the truth.'

Woody reckoned he was probably the more discreet of the two of them, but he did not say so. 'I'll call you later.'

'Okay.' She got out of the taxi and walked up the driveway with a perfunctory wave.

'She's a doll,' said the driver. 'Too old for you, though.'

'Take me to Delaware Avenue,' Woody said. He gave the number and the cross street. He was not going to talk about Joanne to a goddamn cabby.

He pondered his rejection. He should not have been surprised: everyone from his brother to the taxi driver said he was too young for her. All the same it hurt. He felt as if he did not know what to do with his life now. How would he get through the rest of the day?

Back at home, his parents were taking their ritual Sunday afternoon nap. Chuck believed that was when they had sex. Chuck himself had gone swimming with a bunch of friends, according to Betty.

Woody went into the darkroom and developed the film from his camera. He ran warm water into the basin to bring the chemicals to the ideal temperature, then put the film into a black bag to transfer it into a light-trap tank.

It was a lengthy process that required patience, but he was happy to sit in the dark and think about Joanne. Their being together during a riot had not made her fall in love with him, but it had certainly brought them closer. He felt sure she was at least growing to like him more and more. Maybe her rejection was not final. Perhaps he should keep trying. He certainly had no interest in any other girls.

When his timer rang, he transferred the film into a stop bath to halt the chemical reaction, then to a bath of fixer to make the image permanent. Finally, he washed and dried his film and looked at the negative black-and-white images on the reel.

He thought they were pretty good.

He cut the film into frames, then put the first into the enlarger. He laid a sheet of ten-by-eight photographic paper on the base of the enlarger, turned on the light, and exposed the paper to the negative image while he counted seconds. Then he put the print into an open bath of developer.

This was the best part of the process. Slowly the white paper began to show patches of grey, and the image he had photographed began to appear. It always seemed to him like a miracle. The first print showed a Negro and a white man, both in Sunday suits and hats, holding a banner that said BROTHERHOOD in large letters. When the image was clear he moved the paper to a bath of fixer, then washed it and dried it.

He printed all the shots he had taken, took them out into the light, and laid them out on the dining-room table. He was pleased: they were vivid, active pictures that clearly showed a sequence of events. When he heard his parents moving about upstairs he called his mother. She had been a journalist before she married, and she still wrote books and magazine articles. 'What do you think?' he asked her.

She studied them thoughtfully with her one eye. After a while she said: 'I think they're good. You should take them to a newspaper.'

'Really?' he said. He began to feel excited. 'Which paper?'

'They're all conservative, unfortunately. Maybe the *Buffalo Sentinel*. The editor is Peter Hoyle – he's been there since God was a boy. He knows your father well, he'll probably see you.'

'When should I show him the photos?'

'Now. The march is hot news. It will be in all tomorrow's papers. They need the pictures tonight.'

Woody was energized. 'All right,' he said. He picked up the glossy sheets and shuffled them into a neat stack. His mother produced a cardboard folder from Papa's study. Woody kissed her and left the house.

He caught a bus downtown.

The front entrance of the *Sentinel* office was closed, and he suffered a moment of dismay, but he reasoned that reporters must be able to get in and out today if they were to produce a Monday morning paper and, sure enough, he found a side entrance. 'I have some photographs for Mr Hoyle,' he said to a man sitting inside the door, and he was directed upstairs.

He found the editor's office, a secretary took his name, and a minute later he was shaking hands with Peter Hoyle. The editor was a tall, imposing man with white hair and a black moustache. He appeared to be finishing a meeting with a younger colleague. He spoke loudly, as if shouting over the noise of a printing press. 'The hit-and-run driver's story is fine, but the intro stinks, Jack,' he said, with a dismissive hand on the man's shoulder, moving him to the door. 'Put a new nose on it. Move the Mayor's statement to later and start with crippled children.' Jack left, and Hoyle turned to Woody. 'What have you got, kid?' he said without preamble.

'I was at the march today.'

'You mean the riot.'

'It wasn't a riot until the factory guards started hitting women with their clubs.'

'I hear the marchers tried to break into the factory, and the guards repelled them.'

'It's not true, sir, and the photos prove it.'

'Show me.'

Woody had arranged them in order while sitting on the bus. He put the first down on the editor's desk. 'It started peacefully.'

Hoyle pushed the photograph aside. 'That's nothing,' he said.

Woody brought out a picture taken at the factory. 'The guards were waiting at the gate. You can see their nightsticks.' His next picture had been taken when the shoving started. 'The marchers were at least ten yards from the gate, so there was no need for the guards to try to move them back. It was a deliberate provocation.'

'Okay,' said Hoyle, and he did not push the pictures aside.

Woody brought out his best shot: a guard using a truncheon to beat a woman. 'I saw this whole incident,' Woody said. 'All the woman did was tell him to stop shoving her, and he hit her like this.'

'Good picture,' said Hoyle. 'Any more?'

'One,' said Woody. 'Most of the marchers ran away as soon as the fighting began, but a few fought back.' He showed Hoyle the photograph of two demonstrators kicking a guard on the ground. 'These men retaliated against the guard who hit the woman.'

'You did a good job, young Dewar,' said Hoyle. He sat at his desk and pulled a form from a tray. 'Twenty bucks okay?'

'You mean you're going to print my photographs?'

'I assume that's why you brought them here.'

'Yes, sir, thank you, twenty dollars is okay, I mean fine. I mean plenty.'

Hoyle scribbled on the form and signed it. 'Take this to the cashier. My secretary will tell you where to go.'

The phone on the desk rang. The editor picked it up and barked: 'Hoyle.' Woody gathered he was dismissed, and left the room.

He was elated. The payment was amazing, but he was more thrilled that the newspaper would use his photos. He followed the secretary's directions to a little room with a counter and a teller's window, and got his twenty bucks. Then he went home in a taxi.

His parents were delighted by his coup, and even his brother seemed pleased. Over dinner, Grandmama said: 'As long as you don't consider journalism as a career. That would be lowering.'

In fact, Woody had been thinking that he might take up news photography instead of politics, and he was surprised to learn that his grandmother disapproved.

His mother smiled and said: 'But, Ursula dear, I was a journalist.'

'That's different, you're a girl,' Grandmama replied. 'Woodrow must become a man of distinction, like his father and grandfather before him.'

Mother did not take offence at this. She was fond of Grandmama and listened with amused tolerance to her pronouncements of orthodoxy.

However, Chuck resented the traditional focus on the elder son. He said: 'And what must I become, chopped liver?'

'Don't be vulgar, Charles,' said Grandmama, having the last word as usual.

That night Woody lay awake a long time. He could hardly wait to see his photos in the paper. He felt the way he had as a kid on Christmas Eve: his longing for the morning kept him from sleep.

He thought about Joanne. She was wrong to think him too young. He was right for her. She liked him, they had a lot in common, and she had enjoyed the kiss. He still thought he might win her heart.

He fell asleep at last, and when he woke it was daylight. He put on a dressing gown over his pyjamas and ran downstairs. Joe, the butler, always went out early to buy the newspapers, and they were already laid out on the side table in the breakfast room. Woody's parents were there, his father eating scrambled eggs, his mother sipping coffee.

Woody picked up the *Sentinel*. His work was on the front page.

But it was not what he expected.

They had used only one of his shots – the last. It showed a factory guard lying on the ground being kicked by two workers. The headline was: METAL STRIKERS RIOT.

'Oh, no!' he said.

He read the report with incredulity. It said that marchers had attempted to break into the factory and had been bravely repelled by the factory police, several of whom had suffered minor injuries. The behaviour of the workers was condemned by the Mayor, the Chief of Police, and Lev Peshkov. At the foot of the article, like an afterthought, union spokesman Brian Hall was quoted as denying the story and blaming the guards for the violence.

Woody put the newspaper in front of his mother. 'I told Hoyle that the guards started the riot – and I gave him the pictures to prove it!' he said angrily. 'Why would he print the opposite of the truth?'

'Because he's a conservative,' she said.

'Newspapers are supposed to tell the truth!' Woody said, his voice rising with furious indignation. 'They can't just make up lies!'

'Yes, they can,' she said.

'But it's not fair!'

'Welcome to the real world,' said his mother.

(vi)

Greg Peshkov and his father were in the lobby of the Ritz-Carlton hotel in Washington, DC, when they ran into Dave Rouzrokh.

Dave was wearing a white suit and a straw hat. He glared at them with hatred. Lev greeted him, but he turned away contemptuously without answering.

Greg knew why. Dave had been losing money all summer, because Roseroque Theatres was not able to get first-run hit movies. And Dave must have guessed that Lev was somehow responsible.

Last week Lev had offered Dave four million dollars for his movie houses – half the original bid – and Dave had again refused. 'The price is dropping, Dave,' Lev had warned.

Now Greg said: 'I wonder what he's doing here?'

'He's meeting with Sol Starr. He's going to ask why Sol won't give him good movies.' Lev obviously knew all about it.

'What will Mr Starr do?'

'String him along.'

Greg marvelled at his father's ability to know everything and stay on top of a changing situation. He was always ahead of the game.

They rode up in the elevator. This was the first time Greg had visited his father's permanent suite at the hotel. His mother, Marga, had never been here.

Lev spent a lot of time in Washington because the government was forever interfering with the movie business. Men who considered themselves to be moral leaders got very agitated about what was shown on the big screen, and they put pressure on the government to censor pictures. Lev saw this as a negotiation – he saw life as a negotiation –

and his constant aim was to avoid formal censorship by adhering to a voluntary code, a strategy backed by Sol Starr and most other Hollywood big shots.

They entered a living room that was extremely fancy, much more so than the spacious apartment in Buffalo where Greg and his mother lived, and which Greg had always thought to be luxurious. This room had spindly legged furniture that Greg imagined to be French, rich chestnut-brown velvet drapes at the windows, and a large phonograph.

In the middle of the room he was stunned to see, sitting on a yellow silk sofa, the movie star Gladys Angelus.

People said she was the most beautiful woman in the world.

Greg could see why. She radiated sex appeal, from her dark-blue inviting eyes to the long legs crossed under her clinging skirt. As she put out a hand to shake his, her red lips smiled and her round breasts moved alluringly inside a soft sweater.

He hesitated a split second before shaking her hand. He felt disloyal to his mother, Marga. She never mentioned the name of Gladys Angelus, a sure sign that she knew what people were saying about Gladys and Lev. Greg felt he was making friends with his mother's enemy. If Mom knew about this she would cry, he thought.

But he had been taken by surprise. If he had been forewarned, if he had had time to think about his reaction, he might have prepared, and rehearsed a gracious withdrawal. But he could not bring himself to be clumsily rude to this overwhelmingly lovely woman.

So he took her hand, looked into her amazing eyes, and gave what people called a shit-eating grin.

She kept hold of his hand as she said: 'I'm so happy to meet you at long last. Your father has told me all about you – but he didn't say how handsome you are!'

There was something unpleasantly proprietorial about this, as if she were a member of the family, rather than a whore who had usurped his mother. All the same he found himself falling under her spell. 'I love your films,' he said awkwardly.

'Oh, stop it, you don't have to say that,' she said, but Greg thought she liked to hear it all the same. 'Come and sit by me,' she went on. 'I want to get to know you.'

He did as he was told. He could not help himself. Gladys asked him what school he attended, and while he was telling her, the phone rang. He vaguely heard his father say into the phone: 'It was supposed to be tomorrow ... okay, if we have to, we can rush it ... leave it with me, I'll handle it.'

Lev hung up and interrupted Gladys. 'Your room is down the hall, Greg,' he said. He handed over a key. 'And you'll find a gift from me. Settle in and enjoy yourself. We'll meet for dinner at seven.'

This was abrupt, and Gladys looked put out, but Lev could be peremptory sometimes, and it was best just to obey. Greg took the key and left.

In the corridor was a broad-shouldered man in a cheap suit. He reminded Greg of Joe Brekhunov, head of security at the Buffalo Metal Works. Greg nodded, and the man said: 'Good afternoon, sir.' Presumably he was a hotel employee.

Greg entered his room. It was pleasant enough, though not as swanky as his father's suite. He did not see the gift his father had mentioned, but his suitcase was there, and he began to unpack, thinking about Gladys. Was he being disloyal to his mother by shaking hands with his father's mistress? Of course, Gladys was only doing what Marga herself had done, sleeping with a married man. All the same, he felt painfully uncomfortable. Was he going to tell his mother that he had met Gladys? Hell, no.

As he was hanging up his shirts, he heard a knock. It came from a door that looked as if it might lead to the neighbouring room. Next moment, the door opened and a girl walked through.

She was older than Greg, but not much. Her skin was the colour of dark chocolate, and she wore a polka-dot dress and carried a clutch bag. She smiled broadly, showing white teeth, and said: 'Hello, I've got the room next door.'

'I figured that out,' he said. 'Who are you?'

'Jacky Jakes.' She held out her hand. 'I'm an actress.'

Greg shook hands with the second beautiful actress in an hour. Jacky had a playful look that Greg found more attractive than Gladys's overpowering magnetism. Her mouth was a dark-pink bow. He said: 'My dad said he had got me a gift – are you it?'

She giggled. 'I guess I am. He said I would like you. He's going to get me into the movies.'

Greg got the picture. His father had guessed that he might feel bad about being friendly with Gladys. Jacky was his reward for not making a fuss. He thought he probably ought to reject such a bribe, but he could not resist. 'You're a very nice gift,' he said.

'Your father's real good to you.'

'He's wonderful,' Greg said. 'And so are you.'

'Aren't you sweet?' She put her purse down on the dresser, stepped closer to Greg, stood on tiptoe, and kissed his mouth. Her lips were soft and warm. 'I like you,' she said. She felt his shoulders. 'You're strong.'

'I play ice hockey.'

'Makes a girl feel safe.' She put both hands on his cheeks and kissed him again, longer, then she sighed and said: 'Oh, boy, I think we're going to have fun.'

'Are we?' Washington was a Southern city, still largely segregated. In Buffalo, white and black people could eat in the same restaurants and drink in the same bars, mostly, but here it was different. Greg was not sure what the laws were, but he felt certain that in practice a white man with a black woman would cause trouble. It was surprising to find Jacky occupying a room in this hotel: Lev must have fixed it. But certainly there was no question of Greg and Jacky swanning around town with Lev and Gladys in a foursome. So what did Jacky think they were going to do to have fun together? The amazing notion crossed his mind that she might be willing to go to bed with him.

He put his hands on her waist, to draw her to him for another kiss, but she pulled back. 'I need to take a shower,' she said. 'Give me a few minutes.' She turned and disappeared through the communicating door, closing it behind her.

He sat on the bed, trying to take it all in. Jacky wanted to act in movies, and it seemed she was willing to use sex to advance her career. She certainly was not the first actress, black or white, to use that strategy. Gladys was doing the same by sleeping with Lev. Greg and his father were the lucky beneficiaries.

He saw that she had left her clutch bag behind. He picked it up and tried the door. It was not locked. He stepped through.

She was on the phone, wearing a pink bathrobe. She said: 'Yes, hunky-dory, no problem.' Her voice seemed different, more mature, and he realized that with him she had been using a sexy-little-girl tone that was not natural. Then she saw him, smiled, and reverted to the girly voice as she said into the phone: 'Please hold my calls. I don't want to be disturbed. Thank you. Goodbye.'

'You left this,' said Greg, and handed her the purse.

'You just wanted to see me in my bathrobe,' she said coquettishly.

The front of the robe did not entirely hide her breasts, and he could see an enchanting curve of flawless brown skin.

He grinned. 'No, but I'm glad I did.'

'Go back to your room. I have to shower. I might let you see more later.'

'Oh, my God,' he said.

He returned to his room. This was astonishing. 'I might let you see more later,' he repeated to himself aloud. What a thing for a girl to say!

He had a hard-on, but he did not want to jerk off when the real thing seemed so close. To take his mind off it, he went on unpacking. He had an expensive shaving kit, razor and brush with pearl handles, a present from his mother. He laid the things out in the bathroom, wondering whether they would impress Jacky if she saw them.

The walls were thin, and he heard the sound of running water from the next room. The thought of her body naked and wet possessed him. He tried to concentrate on arranging his underwear and socks in a drawer.

Then he heard her scream.

He froze. For a moment he was too surprised to move. What did it mean? Why would she yell out like that? Then she screamed again, and he was shocked into action. He threw open the communicating door and stepped into her room.

She was naked. He had never seen a naked woman in real life. She had pointed breasts with dark-brown tips. At her groin was a thatch of wiry black hair. She was cowering back against the wall, trying ineffectually to cover her nakedness with her hands.

Standing in front of her was Dave Rouzrokh, with twin scratches down his aristocratic cheek, presumably caused by Jacky's pink-varnished nails. There was blood on the broad lapel of Dave's double-breasted white jacket.

Jacky screamed: 'Get him away from me!'

Greg swung a fist. Dave was an inch taller, but he was an old man, and Greg was an athletic teenager. The blow connected with Dave's chin – more by luck than by judgement – and Dave staggered back then fell to the floor.

The room door opened.

The broad-shouldered hotel employee Greg had seen earlier came in. He must have a master key, Greg thought. 'I'm Tom Cranmer, house detective,' the man said. 'What's going on here?'

Greg said: 'I heard her scream and came in to find him here.'

Jacky said: 'He tried to rape me!'

Dave struggled to his feet. 'That's not true,' he said. 'I was asked to come to this room for a meeting with Sol Starr.'

Jacky began to sob. 'Oh, now he's going to lie about it!'

Cranmer said: 'Put something on, please, miss.'

Jacky put on her pink bathrobe.

The detective picked up the room phone, dialled a number, and said: 'There's usually a cop on the corner. Get him into the lobby, right now.'

Dave was staring at Greg. 'You're Peshkov's bastard, aren't you?'

Greg was about to hit him again.

Dave said: 'Oh, my God, this is a set-up.'

Greg was thrown by this remark. He felt intuitively that Dave was telling the truth. He dropped his fist. This whole scene must have been scripted by Lev, he realized. Dave Rouzrokh was no rapist. Jacky was faking. And Greg himself was just an actor in the movie. He felt dazed.

'Please come with me, sir,' said Cranmer, taking Dave firmly by the arm. 'You two as well.'

'You can't arrest me,' said Dave.

'Yes, sir, I can,' said Cranmer. 'And I'm going to hand you over to a police officer.'

Greg said to Jacky: 'Do you want to get dressed?'

She shook her head quickly and decisively. Greg realized it was part of the plan that she would appear in her robe.

He took Jacky's arm and they followed Cranmer and Dave along the corridor and into the elevator. A cop was waiting in the lobby. Both he and the hotel detective must be in on the plot, Greg surmised.

Cranmer said: 'I heard a scream from her room, found the old guy in there. She says he tried to rape her. The kid is a witness.'

Dave looked bewildered, as if he thought this might be a bad dream. Greg found himself feeling sorry for Dave. He had been cruelly trapped. Lev was more pitiless than Greg had imagined. Half of him admired his father; the other half wondered if such ruthlessness was really necessary.

The cop snapped handcuffs on Dave and said: 'All right, let's go.'

'Go where?' Dave said.

'Downtown,' said the cop.

Greg said: 'Do we all have to go?'

'Yeah.'

Cranmer spoke to Greg in a low voice. 'Don't worry, son,' he said.

'You did a great job. We'll go to the precinct house and make our statements, and after that you can fuck her from here to Christmastime.'

The cop led Dave to the door, and the others followed.

As they stepped outside, a photographer popped a flashgun.

(vii)

Woody Dewar got a copy of Freud's *Studies in Hysteria* mailed to him by a bookseller in New York. On the night of the Yacht Club Ball – the climactic social event of the summer season in Buffalo – he wrapped it neatly in brown paper and tied a red ribbon around it. 'Chocolates for a lucky girl?' said his mother, passing him in the hall. She had only one eye but she saw everything.

'A book,' he said. 'For Joanne Rouzrokh.'

'She won't be at the ball.'

'I know.'

Mama stopped and gave him a searching look. After a moment she said: 'You're serious about her.'

'I guess. But she thinks I'm too young.'

'Her pride is probably involved. Her friends would ask why she can't find a guy her own age to go out with. Girls are cruel like that.'

'I'm planning to persist until she grows more mature.'

Mama smiled. 'I bet you make her laugh.'

'I do. It's the best card I hold.'

'Well, heck, I waited long enough for your father.'

'Did you?'

'I loved him from the first time I met him. I pined for years. I had to watch him fall for that shallow cow Olga Vyalov, who wasn't worthy of him but had two working eyes. Thank God she got knocked up by her chauffeur.' Mama's language could be a little coarse, especially when Grandmama was not around. She had picked up bad habits during the years she spent working on newspapers. 'Then he went off to war. I had to follow him to France before I could nail his foot to the goddamn floor.'

Nostalgia was mixed with pain in her reminiscence, Woody could tell. 'But he realized you were the right girl for him.'

'In the end, yes.'

'Maybe that'll happen to me.'

Mama kissed him. 'Good luck, my son,' she said.

The Rouzrokh house was less than a mile away and Woody walked there. None of the Rouzrokhs would be at the Yacht Club tonight. Dave had been all over the papers after a mysterious incident at the Ritz-Carlton Hotel in Washington. A typical headline had read: CINEMA MOGUL ACCUSED BY STARLET. Woody had recently learned to mistrust newspapers. However, gullible people said there must be something in it, otherwise why would the police have arrested Dave?

None of the family had been seen at any social event since.

Outside the house an armed guard stopped Woody. 'The family aren't seeing callers,' he said brusquely.

Woody guessed the man had spent a lot of time repelling reporters, and he forgave the discourteous tone. He recalled the name of the Rouzrokhs' maid. 'Please ask Miss Estella to tell Joanne that Woody Dewar has a book for her.'

'You can leave it with me,' said the guard, holding out his hand.

Woody held on firmly to the book. 'Thanks, but no.'

The guard looked annoyed, but he walked Woody up the drive and rang the doorbell. Estella opened it and said at once: 'Hello, Mr Woody, come in – Joanne will be so glad to see you!' Woody permitted himself a triumphant glance at the guard as he stepped inside.

Estella showed him into an empty drawing room. She offered him milk and cookies, as if he were still a kid, and he declined politely. Joanne came in a minute later. Her face was drawn and her olive skin looked washed-out, but she smiled pleasantly at him and sat down to chat.

She was pleased with the book. 'Now I'll have to read Dr Freud instead of just gabbing about him,' she said. 'You're a good influence on me, Woody.'

'I wish I could be a bad influence.'

She let that pass. 'Aren't you going to the ball?'

'I have a ticket but if you're not there I'm not interested. Would you like to go to a movie instead?'

'No, thanks, really.'

'Or we could just get dinner. Somewhere really quiet. If you don't mind taking the bus.'

'Oh, Woody, of course I don't mind the bus, but you're too young for me. Anyway, the summer's almost over. You'll be back at school soon, and I'm going to Vassar.'

'Where you'll go on dates, I guess.'

'I sure hope so!'

Woody stood up. 'Okay, well, I'm going to take a vow of celibacy and enter a monastery. Please don't come and visit me, you'll distract the other brethren.'

She laughed. 'Thank you for taking my mind off my family's troubles.'

It was the first time she had mentioned what had happened to her father. He had not been planning to raise the subject but, now that she had, he said: 'You know we're all on your side. Nobody believes that actress's story. Everyone in town realizes it was a set-up by that swine Lev Peshkov, and we're furious about it.'

'I know,' she said. 'But the accusation alone is too shameful for my father to bear. I think my parents are going to move to Florida.'

'I'm so sorry.'

'Thank you. Now go to the ball.'

'Maybe I will.'

She walked him to the door.

'May I kiss you goodbye?' he said.

She leaned forward and kissed his lips. This was not like the last kiss, and he knew instinctively not to grab her and press his mouth to hers. It was a gentle kiss, her lips on his for a sweet moment that was over in a breath. Then she pulled away and opened the front door.

'Goodnight,' Woody said as he stepped out.

'Goodbye,' said Joanne.

(viii)

Greg Peshkov was in love.

He knew that Jacky Jakes had been bought for him by his father, as his reward for helping to entrap Dave Rouzrokh, but despite that it was real love.

He had lost his virginity a few minutes after they had returned from the precinct house, and the two of them had then spent most of a week in bed at the Ritz-Carlton. Greg did not need to use birth control, she told him, because she was already 'fixed up'. He had only the vaguest idea what that meant, but he took her at her word.

He had never been so happy in his life, and he adored her, especially when she dropped the little-girl act and revealed a shrewd intelligence and a mordant sense of humour. She admitted that she had seduced Greg on his father's orders, but confessed that against her will she had fallen in love. Her real name was Mabel Jakes and, although she pretended to be nineteen, she was in fact just sixteen, only a few months older than Greg.

Lev had promised her a part in a movie but, he said, he was still looking for just the right role. In a perfect imitation of Lev's vestigial Russian accent she said: 'But I don't guess he's lookin' too fuckin' hard.'

'I guess there aren't many parts written for Negro actors,' Greg said.

'I know, I'll end up playing the maid, rolling my eyes and saying "Lawdy". There are Africans in plays and films – Cleopatra, Hannibal, Othello – but they're usually played by white actors.' Her father, now dead, had been a professor in a Negro college, and she knew more about literature than Greg did. 'Anyway, why should Negroes only play black people? If Cleopatra can be played by a white actress, why can't Juliet be black?'

'People would find it strange.'

'People would get used to it. They get used to anything. Does Jesus have to be played by a Jew? Nobody cares.'

She was right, Greg thought, but, all the same, it was never going to happen.

When Lev had announced their return to Buffalo – leaving it until the last minute, as usual – Greg had been devastated. He had asked his father if Jacky could come to Buffalo, but Lev had laughed and said: 'Son, you don't shit where you eat. You can see her next time you come to Washington.'

Despite that, Jacky had followed him to Buffalo a day later and moved into a cheap apartment near Canal Street.

Lev and Greg had been busy for the next couple of weeks with the takeover of Roseroque Theatres. Dave had sold for two million in the end, a quarter of the original offer, and Greg's admiration for his father went up another notch. Jacky had withdrawn her charges and hinted to the newspapers that she had accepted a cash settlement. Greg was awestruck by his father's callous nerve.

And he had Jacky. He told his mother he was out every night with male friends but, in fact, he spent all his spare time with Jacky. He showed her around town, picnicked with her at the beach, even managed to take her out in a borrowed speedboat. No one connected her with the rather blurred newspaper photograph of a girl walking out of the Ritz-Carlton hotel in a bathrobe. But mostly they spent the warm summer evenings having sweaty, deliriously happy sex, tangling the worn sheets on the narrow bed in her small apartment. They decided to get married as soon as they were old enough.

Tonight he was taking her to the Yacht Club Ball.

It had been extraordinarily difficult to get tickets, but Greg had bribed a school friend.

He had bought Jacky a new dress, pink satin. He got a generous allowance from Marga, and Lev loved to slip him fifty bucks now and again, so he always had more money than he needed.

In the back of his mind a warning was sounding. Jacky would be the only Negro at the ball not serving drinks. She was very reluctant to go, but Greg had talked her round. The young men would envy him but the older ones might be hostile, he knew. There would be some muttering. Jacky's beauty and charm would overcome much prejudice, he felt: how could anyone resist her? But if some fool got drunk and insulted her, Greg would teach him a lesson with both fists.

Even as he thought this, he heard his mother telling him not to be a love-struck fool. But a man could not go through life listening to his mother.

As he walked along Canal Street in white tie and tails, he looked forward to seeing her in the new dress, and maybe kneeling to lift the hem up until he could see her panties and garter belt.

He entered her building, an old house now subdivided. There was a threadbare red carpet on the stairs and a smell of spicy cooking. He let himself into the apartment with his own key.

The place was empty.

That was odd. Where would she go without him?

With fear in his heart, he opened the closet. The pink satin ball dress hung there on its own. Her other clothes were gone.

'No!' he said aloud. How could this happen?

On the rickety pine table was an envelope. He picked it up and saw his name on the front in Jacky's neat, schoolgirl handwriting. A feeling of dread came over him.

He tore open the envelope with shaky hands and read the short message.

My darling Greg,

The last three weeks have been the happiest time of my entire life. I knew in my heart that we couldn't ever get married but it was nice to pretend. You are a lovely boy and will grow into a fine man, if you don't take after your father too much.

Had Lev found out that Jacky was living here, and somehow made her leave? He would not do that – would he?

Goodbye and don't forget me.
    Your Gift,
        Jacky

Greg crumpled the paper and wept.

(ix)

'You look wonderful,' Eva Rothmann said to Daisy Peshkov. 'If I was a boy, I'd fall in love with you in a minute.'

Daisy smiled. Eva was already a little bit in love with her. And Daisy did look wonderful, in an ice-blue silk organdie ball gown that deepened the blue of her eyes. The skirt of the dress had a frilled hem that was ankle length in front but rose playfully to mid-calf behind, giving a tantalizing glimpse of Daisy's legs in sheer stockings.

She wore a sapphire necklace of her mother's. 'Your father bought me that, back in the days when he was still occasionally nice to me,' Olga said. 'But hurry up, Daisy, you're making us all late.'

Olga was wearing matronly navy blue, and Eva was in red, which suited her dark colouring.

Daisy walked down the stairs on a cloud of happiness.

They stepped out of the house. Henry, the gardener, doubling as chauffeur tonight, opened the doors of the shiny old black Stutz.

This was Daisy's big night. Tonight Charlie Farquharson would formally propose to her. He would offer her a diamond ring that was a family heirloom – she had seen and approved it, and it had been altered to fit her. She would accept his proposal, and then they would announce their engagement to everyone at the ball.

She got into the car feeling like Cinderella.

Only Eva had expressed doubts. 'I thought you'd go for someone who was more of a match for you,' she had said.

'You mean a man who won't let me boss him around,' Daisy had replied.

'No, but someone more like you, good-looking and charming and sexy.'

This was unusually sharp for Eva: it implied that Charlie was homely and charmless and unglamorous. Daisy had been taken aback, and did not know how to reply.

Her mother had saved her. Olga had said: 'I married a man who was good-looking and charming and sexy, and he made me utterly miserable.'

Eva had said no more.

As the car approached the Yacht Club, Daisy vowed to restrain herself. She must not show how triumphant she felt. She must act as if there was nothing unexpected about her mother being asked to join the Buffalo Ladies Society. As she showed the other girls her enormous diamond, she would be so gracious as to declare that she did not deserve someone as wonderful as Charlie.

She had plans to make him even more wonderful. As soon as the honeymoon was over she and Charlie would start building their stable of racehorses. In five years they would be entering the most prestigious races around the world: Saratoga Springs, Longchamps, Royal Ascot.

Summer was turning to fall, and it was dusk when the car drew up at the pier. 'I'm afraid we may be very late tonight, Henry,' Daisy said gaily.

'Quite all right, Miss Daisy,' he replied. He adored her. 'You have a wonderful time, now.'

At the door, Daisy noticed Victor Dixon following them in. Feeling well disposed towards everyone, she said: 'So, Victor, your sister met the King of England. Congratulations!'

'Mm, yes,' he said, looking embarrassed.

They entered the club. The first person they saw was Ursula Dewar, who had agreed to accept Olga into her snobby club. Daisy smiled warmly at her and said: 'Good evening, Mrs Dewar.'

Ursula seemed distracted. 'Excuse me, just a moment,' she said, and moved away across the lobby. She thought herself a queen, Daisy reflected, but did that mean she had no need of good manners? One day Daisy would rule over Buffalo society, but she would be unfailingly gracious to all, she vowed.

The three women went into the ladies' room, where they checked their appearance in the mirrors, in case anything had gone wrong in the twenty minutes since they had left home. Dot Renshaw came in, looked at them, and went out again. 'Stupid girl,' Daisy said.

But her mother looked worried. 'What's happening?' she said. 'We've been here five minutes, and already three people have snubbed us!'

'Jealousy,' Daisy said. 'Dot would like to marry Charlie herself.'

Olga said: 'At this point Dot Renshaw would like to marry more or less anybody, I guess.'

'Come on, let's enjoy ourselves,' said Daisy, and she led the way out.

As she entered the ballroom, Woody Dewar greeted her. 'At last, a gentleman!' Daisy said.

In a lowered voice he said: 'I just want to say that I think it's wrong of people to blame you for anything your father might have done.'

'Especially when they all bought their booze from him!' she replied.

Then she saw her future mother-in-law, in a ruched pink gown that did nothing for her angular figure. Nora Farquharson was not ecstatic about her son's choice of bride, but she had accepted Daisy and had been charming to Olga when they had exchanged visits. 'Mrs Farquharson!' Daisy said. 'What a lovely dress!'

Nora Farquharson turned her back and walked away.

Eva gasped.

A feeling of horror came over Daisy. She turned back to Woody. 'This isn't about bootlegging, is it?'

'No.'

'What, then?'

'You must ask Charlie. Here he comes.'

Charlie was perspiring, though it was not warm. 'What's going on?' Daisy asked him. 'Everyone's giving me the cold shoulder!'

He was terribly nervous. 'People are so angry at your family,' he said.

'What for?' she cried.

Several people nearby heard her raised voice and looked around. She did not care.

Charlie said: 'Your father ruined Dave Rouzrokh.'

'Are you talking about that incident in the Ritz-Carlton? What has that got to do with me?'

'Everyone likes Dave, even though he's Persian or something. And they don't believe he would rape anybody.'

'I never said he did!'

'I know,' Charlie said. He was clearly in agony.

People were frankly staring, now: Victor Dixon, Dot Renshaw, Chuck Dewar.

Daisy said to Charlie: 'But I'm going to be blamed. Is that so?'

'Your father did a terrible thing.'

Daisy was cold with fear. Surely she could not lose her triumph at the last minute? 'Charlie,' she said. 'What are you telling me? Talk straight, for the love of God.'

Eva put her arm around Daisy's waist in a gesture of support.

Charlie replied: 'Mother says it's unforgivable.'

'What does that mean, unforgivable?'

He stared miserably at her. He could not bring himself to speak.

But there was no need. She knew what he was going to say. 'It's over, isn't it?' she said. 'You're jilting me.'

He nodded.

Olga said: 'Daisy, we must leave.' She was in tears.

Daisy looked around. She tilted her chin as she stared them all down: Dot Renshaw looking maliciously pleased, Victor Dixon admiring, Chuck Dewar with his mouth open in adolescent shock, and his brother Woody looking sympathetic.

'To hell with you all,' Daisy said loudly. 'I'm going to London to dance with the King!'

3

1936

It was a sunny Saturday afternoon in May, 1936, and Lloyd Williams was at the end of his second year at Cambridge, when Fascism reared its vile head among the white stone cloisters of the ancient university.

Lloyd was at Emmanuel College – known as 'Emma' – doing Modern Languages. He was studying French and German, but he preferred German. As he immersed himself in the glories of German culture, reading Goethe, Schiller, Heine and Thomas Mann, he looked up occasionally from his desk in the quiet library to watch with sadness as today's Germany descended into barbarism.

Then the local branch of the British Union of Fascists announced that their leader, Sir Oswald Mosley, would address a meeting in Cambridge. The news took Lloyd back to Berlin three years earlier. He saw again the Brownshirt thugs wrecking Maud von Ulrich's magazine office; heard again the grating sound of Hitler's hate-filled voice as he stood in the parliament and poured scorn on democracy; shuddered anew at the memory of the dogs' bloody muzzles savaging Jörg with a bucket over his head.

Now Lloyd stood on the platform at Cambridge railway station, waiting to meet his mother off the train from London. With him was Ruby Carter, a fellow activist in the local Labour Party. She had helped him organize today's meeting on the subject of 'The Truth about Fascism'. Lloyd's mother, Eth Leckwith, was to speak. Her book about Germany had been a big success; she had stood for Parliament again in the 1935 election; and she was once again the Member for Aldgate.

Lloyd was tense about the meeting. Mosley's new political party had gained many thousands of members, due in part to the enthusiastic support of the *Daily Mail*, which had run the infamous headline HURRAH FOR THE BLACKSHIRTS! Mosley was a charismatic speaker, and would undoubtedly recruit new members today. It was vital that there should be a bright beacon of reason to contrast with his seductive lies.

However, Ruby was chatty. She was complaining about the social life of Cambridge. 'I'm so bored with local boys,' she said. 'All they want to do is go to a pub and get drunk.'

Lloyd was surprised. He had imagined that Ruby had a well-developed social life. She wore inexpensive clothes that were always a bit tight, showing off her plump curves. Most men would find her attractive, he thought. 'What do you like to do?' he asked. 'Apart from organize Labour Party meetings.'

'I love dancing.'

'You can't be short of partners. There are twelve men for every woman at the university.'

'No offence intended, but most of the university men are pansies.'

There were a lot of homosexual men at Cambridge University, Lloyd knew, but it startled him to hear her mention the subject. Ruby was famously blunt, but this was shocking, even from her. He had no idea how to respond, so he said nothing.

Ruby said: 'You're not one of them, are you?'

'No! Don't be ridiculous.'

'No need to be insulted. You're handsome enough for a pansy, except for that squashed nose.'

He laughed. 'That's what they call a backhanded compliment.'

'You are, though. You look like Douglas Fairbanks Junior.'

'Well, thanks, but I'm not a pansy.'

'Have you got a girlfriend?'

This was becoming embarrassing. 'No, not at the moment.' He made a show of checking his watch and looking for the train.

'Why not?'

'I just haven't met Miss Right.'

'Oh, thank you very much, I'm sure.'

He looked at her. She was only half joking. He felt mortified that she had taken his remark personally. 'I didn't mean . . .'

'Yes, you did. But never mind. Here's the train.'

The locomotive drew into the station and came to a halt in a cloud of steam. The doors opened and passengers stepped out on to the platform: students in tweed jackets, farmers' wives going shopping, working men in flat caps. Lloyd scanned the crowd for his mother. 'She'll be in a third-class carriage,' he said. 'Matter of principle.'

Ruby said: 'Would you come to my twenty-first birthday party?'

'Of course.'

'My friend's got a little flat in Market Street, and a deaf landlady.'

Lloyd was not comfortable about this invitation, and hesitated over his reply; then his mother appeared, as pretty as a songbird in a red summer coat and a jaunty little hat. She hugged and kissed him. 'You look very well, my lovely,' she said. 'But I must buy you a new suit for next term.'

'This one is fine, Mam.' He had a scholarship that paid his university fees and basic living expenses, but it did not run to suits. When he had started at Cambridge his mother had dipped into her savings and bought him a tweed suit for daytime and an evening suit for formal dinners. He had worn the tweed every day for two years, and it showed. He was particular about his appearance, and made sure that he always had a clean white shirt, a perfectly knotted tie, and a folded white handkerchief in his breast pocket: there had to be a dandy somewhere in his ancestry. The suit was carefully pressed, but it was beginning to look shabby, and in truth he longed for a new one, but he did not want his mother to spend her savings.

'We'll see,' she said. She turned to Ruby, smiled warmly, and held out her hand. 'I'm Eth Leckwith,' she said with the easy grace of a visiting duchess.

'Pleased to meet you. I'm Ruby Carter.'

'Are you a student, too, Ruby?'

'No, I'm a maid at Chimbleigh, a big country house.' Ruby looked a bit ashamed as she made this confession. 'It's five miles out of town, but I can usually borrow a bike.'

'Fancy that!' said Ethel. 'When I was your age, I was a maid at a country house in Wales.'

Ruby was amazed. 'You, a housemaid? And now you're a Member of Parliament!'

'That's what democracy means.'

Lloyd said: 'Ruby and I organized today's meeting together.'

His mother said: 'And how is it going?'

'Sold out. In fact, we had to move to a bigger hall.'

'I told you it would work.'

The meeting had been Ethel's idea. Ruby Carter and many others in the Labour Party had wanted to mount a protest demonstration, marching through the town. Lloyd had agreed at first. 'Fascism must be publicly opposed at every opportunity,' he had said.

Ethel had counselled otherwise. 'If we march and shout slogans, we look just like them,' she had said. 'Show that we're different. Hold a quiet, intelligent meeting to discuss the reality of Fascism.' Lloyd had been dubious. 'I'll come and speak, if you like,' she had said.

Lloyd had put that to the Cambridge party. There had been a lively discussion, with Ruby leading the opposition to Ethel's plan; but in the end the prospect of having an MP and famous feminist to speak had clinched it.

Lloyd was still not sure that it had been the right decision. He recalled Maud von Ulrich in Berlin saying: 'We must *not* meet violence with violence.' That had been the policy of the German Social Democratic Party. For the von Ulrich family, and for Germany, the policy had been a catastrophe.

They walked out through the yellow-brick Romanesque arches of the station and hurried along leafy Station Road, a street of smug middle-class houses made of the same yellow brick. Ethel put her arm through Lloyd's. 'How's my little undergraduate, then?' she said.

He smiled at the word 'little'. He was four inches taller than her, and muscular because of his training with the university boxing team: he could have picked her up with one hand. She was bursting with pride, he knew. Few things in life had pleased her as much as his coming to this place. That was probably why she wanted to buy him suits.

'I love it here, you know that,' he said. 'I'll love it more when it's full of working-class boys.'

'And girls,' Ruby put in.

They turned into Hills Road, the main thoroughfare leading to the town centre. Since the coming of the railway, the town had expanded south towards the station, and churches had been built along Hills Road to serve the new suburb. Their destination was a Baptist chapel whose left-wing pastor had agreed to loan it free of charge.

'I made a bargain with the Fascists,' Lloyd said. 'I said we'd refrain from marching if they would promise to do the same.'

'I'm surprised they agreed,' said Ethel. 'Fascists love marching.'

'They were reluctant. But I told the university authorities and the police what I was proposing, and the Fascists pretty much had to go along with it.'

'That was clever.'

'But Mam, guess who is their local leader? Viscount Aberowen, otherwise known as Boy Fitzherbert, the son of your former employer Earl Fitzherbert!' Boy was twenty-one, the same age as Lloyd. He was at Trinity, the aristocratic college.

'What? My God!'

She seemed more shaken than he had expected, and he glanced at her. She had gone pale. 'Are you shocked?'

'Yes!' She seemed to recover her composure. 'His father is a junior minister in the Foreign Office.' The government was a Conservative-dominated coalition. 'Fitz must be embarrassed.'

'Most Conservatives are soft on Fascism, I imagine. They see little wrong with killing Communists and persecuting Jews.'

'Some of them, perhaps, but you exaggerate.' She gave Lloyd a sideways look. 'So you went to see Boy?'

'Yes.' Lloyd thought this seemed to have special significance for Ethel, but he could not imagine why. 'I thought him perfectly frightful. In his room at Trinity he had a whole case of Scotch – twelve bottles!'

'You met him once before – do you remember?'

'No, when was that?'

'You were nine years old. I took you to the Palace of Westminster, shortly after I was elected. We met Fitz and Boy on the stairs.'

Lloyd did vaguely remember. Then, as now, the incident seemed to be mysteriously important to his mother. 'That was him? How funny.'

Ruby put in: 'I know him. He's a pig. He paws maids.'

Lloyd was shocked, but his mother seemed unsurprised. 'Very unpleasant, but it happens all the time.' Her grim acceptance made it more horrifying to him.

They reached the chapel and went in through the back door. There, in a kind of vestry, was Robert von Ulrich, looking startlingly British in a bold green-and-brown check suit and a striped tie. He stood up and Ethel hugged him. In faultless English, Robert said: 'My dear Ethel, what a perfectly charming hat.'

Lloyd introduced his mother to the local Labour Party women, who were preparing urns of tea and plates of biscuits to be served after the meeting. Having heard Ethel complain, many times, that people who organized political events seemed to think that an MP never needed to go to the toilet, he said: 'Ruby, before we start, would you show my mother where the ladies' facilities are?' The two women went off.

Lloyd sat down next to Robert and said conversationally: 'How's business?'

Robert was now the proprietor of a restaurant much favoured by the homosexuals about whom Ruby had been complaining. Somehow he had known that Cambridge in the 1930s was congenial to such men, just as Berlin had been in the 1920s. His new place had the same name as the old, Bistro Robert. 'Business is good,' he answered. A shadow crossed his face, a brief but intense look of real fear. 'This time, I hope I can keep what I've built up.'

'We're doing our best to fight off the Fascists, and meetings such as this are the way to do it,' Lloyd said. 'Your talk will be a big help – it will open people's eyes.' Robert was going to speak about his personal experience of life under Fascism. 'A lot of them say it couldn't happen here, but they're wrong.'

Robert nodded grim agreement. 'Fascism is a lie, but an alluring one.'

Lloyd's visit to Berlin three years ago was vivid in his mind. 'I often wonder what happened to the old Bistro Robert,' he said.

'I had a letter from a friend,' Robert said in a voice full of sadness. 'None of the old crowd go there any more. The Macke brothers auctioned off the wine cellar. Now the clientele is mostly middle-ranking cops and bureaucrats.' He looked even more pained as he added: 'They no longer use tablecloths.' He changed the subject abruptly. 'Do you want to go to the Trinity Ball?'

Most of the colleges held summer dances to celebrate the end of exams. The balls, plus associated parties and picnics, constituted May Week, which illogically took place in June. The Trinity Ball was famously lavish. 'I'd love to go, but I can't afford it,' Lloyd said. 'Tickets are two guineas, aren't they?'

'I've been given one. But you can have it. Several hundred drunk students dancing to a jazz band is actually my idea of hell.'

Lloyd was tempted. 'But I haven't got a tailcoat.' College balls required white-tie-and-tails.

'Borrow mine. It'll be too big at the waist, but we're the same height.'

'Then I will. Thank you!'

Ruby reappeared. 'Your mother is wonderful,' she said to Lloyd. 'I never knew she used to be a maid!'

Robert said: 'I have known Ethel for more than twenty years. She is truly extraordinary.'

'I can see why you haven't met Miss Right,' Ruby said to Lloyd. 'You're looking for someone like her, and there aren't many.'

'You're right about the last part, anyway,' Lloyd said. 'There's no one like her.'

Ruby winced, as if in pain.

Lloyd said: 'What's wrong?'

'Toothache.'

'You must go to the dentist.'

She looked at him as if he had said something stupid, and he realized that on a housemaid's wage she could not afford to pay a dentist. He felt foolish.

He went to the door and peeped through to the main hall. Like many nonconformist churches, this was a plain, rectangular room with walls painted white. It was a warm day, and the clear-glass windows were open. The rows of chairs were full and the audience was waiting expectantly.

When Ethel reappeared, Lloyd said: 'If it's all right with everyone, I'll open the meeting. Then Robert will tell his personal story, and my mother will draw out the political lessons.'

They all agreed.

'Ruby, will you keep an eye on the Fascists? Let me know if anything happens.'

Ethel frowned. 'Is that really necessary?'

'We probably shouldn't trust them to keep their promise.'

Ruby said: 'They're meeting a quarter of a mile up the road. I don't mind running in and out.'

She left by the back door, and Lloyd led the others into the church. There was no stage, but a table and three chairs stood at the near end, with a lectern to one side. As Ethel and Robert took their seats, Lloyd went to the lectern. There was a brief round of subdued applause.

'Fascism is on the march,' Lloyd began. 'And it is dangerously attractive. It gives false hope to the unemployed. It wears a spurious patriotism, as the Fascists themselves wear imitation military uniforms.'

The British government was keen to appease Fascist regimes, to Lloyd's dismay. It was a coalition dominated by Conservatives, with a few Liberals and a sprinkling of renegade Labour ministers who had split with their party. Only a few days after it was re-elected last November, the Foreign Secretary had proposed to yield much of Abyssinia to the conquering Italians and their Fascist leader, Benito Mussolini.

Worse still, Germany was rearming and aggressive. Just a couple of months ago, Hitler had violated the Versailles Treaty by sending troops into the demilitarized Rhineland – and Lloyd had been horrified to see that no country had been willing to stop him.

Any hope he had that Fascism might be a temporary aberration had now vanished. Lloyd believed that democratic countries such as France and Britain must get ready to fight. But he did not say so in his speech today, for his mother and most of the Labour Party opposed a build-up in British armaments and hoped that the League of Nations would be able to deal with the dictators. They wanted at all costs to avoid repeating the dreadful slaughter of the Great War. Lloyd sympathized with that hope, but feared it was not realistic.

He was preparing himself for war. He had been an officer cadet at school and, when he came up to Cambridge, he had joined the Officer Training Corps – the only working-class boy and certainly the only Labour Party member to do so.

He sat down to muted applause. He was a clear and logical speaker, but he did not have his mother's ability to touch hearts – not yet, anyway.

Robert stepped to the lectern. 'I am Austrian,' he said. 'In the war I was wounded, captured by the Russians, and sent to a prison camp in Siberia. After the Bolsheviks made peace with the Central Powers, the guards opened the gates and told us we were free to go. Getting home was our problem, not theirs. It is a long way from Siberia to Austria – more than three thousand miles. There was no bus, so I walked.'

Surprised laughter rippled around the room, with a few appreciative handclaps. Robert had already charmed them, Lloyd saw.

Ruby came up to him, looking annoyed, and spoke in his ear. 'The Fascists just went by. Boy Fitzherbert was driving Mosley to the railway station, and a bunch of hotheads in black shirts were running after the car, cheering.'

Lloyd frowned. 'They promised they wouldn't march. I suppose they'll say that running behind a car doesn't count.'

'What's the difference, I'd like to know?'

'Any violence?'

'No.'

'Keep a lookout.'

Ruby retired. Lloyd was bothered. The Fascists had certainly broken the spirit of the agreement, if not the letter. They had appeared on the street in their uniforms – and there had been no counter-demonstration. The socialists were here, inside the church, invisible. All there was to show for their stand was a banner outside the church saying THE TRUTH ABOUT FASCISM in large red letters.

Robert was saying: 'I am pleased to be here, honoured to have been invited to address you, and delighted to see several patrons of Bistro Robert in the audience. However, I must warn you that the story I have to tell is most unpleasant, and indeed gruesome.'

He related how he and Jörg had been arrested after refusing to sell the Berlin restaurant to a Nazi. He described Jörg as his chef and long-time business partner, saying nothing of their sexual relationship, though the more knowing people in the church probably guessed.

The audience became very quiet as he began to describe events in the concentration camp. Lloyd heard gasps of horror when he got to the part where the starving dogs appeared. Robert described the torture of Jörg in a low, clear voice that carried across the room. By the time he came to Jörg's death, several people were weeping.

Lloyd himself relived the cruelty and anguish of those moments, and he was possessed by rage against such fools as Boy Fitzherbert, whose infatuation with marching songs and smart uniforms threatened to bring the same torment to England.

Robert sat down and Ethel went to the lectern. As she began to speak, Ruby reappeared, looking furious. 'I told you this wouldn't work!' she hissed in Lloyd's ear. 'Mosley has gone, but the boys are singing "Rule Britannia" outside the station.'

That certainly was a breach of the agreement, Lloyd thought angrily. Boy had broken his promise. So much for the word of an English gentleman.

Ethel was explaining how Fascism offered false solutions, simplistically blaming groups such as Jews and Communists for complex problems such as unemployment and crime. She made merciless fun of the concept of the triumph of the will, likening the Führer and the Duce to playground bullies. They claimed popular support, but banned all opposition.

Lloyd realized that when the Fascists returned from the railway station to the centre of town they would have to pass this church. He began to listen to the sounds coming through the open windows. He could hear cars and lorries growling along Hills Road, punctuated now and again by the trill of a bicycle bell or the cry of a child. He thought he heard a distant shout, and it sounded ominously like the noise made by rowdy boys young enough still to be proud of their new, deep voices.

He tensed, straining to hear, and there were more shouts. The Fascists were marching.

Ethel raised her own voice as the bellowing outside got louder. She argued that working people of all kinds needed to band together in trade unions and the Labour Party to build a fairer society step by democratic step, not through the kind of violent upheaval that had gone so badly wrong in Communist Russia and Nazi Germany.

Ruby re-entered. 'They're marching up Hills Road now,' she said in a low, urgent murmur. 'We have to go out there and confront them!'

'No!' Lloyd whispered. 'The party made a collective decision – no demonstration. We must stick to that. We must be a disciplined movement!' He knew the reference to party discipline would carry weight with her.

The Fascists were nearby now, raucously chanting. Lloyd guessed there must be fifty or sixty of them. He itched to go out there and face them. Two young men near the back stood up and went to the windows to look out. Ethel urged caution. 'Don't react to hooliganism by becoming a hooligan,' she said. 'That will only give the newspapers an excuse to say that one side is as bad as the other.'

There was a crash of breaking glass, and a stone came through the window. A woman screamed, and several people got to their feet. 'Please remain seated,' Ethel said. 'I expect they will go away in a minute.' She talked on in a calm and reassuring voice. Few people attended to her speech. Everyone was looking backwards, towards the church door, and listening to the hoots and jeers of the ruffians outside. Lloyd had to struggle to sit still. He looked towards his mother with a neutral expression fixed like a mask on his face. Every bone in his body wanted to rush outside and punch heads.

After a minute the audience quietened somewhat. They returned their attention to Ethel, though still fidgeting and looking back over their shoulders. Ruby muttered: 'We're like a pack of rabbits, shaking in our burrow while the fox barks outside.' Her tone was contemptuous, and Lloyd felt she was right.

But his mother's forecast proved true, and no more stones were thrown. The chanting receded.

'Why do the Fascists want violence?' Ethel asked rhetorically. 'Those out there in Hills Road might be mere hooligans, but someone is directing them, and their tactics have a purpose. When there is fighting in the streets, they can claim that public order has broken down, and drastic measures are needed to restore the rule of law. Those emergency measures will include banning democratic political parties such as Labour, prohibiting trade union action, and jailing people without trial – people such as us, peaceful men and women whose only crime is to disagree with the government. Does this sound fantastic to you, unlikely, something that could never happen? Well, they used exactly those tactics in Germany – and it worked.'

She went on to talk about how Fascism should be opposed: in discussion groups, at meetings such as this one, by writing letters to the newspapers, by using every opportunity to alert others to the danger. But even Ethel had trouble making this sound courageous and decisive.

Lloyd was cut to the quick by Ruby's talk of rabbits. He felt like a coward. He was so frustrated that he could hardly sit still.

Slowly the atmosphere in the hall returned to normal. Lloyd turned to Ruby. 'The rabbits are safe, anyhow,' he said.

'For now,' she said. 'But the fox will be back.'

(ii)

'If you like a boy, you can let him kiss you on the mouth,' said Lindy Westhampton, sitting on the lawn in the sunshine.

'And if you really like him, he can feel your breasts,' said her twin sister, Lizzie.

'But nothing below the waist.'

'Not until you're engaged.'

Daisy was intrigued. She had expected English girls to be inhibited, but she had been wrong. The Westhampton twins were sex mad.

Daisy was thrilled to be a guest at Chimbleigh, the country house of Sir Bartholomew 'Bing' Westhampton. It made her feel that she had been accepted into English society. But she still had not met the King.

She recalled her humiliation at the Buffalo Yacht Club with a sense of shame that was still like a burn on her skin, continuing to give her agonizing pain long after the flame had gone away. But whenever she felt that pain she thought about how she was going to dance with the King, and she imagined them all – Dot Renshaw, Nora Farquharson, Ursula Dewar – poring over her picture in the *Buffalo Sentinel*, reading every word of the report, envying her, and wishing that they could honestly say they had always been her friends.

Things had been difficult at first. Daisy had arrived three months ago with her mother and her friend Eva. Her father had given them a handful of introductions to people who turned out not to be the crème de la crème of London's social scene. Daisy had begun to regret her overconfident exit from the Yacht Club Ball: what if it all came to nothing?

But Daisy was determined and resourceful, and she needed no more than a foot in the door. Even at entertainments that were more or less public, such as horse races and operas, she met high-ranking people. She flirted with the men, and she piqued the curiosity of the matrons by letting them know she was rich and single. Many aristocratic English families had been ruined by the Depression, and an American heiress would have been welcome even if she were not pretty and charming. They liked her accent, they tolerated her holding her fork in her right hand, and they were amused that she could drive a car – in England men did the driving. Many English girls could ride a horse as well as Daisy, but few looked so pertly assured in the saddle. Some older women still viewed Daisy with suspicion, but she would win them around eventually, she felt sure.

Bing Westhampton had been easy to flirt with. An elfin man with a winning smile, he had an eye for a pretty girl; and Daisy knew instinctively that more than his eye would be involved if he got the chance of a twilight fumble in the garden. Clearly his daughters took after him.

The Westhamptons' house party was one of several in Cambridgeshire held to coincide with May Week. The guests included Earl Fitzherbert, known as Fitz, and his wife, Bea. She was Countess Fitzherbert, of course, but she preferred her Russian title of Princess. Their elder son, Boy, was at Trinity College.

Princess Bea was one of the social matriarchs who were doubtful about Daisy. Without actually telling a lie, Daisy had let people assume that her father was a Russian nobleman who had lost everything in the revolution, rather than a factory worker who had fled to America one step ahead of the police. But Bea was not taken in. 'I can't recall a family called Peshkov in St Petersburg or Moscow,' she had said, hardly pretending to be puzzled; and Daisy had forced herself to smile as if it was of no consequence what the princess could remember.

There were three girls the same age as Daisy and Eva: the Westhampton twins plus May Murray, the daughter of a general. The balls went on all night, so everyone slept until midday, but the afternoons were dull. The five girls lazed in the garden or strolled in the woods. Now, sitting up in her hammock, Daisy said: 'What can you do *after* you're engaged?'

Lindy said: 'You can rub his thing.'

'Until it squirts,' said her sister.

May Murray, who was not as daring as the twins, said: 'Oh, disgusting!'

That only encouraged the twins. 'Or you can suck it,' said Lindy. 'They like that best of all.'

'Stop it!' May protested. 'You're just making this up.'

They stopped, having teased May enough. 'I'm bored,' said Lindy. 'What shall we do?'

An imp of mischief seized Daisy, and she said: 'Let's come down to dinner in men's clothes.'

She regretted it immediately. A stunt like that could ruin her social career when it had only just got started.

Eva's German sense of propriety was upset. 'Daisy, you don't mean it!'

'No,' she said. 'Silly notion.'

The twins had their mother's fine blonde hair, not their father's dark curls, but they had inherited his streak of naughtiness, and they both loved the idea. 'They'll all be in tailcoats tonight, so we can steal their dinner jackets,' said Lindy.

'Yes!' said her twin. 'We'll do it while they're having tea.'

Daisy saw that it was too late to back out.

May Murray said: 'We couldn't go to the ball like that!' The whole party was to attend the Trinity Ball after dinner.

'We'll change again before leaving,' said Lizzie.

May was a timid creature, probably cowed by her military father, and she always went along with whatever the other girls decided. Eva as the only dissident was overruled, and the plan went ahead.

When the time came to dress for dinner, a maid brought two evening suits into the bedroom Daisy was sharing with Eva. The maid's name was Ruby. Yesterday she had been miserable with toothache, so Daisy had given her the money for a dentist, and she had had the tooth pulled out. Now Ruby was bright-eyed with excitement, toothache forgotten. 'Here you are, ladies!' she said. 'Sir Bartholomew's should be small enough for you, Miss Peshkov, and Mr Andrew Fitzherbert's for Miss Rothmann.'

Daisy took off her dress and put on the shirt. Ruby helped her with the unfamiliar studs and cufflinks. Then she climbed into Bing Westhampton's trousers, black with a satin stripe. She tucked her slip in and pulled the suspenders over her shoulders. She felt a bit daring as she buttoned the fly.

None of the girls knew how to knot a tie, so the results were distinctly limp. But Daisy came up with the winning touch. Using an eyebrow pencil, she gave herself a moustache. 'It's marvellous!' said Eva. 'You look even prettier!' Daisy drew side-whiskers on Eva's cheeks.

The five girls met up in the twins' bedroom. Daisy walked in with a mannish swagger that made the others giggle hysterically.

May voiced the concern that remained in the back of Daisy's mind. 'I hope we're not going to get into trouble over this.'

Lindy said: 'Oh, who cares if we do?'

Daisy decided to forget her misgivings and enjoy herself, and she led the way down to the drawing room.

They were the first to arrive, and the room was empty. Repeating something she had heard Boy Fitzherbert say to the butler, Daisy put on a man's voice and drawled: 'Pour me a whisky, Grimshaw, there's a good chap – this champagne tastes like piss.' The others squealed with shocked laughter.

Bing and Fitz came in together. Bing in his white waistcoat made Daisy think of a pied wagtail, a cheeky black-and-white bird. Fitz was a good-looking middle-aged man, his dark hair touched with grey. As a result of war wounds he walked with a slight limp, and one eyelid drooped; but this evidence of his courage in battle only made him more dashing.

Fitz saw the girls, looked twice, and said: 'Good God!' His tone was sternly disapproving.

Daisy suffered a moment of sheer panic. Had she spoiled everything? The English could be frightfully straight-laced, everyone knew that. Would she be asked to leave the house? How terrible that would be. Dot Renshaw and Nora Farquharson would crow if she went home in disgrace. She would rather die.

But Bing burst out laughing. 'I say, that's terribly good,' he said. 'Look at this, Grimshaw.'

The elderly butler, coming in with a bottle of champagne in a silver ice bucket, observed them bleakly. In a tone of withering insincerity he said: 'Most amusing, Sir Bartholomew.'

Bing continued to regard them all with a delight mingled with lasciviousness, and Daisy realized – too late – that dressing like the opposite sex might misleadingly suggest, to some men, a degree of sexual freedom and a willingness to experiment – a suggestion that could obviously lead to trouble.

As the party assembled for dinner, most of the other guests followed the lead of their host in treating the girls' prank as an amusing piece of tomfoolery, though Daisy could tell that they were not all equally charmed. Daisy's mother went pale with fright when she saw them, and sat down quickly as if she felt shaky. Princess Bea, a heavily corseted woman in her forties who might once have been pretty, wrinkled her powdered brow in a censorious frown. But Lady Westhampton was a jolly woman who reacted to life, as to her wayward husband, with a tolerant smile: she laughed heartily and congratulated Daisy on her moustache.

The boys, coming last, were also delighted. General Murray's son, Lieutenant Jimmy Murray, not as straight-laced as his father, roared with pleased laughter. The Fitzherbert sons, Boy and Andy, came in together, and it was Boy's reaction that was the most interesting of all. He stared at the girls with mesmeric fascination. He tried to cover up with jollity, haw-hawing like the other men, but it was clear he was weirdly captivated.

At dinner the twins picked up Daisy's joke and talked like men, in deep voices and hearty tones, making the others laugh. Lindy held up her wine glass and said: 'How do you like this claret, Liz?'

Lizzie replied: 'I think it's a bit thin, old boy. I've a notion Bing's been watering it, don't you know.'

All through dinner Daisy kept catching Boy staring at her. He did not resemble his handsome father, but, all the same, he was good-looking, with his mother's blue eyes. She began to feel embarrassed, as if he were ogling her breasts. To break the spell she said: 'And have you been taking exams, Boy?'

'Good Lord, no,' he said.

His father said: 'Too busy flying his plane to study much.' This was phrased as a criticism, but it sounded as if Fitz was actually proud of his elder son.

Boy pretended to be outraged. 'A slander!' he said.

Eva was mystified. 'Why are you at the university if you don't wish to study?'

Lindy explained: 'Some of the boys don't bother to graduate, especially if they're not academic types.'

Lizzie added: 'Especially if they're rich and lazy.'

'I do study!' Boy protested. 'But I don't intend actually to sit the exams. It's not as if I'm hoping to make a living as a doctor, or something.' Boy would inherit one of the largest fortunes in England when Fitz died.

And his lucky wife would be Countess Fitzherbert.

Daisy said: 'Wait a minute. Do you really have your own airplane?'

'Yes, I do. A Hornet Moth. I belong to the University Aero Club. We use a little airfield outside the town.'

'But that's wonderful! You must take me up!'

Daisy's mother said: 'Oh, dear, no!'

Boy said to Daisy: 'Wouldn't you be nervous?'

'Not a bit!'

'Then I will take you.' He turned to Olga. 'It's perfectly safe, Mrs Peshkov. I promise I'll bring her back in one piece.'

Daisy was thrilled.

The conversation moved on to this summer's favourite topic: England's stylish new King, Edward VIII, and his romance with Wallis Simpson, an American woman separated from her second husband. The London newspapers said nothing about it, except to include Mrs Simpson on lists of guests at royal events; but Daisy's mother got the American papers sent over, and they were full of speculation that Wallis would divorce Mr Simpson and marry the King.

'Completely out of the question,' said Fitz severely. 'The King is the head of the Church of England. He cannot possibly marry a divorcée.'

When the ladies retired, leaving the men to port and cigars, the girls hurried to change. Daisy decided to emphasize how very feminine she really was, and chose a ball dress of pink silk patterned with tiny flowers that had a matching jacket with puffed short sleeves.

Eva wore a dramatically simple black silk gown with no sleeves. In the past year she had lost weight, changed her hair, and learned – under Daisy's tuition – to dress in an unfussy tailored style that flattered her. Eva had become like one of the family, and Olga delighted in buying clothes for her. Daisy regarded her as the sister she had never had.

WINTER OF THE WORLD

It was still light when they all climbed into cars and carriages and drove the five miles into the town centre.

Daisy thought Cambridge was the quaintest place she had ever seen, with its winding little streets and elegant college buildings. They got out at Trinity and Daisy gazed up at the statue of its founder, King Henry VIII. When they passed through the sixteenth-century brick gatehouse, Daisy gasped with pleasure at the sight that met her eyes: a large quadrangle, its trimmed green lawn crossed by cobbled paths, with an elaborate architectural fountain in the middle. On all four sides, timeworn buildings of golden stone formed the backdrop against which young men in tailcoats danced with gorgeously dressed girls, and dozens of waiters in evening dress offered trays crowded with glasses of champagne. Daisy clapped her hands with joy: this was just the kind of thing she loved.

She danced with Boy, then Jimmy Murray, then Bing, who held her close and let his right hand drift from the small of her back down to the swell of her hips. She decided not to protest. The English band played a watery imitation of American jazz, but they were loud and fast, and they knew all the latest hits.

Night fell, and the quadrangle was illuminated with blazing torches. Daisy took a break to check on Eva, who was not so self-confident and sometimes needed to be introduced around. However, she need not have worried: she found Eva talking to a strikingly handsome student in a suit too big for him. Eva introduced him as Lloyd Williams. 'We've been talking about Fascism in Germany,' Lloyd said, as if Daisy might want to join in the discussion.

'How extraordinarily dull of you,' Daisy said.

Lloyd seemed not to hear that. 'I was in Berlin three years ago, when Hitler came to power. I didn't meet Eva then, but it turns out we have some acquaintances in common.'

Jimmy Murray appeared and asked Eva to dance. Lloyd was visibly disappointed to see her go, but summoned his manners and graciously asked Daisy, and they moved closer to the band. 'What an interesting person your friend Eva is,' he said.

'Why, Mr Williams, that's what every girl longs to hear from her dancing partner,' Daisy replied. As soon as the words were out of her mouth she regretted sounding shrewish.

But he was amused. He grinned and said: 'Dear me, you're so right. I am justly reproved. I must try to be more gallant.'

She immediately liked him better for being able to laugh at himself. It showed confidence.

He said: 'Are you staying at Chimbleigh, like Eva?'

'Yes.'

'Then you must be the American who gave Ruby Carter the money for the dentist.'

'How on earth do you know about that?'

'She's a friend of mine.'

Daisy was surprised. 'Do many undergraduates befriend housemaids?'

'My goodness, what a snobbish thing to say! My mother was a housemaid, before she became a Member of Parliament.'

Daisy felt herself blush. She hated snobbery and often accused others of it, especially in Buffalo. She thought she was totally innocent of such unworthy attitudes. 'I've got off on the wrong foot with you, haven't I?' she said as the dance came to an end.

'Not really,' he said. 'You think it's dull to talk about Fascism, yet you take a German refugee into your home and even invite her to travel to England with you. You think housemaids have no right to be friends with undergraduates, yet you pay for Ruby to see the dentist. I don't suppose I'll meet another girl half as intriguing as you tonight.'

'I'll take that as a compliment.'

'Here comes your Fascist friend, Boy Fitzherbert. Do you want me to scare him off?' Daisy sensed that Lloyd would relish the chance of a quarrel with Boy.

'Certainly not!' she said, and turned to smile at Boy.

Boy nodded curtly to Lloyd. 'Evening, Williams.'

'Good evening,' said Lloyd. 'I was disappointed that your Fascists marched along Hills Road last Saturday.'

'Ah, yes,' Boy said. 'They got a bit over-enthusiastic.'

'It surprised me, when you had given your word that they would not.' Daisy saw that Lloyd was angry about this, underneath his mask of cool courtesy.

Boy refused to take it seriously. 'Sorry about that,' he said lightly. He turned to Daisy. 'Come and see the library,' he said to her. 'It's by Christopher Wren.'

'With pleasure!' Daisy said. She waved goodbye to Lloyd and let Boy take her arm. Lloyd looked disappointed to see her go, which pleased her.

On the west side of the quadrangle a passage led to a courtyard with a single elegant building at the far end. Daisy admired the cloisters on the ground floor. Boy explained that the books were on the upper floor, because the River Cam was liable to flood. 'Let's go and look at the river,' he said. 'It's pretty at night.'

Daisy was twenty years old and, though she was inexperienced, she knew that Boy did not really care for gazing on rivers at night. But she wondered, after his reaction to seeing her in men's clothing, whether he might really prefer boys to girls. She guessed she was about to find out.

'Do you actually know the King?' she asked as he led her across a second courtyard.

'Yes. He's more my father's friend, obviously, but he comes to our house sometimes. And he's jolly keen on some of my political ideas, I can tell you.'

'I'd love to meet him.' She was sounding naive, she knew, but this was her chance and she was not going to miss it.

They passed through a gateway and emerged on to a smooth lawn sloping down to a narrow walled-in river. 'This area is called the Backs,' Boy said. 'Most of the older colleges own the fields on the other side of the water.' He put his arm around her waist as they approached a little bridge. His hand moved up, as if accidentally, until his forefinger lay along the underside of her breast.

At the far end of the little bridge two college servants in uniform stood guard, presumably to repel gatecrashers. One of the men murmured: 'Good evening, Viscount Aberowen,' and the other smothered a grin. Boy responded with a barely perceptible nod. Daisy wondered how many other girls he had led across this bridge.

She knew Boy had a motive for giving her this tour and, sure enough, he stopped in the darkness and put his hands on her shoulders. 'I say, you looked jolly fetching in that outfit at dinner.' His voice was throaty with excitement.

'I'm glad you thought so.' She knew the kiss was coming, and she felt aroused at the prospect, but she was not quite ready. She put a hand on his shirt front, palm flat, holding him at a distance. 'I really want to be presented at the royal court,' she said. 'Is it difficult to arrange?'

'Not difficult at all,' he said. 'Not for my family, at least. And not for someone as pretty as you.' He dipped his head eagerly towards hers.

She leaned away. 'Would you do that for me? Will you fix it for me to be presented?'

'Of course.'

She moved in closer, and felt the erection bulging at the front of his trousers. No, she thought, he doesn't prefer boys. 'Promise?' she said.

'I promise,' he said breathlessly.

'Thank you,' she said, then she let him kiss her.

(iii)

The little house in Wellington Row, Aberowen, South Wales, was crowded at one o'clock on Saturday afternoon. Lloyd's grandfather sat at the kitchen table looking proud. On one side he had his son, Billy Williams, a coal miner who had become Member of Parliament for Aberowen. On the other was his grandson, Lloyd, the Cambridge University student. Absent was his daughter, also a Member of Parliament. It was the Williams dynasty. No one here would ever say that – the notion of a dynasty was undemocratic, and these people believed in democracy the way the Pope believed in God – but, just the same, Lloyd suspected Granda was thinking it.

Also at the table was Uncle Billy's lifelong friend and agent, Tom Griffiths. Lloyd was honoured to sit with such men. Granda was a veteran of the miners' union; Uncle Billy had been court-martialled in 1919 for revealing Britain's secret war against the Bolsheviks; Tom had fought alongside Billy at the Battle of the Somme. This was more impressive than dining with royalty.

Lloyd's grandmother, Cara Williams, had served them stewed beef with home-made bread, and now they sat drinking tea and smoking. Friends and neighbours had come in, as they always did when Billy was here, and half a dozen of them stood leaning against the walls, smoking pipes and hand-rolled cigarettes, filling the little kitchen with the smell of men and tobacco.

Billy had the short stature and broad shoulders of many miners but, unlike the others, he was well dressed, in a navy-blue suit with a clean white shirt and a red tie. Lloyd noticed that they all used his first name often, as if to emphasize that he was one of them, empowered by their votes. They called Lloyd 'boyo', making it clear that they were not over-impressed by a university student. But they addressed Granda as Mr Williams: he was the one they truly respected.

Through the open back door Lloyd could see the slag heap from the mine, an ever-growing mountain which had now reached the lane behind the house.

Lloyd was spending the summer vacation as a low-paid organizer at a camp for unemployed colliers. Their project was to refurbish the Miners' Institute Library. Lloyd found the physical work of sanding and painting and building shelves a refreshing change from reading Schiller in German and Molière in French. He enjoyed the banter among the men: he had inherited from his mother a love of the Welsh sense of humour.

It was great, but it was not fighting Fascism. He winced every time he remembered how he had skulked in the Baptist chapel while Boy Fitzherbert and the other bullies chanted in the street and threw stones through the window. He wished he had gone outside and punched someone. It might have been stupid but he would have felt better. He thought about it every night before falling asleep.

He also thought about Daisy Peshkov in a pink silk jacket with puffed sleeves.

He had seen Daisy a second time in May Week. He had gone to a recital in the chapel of King's College, because the student in the room next to his at Emmanuel was playing the cello; and Daisy had been in the audience with the Westhamptons. She had been wearing a straw hat with a turned-up brim that made her look like a naughty schoolgirl. He had sought her out afterwards, and asked her questions about America, where he had never been. He wanted to know about President Roosevelt's administration, and whether it had any lessons to teach Britain, but all Daisy talked about was tennis parties and polo matches and yacht clubs. Despite that, he had been captivated by her all over again. He liked her gay chatter all the more because it was punctuated, now and again, by unexpected darts of sarcastic wit. He had said: 'I don't want to keep you from your friends – I just wanted to ask about the New Deal', and she had replied: 'Oh, boy, you really know how to flatter a girl.' But then, as they parted, she had said: 'Call me when you come to London – Mayfair two four three four.'

Today he had come to his grandparents' house for the midday meal, on his way to the railway station. He had a few days off from the work camp, and he was taking the train to London for a short break. He was vaguely hoping he might run into Daisy, as if London were a little town like Aberowen.

At the camp he was in charge of political education, and he told his grandfather he had organized a series of lectures by left-wing dons from Cambridge. 'I tell them it's their chance to get out of the ivory tower and meet the working class, and they find it hard to refuse me.'

Granda's pale-blue eyes looked down his long, sharp nose. 'I hope our lads teach them a thing or two about the real world.'

Lloyd pointed to Tom Griffiths's son, standing in the open back door and listening. At sixteen, Lenny already had the characteristic Griffiths shadow of a black beard that never went away even when his cheeks were freshly shaved. 'Lenny had an argument with a Marxist lecturer.'

'Good for you, Len,' said Granda. Marxism was popular in South Wales, which was sometimes jokingly called Little Moscow, but Granda had always been fiercely anti-Communist.

Lloyd said: 'Tell Granda what you said, Lenny.'

Lenny grinned and said: 'In 1872 the anarchist leader Mikhail Bakunin warned Karl Marx that Communists in power would be as oppressive as the aristocracy they replaced. After what has happened in Russia, can you honestly say Bakunin was wrong?'

Granda clapped his hands. A good debating point had always been relished around his kitchen table.

Lloyd's grandmother poured him a fresh cup of tea. Cara Williams was grey, lined and bent, like all the women of her age in Aberowen. She asked Lloyd: 'Are you courting yet, my lovely?'

The men grinned and winked.

Lloyd blushed. 'Too busy studying, Grandmam.' But an image of Daisy Peshkov came into his mind, together with the phone number: Mayfair two four three four.

His grandmother said: 'Who's this Ruby Carter, then?'

The men laughed, and Uncle Billy said: 'Caught out, boyo!'

Lloyd's mother had obviously been talking. 'Ruby is membership officer of my local Labour Party in Cambridge, that's all,' Lloyd protested.

Billy said sarcastically: 'Oh, aye, very convincing', and the men laughed again.

'You wouldn't want me to go out with Ruby, Grandmam,' Lloyd said. 'You'd think she wears her clothes too tight.'

'She doesn't sound very suitable,' Cara said. 'You're a university man, now. You must set your sights higher.'

She was just as snobbish as Daisy, Lloyd perceived. 'There's nothing wrong with Ruby Carter,' he said. 'But I'm not in love with her.'

'You must marry an educated woman, a schoolteacher or a trained nurse.'

The trouble was that she was right. Lloyd liked Ruby, but he would never love her. She was pretty enough, and intelligent, too, and Lloyd was as vulnerable as the next man to a curvy figure, but still he knew she was not right for him. Worse, Grandmam had put her wrinkled old finger precisely on the reason: Ruby's outlook was restricted, her horizons narrow. She was not exciting. Not like Daisy.

'That's enough women's chatter,' Granda said. 'Billy, tell us the news from Spain.'

'It's bad,' said Billy.

All Europe was watching Spain. The left-wing government elected last February had suffered an attempted military coup backed by Fascists and conservatives. The rebel general, Franco, had won support from the Catholic Church. The news had struck the rest of the continent like an earthquake. After Germany and Italy, would Spain, too, fall under the curse of Fascism?

'The revolt was botched, as you probably know, and it almost failed,' Billy went on. 'But Hitler and Mussolini came to the rescue, and saved the insurrection by airlifting thousands of rebel troops from north Africa as reinforcements.'

Lenny put in: 'And the unions saved the government!'

'That's true,' Billy said. 'The government was slow to react, but the trade unions led the way in organizing workers and arming them with weapons they had seized from military arsenals, ships, gun shops, and anywhere else they could find them.'

Granda said: 'At least someone is fighting back. Until now the Fascists have had it all their own way. In the Rhineland and Abyssinia, they just walked in and took what they wanted. Thank God for the Spanish people, I say. They've got the guts to say no.'

There was a murmur of agreement from the men around the walls.

Lloyd again recalled that Saturday afternoon in Cambridge. He, too, had let the Fascists have it all their own way. He seethed with frustration.

'But can they win?' said Granda. 'Weapons seem to be the issue now, isn't it?'

'Aye,' said Billy. 'The Germans and the Italians are supplying the rebels with guns and ammunition, as well as fighter planes and pilots. But no one is helping the elected Spanish government.'

'And why the bloody hell not?' said Lenny angrily.

Cara looked up from the cooking range. Her dark Mediterranean eyes flashed disapproval, and Lloyd thought he glimpsed the beautiful girl she had once been. 'None of that language in my kitchen!' she said.

'Sorry, Mrs Williams.'

'I can tell you the inside story,' Billy said, and the men went quiet, listening. 'The French Prime Minister, Leon Blum – a socialist, as you know – was all set to help. He's already got one Fascist neighbour, Germany, and the last thing he wants is a Fascist regime on his southern border too. Sending arms to the Spanish government would enrage the French right wing, and French Catholic socialists too, but Blum could withstand that, especially if he had British support and could say that arming the government was an international initiative.'

Granda said: 'So what went wrong?'

'Our government talked him out of it. Blum came to London and our Foreign Secretary, Anthony Eden, told him we would not support him.'

Granda was angered. 'Why does he need support? How can a socialist prime minister let himself be bullied by the conservative government of another country?'

'Because there's a danger of a military coup in France, too,' said Billy. 'The press there is rabidly right wing, and they're whipping their own Fascists into a frenzy. Blum can fight them off with British support – but perhaps not without.'

'So it's our Conservative government being soft on Fascism again!'

'All those Tories have investments in Spain – wine, textiles, coal, steel – and they're afraid the left-wing government will expropriate them.'

'What about America? They believe in democracy. Surely they'll sell guns to Spain?'

'You'd think so, wouldn't you? But there's a well-financed Catholic lobby, led by a millionaire called Joseph Kennedy, opposing any help to the Spanish government. And a Democratic president needs Catholic support. Roosevelt won't do anything to jeopardize his New Deal.'

'Well, there's something we can do,' said Lenny Griffiths, and a look of adolescent defiance came over his face.

'What's that, Len boy?' said Billy.

'We can go to Spain and fight.'

His father said: 'Don't talk daft, Lenny.'

'Lots of people are talking about going, all over the world, even in

America. They want to form volunteer units to fight alongside the regular army.'

Lloyd sat upright. 'Do they?' This was the first he had heard of it. 'How do you know?'

'I read about it in the *Daily Herald*.'

Lloyd was electrified. Volunteers going to Spain to fight the Fascists!

Tom Griffiths said to Lenny: 'Well, you're not going, and that's that.'

Billy said: 'Remember those boys who lied about their age to fight in the Great War? Thousands of them.'

'And totally useless, most of them,' Tom said. 'I recall that kid who cried before the Somme. What was his name, Billy?'

'Owen Bevin. He ran away, didn't he?'

'Aye – to a firing squad. The bastards shot him for desertion. Fifteen, he was, poor little tyke.'

Lenny said: 'I'm sixteen.'

'Aye,' said his father. 'Big difference, that.'

Granda said: 'Lloyd here is going to miss the train to London in about ten minutes.'

Lloyd had been so struck by Lenny's revelation that he had not kept an eye on the clock. He jumped up, kissed his grandmother, and picked up his small suitcase.

Lenny said: 'I'll walk with you to the station.'

Lloyd said his goodbyes and hurried down the hill. Lenny said nothing, seeming preoccupied. Lloyd was glad not to have to talk: his mind was in turmoil.

The train was in. Lloyd bought a third-class ticket to London. As he was about to board, Lenny said: 'Tell me, now, Lloyd, how do you get a passport?'

'You're serious about going to Spain, aren't you?'

'Come on, man, don't muck about, I want to know.'

The whistle blew. Lloyd climbed aboard, closed the door, and let down the window. 'You go to the Post Office and ask for a form,' he said.

Lenny said despondently: 'If I went to the Aberowen post office and asked for a passport form, my mother would hear of it about thirty seconds later.'

'Then go to Cardiff,' said Lloyd; and the train pulled away.

He settled in his seat and took from his pocket a copy of *Le Rouge et le Noir* by Stendhal in French. He stared at the page without taking anything in. He could think of only one idea: going to Spain.

He knew he should be scared, but all he felt was excitement at the prospect of fighting – really fighting, not just holding meetings – against the kind of men who had set the dogs on Jörg. No doubt fear would come later. Before a boxing match he was not scared in the dressing room. But when he entered the ring and saw the man who wanted to beat him unconscious, looked at the muscular shoulders and the hard fists and the vicious face, then his mouth went dry and his heart pounded and he had to suppress the impulse to turn and run away.

Right now he was mainly worried about his parents. Bernie was so proud of having a stepson at Cambridge – he had told half the East End – and he would be devastated if Lloyd left before getting his degree. Ethel would be frightened that her son might be wounded or killed. They would both be terribly upset.

There were other issues. How would he get to Spain? What city would he go to? How would he pay the fare? But only one snag really gave him pause.

Daisy Peshkov.

He told himself not to be ridiculous. He had met her twice. She was not even very interested in him. That was smart of her, because they were ill-suited. She was a millionaire's daughter and a shallow socialite who thought talking about politics was dull. She liked men such as Boy Fitzherbert: that alone proved she was wrong for Lloyd. Yet he could not get her out of his mind, and the thought of going to Spain and losing all chance of seeing her again filled him with sadness.

Mayfair two four three four.

He felt ashamed of his hesitation, especially when he recalled Lenny's simple determination. Lloyd had been talking about fighting Fascism for years. Now there was a chance to do it. How could he not go?

He reached London's Paddington Station, took the Tube to Aldgate, and walked to the row house in Nutley Street where he had been born. He let himself in with his own key. The place had not changed much since he was a child, but one innovation was the telephone on a little table next to the hat stand. It was the only phone in the street, and the neighbours treated it as public property. Beside the phone was a box in which they placed the money for their calls.

His mother was in the kitchen. She had her hat on, ready to go out to address a Labour Party meeting – what else? – but she put the kettle on and made him tea. 'How are they all in Aberowen?' she asked.

'Uncle Billy is there this weekend,' he said. 'All the neighbours came into Granda's kitchen. It's like a medieval court.'

'Are your grandparents well?'

'Granda is the same as ever. Grandmam looks older.' He paused. 'Lenny Griffiths wants to go to Spain, to fight the Fascists.'

She pursed her lips in disapproval. 'Does he, now?'

'I'm considering going with him. What do you think?'

He was expecting opposition, but even so her reaction surprised him. 'Don't you bloody dare,' she said savagely. She did not share her mother's aversion to swear words. 'Don't even speak of it!' She slammed the teapot down on the kitchen table. 'I bore you in pain and suffering, and raised you, and put shoes on your feet and sent you to school, and I didn't go through all that for you to throw your life away in a bloody war!'

He was taken aback. 'I wasn't thinking of throwing my life away,' he said. 'But I might risk it in a cause you brought me up to believe in.'

To his astonishment she began to sob. She rarely cried – in fact, Lloyd could not remember the last time.

'Mother, don't.' He put his arm around her shaking shoulders. 'It hasn't happened yet.'

Bernie came into the kitchen, a stocky middle-aged man with a bald dome. 'What's all this?' he said. He looked a bit scared.

Lloyd said: 'I'm sorry, Dad, I've upset her.' He stepped back and let Bernie put his arms around Ethel.

She wailed: 'He's going to Spain! He'll be killed!'

'Let's all calm down and discuss it sensibly,' Bernie said. He was a sensible man wearing a sensible dark suit and much-repaired shoes with sensible thick soles. No doubt that was why people voted for him: he was a local politician, representing Aldgate on the London County Council. Lloyd had never known his own father, but he could not imagine loving a real father more than he loved Bernie, who had been a gentle stepfather, quick to comfort and advise, slow to command or punish. He treated Lloyd no differently from his daughter, Millie.

Bernie persuaded Ethel to sit at the kitchen table, and Lloyd poured her a cup of tea.

'I thought my brother was dead, once,' Ethel said, her tears still flowing. 'The telegrams came to Wellington Row, and the wretched boy from the post office had to go from one house to the next, giving men and women the bits of paper that said their sons and husbands were dead. Poor lad, what was his name? Geraint, I think. But he didn't have a telegram for our house and, wicked woman that I am, I thanked God it was others that had died and not our Billy!'

'You're not a wicked woman,' Bernie said, patting her.

Lloyd's half-sister, Millie, appeared from upstairs. She was sixteen, but looked older, especially dressed as she was this evening, in a stylish black outfit and small gold earrings. For two years she had worked in a women's wear shop in Aldgate, but she was bright and ambitious, and in the last few days she had got a job in a swanky West End department store. She looked at Ethel and said: 'Mam, what's the matter?' She spoke with a Cockney accent.

'Your brother wants to go to Spain and get himself killed!' Ethel cried.

Millie looked accusingly at Lloyd. 'What have you been saying to her?' Millie was always quick to find fault with her older brother, whom she felt was undeservedly adored.

Lloyd responded with fond tolerance. 'Lenny Griffiths from Aberowen is going to fight the Fascists, and I told Mam I was thinking about going with him.'

'Trust you,' Millie said disgustedly.

'I doubt if you can get there,' said Bernie, ever practical. 'After all, the country is in the middle of a civil war.'

'I can get a train to Marseilles. Barcelona's not far from the French border.'

'Eighty or ninety miles. And it's a cold walk over the Pyrenees.'

'There must be ships going from Marseilles to Barcelona. It's not so far by sea.'

'True.'

'Stop it, Bernie!' Ethel cried. 'You sound as if you're discussing the quickest way to Piccadilly Circus. He's talking about going to war! I won't allow it.'

'He's twenty-one, you know,' Bernie said. 'We can't stop him.'

'I know how bloody old he is!'

Bernie looked at his watch. 'We need to get to the meeting. You're the main speaker. And Lloyd's not going to Spain tonight.'

'How do you know?' she said. 'We might get home and find a note saying he's caught the boat train to Paris!'

'I tell you what,' said Bernie. 'Lloyd, promise your mother you won't go for a month at least. It's not a bad idea anyway – you need to check the lie of the land before you rush off. Set her mind at ease, just temporarily. Then we can talk about it again.'

It was a typical Bernie compromise, calculated to let everyone back off without backing down; but Lloyd was reluctant to make a commitment. On the other hand, he probably could not simply jump on a train. He had to find out what arrangements the Spanish government might be making to receive volunteers. Ideally, he would go in company with Lenny and others. He would need visas, foreign currency, a pair of boots . . . 'All right,' he said. 'I won't go for a month.'

'Promise,' his mother said.

'I promise.'

Ethel became calm. After a minute she powdered her face and looked more normal. She drank her tea.

Then she put her coat on, and she and Bernie left.

'Right, I'm off too,' said Millie.

'Where are you going?' Lloyd asked her.

'The Gaiety.'

It was a music hall in the East End. 'Do they let sixteen-year-olds in?'

She gave him an arch look. 'Who's sixteen? Not me. Anyway, Dave's going and he's only fifteen.' She was speaking of their cousin David Williams, son of Uncle Billy and Aunt Mildred.

'Well, enjoy yourselves.'

She went to the door and came back. 'Just don't get killed in Spain, you stupid sod.' She put her arms around him and hugged him hard, then went out without saying any more.

When he heard the front door slam, he went to the phone.

He did not have to think to recall the number. He could see Daisy in his mind's eye, turning as she left him, smiling winningly under the straw hat, saying: 'Mayfair two four three four.'

He picked up the phone and dialled.

What was he going to say? 'You told me to phone, so here I am.' That was feeble. The truth? 'I don't admire you at all, but I can't get you off my mind.' He should invite her to something, but what? A Labour Party meeting?

A man answered. 'This is Mrs Peshkov's residence. Good evening.'

The deferential tone made Lloyd think he was a butler. No doubt Daisy's mother had rented a London house complete with staff.

'This is Lloyd Williams . . .' He wanted to say something that would explain or justify his call, and he added the first thing that came to mind: '. . . of Emmanuel College.' It meant nothing but he hoped it sounded impressive. 'May I speak to Miss Daisy Peshkov?'

'No, I'm sorry, Professor Williams,' said the butler, assuming Lloyd must be a don. 'They've all gone to the opera.'

Of course, Lloyd thought with disappointment. No socialite was home at this time of the evening, especially on a Saturday. 'I remember,' he lied. 'She told me she was going, and I forgot. Covent Garden, isn't it?' He held his breath.

But the butler was not suspicious. 'Yes, sir. *The Magic Flute*, I believe.'

'Thank you.' Lloyd hung up.

He went to his room and changed. In the West End most people wore evening dress, even to go to the cinema. But what would he do when he got there? He could not afford a ticket to the opera, and anyway it would be over soon.

He took the Tube. The Royal Opera House was incongruously located next to Covent Garden, London's wholesale fruit and vegetable market. The two institutions got along well because they kept different hours: the market opened for business at three or four o'clock in the morning, when London's most determined revellers were beginning to head for home; and it closed before the matinee.

Lloyd walked past the shuttered stalls of the market and looked through glazed doors into the opera house. Its bright lobby was empty, and he could hear muffled Mozart. He stepped inside. Adopting a careless upper-class manner, he said to an attendant: 'What time does the curtain come down?'

If he had been wearing his tweed suit he would probably have been told that it was none of his business, but the dinner jacket was the uniform of authority, and the attendant said: 'In about five minutes, sir.'

Lloyd nodded curtly. To say 'Thank you' would have given him away.

He left the building and walked around the block. It was a moment of quiet. In the restaurants, people were ordering coffee; in the cinemas, the big feature was approaching its melodramatic climax. Everything would change soon, and the streets would be thronged with people shouting for taxis, heading to nightclubs, kissing goodbye at bus stops, and hurrying for the last train back to the suburbs.

He returned to the opera house and went inside. The orchestra was silent, and the audience was just beginning to emerge. Released from long imprisonment in their seats they were talking animatedly, praising the singers, criticizing the costumes, and making plans for late suppers.

He saw Daisy almost immediately.

She was wearing a lavender dress with a little cape of champagne-coloured mink over her bare shoulders, and she looked ravishing. She emerged from the auditorium at the head of a small clutch of people her own age. Lloyd was sorry to recognize Boy Fitzherbert beside her, and to see her laugh gaily at something he murmured to her as they stepped down the red-carpeted stairs. Behind her was the interesting German girl, Eva Rothmann, escorted by a tall young man in the kind of military evening dress known as a mess kit.

Eva recognized Lloyd and smiled, and he spoke to her in German. 'Good evening, Fräulein Rothmann, I hope you enjoyed the opera.'

'Very much, thank you,' she replied in the same language. 'I didn't realize you were in the audience.'

Boy said amiably: 'I say, speak English, you lot.' He sounded slightly drunk. He was good-looking in a dissipated way, like a sulkily handsome adolescent, or a pedigree dog that is fed too many scraps. He had a pleasant manner, and probably could be devastatingly charming when he chose.

Eva said in English: 'Viscount Aberowen, this is Mr Williams.'

'We know each other,' said Boy. 'He's at Emma.'

Daisy said: 'Hello, Lloyd. We're going slumming.'

Lloyd had heard this word before. It meant going to the East End to visit low pubs and watch working-class entertainment such as dog fights.

Boy said: 'I bet Williams knows some places.'

Lloyd hesitated only a fraction of a second. Was he willing to put up with Boy in order to be with Daisy? Of course he was. 'As a matter of fact, I do,' he said. 'Do you want me to show you?'

'Splendid!'

An older woman appeared and wagged a finger at Boy. 'You must have these girls home by midnight,' she said in an American accent. 'Not a second later, please.' Lloyd guessed she must be Daisy's mother.

The tall man in the military outfit replied: 'Leave it to the army, Mrs Peshkov. We'll be on time.'

Behind Mrs Peshkov came Earl Fitzherbert with a fat woman who must be his wife. Lloyd would have liked to question the earl about his government's policy on Spain.

Two cars were waiting for them outside. The earl, his wife, and Daisy's mother got into a black-and-cream Rolls-Royce Phantom III. Boy and his group piled into the other car, a dark-blue Daimler E20 limousine, the royal family's favourite car. There were seven young people including Lloyd. Eva seemed to be with the soldier, who introduced himself to Lloyd as Lieutenant Jimmy Murray. The third girl was his sister, May, and the other boy – a slimmer, quieter version of Boy – turned out to be Andy Fitzherbert.

Lloyd gave the chauffeur directions to the Gaiety.

He noticed that Jimmy Murray discreetly slipped his arm around Eva's waist. Her reaction was to move slightly closer to him: obviously they were courting. Lloyd was happy for her. She was not a pretty girl, but she was intelligent and charming. He liked her, and he was glad she had found herself a tall soldier. He wondered, though, how others in this upper-class social set would react if Jimmy announced he was going to marry a half-Jewish German girl.

It occurred to him that the others formed two more couples: Andy and May, and – annoyingly – Boy and Daisy. Lloyd was the odd one out. Not wanting to stare at them, he studied the polished mahogany window surrounds.

The car went up Ludgate Hill to St Paul's Cathedral. 'Take Cheapside,' Lloyd said to the driver.

Boy took a long pull from a silver hip flask. Wiping his mouth, he said: 'You know your way around, Williams.'

'I live here,' said Lloyd. 'I was born in the East End.'

'How splendid,' said Boy; and Lloyd was not sure whether he was being thoughtlessly polite or unpleasantly sarcastic.

All the seats were taken at the Gaiety, but there was plenty of standing room, and the audience moved around constantly, greeting friends and going to the bar. They were dressed up, the women in brightly coloured frocks, the men in their best suits. The air was warm and smoky, and there was a powerful odour of spilled beer. Lloyd found a place for his group near the back. Their clothes identified them as visitors from the West End, but they were not the only ones: music halls were popular with all classes.

On stage a middle-aged performer in a red dress and blonde wig was doing a double-entendre routine. 'I said to him, "I'm not letting you into my passage."' The audience roared with laughter. 'He said to me, "I can see it from here, love." I told him, "You keep your nose out."' She was pretending indignation. 'He said, "It looks to me like it needs a good clean-out." Well! I ask you.'

Lloyd saw that Daisy was grinning widely. He leaned over and murmured in her ear: 'Do you realize it's a man?'

'No!' she said.

'Look at the hands.'

'Oh, my God!' she said. 'She's a man!'

Lloyd's cousin David walked past, spotted Lloyd, and came back. 'What are you all dressed up for?' he said in a Cockney accent. He was wearing a knotted scarf and a cloth cap.

'Hello, Dave, how's life?'

'I'm going to Spain with you and Lenny Griffiths,' Dave said.

'No, you're not,' said Lloyd. 'You're fifteen.'

'Boys my age fought in the Great War.'

'But they were no use – ask your father. Anyway, who says I'm going?'

'Your sister, Millie,' Dave said, and he walked on.

Boy said: 'What do people usually drink in this place, Williams?'

Lloyd thought Boy did not need any more alcohol, but he replied: 'Pints of best bitter for the men and port-and-lemon for the girls.'

'Port-and-lemon?'

'It's port diluted with lemonade.'

'How perfectly ghastly.' Boy disappeared.

The comedian reached the climax of the act. 'I said to him, "You fool, *that's the wrong passage!*"' She, or he, went off to gales of applause.

Millie appeared in front of Lloyd. 'Hello,' she said. She looked at Daisy. 'Who's your friend?'

Lloyd was glad Millie looked so pretty, in her sophisticated black dress, with a row of fake pearls and a discreet touch of make-up. He said: 'Miss Peshkov, allow me to present my sister, Miss Leckwith. Millie, this is Daisy.'

They shook hands. Daisy said: 'I'm very glad to meet Lloyd's sister.'

'Half-sister, to be exact,' said Millie.

Lloyd explained: 'My father was killed in the Great War. I never knew him. My mother married again when I was still a baby.'

'Enjoy the show,' Millie said, turning away; then, as she left, she murmured to Lloyd: 'Now I see why Ruby Carter has no chance.'

Lloyd groaned inwardly. His mother had obviously told the whole family that he was romancing Ruby.

Daisy said: 'Who's Ruby Carter?'

'She's a maid at Chimbleigh. You gave her the money to see a dentist.'

'I remember. So her name is being romantically linked with yours.'

'In the imagination of my mother, yes.'

Daisy laughed at his discomfiture. 'So you're not going to marry a housemaid.'

'I'm not going to marry Ruby.'

'She might suit you very well.'

Lloyd gave her a direct look. 'We don't always fall in love with the most suitable people, do we?'

She looked at the stage. The show was approaching its end, and the entire cast was beginning a familiar song. The audience joined in enthusiastically. The standing customers at the back linked arms and swayed in time, and Boy's party did likewise.

When the curtain came down, Boy still had not reappeared. 'I'll look for him,' Lloyd said. 'I think I know where he might be.' The Gaiety had a ladies' toilet, but the men's was a back yard with an earth closet and several halved oil drums. Lloyd found Boy puking into one of the drums.

He gave Boy a handkerchief to wipe his mouth, then took his arm and led him through the emptying theatre and outside to the Daimler limousine. The others were waiting. They all got in and Boy immediately fell asleep.

When they got back to the West End, Andy Fitzherbert told the driver to go first to the Murray house, in a modest street near Trafalgar Square. Getting out of the car with May, he said: 'You lot go on. I'll see May to her door then walk home.' Lloyd presumed that Andy was planning a romantic goodnight on May's doorstep.

They drove on to Mayfair. As the car was approaching Grosvenor Square, where Daisy and Eva were living, Jimmy told the chauffeur: 'Just stop at the corner, please.' Then he said quietly to Lloyd: 'I say, Williams, would you mind taking Miss Peshkov to the door, and I'll follow with Fräulein Rothmann in half a minute?'

'Of course.' Jimmy wanted to kiss Eva goodnight in the car, obviously. Boy would know nothing about it: he was snoring. The chauffeur would pretend to be oblivious in the expectation of a tip.

Lloyd got out of the car and handed Daisy out. When she grasped his hand he got a thrill like a mild electric shock. He took her arm and they walked slowly along the pavement. At the midpoint between two street lamps, where the light was dimmest, Daisy stopped. 'Let's give them time,' she said.

Lloyd said: 'I'm so glad Eva has a paramour.'

'Me, too.'

He took a breath. 'I can't say the same about you and Boy Fitzherbert.'

'He got me presented at court!' Daisy said. 'And I danced with the King in a nightclub – it was in all the American newspapers.'

'And that's why you're courting him?' Lloyd said incredulously.

'Not only. He likes all the things I do – parties and racehorses and beautiful clothes. He's such fun! He even has his own airplane.'

'None of that means anything,' Lloyd said. 'Give him up. Be my girlfriend instead.'

She looked pleased, but she laughed. 'You're crazy,' she said. 'But I like you.'

'I mean it,' he said desperately. 'I can't stop thinking about you, even though you're the last person in the world I should marry.'

She laughed again. 'You say the rudest things! I don't know why I talk to you. I guess I think you're nice under your clumsy manners.'

'I'm not really clumsy – only with you.'

'I believe you. But I'm not going to marry a penniless socialist.'

Lloyd had opened his heart only to be charmingly rejected, and now he felt miserable. He looked back at the Daimler. 'I wonder how long they're going to be,' he said disconsolately.

Daisy said: 'I might kiss a socialist, though, just to see what it's like.'

For a moment he did not react. He assumed she was speaking theoretically. But a girl would never say something like that theoretically. It was an invitation. He had almost been stupid enough to miss it.

He moved closer, putting his hands on her small waist. She tilted her face up, and her beauty took his breath away. He bent his head and kissed her mouth softly. She did not close her eyes, and neither did he. He felt tremendously aroused, staring into her blue eyes as he moved his lips against hers. She opened her mouth slightly, and he touched her parted lips with the tip of his tongue. A moment later he felt her tongue respond. She was still looking at him. He was in paradise, and he wanted to stay locked in this embrace for all eternity. She pressed her body to his. He had an erection, and he was embarrassed in case she might feel it, so he eased back — but she pushed forward again, and he understood, looking into her eyes, that she wanted to feel his penis pressed against her soft body. The realization heated him unbearably. He felt as if he was going to ejaculate, and it occurred to him that she might even want him to.

Then he heard the door of the Daimler open, and Jimmy Murray speaking with slightly unnatural loudness, as if giving a warning. Lloyd broke the embrace with Daisy.

'Well,' she murmured in a surprised tone, 'that was an unexpected pleasure.'

Lloyd said hoarsely: 'More than a pleasure.'

Then Jimmy and Eva were beside them, and they all walked to the door of Mrs Peshkov's house. It was a grand building with steps up to a covered porch. Lloyd wondered if the porch might give shelter enough for another kiss, but as they climbed the steps the door was opened from the inside by a man in evening dress, probably the butler Lloyd had spoken to earlier. How glad he was that he had made that phone call!

The two girls said goodnight demurely, giving no hint that only seconds ago they both had been locked in passionate embraces; then the door closed and they were gone.

Lloyd and Jimmy went back down the steps.

'I'm going to walk from here,' Jimmy said. 'Shall I tell the chauffeur to drive you back to the East End? You must be three or four miles from home. And Boy won't care — he'll sleep until breakfast-time, I should think.'

'That's thoughtful of you, Murray, and I appreciate it; but, believe it or not, I feel like walking. Lots to think about.'

'As you wish. Goodnight, then.'

'Goodnight,' said Lloyd; and, with his mind in a whirl and his erection slowly deflating, he turned east and headed for home.

(iv)

London's social season ended in the middle of August, and still Boy Fitzherbert had not proposed marriage to Daisy Peshkov.

Daisy was hurt and puzzled. Everyone knew they were courting. They saw one another almost every day. Earl Fitzherbert talked to Daisy like a daughter, and even the suspicious Princess Bea had warmed to her. Boy kissed her whenever he got the chance, but said nothing about the future.

The long series of lavish lunches and dinners, glittering parties and balls, traditional sporting events and champagne picnics that made up the London season came to an abrupt end. Many of the new friends Daisy had made suddenly left town. Most of them went to country houses where, as far as she could gather, they would spend their time hunting foxes, stalking deer, and shooting birds.

Daisy and Olga stayed for Eva Rothmann's wedding. Unlike Boy, Jimmy Murray was in a rush to marry the woman he loved. The ceremony was held at his parents' parish church in Chelsea.

Daisy felt she had done a great job with Eva. She had taught her friend how to choose clothes that suited her, smart styles without frills, in plain, strong colours that flattered her dark hair and brown eyes. Gaining in confidence, Eva had learned how to use her natural warmth and quick intelligence to charm men and women. And Jimmy had fallen in love with her. He was no movie star, but he was tall and craggily attractive. He came from a military family with a modest fortune, so Eva would be comfortable, though not rich.

The British were as prejudiced as anyone else, and at first General Murray and Mrs Murray had not been thrilled at the prospect of their son marrying a half-Jewish German refugee. Eva had won them over quickly, but many of their friends still expressed coded doubts. At the wedding Daisy had been told that Eva was 'exotic', Jimmy was 'courageous', and the Murrays were 'marvellously broad-minded', all ways of making the best of an unsuitable match.

Jimmy had written formally to Dr Rothmann in Berlin, and had received permission to ask Eva for her hand in marriage; but the German authorities had refused to let the Rothmann family come to the wedding. Eva had said tearfully: 'They hate Jews so much, you'd think they'd be happy to see them leave the country!'

Boy's father, Fitz, had heard this remark, and had later spoken to Daisy about it. 'Tell your friend Eva not to say too much about Jews, if she can avoid it,' he had said, in the tone of one who gives a friendly warning. 'Having a half-Jewish wife is not going to help Jimmy's army career, you know.' Daisy had not passed on this unpleasant counsel.

The happy couple went off to Nice for their honeymoon. Daisy realized with a pang of guilt that she was relieved to get Eva off her hands. Boy and his political pals disliked Jews so much that Eva was becoming a problem. Already the friendship between Boy and Jimmy had ended – Boy had refused to be Jimmy's best man.

After the wedding, Daisy and Olga were invited by the Fitzherberts to a shooting party at their country house in Wales. Daisy's hopes rose. Now that Eva was out of the way, there was nothing to stop Boy proposing. The earl and princess must surely assume he was on the point of it. Perhaps they planned for him to do so this weekend.

Daisy and Olga went to Paddington station on a Friday morning and took a train west. They crossed the heart of England, rich rolling farmland dotted with hamlets, each with its stone church spire rising from a stand of ancient trees. They had a first-class carriage to themselves, and Olga asked Daisy what she thought Boy might do. 'He must know I like him,' Daisy said. 'I've let him kiss me enough times.'

'Have you shown any interest in anyone else?' her mother asked shrewdly.

Daisy suppressed the guilty memory of that brief moment of foolishness with Lloyd Williams. Boy could not possibly know about that and, anyway, she had not seen Lloyd again, nor had she replied to the three letters he had sent her. 'No one,' she said.

'Then it's because of Eva,' said Olga. 'And now she's gone.'

The train went through a long tunnel under the estuary of the River Severn, and when it emerged, they were in Wales. Bedraggled sheep grazed the hills, and in the cleft of each valley was a small mining town, its pithead winding gear rising from a scatter of ugly industrial buildings.

Earl Fitzherbert's black-and-cream Rolls-Royce was waiting for them at Aberowen station. The town was dismal, Daisy thought, with small grey stone houses in rows along the steep hillsides. They drove a mile or so out of town to the house, Tŷ Gwyn.

Daisy gasped with pleasure as they passed through the gates. Tŷ Gwyn was enormous and elegant, with long rows of tall windows in a perfectly classical façade. It was set in elaborate gardens of flowers, shrubs and specimen trees that clearly were the pride of the earl himself. What a joy it would be to be mistress of this house, she thought. The British aristocracy might no longer rule the world, but they had perfected the art of living, and Daisy longed to be one of them.

Tŷ Gwyn meant White House, but the place was actually grey, and Daisy learned why when she touched the stonework with her hand and got coal dust on her fingertips.

She was given a room called the Gardenia Suite.

That evening, she and Boy sat on the terrace before dinner and watched the sun go down over the purple mountaintop, Boy smoking a cigar and Daisy sipping champagne. They were alone for a while, but Boy said nothing about marriage.

Over the weekend her anxiety grew. Boy had plenty more chances to speak to her alone – she made sure of that. On Saturday the men went shooting, but Daisy went out to meet them at the end of the afternoon, and she and Boy walked back through the woods together. On Sunday morning the Fitzherberts and most of their guests went to the Anglican church in the town. After the service, Boy took Daisy to a pub called The Two Crowns, where squat, broad-shouldered miners in flat caps stared at her in her lavender cashmere coat as if Boy had brought in a leopard on a leash.

She told him that she and her mother would soon have to go back to Buffalo, but he did not take the hint.

Could it simply be that he liked her, but not enough to marry her?

By lunch on Sunday she was desperate. Tomorrow she and her mother were to return to London. If Boy had not proposed by then, his parents would begin to think he was not serious, and there would be no more invitations to Tŷ Gwyn.

That prospect frightened Daisy. She had made up her mind to marry Boy. She wanted to be Viscountess Aberowen, and then one day Countess Fitzherbert. She had always been rich, but she craved the respect and deference that went with social status. She longed to be addressed as 'Your Ladyship'. She coveted Princess Bea's diamond tiara. She wanted to count royalty among her friends.

She knew Boy liked her, and there was no doubt about his desire when he kissed her. 'He needs something to spur him on,' Olga murmured to Daisy as they drank their after-lunch coffee with the other ladies in the morning room.

'But what?'

'There is one thing that never fails with men.'

Daisy raised her eyebrows. 'Sex?' She and her mother talked about most things, but generally skirted around this subject.

'Pregnancy would do it,' Olga said. 'But that only happens for sure when you *don't* want it.'

'What, then?'

'You need to give him a glimpse of the Promised Land, but not let him in.'

Daisy shook her head. 'I'm not certain, but I think he may have already been to the Promised Land with someone else.'

'Who?'

'I don't know – a maid, an actress, a widow ... I'm guessing, but he just doesn't have that virginal air.'

'You're right, he doesn't. That means you have to offer him something he can't get from the others. Something he'd do anything for.'

Daisy wondered briefly where her mother had got this wisdom, having spent her life in a cold marriage. Perhaps she had done a lot of thinking about how her husband, Lev, had been stolen from her by his mistress, Marga. Anyway, there was nothing Daisy could offer Boy that he couldn't get from another girl, was there?

The women were finishing their coffee and heading to their bedrooms for the afternoon nap. The men were still in the dining room, smoking their cigars, but they would follow in a quarter of an hour. Daisy stood up.

Olga said: 'What are you going to do?'

'I'm not sure,' she said. 'I'll think of something.'

She left the room. She was going to go to Boy's room, she had decided, but she did not want to say so in case her mother objected. She would be waiting for him when he came for his nap. The servants also took a break at that time of day, so it was unlikely that anyone would come into the room.

She would have Boy on his own, then. But what would she say or do? She did not know. She would have to improvise.

She went to the Gardenia Suite, brushed her teeth, dabbed Jean Naté cologne on her neck, and walked quietly along the corridor to Boy's room.

No one saw her go in.

He had a spacious bedroom with a view of misty mountaintops. It felt as if it might have been his for many years. There were masculine leather chairs, pictures of airplanes and racehorses on the wall, a cedar wood humidor full of fragrant cigars, and a side table with decanters of whisky and brandy and a tray of crystal glasses.

She pulled open a drawer and saw Tŷ Gwyn writing paper, a bottle of ink, and pens and pencils. The paper was blue with the Fitzherbert crest. Would that one day be her crest?

She wondered what Boy would say when he found her here. Would he be pleased, take her in his arms, and kiss her? Or would he be angry that his privacy had been invaded, and accuse her of snooping? She had to take the risk.

She went into the adjoining dressing room. There was a small washbasin with a mirror over it. His shaving tackle was on the marble surround. Daisy thought she would like to learn to shave her husband. How intimate that would be.

She opened the wardrobe doors and looked at his clothes: formal morning dress, tweed suits, riding clothes, a leather pilot's jacket with a fur lining, and two evening suits.

That gave her an idea.

She recalled how aroused Boy had been, at Bing Westhampton's house back in June, by the sight of her and the other girls dressed as men. That evening had been the first time he had kissed her. She was not sure why he had been so excited – such things were generally inexplicable. Lizzie Westhampton said some men liked women to spank their bottoms: how could you account for that?

Perhaps she should dress in his clothes now.

Something he'd do anything for, her mother had said. Was this it?

She stared at the row of suits on hangers, the stack of folded white shirts, the polished leather shoes each with its wooden tree inside. Would it work? Did she have time?

Did she have anything to lose?

She could pick the clothes she needed, take them to the Gardenia Suite, change there, and then hurry back, hoping that no one saw her on the way . . .

No. There was no time for that. His cigar was not long enough. She had to change here, and fast – or not at all.

She made up her mind.

She pulled her dress off.

She was in danger now. Until this moment, she might have explained her presence here, just about plausibly, by pretending that she had lost her bearings in Tŷ Gwyn's miles of corridors and gone into the wrong room by mistake. But no girl's reputation could survive being found in a man's room in her underwear.

She took the top shirt off the pile. The collar had to be attached with a stud, she saw with a groan. She found a dozen starched collars in a drawer with a box of studs, and fixed one to the shirt, then pulled the shirt over her head.

She heard a man's heavy footsteps in the corridor outside, and froze, her heart beating like a big drum; but the steps went by.

She decided to wear formal morning dress. The striped trousers had no suspenders attached, but she found some in another drawer. She figured out how to button the suspenders to the trousers, then pulled the trousers on. The waist was big enough for two of her.

She pushed her stockinged feet into a pair of shiny black shoes and laced them.

She buttoned the shirt and put on a silver tie. The knot was wrong, but it did not matter, and, anyway, she did not know the correct way to tie it, so she left it as it was.

She put on a fawn double-breasted waistcoat and a black tailcoat, then she looked in the full-length mirror on the inside of the wardrobe door.

The clothes were baggy but she looked cute anyway.

Now that she had time, she put gold links in the shirt cuffs and a white handkerchief in the breast pocket of the coat.

Something was missing. She stared at herself in the mirror until she figured out what else she needed.

A hat.

She opened another cupboard and saw a row of hatboxes on a high shelf. She found a grey top hat and perched it on the back of her head.

She remembered the moustache.

She did not have an eyebrow pencil with her. She returned to Boy's bedroom and bent over the fireplace. It was still summer, and there was no fire. She got some soot on her fingertip, returned to the mirror, and carefully drew a moustache on her upper lip.

She was ready.

She sat in one of the leather armchairs to wait for him.

Her instinct told her she was doing the right thing, but rationally it seemed bizarre. However, there was no accounting for arousal. She herself had got wet inside when he took her up in his plane. It had been impossible for them to canoodle while he was concentrating on flying the little aircraft, and that was just as well, for soaring through the air had been so exciting that probably she would have let him do anything he wanted.

However, boys could be unpredictable, and she feared he might be angry. When that happened, his handsome face would twist into an unattractive grimace, he would tap his foot very quickly, and he could become quite cruel. Once, when a waiter with a limp had brought him the wrong drink, he had said: 'Just hobble back to the bar and bring me the Scotch I ordered – being a cripple doesn't make you deaf, does it?' The wretched man had flushed with shame.

She wondered what Boy would say to her if he was angered by her being in his room.

He arrived five minutes later.

She heard his tread outside, and realized she already knew him well enough to recognize his step.

The door opened and he came in without seeing her.

She put on a deep voice and said: 'Hello, old chap, how are you?'

He started and said: 'Good God!' Then he looked again. 'Daisy?'

She stood up. 'The same,' she said in her normal voice. He was still staring at her in surprise. She doffed the top hat, gave a little bow, and said: 'At your service.' She replaced the hat on her head at an angle.

After a long moment, he recovered from the shock and grinned.

Thank God, she thought.

He said: 'I say, that topper does suit you.'

She came closer. 'I put it on to please you.'

'Jolly nice of you, I must say.'

She turned her face up invitingly. She liked kissing him. In truth, she liked kissing most men. She was secretly embarrassed by how much she liked it. She had even enjoyed kissing girls, at her boarding school where they did not see a boy for weeks on end.

He bent his head and touched his lips to hers. Her hat fell off, and they both giggled. Quickly he thrust his tongue into her mouth. She relaxed and enjoyed it. He was enthusiastic about all sensual pleasures, and she was excited by his eagerness.

She reminded herself that she had a purpose. Things were progressing nicely, but she wanted him to propose. Would he be satisfied with just a kiss? She needed him to want more. Often, if they had more than a few hasty moments, he would fondle her breasts.

A lot depended on how much wine he had drunk with lunch. He had a large capacity, but there came a point when he lost the urge.

She moved her body, pressing herself to him. He put a hand on her chest, but she was wearing a baggy waistcoat of woollen cloth and he could not find her small breasts. He grunted in frustration.

Then his hand roamed across her stomach and inside the waistband of the loose-fitting trousers.

She had never before let him touch her down there.

She still had on a silk petticoat and substantial cotton underdrawers, so he surely could not feel much, but his hand went to the fork of her thighs and pressed firmly against her through the layers. She felt a twinge of pleasure.

She pulled away from him.

Panting, he said: 'Have I gone too far?'

'Lock the door,' she said.

'Oh, my goodness.' He went to the door, turned the key in the lock, and came back. They embraced again, and he resumed where he had left off. She touched the front of his trousers, found his erect penis through the cloth, and grasped it firmly. He groaned with pleasure.

She pulled away again.

The shadow of anger crossed his face. An unpleasant memory came back to her. Once, when she had made a boy called Theo Coffman take his hand off her breasts, he had turned nasty and called her a prick-teaser. She had never seen that boy again, but the insult had made her feel irrationally ashamed. Momentarily she feared that Boy might be about to make a similar accusation.

Then his face softened and he said: 'I am dreadfully keen on you, y'know.'

This was her moment. Sink or swim, she told herself. 'We shouldn't be doing this,' she said with a regret that was not greatly exaggerated.

'Why not?'

'We're not even engaged.'

The word hung in the air for a long moment. For a girl to say that was tantamount to a proposal. She watched his face, terrified that he would take fright, turn away, mumble excuses, and ask her to leave.

He said nothing.

'I want to make you happy,' she said. 'But . . .'

'I do love you, Daisy,' he said.

That was not enough. She smiled at him and said: 'Do you?'

'Ever such a lot.'

She said nothing, but looked at him expectantly.

At last he said: 'Will you marry me?'

'Oh, yes,' she said, and she kissed him again. With her mouth pressed to his she unbuttoned his fly, burrowed through his underclothing, found his penis, and took it out. The skin was silky and hot. She stroked it, remembering a conversation with the Westhampton twins. 'You can rub his thing,' Lindy had said, and Lizzie had added: 'Until it squirts.' Daisy was intrigued and excited by the idea of making a man do that. She grasped a bit harder.

Then she remembered Lindy's next remark. 'Or you can suck it – they like that best of all.'

She moved her lips away from Boy's and spoke into his ear. 'I'll do anything for my husband,' she said.

Then she knelt down.

(v)

It was the wedding of the year. Daisy and Boy were married at St Margaret's Church, Westminster, on Saturday 3 October 1936. Daisy was disappointed it was not Westminster Abbey, but she was told that was for the royal family only.

Coco Chanel made her wedding dress. Depression fashion was for simple lines and minimal extravagance. Daisy's floor-length bias-cut satin gown had pretty butterfly sleeves and a short train that could be carried by one pageboy.

Her father, Lev Peshkov, came across the Atlantic for the ceremony. Her mother, Olga, agreed for the sake of appearances to sit beside him in church and generally pretend that they were a more or less happily married couple. Daisy's nightmare was that at some point Marga would show up with Lev's illegitimate son Greg on her arm; but it did not happen.

The Westhampton twins and May Murray were bridesmaids, and Eva

Murray was matron of honour. Boy had been grumpy about Eva's being half Jewish – he had not wanted to invite her at all – but Daisy had insisted.

She stood in the ancient church, conscious that she looked heartbreakingly beautiful, and happily gave herself to Boy Fitzherbert body and soul.

She signed the register 'Daisy Fitzherbert, Viscountess Aberowen.' She had been practising that signature for weeks, carefully tearing the paper into unreadable shreds afterwards. Now she was entitled to it. It was her name.

Processing out of the church, Fitz took Olga's arm amiably, but Princess Bea put a yard of empty space between herself and Lev.

Princess Bea was not a nice person. She was friendly enough towards Daisy's mother, and if there was a heavy strain of condescension in her tone, Olga did not notice it, so relations were amiable. But Bea did not like Lev.

Daisy now realized that Lev lacked the veneer of social respectability. He walked and talked, ate and drank, smoked and laughed and scratched like a gangster, and he did not care what people thought. He did what he liked because he was an American millionaire, just as Fitz did what he liked because he was an English earl. Daisy had always known this, but it struck her with extra force when she saw her father with all these upper-class English people, at the wedding breakfast in the grand ballroom of the Dorchester Hotel.

But it did not matter now. She was Lady Aberowen, and that could not be taken away from her.

Nevertheless, Bea's constant hostility to Lev was an irritant, like a slightly bad smell or a distant buzzing noise, giving Daisy a feeling of dissatisfaction. Sitting beside Lev at the top table, Bea constantly turned slightly away. When he spoke to her she replied briefly without meeting his eye. He seemed not to notice, smiling and drinking champagne, but Daisy, seated on Lev's other side, knew that he had not failed to read the signs. He was uncouth, not stupid.

When the toasts were over and the men began to smoke, Lev, who as the father of the bride was paying the bill, looked along the table and said: 'Well, Fitz, I hope you enjoyed your meal. Were the wines up to your standards?'

'Very good, thank you.'

'I must say, I thought it was a damn fine spread.'

Bea tutted audibly. Men were not supposed to say 'damn' in her hearing.

Lev turned to her. He was smiling, but Daisy knew the dangerous look in his eye. 'Why, Princess, have I offended you?'

She did not want to reply, but he looked expectantly at her, and did not turn his gaze aside. At last she spoke. 'I prefer not to hear coarse language,' she said.

Lev took a cigar from his case. He did not light it at once, but sniffed it and rolled it between his fingers. 'Let me tell you a story,' he said, and he looked up and down the table to make sure they were all listening: Fitz, Olga, Boy, Daisy, and Bea. 'When I was a kid my father was accused of grazing livestock on someone else's land. No big deal, you might think, even if he was guilty. But he was arrested, and the land agent built a scaffold in the north meadow. Then the soldiers came and grabbed me and my brother and our mother and took us there. My father was on the scaffold with a noose around his neck. Then the landlord arrived.'

Daisy had never heard this story. She looked at her mother. Olga seemed equally surprised.

The little group at the table were all silent now.

'We were forced to watch while my father was hanged,' Lev said. He turned to Bea. 'And you know something strange? The landlord's sister was there as well.' He put the cigar in his mouth, wetting the end, and took it out again.

Daisy saw that Bea had turned pale. Was this about her?

'The sister was about nineteen years old, and she was a princess,' Lev said, looking at his cigar. Daisy heard Bea let out a small cry, and realized that this story *was* about her. 'She stood there and watched the hanging, cold as ice,' Lev said.

Then he looked directly at Bea. 'Now that's what I call coarse,' he said.

There was a long moment of silence.

Then Lev put the cigar back in his mouth and said: 'Has anyone got a light?'

(vi)

Lloyd Williams sat at the table in the kitchen of his mother's house in Aldgate, anxiously studying a map.

It was Sunday 4 October 1936, and today there was going to be a riot.

The old Roman town of London, built on a hill beside the river Thames, was now the financial district, called the City. West of this hill were the palaces of the rich, and the theatres and shops and cathedrals that catered to them. The house in which Lloyd sat was to the east of the hill, near the docks and the slums. Here, for centuries, waves of immigrants had landed, determined to work their fingers to the bone so that their grandchildren could one day move from the East End to the West End.

The map Lloyd was looking at so intently was in a special edition of the *Daily Worker*, the Communist Party newspaper, and it showed the route of today's march by the British Union of Fascists. They planned to assemble outside the Tower of London, on the border between the City and the East End, then march east.

Straight into the overwhelmingly Jewish borough of Stepney.

Unless Lloyd and people who thought as he did could stop them.

There were 330,000 Jews in Britain, according to the newspaper, and half of them lived in the East End. Most were refugees from Russia, Poland and Germany, where they had lived in fear that on any day the police, the army or the Cossacks might ride into town, robbing families, beating old men and outraging young women, lining fathers and brothers up against the wall to be shot.

Here in the London slums those Jews had found a place where they had as much right to live as anyone else. How would they feel if they looked out of their windows to see, marching down their own streets, a gang of uniformed thugs sworn to wipe them all out? Lloyd felt that it just could not be allowed to happen.

The *Worker* pointed out that from the Tower there were really only two routes the marchers could take. One went through Gardiner's Corner, a five-way junction known as the Gateway to the East End; the other led along Royal Mint Street and the narrow Cable Street. There were a dozen other routes for an individual using side streets, but not for a march. St George Street led to Catholic Wapping rather than Jewish Stepney, and was therefore no use to the Fascists.

The *Worker* called for a human wall to block Gardiner's Corner and Cable Street, and stop the march.

The paper often called for things that did not happen: strikes, revolutions, or – most recently – an alliance of all left parties to form a People's Front. The human wall might be just another fantasy. It would take many thousands of people to effectively close off the East End. Lloyd did not know whether enough would show up.

All he knew for sure was that there would be trouble.

At the table with Lloyd were his parents, Bernie and Ethel; his sister, Millie; and sixteen-year-old Lenny Griffiths from Aberowen, in his Sunday suit. Lenny was part of a small army of Welsh miners who had come to London to join the counter-demonstration.

Bernie looked up from his newspaper and said to Lenny: 'The Fascists claim that the train fares for all you Welshmen to come to London have been paid by the big Jews.'

Lenny swallowed a mouthful of fried egg. 'I don't know any big Jews,' he said. 'Unless you count Mrs Levy Sweetshop, she's quite big. Anyway, I came to London on the back of a lorry with sixty Welsh lambs going to Smithfield meat market.'

Millie said: 'That accounts for the smell.'

Ethel said: 'Millie! How rude.'

Lenny was sharing Lloyd's bedroom, and he had confided that after the demonstration he was not planning to return to Aberowen. He and Dave Williams were going to Spain to join the International Brigades being formed to fight the Fascist insurrection.

'Did you get a passport?' Lloyd had asked. Getting a passport was not difficult, but the applicant did have to provide a reference from a clergyman, doctor, lawyer, or other person of status, so a young person could not easily keep it secret.

'No need,' Lenny said. 'We go to Victoria Station and get a weekend return ticket to Paris. You can do that without a passport.'

Lloyd had vaguely known that. It was a loophole intended for the convenience of the prosperous middle class. Now the anti-Fascists were taking advantage of it. 'How much is the ticket?'

'Three pounds fifteen shillings.'

Lloyd had raised his eyebrows. That was more money than an unemployed coal miner was likely to have.

Lenny had added: 'But the Independent Labour Party is paying for my ticket, and the Communist Party for Dave's.'

They must have lied about their ages. 'Then what happens when you get to Paris?' Lloyd had asked.

'We'll be met by the French Communists at the Gare du Nord.' He pronounced it *gair duh nord*. He did not speak a word of French. 'From there we'll be escorted to the Spanish border.'

Lloyd had delayed his own departure. He told people he wanted to soothe his parents' worries, but the truth was he could not give up on Daisy. He still dreamed of her throwing Boy over. It was hopeless – she did not even answer his letters – but he could not forget her.

Meanwhile Britain, France, and the USA had agreed with Germany and Italy to adopt a policy of non-intervention in Spain, which meant that none of them would supply weapons to either side. This in itself was infuriating to Lloyd: surely the democracies should support the elected government? But what was worse, Germany and Italy were breaching the agreement every day, as Lloyd's mother and Uncle Billy pointed out at many public meetings held that autumn in Britain to discuss Spain. Earl Fitzherbert, as the government minister responsible, defended the policy stoutly, saying that the Spanish government should not be armed for fear it would go Communist.

This was a self-fulfilling prophecy, as Ethel had argued in a scathing speech. The one nation willing to support the government of Spain was the Soviet Union, and the Spaniards would naturally gravitate towards the only country in the world that helped them.

The truth was that the Conservatives felt Spain had elected people who were dangerously left-wing. Men such as Fitzherbert would not be unhappy if the Spanish government was violently overthrown and replaced by right-wing extremists. Lloyd seethed with frustration.

Then had come this chance to fight Fascism at home.

'It's ridiculous,' Bernie had said a week ago, when the march had been announced. 'The Metropolitan Police must force them to change the route. They have the right to march, of course; but not in Stepney.' However, the police said they did not have the power to interfere with a perfectly legal demonstration.

Bernie and Ethel and the mayors of eight London boroughs had been in a delegation that begged the Home Secretary, Sir John Simon, to ban the march or at least divert it; but he, too, claimed he had no power to act.

The question of what to do next had split the Labour Party, the Jewish community, and the Williams family.

The Jewish People's Council against Fascism and Anti-Semitism, founded by Bernie and others three months ago, had called for a massive counter-demonstration that would keep the Fascists out of Jewish streets. Their slogan was the Spanish phrase *No pasaran*, meaning 'They shall not pass', the cry of the anti-Fascist defenders of Madrid. The Council was a small organization with a grand name. It occupied two upstairs rooms in a building on Commercial Road, and it owned a Gestetner duplicating machine and a couple of old typewriters. But it commanded huge support in the East End. In forty-eight hours it had collected an incredible hundred thousand signatures on a petition calling for the march to be banned. Still the government did nothing.

Only one major political party supported the counter-demonstration, and that was the Communists. The protest was also backed by the fringe Independent Labour Party, to which Lenny belonged. The other parties were against.

Ethel said: 'I see the *Jewish Chronicle* has advised its readers to stay off the streets today.'

This was the problem, in Lloyd's opinion. A lot of people were taking the view that it was best to keep out of trouble. But that would give the Fascists a free hand.

Bernie, who was Jewish though not religious, said to Ethel: 'How can you quote the *Jewish Chronicle* at me? It believes Jews should not be against Fascism, just anti-Semitism. What kind of political sense does that make?'

'I hear that the Board of Deputies of British Jews says the same as the *Chronicle*,' Ethel persisted. 'Apparently there was an announcement yesterday in all the synagogues.'

'Those so-called deputies are alrightniks from Golders Green,' Bernie said with contempt. 'They've never been insulted on the streets by Fascist hooligans.'

'You're in the Labour Party,' Ethel said accusingly. 'Our policy is not to confront the Fascists on the streets. Where's your solidarity?'

Bernie said: 'What about solidarity with my fellow Jews?'

'You're only Jewish when it suits you. And you've never been abused on the street.'

'All the same, the Labour Party has made a political mistake.'

'Just remember, if you allow the Fascists to provoke violence, the press will blame the Left for it, regardless of who really started it.'

Lenny said rashly: 'If Mosley's boys start a fight, they'll get what's coming to them.'

Ethel sighed. 'Think about it, Lenny: in this country, who's got the most guns – you and Lloyd and the Labour Party, or the Conservatives with the army and the police on their side?'

'Oh,' said Lenny. Clearly he had not considered that.

Lloyd said angrily to his mother: 'How can you talk like that? You were in Berlin three years ago – you saw how it was. The German Left tried to oppose Fascism peacefully, and look what happened to them.'

Bernie put in: 'The German Social Democrats failed to form a popular front with the Communists. That allowed them to be picked off separately. Together they might have won.' Bernie had been angry when the local Labour Party branch had refused an offer from the Communists to form a coalition against the march.

Ethel said: 'An alliance with Communists is a dangerous thing.'

She and Bernie disagreed on this. In fact, it was an issue that split the Labour Party. Lloyd thought that Bernie was right and Ethel wrong. 'We have to use every resource we've got to defeat Fascism,' he said; then he added diplomatically: 'But Mam's right, it will be best for us if today goes off without violence.'

'It will be best if you all stay home, and oppose the Fascists through the normal channels of democratic politics,' Ethel said.

'You tried to get equal pay for women through the normal channels of democratic politics,' Lloyd said. 'You failed.' Only last April, women Labour MPs had promoted a parliamentary bill to guarantee female government employees equal pay for equal work. It had been voted down by the male-dominated House of Commons.

'You don't give up on democracy every time you lose a vote,' Ethel said crisply.

The trouble was, Lloyd knew, that these divisions could fatally weaken the anti-Fascist forces, as had happened in Germany. Today would be a harsh test. Political parties could try to lead, but the people would choose whom to follow. Would they stay at home, as urged by the timid Labour Party and the *Jewish Chronicle*? Or would they come out on to the streets in their thousands and say No to Fascism? By the end of the day he would know the answer.

There was a knock at the back door and their neighbour, Sean

Dolan, came in dressed in his churchgoing suit. 'I'll be joining you after Mass,' he said to Bernie. 'Where should we meet up?'

'Gardiner's Corner, not later than two o'clock,' said Bernie. 'We're hoping to have enough people to stop the Fascists there.'

'You'll have every dock worker in the East End with you,' said Sean enthusiastically.

Millie asked: 'Why is that? The Fascists don't hate you, do they?'

'You're too young to remember, you darlin' girl, but the Jews have always supported us,' Sean explained. 'In the dock strike of 1912, when I was only nine years old, my father couldn't feed us, and me and my brother were taken in by Mrs Isaacs the baker's wife in New Road, may God bless her great big heart. Hundreds of dockers' children were looked after by Jewish families then. It was the same in 1926. We're not going to let the bloody Fascists come down our streets – excuse my language, Mrs Leckwith.'

Lloyd was heartened. There were thousands of dockers in the East End: if they showed up en masse it would hugely swell the ranks.

From outside the house came the sound of a loudspeaker. 'Keep Mosley out of Stepney,' said a man's voice. 'Assemble at Gardiner's Corner at two o'clock.'

Lloyd drank his tea and stood up. His role today was to be a spy, checking the position of the Fascists and calling in updates to Bernie's Jewish People's Council. His pockets were heavy with big brown pennies for public phones. 'I'd better get started,' he said. 'The Fascists are probably assembling already.'

His mother got up and followed him to the door. 'Don't get into a fight,' she said. 'Remember what happened in Berlin.'

'I'll be careful,' Lloyd said.

She tried a light tone. 'Your rich American girl won't like you with no teeth.'

'She doesn't like me anyway.'

'I don't believe it. What girl could resist you?'

'I'll be all right, Mam,' Lloyd said. 'Really I will.'

'I suppose I should be glad you're not going to bloody Spain.'

'Not today, anyway.' Lloyd kissed his mother and went out.

It was a bright autumn morning, the sun unseasonably warm. In the middle of Nutley Street a temporary platform had been set up by a group of men, one of whom was speaking through a megaphone. 'People of the East End, we do not have to stand quiet while a crowd of strutting anti-Semites insults us!' Lloyd recognized the speaker as a local official of the National Unemployed Workers' Movement. Because of the Depression there were thousands of unemployed Jewish tailors. They signed on every day at the Settle Street Labour Exchange.

Before Lloyd had gone ten yards, Bernie came after him and handed him a paper bag of the little glass balls that children called marbles. 'I've been in a lot of demonstrations,' he said. 'If the mounted police charge the crowd, throw these under the horses' hooves.'

Lloyd smiled. His stepfather was a peacemaker, almost all the time, but he was no softie.

All the same, Lloyd was dubious about the marbles. He had never had much to do with horses, but they seemed to him to be patient, harmless beasts, and he did not like the idea of causing them to crash to the ground.

Bernie read the look on his face and said: 'Better a horse should fall than my boy should be trampled.'

Lloyd put the marbles in his pocket, thinking that it did not commit him to using them.

He was pleased to see many people already on the streets. He noted other encouraging signs. The slogan 'They shall not pass' in English and Spanish had been chalked on walls everywhere he looked. The Communists were out in force, handing out leaflets. Red flags draped many windowsills. A group of men wearing medals from the Great War carried a banner that read: 'Jewish Ex-Servicemen's Association.' Fascists hated to be reminded how many Jews had fought for Britain. Five Jewish soldiers had won the country's highest medal for bravery, the Victoria Cross.

Lloyd began to think that perhaps there would be enough people to stop the march after all.

Gardiner's Corner was a broad five-way junction, named for the Scottish clothing store, Gardiner and Company, which occupied a corner building with a distinctive clock tower. When he got there, Lloyd saw that trouble was expected. There were several first aid stations and hundreds of St John Ambulance volunteers in their uniforms. Ambulances were parked in every side street. Lloyd hoped there would be no fighting; but better to risk violence, he thought, than to let the Fascists march unhindered.

He took a roundabout route and came towards the Tower of London from the north-west, in order not to be identified as an East Ender. Some minutes before he got there he could hear the brass bands.

The Tower was a riverside palace that had symbolized authority and repression for eight hundred years. It was surrounded by a long wall of pale old stone that looked as if the colour had been washed out of it by centuries of London rain. Outside the walls, on the landward side, was a park called Tower Gardens, and here the Fascists were assembling. He estimated that there were already a couple of thousand of them, in a line that stretched back westward into the financial district. Every now and again they broke into a rhythmic chant:

> *One, two, three, four,*
> *We're gonna get rid of the Yids!*
> *The Yids! The Yids!*
> *We're gonna get rid of the Yids!*

They carried Union Jack flags. Why was it, Lloyd wondered, that the people who wanted to destroy everything good about their country were the quickest to wave the national flag?

They looked impressively military, in their wide black leather belts and black shirts, as they formed neat columns across the grass. Their officers wore a smart uniform: a black military-cut jacket, grey riding breeches, jackboots, a black cap with a shiny peak, and a red-and-white armband. Several motorcyclists in uniform roared around ostentatiously, delivering messages with Fascist salutes. More marchers were arriving, some of them in armoured vans with wire mesh at the windows.

This was not a political party. It was an army.

The purpose of the display, Lloyd figured, was to give them false authority. They wanted to look as if they had the right to close meetings and empty buildings, to burst into homes and offices and arrest people, to drag them to jails and camps and beat them up, interrogate and torture them, as the Brownshirts did in Germany under the Nazi regime so admired by Mosley and the *Daily Mail*'s proprietor, Lord Rothermere.

They would terrify the people of the East End, people whose parents and grandparents had fled from repression and pogroms in Ireland and Poland and Russia.

Would East Enders come out on the streets and fight them? If not – if today's march went ahead as planned – what might the Fascists dare tomorrow?

He walked around the edge of the park, pretending to be one of the hundred or so casual onlookers. Side streets radiated from the hub-like spokes. In one of them he noticed a familiar-looking black-and-cream Rolls-Royce drawing up. The chauffeur opened the rear door and, to Lloyd's shock and dismay, Daisy Peshkov got out.

There was no doubt why she was here. She was wearing a beautifully tailored female version of the uniform, with a long grey skirt instead of the breeches, her fair curls escaping from under the black cap. Much as he hated the outfit, Lloyd could not help finding her irresistibly alluring.

He stopped and stared. He should not have been surprised: Daisy had told him she liked Boy Fitzherbert, and Boy's politics clearly made no difference to that. But to see her obviously supporting the Fascists in their attack on Jewish Londoners rammed home to him how utterly alien she was from everything that mattered in his life.

He should simply have turned away, but he could not. As she hurried along the pavement, he blocked her way. 'What the devil are you doing here?' he said brusquely.

She was cool. 'I might ask you the same question, Mr Williams,' she said. 'I don't suppose you're intending to march with us.'

'Don't you understand what these people are like? They break up peaceful political meetings, they bully journalists, they imprison their political rivals. You're an American – how can you be against democracy?'

'Democracy is not necessarily the most appropriate political system for every country in all times.' She was quoting Mosley's propaganda, Lloyd guessed.

He said: 'But these people torture and kill everyone who disagrees with them!' He thought of Jörg. 'I've seen it for myself, in Berlin. I was in one of their camps, briefly. I was forced to watch while a naked man was savaged to death by starving dogs. That's the kind of thing your Fascist friends do.'

She was unintimidated. 'And who, exactly, has been killed by Fascists here in England recently?'

'The British Fascists haven't got the power yet – but your Mosley admires Hitler. If they ever get the chance, they'll do exactly the same as the Nazis.'

'You mean they will eliminate unemployment and give the people pride and hope.'

Lloyd was drawn to her so powerfully that it broke his heart to hear her spouting this rubbish. 'You know what the Nazis have done to the family of your friend Eva.'

'Eva got married, did you know?' Daisy said, in the determinedly cheerful tone of one who tries to switch a dinner-table conversation to a more agreeable topic. 'To nice Jimmy Murray. She's an English wife, now.'

'And her parents?'

Daisy looked away. 'I don't know them.'

'But you know what the Nazis have done to them.' Eva had told Lloyd all about it at the Trinity Ball. 'Her father is no longer allowed to practise medicine – he's working as an assistant in a pharmacy. He can't enter a park or a public library. *His* father's name has been scraped off the war memorial in his home village!' Lloyd realized he had raised his voice. More quietly he said: 'How can you possibly stand side by side with people who do such things?'

She looked troubled, but she did not answer his question. Instead she said: 'I'm late already. Please excuse me.'

'What you're doing can't be excused.'

The chauffeur said: 'All right, sonny, that's enough.'

He was a heavy middle-aged man who evidently took little exercise, and Lloyd was not in the least intimidated, but he did not want to start a fight. 'I'm leaving,' he said in a mild tone. 'But don't call me sonny.'

The chauffeur took his arm.

Lloyd said: 'You'd better take your hand off me, or I'll knock you down before I go.' He looked into the chauffeur's face.

The chauffeur hesitated. Lloyd tensed, preparing to react, watching for warning signs, as he would in the boxing ring. If the chauffeur tried to hit him, it would be a great swinging haymaker of a blow, easily dodged.

But the man either sensed Lloyd's readiness or felt the well-developed muscle in the arm he was holding; for one reason or the other he backed off and released his grip, saying: 'No need for threats.'

Daisy walked away.

Lloyd looked at her back in the perfectly fitting uniform as she hurried towards the ranks of the Fascists. With a deep sigh of frustration he turned and went in the other direction.

He tried to concentrate on the job at hand. What a fool he had been to threaten the chauffeur. If he had got into a fight he would probably have been arrested, then he would have spent the day in a police cell – and how would that have helped defeat Fascism?

It was now half past twelve. He left Tower Hill, found a telephone box, called the Jewish People's Council, and spoke to Bernie. After he had reported what he had seen, Bernie told him to make an estimate of the number of policemen in the streets between the Tower and Gardiner's Corner.

He crossed to the east side of the park and explored the radiating side streets. What he saw astonished him.

He had expected a hundred or so police. In fact, there were thousands.

They stood lining the pavements, waited in dozens of parked buses, and sat astride huge horses in remarkably neat rows. Only a narrow gap was left for people who wanted to walk along the streets. There were more police than Fascists.

From inside one of the buses, a uniformed constable gave him the Hitler salute.

Lloyd was dismayed. If all these policemen sided with the Fascists, how could the counter-demonstrators resist them?

This was worse than a Fascist march: it was a Fascist march with police authority. What kind of message did that send to the Jews of the East End?

In Mansell Street he saw a beat policeman he knew, Henry Clark. 'Hello, Nobby,' he said. For some reason all Clarks were called Nobby. 'A copper just gave me the Hitler salute.'

'They're not from round here,' Nobby said quietly, as if revealing a confidence. 'They don't live with Jews like I do. I tell them Jews are the same as everyone else, mostly decent law-abiding people, a few villains and troublemakers. But they don't believe me.'

'All the same . . . the Hitler salute?'

'Might have been a joke.'

Lloyd did not think so.

He left Nobby and moved on. The police were forming cordons where the side streets entered the area around Gardiner's Corner, he saw.

He went into a pub with a phone – he had scouted all the available telephones the day before – and told Bernie there were at least five thousand policemen in the neighbourhood. 'We can't resist that many coppers,' he said gloomily.

'Don't be so sure,' Bernie said. 'Have a look at Gardiner's Corner.'

Lloyd found a way around the police cordon and joined the counter-demonstration. It was not until he got into the middle of the street outside Gardiner's that he could appreciate the full extent of the crowd.

It was the largest gathering of people he had ever seen.

The five-way junction was jammed, but that was the least of it. The crowd stretched east along Whitechapel High Street as far as the eye could see. Commercial Road, which ran south-east, was also crammed. Leman Street, where the police station stood, was impenetrable.

There must be a hundred thousand people here, Lloyd thought. He wanted to throw his hat in the air and cheer. East Enders had come out in force to repel the Fascists. There could be no doubt about their feelings now.

In the middle of the junction stood a stationary tram, abandoned by its driver and passengers.

Nothing could pass through this crowd, Lloyd realized with mounting optimism.

He saw his neighbour Sean Dolan climb a lamp post and fix a red flag to its top. The Jewish Lads' Brigade brass band was playing – probably without the knowledge of the respectable conservative organizers of the club. A police aircraft flew overhead, an autogyro of some kind, Lloyd thought.

Near the windows of Gardiner's he ran into his sister Millie and her friend, Naomi Avery. He did not want Millie to become involved in any rough stuff: the thought chilled his heart. 'Does Dad know you've come?' he said in a tone of reproof.

She was insouciant. 'Don't be daft,' she replied.

He was surprised she was there at all. 'You're not usually very political,' he said. 'I thought you were more interested in making money.'

'I am,' she said. 'But this is special.'

Lloyd could imagine how upset Bernie would be if Millie got hurt. 'I think you should go home.'

'Why?'

He looked around. The crowd was amiable and peaceful. The police were some distance away, the Fascists nowhere to be seen. There would be no march today, that was clear. Mosley's people could not force their way through a crowd of a hundred thousand people determined to stop them, and the police would be insane to let them try. Millie was probably quite safe.

Just as he was thinking this, everything changed.

Several whistles shrilled. Looking in the direction of the sound, Lloyd saw the mounted police drawn up in an ominous line. The horses were stamping and blowing in agitation. The police had drawn long clubs shaped like swords.

They seemed to be getting ready to attack – but surely that could not be so.

Next moment, they charged.

There were angry shouts and terrified screams from the people. Everyone scrambled to get out of the way of the giant horses. The crowd made a path, but those at the edge fell under the pounding hooves. The police lashed out left and right with their long clubs. Lloyd was pushed helplessly backwards.

He felt furious: what did the police think they were doing? Were they stupid enough to believe they could clear a path for Mosley to march along? Did they really imagine that two or three thousand Fascists chanting insults could pass through a crowd of a hundred thousand of their victims without starting a riot? Were the police led by idiots, or out of control? He was not sure which would be worse.

They backed away, wheeling their panting horses, and regrouped, forming a ragged line; then a whistle blew and they heeled the flanks of their mounts, urging them into another reckless charge.

Millie was scared now. She was only sixteen, and her bravado had gone. She screamed with fear as the crowd squeezed her up against the plate-glass window of Gardiner and Company. Tailor's dummies in cheap suits and winter coats stared out at the horrified crowd and the warlike riders. Lloyd was deafened by the roar of thousands of voices yelling in fearful protest. He got in front of Millie and pushed against the press with all his might, trying to protect her, but it was in vain. Despite his efforts he was crushed against her. Forty or fifty screaming people had their backs to the window, and the pressure was building dangerously.

Lloyd realized with rage that the police were determined to make a pathway through the crowd regardless of the cost.

A moment later, there was a terrific crash of breaking glass and the window gave way. Lloyd fell on top of Millie, and Naomi fell on him. Dozens of people cried out in pain and panic.

Lloyd struggled to his feet. Miraculously, he was unhurt. He looked around frantically for his sister. It was maddeningly difficult to distinguish the people from the tailor's dummies. Then he spotted Millie lying in a mess of broken glass. He grasped her arms and pulled her to her feet. She was crying. 'My back!' she said.

He turned her around. Her coat was cut to ribbons and there was blood all over her. He felt sick with anguish. He put his arm around her shoulders protectively. 'There's an ambulance just around the corner,' he said. 'Can you walk?'

They had gone only a few yards when the police whistles blew again. Lloyd was terrified that he and Millie would be shoved back into Gardiner's window. Then he remembered what Bernie had given him. He took the paper bag of marbles from his pocket.

The police charged.

Drawing back his arm, Lloyd threw the paper bag over the heads of the crowd to land in front of the horses. He was not the only one so equipped, and several other people threw marbles. As the horses came at them there was the sound of firecrackers. A police horse slipped on marbles and went down. Others stopped and reared at the banging of the fireworks. The police charge turned into chaos. Naomi Avery had somehow pushed to the front of the crowd, and he saw her burst a bag of pepper under the nose of a horse, causing it to veer away, shaking its head frantically.

The crush eased, and Lloyd led Millie around the corner. She was still in pain, but she had stopped crying.

A line of people were waiting for attention from the St John's Ambulance volunteers: a weeping girl whose hand appeared to have been crushed; several young men with bleeding heads and faces; a middle-aged woman sitting on the ground nursing a swollen knee. As Lloyd and Millie arrived, Sean Dolan walked away with a bandage around his head and went straight back into the crowd.

A nurse looked at Millie's back. 'This is bad,' she said. 'You need to go to the London Hospital. We'll take you in an ambulance.' She looked at Lloyd. 'Do you want to go with her?'

Lloyd did, but he was supposed to be phoning in reports, and he hesitated.

Millie solved the dilemma for him with characteristic spunk. 'Don't you dare come,' she said. 'You can't do anything for me, and you've got important work to do here.'

She was right. He helped her into a parked ambulance. 'Are you sure?'

'Yes, I'm sure. Try not to end up in hospital yourself.'

He was leaving her in the best hands, he decided. He kissed her cheek and returned to the fray.

The police had changed their tactics. The people had repelled the horse charges, but the police were still determined to make a path through the crowd. As Lloyd pushed his way to the front they charged on foot, attacking with their batons. The unarmed demonstrators cowered back from them, liked piled leaves in a wind, then surged forward in a different part of the line.

The police started to arrest people, perhaps hoping to weaken the crowd's determination by taking ringleaders away. In the East End, being arrested was no legal formality. Few people came back without a black eye or a few gaps in their teeth. Leman Street police station had a particularly bad reputation.

Lloyd found himself behind a vociferous young woman carrying a red flag. He recognized Olive Bishop, a neighbour in Nutley Street. A policeman hit her over the head with his truncheon, screaming: 'Jewish whore!' She was not Jewish, and she certainly was not a whore; in fact, she played the piano at the Calvary Gospel Hall. But she had forgotten the admonition of Jesus to turn the other cheek, and she scratched the cop's face, drawing parallel red lines on his skin. Two more officers grabbed her arms and held her while the scratched man hit her on the head again.

The sight of three strong men attacking one girl maddened Lloyd. He stepped forward and hit the woman's assailant with a right hook that had all of his rage behind it. The blow landed on the policeman's temple. Dazed, the man stumbled and fell.

More officers converged on the scene, lashing out randomly with their clubs, hitting arms and legs and heads and hands. Four of them picked up Olive, each taking an arm or a leg. She screamed and wriggled desperately but she could not get free.

But the bystanders were not passive. They attacked the police carrying the girl off, trying to pull the uniformed men away from her. The police turned on their attackers, yelling: 'Jew bastards!' even though not all their assailants were Jews and one was a black-skinned Somali sailor.

The police let go of Olive, dropping her to the road, and began to defend themselves. Olive pushed through the crowd and vanished. The cops retreated, hitting out at anyone within reach as they backed away.

Lloyd saw with a thrill of triumph that the police strategy was not working. For all their brutality, the attacks had completely failed to make a way through the crowd. Another baton charge began, but the angry crowd surged forward to meet it, eager now for combat.

Lloyd decided it was time for another report. He worked his way backwards through the crush and found a phone box. 'I don't think they're going to succeed, Dad,' he told Bernie excitedly. 'They're trying to beat a path through us but they're making no progress. We're too many.'

'We're redirecting people to Cable Street,' Bernie said. 'The police may be about to switch their thrust, thinking they have more chance there, so we're sending reinforcements. Go along there, see what's happening, and let me know.'

'Right,' said Lloyd, and he hung up before realizing he had not told his stepfather that Millie had been taken to hospital. But perhaps it was better not to worry him right now.

Getting to Cable Street was not going to be easy. From Gardiner's Corner, Leman Street led directly south to the near end of Cable Street, a distance of less than half a mile, but the road was jammed by demonstrators fighting with police. Lloyd had to take a less direct route. He struggled eastward through the crowd into Commercial Road. Once there, further progress was not much easier. There were no police, therefore there was no violence, but the crowd was almost as dense. It was frustrating, but Lloyd was consoled for his difficulties by the reflection that the police would never force a way through so many.

He wondered what Daisy Peshkov was doing. Probably she was sitting in the car, waiting for the march to begin, tapping the toe of her expensive shoe impatiently on the Rolls-Royce's carpet. The thought that he was helping to frustrate her purpose gave him an oddly spiteful sense of satisfaction.

With persistence and a slightly ruthless attitude to those in his way, Lloyd pushed through the throng. The railway that ran along the north side of Cable Street obstructed his route, and he had to walk some distance before reaching a side road that tunnelled beneath the line. He passed under the tracks and entered Cable Street.

The crowd here was not so closely packed, but the street was narrow, and passage was still difficult. That was a good thing: it would be even more difficult for the police to get through. But there was another obstruction, he saw. A lorry had been parked across the road and turned on its side. At either end of the vehicle, the barricade had been extended the full width of the street with old tables and chairs, odd lengths of timber, and other assorted rubbish piled high.

A barricade! It made Lloyd think of the French revolution. But this was no revolution. The people of the East End did not want to overthrow the British government. On the contrary, they were deeply attached to their elections and their borough councils and their Houses of Parliament. They liked their system of government so much that they were determined to defend it against Fascism, even if it would not defend itself.

He had emerged behind the barrier, and now he moved towards it to see what was happening. He stood on a wall to get a better view. He saw a lively scene. On the far side, police were trying to dismantle the blockage, picking up broken furniture and dragging old mattresses away. But they were not having an easy time of it. A hail of missiles fell on their helmets, some hurled from behind the barricade, some thrown from the upstairs windows of the houses packed closely on either side of the street: stones, milk bottles, broken pots, and bricks that came, Lloyd saw, from a nearby builder's yard. A few daring young men stood on top of the barricade, lashing out at the police with sticks, and occasionally a fight broke out as the police tried to pull one down and give him a kicking. With a start, Lloyd recognized two of the figures standing on the barricade as Dave Williams, his cousin, and Lenny Griffiths, from Aberowen. Side by side they were fighting policemen off with shovels.

But as the minutes passed, Lloyd saw that the police were winning. They were working systematically, picking up the components of the barricade and taking them away. On this side a few people reinforced the wall, replacing what the police removed, but they were less organized and did not have an infinite supply of materials. It looked to Lloyd as if the police would soon prevail. And if they could clear Cable Street, they would let the Fascists march down here, past one Jewish shop after another.

Then, looking behind him, he saw that whoever was organizing the defence of Cable Street was thinking ahead. Even while the police dismantled the barricade, another was going up a few hundred yards farther along the street.

Lloyd retreated and began enthusiastically to help build the second wall. Dockers with pickaxes were prising up paving stones, housewives dragged dustbins from their yards, and shopkeepers brought empty crates and boxes. Lloyd helped carry a park bench, then pulled down a noticeboard from outside a municipal building. Learning from experience, the builders did a better job this time, using their materials economically and making sure the structure was sturdy.

Looking behind him again, Lloyd saw that a third barricade was beginning to rise farther east.

The people began to retreat from the first one and regroup behind the second. A few minutes later the police at last made a gap in the first barricade and poured through it. The first of them went after the few young men remaining, and Lloyd saw Dave and Lenny chased down an alley. The houses on either side were swiftly shut up, doors slamming and windows closing.

Then, Lloyd saw, the police did not know what to do next. They had broken through the barricade only to be confronted with another, stronger one. They seemed not to have the heart to begin dismantling the second. They milled around in the middle of Cable Street, talking desultorily, looking resentfully at the residents watching them from upstairs windows.

It was too early to proclaim victory but, all the same, Lloyd could not suppress a happy feeling of success. It was beginning to look as if the anti-Fascists were going to win the day.

He remained at his post for another quarter of an hour, but the police did nothing more, so he left the scene, found a telephone kiosk, and called in.

Bernie was cautious. 'We don't know what's happening,' he said. 'There seems to be a lull everywhere, but we need to find out what the Fascists are up to. Can you get back to the Tower?'

Lloyd certainly could not fight his way through the massed police, but perhaps there was another way. 'I could try going via St George Street,' he said doubtfully.

'Do the best you can. I want to know their next move.'

Lloyd worked his way south through a maze of alleys. He hoped he was right about St George Street. It was outside the contested area, but the crowds might have spilled over.

However, as he had hoped, there were no crowds here, even though he was still within earshot of the counter-demonstration, and could hear shouting and police whistles. A few women stood in the street talking, and a gaggle of little girls skipped a rope in the middle of the road. Lloyd headed west, breaking into a jog-trot, expecting to see crowds of demonstrators or police around every bend. He came across a few people who had strayed from the fracas – two men with bandaged heads, a woman in a ripped coat, a bemedalled veteran with his arm in a sling – but no crowds. He ran all the way to where the street ended at the Tower. He was able to walk unhindered into Tower Gardens.

The Fascists were still here.

That in itself was an achievement, Lloyd felt. It was now half past three: the marchers had been kept waiting here, not marching, for hours. He saw that their high spirits had evaporated. They were no longer singing or chanting, but stood quiet and listless, lined up but not so neatly, their banners drooping, their bands silent. They already looked beaten.

However, there was a change a few minutes later. An open car emerged from a side street and drove alongside the Fascist lines. Cheers went up. The lines straightened, the officers saluted, the Fascists stood to attention. In the back seat of the car sat their leader, Sir Oswald Mosley, a handsome man with a moustache, wearing the uniform complete with cap. Rigidly straight-backed, he saluted repeatedly as his car went by at walking pace, as if he were a monarch inspecting his troops.

His presence reinvigorated his forces and worried Lloyd. This probably meant that they were going to march as planned – otherwise, why was he here? The car followed the Fascist line along a side street into the financial district. Lloyd waited. Half an hour later Mosley returned, this time on foot, again saluting and acknowledging cheers.

When he reached the head of the line, he turned and, accompanied by one of his officers, entered a side street.

Lloyd followed.

Mosley approached a group of older men standing in a huddle on the pavement. Lloyd was surprised to recognize Sir Philip Game, the Commissioner of Police, in a bow tie and trilby hat. The two men began an intense conversation. Sir Philip must surely be telling Sir Oswald that the crowd of counter-demonstrators was too huge to be dispersed. But what then would be his advice to the Fascists? Lloyd longed to get close enough to eavesdrop, but he decided not to risk arrest, and remained at a discreet distance.

The police commissioner did most of the talking. The Fascist leader nodded briskly several times and asked a few questions. Then the two men shook hands and Mosley walked away.

He returned to the park and conferred with his officers. Among them Lloyd recognized Boy Fitzherbert, wearing the same uniform as Mosley.

Boy did not look so well in it: the trim military outfit did not suit his soft body and the lazy sensuality of his stance.

Mosley seemed to be giving orders. The other men saluted and moved away, no doubt to carry out his commands. What had he told them to do? Their only sensible option was to give up and go home. But if they had been sensible they would not have been Fascists.

Whistles blew, orders were shouted, bands began to play, and the men stood to attention. They were going to march, Lloyd realized. The police must have assigned them a route. But what route?

Then the march began – and they went in the opposite direction. Instead of heading into the East End, they went west, into the financial district, which was deserted on a Sunday afternoon.

Lloyd could hardly believe it. 'They've given up!' he said aloud, and a man standing near him said: 'Looks like it, don't it?'

He watched for five minutes as the columns slowly moved off. When there was absolutely no doubt what was happening, he ran to a phone box and called Bernie. 'They're marching away!' he said.

'What, into the East End?'

'No, the other way! They're going west, into the City. We've won!'

'Good God!' Bernie spoke to the other people with him. 'Everybody! The Fascists are marching west. They've given up!'

Lloyd heard a burst of wild cheering in the room.

After a minute Bernie said: 'Keep an eye on them, let us know when they've all left Tower Gardens.'

'Absolutely.' Lloyd hung up.

He walked around the perimeter of the park in high spirits. It became clearer every minute that the Fascists were defeated. Their bands played, and they marched in time, but there was no spring in their step, and they no longer chanted that they were going to get rid of the Yids. The Yids had got rid of them.

As he passed the end of Byward Street he saw Daisy again.

She was heading towards the distinctive black-and-cream Rolls-Royce, and she had to walk past Lloyd. He could not resist the temptation to gloat. 'The people of the East End have rejected you and your filthy ideas,' he said.

She stopped and looked at him, cool as ever. 'We've been obstructed by a gang of thugs,' she said with disdain.

'Still, you're marching in the other direction now.'

'One battle doesn't make a war.'

That might be true, Lloyd thought; but it was a pretty big battle. 'You're not marching home with your boyfriend?'

'I prefer to drive,' she said. 'And he's not my boyfriend.'

Lloyd's heart leaped in hope.

Then she said: 'He's my husband.'

Lloyd stared at her. He had never really believed that she would be so stupid. He was speechless.

'It's true,' she said, reading the disbelief in his face. 'Didn't you see our engagement reported in the newspapers?'

'I don't read the society pages.'

She showed him her left hand, with a diamond engagement ring and a gold wedding band. 'We were married yesterday. We postponed our honeymoon to join the march today. Tomorrow we're flying to Deauville in Boy's plane.'

She walked the few steps to the car and the chauffeur opened the door. 'Home, please,' she said.

'Yes, my lady.'

Lloyd was so angry he wanted to hit someone.

Daisy looked back over her shoulder. 'Goodbye, Mr Williams.'

He found his voice. 'Goodbye, Miss Peshkov.'

'Oh, no,' she said. 'I'm Viscountess Aberowen now.'

She just loved saying it, Lloyd could tell. She was a titled lady, and it meant the world to her.

She got into the car and the chauffeur closed the door.

Lloyd turned away. He was ashamed to realize that he had tears in his eyes. 'Hell,' he said aloud.

He sniffed, swallowing tears. He squared his shoulders and headed back towards the East End at a brisk walk. Today's triumph had been soured. He knew he was a fool to care about Daisy – clearly she did not care about him – but, all the same, it broke his heart that she was throwing herself away on Boy Fitzherbert.

He tried to put her out of his mind.

The police were getting back into their buses and leaving the scene. Lloyd had not been surprised by their brutality – he had lived in the East End all his life, and it was a rough neighbourhood – but their anti-Semitism had shocked him. They had called every woman a Jewish whore, every man a Jew bastard. In Germany the police had supported the Nazis and sided with the Brownshirts. Would they do the same here? Surely not!

The crowd at Gardiner's Corner had begun to rejoice. The Jewish Lads' Brigade band was playing a jazz tune for men and women to dance to, and bottles of whisky and gin were passed from hand to hand. Lloyd decided to go to the London Hospital and check on Millie. Then he should probably go to the Jewish Council headquarters and break the news to Bernie that Millie had been hurt.

Before he got any further he ran into Lenny Griffiths. 'We sent the buggers packing!' Lenny said excitedly.

'We did, too.' Lloyd grinned.

Lenny lowered his voice. 'We beat the Fascists here, and we're going to beat them in Spain, too.'

'When are you leaving?'

'Tomorrow. Me and Dave are catching a train to Paris in the morning.'

Lloyd put his arm around Lenny's shoulders. 'I'll come with you,' he said.

4

1937

Volodya Peshkov bent his head against the driving snow as he walked across the bridge over the Moscow River. He wore a heavy greatcoat, a fur hat, and a stout pair of leather boots. Few Muscovites were so well dressed. Volodya was lucky.

He always had good boots. His father, Grigori, was an army commander. Grigori was not a high-flyer: although he was a hero of the Bolshevik revolution and a personal acquaintance of Stalin, his career had stalled at some point in the twenties. All the same, the family had always lived comfortably.

Volodya himself *was* a high-flyer. After university he had got into the prestigious Military Intelligence Academy. A year later he had been posted to Red Army Intelligence headquarters.

His greatest piece of luck had been meeting Werner Franck in Berlin, while his father had been a military attaché at the Soviet Embassy there. Werner had been at the same school in a more junior class. Learning that young Werner hated Fascism, Volodya had suggested to him that he could best oppose the Nazis by spying for the Russians.

Werner had been only fourteen years old then, but he was now eighteen, he worked at the Air Ministry, he hated the Nazis even more, and he had a powerful radio transmitter and a code book. He was resourceful and courageous, taking dreadful risks and gathering priceless information. And Volodya was his contact.

Volodya had not seen Werner for four years, but he remembered him vividly. Tall with striking red-blond hair, Werner looked and acted older than he was, and even at fourteen he had been enviably successful with women.

Werner had recently tipped him off about Markus, a diplomat at the German embassy in Moscow who was secretly a Communist. Volodya had sought Markus out and recruited him as a spy. For some months now Markus had been supplying a stream of reports which Volodya translated into Russian and passed to his boss. The latest was a fascinating account of how pro-Nazi American business leaders were supplying the right-wing Spanish rebels with trucks, tyres and oil. Texaco's chairman, the Hitler-admiring Torkild Rieber, was using the company's tankers to smuggle oil to the rebels in defiance of a specific request from President Roosevelt.

Volodya was on his way to meet Markus now.

He walked along Kutuzovsky Prospekt and turned towards the Kiev Station. Their rendezvous today was a workingmen's bar near the station. They never used the same place twice, but finished each meeting by arranging the next one: Volodya was meticulous about tradecraft. They always used cheap bars or cafés where Markus's diplomatic colleagues would never dream of going. If somehow Markus were to fall under suspicion and be followed by a German counter-espionage agent, Volodya would know, for such a man would stand out from the other customers.

This place was called the Ukraine Bar. Like most buildings in Moscow, it was a timber structure. The windows were steamed up, so at least it would be warm inside. But Volodya did not go in immediately. There were further precautions to be taken. He crossed the street and ducked into the entrance of an apartment house. He stood in the cold hallway, looking out through a small window, watching the bar.

He wondered if Markus would show up. He always had, in the past, but Volodya could not feel sure. If he did show up, what information would he bring? Spain was the hot issue in international politics, but Red Army Intelligence was also passionately interested in German armaments. How many tanks were they producing per month? How many Mauser M34 machine guns per day? How good was the new Heinkel He 111 bomber? Volodya longed for such information to pass to his boss, Major Lemitov.

Half an hour went by, and Markus did not come.

Volodya began to worry. Had Markus been found out? He worked as assistant to the ambassador, and therefore saw everything that crossed the ambassador's desk; but Volodya had been urging him to seek access to other documents, especially the correspondence of military attachés. Had that been a mistake? Had someone noticed Markus sneaking a peek at cables that were none of his business?

Then Markus came along the street, a professorial figure in spectacles and an Austrian-style loden coat, white snowflakes spotting the green felt cloth. He turned into the Ukraine Bar. Volodya waited, watching.

Another man followed Markus in, and Volodya frowned anxiously; but the second man was obviously a Russian worker, not a German counter-espionage agent. He was a small, rat-faced man in a threadbare coat, his boots wrapped in rags, and he wiped the wet end of his pointed nose with his sleeve.

Volodya crossed the street and went into the bar.

It was a smoky place, none too clean, and it smelled of men who did not often bathe. On the walls were fading watercolours of Ukrainian scenery in cheap frames. It was mid-afternoon, and there were not many customers. The only woman in the place looked like an aging prostitute recovering from a hangover.

Markus was at the back of the room, hunched over an untasted glass of beer. He was in his thirties but looked older, with a neat fair beard and moustache. He had thrown open his coat, revealing a fur lining. The rat-faced Russian sat two tables away, rolling a cigarette.

As Volodya approached, Markus stood up and punched him in the mouth.

'You cowfucker!' he screamed in German. 'You pig's cunt!'

Volodya was so shocked that for a moment he did nothing. His lips hurt and he tasted blood. Reflexively, he raised his arm to hit back. But he restrained himself.

Markus swung at him again, but this time Volodya was ready, and he easily dodged the wild blow.

'Why did you do it?' Markus yelled. 'Why?'

Then, just as suddenly, he crumpled, falling back into his chair, burying his face in his hands, and beginning to sob.

Volodya spoke through bleeding lips. 'Shut up, you fool,' he said. He turned around and spoke to the other customers, who were all staring. 'It's nothing, he's upset.'

They all looked away, and one man left. Muscovites never voluntarily got involved in trouble. It was dangerous even to separate two scrapping drunks, in case one of them was powerful in the Party. And they knew that Volodya was such a man: they could tell by his good coat.

Volodya turned back to Markus. In a lowered voice he said angrily: 'What the hell was that for?' He spoke German: Markus's Russian was poor.

'You arrested Irina,' the man replied, weeping. 'You fucking bastard, you burned her nipples with a cigarette.'

Volodya winced. Irina was Markus's Russian girlfriend. Volodya began to see what this might be about and he had a bad feeling. He sat down opposite Markus. 'I didn't arrest Irina,' he said. 'And I'm sorry if she's been hurt. Just tell me what happened.'

'They came for her in the middle of the night. Her mother told me. They wouldn't say who they were, but they weren't regular police detectives – they had better clothes. She doesn't know where they took her. They questioned her about me and accused her of being a spy. They tortured her and raped her, then they threw her out.'

'Fuck,' said Volodya. 'I'm really sorry.'

'You're sorry? It must have been you that did it – who else?'

'This is nothing to do with Army Intelligence, I swear.'

'Makes no difference,' Markus said. 'I'm finished with you, and I'm finished with Communism.'

'There are sometimes casualties in the war against capitalism.' It sounded glib even to Volodya as he said it.

'You young fool,' Markus said savagely. 'Don't you understand that socialism means freedom from this kind of shit?'

Volodya glanced up and saw a burly man in a leather coat come through the door. He was not here for a drink, Volodya knew instinctively.

Something was going on, and Volodya did not know what it was. He was new to this game, and right now he felt his lack of experience like a missing limb. He thought he might be in danger but he did not know what to do.

The newcomer approached the table where Volodya sat with Markus.

Then the rat-faced man stood up. He was about the same age as Volodya. Surprisingly, he spoke with an educated accent. 'You two are under arrest.'

Volodya cursed.

Markus jumped to his feet. 'I am commercial attaché at German Embassy!' he screamed in ungrammatical Russian. 'You cannot arrest! I have diplomatic immunity!'

The other customers left the bar in a rush, shoving at each other as they squeezed through the door. Only two people remained: the bartender, nervously swiping the counter with a filthy rag, and the prostitute, smoking a cigarette and staring into an empty vodka glass.

'You can't arrest me, either,' Volodya said calmly. He took his identification card from his pocket. 'I'm Lieutenant Peshkov, Army Intelligence. Who the fuck are you?'

'Dvorkin, NKVD.'

The man in the leather coat said: 'Berezovsky, NKVD.'

The secret police. Volodya groaned: he might have known. The NKVD overlapped with Army Intelligence. He had been warned that the two organizations were always treading on each other's toes, but this was his first experience of it. He said to Dvorkin: 'I suppose it was you who tortured this man's girlfriend.'

Dvorkin wiped his nose on his sleeve: apparently that unpleasant habit was not part of his disguise. 'She had no information.'

'So you burned her nipples for nothing.'

'Lucky for her. If she had been a spy it would have been worse.'

'It didn't occur to you to check with us first?'

'When did you ever check with us?'

Markus said: 'I'm leaving.'

Volodya felt desperate. He was about to lose a valuable asset. 'Don't go,' he pleaded. 'We'll make this up to Irina somehow. We'll get her the best hospital treatment—'

'Fuck you,' said Markus. 'You'll never see me again.' He walked out of the bar.

Dvorkin evidently did not know what to do. He did not want to let Markus go, but clearly he could not arrest him without looking foolish. In the end he said to Volodya: 'You shouldn't let people speak to you that way. It makes you look weak. They should respect you.'

'You prick,' Volodya said. 'Can't you see what you've done? That man was a good source of reliable intelligence – but now he'll never work for us again, thanks to your blundering.'

Dvorkin shrugged. 'As you said to him, sometimes there are casualties.'

'God spare me,' Volodya said, and he went out.

He felt vaguely nauseated as he walked back across the river. He was sickened by what the NKVD had done to an innocent woman, and downcast by the loss of his source. He boarded a tram: he was too junior to have a car. He brooded as the vehicle trundled through the snow to his place of work. He had to report to Major Lemitov, but he hesitated, wondering how to tell the story. He needed to make it clear that he was not to blame, yet avoid seeming to make excuses.

Army Intelligence headquarters stood on one edge of the Khodynka airfield, where a patient snowplough crawled up and down keeping the runway clear. The architecture was peculiar: a two-storey building with no windows in its outer walls surrounded a courtyard in which stood the nine-storey head office, sticking up like a pointed finger out of a brick fist. Cigarette lighters and fountain pens could not be brought in, as they might set off the metal detectors at the entrance, so the army provided its staff with one of each inside. Belt buckles were a problem, too, so most people wore suspenders. The security was superfluous, of course. Muscovites would do anything to stay out of such a building: no one was mad enough to want to sneak inside.

Volodya shared an office with three other subalterns, their steel desks side by side on opposite walls. There was so little space that Volodya's desk prevented the door from opening fully. The office wit, Kamen, looked at his swollen lips and said: 'Let me guess – her husband came home early.'

'Don't ask,' said Volodya.

On his desk was a decrypt from the radio section, the German words pencilled letter by letter under the code groups.

The message was from Werner.

Volodya's first reaction was fear. Had Markus already reported what had happened to Irina, and persuaded Werner, too, to withdraw from espionage? Today seemed a sufficiently unlucky day for such a disaster.

But the message was the opposite of disastrous.

Volodya read with growing amazement. Werner explained that the German military had decided to send spies to Spain posing as anti-Fascist volunteers wanting to fight for the government side in the civil war. They would report clandestinely from behind the lines to German-manned listening stations in the rebel camp.

That in itself was red-hot information.

But there was more.

Werner had the names.

Volodya had to restrain himself from whooping with joy. A coup like this could happen only once in the lifetime of an intelligence man, he thought. It more than made up for losing Markus. Werner was solid gold. Volodya dreaded to think what risks he must have taken to purloin this list of names and smuggle it out of Air Ministry headquarters in Berlin.

He was tempted to run upstairs to Lemitov's office right away, but he restrained himself.

The four subalterns shared a typewriter. Volodya lifted the heavy old machine off Kamen's desk and put it on his own. Using the forefinger of each hand, he typed out a Russian translation of the message from Werner. While he was doing so the daylight faded and powerful security lights came on outside the building.

Leaving a carbon copy in his desk drawer, he took the top copy and went upstairs. Lemitov was in. A good-looking man of about forty, he had dark hair slicked down with brilliantine. He was shrewd, and had a knack of thinking one step ahead of Volodya, who strove to emulate his forethought. He did not subscribe to the orthodox military view that army organization was about shouting and bullying, yet he was merciless with incompetent people. Volodya respected him and feared him.

'This might be tremendously useful information,' Lemitov said when he had read the translation.

'Might be?' Volodya did not see any reason for doubt.

'It could be disinformation,' Lemitov pointed out.

Volodya did not want to believe that, but he realized with a surge of disappointment that he had to acknowledge the possibility that Werner had been caught and turned into a double agent. 'What kind of disinformation?' he asked dispiritedly. 'Are these false names, to send us on a wild goose chase?'

'Perhaps. Or they might be the real names of genuine volunteers, Communists and socialists who have escaped from Nazi Germany and gone to Spain to fight for freedom. We could end up arresting real anti-Fascists.'

'Hell.'

Lemitov smiled. 'Don't look so miserable! The information is still very good. We have our own spies in Spain – young Russian soldiers and officers who have "volunteered" to join the International Brigades. They can investigate.' He picked up a red pencil and wrote on the sheet of paper in small, neat handwriting. 'Well done,' he said.

Volodya took that for dismissal and went to the door.

Lemitov said: 'Did you meet Markus today?'

Volodya turned back. 'There was a problem.'

'I guessed, by your mouth.'

Volodya told the story. 'So I lost a good source,' he finished. 'But I don't know what I could have done differently. Should I have told the NKVD about Markus and warned them off?'

'Fuck, no,' said Lemitov. 'They're completely untrustworthy. Never tell them anything. But don't worry, you haven't lost Markus. You can get him back easily.'

'How?' Volodya said uncomprehendingly. 'He hates us all now.'

'Arrest Irina again.'

'What?' Volodya was horrified. Had she not suffered enough? 'Then he'll hate us even more.'

'Tell him that if he doesn't continue to co-operate with us, we'll interrogate her all over again.'

Volodya desperately tried to hide his revulsion. It was important not to appear squeamish. And he could see that Lemitov's plan would work. 'Yes,' he managed to say.

'Only this time,' Lemitov went on, 'tell him we'll put the lighted cigarettes up her cunt.'

Volodya felt as if he might vomit. He swallowed hard and said: 'Good idea. I'll pick her up now.'

'Tomorrow is soon enough,' said Lemitov. 'Four in the morning. Maximum shock.'

'Yes, sir.' Volodya went out and closed the door behind him.

He stood in the corridor for a moment, feeling unsteady. Then a passing clerk looked strangely at him and he forced himself to walk away.

He was going to have to do this. He would not torture Irina, of course: the threat would be enough. But she would surely *think* she was going to be tortured all over again, and that would terrify her out of her wits. Volodya felt that in her place he might go insane. He had never imagined, when he joined the Red Army, that he might have to do such things. Of course the army was about killing people, he knew that; but torturing girls?

The building was emptying, lights were being switched off in offices, men with hats on were in the corridors. It was time to go home. Returning to his office, Volodya called the military police and arranged to meet a squad at three-thirty in the morning to arrest Irina. Then he put on his coat and went to catch a tram home.

Volodya lived with his parents, Grigori and Katerina, and his sister Anya, nineteen, who was still at university. On the tram he wondered if he could talk to his father about this. He imagined saying: 'Do we have to torture people in Communist society?' But he knew what the answer would be. It was a temporary necessity, essential to defend the revolution against spies and subversives in the pay of the capitalist imperialists. Perhaps he could ask: 'How long will it be before we can abandon such dreadful practices?' Of course his father would not know, nor would anyone else.

On their return from Berlin, the Peshkov family had moved into Government House, sometimes called The House on the Embankment, an apartment block across the river from the Kremlin, occupied by members of the Soviet elite. It was a huge building in the Constructivist style, with more than five hundred flats.

Volodya nodded at the military policeman at the door, then passed through the grand lobby – so large that, some evenings, there was dancing to a jazz band – and went up in the elevator. The apartment was luxurious by Soviet standards, with constant hot water and a phone, but it was not as pleasant as their home in Berlin.

His mother was in the kitchen. Katerina was an indifferent cook and an unenthusiastic housekeeper, but Volodya's father adored her. Back in 1914, in St Petersburg, he had rescued her from the unwelcome attentions of a bullying policeman, and he had been in love with her ever since. She was still attractive at forty-three, Volodya guessed, and while the family had been on the diplomatic circuit she had learned how to dress more stylishly than most Russian women – though she was careful not to look Western, a serious offence in Moscow.

'Did you hurt your mouth?' she said to him after he kissed her hello.

'It's nothing.' Volodya smelled chicken. 'Special dinner?'

'Anya is bringing a boyfriend home.'

'Ah! A fellow student?'

'I don't think so. I'm not sure what he does.'

Volodya was pleased. He was fond of his sister, but he knew she was not beautiful. She was short and stumpy, and wore dull clothes in drab colours. She had not had many boyfriends, and it was good news that one liked her enough to come home with her.

He went to his room, took off his jacket, and washed his face and hands. His lips were almost back to normal: Markus had not hit him very hard. While he was drying his hands he heard voices, and gathered that Anya and her boyfriend had arrived.

He put on a knitted cardigan, for comfort, and left his room. He went into the kitchen. Anya was sitting at the table with a small, rat-faced man Volodya recognized. 'Oh, no!' Volodya said. 'You!'

It was Ilya Dvorkin, the NKVD agent who had arrested Irina. His disguise had gone, and he was dressed in a normal dark suit and decent boots. He stared at Volodya in surprise. 'Of course – Peshkov!' he said. 'I didn't make the connection.'

Volodya turned to his sister. 'Don't tell me this is your boyfriend.'

Anya said in dismay: 'What's the matter?'

Volodya said: 'We met earlier today. He screwed up an important Army operation by sticking his nose in where it didn't belong.'

'I was doing my job,' said Dvorkin. He wiped the end of his nose on his sleeve.

'Some job!'

Katerina stepped in to rescue the situation. 'Don't bring your work home,' she said. 'Volodya, please pour a glass of vodka for our guest.'

Volodya said: 'Really?'

His mother's eyes flashed anger. 'Really!'

'Okay.' Reluctantly, he took the bottle from the shelf. Anya got glasses from a cupboard and Volodya poured.

Katerina took a glass and said: 'Now, let's start again. Ilya, this is my son Vladimir, whom we always call Volodya. Volodya, this is Anya's friend Ilya, who has come to dinner. Why don't you shake hands.'

Volodya had no option but to shake the man's hand.

Katerina put snacks on the table: smoked fish, pickled cucumber, sliced sausage. 'In summer we have salad that I grow at the dacha, but at this time of year, of course, there is nothing,' she said apologetically. Volodya realized that she was keen to impress Ilya. Did his mother really want Anya to marry this creep? He supposed she must.

Grigori came in, wearing his army uniform, all smiles, sniffing the chicken and rubbing his hands together. At forty-eight he was red-faced and corpulent: it was hard to imagine him storming the Winter Palace as he had in 1917. He must have been thinner then.

He kissed his wife with relish. Volodya thought his mother was thankful for his father's unabashed lust without actually returning it. She would smile when he patted her bottom, hug him when he embraced her, and kiss him as often as he wanted, but she was never the initiator. She liked him, respected him, and seemed happy being married to him; but clearly she did not burn with desire. Volodya would want more than that from marriage.

The matter was purely hypothetical: Volodya had had a dozen or so short-term girlfriends but had not yet met a woman he wanted to marry.

Volodya poured his father a shot of vodka, and Grigori tossed it back with relish, then took some smoked fish. 'So, Ilya, what work do you do?'

'I'm with the NKVD,' Ilya said proudly.

'Ah! A very good organization to belong to!'

Grigori did not really think this, Volodya suspected; he was just trying to be friendly. Volodya thought the family should be unfriendly, in the hope that they could drive Ilya away. He said: 'I suppose, Father, that when the rest of the world follows the Soviet Union in adopting the Communist system, there will no longer be a need for the secret police, and the NKVD can then be abolished.'

Grigori chose to treat the question lightly. 'No police at all!' he said jovially. 'No criminal trials, no prisons. No counter-espionage department, as there will be no spies. No army either, since we will have no enemies! What will we all do for a living?' He laughed heartily. 'This, however, may still be some distance in the future.'

Ilya looked suspicious, as if he felt something subversive was being said but he could not put his finger on it.

Katerina brought to the table a plate of black bread and five bowls of hot borscht, and they all began to eat. 'When I was a boy in the countryside,' Grigori said, 'all winter long my mother would save vegetable peelings, apple cores, the discarded outer leaves of cabbages, the hairy part of the onion, anything like that, in a big old barrel outside the house, where it all froze. Then, in the spring, when the snow melted, she would use it to make borscht. That's what borscht really is, you know – soup made from peelings. You youngsters have no idea how well off you are.'

There was a knock at the door. Grigori frowned, not expecting anyone; but Katerina said: 'Oh, I forgot! Konstantin's daughter is coming.'

Grigori said: 'You mean Zoya Vorotsyntsev? The daughter of Magda the midwife?'

'I remember Zoya,' said Volodya. 'Skinny kid with blonde ringlets.'

'She's not a kid any more,' Katerina said. 'She's twenty-four and a scientist.' She stood up to go to the door.

Grigori frowned. 'We haven't seen her since her mother died. Why has she suddenly made contact?'

'She wants to talk to you,' Katerina replied.

'To me? About what?'

'Physics.' Katerina went out.

Grigori said proudly: 'Her father, Konstantin, and I were delegates to the Petrograd Soviet in 1917. We issued the famous Order Number One.' His face darkened. 'He died, sadly, after the Civil War.'

Volodya said: 'He must have been young – what did he die of?'

Grigori glanced at Ilya and quickly looked away. 'Pneumonia,' he said; and Volodya knew he was lying.

Katerina returned, followed by a woman who took Volodya's breath away.

She was a classic Russian beauty, tall and slim, with light-blonde hair, blue eyes so pale they were almost colourless, and perfect white skin. She wore a simple Nile-green dress whose plainness only drew attention to her slender figure.

She was introduced all around, then she sat at the table and accepted a bowl of borscht. Grigori said: 'So, Zoya, you're a scientist.'

'I'm a graduate student, doing my doctorate, and I teach undergraduate classes,' she said.

'Volodya here works in Red Army Intelligence,' Grigori said proudly.

'How interesting,' she said, obviously meaning the opposite.

Volodya realized that Grigori saw Zoya as a potential daughter-in-law. He hoped his father would not hint at this too heavily. He had already made up his mind to ask her for a date before the end of the evening. But he could manage that by himself. He did not need his father's help. On the contrary: unsubtle parental boasting might put her off.

'How is the soup?' Katerina asked Zoya.

'Delicious, thank you.'

Volodya was already getting the impression of a matter-of-fact personality behind the gorgeous exterior. It was an intriguing combination: a beautiful woman who made no attempt to charm.

Anya cleared away the soup bowls while Katerina brought the main course, chicken and potatoes cooked in a pot. Zoya tucked in, stuffing the food into her mouth, chewing and swallowing and eating more. Like most Russians, she did not often see food this good.

Volodya said: 'What kind of science do you do, Zoya?'

With evident regret she stopped eating to answer. 'I'm a physicist,' she said. 'We're trying to understand the atom: what its components are, what holds them together.'

'Is that interesting?'

'Completely fascinating.' She put down her fork. 'We're finding out what the universe is really made of. There's nothing so exciting.' Her eyes lit up. Apparently physics was the one thing that could distract her from her dinner.

Ilya spoke up for the first time. 'Ah, but how does all this theoretical stuff help the revolution?'

Zoya's eyes blazed anger, and Volodya liked her even more. 'Some comrades make the mistake of undervaluing pure science, preferring practical research,' she said. 'But technical developments, such as improved aircraft, are ultimately based on theoretical advances.'

Volodya concealed a grin. Ilya had been demolished with one casual swipe.

But Zoya had not finished. 'This is why I wanted to talk to you, sir,' she said to Grigori. 'We physicists read all the scientific journals published in the West – they foolishly reveal their results to the whole world. And lately we have realized that they are making alarming forward leaps in their understanding of atomic physics. Soviet science is in grave danger of falling behind. I wonder if Comrade Stalin is aware of this.'

The room went quiet. The merest hint of a criticism of Stalin was dangerous. 'He knows most things,' Grigori said.

'Of course,' Zoya said automatically. 'But perhaps there are times when loyal comrades such as yourself need to draw important matters to his attention.'

'Yes, that's true.'

Ilya said: 'Undoubtedly Comrade Stalin believes that science should be consistent with Marxist-Leninist ideology.'

Volodya saw a flash of defiance in Zoya's eyes, but she dropped her gaze and said humbly: 'There can be no question that he is right. We scientists must clearly redouble our efforts.'

This was horseshit, and everyone in the room knew it, but no one would say so. The proprieties had to be observed.

'Indeed,' said Grigori. 'Nevertheless, I will mention it next time I get a chance to talk to the Comrade General Secretary of the Party. He may wish to look into it further.'

'I hope so,' said Zoya. 'We want to be ahead of the West.'

'And how about after work, Zoya?' said Grigori cheerily. 'Do you have a boyfriend, a fiancé perhaps?'

Anya protested: 'Dad! That's none of our business.'

Zoya did not seem to mind. 'No fiancé,' she said mildly. 'No boyfriend.'

'As bad as my son, Volodya! He, too, is single. He is twenty-three years old, well educated, tall and handsome – yet he has no fiancée!'

Volodya squirmed at the heavy-handedness of this hint.

'Hard to believe,' Zoya said, and as she glanced at Volodya he saw a gleam of humour in her eyes.

Katerina put a hand on her husband's arm. 'Enough,' she said. 'Stop embarrassing the poor girl.'

The doorbell rang.

'Again?' said Grigori.

'This time I have no idea who it might be,' said Katerina as she left the kitchen.

She returned with Volodya's boss, Major Lemitov.

Startled, Volodya jumped to his feet. 'Good evening, sir,' he said. 'This is my father, Grigori Peshkov. Dad, may I present Major Lemitov?'

Lemitov saluted smartly.

Grigori said: 'At ease, Lemitov. Sit down and have some chicken. Has my son done something wrong?'

That was precisely the thought that was making Volodya's hands shake.

'No, sir – rather the contrary. But ... I was hoping for a private word with you and him.'

Volodya relaxed a little. Perhaps he was not in trouble after all.

'Well, we've just about finished dinner,' Grigori said, standing up. 'Let's go into my study.'

Lemitov looked at Ilya. 'Aren't you with the NKVD?' he said.

'And proud of it. Dvorkin is the name.'

'Oh! You tried to arrest Volodya this afternoon.'

'I thought he was behaving like a spy. I was right, wasn't I?'

'You must learn to arrest enemy spies, not our own.' Lemitov went out.

Volodya grinned. That was the second time Dvorkin had been put down.

Volodya, Grigori and Lemitov crossed the hallway. The study was a small room, sparsely furnished. Grigori took the only easy chair. Lemitov sat at a small table. Volodya closed the door and remained standing.

Lemitov said to Volodya: 'Does your comrade father know about this afternoon's message from Berlin?'

'No, sir.'

'You'd better tell him.'

Volodya related the story of the spies in Spain. His father was delighted. 'Well done!' he said. 'Of course this might be disinformation, but I doubt it: the Nazis aren't that imaginative. However, we are. We can arrest the spies and use their radios to send misleading messages to the right-wing rebels.'

Volodya had not thought of that. Dad might play the fool with Zoya, he thought, but he still has a sharp mind for intelligence work.

'Exactly,' said Lemitov.

Grigori said to Volodya: 'Your school friend, Werner, is a brave man.' He turned back to Lemitov. 'How do you plan to handle this?'

'We'll need some good intelligence men in Spain to investigate these Germans. It shouldn't be too difficult. If they really are spies, there will be evidence: code books, wireless sets, and so on.' He hesitated. 'I've come here to suggest we send your son.'

Volodya was astonished. He had not seen that coming.

Grigori's face fell. 'Ah,' he said thoughtfully. 'I must confess, the prospect fills me with dismay. We would miss him so much.' Then a look of resignation came over his face, as if he realized he did not really have a choice. 'The defence of the revolution must come first, of course.'

'An intelligence man needs field experience,' Lemitov said. 'You and I have seen action, sir, but the younger generation have never been on the battlefield.'

'True, true. How soon would he go?'

'In three days' time.'

Volodya could see that his father was trying desperately to think of a reason to keep him at home, but finding none. Volodya himself was excited. Spain! He thought of blood-red wine, black-haired girls with strong brown legs, and hot sunshine instead of Moscow snow. It would be dangerous, of course, but he had not joined the army to be safe.

Grigori said: 'Well, Volodya, what do you think?'

Volodya knew his father wanted him to come up with an objection. The only drawback he could think of was that he would not have time to get to know the stunning Zoya. 'It is a wonderful opportunity,' he said. 'I'm honoured to have been chosen.'

'Very well,' said his father.

'There is one small problem,' Lemitov said. 'It has been decided that Army Intelligence will investigate but not actually carry out the arrests.

That will be the prerogative of the NKVD.' His smile was humourless. 'I'm afraid you will be working with your friend Dvorkin.'

(ii)

It was amazing, Lloyd Williams thought, how quickly you could come to love a place. He had been in Spain for only ten months, but already his passion for the country was almost as strong as his attachment to Wales. He loved to see a rare flower blooming in the scorched landscape; he enjoyed sleeping in the afternoon; he liked the way there was wine to drink even when there was nothing to eat. He had experienced flavours he had never tasted before: olives, paprika, chorizo, and the fiery spirit they called orujo.

He stood on a rise, staring across a heat-hazed landscape with a map in his hand. There were a few meadows beside a river, and some trees on distant mountainsides, but in between was a barren, featureless desert of dusty soil and rock. 'Not much cover for our advance,' he said anxiously.

Beside him, Lenny Griffiths said: 'It's going to be a bloody hard battle.'

Lloyd looked at his map. Saragossa straddled the Ebro River about a hundred miles from its Mediterranean end. The town dominated communications in the Aragon region. It was a major crossroads, a rail junction, and the meeting of three rivers. Here the Spanish army confronted the anti-democratic rebels across an arid no-man's-land.

Some people called the government forces Republicans and the rebels Nationalists, but these were misleading names. Many people on both sides were republicans, in that they did not want to be ruled by a king. And they were all nationalist, in that they loved their country and were willing to die for it. Lloyd thought of them as the government and the rebels.

Right now Saragossa was held by Franco's rebels, and Lloyd was looking towards the town from a vantage point fifty miles south. 'Still, if we can take the town, the enemy will be bottled up in the north for another winter,' he said.

'If,' said Lenny.

It was a grim prognosis, Lloyd thought gloomily, when the best he could wish for was that the rebel advance might be halted. But no victory was in sight this year for the government.

All the same, a part of Lloyd was looking forward to the fight. He had been in Spain for ten months, and this would be his first taste of action. Until now he had been an instructor in a base camp. As soon as the Spaniards had discovered that he had been in Britain's Officer Training Corps, they had sped him through his induction, made him a lieutenant, and put him in charge of new arrivals. He had to drill them until obeying orders became a reflex, march them until their feet stopped bleeding and their blisters turned to calluses, and show them how to strip down and clean what few rifles were available.

But the flood of volunteers had now slowed to a trickle, and the instructors had been moved to fighting battalions.

Lloyd wore a beret, a zipped blouson with his badge of rank roughly hand-sewn to the sleeve, and corduroy breeches. He carried a short Spanish Mauser rifle, firing 7mm ammunition that had presumably been stolen from some Civil Guard arsenal.

Lloyd, Lenny and Dave had been split up for a while, but the three had been reunited in the British battalion of the 15th International Brigade for the coming battle. Lenny now had a black beard and looked a decade older than his seventeen years. He had been made a sergeant, though he had no uniform, just blue dungarees and a striped bandana. He looked more like a pirate than a soldier.

Now Lenny said: 'Anyway, this attack has nothing to do with bottling up the rebels. It's political. This region has always been dominated by the anarchists.'

Lloyd had seen anarchism in action during a brief spell in Barcelona. It was a cheerfully fundamentalist form of communism. Officers and men got the same pay. The dining rooms of the grand hotels had been turned into canteens for the workers. Waiters would hand back a tip, explaining amiably that the practice of tipping was demeaning. Posters everywhere condemned prostitution as exploitation of female comrades. There had been a wonderful atmosphere of liberation and camaraderie. The Russians hated it.

Lenny went on: 'Now the government has brought Communist troops from the Madrid area and amalgamated us all into the new Army of the East – under overall Communist command, of course.'

This kind of talk made Lloyd despair. The only way to win was for all the left-wing factions to work together, as they had – in the end, at least – at the Battle of Cable Street. But anarchists and Communists had been fighting each other in the streets of Barcelona. He said: 'Prime Minister Negrín isn't a Communist.'

'He might as well be.'

'He understands that without the support of the Soviet Union we're finished.'

'But does that mean we abandon democracy and let the Communists take over?'

Lloyd nodded. Every discussion about the government ended the same way: do we have to do everything the Soviets want just because they are the only people who will sell us guns?

They walked down the hill. Lenny said: 'We'll have a nice cup of tea, now, is it?'

'Yes, please. Two lumps of sugar in mine.'

It was a standing joke. Neither of them had had tea for months.

They came to their camp by the river. Lenny's platoon had taken over a little cluster of crude stone buildings that had probably been cowsheds until the war drove the farmers away. A few yards upriver a boathouse had been occupied by some Germans from the 11th International Brigade.

Lloyd and Lenny were met by Lloyd's cousin Dave Williams. Like Lenny, Dave had aged ten years in one. He looked thin and hard, his skin tanned and dusty, his eyes wrinkled with squinting into the sun. He wore the khaki tunic and trousers, leather belt pouches and ankle-buckled boots that formed the standard-issue uniform – though few soldiers had a complete set. He had a red cotton scarf around his neck. He carried a Russian Mosin-Nagant rifle with the old-fashioned spike bayonet reversed, making the weapon less clumsy. At his belt he had a German 9mm Luger that he must have taken from the corpse of a rebel officer. Apparently he was very accurate with rifle or pistol.

'We've got a visitor,' he said excitedly.

'Who is he?'

'She!' said Dave, and pointed.

In the shade of a misshapen black poplar tree, a dozen British and German soldiers were talking to a startlingly beautiful woman.

'Oh, *Duw*,' said Lenny, using the Welsh word for God. 'She's a sight for sore eyes.'

She looked about twenty-five, Lloyd thought, and she was petite, with big eyes and a mass of black hair pinned up and topped by a fore-and-aft army cap. Somehow her baggy uniform seemed to cling to her like an evening gown.

A volunteer called Heinz, who knew that Lloyd understood German, spoke to him in that language. 'This is Teresa, sir. She has come to teach us to read.'

Lloyd nodded understanding. The International Brigades consisted of foreign volunteers mixed with Spanish soldiers, and literacy was a problem with the Spanish. They had spent their childhood chanting the catechism in village schools run by the Catholic Church. Many priests did not teach the children to read, for fear that in later life they would get hold of socialist books. As a result, only about half the population had been literate under the monarchy. The republican government elected in 1931 had improved education, but there remained millions of Spaniards who could not read or write, and classes for soldiers continued even in the front line.

'I'm illiterate,' said Dave, who was not.

'Me, too,' said Joe Eli, who taught Spanish literature at Columbia University in New York.

Teresa spoke in Spanish. Her voice was low and calm and very sexy. 'How many times do you think I have heard this joke?' she said, but she did not seem very cross.

Lenny moved closer. 'I'm Sergeant Griffiths,' he said. 'I'll do anything I can to help you, of course.' His words were practical, but his tone of voice made them sound like an amorous invitation.

She gave him a dazzling smile. 'That would be most helpful,' she said.

Lloyd spoke formally to her in his best Spanish. 'I'm so very glad you're here, Señorita.' He had spent much of the last ten months studying the language. 'I am Lieutenant Williams. I can tell you exactly which members of the group require lessons . . . and which do not.'

Lenny said dismissively: 'But the lieutenant has to go to Bujaraloz to get our orders.' Bujaraloz was the small town where government forces had set up headquarters. 'Perhaps you and I should look around here for a suitable place to hold classes.' He might have been suggesting a walk in the moonlight.

Lloyd smiled and nodded agreement. He was happy to let Lenny romance Teresa. He himself was in no mood for flirting, whereas Lenny seemed in love already. In Lloyd's opinion Lenny's chances were close to zero. Teresa was an educated twenty-five-year-old who probably got a dozen propositions a day, and Lenny was a seventeen-year-old coal miner who had not taken a bath for a month. But he said nothing: Teresa seemed capable of looking after herself.

A new figure appeared, a man of Lloyd's age who looked vaguely familiar. He was dressed better than the soldiers, in wool breeches and a cotton shirt, and had a handgun in a buttoned holster. His hair was cut so short that it looked like stubble, a style favoured by Russians. He was only a lieutenant, but had an air of authority, even power. He said in fluent German: 'I am looking for Lieutenant Garcia.'

'He's not here,' said Lloyd in the same language. 'Where have you and I met before?'

The Russian seemed shocked and irritated at the same time, like one who finds a snake in his bedroll. 'We have never met,' he said firmly. 'You are mistaken.'

Lloyd snapped his fingers. 'Berlin,' he said. 'Nineteen thirty-three. We were attacked by Brownshirts.'

A look of relief came briefly over the man's face, as if he had been expecting something worse. 'Yes, I was there,' he said. 'My name is Vladimir Peshkov.'

'But we called you Volodya.'

'Yes.'

'At that scrap in Berlin you were with a boy called Werner Franck.'

Volodya looked panicked for a moment, then hid his feelings with an effort. 'I know no one of that name.'

Lloyd decided not to press the point. He could guess why Volodya was jumpy. The Russians were as terrified as everyone else of their secret police, the NKVD, who were operating in Spain and had a reputation for brutality. To them, any Russian who was friendly with foreigners might be a traitor. 'I'm Lloyd Williams.'

'I do remember.' Volodya looked at him with a penetrating blue-eyed stare. 'How strange that we should meet again here.'

'Not so strange, really,' Lloyd said. 'We fight the Fascists wherever we can.'

'Can I have a quiet word?'

'Of course.'

They walked a few yards away from the others. Peshkov said: 'There is a spy in Garcia's platoon.'

Lloyd was astonished. 'A spy? Who?'

'A German called Heinz Bauer.'

'Why, that's him in the red shirt. A spy? Are you sure?'

Peshkov did not bother to answer that question. 'I'd like you to summon him to your dugout, if you have one, or some other private place.' Peshkov looked at his wristwatch. 'In one hour, an arrest unit will be here to pick him up.'

'I'm using that little shed as my office,' said Lloyd, pointing. 'But I need to speak to my commanding officer about this.' The C.O. was a Communist, and unlikely to interfere, but Lloyd wanted time to think.

'If you wish.' Volodya clearly did not care what Lloyd's commanding officer thought. 'I want the spy taken quietly, without any fuss. I have explained to the arrest unit the importance of discretion.' He sounded as if he was not sure his wishes would be obeyed. 'The fewer people who know, the better.'

'Why?' said Lloyd, but before Volodya could reply he figured out the answer for himself. 'You're hoping to turn him into a double agent, sending misleading reports to the enemy. But, if too many people know he has been caught, then other spies may warn the rebels, and they will not believe the disinformation.'

'It is better not to speculate about such matters,' Peshkov said severely. 'Now let us go to your shed.'

'Wait a minute,' said Lloyd. 'How do you know he is a spy?'

'I can't tell you without compromising security.'

'That's a bit unsatisfactory.'

Peshkov looked exasperated. Clearly he was not used to being told that his explanations were unsatisfactory. Discussion of orders was a feature of the Spanish Civil War that the Russians particularly detested.

Before Peshkov could say anything further, two more men appeared and approached the group under the tree. One of the newcomers wore a leather jacket despite the heat. The other, who seemed to be in charge, was a scrawny man with a long nose and a receding chin.

Peshkov let out an exclamation of anger. 'Too early!' he said, then he called out something indignant in Russian.

The scrawny man made a dismissive gesture. In rough Spanish he said: 'Which one is Heinz Bauer?'

No one answered. The scrawny man wiped the end of his nose with his sleeve.

Then Heinz moved. He did not immediately flee, but cannoned into the man in the leather jacket, knocking him down. Then he dashed away – but the scrawny man stuck out a leg and tripped him up.

Heinz fell hard, his body skidding on the dry soil. He lay stunned – only for a moment, but it was a moment too long. As he got to his knees the two men pounced on him and knocked him down again.

He lay still, but all the same they started to beat him up. They drew wooden clubs. Standing either side of him they took turns to hit his head and body, raising their arms above their heads and striking down in a vicious ballet. In a few seconds there was blood all over Heinz's face. He tried desperately to escape, but when he got to his knees they pushed him down again. Then he curled up in a ball, whimpering. He was clearly finished, but they were not. They clubbed the helpless man again and again.

Lloyd found himself shouting a protest and pulling the scrawny man off. Lenny did the same to the other one. Lloyd grabbed his man in a bear hug and lifted him; Lenny knocked his man to the ground. Then Lloyd heard Volodya say in English: 'Stand still, or I'll shoot!'

Lloyd let go of his man and turned, incredulous. Volodya had drawn his sidearm, a standard-issue Russian Nagant M1895 revolver, and cocked it. 'Threatening an officer with a weapon is a court-martial offence in every army in the world,' Lloyd said. 'You're in deep trouble, Volodya.'

'Don't be a fool,' said Volodya. 'When was the last time a Russian was in trouble in this army?' But he lowered the gun.

The man in the leather jacket raised his club as if to hit Lenny, but Volodya barked: 'Back off, Berezovsky!' and the man obeyed.

Other soldiers appeared, drawn by the mysterious magnetism that attracts men to a fight, and in seconds there were twenty of them.

The scrawny man pointed a finger at Lloyd. Speaking English with a heavy accent, he said: 'You have interfered in matters that do not concern you!'

Lloyd helped Heinz to his feet. He was groaning in pain and covered in blood.

'You people can't just march in and start beating people up!' Lloyd said to the scrawny man. 'Where's your authority?'

'This German is a Trotsky-Fascist spy!' the man screeched.

Volodya said: 'Shut up, Ilya.'

Ilya took no notice. 'He has been photographing documents!' he said.

'Where is your evidence?' Lloyd said calmly.

Ilya clearly did not know or care about the evidence. But Volodya sighed and said: 'Look in his kitbag.'

Lloyd nodded to Mario Rivera, a corporal. 'Go and check,' he said.

Corporal Rivera ran to the boathouse and disappeared inside.

But Lloyd had a dreadful feeling Volodya was telling the truth. He said: 'Even if you're right, Ilya, you could use a little courtesy.'

Ilya said: 'Courtesy? This is a war, not an English tea party.'

'It might save you from getting into unnecessary fights.'

Ilya said something contemptuous in Russian.

Rivera emerged from the building carrying a small, expensive-looking camera and a sheaf of official papers. He showed them to Lloyd. The top document was yesterday's general order for deployment of troops ahead of the coming assault. The paper bore a wine stain of familiar shape, and Lloyd realized with a shock that it was his own copy, and must have been purloined from his shed.

He looked at Heinz, who straightened, gave the Fascist salute, and said: '*Heil Hitler!*'

Ilya looked triumphant.

Volodya said: 'Well, Ilya, you have now ruined the prisoner's value as a double agent. Another coup for the NKVD. Congratulations.' And he walked off.

(iii)

Lloyd went into battle for the first time on Tuesday 24 August.

His side, the elected government, had 80,000 men. The anti-democratic rebels had fewer than half that. The government also had two hundred aircraft against the rebels' fifteen.

To make the most of this superiority, the government advanced over a wide front, a north–south line sixty miles long, so that the rebels could not concentrate their limited numbers.

It was a good plan – so why, Lloyd asked himself two days later, was it not working?

It had started well enough. On the first day, the government had taken two villages north of Saragossa and two to the south. Lloyd's group, in the south, had overcome fierce resistance to take a village called Codo. The only failure was the central push, up the river valley, which had stalled at a place called Fuentes de Ebro.

Before the battle, Lloyd had been scared, and he spent the night awake, imagining what was to come, as he sometimes did before a boxing match. But once the fighting started he was too busy to worry. The worst moment was advancing across the barren scrubland, with no cover but stunted bushes, while the defenders fired from inside stone buildings. Even then, what he had felt was not fear but a kind of desperate cunning, zigzagging as he ran, crawling and rolling when the bullets came too near, then getting up and running, bent double, a few more yards. The main problem was shortage of ammunition: they had to make every shot count. They took Codo by force of numbers, and Lloyd, Lenny and Dave ended the day unhurt.

The rebels were tough and brave – but so were the government forces. The foreign brigades were made up of idealistic volunteers who had come to Spain knowing they might have to give their lives. Because of their reputation for courage they were often chosen to spearhead attacks.

The assault began to go wrong on the second day. The northern forces had stayed put, reluctant to advance because of lack of intelligence about rebel defences – a feeble excuse, Lloyd thought. The central group still could not take Fuentes de Ebro, despite being reinforced on the third day, and Lloyd was appalled to hear that they had lost nearly all their tanks to devastating defensive fire. In the south, Lloyd's group, instead of pushing forward, was directed to make a sideways move, to the riverside village of Quinto. Once again, they had to overcome determined defenders in house-to-house fighting. When the enemy surrendered, Lloyd's group took a thousand prisoners.

Now Lloyd sat in the evening light outside a church that had been wrecked by artillery fire, surrounded by the smoking ruins of houses and the strangely still bodies of the recently dead. A group of exhausted men gathered around him: Lenny, Dave, Joe Eli, Corporal Rivera, and a Welshman called Muggsy Morgan. There were so many Welshmen in Spain that someone had made up a limerick poking fun at the similarity in their names:

> There was a young fellow named Price
> And another young fellow named Price
> And a fellow named Roberts
> And a fellow named Roberts
> And another young fellow named Price.

The men were smoking, waiting quietly to see whether there would be any dinner, too weary even to banter with Teresa, who was, remarkably,

222

still with them, as the transport due to take her to the rear had failed to appear. They could hear occasional bursts of shooting as mopping-up continued a few streets away.

'What have we gained?' Lloyd said to Dave. 'We used scarce ammunition, we lost a lot of men, and we're no farther forward. Worse, we've given the Fascists time to bring up reinforcements.'

'I can tell you the fucking reason,' Dave said in his East End accent. His soul had hardened even more than his body, and he had become cynical and contemptuous. 'Our officers are more afraid of their commissars than of the fucking enemy. At the least excuse they can be branded as Trotsky-Fascist spies and tortured to death, so they're terrified of sticking their necks out. They'd rather sit still than move, they won't do anything on their own initiative, and they never take risks. I bet they don't shit without an order in writing.'

Lloyd wondered whether Dave's scornful analysis was right. The Communists never ceased to talk about the need for a disciplined army with a clear chain of command. By that they meant an army following Russian orders, but, all the same, Lloyd saw their point. However, too much discipline could stifle thinking. Was that what was going wrong?

Lloyd did not want to believe it. Surely Social Democrats, Communists and anarchists could fight in a common cause without one group tyrannizing the others: they all hated Fascism, and they all believed in a future society that was fairer to everyone.

He wondered what Lenny thought, but Lenny was sitting next to Teresa, talking to her in a low voice. She giggled at something he said, and Lloyd guessed that he must be making progress. It was a good sign when you could make a girl laugh. Then she touched his arm, said a few words, and stood up. Lenny said: 'Hurry back.' She smiled over her shoulder.

Lucky Lenny, thought Lloyd, but he felt no envy. A passing romance held no appeal for him: he did not see the point. He was an all-or-nothing man, he supposed. The only girl he had ever really wanted had been Daisy. She was now Boy Fitzherbert's wife, and Lloyd still had not met the girl who might take her place in his heart. He would, one day, he felt sure; but, meanwhile, he was not much attracted to temporary substitutes, even when they were as alluring as Teresa.

Someone said: 'Here come the Russians.' The speaker was Jasper Johnson, a black American electrician from Chicago. Lloyd looked up to see a dozen or so military advisors walking through the village like conquerors. The Russians were recognizable by their leather jackets and buttoned holsters. 'Strange thing, I didn't see them while we were fighting,' Jasper went on sarcastically. 'I guess they must have been in a different part of the battlefield.'

Lloyd looked around, making sure that no political commissars were nearby to hear this subversive talk.

As the Russians passed through the graveyard of the ruined church, Lloyd spotted Ilya Dvorkin, the weaselly secret policeman he had clashed with a week ago. The Russian crossed paths with Teresa and stopped to speak to her. Lloyd heard him say something in bad Spanish about dinner.

She replied, he spoke again, and she shook her head, evidently refusing. She turned to walk away, but he took hold of her arm, detaining her.

Lloyd saw Lenny sit upright, looking alertly at the tableau, the two figures framed by a stone archway that no longer led anywhere.

'Oh, shit,' said Lloyd.

Teresa tried again to move away, and Ilya seemed to tighten his grip.

Lenny moved to get up, but Lloyd put a hand on his shoulder and pushed him down. 'Let me deal with this,' he said.

Dave murmured a low warning. 'Careful, mate – he's in the NKVD. Best not to mess with those fucking bastards.'

Lloyd walked over to Teresa and Ilya.

The Russian saw him and said in Spanish: 'Get lost.'

Lloyd said: 'Hello, Teresa.'

She said: 'I can handle this, don't worry.'

Ilya looked more closely at Lloyd. 'I know you,' he said. 'You tried to prevent the arrest of a dangerous Trotsky-Fascist spy last week.'

Lloyd said: 'And is this young lady also a dangerous Trotsky-Fascist spy? I thought I just heard you ask her to have dinner with you.'

Ilya's sidekick Berezovsky appeared and stood aggressively close to Lloyd.

Out of the corner of his eye, Lloyd saw Dave draw the Luger from his belt.

This was getting out of control.

Lloyd said: 'I came to tell you, Señorita, that Colonel Bobrov wants to see you in his headquarters immediately. Please follow me and I'll take you to him.' Bobrov was a senior Russian military 'advisor'. He had not invited Teresa, but it was a plausible story, and Ilya did not know it was a lie.

For a frozen moment Lloyd could not tell which way it was going to go. Then the bang of a nearby gunshot was heard, perhaps from the next street. It seemed to return the Russians to reality. Teresa again moved away from Ilya, and this time he let her go.

Ilya pointed a finger aggressively at Lloyd's face. 'I'll see you again,' he said, and he made a dramatic exit, followed dog-like by Berezovsky.

Dave said: 'Stupid prick.'

Ilya pretended not to hear.

They all sat down. Dave said: 'You've made a bad enemy, Lloyd.'

'I didn't have much choice.'

'All the same, watch your back from now on.'

'An argument about a girl,' Lloyd said dismissively. 'Happens a thousand times a day.'

As darkness fell, a handbell summoned them to a field kitchen. Lloyd got a bowl of thin stew, a slab of dry bread, and a big cup of red wine so harsh-tasting that he imagined it taking the enamel off his teeth. He dipped his bread in the wine, improving both.

When the food was gone he was still hungry, as usual. He said: 'We'll have a nice cup of tea, shall we?'

'Aye,' said Lenny. 'Two lumps of sugar, please.'

They unrolled their thin blankets and prepared to sleep. Lloyd went in search of a latrine, found none, and relieved himself in a small orchard on the edge of the village. There was a three-quarter moon, and he could see the dusty leaves on olive trees that had survived the shelling.

As he buttoned up he heard a footstep. He turned around slowly – too slowly. By the time he saw Ilya's face, the club was coming down on his head. He felt an agonizing pain and fell to the ground. Dazed, he looked up. Berezovsky held a short-barrelled revolver pointed at his head. Beside him, Ilya said: 'Don't move or you'll be dead.'

Lloyd was terrified. Desperately he shook his head to clear it. This was insane. 'Dead?' he said incredulously. 'And how will you explain the murder of a lieutenant?'

'Murder?' said Ilya. He smiled. 'This is the front line. A stray bullet got you.' He switched to English. 'Jolly bad luck.'

Lloyd realized with despair that Ilya was right. When his body was found, it would look as if he had been killed in the battle.

What a way to die.

Ilya said to Berezovsky: 'Finish him off.'

There was a bang.

Lloyd felt nothing. Was this death? Then Berezovsky crumpled and fell to the ground. At the same moment Lloyd realized that the shot had come from behind him. He turned, incredulous, to look. In the moonlight he saw Dave holding his stolen Luger. Relief swamped him like a tidal wave. He was alive!

Ilya, too, had seen Dave, and he ran like a startled rabbit.

Dave tracked him with the pistol for several seconds, and Lloyd willed him to shoot, but Ilya dodged frantically between the olive trees, like a rat in a maze, then disappeared into the darkness.

Dave lowered the gun.

Lloyd looked down at Berezovsky. He was not breathing. Lloyd said: 'Thanks, Dave.'

'I told you to watch your back.'

'You watched it for me. But it's a pity you didn't get Ilya too. Now you're in trouble with the NKVD.'

'I wonder,' said Dave. 'Will Ilya want people to know that he got his sidekick killed in a squabble over a woman? Even the NKVD people are frightened of the NKVD. I think he'll keep it quiet.'

Lloyd looked again at the body. 'How do we explain this?'

'You heard the man,' Dave said. 'This is the front line. No explanation needed.'

Lloyd nodded. Dave and Ilya were both right. No one would ask how Berezovsky had died. A stray bullet got him.

They walked away, leaving the body where it lay.

'Jolly bad luck,' said Dave.

(iv)

Lloyd and Lenny spoke to Colonel Bobrov and complained that the attack on Saragossa was stalemated.

Bobrov was an older Russian with a cropped fuzz of white hair, nearing retirement and rigidly orthodox. In theory he was there only to help and advise the Spanish commanders. In practice the Russians called the shots.

'We're wasting time and energy on these little villages,' Lloyd said, translating into German what Lenny and all the experienced men were saying. 'Tanks are supposed to be armoured fists, used for deep penetration, striking far into enemy territory. The infantry should follow, mopping up and securing after the enemy has been scattered.'

Volodya was standing nearby, listening, and seemed by his expression to agree, though he said nothing.

'Small strongpoints like this wretched one-horse town should not be allowed to delay the advance, but should be bypassed and dealt with later by a second line,' Lloyd finished.

Bobrov looked shocked. 'This is the theory of the discredited Marshal Tuchachevsky!' he said in hushed tones. It was as if Lloyd had told a bishop to pray to Buddha.

'So what?' said Lloyd.

'He has confessed to treason and espionage, and has been executed.'

Lloyd stared incredulously. 'Are you telling me that the Spanish government cannot use modern tank tactics because some general has been purged in Moscow?'

'Lieutenant Williams, you are becoming disrespectful.'

Lloyd said: 'Even if the charges against Tuchachevsky are true, that doesn't mean his methods are wrong.'

'That will do!' Bobrov thundered. 'This conversation is over.'

Any hope that Lloyd might have had remaining was crushed when his battalion was moved from Quinto back in the direction they had come, another sideways manoeuvre. On 1 September, they were part of the attack on Belchite, a well-defended but strategically worthless small town twenty-five miles wide of their objective.

It was another hard battle.

Some seven thousand defenders were well dug in at the town's largest church, San Agustin, and atop a nearby hill, with trenches and earthworks. Lloyd and his platoon reached the outskirts of town without casualties, but then came under withering fire from windows and rooftops.

Six days later they were still there.

The corpses were stinking in the heat. As well as humans, there were dead animals, for the town's water supply had been cut off and livestock were dying of thirst. Whenever they could, the engineers stacked the bodies up, doused them with gasoline, and set fire to them; but the smell of roasting humans was worse than the stink of corruption. It seemed hard to breathe, and some of the men wore their gas masks.

The narrow streets around the church were killing fields, but Lloyd had devised a way to make progress without going outside. Lenny had found some tools in a workshop. Now two men were making a hole in the wall of the house in which they were sheltering. Joe Eli was using a pickaxe, sweat gleaming on his bald head. Corporal Rivera, who wore a striped shirt in the anarchist colours of red and black, wielded a sledgehammer. The wall was made of flat, yellow local bricks, roughly mortared. Lenny directed the operation to make sure that they did not bring the entire house down: as a miner, he had an instinct for the trustworthiness of a roof.

When the hole was big enough for a man to pass through, Lenny nodded to Jasper, also a corporal. Jasper took one of his few remaining grenades from his belt pouch, drew the pin, and threw it into the next house, just in case there was an ambush. As soon as it had exploded, Lloyd crawled quickly through the hole, rifle at the ready.

He found himself in another poor Spanish home, with whitewashed walls and a floor of beaten earth. There was no one here, dead or alive.

The thirty-five men of his platoon followed him through the hole and ran through the place to flush out any defenders. The house was small and empty.

In this way they were moving slowly but safely through a row of cottages towards the church.

They started work on the next hole but, before they broke through, they were halted by a major called Marquez, who came along the row of houses by the route they had made through the walls. 'Forget all that,' he said in Spanish-accented English. 'We're going to rush the church.'

Lloyd went cold. It was suicidal. He said: 'Is that Colonel Bobrov's idea?'

'Yes,' said Major Marquez non-committally. 'Wait for the signal: three sharp blows on the whistle.'

'Can we get more ammunition?' Lloyd said. 'We're low, especially for this kind of action.'

'No time,' said the major, and he went away.

Lloyd was horrified. He had learned a lot in a few days of battle, and he knew that the only way to rush a well-defended position was under a hail of covering fire. Otherwise the defenders would just mow the attackers down.

The men looked mutinous, and Corporal Rivera said: 'It is impossible.'

Lloyd was responsible for maintaining their morale. 'No complaints, you lot,' he said breezily. 'You're all volunteers. Did you think war wasn't dangerous? If it was safe, your sisters could do it for you.' They laughed, and the moment of danger passed, for now.

He moved to the front of the house, opened the door a crack, and peeped out. The sun glared down on a narrow street with houses and shops on both sides. The buildings and the ground were the same pale tan colour, like undercooked bread, except where shelling had gouged up red earth. Right outside the door a militiaman lay dead, a cloud of flies feasting on the hole in his chest. Looking towards the square, Lloyd saw that the street widened towards the church. The gunmen in the high twin towers had a clear view and an easy shot at anyone approaching. On the ground there was only minimal cover: some rubble, a dead horse, a wheelbarrow.

We're all going to die, he thought.

But why else did we come here?

He turned back to his men, wondering what to say. He had to keep them thinking positively. 'Just hug the sides of the street, close to the houses,' he said. 'Remember, the slower you go, the longer you're exposed – so wait for the whistle, then run like fuck.'

Sooner than expected, he heard the three sharp chirrups of Major Marquez's whistle.

'Lenny, you're last out,' he said.

'Who's first?' said Lenny.

'I am, of course.'

Goodbye, world, Lloyd thought. At least I'll die fighting Fascists.

He threw the door wide. 'Let's go!' he yelled, and he ran out.

Surprise gave him a few seconds' grace, and he ran freely along the street towards the church. He felt the scorch of the midday sun on his face and heard the pounding of his men's boots behind him, and noted with a weird sentiment of gratitude that such sensations meant he was still alive. Then gunfire broke out like a hailstorm. For a few more heartbeats he ran, hearing the zip and thwack of bullets, then there was a feeling in his left arm as if he had banged it against something, and inexplicably he fell down.

He realized he had been hit. There was no pain, but his arm was numb and hung lifeless. He managed to roll sideways until he hit the wall of the nearest building. Shots continued to fly, and he was terribly vulnerable, but a few feet ahead he saw a dead body. It was a rebel soldier, propped against the house. He looked as if he had been sitting on the ground, resting with his back against the wall, and had gone to sleep; except that there was a bullet wound in his neck.

Lloyd wriggled forward, moving awkwardly, rifle in his right hand, left arm dragging behind, then crouched behind the body, trying to make himself small.

He rested his rifle barrel on the dead man's shoulder and took aim at a high window in the church tower. He fired all five rounds in his magazine in rapid succession. He could not tell whether he had hit anyone.

He looked back. To his horror he saw the street littered with the corpses of his platoon. The still body of Mario Rivera in his red and black shirt looked like a crumpled anarchist flag. Next to Mario was Jasper Johnson, his black curls soaked in blood. All the way from a factory in Chicago, Lloyd thought, to die on the street in a small town in Spain, because he believed in a better world.

Worse were those who still lived, moaning and crying on the ground. Somewhere a man was screaming in agony, but Lloyd could not see who or where. A few of his men were still running, but, as he watched, more fell and others threw themselves down. Seconds later no one was moving except the writhing wounded.

What a slaughter, he thought, and a bile of anger and sorrow rose chokingly in his throat.

Where were the other units? Surely Lloyd's platoon was not the only one involved in the attack? Perhaps others were advancing along parallel streets leading to the square. But a rush required overwhelming numbers. Lloyd and his thirty-five were obviously too few. The defenders had been able to kill and wound nearly all of them, and the few who remained of Lloyd's platoon had been forced to take cover before reaching the church.

He caught the eye of Lenny, peering from behind the dead horse. At least he was still alive. Lenny held up his rifle and made a helpless gesture, pantomiming 'no ammunition'. Lloyd was out, too. In the next minute, firing from the street died away as the others also ran out of bullets.

That was the end of the attack on the church. It had been impossible anyway. With no ammunition it would have been pointless suicide.

The hail of fire from the church had lessened as the easier targets were eliminated, but sporadic sniping continued at those remaining behind cover. Lloyd realized that all his men would be killed eventually. They had to withdraw.

They would probably all be killed in the retreat.

He caught Lenny's eye again and waved emphatically towards the rear, away from the church. Lenny looked around, repeating the gesture to the few others left alive. They would have a better chance if they all moved at the same time.

When as many as possible had been forewarned, Lloyd struggled to his feet.

'Retreat!' he yelled at the top of his voice.

Then he began to run.

It was no more than two hundred yards, but it was the longest journey of his life.

The rebels in the church opened fire as soon as they saw the government troops move. Out of the corner of his eye, Lloyd thought he saw five or six of his men retreating. He ran with a ragged gait, his wounded arm putting him off balance. Lenny was ahead of him, apparently unhurt. Bullets scored the masonry of the buildings that Lloyd staggered past. Lenny made it to the house they had come from, dashed in, and held the door open. Lloyd ran in, panting hoarsely, and collapsed on the floor. Three more followed them in.

Lloyd stared at the survivors: Lenny, Dave, Muggsy Morgan and Joe Eli. 'Is that all?' he said.

Lenny said: 'Yes.'

'Jesus. Five of us left, out of thirty-six.'

'What a great military advisor Colonel Bobrov is.'

They stood panting, catching their breath. The feeling returned to Lloyd's arm and it hurt like hell. He found he could move it, painfully, so perhaps it was not broken. Looking down, he saw that his sleeve was soaked with blood. Dave took off his red scarf and improvised a sling.

Lenny had a head wound. There was blood on his face, but he said it was a scratch, and he seemed all right.

Dave, Muggsy and Joe were miraculously unhurt.

'We'd better go back for fresh orders,' Lloyd said when they had lain down a few minutes. 'We can't accomplish anything without ammunition, anyway.'

'Let's have a nice cup of tea first, is it?' said Lenny.

Lloyd said: 'We can't, we haven't got teaspoons.'

'Oh, all right, then.'

Dave said: 'Can't we rest here a bit longer?'

'We'll rest in the rear,' Lloyd said. 'It's safer.'

They made their way back along the row of houses, using the holes they had made in the walls. The repeated bending made Lloyd dizzy. He wondered if he was weak from loss of blood.

They emerged out of sight of the church of San Agustin, and hurried along a side street. Lloyd's relief at still being alive was rapidly giving way to a feeling of rage at the waste of the lives of his men.

They came to the barn on the outskirts where the government forces had made their headquarters. Lloyd saw Major Marquez behind a stack of crates, giving out ammunition. 'Why couldn't we have had some of that?' he said furiously.

Marquez just shrugged.

'I'm reporting this to Bobrov,' Lloyd said.

Colonel Bobrov was outside the barn, sitting on a chair at a table, both of which items of furniture looked as if they had been taken from a village house. His face was reddened with sunburn. He was talking to Volodya Peshkov. Lloyd went straight up to them. 'We rushed the church, but we had no support,' he said. 'And we ran out of ammunition because Marquez refused to supply us!'

Bobrov looked coldly at Lloyd. 'What are you doing here?' he said.

Lloyd was puzzled. He expected Bobrov to congratulate him for a brave effort and at least commiserate with him over the lack of support. 'I just told you,' he said. 'There was no support. You can't rush a fortified building with one platoon. We did our best, but we were slaughtered. I've lost thirty-one of my thirty-six men.' He pointed at his four companions. 'This is all that's left of my platoon!'

'Who ordered you to retreat?'

Lloyd was fighting off dizziness. He felt close to collapse, but he had to explain to Bobrov how bravely his men had fought. 'We came back for fresh orders. What else could we do?'

'You should have fought to the last man.'

'What should we have fought with? We had no bullets!'

'Silence!' Bobrov barked. 'Stand to attention!'

Automatically, they all stood to attention: Lloyd, Lenny, Dave, Muggsy and Joe in a line. Lloyd feared he was about to faint.

'About face!'

They turned their backs. Lloyd thought: What now?

'Those who are wounded, fall out.'

Lloyd and Lenny stepped back.

Bobrov said: 'The walking wounded are transferred to prisoner escort duty.'

Dimly, Lloyd perceived that this meant he would probably be guarding prisoners of war on a train to Barcelona. He swayed on his feet. Right now I couldn't guard a flock of sheep, he thought.

Bobrov said: 'Retreating under fire without orders is desertion.'

Lloyd turned and looked at Bobrov. To his astonishment and horror he saw that Bobrov had drawn his revolver from its buttoned holster.

Bobrov stepped forward so that he was immediately behind the three men standing to attention. 'You three are found guilty and sentenced to death.' He raised the gun until the barrel was three inches from the back of Dave's head.

Then he fired.

There was a bang. A bullet hole appeared in Dave's head, and blood and brains exploded from his brow.

Lloyd could not believe what he was seeing.

Next to Dave, Muggsy began to turn, his mouth open to shout; but Bobrov was quicker. He swung the gun to Muggsy's neck and fired again. The bullet entered behind Muggsy's right ear and came out through his left eye, and he crumpled.

At last Lloyd's voice came, and he shouted: 'No!'

Joe Eli turned, roaring with shock and rage, and raised his hands to grab Bobrov. The gun banged again and Joe got a bullet in the throat. Blood spurted like a fountain from his neck and splashed Bobrov's Red Army uniform, causing the colonel to curse and jump back a pace. Joe fell to the ground but did not die immediately. Lloyd watched, helpless, as the blood pumped out of Joe's carotid artery into the parched Spanish earth. Joe seemed to try to speak, but no words came; and then his eyes closed and he went limp.

'There's no mercy for cowards,' Bobrov said, and he walked away.

Lloyd looked at Dave on the ground: thin, grimy, brave as a lion, sixteen years old and dead. Killed not by the Fascists but by a stupid and brutal Soviet officer. What a waste, Lloyd thought, and tears came to his eyes.

A sergeant came running out of the barn. 'They've given up!' he shouted joyfully. 'The town hall has surrendered – they've raised the white flag. We've taken Belchite!'

The dizziness overwhelmed Lloyd at last, and he fainted.

(v)

London was cold and wet. Lloyd walked along Nutley Street in the rain, heading for his mother's house. He still wore his zipped Spanish army blouson and corduroy breeches, and boots with no socks. He carried a small backpack containing his spare underwear, a shirt, and a tin cup. Around his neck he had the red scarf Dave had turned into an improvised sling for his wounded arm. The arm still hurt, but he no longer needed the sling.

It was late on an October afternoon.

As expected, he had been put on a supply train returning to Barcelona crammed with rebel prisoners. The journey was not much more than a hundred miles, but it had taken three days. In Barcelona he had been separated from Lenny and lost contact with him. He had got a lift in a lorry going north. After the trucker dropped him off he had walked, hitch-hiked, and ridden in railway wagons full of coal or gravel or – on one lucky occasion – cases of wine. He had slipped across the border into France at night. He had slept rough, begged food, done odd jobs for a few coins and, for two glorious weeks, earned his cross-Channel boat fare picking grapes in a Bordeaux vineyard. Now he was home.

He inhaled the damp, soot-smelling Aldgate air as if it were perfume. He stopped at the garden gate and looked up at the terraced house in which he had been born more than twenty-two years ago. Lights glowed behind the rain-streaked windows: someone was at home. He walked up to the front door. He still had his key: he had kept it with his passport. He let himself in.

He dropped his backpack on the floor in the hall, by the hatstand.

From the kitchen he heard: 'Who's that?' It was the voice of his stepfather, Bernie.

Lloyd found he could not speak.

Bernie came into the hall. 'Who . . . ?' Then he recognized Lloyd. 'My life!' he said. 'It's you.'

Lloyd said: 'Hello, Dad.'

'My boy,' said Bernie. He put his arms around Lloyd. 'Alive,' Bernie said. Lloyd could feel him shaking with sobs.

After a minute Bernie rubbed his eyes with the sleeve of his cardigan then went to the bottom of the stairs. 'Eth!' he called.

'What?'

'Someone to see you.'

'Just a minute.'

She came down the stairs a few seconds later, pretty as ever in a blue dress. Halfway down she saw his face and turned pale. 'Oh, *Duw*,' she said. 'It's Lloyd.' She came down the rest of the stairs in a rush and threw her arms around him. 'You're alive!' she said.

'I wrote to you from Barcelona—'

'We never got that letter.'

'Then you don't know . . .'

'What?'

'Dave Williams died.'

'Oh, no!'

'Killed at the Battle of Belchite.' Lloyd had decided not to tell the truth about how Dave had died.

'What about Lenny Griffiths?'

'I don't know. I lost touch with him. I was hoping he might have got home before me.'

'No, there's no word.'

Bernie said: 'What was it like over there?'

'The Fascists are winning. And it's mainly the fault of the Communists, who are more interested in attacking the other left parties.'

Bernie was shocked. 'Surely not.'

'It's true. If I've learned one thing in Spain, it's that we have to fight the Communists just as hard as the Fascists. They're both evil.'

His mother smiled wryly. 'Well, just fancy that.' She had figured out the same thing long ago, Lloyd realized.

'Enough politics,' he said. 'How are you, Mam?'

'Oh, I'm the same, but look at you – you're so thin!'

'Not much to eat in Spain.'

'I'd better make you something.'

'No rush. I've been hungry for twelve months – I can keep going a few more minutes. I tell you what would be nice, though.'

'What? Anything!'

'I'd love a nice cup of tea.'

5

1939

Thomas Macke was watching the Soviet Embassy in Berlin when Volodya Peshkov came out.

The Prussian secret police had been transformed into the new, more efficient Gestapo six years ago, but Commissar Macke was still in charge of the section that monitored traitors and subversives in the city of Berlin. The most dangerous of them were undoubtedly getting their orders from this building at 63–65 Unter den Linden. So Macke and his men watched everyone who went in and came out.

The embassy was an art deco fortress made of a white stone that painfully reflected the glare of the August sun. A pillared lantern stood watchful above the central block, and to either side the wings had rows of tall, narrow windows like guardsmen at attention.

Macke sat at a pavement café opposite. Berlin's most elegant boulevard was busy with cars and bicycles; the women shopped in their summer dresses and hats; the men walked briskly by in suits or smart uniforms. It was hard to believe there were still German Communists. How could anyone possibly be against the Nazis? Germany was transformed. Hitler had wiped out unemployment – something no other European leader had achieved. Strikes and demonstrations were a distant memory of the bad old days. The police had no-nonsense powers to stamp out crime. The country was prospering: many families had a radio, and soon they would have people's cars to drive on the new autobahns.

And that was not all. Germany was strong again. The military was well armed and powerful. In the last two years both Austria and Czechoslovakia had been absorbed into Greater Germany, which was now the dominant power in Europe. Mussolini's Italy was allied with Germany in the Pact of Steel. Earlier this year Madrid had at last fallen to Franco's rebels, and Spain now had a Fascist-friendly government. How could any German wish to undo all that and bring the country under the heel of the Bolsheviks?

236

In Macke's eyes such people were scum, vermin, filth that had to be ruthlessly sought out and utterly destroyed. As he thought about them his face twisted into a scowl of anger, and he tapped his foot on the pavement as if preparing to stomp a Communist.

Then he saw Peshkov.

He was a young man in a blue serge suit, carrying a light coat over his arm as if expecting a change in the weather. His close-cropped hair and quick march indicated the army, despite his civilian clothes, and the way he scanned the street, deceptively casual but thorough, suggested either Red Army Intelligence or the NKVD, the Russian secret police.

Macke's pulse quickened. He and his men knew everyone at the embassy by sight, of course. Their passport photographs were on file and the team watched them all the time. But he did not know much about Peshkov. The man was young – twenty-five, according to his file, Macke recalled – so he might be a junior staffer of no importance. Or he could be good at seeming unimportant.

Peshkov crossed Unter den Linden and walked towards where Macke sat, near the corner of Friedrich Strasse. As Peshkov came closer, Macke noted that the Russian was quite tall, with the build of an athlete. He had an alert look and an intense gaze.

Macke looked away, suddenly nervous. He picked up his cup and sipped the cold dregs of his coffee, partly covering his face. He did not want to meet those blue eyes.

Peshkov turned into Friedrich Strasse. Macke nodded to Reinhold Wagner, standing on the opposite corner, and Wagner followed Peshkov. Macke then got up from his table and followed Wagner.

Not everyone in Red Army Intelligence was a cloak-and-dagger spy, of course. They got most of their information legitimately, mainly by reading the German newspapers. They did not necessarily believe everything they read, but they took note of clues such as an advertisement by a gun factory needing to recruit ten skilled lathe operators. Furthermore, Russians were free to travel Germany and look around – unlike diplomats in the Soviet Union, who were not allowed to leave Moscow unescorted. The young man whom Macke and Wagner were now tailing might be the tame, newspaper-reading kind of intelligence gatherer: all that was required for such a job was fluent German and the ability to summarize.

They followed Peshkov past Macke's brother's restaurant. It was still called Bistro Robert, but it had a different clientele. Gone were the wealthy homosexuals, the Jewish businessmen with their mistresses, and the overpaid actresses calling for pink champagne. Such people kept their heads down nowadays, if they were not already in concentration camps. Some had left Germany – and good riddance, Macke thought, even if it did, unfortunately, mean that the restaurant no longer made much money.

He wondered idly what had become of the former owner, Robert von Ulrich. He vaguely remembered that the man had gone to England. Perhaps he had opened a restaurant for perverts there.

Peshkov went into a bar.

Wagner followed him in a minute or two later, while Macke watched the outside. It was a popular place. While Macke waited for Peshkov to reappear, he saw a soldier and a girl enter, and a couple of well-dressed women and an old man in a grubby coat come out and walk away. Then Wagner came out alone, looked directly at Macke, and spread his arms in a gesture of bewilderment.

Macke crossed the street. Wagner was distressed. 'He's not there!'

'Did you look everywhere?'

'Yes, including the toilets and the kitchen.'

'Did you ask if anyone had gone out the back way?'

'They said not.'

Wagner was scared with reason. This was the new Germany, and errors were no longer dealt with by a slap on the wrist. He could be severely punished.

But not this time. 'That's all right,' said Macke.

Wagner could not hide his relief. 'Is it?'

'We've learned something important,' Macke said. 'The fact that he shook us off so expertly tells us that he's a spy – and a very good one.'

(ii)

Volodya entered the Friedrich Strasse Station and boarded a U-bahn train. He took off the cap, glasses and dirty raincoat that had helped him look like an old man. He sat down, took out a handkerchief, and wiped away the powder he had put on his shoes to make them appear shabby.

He had been unsure about the raincoat. It was such a sunny day that he feared the Gestapo might have noticed it and realized what he was up to. But they had not been that clever, and no one had followed him from the bar after he had done his quick change in the men's room.

He was about to do something highly dangerous. If they caught him contacting a German dissident, the best that he could expect was to be deported back to Moscow with his career in ruins. If he were less lucky, he and the dissident would both vanish into the basement of Gestapo headquarters in Prinz Albrecht Strasse, never to be seen again. The Soviets would complain that one of their diplomats had disappeared, and the German police would pretend to do a missing-persons search then regretfully report no success.

Volodya had never been to Gestapo headquarters, of course, but he knew what it would be like. The NKVD had a similar facility in the Soviet Trade Mission at 11 Lietsenburger Strasse: steel doors, an interrogation room with tiled walls so that the blood could be washed off easily, a tub for cutting up the bodies, and an electrical furnace for burning the parts.

Volodya had been sent to Berlin to expand the network of Soviet spies here. Fascism was triumphant in Europe, and Germany was more of a threat to the USSR now than ever. Stalin had fired his foreign minister, Litvinov, and replaced him with Vyacheslav Molotov. But what could Molotov do? The Fascists seemed unstoppable. The Kremlin was haunted by the humiliating memory of the Great War, in which the Germans had defeated a Russian army of six million men. Stalin had taken steps to form a pact with France and Britain to restrain Germany, but the three powers had been unable to agree, and the talks had broken down in the last few days.

Sooner or later, war was expected between Germany and the Soviet Union, and it was Volodya's job to gather military intelligence that would help the Soviets win that war.

He got off the train in the poor working-class district of Wedding, north of Berlin's centre. Outside the station he stood and waited, watching the other passengers as they left, pretending to study a timetable pasted on the wall. He did not move off until he was quite sure no one had followed him here.

Then he made his way to the cheap restaurant that was his chosen rendezvous. As was his regular practice, he did not go in, but stood at a bus stop on the other side of the road and watched the entrance. He was confident he had shaken off any tail, but now he needed to make sure Werner had not been followed.

He was not sure that he would recognize Werner Franck, who had been a fourteen-year-old boy when Volodya had last seen him, and was now twenty. Werner felt the same, so they had agreed they would both carry today's edition of the *Berliner Morgenpost* open to the sports page. Volodya read a preview of the new soccer season as he waited, glancing up every few seconds to look for Werner. Ever since being a schoolboy in Berlin, Volodya had followed the city's top team, Hertha. He had often chanted: 'Ha! Ho! He! Hertha B-S-C!' He was interested in the team's prospects, but anxiety spoiled his concentration, and he read the same report over and over again without taking anything in.

His two years in Spain had not boosted his career in the way he had hoped – rather the reverse. Volodya had uncovered numerous Nazi spies like Heinz Bauer among the German 'volunteers'. But then the NKVD had used that as an excuse to arrest genuine volunteers who had merely expressed mild disagreement with the Communist line. Hundreds of idealistic young men had been tortured and killed in the NKVD's prisons. At times it had seemed as if the Communists were more interested in fighting their anarchist allies than their Fascist enemies.

And all for nothing. Stalin's policy was a catastrophic failure. The upshot was a right-wing dictatorship, the worst imaginable outcome for the Soviet Union. But the blame was put on those Russians who had been in Spain, even though they had faithfully carried out Kremlin instructions. Some of them had disappeared soon after returning to Moscow.

Volodya had gone home in fear after the fall of Madrid. He had found many changes. In 1937 and 1938 Stalin had purged the Red Army. Thousands of commanders had disappeared, including many of the residents of Government House where his parents lived. But previously neglected men such as Grigori Peshkov had been promoted to take the places of those purged, and Grigori's career had a new impetus. He was in charge of the defence of Moscow against air raids, and was frantically busy. His enhanced status was probably the reason why Volodya was not among those scapegoated for the failure of Stalin's Spanish policy.

The unpleasant Ilya Dvorkin had also somehow avoided punishment. He was back in Moscow and married to Volodya's sister, Anya, much to Volodya's regret. There was no accounting for women's choices in such matters. She was already pregnant, and Volodya could not repress a nightmare image of her nursing a baby with the head of a rat.

After a brief leave, Volodya had been posted to Berlin, where he had to prove his worth all over again.

He looked up from his paper to see Werner walking along the street.

Werner had not changed much. He was a little taller and broader, but he had the same strawberry-blond hair falling over his forehead in a way girls had found irresistible, the same look of tolerant amusement in his blue eyes. He wore an elegant light-blue summer suit, and gold links glinted at his cuffs.

There was no one following him.

Volodya crossed the road and intercepted him before he reached the café. Werner smiled broadly, showing white teeth. 'I wouldn't have recognized you with that army haircut,' he said. 'It's good to see you, after all these years.'

He had not lost any of his warmth and charm, Volodya noted. 'Let's go inside.'

'You don't really want to go into that dump, do you?' Werner said. 'It will be full of plumbers eating sausages with mustard.'

'I want to get off the street. Here we could be seen by anyone passing.'

'There's an alley three doors down.'

'Good.'

They walked a short distance and turned into a narrow passage between a coal yard and a grocery store. 'What have you been doing?' Werner said.

'Fighting the Fascists, just like you.' Volodya considered whether to tell him more. 'I was in Spain.' It was no secret.

'Where you had no more success than we did here in Germany.'

'But it's not over yet.'

'Let me ask you something,' Werner said, leaning against the wall. 'If you thought Bolshevism was wicked, would you be a spy working against the Soviet Union?'

Volodya's instinct was to say *No, absolutely not!* But before the words came out he realized how tactless that would be – for the prospect that revolted him was precisely what Werner was doing, betraying his country for the sake of a higher cause. 'I don't know,' he said. 'I think it must be very difficult for you to work against Germany, even though you hate the Nazis.'

'You're right,' Werner said. 'And what happens if war breaks out? Am I going to help you kill our soldiers and bomb our cities?'

Volodya was worried. It seemed that Werner was weakening. 'It's the only way to defeat the Nazis,' he said. 'You know that.'

'I do. I made my decision a long time ago. And the Nazis have done nothing to change my mind. It's hard, that's all.'

'I understand,' Volodya said sympathetically.

Werner said: 'You asked me to suggest other people who might do for you what I am doing.'

Volodya nodded. 'People like Willi Frunze. Remember him? Cleverest boy in school. He was a serious socialist – he chaired that meeting the Brownshirts broke up.'

Werner shook his head. 'He went to England.'

Volodya's heart sank. 'Why?'

'He's a brilliant physicist and he's studying in London.'

'Shit.'

'But I've thought of someone else.'

'Good!'

'Did you ever know Heinrich von Kessel?'

'I don't think so. Was he at our school?'

'No, he went to a Catholic school. And in those days he didn't share our politics, either. His father was a big shot in the Centre Party—'

'Which put Hitler in power in 1933!'

'Correct. Heinrich was then working for his father. The father has now joined the Nazis, but the son is wracked by guilt.'

'How do you know?'

'He got drunk and told my sister, Frieda. She's seventeen. I think he fancies her.'

This was promising. Volodya's spirits lifted. 'Is he a Communist?'

'No.'

'What makes you think he'll work for us?'

'I asked him, straight out. "If you got a chance to fight against the Nazis by spying for the Soviet Union, would you do it?" He said he would.'

'What's his job?'

'He's in the army, but he has a weak chest, so they made him a pen-pusher – which is lucky for us, because now he works for the Supreme High Command in the economic planning and procurement department.'

Volodya was impressed. Such a man would know exactly how many trucks and tanks and machine guns and submarines the German military was acquiring month by month – and where they were being deployed. He began to feel excited. 'When can I meet him?'

'Now. I've arranged to have a drink with him in the Adlon Hotel after work.'

Volodya groaned. The Adlon was Berlin's swankiest hotel. It was located on Unter den Linden. Because it was in the government and diplomatic district, the bar was a favourite haunt of journalists hoping to pick up gossip. It would not have been Volodya's choice of rendezvous. But he could not afford to miss this chance. 'All right,' he said. 'But I'm not going to be seen talking to either of you in that place. I'll follow you in, identify Heinrich, then follow him out and accost him later.'

'Okay. I'll drive you there. My car's around the corner.'

As they walked to the other end of the alley, Werner told Volodya Heinrich's work and home addresses and phone numbers, and Volodya committed them to memory.

'Here we are,' said Werner. 'Jump in.'

The car was a Mercedes 540K Autobahn Kurier, a model that was head-turningly beautiful, with sensually curved fenders, a bonnet longer than an entire Ford Model T, and a sloping fastback rear end. It was so expensive that only a handful had ever been sold.

Volodya stared aghast. 'Shouldn't you have a less ostentatious car?' he said incredulously.

'It's a double bluff,' Werner said. 'They think no real spy would be so flamboyant.'

Volodya was going to ask how he could afford it, but then he recalled that Werner's father was a wealthy manufacturer.

'I'm not getting into that thing,' Volodya said. 'I'll go by train.'

'As you wish.'

'I'll see you at the Adlon, but don't acknowledge me.'

'Of course.'

Half an hour later, Volodya saw Werner's car carelessly parked in front of the hotel. This cavalier attitude of Werner's seemed foolish to him, but now he wondered whether it was a necessary element of Werner's courage. Perhaps Werner had to pretend to be carefree in order to take the appalling risks required to spy on the Nazis. If he acknowledged the danger he was in, maybe he would not be able to carry on.

The bar of the Adlon was full of fashionable women and well-dressed men, many in smartly tailored uniforms. Volodya spotted Werner right away, at a table with another man who was presumably Heinrich von Kessel. Passing close to them, Volodya heard Heinrich say argumentatively: 'Buck Clayton is a much better trumpeter than Hot Lips Page.' He squeezed in at the counter, ordered a beer, and discreetly studied the new potential spy.

Heinrich had pale skin and thick dark hair that was long by army standards. Although they were talking about the relatively unimportant topic of jazz, he seemed very intense, arguing with gestures and repeatedly running his fingers through his hair. He had a book stuffed into the pocket of his uniform tunic, and Volodya would have bet it contained poetry.

Volodya drank two beers slowly and pretended to read the *Morgenpost* from cover to cover. He tried not to get too keyed up about Heinrich. The man was thrillingly promising, but there was no guarantee he would co-operate.

Recruiting informers was the hardest part of Volodya's work. Precautions were difficult to take because the target was not yet on side. The proposition often had to be made in inappropriate places, usually somewhere public. It was impossible to know how the target would react: he might be angry and shout his refusal, or be terrified and literally run away. But there was not much the recruiter could do to control the situation. At some point he just had to ask the simple, blunt question: 'Do you want to be a spy?'

He thought about how to approach Heinrich. Religion was probably the key to his personality. Volodya recalled his boss, Lemitov, saying: 'Lapsed Catholics make good agents. They reject the total authority of the Church only to accept the total authority of the Party.' Heinrich might need to seek forgiveness for what he had done. But would he risk his life?

At last Werner paid the bill and the two men left. Volodya followed. Outside the hotel they parted company, Werner driving off with a squeal of tyres and Heinrich going on foot across the park. Volodya went after Heinrich.

Night was falling, but the sky was clear and he could see well. There were many people strolling in the warm evening air, most of them in couples. Volodya looked back repeatedly, to make sure no one had followed him or Heinrich from the Adlon. When he was satisfied he took a deep breath, steeled his nerve, and caught up with Heinrich.

Walking alongside him, Volodya said: 'There is atonement for sin.'

Heinrich looked at him warily, as at someone who might be mad. 'Are you a priest?'

'You could strike back at the wicked regime you helped to create.'

Heinrich kept walking, but he looked worried. 'Who are you? What do you know about me?'

Volodya continued to ignore Heinrich's questions. 'The Nazis will be defeated, one day. That day could come sooner, with your help.'

'If you're a Gestapo agent hoping to entrap me, don't bother. I'm a loyal German.'

'Do you notice my accent?'

'Yes – you sound Russian.'

'How many Gestapo agents speak German with a Russian accent? Or have the imagination to fake it?'

Heinrich laughed nervously. 'I know nothing about Gestapo agents,' he said. 'I shouldn't have mentioned the subject – very foolish of me.'

'Your office produces reports of the quantities of armaments and other supplies ordered by the military. Copies of those reports could be immeasurably useful to the enemies of the Nazis.'

'To the Red Army, you mean.'

'Who else is going to destroy this regime?'

'We keep careful track of all copies of such reports.'

Volodya suppressed a surge of triumph. Heinrich was thinking about practical difficulties. That meant he was inclined to agree in principle. 'Make an extra carbon,' Volodya said. 'Or write out a copy in longhand. Or take someone's file copy. There are ways.'

'Of course there are. And any of them could get me killed.'

'If we do nothing about the crimes that are being committed by this regime . . . is life worth living?'

Heinrich stopped and stared at Volodya. Volodya could not guess what the man was thinking, but instinct told him to remain quiet. After a long pause, Heinrich sighed and said: 'I'll think about it.'

I have him, Volodya thought exultantly.

Heinrich said: 'How do I contact you?'

'You don't,' Volodya said. 'I will contact you.' He touched the brim of his hat, then walked back the way he had come.

He felt exultant. If Heinrich had not meant to accept the proposition he would have rejected it firmly. His promising to think about it was almost as good as acceptance. He would sleep on it. He would run over the dangers. But he would do it, eventually. Volodya felt almost certain.

He told himself not to be overconfident. A hundred things could go wrong.

All the same, he was full of hope as he left the park and walked in bright lights past the shops and restaurants of Unter den Linden. He had had no dinner, but he could not afford to eat on this street.

He took a tram eastwards into the low-rent neighbourhood called Friedrichshain and made his way to a small apartment in a tenement. The door was opened by a short, pretty girl of eighteen with fair hair. She wore a pink sweater and dark slacks, and her feet were bare. Although she was slim, she had delightfully generous breasts.

'I'm sorry to call unexpectedly,' Volodya said. 'Is it inconvenient?'

She smiled. 'Not at all,' she said. 'Come in.'

He stepped inside. She closed the door, then threw her arms around him. 'I'm always happy to see you,' she said, and kissed him eagerly.

Lili Markgraf was a girl with a lot of affection to give. Volodya had been taking her out about once a week since he got back to Berlin. He was not in love with her, and he knew that she dated other men, including Werner; but when they were together she was passionate.

After a moment she said: 'Have you heard the news? Is that why you've come?'

'What news?' Lili worked as a secretary in a press agency, and always heard things first.

'The Soviet Union has made a pact with Germany!' she said.

That made no sense. 'You mean with Britain and France, against Germany.'

'No, I don't! That's the surprise – Stalin and Hitler have made friends.'

'But...' Volodya tailed off, baffled. Friends with Hitler? It seemed crazy. Was this the solution devised by the new Soviet foreign minister, Molotov? We have failed to stop the tide of world Fascism – so we give up trying? Did my father fight a revolution for that?

(iii)

Woody Dewar saw Joanne Rouzrokh again after four years.

No one who knew her father actually believed he had tried to rape a starlet in the Ritz-Carlton Hotel. The girl had dropped the charges; but that was dull news, and the papers had given it little prominence.

Consequently, Dave was still a rapist in the eyes of Buffalo people. So Joanne's parents moved to Palm Beach and Woody lost touch.

Next time he saw her it was in the White House.

Woody was with his father, Senator Gus Dewar, and they were going to see the President. Woody had met Franklin D. Roosevelt several times. His father and the President had been friends for many years. But those had been social occasions, when FDR had shaken Woody's hand and asked him how he was getting along at school. This would be the first time Woody attended a real political meeting with the President.

They went in through the main entrance of the West Wing, passed through the entrance lobby, and stepped into a large waiting room; and there she was.

Woody stared at her in delight. She had hardly changed. With her narrow, haughty face and curved nose she still looked like the high priestess of an ancient religion. As ever, she wore simple clothes to dramatic effect: today she had on a dark-blue suit of some cool fabric and a straw hat the same colour with a big brim. Woody was glad he had put on a clean white shirt and his new striped tie this morning.

She seemed pleased to see him. 'You look great!' she said. 'Are you working in DC now?'

'Just helping out in my father's office for the summer,' he replied. 'I'm still at Harvard.'

She turned to his father and said deferentially: 'Good afternoon, Senator.'

'Hello, Joanne.'

Woody was thrilled to run into her. She was as alluring as ever. He wanted to keep the conversation going. 'What are you doing here?' Woody said.

'I work at the State Department.'

Woody nodded. That explained her deference to his father. She had joined a world in which people kowtowed to Senator Dewar. Woody said: 'What's your job?'

'I'm assistant to an assistant. My boss is with the President now, but I'm too lowly to go in with him.'

'You were always interested in politics. I recall an argument about lynching.'

'I miss Buffalo. What fun we used to have!'

Woody remembered kissing her at the Racquet Club Ball, and he felt himself blush.

His father said: 'Please give my best regards to your father,' indicating that they needed to move on.

Woody considered asking for her phone number, but she pre-empted him. 'I'd love to see you again, Woody,' she said.

He was delighted. 'Sure!'

'Are you free tonight? I'm having a few friends for cocktails.'

'Sounds great!'

She gave him the address, an apartment building not far away, then his father hurried him out of the other end of the room.

A guard nodded familiarly to Gus, and they stepped into another waiting room.

Gus said: 'Now, Woody, don't say anything unless the President addresses you directly.'

Woody tried to concentrate on the imminent meeting. There had been a political earthquake in Europe: the Soviet Union had signed a peace pact with Nazi Germany, upsetting everyone's calculations. Woody's father was a key member of the Senate Foreign Relations Committee, and the President wanted to know what he thought.

Gus Dewar had another subject to discuss. He wanted to persuade Roosevelt to revive the League of Nations.

It would be a tough sell. The USA had never joined the League and Americans did not much like it. The League had failed dismally to deal with the crises of the 1930s: Japanese aggression in the Far East, Italian imperialism in Africa, Nazi takeovers in Europe, the ruin of democracy in Spain. But Gus was determined to try. It had always been his dream, Woody knew: a world council to resolve conflicts and prevent war.

Woody was 100 per cent behind him. He had made a speech about this in a Harvard debate. When two nations had a quarrel, the worst possible procedure was for men to kill people on the other side. That seemed to him pretty obvious. 'I understand why it happens, of course,' he had said in the debate. 'Just like I understand why drunks get into fistfights. But that doesn't make it any less irrational.'

But now Woody found it hard to think about the threat of war in Europe. All his old feelings about Joanne came back in a rush. He wondered if she would kiss him again – maybe tonight. She had always liked him, and it seemed she still did – why else would she have invited him to her party? She had refused to date him, back in 1935, because he had been fifteen and she eighteen, which was understandable, though he had not thought so at the time. But now that they were both four years older, the age difference would not seem so stark – would it? He hoped not. He had dated girls in Buffalo and at Harvard, but he had not felt for any of them the overwhelming passion he had had for Joanne.

'Have you got that?' his father said.

Woody felt foolish. His father was about to make a proposal to the President that could bring world peace, and all Woody could think about was kissing Joanne. 'Sure,' he said. 'I won't say anything unless he speaks to me first.'

A tall, slim woman in her early forties came into the room, looking relaxed and confident, as if she owned the place; and Woody recognized Marguerite LeHand, nicknamed Missy, who managed Roosevelt's office. She had a long, masculine face with a big nose, and there was a touch of grey in her dark hair. She smiled warmly at Gus. 'What a pleasure to see you again, Senator.'

'How are you, Missy? You remember my son, Woodrow.'

'I do. The President is ready for you both.'

Missy's devotion to Roosevelt was famous. FDR was more fond of her than a married man was entitled to be, according to Washington gossip. Woody knew, from guarded but revealing remarks his parents made to one another, that Roosevelt's wife, Eleanor, had refused to sleep with him since she gave birth to their sixth child. The paralysis that had struck him five years later did not extend to his sexual equipment. Perhaps a man who had not slept with his wife for twenty years was entitled to an affectionate secretary.

She showed them through another door and across a narrow corridor, then they were in the Oval Office.

The President sat at a desk with his back to three tall windows in a curving bay. The blinds were drawn to filter the August sun coming through the south-facing glass. Roosevelt used an ordinary office chair, Woody saw, not his wheelchair. He wore a white suit and he was smoking a cigarette in a holder.

He was not really handsome. He had receding hair and a jutting chin, and he wore pince-nez glasses that made his eyes seem too close together. All the same, there was something immediately attractive about his engaging smile, his hand extended to shake, and the amiable tone of voice in which he said: 'Good to see you, Gus, come on in.'

'Mr President, you remember my elder son, Woodrow.'

'Of course. How's Harvard, Woody?'

'Just fine, sir, thank you. I'm on the debating team.' He knew that politicians often had the knack of seeming to know everyone intimately. Either they had remarkable memories, or their secretaries reminded them efficiently.

'I was at Harvard myself. Sit down, sit down.' Roosevelt removed the end of his cigarette from the holder and stubbed it in a full ashtray. 'Gus, what the heck is happening in Europe?'

The President knew what was happening in Europe, of course, thought Woody. He had an entire State Department to tell him. But he wanted Gus Dewar's analysis.

Gus said: 'Germany and Russia are still mortal enemies, in my opinion.'

'That's what we all thought. But then why have they signed this pact?'

'Short-term convenience for both. Stalin needs time. He wants to build up the Red Army, so they can defeat the Germans if it comes to that.'

'And the other guy?'

'Hitler is clearly on the point of doing something to Poland. The German press is full of ridiculous stories about how the Poles are mistreating their German-speaking population. Hitler doesn't stir up hatred without a purpose. Whatever he's planning, he doesn't want the Soviets to stand in his way. Hence the pact.'

'That's pretty much what Hull says.' Cordell Hull was Secretary of State. 'But he doesn't know what will happen next. Will Stalin let Hitler do anything he wants?'

'My guess is they'll carve up Poland between them in the next couple of weeks.'

'And then what?'

'A few hours ago the British signed a new treaty with the Poles promising to come to their aid if Poland is attacked.'

'But what can they do?'

'Nothing, sir. The British army, navy and air force have no power to prevent the Germans overrunning Poland.'

'What do you think we should do, Gus?' said the President.

Woody knew that this was his father's chance. He had the President's attention for a few minutes. It was a rare opportunity to make something happen. Woody discreetly crossed his fingers.

Gus leaned forward. 'We don't want our sons to go to war as we did.' Roosevelt had four boys in their twenties and thirties. Woody suddenly understood why he was here: he had been brought to the meeting to remind the President of his own sons. Gus said quietly: 'We can't send American boys to be slaughtered in Europe again. The world needs a police force.'

'What do you have in mind?' Roosevelt said non-committally.

'The League of Nations isn't such a failure as people think. In the 1920s it resolved a border dispute between Finland and Sweden, and another between Turkey and Iraq.' Gus was ticking items off on his fingers. 'It stopped Greece and Yugoslavia from invading Albania, and persuaded Greece to pull out of Bulgaria. And it sent a peacekeeping force to keep Colombia and Peru from hostilities.'

'All true. But in the thirties . . .'

'The League was not strong enough to deal with Fascist aggression. It's not surprising. The League was crippled from the start because Congress refused to ratify the Covenant, so the United States was never a member. We need a new, American-led version, with teeth.' Gus paused. 'Mr President, it's too soon to give up on a peaceful world.'

Woody held his breath. Roosevelt nodded, but then he always nodded, Woody knew. It was rare for him to disagree openly. He hated confrontation. You had to be careful, Woody had heard his father say, not to take his silence for consent. Woody did not dare look at his father, sitting beside him, but he could sense the tension.

At last the President said: 'I believe you're right.'

Woody had to restrain himself from whooping aloud. The President had consented! He looked at his father. The normally imperturbable Gus was barely concealing his surprise. It had been such a quick victory.

Gus moved rapidly to consolidate it. 'In that case, may I suggest that Cordell Hull and I draft a proposal for your consideration?'

'Hull has a lot on his plate. Talk to Welles.'

Sumner Welles was Undersecretary of State. He was both ambitious and flamboyant, and Woody knew he would not have been Gus's first choice. But he was a long-time friend of the Roosevelt family – he had been a pageboy at FDR's wedding.

Anyway, Gus was not going to make difficulties at this point. 'By all means,' he said.

'Anything else?'

That was clearly dismissal. Gus stood up, and Woody followed suit. Gus said: 'What about Mrs Roosevelt, your mother, sir? Last I heard, she was in France.'

'Her ship left yesterday, thank goodness.'

'I'm glad to hear it.'

'Thank you for coming in,' Roosevelt said. 'I really value your friendship, Gus.'

Gus said: 'Nothing could give me more pleasure, sir.' He shook hands with the President, and Woody did the same.

Then they left.

Woody half hoped that Joanne would still be hanging around, but she had gone.

As they made their way out of the building, Gus said: 'Let's go for a celebratory drink.'

Woody looked at his watch. It was five o'clock. 'Sure,' he said.

They went to Old Ebbitt's, on F Street near 15th: stained glass, green velvet, brass lamps and hunting trophies. The place was full of congressmen, senators and the people who followed them around: aides, lobbyists and journalists. Gus ordered a dry martini straight up with a twist for himself and a beer for Woody. Woody smiled: maybe he would have liked a martini. In fact, he would not – to him it just tasted like cold gin – but it would have been nice to be asked. However, he raised his glass and said: 'Congratulations. You got what you wanted.'

'What the world needs.'

'You argued brilliantly.'

'Roosevelt hardly needed convincing. He's a liberal, but a pragmatist. He knows you can't do everything, you have to pick the battles you can win. The New Deal is his number one priority – getting unemployed men back to work. He won't do anything that interferes with the main mission. If my plan becomes controversial enough to upset his supporters, he'll drop it.'

'So we haven't won anything yet.'

Gus smiled. 'We've taken the important first step. But no, we haven't won anything.'

'A pity he forced Welles on you.'

'Not entirely. Sumner strengthens the project. He's closer to the President than I am. But he's unpredictable. He might pick it up and run in a different direction.'

Woody looked across the room and saw a familiar face. 'Guess who's here. I might have known.'

His father looked in the same direction.

'Standing at the bar,' Woody said. 'With a couple of older guys in hats, and a blonde girl. It's Greg Peshkov.' As usual, Greg looked a mess despite his expensive clothes: his silk tie was awry, his shirt was coming out of his waistband, and there was a smear of cigarette ash on his ice-cream-coloured trousers. Nevertheless, the blonde was looking adoringly at him.

'So it is,' said Gus. 'Do you see much of him at Harvard?'

'He's a physics major, but he doesn't hang around with the scientists – too dull for him, I guess. I run into him at the *Crimson*.' The *Harvard Crimson* was the student newspaper. Woody took photographs for the paper and Greg wrote articles. 'He's doing an internship at the State Department this summer, that's why he's here.'

'In the press office, I imagine,' said Gus. 'The two men he's with are reporters, the one in the brown suit for the *Chicago Tribune* and the pipe smoker for the Cleveland *Plain Dealer*.'

Woody saw that Greg was talking to the journalists as if they were old friends, taking the arm of one as he leaned forward to say something in a low voice, patting the other on the back in mock congratulation. They seemed to like him, Woody thought, as they laughed loudly at something he said. Woody envied that talent. It was useful to politicians – though perhaps not essential: his father did not have that hail-fellow-well-met quality, and he was one of the most senior statesmen in America.

Woody said: 'I wonder how his half-sister Daisy feels about the threat of war. She's over there in London. She married some English lord.'

'To be exact, she married the elder son of Earl Fitzherbert, whom I used to know quite well.'

'She's the envy of every girl in Buffalo. The King went to her wedding.'

'I also knew Fitzherbert's sister, Maud – a wonderful woman. She married Walter von Ulrich, a German. I would have married her myself if Walter hadn't got to her first.'

Woody raised his eyebrows. It was not like Papa to talk this way.

'That was before I fell in love with your mother, of course.'

'Of course.' Woody smothered a grin.

'Walter and Maud dropped out of sight after Hitler banned the Social Democrats. I hope they're all right. If there's a war . . .'

Woody saw that talk of war had put his father in a reminiscent mood. 'At least America isn't involved.'

'That's what we thought last time.' Gus changed the subject. 'What do you hear from your kid brother?'

Woody sighed. 'He's not going to change his mind, Papa. He won't go to Harvard, or any other university.'

This was a family crisis. Chuck had announced that as soon as he was eighteen he was going to join the navy. Without a college degree he would be an enlisted man, with no prospect of ever becoming an officer. This horrified his high-achieving parents.

'He's bright enough for college, damn it,' said Gus.

'He beats me at chess.'

'He beats me, too. So what's his problem?'

'He hates to study. And he loves boats. Sailing is the only thing he cares about.' Woody looked at his wristwatch.

'You've got a party to go to,' his father said.

'There's no hurry—'

'Sure there is. She's a very attractive girl. Get the hell out of here.'

Woody grinned. His father could be surprisingly smart. 'Thanks, Papa.' He got up.

Greg Peshkov was leaving at the same time, and they went out together. 'Hello, Woody, how are things?' Greg said amiably, turning in the same direction.

There had been a time when Woody wanted to punch Greg for his part in what had been done to Dave Rouzrokh. His feelings had cooled over the years, and in truth it was Lev Peshkov who had been responsible, not his son, who had then been only fifteen. All the same, Woody was no more than polite. 'I'm enjoying Washington,' he said, walking along one of the city's wide Parisian boulevards. 'How about you?'

'I like it. They soon get over their surprise at my name.' Seeing Woody's enquiring look, Greg explained: 'The State Department is all Smiths, Fabers, Jensens and McAllisters. No one called Kozinsky or Cohen or Papadopoulos.'

Woody realized it was true. Government was carried on by a rather exclusive little ethnic group. Why had he not noticed that before? Perhaps because it had been the same in school, in church, and at Harvard.

Greg went on: 'But they're not narrow-minded. They'll make an exception for someone who speaks fluent Russian and comes from a wealthy family.'

Greg was being flippant, but there was an undertone of real resentment, and Woody saw that the guy had a serious chip on his shoulder.

'They think my father is a gangster,' Greg said. 'But they don't really mind. Most rich people have a gangster somewhere in their ancestry.'

'You sound as if you hate Washington.'

'On the contrary! I wouldn't be anywhere else. The power is here.'

Woody felt he was more high-minded. 'I'm here because there are things I want to do, changes I want to make.'

Greg grinned. 'Same thing, I guess – power.'

'Hmm.' Woody had not thought of it that way.

Greg said: 'Do you think there will be war in Europe?'

'You should know, you're in the State Department!'

'Yeah, but I'm in the press office. All I know is the fairy tales we tell reporters. I have no idea what the truth is.'

'Heck, I don't know, either. I've just been with the President and I don't think even he knows.'

'My sister, Daisy, is over there.'

Greg's tone had changed. His worry was evidently genuine, and Woody warmed to him. 'I know.'

'If there's bombing, even women and children won't be safe. Do you think the Germans will bomb London?'

There was only one honest answer. 'I guess they will.'

'I wish she'd come home.'

'Maybe there won't be a war. Chamberlain, the British premier, made a last-minute deal with Hitler over Czechoslovakia last year—'

'A last-minute sell-out.'

'Right. So perhaps he'll do the same over Poland – although time is running out.'

Greg nodded glumly and changed the subject. 'Where are you headed?'

'To Joanne Rouzrokh's apartment. She's giving a party.'

'I heard about it. I know one of her room-mates. But I'm not invited, as you could probably guess. Her building is— good God!' Greg stopped in mid-sentence.

Woody stopped, too. Greg was staring ahead. Following his gaze, Woody saw that he was looking at an attractive black woman walking towards them on E Street. She was about their age, and pretty, with wide pinky-brown lips that made Woody think about kissing. She had on a plain black dress that might have been part of a waitress uniform, but she wore it with a cute hat and fashionable shoes that gave her a stylish look.

She saw the two of them, caught Greg's eye, and looked away.

Greg said: 'Jacky? Jacky Jakes?'

The girl ignored him and kept walking, but Woody thought she looked troubled.

Greg said: 'Jacky, it's me, Greg Peshkov.'

Jacky – if it were she – did not respond, but she looked as if she might be about to burst into tears.

'Jacky – real name Mabel. You know me!' Greg stood in the middle of the sidewalk with his arms spread in a gesture of appeal.

She deliberately went around him, not speaking or meeting his eye, and walked on.

Greg turned. 'Wait a minute!' he called after her. 'You ran out on me, four years ago – you owe me an explanation!'

This was uncharacteristic of Greg, Woody thought. He had always been such a smooth operator with girls, at school and at Harvard. Now he seemed genuinely upset: bewildered, hurt, almost desperate.

Four years ago, Woody reflected. Could this be the girl in the scandal? It had taken place here in Washington. No doubt she lived here.

Greg ran after her. A cab had stopped at the corner and the passenger, a man in a tuxedo, was standing at the kerb paying the driver. Jacky jumped in, slamming the door.

Greg went to the window and shouted through it: 'Talk to me, please!'

The man in the tuxedo said: 'Keep the change,' and walked away.

The cab moved off, leaving Greg staring after it.

He slowly returned to where Woody stood waiting, intrigued. 'I don't understand it,' Greg said.

Woody said: 'She looked frightened.'

'What of? I never did her any harm. I was crazy about her.'

'Well, she was scared of something.'

Greg seemed to shake himself. 'Sorry,' he said. 'Not your problem, anyway. My apologies.'

'Not at all.'

Greg pointed to an apartment block a few steps away. 'That's Joanne's building,' he said. 'Have a good time.' Then he walked away.

Somewhat bemused, Woody went to the entrance. But he soon forgot about Greg's romantic life and started to think about his own. Did Joanne really like him? She might not kiss him this evening, but maybe he could ask her for a date.

This was a modest apartment house, with no doorman or hall porter. A list in the lobby revealed that Rouzrokh shared her place with Stewart and Fisher, presumably two other girls. Woody went up in the elevator. He realized he was empty-handed: he should have brought candy or flowers. He thought about going back to buy something, then decided that would be taking good manners too far. He rang the bell.

A girl in her early twenties opened the door.

Woody said: 'Hello, I'm—'

'Come on in,' she said, not waiting to hear his name. 'The drinks are in the kitchen, and there's food on the table in the living room, if there's any left.' She turned away, clearly thinking she had given him sufficient welcome.

The small apartment was packed with people drinking, smoking, and shouting at one another over the noise of the phonograph. Joanne had said 'a few friends' and Woody had imagined eight or ten young people sitting around a coffee table discussing the crisis in Europe. He was disappointed: this overcrowded bash would give him little opportunity to demonstrate to Joanne how much he had grown up.

He looked around for her. He was taller than most people and could see over their heads. She was not in sight. He pushed through the crowd, searching for her. A girl with plump breasts and nice brown eyes looked up at him as he squeezed past and said: 'Hello, big guy. I'm Diana Taverner. What's your name?'

'I'm looking for Joanne,' he said.

She shrugged. 'Good luck with that.' She turned away.

He made his way into the kitchen. The noise level dropped a fraction. Joanne was nowhere to be seen, but he decided to get a drink while he was there. A broad-shouldered man of about thirty was rattling a cocktail shaker. Well dressed in a tan suit, pale-blue shirt and dark-blue tie, he clearly was not a barman, but was acting like a host. 'Scotch is over there,' he said to another guest. 'Help yourself. I'm making martinis, for anyone who's interested.'

Woody said: 'Got any bourbon?'

'Right here.' The man passed him a bottle. 'I'm Bexforth Ross.'

'Woody Dewar.' Woody found a glass and poured bourbon.

'Ice in that bucket,' said Bexforth. 'Where are you from, Woody?'

'I'm an intern in the Senate. You?'

'I work in the State Department. I'm in charge of the Italy desk.' He started passing martinis around.

Clearly a rising star, Woody thought. The man had so much self-confidence it was irritating. 'I was looking for Joanne.'

'She's somewhere around. How do you know her?'

Here Woody felt he could show clear superiority. 'Oh, we're old friends,' he said airily. 'In fact, I've known her all my life. We were kids together in Buffalo. How about you?'

Bexforth took a long sip of martini and gave a satisfied sigh. Then he looked speculatively at Woody. 'I haven't known Joanne as long as you have,' he said. 'But I guess I know her better.'

'How so?'

'I'm planning to marry her.'

Woody felt as if he had been slapped. 'Marry her?'

'Yes. Isn't that great?'

Woody could not hide his dismay. 'Does she know about this?'

Bexforth laughed, and patted Woody's shoulder condescendingly. 'She sure does, and she's all for it. I'm the luckiest guy in the world.'

Clearly Bexforth had divined that Woody was attracted to Joanne. Woody felt a fool. 'Congratulations,' he said dispiritedly.

'Thank you. And now I must circulate. Good talking to you, Woody.'

'My pleasure.'

Bexforth moved away.

Woody put his drink down untasted. 'Fuck it,' he said quietly. Then he left.

(iv)

The first day of September was sultry in Berlin. Carla von Ulrich woke up sweaty and uncomfortable, her bedsheets thrown off during the warm night. She looked out of her bedroom window to see low grey clouds hanging over the city, keeping heat in like a saucepan lid.

Today was a big day for her. In fact, it would determine the course of her life.

She stood in front of the mirror. She had her mother's colouring, the dark hair and green eyes of the Fitzherberts. She was prettier than Maud, who had an angular face, striking rather than beautiful. Yet there was a bigger difference. Her mother attracted just about every man she met. Carla, by contrast, could not flirt. She watched other girls her age doing it: simpering, pulling their sweaters tight over their breasts, tossing their hair, and batting their eyelashes, and she just felt embarrassed. Her mother was more subtle, of course, so that men hardly knew they were being enchanted, but it was essentially the same game.

Today, however, Carla did not want to appear sexy. On the contrary, she needed to look practical, sensible, and capable. She put on a plain stone-coloured cotton dress that came to mid-calf, stepped into her flat, unglamorous school sandals, and wove her hair into two plaits in the approved German-maiden fashion. The mirror showed her an ideal girl student: conservative, dull, sexless.

She was up and dressed before the rest of the family. The maid, Ada, was in the kitchen, and Carla helped her set out the breakfast things.

Her brother appeared next. Erik, nineteen and sporting a clipped black moustache, supported the Nazis, infuriating the rest of his family. He was a student at the Charité, the medical school of the University of Berlin, as was his best friend and fellow-Nazi, Hermann Braun. The von Ulrichs could not afford tuition fees, of course, but Erik had won a scholarship.

Carla had applied for the same scholarship to study at the same institution. Her interview was today. If she was successful, she would study and become a doctor. If not . . .

She had no idea what else she would do.

The coming to power of the Nazis had ruined her parents' lives. Her father was no longer a deputy in the Reichstag, having lost his job when the Social-Democratic Party became illegal, along with all other parties except for the Nazis. There was no work her father could do that would use his expertise as a politician and a diplomat. He scraped a living translating German newspaper articles for the British Embassy, where he still had a few friends. Mother had once been a famous left-wing journalist, but newspapers were no longer allowed to publish her articles.

Carla found it heartbreaking. She was deeply devoted to her family, which included Ada. She was saddened by the decline in her father, who in her childhood had been a hard-working and politically powerful man, and was now simply defeated. Even worse was the brave face put on by her mother, a famous suffragette leader in England before the war, now scraping a few marks by giving piano lessons.

But they said they could bear anything as long as their children grew up to lead happy and fulfilled lives.

Carla had always taken it for granted that she would spend her life making the world a better place, as her parents had. She did not know whether she would have followed her father into politics or her mother into journalism, but both were out of the question now.

What else was she to do, under a government that prized ruthlessness and brutality above all else? Her brother had given her the clue. Doctors made the world a better place regardless of the government. So she had made it her ambition to go to medical school. She had studied harder than any other girl in her class, and she had passed every exam with top marks, especially the sciences. She was better qualified than her brother to win a scholarship.

'There are no girls at all in my year,' Erik said. He sounded grumpy. Carla thought he disliked the idea of her following in his footsteps. Their parents were proud of his achievements, despite his repellent politics. Perhaps he was afraid of being outshone.

Carla said: 'All my grades are better than yours: biology, chemistry, maths—'

'All right, all right.'

'And the scholarship is available to female students, in principle – I checked.'

Their mother came in at the end of this exchange, dressed in a grey watered-silk bathrobe with the cord doubled around her narrow waist. 'They should follow their own rules,' she said. 'This is Germany, after all.' Mother said she loved her adopted country, and perhaps she did, but since the coming of the Nazis she had taken to making wearily ironic remarks.

Carla dipped bread into milky coffee. 'How will you feel, Mother, if England attacks Germany?'

'Miserably unhappy, as I felt last time,' she replied. 'I was married to your father throughout the Great War, and every day for more than four years I was terrified that he would be killed.'

Erik said in a challenging tone: 'But whose side will you take?'

'I'm German,' she said. 'I married for better or worse. Of course, we never foresaw anything as wicked and oppressive as this Nazi regime.

No one did.' Erik grunted in protest and she ignored him. 'But a vow is a vow, and, anyway, I love your father.'

Carla said: 'We're not at war yet.'

'Not quite,' said Mother. 'If the Poles have any sense, they will back down and give Hitler what he asks for.'

'They should,' said Erik. 'Germany is strong now. We can take what we want, whether they like it or not.'

Mother rolled her eyes. 'God spare us.'

A car horn sounded outside. Carla smiled. A minute later her friend Frieda Franck entered the kitchen. She was going to accompany Carla to the interview, just to give moral support. She, too, was dressed in sober-schoolgirl fashion, though she, unlike Carla, had a wardrobe full of stylish clothes.

She was followed in by her older brother. Carla thought Werner Franck was wonderful. Unlike so many handsome boys he was kind and thoughtful and funny. He had once been very left wing, but all that seemed to have faded away, and he was non-political now. He had had a string of beautiful and stylish girlfriends. If Carla had known how to flirt she would have started with him.

Mother said: 'I'd offer you coffee, Werner, but ours is ersatz, and I know you have the real thing at home.'

'Shall I steal some from our kitchen for you, Frau von Ulrich?' he said. 'I think you deserve it.'

Mother blushed slightly, and Carla realized, with a twinge of disapproval, that even at forty-eight Mother was susceptible to Werner's charm.

Werner glanced at a gold wristwatch. 'I have to go,' he said. 'Life is completely frantic at the Air Ministry these days.'

Frieda said: 'Thank you for the lift.'

Carla said to Frieda: 'Wait a minute – if you came in Werner's car, where's your bike?'

'Outside. We strapped it to the back of the car.'

The two girls belonged to the Mercury Cycling Club and went everywhere by bike.

Werner said: 'Best wishes for the interview, Carla. Bye, everyone.'

Carla swallowed the last of her bread. As she was about to leave, her father came down. He had not shaved or put on a tie. He had been quite plump, when Carla was a girl, but now he was thin. He kissed Carla affectionately.

Mother said: 'We haven't listened to the news!' She turned on the radio that stood on the shelf.

While the set was warming up, Carla and Frieda left the house, so they did not hear the news.

The University Hospital was in Mitte, the central area of Berlin where the von Ulrichs lived, so Carla and Frieda had a short bicycle ride. Carla began to feel nervous. The fumes from car exhausts nauseated her, and she wished she had not eaten breakfast. They reached the hospital, a new building put up in the twenties, and found their way to the room of Professor Bayer, who had the job of recommending a student for the scholarship. A haughty secretary said they were early and told them to wait.

Carla wished she had worn a hat and gloves. That would have made her look older and more authoritative, like someone sick people would trust. The secretary might have been polite to a girl in a hat.

The wait was long, but Carla was sorry when it came to an end and the secretary said the professor was ready to see her.

Frieda whispered: 'Good luck!'

Carla went in.

Bayer was a thin man in his forties with a small grey moustache. He sat behind a desk, wearing a tan linen jacket over the waistcoat of a grey business suit. On the wall was a photograph of him shaking hands with Hitler.

He did not greet Carla, but barked: 'What is an imaginary number?'

She was taken aback by his abruptness, but at least it was an easy question. 'The square root of a negative real number; for example, the square root of minus one,' she said in a shaky voice. 'It cannot be assigned a real numerical value but can, nevertheless, be used in calculations.'

He seemed a bit surprised. Perhaps he had expected to floor her completely. 'Correct,' he said after a momentary hesitation.

She looked around. There was no chair for her. Was she to be interviewed standing up?

He asked her some questions on chemistry and biology, all of which she answered easily. She began to feel a bit less nervous. Then he suddenly said: 'Do you faint at the sight of blood?'

'No, sir.'

'Aha!' he said triumphantly. 'How do you know?'

'I delivered a baby when I was eleven years old,' she said. 'That was quite bloody.'

'You should have sent for a doctor!'

'I did,' she said indignantly. 'But babies don't wait for doctors.'

'Hmm.' Bayer stood up. 'Wait there.' He left the room.

Carla stayed where she was. She was being subjected to a harsh test, but so far she thought she was doing all right. Fortunately, she was used to give-and-take arguments with men and women of all ages: combative discussions were commonplace in the von Ulrich house, and she had been holding her own with her parents and brother for as long as she could remember.

Bayer was gone for several minutes. What was he doing? Had he gone to fetch a colleague to meet this unprecedentedly brilliant girl applicant? That seemed too much to hope for.

She was tempted to pick up one of the books on his shelf and read, but she was scared of offending him, so she stood still and did nothing.

He came back after ten minutes with a pack of cigarettes. Surely he had not kept her standing in the middle of the room all this time while he went to the tobacconist's shop? Or was that another test? She began to feel angry.

He took his time lighting up, as if he needed to collect his thoughts. He blew out smoke and said: 'How would you, as a woman, deal with a man who had an infection of the penis?'

She was embarrassed, and felt herself blush. She had never discussed the penis with a man. But she knew she had to be robust about such things if she wanted to be a doctor. 'In the same way that you, as a man, would deal with a vaginal infection,' she said. He looked horrified, and she feared she had been insolent. Hastily she went on: 'I would examine the infected area carefully, try to establish the nature of the infection, and probably treat it with sulphonamide, although I have to admit we did not cover this in my school biology course.'

He said sceptically: 'Have you ever seen a naked man?'

'Yes.'

He affected to be outraged. 'But you are a single girl!'

'When my grandfather was dying he was bedridden and incontinent. I helped my mother keep him clean – she could not manage on her own, he was too heavy.' She tried a smile. 'Women do these things all the time, Professor, for the very young and the very old, the sick and the helpless. We're used to it. It's only men who find such tasks embarrassing.'

He was looking more and more cross, even though she was answering well. What was going wrong? It was almost as if he would have been happier for her to be intimidated by his manner and to give stupid replies.

He put out his cigarette thoughtfully in the ashtray on his desk. 'I'm afraid you are not suitable as a candidate for this scholarship,' he said.

She was astonished. How had she failed? She had answered every question! 'Why not?' she said. 'My qualifications are irreproachable.'

'You are unwomanly. You talk freely of the vagina and the penis.'

'It was you who started that! I merely answered your question.'

'You have clearly been brought up in a coarse environment where you saw the nakedness of your male relatives.'

'Do you think old people's diapers should be changed by men? I'd like to see you do it!'

'Worst of all, you are disrespectful and insolent.'

'You asked me challenging questions. If I had given you timid replies you would have said I wasn't tough enough to be a doctor – wouldn't you?'

He was momentarily speechless, and she realized that was exactly what he would have done.

'You've wasted my time,' she said, and she went to the door.

'Get married,' he said. 'Produce children for the Führer. That's your role in life. Do your duty!'

She went out and slammed the door.

Frieda looked up in alarm. 'What happened?'

Carla headed for the exit without replying. She caught the eye of the secretary, who looked pleased, clearly knowing what had happened. Carla said to her: 'You can wipe that smirk off your face, you dried-up old bitch.' She had the satisfaction of seeing the woman's shock and horror.

Outside the building she said to Frieda: 'He had no intention of recommending me for the scholarship, because I'm a woman. My qualifications were irrelevant. I did all that work for nothing.' Then she burst into tears.

Frieda put her arms around her.

After a minute she felt better. 'I'm not going to raise children for the damned Führer,' she muttered.

'What?'

'Let's go home. I'll tell you when we get there.' They climbed on to their bikes.

There was a strange air in the streets, but Carla was too full of her own woes to wonder what was going on. People were gathering around the loudspeakers that sometimes broadcast Hitler's speeches from the Kroll Opera, the building that was being used instead of the burned-out Reichstag. Presumably he was about to speak.

When they got back to the von Ulrich town house, Mother and Father were still in the kitchen, Father sitting next to the radio with a frown of concentration.

'They turned me down,' Carla said. 'Regardless of what their rules say, they don't want to give a scholarship to a girl.'

'Oh, Carla, I'm so sorry,' said Mother.

'What's on the radio?'

'Haven't you heard?' said Mother. 'We invaded Poland this morning. We're at war.'

(v)

The London season was over, but most people were still in town because of the crisis. Parliament, normally in recess at this time of year, had been specially recalled. But there were no parties, no royal receptions, no balls. It was like being at a seaside resort in February, Daisy thought. Today was Saturday, and she was getting ready to go to dinner at the home of her father-in-law, Earl Fitzherbert. What could be more dull?

She sat at her dressing table wearing an evening gown in eau-de-nil silk with a V-neck and a pleated skirt. She had silk flowers in her hair and a fortune in diamonds round her neck.

Her husband, Boy, was getting ready in his dressing room. She was pleased he was here. He spent many nights elsewhere. Although they lived in the same Mayfair house, sometimes several days would go by without their meeting. But he was at home tonight.

She held in her hand a letter from her mother in Buffalo. Olga had divined that Daisy was discontented in her marriage. There must have been hints in Daisy's letters home. Mother had good intuition. 'I only want you to be happy,' she wrote. 'So listen when I tell you not to give up too soon. You're going to be Countess Fitzherbert one day, and your son, if you have one, will be the earl. You might regret throwing all that away just because your husband didn't pay you enough attention.'

She might be right. People had been addressing Daisy as 'My lady' for almost three years, yet it still gave her a little jolt of pleasure every time, like a puff on a cigarette.

But Boy seemed to think that marriage need make no great difference to his life. He spent evenings with his men friends, travelled all over the country to go horse racing, and rarely told his wife what his plans were. Daisy found it embarrassing to go to a party and be surprised to meet her husband there. But if she wanted to know where he was going, she had to ask his valet, and that was too demeaning.

Would he gradually grow up, and start to behave as a husband should, or would he always be like this?

He put his head around her door. 'Come on, Daisy, we're late.'

She put Mother's letter in a drawer, locked it, and went out. Boy was waiting in the hall, wearing a tuxedo. Fitz had at last succumbed to fashion and permitted informal short dinner jackets for family dinners at home.

They could have walked to Fitz's house, but it was raining, so Boy had had the car brought round. It was a Bentley Airline saloon, cream-coloured with whitewall tyres. Boy shared his father's love of beautiful cars.

Boy drove. Daisy hoped he would let her drive back. She enjoyed it, and, anyway, he was not safe after dinner, especially on wet roads.

London was preparing for war. Barrage balloons floated over the city at a height of two thousand feet, to impede bombers. In case that failed, sandbags were stacked outside important buildings. Alternate kerbstones had been painted white, for the benefit of drivers in the blackout, which had begun yesterday. There were white stripes on large trees, street statues, and other obstacles that might cause accidents.

Princess Bea welcomed Boy and Daisy. In her fifties she was quite fat, but she still dressed like a girl. Tonight she wore a pink gown embroidered with beads and sequins. She never spoke about the story Daisy's father had told at the wedding, but she had stopped hinting that Daisy was socially inferior, and now always spoke to Daisy with courtesy, if not warmth. Daisy was cautiously friendly, and treated Bea like a slightly dotty aunt.

Boy's younger brother, Andy, was there. He and May had two

children and May looked, to Daisy's interested eye, as if she might be expecting a third.

Boy wanted a son, of course, to be heir to the Fitzherbert title and fortune, but so far Daisy had failed to get pregnant. It was a sore point, and the evident fecundity of Andy and May made it worse. Daisy would have had a better chance if Boy spent more nights at home.

She was delighted to see her friend Eva Murray there – but without her husband: Jimmy Murray, now a captain, was with his unit and had not been able to get away, for most troops were in barracks and their officers were with them. Eva was family, now, because Jimmy was May's brother and therefore an in-law. So Boy had been forced to overcome his prejudice against Jews and be polite to Eva.

Eva adored Jimmy as much now as she had three years ago when she had married him. They, too, had produced two children in three years. But Eva looked worried tonight, and Daisy could guess why. 'How are your parents?' she said.

'They can't get out of Germany,' Eva said miserably. 'The government won't give them exit visas.'

'Can't Fitz help?'

'He's tried.'

'What have they done to deserve this?'

'It's not them, particularly. There are thousands of German Jews in the same position. Only a few get visas.'

'I'm so sorry.' Daisy was more than sorry. She squirmed with embarrassment when she recalled how she and Boy had supported the Fascists in the early days. Her doubts had grown rapidly as the brutality of Fascism at home and abroad had become more and more obvious, and in the end she had been relieved when Fitz had complained that they were embarrassing him and had begged them to leave Mosley's party. Now Daisy felt she had been an utter fool ever to have joined in the first place.

Boy was not quite so repentant. He still thought that upper-class white Europeans formed a superior species, chosen by God to rule the earth. But he no longer believed that was a practical political philosophy. He was often infuriated by British democracy, but he did not advocate abolishing it.

They sat down to dinner early. 'Neville is making a statement in the House of Commons at half past seven,' Fitz said. Neville Chamberlain was Prime Minister. 'I want to see it – I shall sit in the Peers' Gallery. I may have to leave you before dessert.'

Andy said: 'What do you think will happen, Papa?'

'I really don't know,' Fitz said with a touch of exasperation. 'Of course we would all like to avoid a war, but it's important not to give an impression of indecision.'

Daisy was surprised: Fitz believed in loyalty and rarely criticized his government colleagues, even as obliquely as this.

Princess Bea said: 'If there is a war, I shall go and live in Tŷ Gwyn.'

Fitz shook his head. 'If there is a war, the government will ask owners of large country houses to put them at the disposal of the military for the duration. As a member of the government, I must set an example. I shall have to lend Tŷ Gwyn to the Welsh Rifles for use as a training centre, or possibly a hospital.'

Bea was outraged. 'But it is my country house!'

'We may reserve a small part of the premises for private use.'

'I don't choose to live in a small part of the premises – I am a princess!'

'It might be cosy. We could use the butler's pantry as a kitchen, and the breakfast room as a dining room, plus three or four of the smaller bedrooms.'

'Cosy!' Bea looked disgusted, as if something unpleasant had been set before her, but she said no more.

Andy said: 'Presumably Boy and I will have to join the Welsh Rifles.'

May made a noise in her throat like a sob.

Boy said: 'I shall join the Air Force.'

Fitz was shocked. 'But you can't. The Viscount Aberowen has always been in the Welsh Rifles.'

'They haven't got any planes. The next war will be an air war. The RAF will be desperate for pilots. And I've been flying for years.'

Fitz was about to argue, but the butler came in and said: 'The car is ready, my lord.'

Fitz looked at the clock on the mantelpiece. 'Dash it, I've got to go. Thank you, Grout.' He looked at Boy. 'Don't make a final decision until we've talked some more. This is not right.'

'Very well, Papa.'

Fitz looked at Bea. 'Forgive me, my dear, for leaving in the middle of dinner.'

'Of course,' she said.

Fitz got up from the table and walked to the door. Daisy noticed his limp, a grim reminder of what the last war had done.

The rest of dinner was gloomy. They were all wondering whether the Prime Minister would declare war.

When the ladies got up to withdraw, May asked Andy to take her arm. He excused himself to the two remaining men, saying: 'My wife is in a delicate condition.' It was the usual euphemism for pregnancy.

Boy said: 'I wish my wife were as quick to get delicate.'

It was a cheap shot, and Daisy felt herself blush bright red. She repressed a retort, then asked herself why she should be silent. 'You know what footballers say, Boy,' she said loudly. 'You have to shoot to score.'

It was Boy's turn to blush. 'How dare you!' he said furiously.

Andy laughed. 'You asked for it, brother.'

Bea said: 'Stop it, both of you. I expect my sons to wait until the ladies are out of earshot before indulging in such disgusting talk.' She swept out of the room.

Daisy followed, but she parted company from the other women on the landing and went on upstairs, still feeling angry, wanting to be alone. How could Boy say such a thing? Did he really believe it had to be her fault that she was not pregnant? It could just as easily be his! Perhaps he knew that, and tried to blame her because he was afraid people would think he was infertile. That was probably the truth, but it was no excuse for a public insult.

She went to his old room. After they had married the two of them had lived here for three months while their own house was being redecorated. They had used Boy's old bedroom and the one next door, although in those days they had slept together every night.

She went in and turned on the light. To her surprise she saw that Boy appeared not to have completely moved out. There was a razor on the wash stand and a copy of *Flight* magazine on the bedside table. She opened a drawer and found a tin of Leonard's Liver-Aid, which he took every morning before breakfast. Did he sleep here when he was too disgustingly drunk to face his wife?

The lower drawer was locked, but she knew he kept the key in a pot on the mantelpiece. She had no qualms about prying: in her view a husband should have no secrets from his wife. She opened the drawer.

The first thing she found was a book of photographs of naked women. In artistic paintings and photographs, the women generally posed to half conceal their private parts, but these girls were doing the opposite: legs akimbo, buttocks held open, even the lips of their vaginas spread to show the inside. Daisy would pretend to be shocked if anyone caught her, but in truth she was fascinated. She looked through the entire book with great interest, comparing the women with herself: the size and shape of their breasts, the amount of hair, their sexual organs. What a wonderful variety there was in women's bodies!

Some of the girls were stimulating themselves, or pretending to, and some were photographed in pairs, doing it to each other. Daisy was not really surprised that men liked this sort of thing.

She felt like an eavesdropper. It reminded her of the time she had gone to his room at Tŷ Gwyn, before they were married. Then she had been desperate to learn more about him, to gain intimate knowledge of the man she loved, to find a way to make him her own. What was she doing now? Spying on a husband who seemed no longer to love her, trying to understand where she had failed.

Beneath the book was a brown paper bag. Inside were several small, square paper envelopes, white with red lettering on the front. She read:

'Prentif' Reg. Trade Mark

SERVISPAK

NOTICE
Do not leave the envelope
or contents in public places
as this is likely to cause offence

British made
Latex rubber
Withstands all climates

None of it made any sense. Nowhere did it say what the package actually contained. So she opened it.

Inside was a piece of rubber. She unfolded it. It was shaped like a tube, closed at one end. She took a few seconds to figure out what it was.

She had never seen one, but she had heard people talk about such things. Americans called it a Trojan, the British a rubber johnny. The correct term was condom, and it was to stop you getting pregnant.

Why did her husband have a bag of them? There could be only one answer. They were to be used with another woman.

She felt like crying. She had given him everything he wanted. She had never told him she was too tired to make love – even when she was – nor had she refused anything he suggested in bed. She would even have posed like the women in the book of photographs, if he had asked her to.

What had she done wrong?

She decided to ask him.

Sorrow turned to anger. She stood up. She would take the paper packets down to the dining room and confront him with them. Why should she protect his feelings?

At that moment he walked in.

'I saw the light from the hall,' he said. 'What are you doing in here?' He looked at the open drawers of the bedside cupboard and said: 'How dare you spy on me?'

'I suspected you of being unfaithful,' she said. She held up the condom. 'And I was right.'

'Damn you for a sneak.'

'Damn you for an adulterer.'

He raised his hand. 'I should beat you like a Victorian husband.'

She snatched a heavy candlestick from the mantelpiece. 'Try it, and I'll bop you like a twentieth-century wife.'

'This is ridiculous.' He sat down heavily on a chair by the door, looking defeated.

His evident unhappiness deflated Daisy's rage, and she just felt sad. She sat on the bed. But she had not lost her curiosity. 'Who is she?'

He shook his head. 'Never mind.'

'I want to know!'

He shifted uncomfortably. 'Does it matter?'

'It sure does.' She knew she would get it out of him eventually.

He would not meet her eye. 'Nobody you know, or would ever know.'

'A prostitute?'

He was stung by this suggestion. 'No!'

She goaded him further. 'Do you pay her?'

'No. Yes.' He was clearly ashamed enough to wish to deny it. 'Well, an allowance. It's not the same thing.'

'Why do you pay, if she's not a prostitute?'

'So they don't have to see anyone else.'

'They? You have several mistresses?'

'No! Only two. They live in Aldgate. Mother and daughter.'

'What? You can't be serious.'

'Well, one day Joanie was . . . the French say *Elle avait les fleurs*.'

'American girls call it the curse.'

'So Pearl offered to . . .'

'Act as a substitute? This is the most sordid arrangement imaginable! So you go to bed with them both?'

'Yes.'

She thought of the book of photographs, and an outrageous possibility occurred to her. She had to ask. 'Not at the same time?'

'Occasionally.'

'How utterly foul.'

'You don't need to worry about disease.' He pointed to the condom in her hand. 'Those things prevent infection.'

'I'm overwhelmed by your thoughtfulness.'

'Look, most men do this sort of thing, you know. At least, most men of our class.'

'No, they don't,' she said, but she thought of her father, who had a wife and a long-time mistress and still felt the need to romance Gladys Angelus.

Boy said: 'My father isn't a faithful husband. He has bastards all over the place.'

'I don't believe you. I think he loves your mother.'

'He has one bastard for certain.'

'Where?'

'I don't know.'

'Then you can't be sure.'

'I heard him say something to Bing Westhampton once. You know what Bing is like.'

'I do,' said Daisy. This seemed a moment for telling the truth, so she added: 'He feels my bottom every chance he gets.'

'Dirty old man. Anyway, we were all a bit drunk, and Bing said: "Most of us have got one or two bastards hidden away, haven't we?" and Papa said: "I'm pretty sure I've only got one." Then he seemed to realize what he'd said, and he coughed and looked foolish and changed the subject.'

'Well, I don't care how many bastards your father has, I'm a modern American girl and I won't live with an unfaithful husband.'

'What can you do about it?'

'I'll leave you.' She put on a defiant expression, but she felt in pain, as if he had stabbed her.

'And go back to Buffalo with your tail between your legs?'

'Perhaps. Or I could do something else. I've got plenty of money.' Her father's lawyers had made sure Boy did not get his hands on the Vyalov-Peshkov fortune when they married. 'I could go to California. Act in one of Father's movies. Become a film star. I bet you I could.' This was all pretence. She wanted to burst into tears.

'Leave me, then,' he said. 'Go to hell, for all I care.' She wondered if that was true. Looking at his face, she thought not.

They heard a car. Daisy pulled the blackout curtain aside an inch and saw Fitz's black-and-cream Rolls-Royce outside, its headlights dimmed by slit masks. 'Your father's back,' she said. 'I wonder if we're at war.'

'We'd better go down.'

'I'll follow you.'

Boy went out and Daisy looked in the mirror. She was surprised to see that she looked no different from the woman who had walked in here half an hour ago. Her life had been turned upside-down, but there was no sign of it on her face. She felt terribly sorry for herself, and wanted to cry, but she repressed the urge. Steeling herself, she went downstairs.

Fitz was in the dining room, with raindrops on the shoulders of his dinner jacket. Grout, the butler, had set out cheese and fruit, as Fitz had skipped dessert. The family sat around the table as Grout poured a glass of claret for Fitz. He drank some and said: 'It was absolutely dreadful.'

Andy said: 'What on earth happened?'

Fitz ate a corner of cheddar cheese before answering. 'Neville spoke for four minutes. It was the worst performance by a prime minister that I have ever seen. He mumbled and prevaricated and said Germany might withdraw from Poland, which no one believes. He said nothing about war, or even an ultimatum.'

Andy said: 'But why?'

'Privately, Neville says he's waiting for the French to stop dithering and declare war simultaneously with us. But a lot of people suspect that's just a cowardly excuse.'

Fitz took another draught of wine. 'Arthur Greenwood spoke next.'

Greenwood was deputy leader of the Labour Party. 'As he stood up, Leo Amery – a Conservative Member of Parliament, mind you – shouted out: "Speak for England, Arthur!" To think that a damned socialist might speak for England where a Conservative Prime Minister has failed! Neville looked as sick as a dog.'

Grout refilled Fitz's glass.

'Greenwood was quite mild, but he did say: "I wonder how long we are prepared to vacillate?" and at that, MPs on both sides of the house roared their approval. I should think Neville wanted the earth to swallow him up.' Fitz took a peach and sliced it with a knife and fork.

Andy said: 'How were things left?'

'Nothing is resolved! Neville has gone back to Number Ten Downing Street. But most of the Cabinet is holed up in Simon's room at the Commons.' Sir John Simon was Chancellor of the Exchequer. 'They're saying they won't leave the room until Neville sends the Germans an ultimatum. Meanwhile, Labour's National Executive Committee is in session, and discontented backbenchers are meeting in Winston's flat.'

Daisy had always said she did not like politics, but since becoming part of Fitz's family, and seeing everything from the inside, she had become interested, and she found this drama fascinating and scary. 'Then the Prime Minister must act!' she said.

'Oh, certainly,' said Fitz. 'Before Parliament meets again – which should be at noon tomorrow – I think Neville must either declare war or resign.'

The phone rang in the hall and Grout went out to answer it. A minute later he came back and said: 'That was the Foreign Office, my lord. The gentleman would not wait for you to come to the telephone, but insisted on giving a message.' The old butler looked disconcerted, as if he had been spoken to rather sharply. 'The Prime Minister has called an immediate meeting of the Cabinet.'

'Movement!' said Fitz. 'Good.'

Grout went on: 'The Foreign Secretary would like you to be in attendance, if convenient.' Fitz was not in the Cabinet, but junior ministers were sometimes asked to attend meetings on their area of specialization, sitting at the side of the room rather than at the central table, so that they could answer questions of detail.

Bea looked at the clock. 'It's almost eleven. I suppose you must go.'

'Indeed I must. The phrase "if convenient" is an empty courtesy.' He patted his lips with a snowy napkin and limped out again.

Princess Bea said: 'Make some more coffee, Grout, and bring it to the drawing room. We may be up late tonight.'

'Yes, your Highness.'

They all returned to the drawing room, talking animatedly. Eva was in favour of war: she wanted to see the Nazi regime destroyed. She would worry about Jimmy, of course, but she had married a soldier and had always known he might have to risk his life in battle. Bea was pro-war, too, now that the Germans were allied with the Bolsheviks she hated. May feared that Andy would be killed, and could not stop crying. Boy did not see why two great nations such as England and Germany should go to war over a half-barbaric wasteland such as Poland.

As soon as she could, Daisy got Eva to go with her to another room where they could talk privately. 'Boy's got a mistress,' she said immediately. She showed Eva the condoms. 'I found these.'

'Oh, Daisy, I'm so sorry,' Eva said.

Daisy thought of giving Eva the grisly details – they normally told each other everything – but this time Daisy felt too humiliated, so she just said: 'I confronted him, and he admitted it.'

'Is he sorry?'

'Not very. He says all men of his class do it, including his father.'

'Jimmy doesn't,' Eva said decisively.

'No, I'm sure you're right.'

'What will you do?'

'I'm going to leave him. We can get divorced, then someone else can be the viscountess.'

'But you can't if there's a war!'

'Why not?'

'It's too cruel, when he's on the battlefield.'

'He should have thought of that before he slept with a pair of prostitutes in Aldgate.'

'But it would be cowardly, as well. You can't dump a man who is risking his life to protect you.'

Reluctantly, Daisy saw Eva's point. War would transform Boy from a despicable adulterer who deserved rejection into a hero defending his wife, his mother and his country from the terror of invasion and conquest. It was not just that everyone in London and Buffalo would see Daisy as a coward for leaving him; she would feel that way herself. If there was a war, she wanted to be brave, even though she was not sure what that might involve.

'You're right,' she said grudgingly. 'I can't leave him if there's a war.'

There was a clap of thunder. Daisy looked at the clock: it was midnight. The rain altered in sound as a torrential downpour began.

Daisy and Eva returned to the drawing room. Bea was asleep on a couch. Andy had his arm around May, who was still snivelling. Boy was smoking a cigar and drinking brandy. Daisy decided that she would definitely be driving home.

Fitz came in at half past midnight, his evening suit soaking wet. 'The dithering is over,' he said. 'Neville will send the Germans an ultimatum in the morning. If they do not begin to withdraw their troops from Poland by midday – eleven o'clock our time – we will be at war.'

They all got up and prepared to leave. In the hall, Daisy said: 'I'll drive,' and Boy did not argue with her. They got into the cream Bentley and Daisy started the engine. Grout closed the door of Fitz's house. Daisy turned on the windscreen wipers but did not pull away.

'Boy,' she said, 'let's try again.'

'What do you mean?'

'I don't really want to leave you.'

'I certainly don't want you to go.'

'Give up those women in Aldgate. Sleep with me every night. Let's really try for a baby. It's what you want, isn't it?'

'Yes.'

'Then will you do as I ask?'

There was a long pause. Then he said: 'All right.'

'Thank you.'

She looked at him, hoping for a kiss, but he sat still, looking straight ahead through the windscreen, as the rhythmic wipers swept away the relentless rain.

(vi)

On Sunday the rain stopped and the sun came out. Lloyd Williams felt as if London had been washed clean.

During the course of the morning, the Williams family gathered in the kitchen of Ethel's house in Aldgate. There was no prior arrangement: they turned up spontaneously. They wanted to be together, Lloyd guessed, if war was declared.

Lloyd longed for action against the Fascists, and at the same time dreaded the prospect of war. In Spain he had seen enough bloodshed and suffering for a lifetime. He wished never to take part in another battle. He had even given up boxing. Yet he hoped with all his heart that Chamberlain would not back down. He had seen for himself what Fascism meant in Germany, and the rumours coming out of Spain were equally nightmarish: the Franco regime was murdering former supporters of the elected government in their hundreds and thousands, and the priests were in control of the schools again.

This summer, after he had graduated, he had immediately joined the Welsh Rifles, and as a former member of the Officer Training Corps he had been given the rank of lieutenant. The army was energetically preparing for combat: it was only with the greatest difficulty that he had got a twenty-four hour pass to visit his mother this weekend. If the Prime Minister declared war today, Lloyd would be among the first to go.

Billy Williams came to the house in Nutley Street after breakfast on Sunday morning. Lloyd and Bernie were sitting by the radio, newspapers open on the kitchen table, while Ethel prepared a leg of pork for dinner. Uncle Billy almost wept when he saw Lloyd in uniform. 'It makes me think of our Dave, that's all,' he said. 'He'd be a conscript, now, if he'd come back from Spain.'

Lloyd had never told Billy the truth about how Dave had died. He pretended he did not know the details, just that Dave had been killed in action at Belchite and was presumably buried there. Billy had been in the Great War and knew how haphazardly bodies were dealt with on the battlefield, and that probably made his grief worse. His great hope was to visit Belchite one day, when Spain was freed at last, and to pay his respects to the son who died fighting in that great cause.

Lenny Griffiths was another who had never returned from Spain. No one had any idea where he might be buried. It was even possible he was still alive, in one of Franco's prison camps.

Now the radio reported Prime Minister Chamberlain's statement to the House of Commons last night, but nothing further.

'You'd never know what a stink there was afterwards,' said Billy.

'The BBC doesn't report stinks,' said Lloyd. 'They like to sound reassuring.'

Both Billy and Lloyd were members of the Labour Party's National Executive – Lloyd as the representative of the party's youth section. After he had come back from Spain he had managed to gain readmission to

Cambridge University, and while finishing his studies he had toured the country addressing Labour Party groups, telling people how the elected Spanish government had been betrayed by Britain's Fascist-friendly government. It had done no good – Franco's anti-democracy rebels had won anyway – but Lloyd had become a well-known figure, even something of a hero, especially among young left-wingers – hence his election to the Executive.

So both Lloyd and Uncle Billy had been at last night's committee meeting. They knew that Chamberlain had bowed to pressure from the Cabinet and sent the ultimatum to Hitler. Now they were waiting on tenterhooks to see what would happen.

As far as they knew, no response had yet been received from Hitler.

Lloyd recalled his mother's friend Maud and her family in Berlin. Those two little children would be eighteen and nineteen now, he calculated. He wondered if they were sitting around a radio wondering whether they were going to war against England.

At ten o'clock, Lloyd's half-sister, Millie, arrived. She was now nineteen, and married to her friend Naomi Avery's brother Abe, a leather wholesaler. She earned good money as a salesgirl on commission in an expensive dress shop. She had ambitions to open her own shop, and Lloyd had no doubt that she would do it one day. Although it was not the career Bernie would have chosen for her, Lloyd could see how proud he was of her brains and ambition and smart appearance.

But today her poised self-assurance had collapsed. 'It was awful when you were in Spain,' she said tearfully to Lloyd. 'And Dave and Lenny never did come back. Now it will be you and my Abie off somewhere, and us women waiting every day for news, wondering if you're dead yet.'

Ethel put in: 'And your cousin Keir. He's eighteen now.'

Lloyd said to his mother: 'Which regiment was my real father in?'

'Oh, does it matter?' She was never keen to talk about Lloyd's father, perhaps out of consideration for Bernie.

But Lloyd wanted to know. 'It matters to me,' he said.

She threw a peeled potato into a pan of water with unnecessary vigour. 'He was in the Welsh Rifles.'

'The same as me! Why didn't you tell me before?'

'The past is the past.'

There might be another reason for her caginess, Lloyd knew. She had probably been pregnant when she married. This did not bother

Lloyd, but to her generation it was shameful. All the same, he persisted. 'Was my father Welsh?'

'Yes.'

'From Aberowen?'

'No.'

'Where, then?'

She sighed. 'His parents moved around – something to do with his father's job – but I think they were from Swansea originally. Satisfied now?'

'Yes.'

Lloyd's Aunt Mildred came in from church, a stylish middle-aged woman, pretty except for protruding front teeth. She wore a fancy hat – she was a milliner with a small factory. Her two daughters by her first marriage, Enid and Lillian, both in their late twenties, were married with children of their own. Her elder son was the Dave who had died in Spain. Her younger son, Keir, followed her into the kitchen. Mildred insisted on taking her children to church, even though her husband, Billy, would have nothing to do with religion. 'I had a lifetime's worth of that when I was a child,' he often said. 'If I'm not saved, no one is.'

Lloyd looked around. This was his family: mother, stepfather, half-sister, uncle, aunt, cousin. He did not want to leave them and go away to die somewhere.

Lloyd looked at his watch, a stainless-steel model with a square face that Bernie had given him as a graduation present. It was eleven o'clock. On the radio, the fruity voice of newsreader Alvar Lidell said the Prime Minister was expected to make an announcement shortly. Then there was some solemn classical music.

'Hush, now, everyone,' said Ethel. 'I'll make you all a cup of tea after.'

The kitchen went quiet.

Alvar Lidell announced the Prime Minister, Neville Chamberlain.

The appeaser of Fascism, Lloyd thought; the man who gave Czechoslovakia to Hitler; the man who had stubbornly refused to help the elected government of Spain even after it became indisputably obvious that the Germans and Italians were arming the rebels. Was he about to cave in yet again?

Lloyd noticed that his parents were holding hands, Ethel's small fingers digging into Bernie's palm.

He checked his watch again. It was a quarter past eleven.

Then they heard the Prime Minister say: 'I am speaking to you from the Cabinet Room at Ten Downing Street.'

Chamberlain's voice was reedy and over-precise. He sounded like a pedantic schoolmaster. What we need is a warrior, Lloyd thought.

'This morning the British ambassador in Berlin handed the German government a final note, stating that, unless the British government heard from them by eleven o'clock that they were prepared at once to withdraw their troops from Poland, a state of war would exist between us.'

Lloyd found himself feeling impatient with Chamberlain's verbiage. *A state of war would exist between us*: what a strange way to put it. Get on with it, he thought; get to the point. This is life and death.

Chamberlain's voice deepened and became more statesmanlike. Perhaps he was no longer looking at the microphone, but instead seeing millions of his countrymen in their homes, sitting by their radio sets, waiting for his fateful words. 'I have to tell you now that no such undertaking has been received.'

Lloyd heard his mother say: 'Oh, God, spare us.' He looked at her. Her face was grey.

Chamberlain uttered his next, dreadful words quite slowly: '. . . and that, consequently, this country is at war with Germany.'

Ethel began to cry.

Part Two

A SEASON OF BLOOD

6

1940 (I)

Aberowen had changed. There were cars, trucks and buses on the streets. When Lloyd had come here as a child in the 1920s to visit his grandparents, a parked car had been a rarity that would draw a crowd.

But the town was still dominated by the twin towers of the pithead, with their majestically revolving wheels. There was nothing else: no factories, no office blocks, no industry other than coal. Almost every man in town worked down the pit. There were a few dozen exceptions: some shopkeepers, numerous clergymen of all denominations, a town clerk, a doctor. Whenever the demand for coal slumped, as it had in the thirties, and men were laid off, there was nothing else for them to do. That was why the Labour Party's most passionate demand was help for the unemployed, so that such men would never again suffer the agony and humiliation of being unable to feed their families.

Lieutenant Lloyd Williams arrived by train from Cardiff on a Sunday in April 1940. Carrying a small suitcase, he walked up the hill to Tŷ Gwyn. He had spent eight months training new recruits – the same work he had done in Spain – and coaching the Welsh Rifles boxing team, but the army had at last realized that he spoke fluent German, transferred him to intelligence duties, and sent him on a training course.

Training was all the army had done so far. No British forces had yet fought the enemy in an engagement of any significance. Germany and the USSR had overrun Poland and divided it between them, and the Allied guarantee of Polish independence had proved worthless.

British people called it the Phoney War, and they were impatient for the real thing. Lloyd had no sentimental illusions about warfare – he had heard the piteous voices of dying men begging for water on the battlefields of Spain – but even so he was eager to get started on the final showdown with Fascism.

The army was expecting to send more forces to France, assuming the

Germans would invade. It had not happened, and they remained at the ready, but meanwhile, they did a lot of training.

Lloyd's initiation into the mysteries of military intelligence was to take place in the stately home that had featured in his family's destiny for so long. The wealthy and noble owners of many such palaces had loaned them to the armed forces, perhaps for fear that otherwise they might be confiscated permanently.

The army had certainly made Tŷ Gwyn look different. There were a dozen olive-drab vehicles parked on the lawn, and their tyres had chewed up the earl's lush turf. The gracious entrance courtyard, with its curved granite steps, had become a supply dump, and giant cans of baked beans and cooking lard stood in teetering stacks where, formerly, bejewelled women and men in tailcoats had stepped out of their carriages. Lloyd grinned: he liked the levelling effect of war.

Lloyd entered the house. He was greeted by a podgy officer in a creased and stained uniform. 'Here for the intelligence course, Lieutenant?'

'Yes, sir. My name is Lloyd Williams.'

'I'm Major Lowther.'

Lloyd had heard of him. He was the Marquis of Lowther, known to his pals as Lowthie.

Lloyd looked around. The paintings on the walls had been shrouded with huge dust sheets. The ornate carved marble fireplaces had been boxed in with rough planking, leaving only a small space for a grate. The dark old furniture that his mother sometimes mentioned fondly had all disappeared, to be replaced by steel desks and cheap chairs. 'My goodness, the place looks different,' he said.

Lowther smiled. 'You've been here before. Do you know the family?'

'I was up at Cambridge with Boy Fitzherbert. I met the Viscountess there, too, although they weren't married then. But I suppose they've moved out for the duration.'

'Not entirely. A few rooms have been reserved for their private use. But they don't bother us at all. So you came here as a guest?'

'Goodness, no, I don't know them well. No, I was shown around the place as a boy, one day when the family weren't in residence. My mother worked here at one time.'

'Really? What, looking after the earl's library, or something?'

'No, as a housemaid.' As soon as the words were out of Lloyd's mouth he knew he had made a mistake.

Lowther's face changed to an expression of distaste. 'I see,' he said. 'How very interesting.'

Lloyd knew he had instantly been pigeonholed as a proletarian upstart. He would now be treated as a second-class citizen throughout his time here. He should have kept quiet about his mother's past: he knew how snobbish the army was.

Lowthie said: 'Show the lieutenant to his room, sergeant. Attic floor.'

Lloyd had been assigned a room in the old servants' quarters. He did not really mind. It was good enough for my mother, he thought.

As they walked up the back stairs, the sergeant told Lloyd he had no obligations until dinner in the mess. Lloyd asked whether any of the Fitzherberts happened to be in residence right now, but the man did not know.

It took Lloyd two minutes to unpack. He combed his hair, put on a clean uniform shirt, and went to visit his grandparents.

The house in Wellington Row seemed smaller and more drab than ever, though it now had hot water in the scullery and a flushing toilet in the outhouse. The decor had not altered within Lloyd's memory: same rag rug on the floor, same faded paisley curtains, same hard oak chairs in the single ground-floor room that served as living room and kitchen.

His grandparents had changed, though. Both were about seventy now, he guessed, and looking frail. Granda had pains in his legs, and had reluctantly retired from his job with the miners' union. Grandmam had a weak heart: Dr Mortimer had told her to put her feet up for a quarter of an hour after meals.

They were pleased to see Lloyd in his uniform. 'Lieutenant, is it?' said Grandmam. A class warrior all her life, she nevertheless could not conceal her pride that her grandson was an officer.

News travelled fast in Aberowen, and the fact that Dai Union's grandson was visiting probably went halfway round the town before Lloyd had finished his first cup of Grandmam's strong tea. So he was not really surprised when Tommy Griffiths dropped in.

'I expect my Lenny would be a lieutenant, like you, if he'd come back from Spain,' Tommy said.

'I should think so,' Lloyd said. He had never met an officer who had been a coal miner in civilian life, but anything might happen once the war got going properly. 'He was the best sergeant in Spain, I can tell you that.'

'You two went through a lot together.'

'We went through hell,' Lloyd said. 'And we lost. But the Fascists won't win this time.'

'I'll drink to that,' said Tommy, and emptied his mug of tea.

Lloyd went with his grandparents to the evening service at the Bethesda Chapel. Religion was not a big part of his life, and he certainly did not go along with Granda's dogmatism. The universe was mysterious, Lloyd thought, and people might as well admit it. But it pleased his grandparents that he sat with them in chapel.

The extempore prayers were eloquent, knitting biblical phrases seamlessly into colloquial language. The sermon was a bit tedious, but the singing thrilled Lloyd. Welsh chapelgoers automatically sang in four-part harmony, and when they were in the mood they could raise the roof.

As he joined in, Lloyd felt this was the beating heart of Britain, here in this whitewashed chapel. The people around him were poorly dressed and ill-educated, and they lived lives of unending hard work, the men winning the coal underground, the women raising the next generation of miners. But they had strong backs and sharp minds, and all on their own they had created a culture that made life worth living. They gained hope from nonconformist Christianity and left-wing politics, they found joy in rugby football and male voice choirs, and they were bonded together by generosity in good times and solidarity in bad. This was what he would be fighting for, these people, this town. And if he had to give his life for them, it would be well spent.

Granda gave the closing prayer, standing up with his eyes shut, leaning on a walking stick. 'You see among us, O Lord, your young servant Lloyd Williams, sitting by here in his uniform. We ask you, in your wisdom and grace, to spare his life in the conflict to come. Please, Lord, send him back home to us safe and whole. If it be your will, O Lord.'

The congregation gave a heartfelt amen, and Lloyd wiped away a tear.

He walked the old folk home as the sun went down behind the mountain and an evening gloom settled on the rows of grey houses. He refused the offer of supper and hurried back to Tŷ Gwyn, arriving in time for dinner in the mess.

They had braised beef, boiled potatoes and cabbage. It was no better or worse than most army food, and Lloyd tucked in, aware that it had been paid for by people such as his grandparents who were having bread-and-dripping for their supper. There was a bottle of whisky on the table, and Lloyd took some to be convivial. He studied his fellow trainees and tried to remember their names.

On his way up to bed he passed through the Sculpture Room, now empty of art and furnished with a blackboard and twelve cheap desks. There he saw Major Lowther talking to a woman. At a second glance he saw that the woman was Daisy Fitzherbert.

He was so surprised that he stopped. Lowther looked around with an irritated expression. He saw Lloyd and reluctantly said: 'Lady Aberowen, I believe you know Lieutenant Williams.'

If she denies it, Lloyd thought, I shall remind her of the time she kissed me, long and hard, on a Mayfair street in the dark.

'How nice to see you again, Mr Williams,' she said, and put out her hand to shake.

Her skin was warm and soft to his touch. His heart beat faster.

Lowther said: 'Williams tells me his mother worked at this house as a maid.'

'I know,' Daisy said. 'He told me that at the Trinity Ball. He was reproving me for being a snob. I'm sorry to say that he was quite right.'

'You're generous, Lady Aberowen,' said Lloyd, feeling embarrassed. 'I don't know what business I had to say such a thing to you.' She seemed less brittle than he remembered: perhaps she had matured.

Daisy said to Lowther: 'Mr Williams's mother is a Member of Parliament now, though.'

Lowther was taken aback.

Lloyd said to Daisy: 'And how is your Jewish friend Eva? I know she married Jimmy Murray.'

'They have two children now.'

'Did she get her parents out of Germany?'

'How kind of you to remember – but no, sadly, the Rothmanns can't get exit visas.'

'I'm so sorry. It must be hell for her.'

'It is.'

Lowther was visibly impatient with this talk of housemaids and Jews. 'To get back to what I was saying, Lady Aberowen . . .'

Lloyd said: 'I'll bid you goodnight.' He left the room and ran upstairs.

As he got ready for bed he found himself singing the last hymn from the service:

No storm can shake my inmost calm
While to that rock I'm clinging
Since Love is Lord of heaven and earth
How can I keep from singing?

(ii)

Three days later Daisy was finishing writing to her half-brother, Greg. When war broke out he had sent her a sweetly anxious letter, and since then they had corresponded every month or so. He had told her about seeing his old flame, Jacky Jakes, on E street in Washington, and asked Daisy what would make a girl run away like that? Daisy had no idea. She said so, and wished him luck, then signed off.

She looked at the clock. It was an hour before the trainees' dinner time, so lessons had ended and she had a good chance of catching Lloyd in his room.

She went up to the old servants' quarters on the attic floor. The young officers were sitting or lying on their beds, reading or writing. She found Lloyd in a narrow room with an old cheval-glass, sitting by the window, studying an illustrated book. She said: 'Reading something interesting?'

He sprang to his feet. 'Hello, this is a surprise.'

He was blushing. He probably still had a crush on her. It had been very cruel of her to kiss him, when she had no intention of letting the relationship go any further. But that was four years ago, and they had both been kids. He should have gotten over it by now.

She looked at the book in his hands. It was in German, and had colour pictures of badges.

'We have to know German insignia,' he explained. 'A lot of military intelligence comes from interrogation of prisoners of war immediately after their capture. Some won't talk, of course; so the interrogator needs to be able to tell, just by looking at the prisoner's uniform, what his rank is, what army corps he belongs to, whether he is from infantry, cavalry, artillery, or a specialist unit such as veterinarian, and so on.'

'That's what you're learning here?' she said sceptically. 'The meanings of German badges?'

He laughed. 'It's one of the things we're learning. One I can tell you about without giving away military secrets.'

'Oh, I see.'

'Why are you here in Wales? I'm surprised you're not doing something for the war effort.'

'There you go again,' she said. 'Moral reproof. Did someone tell you this was a way to charm women?'

'Pardon me,' he said stiffly. 'I didn't mean to rebuke you.'

'Anyway, there is no war effort. Barrage balloons float in the air as a hazard to German planes that never come.'

'At least you'd have a social life in London.'

'Do you know that used to be the most important thing in the world, and now it's not?' she said. 'I must be getting old.'

There was another reason she had left London, but she was not going to tell him.

'I imagined you in a nurse's uniform,' he said.

'Not likely. I hate sick people. But before you give me another of those disapproving frowns, take a look at this.' She handed him the framed photograph she was carrying.

He studied it, frowning. 'Where did you get it?'

'I was looking through a box of old pictures in the basement junk room.'

It was a group photo taken on the east lawn of Tŷ Gwyn on a summer morning. In the centre was the young Earl Fitzherbert, with a big white dog at his feet. The girl next to him was probably his sister, Maud, whom Daisy had never met. Lined up on either side of them were forty or fifty men and women in a variety of servants' uniforms.

'Look at the date,' she said.

'Nineteen-twelve,' Lloyd read aloud.

She watched him, studying his reactions to the photo he was holding. 'Is your mother in it?'

'Goodness! She might be.' Lloyd looked closer. 'I believe she is,' he said after a minute.

'Show me.'

Lloyd pointed. 'I think that's her.'

Daisy saw a slim, pretty girl of about nineteen, with curly black hair under a maid's white cap, and a smile that had more than a hint of mischief in it. 'Why, she's enchanting!' she said.

'She was then, anyway,' Lloyd said. 'Nowadays people are more likely to call her formidable.'

'Have you ever met Lady Maud? Do you think that's her next to Fitz?'

'I suppose I've known her all my life, off and on. She and my mother were suffragettes together. I haven't seen her since I left Berlin in 1933, but this is definitely her in the picture.'

'She's not so pretty.'

'Perhaps, but she's very poised, and wonderfully well dressed.'

'Anyway, I thought you might like to have the picture.'

'To keep?'

'Of course. No one else wants it – that's why it was in a box in the basement.'

'Thank you!'

'You're welcome.' Daisy went to the door. 'Go back to your studies.'

Going down the back stairs she hoped she had not flirted. She probably should not have gone to see him at all. She had succumbed to a generous impulse. Heaven forbid that he should misinterpret it.

She felt a sharp pain in her tummy, and stopped on the half-landing. She had had a slight backache all day – which she attributed to the cheap mattress she was sleeping on – but this was different. She thought back over what she had eaten today, but could not identify anything that might have made her ill: no undercooked chicken, no unripe fruit. She had not eaten oysters – no such luck! The pain went as quickly as it had come and she told herself to forget about it.

She returned to her quarters in the basement. She was living in what had been the housekeeper's flat: a tiny bedroom, a sitting room, a small kitchen and an adequate bathroom with a tub. An old footman called Morrison was acting as caretaker to the house, and a young woman from Aberowen was her maid. The girl was called Little Maisie Owen, although she was quite big. 'My mother's Maisie too, so I've always been Little Maisie, even though I'm taller than her now,' she had explained.

The phone rang as Daisy entered. She picked it up and heard her husband's voice. 'How are you?' he said.

'I'm fine. What time will you be here?' He had flown to RAF St Athan, a large air base outside Cardiff, on some mission, and he had promised to visit her and spend the night.

'I'm not going to make it, I'm sorry.'

'Oh, how disappointing!'

'There's a ceremonial dinner at the base that I'm required to attend.'

He did not sound particularly dispirited that he would not see her, and she felt spurned. 'How nice for you,' she said.

'It will be boring, but I can't get out of it.'

'Not half as boring as living here on my own.'

'It must be dull. But you're better off there, in your condition.'

Thousands of people had left London after war was declared, but most of them had drifted back when the expected bombing raids and gas attacks did not materialize. However, Bea and May and even Eva were agreed that Daisy's pregnancy meant she should live at Tŷ Gwyn. Many women gave birth safely every day in London, Daisy had pointed out; but of course the heir to the earldom was different.

In truth, she did not mind as much as she had expected. Perhaps pregnancy had made her uncharacteristically passive. But there was a half-hearted quality about London social life since the declaration of war, as if people felt they did not have the right to enjoy themselves. They were like vicars in a pub, knowing it was supposed to be fun but unable to enter into the spirit.

'I wish I had my motorcycle here, though,' she said. 'Then at least I could explore Wales.' Petrol was rationed, but not severely.

'Really, Daisy!' he said censoriously. 'You can't ride a motorcycle – the doctor absolutely forbade it.'

'Anyway, I've discovered literature,' she said. 'The library here is wonderful. A few rare and valuable editions have been packed away, but nearly all the books are still on the shelves. I'm getting the education I worked so hard to avoid at school.'

'Excellent,' he said. 'Well, curl up with a good murder mystery and enjoy your evening.'

'I had a slight tummy pain earlier.'

'Probably indigestion.'

'I expect you're right.'

'Give my regards to that slob Lowthie.'

'Don't drink too much port at your dinner.'

Just as Daisy hung up she got the tummy cramp again. This time it lasted longer. Maisie came in, saw her face, and said: 'Are you all right, my lady?'

'Just a twinge.'

'I have came to ask if you are ready for your supper.'

'I don't feel hungry. I think I'll skip supper tonight.'

'I done you a lovely cottage pie,' Maisie said reproachfully.

'Cover it and put it in the larder. I'll eat it tomorrow.'

'Shall I make you a nice cup of tea?'

Just to get rid of her Daisy said: 'Yes, please.' Even after four years she had not grown to like strong British tea with milk and sugar in it.

The pain went away, and she sat down and opened *The Mill on the Floss*. She forced herself to drink Maisie's tea and felt a little better. When she had finished the drink, and Maisie had washed the cup and saucer, she sent Maisie home. The girl had to walk a mile in the dark, but she carried a flashlight, and said she did not mind.

An hour later the pain returned, and this time it did not go away. Daisy went to the toilet, vaguely hoping to relieve pressure in her abdomen. She was surprised and worried to see spots of dark-red blood in her underwear.

She put on clean panties and, seriously worried now, she went to the phone. She got the number of RAF St Athan and called the base. 'I need to speak to Flight Lieutenant the Viscount Aberowen,' she said.

'We can't connect personal calls to officers,' said a pedantic Welshman.

'This is an emergency. I must speak to my husband.'

'There are no phones in the rooms, this isn't the Dorchester Hotel.' Perhaps it was her imagination, but he sounded quite pleased that he could not help her.

'My husband will be at the ceremonial banquet. Please send an orderly to bring him to the phone.'

'I haven't got any orderlies, and anyway there's no banquet.'

'No banquet?' Daisy was momentarily at a loss.

'Just the usual dinner in the mess,' the operator said. 'And that was finished an hour ago.'

Daisy slammed the phone down. No banquet? Boy had distinctly said he had to attend a ceremonial dinner at the base. He must have lied. She wanted to cry. He had chosen not to see her, preferring to go drinking with his comrades, or perhaps to visit some woman. The reason did not matter. Daisy was not his priority.

She took a deep breath. She needed help. She did not know the phone number of the Aberowen doctor, if there was one. What was she to do?

Last time Boy had left he had said: 'You'll have a hundred or more army officers to look after you if necessary.' But she could not tell the Marquis of Lowther that she was bleeding from her vagina.

The pain was getting worse, and she could feel something warm and sticky between her legs. She went to the bathroom again and washed herself. There were clots in the blood, she saw. She did not have any sanitary towels – pregnant women did not need them, she had thought. She cut a length off a hand towel and stuffed it in her panties.

Then she thought of Lloyd Williams.

He was kind. He had been brought up by a strong-minded feminist woman. He adored Daisy. He would help her.

She went up to the hall. Where was he? The trainees would have finished their dinner by now. He might be upstairs. Her stomach hurt so much that she did not think she could make it all the way to the attic.

Perhaps he was in the library. The trainees used the room for quiet study. She went in. A sergeant was poring over an atlas. 'Would you be very kind,' she said to him, 'and find Lieutenant Lloyd Williams for me?'

'Of course, my lady,' said the man, closing the book. 'What's the message?'

'Ask him if he would come down to the basement for a moment.'

'Are you all right, ma'am? You look a bit pale.'

'I'll be fine. Just fetch Williams as quickly as you can.'

'Right away.'

Daisy returned to her rooms. The effort of seeming normal had exhausted her, and she lay on the bed. Before long she felt the blood soaking through her dress, but she hurt too much to care. She looked at her watch. Why had Lloyd not come? Perhaps the sergeant could not find him. It was such a big house. Perhaps she would just die here.

There was a tap at the door, and then to her immense relief she heard his voice. 'It's Lloyd Williams.'

'Come in,' she called. He was going to see her in a dreadful state. Perhaps it would put him off her for good.

She heard him enter the next room. 'It took me a while to find your quarters,' he said. 'Where are you?'

'Through here.'

He stepped into the bedroom. 'Good God!' he exclaimed. 'What on earth has happened?'

'Get help,' she said. 'Is there a doctor in this town?'

'Of course. Dr Mortimer. He's been here for centuries. But there may not be time. Let me . . .' He hesitated. 'You may be haemorrhaging, but I can't tell unless I look.'

She closed her eyes. 'Go ahead.' She was almost too scared to be embarrassed.

She felt him raise the skirt of her dress. 'Oh, dear,' he said. 'Poor you.' Then he ripped her underpants. 'I'm sorry,' he said. 'Is there some water . . . ?'

'Bathroom,' she said, pointing.

He stepped into the bathroom and ran a tap. A moment later she felt a warm, damp cloth being used to clean her.

Then he said: 'It's just a trickle. I've seen men bleed to death, and you're not in that danger.' She opened her eyes to see him pulling her skirt back down. 'Where's the phone?' he said.

'Sitting room.'

She heard him say: 'Put me through to Dr Mortimer, quick as you can.' There was a pause. 'This is Lloyd Williams. I'm at Tŷ Gwyn. May I speak to the doctor? Oh, hello, Mrs Mortimer, when do you expect him back? . . . It's a woman with abdominal pain and vaginal bleeding . . . Yes, I do realize most women suffer that every month, but this is clearly abnormal . . . she's twenty-three . . . yes, married . . . no children . . . I'll ask.' He raised his voice. 'Could you be pregnant?'

'Yes,' Daisy replied. 'Three months.'

He repeated her answer, then there was a long silence. Eventually he hung up the phone and returned to her.

He sat on the edge of the bed. 'The doctor will come as soon as he can, but he's operating on a miner crushed by a runaway dram. However, his wife is quite sure that you've suffered a miscarriage.' He took her hand. 'I'm sorry, Daisy.'

'Thank you,' she whispered. The pain seemed less, but she felt terribly sad. The heir to the earldom was no more. Boy would be so upset.

Lloyd said: 'Mrs Mortimer says it's quite common, and most women suffer one or two miscarriages between pregnancies. There's no danger, provided the bleeding isn't copious.'

'What if it gets worse?'

'Then I must drive you to Merthyr Hospital. But going ten miles in an army lorry would be quite bad for you, so it's to be avoided unless your life is in danger.'

She was not frightened any more. 'I'm so glad you were here.'

'May I make a suggestion?'

'Of course.'

'Do you think you can walk a few steps?'

'I don't know.'

'Let me run you a bath. If you can manage it, you'll feel so much better when you're clean.'

'Yes.'

'Then perhaps you can improvise a bandage of some kind.'

'Yes.'

He returned to the bathroom, and she heard water running. She sat upright. She felt dizzy, and rested for a minute, then her head cleared. She swung her feet to the floor. She was sitting in congealing blood, and felt disgusted with herself.

The taps were turned off. He came back in and took her arm. 'If you feel faint, just tell me,' he said. 'I won't let you fall.' He was surprisingly strong, and half carried her as he walked her into the bathroom. At some point her ripped underwear fell to the floor. She stood beside the bath and let him undo the buttons at the back of her dress. 'Can you manage the rest?' he said.

She nodded, and he went out.

Leaning on the linen basket, she took off her clothes slowly, leaving them on the floor in a bloodstained heap. Gingerly, she got into the bath. The water was just hot enough. The pain eased as she lay back and relaxed. She felt overwhelmed with gratitude to Lloyd. He was so kind that it made her want to cry.

After a few minutes, the door opened a crack and his hand appeared holding some clothes. 'A nightdress, and so on,' he said. He placed them on top of the linen basket and closed the door.

When the water began to cool she stood up. She felt dizzy again, but only for a moment. She dried herself with a towel then put on the nightdress and underwear he had brought. She placed a hand towel inside her panties to soak up the blood that continued to seep.

When she returned to the bedroom, her bed was made up with clean sheets and blankets. She climbed in and sat upright, pulling the covers up to her neck.

He came in from the sitting room. 'You must be feeling better,' he said. 'You look embarrassed.'

'Embarrassed isn't the word,' she said. 'Mortified, perhaps, though even that seems understated.' The truth was not so simple. She winced when she thought of how he had seen her – but, on the other hand, he had not seemed disgusted.

He went into the bathroom and picked up her discarded clothes. Apparently he was not squeamish about menstrual blood.

She said: 'Where have you put the sheets?'

'I found a big sink in the flower room. I left them to soak in cold water. I'll do the same with your clothes, shall I?'

She nodded.

He disappeared again. Where had he learned to be so competent and self-sufficient? In the Spanish Civil War, she supposed.

She heard him moving around the kitchen. He reappeared with two cups of tea. 'You probably hate this stuff, but it will make you feel better.' She took the tea. He showed her two white pills in the palm of his hand. 'Aspirin? May ease the stomach cramps a bit.'

She took them and swallowed them with hot tea. He had always struck her as being mature beyond his years. She remembered how confidently he had gone off to find the drunken Boy at the Gaiety Theatre. 'You've always been like this,' she said. 'A real grown-up, when the rest of us were just pretending.'

She finished the tea and felt sleepy. He took the cups away. 'I may just close my eyes for a moment,' she said. 'Will you stay here, if I go to sleep?'

'I'll stay as long as you like,' he said. Then he said something else, but his voice seemed to fade away, and she slept.

(iii)

After that Lloyd began to spend his evenings in the little housekeeper's flat.

He looked forward to it all day.

He would go downstairs a few minutes after eight, when dinner in the mess was over and Daisy's maid had left for the night. They would sit opposite one another in the two old armchairs. Lloyd would bring a book to study – there was always 'homework', with tests in the morning – and Daisy would read a novel; but mostly they talked. They related what had happened during the day, discussed whatever they were reading, and told each other the story of their lives.

He recounted his experiences at the Battle of Cable Street. 'Standing there in a peaceful crowd, we were charged by mounted policemen screaming about dirty Jews,' he told her. 'They beat us with their truncheons and pushed us through the plate-glass windows.'

She had been quarantined with the Fascists in Tower Gardens, and had seen none of the fighting. 'That wasn't the way it was reported,' she said. She had believed the newspapers that said it had been a street riot organized by hooligans.

Lloyd was not surprised. 'My mother watched the newsreel at the Aldgate Essoldo a week later,' he recalled. 'That plummy-voiced commentator said: "From impartial observers the police received nothing but praise." Mam said the entire audience burst out laughing.'

Daisy was shocked by his scepticism about the news. He told her that most British papers had suppressed stories of atrocities by Franco's army in Spain, and exaggerated any report of bad behaviour by government forces. She admitted she had swallowed Earl Fitzherbert's view that the rebels were high-minded Christians liberating Spain from the threat of Communism. She knew nothing of mass executions, rape and looting by Franco's men.

It seemed never to have occurred to her that newspapers owned by capitalists might play down news that reflected badly on the Conservative government, the military or businessmen, and would seize upon any incident of bad behaviour by trade unionists or left-wing parties.

Lloyd and Daisy talked about the war. There was action at last. British and French troops had landed in Norway, and were contending for control with the Germans who had done the same. The newspapers could not quite conceal the fact that it was going badly for the Allies.

Her attitude to him had changed. She no longer flirted. She was always pleased to see him, and complained if he was late arriving in the evening, and she teased him sometimes; but she was never coquettish. She told him how disappointed everyone was about the baby she had lost: Boy, Fitz, Bea, her mother in Buffalo, even her father, Lev. She could not shake the irrational feeling that she had done something shameful, and she asked if he thought that was foolish. He did not. Nothing she did was foolish to him.

Their conversation was personal but they kept their distance from one another physically. He would not exploit the extraordinary intimacy of the night she miscarried. Of course, the scene would live in his heart for ever. Wiping the blood from her thighs and her belly had not been sexy – not in the least – but it had been unbearably tender. However, it had been a medical emergency, and it did not give him permission to take liberties later. He was so afraid of giving the wrong impression about this that he was careful never to touch her.

At ten o'clock she would make them cocoa, which he loved and she said she liked, though he wondered if she was just being nice. Then he would say goodnight and go upstairs to his attic bedroom.

They were like old friends. It was not what he wanted, but she was a married woman, and this was the best he was going to get.

He tended to forget Daisy's status. He was startled, one evening, when she announced that she was going to pay a visit to the earl's retired butler, Peel, who was living in a cottage just outside the grounds. 'He's eighty!' she told Lloyd. 'I'm sure Fitz has forgotten all about him. I should check on him.'

Lloyd raised his eyebrows in surprise, and she added: 'I need to make sure he's all right. It's my duty as a member of the Fitzherbert clan. Taking care of your old retainers is an obligation of wealthy families – didn't you know that?'

'It had slipped my mind.'

'Will you come with me?'

'Of course.'

The next day was a Sunday, and they went in the morning, when Lloyd had no lectures. They were both shocked by the state of the little house. The paint was flaking, the wallpaper was peeling, and the curtains were grey with coal dust. The only decoration was a row of photographs cut from magazines and tacked to the wall: the King and Queen, Fitz and Bea, and other assorted members of the nobility. The place had not been properly cleaned for years, and there was a smell of urine and ash and decay. But Lloyd guessed it was not unusual for an old man on a small pension.

Peel had white eyebrows. He looked at Lloyd and said: 'Good morning, my lord – I thought you were dead!'

Lloyd smiled. 'I'm just a visitor.'

'Are you, sir? My poor brain is scrambled eggs. The old earl died, what, thirty-five or forty years ago? Well, then, who are you, young sir?'

'I'm Lloyd Williams. You knew my mother, Ethel, years ago.'

'You're Eth's boy? Well, in that case, of course . . .'

Daisy said: 'In that case, what, Mr Peel?'

'Oh, nothing. My brain's scrambled eggs!'

They asked him if he needed anything, and he insisted he had everything a man could want. 'I don't eat much, and I rarely drink beer. I've got enough money to buy pipe tobacco, and the newspaper. Will Hitler invade us, do you think, young Lloyd? I hope I don't live to see that.'

Daisy cleaned up his kitchen a bit, though housekeeping was not her forte. 'I can't believe it,' she said to Lloyd in a low voice. 'Living here, like this, he says he's got everything – he thinks he's lucky!'

'Many men his age are worse off,' Lloyd said.

They talked to Peel for an hour. Before they left, he thought of something he did want. He looked at the row of pictures on the wall. 'At the funeral of the old earl, there was a photograph took,' he said. 'I was a mere footman, then, not the butler. We all lined up alongside the hearse. There was a big old camera with a black cloth over it, not like the little modern ones. That was in 1906.'

'I bet I know where that photograph is,' said Daisy. 'We'll go and look.'

They returned to the big house and went down to the basement. The junk room, next to the wine cellar, was quite large. It was full of boxes and chests and useless ornaments: a ship in a bottle, a model of Tŷ Gwyn made of matchsticks, a miniature chest of drawers, a sword in an ornate scabbard.

They began to sort through old photographs and paintings. The dust made Daisy sneeze, but she insisted on continuing.

They found the photograph Peel wanted. In the box with it was an even older photo of the previous earl. Lloyd stared at it in some astonishment. The sepia picture was five inches high and three inches wide, and showed a young man in the uniform of a Victorian army officer.

He looked exactly like Lloyd.

'Look at this,' he said, handing the photo to Daisy.

'It could be you, if you had side-whiskers,' she said.

'Perhaps the old earl had a romance with one of my ancestors,' Lloyd said flippantly. 'If she was a married woman, she might have passed off the earl's child as her husband's. I wouldn't be very pleased, I can tell you, to learn that I was illegitimately descended from the aristocracy – a red-hot socialist like me!'

Daisy said: 'Lloyd, how stupid are you?'

He could not tell whether she was serious. Besides, she had a smear of dust on her nose that looked so sweet that he longed to kiss it. 'Well,' he said, 'I've made a fool of myself more than once, but—'

'Listen to me. Your mother was a maid in this house. Suddenly in 1914 she went to London and married a man called Teddy whom no one knows anything about except that his surname was Williams, the same as hers, so she did not have to change her name. The mysterious Mr Williams died before anyone met him and his life insurance bought her the house she still lives in.'

'Exactly,' he said. 'What are you getting at?'

'Then, after Mr Williams died, she gave birth to a son who happens to look remarkably like the late Earl Fitzherbert.'

He began to get a glimmer of what she might be saying. 'Go on.'

'Has it never occurred to you that there might be a completely different explanation for this whole story?'

'Not until now...'

'What does an aristocratic family do when one of their daughters gets pregnant? It happens all the time, you know.'

'I suppose it does, but I don't know how they handle it. You never hear about it.'

'Exactly. The girl disappears for a few months – to Scotland, or Brittany, or Geneva – with her maid. When the two of them reappear, the maid has a little baby which, she says, she gave birth to during the holiday. The family treat her surprisingly kindly, even though she has admitted fornication, and send her to live a safe distance away, with a small pension.'

It seemed like a fairy story, nothing to do with real life; but all the same Lloyd was intrigued and troubled. 'And you think I was the baby in some such pretence?'

'I think Lady Maud Fitzherbert had a love affair with a gardener, or a coal miner, or perhaps a charming rogue in London; and she got pregnant. She went away somewhere to give birth in secret. Your mother agreed to pretend the baby was hers, and in exchange she was given a house.'

Lloyd was struck by a corroborating thought. 'She's always been evasive whenever I've asked about my real father.' That now seemed suspicious.

'There you are! There never was a Teddy Williams. To maintain her respectability, your mother said she was a widow. She called her fictional late husband Williams to avoid the problem of changing her name.'

Lloyd shook his head in disbelief. 'It seems too fantastic.'

'She and Maud continued friends, and Maud helped raise you. In 1933 your mother took you to Berlin because your real mother wanted to see you again.'

Lloyd felt as if he were either dreaming or just waking up. 'You think I'm Maud's child?' he said incredulously.

Daisy tapped the frame of the picture she was still holding. 'And you look just like your grandfather!'

Lloyd was bewildered. It could not be true – yet it made sense. 'I'm used to Bernie not being my real father,' he said. 'Is Ethel not my real mother?'

Daisy must have seen a look of helplessness on his face, for she leaned forward and touched him – something she did not generally do – and said: 'I'm sorry, have I been brutal? I just want you to see what's in front of your eyes. If Peel suspects the truth, don't you think others may too? It's the kind of news you want to hear from someone who . . . from a friend.'

A gong sounded distantly. Lloyd said mechanically: 'I'd better go to the mess for lunch.' He took the photograph out of its frame and slipped it into a pocket of his uniform jacket.

'You're upset,' Daisy said anxiously.

'No, no. Just . . . astonished.'

'Men always deny that they're upset. Please come and see me later.'

'All right.'

'Don't go to bed without talking to me again.'

'I won't.'

He left the junk room and made his way upstairs to the grand dining room, now the mess. He ate his canned beef mince automatically, his mind in turmoil. He took no part in the discussion at table about the battle raging in Norway.

'Having a daydream, Williams?' said Major Lowther.

'Sorry, sir,' Lloyd said mechanically. He improvised an excuse. 'I was trying to remember which was the higher German rank, *Generalleutnant* or *Generalmajor*.'

Lowther said: '*Generalleutnant* is higher.' Then he added quietly: 'Just don't forget the difference between *meine Frau* and *deine Frau*.'

Lloyd felt himself blush. So his friendship with Daisy was not as discreet as he had imagined. It had even come to Lowther's notice. He felt indignant: he and Daisy had done nothing improper. Yet he did not protest. He felt guilty, even though he was not. He could not put his hand on his heart and swear that his intentions were pure. He knew what Granda would say: 'Whosoever looketh on a woman to lust after her hath committed adultery with her already in his heart.' That was the no-bullshit teaching of Jesus and there was a lot of truth in it.

Thinking of his grandparents led him to wonder if they knew about his real parents. Being in doubt about his real father and mother gave him a lost feeling, like a dream about falling from a height. If he had been told lies about that, he might have been misled about anything.

He decided he would question Granda and Grandmam. He could do it today, as it was Sunday. As soon as he could decently excuse himself from the mess, he walked downhill to Wellington Row.

It occurred to him that if he asked them outright whether he was Maud's son they might simply deny everything point-blank. Perhaps a more gradual approach would be more likely to elicit information.

He found them sitting in their kitchen. To them Sunday was the Lord's Day, devoted to religion, and they would not read newspapers or listen to the radio. But they were pleased to see him, and Grandmam made tea, as always.

Lloyd began: 'I wish I knew more about my real father. Mam says that Teddy Williams was in the Welsh Rifles, did you know that?'

Grandmam said: 'Oh, why do you want to go digging up the past? Bernie's your father.'

Lloyd did not contradict her. 'Bernie Leckwith has been everything a father should be to me.'

Granda nodded. 'A Jew, but a good man, there's no doubt.' He imagined he was being magnanimously tolerant.

Lloyd let it pass. 'All the same, I'm curious. Did you meet Teddy Williams?'

Granda looked angry. 'No,' he said. 'And it was a sorrow to us.'

Grandmam said: 'He came to Tŷ Gwyn as a valet to a guest. We never knew your mother was sweet on him till she went to London to marry him.'

'Why didn't you go to the wedding?'

They were both silent. Then Granda said: 'Tell him the truth, Cara. No good ever comes of lies.'

'Your mother yielded to temptation,' Grandmam said. 'After the valet left Tŷ Gwyn, she found she was with child.' Lloyd had suspected that, and thought it might account for her evasiveness. 'Your Granda was very angry,' Grandmam added.

'Too angry,' Granda said. 'I forgot that Jesus said: "Judge not, that ye be not judged." Her sin was lust, but mine was pride.' Lloyd was astonished to see tears in his grandfather's pale-blue eyes. 'God forgave her, but I didn't, not for a long time. By then my son-in-law was dead, killed in France.'

Lloyd was more bewildered than before. Here was another detailed story, somewhat different from what he had been told by his mother and completely different from Daisy's theory. Was Granda weeping for a son-in-law who had never existed?

He persisted. 'And the family of Teddy Williams? Mam said he came from Swansea. He probably had parents, brothers and sisters . . .'

Grandmam said: 'Your mother never talked about his family. I think she was ashamed. Whatever the reason, she didn't want to know them. And it wasn't our place to go against her in that.'

'But I might have two more grandparents in Swansea. And uncles and aunts and cousins I've never met.'

'Aye,' said Granda. 'But we don't know.'

'My mother knows, though.'

'I suppose she does.'

'I'll ask her, then,' said Lloyd.

(iv)

Daisy was in love.

She knew, now, that she had never loved anyone before Lloyd. She had never truly loved Boy, though she had been excited by him. As for poor Charlie Farquharson, she had been at most fond of him. She had believed that love was something she could bestow upon whomever she liked, and that her main responsibility was to choose cleverly. Now she knew that was all wrong. Cleverness had nothing to do with it, and she had no choice. Love was an earthquake.

Life was empty but for the two hours she spent with Lloyd each evening. The rest of the day was anticipation; the night was recollection.

Lloyd was the pillow she put her cheek on. He was the towel with which she patted her breasts when she got out of the bathtub. He was the knuckle she put into her mouth and sucked thoughtfully.

How could she have ignored him for four years? The love of her life had appeared before her at the Trinity Ball, and she had noticed only that he appeared to be wearing someone else's dress clothes! Why had she not taken him in her arms and kissed him and insisted they get married immediately?

He had known all along, she surmised. He must have fallen in love with her from the start. He had begged her to throw Boy over. 'Give him up,' he had said the night they went to the Gaiety music hall. 'Be my girlfriend instead.' And she had laughed at him. But he had seen the truth to which she had been blind.

However, some intuition deep within her had told her to kiss him, there on the Mayfair pavement in the darkness between two street lights. At the time she had regarded it as a self-indulgent whim; but, in fact, it was the smartest thing she had ever done, for it had probably sealed his devotion.

Now, at Tŷ Gwyn, she refused to think about what would happen next. She was living from day to day, walking on air, smiling at nothing. She got an anxious letter from her mother in Buffalo, worrying about her health and her state of mind after the miscarriage, and she sent back a reassuring reply. Olga included titbits of news: Dave Rouzrokh had died in Palm Beach; Muffie Dixon had married Philip Renshaw; Senator Dewar's wife, Rosa, had written a bestseller called *Behind the Scenes at the White House*, with photographs by Woody. A month ago this would have made her homesick; now she was just mildly interested.

She felt sad only when she thought of the baby she had lost. The pain had gone immediately, and the bleeding had stopped after a week, but the loss grieved her. She no longer cried about it, but occasionally she found herself staring into empty space, thinking about whether it would have been a girl or a boy, and what it would have looked like; and then realized with a shock that she had not moved for an hour.

Spring had come, and she walked on the windy mountainside, in waterproof boots and a raincoat. Sometimes, when she was sure there was no one to hear but the sheep, she shouted at the top of her voice: 'I love him!'

She worried about his reaction to her questions about his parentage. Perhaps she had done wrong to raise the issue: it had only made him unhappy. Yet her excuse had been valid: sooner or later the truth would probably come out, and it was better to hear such things from someone who loved you. His pained bafflement touched her heart, and made her love him even more.

Then he told her he had arranged leave. He was going to a south coast resort called Bournemouth for the Labour Party's annual conference on the second weekend in May, which was a British holiday called Whitsun.

His mother would also be at Bournemouth, he said, so he would have a chance to question her about his parentage; and Daisy thought he looked eager and afraid at the same time.

Lowther would certainly have refused to let him go, but Lloyd had spoken to Colonel Ellis-Jones back in March, when he had been assigned to this course, and the colonel either liked Lloyd or sympathized with the party, or both, and gave him permission which Lowther could not countermand. Of course, if the Germans invaded France, then nobody would be able to take leave.

Daisy was strangely frightened by the prospect of Lloyd's leaving Aberowen without knowing that she loved him. She was not sure why, but she had to tell him before he went.

Lloyd was to leave on Wednesday and return six days later. By coincidence, Boy had announced he would come to visit, arriving on Wednesday evening. Daisy was glad, for reasons she could not quite figure out, that the two men would not be there at the same time.

She decided to make her confession to Lloyd on Tuesday, the day before he left. She had no idea what she was going to say to her husband a day later.

Imagining the conversation she would have with Lloyd, she realized that he would surely kiss her, and when they kissed they would be overwhelmed by their feelings, and they would make love. And then they would lie all night in each other's arms.

At this point in her thinking, the need for discretion intruded into her daydream. Lloyd must not be seen emerging from her quarters in the morning, for both their sakes. Lowthie already had his suspicions: she could tell by his attitude towards her, which was both disapproving and roguish, almost as if he felt that he rather than Lloyd should be the one she should fall for.

How much better it would be if she and Lloyd could meet somewhere else for their fateful conversation. She thought of the unused bedrooms in the west wing, and she felt breathless. He could leave at dawn, and if anyone saw him they would not know he had been with her. She could emerge later, fully dressed, and pretend to be looking for some lost piece of family property, a painting perhaps. In fact, she thought, elaborating on the lie she would tell if necessary, she could take some object from the junk room and place it in the bedroom in advance, ready to be used as concrete evidence of her story.

At nine o'clock on Tuesday, when the students were all in classes, she walked along the upper floor, carrying a set of perfume vials with tarnished silver tops and a matching hand mirror. She felt guilty already. The carpet had been taken up, and her footsteps rang loud on the floorboards, as if announcing the approach of a scarlet woman. Fortunately, there was no one in the bedrooms.

She went to the Gardenia Suite, which she vaguely thought was being used for storage of bed linen. There was no one in the corridor as she stepped inside. She closed the door quickly behind her. She was panting. I haven't done anything yet, she told herself.

She had remembered aright: all around the room, piled up against the gardenia-printed wallpaper, were neat stacks of sheets and blankets and pillows, wrapped in covers of coarse cotton and tied with string like large parcels.

The room smelled musty, and she opened a window. The original furniture was still here: a bed, a wardrobe, a chest of drawers, a writing table, and a kidney-shaped dressing table with three mirrors. She put the perfume vials on the dressing table, then she made the bed up with some of the stored linen. The sheets were cold to her touch.

Now I've done something, she thought. I've made a bed for my lover and me.

She looked at the white pillows and the pink blankets with their satin edging, and she saw herself and Lloyd, locked in a clinging embrace, kissing with mad desperation. The thought aroused her so much that she felt faint.

She heard footsteps outside, ringing on the floorboards as hers had. Who could that be? Morrison, perhaps, the old footman, on his way to look at a leaking gutter or a cracked windowpane. She waited, heart pounding with guilt, as the footsteps came nearer then receded.

The scare calmed her excitement and cooled the heat she felt inside. She took one last look around the scene and left.

There was no one in the corridor.

She walked along, her shoes heralding her progress; but she looked perfectly innocent now, she told herself. She could go anywhere she wanted; she had more right to be here than anyone else; she was at home; her husband was heir to the whole place.

The husband she was carefully planning to betray.

She knew she should be paralysed by guilt, but in fact she was eager to do it, consumed by longing.

Next she had to brief Lloyd. He had come to her apartment last night, as usual; but she could not have made this assignation with him then, for he would have expected her to explain herself and then, she knew, she would have told him everything and taken him to her bed and ruined the whole plan. So she had to speak to him briefly today.

She did not normally see him in the daytime, unless she ran into him by accident, in the hall or library. How could she make sure of meeting him? She went up the back stairs to the attic floor. The trainees were not in their rooms, but at any moment one of them might appear, returning to his room for something he had forgotten. So she had to be quick.

She went into Lloyd's room. It smelled of him. She could not say exactly what the fragrance was. She did not see a bottle of cologne in the room, but there was a jar of some kind of hair lotion beside his razor. She opened it and sniffed: yes, that was it, citrus and spice. Was he vain, she asked herself? Perhaps a little bit. He usually looked well dressed, even in his uniform.

She would leave him a note. On top of the dresser was a pad of cheap writing paper. She opened it and tore out a sheet. She looked around for something to write with. He had a black fountain pen with his name engraved on the barrel, she knew, but he would have that with him, for writing notes in class. She found a pencil in the top drawer.

What could she write? She had to be careful in case someone else should read the note. In the end she just wrote: 'Library'. She left the pad open on the dresser where he could hardly fail to see it. Then she left.

No one saw her.

He would probably come to his room at some point, she speculated, perhaps to fill his pen with ink from the bottle on the dresser. Then he would see the note and come to her.

She went to the library to wait.

The morning was long. She was reading Victorian authors – they seemed to understand how she felt right now – but today Mrs Gaskell could not hold her attention, and she spent most of the time looking out of the window. It was May, and normally there would have been a brilliant display of spring flowers in the grounds of Tŷ Gwyn, but most of the gardeners had joined the armed forces, and the rest were growing vegetables, not flowers.

Several trainees came into the library just before eleven, and settled down in the green leather chairs with their notebooks, but Lloyd was not among them.

The last lecture of the morning ended at half past twelve, she knew. At that point the men got up and left the library, but Lloyd did not appear.

Surely he would go to his room now, she thought, just to put down his books and wash his hands in the nearby bathroom.

The minutes passed, and the gong sounded for lunch.

Then he came in, and her heart leaped.

He looked worried. 'I just saw your note,' he said. 'Are you all right?'

His first concern was for her. A problem of hers was not a nuisance to him, but an opportunity to help her, and he would seize it eagerly. No man had cared for her this way, not even her father.

'Everything is all right,' she said. 'Do you know what a gardenia looks like?' She had rehearsed this speech all morning.

'I suppose so. A bit like a rose. Why?'

'In the west wing there's an apartment called the Gardenia Suite. It has a white gardenia painted on the door, and it's full of stored linen. Do you think you could find it?'

'Of course.'

'Meet me there tonight, instead of coming to the flat. Usual time.'

He stared at her, trying to figure out what was going on. 'I will,' he said. 'But why?'

'I want to tell you something.'

'How exciting,' he said, but he looked puzzled.

She could guess what was going through his mind. He was electrified by the thought that she might intend a romantic assignation, and at the same time he was telling himself that was a hopeless dream.

'Go to lunch,' she said.

He hesitated.

She said: 'I'll see you tonight.'

'I can't wait,' he said, and went out.

She returned to her flat. Maisie, who was not much of a cook, had made her a sandwich with two slabs of bread and a slice of canned ham. Daisy's stomach was full of butterflies: she could not have eaten if it had been peach ice cream.

She lay down to rest. Her thoughts about the night to come were so explicit she felt embarrassed. She had learned a lot about sex from Boy, who clearly had much experience with other women, and she knew a great deal about what men liked. She wanted to do everything with Lloyd, to kiss every part of his body, to do what Boy called *soixante-neuf,* to swallow his semen. The thoughts were so arousing that it took all her willpower to resist the temptation to pleasure herself.

She had a cup of coffee at five, then washed her hair and took a long bath, shaving her underarms and trimming her pubic hair, which grew too abundantly. She dried herself and rubbed in a light body lotion all over. She perfumed herself and began to get dressed.

She put on new underwear. She tried on all her dresses. She liked the look of one with fine blue-and-white stripes, but all down the front it had little buttons that would take forever to undo, and she knew she would want to undress quickly. I'm thinking like a whore, she realized, and she did not know whether to be amused or ashamed. In the end, she decided on a simple peppermint-green cashmere knee-length that showed off her shapely legs.

She studied herself in the narrow mirror on the inside of the wardrobe door. She looked good.

She perched on the edge of the bed to put her stockings on, and Boy came in.

Daisy felt faint. If she had not been sitting she would have fallen down. She stared at him in disbelief.

'Surprise!' he said with jollity. 'I came a day early.'

'Yes,' she said when at last she was able to speak. 'Surprise.'

He bent down and kissed her. She had never much liked his tongue in her mouth, because he always tasted of booze and cigars. He did not mind her distaste – in fact, he seemed to enjoy forcing the issue. But now, out of guilt, she tongued him back.

'Gosh!' he said when he ran out of breath. 'You're frisky.'

You have no idea, Daisy thought; at least, I hope you don't.

'The exercise was brought forward by a day,' he explained. 'No time to warn you.'

'So you're here for the night,' she said.

'Yes.'

And Lloyd was leaving in the morning.

'You don't seem very pleased,' Boy said. He looked at her dress. 'Did you have something else planned?'

'Such as what?' she said. She had to regain her composure. 'A night out at the Two Crowns pub, perhaps?' she asked sarcastically.

'Speaking of that, let's have a drink.' He left the room in search of booze.

Daisy buried her face in her hands. How could this be? Her plan was ruined. She would have to find some way of alerting Lloyd. And she could not declare her love for him in a hurried whisper with Boy around the corner.

She told herself that the whole scheme would simply be postponed. It was only for a few days: he was due back next Tuesday. The delay would be agonizing, but she would survive, and so would her love. All the same, she almost cried with disappointment.

She finished putting on her stockings and shoes, then she went into the little sitting room.

Boy had found a bottle of Scotch and two glasses. She took some to be convivial. He said: 'I see that girl is making a fish pie for supper. I'm starving. Is she a good cook?'

'Not really. Her food is edible, if you're hungry.'

'Oh, well, there's always whisky,' he said, and he poured himself another drink.

'What have you been doing?' She was desperate to get him to talk so that she would not have to. 'Did you fly to Norway?' The Germans were winning the first land battle of the war there.

'No, thank God. It's a disaster. There's a big debate in the House of Commons tonight.' He began to talk about the mistakes the British and French commanders had made.

When supper was ready, Boy went down to the cellar to get some wine. Daisy saw a chance to alert Lloyd. But where would he be? She looked at her wristwatch. It was half past seven. He would be having dinner in the mess. She could not walk into that room and whisper in his ear as he sat at the table with his fellow officers: it would be as good as telling everyone they were lovers. Was there some way she could get him out of there? She racked her brains, but before she could think of anything Boy returned, triumphantly carrying a bottle of 1921 Dom Pérignon. 'The first vintage they made,' he said. 'Historic.'

They sat at the table and ate Maisie's fish pie. Daisy drank a glass of the champagne but she found it difficult to eat. She pushed her food around the plate in an attempt to look normal. Boy had a second helping.

For dessert, Maisie served canned peaches with condensed milk. 'War has been bad for British cuisine,' Boy said.

'Not that it was great before,' Daisy commented, still working on seeming normal.

By now Lloyd must be in the Gardenia Suite. What would he do if she were unable to get a message to him? Would he remain there all night, waiting and hoping for her to arrive? Would he give up at midnight and return to his own bed? Or would he come down here looking for her? That might be awkward.

Boy took out a large cigar and smoked it with satisfaction, occasionally dipping the unlit end into a glass of brandy. Daisy tried to think of an excuse to leave him and go upstairs, but nothing came. What pretext could she possibly cite for visiting the trainees' quarters at this time of night?

She still had done nothing when he put out his cigar and said: 'Well, time for bed. Do you want to use the bathroom first?'

Not knowing what else to do, she got up and went into the bedroom. Slowly, she took off the clothes she had put on so carefully for Lloyd. She washed her face and put on her least alluring nightdress. Then she got into bed.

Boy was moderately drunk when he climbed in beside her, but he still wanted sex. The thought appalled her. 'I'm sorry,' she said. 'Dr Mortimer said no marital relations for three months.' This was not true. Mortimer had said it would be all right when the bleeding stopped. She felt horribly dishonest. She had been planning to do it with Lloyd tonight.

'What?' Boy said indignantly. 'Why?'

Improvising, she said: 'If we do it too soon, it might affect my chances of getting pregnant again, apparently.'

That convinced him. He was desperate for an heir. 'Ah, well,' he said, and turned away.

In a minute he was asleep.

Daisy lay awake, her mind buzzing. Could she slip away now? She would have to get dressed – she certainly could not walk around the house in her nightdress. Boy slept heavily, but often woke to go to the bathroom. What if he did that while she was gone, and saw her return with her clothes on? What story could she tell that had a chance of being believed? Everyone knew there was only one reason why a woman went creeping around a country house at night.

Lloyd would have to suffer. And she suffered with him, thinking of him alone and disappointed in that musty room. Would he lie down in his uniform and fall asleep? He would be cold, unless he pulled a blanket around him. Would he assume some emergency, or just think she had carelessly stood him up? Perhaps he would feel let down, and be angry with her.

Tears rolled down her face. Boy was snoring, so he would never know.

She dozed off in the small hours, and dreamed she was catching a train, but silly things kept happening to delay her: the taxi took her to the wrong place, she had to walk unexpectedly far with her suitcase, she could not find her ticket, and when she reached the platform she found waiting for her an old-fashioned stage coach that would take days to get to London.

When she woke from the dream, Boy was in the bathroom, shaving.

She lost heart. She got up and dressed. Maisie prepared breakfast, and Boy had eggs and bacon and buttered toast. By the time they had finished it was nine o'clock. Lloyd had said he was leaving at nine. He might be in the hall now, with his suitcase in his hand.

Boy got up from the table and went into the bathroom, taking the newspaper with him. Daisy knew his morning habits: he would be there five or ten minutes. Suddenly her apathy left her. She went out of the flat and ran up the stairs to the hall.

Lloyd was not there. He must already have left. Her heart sank.

But he would be walking to the railway station: only the wealthy and infirm took taxis to go a mile. Perhaps she could catch him up. She went out through the front door.

She saw him four hundred yards down the drive, walking smartly, carrying his case, and her heart leaped. Throwing caution to the wind, she ran after him.

A light army pickup truck of the kind they called a Tilly was bowling down the drive ahead of her. To her dismay it slowed alongside Lloyd. 'No!' Daisy said, but Lloyd was too far away to hear her.

He threw his suitcase into the back and jumped into the cab beside the driver.

She kept running, but it was hopeless. The little truck pulled away and picked up speed.

Daisy stopped. She stood and watched as the Tilly passed through the gates of Tŷ Gwyn and disappeared from view. She tried not to cry.

After a moment she turned around and went back inside the house.

(v)

On the way to Bournemouth Lloyd spent a night in London; and that evening, Wednesday 8 May, he was in the visitors' gallery of the House of Commons, watching the debate that would decide the fate of the Prime Minister, Neville Chamberlain.

It was like being in the gods at the theatre: the seats were cramped and hard, and you looked vertiginously down on the drama unfolding below. The gallery was full tonight. Lloyd and his stepfather, Bernie, had got tickets only with difficulty, through the influence of his mother, Ethel, who was now sitting with his Uncle Billy among the Labour MPs down in the packed chamber.

Lloyd had had no chance yet to ask about his real father and mother: everyone was too preoccupied with the political crisis. Both Lloyd and Bernie wanted Chamberlain to resign. The appeaser of Fascism had little credibility as a war leader, and the debacle in Norway only underlined that.

The debate had begun the night before. Chamberlain had been furiously attacked, not just by Labour MPs but by his own side, Ethel had reported. The Conservative Leo Amery had quoted Cromwell at him: 'You have sat too long here for any good you have been doing. Depart, I say, and let us have done with you. In the name of God, go!' It was a cruel speech to come from a colleague, and it was made more wounding by the chorus of 'Hear, hear!' that arose from both sides of the chamber.

Lloyd's mother and the other female MPs had got together in their own room in the palace of Westminster and agreed to force a vote. The men could not stop them and so joined them instead. When this was announced on Wednesday, the debate was transformed into a ballot on Chamberlain. The Prime Minister accepted the challenge, and – in what Lloyd felt was a sign of weakness – appealed to his friends to stand by him.

The attacks continued tonight. Lloyd relished them. He hated Chamberlain for his policy on Spain. For two years, from 1937 to 1939, Chamberlain had continued to enforce 'non-intervention' by Britain and France, while Germany and Italy poured arms and men into the rebel army, and American ultra-conservatives sold oil and trucks to Franco. If any one British politician bore guilt for the mass murders now being carried out by Franco, it was Neville Chamberlain.

'And yet,' said Bernie to Lloyd during a lull, 'Chamberlain isn't really to blame for the fiasco in Norway. Winston Churchill is First Lord of the Admiralty, and your mother says he was the one who pushed for this invasion. After all Chamberlain has done – Spain, Austria, Czechoslovakia – it will be ironic if he falls from power because of something that isn't really his fault.'

'Everything is ultimately the Prime Minister's fault,' said Lloyd. 'That's what it means to be the leader.'

Bernie smiled wryly, and Lloyd knew he was thinking that young people saw everything too simply; but, to his credit, Bernie did not say it.

It was a noisy debate, but the House went quiet when the former Prime Minister, David Lloyd George, stood up. Lloyd had been named after him. Seventy-seven years old now, a white-haired elder statesman, he spoke with the authority of the man who had won the Great War.

He was merciless. 'It is not a question of who are the Prime Minister's friends,' he said, stating the obvious with withering sarcasm. 'It is a far bigger issue.'

Once again, Lloyd was heartened to see that the chorus of approval came from the Conservative side as well as the opposition.

'He has appealed for sacrifices,' Lloyd George said, his nasal North Wales accent seeming to sharpen the edge of his contempt. 'There is nothing which can contribute more to victory, in this war, than that he should sacrifice the seals of office.'

The opposition shouted their approval, and Lloyd could see his mother cheering.

Churchill closed the debate. As a speaker he was the equal of Lloyd George, and Lloyd feared that his oratory might rescue Chamberlain. But the House was against him, interrupting and jeering, sometimes so loudly that he could not be heard over the clamour.

He sat down at 11 p.m. and the vote was taken.

The voting system was cumbersome. Instead of raising their hands, or ticking slips of paper, MPs had to leave the chamber and be counted as they walked through one of two lobbies, for Ayes or Noes. The process took fifteen or twenty minutes. It could have been devised only by men who did not have enough to do, Ethel said. She felt sure it would be modernized soon.

Lloyd waited on tenterhooks. The fall of Chamberlain would give him profound satisfaction, but it was by no means certain.

To distract himself he thought about Daisy, always a pleasant occupation. How strange his last twenty-four hours at Tŷ Gwyn had been: first the one-word note 'Library'; then the rushed conversation, with her tantalizing summons to the Gardenia Suite; then a whole night of waiting, cold and bored and bewildered, for a woman who did not show up. He had stayed there until six o'clock in the morning, miserable but unwilling to give up hope until the moment when he was obliged to wash and shave and change his clothes and pack his suitcase for the trip.

Clearly something had gone wrong, or she had changed her mind; but what had she intended in the first place? She had said she wanted to tell him something. Had she planned to say something earth-shaking, to merit all that drama? Or something so trivial that she had forgotten all about it and the rendezvous? He would have to wait until next Tuesday to ask her.

He had not told his family that Daisy had been at Tŷ Gwyn. That would have required him to explain to them what his relationship with Daisy was now, and he could not do that, for he did not really understand it himself. Was he in love with a married woman? He did not know. How did she feel about him? He did not know. Most likely, he thought, Daisy and he were two good friends who had missed their chance at love. And somehow he did not want to admit that to anyone, for it seemed unbearably final.

He said to Bernie: 'Who will take over, if Chamberlain goes?'

'The betting is on Halifax.' Lord Halifax was currently the Foreign Secretary.

'No!' said Lloyd indignantly. 'We can't have an earl for Prime Minister at a time like this. Anyway, he's an appeaser, just as bad as Chamberlain!'

'I agree,' said Bernie. 'But who else is there?'

'What about Churchill?'

'You know what Stanley Baldwin said about Churchill?' Baldwin, a Conservative, had been Prime Minister before Chamberlain. 'When Winston was born, lots of fairies swooped down on his cradle with gifts – imagination, eloquence, industry, ability – and then came a fairy who said: 'No person has a right to so many gifts,' picked him up, and gave him such a shake and a twist that he was denied judgement and wisdom.'

Lloyd smiled. 'Very witty, but is it true?'

'There's something in it. In the last war he was responsible for the Dardanelles campaign, which was a terrible defeat for us. Now he's pushed us into the Norwegian adventure, another failure. He's a fine orator, but the evidence suggests he has a tendency to wishful thinking.'

Lloyd said: 'He was right about the need to rearm in the thirties – when everyone else was against it, including the Labour Party.'

'Churchill will be calling for rearmament in Paradise, when the lion lies down with the lamb.'

'I think we need someone with an aggressive streak. We want a prime minister who will bark, not whimper.'

'Well, you may get your wish. The tellers are coming back.'

The votes were announced. The Ayes had 280, the Noes 200. Chamberlain had won. There was uproar in the chamber. The Prime Minister's supporters cheered, but others yelled at him to resign.

Lloyd was bitterly disappointed. 'How can they want to keep him, after all that?'

'Don't jump to conclusions,' said Bernie as the Prime Minister left and the noise subsided. Bernie was making calculations with a pencil in the margin of the *Evening News*. 'The government usually has a majority of about two hundred and forty. That's dropped to eighty.' He scribbled numbers, adding and subtracting. 'Taking a rough guess at the number of MPs absent, I reckon about forty of the government's supporters voted against Chamberlain, and another sixty abstained. That's a terrible blow to a prime minister – a hundred of his colleagues don't have confidence in him.'

'But is it enough to force him to resign?' Lloyd said impatiently.

Bernie spread his arms in a gesture of surrender. 'I don't know,' he said.

(vi)

Next day Lloyd, Ethel, Bernie and Billy went to Bournemouth by train.

The carriage was full of delegates from all over Britain. They all spent the entire journey discussing last night's debate and the future of the Prime Minister, in accents ranging from the harsh chop of Glasgow to the swerve and swoop of Cockney. Once again Lloyd had no chance to raise with his mother the subject that was haunting him.

Like most delegates, they could not afford the swanky hotels on the clifftops, so they stayed in a boarding house on the outskirts. That evening the four of them went to a pub and sat in a quiet corner, and Lloyd saw his chance.

Bernie bought a round of drinks. Ethel wondered aloud what was happening to her friend Maud in Berlin: she no longer got news, for the war had ended the postal service between Germany and Britain.

Lloyd sipped his pint of beer then said firmly: 'I'd like to know more about my real father.'

Ethel said sharply: 'Bernie is your father.'

Evasion again! Lloyd suppressed the anger that immediately rose in him. 'You don't need to tell me that,' he said. 'And I don't need to tell Bernie that I love him like a father, because he already knows.'

Bernie patted him on the shoulder, an awkward but genuine gesture of affection.

Lloyd made his voice insistent. 'But I'm curious about Teddy Williams.'

Billy said: 'We need to talk about the future, not the past – we're at war.'

'Exactly,' said Lloyd. 'So I want answers to my questions *now*. I'm not willing to wait, because I will be going into battle soon, and I don't want to die in ignorance.' He did not see how they could deny that argument.

Ethel said: 'You know all there is to know,' but she was not meeting his eye.

'No, I don't,' he said, forcing himself to be patient. 'Where are my other grandparents? Do I have uncles and aunts and cousins?'

'Teddy Williams was an orphan,' Ethel said.

'Raised in what orphanage?'

She said irritably: 'Why are you so stubborn?'

Lloyd allowed his voice to rise in reciprocal annoyance. 'Because I'm like you!'

Bernie could not repress a grin. 'That's true, anyway.'

Lloyd was not amused. 'What orphanage?'

'He might have told me, but I don't remember. In Cardiff, I think.'

Billy intervened. 'You're touching a sore place, now, Lloyd, boy. Drink your beer and drop the subject.'

Lloyd said angrily: 'I've got a bloody sore place, too, Uncle Billy, thank you very much, and I'm fed up with lies.'

'Now, now,' said Bernie. 'Let's not have talk of lies.'

'I'm sorry, Dad, but it's got to be said.' Lloyd held up a hand to stave off interruption. 'Last time I asked, Mam told me Teddy Williams's family came from Swansea but they moved around a lot because of his father's job. Now she says he was raised in an orphanage in Cardiff. One of those stories is a lie – if not both.'

At last Ethel looked him in the eye. 'Me and Bernie fed you and clothed you and sent you to school and university,' she said indignantly. 'You've got nothing to complain about.'

'And I'll always be grateful to you, and I'll always love you,' Lloyd said.

Billy said: 'Why have this come up now, anyhow?'

'Because of something somebody said to me in Aberowen.'

His mother did not respond, but there was a flash of fear in her eyes. Someone in Wales knows the truth, Lloyd thought.

He went on relentlessly: 'I was told that perhaps Maud Fitzherbert fell pregnant in 1914, and her baby was passed off as yours, for which you were rewarded with the house in Nutley Street.'

Ethel made a scornful noise.

Lloyd held up a hand. 'That would explain two things,' he said. 'One, the unlikely friendship between you and Lady Maud.' He reached into his jacket pocket. 'Two, this picture of me in side-whiskers.' He showed them the photograph.

Ethel stared at the picture without speaking.

Lloyd said: 'It could be me, couldn't it?'

Billy said testily: 'Yes, Lloyd, it could. But obviously it's not, so stop mucking about and tell us who it is.'

'It's Earl Fitzherbert's father. Now *you* stop mucking about, Uncle Billy, and you, Mam. Am I Maud's son?'

Ethel said: 'The friendship between me and Maud was a political alliance, foremost. It was broken off when we disagreed about strategy for suffragettes, then resumed later. I like her a lot, and she gave me important chances in life, but there is no secret bond. She doesn't know who your father is.'

'All right, Mam,' said Lloyd. 'I could believe that. But this photo . . .'

'The explanation of that resemblance . . .' She choked up.

Lloyd was not going to let her escape. 'Come on,' he said remorselessly. 'Tell me the truth.'

Billy intervened again. 'You're barking up the wrong tree, boyo,' he said.

'Am I? Well, then, set me straight, why don't you?'

'It's not for me to do that.'

That was as good as an admission. 'So you *were* lying before.'

Bernie looked gobsmacked. He said to Billy: 'Are you saying the Teddy Williams story isn't true?' Clearly he had believed it all these years, just as Lloyd had.

Billy did not reply.

They all looked at Ethel.

'Oh, bugger it,' she said. 'My father would say: "Be sure your sins will find you out." Well, you've asked for the truth, so you shall have it, though you won't like it.'

'Try me,' Lloyd said recklessly.

'You're not Maud's child,' she said. 'You're Fitz's.'

(vii)

Next day, Friday 10 May, Germany invaded Holland, Belgium and Luxembourg.

Lloyd heard the news on the radio as he sat down to breakfast with his parents and Uncle Billy in the boarding house. He was not surprised: everyone in the army had believed the invasion was imminent.

He was much more stunned by the revelations of the previous evening. Last night he had lain awake for hours, angry that he had been misled so long, dismayed that he was the son of a right-wing aristocratic appeaser who was also, weirdly, the father-in-law of the enchanting Daisy.

'How could you fall for him?' he had said to his mother in the pub.

Her reply had been sharp. 'Don't be a hypocrite. You used to be crazy about your rich American girl, and she was so right-wing she married a Fascist.'

Lloyd had wanted to argue that that was different, but quickly realized it was the same. Whatever his relationship with Daisy now, there was no doubt that he had once felt in love with her. Love was not logical. If he could succumb to an irrational passion, so could his mother; indeed, they had been the same age, twenty-one, when it had happened.

He had said she should have told him the truth from the start, but she had an argument for that, too. 'How would you have reacted, as a little boy, if I had told you that you were the son of a rich man, an earl?

How long would it have been before you boasted to the other boys at school? Think how they would have mocked your childish fantasy. Think how they would have hated you for being superior to them.'

'But later . . .'

'I don't know,' she had said wearily. 'There never seemed to be a good time.'

Bernie had at first gone white with shock, but soon recovered and became his usual phlegmatic self. He said he understood why Ethel had not told him the truth. 'A secret shared is a secret no more.'

Lloyd wondered about his mother's relationship with the earl now. 'I suppose you must see him all the time, in Westminster.'

'Just occasionally. Peers have a separate section of the Palace, with their own restaurants and bars, and when we see them it's usually by arrangement.'

That night Lloyd was too shocked and bewildered to know how he felt. His father was Fitz – the aristocrat, the Tory, the father of Boy, the father-in-law of Daisy. Should he be sad about it, angry, suicidal? The revelation was so devastating that he felt numbed. It was like an injury so grave that at first there was no pain.

The morning news gave him something else to think about.

In the early hours the German army had made a lightning westward strike. Although it was anticipated, Lloyd knew that the best efforts of Allied intelligence had been unable to discover the date in advance, and the armies of those small states had been taken by surprise. Nevertheless, they were fighting back bravely.

'That's probably true,' said Uncle Billy, 'but the BBC would say it anyway.'

Prime Minister Chamberlain had called a Cabinet meeting that was going on at that very moment. However, the French army, reinforced by ten British divisions already in France, had long ago agreed a plan for dealing with such an invasion, and that plan had automatically gone into operation. Allied troops had crossed the French border into Holland and Belgium from the west and were rushing to meet the Germans.

With the momentous news heavy on their hearts, the Williams family caught the bus into the town centre and made their way to Bournemouth Pavilion, where the party conference was being held.

There they heard the news from Westminster. Chamberlain was clinging to power. Billy learned that the Prime Minister had asked

Labour Party leader Clement Attlee to become a Cabinet Minister, making the government a coalition of the three main parties.

All three of them were aghast at this prospect. Chamberlain the appeaser would remain Prime Minister, and the Labour Party would be obliged to support him in a coalition government. It did not bear thinking about.

'What did Attlee say?' asked Lloyd.

'That he would have to consult his National Executive Committee,' Billy replied.

'That's us.' Both Lloyd and Billy were members of the committee, which had a meeting scheduled for four o'clock that afternoon.

'Right,' said Ethel. 'Let's start canvassing, and find out how much support Chamberlain's plan might have on our executive.'

'None, I should think,' said Lloyd.

'Don't be so sure,' said his mother. 'There will be some who want to keep Churchill out at any price.'

Lloyd spent the next few hours in constant political activity, talking to members of the committee and their friends and assistants, in cafés and bars in the pavilion and along the seafront. He ate no lunch, but drank so much tea that he felt he might have floated.

He was disappointed to find that not everyone shared his view of Chamberlain and Churchill. There were a few pacifists left over from the last war, who wanted peace at any price, and approved of Chamberlain's appeasement. On the other side, Welsh MPs still thought of Churchill as the Home Secretary who sent the troops in to break a strike in Tonypandy. That had been thirty years ago, but Lloyd was learning that memories could be long in politics.

At half past three Lloyd and Billy walked along the seafront in a fresh breeze and entered the Highcliff Hotel, where the meeting was to be held. They thought that a majority of the committee were against accepting Chamberlain's offer, but they could not be completely sure, and Lloyd was still worried about the result.

They went into the room and sat at the long table with the other committee members. Promptly at four the party leader came in.

Clem Attlee was a slim, quiet, unassuming man, neatly dressed, with a bald head and a moustache. He looked like a solicitor – which his father was – and people tended to underestimate him. In his dry, unemotional way he summarized, for the committee, the events of the last twenty-four hours, including Chamberlain's offer of a coalition with Labour.

Then he said: 'I have two questions to ask you. The first is: Would you serve in a coalition government with Neville Chamberlain as Prime Minister?'

There was a resounding 'No!' from the people around the table, more vehement than Lloyd had expected. He was thrilled. Chamberlain, friend of the Fascists, the betrayer of Spain, was finished. There was some justice in the world.

Lloyd also noted how subtly the unassertive Attlee had controlled the meeting. He had not opened the subject for general discussion. His question had not been: What shall we do? He had not given people the chance to express uncertainty or dither. In his understated way he had put them all up against the wall and made them choose. And Lloyd felt sure the answer he got was the one he had wanted.

Attlee said: 'Then the second question is: Would you serve in a coalition under a different prime minister?'

The answer was not so vocal, but it was Yes. As Lloyd looked around the table it was clear to him that almost everyone was in favour. If there were any against, they did not bother to ask for a vote.

'In that case,' said Attlee, 'I shall tell Chamberlain that our party will serve in a coalition but only if he resigns and a new prime minister is appointed.'

There was a murmur of agreement around the table.

Lloyd noted how cleverly Attlee had avoided asking who they thought the new prime minister should be.

Attlee said: 'I shall now go and telephone Number Ten Downing Street.'

He left the room.

(viii)

That evening Winston Churchill was summoned to Buckingham Palace, in accordance with tradition, and the King asked him to become Prime Minister.

Lloyd had high hopes of Churchill, even if the man was a Conservative. Over the weekend Churchill made his dispositions. He formed a five-man War Cabinet including Clem Attlee and Arthur

Greenwood, respectively leader and deputy leader of the Labour Party. Union leader Ernie Bevin became Minister of Labour. Clearly, Lloyd thought, Churchill intended to have a genuine cross-party government.

Lloyd packed his case ready to catch the train back to Aberowen. Once there, he expected to be quickly redeployed, probably to France. But he only needed an hour or two. He was desperate to learn the explanation of Daisy's behaviour last Tuesday. Knowing he was going to see her soon increased his impatience to understand.

Meanwhile, the German army rolled across Holland and Belgium, overcoming spirited opposition with a speed that shocked Lloyd. On Sunday evening Billy spoke on the phone to a contact in the War Office, and afterwards he and Lloyd borrowed an old school atlas from the boarding-house proprietress and studied the map of north-west Europe.

Billy's forefinger drew an east–west line from Dusseldorf through Brussels to Lille. 'The Germans are thrusting at the softest part of the French defences, the northern section of the border with Belgium.' His finger moved down the page. 'Southern Belgium is bordered by the Ardennes Forest, a huge strip of hilly, wooded terrain virtually impassable to modern motorized armies. So my friend in the War Office says.' His finger moved on. 'Yet farther south, the French–German border is defended by a series of heavy fortifications called the Maginot Line, stretching all the way to Switzerland.' His finger returned up the page. 'But there are no fortifications between Belgium and northern France.'

Lloyd was puzzled. 'Did no one think of this until now?'

'Of course we did. And we have a strategy to deal with it.' Billy lowered his voice. 'Called Plan D. It can't be a secret any more, since we're already implementing it. The best part of the French army, plus all of the British Expeditionary Force already over there, are pouring across the border into Belgium. They will form a solid line of defence at the Dyle River. That will stop the German advance.'

Lloyd was not much reassured. 'So we're committing half our forces to Plan D?'

'We need to make sure it works.'

'It better.'

They were interrupted by the proprietress, who brought Lloyd a telegram.

It had to be from the army. He had given Colonel Ellis-Jones this address before going on leave. He was surprised he had not heard sooner. He ripped open the envelope. The cable said:

DO NOT RETURN ABEROWEN STOP REPORT SOUTHAMPTON DOCKS
IMMEDIATELY STOP A BIENTOT SIGNED ELLISJONES

He was not going back to Tŷ Gwyn. Southampton was one of Britain's largest ports, a common embarkation point for the Continent, and it was located just a few miles along the coast from Bournemouth, an hour perhaps by train or bus.

Lloyd would not be seeing Daisy tomorrow, he realized with an ache in his heart. Perhaps he might never learn what she had wanted to tell him.

Colonel Ellis-Jones's *à bientôt* confirmed the obvious inference.

Lloyd was going to France.

7

1940 (II)

Erik von Ulrich spent the first three days of the Battle of France in a traffic jam.

Erik and his friend Hermann Braun were part of a medical unit attached to the 2nd Panzer Division. They saw no action as they passed through southern Belgium, just mile after mile of hills and trees. They were in the Ardennes Forest, they reckoned. They travelled on narrow roads, many not even paved, and a broken-down tank could cause a fifty-mile tailback in no time. They were stationary, stuck in queues, more than they were moving.

Hermann's freckled face was set in a grimace of anxiety, and he muttered to Erik in an undertone no one else could hear: 'This is stupid!'

'You should know better than to say that – you were in the Hitler Youth,' said Erik quietly. 'Have faith in the Führer.' But he was not angry enough to denounce his friend.

When they did move it was painfully uncomfortable. They sat on the hard wooden floor of an army truck as it bounced over tree roots and swerved around potholes. Erik longed for battle just so that he could get out of the damn truck.

Hermann said more loudly: 'What are we doing here?'

Their boss, Dr Rainer Weiss, was sitting on a real seat beside the driver. 'We are following the orders of the Führer, which are of course always correct.' He said it straight-faced, but Erik felt sure he was being sarcastic. Major Weiss, a thin man with black hair and spectacles, often spoke cynically about the government and the military, but always in this enigmatic way, so that nothing could be proved against him. Anyway, the army could not afford to get rid of a good doctor at this point.

There were two other medical orderlies in the truck, both older than Erik and Hermann. One of them, Christof, had a better answer to Hermann's question. 'Perhaps the French aren't expecting us to attack here, because the terrain is so difficult.'

His friend Manfred said: 'We will have the advantage of surprise, and will encounter light defences.'

Weiss said sarcastically: 'Thanks for that lesson in tactics, you two – most enlightening.' But he did not say they were wrong.

Despite all that had happened there were still people who lacked faith in the Führer, to Erik's amazement. His own family continued to close their eyes to the triumphs of the Nazis. His father, once a man of status and power, was now a pathetic figure. Instead of rejoicing in the conquest of barbarian Poland, he just moaned about ill-treatment of the Poles – which he must have heard about by listening illegally to a foreign radio station. Such behaviour could get them all into trouble – including Erik, who was guilty of not reporting it to the local Nazi block supervisor.

Erik's mother was just as bad. Every now and again she disappeared with small packages of smoked fish or eggs. She said nothing in explanation, but Erik felt sure she was taking them to Frau Rothmann, whose Jewish husband was no longer allowed to practise as a doctor.

Despite that, Erik sent home a large slice of his army pay, knowing his parents would be cold and hungry if he did not. He hated their politics, but he loved them. They undoubtedly felt the same about his politics and him.

Erik's sister, Carla, had wanted to be a doctor, like Erik, and had been furious when it was made clear to her that in today's Germany this was a man's job. She was now training as a nurse, a much more appropriate role for a German girl. And she, too, was supporting their parents with her meagre pay.

Erik and Hermann had wanted to join infantry units. Their idea of battle was to run at the enemy firing a rifle, and kill or be killed for the Fatherland. But they were not going to be killing anyone. Both had had one year of medical school, and such training was not to be wasted; so they were made medical orderlies.

The fourth day in Belgium, Monday 13 May, was like the first three until the afternoon. Above the roar and snarl of hundreds of tank and truck engines, they began to hear another, louder sound. Aircraft were flying low over their heads and, not too far away, dropping bombs on someone. Erik's nose twitched with the smell of high explosives.

They stopped for their mid-afternoon break on high ground overlooking a meandering river valley. Major Weiss said the river was the Meuse, and they were west of the city of Sedan. So they had entered

France. The planes of the Luftwaffe roared past them, one after another, diving towards the river a couple of miles away, bombing and strafing the scattered villages on the banks where, presumably, there were French defensive positions. Smoke rose from countless fires among the ruined cottages and farm buildings. The barrage was relentless, and Erik almost felt pity for anyone trapped in that inferno.

This was the first action he had seen. Before long he would be in it, and perhaps some young French soldier would look from a safe vantage point and feel sorry for the Germans being maimed and killed. The thought made Erik's heart thud with excitement like a big drum in his chest.

Looking to the east, where the details of the landscape were obscured by distance, he could nevertheless see aircraft like specks, and columns of smoke rising through the air, and he realized that battle had been joined along several miles of this river.

As he watched, the air bombardment came to an end, the planes turning and heading north, waggling their wings to say 'Good luck' as they passed overhead on their way home.

Nearer to where Erik stood, on the flat plain leading to the river, the German tanks were going into action.

They were two miles from the enemy, but already the French artillery was shelling them from the town. Erik was surprised that so many gunners had survived the air bombardment. But fire flashed in the ruins, the boom of cannon was heard across the fields, and fountains of French soil spurted where the shells landed. Erik saw a tank explode after a direct hit, smoke and metal and body parts spewing out of the volcano's mouth, and he felt sick.

But the French shelling did not stop the advance. The tanks crawled on relentlessly towards the stretch of river to the east of the town, which Weiss said was called Donchery. Behind them followed the infantry, in trucks and on foot.

Hermann said: 'The air attack wasn't enough. Where's our artillery? We need them to take out the big guns in the town, and give our tanks and infantry a chance to cross the river and establish a bridgehead.'

Erik wanted to punch him to shut his whining mouth. They were about to go into action – they had to be positive now!

But Weiss said: 'You're right, Braun – but our artillery ammunition is gridlocked in the Ardennes Forest. We've only got forty-eight shells.'

A red-faced major came running past, yelling: 'Move out! Move out!'

Major Weiss pointed and said: 'We'll set up our field dressing station over to the east, where you see that farmhouse.' Erik made out a low grey roof about eight hundred yards from the river. 'All right, get moving!'

They jumped into the truck and roared down the hill. When they reached level ground they swerved left along a farm track. Erik wondered what they would do with the family that presumably lived in the building that was about to become an army hospital. Throw them out of their home, he guessed, and shoot them if they made trouble. But where would they go? They were in the middle of a battlefield.

He need not have worried: they had already left.

The building was half a mile from the worst of the fighting, Erik observed. He guessed there was no point setting up a dressing station within range of enemy guns.

'Stretcher bearers, get going,' Weiss shouted. 'By the time you get back here we'll be ready.'

Erik and Hermann took a rolled-up stretcher and first aid kit from the medical supply truck and headed towards the battle. Christof and Manfred were just ahead of them, and a dozen of their comrades followed. This is it, Erik thought exultantly; this is our chance to be heroes. Who will keep his nerve under fire, and who will lose control and crawl into a hole and hide?

They ran across the fields to the river. It was a long jog, and it was going to seem longer coming back, carrying a wounded man.

They passed burned-out tanks but there were no survivors, and Erik averted his eyes from the scorched human remains smeared across the twisted metal. Shells fell around them, though not many: the river was lightly defended, and many of the guns had been taken out by the air attack. All the same, it was the first time in his life Erik had been shot at, and he felt the absurd, childish impulse to cover his eyes with his hands; but he kept running forward.

Then a shell landed right in front of them.

There was a terrific thud, and the earth shook as if a giant had stamped his foot. Christof and Manfred were hit directly, and Erik saw their bodies fly up into the air as if weightless. The blast threw Erik off his feet. As he lay on the ground, face up, he was showered with dirt from the explosion, but he was not injured. He struggled to his feet. Right in front of him were the mangled bodies of Christof and Manfred. Christof lay like a broken doll, as if all his limbs were disjointed.

Manfred's head had somehow been severed from his body and lay next to his booted feet.

Erik was paralysed with horror. In medical school he had not had to deal with maimed and bleeding bodies. He was used to corpses in anatomy class – they had had one between two students, and he and Hermann had shared the cadaver of a shrivelled old woman – and he had watched living people being cut open on the operating table. But none of that had prepared him for this.

He wanted nothing but to run away.

He turned around. His mind was blank of every thought but fear. He started to walk back the way they had come, towards the forest, away from the battle, taking long, determined strides.

Hermann saved him. He stood in front of Erik and said: 'Where are you going? Don't be a fool!' Erik kept moving, and tried to walk past him. Hermann punched him in the stomach, really hard, and Erik folded over and fell to his knees.

'Don't run away!' Hermann said urgently. 'You'll be shot for desertion! Pull yourself together!'

While Erik was trying to catch his breath he came to his senses. He could not run away, he must not desert, he had to stay here, he realized. Slowly his willpower overcame his terror. Eventually he got to his feet.

Hermann looked at him warily.

'Sorry,' said Erik. 'I panicked. I'm all right now.'

'Then pick up the stretcher and keep going.'

Erik picked up the rolled stretcher, balanced it on his shoulder, turned around and ran on.

Closer to the river, Erik and Hermann found themselves among infantry. Some were manhandling inflated rubber dinghies out of the backs of trucks and carrying them to the water's edge, while the tanks tried to cover them by firing at the French defences. But Erik, rapidly recovering his mental powers, soon saw that it was a losing battle: the French were behind walls and inside buildings, while the German infantry were exposed on the bank of the river. As soon as they got a dinghy into the water, it came under intense machine-gun fire.

Upstream, the river turned a right-angled bend, so the infantry could not move out of range of the French without retreating a long distance.

There were already many dead and wounded men on the ground.

'Let's pick this one up,' Hermann said decisively, and Erik bent to the task. They unrolled their stretcher on the ground next to a groaning infantryman. Erik gave him water from a flask, as he had learned in training. The man seemed to have numerous superficial wounds on his face and one limp arm. Erik guessed he had been hit by machine-gun fire that luckily had missed his vital areas. He saw no gush of blood, so they did not attempt to staunch his wounds. They lifted the man on to the stretcher, picked it up, and began to jog back to the dressing station.

The wounded man cried out in agony as they moved; then, when they stopped, he shouted: 'Keep going, keep going!' and gritted his teeth.

Carrying a man on a stretcher was not as easy as it might seem. Erik thought his arms would fall off when they were only halfway. But he could see that the patient was in greater pain by far, and he just kept running.

Shells no longer fell around them, he noticed gratefully. The French were concentrating all their fire on the river bank, trying to prevent the Germans crossing.

At last Erik and Hermann reached the farmhouse with their burden. Weiss had the place organized, the rooms cleared of superfluous furniture, places marked on the floor for patients, the kitchen table set up for operations. He showed Erik and Hermann where to put the wounded man. Then he sent them back for another.

The run back to the river was easier. They were unburdened and going slightly downhill. As they approached the bank Erik wondered fearfully whether he would panic again.

He saw with trepidation that the battle was going badly. There were several deflated vessels in midstream and many more bodies on the bank – and still no Germans on the far side.

Hermann said: 'This is a catastrophe. We should have waited for our artillery!' His voice was shrill.

Erik said: 'Then we would have lost the advantage of surprise, and the French would have had time to bring up reinforcements. There would have been no point in that long trek through the Ardennes.'

'Well, this isn't working,' said Hermann.

Deep in his heart Erik was beginning to wonder whether the Führer's plans really were infallible. The thought undermined his resolution and threatened to throw him completely off balance. Fortunately there was no more time for reflection. They stopped beside a man with most of one leg blown off. He was about their age, twenty, with pale, freckled skin and copper-red hair. His right leg ended at mid-thigh in a ragged stump. Amazingly, he was conscious, and he stared at them as if they were angels of mercy.

Erik found the pressure point in his groin and stopped the bleeding while Hermann got out a tourniquet and applied it. Then they put him on the stretcher and began the run back.

Hermann was a loyal German, but he sometimes allowed negative feelings to get the better of him. If Erik ever had such feelings he was careful not to voice them. That way he did not lower anyone else's morale – and he stayed out of trouble.

But he could not help thinking. It seemed the approach through the Ardennes had not given the Germans the walkover victory they had expected. The Meuse defences were light but the French were fighting back fiercely. Surely, he thought, his first experience of battle was not going to destroy his faith in his Führer? The idea made him feel panicky.

He wondered whether the German forces farther east were faring any better. The 1st Panzer and the 10th Panzer had been alongside Erik's division, the 2nd, as they approached the border, and it must be they who were attacking upstream.

His arm muscles were now in constant agony.

They arrived back at the dressing station for the second time. The place was now frantically busy, the floor crowded with men groaning and crying, bloody bandages everywhere, Weiss and his assistants moving quickly from one maimed body to the next. Erik had never imagined there could be so much suffering in one small place. Somehow, when the Führer spoke of war, Erik never thought of this kind of thing.

Then he noticed that his own patient's eyes were closed.

Major Weiss felt for a pulse then said harshly. 'Put him in the barn – and for fuck's sake don't waste time bringing me corpses!'

Erik could have cried with frustration, and with the pain in his arms, which was beginning to afflict his legs, too.

They put the body in the barn, and saw that there were already a dozen dead young men there.

This was worse than anything he had envisaged. When he had thought about battle he had foreseen courage in the face of danger, stoicism in suffering, heroism in adversity. What he saw now was agony, screaming, blind terror, broken bodies, and a complete lack of faith in the wisdom of the mission.

They went back again to the river.

The sun was low in the sky, now, and something had changed on the battlefield. The French defenders in Donchery were being shelled from the far side of the river. Erik guessed that farther upstream the 1st Panzers had had better luck, and had secured a bridgehead on the south bank; and now they were coming to the aid of the comrades on their flanks. Clearly *they* had not lost their ammunition in the forest.

Heartened, Erik and Hermann rescued another wounded man. When they got back to the dressing station this time they were given tin bowls of a tasty soup. Resting for ten minutes while he drank the soup made Erik want to lie down and go to sleep for the night. It took a mighty effort to stand up and pick up his end of the stretcher and jog back to the battlefield.

Now they saw a different scene. Tanks were crossing the river on rafts. The Germans on the far side were coming under heavy fire, but they were shooting back, with the help of reinforcements from the 1st Panzers.

Erik saw that his side had a chance of winning their objective after all. He was heartened, and he began to feel ashamed that he had doubted the Führer.

He and Hermann kept on retrieving the wounded, hour after hour, until they forgot what it was like to be free from pain in their arms and legs. Some of their charges were unconscious; some thanked them, some cursed them; many just screamed; some lived and some died.

By eight o'clock that evening there was a German bridgehead on the far side of the river, and by ten it was secure.

The fighting came to an end at nightfall. Erik and Hermann continued to sweep the battlefield for wounded men. They brought back the last one at midnight. Then they lay down under a tree and fell into a sleep of utter exhaustion.

Next day Erik and Hermann and the rest of the 2nd Panzers turned west and broke through what remained of the French defences.

Two days later they were fifty miles away, at the river Oise, and moving fast through undefended territory.

By 20 May, a week after emerging unexpectedly from the Ardennes Forest, they had reached the coast of the English Channel.

Major Weiss explained their achievement to Erik and Hermann. 'Our attack on Belgium was a feint, you see. Its purpose was to draw the French and British into a trap. We Panzer divisions formed the jaws of the trap, and now we have them between our teeth. Much of the French army and nearly all of the British Expeditionary Force are in Belgium, encircled by the German army. They are cut off from supplies and reinforcements, helpless – and defeated.'

Erik said triumphantly: 'This was the Führer's plan all along!'

'Yes,' said Weiss, and, as ever, Erik could not tell whether he was sincere. 'No one thinks like the Führer!'

(ii)

Lloyd Williams was in a football stadium somewhere between Calais and Paris. With him were another thousand or more British prisoners of war. They had no shelter from the blazing June sun, but they were grateful for the warm nights as they had no blankets. There were no toilets and no water for washing.

Lloyd was digging a hole with his hands. He had organized some of the Welsh miners to make latrines at one end of the soccer pitch, and he was working alongside them to show willing. Other men joined in, having nothing else to do, and soon there were a hundred or so helping. When a guard strolled over to see what was going on, Lloyd explained.

'You speak good German,' said the guard amiably. 'What's your name?'

'Lloyd.'

'I'm Dieter.'

Lloyd decided to exploit this small expression of friendliness. 'We could dig faster if we had tools.'

'What's the hurry?'

'Better hygiene would benefit you as well as us.'

Dieter shrugged and went away.

Lloyd felt awkwardly unheroic. He had seen no fighting. The Welsh Rifles had gone to France as reserves, to relieve other units in what was expected to be a long battle. But it had taken the Germans only ten days to defeat the bulk of the Allied army. Many of the defeated British troops had then been evacuated from Calais and Dunkirk, but thousands had missed the boat, and Lloyd was among them.

Presumably the Germans were now pushing south. As far as he knew, the French were still fighting; but their best troops had been annihilated in Belgium, and there was a triumphant look about the German guards, as if they knew victory was assured.

Lloyd was a prisoner of war, but how long would he remain so? At this point there must be powerful pressure on the British government to make peace. Churchill would never do so, but he was a maverick, different from all other politicians, and he could be deposed. Men such as Lord Halifax would have little difficulty signing a peace treaty with the Nazis. The same was true, Lloyd thought bitterly, of the junior Foreign Office minister Earl Fitzherbert, whom he now shamefully knew to be his father.

If peace came soon, his time as a prisoner of war could be short. He might spend all of it here, in this French arena. He would go home scrawny and sunburnt, but otherwise whole.

But if the British fought on, it would be a different matter. The last war had continued more than four years. Lloyd could not bear the thought of wasting four years of his life in a prisoner-of-war camp. To avoid that, he decided, he would try to escape.

Dieter reappeared carrying half a dozen spades.

Lloyd gave them to the strongest men, and the work went faster.

At some point the prisoners would have to be moved to a permanent camp. That would be the time to make a run for it. Based on experience in Spain, Lloyd guessed that the army would not prioritize the guarding of prisoners. If one tried to get away he might succeed, or he might be shot dead; either way, it was one less mouth to feed.

They spent the rest of the day completing the latrines. Apart from the improvement in hygiene, this project had boosted morale, and Lloyd lay awake that night, looking at the stars, trying to think of other communal activities he might organize. He decided on a grand athletics contest, a prison-camp Olympic Games.

But he did not have the chance to put this into practice, for the next morning they were marched away.

At first he was not sure of the direction they were taking, but before long they got on to a Route Napoléon two-lane road and began to go steadily east. In all probability, Lloyd thought, they were intended to walk all the way to Germany.

Once there, he knew, escape would be much more difficult. He had to seize this opportunity. And the sooner the better. He was scared – those guards had guns – but determined.

There was not much motor traffic other than the occasional German staff car, but the road was busy with people on foot, heading in the opposite direction. With their possessions in handcarts and wheelbarrows, some driving their livestock ahead of them, they were clearly refugees whose homes had been destroyed in battle. That was a heartening sign, Lloyd told himself. An escaped prisoner might hide himself among them.

The prisoners were lightly guarded. There were only ten Germans in charge of this moving column of a thousand men. The guards had one car and a motorcycle; the rest were on foot and on civilian bicycles which they must have commandeered from the locals.

All the same, escape seemed hopeless at first. There were no English-style hedgerows to provide cover, and the ditches were too shallow to hide in. A man running away would provide an easy target for a competent rifleman.

Then they entered a village. Here it was a little harder for the guards to keep an eye on everyone. Local men and women stood at the edges of the column, staring at the prisoners. A small flock of sheep got mixed up with them. There were cottages and shops beside the road. Lloyd watched hopefully for his opportunity. He needed a place to hide instantly, an open door or a passage between houses or a bush to hide behind. And he needed to be passing it at a moment when none of the guards was in sight.

In a couple of minutes he had left the village behind without spotting his opportunity.

He felt annoyed, and told himself to be patient. There would be more chances. It was a long way to Germany. On the other hand, with every day that passed the Germans would tighten their grip on conquered territory, improve their organization, impose curfews and passes and checkpoints, stop the movement of refugees. Being on the run would be easier at first, harder as time went on.

It was hot, and he took off his uniform jacket and tie. He would get rid of them as soon as he could. Close up he probably still looked like a British soldier, in his khaki trousers and shirt, but at a distance he hoped he would not be so conspicuous.

They passed through two more villages then came to a small town. This should present some possible escape routes, Lloyd thought nervously. He realized that a part of him hoped he would not see a good opportunity, would not have to put himself in danger of those rifles. Was he getting accustomed to captivity already? It was too easy to continue marching, footsore but safe. He had to snap out of it.

The road through the town was unfortunately broad. The column kept to the middle of the street, leaving wide aisles either side that would have to be crossed before an escaper could find concealment. Some shops were closed and a few buildings were boarded up, but Lloyd could see promising-looking alleys, cafés with open doors, a church – but he could not get to any of them unobserved.

He studied the faces of the townspeople as they stared at the passing prisoners. Were they sympathetic? Would they remember that these men had fought for France? Or would they be understandably terrified of the Germans, and refuse to put themselves in danger? Half and half, probably. Some would risk their lives to help, others would hand him over to the Germans in a heartbeat. And he would not be able to tell the difference until it was too late.

They reached the town centre. I've lost half my opportunities already, he told himself. I have to act.

Up ahead he saw a crossroads. An oncoming line of traffic was waiting to turn left, its way blocked by the marching men. Lloyd saw a civilian pickup truck in the queue. Dusty and battered, it looked as if it might belong to a builder or a road mender. The back was open, but Lloyd could not see inside, for its sides were high.

He thought he might be able to pull himself up the side and scramble over the edge into the truck.

Once inside he could not be seen by anyone standing or walking on the street, nor by the guards on their bikes. But he would be plainly visible to people looking out of the upstairs windows of the buildings that lined the streets. Would they betray him?

He came closer to the truck.

He looked back. The nearest guard was two hundred yards behind.

He looked ahead. A guard on a bicycle was twenty yards in front.

He said to the man beside him: 'Hold this for me, would you?' and gave him his jacket.

He drew level with the front of the truck. At the wheel was a bored-looking man in overalls and a beret with a cigarette dangling from his lip. Lloyd passed him. Then he was level with the side of the truck. There was no time to check the guards again.

Without breaking step, Lloyd put both hands on the side of the truck, heaved himself up, threw one leg over then the other, and fell inside, hitting the bed of the truck with a crash that seemed terribly loud despite the tramp of a thousand pairs of feet. He flattened himself immediately. He lay still, listening for a clamour of shouted German, the roar of a motorcycle approaching, the crack of a rifle shot.

He heard the irregular snore of the truck's engine, the stamp and shuffle of the prisoners' feet, the background noises of a small town's traffic and people. Had he got away with it?

He looked around him, keeping his head low. In the truck with him were buckets, planks, a ladder and a wheelbarrow. He had been hoping for a few sacks with which to cover himself, but there were none.

He heard a motorcycle. It seemed to come to a halt nearby. Then, a few inches from his head, someone spoke French with a strong German accent. 'Where are you going?' A guard was talking to the truck driver, Lloyd figured with a racing heart. Would the guard try to look into the back?

He heard the driver reply, an indignant stream of fast French that Lloyd could not decipher. The German soldier almost certainly could not understand it either. He asked the question again.

Looking up, Lloyd saw two women at a high window overlooking the street. They were staring at him, mouths open in surprise. One was pointing, her arm sticking out through the open window.

Lloyd tried to catch her eye. Lying still, he moved one hand from side to side in a gesture that meant: 'No.'

She got the message. She withdrew her arm suddenly and covered her mouth with her hand as if realizing, with horror, that her pointing could be a sentence of death.

Lloyd wanted both women to move away from the window, but that was too much to hope for, and they continued to stare.

Then the motorcycle guard seemed to decide not to pursue his enquiry for, a moment later, the motorcycle roared away.

The sound of feet receded. The body of prisoners had passed. Was Lloyd free?

There was a crash of gears and the truck moved. Lloyd felt it turn the corner and pick up speed. He lay still, too scared to move.

He watched the tops of buildings pass by, alert in case anyone else should spot him, though he did not know what he would do if it happened. Every second was taking him away from the guards, he told himself encouragingly.

To his disappointment, the truck came to a halt quite soon. The engine was turned off, then the driver's door opened and slammed shut. Then nothing. Lloyd lay still for a while, but the driver did not return.

Lloyd looked at the sky. The sun was high: it must be after midday. The driver was probably having lunch.

The trouble was, Lloyd continued to be visible from high windows on both sides of the street. If he remained where he was he would be noticed sooner or later. And then there was no telling what might happen.

He saw a curtain twitch in an attic, and that decided him.

He stood up and looked over the side. A man in a business suit walking along the pavement stared in curiosity but did not stop.

Lloyd scrambled over the side of the truck and dropped to the ground. He found himself outside a bar-restaurant. No doubt that was where the driver had gone. To Lloyd's horror there were two men in German army uniforms sitting at a window table with glasses of beer in their hands. By a miracle they did not look at Lloyd.

He walked quickly away.

He looked around alertly as he walked. Everyone he passed stared at him: they knew exactly what he was. One woman screamed and ran away. He realized he needed to change his khaki shirt and trousers for something more French in the next few minutes.

A young man took him by the arm. 'Come with me,' he said in English with a heavy accent. 'I will 'elp you 'ide.'

He turned down a side street. Lloyd had no reason to trust this man, but he had to make a split-second decision, and he went along.

'This way,' the young man said, and steered Lloyd into a small house.

In a bare kitchen was a young woman with a baby. The young man introduced himself as Maurice, the woman as his wife, Marcelle, and the baby as Simone.

Lloyd allowed himself a moment of grateful relief. He had escaped from the Germans! He was still in danger, but he was off the streets and in a friendly house.

The stiffly correct French Lloyd had learned in school and at Cambridge had become more colloquial during his escape from Spain, and especially in the two weeks he spent picking grapes in Bordeaux. 'You're very kind,' he said. 'Thank you.'

Maurice replied in French, evidently relieved not to have to speak English. 'I guess you'd like something to eat.'

'Very much.'

Marcelle rapidly cut several slices off a long loaf and put them on the table with a round of cheese and a wine bottle with no label. Lloyd sat down and tucked in ravenously.

'I'll give you some old clothes,' said Maurice. 'But also, you must try to walk differently. You were striding along looking all around you, so alert and interested, you might as well have a sign around your neck saying "Visitor from England". Better to shuffle with your eyes on the ground.'

With his mouth full of bread and cheese Lloyd said: 'I'll remember that.'

There was a small shelf of books including French translations of Marx and Lenin. Maurice noticed Lloyd looking at them and said: 'I was a Communist – until the Hitler-Stalin pact. Now – it's finished.' He made a swift cutting-off gesture with his hand. 'All the same, we have to defeat Fascism.'

'I was in Spain,' said Lloyd. 'Before that, I believed in a united front of all left parties. Not any more.'

Simone cried. Marcelle lifted a large breast out of her loose dress and began to feed the baby. French women were more relaxed about this than the prudish British, Lloyd remembered.

When he had eaten, Maurice took him upstairs. From a wardrobe that had very little in it he took a pair of dark-blue overalls, a light-blue shirt, underwear and socks, all worn but clean. The kindness of this evidently poor man overwhelmed Lloyd, and he had no idea how to say thank you.

'Just leave your army clothes on the floor,' Maurice said. 'I'll burn them.'

Lloyd would have liked a wash, but there was no bathroom. He guessed it was in the back yard.

He put on the fresh clothes and studied his reflection in a mirror hanging on the wall. French blue suited him better than army khaki, but he still looked British.

He went back downstairs.

Marcelle was burping the baby. 'Hat,' she said.

Maurice produced a typical French beret, dark blue, and Lloyd put it on.

Then Maurice looked anxiously at Lloyd's stout black leather British army boots, dusty but unmistakably good quality. 'They give you away,' he said.

Lloyd did not want to give up his boots. He had a long way to walk. 'Perhaps we can make them look older?' he said.

Maurice looked doubtful. 'How?'

'Do you have a sharp knife?'

Maurice took a clasp knife from his pocket.

Lloyd took his boots off. He cut holes in the toecaps, then slashed the ankles. He removed the laces and re-threaded them untidily. Now they looked like something a down-and-out would wear, but they still fit well and had thick soles that would last many miles.

Maurice said: 'Where will you go?'

'I have two options,' Lloyd said. 'I can head north, to the coast, and hope to persuade a fisherman to take me across the English Channel. Or I can go south-west, across the border into Spain.' Spain was neutral, and still had British consuls in major cities. 'I know the Spanish route — I've travelled it twice.'

'The Channel is a lot nearer than Spain,' Maurice said. 'But I think the Germans will close all the ports and harbours.'

'Where's the front line?'

'The Germans have taken Paris.'

Lloyd suffered a moment of shock. Paris had fallen already!

'The French government has moved to Bordeaux.' Maurice shrugged. 'But we are beaten. Nothing can save France now.'

'All Europe will be Fascist,' Lloyd said.

'Except for Britain. So you must go home.'

Lloyd mused. North or south-west? He could not tell which would be better.

Maurice said: 'I have a friend, a former Communist, who sells cattle feed to farmers. I happen to know he's delivering this afternoon to a place south-west of here. If you decide to go to Spain, he could take you twenty miles.'

That helped Lloyd make up his mind. 'I'll go with him,' he said.

(iii)

Daisy had been on a long journey that had brought her around in a circle.

When Lloyd was sent to France she was heartbroken. She had missed her chance of telling him she loved him — she had not even kissed him!

And now there might never be another opportunity. He was reported missing in action after Dunkirk. That meant his body had not been found and identified, but neither was he registered as a prisoner of war.

Most likely he was dead, blown up into unidentifiable fragments by a shell, or perhaps lying unmarked beneath the debris of a destroyed farmhouse. She cried for days.

For another month she moped about Tŷ Gwyn, hoping to hear more, but no further news came. Then she began to feel guilty. There were many women as badly off as she or worse. Some had to face the prospect of raising two or three children with no man to support the family. She had no right to feel sorry for herself just because the man with whom she had been contemplating an adulterous affair was missing.

She had to pull herself together and do something positive. Fate did not intend her to be with Lloyd, that was clear. She already had a husband, one who was risking his life every day. It was her duty, she told herself, to take care of Boy.

She returned to London. She opened up the Mayfair house, as best she could with limited servants, and made it into a pleasant home for Boy to come to when on leave.

She needed to forget Lloyd and be a good wife. Perhaps she would even get pregnant again.

Many women signed up for war work, joining the Women's Auxiliary Air Force, or doing agricultural labour with the Women's Land Army. Others worked for no pay in the Women's Voluntary Service for Air Raid Precautions. But there was not enough for most such women to do, and *The Times* published letters to the editor complaining that air raid precautions were a waste of money.

The war in Continental Europe appeared to be over. Germany had won. Europe was Fascist from Poland to Sicily and from Hungary to Portugal. There was no fighting anywhere. Rumours said the British government had discussed peace terms.

But Churchill did not make peace with Hitler, and that summer the Battle of Britain began.

At first, civilians were not much affected. Church bells were silenced, their peal reserved to warn of the expected German invasion. Daisy followed government instructions and placed buckets of sand and water on every landing in the house, for firefighting, but they were not needed. The Luftwaffe bombed harbours, hoping to cut Britain's supply lines. Then they started on air bases, trying to destroy the Royal Air Force. Boy was flying a Spitfire, engaging enemy aircraft in sky battles that were watched by open-mouthed farmers in Kent and Sussex. In a rare letter home he said proudly that he had shot down three German planes. He had no leave for weeks on end, and Daisy sat alone in the house she filled with flowers for him.

At last, on the morning of Saturday 7 September, Boy showed up with a weekend pass. The weather was glorious, hot and sunny, a late spell of warmth that people called an Indian summer.

As it happened, that was the day the Luftwaffe changed their tactics.

Daisy kissed her husband and made sure there were clean shirts and fresh underwear in his dressing room.

From what other women said, she believed that fighting men on leave wanted sex, booze, and decent food, in that order.

Boy and she had not slept together since the miscarriage. This would be the first time. She felt guilty that she did not really relish the prospect. But she certainly would not refuse to do her duty.

She half expected him to tumble her into bed the minute he arrived, but he was not that desperate. He took off his uniform, bathed and washed his hair, and dressed again in a civilian suit. Daisy ordered the cook to spare no ration coupons in the preparation of a good lunch, and Boy brought up from the cellar one of his oldest bottles of claret.

She was surprised and hurt after lunch when he said: 'I'm going out for a few hours. I'll be back for dinner.'

She wanted to be a good wife, but not a passive one. 'This is your first leave for months!' she protested. 'Where the heck are you going?'

'To look at a horse.'

That was all right. 'Oh, fine – I'll come with you.'

'No, don't. If I show up with a woman in tow, they'll think I'm a softie and put the price up.'

She could not hide her disappointment. 'I always dreamed this would be something we did together – buying and breeding racehorses.'

'It's not really a woman's world.'

'Oh, stink on that!' she said indignantly. 'I know as much about horseflesh as you do.'

He looked irritated. 'Perhaps you do, but I still don't want you hanging around when I'm bargaining with these blighters – and that's final.'

She gave in. 'As you please,' she said, and she left the dining room.

Her instinct told her that he was lying. Fighting men on leave did not think about buying horses. She intended to find out what he was up to. Even heroes had to be true to their wives.

In her room she put on trousers and boots. As Boy went down the main staircase to the front door, she ran down the back stairs, through the kitchen, across the yard and into the old stables. There she put on a leather jacket, goggles and a crash helmet. She opened the garage door into the mews and wheeled out her motorcycle, a Triumph Tiger 100, so called because its top speed was one hundred miles per hour. She kicked it into life and drove out of the mews effortlessly.

She had taken quickly to motorcycling when petrol rationing was introduced back in September 1939. It was like bicycling, but easier. She loved the freedom and independence it gave her.

She turned into the street just in time to see Boy's cream-coloured Bentley Airline disappear around the next corner.

She followed.

He drove across Trafalgar Square and through the theatre district. Daisy stayed a discreet distance behind, not wanting to be conspicuous. There was still plenty of traffic in Central London, where there were hundreds of cars on official business. In addition, the petrol ration for private vehicles was not unreasonably small, especially for people who only wanted to drive around town.

Boy continued east, through the financial district. There was little traffic here on a Saturday afternoon, and Daisy became more concerned about being noticed. But she was not easily recognizable in her goggles and helmet, and Boy was paying little attention to his surroundings, driving with the window open, smoking a cigar.

He headed into Aldgate, and Daisy had a dreadful feeling she knew why.

He turned into one of the East End's less squalid streets and parked outside a pleasant eighteenth-century house. There were no stables in sight: this was not a place where racehorses were bought and sold. So much for his story.

Daisy stopped her motorcycle at the end of the street and watched. Boy got out of the car and slammed the door. He did not look around, or study the house numbers; clearly he had been here before and knew exactly where he was going. Walking with a jaunty air, cigar in his mouth, he went up to the front door and opened it with a key.

Daisy wanted to cry.

Boy disappeared into the house.

Somewhere to the east, there was an explosion.

Daisy looked in that direction and saw planes in the sky. Had the Germans chosen today to begin bombing London?

If so, she did not care. She was not going to let Boy enjoy his infidelity in peace. She drove up to the house and parked her bike behind his car. She took off her helmet and goggles, marched up to the front door of the house, and knocked.

She heard another explosion, this one closer; then the air raid sirens began their mournful song.

The door came open a crack, and she shoved it hard. A young woman in a maid's black dress cried out and staggered backwards, and Daisy walked in. She slammed the door behind her. She was in the hallway of a standard middle-class London house, but it was decorated in exotic fashion with Oriental rugs, heavy curtains, and a painting of naked women in a bathhouse.

She threw open the nearest door and stepped into the front parlour. It was dimly lit, velvet drapes keeping out the sunlight. There were three people in the room. Standing up, staring at her in shock, was a woman of about forty, dressed in a loose silk wrap, but carefully made up with bright red lipstick: the mother, she assumed. Behind her, sitting on a couch, was a girl of about sixteen wearing only underwear and stockings, smoking a cigarette. Next to the girl sat Boy, his hand on her thigh above the top of the stocking. He snatched his hand away guiltily. It was a ludicrous gesture, as if taking his hand off her could make this tableau look innocent.

Daisy fought back tears. 'You promised me you would give them up!' she said. She wanted to be coldly angry, like the avenging angel, but she could hear that her voice was just wounded and sad.

Boy reddened and looked panicked. 'What the devil are you doing here?'

The older woman said: 'Oh, fuck, it's his wife.'

Her name was Pearl, Daisy recalled, and the daughter was Joanie. How dreadful that she should know the names of such women.

The maid came to the door of the room and said: 'I didn't let the bitch in, she just shoved past me!'

Daisy said to Boy: 'I tried so hard to make our home beautiful and welcoming for you – and yet you prefer this!'

He started to say something, but had trouble finding his words. He sputtered incoherently for a moment or two. Then a big explosion nearby shook the floor and rattled the windows.

The maid said: 'Are you all deaf? There's a fucking air raid on!' No one looked at her. 'I'm going down the basement,' she said, and she disappeared.

They all needed to seek shelter. But Daisy had something to say to Boy before she left. 'Don't come to my bed again, ever, please. I refuse to be contaminated.'

The girl on the couch – Joanie – said: 'It's only a bit of fun, love. Why don't you join in? You might like it.'

Pearl, the older one, looked Daisy up and down. 'She's got a nice little figure.'

Daisy realized they would humiliate her further if she gave them the chance. Ignoring them, she spoke to Boy. 'You've made your choice,' she said. 'And I've made my decision.' She left the room, holding her head high even though she felt debased and spurned.

She heard Boy said: 'Oh, damn, what a mess.'

A mess? she thought. Is that all?

She went out of the front door.

Then she looked up.

The sky was full of planes.

The sight made her shake with fear. They were high, about ten thousand feet, but all the same they seemed to block the sun. There were hundreds of them, fat bombers and waspish fighters, a fleet that seemed twenty miles wide. To the east, in the direction of the docks and Woolwich Arsenal, palls of smoke rose from the ground where the bombs were landing. The explosions ran together into a continuous tidal roar like an angry sea.

Daisy recalled that Hitler had made a speech in the German parliament, just last Wednesday, ranting about the wickedness of RAF bombing raids on Berlin, and threatening to erase British cities in retaliation. Apparently he had meant it. They were intending to flatten London.

This was already the worst day of Daisy's life. Now she realized it might be the last.

But she could not bring herself to go back into that house and share their basement shelter. She had to get away. She needed to be at home where she could cry in private.

Hurriedly, she put on helmet and goggles. She resisted an irrational but nonetheless powerful impulse to throw herself behind the nearest wall. She jumped on her motorcycle and drove away.

She did not get far.

Two streets away, a bomb landed on a house directly in her line of vision, and she braked suddenly. She saw the hole in the roof, felt the thump of the explosion, and a few seconds later saw flames inside, as if kerosene from a heater had spilled and caught fire. A moment later, a girl of about twelve came out, screaming, with her hair on fire, and ran straight at Daisy.

Daisy jumped off the bike, pulled off her leather jacket, and used it to cover the girl's head, wrapping it tightly over the hair, denying oxygen to the flames.

The screaming stopped. Daisy removed the jacket. The girl was sobbing. She was no longer in agony, but she was bald.

Daisy looked up and down the street. A man wearing a steel helmet and an ARP armband came running up carrying a tin case with a white First Aid cross painted on its side.

The girl looked at Daisy, opened her mouth, and screamed: 'My mother's in there!'

The ARP warden said: 'Calm down, love, let's have a look at you.'

Daisy left the girl with him and ran to the front door of the building. It seemed to be an old house subdivided into cheap apartments. The upper floors were burning but she was able to enter the hall. Taking a guess, she ran to the back and found herself in a kitchen. There she saw a woman unconscious on the floor and a toddler in a cot. She picked up the child and ran out again.

The girl with the burned hair yelled: 'That's my sister!'

Daisy thrust the toddler into the girl's arms and ran back inside.

The unconscious woman was too heavy for her to lift. Daisy got behind her, raised her to a sitting position, took hold of her under the arms, and dragged her across the kitchen floor and through the hallway into the street.

An ambulance had arrived, a converted saloon car, its rear bodywork replaced by a canvas roof with a back opening. The ARP warden was helping the burned girl into the vehicle. The driver came running over to Daisy. Between them, they lifted the mother into the ambulance.

The driver said to Daisy: 'Is there anyone else inside?'

'I don't know!'

He ran into the hall. At that moment the entire building sagged. The burning upper storeys crashed through to the ground floor. The ambulance driver disappeared into an inferno.

Daisy heard herself scream.

She covered her mouth with her hand and stared into the flames, searching for him, even though she could not have helped him, and it would have been suicide to try.

The ARP warden said: 'Oh, my God, Alf's been killed.'

There was another explosion as a bomb landed a hundred yards along the street.

The warden said: 'Now I've got no driver, and I can't leave the scene.' He looked up and down the street. There were little knots of people standing outside some of the houses, but most were probably in shelters.

Daisy said: 'I'll drive it. Where should I go?'

'Can you drive?'

Most British women could not drive: it was still a man's job here. 'Don't ask stupid questions,' Daisy said. 'Where am I taking the ambulance?'

'St Bart's. Do you know where it is?'

'Of course.' St Bartholomew's was one of the biggest hospitals in London, and Daisy had been living here for four years. 'West Smithfield,' she added, to make sure he believed her.

'Emergency ward is around the back.'

'I'll find it.' She jumped in. The engine was still running.

The warden shouted: 'What's your name?'

'Daisy Fitzherbert. What's yours?'

'Nobby Clarke. Take care of my ambulance.'

The car had a standard gearshift with a clutch. Daisy put it into first and drove off.

The planes continued to roar overhead, and the bombs fell relentlessly. Daisy was desperate to get the injured people to hospital, and St Bart's was not much more than a mile away, but the journey was maddeningly difficult. She drove along Leadenhall Street, Poultry, and Cheapside, but several times she found the road blocked, and had to reverse away and find another route. There seemed to be at least one destroyed house in every street. Everywhere was smoke and rubble, people bleeding and crying.

With huge relief she reached the hospital and followed another ambulance to the emergency entrance. The place was frantically busy, with a dozen vehicles discharging maimed and burned patients into the care of hurrying porters with bloodstained aprons. Perhaps I've saved the mother of these children, Daisy thought. I'm not completely worthless, even if my husband doesn't want me.

The girl with no hair was still carrying her baby sister. Daisy helped them both out of the back of her ambulance.

A nurse helped Daisy lift the unconscious mother and carry her in.

But Daisy could see that the woman had stopped breathing.

She said to the nurse: 'These two are her children!' She heard the edge of hysteria in her own voice. 'What will happen now?'

'I'll deal with it,' the nurse said briskly. 'You have to go back.'

'Must I?' said Daisy.

'Pull yourself together,' said the nurse. 'There will be a lot more dead and injured before this night is over.'

'All right,' said Daisy; and she got back behind the wheel and drove off.

(iv)

On a warm Mediterranean afternoon in October, Lloyd Williams arrived in the sunlit French town of Perpignan, only twenty miles from the border with Spain.

He had spent the month of September in the Bordeaux area, picking grapes for the wine harvest, just as he had in the terrible year of 1937. Now he had money in his pockets for buses and trams, and could eat in cheap restaurants instead of living on unripe vegetables he dug up in people's gardens or raw eggs stolen from hen-coops. He was going back along the route he had taken when he left Spain three years ago. He had come south from Bordeaux through Toulouse and Béziers, occasionally riding freight trains, mostly begging lifts from truck drivers.

Now he was at a roadside café on the main highway running south-east from Perpignan towards the Spanish border. Still dressed in Maurice's blue overalls and beret, he carried a small canvas bag containing a rusty trowel and a mortar-spattered spirit level, evidence that he was a Spanish bricklayer making his way home. God forbid that anyone should offer him work: he had no idea how to build a wall.

He was worried about finding his way across the mountains. Three months ago, back in Picardy, he had told himself glibly that he could find the route over the Pyrenees along which his guides had led him into Spain in 1936, parts of which he had retraced in the opposite direction when he left a year later. But as the purple peaks and green passes came into distant view on the horizon, the prospect seemed more daunting. He had thought that every step of the journey must be engraved on his memory, but when he tried to recall specific paths and bridges and turning points he found that the pictures were blurred, and the exact details slipped infuriatingly from his mind's grasp.

He finished his lunch – a peppery fish stew – then spoke quietly to a group of drivers at the next table. 'I need a lift to Cerbère.' It was the last village before the Spanish border. 'Anyone going that way?'

They were probably all going that way: it was the only reason for being here on this south-east route. All the same, they hesitated. This was Vichy France, technically an independent zone, in practice under the thumb of the Germans occupying the other half of the country. No one was in a hurry to help a travelling stranger with a foreign accent.

'I'm a mason,' he said, hefting his canvas bag. 'Going home to Spain. Leandro is my name.'

A fat man in an undershirt said: 'I can take you halfway.'

'Thank you.'

'Are you ready now?'

'Of course.'

They went outside and got into a grimy Renault van with the name of an electrical goods store on the side. As they pulled away, the driver asked Lloyd if he was married. A series of unpleasantly personal inquiries followed, and Lloyd realized the man had a fascination with other people's sex lives. No doubt that was why he had agreed to take Lloyd: it gave him the chance to ask intrusive questions. Several of the men who had given Lloyd lifts had had some such creepy motive.

'I'm a virgin,' Lloyd told him, which was true; but that only led to an interrogation about heavy petting with schoolgirls. Lloyd did have considerable experience of that, but he was not going to share it. He refused to give details while trying not to be rude, and eventually the driver despaired. 'I have to turn off here,' he said, and pulled up.

Lloyd thanked him for the ride and walked on.

He had learned not to march like a soldier, and had developed what he thought was a fairly realistic peasant slouch. He never carried a newspaper or a book. His hair had last been cut by a brutally incompetent barber in the poorest quarter of Toulouse. He shaved about once a week, so that he normally had a growth of stubble, which was surprisingly effective in making him look like a nobody. He had stopped washing, and acquired a ripe odour that discouraged people from talking to him.

Few working-class people had watches, in France or Spain, so the steel wristwatch with the square face that Bernie had given him as a graduation present had to go. He could not give it to one of the many French people who had helped him, for a British watch could have incriminated them, too. In the end, with great sadness, he had thrown it into a pond.

His greatest weakness was that he had no identity papers.

He had tried to buy papers from a man who looked vaguely like him, and schemed to steal them from two others, but not surprisingly, people were cautious about such things these days. His strategy was therefore to steer clear of situations in which he might be asked to identify himself. He made himself inconspicuous, he walked across fields rather than take roads when he had the choice, and he never travelled by passenger train because there were often checkpoints at stations. So far he had been lucky. One village gendarme had demanded his papers, and when he explained that they had been stolen from him after he got drunk and passed out in a bar in Marseilles, the policeman had believed him and sent him on his way.

Now, however, his luck ran out.

He was passing through poor agricultural terrain. He was in the foothills of the Pyrenees, close to the Mediterranean, and the soil was sandy. The dusty road ran through struggling smallholdings and poor villages. The landscape was sparsely populated. To his left, through the hills, he got blue glimpses of the distant sea.

The last thing he expected was the green Citroën that pulled up alongside him with three gendarmes inside.

It happened very suddenly. He heard the car approaching – the only car he had heard since the fat man had dropped him off. He carried on shuffling like a tired worker going home. Either side of the road were dry fields with sparse vegetation and stunted trees. When the car stopped, he thought for a second of making a run for it across the fields. He dropped the idea when he saw the holstered pistols of the two gendarmes who jumped out of the car. They were probably not very good shots, but they might get lucky. His chances of talking his way out of this were better. These were country constables, more amiable than the hard-nosed French city police.

'Papers?' said the nearest gendarme in French.

Lloyd spread his hands in a helpless gesture. 'Monsieur, I am so unfortunate, my papers were stolen in Marseilles. I am Leandro, Spanish mason, going—'

'Get in the car.'

Lloyd hesitated, but it was hopeless. The odds against his getting away were now worse than before.

A gendarme took him firmly by the arm, hustled him into the back seat, and got in beside him.

His spirits sank as the car pulled away.

The gendarme next to him said: 'Are you English, or what?'

'I am Spanish mason. My name—'

The gendarme made a waving-away gesture and said: 'Don't bother.'

Lloyd saw that he had been wildly optimistic. He was a foreigner without papers heading for the Spanish border: they simply assumed he was an escaping British soldier. If they had any doubt, they would find proof when they ordered him to strip, for they would see the identity tag around his neck. He had not thrown it away, for without it he would automatically be shot as a spy.

And now he was stuck in a car with three armed men, and the likelihood that he would find a way to escape was zero.

They drove on, in the direction in which he had been heading, as the sun went down over the mountains on their right-hand side. There were no big towns between here and the border, so he assumed they intended to put him in a village jail for the night. Perhaps he could escape from there. Failing that, they would undoubtedly take him back to Perpignan tomorrow and hand him over to the city police. What then? Would he be interrogated? The prospect made him cold with fear. The French police would beat him up, the Germans would torture him. If he survived, he would end up in a prisoner of war camp, where he would remain until the end of the war, or until he died of malnutrition. And yet he was only a few miles from the border!

They drove into a small town. Could he escape between the car and the jail? He could make no plan: he did not know the terrain. There was nothing he could do but remain alert and seize any opportunity.

The car turned off the main street and into an alley behind a row of shops. Were they going to shoot him here and dump his body?

The car stopped at the back of a restaurant. The yard was littered with boxes and giant cans. Through a small window Lloyd could see a brightly lit kitchen.

The gendarme in the front passenger seat got out, then opened Lloyd's door, on the side of the car nearest the building. Was this his chance? He would have to run around the car and along the alley. It was dusk: after the first few yards he would not be an easy target.

The gendarme reached into the car and grasped Lloyd's arm, holding him as he got out and stood up. The second one got out immediately behind Lloyd. The opportunity was not good enough.

But why had they brought him here?

They walked him into the kitchen. A chef was beating eggs in a bowl and an adolescent boy was washing up in a big sink. One of the gendarmes said: 'Here's an Englishman. He calls himself Leandro.'

Without pausing in his work, the chef lifted his head and bawled: 'Teresa! Come here!'

Lloyd remembered another Teresa, a beautiful Spanish anarchist who had taught soldiers to read and write.

The kitchen door swung wide and she walked in.

Lloyd stared at her in astonishment. There was no possibility of mistake: he would never forget those big eyes and that mass of black hair, even though she wore the white cotton cap and apron of a waitress.

At first she did not look at him. She put a pile of plates on the counter next to the young washer-up, then turned to the gendarmes with a smile and kissed each on both cheeks, saying: 'Pierre! Michel! How are you?' Then she turned to Lloyd, stared at him, and said in Spanish: 'No – it's not possible. Lloyd, is it really you?'

He could only nod dumbly.

She put her arms around him, embraced him, and kissed him on both cheeks.

One of the gendarmes said: 'There we are. All is well. We have to go. Good luck!' He handed Lloyd his canvas bag, then they left.

Lloyd found his tongue. 'What's going on?' he said to Teresa in Spanish. 'I thought I was being taken to jail!'

'They hate the Nazis, so they help us,' she said.

'Who is *us*?'

'I'll explain later. Come with me.' She opened a door that gave on to a staircase and led him to an upper storey, where there was a sparsely furnished bedroom. 'Wait here. I'll bring you something to eat.'

Lloyd lay down on the bed and contemplated his extraordinary fortune. Five minutes ago he had been expecting torture and death. Now he was waiting for a beautiful woman to bring him supper.

It could change again just as quickly, he reflected.

She returned half an hour later with an omelette and fried potatoes on a thick plate. 'We've been busy, but we close soon,' she said. 'I'll be back in a few minutes.'

He ate the food quickly.

Night fell. He listened to the chatter of customers leaving and the clang of pots being put away, then Teresa reappeared with a bottle of red wine and two glasses.

Lloyd asked her why she had left Spain.

'Our people are being murdered by the thousand,' she said. 'For those they don't kill, they have passed the Law of Political Responsibilities, making criminals of everyone who supported the government. You can lose all your assets if you opposed Franco even by "grave passivity". You are innocent only if you can prove you supported him.'

Lloyd thought bitterly of Chamberlain's reassurance to the House of Commons, back in March, that Franco had renounced political reprisals. What an evil liar Chamberlain had been.

Teresa went on: 'Many of our comrades are in filthy prison camps.'

'I don't suppose you have any idea what happened to Sergeant Lenny Griffiths, my friend?'

Teresa shook her head. 'I never saw him again after Belchite.'

'And you . . . ?'

'I escaped from Franco's men, came here, got a job as a waitress . . . and found there was other work for me to do.'

'What work?'

'I take escaping soldiers across the mountains. That's why the gendarmes brought you to me.'

Lloyd was heartened. He had been planning to do it alone, and he had been worried about finding the way. Now perhaps he would have a guide.

'I have two others waiting,' she said. 'A British gunner and a Canadian pilot. They are in a farmhouse in the hills.'

'When are you planning to go across?'

'Tonight,' she said. 'Don't drink too much wine.'

She went away again and returned half an hour later carrying an old, ripped brown overcoat for him. 'It's cold where we're going,' she explained.

They slipped out of the kitchen door and threaded their way through the small town by starlight. Leaving the houses behind, they followed a dirt track steadily uphill. After an hour they came to a small group of stone buildings. Teresa whistled then opened the door to a barn, and two men came out.

'We always use false names,' she said in English. 'I am Maria and these two are Fred and Tom. Our new friend is Leandro.' The men shook hands. She went on: 'No talking, no smoking, and anyone who falls behind will be left. Are we ready?'

From here the path was steeper. Lloyd found himself slipping on stones. Now and again he clutched at stunted bushes of heather beside the path and pulled himself upwards with their aid. The petite Teresa set a pace that soon had the three men puffing and blowing. She was carrying a flashlight, but she refused to use it while the stars were bright, saying she had to conserve the battery.

The air got colder. They waded across an icy stream, and Lloyd's feet did not get warm again afterwards.

An hour later, Teresa said: 'Take care to stay in the middle of the path here.' Lloyd looked down and realized he was on a ridge between steep slopes. When he saw how far he could fall, he felt a little giddy, and quickly looked up and ahead at Teresa's swiftly moving silhouette. In normal circumstances he would have enjoyed every minute of walking behind a figure like that, but now he was so tired and cold he did not have the energy even to ogle.

The mountains were not uninhabited. At one point a distant dog barked; at another they heard a tinkling of eerie bells, which spooked the men until Teresa explained that mountain shepherds hung bells on their sheep so that they could find their flocks.

Lloyd thought about Daisy. Was she still at Tŷ Gwyn? Or had she gone back to her husband? Lloyd hoped she had not returned to London, for London was being bombed every night, the French newspapers said. Was she alive or dead? Would he ever see her again? If he did, how would she feel about him?

They stopped every two hours to rest, drink water, and take a few mouthfuls from a bottle of wine Teresa was carrying.

It started to rain around dawn. The ground underfoot instantly became treacherous, and they all stumbled and slipped, but Teresa did not slow down. 'Be glad it's not snow,' she said.

Daylight revealed a landscape of scrubby vegetation in which rocky outcrops stuck up like tombstones. The rain continued, and a cold mist obscured the distance.

After a while, Lloyd realized they were walking downhill. At the next rest stop, Teresa announced: 'We are now in Spain.' Lloyd should have been relieved, but he just felt exhausted.

Gradually the landscape softened, rocks giving way to coarse grass and shrubs.

Suddenly Teresa dropped to the ground and lay flat.

The three men instantly did the same, not needing to be prompted. Following Teresa's gaze, Lloyd saw two men in green uniforms and peculiar hats: Spanish border guards, presumably. He realized that being in Spain did not mean he was out of trouble. If he was caught entering the country illegally he might just be sent back. Worse, he could disappear into one of Franco's prison camps.

The border guards were walking along a mountain track towards the fugitives. Lloyd prepared himself for a fight. He would have to move fast, in order to overcome them before they could draw their guns. He wondered how good the other two men would be in a fracas.

But his trepidation was unnecessary. The two guards reached some unmarked boundary and then turned back. Teresa acted as if she had known this would happen. When the guards disappeared from sight, she stood up and the four of them walked on.

Soon afterwards the mist lifted. Lloyd saw a fishing village around a sandy bay. He had been here before, when he came to Spain in 1936. He even remembered that there was a railway station.

They walked into the village. It was a sleepy place, with no signs of officialdom: no police, no town hall, no soldiers, no checkpoints. Doubtless that was why Teresa had chosen it.

They went to the station and Teresa bought tickets, flirting with the vendor as if they were old friends.

Lloyd sat on a bench on the shady platform, footsore, weary, grateful and happy.

An hour later they caught a train to Barcelona.

(v)

Daisy had never before understood the meaning of work.

Or tiredness.

Or tragedy.

She sat in a school classroom, drinking sweet English tea out of a cup with no saucer. She wore a steel helmet and rubber boots. It was five o'clock in the afternoon, and she was still weary from the night before.

She was part of the Aldgate district Air Raid Precautions sector. Theoretically, she worked an eight-hour shift followed by eight hours on standby and eight hours off duty. In practice, she worked as long as the air raid continued and there were wounded people to be driven to hospital.

London was bombed every single night of October 1940.

Daisy always worked with one other woman, the driver's attendant, and four men forming a first-aid party. Their headquarters was in a school, and now they were sitting at the children's desks, waiting for the planes to come and the sirens to wail and the bombs to fall.

The ambulance she drove was a converted American Buick. They also had a normal car and driver to transport what they called sitting cases – injured people who could nevertheless sit upright without assistance while being transported to hospital.

Her attendant was Naomi Avery, an attractive blonde Cockney who liked men and enjoyed the camaraderie of the team. Now she bantered with the post warden, Nobby Clarke, a retired policeman. 'The Chief Warden is a man,' she said. 'The District Warden is a man. You're a man.'

'I hope so,' Nobby said, and the others chuckled.

'There are plenty of women in ARP,' Naomi went on. 'How come none of them are officials?'

The men laughed. A bald man with a big nose called Gorgeous George said: 'Here we go, women's rights again.' He had a misogynist streak.

Daisy joined in. 'You don't really think all you men are smarter than all of us women, do you?'

Nobby said: 'Matter of fact, there are some women senior wardens.'

'I've never met one,' said Naomi.

'It's tradition, isn't it,' Nobby said. 'Women have always been home-makers.'

'Like Catherine the Great of Russia,' Daisy said sarcastically.

Naomi put in: 'Or Queen Elizabeth of England.'

'Amelia Earhart.'

'Jane Austen.'

'Marie Curie, the only scientist ever to win the Nobel Prize twice.'

'Catherine the Great?' said Gorgeous George. 'Isn't there a story about her and her horse?'

'Now, now, ladies present,' said Nobby in a tone of reproof. 'Anyway, I can answer Daisy's question,' he went on.

Daisy, willing to be his foil, said: 'Go on, then.'

'I grant you that some women may be just as clever as a man,' he said with the air of one who makes a remarkably generous concession. 'But there is one very good reason why almost all ARP officials are men, nevertheless.'

'And what would that reason be, Nobby?'

'It's very simple. Men won't take orders from a woman.' He sat back with a triumphant expression, confident that he had won the argument.

The irony was that when the bombs were falling, and they were digging through the rubble to rescue the injured, they *were* equals. There was no hierarchy then. If Daisy shouted at Nobby to pick up the other end of a roof beam he would do it without demur.

Daisy loved these men, even George. They would give their lives for her, and she for them.

She heard a low hooting sound outside. Slowly it rose in pitch until it became the tiresomely familiar siren of an air raid warning. Seconds later there was the boom of a distant explosion. The warning was often late; sometimes it sounded after the first bombs had fallen.

The phone rang and Nobby picked it up.

They all stood up. George said wearily: 'Don't the Germans ever take a ruddy day off?'

Nobby put the phone down and said: 'Nutley Street.'

'I know where that is,' said Naomi as they all hurried out. 'Our MP lives there.'

They jumped into the cars. As Daisy put the ambulance in gear and drove off, Naomi, sitting beside her, said: 'Happy days.'

Naomi was being ironic but, strangely, Daisy *was* happy. It was very odd, she thought as she careered around a bend. Every night she saw destruction, tragic bereavement, and horribly maimed bodies. There was a good chance she herself would die in a blazing building tonight. Yet she felt wonderful. She was working and suffering for a cause, and, paradoxically, that was better than pleasing herself. She was part of a group that would risk everything to help others, and it was the best feeling in the world.

Daisy did not hate the Germans for trying to kill her. She had been told by her father-in-law, Earl Fitzherbert, why they were bombing London. Until August the Luftwaffe had raided only ports and airfields. Fitz had explained, in an unusually candid moment, that the British were not so scrupulous: the government had approved bombing of targets in German cities back in May, and all through June and July the RAF had dropped bombs on women and children in their homes. The German public had been enraged by this and demanded retaliation. The Blitz was the result.

Daisy and Boy were keeping up appearances, but she locked her bedroom door when he was at home, and he made no objection. Their marriage was a sham, but they were both too busy to do anything about it. When Daisy thought about it, she felt sad; for she had lost both Boy and Lloyd now. Fortunately, she hardly had time to think.

Nutley Street was on fire. The Luftwaffe dropped incendiary bombs and high explosive together. Fire did the most damage, but the high explosive helped the blaze to spread by blowing out windows and ventilating the flames.

Daisy brought the ambulance to a screeching halt and they all went to work.

People with minor injuries were helped to the nearest First Aid station. Those more seriously hurt were driven to St Bart's or the London Hospital in Whitechapel. Daisy made one trip after another. When darkness fell she switched on her headlights. They were masked, with only a slit of light, as part of the blackout, though it seemed a superfluous precaution when London was burning like a bonfire.

The bombing went on until dawn. In full daylight the bombers were too vulnerable to being shot down by the fighter aircraft piloted by Boy and his comrades, so the air raid petered out. As the cold grey light washed over the wreckage, Daisy and Naomi returned to Nutley Street to find that there were no more victims to be taken to hospital.

They sat down wearily on the remains of a brick garden wall. Daisy took off her steel helmet. She was filthy dirty and worn out. I wonder what the girls in the Buffalo Yacht Club would think of me now, she thought; then she realized she no longer cared much what they thought. The days when their approval was all-important to her seemed a long time in the past.

Someone said: 'Would you like a cup of tea, my lovely?'

She recognized the accent as Welsh. She looked up to see an attractive middle-aged woman carrying a tray. 'Oh, boy, that's what I need,' she said, and helped herself. She had now grown to like this beverage. It tasted bitter but it had a remarkable restorative effect.

The woman kissed Naomi, who explained: 'We're related. Her daughter, Millie, is married to my brother, Abie.'

Daisy watched the woman take the tray around the little crowd of ARP wardens and firemen and neighbours. She must be a local dignitary, Daisy decided: she had an air of authority. Yet at the same time she was clearly a woman of the people, speaking to everyone with an easy warmth, making them smile. She knew Nobby and Gorgeous George, and greeted them as old friends.

She took the last cup on the tray for herself and came to sit beside Daisy. 'You sound American,' she said pleasantly.

Daisy nodded. 'I'm married to an Englishman.'

'I live in this street – but my house escaped the bombs last night. I'm the Member of Parliament for Aldgate. My name is Eth Leckwith.'

Daisy's heart skipped a beat. This was Lloyd's famous mother! She shook hands. 'Daisy Fitzherbert.'

Ethel's eyebrows went up. 'Oh!' she said. 'You're the Viscountess Aberowen.'

Daisy blushed and lowered her voice. 'They don't know that in the ARP.'

'Your secret is safe with me.'

Hesitantly, Daisy said: 'I knew your son, Lloyd.' She could not help the tears that came to her eyes when she thought of their time at Tŷ Gwyn, and the way he had looked after her when she had miscarried. 'He was very kind to me, once, when I needed help.'

'Thank you,' said Ethel. 'But don't talk as if he's dead.'

The reproof was mild, but Daisy felt she had been dreadfully tactless. 'I'm so sorry!' she said. 'He's missing in action, I know. How frightfully stupid of me.'

'But he's not missing any longer,' Ethel said. 'He escaped through Spain. He arrived home yesterday.'

'Oh, my God!' Daisy's heart was racing. 'Is he all right?'

'Perfectly. In fact, he looks very well, despite what he's been through.'

'Where . . .' Daisy swallowed. 'Where is he now?'

'Why, he's here somewhere.' Ethel looked around. 'Lloyd?' she called.

Daisy scanned the crowd wildly. Could it be true?

A man in a ripped brown overcoat turned around and said: 'Yes, Mam?'

Daisy stared at him. His face was sunburned, and he was as thin as a stick, but he looked more attractive than ever.

'Come here, my lovely,' said Ethel.

Lloyd took a step forward, then saw Daisy. Suddenly his face was transformed. He smiled happily. 'Hello,' he said.

Daisy sprang to her feet.

Ethel said: 'Lloyd, there's someone here you may remember—'

Daisy could not restrain herself. She ran to Lloyd and threw herself into his arms. She hugged him. She looked into his green eyes then kissed his brown cheeks and his broken nose and then his mouth. 'I love you, Lloyd,' she said madly. 'I love you, I love you, I love you.'

'I love you, too, Daisy,' he said.

Behind her, Daisy heard Ethel's wry voice. 'You do remember, I see.'

(vi)

Lloyd was eating toast and jam when Daisy entered the kitchen of the house in Nutley Street. She sat at the table, looking exhausted, and took off her steel helmet. Her face was smudged and her hair was dirty with ash and dust, and Lloyd thought she looked irresistibly beautiful.

She came in most mornings when the bombing ended and the last victim had been driven to the hospital. Lloyd's mother had told her she did not need an invitation, and Daisy had taken her at her word.

Ethel poured Daisy a cup of tea and said: 'Hard night, my lovely?'

Daisy nodded grimly. 'One of the worst. The Peabody building in Orange Street burned down.'

'Oh, no!' Lloyd was horrified. He knew the place: a big overcrowded tenement full of poor families with numerous children.

Bernie said: 'That's a big building.'

'It was,' said Daisy. 'Hundreds of people were burned and God knows how many children are orphans. Nearly all my patients died on the way to hospital.'

Lloyd reached across the little table and took her hand.

She looked up from her cup of tea. 'You don't get used to it. You think you'll become hardened, but you don't.' She was stricken with sadness.

Ethel put a hand on her shoulder for a moment in a gesture of compassion.

Daisy said: 'And we're doing the same to families in Germany.'

Ethel said: 'Including my old friends Maud and Walter and their children, I presume.'

'Isn't that terrible?' Daisy shook her head despairingly. 'What's wrong with us?'

Lloyd said: 'What's wrong with the human race?'

Bernie, ever practical, said: 'I'll go over to Orange Street later and make sure everything's being done for the children.'

'I'll come with you,' said Ethel.

Bernie and Ethel thought alike and acted together effortlessly, often seeming to read each other's mind. Lloyd had been observing them carefully since he got home, worrying that their marriage might have been affected by the shocking revelation that Ethel had never had a husband called Teddy Williams, and that Lloyd's father was Earl Fitzherbert. He had discussed this at length with Daisy, who now knew the whole truth. How did Bernie feel about having been lied to for twenty years? But Lloyd saw no sign that it had made any difference. In his unsentimental way Bernie adored Ethel, and to him she could do no wrong. He believed she would never do anything to hurt him, and he was right. It made Lloyd hope that he, too, might one day have such a marriage.

Daisy noticed that Lloyd was in uniform. 'Where are you off to this morning?'

'I've had a summons from the War Office.' He looked at the clock on the mantelpiece. 'I'd better get going.'

'I thought you'd already been debriefed.'

'Come to my room and I'll explain while I'm putting on my tie. Bring your tea.'

They went upstairs. Daisy looked around with interest, and he realized she had not been in his bedroom before. He looked at the single bed, the bookshelf of novels in German, French and Spanish, and the writing table with the row of sharpened pencils, and wondered what she thought of it.

'What a nice little room,' she said.

It was not little. It was the same size as the other bedrooms in the house. But she had different standards.

She picked up a framed photograph. It showed the family at the seaside: little Lloyd in shorts, toddling Millie in a swimsuit, young Ethel in a big floppy hat, Bernie wearing a grey suit with a white shirt open at the neck and a knotted handkerchief on his head.

'Southend,' Lloyd explained. He took her cup, put it on the dressing table, and folded her into his arms. He kissed her mouth. She kissed him back with weary tenderness, stroking his cheek, letting her body slump against his.

After a minute he released her. She was really too tired to canoodle, and he had an appointment.

She took off her boots and lay down on his bed.

'The War Office have asked me to go in and see them again,' he said as he tied his tie.

'But you were there for hours last time.'

It was true. He had had to dredge his memory for every last detail of his time on the run in France. They wanted to know the rank and regiment of every German he had encountered. He could not remember them all, of course, but he had done his homework meticulously on the Tŷ Gwyn course and he was able to give them a great deal of information.

That was standard military intelligence debriefing. But they had also asked about his escape, the roads he had taken and who had helped him. They were even interested in Maurice and Marcelle, and reproved him for not knowing their surname. They had got very excited about Teresa, who clearly could be a major asset to future escapers.

'I'm seeing a different lot today.' He glanced at a typed note on his dressing table. 'At the Metropole Hotel in Northumberland Avenue. Room four two four.' The address was off Trafalgar Square in a neighbourhood of government offices. 'Apparently it's a new department dealing with British prisoners of war.' He put on his peaked cap and looked in the mirror. 'Am I smart enough?'

There was no answer. He looked at the bed. She had fallen asleep.

He pulled a blanket over her, kissed her forehead, and went out.

He told his mother that Daisy was asleep on his bed, and she said she would check on her later to make sure she was all right.

He took the Tube to Central London.

He had told Daisy the true story of his parentage, disabusing her of the theory that he was Maud's child. She believed him readily, for she suddenly recalled Boy telling her that Fitz had an illegitimate child somewhere. 'This is creepy,' she had said, looking thoughtful. 'The two Englishmen I've fallen for turn out to be half-brothers.' She had looked appraisingly at Lloyd. 'You inherited your father's good looks. Boy just got his selfishness.'

Lloyd and Daisy had not yet made love. One reason was that she never had a night off. Then, on the single occasion they had had a chance to be alone together, things had gone wrong.

It had been last Sunday, at Daisy's home in Mayfair. Her servants had Sunday afternoon off, and she had taken him to her bedroom in the empty house. But she had been nervy and ill at ease. She had kissed him, then turned her head aside. When he put his hands on her breasts she had pushed them away. He had been confused: if he was not supposed to behave this way, why were they in her bedroom?

'I'm sorry,' she had said at last. 'I love you, but I can't do this. I can't betray my husband in his own house.'

'But he betrayed you.'

'At least he went somewhere else.'

'All right.'

She had looked at him. 'Do you think I'm being silly?'

He shrugged. 'After all we've been through together, this seems overly fastidious of you, yes – but, look, you feel the way you feel. What a rotter I would be if I tried to bully you into doing it when you're not ready.'

She put her arms around him and hugged him hard. 'I said it before,' she said. 'You're a grown-up.'

'Don't let's spoil the whole afternoon,' he said. 'We'll go to the pictures.'

They saw Charlie Chaplin in *The Great Dictator* and laughed their heads off, then she went back on duty.

Pleasant thoughts of Daisy occupied Lloyd all the way to Embankment station, then he walked up Northumberland Avenue to the Metropole. The hotel had been stripped of its reproduction antiques and furnished with utilitarian tables and chairs.

After a few minutes' wait, Lloyd was taken to see a tall colonel with a brisk manner. 'I've read your account, Lieutenant,' he said. 'Well done.'

'Thank you, sir.'

'We expect more people to follow in your footsteps, and we'd like to help them. We're especially interested in downed airmen. They're expensive to train, and we want them back so that they can fly again.'

Lloyd thought that was harsh. If a man survived a crash landing, should he really be asked to risk going through the whole thing again? But wounded men were sent back into battle as soon as they recovered. That was war.

The colonel said: 'We're setting up a kind of underground railroad, all the way from Germany to Spain. You speak German, French and Spanish, I see; but, more importantly, you've been at the sharp end. We'd like to second you to our department.'

Lloyd had not been expecting this, and he was not sure how he felt about it. 'Thank you, sir. I'm honoured. But is it a desk job?'

'Not at all. We want you to go back to France.'

Lloyd's heart raced. He had not thought he would have to face those perils again.

The colonel saw the dismay on his face. 'You know how dangerous it is.'

'Yes, sir.'

In an abrupt tone the colonel said: 'You can refuse if you like.'

Lloyd thought of Daisy in the Blitz, and of the people burned to death in the Peabody tenement, and realized he did not even want to refuse. 'If you think it's important, sir, then I will go back most willingly, of course.'

'Good man,' said the colonel.

Half an hour later Lloyd was dazedly walking back to the Tube station. He was now part of a department called MI9. He would return to France with false papers and large sums in cash. Already dozens of German, Dutch, Belgian and French people in occupied territory had been recruited to the deadly dangerous task of helping British and Commonwealth airmen return home. He would be one of numerous MI9 agents expanding the network.

If he were caught, he would be tortured.

Although he was scared, he was also excited. He was going to fly to Madrid: it would be his first time up in an airplane. He would re-enter France across the Pyrenees and make contact with Teresa. He would be moving in disguise among the enemy, rescuing people under the noses of the Gestapo. He would make sure that men following in his footsteps would not be as alone and friendless as he had been.

He got back to Nutley Street at eleven o'clock. There was a note from his mother: 'Not a peep from Miss America.' After visiting the bomb site, Ethel would have gone to the House of Commons, Bernie to County Hall. Lloyd and Daisy had the house to themselves.

He went up to his room. Daisy was still asleep. Her leather jacket and heavy-duty wool trousers were carelessly tossed on the floor. She was in his bed wearing only her underwear. This had never happened before.

He took off his jacket and tie.

A sleepy voice from the bed said: 'And the rest.'

He looked at her. 'What?'

'Take off your clothes and get into bed.'

The house was empty: no one would disturb them.

He took off his boots, trousers, shirt and socks, then he hesitated.

'You're not going to feel cold,' she said. She wriggled under the blankets, then threw a pair of silk camiknickers at him.

He had expected this to be a solemn moment of high passion, but Daisy seemed to think it should be a matter of laughter and fun. He was willing to be guided by her.

He took off his vest and pants and slipped into bed beside her. She was warm and languid. He felt nervous: he had never actually told her that he was a virgin.

He had always heard that the man should take the initiative, but it seemed that Daisy did not know that. She kissed and caressed him, then she grasped his penis. 'Oh, boy,' she said. 'I was hoping you'd have one of these.'

After that he stopped being nervous.

8

1941 (I)

On a cold winter Sunday, Carla von Ulrich went with the maid, Ada, to visit Ada's son, Kurt, at the Wannsee Children's Nursing Home, by the lake on the western outskirts of Berlin. It took an hour to get there on the train. Carla made a habit of wearing her nurse's uniform on these visits, because the staff at the home talked more frankly about Kurt to a fellow professional.

In summer the lakeside would be crowded with families and children playing on the beach and paddling in the shallows, but today there were just a few walkers, well wrapped up against the chill, and one hardy swimmer with an anxious wife waiting at the waterside.

The home, which specialized in caring for severely handicapped children, was a once-grand house whose elegant reception rooms had been subdivided and painted pale green and furnished with hospital beds and cots.

Kurt was now eight years old. He could walk and feed himself about as well as a two-year-old, but he could not talk and still wore diapers. He had shown no sign of improvement for years. However, there was no doubt of his joy at seeing Ada. He beamed with happiness, burbled excitedly, and held out his arms to be picked up and hugged and kissed.

He recognized Carla, too. Whenever she saw him she remembered the frightening drama of his birth, when she had delivered him while her brother Erik ran to fetch Dr Rothmann.

They played with him for an hour or so. He liked toy trains and cars, and books with highly coloured pictures. Then the time for his afternoon nap drew near, and Ada sang to him until he went to sleep.

On their way out a nurse spoke to Ada. 'Frau Hempel, please come with me to the office of Herr Professor Doctor Willrich. He would like to speak to you.'

Willrich was Director of the home. Carla had never met him and she was not sure Ada had either.

Ada said nervously: 'Is there some problem?'

The nurse said: 'I'm sure the Director just wants to talk to you about Kurt's progress.'

Ada said: 'Fräulein von Ulrich will come with me.'

The nurse did not like that idea. 'Professor Willrich asked only for you.'

But Ada could be stubborn when necessary. 'Fräulein von Ulrich will come with me,' she repeated firmly.

The nurse shrugged and said curtly: 'Follow me.'

They were shown into a pleasant office. This room had not been subdivided. A coal fire burned in the grate, and a bay window gave a view of the Wannsee lake. Someone was sailing, Carla saw, slicing through the wavelets before a stiff breeze. Willrich sat behind a leather-topped desk. He had a jar of tobacco and a rack of different-shaped pipes. He was about fifty, tall and heavily built. All his features seemed large: big nose, square jaw, huge ears, and a domed bald head. He looked at Ada and said: 'Frau Hempel, I presume?' Ada nodded. Willrich turned to Carla. 'And you are Fräulein . . . ?'

'Carla von Ulrich, Professor. I'm Kurt's godmother.'

He raised his eyebrows. 'A little young to be a godmother, surely?'

Ada said indignantly: 'She delivered Kurt! She was only eleven, but she was better than the doctor, because he wasn't there!'

Willrich ignored that. Still looking at Carla, he said disdainfully: 'And hoping to become a nurse, I see.'

Carla wore a beginner's uniform, but she considered herself to be more than just hopeful. 'I am a trainee nurse,' she said. She did not like Willrich.

'Please sit.' He opened a thin file. 'Kurt is eight years old, but has reached the developmental stage of only two years.'

He paused. Neither woman said anything.

'This is unsatisfactory,' he said.

Ada looked at Carla. Carla did not know what he was getting at, and indicated as much with a shrug.

'There is a new treatment available for cases of this type. However, it will necessitate moving Kurt to another hospital.' Willrich closed the file. He looked at Ada and, for the first time, he smiled. 'I'm sure you would like Kurt to undergo a therapy that might improve his condition.'

Carla did not like his smile: it seemed creepy. She said: 'Could you tell us more about the treatment, Professor?'

'I'm afraid it would be beyond your understanding,' he said. 'Even though you are a trainee nurse.'

Carla was not going to let him get away with that. 'I'm sure Frau Hempel would like to know whether it would involve surgery, or drugs, or electricity, for example.'

'Drugs,' he said with evident reluctance.

Ada said: 'Where would he have to go?'

'The hospital is in Akelberg, in Bavaria.'

Ada's geography was weak, and Carla knew she had no sense of how far that was. 'It's two hundred miles,' she said.

'Oh, no!' said Ada. 'How would I visit him?'

'By train,' said Willrich impatiently.

Carla said: 'It would take four or five hours. She would probably have to stay overnight. And what about the cost of the fare?'

'I cannot concern myself with such things!' said Willrich angrily. 'I am a doctor, not a travel agent!'

Ada was close to tears. 'If it means Kurt will get better, and learn to say a few words, and not to soil himself . . . one day we might perhaps bring him home.'

'Exactly,' said Willrich. 'I felt sure you would not wish to deny him the chance of getting better just for your own selfish reasons.'

'Is that what you're telling us?' said Carla. 'That Kurt might be able to live a normal life?'

'Medicine offers no guarantees,' he said. 'Even a trainee nurse should know that.'

Carla had learned, from her parents, to be impatient with prevarication. 'I don't ask you for a guarantee,' she said crisply. 'I ask you for a prognosis. You must have one, otherwise you would not be proposing the treatment.'

He reddened. 'The treatment is new. We hope it will improve Kurt's condition. That is what I am telling you.'

'Is it experimental?'

'All medicine is experimental. All therapies work on some patients but not on others. You must listen to what I tell you: medicine offers no guarantees.'

Carla wanted to oppose him just because he was so arrogant, but she realized that was not the basis on which to make a judgement. Besides, she was not sure that Ada really had a choice. Doctors could go against the wishes of parents if the child's health was at risk: in effect, they could do what they liked. Willrich was not asking Ada's permission – he had no real need of it. He was speaking to her only in order to avoid a fuss.

Carla said: 'Can you tell Frau Hempel how long it might be before Kurt returns from Akelberg to Berlin?'

'Quite soon,' said Willrich.

It was no answer at all, but Carla felt that if she pressed him he would become angry again.

Ada was looking helpless. Carla sympathized: she herself found it difficult to know what to say. They had not been given enough information. Doctors were often like this, Carla had noticed: they seemed to want to hug their knowledge to themselves. They preferred to fob patients off with platitudes, and became defensive when questioned.

Ada had tears in her eyes. 'Well, if there's a chance he could get better . . .'

'That's the attitude,' Willrich said.

But Ada had not finished. 'What do you think, Carla?'

Willrich looked outraged at this appeal to the opinion of a mere nurse.

Carla said: 'I agree with you, Ada. This opportunity must be seized, for Kurt's sake, even though it will be hard for you.'

'Very sensible,' said Willrich, and he got to his feet. 'Thank you for coming to see me.' He went to the door and opened it. Carla felt he could not get rid of them quickly enough.

They left the home and walked back to the station. As their nearly empty train pulled away, Carla picked up a leaflet that had been left on the seat. It was headed How to Oppose the Nazis, and it listed ten things people could do to hasten the end of the regime, starting with slowing down their rate of work.

Carla had seen such flyers before, though not often. They were placed by some underground resistance movement.

Ada snatched it from her, crumpled it, and threw it out of the window. 'You can be arrested for reading such things!' she said. She had been Carla's nanny, and sometimes she behaved as though Carla had not grown up. Carla did not mind her occasional bossiness, for she knew it came from love.

However, in this case Ada was not overreacting. People could be imprisoned not just for reading such things but even for failing to report that they had found one. Ada could be in trouble merely for throwing it out of the window. Fortunately, there was no one else in the carriage to see what she had done.

Ada was still troubled by what she had been told at the home. 'Do you think we did the right thing?' she said to Carla.

'I don't really know,' Carla said candidly. 'I think so.'

'You're a nurse, you understand these things better than I do.'

Carla was enjoying nursing, though she still felt frustrated that she had not been allowed to train as a doctor. Now, with so many young men in the army, the attitude to female medical students had changed, and more women were going to medical school. Carla could have applied again for a scholarship – except that her family was so desperately poor that they depended on her meagre wages. Her father had no work at all, her mother gave piano lessons, and Erik sent home as much as he could afford out of his army pay. The family had not paid Ada for years.

Ada was a naturally stoical person, and by the time they got home she was getting over her upset. She went into the kitchen, put on her apron, and began to prepare dinner for the family, and the comfortable routine seemed to console her.

Carla was not having dinner. She had plans for the evening. She felt she was abandoning Ada to her sadness, and she was a bit guilty; but not guilty enough to sacrifice her night out.

She put on a knee-length tennis dress she had made herself by shortening the frayed hem of an old frock of her mother's. She was not going to play tennis, she was going to dance, and her aim was to look American. She put on lipstick and face powder, and combed out her hair in defiance of the government's preference for braids.

The mirror showed her a modern girl with a pretty face and a defiant air. She knew that her confidence and self-possession put a lot of boys off her. Sometimes she wished she could be seductive as well as capable, a trick her mother had always been able to pull off; but it was not in her nature. She had long ago given up trying to be winsome: it just made her feel silly. Boys had to accept her as she was.

Some boys were scared of her, but others were attracted, and at parties she often ended up with a small cluster of admirers. She, in turn, liked boys, especially when they forgot about trying to impress people and started to talk normally. Her favourites were the ones who made her laugh. So far she had not had a serious boyfriend, though she had kissed quite a few.

To complete her outfit she put on a striped blazer she had bought from a second-hand clothing cart. She knew her parents would disapprove of her appearance, and try to make her change, saying it was dangerous to defy the Nazis' prejudices. So she needed to get out of the house without seeing them. It should be easy enough. Mother was giving a piano lesson: Carla could hear the painfully hesitant playing of her pupil. Father would be reading the newspaper in the same room, for they could not afford to heat more than one room of the house. Erik was away with the army, though he was now stationed near Berlin and due home on leave shortly.

She covered up with a conventional raincoat and put her white shoes in her pocket.

She went down to the hall, opened the front door, shouted: 'Goodbye, back soon!' and hurried out.

She met Frieda at the Friedrich Strasse station. She was dressed similarly with a stripey dress under a plain tan coat, her hair hanging loose; the main difference being that Frieda's clothes were new and expensive. On the platform, two boys in Hitler Youth outfits stared at them with a mixture of disapproval and desire.

They got off the train in the northern suburb of Wedding, a working-class district that had once been a left-wing stronghold. They headed for the Pharus Hall, where in the past Communists had held their conferences. Now there was no political activity at all, of course. Nevertheless, the building had become the centre of the movement called Swing Kids.

Kids of between fifteen and twenty-five were already gathering in the streets around the hall. Swing boys wore check jackets and carried umbrellas, to look English. They let their hair grow long to show their contempt for the military. Swing girls had heavy make-up and American sports clothes. They all thought the Hitler Youth were stupid and boring, with their folk music and community dances.

Carla thought it was ironic. When she was little she had been teased by the other kids and called a foreigner because her mother was English: now the same children, a little older, thought English was the fashionable thing to be.

Carla and Frieda went into the hall. There was a conventional, innocent youth club there, with girls in pleated skirts and boys in short trousers playing table tennis and drinking sticky orange cordial. But the action was in the side rooms.

Frieda quickly led Carla to a large storeroom with stacked chairs around the walls. There her brother, Werner, had plugged in a record player. Fifty or sixty boys and girls were dancing the jitterbug jive. Carla recognized the tune that was playing: 'Ma, He's Making Eyes at Me.' She and Frieda started to dance.

Jazz records were banned because most of the best musicians were Negroes. The Nazis had to denigrate anything that was done well by non-Aryans: it threatened their theories of superiority. Unfortunately for them, Germans loved jazz just as much as everyone else. People who visited other countries brought records home, and you could buy them from American sailors in Hamburg. There was a lively black market.

Werner had lots of discs, of course. He had everything: a car, modern clothes, cigarettes, money. He was still Carla's dream boy, though he always went for girls older than she – women, really. Everyone assumed he went to bed with them. Carla was a virgin.

Werner's earnest friend Heinrich von Kessel immediately came up to them and started to dance with Frieda. He wore a black jacket and waistcoat, which looked dramatic with his longish dark hair. He was devoted to Frieda. She liked him – she enjoyed talking to clever men – but she would not go out with him because he was too old, twenty-five or twenty-six.

Soon a boy Carla did not know came and danced with her, and the evening was off to a good start.

She abandoned herself to the music: the irresistible sexual drumbeat, the suggestively crooned lyrics, the exhilarating trumpet solos, the joyous flight of the clarinet. She whirled and kicked, let her skirt flare outrageously high, fell into the arms of her partner and sprang out again.

When they had danced for an hour or so Werner put on a slow tune. Frieda and Heinrich began dancing cheek to cheek. There was no one available whom Carla liked enough for slow dancing, so she left the room and went to get a Coke. Germany was not at war with America so Coca-Cola syrup was imported and bottled in Germany.

To her surprise, Werner followed her out, leaving someone else to put on records for a while. She was flattered that the most attractive man in the room wanted to talk to her.

She told him about Kurt being moved to Akelberg, and Werner said the same thing had happened to his brother, Axel, who was fifteen. Axel had been born with spina bifida. 'Could the same treatment work for both of them?' he said with a frown.

'I doubt it, but I don't really know,' Carla said.

'Why is it that medical men never explain what they're doing?' Werner said irritably.

She laughed humourlessly. 'They think that if ordinary people understand medicine they won't hero-worship doctors any longer.'

'Same principle as a conjurer: it's more impressive if you don't know how it's done,' said Werner. 'Doctors are as egocentric as anyone else.'

'More so,' said Carla. 'As a nurse, I know.'

She told him about the leaflet she had read on the train. Werner said: 'How did you feel about it?'

Carla hesitated. It was dangerous to speak honestly about such things. But she had known Werner all her life, he had always been left-wing, and he was a Swing Kid. She could trust him. She said: 'I'm pleased someone is opposing the Nazis. It shows that not all Germans are paralysed by fear.'

'There are lots of things you can do against the Nazis,' he said quietly. 'Not just wearing lipstick.'

She assumed he meant she could distribute such leaflets. Could he be involved in such activity? No, he was too much of a playboy. Heinrich might be different: he was very intense.

'No, thanks,' she said. 'I'm too scared.'

They finished their Cokes and returned to the storeroom. It was packed, now, with hardly room enough to dance.

To Carla's surprise, Werner asked her for the last dance. He put on Bing Crosby singing 'Only Forever'. Carla was thrilled. He held her close and they swayed, rather than danced, to the slow ballad.

At the end, by tradition, someone turned off the light for a minute, so that couples could kiss. Carla was embarrassed: she had known Werner since they were children. But she had always been attracted to him, and now she turned her face up eagerly. As she had expected, he kissed her expertly, and she returned the kiss with enthusiasm. To her delight she felt his hand gently grasp her breast. She encouraged him by opening her mouth. Then the light came on and it was all over.

'Well,' she said breathlessly, 'that was a surprise.'

He gave his most charming smile. 'Perhaps I can surprise you again some time.'

(ii)

Carla was passing through the hall, on her way to the kitchen for breakfast, when the phone rang. She picked up the handset. 'Carla von Ulrich.'

She heard Frieda's voice. 'Oh, Carla, my little brother's dead!'

'What?' Carla could hardly believe it. 'Frieda, I'm so sorry! Where did it happen?'

'In that hospital.' Frieda was sobbing.

Carla recalled Werner telling her that Axel had been sent to the same Akelberg hospital as Kurt. 'How did he die?'

'Appendicitis.'

'That's terrible.' Carla was sad for her friend, but also suspicious. She had had a bad feeling when Professor Willrich spoke to them a month ago about the new treatment for Kurt. Had it been more experimental than he had let on? Could it have actually been dangerous? 'Do you know any more?'

'We just got a short letter. My father is enraged. He phoned the hospital but he wasn't able to speak to the senior people.'

'I'll come round to your house. I'll be there in a few minutes.'

'Thanks.'

Carla hung up and went into the kitchen. 'Axel Franck has died at that hospital in Akelberg,' she said.

Her father, Walter, was looking at the morning post. 'Oh!' he said. 'Poor Monika.' Carla recalled that Axel's mother, Monika Franck, had once been in love with Walter, according to family legend. The look of concern on Walter's face was so pained that Carla wondered if he had had a slight tendresse for Monika, despite being in love with Maud. How complicated love was.

Carla's mother, who was now Monika's best friend, said: 'She must be devastated.'

Walter looked down at the post again and said in a tone of surprise: 'Here's a letter for Ada.'

The room went quiet.

Carla stared at the white envelope as Ada took it from Walter.

Ada did not receive many letters.

Erik was home – it was the last day of his short leave – so there were four people watching as Ada opened the envelope.

Carla held her breath.

Ada drew out a typed letter on headed paper. She read the message quickly, gasped, then screamed.

'No!' said Carla. 'It can't be!'

Maud jumped up and put her arms around Ada.

Walter took the letter from Ada's fingers and read it. 'Oh, dear, how terribly sad,' he said. 'Poor little Kurt.' He put the paper down on the breakfast table.

Ada began to sob. 'My little boy, my dear little boy, and he died without his mother – I can't bear it!'

Carla fought back tears. She felt bewildered. 'Axel *and* Kurt?' she said. 'At the same time?'

She picked up the letter. It was printed with the name of the hospital and its address in Akelberg. It read:

Dear Mrs Hempel,

I regret to inform you of the sad death of your son, Kurt Walter Hempel, age eight years. He passed away on 4 April at this hospital as a result of a burst appendix. Everything possible was done for him but to no avail. Please accept my deepest condolences.

It was signed by the Senior Physician.

Carla looked up. Her mother was sitting next to Ada, arm around her, holding her hand as she sobbed.

Carla was grief-stricken, but more alert than Ada. She spoke to her father in a shaky voice. 'There's something wrong.'

'What makes you say that?'

'Look again.' She handed him the letter. 'Appendicitis.'

'What is the significance?'

'Kurt had had his appendix removed.'

'I remember,' her father said. 'He had an emergency operation, just after his sixth birthday.'

Carla's sorrow was mixed with angry suspicion. Had Kurt been killed by a dangerous experiment which the hospital was now trying to cover up? 'Why would they lie?' she said.

Erik banged his fist on the table. 'Why do you say it is a lie?' he cried. 'Why do you always accuse the establishment? This is obviously a mistake! Some typist has made a copying error!'

Carla was not so sure. 'A typist working in a hospital is likely to know what an appendix is.'

Erik said furiously: 'You will seize upon even this personal tragedy as a way of attacking those in authority!'

'Be quiet, you two,' said their father.

They looked at him. There was a new tone in his voice. 'Erik may be right,' he said. 'If so, the hospital will be perfectly happy to answer questions and give further details of how Kurt and Axel died.'

'Of course they will,' said Erik.

Walter went on: 'And if Carla is right, they will try to discourage inquiries, withhold information and intimidate the parents of the dead children by suggesting that their questions are somehow illegitimate.'

Erik looked less comfortable about that.

Half an hour ago Walter had been a shrunken man. Now somehow he seemed to fill his suit again. 'We will find out as soon as we start asking questions.'

Carla said: 'I'm going to see Frieda.'

Her mother said: 'Don't you have to go to work?'

'I'm on the late shift.'

Carla phoned Frieda, told her that Kurt was dead too, and said she was coming to talk about it. She put on her coat, hat and gloves then wheeled her bicycle outside. She was a fast rider and it took her only a quarter of an hour to get to the Francks' villa in Schöneberg.

The butler let her in and told her the family were still in the dining room. As soon as she walked in, Frieda's father, Ludwig Franck, bellowed at her: 'What did they tell you at the Wannsee Children's Home?'

Carla did not much like Ludwig. He was a right-wing bully and he had supported the Nazis in the early days. Perhaps he had changed his views: many businessmen had, by now, though they showed little sign of the humility that ought to go with having been so wrong.

She did not answer immediately. She sat down at the table and looked at the family: Ludwig, Monika, Werner and Frieda, and the butler hovering in the background. She collected her thoughts.

'Come on, girl, answer me!' Ludwig demanded. He had in his hand a letter that looked very like Ada's, and he was waving it angrily.

Monika put a restraining hand on her husband's arm. 'Take it easy, Ludi.'

'I want to know!' he said.

Carla looked at his pink face and little black moustache. He was in an agony of grief, she saw. In other circumstances she would have refused to speak to someone so rude. But he had an excuse for his bad manners, and she decided to overlook them. 'The Director, Professor Willrich, told us there was a new treatment for Kurt's condition.'

'The same as he told us,' said Ludwig. 'What kind of treatment?'

'I asked him that question. He said I would not be able to understand it. I persisted, and he said it involved drugs, but he did not give any further information. May I see your letter, Herr Franck?'

Ludwig's expression said he was the one who should be asking questions; but he handed the sheet of paper to Carla.

It was exactly the same as Ada's, and Carla had a queer feeling that the typist had done several of them, just changing the names.

Franck said: 'How can two boys have died of appendicitis at the same time? It's not a contagious illness.'

Carla said: 'Kurt certainly did not die of appendicitis, for he had no appendix. It was removed two years ago.'

'Right,' said Ludwig. 'That's enough talk.' He snatched the letter from Carla's hand. 'I'm going to see someone in the government about this.' He went out.

Monika followed him, and so did the butler.

Carla went over to Frieda and took her hand. 'I'm so sorry,' she said.

'Thank you,' Frieda whispered.

Carla went to Werner. He stood up and put his arms around her. She felt a tear fall on her forehead. She was gripped by she did not know what intense emotion. Her heart was full of grief, yet she thrilled to the pressure of his body against hers, and the gentle touch of his hands.

After a long moment Werner stepped back. He said angrily: 'My father has phoned the hospital twice. The second time, they told him they had no more information and hung up on him. But I'm going to find out what happened to my brother, and I won't be brushed off.'

Frieda said: 'Finding out won't bring him back.'

'I still want to know. If necessary, I'll go to Akelberg.'

Carla said: 'I wonder if there's anyone in Berlin who could help us.'

'It would have to be someone in the government,' Werner said.

Frieda said: 'Heinrich's father is in the government.'

Werner snapped his fingers. 'The very man. He used to belong to the Centre Party, but he's a Nazi now, and something important in the Foreign Office.'

Carla said: 'Will Heinrich take us to see him?'

'He will if Frieda asks him,' said Werner. 'Heinrich will do anything for Frieda.'

Carla could believe that. Heinrich had always been intense about everything he did.

'I'll phone him now,' said Frieda.

She went into the hall, and Carla and Werner sat down side by side. He put his arm around her, and she leaned her head on his shoulder. She did not know whether these signs of affection were merely a side-effect of the tragedy, or something more.

Frieda came back in and said: 'Heinrich's father will see us right away if we go over there now.'

They all got into Werner's sports car, squeezing on to the front seat. 'I don't know how you keep this car going,' Frieda said as he pulled away. 'Even Father can't get petrol for private use.'

'I tell my boss it's for official business,' he said. Werner worked for an important general. 'But I don't know how much longer I can get away with it.'

The von Kessel family lived in the same suburb. Werner drove there in five minutes.

The house was luxurious, though smaller than the Francks'. Heinrich met them at the door and showed them into a living room with leather-bound books and an old German woodcarving of an eagle.

Frieda kissed him. 'Thank you for doing this,' she said. 'It probably wasn't easy – I know you don't get on so well with your father.'

Heinrich beamed with pleasure.

His mother brought them coffee and cake. She seemed a warm, simple person. When she had served them she left, like a maid.

Heinrich's father, Gottfried, came in. He had the same thick straight hair, but it was silver instead of black.

Heinrich said: 'Father, here are Werner and Frieda Franck, whose father manufactures People's Radios.'

'Ah, yes,' said Gottfried. 'I have seen your father in the Herrenklub.'

'And this is Carla von Ulrich – I believe you know her father, too.'

'We were colleagues at the German embassy in London,' Gottfried said carefully. 'That was in 1914.' Clearly he was not so pleased to be reminded of his association with a social democrat. He took a piece of cake, clumsily dropped it on the rug, tried ineffectually to pick up the crumbs, then abandoned the effort and sat back.

Carla thought: What is he afraid of?

Heinrich got straight down to the purpose of the visit. 'Father, I expect you've heard of Akelberg.'

Carla was watching Gottfried closely. There was a split-second flash of something in his expression, but he quickly adopted a pose of indifference. 'A small town in Bavaria?' he said.

'There is a hospital there,' said Heinrich. 'For mentally handicapped people.'

'I don't think I was aware of that.'

'We think something strange is going on there, and we wondered if you might know about it.'

'I certainly don't. What seems to be happening?'

Werner broke in. 'My brother died there, apparently of appendicitis. Herr von Ulrich's maid's child died at the same time in the same hospital of the same illness.'

'Very sad – but a coincidence, surely?'

Carla said: 'My maid's child did not have an appendix. It was removed two years ago.'

'I understand why you are keen to ascertain the facts,' said Gottfried. 'This is deeply unsatisfactory. However, the likeliest explanation would seem to be clerical error.'

Werner said: 'If so, we would like to know.'

'Of course. Have you written to the hospital?'

Carla said: 'I wrote to ask when my maid could visit her son. They never replied.'

Werner said: 'My father telephoned the hospital this morning. The Senior Physician slammed the phone down on him.'

'Oh, dear. Such bad manners. But, you know, this is hardly a Foreign Office matter.'

Werner leaned forward. 'Herr von Kessel, is it possible that both boys were involved in a secret experiment that went wrong?'

Gottfried sat back. 'Quite impossible,' he said, and Carla had a feeling he was telling the truth. 'That is definitely not happening.' He sounded relieved.

Werner looked as if he had run out of questions, but Carla was not satisfied. She wondered why Gottfried seemed so happy about the assurance he had just given. Was it because he was concealing something worse?

She was struck by a possibility so appalling that she could hardly contemplate it.

Gottfried said: 'Well, if that's all ...'

Carla said: 'You're very sure, sir, that they were not killed by an experimental therapy that went wrong?'

'Very sure.'

'To know for certain that is *not* true, you must have some knowledge of what *is* being done at Akelberg.'

'Not necessarily,' he said, but all his tension had returned, and she knew she was on to something.

'I remember seeing a Nazi poster,' she went on. It was this memory that had triggered her dreadful thought. 'There was a picture of a male nurse and a mentally handicapped man. The text said something like: 'Sixty thousand Reichsmarks is what this person suffering from hereditary defects costs the people's community during his lifetime. Comrade, that is your money too!' It was an advertisement for a magazine, I think.'

'I have seen some of that propaganda,' Gottfried said disdainfully, as if it were nothing to do with him.

Carla stood up. 'You're a Catholic, Herr von Kessel, and you brought up Heinrich in the Catholic faith.'

Gottfried made a scornful noise. 'Heinrich says he's an atheist now.'

'But you're not. And you believe that human life is sacred.'

'Yes.'

'You say that the doctors at Akelberg are not testing dangerous new therapies on handicapped people, and I believe you.'

'Thank you.'

'But are they doing something else? Something worse?'

'No, no.'

'Are they deliberately *killing* the handicapped?'

Gottfried shook his head silently.

Carla moved closer to Gottfried and lowered her voice, as if they were the only two people in the room. 'As a Catholic who believes that human life is sacred, will you put your hand on your heart and tell me that mentally ill children are not being murdered at Akelberg?'

Gottfried smiled, made a reassuring gesture, and opened his mouth to speak, but no words came out.

Carla knelt on the rug in front of him. 'Would you do that, please? Right now? Here in your house with you are four young Germans, your son and his three friends. Just tell us the truth. Look me in the eye and say that our government does not kill handicapped children.'

The silence in the room was total. Gottfried seemed about to speak, but changed his mind. He squeezed his eyes shut, twisted his mouth into a grimace, and bowed his head. The four young people watched his facial contortions in amazement.

At last he opened his eyes. He looked at them one by one, ending with his gaze on his son.

Then he stood up and walked out of the room.

(iii)

The next day, Werner said to Carla: 'This is awful. We've talked of the same thing for more than twenty-four hours. We'll go mad if we don't do something else. Let's see a movie.'

They went to the Kurfürstendamm, a street of theatres and shops, always called the Ku'damm. Most of the good German film-makers had gone to Hollywood years ago, and the domestic movies were now second-rate. They saw *Three Soldiers*, set during the invasion of France.

The three soldiers were a tough Nazi sergeant, a snivelling complainer who looked a bit Jewish, and an earnest young man. The earnest one asked naive questions such as: 'Do the Jews really do us any harm?' and in answer received long, stern lectures from the sergeant. When battle was joined the sniveller admitted to being a Communist, deserted, and was blown up in an air raid. The earnest young man fought bravely, was promoted to sergeant, and became an admirer of the Führer. The script was dire but the battle scenes were exciting.

Werner held Carla's hand all the way through. She hoped he would kiss her in the dark, but he did not.

As the lights came up he said: 'Well, it was terrible, but it took my mind off things for a couple of hours.'

They went outside and found his car. 'Shall we go for a drive?' he said. 'It could be our last chance. This car goes up on blocks next week.'

He drove out to the Grunewald. On the way, Carla's thoughts inevitably returned to yesterday's conversation with Gottfried von Kessel. No matter how many times she went over it in her mind, there was no way she could escape the terrible conclusion all four of them had reached at the end of it. Kurt and Axel had not been accidental victims of a dangerous medical experiment, as she had at first thought. Gottfried had denied that convincingly. But he had not been able to bring himself to deny that the government was deliberately killing the handicapped, and lying to their families about it. It was hard to believe, even of people as ruthless and brutal as the Nazis. Yet Gottfried's response had been the clearest example of guilty behaviour that Carla had ever witnessed.

When they were in the forest Werner pulled off the road and drove along a track until the car was hidden by shrubbery. Carla guessed he had brought other girls to this spot.

He turned out the lights, and they were in deep darkness. 'I'm going to speak to General Dorn,' he said. Dorn was his boss, an important officer in the Air Force. 'What about you?'

'My father says there's no political opposition left, but the churches are still strong. No one who is sincere about their religious beliefs could condone what's being done.'

'Are you religious?' Werner asked.

'Not really. My father is. For him, the Protestant faith is part of the German heritage he loves. Mother goes to church with him, though I suspect her theology might be a bit unorthodox. I believe in God, but I can't imagine He cares whether people are Protestant or Catholic or Muslim or Buddhist. And I like singing hymns.'

Werner's voice fell to a whisper. 'I can't believe in a God who allows the Nazis to murder children.'

'I don't blame you.'

'What is your father going to do?'

'Speak to the pastor of our church.'

'Good.'

They were silent for a while. He put his arm around her. 'Is this all right?' he said in a half-whisper.

She was tense with anticipation, and her voice seemed to fail. Her reply came out as a grunt. She tried again, and managed to say: 'If it stops you feeling so sad . . . yes.'

Then he kissed her.

She kissed him back eagerly. He stroked her hair, then her breasts. At this point, she knew, a lot of girls would call a halt. They said if you went any further you would lose control of yourself.

Carla decided to risk it.

She touched his cheek while he was kissing her. She caressed his throat with her fingertips, enjoying the feel of the warm skin. She put her hand under his jacket and explored his body, her hand on his shoulder blades and his ribs and his spine.

She sighed when she felt his hand on her thigh, under her skirt. As soon as he touched her between her legs she parted her knees. Girls said a boy would think you cheap for doing that, but she could not help herself.

He touched her in just the right place. He did not try to put his hand inside her underwear, but stroked her lightly through the cotton. She heard herself making noises in her throat, quietly at first but then louder. Eventually she cried out with pleasure, burying her face in his neck to muffle the sound. Then she had to push his hand away because she felt too sensitive.

She was panting. As she began to get her breath back she kissed his neck. He touched her cheek lovingly.

After a minute she said: 'Can I do something for you?'

'Only if you want to.'

She was embarrassed by how much she wanted to. 'The only thing is, I've never . . .'

'I know,' he said. 'I'll show you.'

(iv)

Pastor Ochs was a portly, comfortable clergyman with a large house, a nice wife and five children, and Carla feared he would refuse to get involved. But she underestimated him. He had already heard rumours that were troubling his conscience, and he agreed to go with Walter to the Wannsee Children's Home. Professor Willrich could hardly refuse a visit from an interested clergyman.

They decided to take Carla with them, because she had witnessed the interview with Ada. The Director might find it more difficult to change his story in front of her.

On the train, Ochs suggested he should do the talking. 'The Director is probably a Nazi,' he said. Most people in senior jobs nowadays were party members. 'He will naturally see a former social-democrat deputy as an enemy. I will play the role of unbiased arbitrator. That way, I believe, we may learn more.'

Carla was not sure about that. She felt her father would be a more expert questioner. But Walter went along with the pastor's suggestion.

It was spring, and the weather was warmer than on Carla's last visit. There were boats on the lake. Carla decided to ask Werner to come out here for a picnic. She wanted to make the most of him before he drifted off to another girl.

Professor Willrich had a fire blazing, but a window was open, letting in a fresh breeze off the water.

The Director shook hands with Pastor Ochs and Walter. He gave

Carla a brief glance of recognition then ignored her. He invited them to sit down, but Carla saw there was angry hostility behind his superficial courtesy. Clearly he did not relish being questioned. He picked up one of his pipes and played with it nervously. He was less arrogant today, confronted by two mature men rather than a couple of young women.

Ochs opened the discussion. 'Herr von Ulrich and others in my congregation are concerned, Professor Willrich, about the mysterious deaths of several handicapped children known to them.'

'No children have died mysteriously here,' Willrich shot back. 'In fact, no child has died here in the last two years.'

Ochs turned to Walter. 'I find that very reassuring, Walter, don't you?'

'Yes,' said Walter.

Carla did not, but she kept her mouth shut for the moment.

Ochs went on unctuously: 'I feel sure that you give your charges the best possible care.'

'Yes.' Willrich looked a little less anxious.

'But you do send children from here to other hospitals?'

'Of course, if another institution can offer a child some treatment not available here.'

'And when a child is transferred, I suppose you are not necessarily kept informed about his treatment or his condition thereafter.'

'Exactly!'

'Unless they come back.'

Willrich said nothing.

'Have any come back?'

'No.'

Ochs shrugged. 'Then you cannot be expected to know what happened to them.'

'Precisely.'

Ochs sat back and spread his hands in a gesture of openness. 'So you have nothing to hide!'

'Nothing at all.'

'Some of those transferred children have died.'

Willrich said nothing.

Ochs gently persisted. 'That's true, isn't it?'

'I cannot answer you with any certain knowledge, Herr Pastor.'

'Ah!' said Ochs. 'Because even if one of those children died, you would not be notified.'

'As we said before.'

'Forgive me the repetition, but I simply want to establish beyond doubt that you cannot be asked to shed light on those deaths.'

'Not at all.'

Once again Ochs turned to Walter. 'I think we're clearing matters up splendidly.'

Walter nodded.

Carla wanted to say *Nothing has been cleared up!*

But Ochs was speaking again. 'Approximately how many children have you transferred in, say, the last twelve months.'

'Ten,' said Willrich. 'Exactly.' He smiled complacently. 'We scientific men prefer not to deal in approximations.'

'Ten patients, out of . . . ?'

'Today we have one hundred and seven children here.'

'A very small proportion!' said Ochs.

Carla was getting angry. Ochs was obviously on Willrich's side! Why was her father swallowing this?

Ochs said: 'And did those children suffer from one common condition, or a variety?'

'A variety.' Willrich opened a folder on his desk. 'Idiocy, Down's syndrome, microcephaly, hydrocephaly, malformations of limbs, head and spinal column, and paralysis.'

'These are the types of patient you were instructed to send to Akelberg.'

That was a jump. It was the first mention of Akelberg, and the first suggestion that Willrich had received instructions from a higher authority. Perhaps Ochs was more subtle than he had seemed.

Willrich opened his mouth to say something, but Ochs forestalled him with another question. 'Were they all to receive the same special treatment?'

Willrich smiled. 'Again, I was not informed, so I cannot tell you.'

'You simply complied . . .'

'With my instructions, yes.'

Ochs smiled. 'You're a judicious man. You choose your words carefully. Were the children all ages?'

'Initially the programme was restricted to children under three, but later it was expanded to benefit all ages, yes.'

Carla noted the mention of a 'programme'. That had not been admitted before. She began to realize that Ochs was cleverer than he might at first appear.

Ochs spoke his next sentence as if confirming something already stated. 'And all handicapped Jewish children were included, irrespective of their particular disability.'

There was a moment of silence. Willrich looked shocked. Carla wondered how Ochs knew that about Jewish children. Perhaps he did not: he might have been guessing.

After a pause, Ochs added: 'Jewish children, and those of mixed race, I should have said.'

Willrich did not speak, but gave a slight nod.

Ochs went on: 'It's unusual, in this day and age, for Jewish children to be given preference, isn't it?'

Willrich looked away.

The pastor stood up, and when he spoke again his voice rang with anger. 'You have told me that ten children suffering from a range of illnesses, who could not possibly all benefit from the same treatment, were sent away to a special hospital from which they never returned; and that Jews got priority. What did you think happened to them, Herr Professor Doctor Willrich? In God's name, *what did you think?*'

Willrich looked as if he would cry.

'You may say nothing, of course,' Ochs said more quietly. 'But one day you will be asked the same question by a higher authority – in fact, by the highest of all authorities.'

He stretched out his arm and pointed a condemning finger.

'And on that day, my son, you *will* answer.'

With that he turned around and left the room.

Carla and Walter followed him out.

(v)

Inspector Thomas Macke smiled. Sometimes the enemies of the state did his job for him. Instead of working in secret, and hiding away where they were difficult to find, they identified themselves to him and generously provided irrefutable evidence of their crimes. They were like fish that did not require bait and a hook but simply jumped out of the river into the fisherman's basket and begged to be fried.

Pastor Ochs was one such.

Macke read his letter again. It was addressed to the Justice Minister, Franz Gürtner.

1941 (I)

Dear Minister,

*Is the government killing handicapped children? I ask you this question
bluntly because I must have a plain answer.*

What a fool! If the answer was No, this was a criminal libel; if Yes,
Ochs was guilty of revealing state secrets. Could he not figure that out
for himself?

*After it became impossible to ignore rumours circulating in my congregation,
I visited the Wannsee Children's Nursing Home and spoke to its director,
Professor Willrich. His responses were so unsatisfactory that I became
convinced something terrible is going on, something that is presumably
a crime and unquestionably a sin.*

The man had the nerve to write of crimes! Did it not occur to him
that accusing government agencies of illegal acts was itself an illegal act?
Did he imagine he was living in a degenerate liberal democracy?

Macke knew what Ochs was complaining about. The programme
was called Aktion T4 after its address, Tiergarten Strasse 4. The agency
was officially the Charitable Foundation for Cure and Institutional Care,
though it was supervised by Hitler's personal office, the Chancellery of
the Führer. Its job was to arrange the painless deaths of handicapped
people who could not survive without costly care. It had done splendid
work in the last couple of years, disposing of tens of thousands of useless
people.

The problem was that German public opinion was not yet
sophisticated enough to understand the need for such deaths, so the
programme had to be kept quiet.

Macke was in on the secret. He had been promoted to Inspector
and had at last been admitted to the Nazi party's elite paramilitary
Schutzstaffel, the SS. He had been briefed on Aktion T4 when he was
assigned to the Ochs case. He felt proud: he was a real insider now.

Unfortunately, people had been careless, and there was a danger that
the secret of Aktion T4 would get out.

It was Macke's job to plug the leak.

Preliminary inquiries had swiftly revealed that there were three men
to be silenced: Pastor Ochs, Walter von Ulrich, and Werner Franck.

Franck was the elder son of a radio manufacturer who had been an
important early supporter of the Nazis. The manufacturer himself, Ludwig
Franck, had initially made furious demands for information about the

death of his disabled younger son, but had quickly fallen silent after a threat to close his factories. Young Werner, a fast-rising officer in the Air Ministry, had persisted in asking awkward questions, trying to involve his influential boss, General Dorn.

The Air Ministry, said to be the largest office building in Europe, was an ultra-modern edifice occupying an entire block of Wilhelm Strasse, just around the corner from Gestapo headquarters in Prinz Albrecht Strasse. Macke walked there.

In his SS uniform he was able to ignore the guards. At the reception desk he barked: 'Take me to Lieutenant Werner Franck immediately.'

The receptionist took him up in an elevator and along a corridor to an open door leading into a small office. The young man at the desk did not at first look up from the papers in front of him. Observing him, Macke guessed he was about twenty-two years old. Why was he not with a front-line unit, bombing England? The father had probably pulled strings, Macke thought resentfully. Werner looked like a son of privilege: tailored uniform, gold rings, and over-long hair that was distinctly un-military. Macke despised him already.

Werner wrote a note with a pencil then looked up. The amiable expression on his face died quickly when he saw the SS uniform, and Macke noted with interest a flash of fear. The boy immediately tried to cover up with a show of bonhomie, standing up deferentially and smiling a welcome, but Macke was not fooled.

'Good afternoon, Inspector,' said Werner. 'Please be seated.'

'*Heil Hitler*,' said Macke.

'*Heil Hitler*. How can I help you?'

'Sit down and shut up, you foolish boy,' Macke spat.

Werner struggled to hide his fear. 'My goodness, what can I have done to incur such wrath?'

'Don't presume to question me. Speak when you're spoken to.'

'As you wish.'

'From this moment on you will ask no further questions about your brother Axel.'

Macke was surprised to see a momentary look of relief pass over Werner's face. That was puzzling. Had he been afraid of something else, something more frightening than the simple order to stop asking questions about his brother? Could Werner be involved in other subversive activities?

Probably not, Macke thought on reflection. Most likely Werner was relieved he was not being arrested and taken to the basement in Prinz Albrecht Strasse.

Werner was not yet completely cowed. He summoned the nerve to say: 'Why should I not ask how my brother died?'

'I told you not to question me. Be aware that you are being treated gently only because your father has been a valued friend of the Nazi party. Were it not for that, *you* would be in *my* office.' That was a threat everyone understood.

'I'm grateful for your forbearance,' Werner said, struggling to retain a shred of dignity. 'But I want to know who killed my brother, and why.'

'You will learn no more, regardless of what you do. But any further inquiries will be regarded as treason.'

'I hardly need to make further inquiries, after this visit from you. It is now clear that my worst suspicions were right.'

'I require you to drop your seditious campaign immediately.'

Werner stared defiantly back but said nothing.

Macke said: 'If you do not, General Dorn will be informed that there are questions about your loyalty.' Werner could be in no doubt about what that meant. He would lose his cosy job here in Berlin and be dispatched to a barracks on an airstrip in northern France.

Werner looked less defiant, more thoughtful.

Macke stood up. He had spent enough time here. 'Apparently General Dorn finds you a capable and intelligent assistant,' he said. 'If you do the right thing, you may continue in that role.' He left the room.

He felt edgy and dissatisfied. He was not sure he had succeeded in crushing Werner's will. He had sensed a bedrock defiance that remained untouched.

He turned his mind to Pastor Ochs. A different approach would be required for him. Macke returned to Gestapo headquarters and collected a small team: Reinhold Wagner, Klaus Richter and Günther Schneider. They took a black Mercedes 260D, the Gestapo's favourite car, unobtrusive because many Berlin taxis were the same model and colour. In the early days, the Gestapo had been encouraged to make themselves visible and let the public see the brutal way they dealt with opposition. However, the terrorization of the German people had been accomplished long ago, and open violence was no longer necessary. Nowadays the Gestapo acted discreetly, always with a cloak of legality.

They drove to Ochs's house next to the large Protestant church in

Mitte, the central district. In the same way that Werner might think he was protected by his father, so Ochs probably imagined his church made him safe. He was about to learn otherwise.

Macke rang the bell: in the old days they would have kicked the door down, just for effect.

A maid opened the door, and he walked into a broad, well-lit hallway with polished floorboards and heavy rugs. The other three followed him in. 'Where is your master?' Macke said pleasantly to the maid.

He had not threatened her, but all the same she was frightened. 'In his study, sir,' she said, and she pointed to a door.

Macke said to Wagner: 'Get the women and children together in the next room.'

Ochs opened the study door and looked into the hall, frowning. 'What on earth is going on?' he said indignantly.

Macke walked directly towards him, forcing him to step back and allow Macke to enter the room. It was a small, well-appointed den, with a leather-topped desk and shelves of biblical commentaries. 'Close the door,' said Macke.

Reluctantly, Ochs did as he was told; then he said: 'You'd better have a very good explanation for this intrusion.'

'Sit down and shut up,' said Macke.

Ochs was dumbfounded. Probably he had not been told to shut up since he was a boy. Clergymen were not normally insulted, even by policemen. But the Nazis ignored such enfeebling conventions.

'This is an outrage!' Ochs managed at last. Then he sat down.

Outside the room, a woman's voice was raised in protest: the wife, presumably. Ochs paled when he heard it, and rose from his chair.

Macke pushed him back down. 'Stay where you are.'

Ochs was a heavy man, and taller than Macke, but he did not resist.

Macke loved to see these pompous types deflated by fear.

'Who are you?' said Ochs.

Macke never told them. They could guess, of course, but it was more frightening if they did not know for sure. Afterwards, in the unlikely event that anyone asked questions, the whole team would swear that they had begun by identifying themselves as police officers and showing their badges.

He went out. His men were hustling several children into the parlour.

Macke told Reinhold Wagner to go into the study and keep Ochs there. Then he followed the children into the other room.

There were flowered curtains, family photographs on the mantelpiece, and a set of comfortable chairs upholstered in a checked fabric. It was a nice home and a nice family. Why could they not be loyal to the Reich and mind their own business?

The maid was by the window, hand over her mouth as if to stop herself crying out. Four children clustered around Ochs's wife, a plain, heavy-breasted woman in her thirties. She held a fifth child in her arms, a girl of about two years with blonde ringlets.

Macke patted the girl's head. 'And what is this one's name?' he said.

Frau Ochs was terrified. She whispered: 'Lieselotte. What do you want with us?'

'Come to Uncle Thomas, little Lieselotte,' said Macke, holding out his arms.

'No!' Frau Ochs cried. She clutched the child closer and turned away.

Lieselotte began to cry loudly.

Macke nodded to Klaus Richter.

Richter grabbed Frau Ochs from behind, pulling her arms back, forcing her to let go of the child. Macke took Lieselotte before she fell. The child wriggled like a fish, but he just held her tighter, as he would have held a cat. She wailed louder.

A boy of about twelve flung himself at Macke, small fists pounding ineffectually. It was about time he learned to respect authority, Macke decided. He put Lieselotte on his left hip then, with his right hand, picked the boy up by his shirt front and threw him across the room, making sure he landed in an upholstered chair. The boy yelled in fear and Frau Ochs screamed. The chair went over backwards and the boy tumbled to the floor. He was not really hurt but he began to cry.

Macke took Lieselotte out into the hall. She screamed at the top of her voice for her mother. Macke put her down. She ran to the parlour door and banged on it, screeching in terror. She had not yet learned to turn doorknobs, Macke noted.

Leaving the child in the hallway, Macke re-entered the study. Wagner was by the door, guarding it; Ochs was standing in the middle of the room, white with fear. 'What are you doing to my children?' he said. 'Why is Lieselotte screaming?'

'You will write a letter,' Macke said.

'Yes, yes, anything,' Ochs said, going to the leather-topped desk.
'Not now, later.'
'All right.'
Macke was enjoying this. Ochs's collapse was complete, unlike Werner's. 'A letter to the Justice Minister,' he went on.
'So that's what this is about.'
'You will say you now realize there is no truth in the allegations you made in your first letter. You were misled by secret Communists. You will apologize to the minister for the trouble you have caused by your incautious actions, and assure him that you will never again speak of the matter to anyone.'
'Yes, yes, I will. What are they doing to my wife?'
'Nothing. She is screaming because of what will happen to her if you fail to write the letter.'
'I want to see her.'
'It will be worse for her if you annoy me with stupid demands.'
'Of course, I'm sorry, I beg your pardon.'
The opponents of Nazism were so weak. 'Write the letter this evening, and mail it in the morning.'
'Yes. Should I send you a copy?'
'It will come to me anyway, you idiot. Do you think the minister himself reads your insane scribbling?'
'No, no, of course not, I see that.'
Macke went to door. 'And stay away from people like Walter von Ulrich.'
'I will, I promise.'
Macke went out, beckoning Wagner to follow. Lieselotte was sitting on the floor screaming hysterically. Macke opened the parlour door and summoned Richter and Schneider.
They left the house.
'Sometimes violence is quite unnecessary,' Macke said reflectively as they got into the car.
Wagner took the wheel and Macke gave him the address of the von Ulrich house.
'And then again, sometimes it's the simplest way,' he added.
Von Ulrich lived in the neighbourhood of the church. His house was a spacious old building that he evidently could not afford to maintain. The paint was peeling, the railings were rusty, and a broken window

had been patched with cardboard. This was not unusual: wartime austerity meant that many houses were not kept up.

The door was opened by a maid. Macke presumed this was the woman whose handicapped child had started the whole problem – but he did not bother to enquire. There was no point in arresting girls.

Walter von Ulrich stepped into the hall from a side room.

Macke remembered him. He was the cousin of the Robert von Ulrich whose restaurant Macke and his brother had bought eight years ago. In those days he had been proud and arrogant. Now he wore a shabby suit, but his manner was still bold. 'What do you want?' he said, attempting to sound as if he still had the power to demand explanations.

Macke did not intend to waste much time here. 'Cuff him,' he said.

Wagner stepped forward with the handcuffs.

A tall, handsome woman appeared and stood in front of von Ulrich. 'Tell me who you are and what you want,' she demanded. She was obviously the wife. She had the hint of a foreign accent. No surprise there.

Wagner slapped her face, hard, and she staggered back.

'Turn around and put your wrists together,' Wagner said to von Ulrich. 'Otherwise I'll knock her teeth down her throat.'

Von Ulrich obeyed.

A pretty young woman dressed in a nurse's uniform came rushing down the stairs. 'Father!' she said. 'What's happening?'

Macke wondered how many more people there might be in the house. He felt a twinge of anxiety. An ordinary family could not overcome trained police officers, but a crowd of them might create enough of a fracas for von Ulrich to slip away.

However, the man himself did not want a fight. 'Don't confront them!' he said to his daughter in a voice of urgency. 'Stay back!'

The nurse looked terrified and did as she was told.

Macke said: 'Put him in the car.'

Wagner walked von Ulrich out of the door.

The wife began to sob.

The nurse said: 'Where are you taking him?'

Macke went to the door. He looked at the three women: the maid, the wife and the daughter. 'All this trouble,' he said, 'for the sake of an eight-year-old moron. I will never understand you people.'

He went out and got into the car.

They drove the short distance to Prinz Albrecht Strasse. Wagner parked at the back of the Gestapo headquarters building alongside a dozen identical black cars. They all got out.

They took von Ulrich in through a back door and down the stairs to the basement, and put him in a white-tiled room.

Macke opened a cupboard and took out three long, heavy clubs like American baseball bats. He gave one to each of his assistants.

'Beat the shit out of him,' he said; and he left them to it.

(vi)

Captain Volodya Peshkov, head of the Berlin section of Red Army Intelligence, met Werner Franck at the Invalids' Cemetery beside the Berlin-Spandau Ship Canal.

It was a good choice. Looking around the graveyard carefully, Volodya was able to confirm that no one followed him or Werner in. The only other person present was an old woman in a black headscarf, and she was on her way out.

Their rendezvous was the tomb of General von Scharnhorst, a large pedestal bearing a slumbering lion made of melted-down enemy cannons. It was a sunny day in spring, and the two young spies took off their jackets as they walked among the graves of German heroes.

After the Hitler–Stalin pact almost two years ago, Soviet espionage had continued in Germany, and so had surveillance of Soviet Embassy staff. Everyone saw the treaty as temporary, though no one knew how temporary. So counter-intelligence agents were still tailing Volodya everywhere.

They ought to be able to tell when he was going out on a genuine secret intelligence mission, he thought, for that was when he shook them off. If he went out to buy a frankfurter for lunch he let them shadow him. He wondered whether they were smart enough to figure that out.

'Have you seen Lili Markgraf lately?' said Werner.

She was a girl they had both dated at different times in the past. Volodya had now recruited her, and she had learned to encode and decode messages in the Red Army Intelligence cipher. Of course Volodya would not tell Werner that. 'I haven't seen her for a while,' he lied. 'How about you?'

Werner shook his head. 'Someone else has won my heart.' He seemed bashful. Perhaps he was embarrassed about belying his playboy reputation. 'Anyway, why did you want to see me?'

'We have received devastating information,' Volodya said. 'News that will change the course of history – if it is true.'

Werner looked sceptical.

Volodya went on: 'A source has told us that Germany will invade the Soviet Union in June.' He thrilled again as he said it. It was a huge triumph for Red Army Intelligence, and a terrible threat to the USSR.

Werner pushed a lock of hair out of his eyes in a gesture that probably made girls' hearts beat faster. He said: 'A reliable source?'

It was a journalist in Tokyo who was in the confidence of the German ambassador there, but was in fact a secret Communist. Everything he had said so far had turned out to be true. But Volodya could not tell Werner that. 'Reliable,' he said.

'So you believe it?'

Volodya hesitated. That was the problem. Stalin did not believe it. He thought it was Allied disinformation intended to sow mistrust between himself and Hitler. Stalin's scepticism about this intelligence coup had devastated Volodya's superiors, souring their jubilation. 'We seek verification,' he said.

Werner looked around at the trees in the graveyard coming into leaf. 'I hope to God it's true,' he said with sudden savagery. 'It will finish the damned Nazis.'

'Yes,' said Volodya. 'If the Red Army is prepared.'

Werner was surprised. 'Are you not prepared?'

Once again Volodya was not able to tell Werner the whole truth. Stalin believed the Germans would not attack before they had defeated the British, fearing a war on two fronts. While Britain continued to defy Germany, the Soviet Union was safe, he thought. In consequence the Red Army was nowhere near prepared for a German invasion.

'We *will be* prepared,' Volodya said, 'if you can get me verification of the invasion plan.'

He could not help enjoying a moment of self-importance. His spy could be the key.

Werner said: 'Unfortunately, I can't help you.'

Volodya frowned. 'What do you mean?'

'I can't get verification, or otherwise, of this information, nor can I get you anything else. I'm about to be fired from my job at the Air Ministry. I'll probably be posted to France – or, if your intelligence is correct, sent to invade the Soviet Union.'

Volodya was horrified. Werner was his best spy. It was Werner's information that had won Volodya promotion to captain. He found he could hardly breathe. With an effort he said: 'What the hell happened?'

'My brother died in a home for the handicapped, and the same thing happened to my girlfriend's godson; and we're asking too many questions.'

'Why would you be demoted for that?'

'The Nazis are killing off handicapped people, but it's a secret programme.'

Volodya was momentarily diverted from his mission. 'What? They just murder them?'

'So it seems. We don't know the details yet. But if they had nothing to hide they wouldn't have punished me – and others – for asking questions.'

'How old was your brother?'

'Fifteen.'

'God! Still a child!'

'They're not going to get away with it. I refuse to shut up.'

They stopped in front of the tomb of Manfred von Richthofen, the air ace. It was a huge slab, six feet high and twice as wide. On it was carved, in elegant capital letters, the single word RICHTHOFEN. Volodya always found its simplicity moving.

He tried to recover his composure. He told himself that the Soviet secret police murdered people, after all, especially anyone suspected of disloyalty. The head of the NKVD, Lavrentiy Beria, was a torturer whose favourite trick was to have his men pull a couple of pretty girls off the street for him to rape as his evening's entertainment, according to rumour. But the thought that Communists could be as bestial as Nazis was no consolation. One day, he reminded himself, the Soviets would get rid of Beria and his kind, then they could begin to build true Communism. Meanwhile, the priority was to defeat the Nazis.

They came to the canal wall and stood there, watching a barge make its slow progress along the waterway, belching oily black smoke. Volodya mulled over Werner's alarming confession. 'What would happen if you stopped investigating these deaths of handicapped children?' he asked.

'I'd lose my girlfriend,' Werner said. 'She's as angry about it as I am.'

Volodya was struck by the scary thought that Werner might reveal the truth to his girlfriend. 'You certainly couldn't tell her the real reason for your change of mind,' he said emphatically.

Werner looked stricken, but he did not argue.

Volodya realized that by persuading Werner to abandon his campaign he would be helping the Nazis hide their crimes. He pushed the uncomfortable thought aside. 'But would you be allowed to keep your job with General Dorn if you promised to drop the matter?'

'Yes. That's what they want. But I'm not letting them murder my brother then cover it up. They'll send me to the front line, but I won't shut up.'

'What do you think they'll do to you when they realize how determined you are?'

'They'll throw me in some camp.'

'And what good will that do?'

'I just can't lie down for this.'

Volodya had to get Werner back on side, but so far he had failed to get through. Werner had an answer for everything. He was a smart guy. That was why he was such a valuable spy.

'What about the others?' Volodya said.

'What others?'

'There must be thousands more handicapped adults and children. Are the Nazis going to kill them all?'

'Probably.'

'You certainly won't be able to stop them if you're in a prison camp.'

For the first time, Werner did not have a comeback.

Volodya turned away from the water and surveyed the cemetery. A young man in a suit was kneeling at a small tombstone. Was he a tail? Volodya watched carefully. The man was shaking with sobs. He seemed genuine: counter-intelligence agents were not good actors.

'Look at him,' Volodya said to Werner.

'Why?'

'He's grieving. Which is what you're doing.'

'So what?'

'Just watch.'

After a minute the man got up, wiped his face with a handkerchief, and walked away.

Volodya said: 'Now he's happy. That's what grieving is about. It doesn't achieve anything, it just makes you feel better.'

'You think my asking questions is just to make me feel better.'

Volodya turned and looked him in the eye. 'I don't criticize you,' he said. 'You want to discover the truth, and shout it out loud. But think about it logically. The only way to end this is to bring down the regime. And the only way that's going to happen is if the Nazis are defeated by the Red Army.'

'Maybe.'

Werner was weakening, Volodya perceived with a surge of hope. 'Maybe?' he said. 'Who else is there? The British are on their knees, desperately trying to fight off the Luftwaffe. The Americans are not interested in European squabbles. Everyone else supports the Fascists.' He put his hands on Werner's shoulders. 'The Red Army is your only hope, my friend. If we lose, those Nazis will be murdering handicapped children – and Jews, and Communists, and homosexuals – for a thousand more blood-soaked years.'

'Hell,' said Werner. 'You're right.'

(vii)

Carla and her mother went to church on Sunday. Maud was distraught about Walter's arrest and desperate to find out where he had been taken. Of course the Gestapo refused to give out any information. But Pastor Ochs's church was a fashionable one, people came in from the wealthier suburbs to attend, and the congregation included some powerful men, one or two of whom might be able to make inquiries.

Carla bowed her head and prayed that her father might not be beaten or tortured. She did not really believe in prayer but she was desperate enough to try anything.

She was glad to see the Franck family, sitting a few rows in front. She studied the back of Werner's head. His hair curled a little at the neck, in contrast with most of the men who were close-cropped. She had touched his neck and kissed his throat. He was adorable. He was easily the nicest boy who had ever kissed her. Every night before sleeping she relived that evening when they had driven to the Grunewald.

But she was not in love with him, she told herself.

Not yet.

When Pastor Ochs entered, she saw at once that he had been crushed. The change in him was horrifying. He walked slowly to the lectern, head bent and shoulders slumped, causing a few in the congregation to exchange concerned whispers. He recited the prayers without expression then read the sermon from a book. Carla had been a nurse for two years now and she recognized in him the symptoms of depression. She guessed that he, too, had received a visit from the Gestapo.

She noticed that Frau Ochs and the five children were not in their usual places in the front pew.

As they sang the last hymn Carla vowed that she would not give up, scared though she was. She still had allies: Frieda and Werner and Heinrich. But what could they do?

She wished she had solid proof of what the Nazis were doing. She had no doubts, herself, that they were exterminating the handicapped – this Gestapo crackdown made it obvious. But she could not convince others without concrete evidence.

How could she get it?

After the service she walked out of the church with Frieda and Werner. Drawing them away from their parents, she said: 'I think we have to get evidence of what's going on.'

Frieda immediately saw what she meant. 'We should go to Akelberg,' she said. 'Visit the hospital.'

Werner had proposed that, right at the start, but they had decided to begin their inquiries here in Berlin. Now Carla considered the idea afresh. 'We'd need permits to travel.'

'How could we manage that?'

Carla snapped her fingers. 'We both belong to the Mercury Cycling Club. They can get permits for bicycle holidays.' It was just the kind of thing the Nazis were keen on, healthy outdoor exercise for young people.

'Could we get inside the hospital?'

'We could try.'

Werner said: 'I think you should drop the whole thing.'

Carla was startled. 'What do you mean?'

'Pastor Ochs has obviously been scared half to death. This is a very dangerous business. You could be imprisoned, tortured. And it won't bring back Axel or Kurt.'

She stared at him incredulously. 'You want us to give it up?'

'You must give it up. You're talking as if Germany were a free country! You'll get yourselves killed, both of you.'

'We have to take risks!' Carla said angrily.

'Leave me out of this,' he said. 'I've had a visit from the Gestapo, too.'

Carla was immediately concerned. 'Oh, Werner – what happened?'

'Just threats, so far. If I ask any more questions I'll be sent to the front line.'

'Oh, well, thank God it's not worse.'

'It's bad enough.'

The girls were silent for a few moments, then Frieda said what Carla was thinking. 'This is more important than your job, you must see that.'

'Don't tell me what I must see,' Werner replied. He was superficially angry but, underneath that, Carla could tell he was in fact ashamed. 'It's not your career that's at stake,' he went on. 'And you haven't met the Gestapo yet.'

Carla was astonished. She thought she knew Werner. She would have been sure he would see this the way she did. 'Actually, I have met them,' she said. 'They arrested my father.'

Frieda was appalled. 'Oh, Carla!' she said, and put her arm around Carla's shoulders.

'We can't find out where he is,' Carla added.

Werner showed no sympathy. 'Then you should know better than to defy them!' he said. 'They would have arrested you, too, except that Inspector Macke thinks girls aren't dangerous.'

Carla wanted to cry. She had been on the point of falling in love with Werner, and now he turned out to be a coward.

Frieda said: 'Are you saying you won't help us?'

'Yes.'

'Because you want to keep your job?'

'It's pointless – you can't beat them!'

Carla was furious with him for his cowardice and defeatism. 'We can't just let this happen!'

'Open confrontation is insane. There are other ways to oppose them.'

Carla said: 'How, by working slowly, like those leaflets say? That won't stop them killing handicapped children!'

'Defying the government is suicidal!'

'Anything else is cowardice!'

'I refused to be judged by two girls!' With that he stalked off.

Carla fought back tears. She could not cry in front of two hundred people standing outside the church in the sunshine. 'I thought he was different,' she said.

Frieda was upset, but baffled too. 'He *is* different,' she said. 'I've known him all my life. Something else is going on, something he's not telling us about.'

Carla's mother approached. She did not notice Carla's distress, which was unusual. 'Nobody knows anything!' she said despairingly. 'I can't find out where you father might be.'

'We'll keep trying,' Carla said. 'Didn't he have friends at the American Embassy?'

'Acquaintances. I've asked them already, but they haven't come up with any information.'

'We'll ask them again tomorrow.'

'Oh, God, I suppose there are a million German wives in the same situation as me.'

Carla nodded. 'Let's go home, Mother.'

They walked back slowly, not talking, each with her own thoughts. Carla was angry with Werner, the more so because she had badly mistaken his character. How could she have fallen for someone so weak?

They reached their street. 'I shall go to the American Embassy in the morning,' Maud said as they approached the house. 'I'll wait in the lobby all day if necessary. I'll beg them to do something. If they really want to they can make a semi-official inquiry about the brother-in-law of a British government minister. Oh! Why is our front door open?'

Carla's first thought was that the Gestapo had paid them a second visit. But there was no black car parked at the kerb. And a key was sticking out of the lock.

Maud stepped into the hall and screamed.

Carla rushed in after her.

There was a man lying on the floor covered in blood.

Carla managed to stop herself screaming. 'Who is it?' she said.

Maud knelt beside the man. 'Walter,' she said. 'Oh, Walter, what have they done to you?'

Then Carla saw that it was her father. He was so badly injured he was almost unrecognizable. One eye was closed, his mouth was swollen into a single huge bruise, and his hair was covered with congealed blood. One arm was twisted oddly. The front of his jacket was stained with vomit.

Maud said: 'Walter, speak to me, speak to me!'

He opened his ruined mouth and groaned.

Carla suppressed the hysterical grief that bubbled up inside her by shifting into professional gear. She fetched a cushion and propped up his head. She got a cup of water from the kitchen and dribbled a little on his lips. He swallowed and opened his mouth for more. When he seemed to have had enough, she went into his study and got a bottle of schnapps and gave him a few drops. He swallowed them and coughed.

'I'm going for Dr Rothmann,' Carla said. 'Wash his face and give him more water. Don't try to move him.'

Maud said: 'Yes, yes – hurry!'

Carla wheeled her bike out of the house and pedalled away. Dr Rothmann was not allowed to practise any longer – Jews could not be doctors – but, unofficially, he still attended poor people.

Carla pedalled furiously. How had her father got home? She guessed they had brought him in a car, and he had managed to stagger from the kerbside into the house, then collapsed.

She reached the Rothmann house. Like her own home, it was in bad repair. Most of the windows had been broken by Jew-haters. Frau Rothmann opened the door. 'My father has been beaten,' Carla said breathlessly. 'The Gestapo.'

'My husband will come,' said Frau Rothmann. She turned and called up the stairs. 'Isaac!'

The doctor came down.

'It's Herr von Ulrich,' said Frau Rothmann.

The doctor picked up a canvas shopping bag that stood near the door. Because he was banned from practising medicine, Carla guessed he could not carry anything that looked like an instrument case.

They left the house. 'I'll cycle on ahead,' Carla said.

When she got home she found her mother sitting on the doorstep, weeping.

'The doctor's on his way!' Carla said.

'He is too late,' said Maud. 'Your father's dead.'

(viii)

Volodya was outside the Wertheim department store, just off the Alexander Platz, at half past two in the afternoon. He patrolled the area several times, looking for men who might be plain-clothes police officers. He was sure he had not been followed here, but it was not impossible that a passing Gestapo agent might recognize him and wonder what he was up to. A busy place with crowds was the best camouflage, but it was not perfect.

Was the invasion story true? If so, Volodya would not be in Berlin much longer. He would kiss goodbye to Gerda and Sabine. He would presumably return to Red Army Intelligence headquarters in Moscow. He looked forward to spending some time with his family. His sister, Anya, had twin babies whom he had never seen. And he felt he could do with a rest. Undercover work meant continual stress: losing Gestapo shadows, holding clandestine meetings, recruiting agents, and worrying about betrayal. He would welcome a year or two at headquarters, assuming the Soviet Union survived that long. Alternatively, he might be sent on another foreign posting. He fancied Washington. He had always had a yen to see America.

He took from his pocket a ball of crumpled tissue paper and dropped it into a litter bin. At one minute to three he lit a cigarette, although he did not smoke. He dropped the lighted match carefully into the bin so that it landed in the nest of tissue paper. Then he walked away.

Seconds later, someone cried: 'Fire!'

Just when everyone in the vicinity was looking at the fire in the litter bin, a taxi drew up at the entrance to the store, a regular black Mercedes 260D. A handsome young man in the uniform of an air force lieutenant jumped out. As the lieutenant was paying the driver, Volodya jumped into the cab and slammed the door.

On the floor of the cab, where the driver could not see it, was a copy of *Neues Volk*, the Nazi magazine of racial propaganda. Volodya picked it up, but did not read it.

'Some idiot has set fire to a litter bin,' said the driver.

'Adlon Hotel,' Volodya said, and the car pulled away.

He riffled the pages of the magazine and verified that a buff-coloured envelope was concealed within.

He longed to open it, but he waited.

He got out of the cab at the hotel, but did not go inside. Instead, he walked through the Brandenburg Gate and into the park. The trees were showing bright new leaves. It was a warm spring day and there were plenty of afternoon strollers.

The magazine seemed to burn the skin of Volodya's hand. He found an unobtrusive bench and sat down.

He unfolded the magazine and, behind its screen, he opened the buff-coloured envelope.

He drew out a document. It was a carbon copy, typed and a bit faint, but legible. It was headed:

DIRECTIVE NO. 21: CASE 'BARBAROSSA'

Friedrich Barbarossa was the German Emperor who had led the Third Crusade in the year 1189.

The text began: 'The German Wehrmacht must be prepared, even before the completion of the war against England, to overthrow Russia in a rapid campaign.'

Volodya found himself gasping for breath. This was dynamite. The Tokyo spy had been right, and Stalin wrong. And the Soviet Union was in mortal danger.

Heart pounding, Volodya looked at the end of the document. It was signed: 'Adolf Hitler.'

He scanned the pages, looking for a date, and found one. The invasion was scheduled for 15 May 1941.

Next to this was a pencilled note in Werner Franck's handwriting: 'The date has now been changed to 22 June.'

'Oh, my God, he's done it,' Volodya said aloud. 'He's confirmed the invasion.'

He put the document back into the envelope and the envelope into the magazine.

This changed everything.

He got up from the bench and walked back to the Soviet Embassy to give them the news.

(ix)

There was no railway station at Akelberg, so Carla and Frieda got off at the nearest stop, ten miles away, and wheeled their bicycles off the train.

They wore shorts, sweaters, and utilitarian sandals, and they had put their hair up in plaits. They looked like members of the League of German Girls, the Bund Deutscher Mädel or BDM. Such girls often took cycling holidays. Whether they did anything other than cycle, especially during the evenings in the spartan hostels at which they stayed, was the subject of much speculation. Boys said BDM stood for *Bubi Drück Mir*, Baby Do Me.

Carla and Frieda consulted their map then rode out of town in the direction of Akelberg.

Carla thought about her father every hour of every day. She knew she would never get over the horror of finding him savagely beaten and dying. She had cried for days. But alongside her grief was another emotion: rage. She was not merely going to be sad. She was going to do something about it.

Maud, distraught with grief, had at first tried to persuade Carla not to go to Akelberg. 'My husband is dead, my son is in the army, I don't want my daughter to put her life on the line too!' she had wailed.

After the funeral, when horror and hysteria gave way to a calmer, more profound mourning, Carla had asked her what Walter would have wanted. Maud had thought for a long time. It was not until the next day that she answered. 'He would have wanted you to carry on the fight.'

It was hard for Maud to say it, but they both knew it was true.

Frieda had had no such discussion with her parents. Her mother, Monika, had once loved Walter, and was devastated by his death; nonetheless, she would have been horrified if she knew what Frieda was doing. Her father, Ludi, would have locked her in the cellar. But they believed she was going bicycling. If anything, they might have suspected she was meeting some unsuitable boyfriend.

The countryside was hilly, but they were both in good shape, and an hour later they coasted down a slope into the small town of Akelberg. Carla felt apprehensive: they were entering enemy territory.

They went into a café. There was no Coca-Cola. 'This isn't Berlin!' said the woman behind the counter, with as much indignation as if they had asked to be serenaded by an orchestra. Carla wondered why someone who disliked strangers would run a café.

They got glasses of Fanta, a German product, and took the opportunity to refill their water bottles.

They did not know the precise location of the hospital. They needed to ask directions, but Carla was concerned about arousing suspicion. The local Nazis might take an interest in strangers asking questions. As they were paying, Carla said: 'We're supposed to meet the rest of our group at the crossroads by the hospital. Which way is that?'

The woman would not meet her eye. 'There's no hospital here.'

'The Akelberg Medical Institution,' Carla persisted, quoting from the letterhead.

'Must be another Akelberg.'

Carla thought she was lying. 'How strange,' she said, keeping up the pretence. 'I hope we're not in the wrong place.'

They wheeled their bikes along the high street. There was nothing else for it, Carla thought: she had to ask the way.

A harmless-looking old man was sitting on a bench outside a bar, enjoying the afternoon sunshine. 'Where's the hospital?' Carla asked him, covering her anxiety with a cheery veneer.

'Through the town and up the hill on your left,' he said. 'Don't go inside, though – not many people come out!' He cackled as if he had made a joke.

The directions were a bit vague, but might suffice, Carla thought. She decided she would not draw further attention by asking again.

A woman in a headscarf took the arm of the old man. 'Pay no attention to him – he doesn't know what he's saying,' she said, looking worried. She jerked him to his feet and hustled him along the sidewalk. 'Keep your mouth shut, you old fool,' she muttered.

It seemed these people had an inkling of what was going on in their neighbourhood. Fortunately their main reaction was to act surly and not get involved. Perhaps they would not be in a hurry to give information to the police or the Nazi party.

Carla and Frieda went farther along the street and found the youth hostel. There were thousands of such places in Germany, designed to cater for exactly such people as they were pretending to be, athletic youngsters on a vigorous open-air holiday. They checked in. The facilities were primitive, with three-tiered bunk beds, but the place was cheap.

It was late afternoon when they cycled out of town. After a mile they came to a left turn. There was no signpost, but the road led uphill, so they took it.

Carla's apprehension intensified. The nearer they got, the harder it would be to seem innocent under questioning.

A mile later they saw a large house in a park. It did not seem to be walled or fenced, and the road led up to the door. Once again there were no signs.

Unconsciously, Carla had been expecting a hilltop castle of forbidding grey stone, with barred windows and ironbound oak doors. But this was a Bavarian country house, with steep overhanging roofs, wooden balconies, and a little bell tower. Surely nothing as horrible as child murder could go on here? It also seemed small, for a hospital. Then she saw that a modern extension had been added to one side, with a tall chimney.

They dismounted and leaned their bikes against the side of the building. Carla's heart was in her mouth as they walked up the steps to the entrance. Why were there no guards? Because no one would be so foolhardy as to try to investigate the place?

There was no bell or knocker, but when Carla pushed the door it opened. She stepped inside, and Frieda followed. They found themselves in a cool hall with a stone floor and bare white walls. There were several rooms off the hall, but all the doors were closed. A middle-aged woman in spectacles was coming down a broad staircase. She wore a smart grey dress. 'Yes?' she said.

'Hello,' said Frieda casually.

'What are you doing? You can't come in here.'

Frieda and Carla had prepared a story. 'I just wanted to visit the place where my brother died,' Frieda said. 'He was fifteen—'

'This isn't a public facility!' the woman said indignantly.

'Yes, it is.' Frieda had been brought up in a wealthy family, and was not cowed by minor functionaries.

A nurse of about nineteen appeared from a side door and stared at them. The woman in the grey dress spoke to her. 'Nurse König, fetch Herr Römer immediately.'

The nurse hurried away.

The woman said: 'You should have written in advance.'

'Did you not get my letter?' said Frieda. 'I wrote to the Senior Physician.' This was not true: Frieda was improvising.

'No such letter has been received!' Clearly the woman felt that Frieda's outrageous request could not possibly have gone unnoticed.

Carla was listening. The place was strangely quiet. She had dealt with physically and mentally handicapped people, adults and children, and they were not often silent. Even through these closed doors she should have been able to hear shouts, laughter, crying, voices raised in protest, and nonsensical ravings. But there was nothing. It was more like a morgue.

Frieda tried a new tack. 'Perhaps you can tell me where my brother's grave is. I'd like to visit it.'

'There are no graves. We have an incinerator.' She immediately corrected herself. 'A cremation facility.'

Carla said: 'I noticed the chimney.'

Frieda said: 'What happened to my brother's ashes?'

'They will be sent to you in due course.'

'Don't mix them up with anyone else's, will you?'

The woman's neck reddened in a blush, and Carla guessed they did mix up the ashes, figuring that no one would know.

Nurse König reappeared, followed by a burly man in the white uniform of a male nurse. The woman said: 'Ah, Römer. Please escort these girls off the premises.'

'Just a minute,' said Frieda. 'Are you quite sure you're doing the right thing? I only wanted to see the place where my brother died.'

'Quite sure.'

'Then you won't mind letting me know your name.'

There was a second's hesitation. 'Frau Schmidt. Now please leave us.'

Römer moved towards them in a menacing way.

'We're going,' Frieda said frostily. 'We have no intention of giving Herr Römer an excuse to molest us.'

The man changed course and opened the door for them.

They went out, climbed on their bikes, and rode down the drive. Frieda said: 'Do you think she believed our story?'

'Totally,' said Carla. 'She didn't even ask our names. If she had suspected the truth she would have called the police right away.'

'But we didn't learn much. We saw the chimney. But we didn't find anything we could call proof.'

Carla felt a bit down. Getting evidence was not as easy as it sounded.

They returned to the hostel. They washed and changed and went out in search of something to eat. The only café was the one with the grumpy proprietress. They ate potato pancakes with sausage. Afterwards they went to the town's bar. They ordered beers and spoke cheerfully to the other customers, but no one wanted to talk to them. This in itself was suspicious. People everywhere were wary of strangers, for anyone might be a Nazi snitch, but even so Carla wondered how many towns there were where two young girls could spend an hour in a bar without anyone even trying to flirt with them.

They returned to the hostel for an early night. Carla could not think what else to do. Tomorrow they would return home empty-handed. It seemed incredible that she should know about these awful killings yet be unable to stop them. She felt so frustrated she wanted to scream.

It occurred to her that Frau Schmidt – if that really was her name – might have further thoughts about her visitors. At the time, she had taken Carla and Frieda for what they claimed to be, but she might develop suspicions later, and call the police just to be safe. If that happened, Carla and Frieda would not be hard to find. There were just five people at the hostel tonight and they were the only girls. She listened in fear for the fatal knock on the door.

If they were questioned, they would tell part of the truth, saying that Frieda's brother and Carla's godson had died at Akelberg, and they wanted to visit their graves, or at least see the place where they died and spend a few minutes in remembrance. The local police might buy that story. But if they checked with Berlin they would swiftly learn the connection with Walter von Ulrich and Werner Franck, two men who had been investigated by the Gestapo for asking disloyal questions about Akelberg. Then Carla and Frieda would be deep in trouble.

As they were getting ready to go to bed in the uncomfortable-looking bunks, there was a knock at the door.

Carla's heart stopped. She thought of what the Gestapo had done to her father. She knew she could not withstand torture. In two minutes she would name every Swing Kid she knew.

Frieda, who was less imaginative, said: 'Don't look so scared!' and opened the door.

It was not the Gestapo but a small, pretty, blonde girl. It took Carla a moment to recognize her as Nurse König, out of uniform.

'I have to speak to you,' she said. She was distressed, breathless and tearful.

Frieda invited her in. She sat on a bunk bed and wiped her eyes on the sleeve of her dress. Then she said: 'I can't keep it inside any longer.'

Carla glanced at Frieda. They were thinking the same thing. Carla said: 'Keep what inside, Nurse König?'

'My name is Ilse.'

'I'm Carla and this is Frieda. What's on your mind, Ilse?'

Ilse spoke in a voice so low they could hardly hear her. She said: 'We kill them.'

Carla could hardly breathe. She managed to say: 'At the hospital?'

Ilse nodded. 'The poor people who come in on the grey buses. Children, even babies, and old people, grandmothers. They're all more or less helpless. Sometimes they're horrid, dribbling and soiling themselves, but they can't help it, and some of them are really sweet and innocent. It makes no difference – we kill them all.'

'How do you do it?'

'An injection of morphium-scopolamine.'

Carla nodded. It was a common anaesthetic, fatal in overdose. 'What about the special treatments they're supposed to have?'

Ilse shook her head. 'There are no special treatments.'

Carla said: 'Ilse, let me get this clear. Do they kill every patient that comes here?'

'Every one.'

'As soon as they arrive?'

'Within a day, no more than two.'

It was what Carla had suspected but, even so, the stark reality was horrifying, and she felt nauseated.

After a minute she said: 'Are there any patients there now?'

'Not alive. We were giving injections this afternoon. That's why Frau Schmidt was so frightened when you walked in.'

'Why don't they make it harder for strangers to get into the building?'

'They think guards and barbed wire around a hospital would make it obvious that something sinister was going on. Anyway, no one ever tried to visit before you.'

'How many people died today?'

'Fifty-two.'

Carla's skin crawled. 'The hospital killed fifty-two people this afternoon, around the time we were there?'

'Yes.'

'So they're all dead, now?'

Ilse nodded.

An intention had been germinating in Carla's mind, and now she resolved to carry it out. 'I want to see,' she said.

Ilse looked frightened. 'What do you mean?'

'I want to go inside the hospital and see those corpses.'

'They're burning them already.'

'Then I want to see that. Can you sneak us in?'

'Tonight?'

'Right now.'

'Oh, God.'

Carla said: 'You don't have to do anything. You've already been brave, just by talking to us. If you don't want to do any more, it's okay. But if we're going to put a stop to this we need proof.'

'Proof.'

'Yes. Look, the government is ashamed of this project – that's why it's secret. The Nazis know that ordinary Germans won't tolerate the killing of children. But people prefer to believe it's not happening, and it's easy for them to dismiss a rumour, especially if they hear it from a young girl. So we have to prove it to them.'

'I see.' Ilse's pretty face took on a look of grim determination. 'All right, then. I'll take you.'

Carla stood up. 'How do you normally get there?'

'Bicycle. It's outside.'

'Then we'll all ride.'

They went out. Darkness had fallen. The sky was partly cloudy, and the starlight was faint. They used their cycle lights as they rode out of town and up the hill. When they came in sight of the hospital they switched off their lights and continued on foot, pushing their bikes. Ilse took them by a forest path that led to the rear of the building.

Carla smelled an unpleasant odour, somewhat like a car's exhaust. She sniffed.

Ilse whispered: 'The incinerator.'

'Oh, no!'

They hid the bikes in a shrubbery and walked silently to the back door. It was unlocked. They went in.

The corridors were bright. There were no shadowy corners: the place was lit like the hospital it pretended to be. If they met someone they would be seen clearly. Their clothes would give them away immediately as intruders. What would they do then? Run, probably.

Ilse walked quickly along a corridor, turned a corner, and opened a door. 'In here,' she whispered.

They walked in.

Frieda let out a squeal of horror and covered her mouth.

Carla whispered: 'Oh, my soul.'

In a large, cold room were about thirty dead people, all lying face up on tables, naked. Some were fat, some thin; some old and withered, some children, and one baby of about a year. A few were bent and twisted, but most appeared physically normal.

Each one had a small sticking-plaster on the upper left arm, where the needle had gone in.

Carla heard Frieda crying softly.

She steeled her nerves. 'Where are the others?' she whispered.

'Already gone to the furnace,' Ilse replied.

They heard voices coming from behind the double door at the far end of the room.

'Back outside,' Ilse said.

They stepped into the corridor. Carla closed the door all but a crack, and peeped through. She saw Herr Römer and another man push a hospital trolley through the doors.

The men did not look in Carla's direction. They were arguing about soccer. She heard Römer say: 'It's only nine years ago that we won the national championship. We beat Eintracht Frankfurt two-nil.'

'Yes, but half your best players were Jews, and they've all gone.'

Carla realized they were talking about the Bayern Munich team.

Römer said: 'The old days will come back, if only we play the right tactics.'

Still arguing, the two men went to a table where a fat woman lay dead. They took her by the shoulders and knees, then unceremoniously swung her on to the trolley, grunting with the effort.

They moved the trolley to another table and put a second corpse on top of the first.

When they had three they wheeled the trolley out.

Carla said: 'I'm going to follow them.'

She crossed the morgue to the double doors, and Frieda and Ilse followed her. They passed into an area that felt more industrial than medical: the walls were painted brown, the floor was concrete, and there were store cupboards and tool racks.

They looked around a corner.

They saw a large room like a garage, with harsh lighting and deep shadows. The atmosphere was warm, and there was a faint smell of cooking. In the middle of the space was a steel box large enough to hold a motor car. A metal canopy led from the top of the box through the roof. Carla realized she was looking at a furnace.

The two men lifted a body off the trolley and shifted it to a steel conveyor belt. Römer pushed a button on the wall. The belt moved, a door opened, and the corpse passed into the furnace.

They put the next corpse on the belt.

Carla had seen enough.

She turned and motioned the others back. Frieda bumped into Ilse, who let out an involuntary cry. They all froze.

They heard Römer say: 'What was that?'

'A ghost,' the other replied.

Römer's voice was shaky. 'Don't joke about such things!'

'Are you going to pick up the other end of this stiff, or what?'

'All right, all right.'

The three girls hurried back to the morgue. Seeing the remaining bodies, Carla suffered a wave of grief about Ada's Kurt. He had lain here, with a sticking-plaster on his arm, and had been thrown on to the conveyor belt and disposed of like a bag of garbage. But you're not forgotten, Kurt, she thought.

They went out into the corridor. As they turned towards the back door, they heard footsteps and the voice of Frau Schmidt. 'What is taking those two men so long?'

They hurried along the corridor and through the door. The moon was out, and the park was brightly lit. Carla could see the shrubbery where they had hidden the bikes, two hundred yards away across the grass.

Frieda came out last, and in her rush she let the door bang.

Carla thought fast. Frau Schmidt was likely to investigate the noise. The three girls might not reach the shrubbery before she opened the door. They had to hide. 'This way!' Carla hissed, and she ran around the corner of the building. The others followed.

They flattened themselves against the wall. Carla heard the door open. She held her breath.

There was a long pause. Then Frau Schmidt muttered something unintelligible, and the door banged again.

Carla peeped around the corner. Frau Schmidt had gone.

The three girls ran across the lawn and retrieved their bicycles.

They pushed the bikes along the forest path and emerged on to the road. They switched on their lights, mounted up, and pedalled away. Carla felt euphoric. They had got away with it!

As they approached the town, triumph gave way to more practical considerations. What had they achieved, exactly? What would they do next?

They must tell someone what they had seen. She was not sure who. In any event, they had to convince someone. Would they be believed? The more she thought about it, the less sure she was.

When they reached the hostel and dismounted, Ilse said: 'Thank goodness that's over. I've never been so scared in all my life.'

'It's not over,' said Carla.

'What do you mean?'

'It won't be over until we've closed that hospital, and any others like it.'

'How can you do that?'

'We need you,' Carla said to her. 'You're the proof.'

'I was afraid you were going to say that.'

'Will you come with us, tomorrow, when we go back to Berlin?'

There was a long pause, then Ilse said: 'Yes, I will.'

(x)

Volodya Peshkov was glad to be home. Moscow was at its summery best, sunny and warm. On Monday 30 June he returned to Red Army Intelligence headquarters beside the Khodynka airfield.

Both Werner Franck and the Tokyo spy had been right: Germany had invaded the Soviet Union on 22 June. Volodya and all the personnel at the Soviet Embassy in Berlin had returned to Moscow, by ship and train. Volodya had been prioritized, and made it back faster than most: some were still travelling.

Volodya now realized how much Berlin had been getting him down. The Nazis were tedious in their self-righteousness and triumphalism. They were like a winning soccer team at the after-match party, getting drunker and more boring and refusing to go home. He was sick of them.

Some people might say that the USSR was similar, with its secret police, its rigid orthodoxy, and its puritan attitudes to such pleasures as abstract painting and fashion. They were wrong. Communism was a work in progress, with mistakes being made on the road to a fair society.

The NKVD with its torture chambers was an aberration, a cancer in the body of Communism. One day it would be surgically removed. But probably not in wartime.

Anticipating the outbreak of war, Volodya had long ago equipped his Berlin spies with clandestine radios and code books. Now it was more vital than ever that the handful of brave anti-Nazis should continue to pass information to the Soviets. Before leaving he had destroyed all records of their names and addresses, which now existed only in his head.

He had found both his parents fit and well, although his father looked harassed: it was his responsibility to prepare Moscow for air raids. Volodya had gone to see his sister, Anya, her husband, Ilya Dvorkin, and the twins, now eighteen months old: Dmitriy, called Dimka, and Tatiana, called Tania. Unfortunately their father struck Volodya as being just as rat-like and contemptible as ever.

After a pleasant day at home, and a good night's sleep in his old room, he was ready to start work again.

He passed through the metal detector at the entrance to the Intelligence building. The familiar corridors and staircases touched a nostalgic chord, even if they were drab and utilitarian. Walking through the building he half expected people to come up and congratulate him: many of them must know he had been the one to confirm Barbarossa. But no one did: perhaps they were being discreet.

He entered a large open area of typists and file clerks and spoke to the middle-aged woman receptionist. 'Hello, Nika – are you still here?'

'Good morning, Captain Peshkov,' she said, not as warmly as he might have hoped. 'Colonel Lemitov would like to see you right away.'

Like Volodya's father, Lemitov had not been important enough to suffer in the great purge of the late thirties, and now he had been promoted to fill the place of an unlucky former superior. Volodya did not know much about the purge, but he found it hard to believe that so many senior men had been disloyal enough to merit such punishment. Not that Volodya knew exactly what the punishment was. They could be in exile in Siberia, or in prison somewhere, or dead. All he knew was that they had vanished.

Nika added: 'He has the big office at the end of the main corridor now.'

Volodya walked through the open room, nodding and smiling at one or two acquaintances, but again he got feeling that he was not the hero he had expected to be. He tapped on Lemitov's door, hoping the boss might shed some light.

'Come in.'

Volodya entered, saluted, and closed the door behind him.

'Welcome back, Captain.' Lemitov came around his desk. 'Between you and me, you did a great job in Berlin. Thank you.'

'I'm honoured, sir,' said Volodya. 'But why is this between you and me?'

'Because you contradicted Stalin.' He held up a hand to forestall protest. 'Stalin doesn't know it was you, of course. But all the same, people around here are nervous, after the purge, of associating with anyone who takes the wrong line.'

'What should I have done?' Volodya said incredulously. 'Faked wrong intelligence?'

Lemitov shook his head emphatically. 'You did exactly the right thing, don't get me wrong. And I've protected you. But just don't expect people around here to treat you like a champion.'

'Okay,' said Volodya. Things were worse than he had imagined.

'You have your own office, now, at least – three doors down. You'll need to spend a day or so catching up.'

Volodya took that for dismissal. 'Yes, sir,' he said. He saluted and left.

His office was not luxurious – a small room with no carpet – but he had it to himself. He was out of touch with the progress of the German invasion, having been busy trying to get home as fast as possible. Now he put his disappointment aside and began to read the reports of the battlefield commanders for the first week of the war.

As he did so, he became more and more desolate.

The invasion had taken the Red Army by surprise.

It seemed impossible, but the evidence covered his desk.

On 22 June, when the Germans attacked, many forward units of the Red Army had had *no live ammunition.*

That was not all. Planes had been lined up neatly on airstrips with no camouflage, and the Luftwaffe had destroyed 1,200 Soviet aircraft in the first few hours of the war. Army units had been thrown at the advancing Germans without adequate weapons, with no air cover, and lacking intelligence about enemy positions; and in consequence had been annihilated.

Worst of all, Stalin's standing order to the Red Army was that retreat was forbidden. Every unit had to fight to the last man, and officers were expected to shoot themselves to avoid capture. Troops were never allowed to regroup at a new, stronger defensive position. This meant that every defeat turned into a massacre.

Consequently, the Red Army was haemorrhaging men and equipment.

The warning from the Tokyo spy, and Werner Franck's confirmation, had been ignored by Stalin. Even when the attack began, Stalin had at first insisted it was a limited act of provocation, done by German army officers without the knowledge of Hitler, who would put a stop to it as soon as he found out.

By the time it became undeniable that it was not a provocation but the largest invasion in the history of warfare, the Germans had overwhelmed the Soviets' forward positions. After a week they had pushed three hundred miles inside Soviet territory.

It was a catastrophe – but what made Volodya want to scream out loud was that it could have been avoided.

There was no doubt whose fault it was. The Soviet Union was an autocracy. Only one person made the decisions: Josef Stalin. He had been stubbornly, stupidly, disastrously wrong. And now his country was in mortal danger.

Until now Volodya had believed that Soviet Communism was the true ideology, marred only by the excesses of the secret police, the NKVD. Now he saw that the failure was at the very top. Beria and the NKVD existed only because Stalin permitted them. It was Stalin who was preventing the march to true Communism.

Late that afternoon, as Volodya was staring out of the window over the sunlit airstrip, brooding over what he had learned, he was visited by Kamen. They had been lieutenants together four years ago, fresh out of the Military Intelligence Academy, and had shared a room with two others. In those days Kamen had been the clown, making fun of everyone, daringly mocking pious Soviet orthodoxy. Now he was heavier and seemed more serious. He had grown a small black moustache like that of the Foreign Minister, Molotov, perhaps to make himself look more mature.

Kamen closed the door behind him and sat down. He took from his pocket a toy, a tin soldier with a key in its back. He wound up the key and placed the toy on Volodya's desk. The soldier swung his arms as if marching, and the clockwork mechanism made a loud ratcheting sound as it wound down.

In a lowered voice Kamen said: 'Stalin has not been seen for two days.'

Volodya realized that the clockwork soldier was there to swamp any listening device that might be hidden in his office.

He said: 'What do you mean, he hasn't been seen?'

'He has not come to the Kremlin, and he is not answering the phone.'

Volodya was baffled. The leader of a nation could not just disappear. 'What's he doing?'

'No one knows.' The soldier ran down. Kamen wound it up and set it going again. 'On Saturday night, when he heard that the Soviet Western Army Group had been encircled by the Germans, he said: "Everything's lost. I give up. Lenin founded our state and we've fucked it up." Then he went to Kuntsevo.' Stalin had a country house near the town of Kuntsevo on the outskirts of Moscow. 'Yesterday he didn't show up at the Kremlin at his usual time of midday. When they phoned Kuntsevo, no one answered. Today, the same.'

Volodya leaned forward. 'Is he suffering . . .' his voice fell to a whisper, 'a mental breakdown?'

Kamen made a helpless gesture. 'It wouldn't be surprising. He insisted, against all the evidence, that Germany would not attack us this year, and now look.'

Volodya nodded. It made sense. Stalin had allowed himself to be officially called Father, Teacher, Great Leader, Transformer of Nature, Great Helmsman, Genius of Mankind, the Greatest Genius of All Times and Peoples. But now it had been proved, even to him, that he had been wrong and everyone else right. Men committed suicide in such circumstances.

The crisis was even worse than Volodya had thought. Not only was the Soviet Union under attack and losing. It was also leaderless. This had to be its most perilous moment since the revolution.

But was it also an opportunity? Could it be a chance to get rid of Stalin?

The last time Stalin had appeared vulnerable was in 1924, when Lenin's Testament had said that Stalin was not fit to hold power. Since Stalin had survived that crisis his power had seemed unassailable, even –

Volodya could now see clearly – when his decisions had verged on madness: the purges, the blunders in Spain, the appointment of the sadist Beria as head of the secret police, the pact with Hitler. Was this emergency the occasion, at last, to break his hold?

Volodya hid his excitement from Kamen and everyone else. He hugged his thoughts to himself as he rode the bus home through the soft light of a summer evening. His journey was delayed by a slow-moving convoy of lorries towing anti-aircraft guns – presumably being deployed by his father, who was in charge of Moscow's air raid defences.

Could Stalin be deposed?

He wondered how many Kremlin insiders were asking themselves the same question.

He entered his parents' apartment building, the ten-storey Government House, across the Moskva River from the Kremlin. They were out, but his sister was there with the twins, Dimka and Tania. The boy, Dimka, had dark eyes and hair. He held a red pencil and was scribbling messily on an old newspaper. The girl had the same intense blue-eyed stare that Grigori had – and so did Volodya, people said. She immediately showed Volodya her doll.

Also there was Zoya Vorotsyntsev, the astonishingly beautiful physicist Volodya had last seen four years earlier when he was about to leave for Spain. She and Anya had discovered a shared interest in Russian folk music: they went to recitals together, and Zoya played the gudok, a three-stringed fiddle. Neither could afford a phonograph, but Grigori had one, and they were listening to a record of a balalaika orchestra. Grigori was not a great music lover but he thought the record sounded jolly.

Zoya was wearing a short-sleeved summer dress the pale colour of her blue eyes. When Volodya asked her the conventional question about how she was, she replied sharply: 'I'm very angry.'

There were lots of reasons for Russians to be angry just now. Volodya asked: 'Why's that?'

'My research into nuclear physics has been cancelled. All the scientists I work with have been reassigned. I myself am working on improvements to the design of bomb sights.'

That seemed very reasonable to Volodya. 'We are at war, after all.'

'You don't understand,' she said. 'Listen. When uranium metal undergoes a process called fission, enormous quantities of energy are released. I mean *enormous*. We know this, and Western scientists do too – we have read their papers in scientific journals.'

'Still, the question of bomb sights seems more immediate.'

Zoya said angrily: 'This process, fission, could be used to create bombs that would be a hundred times more powerful than anything anyone has now. One nuclear explosion could flatten Moscow. What if the Germans make such a bomb and we don't have it? It will be as if they had rifles and we only had swords!'

Volodya said sceptically: 'But is there any reason to believe that scientists in other countries are working on a fission bomb?'

'We're sure they are. The concept of fission leads automatically to the idea of a bomb. We thought of it – why shouldn't they? But there's another reason. They published all their early results in the journals – and then they stopped, suddenly, one year ago. There have been no new scientific papers on fission since this time last year.'

'And you believe the politicians and generals in the West realized the military potential of the research and made it secret?'

'I can't think of another reason. And yet here in the Soviet Union we have not even begun to prospect for uranium.'

'Hmm.' Volodya was pretending to be doubtful, but in truth he found it all too credible. Even Stalin's greatest admirers – a group that included Volodya's father, Grigori – did not claim he understood science. And it was all too easy for an autocrat to ignore anything that made him uncomfortable.

'I've told your father,' Zoya went on. 'He listens to me, but no one listens to him.'

'So what are you going to do?'

'What can I do? I'm going to make a damn good bomb sight for our airmen, and hope for the best.'

Volodya nodded. He liked that attitude. He liked this girl. She was smart and feisty and a joy to look at. He wondered if she would go to a movie with him.

Talk of physics reminded him of Willi Frunze, who had been his friend at the Berlin Boys' Academy. According to Werner Franck, Willi was a brilliant physicist now studying in England. He might know something about the fission bomb Zoya was so exercised about. And if he was still a Communist he might be willing to tell what he knew.

Volodya made a mental note to send a cable to the Red Army Intelligence desk in the London embassy.

His parents came in. Father was in full dress uniform, Mother in a coat and hat. They had been to one of the many interminable ceremonies the army loved: Stalin insisted such rituals continue, despite the German invasion, because they were so good for morale.

They cooed over the twins for a few minutes, but Father looked distracted. He muttered something about a phone call and went immediately to his study. Mother began to make supper.

Volodya talked to the three women in the kitchen, but he was desperate to speak to his father. He thought he could guess the subject of Father's urgent phone call: the overthrow of Stalin was being either planned or prevented right now, probably here in this building.

After a few minutes he decided to risk the old man's wrath and interrupt him. He excused himself and went to the study. But his father was just coming out. 'I have to go to Kuntsevo,' he said.

Volodya longed to know what was going on. 'Why?' he said.

Grigori ignored the question. 'I've called down for my car, but my chauffeur has gone home. You can drive me.'

Volodya was thrilled. He had never been to Stalin's dacha. Now he was going there at a moment of profound crisis.

'Come on,' his father said impatiently.

They shouted goodbyes from the hallway and went out.

Grigori's car was a black ZIS-101A, a Soviet copy of an American Packard, with three-speed automatic transmission. Its top speed was about eighty miles per hour. Volodya got behind the wheel and pulled away.

He drove through the Arbat, a neighbourhood of craftsmen and intellectuals, and out on to the westward Mozhaisk Highway. 'Have you been summoned by Comrade Stalin?' he asked his father.

'No. Stalin has been incommunicado for two days.'

'That's what I heard.'

'Did you? It's supposed to be secret.'

'You can't keep something like that secret. What's happening now?'

'A group of us are going to Kuntsevo to see him.'

Volodya asked the key question. 'For what purpose?'

'Primarily to find out whether he's alive or dead.'

Could he really be dead already, and no one know about it? Volodya wondered. It seemed unlikely. 'And if he's alive?'

'I don't know. But whatever happens, I'd rather be there to see it than find out later.'

Listening devices did not work in moving cars, Volodya knew – the microphone just picked up engine noise – so he was confident he could not be overheard. Nevertheless, he felt fearful as he said the unthinkable. 'Could Stalin be overthrown?'

His father answered irritably: 'I told you, I don't know.'

Volodya was electrified. Such a question demanded a confident negative. Anything else was a Yes. His father had admitted the possibility that Stalin could be finished.

Volodya's hopes rose volcanically. 'Think what that could be like!' he said joyously. 'No more purges! The labour camps will be closed. Young girls will no longer be pulled off the street to be raped by the secret police.' He half expected his father to interrupt, but Grigori just listened with half-closed eyes. Volodya went on: 'The stupid phrase "Trotsky-Fascist spy" will disappear from our language. Army units who find themselves outnumbered and outgunned could retreat, instead of sacrificing themselves uselessly. Decisions will be made rationally, by groups of intelligent men working out what's best for everyone. It's the Communism you dreamed of thirty years ago!'

'Young fool,' his father said contemptuously. 'The last thing we want at this point is to lose our leader. We're at war and retreating! Our sole aim must be to defend the revolution – whatever it takes. We need Stalin now more than ever.'

Volodya felt as if he had been slapped. It was many years since his father had called him a fool.

Was the old man right? Did the Soviet Union need Stalin? The leader had made so many disastrous decisions that Volodya did not see how the country could possibly be worse off with someone else in charge.

They reached their destination. Stalin's home was conventionally called a dacha, but it was not a country cottage. A long, low building with five tall windows each side of a grand entrance, it stood in a pine forest and was painted dull green, as if to hide it. Hundreds of armed troops guarded the gates and the double barbed-wire fence. Grigori pointed to an anti-aircraft battery partly concealed by camouflage netting. 'I put that there,' he said.

The guard at the gate recognized Grigori, but nevertheless asked for their identification documents. Even though Grigori was a general and Volodya a captain in Intelligence, they were both patted down for weapons.

Volodya drove up to the door. There were no other cars in front of the house. 'We'll wait for the others,' his father said.

A few moments later three more ZIS limousines drew up. Volodya recalled that ZIS stood for *Zavod Imeni Stalina*, Factory Called Stalin. Had the executioners arrived in cars named after their victim?

They all got out, eight middle-aged men in suits and hats, holding in their hands the future of their country. Among them Volodya recognized Foreign Minister Molotov and secret-police chief Beria.

'Let's go,' said Grigori.

Volodya was astonished. 'I'm coming in there with you?'

Grigori reached under his seat and handed Volodya a Tokarev TT-33 pistol. 'Put this in your pocket,' he said. 'If that prick Beria tries to arrest me, you shoot the fucker.'

Volodya took it gingerly: the TT-33 had no safety catch. He slipped the gun into his jacket pocket – it was about seven inches long – and got out of the car. There were eight rounds, he recalled, in the magazine of the gun.

They all went inside. Volodya feared he would be patted down again, and his gun discovered, but there was no second check.

The house was painted dark colours and poorly lit. An officer showed the group into what looked like a small dining room. Stalin sat there in an armchair.

The most powerful man in the Eastern Hemisphere appeared haggard and depressed. Looking up at the group entering the room he said: 'Why have you come?'

Volodya gasped. Clearly he thought they were here either to arrest him or to execute him.

There was a long pause, and Volodya realized the group had not planned what to do. How could they, not even knowing whether Stalin was alive?

But what would they do now? Shoot him? There might never be another chance.

At last Molotov stepped forward. 'We're asking you to come back to work,' he said.

Volodya had to suppress the urge to protest.

But Stalin shook his head. 'Can I live up to people's hopes? Can I lead the country to victory?'

Volodya was flabbergasted. Would he really refuse?

Stalin added: 'There may be better candidates.'

He was giving them a second chance to fire him!

Another member of the group spoke up, and Volodya recognized Marshal Voroshilov. 'There's none more worthy,' he said.

How did that help? This was hardly the time for naked sycophancy.

Then his father joined in, saying: 'That's right!'

Were they not going to let Stalin go? How could they be so stupid?

Molotov was the first to say something sensible. 'We propose to form a war cabinet called the State Defence Committee, a kind of ultra-politburo with a very small membership and sweeping powers.'

Stalin quickly interposed: 'Who will be its head?'

'You, Comrade Stalin!'

Volodya wanted to shout: 'No!'

There was another long silence.

At last Stalin spoke. 'Very well,' he said. 'Now, who else shall we have on the committee?'

Beria stepped forward and began to propose the members.

It was all over, Volodya realized, feeling dizzy with frustration and disappointment. They had lost their chance. They could have deposed a tyrant, but they had lacked the nerve. Like the children of a violent father, they feared they could not manage without him.

In fact, it was worse than that, he saw with growing despondency. Perhaps Stalin really had had a breakdown – it had certainly seemed real – but he had also made a brilliant political move. All the men who might replace him were here in this room. At the moment when his catastrophically poor judgement had been exposed for all to see, he had forced his rivals to come out and beg him to be their leader again. He had drawn a line under his appalling mistake and given himself a new start.

Stalin was not just back.

He was stronger than ever.

Who would have the courage to make a public protest about what was going on at Akelberg? Carla and Frieda had seen it with their own eyes, and they had Ilse König as a witness, but now they needed an advocate. There were no elected representatives any more: all Reichstag deputies were Nazis. There were no real journalists, either; just scribbling sycophants. The judges were all Nazi appointees subservient to the government. Carla had never before realized how much she had been protected by politicians, newspapermen and lawyers. Without them, she saw now, the government could do anything it liked, even kill people.

Who could they turn to? Frieda's admirer Heinrich von Kessel had a friend who was a Catholic priest. 'Peter was the cleverest boy in my class,' he told them. 'But he wasn't the most popular. A bit upright and stiff-necked. I think he'll listen to us, though.'

Carla thought it was worth a try. Her Protestant pastor had been sympathetic, until the Gestapo terrified him into silence. Perhaps the same would happen again. But she did not know what else to do.

Heinrich took Carla, Frieda and Ilse to Peter's church in Schöneberg early on a Sunday morning in July. Heinrich was handsome in a black suit; the girls all wore their nurses' uniforms, symbols of trustworthiness. They entered by a side door and went to a small, dusty room with a few old chairs and a large wardrobe. They found Father Peter alone, praying. He must have heard them come in, but he remained on his knees for a minute before getting up and turning to greet them.

Peter was tall and thin, with regular features and a neat haircut. He was twenty-seven, Carla calculated, if he was Heinrich's contemporary. He frowned at them, not troubling to conceal his irritation at being disturbed. 'I am preparing myself for Mass,' he said severely. 'I am pleased to see you in church, Heinrich, but you must leave me now. I will see you afterwards.'

'This is a spiritual emergency, Peter,' said Heinrich. 'Sit down, we have something important to tell you.'

'It could hardly be more important than Mass.'

'Yes, it could, Peter, believe me. In five minutes' time you will agree.'

'Very well.'

'This is my girlfriend, Frieda Franck.'

Carla was surprised. Was Frieda his girlfriend now?

Frieda said: 'I had a younger brother who was born with spina bifida. Earlier this year he was transferred to a hospital at Akelberg in Bavaria for special treatment. Shortly afterwards we got a letter saying he had died of appendicitis.'

She turned to Carla, who took up the tale. 'My maid had a son born brain-damaged. He, too, was transferred to Akelberg. The maid got an identical letter on the same day.'

Peter spread his hands in a so-what gesture. 'I have heard this kind of thing before. It's anti-government propaganda. The Church does not interfere in politics.'

What rubbish that was, Carla thought. The Church was up to its neck in politics. But she let it pass. 'My maid's son did not have an appendix,' she went on. 'He had had it removed two years earlier.'

'Please,' said Peter. 'What does this prove?'

Carla felt discouraged. Peter was obviously biased against them.

Heinrich said: 'Wait, Peter. You haven't heard it all. Ilse here worked at the hospital in Akelberg.'

Peter looked at her expectantly.

'I was raised Catholic, Father,' Ilse said.

Carla had not known that.

'I'm not a good Catholic,' Ilse went on.

'God is good, not us, my daughter,' said Peter piously.

Ilse said: 'But I knew that what I was doing was a sin. Yet I did it, because they told me to, and I was frightened.' She began to cry.

'What did you do?'

'I killed people. Oh, Father, will God forgive me?'

The priest stared at the young nurse. He could not dismiss this as propaganda: he was looking at a soul in torment. He went pale.

The others were silent. Carla held her breath.

Ilse said: 'The handicapped people are brought to the hospital in grey buses. They don't have special treatment. We give them an injection, and they die. Then we cremate them.' She looked up at Peter. 'Will I ever be forgiven for what I have done?'

He opened his mouth to speak. His words caught in his throat, and he coughed. At last he said quietly: 'How many?'

'Usually four. Buses, I mean. There are about twenty-five patients in a bus.'

'A hundred people?'

'Yes. Every week.'

Peter's proud composure had vanished. His face was pale grey, and his mouth hung open. 'A hundred handicapped people a week?'

'Yes, Father.'

'What sort of handicap?'

'All sorts, mental and physical. Some senile old people, some deformed babies, men and women, paralysed or retarded or just helpless.'

He had to keep repeating it. 'And the staff of the hospital kill them all?'

Ilse sobbed. 'I'm sorry, I'm sorry, I knew it was wrong.'

Carla watched Peter. His supercilious air had gone. It was a remarkable transformation. After years of hearing the prosperous Catholics of this sylvan suburb confess their little sins, he had suddenly been confronted with raw evil. And he was shocked to his core.

But what would he do?

Peter stood up. He took Ilse by the hands and raised her from her seat. 'Come back to the church,' he said. 'Confess to your priest. God will forgive you. This much I know.'

'Thank you,' she whispered.

He released her hands and looked at Heinrich. 'It may not be so simple for the rest of us,' he said.

Then he turned his back on them and knelt to pray again.

Carla looked at Heinrich, who shrugged. They got up and left the little room, Carla with her arm around the weeping Ilse.

Carla said: 'We'll stay for the service. Perhaps he'll speak to us again afterwards.'

The four of them walked into the nave of the church. Ilse stopped crying and became calmer. Frieda held Heinrich's arm. They took seats among the gathering congregation, prosperous men and plump women and restless children in their best clothes. People such as these would never kill the handicapped, Carla thought. Yet their government did, on their behalf. How had this happened?

She did not know what to expect of Father Peter. Clearly he had believed what they had told him, in the end. He had wanted to dismiss them as politically motivated, but Ilse's sincerity had convinced him. He had been horrified. But he had not made any promises, except that God would forgive Ilse.

Carla looked around the church. The decoration was more colourful than what she was used to in Protestant churches. There were more statues and paintings, more marble and gilding and banners and candles.

Protestants and Catholics had fought wars about such trivia, she recalled. How strange it seemed, in a world where children could be murdered, that anyone should care about candles.

The service began. The priests entered in their robes, Father Peter the tallest among them. Carla could not read anything in his facial expression except stern piety.

She sat indifferent through the hymns and prayers. She had prayed for her father, and two hours later had found him cruelly beaten and dying on the floor of their home. She missed him every day, sometimes every hour. Praying had not saved him, nor would it protect those deemed useless by the government. Action was needed, not words.

Thinking of her father brought her brother, Erik, to mind. He was somewhere in Russia. He had written a letter home, jubilantly celebrating the rapid progress of the invasion, and angrily refusing to believe that Walter had been murdered by the Gestapo. Their father had obviously been released unharmed by the Gestapo and then attacked in the street by criminals or Communists or Jews, he asserted. He was living in a fantasy, beyond the reach of reason.

Was the same true of Father Peter?

Peter mounted the pulpit. Carla had not known he was due to preach a sermon. She wondered what he would say. Would he be inspired by what he had heard this morning? Would he speak of something irrelevant, the virtue of modesty or the sin of envy? Or would he close his eyes and devoutly thank God for the German army's continuing victories in Russia?

He stood tall in the pulpit and swept the church with a gaze that might have been arrogant, or proud, or defiant.

'The fifth commandment says: "Thou shalt not kill".'

Carla met Heinrich's eye. What was Peter going to say?

His voice rang out between the echoing stones of the nave. 'There is a place in Akelberg, Bavaria, where our government is breaking the commandment a hundred times a week!'

Carla gasped. He was doing it – he was preaching a sermon against the programme! This could change everything.

'It makes no difference that the victims are handicapped, or mentally ill, or incapable of feeding themselves, or paralysed.' Peter was letting his anger show. 'Helpless babies and senile old people are all God's children, and their lives are as sacred as yours and mine.' His voice rose in volume. 'To kill them is a mortal sin!' He lifted his right arm and made a fist, and his voice shook with emotion. 'I say to you that if we do nothing about it, we sin just as much as the doctors and nurses who administer the lethal injections. If we remain silent . . .' He paused. 'If we remain silent, we are murderers too!'

(xii)

Inspector Thomas Macke was furious. He had been made to look a fool in the eyes of Superintendent Kringelein and the rest of his superiors. He had assured them he had plugged the leak. The secret of Akelberg – and hospitals of the same kind in other parts of the country – was safe, he had said. He had tracked down the three troublemakers, Werner Franck, Pastor Ochs and Walter von Ulrich, and in different ways he had silenced each of them.

And yet the secret had come out.

The man responsible was an arrogant young priest called Peter.

Father Peter was in front of Macke now, naked, strapped by wrists and ankles to a specially constructed chair. He was bleeding from the ears, nose, and mouth, and had vomit all down his chest. Electrodes were attached to his lips, his nipples and his penis. A strap around his forehead prevented him from breaking his neck while the convulsions shook him.

A doctor sitting beside the priest checked his heart with a stethoscope and looked dubious. 'He can't stand much more,' he said in a matter-of-fact tone.

Father Peter's seditious sermon had been taken up elsewhere. The Bishop of Münster, a much more important clergyman, had preached a similar sermon, denouncing the T4 programme. The bishop had called upon Hitler to save the people from the Gestapo, cleverly implying that the Führer could not possibly know about the programme, thereby offering Hitler a ready-made alibi.

His sermon had been typed out and duplicated and passed from hand to hand all over Germany.

The Gestapo had arrested every person found in possession of a copy, but to no avail. It was the only time in the history of the Third Reich that there had been a public outcry against any government action.

The clampdown was savage, but it did no good: the duplicates of the sermon continued to proliferate, more clergymen prayed for the handicapped, and there was even a protest march in Akelberg. It was out of control.

And Macke was to blame.

He bent over Peter. The priest's eyes were closed and his breathing was shallow, but he was conscious. Macke shouted in his ear: 'Who told you about Akelberg?'

There was no reply.

Peter was Macke's only lead. Investigations in the town of Akelberg had turned up nothing of significance. Reinhold Wagner had been told a story about two girl cyclists who had visited the hospital, but no one knew who they were; and another story about a nurse who had resigned suddenly, writing a letter saying she was getting married in haste, but not revealing who the husband was. Neither clue led anywhere. In any case, Macke felt sure this calamity could not be the work of a gaggle of girls.

Macke nodded to the technician operating the machine. He turned a knob.

Peter screamed in agony as the electrical current coursed through his body, torturing his nerves. He shook as if in a fit, and the hair on his head stood up.

The operator turned the current off.

Macke screamed: 'Give me his name!'

At last Peter opened his mouth.

Macke leaned closer.

Peter whispered: 'No man.'

'A woman, then! Give me the name!'

'It was an angel.'

'Damn you to hell!' Macke seized the knob and turned it. 'This goes on until you tell me!' he yelled, as Peter shuddered and screamed.

The door opened. A young detective looked in, turned pale, and beckoned to Macke.

The technician turned the current off, and the screaming stopped. The doctor leaned forward to check Peter's heart.

The detective said: 'Excuse me, Inspector Macke, but you're wanted by Superintendent Kringelein.'

'Now?' said Macke irritably.

'That's what he said, sir.'

Macke looked at the doctor, who shrugged. 'He's young,' he said. 'He'll be alive when you get back.'

Macke left the room and went upstairs with the detective. Kringelein's office was on the first floor. Macke knocked and went in. 'The damn priest hasn't talked yet,' he said without preamble. 'I need more time.'

Kringelein was a slight man with spectacles, clever but weak-willed. A late convert to Nazism, he was not a member of the elite SS. He lacked the fervour of enthusiasts such as Macke. 'Don't bother any further with that priest,' he said. 'We're no longer interested in any of the clergymen. Throw them in camps and forget them.'

Macke could not believe his ears. 'But these people have conspired to undermine the Führer!'

'And they have succeeded,' said Kringelein. 'Whereas you have failed.'

Macke suspected that Kringelein was privately pleased about this.

'A decision has been made at the top,' the superintendent went on. 'Aktion T4 has been cancelled.'

Macke was flabbergasted. The Nazis never allowed their decisions to be swayed by the misgivings of the ignorant. 'We didn't get where we are by kowtowing to public opinion!' he said.

'We have this time.'

'Why?'

'The Führer neglected to explain his decision to me personally,' Kringelein said sarcastically. 'But I can guess. The programme has attracted remarkably angry protests from a normally passive public. If we persist with it, we risk an open confrontation with churches of all denominations. That would be a bad thing. We must not weaken the unity and determination of the German people – particularly right now, when we are at war with the Soviet Union, our strongest enemy yet. So the programme is cancelled.'

'Very good, sir,' said Macke, controlling his anger. 'Will there be anything else?'

'Dismissed,' said Kringelein.

Macke went to the door.

'Macke.'

He turned. 'Yes, sir.'

'Change your shirt.'

'My shirt?'

'There's blood on it.'

'Yes, sir. Sorry, sir.'

Macke stamped down the stairs, boiling. He returned to the basement chamber. Father Peter was still alive.

Raging, he yelled again: 'Who told you about Akelberg?'

There was no reply.

He turned the current up to maximum.

Father Peter screamed for a long time; then, at last, he fell into a final silence.

(xiii)

The villa where the Franck family lived was set in a small park. Two hundred yards from the house, on a slight rise, was a little pagoda, open on all sides, with seats. As children Carla and Frieda had pretended it was their country house, and had played for hours pretending to have grand parties where dozens of servants waited on their glamorous guests. Later it became their favourite place to sit and talk where no one could hear them.

'The first time I sat on this bench, my feet didn't reach the floor,' Carla said.

Frieda said: 'I wish we could go back to those days.'

It was a sultry afternoon, overcast and humid, and they both wore sleeveless dresses. They were in sombre mood. Father Peter was dead: he had committed suicide in custody, having become depressed about his crimes, according to the police. Carla wondered if he had been beaten as her father had. It seemed dreadfully likely.

There were dozens more in police cells all over Germany. Some had protested publicly about the killing of the handicapped, others had done no more than pass round copies of Bishop von Galen's sermon. She wondered if all of them would be tortured. She wondered how long she would escape such a fate.

Werner came out of the house with a tray. He carried it across the lawn to the pagoda. Cheerily he said: 'How about some lemonade, girls?'

Carla looked away. 'No, thank you,' she said coldly. She did not understand how he could pretend to be her friend after the cowardice he had shown.

Frieda said: 'Not for me.'

'I hope we're not bad friends,' Werner said, looking at Carla.

How could he say such a thing? Of course they were bad friends.

Frieda said: 'Father Peter is dead, Werner.'

Carla added: 'Probably tortured to death by the Gestapo, because he refused to accept the murder of people such as your brother. My father is dead, too, for the same reason. Lots of other people are in jail or in camps. But you kept your cushy desk job, so that's all right.'

Werner looked hurt. That surprised Carla. She had expected defiance, or at least an effort at insouciance. But he seemed genuinely upset. He said: 'Don't you think we each have our different ways of doing what we can?'

This was feeble. 'You did nothing!' Carla said.

'Perhaps,' he said sadly. 'No lemonade, then?'

Neither girl answered, and he went back to the house.

Carla was indignant and angry, but she could not help also feeling regret. Before she discovered that Werner was a coward she had been embarking on a romance with him. She had liked him a lot, ten times more than any other boy she had kissed. She was not quite heartbroken, but she was deeply disappointed.

Frieda was luckier. This thought was prompted by the sight of Heinrich coming out of the house. Frieda was glamorous and fun-loving, and Heinrich was brooding and intense, but somehow they made a pair. 'Are you in love with him?' Carla said while he was still out of earshot.

'I don't know yet,' Frieda replied. 'He's terribly sweet, though. I kind of adore him.'

That might not be love, Carla thought, but it was well on the way.

Heinrich was bursting with news. 'I had to come and tell you right away,' he said. 'My father told me after lunch.'

'What?' said Frieda.

'The government has cancelled the project. It was called Aktion T4. The killing of the handicapped. They're stopping.'

Carla said: 'You mean we won?'

Heinrich nodded vigorously. 'My father is amazed. He says he has never known the Führer give in to public opinion before.'

Frieda said: 'And we forced him to!'

'Thank God no one knows that,' Heinrich said fervently.

Carla said: 'They're just going to close the hospitals and end the whole programme?'

'Not exactly.'
'What do you mean?'
'My father says all those doctors and nurses are being transferred.'
Carla frowned. 'Where?'
'To Russia,' said Heinrich.

9

1941 (II)

The phone rang on Greg Peshkov's desk on a hot morning in July. He had finished his penultimate year at Harvard, and was once again interning at the State Department for the summer, working in the information office. He was good at physics and math, and passed exams effortlessly, but he had no interest in becoming a scientist. Politics was what excited him. He picked up the phone. 'Greg Peshkov.'

'Morning, Mr Peshkov. This is Tom Cranmer.'

Greg's heart beat a little faster. 'Thank you for returning my call. You obviously remember me.'

'The Ritz-Carlton Hotel, 1935. Only time I ever got my picture in the paper.'

'Are you still the hotel detective?'

'I moved to retail. I'm a store detective now.'

'Do you ever do any freelance work?'

'Sure. What did you have in mind?'

'I'm in my office now. I'd like to talk privately.'

'You work in the Old Executive Office Building, across the street from the White House.'

'How did you know that?'

'I'm a detective.'

'Of course.'

'I'm around the corner, at Aroma Coffee on F Street and Nineteenth.'

'I can't come now.' Greg looked at his watch. 'In fact, I have to hang up right away.'

'I'll wait.'

'Give me an hour.'

Greg hurried down the stairs. He arrived at the main entrance just as a Rolls-Royce motor car came silently to a stop outside. An overweight chauffeur clambered out and opened the rear door. The passenger who emerged was tall, lean and handsome, with a full head of silver hair. He wore a perfectly cut double-breasted suit of pearl-grey flannel that draped him in a style only London tailors could achieve. As he ascended the granite steps to the huge building, his fat chauffeur hurried after him, carrying his briefcase.

He was Sumner Welles, Undersecretary of State, number two at the State Department, and personal friend of President Roosevelt.

The chauffeur was about to hand the briefcase to a waiting State Department usher when Greg stepped forward. 'Good morning, sir,' he said, and he smoothly took the briefcase from the chauffeur and held the door open. Then he followed Welles into the building.

Greg had got into the information office because he was able to show factual, well-written articles he had produced for the *Harvard Crimson*. However, he did not want to end up a press attaché. He had higher ambitions.

Greg admired Sumner Welles, who reminded him of his father. The good looks, the fine clothes and the charm concealed a ruthless operator. Welles was determined to take over from his boss, Secretary of State Cordell Hull, and never hesitated to go behind his back and speak directly to the President – which infuriated Hull. Greg found it exciting to be close to someone who had power and was not afraid to use it. That was what he wanted for himself.

Welles had taken a shine to him. People often did take a shine to Greg, especially when he wanted them to; but in the case of Welles there was another factor. Though Welles was married – apparently happily, to an heiress – he had a fondness for attractive young men.

Greg was heterosexual to a fault. He had a steady girl at Harvard, a Radcliffe student named Emily Hardcastle, who had promised to acquire a birth-control device before September; and here in Washington he was dating Rita, the voluptuous daughter of Congressman Lawrence of Texas. He walked a tightrope with Welles. He avoided all physical contact while being amiable enough to remain in favour. Also, he stayed away from Welles any time after the cocktail hour, when the older man's inhibitions weakened and his hands began to stray.

Now, as the senior staff gathered in the office for the ten o'clock meeting, Welles said: 'You can stay for this, my boy. It will be good for your education.' Greg was thrilled. He wondered if the meeting would give him a chance to shine. He wanted people to notice him and be impressed.

A few minutes later, Senator Dewar arrived with his son Woody. Father and son were lanky and large-headed, and wore similar dark-blue single-breasted linen summer suits. However, Woody differed from his father in being artistic: his photographs for the *Harvard Crimson* had won prizes. Woody nodded to Welles's senior assistant, Bexforth Ross: they must have met before. Bexforth was an excessively self-satisfied guy who called Greg 'Russkie' because of his Russian name.

Welles opened the meeting by saying: 'I now have to tell you all something highly confidential that must not be repeated outside this room. The President is going to meet with the British Prime Minister early next month.'

Greg just stopped himself from saying *Wow*.

'Good!' said Gus Dewar. 'Where?'

'The plan is to rendezvous by ship somewhere in the Atlantic, for security and to reduce Churchill's travel time. The President wants me to attend, while Secretary of State Hull stays here in Washington to mind the store. He also wants you there, Gus.'

'I'm honoured,' said Gus. 'What's the agenda?'

'The British seem to have beaten off the threat of invasion, for now, but they're too weak to attack the Germans on the European continent – unless we help. Therefore Churchill will ask us to declare war on Germany. We will refuse, of course. Once we've got past that, the President wants a joint statement of aims.'

'Not war aims,' Gus said.

'No, because the United States is not at war and has no intention of going to war. But we are non-belligerently allied with the British, we're supplying them with just about everything they need on unlimited credit, and when peace comes at last we expect to have a say in how the post-war world is run.'

'Will that include a strengthened League of Nations?' Gus asked. He was keen on this idea, Greg knew; and so was Welles.

'That's why I wanted to talk to you, Gus. If we want our plan implemented, we need to be prepared. We have to get FDR and Churchill to commit to it as part of their statement.'

Gus said: 'We both know that the President is in favour, theoretically, but he's nervous about public opinion.'

An aide came in and passed a note to Bexforth, who read it and said: 'Oh! My goodness.'

Welles said testily: 'What is it?'

'The Japanese Imperial Council met last week, as you know,' Bexforth said. 'We have some intelligence on their deliberations.'

He was being vague about the source of information, but Greg knew what he meant. The Signal Intelligence Unit of the US Army was able to intercept and decode wireless messages from the Foreign Ministry in Tokyo to its embassies abroad. The data from these decrypts was codenamed MAGIC. Greg knew about this, even though he was not supposed to – in fact, there would have been a hell of a stink if the army found out he was in on the secret.

'The Japanese discussed extending their empire,' Bexforth went on. They had already annexed the vast region of Manchuria, Greg knew, and had moved troops into much of the rest of China. 'They do not favour the option of westward expansion, into Siberia, which would mean war with the Soviet Union.'

'That's good!' said Welles. 'It means the Russians can concentrate on fighting the Germans.'

'Yes, sir. But the Japs are planning instead to extend southwards, by taking full control of Indochina, then the Dutch East Indies.'

Greg was shocked. This was hot news – and he was among the first to hear it.

Welles was indignant. 'Why, that's nothing less than an imperialist war!'

Gus interposed: 'Technically, Sumner, it's not war. The Japanese already have some troops in Indochina, with formal permission from the incumbent colonial power, France, as represented by the Vichy government.'

'Puppets of the Nazis!'

'I did say "technically". And the Dutch East Indies are theoretically ruled by the Netherlands, which is now occupied by the Germans, who are perfectly happy for their Japanese allies to take over a Dutch colony.'

'That's a quibble.'

'It's a quibble that others will raise with us – the Japanese ambassador, for one.'

'You're right, Gus, and thanks for forewarning me.'

Greg was alert for an opportunity to make a contribution to the discussion. He wanted above all else to impress the senior men around him. But they all knew so much more than he did.

Welles said: 'What are the Japanese after, anyway?'

Gus said: 'Oil, rubber and tin. They're securing their access to natural resources. It's hardly surprising, since we keep interfering with their supplies.' The United States had embargoed exports of materials such as oil and scrap iron to Japan, in a failed attempt to discourage the Japanese from taking over ever larger tracts of Asia.

Welles said irritably: 'Our embargoes have never been applied very effectively.'

'No, but the threat is obviously sufficient to panic the Japanese, who have almost no natural resources of their own.'

'Clearly we need to take more effective measures,' Welles snapped. 'The Japanese have a lot of money in American banks. Can we freeze their assets?'

The officials around the room looked disapproving. This was a radical idea. After a moment Bexforth said: 'I guess we could. That would be more effective than any embargoes. They would be unable to buy oil or any other raw materials here in the States because they couldn't pay for them.'

Gus Dewar said: 'The Secretary of State will be concerned, as usual, to avoid any action that might lead to war.'

He was right. Cordell Hull was cautious to the point of timidity, and frequently clashed with his more aggressive deputy, Welles.

'Mr Hull has always followed that course, and very wisely,' said Welles. They all knew he was insincere, but etiquette required it. 'However, the United States must walk tall on the international stage. We're prudent, not cowardly. I'm going to put this idea of an asset freeze to the President.'

Greg was awestruck. This was what power meant. In a heartbeat, Welles could propose something that would rock an entire nation.

Gus Dewar frowned. 'Without imported oil, the Japanese economy will grind to a halt, and their military will be powerless.'

'Which is good!' said Welles.

'Is it? What do you imagine Japan's military government will do, faced with such a catastrophe?'

Welles did not much like to be challenged. He said: 'Why don't you tell me, Senator?'

'I don't know. But I think we should have an answer before we take the action. Desperate men are dangerous. And I do know that the United States is not ready to go to war against Japan. Our navy isn't ready and our air force isn't ready.'

Greg saw his chance to speak and took it. 'Mr Undersecretary, sir, it may help you to know that public opinion favours war with Japan, rather than appeasement, by a factor of two to one.'

'Good point, Greg, thank you. Americans don't want to let Japan get away with murder.'

'They don't really want war, either,' said Gus. 'No matter what the poll says.'

Welles closed the folder on his desk. 'Well, Senator, we agree about the League of Nations and disagree about Japan.'

Gus stood up. 'And in both cases the decision will be made by the President.'

'Good of you to come in to see me.'

The meeting broke up.

Greg left on a high. He had been invited into the briefing, he had learned startling news, and he had made a comment that Welles had thanked him for. It was a great start to the day.

He slipped out of the building and headed for Aroma Coffee.

He had never hired a private detective before. It felt vaguely illegal. But Cranmer was a respectable citizen. And there was nothing criminal about trying to get in contact with an old girlfriend.

At Aroma Coffee there were two girls who looked like secretaries taking a break, an older couple out shopping, and Cranmer, a broad man in a rumpled seersucker suit, dragging on a cigarette. Greg slid into the booth and asked the waitress for coffee.

'I'm trying to reconnect with Jacky Jakes,' he said to Cranmer.

'The black girl?'

She had been a girl, back then, Greg thought nostalgically; sweet sixteen, though she was pretending to be older. 'It's six years ago,' he said to Cranmer. 'She's not a girl any more.'

'It was your father who hired her for that little drama, not me.'

'I don't want to ask him. But you can find her, right?'

'I expect so.' Cranmer took out a little notebook and a pencil. 'I guess Jacky Jakes was an assumed name?'

'Mabel Jakes is her real name.'

'Actress, right?'

'Would-be. I don't know that she made it.' She had had good looks and charm in abundance, but there were not many parts for black actors.

'Obviously she's not in the phone book, or you wouldn't need me.'

'Could be unlisted, but more likely she can't afford a phone.'

'Have you seen her since 1935?'

'Twice. First time two years ago, not far from here, on E Street. Second time, two weeks ago, two blocks away.'

'Well, she sure as hell doesn't live in this swanky neighbourhood, so she must work nearby. You have a photo?'

'No.'

'I remember her vaguely. Pretty girl, dark skin, big smile.'

Greg nodded, remembering that thousand-watt smile. 'I just want her address, so I can write her a letter.'

'I don't need to know what you want the information for.'

'Suits me.' Was it really this easy, Greg thought?

'I charge ten bucks a day, with a two-day minimum, plus expenses.'

It was less than Greg had expected. He took out his billfold and gave Cranmer a twenty.

'Thanks,' said the detective.

'Good luck,' said Greg.

(ii)

Saturday was hot, so Woody went to the beach with his brother, Chuck.

The whole Dewar family was in Washington. They had a nine-room apartment near the Ritz-Carlton Hotel. Chuck was on leave from the navy, Papa was working twelve hours a day planning the summit meeting he referred to as the Atlantic Conference, and Mama was writing a new book, about the wives of presidents.

Woody and Chuck put on shorts and polo shirts, grabbed towels and sunglasses and newspapers, and caught a train to Rehoboth Beach, on the Delaware coast. The journey took a couple of hours, but this was the only place to go on a summer Saturday. There was a wide stretch of sand and a refreshing breeze off the Atlantic Ocean. And there were a thousand girls in swimsuits.

The two brothers were different. Chuck was shorter, with a compact, athletic figure. He had their mother's attractive looks and winning smile. He had been a poor student at school, but he also displayed Mama's quirky intelligence, always taking an off-centre view of life. He was better than Woody at all sports except running, where Woody's long legs gave him speed, and boxing, in which Woody's long arms made him nearly impossible to hit.

At home, Chuck had not said much about the navy, no doubt because their parents were still angry with him for not going to Harvard. But alone with Woody he opened up a bit. 'Hawaii is great, but I'm really disappointed to have a shore job,' he said. 'I joined the navy to go to sea.'

'What are you doing, exactly?'

'I'm part of the Signal Intelligence Unit. We listen to radio messages, mainly from the Imperial Japanese Navy.'

'Aren't they in code?'

'Yes, but you can learn a lot even without breaking the codes. It's called traffic analysis. A sudden increase in the number of messages indicates that some action is imminent. And you learn to recognize patterns in the traffic. An amphibious landing has a distinctive configuration of signals, for example.'

'That's fascinating. And I bet you're good at it.'

Chuck shrugged. 'I'm just a clerk, annotating and filing the transcripts. But you can't help picking up the basics.'

'How's the social life in Hawaii?'

'Lots of fun. Navy bars can get pretty riotous. The Black Cat Café is the best. I have a good pal, Eddie Parry, and we go surfboarding on Waikiki Beach every chance we get. I've had some good times. But I wish I was on a ship.'

They swam in the cold Atlantic, ate hot dogs for lunch, took photos of each other with Woody's camera, and studied the swimsuits until the sun began to go down. As they were leaving, picking their way through the crowd, Woody saw Joanne Rouzrokh.

He did not need to look twice. She was like no other girl on the beach, nor indeed in Delaware. There was no mistaking those high cheekbones, that scimitar nose, the luxuriant dark hair, the skin the colour and smoothness of *café au lait*.

Without hesitation he walked straight towards her.

She looked absolutely sensational. Her black one-piece swimsuit had spaghetti straps that revealed the elegant bones of her shoulders. It was cut straight across her upper thighs, showing almost all of her long, brown legs.

He could hardly believe that he had once taken this fabulous woman in his arms and smooched her like there was no tomorrow.

She looked up at him, shading her eyes from the sun. 'Woody Dewar! I didn't know you were in Washington.'

That was all the invitation he needed. He knelt on the sand beside her. Just being this close made him breathe harder. 'Hello, Joanne.' He glanced briefly at the plump brown-eyed girl beside her. 'Where's your husband?'

She burst out laughing. 'Whatever made you think I was married?'

He was flustered. 'I came to your apartment for a party, a couple of summers back.'

'You did?'

Joanne's companion said: 'I remember. I asked you your name, but you didn't answer.'

Woody had no memory of her at all. 'I'm sorry I was so impolite,' he said. 'I'm Woody Dewar, and this is my brother Chuck.'

The brown-eyed girl shook hands with both of them and said: 'I'm Diana Taverner.' Chuck sat beside her on the sand, which seemed to please her: Chuck was good-looking, much more handsome than Woody.

Woody went on: 'Anyway, I went into the kitchen, looking for you, and a man called Bexforth Ross introduced himself to me as your fiancé. I assumed you'd be married by now. Is it an extraordinarily long engagement?'

'Don't be silly,' she said with a touch of irritation, and he remembered that she did not respond well to teasing. 'Bexforth told people we were engaged, because he was practically living at our apartment.'

Woody was startled. Did that mean that Bexforth had been sleeping there? With Joanne? It was not uncommon, of course, but few girls admitted it.

'He was the one who talked about marriage,' she went on. 'I never agreed to it.'

So she was single. Woody could not have been happier if he had won the lottery.

There might be a boyfriend, he warned himself. He would have to find out. But anyway, a boyfriend was not the same as a husband.

'I was at a meeting with Bexforth a few days back,' Woody said. 'He's a great man in the State Department.'

'He'll go far, and he'll find a woman more suitable than I to be the wife of a great man in the State Department.'

It seemed from her tone that she did not have warm feelings towards her former lover. Woody found that he was pleased about that, although he could not have said why.

He reclined on his elbow. The sand was hot. If she had a serious boyfriend, she would find a reason to mention him before too long, he felt sure. He said: 'Speaking of the State Department, are you still working there?'

'Yes. I'm assistant to the Undersecretary for Europe.'

'Exciting.'

'Right now it is.'

Woody was looking at the line where her swimsuit crossed her thighs, and thinking that no matter how little a girl was wearing, a man was always thinking about the parts of her that were hidden. He began to get an erection, and rolled on to his front to conceal it.

Joanne saw the direction of his gaze and said: 'You like my swimsuit?' She was always frank. It was one of the many things he found attractive about her.

He decided to be equally candid. 'I like *you*, Joanne. I always did.'

She laughed. 'Don't beat about the bush, Woody – come right out with it!'

All around them, people were packing up. Diane said: 'We'd better get going.'

'We were just leaving,' Woody said. 'Shall we travel together?'

This was the moment for her to give him the polite brush-off. She could easily say *Oh, no, thanks, you guys go on ahead.* But instead she said: 'Sure, why not?'

The girls pulled dresses over their swimsuits and threw their stuff into a couple of bags, and they all walked up the beach.

The train was crowded with trippers like them, sunburned and hungry and thirsty. Woody bought four Cokes at the station and produced them as the train pulled out. Joanne said: 'You once bought me a Coke on a hot day in Buffalo, do you remember?'

'On that demonstration. Of course I remember.'

'We were just kids.'

'Buying Cokes is a technique I use with beautiful women.'

She laughed. 'Is it successful?'

'It has never got me a single smooch.'

She raised her bottle in a toast. 'Well, keep trying.'

He thought that was encouraging, so he said: 'When we get back to the city, do you want to get a hamburger, or something, and maybe see a movie?'

This was the moment for her to say *No, thanks, I'm meeting my boyfriend.*

Diana said quickly: 'I'd like that. How about you, Joanne?'

Joanne said: 'Sure.'

No boyfriend – and a date! Woody tried to hide his elation. 'We could see *The Bride Came C.O.D.*,' he said. 'I hear it's pretty funny.'

Joanne said: 'Who's in it?'

'James Cagney and Bette Davis.'

'I'd like to see that.'

Diana said: 'Me, too.'

'That's settled, then,' said Woody.

Chuck said: 'How about you, Chuck? Would you like that? Oh, sure, I'd like it swell, but nice of you to ask, big brother.'

It was not all that funny, but Diana giggled appreciatively.

Soon afterwards, Joanne fell asleep with her head on Woody's shoulder.

Her dark hair tickled his neck, and he could feel her warm breath on his skin below the cuff of his short-sleeved shirt. He felt blissfully contented.

They parted company at Union Station, went home to change, and met up again at a Chinese restaurant downtown.

Over chow mein and beer they talked about Japan. Everyone was talking about Japan. 'Those people have to be stopped,' said Chuck. 'They're Fascists.'

'Maybe,' said Woody.

'They're militaristic and aggressive, and the way they treat the Chinese is racialist. What else do they have to do to be Fascists?'

'I can answer that,' said Joanne. 'The difference is in their vision of the future. Real Fascists want to kill off all their enemies then create a radically new type of society. The Japanese are doing all the same things in defence of traditional power groups, the military caste and the emperor. For the same reason, Spain is not really Fascist: Franco is murdering people for the sake of the Catholic Church and the old aristocracy, not to create a new world.'

'Either way, the Japs must be stopped,' said Diana.

'I see it differently,' said Woody.

Joanne said: 'Okay, Woody, how do you see it?'

She was seriously political, and would appreciate a thoughtful

445

answer, he knew. 'Japan is a trading nation, with no natural resources: no oil, no iron, just some forests. The only way they can make a living is by doing business. For example, they import raw cotton, weave it, and sell it to India and the Philippines. But in the Depression the two great economic empires – Britain and the USA – put up tariff walls to protect our own industries. That was the end of Japanese trade with the British Empire, including India, and the American zone, including the Philippines. It hit them pretty hard.'

Diana said: 'Does that give them the right to conquer the world?'

'No, but it makes them think that the only way to economic security is to have your own empire, as the British do, or at least to dominate your hemisphere, as the US does. Then nobody else can close down your business. So they want the Far East to be their backyard.'

Joanne agreed. 'And the weakness of our policy is that every time we impose economic sanctions to punish the Japanese for their aggression, it only reinforces their feeling that they've got to be self-sufficient.'

'Maybe,' said Chuck. 'But they still have to be stopped.'

Woody shrugged. He did not have an answer to that.

After dinner they went to the cinema. The movie was great. Then Woody and Chuck walked the girls back to their apartment. On the way, Woody took Joanne's hand. She smiled at him and squeezed his hand, and he took that for encouragement.

Outside the girls' building he took her in his arms. Out of the corner of his eye he saw Chuck do the same with Diana.

Joanne kissed Woody's lips briefly, almost chastely, then said: 'The traditional goodnight kiss.'

'There was nothing traditional about it last time I kissed you,' he said. He bent his head to kiss her again.

She put a forefinger on his chin and pushed him away.

Surely, he thought, that little peck was not all he was going to get?

'I was drunk that night,' she said.

'I know.' He saw what the problem was. She was afraid he was going to think she was easy. He said: 'You're even more alluring when you're sober.'

She looked thoughtful for a moment. 'That was the right thing to say,' she said eventually. 'You win the prize.' Then she kissed him again, softly, lingering, not with the urgency of passion but with a concentration that suggested tenderness.

All too soon he heard Chuck sing out: 'Goodnight, Diana!'

Joanne broke the kiss with Woody.

Woody said in dismay: 'My brother was a bit quick!'

She laughed softly. 'Goodnight, Woody,' she said, then she turned and walked to the building.

Diana was already at the door, looking distinctly disappointed.

Woody blurted out: 'Can we have another date?' He sounded needy, even to himself, and he cursed his haste.

But Joanne did not seem to mind. 'Call me,' she said, and went inside.

Woody watched until the two girls disappeared, then he rounded on his brother. 'Why didn't you kiss Diana longer?' he said crossly. 'She seems really nice.'

'Not my type,' said Chuck.

'Really?' Woody was more mystified than annoyed. 'Nice round tits, pretty face – what's not to like? I'd have kissed her, if I wasn't with Joanne.'

'We all have different tastes.'

They started to walk back towards their parents' apartment. 'Well, what is your type, then?' Woody asked Chuck.

'There's something I should probably explain to you, before you plan any more double dates.'

'Okay, what?'

Chuck stopped, forcing Woody to do the same. 'You have to swear never to tell Papa and Mama.'

'I swear.' Woody studied his brother in the yellow light of the street lamps. 'What's the big secret?'

'I don't like girls.'

'A pain in the ass, I agree, but what are you going to do?'

'I mean, I don't like to hug and kiss them.'

'What? Don't be stupid.'

'We're all made differently, Woody.'

'Yeah, but you'd have to be some kind of pansy.'

'Yes.'

'Yes, what?'

'Yes, I'm some kind of pansy.'

'You're such a kidder.'

'I'm not kidding, Woody, I'm dead serious.'

'You're *queer*?'

'That's exactly what I am. I didn't choose to be. When we were kids, and we started jerking off, you used to think about bouncy tits and hairy cunts. I never told you that I used to think about big stiff cocks.'

'Chuck, this is disgusting!'

'No, it's not. It's the way some guys are made. More guys than you think – especially in the navy.'

'There are pansies in the navy?'

Chuck nodded vigorously. 'A lot.'

'Well . . . how do you know?'

'We usually recognize one another. Like Jews always know who's Jewish. For example, the waiter in the Chinese restaurant.'

'He was one?'

'Didn't you hear him say he liked my jacket?'

'Yes, but I didn't think anything of it.'

'There you are.'

'He was attracted to you?'

'I guess.'

'Why?'

'Same reason Diana liked me, probably. Hell, I'm better-looking than you.'

'This is weird.'

'Come on, let's go home.'

They continued on their way. Woody was still reeling. 'You mean there are Chinese pansies?'

Chuck laughed. 'Of course!'

'I don't know, you never think of Chinese guys being that way.'

'Remember, not a word to anyone, especially the parents. God knows what Papa would say.'

After a while, Woody put his arm around Chuck's shoulders. 'Well, what the hell,' he said. 'At least you're not a Republican.'

(iii)

Greg Peshkov sailed with Sumner Welles and President Roosevelt on a heavy cruiser, the *Augusta*, to Placentia Bay, off the coast of Newfoundland. Also in the convoy were the battleship *Arkansas*, the cruiser *Tuscaloosa*, and seventeen destroyers.

They anchored in two long lines, with a broad sea passage down the middle. At nine o'clock in the morning of Saturday 9 August, in bright sunshine, the crews of all twenty vessels mustered at the rails in their dress whites as the British battleship *Prince of Wales* arrived, escorted by three destroyers, and steamed majestically down the middle, bearing Prime Minister Churchill.

It was the most impressive show of power Greg had ever seen, and he was delighted to be part of it.

He was also worried. He hoped the Germans did not know about this rendezvous. If they found out, one U-boat could kill the two leaders of what remained of Western civilization – and Greg Peshkov.

Before leaving Washington, Greg had met with the detective, Tom Cranmer, again. Cranmer had produced an address, a house in a low-rent neighbourhood on the far side of Union Station. 'She's a waitress at the University Women's Club near the Ritz-Carlton, which is why you saw her in that neighbourhood twice,' he had said as he pocketed the balance of his fee. 'I guess acting didn't work out for her – but she still goes by Jacky Jakes.'

Greg had written her a letter.

Dear Jacky,

I just want to know why you ran out on me six years ago. I thought we were so happy, but I must have been wrong. It bugs me, that's all.

You act scared when you see me, but there's nothing to be afraid of. I'm not angry, just curious. I would never do anything to hurt you. You were the first girl I ever loved.

Can we meet, just for a cup of coffee or something, and talk?

Very sincerely,

      Greg Peshkov

He had added his phone number and mailed the note the day he left for Newfoundland.

The President was keen that the conference should result in a joint statement. Greg's boss, Sumner Welles, wrote a draft, but Roosevelt refused to use it, saying it was better to let Churchill produce the first draft.

Greg immediately saw that Roosevelt was a smart negotiator. Whoever produced the first draft would need, in all fairness, to put in some of what the other side wanted alongside his own demands. His statement of the other side's wishes then became an irreducible minimum, while all of his own demands were still up for negotiation. So the drafter always started at a disadvantage. Greg vowed to remember never to write the first draft.

On Saturday, the President and the Prime Minister enjoyed a convivial lunch on board the *Augusta*. On Sunday, they attended a church service on the deck of the *Prince of Wales*, with the Stars and Stripes and the Union Jack draping the altar red, white and blue. On Monday morning, by which time they were firm friends, they got down to brass tacks.

Churchill produced a five-point plan that delighted Sumner Welles and Gus Dewar by calling for an effective international organization to assure the security of all states – in other words, a strengthened League of Nations. But they were disappointed to find that that was too much for Roosevelt. He was in favour, but he feared the isolationists, people who still believed America did not need to get involved with the troubles of the rest of the world. He was extraordinarily sensitive to public opinion, and made ceaseless efforts not to provoke opposition.

Welles and Dewar did not give up, nor did the British. They got together to seek a compromise acceptable to both leaders. Greg took notes for Welles. The group came up with a clause that called for disarmament 'pending the establishment of a wider and more permanent system of general security'.

They put it to the two great men, who accepted it.

Welles and Dewar were jubilant.

Greg could not see why. 'It seems so little,' he said. 'All that effort – the leaders of two great countries brought together across thousands of miles, dozens of staffers, twenty-four ships, three days of talks – and all for a few words that don't quite say what we want.'

'We move by inches, not miles,' said Gus Dewar with a smile. 'That's politics.'

(iv)

Woody and Joanne had been dating for five weeks.

Woody wanted to go out with her every night, but he held back. Nevertheless, he had seen her on four of the last seven days. Sunday they had gone to the beach; Wednesday they had dinner; Friday they saw a movie; and today, Saturday, they were spending the whole day together.

He never tired of talking to her. She was funny and intelligent and sharp-tongued. He loved the way she was so definite about everything. They jawed for hours about the things they liked and hated.

The news from Europe was bad. The Germans were still thrashing the Red Army. East of Smolensk they had wiped out the Russian 16th and 20th Armies, taking 300,000 prisoners, leaving few Soviet forces between the Germans and Moscow. But bad news from afar could not dampen Woody's elation.

Joanne probably was not as crazy about him as he was about her. But she was fond of him, he could tell. They always kissed goodnight, and she seemed to enjoy it, though she did not show the kind of passion he knew she was capable of. Perhaps it was because they always had to kiss in public places, such as the cinema, or a doorway on the street near her building. When they were in her apartment there was always at least one of her two flatmates in the living room, and she had not yet invited him to her bedroom.

Chuck's leave had ended weeks ago, and he was back in Hawaii. Woody still did not know what to think about Chuck's confession. Sometimes he felt as shocked as if the world had turned upside-down; other times he asked himself what difference it made to anything. But he kept his promise not to tell anyone, not even Joanne.

Then Woody's father went off with the President, and his mother went to Buffalo to spend a few days with her parents. So Woody had the Washington apartment – all nine rooms – to himself for a few days. He decided he would look out for an opportunity to invite Joanne Rouzrokh there, in the hope of getting a real kiss.

They had lunch together and went to an exhibition called 'Negro Art', which had been attacked by conservative writers who said there was no such thing as Negro art – despite the unmistakable genius of such people as the painter Jacob Lawrence and the sculptor Elizabeth Catlett.

As they left the exhibition Woody said: 'Would you like to have cocktails while we decide where to go for dinner?'

'No, thanks,' she said in her usual decisive manner. 'I'd really like a cup of tea.'

'Tea?' He was not sure where you could get good tea in Washington. Then he had a brainwave. 'My mother has English tea,' he said. 'We could go to the apartment.'

'Okay.'

The building was a few blocks away on 22nd Street NW, near L Street. They breathed easier as they stepped out of the summer heat into the air-conditioned lobby. A porter took them up in the elevator.

As they entered the apartment Joanne said: 'I see your Papa around Washington all the time, but I haven't talked to your Mama for years. I must congratulate her on her bestseller.'

'She's not here right now,' Woody said. 'Come into the kitchen.'

He filled the kettle from the tap and put it on the heat. Then he put his arms around Joanne and said: 'Alone at last.'

'Where are your parents?'

'Out of town, both of them.'

'And Chuck is in Hawaii.'

'Yes.'

She moved away from him. 'Woody, how could you do this to me?'

'Do what? I'm making you tea!'

'You've got me up here on false pretences! I thought your parents were at home.'

'I never said that.'

'Why didn't you tell me they were away!'

'You didn't ask!' he said indignantly, though there was a grain of truth in her complaint. He would not have lied to her, but he had been hoping he would not have to tell her in advance that the apartment was empty.

'You got me up here to make a pass! You think I'm a cheap broad.'

'I do not! It's just that we're never really private. I was hoping for a kiss, that's all.'

'Don't try to kid me.'

Now she really was being unjust. Yes, he hoped to go to bed with her one day, but no, he had not expected to do so today. 'We'll go,' he said. 'We'll get tea somewhere else. The Ritz-Carlton is right down the street, all the British stay there, they must have tea.'

'Oh, don't be stupid, we don't need to leave. I'm not afraid of you, I can fight you off. I'm just mad at you. I don't want a man who goes out with me because he thinks I'm easy.'

'Easy?' he said, his voice rising. 'Hell! I've waited six years for you to condescend to go out with me. Even now, all I'm asking for is a kiss. If you're easy, I'd hate to be in love with a girl who's difficult!'

To his astonishment, she started to laugh.

'Now what?' he said irritably.

'I'm sorry, you're right,' she said. 'If you wanted a girl who was easy, you would have given up on me long ago.'

'Exactly!'

'After I kissed you like that when I was drunk, I thought you must have a low opinion of me. I assumed you were chasing me for a cheap thrill. I've even been worrying about that in the last few weeks. I misjudged you. I'm sorry.'

He was bewildered by her rapid changes of mood, but he figured this latest phase was an improvement. 'I was crazy about you even before that kiss,' he said. 'I guess you didn't notice.'

'I hardly noticed *you*.'

'I'm pretty tall.'

'It's your only attractive feature, physically.'

He smiled. 'I won't get swollen-headed talking to you, will I?'

'Not if I can help it.'

The kettle boiled. He put tea in a china pot and poured water on top.

Joanne looked thoughtful. 'You said something else a minute ago.'

'What?'

'You said: "I'd hate to be in love with a girl who's difficult." Did you mean it?'

'Did I mean what?'

'The part about being in love.'

'Oh! I didn't intend to say that.' He threw caution to the wind. 'But hell, yes, if you want to know the truth, I'm in love with you. I think I've loved you for years. I adore you. I want—'

She put her arms around his neck and kissed him.

This time it was the real thing, her mouth moving urgently against his, the tip of her tongue touching his lips, her body pressing against his. It was like 1935, except that she did not taste of whisky. This was the girl he loved, the real Joanne, he thought ecstatically: a woman of strong passions. And she was in his arms and kissing him for all she was worth.

She pushed her hands up inside his summer sports shirt and rubbed his chest, pressing her fingers into his ribs, grazing his nipples with her palms, grasping his shoulders, as if she wanted to sink her hands deep into his flesh. He realized that she, too, had a store of frustrated desire that was now overflowing like a busted dam, out of control. He did the same to her, stroking her sides and grasping her breasts, with a feeling of happy liberation, like a child let out of school for an unexpected holiday.

When he pressed his eager hand between her thighs she pulled away.

But what she said surprised him. 'Have you got any birth control?'

'No! I'm sorry—'

'It's okay. In fact, it's good. It proves you really didn't plan to seduce me.'

'I wish I had.'

'Never mind. I know a woman doctor who'll fix me up on Monday. Meanwhile, we'll improvise. Kiss me again.'

As he did so he felt her unbuttoning his pants.

'Oh,' she said a moment later. 'How nice.'

'That's just what I was thinking,' he whispered.

'I may need two hands, though.'

'What?'

'I guess it goes with being so tall.'

'I don't know what you're talking about.'

'Then I'll shut up and kiss you.'

A few minutes later she said: 'Handkerchief.'

Fortunately, he had one.

He opened his eyes, a few moments before the end, and saw her looking at him. In her expression he read desire and excitement and something else that he thought might even be love.

When it was over he felt blissfully calm. I love her, he thought, and I'm happy. How good life is. 'That was wonderful,' he said. 'I'd like to do the same for you.'

'Would you?' she said. 'Really?'

'You bet.'

They were still standing, there in the kitchen, leaning against the door of the refrigerator, but neither of them wanted to move. She took his hand and guided it under her summer dress and inside her cotton underwear. He felt hot skin, crisp hair, and a wet cleft. He tried to push his finger inside, but she said: 'No.' Grasping his fingertip, she guided it between the soft folds. He felt something small and hard, the size of a pea, just under the skin. She moved his finger in a little circle. 'Yes,' she said, closing her eyes. 'Just like that.' He watched her face adoringly as she abandoned herself to the sensation. In a minute or two she gave a little cry, and repeated it two or three times. Then she withdrew his hand and slumped against him.

After a while he said: 'Your tea will be cold.'

She laughed. 'I love you, Woody.'

'Do you really?'

'I hope you're not spooked by me saying that.'

'No.' He smiled. 'It makes me very happy.'

'I know girls aren't supposed to come right out with it, just like that. But I can't pretend to dither. Once I make up my mind, that's it.'

'Yes,' said Woody. 'I'd noticed that.'

(v)

Greg Peshkov was living in his father's permanent apartment at the Ritz-Carlton. Lev came and went, stopping off for a few days between Buffalo and Los Angeles. At present Greg had the place to himself – except that the congressman's curvy daughter, Rita Lawrence, had stayed overnight, and now looked adorably tousled in a man's red silk dressing gown.

A waiter brought them breakfast, the newspapers, and a message envelope.

The joint statement by Roosevelt and Churchill had caused more of a stir than Greg had expected. It was still the main news more than a week later. The press called it the Atlantic Charter. It had seemed, to Greg, to be all cautious phrases and vague commitments, but the world saw it otherwise. It was hailed as a trumpet blast for freedom, democracy and world trade. Hitler was reported to be furious, saying it amounted to a declaration of war by the United States against Germany.

Countries that had not been at the conference nevertheless wanted to sign the charter, and Bexforth Ross had suggested the signatories should be called the United Nations.

Meanwhile, the Germans were overrunning the Soviet Union. In the north they were closing in on Leningrad. In the south the retreating Russians had blown up the Dnieper Dam, the biggest hydro-electric power complex in the world and their pride and joy, in order to deny its power to the conquering Germans – a heartbreaking sacrifice. 'The Red Army has slowed the invasion a bit,' Greg said to Rita, reading from the *Washington Post.* 'But the Germans are still advancing five miles a day. And they claim to have killed three and a half million Soviet soldiers. Is it possible?'

'Do you have any relatives in Russia?'

'As a matter of fact, I do. My father told me, one time when he was a little drunk, that he left a pregnant girl behind.'

Rita made a disapproving face.

'That's him, I'm afraid,' Greg said. 'He's a great man, and great men don't obey the rules.'

She said nothing, but he could read her expression. She disagreed with his view, but was not willing to quarrel with him about it.

'Anyway, I have a Russian half-brother, illegitimate like me,' Greg went on. 'His name is Vladimir, but I don't know anything else about him. He may be dead by now. He's the right age to fight. He's probably one of those three and a half million.' He turned the page.

When he had finished the paper, he read the message the waiter had brought.

It was from Jacky Jakes. It gave a phone number and just said: *Not between 1 and 3.*

Suddenly Greg could not wait to get rid of Rita. 'What time are you expected home?' he asked unsubtly.

She looked at her watch. 'Oh, my gosh, I should be there before my mother starts looking for me.' She had told her parents she was staying over with a girlfriend.

They got dressed together and left in two cabs.

Greg figured the phone number must be Jacky's place of work, and that she would be busy between one o'clock and three. He would phone her around mid-morning.

He wondered why he was so excited. After all, he was only curious. Rita Lawrence was great-looking and very sexy, but with her and several others he had never recaptured the excitement of that first affair with Jacky. No doubt that was because he could never again be fifteen years old.

He got to the Old Executive Office building and began his main task for the day, which was drafting a press release on advice to Americans living in North Africa, where British, Italians and Germans fought backwards and forwards, mostly on a coastal strip two thousand miles long and forty miles wide.

At ten-thirty he phoned the number on the message.

A woman's voice answered: 'University Women's Club.' Greg had never been there: men went only as guests of female members.

He said: 'Is Jacky Jakes there?'

'Yes, she's expecting a call. Please hold on.' She probably had to get special permission to receive a phone call at work, he reflected.

A few moments later he heard 'This is Jacky, who's that?'

'Greg Peshkov.'

'I thought so. How did you get my address?'

'I hired a private detective. Can we meet?'

'I guess we have to. But there's one condition.'

'What?'

'You have to swear by all that's holy not to tell your father. Never, ever.'

'Why?'

'I'll explain later.'

He shrugged. 'Okay.'

'Do you swear?'

'Sure.'

She persisted. 'Say it.'

'I swear it, okay?'

'All right. You can buy me lunch.'

Greg frowned. 'Are there any restaurants in this neighbourhood that will serve a white man and a black woman together?'

'Only one that I know of – the Electric Diner.'

'I've seen it.' He had noticed the name, but he had never been inside: it was a cheap lunch counter used by janitors and messengers. 'What time?'

'Half past eleven.'

'So early?'

'What time do you think waitresses have lunch – one o'clock?'

He grinned 'You're as sassy as ever.'

She hung up.

Greg finished his press release and took the typed sheets into his boss's office. Dropping the draft into the in-tray, he said: 'Would it be convenient for me to take an early lunch, Mike? Around eleven-thirty?'

Mike was reading the op-ed page of the *New York Times*. 'Yeah, no problem,' he said without looking up.

Greg walked past the White House in the sunshine and reached the diner at eleven-twenty. It was empty but for a handful of people taking a mid-morning break. He sat in a booth and ordered coffee.

He wondered what Jacky would have to say. He looked forward to the solution of a puzzle that had mystified him for six years.

She arrived at eleven thirty-five, wearing a black dress and flat shoes – her waitress uniform without the apron, he presumed. Black suited her, and he remembered vividly the sheer pleasure of looking at her, with her bow-shaped mouth and her big brown eyes. She sat opposite him and ordered a salad and a Coke. Greg had more coffee: he was too tense to eat.

Her face had lost the childish plumpness he remembered. She had been sixteen when they met, so she was twenty-two now. They had been kids playing at being grown up; now they really were adults. In her face he read a story that had not been there six years ago: disappointment and suffering and hardship.

'I work the day shift,' she told him. 'Come in at nine, set the tables, dress the room. Wait at lunch, clear away, leave at five.'

'Most waitresses work in the evening.'

'I like to have evenings and weekends free.'

'Still a party girl!'

'No, mostly I stay home and listen to the radio.'

'I guess you have lots of boyfriends.'

'All I want.'

It took him a moment to realize that could mean anything.

Her lunch came. She drank her Coke and picked at the salad.

Greg said: 'So why did you run out, back in 1935?'

She sighed. 'I don't want to tell you this, because you're not going to like it.'

'I have to know.'

'I got a visit from your father.'

Greg nodded. 'I figured he must have had something to do with it.'

'He had a goon with him – Joe something.'

'Joe Brekhunov. He's a thug.' Greg began to feel angry. 'Did he hurt you?'

'He didn't need to, Greg. I was scared to death just looking at him. I was ready to do anything your father wanted.'

Greg suppressed his fury. 'What did he want?'

'He said I had to leave, right then. I could write you a note but he would read it. I had to come back here to Washington. I was so sad to leave you.'

Greg remembered his own anguish. 'Me, too,' he said. He was tempted to reach across the table and take her hand, but he was not sure she would want that.

She went on: 'He said he would give me a weekly allowance just to keep away from you. He's still paying me. It's only a few bucks but it takes care of the rent. I promised – but somehow I managed to summon up the nerve to make one condition.'

'What?'

'That he would never make a pass at me. If he did, I would tell you everything.'

'And he agreed?'

'Yes.'

'Not many people get away with threatening him.'

She pushed her plate away. 'Then he said if I broke my word Joe would cut my face. Joe showed me his straight razor.'

It all fell into place. 'That's why you're still scared.'

Her dark skin was bloodless with fear. 'You bet your goddamn life.'

Greg's voice fell to a whisper. 'Jacky, I'm sorry.'

She forced a smile. 'Are you sure he was so wrong? You were fifteen. It's not a good age to get married.'

'If he had said that to me, it might be different. But he decides what's going to happen and just does it, as if no one else is entitled to an opinion.'

'Still, we had good times.'

'You bet.'

'I was your Gift.'

He laughed. 'Best present I ever got.'

'So what are you doing these days?'

'Working in the press office at the State Department for the summer.'

She made a face. 'Sounds boring.'

'It's the opposite! It's so exciting to watch powerful men make earth-shaking decisions, just sitting there at their desks. They run the world!'

She looked sceptical, but said: 'Well, it probably beats waitressing.'

He began to see how far apart they had moved. 'In September I'm going back to Harvard for my last year.'

'I bet you're a gift to the co-eds.'

'There are lots of men and not many girls.'

'You do all right, though, don't you?'

'I can't lie to you.' He wondered whether Emily Hardcastle had kept her promise and got herself fitted with a contraceptive device.

'You'll marry one of them and have beautiful children and live in a house on the edge of a lake.'

'I'd like to be something in politics, maybe Secretary of State, or a senator like Woody Dewar's father.'

She looked away.

Greg thought about that house on the edge of a lake. It must be her dream. He felt sad for her.

'You'll make it,' she said. 'I know. You have that air about you. Even when you were fifteen you had it. You're like your father.'

'What? Come on!'

She shrugged. 'Think about it, Greg. You knew I didn't want to see you. But you set a private dick on me. *He decides what's going to happen and just does it, as if no one else is entitled to an opinion.* That's what you said about him a minute ago.'

Greg was dismayed. 'I hope I'm not completely like him.'

She gave him an appraising look. 'The jury's still out.'

The waitress took her plate. 'Some dessert?' she said. 'Peach pie's good.'

Neither of them wanted dessert, so the waitress gave Greg the bill.

Jacky said: 'I hope I've satisfied your curiosity.'

'Thank you, I appreciate it.'

'Next time you see me on the street, just walk on by.'

'If that's what you want.'

She stood up. 'Let's leave separately. I'd feel more comfortable.'

'Whatever you say.'

'Good luck, Greg.'

'Good luck to you.'

'Tip the waitress,' she said, and she walked away.

10

1941 (III)

In October the snow fell and melted, and the streets of Moscow were cold and wet. Volodya was searching in the store cupboard for his *valenki*, the traditional felt boots that warmed the feet of Muscovites in winter, when he was astonished to see six cases of vodka.

His parents were not great drinkers. They rarely took more than one small glass. Now and again his father went to one of Stalin's long, boozy dinners with old comrades, and staggered in through the door in the early hours of the morning as drunk as a skunk. But in this house a bottle of vodka lasted a month or more.

Volodya went into the kitchen. His parents were having breakfast, canned sardines with black bread and tea. 'Father,' he said, 'why do we have six years' supply of vodka in the store cupboard?'

His father looked surprised.

Both men looked at Katerina, who blushed. Then she switched on the radio and turned the volume down to a low mutter. Did she suspect their apartment had concealed listening devices, Volodya wondered?

She spoke quietly but angrily. 'What are you going to use for money when the Germans get here?' she said. 'We won't belong to the privileged elite any longer. We'll starve unless we can buy food on the black market. I'm too damn old to sell my body. Vodka will be better than gold.'

Volodya was shocked to hear his mother talking this way.

'The Germans aren't going to get here,' his father said.

Volodya was not so sure. They were advancing again, closing the jaws of a pincer around Moscow. They had reached Kalinin in the north and Kaluga to the south, both cities only about a hundred miles away. Soviet casualties were unimaginably high. A month ago 800,000 Red Army troops had held the line, but only 90,000 were left, according to the estimates reaching Volodya's desk. He said to his father: 'Who the hell is going to stop them?'

'Their supply lines are stretched. They're unprepared for our winter weather. We will counter-attack when they're weakened.'

'So why are you moving the government out of Moscow?'

The bureaucracy was in the process of being transported two thousand miles east, to the city of Kuibyshev. The citizens of the capital had been unnerved by the sight of government clerks carrying boxes of files out of their office buildings and packing them into trucks.

'That's just a precaution,' Grigori said. 'Stalin is still here.'

'There is a solution,' Volodya argued. 'We have hundreds of thousands of men in Siberia. We need them here as reinforcements.'

Grigori shook his head. 'We can't leave the east undefended. Japan is still a threat.'

'Japan is not going to attack us – we know that!' Volodya glanced at his mother. He knew he should not talk about secret intelligence in front of her, but he did anyway. 'The Tokyo source that warned us – correctly – that the Germans were about to invade has now told us that the Japanese will not. Surely we're not going to disbelieve him again!'

'Evaluating intelligence is never easy.'

'We don't have a choice!' Volodya said angrily. 'We have twelve armies in reserve – a million men. If we deploy them, Moscow might survive. If we don't, we're finished.'

Grigori looked troubled. 'Don't speak like that, even in private.'

'Why not? I'll probably be dead soon anyway.'

His mother started to cry.

His father said: 'Now look what you've done.'

Volodya left the room. Putting on his boots, he asked himself why he had shouted at his father and made his mother cry. He saw that it was because he now believed that Germany would defeat the Soviet Union. His mother's stash of vodka to be used as currency during a Nazi occupation had forced him to confront the reality. We're going to lose, he said to himself. The end of the Russian revolution is in sight.

He put on his coat and hat. Then he returned to the kitchen. He kissed his mother and embraced his father.

'What's this for?' said his father. 'You're only going to work.'

'It's just in case we never meet again,' Volodya said. Then he went out.

When he crossed the bridge into the city centre he found that all public transport had stopped. The metro was closed and there were no buses or trams.

It seemed there was nothing but bad news.

This morning's bulletin from SovInformBuro, broadcast on the radio and from black-painted loudspeaker posts on street corners, had been uncharacteristically honest. 'During the night of 14 to 15 October, the position on the Western Front became worse,' it had said. 'Large numbers of German tanks broke through our defences.' Everyone knew that SovInformBuro always lied, so they assumed the real situation was even worse.

The city centre was clogged with refugees. They were pouring in from the west, with their possessions in handcarts, driving herds of skinny cows and filthy pigs and wet sheep through the streets, heading for the countryside east of Moscow, desperate to get as far away as possible from the advancing Germans.

Volodya tried to hitch a lift. There was not much civilian traffic in Moscow these days. Fuel was being saved for the endless military convoys driving around the Garden Ring orbital road. He was picked up by a new GAZ-64 jeep.

Looking from the open vehicle, he saw a good deal of bomb damage. Diplomats returning from England said this was nothing by comparison with the London Blitz, but Muscovites thought it was bad enough. Volodya passed several wrecked buildings and dozens of burned-out wooden houses.

Grigori, in charge of air raid defence, had mounted anti-aircraft guns on the tops of the tallest buildings, and launched barrage balloons to float below the snow clouds. His most bizarre decision had been to order the golden onion domes of the churches to be painted in camouflage green and brown. He had admitted to Volodya that this would make no difference to the accuracy – or otherwise – of the bombing but, he said, it gave citizens the feeling that they were being protected.

If the Germans won, and the Nazis ruled Moscow, then Volodya's nephew and niece, the twin children of his sister, Anya, would be brought up not as patriotic Communists but as slavish Nazis, saluting Hitler. Russia would be like France, a country in servitude, perhaps partly ruled by an obedient pro-Fascist government that would round up Jews to be sent to concentration camps. It hardly bore thinking about. Volodya wanted a future in which the Soviet Union could free itself from the malign rule of Stalin and the brutality of the secret police and begin to build true Communism.

When Volodya reached the headquarters building at the Khodynka airfield, he found the air full of greyish flakes that were not snow but ash. Red Army Intelligence was burning its records to prevent them falling into enemy hands.

Shortly after he arrived, Colonel Lemitov came into his office. 'You sent a memo to London about a German physicist called Wilhelm Frunze. That was a very smart move. It turned out to be a great lead. Well done.'

What does it matter, Volodya thought? The Panzers were only a hundred miles away. It was too late for spies to help. But he forced himself to concentrate. 'Frunze, yes. I was at school with him in Berlin.'

'London contacted him and he is willing to talk. They met at a safe house.' As Lemitov talked, he fiddled with his wristwatch. It was unusual for him to fidget. He was clearly tense. Everyone was tense.

Volodya said nothing. Obviously some information had come out of the meeting, otherwise Lemitov would not be talking about it.

'London say that Frunze was wary at first, and suspected our man of belonging to the British secret police,' Lemitov said with a smile. 'In fact, after the initial meeting he went to Kensington Palace Gardens and knocked on the door of our embassy and demanded confirmation that our man was genuine!'

Volodya smiled. 'A real amateur.'

'Exactly,' said Lemitov. 'A disinformation decoy wouldn't do anything so stupid.'

The Soviet Union was not finished yet, not quite; so Volodya had to carry on as if Willi Frunze mattered. 'What did he give us, sir?'

'He says he and his fellow scientists are collaborating with the Americans to make a super-bomb.'

Volodya, startled, recalled what Zoya Vorotsyntsev had told him. This confirmed her worst fears.

Lemitov went on: 'There's a problem with the information.'

'What?'

'We've translated it, but we still can't understand a word.' Lemitov handed Volodya a sheaf of typewritten sheets.

Volodya read a heading aloud. 'Isotope separation by gaseous diffusion.'

'You see what I mean.'

'I did languages at university, not physics.'

'But you once mentioned a physicist you know.' Lemitov smiled. 'A gorgeous blonde who declined to go to a movie with you, if I remember.'

Volodya blushed. He had told Kamen about Zoya, and Kamen must have repeated the gossip. The trouble with having a spy for a boss was that he knew everything. 'She's a family friend. She told me about an explosive process called fission. Do you want me to question her?'

'Unofficially and informally. I don't want to make a big thing of this until I understand it. Frunze may be a crackpot, and he could make us look foolish. Find out what the reports are about, and whether Frunze is making scientific sense. If he's genuine, can the British and Americans really make a super-bomb? And the Germans too?'

'I haven't seen Zoya for two or three months.'

Lemitov shrugged. It did not really matter how well Volodya knew Zoya. In the Soviet Union, answering questions put by the authorities was never optional.

'I'll track her down.'

Lemitov nodded. 'Do it today.' He went out.

Volodya frowned thoughtfully. Zoya was sure the Americans were making a super-bomb, and she had been convincing enough to persuade Grigori to mention it to Stalin, but Stalin had scorned the idea. Now a spy in England was saying what Zoya had said. It looked as if she had been right. And Stalin had been wrong – again.

The leaders of the Soviet Union had a dangerous tendency to deny the truth of bad news. Only last week, an air reconnaissance mission had spotted German armoured vehicles just eighty miles from Moscow. The General Staff had refused to believe it until the sighting had been confirmed twice. Then they had ordered the reporting air officer to be arrested and tortured by the NKVD for 'provocation'.

It was difficult to think long term when the Germans were so close, but the possibility of a bomb that could flatten Moscow could not be disregarded, even at this moment of extreme peril. If the Soviets beat the Germans, they might afterwards be attacked by Britain and America: something similar had happened after the 1914–18 war. Would the USSR find itself helpless against a capitalist-imperialist super-bomb?

Volodya detailed his assistant, Lieutenant Belov, to find out where Zoya was.

While waiting for the address Volodya studied Frunze's reports, in the original English and in translation, memorizing what seemed to be key phrases, as he could not take the papers out of the building. At the end of an hour he understood enough to ask further questions.

Belov discovered that Zoya was not at the university, nor at the nearby apartment building for scientists. However, the building administrator told him that all the younger residents had been requested to help with the construction of new inner defences for the city, and gave him the location where Zoya was working.

Volodya put on his coat and went out.

He felt excited, but he was not sure whether that was on account of Zoya or the super-bomb. Maybe both.

He was able to get an army ZIS and driver.

Passing the Kazan station – for trains to the east – he saw what looked like a full-blown riot. It seemed that people could not get into the station, let alone board the trains. Affluent men and women were struggling to reach the entrance doors with their children and pets and suitcases and trunks. Volodya was shocked to see some of them punching and kicking one another shamelessly. A few policemen looked on, helpless: it would have taken an army to impose order.

Military drivers were normally taciturn, but this one was moved to comment. 'Fucking cowards,' he said. 'Running away, leaving us to fight the Nazis. Look at them, in their fur fucking coats.'

Volodya was surprised. Criticism of the ruling elite was dangerous. Such remarks could cause a man to be denounced. Then he would spend a week or two in the basement of the NKVD's headquarters in Lubyanka Square. He might come out crippled for life.

Volodya had an unnerving sense that the rigid system of hierarchy and deference that sustained Soviet Communism was beginning to weaken and disintegrate.

They found the barricade party just where the building administrator had predicted. Volodya got out of the car, told the driver to wait, and studied the work.

A main road was strewn with anti-tank 'hedgehogs'. A hedgehog consisted of three pieces of steel railway track, each a yard long, welded together at their centres, forming an asterisk that stood on three feet and stuck three arms up. Apparently they wreaked havoc with caterpillar tracks.

Behind the hedgehog field an anti-tank ditch was being dug with pickaxes and shovels, and beyond that a sandbag wall was going up, with gaps for defenders to shoot through. A narrow zigzag path had been left between the obstacles so that the road could continue to be used by Muscovites until the Germans arrived.

Almost all the workers digging and building were women.

Volodya found Zoya beside a sand mountain, filling sacks with a shovel. For a minute he watched her from a distance. She wore a dirty coat, woollen mittens and felt boots. Her blonde hair was pulled back and covered with a colourless rag tied under her chin. Her face was smeared with mud, but she still looked sexy. She wielded the shovel in a steady rhythm, working efficiently. Then the supervisor blew a whistle and work stopped.

Zoya sat on a stack of sandbags and took from her coat pocket a small packet wrapped in newspaper. Volodya sat beside her and said: 'You could have got exemption from this work.'

'It's my city,' she said. 'Why wouldn't I help to defend it?'

'So you're not fleeing to the east.'

'I'm not running away from the motherfucking Nazis.'

Her vehemence surprised him. 'Plenty of people are.'

'I know. I thought you'd be long gone.'

'You have a low opinion of me. You think I belong to a selfish elite.' She shrugged. 'Those who are able to save themselves generally do.'

'Well, you're wrong. All my family are still here in Moscow.'

'Perhaps I misjudged you. Would you like a pancake?' She opened her packet to reveal four pale-coloured patties wrapped in cabbage leaves. 'Try one.'

He accepted and took a bite. It was not very tasty. 'What is it?'

'Potato peelings. You can get a bucketful free at the back door of any Party canteen or officers' mess. You mince them small in the kitchen grinder, boil them until they're soft, mix them with a little flour and milk, add salt if you've got any, and fry them in lard.'

'I didn't know you were so badly off,' he said, feeling embarrassed. 'You can always get a meal at our place, you know.'

'Thank you. What brings you here?'

'A question. What is isotope separation by gaseous diffusion?'

She stared at him. 'Oh, my God – what's happened?'

'Nothing has happened. I'm simply trying to evaluate some dubious information.'

'Are we building a fission bomb at last?'

Her reaction told him that the information from Frunze was probably sound. She had immediately understood the significance of what he'd said. 'Please answer the question,' Volodya said sternly. 'Even though we're friends, this is official business.'

'Okay. Do you know what an isotope is?'

'No.'

'Some elements exist in slightly different forms. Carbon atoms, for example, always have six protons, but some have six neutrons and others have seven or eight. The different types are isotopes, called carbon-12, carbon-13 and carbon-14.'

'Simple enough, even for a student of languages,' Volodya said. 'Why is it important?'

'Uranium has two isotopes, U-235 and U-238. In natural uranium the two are mixed up. But only U-235 is explosive.'

'So we need to separate them.'

'Gaseous diffusion would be one way, theoretically. When a gas is diffused through a membrane, the lighter molecules pass through faster, so the emerging gas is richer in the lower isotope. Of course I've never seen it done.'

Frunze's report said that the British were building a gaseous diffusion plant in Wales, in the west of the United Kingdom. The Americans were also building one. 'Would there be any other purpose for such a plant?'

She shook her head. 'Figure the odds,' she said. 'Anyone who prioritizes this kind of process in wartime is either going crazy or building a weapon.'

Volodya saw a car approach the barricade and begin to negotiate the zigzag passage. It was a KIM-10, a small two-door car designed for affluent families. It had a top speed of sixty miles per hour, but this one was so overloaded it probably would not do forty.

A man in his sixties was at the wheel, wearing a hat and a Western-style cloth coat. Beside him was a young woman in a fur hat. The back seat of the car was piled with cardboard boxes. There was a piano strapped precariously to the roof.

This was clearly a senior member of the ruling elite trying to get out of town with his wife, or mistress, and as many of his valuables as he could take; the kind of person Zoya assumed Volodya to be, which was perhaps why she had declined to go out with him. He wondered if she might be revising her opinion of him.

One of the barricade volunteers moved a hedgehog in front of the KIM-10, and Volodya saw that there was going to be trouble.

The car inched forward until its bumper touched the hedgehog. Perhaps the driver thought he could nudge it out of the way. Several more women came closer to watch. The device was designed to resist being pushed out of the way. Its legs dug into the ground, jamming, and it stuck fast. There was a sound of bending metal as the car's front bumper deformed. The driver put it in reverse and backed off.

He stuck his head out of the window and yelled: 'Move that thing, right now!' He sounded as if he was used to being obeyed.

The volunteer, a chunky middle-aged woman wearing a man's checked cap, folded her arms. She shouted: 'Move it yourself – deserter!'

The driver got out, red-faced with anger, and Volodya was surprised to recognize Colonel Bobrov, whom he had known in Spain. Bobrov had been famous for shooting his own men in the back of the head if they retreated. 'No mercy for cowards' had been his slogan. At Belchite, Volodya had personally seen him kill three International Brigade troops for retreating when they ran out of ammunition. Now Bobrov was in civilian clothes. Volodya wondered if he would shoot the woman who had blocked his way.

Bobrov walked to the front of the car and took hold of the hedgehog. It was heavier than he had expected, but with an effort he was able to drag it out of the way.

As he was walking back to his car, the woman in the cap replaced the hedgehog in front of the car.

The other volunteers were now crowding around, watching the confrontation, grinning and making jokes.

Bobrov walked up to the woman, taking from his coat pocket an identification card. 'I am General Bobrov!' he said. He must have been promoted since returning from Spain. 'Let me pass!'

'You call yourself a soldier?' the woman sneered. 'Why aren't you fighting?'

Bobrov flushed. He knew her contempt was justified. Volodya wondered if the brutal old soldier had been talked into fleeing by his younger wife.

'I call you a traitor,' said the volunteer in the cap. 'Trying to run away with your piano and your young tart.' Then she knocked his hat off.

Volodya was flabbergasted. He had never seen such defiance of authority in the Soviet Union. Back in Berlin, before the Nazis came to power, he had been surprised by the sight of ordinary Germans fearlessly arguing with police officers; but it did not happen here.

The crowd of women cheered.

Bobrov still had short-cropped white hair all over his head. He looked at his hat as it rolled across the wet road. He took one step in pursuit, then thought better of it.

Volodya was not tempted to intervene. There was nothing he could do against the mob, and anyway he had no sympathy for Bobrov. It seemed just that Bobrov should be treated with the brutality he had always shown to others.

Another volunteer, an older woman wrapped in a filthy blanket, opened the car's trunk. 'Look at all this!' she said. The trunk was full of leather luggage. She pulled out a suitcase and thumbed its catches. The lid came open, and the contents fell out: lacy underwear, linen petticoats and nightdresses, silk stockings and camisoles, all obviously made in the West, finer than anything ordinary Russian women ever saw, let alone bought. The filmy garments dropped into the filthy slush of the street and stuck there like petals on a dunghill.

Some of the women started to pick them up. Others seized more suitcases. Bobrov ran to the back of his car and started to shove the women away. This was turning very nasty, Volodya thought. Bobrov probably carried a gun, and he would draw it any second now. But then the woman in the blanket lifted a spade and hit Bobrov hard over the head. A woman who could dig a trench with a spade was no weakling, and the blow made a sickeningly loud thud as it connected. The general fell to the ground, and the woman kicked him.

The young mistress got out of the car.

The woman in the cap shouted: 'Coming to help us dig?' and the others laughed.

The general's girlfriend, who looked about thirty, put her head down and walked back along the road the way the car had come. The volunteer in the checked cap shoved her, but she dodged between the hedgehogs and started to run. The volunteer ran after her. The mistress was wearing tan suede shoes with a high heel, and she slipped in the wet and fell down. Her fur hat came off. She struggled to her feet and started to run again. The volunteer went after the hat, letting the mistress go.

All the suitcases now lay open around the abandoned car. The workers pulled the boxes from the back seat and turned them upside down, emptying the contents on to the road. Cutlery spilled out, china broke, and glassware smashed. Embroidered bedsheets and white towels were dragged through the slush. A dozen pretty pairs of shoes were scattered across the tarmac.

Bobrov got to his knees and tried to stand. The woman in the blanket hit him with the spade again. Bobrov collapsed on the ground. She unbuttoned Bobrov's fine wool coat and tried to pull it off him. Bobrov struggled, resisting. The woman became furious and hit Bobrov again and again until he lay still, his cropped white head covered with blood. Then she discarded her old blanket and put Bobrov's coat on.

Volodya walked across to Bobrov's unmoving body. The eyes stared lifelessly. Volodya knelt down and checked for breathing, a heartbeat or a pulse. There was none. The man was dead.

'No mercy for cowards,' Volodya said; but he closed Bobrov's eyes.

Some of the women unstrapped the piano. The instrument slid off the car roof and hit the ground with a discordant clang. They began gleefully to smash it up with picks and shovels. Others were quarrelling over the scattered valuables, snatching up the cutlery, bundling the bedsheets, tearing the fine underwear as they struggled for possession. Fights broke out. A china teapot came flying through the air and just missed Zoya's head.

Volodya hurried back to her. 'This is developing into a full-scale riot,' he said. 'I've got an army car and a driver. I'll get you out of here.'

She hesitated only for a second. 'Thanks,' she said; and they ran to the car, jumped in, and drove away.

(ii)

Erik von Ulrich's faith in the Führer was vindicated by the invasion of the Soviet Union. As the German armies raced across the vastness of Russia, sweeping the Red Army aside like chaff, Erik rejoiced in the strategic brilliance of the leader to whom he had given his allegiance.

Not that it was easy. During rainy October the countryside had been a mud bath: they called it the *rasputitsa*, the time of no roads. Erik's ambulance had ploughed through a quagmire. A wave of mud built up in front of the vehicle, gradually slowing it, until he and Hermann had to get out and clear it away with shovels before they could drive any farther. It was the same for the entire German army, and the dash for Moscow had slowed to a crawl. Furthermore, the swamped roads meant that supply trucks never caught up. The army was low on ammunition, fuel and food, and Erik's unit was dangerously short of drugs and other medical necessities.

So Erik had at first rejoiced when the frost had set in at the beginning of November. The freeze seemed a blessing, making the roads hard again and allowing the ambulance to move at normal speed. But Erik shivered in his summer coat and cotton underwear – winter uniforms had not yet arrived from Germany. Nor had the low-temperature lubricants needed to keep the engine of his ambulance operating – and the engines of all the army's trucks, tanks and artillery. While on the road, Erik got up every two hours in the night to start his engine and run it for five minutes, the only way to keep the oil from congealing and the coolant from freezing solid. Even then he cautiously lit a fire under the vehicle every morning an hour before moving off.

Hundreds of vehicles broke down and were abandoned. The planes of the Luftwaffe, left outside all night on makeshift airfields, froze solid and refused to start, and air cover for the troops simply disappeared.

Despite all that, the Russians were retreating. They fought hard, but they were always pushed back. Erik's unit stopped continually to clear away Russian bodies, and the frozen dead stacked by the roadside made a grisly embankment. Relentlessly, remorselessly, the German army was closing in on Moscow.

Soon, Erik felt sure, he would see Panzers majestically rolling across Red Square, while swastika banners fluttered jubilantly from the towers of the Kremlin.

Meanwhile, the temperature was minus ten degrees Centigrade, and falling.

Erik's field hospital unit was in a small town beside a frozen canal, surrounded by spruce forest. Erik did not know the name of the place. The Russians often destroyed everything as they retreated, but this town had survived more or less intact. It had a modern hospital, which the Germans had taken over. Dr Weiss had briskly instructed the local doctors to send their patients home, regardless of condition.

Now Erik studied a frostbite patient, a boy of about eighteen. The skin of his face was a waxy yellow, and frozen hard to the touch. When Erik and Hermann cut away the flimsy summer uniform, they saw that his arms and legs were covered with purple blisters. His torn and broken boots had been stuffed with newspaper in a pathetic attempt to keep out the cold. When Erik took them off he smelled the characteristic rotting stink of gangrene.

Nevertheless, he thought they might yet save the boy from amputation.

They knew what to do. They were treating more men for frostbite than for combat wounds.

He filled a bathtub, then he and Hermann Braun lowered the patient into the warm water.

Erik studied the body as it thawed. He saw the black colour of gangrene on one foot and the toes of the other.

When the water began to cool they took him out, patted him dry, put him in a bed and covered him with blankets. Then they surrounded him with hot stones wrapped in towels.

The patient was conscious and alert. He said: 'Am I going to lose my foot?'

'That's up to the doctor,' Erik said automatically. 'We're just orderlies.'

'But you see a lot of patients,' he persisted. 'What's your best guess?'

'I think you might be all right,' Erik said. If not, he knew what would happen. On the foot less badly affected, Weiss would amputate the toes, cutting them off with a big pair of clippers like bolt cutters. The other leg would be amputated below the knee.

Weiss came a few minutes later and examined the boy's feet. 'Prepare the patient for amputation,' he said brusquely.

Erik was desolate. Another strong young man was going to spend the rest of his life a cripple. What a shame.

But the patient saw it differently. 'Thank God,' he said. 'I won't have to fight any more.'

As they got the boy ready for surgery, Erik reflected that the patient was one of many who persisted in a defeatist attitude – his own family among them. He thought a lot about his late father, and felt deep rage mingled with his grief and loss. The old man would not have joined in with the majority and celebrated the triumph of the Third Reich, he thought bitterly. He would have complained about something, questioned the Führer's judgement, undermined the morale of the armed forces. Why had he had to be such a rebel? Why had he been so attached to the outdated ideology of democracy? Freedom had done nothing for Germany, whereas Fascism had saved the country!

He was angry with his father, yet hot tears came to his eyes when he thought about how he had died. Erik had at first denied that the Gestapo had killed him, but he soon realized it was probably true. They were not Sunday School teachers: they beat people who told wicked lies about the government. Father had persisted in asking whether the government was killing handicapped children. He had been foolish to listen to his English wife and his over-emotional daughter. Erik loved them, which made it all the more painful to him that they were so misguided and obstinate.

While on leave in Berlin, Erik had gone to see Hermann's father, the man who had first revealed the exciting Nazi philosophy to him when he and Hermann were boys. Herr Braun was in the SS now. Erik said he had met a man in a bar who claimed the government killed disabled people in special hospitals. 'It is true that the handicapped are a costly drag on the forward march to the new Germany,' Herr Braun had said to Erik. 'The race must be purified, by repressing Jews and other degenerate types, and preventing mixed marriages that produce mongrel people. But euthanasia has never been Nazi policy. We are determined, tough, even brutal sometimes, but we do not murder people. That is a Communist lie.'

Father's accusations had been wrong. Still Erik wept sometimes.

Fortunately, he was frantically busy. There was always a morning rush of patients, mostly men injured the day before. Then there was a short lull before the first new casualties of the day. When Weiss had operated on the frostbitten boy, he and Erik and Hermann took a mid-morning break in the cramped staff room.

Hermann looked up from a newspaper. 'In Berlin they're saying we've already won!' he exclaimed. 'They ought to come here and see for themselves.'

Dr Weiss spoke with his usual cynicism. 'The Führer made a most interesting speech at the Sportpalast,' he said. 'He spoke of the bestial inferiority of the Russians. I find that reassuring. I had the impression that the Russians were the toughest fighters we have yet come across. They have fought longer and harder than the Poles, the Belgians, the Dutch, the French, or the British. They may be underequipped and badly led and half starved, but they come running at our machine guns, waving their obsolete rifles, as if they don't care whether they live or die. I'm glad to hear that this is no more than a sign of their bestiality. I was beginning to fear that they might be courageous and patriotic.'

As always, Weiss pretended to agree with the Führer, while meaning the opposite. Hermann just looked confused, but Erik understood and was infuriated. 'Whatever the Russians may be, they're losing,' he said. 'We're forty miles from Moscow. The Führer has been proved right.'

'And he is much smarter than Napoléon,' said Dr Weiss.

'In Napoléon's time nothing could move faster than a horse,' said Erik. 'Today we have motor vehicles and wireless telegraphy. Modern communications have enabled us to succeed where Napoléon failed.'

'Or they will have, when we take Moscow.'

'Which we will do in a few days, if not hours. You can hardly doubt that!'

'Can I not? I believe some of our own generals have suggested we halt where we are and build a defence line. We could secure our positions, resupply over the winter, and go back on the offensive when the spring comes.'

'That sounds to me like treacherous cowardice!' Erik said hotly.

'You are right – you must be, because that is exactly what Berlin told the generals, I understand. Headquarters people obviously have a better perspective than the men on the front line.'

'We have almost wiped out the Red Army!'

'But Stalin seems to produce more armies from nowhere, like a magician. At the beginning of this campaign we thought he had two hundred divisions. Now we think he has more than three hundred. Where did he find another hundred divisions?'

'The Führer's judgement will be proved right – again.'

'Of course it will, Erik.'

'He has never yet been wrong!'

'A man thought he could fly, so he jumped off the top of a ten-storey building, and as he fell past the fifth floor, flapping his arms uselessly in the air, he was heard to say: So far, so good.'

A soldier rushed into the staff room. 'There's been an accident,' he said. 'At the quarry north of the town. A collision, three vehicles. Some SS officers are injured.'

The SS, or *Schutzstaffel*, had originally been Hitler's personal guard, and now formed a powerful elite. Erik admired their superb discipline, their ultra-smart uniforms, and their specially close relationship with Hitler.

'We'll send an ambulance,' said Weiss.

The soldier said: 'It's the *Einsatzgruppe*, the Special Group.'

Erik had heard of the Special Groups, vaguely. They followed the army into conquered territory and rounded up troublemakers and potential saboteurs such as Communists. They were probably setting up a prison camp outside the town.

'How many hurt?' asked Weiss.

'Six or seven. They're still getting people out of the cars.'

'Okay. Braun and von Ulrich, you go.'

Erik was pleased. He would be glad to rub shoulders with the Führer's most fervent supporters, even happier if he could be of service to them.

The soldier handed him a message slip with directions.

Erik and Hermann gulped their tea, stubbed their cigarettes, and left the room. Erik put on a fur coat he had taken from a dead Russian officer, but left it open to show his uniform. They hurried down to the garage, and Hermann drove the ambulance out into the street. Erik read out the directions, peering through a light snowfall.

The road led out of town and snaked through the forest. They passed several buses and trucks coming the other way. The snow on the road was packed hard, and Hermann could not go fast on the glossy surface. Erik could easily imagine how there had been a collision.

It was the afternoon of the short day. At this time of year, daylight began at ten and ended at five. A grey light came through the snow clouds. The tall pine trees crowding in on either side darkened the road further. Erik felt as if he were in one of the fairy tales of the Brothers Grimm, following the path into the deep wood where evil lurked.

They looked out for a turning to the left, and found it guarded by a soldier who pointed the way. They bumped along a treacherous path between the trees until they were waved down by a second guard, who said: 'Don't go faster than walking pace. That's how the crash happened.'

A minute later they came upon the accident. Three damaged vehicles stood as if welded together: a bus, a jeep and a Mercedes limousine with snow chains on the tyres. Erik and Hermann jumped out of their ambulance.

The bus was empty. There were three men on the ground, perhaps the occupants of the jeep. Several soldiers gathered around the car sandwiched between the other two vehicles, apparently trying to get the people out of it.

Erik heard a volley of rifle fire, and wondered for a moment who was shooting, but he put the thought aside and concentrated on the job.

He and Hermann went from one man to the next, assessing the

gravity of the injuries. Of the three people on the ground one was dead, another had a broken arm, and the third appeared to be no worse than bruised. In the car, one man had bled to death, another was unconscious, and a third was screaming.

Erik gave the screamer a shot of morphine. When the drug took effect, he and Hermann were able to get the patient out of the car and into the ambulance. With him out of the way, the soldiers could begin to free the unconscious man, who was trapped by the deformed bodywork of the Mercedes. The man had a head injury that was going to kill him anyway, Erik thought, but he did not tell them that. He turned his attention to the men from the jeep. Hermann put a splint on the broken arm, and Erik walked the bruised man to the ambulance and sat him inside.

He returned to the Mercedes. 'We'll have him out in five to ten minutes,' said a captain. 'Just hold on.'

'Okay,' said Erik.

He heard shooting again, and walked a little farther into the forest, curious about what the Special Group might be doing here. The snow on the ground between the trees was heavily trodden and littered with cigarette ends, apple cores, discarded newspapers and other litter, as if a factory outing had passed this way.

He entered a clearing where lorries and buses were parked. A lot of people had been brought here. Some buses were leaving, skirting the accident; another arrived as Erik passed through. Beyond the car park, he came upon a hundred or so Russians of all ages, apparently prisoners, though many had suitcases, boxes and sacks that they clutched as if guarding precious possessions. One man held a violin. A little girl with a doll caught Erik's eye, and he felt in his guts a sensation of sick foreboding.

The prisoners were being guarded by local policemen armed with truncheons. Clearly the Special Group had collaborators for whatever they were doing. The policemen looked at him, noted the German army uniform visible beneath the unbuttoned coat, and said nothing.

As he walked by, a well-dressed Russian prisoner spoke to him in German. 'Sir, I am the director of the tyre factory in this town. I have never believed in Communism, but only paid lip service, as all managers had to. I can help you – I know where everything is. Please take me away from here.'

Erik ignored him and walked in the direction of the shooting.

He came upon the quarry. It was a large, irregular hole in the ground, its edge fringed by tall spruce trees like guardsmen in dark-green uniforms laden with snow. At one end a long slope led into the pit. As he watched, a dozen prisoners began to walk down, two by two, marshalled by soldiers, into the shadowed valley.

Erik noticed three women and a boy of about eleven among them. Was their prison camp somewhere in that quarry? But they were no longer carrying luggage. Snow fell on their bare heads like a benison.

Erik spoke to an SS sergeant standing nearby. 'Who are these prisoners, Sarge?'

'Communists,' said the man. 'From the town. Political commissars, and so on.'

'What, even that little boy?'

'Jews, too,' said the sergeant.

'Well, what are they, Communists or Jews?'

'What's the difference?'

'It's not the same thing.'

'Balls. Most Communists are Jews. Most Jews are Communists. Don't you know anything?'

The tyre factory director who had spoken to Erik seemed to be neither, he thought.

The prisoners reached the rocky floor of the quarry. Until this moment they had shuffled along like sheep in a herd, not speaking or looking around, but now they became animated, pointing at something on the ground. Peering through the snowflakes, Erik saw what looked like bodies scattered among the rocks, snow dusting their garments.

For the first time Erik noticed twelve riflemen standing on the lip of the ravine, among the trees. Twelve prisoners, twelve riflemen: he realized what was happening here, and incredulity mixed with horror rose like bile inside him.

They raised their guns and aimed at the prisoners.

'No,' Erik said. 'No, you can't.' Nobody heard him.

A woman prisoner screamed. Erik saw her grab the eleven-year-old boy and clasp him to herself, as if her arms around him could stop bullets. She seemed to be his mother.

An officer said: 'Fire.'

The rifles cracked. The prisoners staggered and fell. The noise dislodged a little snow from the pines, and it fell on the riflemen, a sprinkling of pure white.

Erik saw the boy and his mother drop, still locked together in an embrace. 'No,' he said. 'Oh, no!'

The sergeant looked at him. 'What's the matter with you?' he said irritably. 'Who are you, anyway?'

'Medical orderly,' said Erik, without taking his eyes off the dread scene in the pit.

'What are you doing here?'

'I brought an ambulance for the officers hurt in the collision.' Erik saw that another twelve prisoners were already being marched down the slope into the quarry. 'Oh, God, my father was right,' he moaned. 'We're murdering people.'

'Stop whining and fuck off back to your ambulance.'

'Yes, Sergeant,' said Erik.

(iii)

At the end of November Volodya asked for a transfer to a fighting unit. His intelligence work no longer seemed important: the Red Army did not need spies in Berlin to discover the intentions of a German army that was already on the outskirts of Moscow. And he wanted to fight for his city.

His misgivings about the government came to seem trivial. Stalin's stupidity, the brutishness of the secret police, the way nothing in the Soviet Union worked the way it was supposed to work – all that faded away. He felt nothing but a blazing need to repel the invader who threatened to bring violence, rape, starvation and death to his mother, his sister, the twins Dimka and Tania, and Zoya.

He was sharply aware that if everyone thought that way he would have no spies. His German informants were people who had decided that patriotism and loyalty were outweighed by the terrible wickedness of the Nazis. He was grateful to them for their courage and the stern morality that drove them. But he felt differently.

So did many of the younger men in Red Army Intelligence, and a small company of them joined a rifle battalion at the beginning of December. Volodya kissed his parents, wrote a note to Zoya saying he hoped to survive to see her again, and moved into barracks.

At long last, Stalin brought reinforcements from the east to Moscow. Thirteen Siberian divisions were deployed against the ever-nearer

Germans. On their way to the front line some of them stopped briefly in Moscow, and Muscovites on the streets stared at them in their white padded coats and warm sheepskin boots, with their skis and goggles and hardy steppe ponies. They arrived in time for the Russian counter-attack.

This was the Red Army's last chance. Time and time again, in the last five months, the Soviet Union had hurled hundreds of thousands of men at the invaders. Each time the Germans had paused, dealt with the attack, and continued their relentless advance. But if this attempt failed there would be no more. The Germans would have Moscow; and when they had Moscow they would have the USSR. And then his mother would be trading vodka for black-market milk for Dimka and Tania.

On the fourth day of December the Soviet forces moved out of the city to the north, west and south and took up their positions for the last effort. They went without lights, to avoid alerting the enemy. They were not allowed to have fires or smoke tobacco.

That evening the front line was visited by NKVD agents. Volodya did not see his rodent-faced brother-in-law Ilya Dvorkin, who must have been among them. A pair he did not recognize came to the bivouac where Volodya and a dozen men were cleaning their rifles. Have you heard anyone criticizing the government? they asked. What do the fellows say about Comrade Stalin? Who among your comrades questions the wisdom of the army's strategy and tactics?

Volodya was incredulous. What did it matter at this point? In the next few days Moscow would be saved or lost. Who cared if soldiers bitched about their officers? He cut the questioning short, saying that he and his men were under a rule of silence, and he had orders to shoot anyone who broke it, but – he added recklessly – he would let the secret policemen off if they left immediately.

That worked, but Volodya had no doubt that the NKVD was undermining the morale of the troops all along the line.

On Friday 5 December in the evening the Russian artillery thundered into action. Next morning at dawn, Volodya and his battalion moved off in a blizzard. Their orders were to take a small town on the far side of a canal.

Volodya ignored orders to attack the German defences frontally – that was the old-fashioned Russian tactic, and this was no moment to stick obstinately to wrong-headed ideas. With his company of a hundred men he went upstream and crossed the ice to the north of the town, then moved in on the Germans' flank. He could hear the crash and roar of battle off to his left, so he knew he was behind the enemy's front line.

Volodya was almost blinded by the blizzard. The occasional blaze of gunfire lit up the clouds for a moment, but at ground level visibility was only a few yards. However, he thought optimistically, that would help the Russians creep up on the Germans and take them by surprise.

It was viciously cold, down to minus 35 Centigrade in places; and while this was bad for both sides, it was worse for the Germans, who lacked cold-weather supplies.

Somewhat to his surprise Volodya found that the normally efficient Germans had not consolidated their line. There were no trenches, no anti-tank ditches, no dugouts. Their front was no more than a series of strongpoints. It was easy to slip through the gaps into the town and look for soft targets: barracks and canteens and ammunition dumps.

His men shot three sentries to take a soccer field in which were parked fifty tanks. Could it be so easy, Volodya wondered? Was the force that had conquered half Russia now depleted and spent?

The corpses of Soviet soldiers, killed in previous skirmishes and left to freeze where they had died, were without their boots and coats, which had presumably been taken by shivering Germans.

The streets of the town were littered with abandoned vehicles – empty trucks with open doors, snow-covered tanks with cold engines, and jeeps with their bonnet lids propped up as if to show that mechanics had tried to fix them but had given up in despair.

Crossing a main road, Volodya heard a car engine and made out, through the snowfall, a pair of headlights approaching on his left. At first he assumed it was a Soviet vehicle that had pushed through the German lines. Then he and his group were fired on, and he yelled at them to take cover. The car turned out to be a Kubelwagen, a Volkswagen jeep with the spare wheel on the hood in front. It had an air-cooled engine, which was why it had not frozen up. It rattled past them at top speed, the Germans firing from their seats.

Volodya was so surprised that he forgot to fire back. Why was a vehicle full of armed Germans driving away from the battle?

He took his company across the road. He had expected that by now they would be fighting their way from house to house, but they met little opposition. The buildings of the occupied town were locked up, shuttered, dark. Any Russians inside were hiding under their beds, if they had any sense.

More cars came along the road, and Volodya decided that officers must be fleeing the battlefield. He detailed a section with a Degtyarev DP-28 light machine gun to take cover in a café and fire on them. He did not want them to live to kill Russians tomorrow.

Just off the main road he spotted a low brick building with bright lights behind skimpy curtains. Creeping past a sentry who could not see far in the snowstorm, he was able to peer in and discern officers inside. He guessed he was looking at a battalion headquarters.

He gave whispered instructions to his sergeants. They shot out the windows then tossed grenades through. A few Germans came out with their hands on their heads. A minute later, Volodya had taken the building.

He heard a new noise. He listened, frowning in puzzlement. More than anything else, it sounded like a football crowd. He stepped out of the headquarters building. The sound was coming from the front line, and it was growing louder.

There was a rattle of machine-gun fire then, a hundred yards away on the main road, a truck slewed sideways and careered off the road into a brick wall, then burst into flames – hit, presumably, by the DP-28 Volodya had deployed. Two more vehicles followed immediately behind it and escaped.

Volodya ran to the café. The machine gun stood on its bipod on a dining table. This model was nicknamed Record Player because of the disc-shaped magazine that sat atop the barrel. The men were enjoying themselves. 'It's like shooting pigeons in the yard, sir!' said a gunner. 'Easy!' One of the men had raided the kitchen and found a big canister of ice cream, miraculously unspoiled, and they were taking turns to scoff it.

Volodya looked out through the smashed window of the café. He saw another vehicle coming, a jeep he thought, and behind it some men running. As they got nearer he recognized German uniforms. More followed behind, dozens, perhaps hundreds. They were responsible for the football-crowd sound.

The gunner trained the barrel on the oncoming car, but Volodya put a hand on his shoulder. 'Wait,' he said.

He stared into the blizzard, making his eyes sting. All he could see was more vehicles and more running men, plus a few horses.

A soldier raised a rifle. 'Don't shoot,' Volodya said. The crowd came closer. 'We can't stop this lot – we'd be overrun in a minute,' he said. 'Let them pass. Take cover.' The men lay down. The gunner lifted the DP-28 off the table. Volodya sat on the floor and peered over the windowsill.

The noise rose to a roar. The leading men drew level with the café and passed. They were running, stumbling and limping. Some carried rifles, most seemed to have lost their weapons; some had coats and hats, others nothing but their uniform tunics. Many were wounded. Volodya saw a man with a bandaged head fall down, crawl a few yards, and collapse. No one took any notice. A cavalryman on horseback trampled an infantryman and galloped on, heedless. Jeeps and staff cars drove dangerously through the crowd, skidding on the ice, honking madly and scattering men to both sides.

It was a rout, Volodya realized. They went by in their thousands. It was a stampede. They were on the run.

At last, the Germans were in retreat.

11

1941 (IV)

Woody Dewar and Joanne Rouzrokh flew from Oakland, California, to Honolulu on a Boeing B-314 flying boat. The Pan Am flight took fourteen hours. Just before arriving they had a massive row.

Perhaps it was spending so long in a small space. The flying boat was one of the biggest planes in the world, but passengers sat in one of six small cabins, each of which had two facing rows of four seats. 'I prefer the train,' said Woody, awkwardly crossing his long legs, and Joanne had the grace not to point out that you could not go to Hawaii by train.

The trip was Woody's parents' idea. They had decided to take a vacation in Hawaii so they could see Woody's younger brother, Chuck, who was stationed there. Then they invited Woody and Joanne to join them for the second week of the holiday.

Woody and Joanne were engaged. Woody had proposed at the end of the summer, after four weeks of hot weather and passionate love in Washington. Joanne had said it was too soon, but Woody had pointed out that he had been in love with her for six years, and asked how long would be enough? She had given in. They would get married next June, as soon as Woody graduated from Harvard. Meanwhile, their engaged status entitled them to go on family holidays together.

She called him Woods, and he called her Jo.

The plane began to lose height as they approached Oahu, the main island. They could see forested mountains, a sparse scatter of villages in the lowlands, and a fringe of sand and surf. 'I bought a new swimsuit,' Joanne said. They were sitting side by side, and the roar of the four Wright Twin Cyclone 14-cylinder engines was too loud for her to be overheard.

Woody was reading *The Grapes of Wrath* but he put it down willingly. 'I can't wait to see you in it.' He meant it. She was a swimsuit manufacturer's dream, making all their products look sensational.

She glanced at him from under half-closed eyelids. 'I wonder if your parents booked us adjoining rooms at the hotel?' Her dark-brown eyes seemed to smoulder.

Their engaged status did not allow them to sleep together, at least not officially; though Woody's mother did not miss much and she might have guessed they were lovers.

Woody said: 'I'll find you, wherever you are.'

'You'd better.'

'Don't talk like that. I'm already uncomfortable enough in this seat.'

She smiled contentedly.

The American naval base came into view. A lagoon shaped like a palm leaf formed a large natural harbour. Half the Pacific Fleet was here, about a hundred ships. The rows of fuel storage tanks looked like checkers on a board.

In the middle of the lagoon was an island with an airstrip. At the western end of the island, Woody saw a dozen or more seaplanes moored.

Right next to the lagoon was Hickam air base. Several hundred aircraft were parked with military precision, wingtip to wingtip, on the tarmac.

Banking for its approach, the plane flew over a beach with palm trees and gaily striped umbrellas – which Woody guessed must be Waikiki – then a small town that had to be Honolulu, the capital.

Joanne was owed some leave by the State Department, but Woody had had to skip a week of classes in order to take this vacation. 'I'm kind of surprised at your father,' Joanne said. 'He's usually against anything that interrupts your education.'

'I know,' said Woody. 'But you know the real reason for this trip, Jo? He thinks it could be the last time we see Chuck alive.'

'Oh, my God, really?'

'He thinks there's going to be a war, and Chuck is in the navy.'

'I think he's right. There will be a war.'

'What makes you so sure?'

'The whole world is hostile to freedom.' She pointed to the book in her lap, a bestseller called *Berlin Diary* by the radio broadcaster William Shirer. 'The Nazis have Europe,' she said. 'The Bolsheviks have Russia. And now the Japanese are taking control of the Far East. I don't see how America can survive in such a world. We have to trade with somebody!'

'That's pretty much what my father thinks. He believes we'll go to war against Japan next year.' Woody frowned thoughtfully. 'What's happening in Russia?'

'The Germans don't seem quite able to take Moscow. Just before I left there was a rumour of a massive Russian counter-attack.'

'Good news!'

Woody looked out. He could see Honolulu airport. The plane would splash down in a sheltered inlet alongside the runway, he presumed.

Joanne said: 'I hope nothing major happens while I'm away.'

'Why?'

'I want a promotion, Woods – so I don't want someone bright and promising to shine in my absence.'

'Promotion? You didn't say.'

'I don't have it yet, but I'm aiming for Research Officer.'

He smiled. 'How high do you want to go?'

'I'd like to be ambassador to someplace fascinating and complex, Nanking or Addis Ababa.'

'Really?'

'Don't look sceptical. Frances Perkins is the first woman Secretary of Labour – and a damn good one.'

Woody nodded. Perkins had been Labor Secretary from the start of Roosevelt's presidency eight years ago, and had won union support for the New Deal. An exceptional woman could aspire to almost anything nowadays. And Joanne was truly exceptional. But somehow it came as a shock to him that she was so ambitious. 'But an ambassador has to live overseas,' he said.

'Wouldn't it be great? Foreign culture, weird weather, exotic customs.'

'But . . . how does that fit in with marriage?'

'Excuse me?' she said with asperity.

He shrugged. 'It's a natural question, don't you think?'

Her expression did not change, except that her nostrils flared – a sign, he knew, that she was getting angry. 'Have I asked *you* that question?' she said.

'No, but . . .'

'Well?'

'I'm just wondering, Jo – do you expect me to live wherever your career takes you?'

'I'll try to fit in with your needs, and I think you should try to fit in with mine.'

'But it's not the same.'

'Isn't it?' She was openly annoyed now. 'This is news to me.'

He wondered how the conversation had become so acrimonious so quickly. With an effort at making his tone of voice reasonable and amiable, he said: 'We've talked about having children, haven't we.'

'You'll have them, as well as me.'

'Not in exactly the same way.'

'If children are going to make me a second-class citizen in this marriage, then we're not having any.'

'That's not what I mean!'

'What the heck do you mean?'

'If you're appointed ambassador somewhere, do you expect me to drop everything and go with you?'

'I expect you to say: 'My darling, this is a wonderful opportunity for you, and I'm certainly not going to stand in your way.' Is that unreasonable?'

'Yes!' Woody was baffled and angry. 'What's the point of being married, if we're not together?'

'If war breaks out, will you volunteer?'

'I guess I might.'

'And the army would send you wherever they needed you – Europe, the Far East.'

'Well, yes.'

'So you'll go where your duty takes you, and leave me at home.'

'If I have to.'

'But I can't do that.'

'It's not the same! Why are you pretending it is?'

'Strangely enough, my career and my service to my country seem important to me – just as important as yours to you.'

'You're just being perverse!'

'Well, Woods, I'm really sorry you think that, because I've been talking very seriously about our future together. Now I have to ask myself whether we even have one.'

'Of course we do!' Woody could have screamed with frustration. 'How did this happen? How did we get to this?'

There was a bump, and the plane splashed down in Hawaii.

(ii)

Chuck Dewar was terrified that his parents would learn his secret.

Back home in Buffalo he had never had a real love affair, just a few hasty fumbles in dark alleys with boys he hardly knew. Half the reason he had joined the navy was to go places where he could be himself without his parents finding out.

Since he got to Hawaii it had been different. Here he was part of an underground community of similar people. He went to bars and restaurants and dance halls where he did not have to pretend to be heterosexual. He had had some affairs, and then he had fallen in love. A lot of people knew his secret.

And now his parents were here.

His father was invited to visit the signal intelligence unit at the naval base, known as Station HYPO. As a member of the Senate Foreign Relations Committee, Senator Dewar was let into many military secrets, and he had already been shown around signals intelligence headquarters, called Op-20-G, in Washington.

Chuck picked him up at his hotel in Honolulu in a navy car, a Packard LeBaron limousine. Papa was wearing a white straw hat. As they drove around the rim of the harbour, he whistled. 'The Pacific Fleet,' he said. 'A beautiful sight.'

Chuck agreed. 'Quite something, isn't it?' he said. Ships were beautiful, especially in the US Navy, where they were painted and scrubbed and shined. Chuck thought the navy was great.

'All those battleships in a perfect straight line,' Gus marvelled.

'We call it Battleship Row. Moored off the island are *Maryland, Tennessee, Arizona, Nevada, Oklahoma* and *West Virginia.*' Battleships were named after states. 'We also have *California* and *Pennsylvania* in harbour, but you can't see them from here.'

At the main gate to the Navy Yard, the marine on sentry duty recognized the official car and waved them in. They drove to the submarine base and stopped in the parking lot behind headquarters, the Old Administration Building. Chuck took his father into the recently opened new wing.

Captain Vandermeier was waiting for them.

Vandermeier was Chuck's greatest fear. He had taken a dislike to

Chuck, and he had guessed the secret. He was always calling Chuck a powder puff or a pantywaist. If he could, he would spill the beans.

Vandermeier was a short, stocky man with a gravelly voice and bad breath. He saluted Gus and shook hands. 'Welcome, Senator. It'll be my privilege to show you the Communications Intelligence Unit of the fourteenth Naval District.' This was the deliberately vague title for the group monitoring the radio signals of the Imperial Japanese Navy.

'Thank you, Captain,' said Gus.

'A word of warning, first, sir. It's an informal group. This kind of work is often done by eccentric people, and correct naval uniform is not always worn. The officer in charge, Commander Rochefort, wears a red velvet jacket.' Vandermeier gave a man-to-man grin. 'You may think he looks like a goddamn homo.'

Chuck tried not to wince.

Vandermeier said: 'I won't say any more until we're in the secure zone.'

'Very good,' said Gus.

They went down the stairs and into the basement, passing through two locked doors on the way.

Station HYPO was a windowless neon-lit cellar housing thirty men. As well as the usual desks and chairs, it had oversized chart desks, racks of exotic IBM machine printers, sorters and collators, and two cots where the cryptanalysts took naps during their marathon codebreaking sessions. Some of the men wore neat uniforms but others, as Vandermeier had warned, were in scruffy civilian clothing, unshaven, and – to judge by the smell – unwashed.

'Like all navies, the Japanese have many different codes, using the simplest for less secret signals, such as weather reports, and saving the complex ones for the most highly sensitive messages,' Vandermeier said. 'For example, call signs identifying the sender of a message and its destination are in a primitive cipher, even when the text itself is in a high-grade cipher. They recently changed the code for call signs, but we cracked the new one in a few days.'

'Very impressive,' said Gus.

'We can also figure out where the signal originated, by triangulation. Given locations and the call signs, we can build up a pretty good picture of where most of the ships of the Japanese navy are, even if we can't read the messages.'

'So we know where they are, and what direction they're taking, but not what their orders are,' said Gus.

'Frequently, yes.'

'But if they wanted to hide from us, all they would have to do is impose radio silence.'

'True,' said Vandermeier. 'If they go quiet, this whole operation becomes useless, and we are well and truly fucked up the ass.'

A man in a smoking jacket and carpet slippers approached, and Vandermeier introduced the head of the unit. 'Commander Rochefort is fluent in Japanese, as well as being a master cryptanalyst,' Vandermeier said.

'We were making good progress decrypting the main Japanese cipher until a few days ago,' Rochefort said. 'Then the bastards changed it and undid all our work.'

Gus said: 'Captain Vandermeier was telling me you can learn a lot without actually reading the messages.'

'Yes.' Rochefort pointed to a wall chart. 'Right now, most of the Japanese fleet has left home waters and is heading south.'

'Ominous.'

'It sure is. But tell me, Senator, what's your reading of Japanese intentions?'

'I believe they will declare war on the United States. Our oil embargo is really hurting them. The British and the Dutch are refusing to supply them, and right now they're trying to ship it from South America. They can't survive like this indefinitely.'

Vandermeier said: 'But what would they achieve by attacking us? A little country such as Japan can't invade the USA!'

Gus said: 'Great Britain is a little country, but they achieved world domination just by ruling the seas. The Japanese don't have to conquer America, they just need to defeat us in a naval war, so that they can control the Pacific, and no one can stop them trading.'

'So, in your opinion, what might they be doing, heading south?'

'Their likeliest target has to be the Philippines.'

Rochefort nodded agreement. 'We've already reinforced our base there. But one thing bothers me: the commander of the Japanese aircraft carrier fleet hasn't received any signals for several days.'

Gus frowned. 'Radio silence. Has that ever happened before?'

'Yes. Aircraft carriers go quiet when they return to home waters. So we assume that's the explanation this time.'

Gus nodded. 'It sounds reasonable.'

'Yes,' said Rochefort. 'I just wish I could be sure.'

(iii)

The Christmas lights were ablaze on Fort Street in Honolulu. It was Saturday night, 6 December, and the street was thronged with sailors in white tropical uniform, each with a round white cap and a crossed black scarf, all out for a good time.

The Dewar family strolled along enjoying the atmosphere, Rosa on Chuck's arm and Gus and Woody on either side of Joanne.

Woody had patched up his quarrel with his fiancée. He apologized for making wrong assumptions about what Joanne expected in their marriage. Joanne admitted she had flown off the handle. Nothing was truly resolved, but it was enough of a rapprochement for them to tear off their clothes and jump into bed.

Afterwards, the quarrel seemed less important, and nothing really mattered except how much they loved each other. Then they vowed that in future they would discuss such agreements in a loving and tolerant way. As they got dressed Woody felt that they had passed a milestone. They had had an acrimonious quarrel about a serious difference of view, but they had survived it. It could even be a good sign.

Now they were heading out for dinner, Woody carrying his camera, snapping photos of the scene as they walked along. Before they had gone far Chuck stopped and introduced another sailor. 'This is my pal, Eddie Parry. Eddie, meet Senator Dewar, Mrs Dewar, my brother Woody, and Woody's fiancée, Miss Joanne Rouzrokh.'

Rosa said: 'I'm pleased to meet you, Eddie. Chuck has mentioned you several times in his letters home. Won't you join us for dinner? We're only going to eat Chinese.'

Woody was surprised. It was not like his mother to invite a stranger to a family meal.

Eddie said: 'Thank you, ma'am. I'd be honoured.' He had a southern accent.

They went into the Heavenly Delight restaurant and sat down at a table for six. Eddie had formal manners, calling Gus 'sir' and the women 'ma'am', but he seemed relaxed. After they had ordered he said: 'I've heard so much about this family, I feel as if I know y'all.' He had a freckled face and a big smile, and Woody could tell that everyone liked him.

Eddie asked Rosa how she liked Hawaii. 'To tell you the truth, I'm a little disappointed,' she said. 'Honolulu is just like any small American town. I expected it to be more Asian.'

'I agree,' said Eddie. 'It's all diners and motor courts and jazz bands.'

He asked Gus if there was going to be a war. Everyone asked Gus that question. 'We've tried our darnedest to reach a modus vivendi with Japan,' Gus said. Woody wondered if Eddie knew what a modus vivendi was. 'Secretary of State Hull had a whole series of talks with Ambassador Nomura that lasted all summer long. But we can't seem to agree.'

'What's the problem?' said Eddie.

'American business needs a free trade zone in the Far East. Japan says okay, fine, we love free trade, let's have it, not just in our backyard, but all over the world. The United States can't deliver that, even if we wanted it. So Japan says that as long as other countries have their own economic zone, they need one too.'

'I still don't see why they had to invade China.'

Rosa, who always tried to see the other side, said: 'The Japanese want troops in China and Indochina and the Dutch East Indies to protect their interests, just as we Americans have troops in the Philippines, and the British have theirs in India, and the French in Algeria, and so on.'

'When you put it that way, the Japs don't seem so unreasonable!'

Joanne said firmly: 'They're not unreasonable, but they're wrong. Conquering an empire is the nineteenth-century solution. The world is changing. We're moving away from empires and closed economic zones. To give them what they want would be a backward step.'

Their food arrived. 'Before I forget,' Gus said, 'we're having breakfast tomorrow morning aboard the *Arizona*. Eight o'clock sharp.'

Chuck said: 'I'm not invited, but I've been detailed to get you there. I'll pick you up at seven-thirty and drive you to the Navy Yard, then take you across the harbour in a launch.'

'Fine.'

Woody tucked in to fried rice. 'This is great,' he said. 'We should have Chinese food at our wedding.'

Gus laughed. 'I don't think so.'

'Why not? It's cheap, and it tastes good.'

'A wedding is more than a meal, it's an occasion. Speaking of which, Joanne, I must call your mother.'

Joanne frowned. 'About the wedding?'

'About the guest list.'

Joanne put down her chopsticks. 'Is there a problem?' Woody saw her nostrils flare, and knew there was going to be trouble.

'Not really a problem,' said Gus. 'I have a rather large number of friends and allies in Washington who would be offended if they were not invited to the wedding of my son. I'm going to suggest that your mother and I share the cost.'

Papa was being thoughtful, Woody guessed. Because Dave had sold his business for a bargain price before he died, Joanne's mother might not have a lot of money to spare for a swanky wedding. But Joanne disliked the idea of the two parents making wedding arrangements over her head.

'Who are the friends and allies you're thinking about?' Joanne said coolly.

'Senators and congressmen, mostly. We must invite the President, but he won't come.'

'Which senators and congressmen?' Joanne asked.

Woody saw his mother hide a grin. She was amused at Joanne's insistence. Not many people had the nerve to push Gus up against the wall like this.

Gus began a list of names.

Joanne interrupted him. 'Did you say Congressman Cobb?'

'Yes.'

'He voted against the anti-lynching law!'

'Peter Cobb is a good man. But he's a Mississippi politician. We live in a democracy, Joanne: we have to represent our voters. Southerners won't support an anti-lynching law.' He looked at Chuck's friend. 'I hope I'm not treading on any toes here, Eddie.'

'Don't mince your words on my account, sir,' Eddie said. 'I'm from Texas, but I feel ashamed when I think of southern politics. I hate prejudice. A man's a man, whatever his colour.'

Woody glanced at Chuck. He looked so proud of Eddie he might have burst.

At that moment, Woody realized that Eddie was more than just Chuck's pal.

That was weird.

There were three loving couples around the table: Papa and Mama, Woody and Joanne, and Chuck and Eddie.

He stared at Eddie. Chuck's lover, he thought.

Damn weird.

Eddie caught him staring, and smiled amiably.

Woody tore his gaze away. Thank God Papa and Mama haven't figured it out, he thought.

Unless that was why Mama had invited Eddie to join in a family dinner. Did she know? Did she even approve? No, that was beyond the bounds of possibility.

'Anyway, Cobb has no choice,' Papa was saying. 'And in everything else he's a liberal.'

'There's nothing democratic about it,' Joanne said hotly. 'Cobb doesn't represent the people of the south. Only white people are allowed to vote there.'

Gus said: 'Nothing is perfect in this life. Cobb supported Roosevelt's New Deal.'

'That doesn't mean I have to invite him to my wedding.'

Woody put in: 'Papa, I don't want him either. He has blood on his hands.'

'That's unfair.'

'It's how we feel.'

'Well, the decision is not entirely up to you. Joanne's mother will be throwing the party, and if she'll let me I'll share the cost. I guess that gives us at least a say in the guest list.'

Woody sat back. 'Heck, it's our wedding.'

Joanne looked at Woody. 'Maybe we should have a quiet town hall wedding, with just a few friends.'

Woody shrugged. 'Suits me.'

Gus said severely: 'That would upset a lot of people.'

'But not us,' said Woody. 'The most important person of the day is the bride. I just want her to have what she wants.'

Rosa spoke up. 'Listen to me, everyone,' she said. 'Don't let's go overboard. Gus, my darling, you may have to take Peter Cobb aside and explain to him, gently, that you are lucky enough to have an idealistic son, who is marrying a wonderful and equally idealistic girl, and they have stubbornly refused your impassioned request to invite Congressman Cobb to the wedding. You're sorry, but you cannot follow your own inclinations in this any more than Peter can follow his when voting on anti-lynching bills. He will smile and say he understands, and he has always liked you because you're as straight as a die.'

Gus hesitated for a long moment, then decided to give in graciously. 'I guess you're right, my dear,' he said. He smiled at Joanne. 'Anyway, I'd be a fool to quarrel with my delightful daughter-in-law on account of Pete Cobb.'

Joanne said: 'Thank you . . . Should I start calling you Papa yet?'

Woody almost gasped. It was the perfect thing to say. She was so damn smart!

Gus said: 'I would really like that.'

Woody thought he saw the glint of a tear in his father's eye.

Joanne said: 'Then thank you, Papa.'

How about that? thought Woody. She stood up to him – and she won.

What a girl!

(iv)

On Sunday morning, Eddie wanted to go with Chuck to pick up the family at their hotel.

'I don't know, baby,' said Chuck. 'You and I are supposed to be friendly, not inseparable.'

They were in bed in a motel at dawn. They had to sneak back into barracks before sunup.

'You're ashamed of me,' said Eddie.

'How can you say that? I took you to dinner with my family!'

'That was your Mama's idea, not yours. But your Papa liked me, didn't he?'

'They all adored you. Who wouldn't? But they don't know you're a filthy homo.'

'I am not a filthy homo. I'm a very clean homo.'

'True.'

'Please take me. I want to know them better. It's really important to me.'

Chuck sighed. 'Okay.'

'Thank you.' Eddie kissed him. 'Do we have time . . . ?'

Chuck grinned. 'If we're quick.'

Two hours later they were outside the hotel in the navy's Packard. Their four passengers appeared at seven-thirty. Rosa and Joanne wore hats and gloves, Gus and Woody white linen suits. Woody had his camera.

Woody and Joanne were holding hands. 'Look at my brother,' Chuck murmured to Eddie. 'He's so happy.'

'She's a beautiful girl.'

They held the doors open and the Dewars climbed into the back of the limousine. Woody and Joanne folded down the jump seats. Chuck pulled away and headed for the naval base.

It was a fine morning. On the car radio, station KGMB was playing hymns. The sun shone over the lagoon and glinted off the glass portholes and polished brass rails of a hundred ships. Chuck said: 'Isn't that a pretty sight?'

They entered the base and drove to the Navy Yard, where a dozen ships were in floating docks and dry docks for repair, maintenance and refuelling. Chuck pulled up at the Officers Landing. They all got out and looked across the lagoon at the mighty battleships standing proud in the morning light. Woody took a photo.

It was a few minutes before eight o'clock. Chuck could hear the tolling of church bells in nearby Pearl City. On the ships, the forenoon watch was being piped to breakfast, and colour parties were assembling to hoist ensigns at eight precisely. A band on the deck of the *Nevada* was playing 'The Star-Spangled Banner'.

They walked to the jetty, where a launch was tied up ready for them. The boat was big enough to take a dozen passengers and had an inboard motor under a hatch in the stern. Eddie started the engine while Chuck handed the guests into the boat. The small motor burbled cheerfully. Chuck stood in the bows while Eddie eased the launch away from the dockside and turned towards the battleships. The prow lifted as the launch picked up speed, throwing off twin curves of foam like a seagull's wings.

Chuck heard a plane and looked up. It was coming in from the west, so low it looked as if it might be in danger of crashing. He assumed it was about to land at the naval airstrip on Ford Island.

Woody, sitting near Chuck in the bows, frowned and said: 'What kind of plane is that?'

Chuck knew every aircraft of both the army and the navy, but he had trouble identifying this one. 'It almost looks like a Type Ninety-seven,' he said. That was the carrier-based torpedo bomber of the Imperial Japanese Navy.

Woody pointed his camera.

As the plane came nearer, Chuck saw large red suns painted on its wings. 'It *is* a Jap plane!' he said.

Eddie, steering the boat from the stern, heard him. 'They must have faked it up for an exercise,' he said. 'A surprise drill to spoil everyone's Sunday morning.'

'I guess so,' said Chuck.

Then he saw a second plane behind the first.

And another.

He heard his father say anxiously: 'What the heck is going on?'

The planes banked over the Navy Yard and passed low over the launch, their noise rising to a roar like Niagara Falls. There were about ten of them, Chuck saw; no, twenty; no, more.

They headed straight for Battleship Row.

Woody stopped taking pictures to say: 'It can't be a real attack, can it?' There was fear as well as doubt in his voice.

'How could they be Japanese?' Chuck said incredulously. 'Japan is nearly four thousand miles away! No plane can fly that far.'

Then he remembered that the aircraft carriers of the Japanese navy had gone into radio silence. The signal intelligence unit had assumed they were in home waters, but had never been able to confirm that.

He caught his father's eye, and guessed he was remembering the same conversation.

Everything suddenly became clear, and incredulity turned to fear.

The lead plane flew low over the *Nevada*, the stern marker in Battleship Row. There was a burst of cannon fire. On deck, seamen scattered and the band left off in a ragged diminuendo of abandoned notes.

In the launch, Rosa screamed.

Eddie said: 'Christ Jesus in heaven, it's an attack.'

Chuck's heart pounded. The Japanese were bombing Pearl Harbor, and he was in a small boat in the middle of the lagoon. He looked at the scared faces of the others – both parents, his brother, and Eddie – and realized that all the people he loved were in the boat with him.

Long bullet-shaped torpedoes began to fall from the underbellies of the planes and splash into the tranquil waters of the lagoon.

Chuck yelled: 'Turn back, Eddie!' But Eddie was already doing it, swinging the launch around in a tight arc.

As it turned, Chuck saw, over Hickam air base, another flight of aircraft with the big red discs on their wings. These were dive bombers, and they were streaming down like birds of prey on the rows of American aircraft perfectly lined up on the runways.

How the hell many of the bastards were there? Half the Japanese air force seemed to be in the sky over Pearl.

Woody was still taking pictures.

Chuck heard a deep bang like an underground explosion, then another immediately after. He spun around. There was a flash of flame aboard the *Arizona*, and smoke began to rise from her.

The stern of the launch squatted farther into the water as Eddie opened the throttle. Chuck said unnecessarily: 'Hurry, hurry!'

From one of the ships Chuck heard the insistent rhythmic hoot of a klaxon sounding General Quarters, calling the crew to battle stations, and he realized that this *was* a battle, and his family was in the middle of it. A moment later on Ford Island the air-raid siren began with a low moan and wailed higher in pitch until it struck its frantic top note.

There was a long series of explosions from Battleship Row as torpedoes found their targets. Eddie yelled: 'Look at the Wee Vee!' It was what they called the *West Virginia*. 'She's listing to port!'

He was right, Chuck saw. The ship had been holed on the side nearest the attacking planes. Millions of tons of water must have poured into her in a few seconds to make such a huge vessel tilt sideways.

Next to her, the same fate was overtaking the *Oklahoma*, and to his horror Chuck could see sailors slipping helplessly, sliding across the tilted deck and falling over the side into the water.

Waves from the explosions rocked the launch. Everyone clung to the sides.

Chuck saw bombs rain down on the seaplane base at the near end of Ford Island. The planes were moored close together, and the fragile aircraft were blown to pieces, fragments of wings and fuselages flying into the air like leaves in a hurricane.

Chuck's intelligence-trained mind was trying to identify aircraft types, and now he spotted a third model among the Japanese attackers, the deadly Mitsubishi 'Zero', the best carrier-based fighter in the world. It had only two small bombs, but was armed with twin machine guns and a pair of 20mm cannon. Its role in this attack must be to escort the bombers, defending them from American fighters – but all the American fighters were still on the ground, where many of them had already been destroyed. That left the Zeroes free to strafe buildings, equipment and troops.

Or, Chuck thought fearfully, to strafe a family crossing the lagoon, desperately trying to get to shore.

At last the United States began to shoot back. On Ford Island, and on the decks of the ships that had not yet been hit, anti-aircraft guns and regular machine guns came to life, adding their rattle to the cacophony of lethal noise. Anti-aircraft shells burst in the sky like black flowers blossoming. Almost immediately, a machine-gunner on the island scored a direct hit on a dive-bomber. The cockpit burst into flames and the plane hit the water with a mighty splash. Chuck found himself cheering savagely, shaking his fists in the air.

The listing *West Virginia* began to return to the vertical, but continued to sink, and Chuck realized that the commander must have opened the starboard seacocks, to ensure that she remained upright while she went down, giving the crew a better chance of survival. But the *Oklahoma* was not so fortunate, and they all watched in terrified awe as the great ship began to turn over. Joanne said: 'Oh, God, look at the crew.' The sailors were frantically scrambling up the steeply banked deck and over the starboard rail in a desperate attempt to save themselves. But they were the lucky ones, Chuck realized, as at last the mighty vessel turned turtle with a terrible crash and began to sink, for how many hundreds of men were trapped below decks?

'Hold on, everyone!' Chuck yelled. A huge wave created by the capsize of the *Oklahoma* was approaching. Papa grabbed Mama and Woody held on to Joanne. The wave reached them and lifted the launch impossibly high. Chuck staggered but kept hold of the rail. The launch stayed afloat. Smaller waves followed, rocking them, but everyone was safe.

They were still a long quarter of a mile offshore, Chuck saw with consternation.

Astonishingly the *Nevada*, which had been strafed at the start, began to move off. Someone must have had the presence of mind to signal all ships to sail. If they could get out of the harbour they could scatter and present less easy targets.

Then from Battleship Row came a bang ten times bigger than anything that had gone before. The explosion was so violent that Chuck felt the blast like a blow to his chest, though he was now almost half a mile away. A spurt of flame spewed out of the No. 2 gun turret of the *Arizona*. A split-second later the forward half of the ship seemed to burst. Debris flew into the air, twisted steel girders and warped plates drifting up through the smoke with a nightmare slowness, like scraps of charred paper from a bonfire. Flames and smoke enveloped the front of the ship. The lofty mast tipped forward drunkenly.

Woody said: 'What was *that*?'

'The ship's ammunition store must have gone up,' Chuck said, and he realized with heartfelt grief that hundreds of his fellow seamen must have been killed in that mammoth detonation.

A column of dark-red smoke rose into the air as from a funeral pyre.

There was a crash and the boat lurched as something hit it. Everyone ducked. Falling to his knees, Chuck thought it must be a bomb, then realized it could not be, for he was still alive. When he recovered, he saw that a heavy scrap of metal debris a yard long had pierced the deck over the engine. It was a miracle it had not hit anyone.

However, the engine died.

The boat slowed and was becalmed. It wallowed in the choppy waves while Japanese planes rained hell fire on the lagoon.

Gus said tightly: 'Chuck, we have to get out of here right now.'

'I know.' Chuck and Eddie examined the damage. They grabbed the metal scrap and tried to wrestle it out of the teak deck, but it was firmly stuck.

'We don't have time for this!' Gus said.

Woody said: 'The engine is blitzed anyway, Chuck.'

They were still a quarter of a mile from shore. However, the launch was equipped for an emergency such as this. Chuck unshipped a pair of oars. He took one and Eddie took the other. The boat was large, for rowing, and their progress was slow.

Luckily for them there was a lull in the attack. The sky was no longer swarming with planes. Vast billows of smoke rose from the damaged ships, including a column a thousand feet high from the fatally wounded *Arizona*, but there were no new explosions. The amazingly plucky *Nevada* was now heading for the mouth of the harbour.

The water around the ships was crowded with life rafts, motor launches, and seamen swimming or clinging to floating wreckage. Drowning was not their only fear: oil from the holed ships had spread across the surface and caught fire. The cries for help of those who could not swim mingled horrifyingly with the screams of the burned.

Chuck stole a glance at his watch. He thought the attack had been going on for hours but, amazingly, it was only thirty minutes.

Just as he was thinking that, the second wave began.

This time the planes came from the east. Some of them chased the escaping *Nevada*; others targeted the Navy Yard where the Dewars had boarded the launch. Almost immediately the destroyer *Shaw* in a floating dock exploded with great gouts of flame and billows of smoke. Oil spread across the water and caught fire. Then in the largest dry dock the battleship *Pennsylvania* was hit. Two destroyers in the same dry dock blew up as their ammunition stores were ignited.

Chuck and Eddie strained at the oars, sweating like racehorses.

At the Navy Yard, Marines appeared – presumably from the nearby barracks – and broke out firefighting gear.

At last the launch reached the Officers' Landing. Chuck leaped out and swiftly tied up while Eddie helped the passengers out. They all ran to the car.

Chuck jumped into the driving seat and started the engine. The car radio came on automatically, and he heard the KGMB announcer say: 'All Army, Navy and Marine personnel report for duty immediately.' Chuck had not had a chance to report to anyone, but he felt sure that his orders would be first to ensure the safety of the four civilians in his care, especially as two were women and one was a senator.

As soon as everyone was in the car he pulled away.

The second wave of the attack seemed to be ending. Most of the Japanese planes were heading away from the harbour. All the same, Chuck drove fast: there might be a third wave.

The main gate was open. If it had been shut he would have been tempted to crash it.

There was no other traffic.

He raced away from the harbour along Kamehameha Highway. The farther he got from Pearl Harbor, the safer his family would be, he figured.

Then he saw a lone Zero coming towards him.

It was flying low and following the highway, and after a moment he realized it was targeting the car.

The cannon were in the wings, and there was a good chance they would miss the narrow target of the car; but the machine guns were set close together, either side of the engine cowling. That was what the pilot would use if he was smart.

Chuck looked frantically at both sides of the road. There was no hiding place, nothing but cane fields.

He began to zigzag. The approaching pilot sensibly did not attempt to track him. The road was not wide, and if Chuck drove into the cane field the car would be slowed to a walking pace. He stepped on the gas, realizing that the faster he was going the better his chances of not being hit.

Then it was too late for forethought. The plane was so close Chuck could see the round black holes in the wings through which the cannon fired. But, as he had guessed, the pilot opened up with machine guns, and bullets spat dust from the road ahead.

Chuck moved left, to the crown of the road, then instead of continuing left he swerved right. The pilot corrected. Bullets hit the hood. The windscreen smashed. Eddie roared with pain, and in the back one of the women screamed.

Then the Zero was gone.

The car began to zigzag of its own accord. A forward wheel must have been damaged. Chuck fought with the steering wheel, trying to stay on the road. The car slewed sideways, skidded across the tarmac, crashed into the field at the side of the road, and bumped to a stop.

Flames rose from the engine, and Chuck smelled gasoline.

'Everybody out!' Chuck yelled. 'Before the fuel tank blows!' He opened his door and leaped out. He yanked open the rear door and his father jumped out, pulling his mother along. Chuck could see the others getting out on the far side. 'Run!' he shouted, but it was superfluous. Eddie was already heading into the cane field, limping as though wounded. Woody was half pulling, half carrying Joanne, who also seemed to have been hit. His parents charged into the field, apparently unhurt. He joined them. They all ran a hundred yards then threw themselves flat.

There was a moment of stillness. The sounds of planes had become a distant buzz. Glancing up, Chuck saw oily smoke from the harbour rising thousands of feet into the air. Above that, the last few high-level bombers were heading away to the north.

Then there was a bang that stunned his eardrums. Even with closed eyes he saw the bright flash of exploding gasoline. A wave of heat passed over him.

He lifted his head and looked back. The car was ablaze.

He jumped to his feet. 'Mama! Are you okay?'

'Miraculously unhurt,' she said coolly as his father helped her up.

He scanned the field and spotted the others. He ran to Eddie, who was sitting upright, clutching his thigh. 'Are you hit?'

'Hurts like fuck,' Eddie said. 'But there's not much blood.' He managed a grin. 'Top of my thigh, I think, but no vital organs damaged.'

'We'll get you to hospital.'

At that moment Chuck heard a terrible noise.

His brother was crying.

Woody was weeping not like a baby but like a lost child: a loud, sobbing noise of utter wretchedness.

Chuck knew immediately that it was the sound of a broken heart.

He ran to his brother. Woody was on his knees, his chest shaking, his mouth open, his eyes running with tears. There was blood all over his white linen suit, but he was not wounded. Between sobs he moaned: 'No, no.'

Joanne lay on the ground in front of him, face up.

Chuck could see right away that she was dead. Her body was still and her eyes were open, staring at nothing. The front of her gaily striped cotton dress was soaked with bright red arterial blood, already darkening in patches. Chuck could not see the wound but he guessed she had taken a bullet to the shoulder that had opened her axillary artery. She would have bled to death in minutes.

He did not know what to say.

The others came and stood by him: Mama, Papa, and Eddie. Mama knelt on the ground beside Woody and put her arms around him. 'My poor boy,' she said, as if he was a child.

Eddie put his arm around Chuck's shoulders and gave him a discreet hug.

Papa knelt by the body. He reached out and took Woody's hand.

Woody's sobs quieted a little.

Papa said: 'Close her eyes, Woody.'

Woody's hand was shaking. With an effort, he steadied it.

He stretched out his fingertips to her eyelids.

Then, with infinite gentleness, he closed her eyes.

12

1942 (I)

On the first day of 1942 Daisy got a letter from her former fiancé, Charlie Farquharson.

When she opened it she was at the breakfast table in the Mayfair house, alone except for the aged butler who poured her coffee and the fifteen-year-old maid who brought her hot toast from the kitchen.

Charlie wrote not from Buffalo but from RAF Duxford, an air base in the east of England. Daisy had heard of the place: it was near Cambridge, where she had met both her husband, Boy Fitzherbert, and the man she loved, Lloyd Williams.

She was pleased to hear from Charlie. He had jilted her, of course, and she had hated him then; but it was a long time ago. She felt like a different person now. In 1935 she had been an American heiress called Miss Peshkov; today she was Viscountess Aberowen, an English aristocrat. All the same, she was pleased she was still in Charlie's mind. A woman would always prefer to be remembered than forgotten.

Charlie wrote with a heavy black pen. His handwriting was untidy, the letters large and jagged. Daisy read:

Before anything else, I need, of course, to apologize for the way I treated you back in Buffalo. I shudder with mortification every time I think of it.

Good Lord, thought Daisy, he seems to have grown up.

What snobs we all were, and how weak I was to allow my late mother to bully me into behaving shabbily.

Ah, she thought, his *late* mother. So the old bitch is dead. That might explain the change.

I have joined No. 133 Eagle Squadron. We fly Hurricanes, but we're getting Spitfires any day now.

There were three Eagle squadrons, Royal Air Force units manned by
American volunteers. Daisy was surprised: she would not have expected
Charlie to go to war voluntarily. When she knew him he had been
interested in nothing but dogs and horses. He really had grown up.

*If you can find it in your heart to forgive me, or at least to put the past
behind you, I would love to see you and meet your husband.*

The mention of a husband was a tactful way of saying he had no
romantic intentions, Daisy guessed.

*I will be in London on leave next weekend. May I take the two of you to
dinner? Do say yes.*

With affectionate good wishes,

Charles H.B. Farquharson

Boy was not at home that weekend, but Daisy accepted for herself.
She was starved of male companionship, like many women in wartime
London. Lloyd had gone to Spain and disappeared. He said he was
going to be a military attaché at the British embassy in Madrid. Daisy
wished it might be true that he had such a safe job, but she did not
believe it. When she asked why the government would send an able-
bodied young officer to do a desk job in a neutral country, he had
explained how important it was to discourage Spain from joining in
the war on the Fascist side. But he said it with a rueful smile that told
her plainly she was not to be fooled. She feared that in reality he was
slipping across the border to work with the French Resistance, and she
had nightmares about him being captured and tortured.

She had not seen him for more than a year. His absence was like an
amputation: she felt it every hour of the day. But she was glad of the
chance to spend an evening out with a man, even if it was the awkward,
unglamorous, overweight Charlie Farquharson.

Charlie booked a table in the Grill Room of the Savoy Hotel.

In the lobby of the hotel, as a waiter was helping her take off her
mink coat, she was approached by a tall man in a well-cut dinner jacket
who looked vaguely familiar. He stuck out his hand and said shyly:
'Hello, Daisy. What a pleasure to see you after all these years.'

When she heard his voice she realized it was Charlie. 'Good Lord!'
she said. 'You've changed!'

'I lost a little weight,' he admitted.

'You sure did.' Forty or fifty pounds, she guessed. It made him better-looking. His features now seemed craggy rather than ugly.

'But you haven't changed at all,' he said, looking her up and down.

She had made an effort with her clothes. She had bought nothing new for years, because of wartime austerity, but for tonight she had exhumed an off-the-shoulder sapphire-blue silk evening gown by Lanvin that she had acquired on her last pre-war trip to Paris. 'In a couple of months I'll be twenty-six,' she said. 'I can't believe I look the same as I did when I was eighteen.'

He glanced down at her décolletage, blushed, and said: 'Believe me, you do.'

They went into the restaurant and sat down. 'I was afraid you weren't coming,' he said.

'My watch stopped. I'm sorry I'm late.'

'Only by twenty minutes. I would have waited an hour.'

A waiter asked if they would like a drink. Daisy said: 'This is one of the few places in England where you can get a decent martini.'

'Two of those, please,' Charlie said.

'I like mine straight up with an olive.'

'So do I.'

She studied him, intrigued by the way he had altered. His old awkwardness had softened to a charming shyness. It was still hard to imagine him as a fighter pilot, shooting down German planes. Anyway, the Blitz on London had come to an end half a year ago, and there were no longer air battles in the skies over southern England. 'What kind of flying do you do?' she said.

'Mainly daytime circus operations over northern France.'

'What's a circus operation?'

'A bomber attack with a heavy escort of fighters, the main object being to lure enemy planes into an air battle in which they're outnumbered.'

'I hate bombers,' she said. 'I lived through the Blitz.'

He was surprised. 'I would have thought you'd want to give the Germans a taste of their own medicine.'

'Not at all.' Daisy had thought about this a lot. 'I could weep for all the innocent women and children who were burned and maimed in London – and it doesn't help at all to know that German women and children are suffering the same.'

'I never looked at it that way.'

They ordered dinner. Wartime regulations restricted them to three courses, and their meal could not cost more than five shillings. On the menu were special austerity dishes such as Mock Duck – made out of pork sausages – and Woolton Pie, which contained no meat at all.

Charlie said: 'I can't tell you how good it is to hear a girl speak real American. I like English girls, and I've even dated one, but I miss American voices.'

'Me, too,' she said. 'This is my home now, and I don't guess I'll ever go back, but I know how you feel.'

'I'm sorry I missed meeting Viscount Aberowen.'

'He's in the air force, like you. He's a pilot trainer. He gets home now and again – but not this weekend.'

Daisy was sleeping with Boy again, on his occasional visits home. She had sworn she never would after catching him with those awful women in Aldgate. But he had put pressure on her. He said that fighting men needed consolation when they came home, and he had promised never to visit prostitutes again. She did not really believe his promises, but all the same she gave in, albeit against her inclination. After all, she told herself, I did marry him for better or worse.

However, she no longer took any pleasure in sex with him, unfortunately. She could go to bed with Boy but she could not fall back in love with him. She had to use cream for lubrication. She had tried to summon again the fond feelings she had once had for him, when she had found him an exciting young aristocrat with the world at his feet, full of fun and capable of enjoying life thoroughly. But he was not really exciting, she now realized: he was just a selfish and rather limited man with a title. When he was on top of her, all she could think about was that he might be passing her some disgusting infection.

Charlie said carefully: 'I'm sure you don't want to talk too much about the Rouzrokh family . . .'

'No.'

'. . . but did you hear that Joanne died?'

'No!' Daisy was shocked. 'How?'

'At Pearl Harbor. She was engaged to Woody Dewar, and she went with him to visit his brother, Chuck, who is stationed there. They were in a car that was strafed by a Zero – that's a Jap fighter plane – and she was hit.'

'I'm so sorry. Poor Joanne. Poor Woody.'

Their food came, and a bottle of wine. They ate in silence for a while. Daisy discovered that Mock Duck did not taste much like duck.

Charlie said: 'Joanne was one of two thousand, four hundred people killed at Pearl Harbor. We lost eight battleships and ten other vessels. Goddamn sneaky Japs.'

'People here are secretly pleased, because the US is in the fight now. God alone knows why Hitler was dumb enough to declare war on the States. But the British think they have a chance of winning at last, with the Russians and us on their side.'

'Americans are very angry about Pearl Harbor.'

'People here don't see why.'

'The Japanese kept on negotiating right up until the last minute – long after they must have made the decision. That's deceitful!'

Daisy frowned. 'It seems sensible to me. If agreement had been reached at the last minute, they could have called off the attack.'

'But they didn't declare war!'

'Would that have made any difference? We were expecting them to attack the Philippines. Pearl Harbor would have taken us by surprise even after a declaration of war.'

Charlie spread his hands in a gesture of bafflement. 'Why did they have to attack us anyway?'

'We stole their money.'

'Froze their assets.'

'They can't see the difference. And we cut off their oil. We had them up against the wall. They were facing ruin. What were they to do?'

'They should have given in, and agreed to withdraw from China.'

'Yes, they should. But if it was America that was being pushed around and told what to do by some other country, would you want us to give in?'

'Maybe not.' He grinned. 'I said you hadn't changed. I'd like to take that back.'

'Why?'

'You never used to talk like this. In the old days you wouldn't discuss politics at all.'

'If you don't take an interest, then what happens is your fault.'

'I guess we've all learned that.'

They ordered dessert. Daisy said: 'What's going to happen to the world, Charlie? All Europe is Fascist. The Germans have conquered much of Russia. The USA is an eagle with a broken wing. Sometimes I'm glad I don't have children.'

'Don't underestimate the USA. We're wounded, not crushed. Japan is cock of the walk now, but the day will come when the Japanese people shed bitter tears of regret for Pearl Harbor.'

'I hope you're right.'

'And the Germans aren't having things all their own way any longer. They failed to take Moscow, and they're on the retreat. Do you realize the battle of Moscow was Hitler's first real defeat?'

'Is it a defeat, or just a setback?'

'Either way, it's the worst military result he's ever had. The Bolsheviks gave the Nazis a bloody nose.'

Charlie had discovered vintage port, a British taste. In London men drank it after the ladies had retired from the dinner table, a tiresome practice that Daisy had tried to abolish in her own house, without success. They had a glass each. On top of the martini and the wine, it made Daisy feel a little drunk and happy.

They reminisced about their adolescence in Buffalo, and laughed about the foolish things they and others had done. 'You told us all you were going to London to dance with the King,' Charlie said. 'And you did!'

'I hope they were jealous.'

'And how! Dot Renshaw went into spasm.'

Daisy laughed happily.

'I'm glad we got back in contact,' Charlie said. 'I like you so much.'

'I'm glad, too.'

They left the restaurant and got their coats. The doorman summoned a taxi. 'I'll take you home,' Charlie said.

As they drove along the Strand, he put his arm around her. She was about to protest, then she thought: What the hell. She snuggled up to him.

'What a fool I am,' he said. 'I wish I'd married you when I had the chance.'

'You would have made a better husband than Boy Fitzherbert,' she said. But then she would never have met Lloyd.

She realized she had not said anything to Charlie about Lloyd.

As they turned into her street, Charlie kissed her.

It felt nice to be wrapped in a man's arms and kissing his lips, but she knew it was the booze making her feel that way, and in truth the only man she wanted to kiss was Lloyd. All the same she did not push him away until the cab came to a halt.

'How about a nightcap,' he said.

For a moment she was tempted. It was a long time since she had touched a man's hard body. But she did not really want Charlie. 'No,' she said. 'I'm sorry, Charlie, but I love someone else.'

'We don't have to go to bed together,' he whispered. 'But if we could just, you know, smooch a while . . .'

She opened the door and stepped out. She felt like a heel. He was risking his life for her every day, and she would not even give him a cheap thrill. 'Goodnight, Charlie, and good luck,' she said. Before she could change her mind, she slammed the car door and went into her house.

She went straight upstairs. A few minutes later, alone in bed, she felt wretched. She had betrayed two men: Lloyd, because she had kissed Charlie; and Charlie, because she had sent him away dissatisfied.

She spent most of Sunday in bed with a hangover.

On Monday evening she got a phone call. 'I'm Hank Bartlett,' said a young American voice. 'Friend of Charlie Farquharson, at Duxford. He talked to me about you, and I found your number in his book.'

Her heart stopped. 'Why are you calling me?'

'Bad news, I'm afraid,' he said. 'Charlie died today, shot down over Abbeville.'

'No!'

'It was his first mission in his new Spitfire.'

'He talked about that,' she said dazedly.

'I thought you might like to know.'

'Thank you, yes,' she whispered.

'He just thought you were the bee's knees.'

'Did he?'

'You should have heard him go on about how great you are.'

'I'm sorry,' she said. 'I'm so sorry.' Then she could no longer speak, and she hung up the phone.

(ii)

Chuck Dewar looked over the shoulder of Lieutenant Bob Strong, one of the cryptanalysts. Some of them were chaotic but Strong was the tidy kind, and he had nothing on his desk but a single sheet of paper on which he had written:

YO—LO—KU—TA—WA—NA

'I can't get it,' Strong said in frustration. 'If the decrypt is right, it says they have struck yolokutawana. But it doesn't mean anything. There's no such word.'

Chuck stared at the six Japanese syllables. He felt sure they ought to mean something to him, even though he knew only a smattering of the language. But he could not figure it out, and he got on with his work.

The atmosphere in the Old Administration Building was grim.

For weeks after the raid, Chuck and Eddie saw bloated bodies from sunk ships floating on the oily surface of Pearl Harbor. At the same time, the intelligence they were handling reported more devastating attacks by the Japanese. Only three days after Pearl Harbor, Japanese planes hit the American base at Luzon in the Philippines and destroyed the Pacific Fleet's entire stock of torpedoes. The same day in the South China Sea they sank two British battleships, the *Repulse* and the *Prince of Wales*, leaving the British helpless in the Far East.

They seemed unstoppable. Bad news just kept coming. In the first few months of the New Year Japan defeated US forces in the Philippines and beat the British in Hong Kong, Singapore, and Rangoon, the capital of Burma.

Many of the place names were unfamiliar even to seamen such as Chuck and Eddie. To the American public they sounded like distant planets in a science-fiction yarn: Guam, Wake, Bataan. But everyone knew the meaning of retreat, submit and surrender.

Chuck felt bewildered. Could Japan really beat America? He could hardly believe it.

By May, the Japanese had what they wanted: an empire that gave them rubber, tin, and – most important of all – oil. Information leaking out indicated that they were ruling their empire with a brutality that would have made Stalin blush.

But there was a fly in their ointment, and it was the US Navy. The thought made Chuck proud. The Japanese had hoped to destroy Pearl Harbor completely, and gain control of the Pacific Ocean; but they had failed. American aircraft carriers and heavy cruisers were still afloat. Intelligence suggested the Japanese commanders were infuriated that the Americans refused to lie down and die. After their losses at Pearl Harbor the Americans were outnumbered and outgunned, but they did not flee and hide. Instead they launched hit-and-run raids on Japanese ships, doing minor damage but boosting American morale and giving the Japanese the unshakable feeling that they had not yet won. Then, on 25 April, planes launched from a carrier bombed the centre of Tokyo, inflicting a terrible wound on the pride of the Japanese military. The celebrations in Hawaii were ecstatic. Chuck and Eddie got drunk that night.

But there was a showdown coming. Every man Chuck spoke to in the Old Administration Building said the Japanese would launch a major attack early in the summer to tempt American ships to come out in force for a final battle. The Japanese hoped the superior strength of their navy would be decisive, and the American Pacific Fleet would be wiped out. The only way the Americans could win was to be better prepared and have better intelligence, to move faster and be smarter.

During those months, Station HYPO worked day and night to crack JN-25b, the new code of the Imperial Japanese Navy. By May they had made progress.

The US Navy had wireless intercept stations all around the Pacific Rim, from Seattle to Australia. There, men known as the On The Roof Gang sat with headsets and radio receivers listening to Japanese radio traffic. They scanned the airwaves and wrote what they heard on message pads.

The signals were in Morse Code, but the dots and dashes of naval signals translated into five-digit number groups, each representing a letter, word or phrase in a code book. The apparently random numbers were relayed by secure cable to teleprinters in the basement of the Old Administration Building. Then the difficult part began: cracking the code.

They always started with small things. The last word of any signal was often OWARI, meaning end. The cryptanalyst would look for other appearances of that number group in the same signal, and write 'END?' above any he found.

The Japanese helped them by making an uncharacteristically careless mistake.

Delivery of the new code books for JN-25b was delayed to some far-flung units. So, for a fatal few weeks, the Japanese high command sent out some messages *in both codes*. Since the Americans had broken much of the original JN-25, they were able to translate the message in the old code, set the decrypt alongside the message in the new code, and figure out the meanings of the five-digit groups of the new code. For a while they progressed by leaps and bounds.

The original eight cryptanalysts were supplemented, after Pearl Harbor, by some of the musicians from the band of the sunk battleship *California.* For reasons no one understood, musicians were good at decoding.

Every signal was kept and every decrypt filed. Comparison of one with another was crucial to the work. An analyst might ask for all the signals from a particular day, or all the signals to one ship, or all the signals that mentioned Hawaii. Chuck and the other clerical staff developed ever-more-complex systems of cross-indexing to help them find whatever the analysts needed.

The unit predicted that in the first week of May the Japanese would attack Port Moresby, the Allied base in Papua. They were right, and the US Navy intercepted the invasion fleet in the Coral Sea. Both sides claimed victory, but the Japanese did not take Port Moresby. And Admiral Nimitz, Commander-in-Chief of the Pacific, began to trust his codebreakers.

The Japanese did not use regular names for locations in the Pacific Ocean. Every important place had a designation consisting of two letters – in fact, two characters or kanas of the Japanese alphabet, although the codebreakers usually used equivalents from the Roman A to Z. The men in the basement struggled to figure out the meaning of each of these two-kana designators. They made slow progress: MO was Port Moresby, AH was Oahu, but many were unknown.

In May, evidence was fast building up of a major Japanese assault at a location they called AF.

The best guess of the unit was that AF meant Midway, the atoll at the western end of the fifteen-hundred-mile-long chain of islands that started at Hawaii. Midway was halfway between Los Angeles and Tokyo.

A guess was not enough, of course. Given the numerical superiority of the Japanese navy, Admiral Nimitz had to *know*.

Day by day, the men Chuck was working with built up an ominous picture of the Japanese order of battle. New planes were delivered to aircraft carriers. An 'occupation force' was embarked: the Japanese were planning to hold on to whatever territory they won.

It looked as if this was the big one. But where would the attack come?

The men in the basement were particularly proud of decoding a signal from the Japanese fleet urging Tokyo: 'Expedite delivery of fuelling hose.' They were pleased partly because of the specialized language but mainly because the signal proved that a long-range mid-ocean manoeuvre was imminent.

But the American high command thought the attack might come at Hawaii, and the army feared an invasion of the west coast of the United States. Even the team at Pearl Harbor had a nagging suspicion it could be Johnston Island, an airstrip a thousand miles south of Midway.

They had to be one hundred per cent certain.

Chuck had a notion how it might be done, but he hesitated to say anything. The cryptanalysts were so clever, and he was not. He had never done well in school. In third grade a classmate had called him Chucky the Chump. He had cried, and that had guaranteed that the nickname would stick. He still thought of himself as Chucky the Chump.

At lunchtime he and Eddie got sandwiches and coffee from the commissary and sat on the dockside, looking across the harbour. It was returning to normal. Most of the oil had gone, and some of the wrecks had been raised.

While they were eating, a wounded aircraft carrier appeared around Hospital Point and steamed slowly into harbour, trailing an oil slick that stretched all the way out to sea. Chuck identified the vessel as the *Yorktown*. Her hull was blackened with soot and she had a huge hole in the flight deck, presumably caused by a Japanese bomb in the Battle of the Coral Sea. Sirens and hooters sounded a congratulatory fanfare as she approached the Navy Yard, and tugs assembled to nudge her through the open gates of No. 1 Dry Dock.

'She needs three months' work, I hear,' Eddie said. He was based in the same building as Chuck, but in the naval intelligence office upstairs, so he got to hear more gossip. 'But she's putting to sea again in three days.'

'How are they going to manage that?'

'They've started already. The master shipfitter flew to meet her – he's on board already, with a team. And look at the dry dock.'

Chuck saw that the vacant dock was already swarming with men and equipment: he could not count the number of welding machines waiting at the quayside.

'All the same,' Eddie said, 'they'll just be patching her up. They'll repair the deck and make her seaworthy, and everything else will have to wait.'

Something about the name of the ship bugged Chuck. He could not shake the nagging feeling. What did Yorktown mean? The siege of Yorktown was the last big battle of the War of Independence. Did that have some significance?

Captain Vandermeier walked by. 'Get back to work, you two girlieboys,' he said.

Eddie said under his breath: 'One of these days I'm going to punch him out.'

'After the war, Eddie,' said Chuck.

When he returned to the basement and saw Bob Strong at his desk, Chuck realized he had solved Strong's problem.

Looking over the cryptanalyst's shoulder again, he saw the same sheet of paper with the same six Japanese syllables:

YO—LO—KU—TA—WA—NA

He tactfully tried to make it sound as if Strong himself had solved it. 'But you have got it, Lieutenant!' he said.

Strong was disconcerted. 'Do I?'

'It's an English name, so the Japanese have spelt it out phonetically.'

'Yolokutawana is an English name?'

'Yes, sir. That's how the Japanese pronounce Yorktown.'

'What?' Strong looked baffled.

For a dreadful moment, Chucky the Chump wondered if he was completely wrong.

Then Strong said: 'Oh, my God, you're right! Yolokutawana – Yorktown, with a Japanese accent!' He laughed delightedly. 'Thank you!' he enthused. 'Well done!'

Chuck hesitated. He had another idea. Should he say what was on his mind? It was not his job to solve codes. But America was an inch away from defeat. Maybe he should take a chance. 'Can I make another suggestion?' he said.

'Fire away.'

'It's about the designator AF. We need definite confirmation that it's Midway, right?'

'Yup.'

'Couldn't we write a message about Midway that the Japanese would want to rebroadcast in code? Then when we intercepted the broadcast we could find out how they encode the name.'

Strong looked thoughtful. 'Maybe,' he said. 'We might have to send our message in clear, to be sure they understood it.'

'We could do that. It would have to be something not very confidential – like, say: 'There is an outbreak of venereal disease on Midway, please send medicine,' or something like that.'

'But why would the Japs rebroadcast that?'

'Okay, so it has to be something of military significance, but not top secret; something like the weather.'

'Even weather forecasts are secret nowadays.'

The cryptanalyst at the next desk put in: 'How about a water shortage? If they're planning to occupy the place, that would be important information.'

'Hell, this could work.' Strong was getting excited. 'Suppose Midway sends a message in clear to Hawaii, saying their desalination plant has broken down.'

Chuck said: 'And Hawaii replies, saying we're sending a water barge.'

'The Japanese would be sure to rebroadcast that, if they're planning to attack Midway. They would need to make plans to ship fresh water there.'

'And they would broadcast in code to avoid alerting us to their interest in Midway.'

Strong stood up. 'Come with me,' he said to Chuck. 'Let's put this to the boss, see what he thinks of the idea.'

The signals were exchanged that day.

Next day, a Japanese radio signal reported a water shortage at AF.

The target was Midway.

Admiral Nimitz commenced to set a trap.

(iii)

That evening, while more than a thousand workmen swarmed over the crippled aircraft carrier *Yorktown*, repairing the damage under arc lights, Chuck and Eddie went to The Band Round The Hat, a bar down a dark alley in Honolulu. It was packed, as always, with sailors and locals. Almost all the customers were men, though there were a few nurses in pairs. Chuck and Eddie liked the place because the other men were their kind. The lesbians liked it because the men did not hit on them.

There was nothing overt, of course. You could be thrown out of the navy and put in jail for homosexual acts. All the same the place was congenial. The bandleader wore make-up. The Hawaiian singer was in drag, although he was so convincing that some people did not realize he was a man. The owner was as queer as a three-dollar bill. Men could dance together. And no one would call you a wimp for ordering vermouth.

Since the death of Joanne, Chuck felt he loved Eddie even more. Of course he had always known that Eddie could be killed, in theory; but the danger had never seemed real. Now, after the attack on Pearl Harbor, Chuck never passed a day without visualizing that beautiful girl lying on the ground covered in blood, and his brother sobbing his heart out beside her. It could so easily have been Chuck kneeling next to Eddie, and feeling the same unbearable grief. Chuck and Eddie had cheated death on 7 December, but they were at war now, and life was cheap. Every day together was precious because it might be the last.

Chuck was leaning on the bar with a beer in his hand, and Eddie was sitting on a high stool. They were laughing at a navy pilot called Trevor Paxman – known as Trixie – who was talking about the time he tried to have sex with a girl. 'I was horrified!' Trixie said. 'I thought it would be all tidy down there, and kind of sweet, like girls in paintings, but she had more hair than me!' They roared with laughter. 'She was like a gorilla!' At that point Chuck saw, out of the corner of his eye, the stocky figure of Captain Vandermeier entering the bar.

Few officers went into enlisted men's bars. It was not forbidden, merely thoughtless and inconsiderate, like wearing muddy boots in the restaurant of the Ritz-Carlton. Eddie turned his back, hoping Vandermeier would not see him.

No such luck. Vandermeier came right up to them and said: 'Well, well, all girls together, are we?'

Trixie turned away and melted into the crowd. Vandermeier said: 'Where did he go?' He was already drunk enough to slur his words.

Chuck saw Eddie's face darken. Chuck said stiffly: 'Good evening, Captain, may I buy you a beer?'

'Scotch onna rocks.'

Chuck got him a drink. Vandermeier took a swallow and said: 'So, I hear the action in this place is out the back – is that right?' He looked at Eddie.

'No idea,' Eddie said coldly.

'Aw, come on,' said Vandermeier. 'Off the record.' He patted Eddie's knee.

Eddie stood up abruptly and pushed his stool back. 'Don't you touch me,' he said.

Chuck said: 'Take it easy, Eddie.'

'There's no rule in the navy says I have to be pawed by this old queen!'

Vandermeier said drunkenly: 'What did you call me?'

Eddie said: 'If he touches me again, I swear I'll knock his ugly head off.'

Chuck said: 'Captain Vandermeier, sir, I know a much better place than this. Would you like to go there?'

Vandermeier looked confused. 'What?'

Chuck improvised: 'A smaller, quieter place – like this, but more intimate. Do you know what I mean?'

'Sounds good!' The captain drained his glass.

Chuck took Vandermeier's right arm and gestured to Eddie to take the left. They led the drunk captain outside.

Luckily, a taxi was waiting in the gloom of the alley. Chuck opened the car door.

At that point, Vandermeier kissed Eddie.

The captain threw his arms around him, pressed his lips to Eddie's, then said: 'I love you.'

Chuck's heart filled with fear. There was no good ending to this now.

Eddie punched Vandermeier in the stomach, hard. The captain grunted and gasped. Eddie hit him again, in the face this time. Chuck stepped between them. Before Vandermeier could fall down, Chuck bundled him into the back seat of the taxi.

He leaned through the window and gave the driver a ten-dollar bill. 'Take him home, and keep the change,' he said.

The taxi pulled away.

Chuck looked at Eddie. 'Oh, boy,' he said. 'Now we're in trouble.'

(iv)

But Eddie Parry was never charged with the crime of assaulting an officer.

Captain Vandermeier showed up at the Old Administration Building next morning with a black eye, but he made no accusation. Chuck figured it would ruin the man's career if he admitted he had got into a fight at The Band Round The Hat. All the same everyone was talking about his bruise. Bob Strong said: 'Vandermeier claims he slipped on a patch of oil in his garage, and hit his face on the lawn mower, but I think his wife socked him. Have you seen her? She looks like Jack Dempsey.'

That day, the cryptanalysts in the basement told Admiral Nimitz that the Japanese would attack Midway on 4 June. More specifically, the Japanese force would be 175 miles north of the atoll at 7 a.m.

They were almost as confident as they sounded.

Eddie was gloomy. 'What can we do?' he said when he and Chuck met for lunch. He worked in naval intelligence too, and he knew the Japanese strength as revealed by the codebreakers. 'The Japs have two hundred ships at sea – practically their entire navy – and how many do we have? Thirty-five!'

Chuck was not so glum. 'But their strike force is only a quarter of their strength. The rest are the occupation force, the diversion force and the reserves.'

'So? A quarter of their strength is still more than our entire Pacific Fleet!'

'The actual Japanese strike force has only four aircraft carriers.'

'But we have just three.' Eddie pointed with his ham sandwich at the smoke-blackened carrier in the dry dock, with workmen swarming all over her. 'And that includes the broken-down *Yorktown*.'

'Well, we know they're coming, and they don't know we're lying in wait.'

'I sure hope that makes as much difference as Nimitz thinks.'

'Yeah, so do I.'

When Chuck returned to the basement, he was told that he no longer worked there. He had been reassigned – to the *Yorktown*.

'It's Vandermeier's way of punishing me,' Eddie said tearfully that evening. 'He thinks you'll die.'

'Don't be pessimistic,' Chuck said. 'We might win the war.'

A few days before the attack, the Japanese changed to new code books. The men in the basement sighed and started again from scratch, but they produced little new intelligence before the battle. Nimitz had to make do with what he already had, and hope the Japanese did not revise the whole plan at the last minute.

The Japanese expected to take Midway by surprise and overwhelm it easily. They hoped the Americans would then attack in full force in a bid to win the atoll back. At that point, the Japanese reserve fleet would pounce and wipe out the entire American fleet. Japan would rule the Pacific.

And the USA would ask for peace talks.

Nimitz planned to nip the scheme in the bud by ambushing the strike force before they could take Midway.

Chuck was now part of the ambush.

He packed his kitbag and kissed Eddie goodbye, then they went together to the dockside.

There they ran into Vandermeier.

'There was no time to repair the watertight compartments,' he told them. 'If she's holed, she'll go down like a lead coffin.'

Chuck put a restraining hand on Eddie's shoulder and said: 'How's your eye, Captain?'

Vandermeier's mouth twisted in a grimace of malice. 'Good luck, faggot.' He walked away.

Chuck shook hands with Eddie and went on board.

He forgot about Vandermeier instantly, for at long last he had his wish: he was at sea – and on one of the greatest ships ever made.

The *Yorktown* was the lead ship of the carrier class. She was longer than two football pitches and had a crew of more than two thousand. She carried ninety aircraft: elderly Douglas Devastator torpedo bombers with folding wings; newer Douglas Dauntless dive bombers; and Grumman Wildcat fighters to escort the bombers.

Almost everything was below, apart from the island structure, which stood up thirty feet from the flight deck. It contained the ship's command and communications heart, with the bridge, the radio room just below it, the chart house and the aviators' ready room. Behind these was a huge smokestack containing three funnels in a row.

Some of the repairmen were still aboard, finishing their work, when she left the dry dock and steamed out of Pearl Harbor. Chuck thrilled to the throb of her colossal engines as she put to sea. When she reached deep water and began to rise and fall with the swell of the Pacific Ocean, he felt as if he were dancing.

Chuck was assigned to the radio room, a sensible posting that made use of his experience in handling signals.

The carrier steamed to a rendezvous north-east of Midway, her welded patches creaking like new shoes. The ship had a soda fountain, known as the Gedunk, that served freshly made ice cream. There on the first afternoon Chuck ran into Trixie Paxman, whom he had last seen at The Band Round The Hat. He was glad to have a friend aboard.

On Wednesday 3 June, the day before the predicted attack, a navy flying boat on reconnaissance west of Midway spotted a convoy of Japanese transport ships – presumably carrying the occupation force that was to take over the atoll after the battle. The news was broadcast to all US ships, and Chuck in the radio room of the *Yorktown* was among the first to know. It was hard confirmation that his comrades in the basement had been right, and he felt a sense of relief that they had been vindicated. That was ironic, he realized: he would not be in such danger if they had been wrong and the Japanese were elsewhere.

He had been in the navy for a year and a half, but until now he had never gone into battle. The hastily repaired *Yorktown* was going to be the target of Japanese torpedoes and bombs. She was steaming towards people who would do everything in their power to sink her, and sink Chuck too. It was a weird feeling. Most of the time he was strangely calm, but every now and again he felt an impulse to dive over the side and start swimming back towards Hawaii.

That night he wrote to his parents. If he died tomorrow, he and the letter would probably go down with the ship, but he wrote it anyway. He said nothing about why he had been reassigned. It crossed his mind to confess that he was queer, but he quickly dismissed that idea. He told them he loved them and was grateful for everything they had done for him. 'If I die fighting for a democratic country against a cruel military dictatorship, my life will not have been wasted,' he wrote. When he read it over it sounded a bit pompous, but he left it as it was.

It was a short night. Aircrew were piped to breakfast at 1.30 a.m. Chuck went to wish Trixie Paxman good luck. In recompense for the early start, the airmen were eating steak and eggs.

Their planes were brought up from the below-decks hangars in the ship's huge elevators, then manoeuvred by hand to their parking slots on deck to be fuelled and armed. A few pilots took off and went looking for the enemy. The rest sat in the briefing room, wearing their flying gear, waiting for news.

Chuck went on duty in the radio room. Just before six he picked up a signal from a reconnaissance flying boat:

MANY ENEMY PLANES HEADING MIDWAY

A few minutes later he got a partial signal:

ENEMY CARRIERS

It had started.

When the full report came in a minute later, it placed the Japanese strike force almost exactly where the cryptanalysts had forecast. Chuck felt proud – and scared.

The three American aircraft carriers – *Yorktown, Enterprise* and *Hornet* – set a course that would bring their planes within striking distance of the Japanese ships.

On the bridge was the long-nosed Admiral Frank Fletcher, a fifty-seven-year-old veteran who had won the Navy Cross in the First World War. Carrying a signal to the bridge, Chuck heard him say: 'We haven't seen a Japanese plane yet. That means they still don't know we're here.'

That was all the Americans had going for them, Chuck knew: the advantage of better intelligence.

The Japanese undoubtedly hoped to catch Midway napping, in a repeat of the Pearl Harbor scenario, but it was not going to happen, thanks to the cryptanalysts. The American planes at Midway were not sitting targets parked on their runways. By the time the Japanese bombers arrived they were all in the air and spoiling for a fight.

Tensely listening to the crackling wireless traffic from Midway and the Japanese ships, the officers and men in the radio room of the *Yorktown* had no doubt that there was a terrific air battle going on over the tiny atoll; but they did not know who was winning.

Soon afterwards, American planes from Midway took the fight to the enemy and attacked the Japanese aircraft carriers.

In both battles, as far as Chuck could make out, the anti-aircraft guns had the best of it. Only moderate damage was done to the base at Midway, and almost all the bombs and torpedoes aimed at the Japanese fleet missed; but in both encounters a lot of aircraft were shot down.

The score seemed even – but that bothered Chuck, for the Japanese had more in reserve.

Just before seven the *Yorktown*, the *Enterprise* and the *Hornet* swung around to the south-east. It was a course that unfortunately took them away from the enemy, but their planes had to take off into the south-easterly wind.

Every corner of the mighty *Yorktown* trembled to the thunder of the aircraft as their engines rose to full throttle and they powered along the deck, one after another, and shot up into the air. Chuck noticed the tendency of the Wildcat to lift its right wing and wander left as it accelerated along the deck, a characteristic much complained of by pilots.

By half past eight the three carriers had sent 155 American planes to attack the enemy strike force.

The first planes arrived in the target area, with perfect timing, when the Japanese were busy refuelling and rearming their own planes returning from Midway. The flight decks were littered with ammunition cases scattered in a snakes' nest of fuel hoses, all ready to blow up in an instant. There should have been carnage.

But it did not happen.

Almost all the American aircraft in the first wave were destroyed.

The Devastators were obsolete. The Wildcats that escorted them were better, but no match for the fast, manoeuvrable Japanese Zeroes. Those planes that survived to deliver their ordnance were decimated by devastating anti-aircraft fire from the carriers.

Dropping a bomb from a moving aircraft on to a moving ship, or dropping a torpedo where it would hit a ship, was extraordinarily difficult, especially for a pilot who was under fire from above and below.

Most of the airmen gave their lives in the attempt.

And not one of them scored a hit.

No American bomb or torpedo found its target. The first three waves of attacking planes, one from each American carrier, did no damage at all to the Japanese strike force. The ammunition on their decks did not explode, and their fuel lines did not catch fire. They were unharmed.

Listening to the radio chatter, Chuck despaired.

He saw with new vividness the genius of the attack on Pearl Harbor seven months earlier. The American ships had been at anchor, static targets crowded together, relatively easy to hit. The fighter planes that might have protected them were destroyed on their airstrips. And by the time the Americans had armed and deployed their anti-aircraft guns, the attack was almost over.

However, this battle was still going on, and not all the American planes had yet reached the target area. He heard an air officer on the *Enterprise* radio shout: 'Attack! Attack!' and the laconic response from a pilot: 'Wilco, as soon as I can find the bastards.'

The good news was that the Japanese commander had not yet sent aircraft to attack the American ships. He was sticking to his plan and concentrating on Midway. He might by now have figured out that he must be under attack from carrier-borne planes, but perhaps he was not sure where the American ships were located.

Despite this advantage, the Americans were not winning.

Then the picture changed. A flight of thirty-seven Dauntless dive bombers from the *Enterprise* sighted the Japanese. The Zeroes protecting the ships had come down almost to sea level in their dog-fights with previous attackers, so the bombers found themselves fortunately above the fighters, and able to come down at them out of the sun. Just minutes later another eighteen Dauntlesses from the *Yorktown* reached the target area. One of the pilots was Trixie.

The radio exploded with excited chatter. Chuck closed his eyes and concentrated, trying to make sense of the distorted sounds. He could not identify Trixie's voice.

Then, behind the talk, he began to hear the characteristic scream of bombers diving. The attack had begun.

Suddenly, for the first time, there were cries of triumph from the pilots.

'Got you, you bastard!'

'Shit, I felt that go up!'

'Eat that, you sons of bitches!'

'Bullseye!'

'Look at her burn!'

The men in the radio room cheered wildly, but they were not sure what was happening.

It was over in a few minutes, but it took a long time to get a clear report. The pilots were incoherent with the joy of victory. Gradually, as they calmed down and headed back towards their ships, the picture emerged.

Trixie Paxman was among the survivors.

Most of their bombs had missed, as previously, but about ten had scored direct hits, and those few had done tremendous damage. Three mighty Japanese aircraft carriers were burning out of control: *Kaga*, *Soryu* and the flagship *Akagi*. The enemy had only one left, the *Hiryu*.

'Three out of the four!' Chuck said elatedly. 'And they still haven't come anywhere near our ships!'

That soon changed.

Admiral Fletcher sent out ten Dauntlesses to scout for the surviving Japanese carrier. But it was the *Yorktown*'s radar that picked up a flight of planes, presumably from the *Hiryu*, fifty miles away and approaching. At noon, Fletcher sent up twelve Wildcats to meet the attackers. The rest of the planes were also ordered up so they would not be on deck and vulnerable when the attack came. Meanwhile the *Yorktown*'s fuel lines were flooded with carbon dioxide as a fire precaution.

The attacking flight included fourteen 'Vals', Aichi D3A dive bombers, plus escorting Zeroes.

Here it comes, Chuck thought; my first action. He wanted to throw up. He swallowed hard.

Before the attackers could be seen, the *Yorktown*'s gunners opened up. The ship had four pairs of large anti-aircraft guns with five-inch-diameter barrels that could send their shells several miles. Plotting the enemy's position with the aid of radar, gunnery officers sent a salvo of giant fifty-four-pound shells towards the approaching aircraft, setting the timers to explode when they reached their target.

The Wildcats got above the attackers and, according to the pilots' radio reports, shot down six bombers and three fighters.

Chuck ran to the flag bridge with a signal to say the remainder of the attack force were diving in. Admiral Fletcher said coolly: 'Well, I've got my tin hat on – I can't do anything else.'

Chuck looked out of the window and saw the dive bombers screaming out of the sky towards him at an angle so steep they seemed to be falling straight down. He resisted the impulse to throw himself to the floor.

The ship made a sudden full-rudder turn to port. Anything that might throw the attacking aircraft off course was worth a try.

The *Yorktown* deck also had four Chicago pianos – smaller, short-range anti-aircraft guns with four barrels each. Now these opened up, and so did the guns of *Yorktown*'s escort of cruisers.

As Chuck stared forward from the bridge, terrified and helpless to do anything to defend himself, a deck gunner found his range and hit a Val. The plane seemed to break into three pieces. Two fell into the sea and one crashed into the side of the ship. Then another Val blew up. Chuck cheered.

But that left six.

The *Yorktown* made a sudden turn to starboard.

The Vals braved the hail of death from the deck guns to chase after the ship.

As they got closer, the machine guns on the catwalks either side of the flight deck also opened up. Now the *Yorktown*'s guns played a lethal symphony, with deep booms from the five-inch barrels, mid-range sounds from the Chicago pianos, and the urgent rattle of machine guns.

Chuck saw the first bomb.

Many Japanese bombs had a delayed fuse. Instead of exploding on impact, they went off a second or so later; the idea being that they would crash through the deck and explode deep in the interior, causing maximum devastation.

But this bomb rolled along the *Yorktown*'s deck.

Chuck watched in mesmerized horror. For a moment it looked as if it might do no harm. Then it went off with a boom and a flash of flame. The two Chicago pianos aft were destroyed in an instant. Small fires appeared on deck and in the towers.

To Chuck's amazement the men around him remained as cool as if they were attending a war game in a conference room. Admiral Fletcher issued orders even as he staggered across the shuddering deck of the flag bridge. Moments later, damage control teams were dashing across the flight deck with fire hoses, and stretcher parties were picking up the wounded and carrying them down steep companionways to dressing stations below.

There were no major fires: the carbon dioxide in the fuel lines had prevented that. And there were no bomb-loaded planes on deck to blow up.

A moment later another Val screamed down at the *Yorktown* and a bomb hit the smokestack. The explosion rocked the mighty ship. A huge pall of oily black smoke gouted from the funnels. The bomb must have damaged the engines, Chuck realized, because the ship lost speed immediately.

More bombs missed their targets, landing in the sea, sending up geysers that splashed on to the deck, where sea water mingled with the blood of the wounded.

The *Yorktown* slowed to a halt. When the crippled ship was dead in the water, the Japanese scored a third hit, and a bomb crashed through the forward elevator and exploded somewhere below.

Then, suddenly, it was over, and the surviving Vals climbed into the clear blue Pacific sky.

I'm still alive, Chuck thought.

The ship was not lost. Fire-control parties were at work before the Japanese were out of sight. Down below, the engineers said they could get the boilers going within an hour. Repair crews patched the hole in the flight deck with six-by-four planks of Douglas fir.

But the radio gear had been destroyed, so Admiral Fletcher was deaf and blind. With his personal staff he transferred to the cruiser *Astoria*, and he handed over tactical command to Spruance on the *Enterprise*.

Under his breath, Chuck said: 'Fuck you, Vandermeier – I survived.'

He spoke too soon.

The engines throbbed back to life. Now under the command of Captain Buckmaster, the *Yorktown* began once again to cut through the Pacific waves. Some of her planes had already taken refuge on the *Enterprise*, but others were still in the air, so she turned into the wind, and they began to touch down and refuel. As she had no working radio, Chuck and his colleagues became a semaphore team to communicate with other ships using old-fashioned flags.

At half past two, the radar of a cruiser escorting the *Yorktown* revealed planes coming in low from the west – an attack flight from the *Hiryu*, presumably. The cruiser signalled the news to the carrier. Buckmaster sent up twelve Wildcats to intercept.

The Wildcats must have been unable to stop the attack, for ten torpedo bombers appeared, skimming the waves, heading straight for the *Yorktown*.

Chuck could see the planes clearly. They were Nakajima B5Ns, called Kates by the Americans. Each carried a torpedo slung under its fuselage, the weapon almost half the length of the entire plane.

The four heavy cruisers escorting the carrier shelled the sea around her, throwing up a screen of foamy water, but the Japanese pilots were not so easily deterred, and they flew straight through the spray.

Chuck saw the first plane drop its torpedo. The long bomb splashed into the water, pointed at the *Yorktown*.

The plane flashed past the ship so close that Chuck saw the pilot's face. He was wearing a white-and-red headband as well as his flight helmet. He shook a triumphant fist at the crew on deck. Then he was gone.

More planes roared by. Torpedoes were slow, and ships could

sometimes dodge them, but the crippled *Yorktown* was too cumbersome to zigzag. There was a tremendous bang, shaking the ship: torpedoes were several times more powerful than regular bombs. It felt to Chuck as if she had been struck on the port stern. Another explosion followed close behind, and this one actually lifted the ship, throwing half the crew to the deck. Immediately afterwards, the mighty engines faltered.

Once again the damage parties were at work before the attacking planes were out of sight. But this time the men could not cope. Chuck joined the teams manning the pumps, and saw that the steel hull of the great ship was ripped like a tin can. A Niagara of sea water poured through the gash. Within minutes Chuck could feel that the deck had tilted. The *Yorktown* was listing to port.

The pumps could not cope with the inward rush of water, especially as the ship's watertight compartments had been damaged at the Battle of the Coral Sea and not fixed during her rush repairs.

How long could it be before she capsized?

At three o'clock Chuck heard the order: 'Abandon Ship!'

Sailors dropped ropes over the high edge of the sloping deck. On the hangar deck, by jerking a few strings crewmen released thousands of life jackets from overhead stowage to fall like rain. The escort vessels moved closer and launched their boats. The crew of the *Yorktown* took off their shoes and swarmed over the side. For some reason, they put their shoes on the deck in neat lines, hundreds of pairs, like some ritual sacrifice. Wounded men were lowered on stretchers to waiting whaleboats. Chuck found himself in the water, swimming as fast as he could to get away from the *Yorktown* before she turned over. A wave took him by surprise and washed away his cap. He was glad he was in the warm Pacific: the Atlantic might have killed him with cold while he was waiting to be rescued.

He was picked up by a lifeboat, which continued to retrieve men from the sea. Dozens of other boats were doing the same. Many crew climbed down from the main deck, which was lower than the flight deck. The *Yorktown* somehow managed to stay afloat.

When all the crew were safe they were taken aboard the escorting vessels.

Chuck stood on deck, looking across the water as the sun went down behind the slowly sinking *Yorktown*. It occurred to him that during the whole day he had not seen a Japanese ship. The entire battle had been fought by aircraft. He wondered if this was the first of a new kind of naval battle. If so, aircraft carriers would be the key vessels in future. Nothing else would count for much.

Trixie Paxman appeared beside him. Chuck was so pleased to see him alive that he hugged him.

Trixie told Chuck that the last flight of Dauntless dive bombers, from the *Enterprise* and the *Yorktown,* had set alight the *Hiryu,* the surviving Japanese carrier, and destroyed her.

'So all four Japanese carriers are out of action,' Chuck said.

'That's right. We got them all, and lost only one of our own.'

'So,' said Chuck, 'does that mean we won?'

'Yes,' said Trixie. 'I guess it does.'

(v)

After the Battle of Midway it was clear that the Pacific war would be won by planes launched from ships. Both Japan and the United States began crash programmes to build aircraft carriers as fast as possible.

During 1943 and 1944, Japan produced seven of these huge, costly vessels.

In the same period, the United States produced ninety.

13

1942 (II)

Nursing Sister Carla von Ulrich wheeled a cart into the supply room and closed the door behind her.

She had to work quickly. What she was about to do would get her sent to a concentration camp if she were caught.

She took a selection of wound dressings from a cupboard, plus a roll of bandage and a jar of antiseptic cream. Then she unlocked the drugs cabinet. She took morphine for pain relief, sulphonamide for infections, and aspirin for fever. She added a new hypodermic syringe, still in its box.

She had already falsified the register, over a period of weeks, to look as if what she was stealing had been used legitimately. She had rigged the register before taking the stuff, rather than afterwards, so that any spot check would reveal a surplus, suggesting mere carelessness, instead of a deficit, which indicated theft.

She had done all this twice before, but she felt no less frightened.

As she wheeled the cart out of the store, she hoped she looked innocent: a nurse bringing medical necessities to a patient's bedside.

She walked into the ward. To her dismay she saw Dr Ernst there, sitting beside a bed, taking a patient's pulse.

All the doctors should have been at lunch.

It was now too late to change her mind. Trying to assume an air of confidence that was the opposite of what she felt, she held her head high and walked through the ward, pushing her cart.

Dr Ernst glanced up at her and smiled.

Berthold Ernst was the nurses' dreamboat. A talented surgeon with a warm bedside manner, he was tall, handsome and single. He had romanced most of the attractive nurses, and had slept with many of them, if hospital gossip could be credited.

She nodded to him and went briskly past.

She pushed the trolley out of the ward then suddenly turned into the nurses' cloakroom.

Her outdoor coat was on a hook. Beneath it was a basketwork shopping bag containing an old silk scarf, a cabbage and a box of sanitary towels in a brown paper bag. Carla removed the contents, then swiftly transferred the medical supplies from the trolley to the bag. She covered the supplies with the scarf, a blue and gold geometric design that her mother must have bought in the twenties. Then she put the cabbage and the sanitary towels on top, hung the bag on a hook, and arranged her coat to cover it.

I got away with it, she thought. She realized she was trembling a little. She took a deep breath, got herself under control, opened the door – and saw Dr Ernst standing just outside.

Had he been following her? Was he about to accuse her of stealing? His manner was not hostile; in fact, he looked friendly. Perhaps she had got away with it.

She said: 'Good afternoon, Doctor. Can I help you with something?'

He smiled. 'How are you, Sister? Is everything going well?'

'Perfectly, I think.' Guilt made her add ingratiatingly: 'But it is you, Doctor, who must say whether things are going well.'

'Oh, I have no complaints,' he said dismissively.

Carla thought: So what is this about? Is he toying with me, sadistically delaying the moment when he makes his accusation?

She said nothing, but stood waiting, trying not to shake with anxiety.

He looked down at the cart. 'Why did you take that into the cloakroom?'

'I wanted something,' she said, improvising desperately. 'Something from my raincoat.' She tried to suppress the frightened tremor in her voice. 'A handkerchief, from my pocket.' Stop gabbling, she told herself. He's a doctor, not a Gestapo agent. But he scared her all the same.

He looked amused, as if he enjoyed her nervousness. 'And the trolley?'

'I'm returning it to its place.'

'Tidiness is essential. You're a very good nurse ... Fräulein von Ulrich ... or is it Frau?'

'Fräulein.'

'We should talk some more.'

The way he smiled told her this was not about stealing medical supplies. He was about to ask her to go out with him. She would be the envy of dozens of nurses if she said yes.

But she had no interest in him. Perhaps it was because she had loved one dashing Lothario, Werner Franck, and he had turned out to be a self-centred coward. She guessed that Berthold Ernst was similar.

However, she did not want to risk annoying him, so she just smiled and said nothing.

'Do you like Wagner?' he said.

She could see where this was going. 'I have no time for music,' she said firmly. 'I take care of my elderly mother.' In fact Maud was fifty-one and enjoyed robust good health.

'I have two tickets for a recital tomorrow evening. They're playing the Siegfried Idyll.'

'A chamber piece!' she said. 'Unusual.' Most of Wagner's work was on a grand scale.

He looked pleased. 'You know about music, I see.'

She wished she had not said it. She had just encouraged him. 'My family is musical – my mother gives piano lessons.'

'Then you must come. I'm sure someone else could take care of your mother for an evening.'

'It's really not possible,' Carla said. 'But thank you very much for the invitation.' She saw anger in his eyes: he was not used to rejection. She turned and started to push the cart away.

'Another time, perhaps?' he called after her.

'You're very kind,' she replied, without slowing her pace.

She was afraid he would come after her, but her ambiguous reply to his last question seemed to have mollified him. When she looked back over her shoulder he had gone.

She stowed the trolley and breathed more easily.

She returned to her duties. She checked on all the patients in her ward and wrote her reports. Then it was time to hand over to the evening shift.

She put on her raincoat and slung her bag over her arm. Now she had to walk out of the building with stolen property, and her fear mounted again.

Frieda Franck was going at the same time, and they left together. Frieda had no idea Carla was carrying contraband. They walked in June sunshine to the tram stop. Carla wore a coat mainly to keep her uniform clean.

She thought she was giving a convincing impression of normality until Frieda said: 'Are you worried about something?'

'No, why?'

'You seem nervous.'

'I'm fine.' To change the subject, she pointed at a poster. 'Look at that.'

The government had opened an exhibition in Berlin's Lustgarten, the park in front of the cathedral. 'The Soviet Paradise' was the ironic title of a show about life under Communism, portraying Bolshevism as a Jewish trick and the Russians as subhuman Slavs. But even today the Nazis did not have everything their own way, and someone had gone around Berlin pasting up a spoof poster that read:

Permanent Installation

The NAZI PARADISE

WAR HUNGER LIES GESTAPO

How much longer?

There was one such poster stuck to the tram shelter, and it warmed Carla's heart. 'Who puts these things up?' she said.

Frieda shrugged.

Carla said: 'Whoever they are, they're brave. They would be killed if caught.' Then she remembered what was in her bag. She, too, could be killed if caught.

Frieda just said: 'I'm sure.'

Now it was Frieda who seemed a little jumpy. Could she be one of those who put up the posters? Probably not. Maybe her boyfriend, Heinrich, was. He was the intense, moralistic type who would do that sort of thing. 'How's Heinrich?' said Carla.

'He wants to get married.'

'Don't you?'

Frieda lowered her voice. 'I don't want to have children.' This was a seditious remark: young women were supposed to produce children gladly for the Führer. Frieda nodded at the illegal poster. 'I wouldn't like to bring a child into this paradise.'

'I guess I wouldn't, either,' said Carla. Maybe that was why she had turned down Dr Ernst.

A tram arrived and they got on. Carla perched the basket on her lap nonchalantly, as if it contained nothing more sinister than cabbage. She scanned the other passengers. She was relieved to see no uniforms.

Frieda said: 'Come home with me. Let's have a jazz night. We can play Werner's records.'

'I'd love to, but I can't,' Carla said 'I've got a call to pay. Remember the Rothmann family?'

Frieda looked around warily. Rothmann might or might not be a Jewish name. But no one was near enough to hear them. 'Of course – he used to be our doctor.'

'He's not supposed to practise any more. Eva Rothmann went to London before the war and married a Scottish soldier. But the parents can't get out of Germany, of course. Their son, Rudi, was a violin maker – quite brilliant, apparently – but he lost his job, and now he repairs instruments and tunes pianos.' He came to the von Ulrich house four times a year to tune the Steinway grand. 'Anyway, I said I'd go round there this evening and see them.'

'Oh,' said Frieda. It was the long drawn-out 'oh' of someone who has just seen the light.

'Oh, what?' said Carla.

'Now I understand why you're clutching that basket as if it contained the Holy Grail.'

Carla was thunderstruck. Frieda had guessed her secret! 'How did you know?'

'You said he's not *supposed* to practise. That suggests he does.'

Carla saw that she had given Dr Rothmann away. She should have said that he was not *allowed* to practise. Fortunately, it was only to Frieda that she had betrayed him. She said: 'What is he to do? They come to his door and beg him to help them. He can't turn sick people away! It's not as if he makes any money – all his patients are Jews and other poor folk who pay him with a few potatoes or an egg.'

'You don't have to defend him to me,' said Frieda. 'I think he's brave. And you're heroic, stealing supplies from the hospital to give to him. Is this the first time?'

Carla shook her head. 'Third. But I feel such a fool for letting you find out.'

'You're not a fool. It's just that I know you too well.'

The tram approached Carla's stop. 'Wish me luck,' she said, and she got off.

When she entered her house she heard hesitant notes on the piano upstairs. Maud had a pupil. Carla was glad. It would cheer her mother up as well as providing a little money.

Carla took off her raincoat then went into the kitchen and greeted Ada. When Maud had announced that she could no longer pay Ada's wages, Ada had asked if she could stay on anyway. Now she had a job cleaning an office in the evening, and she did housework for the von Ulrich family in exchange for her room and board.

Carla kicked off her shoes under the table and rubbed her feet together to ease their ache. Ada made her a cup of grain coffee.

Maud came into the kitchen, eyes sparkling. 'A new pupil!' she said. She showed Carla a handful of banknotes. 'And he wants a lesson every day!' She had left him practising scales, and his novice fingering sounded in the background like a cat walking along the keyboard.

'That's great,' said Carla. 'Who is he?'

'A Nazi, of course. But we need the money.'

'What's his name?'

'Joachim Koch. He's quite young and shy. If you meet him, for goodness' sake bite your tongue and be polite.'

'Of course.'

Maud disappeared.

Carla drank her coffee gratefully. She had got used to the taste of burnt acorns, as most people had.

She chatted idly to Ada for a few minutes. Ada had once been plump, but now she was thin. Few people were fat in today's Germany, but there was something wrong with Ada. The death of her handicapped son, Kurt, had hit her hard. She had a lethargic air. She did her job competently, but then she sat staring out of the window for hours, her expression blank. Carla was fond of her, and felt her anguish, but did not know what to do to help her.

The sound of the piano ceased and, a little later, Carla heard two voices in the hallway, her mother's and a man's. She assumed Maud was seeing Herr Koch out, and she was horrified, a moment later, when her mother entered the kitchen, closely followed by a man in an immaculate lieutenant's uniform.

'This is my daughter,' Maud said cheerfully. 'Carla, this is Lieutenant Koch, a new pupil.'

Koch was an attractive, shy-looking man in his twenties. He had a fair moustache, and reminded Carla of pictures of her father when young.

Carla's heart raced with fear. The basket containing the stolen medical supplies was on the kitchen chair next to her. Would she accidentally betray herself to Lieutenant Koch, as she had to Frieda?

She could hardly speak. 'I–I–I am pleased to make your acquaintance,' she said.

Maud looked at her with curiosity, surprised at her nervousness. All Maud wanted was for Carla to be nice to the new pupil in the hope that he would continue his studies. She saw no harm in bringing an army officer into the kitchen. She had no idea that Carla had stolen medicines in her shopping basket.

Koch made a formal bow and said: 'The pleasure is mine.'

'And Ada is our maid.'

Ada shot him a hostile look, but he did not see it: maids were beneath his notice. He put his weight on one leg and stood lopsided, trying to seem at ease but giving the opposite impression.

He acted younger than he looked. There was an innocence about him that suggested an over-protected child. All the same he was a danger.

Changing his stance, he rested his hands on the back of the chair on which Carla had put her basket. 'I see you are a nurse,' he said to her.

'Yes.' Carla tried to think calmly. Did Koch have any idea who the von Ulrichs were? He might be too young to know what a social democrat was. The party had been illegal for nine years. Perhaps the infamy of the von Ulrich family had faded away with the death of Walter. At any rate, Koch seemed to take them for a respectable German family who were poor simply because they had lost the man who had supported them, a situation in which many well-bred women found themselves.

There was no reason he should look in the basket.

Carla made herself speak pleasantly to him. 'How are you getting on with the piano?'

'I believe I am making rapid progress!' He glanced at Maud. 'So my teacher tells me.'

Maud said: 'He shows evidence of talent, even at this early stage.' She always said that, to encourage them to pay for a second lesson; but it seemed to Carla that she was being more charming than usual. She was entitled to flirt, of course; she had been a widow for more than a year. But she could not possibly have romantic feelings for someone half her age.

'However, I have decided not to tell my friends until I have mastered the instrument,' Koch added. 'Then I will astonish them with my skill.'

'Won't that be fun?' said Maud. 'Please sit down, Lieutenant, if you have a few minutes to spare.' She pointed to the chair on which Carla's basket stood.

Carla reached out to grab the basket, but Koch beat her to it. He picked it up, saying: 'Allow me.' He glanced inside. Seeing the cabbage, he said: 'Your supper, I presume?'

Carla said: 'Yes.' Her voice came out as a squeak.

He sat on the chair and placed the basket on the floor by his feet, on the side away from Carla. 'I always fancied I might be musical. Now I have decided it is time to find out.' He crossed his legs, then uncrossed them.

Carla wondered why he was so fidgety. He had nothing to fear. The thought crossed her mind that his unease might be sexual. He was alone with three single women. What was going through his mind?

Ada put a cup of coffee in front of him. He took out cigarettes. He smoked like a teenager, as if he was trying it out. Ada gave him an ashtray.

Maud said: 'Lieutenant Koch works at the Ministry of War on Bendler Strasse.'

'Indeed!' That was the headquarters of the Supreme Staff. It was just as well Koch was telling no one there about learning the piano. All the greatest secrets of the German military were in that building. Even if Koch himself was ignorant, some of his colleagues might remember that Walter von Ulrich had been an anti-Nazi. And that would be the end of his lessons with Frau von Ulrich.

'It is a great privilege to work there,' said Koch.

Maud said: 'My son is in Russia. We're terribly worried about him.'

'That is natural in a mother, of course,' Koch said. 'But please do not be pessimistic! The recent Russian counter-offensive has been decisively beaten back.'

That was rubbish. The propaganda machine could not conceal the fact that the Russians had won the battle of Moscow and pushed the German line back a hundred miles.

Koch went on: 'We are now in a position to resume our advance.'

'Are you sure?' Maud looked anxious. Carla felt the same. They were both tortured by fear of what might happen to Erik.

Koch tried a superior smile. 'Believe me, Frau von Ulrich, I am certain. Of course I cannot reveal all that I know. However, I can assure you that a very aggressive new operation is being planned.'

'I am sure our troops have everything they need – enough food, and so on.' She put a hand on Koch's arm. 'All the same, I worry. I shouldn't say that, I know, but I feel I can trust you, Lieutenant.'

'Of course.'

'I haven't heard from my son for months. I don't know if he's dead or alive.'

Koch reached into his pocket and took out a pencil and a small notebook. 'I can certainly find out for you,' he said.

'Could you?' said Maud, wide-eyed.

Carla thought this might be her reason for flirting.

Koch said: 'Oh, yes. I am on the General Staff, you know – albeit in a humble role.' He tried to look modest. 'I can inquire about . . .'

'Erik.'

'Erik von Ulrich.'

'That would be wonderful. He's a medical orderly. He was studying to be a doctor, but he was impatient to fight for the Führer.'

It was true. Erik had been a gung-ho Nazi – although his last few letters home had taken a more subdued tone.

Koch wrote down the name.

Maud said: 'You're a wonderful man, Lieutenant Koch.'

'It is nothing.'

'I'm so glad we're about to counter-attack on the Eastern Front. But you mustn't tell me when the attack will begin. Though I'm desperate to know.'

Maud was fishing for information. Carla could not imagine why. She had no use for it.

Koch lowered his voice, as if there might be a spy outside the open kitchen window. 'It will be very soon,' he said. He looked around at the three women. Carla saw that he was basking in their attention. Perhaps it was unusual for him to have women hanging on his words. Prolonging the moment, he said: 'Case Blue will begin very soon.'

Maud flashed her eyes at him. 'Case Blue – how tremendously thrilling!' she said in the tone a woman might use if a man offered to take her to the Ritz in Paris for a week.

He whispered: 'The twenty-eighth of June.'

Maud put her hand on her heart. 'So soon! That's marvellous news.'

'I should not have said anything.'

Maud put her hand over his. 'I'm so glad you did, though. You've made me feel so much better.'

He stared at her hand. Carla realized that he was not used to being touched by women. He looked up from her hand to her eyes. She smiled warmly – so warmly that Carla could hardly believe it was 100 per cent faked.

Maud withdrew her hand. Koch stubbed out his cigarette and stood up. 'I must go,' he said.

Thank God, Carla thought.

He bowed to her. 'A pleasure to meet you, Fräulein.'

'Goodbye, Lieutenant,' she replied neutrally.

Maud saw him to the door, saying: 'Same time tomorrow, then.'

When she came back into the kitchen she said: 'What a find – a foolish boy who works for the General Staff!'

Carla said: 'I don't understand why you're so excited.'

Ada said: 'He's very handsome.'

Maud said: 'He gave us secret information!'

'What good is it to us?' Carla asked. 'We're not spies.'

'We know the date of the next offensive – surely we can find a way to pass it to the Russians?'

'I don't know how.'

'We're supposed to be surrounded by spies.'

'That's just propaganda. Everything that goes wrong is blamed on subversion by Jewish-Bolshevik secret agents, instead of Nazi bungling.'

'All the same, there must be some real spies.'

'How would we get in touch with them?'

Mother looked thoughtful. 'I'd speak to Frieda.'

'What makes you say that?'

'Intuition.'

Carla recalled the moment at the bus stop, when she had wondered aloud who put up the anti-Nazi posters, and Frieda had gone quiet. Carla's intuition agreed with her mother's.

But that was not the only problem. 'Even if we could, do we want to betray our country?'

Maud was emphatic. 'We have to defeat the Nazis.'

'I hate the Nazis more than anyone, but I'm still German.'

'I know what you mean. I don't like the idea of turning traitor, even though I was born English. But we aren't going to get rid of the Nazis unless we lose the war.'

'But suppose we could give the Russians information that would ensure we lost a battle. Erik might die in that battle! Your son – my brother! We might be the cause of his death.'

Maud opened her mouth to answer, but found she could not speak. Instead, she began to cry. Carla stood up and put her arms around her.

After a minute, Maud whispered: 'He might die anyway. He might die fighting for Nazism. Better he should be killed losing a battle than winning it.'

Carla was not sure about that.

She released her mother. 'Anyway, I wish you'd warn me before bringing someone like that into the kitchen,' she said. She picked up her basket from the floor. 'It's a good thing Lieutenant Koch didn't look any further into this.'

'Why, what have you got in there?'

'Medicines stolen from the hospital for Dr Rothmann.'

Maud smiled proudly through her tears. 'That's my girl.'

'I nearly died when he picked up the basket.'

'I'm sorry.'

'You couldn't know. But I'm going to get rid of the stuff right now.'

'Good idea.'

Carla put her raincoat back on over her uniform and went out.

She walked quickly to the street where the Rothmanns lived. Their house was not as big as the von Ulrich place, but it was a well-proportioned town dwelling with pleasant rooms. However, the windows were now boarded up and there was a crude sign on the front door that said: 'Surgery closed'.

The Rothmanns had once been prosperous. Dr Rothmann had had a flourishing practice with many wealthy patients. He had also treated poor people at cheaper prices. Now only the poor were left.

Carla went around the back, as the patients did.

She knew immediately that something was wrong. The back door was open, and when she stepped into the kitchen she saw a guitar with a broken neck lying on the tiled floor. The room was empty, but she could hear sounds from elsewhere in the house.

She crossed the kitchen and entered the hall. There were two main rooms on the ground floor. They had been the waiting room and the consulting room. Now the waiting room was disguised as a family sitting room, and the surgery had become Rudi's workshop, with a bench and woodworking tools, and usually half a dozen mandolins, violins and cellos in various states of repair. All medical equipment was stashed out of sight in locked cupboards.

But not any more, she saw when she walked in.

The cupboards had been opened and their contents thrown out. The floor was littered with smashed glass and assorted pills, powders and liquids. In the debris Carla saw a stethoscope and a blood pressure gauge. Parts of several instruments were strewn around, evidently having been thrown on the floor and stamped upon.

Carla was shocked and disgusted. All that waste!

Then she looked into the other room. Rudi Rothmann lay in a corner. He was twenty-two years old, a tall man with an athletic build. His eyes were closed, and he was moaning in agony.

His mother, Hannelore, knelt beside him. Once a handsome blonde, Hannelore was now grey and gaunt.

'What happened?' said Carla, fearing the answer.

'The police,' said Hannelore. 'They accused my husband of treating Aryan patients. They have taken him away. Rudi tried to stop them smashing the place up. They have . . .' She choked up.

Carla put down her basket and knelt beside Hannelore. 'What have they done?'

Hannelore recovered the power of speech. 'They broke his hands,' she whispered.

Carla saw it at once. Rudi's hands were red and horribly twisted. The police seemed to have broken his fingers one by one. No wonder he was moaning. She was sickened. But she saw horror every day, and she knew how to suppress her personal feelings and give practical help. 'He needs morphine,' she said.

Hannelore indicated the mess on the floor. 'If we had any, it's gone.'

Carla felt a spasm of pure rage. Even the hospitals were short of supplies – and yet the police had wasted precious drugs in an orgy of destruction. 'I brought you morphine.' She took from her basket a vial of clear fluid and the new syringe. Swiftly, she took the syringe from its box and charged it with the drug. Then she injected Rudi.

The effect was almost instant. The moaning stopped. He opened his eyes and looked at Carla. 'You angel,' he said. Then he closed his eyes and seemed to sleep.

'We must try to set his fingers,' Carla said. 'So that the bones heal straight.' She touched Rudi's left hand. There was no reaction. She grasped the hand and lifted it. Still he did not stir.

'I've never set bones,' said Hannelore. 'Though I've seen it done often enough.'

'Same here,' said Carla. 'But we'd better try. I'll do his left hand, you do the right. We must finish before the drug wears off. God knows he'll be in enough pain.'

'All right,' said Hannelore.

Carla paused a moment longer. Her mother was right. They had to do anything they could to end this Nazi regime, even if it meant betraying their own country. She was no longer in any doubt.

'Let's get it done,' Carla said.

Gently, carefully, the two women began to straighten Rudi's broken hands.

(ii)

Thomas Macke went to the Tannenberg Bar every Friday afternoon.

It was not much of a place. On one wall was a framed photograph of the proprietor, Fritz, in a First World War uniform, twenty-five years younger and without a beer belly. He claimed to have killed nine Russians at the Battle of Tannenberg. There were a few tables and chairs, but the regulars all sat at the bar. A menu in a leather cover was almost entirely fantasy: the only dishes served were sausages with potatoes or sausages without potatoes.

But the place stood across the street from the Kreuzberg police station, so it was a cop bar. That meant it was free to break all the rules. Gambling was open, street girls gave blow jobs in the toilet, and the food inspectors of the Berlin city government never entered the kitchen. It opened when Fritz got up and closed when the last drinker went home.

Macke had been a lowly police officer at the Kreuzberg station years ago, before the Nazis took over and men such as he were suddenly given a break. Some of his former colleagues still drank at the Tannenberg, and he could be sure of seeing a familiar face or two. He still liked to talk to old friends, even though he had risen so far above them, becoming an inspector and a member of the SS.

'You've done well, Thomas, I'll give you that,' said Bernhardt Engel, who had been a sergeant over Macke in 1932 and was still a sergeant. 'Good luck to you, son.' He raised to his lips the stein of beer that Macke had bought him.

'I won't argue with you,' Macke replied. 'Though I will say, Superintendent Kringelein is a lot worse to work for than you were.'

'I was too soft on you boys,' Bernhardt admitted.

Another old comrade, Franz Edel, laughed scornfully. 'I wouldn't say soft!'

Glancing out of the window, Macke saw a motorcycle pull up outside driven by a young man in the light-blue belted jacket of an air force officer. He looked familiar: Macke had seen him somewhere before. He had over-long red-blond hair flopping on to a patrician forehead. He crossed the pavement and came into the Tannenberg.

Macke remembered the name. He was Werner Franck, spoiled son of the radio manufacturer Ludi Franck.

Werner came to the bar and asked for a pack of Kamel cigarettes. How predictable, Macke thought, that the playboy should smoke American-style cigarettes, even if they were a German imitation.

Werner paid, opened the pack, took out a cigarette, and asked Fritz for a light. Turning to leave, cigarette in his mouth tilted at a rakish angle, he caught Macke's eye and, after a moment's thought, said: 'Inspector Macke.'

The men in the bar all stared at Macke to see what he would say.

He nodded casually. 'How are you, young Werner?'

'Very well, sir, thank you.'

Macke was pleased, but surprised, by the respectful tone. He recalled Werner as an arrogant whippersnapper with insufficient respect for authority.

'I'm just back from a visit to the Eastern Front with General Dorn,' Werner added.

Macke sensed the cops in the bar become alert to the conversation. A man who had been to the Eastern Front merited respect. Macke could not help feeling pleased that they were all impressed that he moved in such elevated circles.

Werner offered Macke the cigarette pack, and Macke took one. 'A beer,' Werner said to Fritz. Turning back to Macke, he said: 'May I buy you a drink, Inspector?'

'The same, thank you.'

Fritz filled two steins. Werner raised his glass to Macke and said: 'I want to thank you.'

That was another surprise. 'For what?' said Macke.

His friends were all listening intently.

Werner said: 'A year ago you gave me a good telling-off.'

'You didn't seem grateful at the time.'

'And for that I apologize. But I thought very hard about what you said to me, and eventually I realized you were right. I had allowed personal emotion to cloud my judgement. You set me straight. I'll never forget that.'

Macke was touched. He had disliked Werner, and had spoken

harshly to him; but the young man had taken his words to heart, and changed his ways. It gave Macke a warm glow to feel that he had made such a difference in a young man's life.

Werner went on: 'In fact, I thought of you the other day. General Dorn was talking about catching spies, and asking if we could track them down by their radio signals. I'm afraid I couldn't tell him much.'

'You should have asked me,' said Macke. 'It's my specialty.'

'Is that so?'

'Come and sit down.'

They carried their drinks to a grubby table.

'These men are all police officers,' Macke said. 'But still, one should not talk publicly about such matters.'

'Of course.' Werner lowered his voice. 'But I know I may confide in you. You see, some of the battlefield commanders told Dorn they believe the enemy often knows our intentions in advance.'

'Ah!' said Macke. 'I feared as much.'

'What can I tell Dorn about radio signal detection?'

'The correct term is goniometry.' Macke collected his thoughts. This was an opportunity to impress an influential general, albeit indirectly. He needed to be clear, and emphasize the importance of what he was doing without exaggerating its success. He imagined General Dorn saying casually to the Führer: 'There's a very good man in the Gestapo – name of Macke – only an inspector, at the moment, but most impressive . . .'

'We have an instrument that tells us the direction from which the signal is coming,' he began. 'If we take three readings from widely separated locations, we can draw three lines on the map. Where they intersect is the address of the transmitter.'

'That's fantastic!'

Macke raised a cautionary hand. 'In theory,' he said. 'In practice, it's more difficult. The pianist – that's what we call the radio operator – does not usually stay in the location long enough for us to find him. A careful pianist never broadcasts from the same place twice. And our instrument is housed in a van with a conspicuous aerial on its roof, so they can see us coming.'

'But you have had some success.'

'Oh, yes. But perhaps you should come out in the van with us one evening. Then you could see the whole process for yourself – and make a first-hand report to General Dorn.'

'That's a good idea,' said Werner.

(iii)

Moscow in June was sunny and warm. At lunchtime Volodya waited for Zoya at a fountain in the Alexander Gardens behind the Kremlin. Hundreds of people strolled by, many in pairs, enjoying the weather. Life was hard, and the water in the fountain had been turned off to save power, but the sky was blue, the trees were in leaf and the German army was a hundred miles away.

Volodya was full of pride every time he thought back to the Battle of Moscow. The dreaded German army, master of blitzkrieg attack, had been at the gates of the city – and had been thrown back. Russian soldiers had fought like lions to save their capital.

Unfortunately the Russian counter-attack had petered out in March. It had won back much territory, and made Muscovites feel safer; but the Germans had licked their wounds and were now preparing to try again.

And Stalin was still in charge.

Volodya spotted Zoya walking through the crowd towards him. She was wearing a red-and-white check dress. There was a spring in her step, and her pale-blonde hair seemed to bounce with her stride. Every man stared at her.

Volodya had dated some beautiful women, but he was surprised to find himself courting Zoya. For years she had treated him with cool indifference, and talked to him about nothing but nuclear physics. Then one day, to his astonishment, she had asked him to go to a movie.

It was shortly after the riot in which General Bobrov had been killed. Her attitude to him had changed that day; he was not sure he understood why; somehow the shared experience had created an intimacy. Anyway, they had gone to see *George's Dinky Jazz Band*, a knockabout comedy starring an English banjolele player called George Formby. It was a popular movie, and had been running for months in Moscow. The plot was about as unrealistic as could be: unknown to George, his instrument was sending messages to German U-boats. It was so silly that they had both laughed their socks off.

Since then they had been dating regularly.

Today they were to have lunch with his father. He had arranged to meet her beforehand at the fountain in order to have a few minutes alone with her.

Zoya gave him her thousand-candlepower smile and stood on tiptoe to kiss him. She was tall, but he was taller. He relished the kiss. Her lips were soft and moist on his. It was over too soon.

Volodya was not completely sure of her yet. They were still 'walking out', as the older generation termed it. They kissed a lot, but they had not yet gone to bed together. They were not too young: he was twenty-seven, she twenty-eight. All the same, Volodya sensed that Zoya was not going to sleep with him until she was ready.

Half of him did not believe he would ever spend a night with this dream girl. She seemed too blonde, too intelligent, too tall, too self-possessed, too sexy ever to give herself to a man. Surely he would never be allowed to watch her take off her clothes, to gaze at her naked body, to touch her all over, to lie on top of her . . . ?

They walked through the long, narrow park. On one side was a busy road. All along the other side, the towers of the Kremlin loomed over a high wall. 'To look at it, you'd think our leaders in there were being held prisoner by the Russian people,' Volodya said.

'Yes,' Zoya agreed. 'Instead of the other way round.'

He looked behind them, but no one had heard. All the same it was foolhardy to talk like that. 'No wonder my father thinks you're dangerous.'

'I used to think you were like your father.'

'I wish I was. He's a hero. He stormed the Winter Palace! I don't suppose I'll ever change the course of history.'

'Oh, I know, but he's so narrow-minded and conservative. You're not like that.'

Volodya thought he was pretty much like his father, but he was not going to argue.

'Are you free this evening?' she said. 'I'd like to cook for you.'

'You bet!' She had never invited him to her place.

'I've got a piece of steak.'

'Great!' Good beef was a treat even in Volodya's privileged home.

'And the Kovalevs are out of town.'

That was even better news. Like many Muscovites, Zoya lived in someone else's apartment. She had two rooms and shared the kitchen and bathroom with another scientist, Dr Kovalev, and his wife and child. But the Kovalevs had gone away, so Zoya and Volodya would have the place to themselves. His pulse quickened. 'Should I bring my toothbrush?' he said.

She gave him an enigmatic smile and did not answer the question.

They left the park and crossed the road to a restaurant. Many were closed, but the city centre was full of offices whose workers had to eat lunch somewhere, and a few cafés and bars survived.

Grigori Peshkov was at a pavement table. There were better restaurants inside the Kremlin, but he liked to be seen in places used by ordinary Russians. He wanted to show that he was not above the common people just because he wore a general's uniform. All the same, he had chosen a table well away from the rest, so that he could not be overheard.

He disapproved of Zoya, but he was not immune to her enchantment, and he stood up and kissed her on both cheeks.

They ordered potato pancakes and beer. The only alternatives were pickled herrings and vodka.

'Today I am not going to speak to you about nuclear physics, General,' said Zoya. 'Please take it as read that I still believe everything I said last time we talked about the subject. I don't want to bore you.'

'That's a relief,' he said.

She laughed, showing white teeth. 'Instead, you can tell me how much longer we will be at war.'

Volodya shook his head in mock despair. She always had to challenge his father. If she had not been a beautiful young woman, Grigori would have had her arrested long ago.

'The Nazis are beaten, but they won't admit it,' Grigori said.

Zoya said: 'Everyone in Moscow is wondering what will happen this summer – but you two probably know.'

Volodya said: 'If I did, I certainly could not tell my girlfriend, no matter how crazy I am about her.' Apart from anything else, it could get her shot, he thought, but he did not say it.

The potato pancakes came and they began to eat. As always, Zoya tucked in hungrily. Volodya loved the relish with which she attacked food. But he did not much like the pancakes. 'These potatoes taste suspiciously like turnips,' he said.

His father shot him a disapproving look.

'Not that I'm complaining,' Volodya added hastily.

When they had finished, Zoya went to the ladies' room. As soon as she was out of earshot, Volodya said: 'We think the German summer offensive is imminent.'

'I agree,' said his father.

'Are we ready?'

'Of course,' said Grigori, but he looked anxious.

'They will attack in the south. They want the oilfields of the Caucasus.'

Grigori shook his head. 'They will come back to Moscow. It's all that matters.'

'Stalingrad is equally symbolic. It bears the name of our leader.'

'Fuck symbolism. If they take Moscow, the war is over. If they don't, they haven't won, no matter what else they gain.'

'You're just guessing,' Volodya said with irritation.

'So are you.'

'On the contrary, I have evidence.' He looked around, but there was no one nearby. 'The offensive is codenamed Case Blue. It will start on 28 June.' He had learned that much from Werner Franck's network of spies in Berlin. 'And we found partial details in the briefcase of a German officer who crash-landed a reconnaissance plane near Kharkov.'

'Officers on reconnaissance do not carry battle plans in briefcases,' Grigori said. 'Comrade Stalin thinks that was a ruse to deceive us, and I agree. The Germans want us to weaken our central front by sending forces south to deal with what will turn out to be no more than a diversion.'

This was the problem with intelligence, Volodya thought with frustration. Even when you had the information, stubborn old men would believe what they wanted.

He saw Zoya coming back, all eyes on her as she walked across the plaza. 'What would convince you?' he said to his father before she arrived.

'More evidence.'

'Such as?'

Grigori thought for a moment, taking the question seriously. 'Get me the battle plan.'

Volodya sighed. Werner Franck had not yet succeeded in obtaining the document. 'If I get it, will Stalin reconsider?'

'If you get it, I'll ask him to.'

'It's a deal,' said Volodya.

He was being rash. He had no idea how he was going achieve this. Werner, Heinrich, Lili, and the others already took horrendous risks. Yet he would have to put even more pressure on them.

Zoya reached their table and Grigori stood up. They were going in three different directions, so they said goodbye.

'I'll see you tonight,' Zoya said to Volodya.

He kissed her. 'I'll be there at seven.'

'Bring your toothbrush,' she said.

He walked away a happy man.

(iv)

A girl knows when her best friend has a secret. She may not know what the secret is, but she knows it is there, like an unidentifiable piece of furniture under a dust sheet. She realizes, from guarded and unforthcoming answers to innocent questions, that her friend is seeing someone she shouldn't; she just doesn't know the name, although she may guess that the forbidden lover is a married man, or a dark-skinned foreigner, or another woman. She admires that necklace, and knows from her friend's muted reaction that it has shameful associations, though it may not be until years later that she discovers it was stolen from a senile grandmother's jewel box.

So Carla thought when she reflected on Frieda.

Frieda had a secret, and it was connected with resistance to the Nazis. She might be deeply, criminally involved: perhaps she went through her brother Werner's briefcase every night, copied secret papers, and handed the copies to a Russian spy. More likely it was not so dramatic: she probably helped print and distribute those illegal posters and leaflets that criticized the government.

So Carla was going to tell Frieda about Joachim Koch. However, she did not immediately get a chance. Carla and Frieda were nurses in different departments of a large hospital, and had different rotas, so they did not necessarily meet every day.

Meanwhile, Joachim came to the house daily for lessons. He made no more indiscreet revelations, but Maud continued to flirt with him. 'You do realize that I'm almost forty years old?' Carla heard her say one day, although she was in fact fifty-one. Joachim was completely infatuated. Maud was enjoying the power she still had to fascinate an attractive young man, albeit a very naive one. The thought crossed Carla's mind that her mother might be developing deeper feelings for this boy with a fair moustache who looked a bit like the young Walter; but that seemed ridiculous.

Joachim was desperate to please her, and soon brought news of her son. Erik was alive and well. 'His unit is in the Ukraine,' Joachim said. 'That's all I can tell you.'

'I wish he could get leave to come home,' Maud said wistfully.

The young officer hesitated.

She said: 'A mother worries so much. If I could just see him, even for only a day, it would be such a comfort to me.'

'I *might* be able to arrange that.'

Maud pretended to be astonished. 'Really? You're that powerful?'

'I'm not sure. I could try.'

'Thank you for even trying.' She kissed his hand.

It was a week before Carla saw Frieda again. When she did, she told her all about Joachim Koch. She told the story as if simply retailing an interesting piece of news, but she felt sure Frieda would not regard it in that innocent light. 'Just imagine,' she said. 'He told us the code name of the operation and the date of the attack!' She waited to see how Frieda would respond.

'He could be executed for that,' Frieda said.

'If we knew someone who could get in touch with Moscow, we might turn the course of the war,' Carla went on, as if still talking about the gravity of Joachim's crime.

'Perhaps,' said Frieda.

That proved it. Frieda's normal reaction to such a story would include expressions of surprise, lively interest, and further questions. Today she offered nothing but neutral phrases and noncommittal grunts. Carla went home and told her mother that her intuition had been correct.

Next day at the hospital, Frieda appeared in Carla's ward looking frantic. 'I have to talk to you urgently,' she said.

Carla was changing a dressing for a young woman who had been badly burned in a munitions factory explosion. 'Go to the cloakroom,' she said. 'I'll be there as soon as I can.'

Five minutes later she found Frieda in the little room, smoking by an open window. 'What is it?' she said.

Frieda put out the cigarette. 'It's about your Lieutenant Koch.'

'I thought so.'

'You have to find out more from him.'

'I *have* to? What are you talking about?'

'He has access to the entire battle plan for Case Blue. We know something about it, but Moscow needs the details.'

Frieda was making a bewildering set of assumptions, but Carla went along with it. 'I can ask him . . .'

'No. You have to *make* him bring you the battle plan.'

'I'm not sure that's possible. He's not completely stupid. Don't you think—'

Frieda was not even listening. 'Then you have to photograph it,' she interrupted. She produced from the pocket of her uniform a stainless-steel box about the size of a pack of cigarettes, but longer and narrower. 'This is a miniature camera specially designed for photographing documents.' Carla noticed the name 'Minox' on the side. 'You'll get eleven pictures on one film. Here are three films.' She brought out three cassettes, the shape of dumbbells but small enough to fit into the little camera. 'This is how you load the film.' Frieda demonstrated. 'To take a picture, you look through this window. If you're not sure, read this manual.'

Carla had never known Frieda to be so domineering. 'I really need to think about this.'

'There's no time. This is your raincoat, isn't it?'

'Yes, but—'

Frieda stuffed the camera, films and booklet into the pockets of the coat. She seemed relieved they were out of her hands. 'I've got to go.' She went to the door.

'But, Frieda!'

At last Frieda stopped and looked directly at Carla. 'What?'

'Well . . . you're not behaving like a friend.'

'This is more important.'

'You've backed me into a corner.'

'You created this situation when you told me about Joachim Koch. Don't pretend you didn't expect me to do something with the information.'

It was true. Carla had triggered this emergency herself. But she had not envisaged things turning out this way. 'What if he says no?'

'Then you'll probably be living under the Nazis for the rest of your life.' Frieda went out.

'Hell,' said Carla.

She stood alone in the cloakroom, thinking. She could not even get rid of the little camera without risk. It was in her raincoat, and she could hardly throw it into a hospital rubbish bin. She would have to leave the building with it in her pocket, and try to find a place where she could dispose of it secretly.

But did she want to?

It seemed unlikely that Koch, naive though he was, could be talked into smuggling a copy of a battle plan out of the War Ministry and bringing it to show his inamorata. However, if anyone could persuade him, Maud could.

But Carla was scared. There would be no mercy for her if she were caught. She would be arrested and tortured. She thought of Rudi Rothmann, moaning in the agony of broken bones. She recalled her father after they released him, so brutally beaten that he had died. Her crime would be worse than theirs; her punishment correspondingly bestial. She would be executed, of course – but not for a long time.

She told herself she was willing to risk that.

What she could not accept was the danger that she would help kill her brother.

He was there, on the Eastern Front, Joachim had confirmed it. He would be involved in Case Blue. If Carla enabled the Russians to win that battle, Erik could die as a result. She could not bear that.

She went back to her work. She was distracted and made mistakes, but fortunately the doctors did not notice and the patients could not tell. When at last her shift ended, she hurried away. The camera was burning a hole in her pocket but she did not see a safe place to dump it.

She wondered where Frieda had got it. Frieda had plenty of money, and could easily have bought it, though she would have had to come up with a story about why she needed such a thing. More likely she could have got it from the Russians before they closed their embassy a year ago.

The camera was still in Carla's coat pocket when she arrived home.

There was no sound from the piano upstairs: Joachim was having his lesson later today. Her mother was sitting at the kitchen table. When Carla walked in, Maud beamed and said: 'Look who's here!'

It was Erik.

Carla stared at him. He was painfully thin, but apparently uninjured. His uniform was grimy and ripped, but he had washed his face and hands. He stood up and put his arms around her.

She hugged him hard, careless of dirtying her spotless uniform. 'You're safe,' she said. There was so little flesh on him that she could feel his bones, his ribs and hips and shoulders and spine, through the thin material.

'Safe for the moment,' he said.

She released her hold. 'How are you?'

'Better than most.'

'You weren't wearing this flimsy uniform in the Russian winter?'

'I stole a coat from a dead Russian.'

She sat down at the table. Ada was there too. Erik said: 'You were right. About the Nazis, I mean. You were right.'

She was pleased, but not sure exactly what he meant. 'In what way?'

'They murder people. You told me that. Father told me, too, and Mother. I'm sorry I didn't believe you. I'm sorry, Ada, that I didn't believe they killed your poor little Kurt. I know better now.'

This was a big reversal. Carla said: 'What changed your mind?'

'I saw them doing it, in Russia. They round up all the important people in town, because they must be Communists. And they get the Jews, too. Not just men, but women and children. And old people too frail to do anyone any harm.' Tears were streaming down his face now. 'Our regular soldiers don't do it – there are special groups. They take the prisoners out of town. Sometimes there's a quarry, or some other kind of pit. Or they make the younger ones dig a great hole. Then—'

He choked up, but Carla had to hear him say it. 'Then what?'

'They do them twelve at a time. Six pairs. Sometimes the husbands and wives hold hands as they walk down the slope. The mothers carry the babies. The riflemen wait until the prisoners are in the right spot. Then they shoot.' Erik wiped his tears with his dirty uniform sleeve. 'Bang,' he said.

There was a long silence in the kitchen. Ada was crying. Carla was aghast. Only Maud was stony-faced.

Eventually Erik blew his nose, then took out cigarettes. 'I was surprised to get leave and a ticket home,' he said.

Carla said: 'When do you have to go back?'

'Tomorrow. I have only twenty-four hours here. All the same I'm the envy of all my comrades. They'd give anything for a day at home. Dr Weiss said I must have friends in high places.'

'You do,' said Maud. 'Joachim Koch, a young lieutenant who works at the War Ministry and comes to me for piano lessons. I asked him to arrange leave for you.' She glanced at her watch. 'He'll be here in a few minutes. He has grown fond of me – he's in need of a mother figure, I think.'

Mother, hell, Carla thought. There was nothing maternal about Maud's relationship with Joachim.

Maud went on: 'He's very innocent. He told us there's going to be a new offensive on the Eastern Front starting on 28 June. He even mentioned the code name: Case Blue.'

Erik said: 'He's going to get himself shot.'

Carla said: 'Joachim is not the only one who might be shot. I told someone what I learned. Now I've been asked to persuade Joachim, somehow, to get me the battle plan.'

'Good God!' Erik was rocked. 'This is serious espionage – you're in more danger than I am on the Eastern Front!'

'Don't worry, I can't imagine Joachim would do it,' Carla said.

'Don't be so sure,' said Maud.

They all looked at her.

'He might do it for me,' she said. 'If I asked him the right way.'

Erik said: 'He's *that* naive?'

She looked defiant. 'He's in love with me.'

'Oh.' Erik was embarrassed at the idea of his mother being involved in a romance.

Carla said: 'All the same, we can't do it.'

Erik said: 'Why not?'

'Because if the Russians win the battle you might die!'

'I'll probably die anyway.'

Carla heard her own voice rise in pitch agitatedly. 'But we'd be helping the Russians kill you!'

'I still want you to do it,' Erik said fiercely. He looked down at the chequered oilcloth on the kitchen table, but what he was seeing was a thousand miles away.

Carla felt torn. If he *wanted* her to ... She said: 'But why?'

'I think of those people walking down the slope into the quarry, holding hands.' His own hands on the table grasped each other hard enough to bruise. 'I'll risk my life, if we can put a stop to that. I *want* to risk my life – I'll feel better about myself, and my country, if I do. Please, Carla, if you can, send the Russians that battle plan.'

Still she hesitated. 'Are you sure?'

'I'm begging you.'

'Then I will,' said Carla.

(v)

Thomas Macke told his men – Wagner, Richter and Schneider – to be on their best behaviour. 'Werner Franck is only a lieutenant, but he works for General Dorn. I want him to have the best possible impression of our team and our work. No swearing, no jokes, no eating, and no rough stuff unless it's really necessary. If we catch a Communist spy, you can give him a good kicking. But if we fail, I don't want you to pick on someone else just for fun.' Normally he would turn a blind eye to that sort of thing. It all helped to keep people in fear of the displeasure of the Nazis. But Franck might be squeamish.

Werner turned up punctually at Gestapo headquarters in Prinz Albrecht Strasse on his motorcycle. They all got into the surveillance van with the revolving aerial on the roof. With so much radio equipment inside it was cramped. Richter took the wheel and they drove around the city in the early evening, the favoured time for spies to send messages to the enemy.

'Why is that, I wonder?' said Werner.

'Most spies have a regular job,' Macke explained. 'It's part of their cover story. So they go to an office or a factory in the daytime.'

'Of course,' said Werner. 'I never thought of that.'

Macke was worried they might not pick up anything at all tonight. He was terrified that he would get the blame for the reverses the German army was suffering in Russia. He had done his best, but there were no prizes for effort in the Third Reich.

It sometimes happened that the unit picked up no signals. On other occasions there would be two or three, and Macke would have to choose which to follow up and which to ignore. He felt sure there was more than one spy network in the city, and they probably did not know of each other's existence. He was trying to do an impossible job with inadequate tools.

They were near the Potsdamer Platz when they heard a signal. Macke recognized the characteristic sound. 'That's a pianist,' he said with relief. At least he could prove to Werner that the equipment worked. Someone was broadcasting five-digit numbers, one after the other. 'Soviet Intelligence uses a code in which pairs of numbers stand for letters,' Macke explained to Werner. 'So, for example, 11 might stand for A. Transmitting them in groups of five is just a convention.'

The radio operator, an electrical engineer named Mann, read off a set of co-ordinates, and Wagner drew a line on a map with a pencil and rule. Richter put the van in gear and set off again.

The pianist continued to broadcast, his beeps sounding loud in the van. Macke hated the man, whoever he was. 'Bastard Communist swine,' he said. 'One day he'll be in our basement, begging me to let him die so the pain will come to an end.'

Werner looked pale. He was not used to police work, Macke thought.

After a moment the young man pulled himself together. 'The way you describe the Soviet code, it sounds as if it might not be too difficult to break,' he said thoughtfully.

'Correct!' Macke was pleased that Werner caught on so fast. 'But I was simplifying. They have refinements. After encoding the message as a series of numbers, the pianist then writes a key word underneath it repeatedly – it might be Kurfürstendamm, say – and encodes that. Then he subtracts the second numbers from the first and broadcasts the result.'

'Almost impossible to decipher if you don't know the key word!'

'Exactly.'

They stopped again near the burned-out Reichstag building and drew another line on the map. The two met in Friedrichshain, to the east of the city centre.

Macke told the driver to swing north-east, taking them nearer to the likely spot while giving them a third line from a different angle. 'Experience shows that it's best to take three bearings,' Macke told Werner. 'The equipment is only approximate, and the extra measurement reduces error.'

'Do you always catch him?' said Werner.

'By no means. In most cases we don't. Often we're just not quick enough. He may change frequency halfway through, so that we lose him. Sometimes he breaks off in mid-transmission and resumes at another location. He may have lookouts who see us coming and warn him to flee.'

'A lot of snags.'

'But we catch them, sooner or later.'

Richter stopped the van and Mann took the third bearing. The three pencil lines on Wagner's map met to form a small triangle near the East Station. The pianist was somewhere between the railway line and the canal.

Macke gave Richter the location and added: 'Quick as you can.'

Werner was perspiring, Macke noticed. Perhaps it was rather hot in the van. And the young lieutenant was not accustomed to action. He was learning what life was like in the Gestapo. All the better, Macke thought.

Richter headed south on Warschauer Strasse, crossed the railway, then turned into a cheap industrial neighbourhood of warehouses, yards and small factories. There was a group of soldiers toting kitbags outside a back entrance to the station, no doubt embarking for the Eastern Front. And a fellow-countryman somewhere in this neighbourhood doing his best to betray them, Macke thought angrily.

Wagner pointed down a narrow street leading away from the station. 'He's in the first few hundred yards, but he could be on either side,' he said. 'If we take the van any closer he'll see us.'

'All right, men, you know the drill,' Macke said. 'Wagner and Richter take the left-hand side. Schneider and I will take the right.' They all picked up long-handled sledgehammers. 'Come with me, Franck.'

There were few people on the street – a man in a worker's cap walking briskly towards the railway station, an older woman in shabby clothes probably on her way to clean offices – and they hurried quickly past, not wanting to attract the attention of the Gestapo.

Macke's team entered each building, one man leapfrogging his partner. Most businesses were closed for the day so they had to rouse a janitor. If he took more than a minute to come to the door they knocked it down. Once inside they raced through the building checking every room.

The pianist was not in the first block.

The first building on the right-hand side of the next block had a fading sign that said: 'Fashion Furs'. It was a two-storey factory that stretched along the side street. It looked disused, but the front door was steel and the windows were barred: a fur coat factory naturally had heavy security.

Macke led Werner down the side street, looking for a way in. The adjacent building was bomb-damaged and derelict. The rubble had been cleared from the street and there was a hand-painted sign saying: 'Danger – No Entry'. The remains of a name board identified it as a furniture warehouse.

They stepped over a pile of stones and splintered timbers, going as fast as they could but forced to tread carefully. A surviving wall concealed the rear of the building. Macke went behind it and found a hole through to the factory next door.

He had a strong feeling the pianist was in here.

He stepped through the hole, and Werner followed.

They found themselves in an empty office. There was an old steel desk with no chair, and a filing cabinet opposite. The calendar pinned to the wall was for 1939, probably the last year during which Berliners could afford such frivolities as fur coats.

Macke heard a footstep on the floor above.

He drew his gun.

Werner was unarmed.

They opened the door and stepped into a corridor.

Macke noted several open doors, a staircase up, and a door under the staircase that might lead to a basement.

Macke crept along the corridor towards the foot of the stairs, then noticed that Werner was checking the door to the basement.

'I thought I heard a noise from below,' Werner said. He turned the handle but the door had a flimsy lock. He stepped back and raised his right foot.

Macke said: 'No—'

'Yes – I hear them!' Werner said, and he kicked the door open.

The crash resounded throughout the empty factory.

Werner burst through the door and disappeared. A light came on, showing a stone staircase. 'Don't move!' Werner yelled. 'You are under arrest!'

Macke went down the stairs after him.

He reached the basement. Werner stood at the foot of the stairs, looking baffled.

The room was empty.

Suspended from the ceiling were rails on which coats had probably been hung. An enormous roll of brown paper stood on end in one corner, probably intended for wrapping. But there was no radio and no spy tapping messages to Moscow.

'You fucking idiot,' Macke said to Werner.

He turned and ran back up the stairs. Werner ran after him. They traversed the hallway and went up to the next floor.

There were rows of workbenches under a glass roof. At one time the place must have been full of women working at sewing machines. Now there was nobody.

A glass door led to a fire escape, but the door was locked. Macke looked out and saw nobody.

He put his gun away. Breathing hard, he leaned on a workbench.

On the floor he noticed a couple of cigarette ends, one with lipstick on. They did not look very old. 'They were here,' he said to Werner, pointing at the floor. 'Two of them. Your shout warned them, and they escaped.'

'I was a fool,' Werner said. 'I'm sorry, but I'm not used to this kind of thing.'

Macke went to the corner window. Along the street he saw a young man and woman walking briskly away. The man was carrying a tan leather suitcase. As he watched, they disappeared into the railway station. 'Shit,' he said.

'I don't think they were spies,' Werner said. He pointed to something on the floor, and Macke saw a crumpled condom. 'Used, but empty,' Werner said. 'I think we caught them in the act.'

'I hope you're right,' said Macke.

(vi)

The day Joachim Koch promised to bring the battle plan, Carla did not go to work.

She probably could have done her usual morning shift and been home in time – but 'probably' was not enough. There was always a risk that there might be a major fire or a road accident obliging her to work after the end of her shift to deal with an inrush of injured people. So she stayed at home all day.

In the end Maud had not had to ask Joachim to bring the plan. He had said he needed to cancel his lesson; then, unable to resist the temptation to boast, he had explained that he had to carry a copy of the plan across town. 'Come for your lesson on the way,' Maud had said; and he had agreed.

Lunch was strained. Carla and Maud ate a thin soup made with a ham bone and dried peas. Carla did not ask what Maud had done, or promised to do, to persuade Koch. Perhaps she had told him he was making marvellous progress on the piano but could not afford to miss a lesson. She might have asked whether he was so junior that he was monitored every minute: such a remark would sting him, for he pretended constantly to be more important than he was, and it might easily provoke him into showing up just to prove her wrong. However, the ploy most likely to have succeeded was the one Carla did not want to think about: sex. Her mother flirted outrageously with Koch, and he responded with slavish devotion. Carla suspected that this was the irresistible temptation that had made Joachim ignore the voice in his head saying: 'Don't be so damn stupid.'

Or perhaps not. He might see sense. He could show up this afternoon, not with a carbon copy in his bag, but with a Gestapo squad and a set of handcuffs.

Carla loaded a film cassette into the Minox camera, then put the camera and the two remaining cassettes in the top drawer of a low kitchen cupboard, under some towels. The cupboard stood next to the window, where the light was bright. She would photograph the document on the cupboard top.

She did not know how the exposed film would reach Moscow, but Frieda had assured her it would, and Carla imagined a travelling salesman – in pharmaceuticals, perhaps, or German-language Bibles – who had permission to sell his wares in Switzerland and could discreetly pass the film to someone from the Soviet Embassy in Bern.

The afternoon was long. Maud went to her room to rest. Ada did laundry. Carla sat in the dining room, which they rarely used nowadays, and tried to read, but she could not concentrate. The newspaper was all lies. She needed to cram for her next nursing exam, but the medical terms in her textbook swam before her eyes. She was reading an old copy of *All Quiet on the Western Front*, a German bestseller about the First World War, now banned because it was too honest about the hardships of soldiers; but she found herself holding the book in her hand and gazing out of the window at the June sunlight beating down on the dusty city.

At last he came. Carla heard a footstep on the path and jumped up to look out. There was no Gestapo squad, just Joachim Koch in his pressed uniform and shiny boots, his movie-star face as full of eager anticipation as that of a child arriving for a birthday party. He had his canvas bag over his shoulder as usual. Had he kept his promise? Did that bag hold a copy of the battle plan for Case Blue?

He rang the bell.

Carla and Maud had premeditated every move from now on. In accordance with their plan, Carla did not answer the door. A few moments later she saw her mother walk across the hall wearing a purple silk dressing gown and high-heeled slippers – almost like a prostitute,

Carla thought with shame and embarrassment. She heard the front door
open, then close again. From the hall there was a whisper of silk and a
murmured endearment that suggested an embrace. Then the purple robe
and the field-grey uniform passed the dining-room door and disappeared
upstairs.

Maud's first priority was to make sure he had the document. She was
to look at it, say something admiring, then put it down. She would lead
Joachim to the piano. Then she would find some pretext – Carla tried
not to think what – for taking the young man through the double doors
that led from the drawing room into the neighbouring study, a smaller,
more intimate room with red velvet curtains and a big, sagging old
couch. As soon as they were there, Maud would give the signal.

Because it was hard to know in advance the exact choreography of
their movements, there were several possible signals, all of which meant
the same thing. The simplest was that she would slam the door loud
enough to be heard throughout the house. Alternatively, she would use
the bell-push beside the fireplace that sounded a ring in the kitchen, part
of the obsolete system for summoning servants. But any other noise
would do, they had decided: in desperation she would knock the marble
bust of Goethe to the floor or 'accidentally' smash a vase.

Carla stepped out of the dining room and stood in the hall, looking
up the stairs. There was no sound.

She looked into the kitchen. Ada was washing the iron pot in which
she had made the soup, scrubbing with an energy that was undoubtedly
fuelled by tension. Carla gave her what she hoped was an encouraging
smile. Carla and Maud would have liked to keep this whole affair secret
from Ada, not because they did not trust her – quite the contrary, her
hostility to the Nazis was fanatical – but because the knowledge made
her complicit in treachery, and liable to the most extreme punishment.
However, they lived too much together for secrecy to be possible, and
Ada knew everything.

Carla faintly heard Maud give a tinkling laugh. She knew that sound.
It struck an artificial note, and indicated that she was straining her
powers of fascination to the limit.

Did Joachim have the document, or not?

A minute or two later Carla heard the piano. It was undoubtedly
Joachim playing. The tune was a simple children's song about a cat in
the snow: 'A.B.C., Die Katze lief im Schnee'. Carla's father had sung it to

her a hundred times. She felt a lump in her throat now when she thought of that. How dare the Nazis play such songs when they had made orphans of so many children?

The song stopped abruptly in the middle. Something had happened. Carla strained to hear – voices, footsteps, anything – but there was nothing.

A minute went by, then another.

Something had gone wrong – but what?

She looked through the kitchen doorway at Ada, who stopped scrubbing to spread her hands in a gesture that signified: *I have no idea.*

Carla had to find out.

She went quietly up the stairs, treading noiselessly on the threadbare carpet.

She stood outside the drawing room. Still she could hear nothing: no piano music, no movement, no voices.

She opened the door as quietly as possible.

She peeped in. She could see no one. She stepped inside and looked all around. The room was empty.

There was no sign of Joachim's canvas bag.

She looked at the double door that led to the study. One of the two doors stood half open.

Carla tiptoed across the room. There was no carpet here, just polished wood blocks, and her footsteps were not completely silent; but she had to take the risk.

As she got nearer, she heard whispers.

She reached the doorway. She flattened herself against the wall then risked a look inside.

They were standing up, embracing, kissing. Joachim had his back to the door and to Carla: no doubt Maud had taken care to move him into that position. As Carla watched, Maud broke the kiss, looked over his shoulder, and caught Carla's eye. She took her hand away from Joachim's neck and made an urgent pointing gesture.

Carla saw the canvas bag on a chair.

She understood immediately what had gone wrong. When Maud had inveigled Joachim into the study, he had not obliged them by leaving his bag in the drawing room, but had nervously taken it with him.

Now Carla had to retrieve it.

Heart thudding, she stepped into the room.

Maud murmured: 'Oh, yes, keep doing that, my sweet boy.'

Joachim groaned: 'I love you, my darling.'

Carla took two paces forward, picked up the canvas bag, turned around, and stepped silently out of the room.

The bag was light.

She walked quickly across the drawing room and ran down the stairs, breathing hard.

In the kitchen she put the bag on the table and unbuckled its straps. Inside were today's edition of the Berlin newspaper *Der Angriff,* a fresh pack of Kamel cigarettes, and a plain buff-coloured cardboard folder. With trembling hands she took out the folder and opened it. It contained a carbon copy of a document.

The first page was headed:

DIRECTIVE NO. 41

On the last page was a dotted line for a signature. Nothing was penned there, no doubt because this was a copy, but the name typed beside the line was Adolf Hitler.

In between was the plan for Case Blue.

Exultation rose in her heart, mingled with the tension she already felt and the terrible dread of discovery.

She put the document on the low cupboard next to the kitchen window. She jerked open the drawer and took out the Minox camera and the two spare films. She positioned the document carefully, then began to photograph it page by page.

It did not take long. There were just ten pages. She did not even have to reload film. She was done. She had stolen the battle plan.

That was for you, Father.

She put the camera back in the drawer, closed the drawer, slipped the document into the cardboard folder, put the folder back in the canvas bag, and closed the bag, fastening the straps.

Moving as quietly as she could, she carried the bag back upstairs.

As she crept into the drawing room she heard her mother's voice. Maud was speaking clearly and emphatically, as if she wanted to be overheard, and Carla immediately sensed a warning. 'Please don't worry,' she was saying. 'It's because you were so excited. We were both excited.'

Joachim's voice came in reply, low and embarrassed. 'I feel a fool,' he said. 'You only touched me, and it was all over.'

Carla could guess what had happened. She had no experience of it, but girls talked, and nurses' conversations were brutally detailed. Joachim must have ejaculated prematurely. Frieda had told her that Heinrich had done the same, several times, when they were first together, and had been mortified with embarrassment, though he had soon got over it. It was a sign of nervousness, she said.

The fact that Maud and Joachim's embraces were over so early created a difficulty for Carla. Joachim would be more alert now, no longer blind and deaf to everything going on around him.

All the same, Maud must be doing her best to keep his back to the doorway. If Carla could just slip in for a second and replace the bag on the chair without being seen by Joachim, they could still get away with it.

Heart pounding, Carla crossed the drawing room and paused at the open door.

Maud said reassuringly: 'It happens often – the body becomes impatient. It's nothing.'

Carla put her head around the door.

The two of them were still standing in the same place, still close together. Maud looked past Joachim and saw Carla. She put her hand on Joachim's cheek, keeping his gaze away from Carla, and said: 'Kiss me again, and tell me you don't hate me for this little accident.'

Carla stepped inside.

Joachim said: 'I need a cigarette.'

As he turned around, Carla stepped back outside.

She waited by the door. Did he have cigarettes in his pocket, or would he look for the new pack in his bag?

The answer came a second later. 'Where's my bag?' he said.

Carla's heart stopped.

Maud's voice came clearly. 'You left it in the drawing room.'

'No, I didn't.'

Carla crossed the room, dropped the bag on a chair, and stepped outside. Then she paused on the landing, listening.

She heard them move from the study to the drawing room.

Maud said: 'There it is, I told you so.'

'I did not leave it there,' he said stubbornly. 'I vowed I would not let it out of my sight. But I did – when I was kissing you.'

'My darling, you're upset about what happened between us. Try to relax.'

'Someone must have come into the room, while I was distracted . . .'

'How absurd.'

'I don't think so.'

'Let's sit at the piano, side by side, the way you like to,' she said, but she was beginning to sound desperate.

'Who else is in this house?'

Guessing what would happen next, Carla ran down the stairs and into the kitchen. Ada stared at her in alarm, but there was no time to explain.

She heard Joachim's boots on the stairs.

A moment later he was in the kitchen. He had the canvas bag in his hand. His face was angry. He looked at Carla and Ada. 'One of you has been looking inside this bag!' he said.

Carla spoke as calmly as she could. 'I don't know why you should think that, Joachim,' she said.

Maud appeared behind Joachim and came past him into the kitchen. 'Let's have coffee, please, Ada,' she said brightly. 'Joachim, do sit down, please.'

He ignored her and scrutinized the kitchen. His eye lit upon the top of the low cupboard by the window. Carla saw, to her horror, that although she had put the camera away, she had left the two spare film cassettes out.

'Those are eight-millimetre film cassettes, aren't they?' Joachim said. 'Have you got a miniature camera?'

Suddenly he did not seem such a little boy.

'Is that what those things are?' said Maud. 'I've been wondering. They were left behind by another pupil, a Gestapo officer, in fact.'

It was a clever improvisation, but Joachim was not buying it. 'And did he also leave behind his camera, I wonder?' he said. He pulled open the drawer.

The neat little stainless-steel camera lay there on a white towel, guilty as a bloodstain.

Joachim looked shocked. Perhaps he had not really believed he was the victim of treachery, but had been blustering to compensate for his sexual failure; and now he was facing the truth for the first time. Whatever the reason, he was momentarily stunned. Still holding the knob of the drawer, he stared at the camera as if hypnotized. In that short moment Carla saw that a young man's dream of love had been defiled, and his rage was going to be terrible.

At last he raised his eyes. He looked at the three women around him, and his gaze rested on Maud. 'You have done this,' he said. 'You tricked me. But you will be punished.' He picked up the camera and films and put them in his pocket. 'You are under arrest, Frau von Ulrich.' He took a step forward and grabbed her arm. 'I am taking you to Gestapo headquarters.'

Maud jerked her arm free of his grasp and took a step back.

Joachim drew back his arm and punched her with all his might. He was tall, strong and young. The blow landed on her face and knocked her down.

Joachim stood over her. 'You made a fool of me!' he screeched. 'You lied, and I believed you!' He was hysterical now. 'We will both be tortured by the Gestapo, and we both deserve it!' He began to kick her where she lay. She tried to roll away, but came up against the cooker. His right boot thudded into her ribs, her thigh, her belly.

Ada rushed at him and scratched his face with her nails. He batted her away with a swipe. Then he kicked Maud in the head.

Carla moved.

She knew that people recovered from all kinds of trauma to the body, but a head injury often did irreparable damage. However, the reasoning was barely conscious. She acted without forethought. She picked up from the kitchen table the iron soup pot that Ada had so energetically scrubbed clean. Holding it by its long handle, she raised it high then brought it down with all her might on top of Joachim's head.

He staggered, stunned.

She hit him again, even harder.

He slumped to the floor, unconscious. Maud moved out of the way of his falling body, and sat upright against the wall, holding her chest.

Carla raised the pot again.

Maud screamed: 'No! Stop!'

Carla put the pot down on the kitchen table.

Joachim moved, trying to rise.

Ada seized the pot and hit him again, furiously. Carla tried to grab her arm but she was in a mad rage. She battered the unconscious man's head again and again until she was exhausted, and then she dropped the pot to the floor with a clang.

Maud struggled to her knees and stared at Joachim. His eyes were wide and staring. His nose was twisted sideways. His skull seemed to be out of shape. Blood came from his ear. He did not appear to be breathing.

Carla knelt beside him, put her fingertips to his neck and felt for a pulse. There was none. 'He's dead,' she said. 'We've killed him. Oh, my God.'

Maud said: 'You poor, stupid boy.' She was crying.

Ada, panting with effort, said: 'What do we do now?'

Carla realized they had to get rid of the body.

Maud struggled to her feet with difficulty. The left side of her face was swelling. 'Dear God, it hurts,' she said, holding her side. Carla guessed she had a cracked rib.

Looking down at Joachim, Ada said: 'We could hide him in the attic.'

Carla said: 'Yes, until the neighbours start to complain about the smell.'

'Then we'll bury him in the back garden.'

'And what will people think when they see three women digging a hole six feet long in the yard of a Berlin town house? That we are prospecting for gold?'

'We could dig at night.'

'Would that seem less suspicious?'

Ada scratched her head.

Carla said: 'We have to take the body somewhere and dump it. A park, or a canal.'

'But how will we carry it?' said Ada.

'He doesn't weigh much,' said Maud sadly. 'So slim and strong.'

Carla said: 'It's not the weight that's the problem. Ada and I can carry him. But somehow we have to do it without arousing suspicion.'

Maud said: 'I wish we had a car.'

Carla shook her head. 'No one can get petrol anyway.'

They were silent. Outside, dusk was falling. Ada got a towel and wrapped it around Joachim's head, to prevent his blood staining the floor. Maud cried silently, the tears rolling down a face twisted in anguish. Carla wanted to sympathize but first she had to solve this problem.

'We could put him in a box,' she said.

Ada said: 'The only box that size is a coffin.'

'How about a piece of furniture? A sideboard?'

'Too heavy.' Ada looked thoughtful. 'But the wardrobe in my room is not so weighty.'

Carla nodded. A maid was assumed not to have many clothes, nor to need mahogany furniture, she realized with a touch of embarrassment; so

Ada's room had a narrow hanging cupboard made of flimsy deal wood. 'Let's get it,' she said.

Ada had originally lived in the basement, but that was now an air raid shelter, and her room was upstairs. Carla and Ada went up. Ada opened her cupboard and pulled all the clothes off the rail. There were not many: two sets of uniform, a few dresses, one winter coat, all old. She laid them neatly on the single bed.

Carla tilted the wardrobe and took its weight, then Ada picked up the other end. It was not heavy, but it was awkward, and it took them some time to manhandle it out of the door and down the stairs.

At last they laid it on its back in the hall. Carla opened the door. Now it looked like a coffin with a hinged lid.

Carla went back into the kitchen and bent over the body. She took the camera and films from Joachim's pocket, and replaced them in the kitchen drawer.

Carla took his arms, Ada took his legs, and they lifted the body. They carried it out of the kitchen into the hall and lowered it into the wardrobe. Ada rearranged the towel about the head, though the bleeding had stopped.

Should they take off his uniform, Carla wondered? It would make the body harder to identify – but it would give her two problems of disposal instead of one. She decided against.

She picked up the canvas bag and dropped it into the wardrobe with the corpse.

She closed the wardrobe door and turned the key, to make sure it did not fall open by accident. She put the key in the pocket of her dress.

She went into the dining room and looked out through the window. 'It's getting dark,' she said. 'That's good.'

Maud said: 'What will people think?'

'That we're moving a piece of furniture – selling it, perhaps, to get money for food.'

'Two women, moving a wardrobe?'

'Women do this sort of thing all the time, now that so many men are in the army or dead. It's not as if we could get a removal van – they can't buy petrol.'

'Why would you be doing it in the half-dark?'

Carla let her frustration show. 'I don't know, Mother. If we're asked, I'll have to make something up. But the body can't stay here.'

'They'll know he's been murdered, when they find the body. They'll examine the injuries.'

Carla, too, was worried about that. 'Nothing we can do.'

'They may try to investigate where he went today.'

'He said he had not told anyone about his piano lessons. He wanted to astonish his friends with his skill. With luck, no one knows he came here.'

And without luck, Carla thought, we're all dead. 'What will they guess to be the motive for the murder?'

'Will they find traces of semen in his underwear?'

Maud looked away, embarrassed. 'Yes.'

'Then they will imagine a sexual encounter, perhaps with another man, that ended in a quarrel.'

'I hope you're right.'

Carla was not at all sure, but she could not think of anything they could do about it. 'The canal,' she said. The body would float, and be found sooner or later; and there would be a murder investigation. They would just have to hope it did not lead to them.

Carla opened the front door.

She stood at the front of the wardrobe on its left, and Ada positioned herself at the back on the right. They bent down.

Ada, who undoubtedly had more experience of heavy lifting than her employers, said: 'Tilt it sideways and get your hands under it.'

Carla did as she said.

'Now lift your end a little.'

Carla did so.

Ada got her hands underneath her end and said: 'Bend your knees. Take the weight. Straighten up.'

They raised the wardrobe to hip height. Ada bent down and got her shoulder underneath. Carla did the same.

The two women straightened up.

The weight tilted to Carla as they went down the steps from the front door, but she could bear it. When they reached the street, she turned towards the canal, a few blocks away.

It was now full dark, with no moon but a few stars shedding a faint light. With the blackout, there was a good chance no one would see them tip the wardrobe into the water. The disadvantage was that Carla could hardly see where she was going. She was terrified she would stumble and fall, and the wardrobe would smash to splinters, revealing the murdered man inside.

An ambulance drove by, its headlights covered by slit masks. It was probably hurrying to a road accident. There were many during the blackout. That meant there would be police cars in the vicinity.

Carla recalled a sensational murder case from the beginning of the blackout. A man had killed his wife, forced her body into a packing-case, and carried it across town on the seat of his bicycle in the dark before dropping it in the Havel river. Would the police remember the case and suspect anyone transporting a large object?

As she thought that, a police car drove by. A cop stared out at the two women with their wardrobe, but the car did not stop.

The burden seemed to get heavier. It was a warm night, and soon Carla was running with perspiration. The wood hurt her shoulder, and she wished she had thought of putting a folded handkerchief inside her blouse as a cushion.

They turned a corner and came upon the accident.

An eight-wheeler articulated truck carrying timber had collided head-on with a Mercedes saloon car which had been badly crushed. The police car and the ambulance were shining their headlights on to the wreckage. In a little pool of faint light, a group of men gathered around the car. The crash must have happened in the last few minutes, for there were still people inside the car. An ambulance man was leaning in at the back door, probably examining the injuries to see whether the passengers could be moved.

Carla was momentarily terrified. Guilt froze her and she stopped in her tracks. But no one had noticed her and Ada and the wardrobe, and after a moment she realized she just needed to steal away, double back, and take a different route to the canal.

She began to turn; but just then an alert policeman shone a flashlight her way.

She was tempted to drop the wardrobe and run, but she held her nerve.

The cop said: 'What are you up to?'

'Moving a wardrobe, officer,' she said. Recovering her presence of mind, she faked a grisly curiosity to cover her guilty nervousness. 'What happened here?' she said. For good measure she added: 'Is anyone dead?'

Professionals disliked this kind of vampire inquisitiveness, she knew – she was a professional herself. As she expected, the policeman reacted dismissively. 'None of your business,' he said. 'Just keep out of the way.' He turned back and shone his light into the crashed car.

The pavement on this side of the street was clear. Carla made a snap decision and walked straight on. She and Ada carried the wardrobe containing the dead man towards the wreckage.

She kept her eyes on the little knot of emergency workers in the small circle of light. They were intensely focused on their task and no one looked up as Carla passed the car.

It seemed to take for ever to pass along the length of the eight-wheel trailer. Then, when at last she drew level with the back end, she had a flash of inspiration.

She stopped.

Ada hissed: 'What is it?'

'This way.' Carla stepped into the road at the back of the truck. 'Put the wardrobe down,' she hissed. 'No noise.'

They placed the wardrobe gently on the pavement.

Ada whispered: 'Are we leaving it here?'

Carla drew the key from her pocket and unlocked the wardrobe door. She looked up: as far as she could tell, the men were still gathered around the car, twenty feet away on the other side of the truck.

She opened the wardrobe door.

Joachim Koch stared up sightlessly, his head wrapped in a bloody towel.

'Tip him out,' Carla said. 'By the wheels.'

They tilted the wardrobe, and the body rolled out, coming to rest up against the tyres.

Carla retrieved the bloody towel and threw it into the wardrobe. She left the canvas bag lying beside the corpse: she was glad to get rid of it. She closed and locked the wardrobe door, then they picked it up and walked away.

It was easy to carry now.

When they were fifty yards away in the dark, Carla heard a distant voice say: 'My God, there's another casualty – looks like a pedestrian was run over!'

Carla and Ada turned a corner, and relief washed over Carla like a tidal wave. She had got rid of the corpse. If only she could get home without attracting further attention – and without anyone looking inside the wardrobe and seeing the bloody towel – she would be safe. There would be no murder investigation. Joachim had become a pedestrian killed in a blackout accident. If he had really been dragged along the cobbled street by the wheels of the truck, he might have received injuries similar to those caused by the heavy base of Ada's soup pot. Perhaps a skilled autopsy doctor could tell the difference – but no one would consider an autopsy necessary.

Carla thought about dumping the wardrobe, and decided against it. Even without the towel it had bloodstains inside, and might spark a police investigation on its own. They had to take it home and scrub it clean.

They got home without meeting anyone else.

They put the wardrobe down in the hall. Ada took out the towel, put it in the kitchen sink, and ran the cold tap. Carla felt a mixture of elation and sadness. She had stolen the Nazis' battle plan, but she had killed a young man who was more foolish than wicked. She would think about that for many days, perhaps years, before she could be sure how she felt about it. For now she was just too tired.

She told her mother what they had done. Maud's left cheek was so puffed up that her eye was almost closed. She was pressing her left side as if to ease a pain. She looked terrible.

Carla said: 'You were terribly brave, Mother. I admire you so much for what you did today.'

Maud said wearily: 'I don't feel admirable. I'm so ashamed. I despise myself.'

'Because you didn't love him?' said Carla.

'No,' said Maud. 'Because I did.'

14

Greg Peshkov graduated from Harvard *summa cum laude*, the highest honour. He could have gone on effortlessly to take a doctorate in physics, his major, and thus have avoided military service. But he did not want to be a scientist. His ambition was to wield a different kind of power. And, after the war was over, a military record would be a huge plus for a rising young politician. So he joined the army.

On the other hand, he did not want actually to have to fight.

He followed the European war with heightened interest at the same time as he pressured everyone he knew in Washington – which was a lot of people – to get him a desk job at War Department headquarters.

The German summer offensive had started on 28 June, and they had swiftly pushed east, meeting relatively light opposition, until they reached the city of Stalingrad, formerly called Tsaritsyn, where they were halted by fierce Russian resistance. Now they were stalled, with overstretched supply lines, and it was looking more and more as if the Red Army had drawn them into a trap.

Greg had not long been in basic training when he was summoned to the colonel's office. 'The Army Corps of Engineers needs a bright young officer in Washington,' the colonel said. 'You've interned in Washington, but all the same you wouldn't have been my first choice – you can't even keep your goddamn uniform clean, look at you – but the job requires a knowledge of physics, and the field is kind of limited.'

Greg said: 'Thank you, sir.'

'Try that kind of sarcasm on your new boss and you'll regret it. You're going to be an assistant to a Colonel Groves. I was at West Point with him. He's the biggest son of a bitch I ever met, in the army or out. Good luck.'

Greg called Mike Penfold in the State Department press office and found out that until recently Leslie Groves had been chief of construction for the entire US Army, and had been responsible for the military's new

Washington headquarters, the vast five-sided building they were beginning to call the Pentagon. But he had been moved to a new project that no one knew much about. Some said he had offended his superiors so often that he had been effectively demoted; others that his new role was even more important but top secret. They all agreed he was egotistical, arrogant and ruthless.

'Does *everybody* hate him?' Greg asked.

'Oh, no,' Mike said. 'Only those who have met him.'

Lieutenant Greg Peshkov was full of trepidation when he arrived at Groves's office in the striking New War Department Building, a pale-tan art deco palace on 21st Street and Virginia Avenue. Right away he learned that he was part of a group called the Manhattan Engineer District. This deliberately uninformative name camouflaged a team who were trying to invent a new kind of bomb using uranium as an explosive.

Greg was intrigued. He knew there was incalculable energy locked up in uranium's lighter isotope, U-235, and he had read several papers on the subject in scientific journals. But news of the research had dried up a couple of years ago, and now Greg knew why.

He learned that President Roosevelt felt the project was moving too slowly, and Groves had been appointed to crack the whip.

Greg arrived six days after Groves had been reassigned. His first task for Groves was to help him pin stars to the collar of his khaki shirt: he had just been promoted to brigadier-general. 'It's mainly to impress all these civilian scientists we have to work with,' Groves growled. 'I have a meeting in the Secretary of War's office in ten minutes. You'd better come with me, it'll serve you for a briefing.'

Groves was heavy. An inch under six feet tall, he had to weigh two hundred and fifty pounds, maybe three hundred. He wore his uniform pants high, and his belly bulged under his webbing belt. He had chestnut-coloured hair that might have curled if it had been grown long enough. He had a narrow forehead, fat cheeks, and a jowly chin. His small moustache was all but invisible. He was an unattractive man in every way, and Greg was not looking forward to working for him.

Groves and his entourage, including Greg, left the building and walked down Virginia Avenue to the National Mall. On the way, Groves said to Greg: 'When they gave me this job, they told me it could win the war. I don't know if that's true, but my plan is to act as if it is. You'd better do the same.'

'Yes, sir,' said Greg.

The Secretary of War had not yet moved into the unfinished Pentagon, and War Department headquarters were still in the old Munitions Building, a long, low, out-of-date 'temporary' structure on Constitution Avenue.

Secretary of War Henry Stimson was a Republican, brought in by the President to keep that party from undermining the war effort by making trouble in Congress. At seventy-five, Stimson was an elder statesman, a dapper old man with a white moustache, but the light of intelligence still gleamed in his grey eyes.

The meeting was a full-dress performance, and the room was full of bigwigs including Army Chief of Staff George Marshall. Greg felt nervous, and he thought admiringly that Groves was remarkably calm for someone who had been a mere colonel yesterday.

Groves began by outlining how he intended to impose order on the hundreds of civilian scientists and dozens of physics laboratories involved in the Manhattan project. He made no attempt to defer to the high-ranking men who might well have thought they were in charge. He outlined his plans without troubling to use such mollifying phrases as 'with your permission' and 'if you agree'. Greg wondered whether the man was trying to get himself fired.

Greg learned so much new information that he wanted to take notes, but no one else did, and he guessed it would not look right.

When Groves had done, one of the group said: 'I believe supplies of uranium are crucial to the project. Do we have enough?'

Groves answered: 'There are 1,250 tons of pitchblende – that's the ore that contains uranium oxide – in a yard on Staten Island.'

'Then we'd better acquire some of that,' said the questioner.

'I bought it all on Friday, sir.'

'Friday? The day after you were appointed?'

'Correct.'

The Secretary of War smothered a smile. Greg's surprise at Groves's arrogance began to turn to admiration of his nerve.

A man in admiral's uniform said: 'What about the priority rating of this project? You need to clear the decks with the War Production Board.'

'I saw Donald Nelson on Saturday, sir,' said Groves. Nelson was the civilian head of the board. 'I asked him to raise our rating.'

'What did he say?

'He said no.'

'That's a problem.'

'Not any longer. I told him I would have to recommend to the President that the Manhattan project be abandoned because the War Production Board was unwilling to co-operate. Then he gave us a triple-A.'

'Good,' said the Secretary of War.

Greg was impressed again. Groves was a real pistol.

Stimson said: 'Now, you'll be supervised by a committee that will report to me. Nine members have been suggested—'

'Hell, no,' said Groves.

The Secretary of War said: 'What did you say?'

Surely, Greg thought, Groves has gone too far this time.

Groves said: 'I can't report to a committee of nine, Mr Secretary. I'll never get 'em off my back.'

Stimson grinned. He was too old a hand to get offended by this kind of talk, it seemed. He said mildly: 'What number would you suggest, General?'

Greg could see that Groves wanted to say 'None,' but what came out was: 'Three would be perfect.'

'All right,' said the Secretary of War, to Greg's amazement. 'Anything else?'

'We're going to need a large site, something like sixty thousand acres, for a uranium enrichment plant and associated facilities. There's a suitable area in Oak Ridge, Tennessee. It's a ridge valley, so that if there should be an accident the explosion will be contained.'

'An accident?' said the admiral. 'Is that likely?'

Groves did not hide his feeling that this was a dumb question. 'We're making an experimental bomb, for Christ's sake,' he said. 'A bomb so powerful that it promises to flatten a medium-size city with one detonation. We'd be pretty goddamn dumb if we ignored the possibility of accidents.'

The admiral looked as if he wanted to protest, but Stimson intervened, saying: 'Carry on, General.'

'Land is cheap in Tennessee,' Groves said. 'So is electricity – and our plant will use huge quantities of power.'

'So you're proposing to buy this land.'

'I'm proposing to view it today.' Groves looked at his watch. 'In fact, I need to leave now to catch my train to Knoxville.' He stood up. 'If you will excuse me, gentlemen, I don't want to lose any time.'

The other men in the room were flabbergasted. Even Stimson looked startled. No one in Washington dreamed of leaving a Secretary's office before he indicated he was through. It was a major breach of etiquette. But Groves seemed not to care.

And he got away with it. 'Very well,' said Stimson. 'Don't let us hold you up.'

'Thank you, sir,' said Groves, and he left the room.

Greg hurried out after him.

(ii)

The most attractive civilian secretary in the New War Office Building was Margaret Cowdry. She had big dark eyes and a wide, sensual mouth. When you saw her sitting behind her typewriter, and she glanced up at you and smiled, you felt as if you were already making love to her.

Her father had turned baking into a mass-production industry: 'Cowdry's Cookies crumble just like Ma's!' She had no need to work, but she was doing her bit for the war effort. Before inviting her to lunch, Greg made sure she knew that he, too, was the child of a millionaire. An heiress usually preferred to date a rich boy: she could feel confident he was not after her money.

It was October and cold. Margaret wore a stylish navy-blue coat with padded shoulders and a nipped-in waist. Her matching beret had a military look.

They went to the Ritz-Carlton, but when they got to the dining room Greg saw his father having lunch with Gladys Angelus. He did not want to make it a foursome. When he explained this to Margaret, she said: 'No problem. We'll have lunch at the University Women's Club around the corner. I'm a member there.'

Greg had never been there, but he had a feeling he knew something about it. For a moment he chased the thought around his memory, but it eluded him, so he put it out of his mind.

At the club Margaret removed her coat to reveal a royal-blue cashmere dress that clung to her alluringly. She kept on her hat and gloves, as all respectable women did when eating out.

As always, Greg loved the sensation of walking into a place with a beautiful woman on his arm. In the dining room of the University Women's Club there were only a handful of men, but they all envied him. Although he might not admit it to anyone else, he enjoyed this as much as sleeping with women.

He ordered a bottle of wine. Margaret mixed hers with mineral water, French style, saying: 'I don't want to spend the afternoon correcting my typing mistakes.'

He told her about General Groves. 'He's a real go-getter. In some ways he's a badly dressed version of my father.'

'Everyone hates him,' Margaret said.

Greg nodded. 'He rubs people up the wrong way.'

'Is your father like that?'

'Sometimes, but mostly he uses charm.'

'Mine's the same! Maybe all successful men are that way.'

The meal went quickly. Service in Washington restaurants had speeded up. The nation was at war and men had urgent work to do.

A waitress brought them the dessert menu. Greg glanced at her and was startled to recognize Jacky Jakes. 'Hello, Jacky!' he said.

'Hi, Greg,' she replied, familiarity overlaying nervousness. 'How have you been?'

Greg recalled the detective telling him that she worked at the University Women's Club. That was the memory that had eluded him before. 'I'm just fine,' he said. 'How about you?'

'Real good.'

'Everything going on just the same?' He was wondering if his father was still paying her an allowance.

'Pretty much.'

Greg guessed that some lawyer was paying out the money and Lev had forgotten all about it. 'That's good,' he said.

Jacky remembered her job. 'Can I offer you some dessert today?'

'Yes, thank you.'

Margaret asked for fruit salad and Greg had ice cream.

When Jacky had gone, Margaret said: 'She's very pretty,' then looked expectant.

'I guess,' he said.

'No wedding ring.'

Greg sighed. Women were so perceptive. 'You're wondering how come I'm friendly with a pretty black waitress who isn't married,' he said. 'I might as well tell you the truth. I had an affair with her when I was fifteen. I hope you're not shocked.'

'Of course I am,' she said. 'I'm morally outraged.' She was neither serious nor joking, but something in between. She was not really scandalized, he felt sure, but perhaps she did not want to give him the impression that she was easygoing about sex – not on their first lunch date, anyway.

Jacky brought the desserts and asked if they wanted coffee. They did not have time – the army did not believe in long lunch breaks – and Margaret asked for the bill. 'Guests aren't allowed to pay here,' she explained.

When Jacky had gone, Margaret said: 'What's nice is that you're so fond of her.'

'Am I?' Greg was surprised. 'I have fond memories, I guess. I wouldn't mind being fifteen again.'

'And yet she's scared of you.'

'She is not!'

'Terrified.'

'I don't think so.'

'Take my word. Men are blind, but a woman sees these things.'

Greg looked hard at Jacky when she brought the bill, and he realized that Margaret was right. Jacky was still scared. Every time she saw Greg she was reminded of Joe Brekhunov and his straight razor.

It made Greg angry. The girl had a right to live in peace.

He was going to have to do something about this.

Margaret, who was as sharp as a tack, said: 'I think you know why she's scared.'

'My father frightened her off. He was worried I might marry her.'

'Is your father scary?'

'He does like to get his own way.'

'My father's the same,' she said. 'Sweet as cherry pie, until you cross him. Then he turns mean.'

'I'm so glad you understand.'

They returned to work. Greg felt angry all afternoon. Somehow his father's curse still lay like a blight over Jacky's life. But what could he do?

What would his father do? That was a good way to look at it. Lev would be completely single-minded about getting his way, and would not care who he hurt in the process. General Groves would be similar. I can be like that, Greg thought; I'm my father's son.

The beginning of a plan began to form in his mind.

He spent the afternoon reading and summarizing an interim report from the University of Chicago Metallurgical Laboratory. The scientists there included Leo Szilard, the man who first conceived of the nuclear chain reaction. Szilard was a Hungarian Jew who had studied at the University of Berlin – until the fatal year of 1933. The research team in Chicago was led by Enrico Fermi, the Italian physicist. Fermi, whose wife was Jewish, had left Italy when Mussolini published his *Manifesto of Race*.

Greg wondered whether the Fascists realized that their racism had brought such a windfall of brilliant scientists to their enemies.

He understood the physics perfectly well. The theory of Fermi and Szilard was that when a neutron struck a uranium atom, the collision could produce two neutrons. Those two neutrons could then collide with further uranium atoms to make four, then eight, and so on. Szilard had called this a chain reaction – a brilliant insight.

That way, a ton of uranium could produce as much energy as three million tons of coal – in theory.

In practice, it had never been done.

Fermi and his team were building a pile of uranium at Stagg Field, a disused football stadium belonging to the University of Chicago. To prevent the stuff exploding spontaneously, they buried the uranium in graphite, which absorbed the neutrons and killed the chain reaction. Their aim was to bring the radioactivity up, very gradually, to the level at which more was being created than absorbed – which would prove that a chain reaction was a reality – then close it down, fast, before it blew up the pile, the stadium, the campus of the university, and quite possibly the city of Chicago.

So far they had not succeeded.

Greg wrote a favourable précis of the report, asked Margaret Cowdry to type it right away, then took it in to Groves.

The general read the first paragraph and said: 'Will it work?'

'Well, sir—'

'You're the goddamn scientist. Will it work?'

'Yes, sir, it will work,' Greg said.

'Good,' said Groves, and threw the summary in his waste-paper bin.

Greg returned to his desk and sat for a while, staring at the representation of the Periodic Table of the Elements on the wall opposite his desk. He was pretty sure the nuclear pile would work. He was more worried about how to force his father to withdraw the threat to Jacky.

Earlier, he had thought about handling the problem as Lev would have done. Now he began to think about practical details. He needed to take a dramatic stand.

His plan began to take shape.

But did he have the guts to confront his father?

At five he left for the day.

On the way home he stopped at a barbershop and bought a straight razor, the folding kind where the blade slid into the handle. The barber said: 'You'll find it better than a safety razor, with your beard.'

Greg was not going to shave with it.

His home was his father's permanent suite at the Ritz-Carlton. When Greg arrived, Lev and Gladys were having cocktails.

He remembered meeting Gladys for the first time in this room seven years ago, sitting on the same yellow silk couch. She was an even bigger star now. Lev had put her in a series of shamelessly gung-ho war movies in which she defied sneering Nazis, outwitted sadistic Japanese, and nursed square-jawed American pilots back to health. She was not quite as beautiful as she had been at twenty, Greg observed. The skin of her face did not have the same perfect smoothness; her hair did not seem so luxuriant; and she was wearing a brassiere, which she would undoubtedly have scorned before. But she still had dark-blue eyes that seemed to issue an irresistible invitation.

Greg accepted a martini and sat down. Was he really going to defy his father? He had not done it in the seven years since he had first shaken Gladys's hand. Perhaps it was time.

I'll do it just the way he would, Greg thought.

He sipped his drink and set it down on a side table with spidery legs. Speaking conversationally, he said to Gladys: 'When I was fifteen, my father introduced me to an actress called Jacky Jakes.'

Lev's eyes widened.

'I don't think I know her,' said Gladys.

Greg took the razor from his pocket, but did not open it. He held it in his hand as if feeling its weight. 'I fell in love with her.'

Lev said: 'Why are you dragging this ancient history up now?'

Gladys sensed the tension and looked anxious.

Greg went on: 'Father was afraid I might want to marry her.'

Lev laughed mockingly. 'That cheap tart?'

'Was she a cheap tart?' Greg said. 'I thought she was an actress.' He looked at Gladys.

Gladys flushed at the implied insult.

Greg said: 'Father paid her a visit, and took with him a colleague, Joe Brekhunov. Have you met him, Gladys?'

'I don't believe so.'

'Lucky you. Joe has a razor like this.' Greg snapped the razor open, showing the gleaming sharp blade.

Gladys gasped.

Lev said: 'I don't know what game you think you're playing—'

'Just a minute,' Greg said. 'Gladys wants to hear the rest of the story.' He smiled at her. She looked terrified. He said: 'My father told Jacky that if she ever saw me again, Joe would cut her face with his razor.'

He jerked the knife, just a little, and Gladys gave a small scream.

'The hell with this,' Lev said, and took a step towards Greg. Greg raised the hand holding the razor. Lev stopped.

Greg did not know whether he would be able to cut his father. But Lev did not know either.

'Jacky lives right here in Washington,' Greg said.

His father said crudely: 'Are you fucking her again?'

'No. I'm not fucking anyone, though I have plans for Margaret Cowdry.'

'The cookie heiress?'

'Why, do you want Joe to threaten her too?'

'Don't be stupid.'

'Jacky is a waitress now – she never got the movie part she was hoping for. I run into her on the street sometimes. Today she served me in a restaurant. Every time she sees my face, she thinks Joe is going to come after her.'

'She's out of her mind,' Lev said. 'I'd forgotten all about her until five minutes ago.'

'Can I tell her that?' Greg said. 'I think by now she's entitled to her peace of mind.'

'Tell her whatever the hell you like. For me she doesn't exist.'

'That's great,' said Greg. 'She'll be pleased to hear it.'

'Now put that damn blade away.'

'One more thing. A warning.'

Lev looked angry. 'You're warning *me*?'

'If anything bad happens to Jacky – anything at all . . .' Greg moved the razor side to side, just a little.

Lev said scornfully: 'Don't tell me you're going to cut Joe Brekhunov.'

'No.'

Lev showed a hint of fear. 'You'd cut me?'

Greg shook his head.

Angrily, Lev said: 'What, then, for Christ's sake?'

Greg looked at Gladys.

She took a second to catch his drift. Then she jerked back in her silk-upholstered chair, put both hands on her cheeks as if to protect them, and gave another little scream, louder this time.

Lev said to Greg: 'You little asshole.'

Greg folded the razor and stood up. 'It's how you would have handled it, Father,' he said.

Then he went out.

He slammed the door and leaned against the wall, breathing as hard as if he had been running. He had never felt so scared in his life. Yet he also felt triumphant. He had stood up to the old man, used his own tactics back on him, even scared him a little.

He walked to the elevator, pocketing the razor. His breathing eased. He looked back along the hotel corridor, half expecting his father to come running after him. But the door of the suite remained closed, and Greg boarded the elevator and went down to the lobby.

He entered the hotel bar and ordered a dry martini.

(iii)

On Sunday Greg decided to visit Jacky.

He wanted to tell her the good news. He remembered the address – the only piece of information he had ever paid a private detective for. Unless she had moved, she lived just the other side of Union Station. He had promised her he would not go there, but now he could explain to her that such caution was no longer necessary.

He went by cab. Crossing town, he told himself he would be glad to draw a line at last under his affair with Jacky. He had a soft spot for his first lover, but he did not want to be involved in her life in any way. It would be a relief to get her off his conscience. Then, next time he ran into her, she would not look scared to death. They could say hello, chat for a while, and walk on.

The cab took him to a poor neighbourhood of one-storey homes with low chain-link fences around small yards. He wondered how Jacky lived these days. What did she do during those evenings she was so keen to have to herself? No doubt she saw movies with her girlfriends. Did she go to Washington Redskins football games, or follow the Nats baseball team? When he had asked her about boyfriends, she had been enigmatic. Perhaps she was married and could not afford a ring. By his calculation she was twenty-four. If she was looking for Mr Right she should have found him by now. But she had never mentioned a husband, nor had the detective.

He paid off the taxi outside a small, neat house with flower pots in a concrete front yard – more domesticated than he had expected. As soon as he opened the gate he heard a dog bark. That made sense: a woman living alone might feel safer with a dog. He stepped on to the porch and rang the doorbell. The barking got louder. It sounded like a big dog, but that could be deceptive, Greg knew.

No one came to the door.

When the dog paused for breath, Greg heard the distinctive silence of an empty house.

There was a wooden bench on the stoop. He sat and waited a few minutes. No one came, and no helpful neighbour appeared to tell him whether Jacky was away for a few minutes, all day, or two weeks.

He walked a few blocks, bought the Sunday edition of the *Washington Post*, and returned to the bench to read it. The dog continued to bark intermittently, knowing he was still there. It was the first of November, and he was glad he had worn his olive-green uniform greatcoat and cap: the weather was wintry. Mid-term elections would be held on Tuesday, and the *Post* was predicting that the Democrats would take a beating because of Pearl Harbor. That incident had transformed America, and it came as a surprise to Greg to realize that it had happened less than a year ago. Now American men of his own age were dying on an island no one had ever heard of called Guadalcanal.

He heard the gate click, and looked up.

At first Jacky did not notice him, and he had a moment to study her. She looked dowdily respectable in a dark coat and a plain felt hat, and she carried a book with a black cover. If he had not known her better, Greg would have thought she was coming home from church.

With her was a little boy. He wore a tweed coat and a cap, and he was holding her hand.

The boy saw Greg first, and said: 'Look, Mommy, there's a soldier!'

Jacky looked at Greg, and her hand flew to her mouth.

Greg stood up as they mounted the steps to the stoop. A child! She had kept that secret. It explained why she needed to be home in the evenings. He had never thought of it.

'I told you never to come here,' she said as she put the key in the lock.

'I wanted to tell you that you need not be afraid of my father any more. I didn't know you had a son.'

She and the boy stepped into the house. Greg stood expectantly at the door. A German shepherd growled at him then looked up at Jacky for guidance. Jacky glared at Greg, evidently thinking about slamming the door in his face; but after a moment she gave an exasperated sigh and turned away, leaving it open.

Greg walked in and offered his left fist to the dog. It sniffed warily and gave him provisional approval. He followed Jacky into a small kitchen.

'It's All Saints' Day,' Greg said. He was not religious, but at his boarding school he had been forced to learn all the Christian festivals. 'Is that why you went to church?'

'We go every Sunday,' she replied.

'This is a day of surprises,' Greg murmured.

She took off the boy's coat, sat him at the table, and gave him a cup of orange juice. Greg sat opposite and said: 'What's your name?'

'Georgy.' He said it quietly, but with confidence: he was not shy. Greg studied him. He was as pretty as his mother, with the same bow-shaped mouth, but his skin was lighter than hers, more like coffee with cream, and he had green eyes, unusual in a Negro face. He reminded Greg a little of his half-sister, Daisy. Meanwhile Georgy looked at Greg with an intense gaze that was almost intimidating.

Greg said: 'How old are you, Georgy?'

He looked at his mother for help. She gave Greg a strange look and said: 'He's six.'

'Six!' said Greg. 'You're quite a big boy, aren't you? Why . . .'

A bizarre thought crossed his mind, and he fell silent. Georgy had been born six years ago. Greg and Jacky had been lovers seven years ago. His heart seemed to falter.

He stared at Jacky. 'Surely not,' he said.

She nodded.

'He was born in 1936,' said Greg.

'May,' she said. 'Eight and a half months after I left that apartment in Buffalo.'

'Does my father know?'

'Heck, no. That would have given him even more power over me.'

Her hostility had vanished, and now she just looked vulnerable. In her eyes he saw a plea, though he was not sure what she was pleading for.

He looked at Georgy with new eyes: the light skin, the green eyes, the odd resemblance to Daisy. Are you mine? he thought. Can it be true?

But he knew it was.

His heart filled with a strange emotion. Suddenly Georgy seemed terribly vulnerable, a helpless infant in a cruel world, and Greg needed to take care of him, make sure he came to no harm. He had an impulse to take the boy in his arms, but he realized that might scare him, so he held back.

Georgy put down his orange juice. He got off his chair and came around the table to stand close to Greg. With a remarkably direct look, he said: 'Who are you?'

Trust a kid to ask the toughest question of all, Greg thought. What the hell was he going to say? The truth was too much for a six-year-old to take. I'm just a former friend of your mother's, he thought; I was just passing the door, thought I'd say hello. Nobody special. May see you again, most likely not.

He looked at Jacky, and saw that pleading expression intensified. He realized what was on her mind: she was desperately afraid he was going to reject Georgy.

'I tell you what,' Greg said, and he lifted Georgy on to his knees. 'Why don't you call me Uncle Greg?'

(iv)

Greg stood shivering in the spectators' gallery of an unheated squash court. Here, under the west stand of the disused stadium on the edge of the University of Chicago campus, Fermi and Szilard had built their atomic pile. Greg was impressed and scared.

The pile was a cube of grey bricks reaching the ceiling of the court, standing just shy of the end wall which still bore the polka-dot marks of hundreds of squash balls. The pile had cost a million dollars, and it could blow up the entire city.

Graphite was the material of which pencil leads were made, and it gave off a filthy dust that covered the floor and walls. Everyone who had been in the room a while was as black-faced as a coal miner. No one had a clean lab coat.

Graphite was not the explosive material – on the contrary, it was there to suppress radioactivity. But some of the bricks in the stack were drilled with narrow holes stuffed with uranium oxide, and this was the material that radiated the neutrons. Running through the pile were ten channels for control rods. These were thirteen-foot strips of cadmium, a metal that absorbed neutrons even more hungrily than graphite. Right now the rods were keeping everything calm. When they were withdrawn from the pile, the fun would start.

The uranium was already throwing off its deadly radiation, but the graphite and the cadmium were soaking it up. Radiation was measured by counters that clicked menacingly and a cylindrical pen recorder that was mercifully silent. The array of controls and meters near Greg in the gallery gave off the only heat in the place.

Greg visited on Wednesday 2 December, a bitterly cold, windy day in Chicago. Today for the first time the pile was supposed to go critical. Greg was there to observe the experiment on behalf of his boss, General Groves. He hinted jovially to anyone who asked that Groves feared an explosion and had deputed Greg to take the risk for him. In fact Greg had a more sinister mission. He was making an initial assessment of the scientists with a view to deciding who might be a security risk.

Security on the Manhattan Project was a nightmare. The top scientists were foreigners. Most of the rest were left-wingers, either Communists themselves or liberals who had Communist friends. If everyone suspicious was fired there would be hardly any scientists left. So Greg was trying to figure out which ones were the worst risks.

Enrico Fermi was about forty. A small, balding man with a long nose, he smiled engagingly while supervising this terrifying experiment. He was smartly dressed in a suit with a waistcoat. It was mid-morning when he ordered the trial to begin.

He instructed a technician to withdraw all but one of the control rods from the pile. Greg said: 'What, all at once?' It seemed frighteningly precipitate.

The scientist standing next to him, Barney McHugh, said: 'We took it this far last night. It worked fine.'

'I'm glad to hear it,' said Greg.

McHugh, bearded and podgy, was low down on Greg's list of suspects. He was American, with no interest in politics. The only black mark against him was a foreign wife: she was British – never a good sign, but not in itself evidence of treachery.

Greg had assumed there would be some sophisticated mechanism for moving the rods in and out, but it was simpler than that. The technician just put a ladder up against the pile, climbed halfway up it, and pulled out the rods by hand.

Speaking conversationally, McHugh said: 'We were originally going to do this in the Argonne Forest.'

'Where's that?'

'Twenty miles south-west of Chicago. Pretty isolated. Fewer casualties.'

Greg shivered. 'So why did you change your minds and decide to do it right here on Fifty-seventh Street?'

'The builders we hired went on strike, so we had to build the damn thing ourselves, and we couldn't be that far away from the laboratories.'

'So you took the risk of killing everyone in Chicago.'

'We don't think that will happen.'

Greg had not thought so, either, but he did not feel so sure now, standing a few feet away from the pile.

Fermi was checking his monitors against a forecast he had prepared of radiation levels at every stage of the experiment. Apparently the initial stage went according to plan, for he now ordered the last rod to be pulled halfway out.

There were some safety measures. A weighted rod hung poised to be dropped into the pile automatically if the radiation rose too high. In case that did not work, a similar rod was tied to the gallery railing with a rope, and a young physicist, looking as if he felt a bit silly, stood holding an axe, ready to cut the rope in an emergency. Finally three more scientists called the suicide squad were positioned near the ceiling, standing on the platform of the elevator used during construction, holding large jugs of cadmium sulphate solution, which they would throw on to the pile, as if dousing a bonfire.

Greg knew that neutron generation multiplied in thousandths of a second. However, Fermi argued that some neutrons took longer, perhaps several seconds. If Fermi was right, there would be no problem. But if he was wrong, the squad with the jugs and the physicist with the axe would be vaporized before they could blink.

Greg heard the clicking become more rapid. He looked anxiously at Fermi, who was doing calculations with a slide-rule. Fermi looked pleased. Anyway, Greg thought, *if things go wrong it will probably happen so fast that we'll never know anything about it.*

The rate of clicking levelled off. Fermi smiled and gave the order for the rod to be pulled out another six inches.

More scientists were arriving, climbing the stairs to the gallery in their heavyweight Chicago-winter clothing, coats and hats and scarves and gloves. Greg was appalled at the lack of security. No one was checking credentials: any one of these men could have been a spy for the Japanese.

Among them Greg recognized the great Szilard, tall and heavy, with a round face and thick curly hair. Leo Szilard was an idealist who had imagined nuclear power liberating the human race from toil. It was with a heavy heart that he had joined the team designing the atom bomb.

Another six inches, another increase in the pace of the clicking.

Greg looked at his watch. It was eleven-thirty.

Suddenly there was a loud crash. Everyone jumped. McHugh said: 'Fuck.'

Greg said: 'What happened?'

'Oh, I see,' said McHugh. 'The radiation level activated the safety mechanism and released the emergency control rod, that's all.'

Fermi announced: 'I'm hungry. Let's go to lunch.' In his Italian accent it came out: 'I'm hungary. Les go to luncha.'

How could they think about food? But no one argued. 'You never know how long an experiment is going to take,' said McHugh. 'Could be all day. Best to eat when you can.' Greg could have screamed.

All the control rods were re-inserted into the pile and locked into position, and everyone left.

Most of them went to a campus canteen. Greg got a grilled-cheese sandwich and sat next to a solemn physicist called Wilhelm Frunze. Most scientists were badly dressed but Frunze was notably so, in a green suit with tan suede trimmings: buttonholes, collar lining, elbow patches, pocket flaps. This guy was high on Greg's suspect list. He was German, though he had left in the mid-1930s and gone to London. He was an anti-Nazi but not a Communist: his politics were social-democrat. He was married to an American girl, an artist. Talking to him over lunch, Greg found no reason for suspicion: he seemed to love living in America and to be interested in little but his work. But with foreigners you could never be quite sure where their ultimate loyalty lay.

After lunch he stood in the derelict stadium, looking at thousands of empty stands, and thought about Georgy. He had told no one he had a son – not even Margaret Cowdry, with whom he was now enjoying delightfully carnal relations – but he longed to tell his mother. He felt proud, for no reason – he had made no contribution to bringing Georgy into the world apart from making love to Jacky, probably about the easiest thing he had ever done. Most of all he felt excited. He was at the beginning of some kind of adventure. Georgy was going to grow, and learn, and change, and one day become a man; and Greg would be there, watching and marvelling.

The scientists reassembled at two o'clock. Now there were about forty people crowded into the gallery with the monitoring equipment. The experiment was carefully reset in the position at which they had left off, Fermi checking his instruments constantly.

Then he said: 'This time, withdraw the rod twelve inches.'

The clicks became rapid. Greg waited for the increase to level off, as it had before, but it did not. Instead the clicking became faster and faster until it was continuous roar.

The radiation level was above the maximum of the counters, Greg realized when he noticed that everyone's attention had switched to the pen recorder. Its scale was adjustable. As the level rose the scale was changed, then changed again, and again.

Fermi raised a hand. They all went silent. 'The pile has gone critical,' he said. He smiled – and did nothing.

Greg wanted to scream: *So turn the fucker off!* But Fermi remained silent and still, watching the pen, and such was his authority that no one challenged him. The chain reaction was happening, but it was under control. He let it run for a minute, then another.

McHugh muttered: 'Jesus Christ.'

Greg did not want to die. He wanted to be a senator. He wanted to sleep with Margaret Cowdry again. He wanted to see Georgy go to college. I haven't had half a life yet, he thought.

At last Fermi ordered the control rods to be pushed in.

The noise of the counters reverted to a clicking that gradually slowed and stopped.

Greg breathed normally.

McHugh was jubilant. 'We proved it!' he said. 'The chain reaction is real!'

'And it's controllable, more importantly,' said Greg.'

'Yes, I suppose that is more important, from the practical point of view.'

Greg smiled. Scientists were like this, he knew from Harvard: for them theory was reality, and the world a rather inaccurate model.

Someone produced a bottle of Italian wine in a straw basket and some paper cups. The scientists all drank a tiny share. This was another reason Greg was not a scientist: they had no idea how to party.

Someone asked Fermi to sign the basket. He did so, then all the others signed it.

The technicians shut down the monitors. Everyone began to drift away. Greg stayed, observing. After a while he found himself alone in the gallery with Fermi and Szilard. He watched as the two intellectual giants shook hands. Szilard was a big, round-faced man; Fermi was elfin; and for a moment Greg was inappropriately reminded of Laurel and Hardy.

Then he heard Szilard speak. 'My friend,' he said, 'I think this will go down as a black day in the history of mankind.'

Greg thought: Now what the hell did he mean by that?

(v)

Greg wanted his parents to accept Georgy.

It would not be easy. No doubt it would be unnerving for them to be told they had a grandson who had been concealed from them for six years. They might be angry. On top of that, they might look down on Jacky. They had no right to take a moralistic attitude, he thought wryly: they themselves had an illegitimate child – himself. But people were not rational.

He was not sure how much difference it would make that Georgy was black. Greg's parents were laid back about race, and never talked viciously about niggers or kikes as some people of their generation did; but they might change when they learned there was a Negro in the family.

His father would be the more difficult one, he guessed; so he spoke to his mother first.

He got a few days leave at Christmas and went home to her place in Buffalo. Marga had a large apartment in the best building in town. She lived mostly alone, but she had a cook, two maids and a chauffeur. She had a safe full of jewellery and a dress closet the size of a two-car garage. But she did not have a husband.

Lev was in town, but traditionally he took Olga out on Christmas Eve. He was still married to her, technically, though he had not spent a night at her house for years. As far as Greg knew, Olga and Lev hated one another; but for some reason they met once a year.

That evening, Greg and his mother had dinner together in the apartment. He put on a tuxedo to please her. 'I love to see my men dressed up,' she often said. They had fish soup, roast chicken, and Greg's boyhood favourite, peach pie.

'I have some news for you, Mother,' he said nervously as the maid poured coffee. He feared she would be angry. He was not frightened for himself, but for Georgy, and he wondered if this was what parenthood was about – worrying about someone else more than you worried about yourself.

'Good news?' she said.

She had become heavier in recent years, but she was still glamorous at forty-six. If there was any grey in her dark hair it had been carefully camouflaged by her hairdresser. Tonight she wore a simple black dress and a diamond choker.

'Very good news, but I guess a little surprising, so please don't fly off the handle.'

She raised a black eyebrow but said nothing.

He reached inside his dinner jacket and took out a photograph. It showed Georgy on a red bicycle with a ribbon around the handlebars. The rear wheel of the bike had a pair of stabilizing wheels so that it would not fall over. The expression on the boy's face was ecstatic. Greg was kneeling beside him, looking proud.

He handed the picture to his mother.

She studied it thoughtfully. After a minute she said: 'I'm assuming you gave this little boy a bicycle for Christmas.'

'That's right.'

She looked up. 'Are you telling me you have a child?'

Greg nodded. 'His name is Georgy.'

'Are you married?'

'No.'

She threw down the photo. 'For God's sake!' she said angrily. 'What is the matter with you Peshkov men!'

Greg was dismayed. 'I don't know what you mean!'

'Another illegitimate child! Another woman bringing him up alone!'

He realized that she saw Jacky as a younger version of herself. 'Mother, I was fifteen . . .'

'Why can't you be normal?' she stormed. 'For the love of Jesus Christ, what's wrong with having a regular family?'

Greg looked down. 'There's nothing wrong with it.'

He felt ashamed. Until this moment he had seen himself as a passive player in this drama, even a victim. Everything that had happened had been done to him by his father and Jacky. But his mother did not view it that way, and now he saw that she was right. He had not thought twice about sleeping with Jacky; he had not questioned her when she had said airily that there was no need to worry about contraception; and he had not confronted his father when Jacky left. He had been very young, yes; but if he was old enough to fuck her, he was old enough to take responsibility for the consequences.

His mother was still raging. 'Don't you remember how you used to carry on? "Where is my Daddy? Why doesn't he sleep here? Why can't we go with him to Daisy's house?" And then later, the fights you had at school when the boys called you a bastard. And you were so angry to be refused membership of that goddamned yacht club.'

'Of course I remember.'

She banged a beringed fist on the table, causing crystal glasses to shake. 'Then how can you put another little boy through the same torture?'

'I didn't know he existed until two months ago. Father scared the mother away.'

'Who is she?'

'Her name is Jacky Jakes. She's a waitress.' He took out another photo.

His mother sighed. 'A pretty Negress.' She was calming down.

'She was hoping to be an actress, but I guess she gave that up when Georgy came along.'

Marga nodded. 'A baby will ruin your career faster than a dose of the clap.'

Mother assumed that an actress had to sleep with the right people to progress, Greg noted. How the hell would she know? But then she had been a nightclub singer when his father met her . . .

He did not want to go down that road.

She said: 'What did you give her for Christmas?'

'Medical insurance.'

'Good choice. Better than a fluffy bear.'

Greg heard a step in the hall. His father was home. Hastily, he said: 'Mother, will you meet Jacky? Will you accept Georgy as your grandson?'

Her hand went to her mouth. 'Oh, my God, I'm a grandmother.' She did not know whether to be shocked or pleased.

Greg leaned forward. 'I don't want Father to reject him. Please!'

Before she could reply, Lev came into the room.

Marga said: 'Hello, darling, how was your evening?'

He sat at the table looking grumpy. 'Well, I've had my shortcomings explained to me in full detail, so I guess I had a great time.'

'You poor thing. Did you get enough to eat? I can make you an omelette in a minute.'

'The food was fine.'

The photographs were on the table, but Lev had not noticed them yet.

The maid came in and said: 'Would you like coffee, Mr Peshkov?'

'No, thank you.'

Marga said: 'Bring the vodka, in case Mr Peshkov would like a drink later.'

'Yes, ma'am.'

Greg noticed how solicitous Marga was about Lev's comfort and pleasure. He guessed that was why Lev was here, not at Olga's, for the night.

The maid brought a bottle and three small glasses on a silver tray. Lev still drank vodka the Russian way, warm and neat.

Greg said: 'Father, you know Jacky Jakes—'

'Her again?' Lev said irritably.

'Yes, because there's something you don't know about her.'

That got his attention. He hated to think other people knew things he did not. 'What?'

'She has a child.' He pushed the photographs across the polished table.

'It it yours?'

'He's six years old. What do you think?'

'She kept this pretty damned quiet.'

'She was scared of you.'

'What did she think I might do, cook the baby and eat it?'

'I don't know, Father – you're the expert at scaring people.'

Lev gave him a hard look. 'You're learning, though.'

He was talking about the scene with the razor. Maybe I am learning to scare people, Greg thought.

Lev said: 'Why are you showing me these photos?'

'I thought you might like to know that you have a grandson.'

'By a goddamn two-bit actress who was hoping to snag herself a rich man!'

Marga said: 'Darling! Please remember that I was a two-bit nightclub singer hoping to snag myself a rich man.'

He looked furious. For a moment he glared at Marga. Then his expression changed. 'You know what?' he said. 'You're right. Who am I to judge Jacky Jakes?'

Greg and Marga stared at him, astonished at this sudden humility.

He said: 'I'm just like her. I was a two-kopek hoodlum from the slums of St Petersburg until I married Olga Vyalov, my boss's daughter.'

Greg caught his mother's eye, and she gave an almost imperceptible shrug that simply said: *You never can tell.*

Lev looked again at the photo. 'Apart from the colour, this kid looks like my brother, Grigori. There's a surprise. Until now I thought all these piccaninnies looked the same.'

Greg could hardly breathe. 'Will you see him, father? Will you come with me and meet your grandson?'

'Hell, yes.' Lev uncorked the bottle, poured vodka into three glasses, and passed them round. 'What's the boy's name, anyway?'

'Georgy.'

Lev raised his glass. 'So here's to Georgy.'

They all drank.

15

Lloyd Williams walked along a narrow uphill path at the tail end of a line of desperate fugitives.

He breathed easily. He was used to this. He had now crossed the Pyrenees several times. He wore rope-soled espadrilles that gave his feet a better grip on the rocky ground. He had a heavy coat on top of his blue overalls. The sun was hot now but later, when the party reached higher altitudes and the sun went down, the temperature would drop below freezing.

Ahead of him were two sturdy ponies, three local people, and eight weary, bedraggled escapers, all loaded with packs. There were three American airmen, the surviving crew of a B-24 Liberator bomber that had crash-landed in Belgium. Two more were British officers who had escaped from the Oflag 65 prisoner-of-war camp in Strasbourg. The others were a Czech Communist, a Jewish woman with a violin, and a mysterious Englishman called Watermill who was probably some kind of spy.

They had all come a long way and suffered many hardships. This was the last leg of their journey, and the most dangerous. If captured now, they would all be tortured until they betrayed the brave men and women who had helped them en route.

Leading the party was Teresa. The climb was hard work for people who were not used to it, but they had to keep up a brisk pace to minimize their exposure, and Lloyd had found that the refugees were less likely to fall behind when they were led by a small, ravishingly pretty woman.

The path levelled and broadened into a small clearing. Suddenly a loud voice rang out. Speaking French with a German accent, it shouted: 'Halt!'

The column came to an abrupt halt.

Two German soldiers emerged from behind a rock. They carried standard Mauser bolt-action rifles, each holding five rounds of ammunition.

Reflexively Lloyd touched the overcoat pocket that contained his loaded 9mm Luger pistol.

Escaping from mainland Europe had become harder, and Lloyd's job had grown even more dangerous. At the end of last year the Germans had occupied the southern half of France, contemptuously ignoring the Vichy French government like the flimsy sham it had always been. A forbidden zone ten miles deep was declared all along the frontier with Spain. Lloyd and his party were in that zone now.

Teresa addressed the soldiers in French. 'Good morning, gentlemen. Is everything all right?' Lloyd knew her well, and he could hear the tremor of fear in her voice, but he hoped it was too faint for the sentries to notice.

Among the French police there were many Fascists and a few Communists, but all of them were lazy, and none wanted to chase refugees across the icy passes of the Pyrenees. However, the Germans did. German troops had moved into border towns and begun to patrol the hill paths and mule trails Lloyd and Teresa used. The occupiers were not crack troops: those were fighting in Russia, where they had recently surrendered Stalingrad after a long and murderous struggle. Many of the Germans in France were old men, boys, and the walking wounded. But that seemed to make them more determined to prove themselves. Unlike the French, they rarely turned a blind eye.

Now the older of the two soldiers, cadaverously thin with a grey moustache, said to Teresa: 'Where are you going?'

'To the village of Lamont. We have groceries for you and your comrades.'

This particular German unit had moved into a remote hill village, kicking out the local inhabitants. Then they had realized how difficult it was to supply troops in that location. It had been a stroke of genius on Teresa's part to undertake to carry food to them – at a healthy profit – and thereby get permission to enter the prohibited zone.

The thin soldier looked suspiciously at the men with their backpacks. 'All this is for German soldiers?'

'I hope so,' Teresa said. 'There's no one else up here to sell it to.' She took a piece of paper from her pocket. 'Here's the order, signed by your Sergeant Eisenstein.'

The man read it carefully and handed it back. Then he looked at Lieutenant-Colonel Will Donelly, a beefy American pilot. 'Is he French?'

Lloyd put his hand on the gun in his pocket.

The appearance of the fugitives was a problem. In this part of the world the local people, French and Spanish, were usually small and dark. And everyone was thin. Both Lloyd and Teresa fitted that description, as did the Czech and the violinist. But the British were pale and fair-haired, and the Americans were huge.

Teresa said: 'Guillaume was born in Normandy. All that butter.'

The younger of the two soldiers, a pale boy with glasses, smiled at Teresa. She was easy to smile at. 'Do you have wine?' he said.

'Of course.'

The two sentries brightened visibly.

Teresa said: 'Would you like some right now?'

The older man said: 'It's thirsty in the sun.'

Lloyd opened a pannier on one of the ponies, took out four bottles of Roussillon white wine, and handed them over. The Germans took two each. Suddenly everyone was smiling and shaking hands. The older sentry said: 'Carry on, friends.'

The fugitives went on. Lloyd had not really expected trouble, but you could never be sure, and he was relieved to have got past the sentry post.

It took them two more hours to reach Lamont. A dirt-poor hamlet with a handful of crude houses and some empty sheep pens, it stood on the edge of a small upland plain where the new spring grass was just beginning to show. Lloyd pitied the people who had lived here. They had had so little, and even that had been taken from them.

The party walked into the centre of the village and gratefully unshouldered their burdens. They were surrounded by German soldiers.

This was the most dangerous moment, Lloyd thought.

Sergeant Eisenstein was in charge of a platoon of fifteen or twenty men. Everyone helped to unload the supplies: bread, sausage, fresh fish, condensed milk, canned food. The soldiers were pleased to get supplies and glad to see new faces. They merrily attempted to engage their benefactors in conversation.

The fugitives had to say as little as possible. This was the moment when they could so easily betray themselves by a slip. Some Germans spoke French well enough to detect an English or American accent. Even those who had passable accents, such as Teresa and Lloyd, could give themselves away with a grammatical error. It was so easy to say *sur le table* instead of *sur la table*, but it was a mistake no French person would ever make.

To compensate, the two genuine Frenchmen in the party went out of their way to be voluble. Any time a soldier began to talk to a fugitive, someone would jump into the conversation.

Teresa presented the sergeant with a bill, and he took a long time to check the numbers then count out the money.

At last they were able to take their leave, with empty backpacks and lighter hearts.

They walked back down the mountain half a mile, then they split up. Teresa went on down with the Frenchmen and the horses. Lloyd and the fugitives turned on to an upward path.

The German sentries at the clearing would probably be too drunk by now to notice that fewer people were coming down than went up. But if they asked questions, Teresa would say some of the party had started a card game with the soldiers, and would be following later. Then there would be a change of shift and the Germans would lose track.

Lloyd made his group walk for two hours, then he allowed them a ten-minute break. They had all been given bottles of water and packets of dried figs for energy. They were discouraged from bringing anything else: Lloyd knew from experience that treasured books, silverware, ornaments and gramophone records would become too heavy and be thrown into a snow-filled ravine long before the footsore travellers crested the pass.

This was the hard part. From now on it would only get darker and colder and rockier.

Just before the snowline, he instructed them to refill their water bottles at a clear cold stream.

When night fell they kept going. It was dangerous to let people sleep: they might freeze to death. They were tired, and they slipped and stumbled on the icy rocks. Inevitably their pace slowed. Lloyd could not let the line spread: stragglers might lose their way, and there were precipitous ravines for the careless to fall into. But he had never lost anyone, yet.

Many of the fugitives were officers, and this was the point where they would sometimes challenge Lloyd, arguing when he ordered them to keep going. Lloyd had been promoted to major to give him more authority.

In the middle of the night, when their morale was at rock bottom, Lloyd announced: 'You are now in neutral Spain!' and they raised a ragged cheer. In truth he did not know exactly where the border was, and always made the announcement when they seemed most in need of a boost.

Their spirits lifted again when dawn broke. They still had some way to go, but the route now led downhill, and their cold limbs gradually thawed.

At sunrise they skirted a small town with a dust-coloured church at the top of a hill. Just beyond, they reached a large barn beside the road. Inside was a green Ford flatbed truck with a grimy canvas cover. The lorry was large enough to carry the whole party. At the wheel was Captain Silva, a middle-aged Englishman of Spanish descent who worked with Lloyd.

Also there, to Lloyd's surprise, was Major Lowther, who had been in charge of the intelligence course at Tŷ Gwyn, and had been snootily disapproving – or perhaps just envious – of Lloyd's friendship with Daisy.

Lloyd knew that Lowthie had been posted to the British Embassy in Madrid, and guessed he worked for MI6, the Secret Intelligence Service, but he would not have expected to see him this far from the capital.

Lowther wore an expensive white flannel suit that was crumpled and grubby. He stood beside the truck looking proprietorial. 'I'll take over from here, Williams,' he said. He looked at the fugitives. 'Which one of you is Watermill?'

Watermill could have been a real name or a code.

The mysterious Englishman stepped forward and shook hands.

'I'm Major Lowther. I'm taking you straight to Madrid.' Turning back to Lloyd he said: 'I'm afraid your party will have to make your way to the nearest railway station.'

'Just a minute,' said Lloyd. 'That truck belongs to my organization.' He had purchased it with his budget from MI9, the department that helped escaping prisoners. 'And the driver works for me.'

'Can't be helped,' Lowther said briskly. 'Watermill has priority.'

The Secret Intelligence Service always thought they had priority. 'I don't agree,' Lloyd said. 'I see no reason why we can't all go to Barcelona in the truck, as planned. Then you can take Watermill on to Madrid by train.'

'I didn't ask for your opinion, laddie. Just do as you're told.'

Watermill himself interjected, in a reasonable tone: 'I'm perfectly happy to share the truck.'

'Leave this to me, please,' Lowther told him.

Lloyd said: 'All these people have just walked across the Pyrenees. They're exhausted.'

'Then they'd better have a rest before going on.'

Lloyd shook his head. 'Too dangerous. The town on the hill has a sympathetic mayor – that's why we rendezvous here. But farther down the valley their politics are different. The Gestapo are everywhere, you know that – and most of the Spanish police are on their side, not ours. My group will be in serious danger of arrest for entering the country illegally. And you know how difficult it is to get people out of Franco's jails, even when they're innocent.'

'I'm not going to waste my time arguing with you. I outrank you,'

'No, you don't.'

'What?'

'I'm a major. So don't call me "laddie" ever again, unless you want a punch on the nose.'

'My mission is urgent!'

'So why didn't you bring your own vehicle?'

'Because this one was available!'

'But it wasn't.'

Will Donelly, the big American, stepped forward. 'I'm with Major Williams,' he drawled. 'He's just saved my life. You, Major Lowther, haven't done shit.'

'That's got nothing to do with it,' said Lowther.

'Well, the situation here seems pretty clear,' Donelly said. 'The truck is under the authority of Major Williams. Major Lowther wants it, but he can't have it. End of story.'

Lowther said: 'You keep out of this.'

'I happen to be a Lieutenant-Colonel, so I guess I outrank you both.'

'But this isn't under your jurisdiction.'

'Nor yours, evidently.' Donelly turned to Lloyd. 'Should we get going?'

'I insist!' spluttered Lowther.

Donelly turned back to him. 'Major Lowther,' he said. 'Shut the fuck up. And that's an order.'

Lloyd said: 'All right, everybody – climb aboard.'

Lowther glared furiously at Lloyd. 'I'll get you for this, you little Welsh bastard,' he said.

(ii)

The daffodils were out in London on the day Daisy and Boy went for their medical.

The visit to the doctor was Daisy's idea. She was fed up with Boy blaming her for not getting pregnant. He constantly compared her to his brother Andy's wife, May, who now had three children. 'There must be something wrong with you,' he had said aggressively.

'I got pregnant once before.' She winced at the remembered pain of her miscarriage; then she recalled how Lloyd had taken care of her, and she felt a different kind of pain.

Boy said: 'Something could have happened since then to make you infertile.'

'Or you.'

'What do you mean?'

'There might just as easily be something wrong with you.'

'Don't be absurd.'

'Tell you what, I'll make a deal.' The thought flashed through her mind that she was negotiating rather as her father, Lev, might have done. 'I'll go for an examination – if you will.'

That had surprised him, and he had hesitated, then said: 'All right. You go first. If they say there's nothing wrong with you, I'll go.'

'No,' she said. 'You go first.'

'Why?'

'Because I don't trust you to keep your promises.'

'All right, then, we'll go together.'

Daisy was not sure why she was bothering. She did not love Boy – had not loved him for a long time. She was in love with Lloyd Williams, still in Spain on a mission he could not say much about. But she was married to Boy. He had been unfaithful to her, of course, with numerous women. But she had committed adultery too, albeit with only one man. She had no moral ground to stand on, and in consequence she was paralysed. She just felt that if she did her duty as a wife she might retain the last shreds of her self-respect.

The doctor's office was in Harley Street, not far from their house though in a less expensive neighbourhood. Daisy found the examination unpleasant. The doctor was a man, and he was grumpy about her being ten minutes late. He asked her a lot of questions about her general health, her menstrual periods, and what he called her 'relations' with her husband, not looking at her but making notes with a fountain pen. Then he put a series of cold metal instruments up her vagina. 'I do this every day, so you don't need to worry,' he said, then he gave her a grin that told her the opposite.

When she came out of the doctor's office she half expected Boy to renege on their deal and refuse to take his turn. He looked sour about it, but he went in.

While she was waiting, Daisy reread a letter from her half-brother, Greg. He had discovered he had a child, from an affair he had with a black girl when he was fifteen. To Daisy's astonishment the playboy Greg was excited about his son and keen to be part of the child's life, albeit as an uncle rather than a father. Even more surprising, Lev had met the child and announced that he was smart.

It was ironic, she thought, that Greg had a son even though he had never wanted one, and Boy had no son even though he longed for one so badly.

Boy came out of the doctor's office an hour later. The doctor promised to give them their results in a week. They left at twelve noon.

'I need a drink after that,' Boy said.

'So do I,' said Daisy.

They looked up and down the street of identical row houses. 'This neighbourhood is a bloody desert. Not a pub in sight.'

'I'm not going to a pub,' said Daisy. 'I want a martini, and they don't know how to make them in pubs.' She spoke from experience. She had asked for a dry martini at the King's Head in Chelsea and had been served a glass of disgustingly warm vermouth. 'Take me to Claridge's hotel, please. It's only five minutes' walk.'

'Now that's a damn good idea.'

The bar at Claridge's was full of people they knew. There were austerity rules about the meals restaurants could sell, but Claridge's had found a loophole: there were no restrictions on giving food away, so they offered a free buffet, charging only their usual high prices for drinks.

Daisy and Boy sat in art deco splendour and sipped perfect cocktails, and Daisy began to feel better.

'The doctor asked me if I'd had mumps,' Boy said.

'But you have.' It was mainly a childhood illness, but Boy had caught it a couple of years back. He had been briefly billeted at a vicarage in

East Anglia, and had picked up the infection from the vicar's three small sons. It had been very painful. 'Did he say why?'

'No. You know what these chaps are like. Never tell you a bloody thing.'

It occurred to Daisy that she was not as happy-go-lucky as she had once been. In the old days she would never have brooded about her marriage this way. She had always liked what Scarlett O'Hara said in *Gone with the Wind*: 'I'll think about that tomorrow.' Not any more. Perhaps she was growing up.

Boy was ordering a second cocktail when Daisy looked towards the door and saw the Marquis of Lowther walking in, dressed in a creased and stained uniform.

Daisy disliked him. Ever since he had guessed at her relationship with Lloyd he had treated her with oily familiarity, as if they shared a secret that made them intimates.

Now he sat at their table uninvited, dropping cigar ash on his khaki trousers, and asked for a manhattan.

Daisy knew at once that he was up to no good. There was a look of malignant relish in his eye that could not be explained merely as anticipation of a good cocktail.

Boy said: 'I haven't seen you for a year or so, Lowthie. Where have you been?'

'Madrid,' Lowthie said. 'Can't say much about it. Hush-hush, you know. How about you?'

'I spend a lot of time training pilots, though I've flown a few missions lately, now that we've stepped up the bombing of Germany.'

'Jolly good thing, too. Give the Germans a taste of their own medicine.'

'You may say that, but there's a lot of muttering among the pilots.'

'Really – why?'

'Because all this stuff about military targets is absolute rubbish. There's no point in bombing German factories because they just rebuild them. So we're targeting large areas of dense working-class housing. They can't replace the workers so fast.'

Lowther looked shocked. 'That would mean it's our policy to kill civilians.'

'Exactly.'

'But the government assures us—'

'The government lies,' Boy said. 'And the bomber crews know it.

Many of them don't give a damn, of course, but some feel bad. They believe that if we're doing the right thing, then we should say so; and if we're doing the wrong thing we should stop.'

Lowther looked uneasy. 'I'm not sure we should be talking like this here.'

'You're probably right,' Boy said.

The second round of cocktails came. Lowther turned to Daisy. 'And what about the little woman?' he said. 'You must have some war work. The devil finds mischief for idle hands, according to the proverb.'

Daisy replied in a neutral matter-of-fact tone. 'Now that the Blitz is over, they don't need women ambulance drivers, so I'm working with the American Red Cross. We have an office in Pall Mall. We do what we can to help American servicemen over here.'

'Men lonely for a bit of feminine company, eh?'

'Mostly they're just homesick. They like to hear an American accent.'

Lowthie leered. 'I expect you're very good at consoling them.'

'I do what I can.'

'I bet you do.'

Boy said: 'Look here, Lowthie, are you a bit drunk? Because this sort of talk is awfully bad form, you know.'

Lowther's expression turned spiteful. 'Oh, come on, Boy, don't tell me you don't know. What are you, blind?'

Daisy said: 'Take me home, please, Boy.'

He ignored her and spoke to Lowther. 'What the devil do you mean?'

'Ask her about Lloyd Williams.'

Boy said: 'Who the hell is Lloyd Williams?'

Daisy said: 'I'm going home alone, if you won't take me.'

'Do you know a Lloyd Williams, Daisy?'

He's your brother, Daisy thought; and she felt a powerful impulse to reveal the secret, and knock him sideways; but she resisted the temptation. 'You know him,' she said. 'He was up at Cambridge with you. He took us to a music hall in the East End, years ago.'

'Oh!' said Boy, remembering. Then, puzzled, he said to Lowther: 'Him?' It was difficult for Boy to see someone such as Lloyd as a rival. With growing incredulity he added: 'A man who can't even afford his own dress clothes?'

Lowther said: 'Three years ago he was on my intelligence course down at Tŷ Gwyn while Daisy was living there. You were risking your life in a Hawker Hurricane over France at the time, I seem to remember. She was dallying with that Welsh weasel – in your family's house!'

Boy was getting red in the face. 'If you're making this up, Lowthie, by God I'll thrash you.'

'Ask your wife!' said Lowther with a confident grin.

Boy turned to Daisy.

She had not slept with Lloyd at Tŷ Gwyn. She had slept with him in his own bed at his mother's house during the Blitz. But she could not explain that to Boy in front of Lowther, and anyway it was a detail. The accusation of adultery was true, and she was not going to deny it. The secret was out. All she wanted now was to retain some semblance of dignity.

She said: 'I will tell you everything you want to know, Boy – but not in front of this leering slob.'

Boy raised his voice in astonishment. 'So you don't deny it?'

The people at the next table looked around, seemed embarrassed, and returned their attention to their drinks.

Daisy raised her own voice. 'I refuse to be cross-examined in the bar of Claridge's Hotel.'

'You admit it, then?' he shouted.

The room went quiet.

Daisy stood up. 'I don't admit or deny anything here. I'll tell you everything in private at home, which is where civilized couples discuss such matters.'

'My God, you did it, you slept with him!' Boy roared.

Even the waiters had paused in their work and were standing still, watching the row.

Daisy walked to the door.

Boy yelled: 'You slut!'

Daisy was not going to exit on that line. She turned around. 'You know about sluts, of course. I had the misfortune to meet two of yours, remember?' She looked around the room. 'Joanie and Pearl,' she said contemptuously. 'How many wives would put up with that?' She went out before he could reply.

She stepped into a waiting taxi. As it pulled away, she saw Boy emerge from the hotel and get into the next cab in line.

She gave the driver her address.

In a way she felt relieved that the truth was out. But she also felt terribly sad. Something had ended, she knew.

The house was only a quarter of a mile away. As she arrived, Boy's taxi pulled up behind hers.

He followed her into the hall.

She could not stay here with him, she realized. That was over. She would never again share his home or his bed. 'Bring me a suitcase, please,' she said to the butler.

'Very good, my lady.'

She looked around. It was an eighteenth-century town house of perfect proportions, with an elegantly curving staircase, but she was not really sorry to leave it.

Boy said: 'Where are you going?'

'To a hotel, I suppose. Probably not Claridge's.'

'To meet your lover!'

'No, he's overseas. But, yes, I do love him. I'm sorry, Boy. You have no right to judge me – your offences are worse – but I judge myself.'

'That's it,' he said. 'I'm going to divorce you.'

Those were the words she had been waiting for, she realized. Now they had been said, and everything was over. Her new life began from this moment.

She sighed. 'Thank God,' she said.

(iii)

Daisy rented an apartment in Piccadilly. It had a large American-style bathroom with a shower. There were two separate toilets, one for guests – a ridiculous extravagance in the eyes of most English people.

Fortunately, money was not an issue for Daisy. Her grandfather Vyalov had left her rich, and she had had control of her own fortune since she was twenty-one. And it was all in American dollars.

New furniture was difficult to buy, so she shopped for antiques, of which there were plenty for sale cheap. She hung modern paintings for a gay, youthful look. She hired an elderly laundress and a girl to clean, and found it was easy to manage the place without a butler or a cook, especially when you did not have a husband to mollycoddle.

The servants at the Mayfair house packed all her clothes and sent them to her in a pantechnicon. Daisy and the laundress spent an afternoon opening the boxes and putting everything away tidily.

She had been both humiliated and liberated. On balance, she thought she was better off. The wound of rejection would heal, but she would be free of Boy for ever.

After a week she wondered what had been the results of the medical examination. The doctor would have reported to Boy, of course, as the husband. She did not want to ask him, and, anyway, it did not seem important any longer, so she forgot about it.

She enjoyed making a new home. For a couple of weeks she was too busy to socialize. When she had fixed up the apartment she decided to see all the friends she had been ignoring.

She had a lot of friends in London. She had been here seven years. For the last four years Boy had been away more than he was home, and she had gone to parties and balls on her own, so being without a husband would not make much difference to her life, she figured. No doubt she would be crossed off the Fitzherbert family's invitation lists, but they were not the only people in London society.

She bought crates of whisky, gin and champagne, scouring London for what little was available legitimately and buying the rest on the black market. Then she sent out invitations to a flatwarming party.

The responses came back with ominous promptness, and they were all declines.

In tears, she phoned Eva Murray. 'Why won't anyone come to my party?' she wailed.

Eva was at her door ten minutes later.

She arrived with three children and a nanny. Jamie was six, Anna four, and baby Karen two.

Daisy showed her around the apartment, then ordered tea while Jamie turned the couch into a tank, using his sisters as crew.

Speaking English with a mixture of German, American and Scots accents, Eva said: 'Daisy, dear, this isn't Rome.'

'I know. Are you sure you're comfortable?'

Eva was heavily pregnant with her fourth child. 'Would you mind if I put my feet up?'

'Of course not.' Daisy fetched a cushion.

'London society is respectable,' Eva went on. 'Don't imagine I approve of it. I have been excluded often, and poor Jimmy is snubbed sometimes for having married a half-Jewish German.'

'That's awful.'

'I wouldn't wish it on anyone, whatever the reason.'

'Sometimes I hate the British.'

'You're forgetting what Americans are like. Don't you remember telling me that all the girls in Buffalo were snobs?'

Daisy laughed. 'What a long time ago it seems.'

'You've left your husband,' Eva said. 'And you did so in undeniably spectacular fashion, hurling insults at him in the bar of Claridge's hotel.'

'And I'd only had one Martini!'

Eva grinned. 'How I wish I'd been there!'

'I kind of wish I hadn't.'

'Needless to say, everyone in London society has talked about little else for the last three weeks.'

'I guess I should have anticipated that.'

'Now, I'm afraid, anyone who appears at your party will be seen as approving of adultery and divorce. Even I wouldn't like my mother-in-law to know I'd come here and had tea with you.'

'But it's so unfair – Boy was unfaithful first!'

'And you thought women were treated equally?'

Daisy remembered that Eva had a great deal more to worry about than snobbery. Her family was still in Nazi Germany. Fitz had made inquiries through the Swiss embassy and learned that her doctor father was now in a concentration camp, and her brother, a violin maker, had been beaten up by the police, his hands smashed. 'When I think about your troubles, I'm ashamed of myself for complaining,' Daisy said.

'Don't be. But cancel the party.'

Daisy did.

But it made her miserable. Her work for the Red Cross filled her days, but in the evenings she had nowhere to go and nothing to do. She went to the movies twice a week. She tried to read *Moby Dick* but found it tedious. One Sunday she went to church. St James's, the Wren church opposite her apartment building in Piccadilly, had been bombed, so she went to St Martin-in-the-Fields. Boy was not there, but Fitz and Bea were, and Daisy spent the service looking at the back of Fitz's head, reflecting that she had fallen in love with two of this man's sons. Boy had his mother's looks and his father's single-minded selfishness. Lloyd had Fitz's good looks and Ethel's big heart. Why did it take me so long to see that, she wondered?

The church was full of people she knew, and after the service none of them spoke to her. She was lonely and almost friendless in a foreign country in the middle of a war.

One evening she took a taxi to Aldgate and knocked at the Leckwith house. When Ethel opened the door, Daisy said: 'I've come to ask for your son's hand in marriage.' Ethel let out a peal of laughter and hugged her.

She had brought a gift, an American tin of ham she had got from a USAF navigator. Such things were luxuries to British families on rations. She sat in the kitchen with Ethel and Bernie, listening to dance tunes on the radio. They all sang along with 'Underneath the Arches' by Flanagan and Allen. 'Bud Flanagan was born right here in the East End,' Bernie said proudly. 'Real name Chaim Reuben Weintrop.'

The Leckwiths were excited about the Beveridge Report, a government paper that had become a bestseller. 'Commissioned under a Conservative Prime Minister and written by a Liberal economist,' said Bernie. 'Yet it proposes what the Labour Party has always wanted! You know you're winning, in politics, when your opponents steal your ideas.'

Ethel said: 'The idea is that everyone of working age should pay a weekly insurance premium, then get benefits when they are sick, unemployed, retired or widowed.'

'A simple proposal, but it will transform our country,' Bernie said enthusiastically. 'Cradle to grave, no one will ever be destitute again.'

Daisy said: 'Has the government accepted it?'

'No,' said Ethel. 'Clem Attlee pressed Churchill very hard, but Churchill won't endorse the report. The Treasury thinks it will cost too much.'

Bernie said: 'We'll have to win an election before we can implement it.'

Ethel and Bernie's daughter, Millie, dropped in. 'I can't stay long,' she said. 'Abie's watching the children for half an hour.' She had lost her job – women were not buying expensive gowns, now, even if they could afford them – but, fortunately, her husband's leather business was flourishing, and they had two babies, Lennie and Pammie.

They drank cocoa and talked about the young man they all adored. They had little real news of Lloyd. Every six or eight months Ethel received a letter on the headed paper of the British embassy in Madrid, saying he was safe and well and doing his bit to defeat Fascism. He had been promoted to major. He had never written to Daisy, for fear Boy might see the letters, but now he could. Daisy gave Ethel the address of her new flat, and took down Lloyd's address, which was a British Forces Post Office number.

They had no idea when he might come home on leave.

Daisy told them about her half-brother, Greg, and his son, Georgy. She knew that the Leckwiths of all people would not be censorious, and would be able to rejoice in such news.

She also told the story of Eva's family in Berlin. Bernie was Jewish, and tears came to his eyes when he heard about Rudi's broken hands. 'They should have fought the bastard Fascists on the street, when they had the chance,' he said. 'That's what we did.'

Millie said: 'I've still got the scars on my back, where the police pushed us through Gardiner's plate-glass window. I used to be ashamed of them – Abie never saw my back until we'd been married six months – but he says they make him proud of me.'

'It wasn't pretty, the fighting in Cable Street,' said Bernie. 'But we put a stop to their bloody nonsense.' He took off his glasses and wiped his eyes with his handkerchief.

Ethel put her arm around his shoulders. 'I told people to stay home that day,' she said. 'I was wrong, and you were right.'

He smiled ruefully. 'Doesn't happen often.'

'But it was the Public Order Act, brought in after Cable Street, that finished the British Fascists,' Ethel said. 'Parliament banned the wearing of political uniforms in public. That finished them. If they couldn't strut up and down in their black shirts they were nothing. The Conservatives did that – credit where credit's due.'

Always a political family, the Leckwiths were planning the post-war reform of Britain by the Labour Party. Their leader, the quietly brilliant Clement Attlee, was now deputy prime minister under Churchill, and union hero Ernie Bevin was Minister of Labour. Their vision made Daisy feel excited about the future.

Millie left and Bernie went to bed. When they were alone Ethel said to Daisy: 'Do you really want to marry my Lloyd?'

'More than anything in the world. Do you think it will be all right?'

'I do. Why not?'

'Because we come from such different backgrounds. You're all such good people. You live for public service.'

'Except for our Millie. She's like Bernie's brother – she wants to make money.'

'Even she has scars on her back from Cable Street.'

'True.'

'Lloyd is like you. Political work isn't something extra he does, like a hobby – it's the centre of his life. And I'm a selfish millionaire.'

'I think there are two kinds of marriage,' Ethel said thoughtfully. 'One is a comfortable partnership, where two people share the same hopes and fears, raise children as a team, and give each other comfort and help.' She was talking about herself and Bernie, Daisy realized. 'The other is a wild passion, madness and joy and sex, possibly with someone completely unsuitable, maybe someone you don't admire or don't even really like.' She was thinking about her affair with Fitz, Daisy felt sure. She held her breath: she knew Ethel was now telling her the raw truth. 'I've been lucky, I've had both,' Ethel said. 'And here's my advice to you. If you get the chance of the mad kind of love, grab it with both hands, and to hell with the consequences.'

'Wow,' said Daisy.

She left a few minutes later. She felt privileged that Ethel had given her a glimpse into her soul. But when she got back to her empty apartment she felt depressed. She made a cocktail and poured it away. She put the kettle on and took it off again. The radio went off the air. She lay between cold sheets and wished Lloyd was there.

She compared Lloyd's family with her own. Both had troubled histories, but Ethel had forged a strong, supportive family out of unfavourable materials, which Daisy's own mother had been unable to do – though that was more Lev's fault than Olga's. Ethel was a remarkable woman, and Lloyd had many of her qualities.

Where was he now, and what was he doing? Whatever the answer, he was sure to be in danger. Would he be killed now, when at last she was free to love him without restraint and, eventually, to marry him? What would she do if he died? Her own life would be at an end, she felt: no husband, no lover, no friends, no country. In the early hours of the morning she cried herself to sleep.

Next day she slept late. At midday she was drinking coffee in her little dining room, dressed in a black silk wrap, when her fifteen-year-old maid came in and said: 'Major Williams is here, my lady.'

'What?' she screeched. 'He can't be!'

Then he came through the door with his kitbag over his shoulder.

He looked tired and had several days' growth of beard, and he had evidently slept in his uniform.

She threw her arms around him and kissed his bristly face. He kissed her back, inhibited somewhat by being unable to stop grinning. 'I must stink,' he said between kisses. 'I haven't changed my clothes for a week.'

'You smell like a cheese factory,' she said. 'I love it.' She pulled him into her bedroom and started to take his clothes off.

'I'll take a quick shower,' he said.

'No,' she said. She pushed him back on the bed. 'I'm in too much of a hurry.' Her longing for him was frantic. And the truth was that she relished the strong smell. It should have repelled her, but it had the opposite effect. It was him, the man she had thought might be dead, and he was filling her nostrils and her lungs. She could have wept with joy.

Taking off his trousers would require removing his boots, and she could see that would be complicated, so she did not bother. She just unbuttoned his fly. She threw off her black silk robe and hiked her nightdress up to her waist, all the time staring with happy lust at the white penis sticking up out of the rough khaki cloth. Then she straddled him, easing herself down, and leaned forward and kissed him. 'Oh, God,' she said. 'I can't tell you how much I've been longing for you.'

She lay on him, not moving much, kissing him again and again. He held her face in his hands and stared at her. 'This is real, isn't it?' he said. 'Not just another happy dream?'

'It's real,' she said.

'Good. I wouldn't like to wake up now.'

'I want to stay like this for ever.'

'Nice idea, but I can't keep still much longer.' He began to move under her.

'If you do that I'll come,' she said.

And she did.

Afterwards they lay on her bed for a long time, talking.

He had two weeks' leave. 'Live here,' she said. 'You can visit your parents every day, but I want you at night.'

'I wouldn't like you to get a bad reputation.'

'That ship has sailed. I've already been shunned by London society.'

'I know.' He had telephoned Ethel from Waterloo Station, and she had told him about Daisy's separation from Boy and given him the address of the flat.

'We must do something about contraception,' he said. 'I'll get some rubber johnnies. But you might want to get fixed up with a device. What do you think?'

'You want to make sure I don't get pregnant?' she said.

There was a note of sadness in her voice, she realized; and he heard it. 'Don't get me wrong,' he said. He raised himself on his elbow. 'I'm illegitimate. I was told lies about my parentage, and when I found out the truth it was a terrible shock.' His voice shook a little with emotion. 'I'll never put my children through that. Never.'

'We wouldn't have to lie to them.'

'Would we tell them that we're not married? That in fact you're married to someone else?'

'I don't see why not.'

'Think how they would be teased at school.'

She was not convinced, but clearly the issue was a profound one for him. 'So, what's your plan?' she said.

'I want us to have children. But not until we're married. To each other.'

'I get that,' she said. 'So . . .'

'We have to wait.'

Men were slow to pick up hints. 'I'm not much of a girl for tradition,' she said. 'But, still, there are some things . . .'

At last he saw what she was getting at. 'Oh! Okay. Just a minute.' He knelt upright on the bed. 'Daisy, dear—'

She burst out laughing. He looked comical, in full uniform with his limp dick hanging out of his fly. 'Can I take a photo of you like that?' she said.

He looked down and saw what she meant. 'Oh, sorry.'

'No – don't you dare put it away! Stay just as you are, and say what you were going to say.'

He grinned. 'Daisy, dear, will you be my wife?'

'In a heartbeat,' she said.

They lay down again, embracing.

Soon the novelty of his odour wore off. They got into the shower together. She soaped him all over, taking merry pleasure in his embarrassment when she washed his most intimate places. She put shampoo on his hair and scrubbed his grimy feet with a brush.

When he was clean he insisted on washing her, but he had only got as far as her breasts when they had to make love again. They did it standing in the shower with the hot water coursing down their bodies. Clearly he had momentarily forgotten his aversion to illegitimate pregnancy, and she did not care.

Afterwards he stood at her mirror shaving. She wrapped a large towel around herself and sat on the lid of the toilet, watching him. He asked: 'How long will it take you to get divorced?'

'I don't know. I'd better speak to Boy.'

'Not today, though. I want you to myself all day.'

'When will you go to see your parents?'

'Tomorrow, maybe.'

'Then I'll go to see Boy at the same time. I want to get this over as soon as possible.'

'Good,' he said. 'That's settled, then.'

(iv)

Daisy felt strange going into the house where she had lived with Boy. A month ago it had been hers. She had been free to come and go as she wished, and enter any room without asking permission. The servants had obeyed her every order without question. Now she was a stranger in the same house. She kept her hat and gloves on, and she had to follow the old butler as he led her to the morning room.

Boy did not shake hands or kiss her cheek. He looked full of righteous indignation.

'I haven't hired a lawyer yet,' Daisy said as she sat down. 'I wanted to talk to you personally first. I'm hoping we can do this without hating one another. After all, there are no children to fight over, and we both have plenty of money.'

'You betrayed me!' he said.

Daisy sighed. Clearly it was not going to go the way she had hoped. 'We both committed adultery,' she said. 'You first.'

'I've been humiliated. Everyone in London knows!'

'I did try to stop you making a fool of yourself in Claridge's – but you were too busy humiliating me! I hope you've thrashed the loathsome marquis.'

'How could I? He did me a favour.'

'He might have done you a bigger favour by having a quiet word at the club.'

'I don't understand how you could fall for such a low-class oik as Williams. I've found out a few things about him. His mother was a housemaid!'

'She's probably the most impressive woman I've ever met.'

'I hope you realize that no one really knows who his father is.'

That was about as ironic as you could get, Daisy thought. 'I know who his father is,' she said.

'Who?'

'I'm certainly not telling you.'

'There you are, then.'

'This isn't getting us anywhere, is it?'

'No.'

'Perhaps I should just have a lawyer write to you.' She stood up. 'I loved you once, Boy,' she said sadly. 'You were fun. I'm sorry I wasn't enough for you. I wish you happiness. I hope you marry someone who suits you better, and that she gives you lots of sons. I would be happy for you if that came about.'

'Well, it won't,' he said.

She had turned towards the door, but now she looked back. 'Why do you say that?'

'I got the report from that doctor we went to.'

She had forgotten about the medical. It had seemed irrelevant after they split. 'What did he say?'

'There's nothing wrong with you – you can have a whole litter of pups. But I can't father children. Mumps in adult men sometimes causes infertility, and I copped it.' He laughed bitterly. 'All those bloody Germans shooting at me for years, and I've been downed by a vicar's three little brats.'

She felt sad for him. 'Oh, Boy, I'm really sorry to hear that.'

'Well, you're going to be sorrier, because I'm not divorcing you.'

She suddenly felt cold. 'What do you mean? Why not?'

'Why should I bother? I don't want to marry again. I can't have children. Andy's son will inherit.'

'But I want to marry Lloyd!'

'Why should I care about that? Why should he have children if I can't?'

Daisy was devastated. Would happiness be snatched away from her just when it seemed to be within her reach? 'Boy, you can't mean this!'

'I've never been more serious in my life.'

Her voice was anguished. 'But Lloyd wants children of his own!'

'He should have thought of that before he f-f-fucked another man's wife.'

'Very well, then,' she said defiantly. 'I'll divorce you.'

'On what grounds?'

'Adultery, of course.'

'But you have no evidence.' She was about to say that that shouldn't be a problem when he grinned maliciously and added: 'And I'll take care you don't get any.'

He could do that, if he was discreet about his liaisons, she realized with growing horror. 'But you threw me out!' she said.

'I shall tell the judge you're welcome to come home any time.'

She tried to stop herself crying. 'I never thought you'd hate me this much,' she said miserably.

'Didn't you?' said Boy. 'Well, now you bloody well know.'

(v)

Lloyd Williams went to Boy Fitzherbert's house in Mayfair at mid-morning, when Boy would be sober, and told the butler he was Major Williams, a distant relative. He thought a man-to-man conversation was worth a try. Surely Boy did not really want to dedicate the rest of his life to revenge? Lloyd was in uniform, hoping to appeal to Boy as one fighting man to another. Good sense must surely prevail.

He was shown into the morning room where Boy sat reading the paper and smoking a cigar. It took Boy a moment to recognize him. 'You!' he said when comprehension dawned. 'You can piss off right away.'

'I've come to ask you to give Daisy a divorce,' Lloyd said.

'Get out.' Boy got to his feet.

Lloyd said: 'I can see that you're toying with the idea of taking a swing at me, so in fairness I should tell you that it won't be as easy as you imagine. I'm a bit smaller than you, but I box at welterweight, and I've won quite a lot of contests.'

'I'm not going to soil my hands on you.'

'Good decision. But will you reconsider the divorce?'

'Absolutely not.'

'There's something you don't know,' Lloyd said. 'I wonder if it might change your mind.'

'I doubt it,' Boy said. 'But go on, now that you're here, give it a shot.' He sat down, but did not offer Lloyd a chair.

Be it on your own head, Lloyd thought.

He took from his pocket a faded sepia photograph. 'If you'd be so kind, glance at this picture of me.' He put it on the side table next to Boy's ashtray.

Boy picked it up. 'This isn't you. It looks like you, but the uniform is Victorian. It must be your father.'

'My grandfather, in fact. Turn it over.'

Boy read the inscription on the back. 'Earl Fitzherbert?' he said scornfully.

'Yes. The previous earl, your grandfather – and mine. Daisy found that photo at Tŷ Gwyn.' Lloyd took a deep breath. 'You told Daisy that no one knows who my father is. Well, I can tell you. It's Earl Fitzherbert. You and I are brothers.' He waited for Boy's response.

Boy laughed. 'Ridiculous!'

'My reaction, exactly, when I was first told.'

'Well, I must say, you have surprised me. I would have thought you could come up with something better than this absurd fantasy.'

Lloyd had been hoping the revelation would shock Boy into a different frame of mind, but so far it was not working. Nevertheless he continued to reason. 'Come on, Boy – how unlikely is it? Doesn't it happen all the time in great houses? Maids are pretty, young noblemen are randy, and nature takes its course. When a baby is born, the matter is hushed up. Please don't pretend you had no idea such things could occur.'

'No doubt it's common enough.' Boy's confidence was shaken, but still he blustered. 'However, lots of people pretend they have connections with the aristocracy.'

'Oh, please,' Lloyd said disparagingly. 'I don't want connections with the aristocracy. I'm not a draper's assistant with daydreams of grandeur. I come from a distinguished family of socialist politicians. My maternal grandfather was one of the founders of the South Wales Miners' Federation. The last thing I need is a wrong-side-of-the-blanket link with a Tory peer. It's highly embarrassing to me.'

Boy laughed again, but with less conviction. '*You're* embarrassed! Talk about inverted snobbery.'

'Inverted? I'm more likely to become prime minister than you are.' Lloyd realized they had got into a pissing contest, which was not what he wanted. 'Never mind that,' he said. 'I'm trying to persuade you that you can't spend the rest of your life taking revenge on me – if only because we're brothers.'

'I still don't believe it,' Boy said, putting the photo down on the side table and picking up his cigar.

'Nor did I, at first.' Lloyd kept trying: his whole future was at stake. 'Then it was pointed out to me that my mother was working at Tŷ Gwyn when she fell pregnant; that she had always been evasive about my father's identity; and that shortly before I was born she somehow acquired the funds to buy a three-bedroom house in London. I confronted her with my suspicions and she admitted the truth.'

'This is laughable.'

'But you know it's true, don't you?'

'I know no such thing.'

'You do, though. For the sake of our brotherhood, won't you do the decent thing?'

'Certainly not.'

Lloyd saw that he was not going to win. He felt downcast. Boy had the power to blight Lloyd's life, and he was determined to use it.

He picked up the photograph and put it back in his pocket. 'You'll ask our father about this. You won't be able to restrain yourself. You'll have to find out.'

Boy made a scornful noise.

Lloyd went to the door. 'I believe he will tell you the truth. Goodbye, Boy.'

He went out and closed the door behind him.

16

Colonel Albert Beck got a Russian bullet in his right lung at Kharkov in March 1943. He was lucky: a field surgeon put in a chest drain and reinflated the lung, saving his life, just. Weakened by blood loss and the almost inevitable infection, Beck was put on a train home and ended up in Carla's hospital in Berlin.

He was a tough, wiry man in his early forties, prematurely bald, with a protruding jaw like the prow of a Viking longboat. The first time he spoke to Carla, he was drugged and feverish and wildly indiscreet. 'We're losing the war,' he said.

She was immediately alert. A discontented officer was a potential source of information. She said lightly: 'The newspapers say we're shortening the line on the Eastern Front.'

He laughed scornfully. 'That means we're retreating.'

She continued to draw him out. 'And Italy looks bad.' The Italian dictator Benito Mussolini – Hitler's greatest ally – had fallen.

'Remember 1939, and 1940?' Beck said nostalgically. 'One brilliant lightning victory after another. Those were the days.'

Clearly he was not ideological, perhaps not even political. He was a normal patriotic soldier who had stopped kidding himself.

Carla led him on. 'It can't be true that the army is short of everything from bullets to underpants.' This kind of mildly risky talk was not unusual in Berlin nowadays.

'Of course we are.' Beck was radically disinhibited but quite articulate. 'Germany simply can't produce as many guns and tanks as the Soviet Union, Great Britain and the United States combined – especially when we're being bombed constantly. And no matter how many Russians we kill, the Red Army seems to have an inexhaustible supply of new recruits.'

'What do you think will happen?'

'The Nazis will never admit defeat, of course. So more people will die. Millions more, just because they're too proud to yield. Insanity. Insanity.' He drifted off to sleep.

You had to be sick – or crazy – to voice such thoughts, but Carla believed that more and more people were thinking that way. Despite relentless government propaganda it was becoming clear that Hitler was losing the war.

There had been no police investigation of the death of Joachim Koch. It had been reported in the newspaper as a road accident. Carla had got over the initial shock, but every now and again the realization hit her that she had killed a man, and she would relive his death in her imagination. It made her shake and she had to sit down. This had happened only once when she was on duty, fortunately, and she had passed that off as a faint due to hunger – highly plausible in wartime Berlin. Her mother was worse. Strange, that Maud had loved Joachim, weak and foolish as he was; but there was no explaining love. Carla herself had completely misjudged Werner Franck, thinking he was strong and brave, only to learn that he was selfish and weak.

She talked to Beck a lot before he was discharged, probing to find out what kind of man he was. Once recovered, he never again spoke indiscreetly about the war. She learned that he was a career soldier, his wife was dead, and his married daughter lived in Buenos Aires. His father had been a Berlin city councillor: he did not say for which party, so clearly it was not the Nazis or any of their allies. He never said anything bad about Hitler, but he never said anything good either, nor did he speak disparagingly of Jews or Communists. These days that in itself was close to insubordination.

His lung would heal, but he would never again be strong enough for active service, and he told her he was being posted to the General Staff. He could become a diamond mine of vital secrets. She would be risking her life if she tried to recruit him – but she had to try.

She knew he would not remember their first conversation. 'You were very candid,' Carla told him in a low voice. There was no one nearby. 'You said we were losing the war.'

His eyes flashed fear. He was no longer a woozy patient in a hospital gown with stubble on his cheeks. He was washed and shaved, sitting upright in dark-blue pyjamas buttoned to the throat. 'I suppose you're going to report me to the Gestapo,' he said. 'I don't think a man should be held to account for what he says when he's sick and raving.'

'You weren't raving,' she said. 'You were very clear. But I'm not going to report you to anyone.'

'No?'

'Because you are right.'

He was surprised. 'Now I should report *you*.'

'If you do, I'll say that you insulted Hitler in your delirium, and when I threatened to report it you made up a story about me in self-defence.'

'If I denounce you, you'll denounce me,' he said. 'Stalemate.'

'But you're not going to denounce me,' she said. 'I know that, because I know you. I've nursed you. You're a good man. You joined the army for love of your country, but you hate the war and you hate the Nazis.' She was 99 per cent sure of this.

'It's very dangerous to talk like that.'

'I know.'

'So this isn't just a casual conversation.'

'Correct. You said that millions of people are going to die just because the Nazis are too proud to surrender.'

'Did I?'

'You can help save some of those millions.'

'How?'

Carla paused. This was where she put her life on the line. 'Any information you have, I can pass it to the appropriate quarters.' She held her breath. If she was wrong about Beck, she was dead.

She read amazement in his look. He could hardly imagine that this briskly efficient young nurse was a spy. But he believed her, she could see that. He said: 'I think I understand you.'

She handed him a green hospital file folder, empty.

He took it. 'What's this for?' he said.

'You're a soldier, you understand camouflage.'

He nodded. 'You're risking your life,' he said, and she saw something like admiration in his eyes.

'So are you, now.'

'Yes,' said Colonel Beck. 'But I'm used to it.'

(ii)

Early in the morning, Thomas Macke took young Werner Franck to the Plötzensee Prison in the western suburb of Charlottenburg. 'You should see this,' he said. 'Then you can tell General Dorn how effective we are.'

He parked in the Königsdamm and led Werner to the rear of the main prison. They entered a room twenty-five feet long and about half as wide. Waiting there was a man dressed in a tailcoat, a top hat and white gloves. Werner frowned at the peculiar costume. 'This is Herr Reichhart,' said Macke. 'The executioner.'

Werner swallowed. 'So we're going to witness an execution?'

'Yes.'

With a casual air that might have been faked, Werner said: 'Why the fancy dress outfit?'

Macke shrugged. 'Tradition.'

A black curtain divided the room in two. Macke drew it back to show eight hooks attached to an iron girder that ran across the ceiling.

Werner said: 'For hanging?'

Macke nodded.

There was also a wooden table with straps for holding someone down. At one end of the table was a high device of distinctive shape. On the floor was a heavy basket.

The young lieutenant was pale. 'A guillotine,' he said.

'Exactly,' said Macke. He looked at his watch. 'We shan't be kept waiting long.'

More men filed into the room. Several nodded in a familiar way to Macke. Speaking quietly into Werner's ear, Macke said: 'Regulations demand that the judges, the court officers, the prison governor and the chaplain all attend.'

Werner swallowed. He was not liking this, Macke could see.

He was not meant to. Macke's motive in bringing him here had nothing to do with impressing General Dorn. Macke was worried about Werner. There was something about him that did not ring true.

Werner worked for Dorn; that was not in question. He had accompanied Dorn on a visit to Gestapo headquarters, and subsequently Dorn had written a note saying that the Berlin counter-espionage effort was most impressive, and mentioning Macke by name. For weeks afterwards Macke had walked around in a miasma of warm pride.

But Macke could not forget Werner's behaviour on that evening, nearly a year ago now, when they had almost caught a spy in a disused fur coat factory near the East Station. Werner had panicked – or had he? Accidentally or otherwise, he had given the pianist enough warning to get away. Macke could not shake the suspicion that the panic had been an act, and Werner had, in fact, been coolly and deliberately sounding the alarm.

Macke did not quite have the nerve to arrest and torture Werner. It could be done, of course, but Dorn might well kick up a fuss, and then Macke would be questioned. His boss, Superintendent Kringelein, who did not much like him, would ask what hard evidence he had against Werner – and he had none.

But this ought to reveal the truth.

The door opened again, and two prison guards entered on either side of a young woman called Lili Markgraf.

He heard Werner gasp. 'What's the matter?' Macke asked.

Werner said: 'You didn't tell me it was going to be a girl.'

'Do you know her?'

'No.'

Lili was twenty-two, Macke knew, though she looked younger. Her fair hair had been cut this morning, and it was now as short as a man's. She was limping, and walked bent over as if she had an abdominal injury. She wore a plain blue dress of heavy cotton with no collar, just a round neckline. Her eyes were red with crying. The guards held her arms firmly, not taking any chances.

'This woman was denounced by a relative who found a code book hidden in her room,' Macke said. 'The five-digit Russian code.'

'Why is she walking like that?'

'The effects of interrogation. But we didn't get anything from her.'

Werner's face was impassive. 'What a shame,' he said. 'She might have led us to other spies.'

Macke saw no sign that he was faking. 'She knew her associate only as Heinrich – no last name – and he may have used a pseudonym anyway. I find we rarely profit by arresting women – they don't know enough.'

'But at least you have her code book.'

'For what it's worth. They change the key word regularly, so we still face a challenge in decrypting their signals.'

'Pity.'

One of the men cleared his throat and spoke loudly enough for everyone to hear. He said he was the President of the Court, then read out the death sentence.

The guards walked Lili to the wooden table. They gave her the chance of lying on it voluntarily, but she took a step backwards, so they picked her up forcibly. She did not struggle. They laid her face down and strapped her in.

The chaplain began a prayer.

Lili began to plead. 'No, no,' she said, without raising her voice. 'No, please, let me go. Let me go.' She spoke coherently, as if she were merely asking someone for a favour.

The man in the top hat looked at the president, who shook his head and said: 'Not yet. The prayer must be finished.'

Lili's voice rose in pitch and urgency. 'I don't want to die! I'm afraid to die! Don't do this to me, please!'

The executioner looked again at the court president. This time the president just ignored him.

Macke studied Werner. He looked sick, but so did everybody else in the room. As a test, this was not really working. Werner's reaction showed that he was sensitive, not that he was a traitor. Macke might have to think of something else.

Lili began to scream.

Even Macke felt impatient.

The pastor hurried through the rest of the prayer.

When he said 'Amen' she stopped screaming, as if she knew it was all over.

The president gave the nod.

The executioner moved a lever, and the weighted blade fell.

It made a whispering sound as it sliced through Lili's pale neck. Her short-cropped head fell forward and there was a gush of blood. The head hit the basket with a loud thump that seemed to resound in the room.

Absurdly, Macke wondered if the head felt any pain.

(iii)

Carla bumped into Colonel Beck in the hospital corridor. He was in uniform. She looked at him in sudden fear. Ever since he had been

discharged, she had lived every day in fear that he had betrayed her, and the Gestapo were on their way.

But he smiled and said: 'I came back for a check-up with Dr Ernst.'

Was that all? Had he forgotten their conversation? Was he pretending to have forgotten it? Was there a black Gestapo Mercedes waiting outside?

Beck was carrying a green hospital file folder.

A cancer specialist in a white coat approached. As he went by, Carla said brightly to Beck: 'How are things?'

'I'm as fit as I'm ever going to be. I'll never lead a battalion into battle again, but aside from athletics I can lead a normal life.'

'I'm glad to hear that.'

People kept walking by. Carla feared Beck would never get the chance to say anything to her privately.

But he remained unruffled. 'I'd just like to thank you for your kindness and professionalism.'

'You're welcome.'

'Goodbye, Sister.'

'Goodbye, Colonel.'

When Beck left, Carla was holding the file folder.

She walked briskly to the nurses' cloakroom. It was empty. She stood with her heel firmly wedged against the door so no one could come in.

Inside the folder was a large envelope made of the cheap buff-coloured paper used in offices everywhere. Carla opened the envelope. It contained several typewritten sheets. She looked at the first without removing it from the envelope. It was headed:

OPERATIONAL ORDER NO. 6
CODE ZITADELLE

It was the battle plan for the summer offensive on the Eastern Front. Her heart raced. This was gold dust.

She had to pass the envelope to Frieda. Unfortunately, Frieda was not working at the hospital today: it was her day off. Carla considered leaving the hospital right away, in the middle of her shift, and going to Frieda's house; but she swiftly rejected that idea. Better to behave normally, not to attract attention.

She slipped the envelope into the shoulder bag hanging on her coat hook. She covered it with the blue-and-gold silk scarf that she always

carried for hiding things. She stood still for a few moments, letting her breathing return to normal. Then she went back to the ward.

She worked the rest of her shift as best she could, then she put on her coat, left the hospital, and walked to the station. Passing a bomb site, she saw graffiti on the remains of the building. A defiant patriot had written: 'Our walls might break, but not our hearts.' But someone else had ironically quoted Hitler's 1933 election slogan: 'Give me four years, and you will not recognize Germany.'

She bought a ticket to the Zoo.

On the train she felt like an alien. All the other passengers were loyal Germans, and she was the one with secrets in her bag to betray to Moscow. She did not like the feeling. No one looked at her, but that only made her think they were all deliberately avoiding her eye. She could hardly wait to hand over the envelope to Frieda.

The Zoo Station was on the edge of the Tiergarten. The trees were dwarfed, now, by a huge flak tower. One of three in Berlin, this square concrete block was more than 100 feet high. At the corners of the roof were four giant 128mm anti-aircraft guns weighing 25 tons each. The raw concrete was painted green in a hopelessly optimistic attempt to make the monstrosity less of an eyesore in the park.

Ugly though it was, Berliners loved it. When the bombs were falling, its thunder reassured them that someone was shooting back.

Still in a state of high tension, Carla walked from the station to Frieda's house. It was mid-afternoon, so the Franck parents would probably be out, Ludi at his factory and Monika seeing a friend, possibly Carla's mother. Werner's motorcycle was parked on the drive.

The manservant opened the door. 'Miss Frieda is out, but she won't be long,' he said. 'She went to KaDeWe to buy gloves. Mr Werner is in bed with a heavy cold.'

'I'll wait for Frieda in her room, as usual.'

Carla took off her coat and went upstairs, still carrying her bag. In Frieda's room she kicked off her shoes and lay on the bed to read the battle plan for Operation Zitadelle. She was as stressed as an overwound clock, but she would feel better when she had given the purloined document to someone else.

From the next room she heard the sound of sobbing.

She was surprised. That was Werner's room. Carla found it hard to imagine the suave playboy in tears.

But the sound definitely came from a man, and he seemed to be trying and failing to suppress his grief.

Against her will, Carla felt pity. She told herself that some feisty woman had thrown Werner over, probably for very good reasons. But she could not help responding to the real distress she was hearing.

She got off the bed, put the battle plan back in her bag, and stepped outside.

She listened at Werner's door. She could hear it even more clearly. She was too soft-hearted to ignore it. She opened the door and went in.

Werner was sitting on the edge of the bed, head in hands. When he heard the door he looked up, startled. His face was red with emotion and wet with tears. His tie was pulled down and his collar undone. He looked at Carla with misery in his eyes. He was bowled over, devastated, and too wretched to care who knew it.

Carla could not pretend to be heartless. 'What is it?' she said.

'I can't do this any more,' he said.

She closed the door behind her. 'What happened?'

'They cut off Lili Markgraf's head – and I had to watch.'

Carla stared open-mouthed. 'What on earth are you talking about?'

'She was twenty-two.' He took a handkerchief from his pocket and wiped his face. 'You're already in danger, but if I tell you this it will be a lot worse.'

Her mind was full of amazing surmises. 'I think I can guess, but tell me,' she said.

He nodded. 'You'll figure it out soon, anyway. Lili helped Heinrich broadcast to Moscow. It's much quicker if someone reads you the code groups. And the faster you go, the less likely you are to be caught. But Lili's cousin stayed at the apartment for a few days and found her code books. Nazi bitch.'

His words confirmed her astonishing suspicions. 'You know about the spying?'

He looked at her with an ironic smile. 'I'm in charge of it.'

'Good God!'

'That's why I had to drop the whole business of the murdered children. Moscow ordered me to. And they were right. If I'd lost my job at the Air Ministry I would have had no access to secret papers, nor to other people who could bring me secrets.'

She needed to sit down. She perched on the edge of the bed beside him. 'Why didn't you tell me?'

'We work on the assumption that everyone talks under torture. Knowing nothing, you can't betray others. Poor Lili was tortured, but she only knew Volodya, who's back in Moscow now, and Heinrich, and she never knew Heinrich's second name or anything else about him.'

Carla was chilled to the bone. *Everyone talks under torture.*

Werner finished: 'I'm sorry I've told you, but after seeing me like this you were on the point of guessing it all anyway.'

'So I've completely misjudged you.'

'Not your fault. I deliberately misled you.'

'I feel a fool just the same. I've despised you for two years.'

'All the while I was desperate to explain to you.'

She put her arm around him.

He took her other hand and kissed it. 'Can you forgive me?'

She was not sure how she felt, but she did not want to reject him when he was so down, so she said: 'Yes, of course.'

'Poor Lili,' he said. His voice fell to a whisper. 'She had been so badly beaten, she could hardly walk to the guillotine. Yet she begged for life, right up to the end.'

'How come you were there?'

'I've befriended a Gestapo man, Inspector Thomas Macke. He took me.'

'Macke? I remember him – he arrested my father.' She vividly recalled a round-faced man with a small black moustache, and she experienced again her rage at the arrogant power Macke had to take her father away, and her grief when he died of the injuries he suffered at Macke's hands.

'I think he suspects me, and taking me to the execution was a test. Perhaps he thought I might lose my self-control and try to intervene. Anyway, I think I passed the test.'

'But if you were arrested . . .'

Werner nodded. 'Everyone talks under torture.'

'And you know everything.'

'Every agent, every code . . . The only thing I don't know is where they broadcast from. I leave it up to them to pick the locations, and they don't tell me.'

They held hands in silence. After a while, Carla said: 'I came to give it to Frieda, but I might as well give it to you.'

'Give what?'

'The battle plan for Operation Zitadelle.'

Werner was electrified. 'But I've been trying to put my hands on that for weeks! Where did you get it?'

'From an officer on the General Staff. Perhaps I shouldn't say his name.'

'Quite right, don't tell me. But is it authentic?'

'You'd better take a look.' She went to Frieda's room and returned with the buff envelope. It had never occurred to her that the document might not be genuine. 'It looks all right to me, but what do I know?'

He took out the typewritten sheets. After a minute he said: 'This is the real thing. Fantastic!'

'I'm so glad.'

He stood up. 'I have to take this to Heinrich right away. We must get this encrypted and broadcast tonight.'

Carla felt disappointed that their moment of intimacy was over so soon, though she could not have said what she had been expecting. She followed him through the door. She picked up her bag from Frieda's room and went downstairs.

With his hand on the front door, Werner said: 'I'm so glad we're friends again.'

'Me, too.'

'Do you think we'll be able to forget this period of estrangement?'

She did not know what he was trying to say. Did he want to be her lover again – or was he telling her that was out of the question? 'I think we can put it behind us,' she said neutrally.

'Good.' He bent and kissed her lips very quickly. Then he opened the door.

They left the house together, and he climbed on his motorcycle.

Carla walked down the driveway to the street and headed for the station. A moment later, Werner drove past her with a honk and a wave.

Now that she was alone, she could begin to think about his revelation. How did she feel? For two years she had hated him. But in that time she had not had a serious boyfriend. Had she remained in love with him all along? At a minimum she had retained, in her heart of hearts, a fondness for him despite everything. Today, when she heard him in such distress, her hostility had melted away. Now she felt a glow of affection.

Did she love him still?

She did not know.

(iv)

Macke sat in the rear seat of the black Mercedes with Werner beside him. Around Macke's neck was a bag like a school satchel, except that he wore it in front instead of behind. It was small enough to be covered by a buttoned overcoat. A thin wire ran from the bag to a small earphone. 'It's the latest thing,' Macke said. 'As you get closer to the broadcaster, the sound gets louder.'

Werner said: 'More discreet than a van with a big aerial on its roof.'

'We have to use both – the van to discover the general area, and this to pinpoint the exact location.'

Macke was in trouble. Operation Zitadelle had been a catastrophe. Even before the offensive opened, the Red Army had attacked the airfields where the Luftwaffe were assembling. Zitadelle had been called off after a week, but even that was too late to prevent irreparable damage to the German army.

Germany's leaders were always quick to blame Jewish-Bolshevik conspirators whenever things went wrong, but in this case they were right. The Red Army had appeared to know the entire battle plan in advance. And that, according to Superintendent Kringelein, was Thomas Macke's fault. He was head of counter-espionage for the city of Berlin. His career was on the line. He faced dismissal and worse.

His only hope now was a tremendous coup, a massive operation to round up the spies who were undermining the German war effort. So tonight he had set a trap for Werner Franck.

If Franck turned out to be innocent, he did not know what he would do.

In the front seat of the car, a walkie-talkie crackled. Macke's pulse quickened. The driver picked up the handset. 'Wagner here.' He started the engine. 'We're on our way,' he said. 'Over and out.'

It had started.

Macke asked him: 'Where are we headed?'

'Kreuzberg.' It was a densely populated low-rent neighbourhood south of the city centre.

As they pulled away, the air raid siren sounded.

That was an unwelcome complication. Macke looked out of the window. The searchlights came on, waving like giant wands. Macke supposed they must find planes sometimes, but he had never seen it happen. When the sirens ceased their howling, he could hear the thunder of approaching bombers. In the early years of the war, a British bombing mission had consisted of a few dozen aircraft – which was bad enough – but now they were sending hundreds at a time. The noise was terrifying even before they dropped their bombs.

Werner said: 'I suppose we'd better call off our mission tonight.'

'Hell, no,' said Macke.

The roar of the planes grew.

Flares and small incendiary bombs began to fall as the car approached Kreuzberg. The neighbourhood was a typical target for the RAF's current strategy of killing as many civilian factory workers as possible. With staggering hypocrisy Churchill and Attlee were claiming they attacked only military targets, and civilian casualties were a regrettable side effect. Berliners knew better.

Wagner drove as fast as he could along streets lit fitfully by flames. There were no people around apart from air raid officials: everyone else was legally obliged to take shelter. The only other vehicles were ambulances, fire engines and police cars.

Macke covertly studied Werner. The boy was edgy, never quite still, staring out of the window anxiously, tapping his foot in unconscious tension.

Macke had not confided his suspicions to anyone but his immediate team. It was going to be difficult for him if he had to admit that he had demonstrated Gestapo operations to someone who he now thought was a spy. He could end up under interrogation in his own basement torture chamber. He was not going to do it until he was sure. The only way he might get away with it would be if at the same time he could present his superiors with a captured spy.

But then, if his suspicion turned out to be true, he would arrest not just Werner but his family and friends, and announce the destruction of a massive spy ring. That would transform the picture. He might even be promoted.

As the raid progressed the type of bombs changed, and Macke heard the profound thudding sound of high explosive. Once the target was illuminated, the RAF liked to drop a mixture of large oil bombs to start fires and high explosive to ventilate the flames and hamper the emergency services. It was cruel, but Macke knew that the Luftwaffe's bombing pattern was similar.

The sound in Macke's earphone started up as they drove cautiously along a street of five-storey tenements. The area was taking a terrific pounding and several buildings were newly demolished. Werner said shakily: 'We're in the middle of the target area, for Christ's sake.'

Macke did not care: tonight was already life or death to him. 'All the better,' he said. 'The pianist will imagine he doesn't need to worry about the Gestapo, in the middle of an air raid.'

Wagner stopped the car next to a burning church and pointed along a side street. 'Down there,' he said.

Macke and Werner jumped out.

Macke walked quickly along the street with Werner beside him and Wagner behind. Werner said: 'Are you sure it's a spy? Could it be anything else?'

'Broadcasting a radio signal?' Macke said. 'What else could it be?'

Macke could still hear his earphone, but only just, for the air raid was cacophonous: the planes, the bombs, the anti-aircraft guns, the crash of falling buildings and the roar of huge fires.

They passed a stable where horses were neighing in terror, the signal growing ever stronger. Werner was glancing from side to side anxiously. If he was a spy, he would now be afraid that one of his colleagues was about to be arrested by the Gestapo – and wondering what the hell he could do about it. Would he repeat the trick he used last time, or think of some new way of giving a warning? If he was not a spy this whole farce was a waste of time.

Macke took out the earpiece and handed it to Werner. 'Listen,' he said, continuing to walk.

Werner nodded. 'Getting stronger,' he said. The look in his eyes was almost frantic. He handed the earpiece back.

I believe I've got you, Macke thought triumphantly.

There was a thunderous crash as a bomb landed in a building they had just passed. They turned to see flames already licking up beyond the smashed windows of a bakery. Wagner said: 'Christ, that was close.'

They came to a school, a low brick building in an asphalt yard. 'In there, I think,' said Macke.

The three men walked up a short flight of stone steps to the entrance. The door was not locked. They went in.

They were at one end of a broad corridor. At its far end was a large door that probably led to the school hall. 'Straight ahead,' said Macke.

He drew his gun, a 9mm Luger pistol.

Werner was not armed.

There was a crash, a thud, and the roar of an explosion, all terrifyingly close. All the windows in the corridor smashed, and shards of glass rained on the tiled floor. A bomb must have landed in the playground.

Werner shouted: 'Clear out, everyone! The building is about to collapse.'

There was no danger of the building collapsing, Macke could see. This was Werner's ruse for giving the alarm to the pianist.

Werner broke into a run, but instead of heading back the way they had come he went on down the corridor towards the hall.

To warn his friends, Macke thought.

Wagner drew his gun, but Macke said: 'No! Don't shoot!'

Werner reached the end of the corridor and flung open the door to the hall. 'Run, everyone!' he yelled. Then he fell silent and stood still.

Inside the hall Macke's colleague Mann, the electrical engineer, was tapping out nonsense on a suitcase radio.

Beside him stood Schneider and Richter, both holding drawn guns.

Macke smiled triumphantly. Werner had fallen straight into his trap.

Wagner walked forward and put his gun to Werner's head.

Macke said: 'You're under arrest, you subhuman Bolshevik.'

Werner acted fast. He jerked his head away from Wagner's gun, seized Wagner's arm, and pulled him into the hall. For a moment Wagner shielded Werner from the guns in the hall. Then he thrust Wagner away from him, causing Wagner to stumble and fall. In the next moment he stepped out of the hall and slammed the door.

For a few seconds it was just Macke and Werner in the corridor.

Werner walked towards Macke.

Macke pointed his Luger. 'Stop, or I'll shoot.'

'No, you won't.' Werner came closer. 'You need to interrogate me, and find out who the others are.'

Macke pointed his gun at Werner's legs. 'I can interrogate you with a bullet in your knee,' he said, and he fired.

The shot missed.

Werner lunged and knocked Macke's gun hand aside. Macke dropped the weapon. As he stooped to retrieve it, Werner ran past.

Macke picked up the gun.

Werner reached the school door. Macke took careful aim at his legs and fired.

His first three shots missed, and Werner went through the door.

1943 (II)

Macke fired one more shot through the still-open door, and Werner cried out and fell down.

Macke ran along the corridor. Behind him, he heard the others coming out of the school hall.

Then the roof opened with a crash, there was another noise like a thud, and liquid fire splashed like a fountain. Macke screamed in terror, then in agony as his clothes caught alight. He fell to the ground, then there was silence, then darkness.

(v)

The doctors were triaging patients in the hospital lobby. Those merely bruised and cut were sent into the out-patients' waiting area where the most junior nurses cleaned their cuts and consoled them with aspirins. The serious cases were given emergency treatment right there in the lobby then sent to specialists upstairs. The dead were taken into the yard and laid on the cold ground until someone claimed them.

Dr Ernst examined a screaming burn victim and prescribed morphine. 'Then get his clothes off and put some gel on those burns,' he said, and moved on to the next one.

Carla loaded a syringe while Frieda cut the patient's blackened clothes away. He had severe burns all down his right side, but the left was not so bad. Carla found an intact patch of skin and flesh on his left thigh. She was about to inject the patient when she looked at his face and froze.

She knew that fat round countenance with the moustache like a dirt mark under the nose. Two years ago he had come into the hall of her house and arrested her father. Next time she saw her father he had been dying. This was Inspector Thomas Macke of the Gestapo.

You killed my father, she thought.

Now I can kill you.

It would be simple. She would give him four times the maximum dose of morphine. No one would notice, especially on a night like tonight. He would fall unconscious immediately and die in a few minutes. A doctor who was almost asleep on his feet would assume his heart had failed. No one would doubt the diagnosis, and no one would ask sceptical questions. He would be one of thousands killed in a massive air raid. Rest in peace.

She knew that Werner feared Macke might be on to him. Any day now Werner could be arrested. *Everyone talks under torture.* Werner would give away Frieda, and Heinrich, and others – and Carla. She could save them all, now, in a minute.

But she hesitated.

She asked herself why. Macke was a torturer and a killer. He deserved to die a thousand deaths.

Carla had killed Joachim, or at least helped to kill him. But Joachim had been kicking Carla's mother to death when she hit him over the head with a soup cauldron. This was different.

Macke was a patient.

Carla was not very religious, but she did believe that some things were sacred. She was a nurse, and patients put their trust in her. She knew that Macke would torture and kill her without hesitation – but she was not like Macke, she was not that kind. This was nothing to do with him: it was about her.

If she killed a patient, she felt, she would have to leave the profession and never again dare to care for sick people. She would be like a banker who steals money, or a politician who takes bribes, or a priest who feels up the young girls who come to him for First Communion classes. She would have betrayed herself.

Frieda said: 'What are you waiting for? I can't gel him until he calms down.'

Carla stuck the needle in Thomas Macke, and he stopped screaming.

Frieda started to put gel on his burned skin.

'This one's only concussed,' Dr Ernst was saying of another patient. 'But he's got a bullet in his backside.' He raised his voice to talk to the patient. 'How did you get shot? Bullets are about the only things the RAF isn't throwing at us tonight.'

Carla turned to look. The patient was lying on his front. His trousers had been cut off, showing his rear. He had white skin and fine, fair hair on the small of his back. He was woozy, but he muttered something.

Ernst said: 'Policeman's gun went off by accident, did you say?'

The patient spoke more clearly. 'Yes.'

'I'm going to take the bullet out. It will hurt, but we're short of morphine, and there are worse cases than you.'

'Go ahead.'

Carla swabbed the wound. Ernst picked up a long, narrow pair of forceps. 'Bite the pillow,' he said.

He inserted the forceps into the wound. A muffled cry of pain came from the patient.

Dr Ernst said: 'Try not to tense your muscles. It makes it worse.'

Carla thought that was a stupid thing to say. No one could relax their muscles while a wound was being probed.

The patient roared: 'Ah, shit!'

'I've got it,' Dr Ernst said. 'Try to keep still!'

The patient lay still, and Ernst drew the slug out and dropped it into a tray.

Carla wiped the blood from the hole and slapped a dressing on the wound.

The patient rolled over.

'No,' Carla said. 'You must lie on your—'

She stopped. The patient was Werner.

'Carla?' he said.

'It's me,' she said happily. 'Putting a bandage on your bum.'

'I love you,' he said.

She threw her arms around him in the most unprofessional way possible and said: 'Oh, my dearest, I love you, too.'

(vi)

Thomas Macke came around slowly. At first he was in a dreamlike state. Then he became more aware, and realized he was in a hospital and drugged. He knew why, too: his skin hurt intensely, especially down his left side. He was able to figure out that the drugs must be reducing the pain but not completely eliminating it.

Slowly he remembered how he had come here. He had been bombed. He would be dead if he had not been running away from the blast, chasing a fugitive. Those behind him were certainly dead: Mann, Schneider, Richter and young Wagner. His whole team.

But he had caught Werner.

Or had he? He had shot Werner, and Werner had fallen; then the bomb had dropped. Macke had survived, so Werner might have too.

Macke was now the only man living who knew that Werner was a spy. He had to speak to his boss, Superintendent Kringelein. He tried to sit upright, but found he did not have the strength to move. He decided to call a nurse, but when he opened his mouth no sound came out. The effort exhausted him and he went back to sleep.

The next time he awoke, he sensed it was night. The place was quiet, no one moving. He opened his eyes to see a face hovering over him.

It was Werner.

'You're leaving here now,' Werner said.

Macke tried to call for help, but found he could not speak.

'You're going to a new place,' Werner said. 'You won't be a torturer any more – in fact, you'll be the one who gets tortured there.'

Macke opened his mouth to scream.

A pillow descended on his face. It was pressed firmly over his mouth and nose. He found he could not breathe. He tried to struggle, but there was no strength in his limbs. He tried to gasp for air, but there was no air. He started to panic. He managed to move his head from side to side, but the pillow was pressed down more firmly. At last he made a noise, but it was only a whimper in his throat.

The universe became a disc of light that shrank slowly until it was a pinpoint.

Then it went out.

17

'Will you marry me?' said Volodya Peshkov, and held his breath.

'No,' said Zoya Vorotsyntsev. 'But thank you.'

She was remarkably matter-of-fact about everything, but this was unusually brisk even for her.

They were in bed at the lavish Hotel Moskva, and they had just made love. Zoya had come twice. Her preferred type of sex was cunnilingus. She liked to recline on a pile of pillows while he knelt worshipfully between her legs. He was a willing acolyte, and she returned the favour with enthusiasm.

They had been a couple for more than a year, and everything seemed to be going wonderfully well. Her refusal baffled him.

He said: 'Do you love me?'

'Yes. I adore you. Thank you for loving me enough to propose marriage.'

That was a bit better. 'So why won't you accept?'

'I don't want to bring children into a world at war,' she said.

'Okay, I can understand that.'

'Ask me again when we've won.'

'By then I may not want to marry you.'

'If that's how inconstant you are, it's a good thing I refused you today.'

'Sorry. For a moment, there, I forgot that you don't understand teasing.'

'I have to pee.' She got off the bed and walked naked across the hotel room. Volodya could hardly believe he was allowed to see this. She had the body of a fashion model or a movie star. Her skin was milk-white and her hair pale blonde – all of it. She sat on the toilet without closing the bathroom door, and he listened to her peeing. Her lack of modesty was a perpetual delight.

He was supposed to be working.

The Moscow intelligence community was thrown into disarray every time Allied leaders visited, and Volodya's normal routine had been disrupted again for the Foreign Ministers' Conference that had opened on 18 October.

The visitors were the American Secretary of State, Cordell Hull, and the British Foreign Secretary, Anthony Eden. They had a hare-brained scheme for a Four-Power Pact including China. Stalin thought it was all nonsense and did not understand why they were wasting time on it. The American, Hull, was seventy-two years old and coughing blood – his doctor had come to Moscow with him – but he was no less forceful for that, and he was insistent on the pact.

There was so much to do during the conference that the NKVD – the secret police – were forced to co-operate with their hated rivals in Red Army Intelligence, Volodya's outfit. Microphones had to be concealed in hotel rooms – there was one in here, only Volodya had disconnected it. The visiting ministers and all their aides had to be kept under minute-by-minute surveillance. Their luggage had to be clandestinely opened and searched. Their phone calls had to be tape-recorded and transcribed and translated into Russian and read and summarized. Most of the people they met, including waiters and chambermaids, were NKVD agents, but anyone else they happened to speak to, in the hotel lobby or on the street, had to be checked out, perhaps arrested and imprisoned and interrogated under torture. It was a lot of work.

Volodya was riding high. His spies in Berlin were producing remarkable intelligence. They had given him the battle plan for the Germans' main summer offensive, Zitadelle, and the Red Army had inflicted a tremendous defeat.

Zoya was happy, too. The Soviet Union had resumed nuclear research, and Zoya was part of the team trying to design a nuclear bomb. They were a long way behind the West, because of the delay caused by Stalin's scepticism, but in compensation they were getting invaluable help from Communist spies in England and America, including Volodya's old school friend Willi Frunze.

Zoya came back to bed. Volodya said: 'When we first met, you didn't seem to like me much.'

'I didn't like men,' she replied. 'I still don't. Most of them are drunks and bullies and fools. It took me a while to figure out that you were different.'

'Thanks, I think,' he said. 'But are men really so bad?'

'Look around you,' she said. 'Look at our country.'

He reached over her and turned on the bedside radio. Even though he had disconnected the listening device behind the headboard, you couldn't be too careful. When the radio had warmed up, a military band played a march. Satisfied that he could not be overheard, Volodya said: 'You're thinking of Stalin and Beria. But they won't always be around.'

'Do you know how my father fell from favour?' she said.

'No. My parents never mentioned it.'

'There's a reason for that.'

'Go on.'

'According to my mother, there was an election at my father's factory for a deputy to attend the Moscow Soviet. A Menshevik candidate stood against the Bolshevik, and my father went to a meeting to hear him speak. He did not support the Menshevik, nor vote for him; but everyone who went to that meeting was sacked, and a few weeks later my father was arrested and taken to the Lubyanka.'

She meant the NKVD headquarters and prison in Lubyanka Square.

She went on: 'My mother went to your father and begged him to help. He immediately went with her to the Lubyanka. They saved my father, but they saw twelve other workers shot.'

'That's terrible,' Volodya said. 'But it was Stalin—'

'No. This was 1920. Stalin was just a Red Army commander fighting in the Soviet–Polish War. Lenin was leader.'

'This happened under Lenin?'

'Yes. So, you see, it's not just Stalin and Beria.'

Volodya's view of Communist history was badly shaken. 'What is it, then?'

The door opened.

Volodya reached for his gun in the bedside-table drawer.

But the person who came in was a girl wearing a fur coat and, as far as he could see, nothing else.

'Sorry, Volodya,' she said. 'I didn't know you had company.'

Zoya said: 'Who the fuck is she?'

Volodya said: 'Natasha, how did you open my door?'

'You gave me a pass key. It opens every door in the hotel.'

'Well, you might have knocked!'

'Sorry. I just came to tell you the bad news.'

'What?'

'I went into Woody Dewar's room, just as you told me. But I didn't succeed.'

'What did you do?'

'This.' Natasha opened her coat to show her naked body. She had a voluptuous figure and a luxuriant bush of dark pubic hair.

'All right, I get the picture, close your coat,' said Volodya. 'What did he say?'

She switched to English. 'He just said: "No." I said: "What do you mean, no?" He said: "It's the opposite of yes." Then he just held the door wide open until I went out.'

'Bugger,' said Volodya. 'I'll have to think of something else.'

(ii)

Chuck Dewar knew there was going to be trouble when Captain Vandermeier came into the enemy land section in the middle of the afternoon, red-faced from a beery lunch.

The intelligence unit at Pearl Harbor had expanded. Formerly called Station HYPO, it now had the grand title of Joint Intelligence Center, Pacific Ocean Area, or JICPOA.

Vandermeier had a marine sergeant in tow. 'Hey, you two powder puffs,' Vandermeier said. 'You got a customer complaint here.'

The operation had grown, everyone began to specialize, and Chuck and Eddie had become experts at mapping the territory where American forces were about to land as they fought their way island by island across the Pacific.

Vandermeier said: 'This is Sergeant Donegan.' The marine was very tall and looked as hard as a rifle. Chuck guessed that the sexually troubled Vandermeier was smitten.

Chuck stood up: 'Good to meet you, Sergeant. I'm Chief Petty Officer Dewar.'

Chuck and Eddie had both been promoted. As thousands of conscripts poured into the US military, there was a shortage of officers, and pre-war enlisted men who knew the ropes rose fast. Chuck and Eddie were now permitted to live off base. They had rented a small apartment together.

Chuck put out his hand, but Donegan did not shake it.

Chuck sat down again. He slightly outranked a sergeant, and he was not going to be polite to one who was rude. 'Something I can do for you, Captain Vandermeier?'

There were many ways a captain could torment petty officers in the navy, and Vandermeier knew them all. He adjusted rotas so that Chuck and Eddie never had the same day off. He marked their reports 'adequate', knowing full well that anything less than 'excellent' was, in fact, a black mark. He sent confusing messages to the pay office, so that Chuck and Eddie were paid late or got less than they should have, and had to spend hours straightening things out. He was a royal pain. And now he had thought up some new mischief.

Donegan pulled from his pocket a grubby sheet of paper and unfolded it. 'Is this your work?' he said aggressively.

Chuck took the paper. It was a map of New Georgia, a group in the Solomon Islands. 'Let me check,' he said. It was his work, and he knew it, but he was playing for time.

He went to a filing cabinet and pulled open a drawer. He took out the file for New Georgia and shut the drawer with his knee. He returned to his desk, sat down, and opened the file. It contained a duplicate of Donegan's map. 'Yes,' Chuck said. 'That's my work.'

'Well, I'm here to tell you it's shit,' said Donegan.

'Is it?'

'Look, right here. You show the jungle coming down to the sea. In fact, there's a beach a quarter of a mile wide.'

'I'm sorry to hear that.'

'Sorry!' Donegan had drunk about the same amount of beer as Vandermeier, and he was spoiling for a fight. 'Fifty of my men died on that beach.'

Vandermeier belched and said: 'How could you make a mistake like that, Dewar?'

Chuck was shaken. If he was responsible for an error that had killed fifty men, he deserved to be shouted at. 'This is what we had to work on,' he said. The file contained an inaccurate map of the islands that might have been Victorian, and a more recent naval chart that showed sea depths but almost no terrain features. There were no on-the-spot reports and no wireless decrypts. The only other item in the file was a blurred black-and-white aerial reconnaissance photograph. Putting his finger on the relevant spot in the photo, Chuck said: 'It sure looks as if the trees come all the way to the waterline. Is there a tide? If not, the sand might have been covered with algae when the photograph was taken. Algae can bloom suddenly, and die off just as fast.'

Donegan said: 'You wouldn't be so goddamn casual about it if you had to fight over the terrain.'

Maybe that was true, Chuck thought. Donegan was aggressive and rude, and he was being egged on by the malicious Vandermeier, but that did not mean he was wrong.

Vandermeier said: 'Yeah, Dewar. Maybe you and your nancy-boy friend should go with the marines on their next assault. See how your maps are used in action.'

Chuck was trying to think of a smart retort when it occurred to him to take the suggestion seriously. Maybe he ought to see some action. It *was* easy to be blasé behind a desk. Donegan's complaint deserved to be taken seriously.

On the other hand, it would mean risking his life.

Chuck looked Vandermeier in the eye. 'That sounds like a good idea, Captain,' he said. 'I'd like to volunteer for that duty.'

Donegan looked startled, as if he was beginning to think he might have misjudged the situation.

Eddie spoke for the first time. 'So would I. I'll go, too.'

'Good,' said Vandermeier. 'You'll come back wiser – or not at all.'

(iii)

Volodya could not get Woody Dewar drunk.

In the bar of the Hotel Moskva he thrust a glass of vodka in front of the young American and said in schoolboy English: 'You'll like this – it's the very best.'

'Thank you very much,' said Woody. 'I appreciate it.' And he left the glass untouched.

Woody was tall and gangly and seemed straightforward to the point of naivety, which was why Volodya had targeted him.

Speaking through the interpreter, Woody said: 'Is Peshkov a common Russian name?'

'Not especially,' Volodya replied in Russian.

'I'm from Buffalo, where there is a well-known businessman called Lev Peshkov. I wonder if you're related.'

Volodya was startled. His father's brother was called Lev Peshkov and had gone to Buffalo before the First World War. But caution made him prevaricate. 'I must ask my father,' he said.

'I was at Harvard with Lev Peshkov's son, Greg. He could be your cousin.'

'Possibly.' Volodya glanced nervously at the police spies around the table. Woody did not understand that any connection with someone in America could bring down suspicion on a Soviet citizen. 'You know, Woody, in this country it's considered an insult to refuse to drink.'

Woody smiled pleasantly. 'Not in America,' he said.

Volodya picked up his own glass and looked around the table at the assorted secret policemen pretending to be civil servants and diplomats. 'A toast!' he said. 'To friendship between the United States and the Soviet Union!'

The others raised their glasses high. Woody did the same. 'Friendship!' they all echoed.

Everyone drank except Woody, who put his glass down untasted.

Volodya began to suspect that he was not as naive as he seemed.

Woody leaned across the table. 'Volodya, you need to understand that I don't know any secrets. I'm too junior.'

'So am I,' said Volodya. It was far from the truth.

Woody said: 'What I'm trying to explain is that you can just ask me questions. If I know the answers, I'll tell you. I can do that, because anything I know can't possibly be secret. So you don't need to get me drunk or send prostitutes to my room. You can just ask me.'

It was some kind of trick, Volodya decided. No one could be so innocent. But he decided to humour Woody. Why not? 'All right,' he said. 'I need to know what you're after. Not you personally, of course. Your delegation, and Secretary Hull, and President Roosevelt. What do you want from this conference?'

'We want you to back the Four-Power Pact.'

It was the standard answer, but Volodya decided to persist. 'This is what we don't understand.' He was being candid now, perhaps more than he should have, but instinct was telling him to take the risk of opening up a little. 'Who cares about a pact with China? We need to defeat the Nazis in Europe. We want you to help us do that.'

'And we will.'

'So you say. But you said you would invade Europe this summer.'

'Well, we did invade Italy.'

'It's not enough.'

'France next year. We've promised that.'

'So why do you need the pact?'

'Well.' Woody paused, collecting his thoughts. 'We have to show the American people how it's in their interests to invade Europe.'

'Why?'

'Why what?'

'Why do you need to explain this to the public? Roosevelt is President, isn't he? He should just do it!'

'Next year is election year. He wants to get re-elected.'

'So?'

'American people won't vote for him if they think he's involved them unnecessarily in the war in Europe. So he wants to put it to them as part of his overall plan for world peace. If we have the Four-Power Pact, showing that we're serious about the United Nations organization, then American voters are more likely to accept that the invasion of France is a step on the road to a more peaceful world.'

'This is amazing,' Volodya said. 'He's the President, yet he has to make excuses all the time for what he does!'

'Something like that,' Woody said. 'We call it democracy.'

Volodya had a sneaking suspicion that this incredible story might actually be the truth. 'So the pact is necessary to persuade American voters to support the invasion of Europe.'

'Exactly.'

'Then why do we need China?' Stalin was particularly scornful of the Allies' insistence that China should be included in the pact.

'China is a weak ally.'

'So ignore China.'

'If the Chinese are left out they will become discouraged, and may fight less enthusiastically against the Japanese.'

'So?'

'So we will have to bolster our forces in the Pacific theatre, and that will take away from our strength in Europe.'

That alarmed Volodya. The Soviet Union did not want Allied forces diverted from Europe to the Pacific. 'So you are making a friendly gesture to China simply in order to conserve more forces for the invasion of Europe.'

'Yes.'

'You make it seem simple.'

'It is,' said Woody.

(iv)

In the early hours of the morning on 1 November, Chuck and Eddie ate a steak breakfast with the US Marine 3rd Division just off the South Sea island of Bougainville.

The island was about 125 miles long. It had two Japanese naval air bases, one in the north and one in the south. The marines were getting ready to land halfway along the lightly defended west coast. Their object was to establish a beachhead and win enough territory to build an airstrip from which to launch attacks on the Japanese bases.

Chuck was on deck at twenty-six minutes past seven when marines in helmets and backpacks began to swarm down the rope nets hanging over the sides of the ship and jump into high-sided landing craft. With them were a small number of war dogs, Dobermann Pinschers that made tireless sentries.

As the boats approached land, Chuck could already see a flaw in the map he had prepared. Tall waves crashed on to a steeply sloping beach. As he watched, a boat turned sideways to the waves and capsized. The marines swam for shore.

'We have to show surf conditions,' Chuck said to Eddie, who was standing beside him on the deck.

'How do we find them out?'

'Reconnaissance aircraft will have to fly low enough for whitecaps to register on their photographs.'

'They can't risk coming that low when there are enemy air bases so close.'

Eddie was right. But there had to be a solution. Chuck filed it away as the first question to be considered as a result of this mission.

For this landing they had benefited from more information than usual. As well as the normal unreliable maps and hard-to-decipher aerial photographs, they had a report from a reconnaissance team landed by submarine six weeks earlier. The team had identified twelve beaches suitable for landing along a four-mile stretch of coast. But they had not warned of the surf. Perhaps it was not so high that day.

In other respects, Chuck's map was right, so far. There was a sandy beach about a hundred yards wide, then a tangle of palm trees and other vegetation. Just beyond the brush line, according to the map, there should be a swamp.

The coast was not completely undefended. Chuck heard the roar of artillery fire, and a shell landed in the shallows. It did no harm, but the gunner's aim would improve. The marines were galvanized with a new urgency as they leaped from the landing craft to the beach and ran for the brush line.

Chuck was glad he had decided to come. He had never been careless or slack about his maps, but it was salutary to see first-hand how correct mapping could save men's lives, and how the smallest errors could be deadly. Even before they embarked, he and Eddie had become a lot more demanding. They asked for blurred photographs to be taken again, they interrogated reconnaissance parties by phone, and they cabled all over the world for better charts.

He was glad for another reason. He was at sea, which he loved. He was on a ship with seven hundred young men, and he relished the camaraderie, the jokes, the songs, and the intimacy of crowded berths and shared showers. 'It's like being a straight guy in a girls' boarding school,' he said to Eddie one evening.

'Except that that never happens, and this does,' Eddie said. He felt the same as Chuck. They loved each other, but they did not mind looking at naked sailors.

Now all seven hundred marines were getting off the ship and on to land as fast as they could. The same was happening at eight other locations along this stretch of coast. As soon as a landing craft emptied out, it lost no time in turning around and coming back for more; but the process still seemed desperately slow.

The Japanese artillery gunner, hidden somewhere in the jungle, found his range at last, and to Chuck's shock a well-aimed shell exploded in a knot of marines, sending men and rifles and body parts flying through the air to litter the beach and stain the sand red.

Chuck was staring in horror at the carnage when he heard the roar of a plane, and looked up to see a Japanese Zero flying low, following the coast. The red suns painted on the wings struck fear into his heart. Last time he saw that sight had been at the Battle of Midway.

The Zero strafed the beach. Marines who were in the process of disembarking from landing craft were caught defenceless. Some threw themselves flat in the shallows, some tried to get behind the hull of the boat, some ran for the jungle. For a few seconds blood spurted and men fell.

Then the plane was gone, leaving the beach scattered with American dead.

Chuck heard it open up a moment later, strafing the next beach.

It would be back.

There were supposed to be US planes in attendance, but he could not see any. Air support was never where you wanted it to be, which was directly above your head.

When all the marines were ashore, alive and dead, the boats transported medics and stretcher parties to the beach. Then they began landing supplies: ammunition, drinking water, food, drugs and dressings. On the return trip the landing craft brought the wounded back to the ship.

Chuck and Eddie, as non-essential personnel, went ashore with the supplies.

The boat skippers had got used to the swell now, and their craft held a stable position, with its ramp on the sand and the waves breaking on its stern, while the boxes were unloaded and Chuck and Eddie jumped into the surf to wade to shore.

They reached the waterline together.

As they did so, a machine gun opened up.

It seemed to be in the jungle about four hundred yards along the beach. Had it been there all along, the gunner biding his time, or had it just been moved into position from another location? Eddie and Chuck bent double and ran for the tree line.

A sailor with a crate of ammunition on his shoulder gave a shout of pain and fell, dropping the box.

Then Eddie cried out.

Chuck ran on two paces before he could stop. When he turned, Eddie was rolling on the sand clutching his knee, yelling: 'Ah, fuck!'

Chuck came back and knelt beside him. 'It's okay, I'm here!' he shouted. Eddie's eyes were closed, but he was alive, and Chuck could see no wounds other than the knee.

He glanced up. The boat that had brought them was still close to shore, being unloaded. He could get Eddie back to the ship in minutes. But the machine gun was still firing.

He got into a crouching position. 'This is going to hurt,' he said. 'Yell as much as you like.'

He got his right arm under Eddie's shoulder, then slid his left under Eddie's thighs. He took the weight and straightened up. Eddie screamed with pain as his smashed leg swung free. 'Hang in there, buddy,' Chuck said. He turned towards the water.

He felt sudden, unbearably sharp pains in his legs, his back and finally his head. In the next fraction of a second he thought he must not drop Eddie. A moment later he knew he was going to. There was a flash of light behind his eyes that rendered him blind.

And then the world came to an end.

(v)

On her day off, Carla worked at the Jewish Hospital.

Dr Rothmann had persuaded her. He had been released from the camp – no one knew why, except the Nazis, and they did not tell anyone. He had lost one eye and he walked with a limp, but he was alive, and capable of practising medicine.

The hospital was in the northern working-class district of Wedding, but there was nothing proletarian about the architecture. It had been built before the First World War, when Berlin's Jews had been prosperous and proud. There were seven elegant buildings set in a large garden. The different departments were linked by tunnels, so that patients and staff could move from one to another without braving the weather.

It was a miracle there was still a Jewish hospital. Very few Jews were left in Berlin. They had been rounded up in their thousands and sent away in special trains. No one knew where they had gone or what happened to them. There were incredible rumours about extermination camps.

The few Jews still in Berlin could not be treated, if they were sick, by Aryan doctors and nurses. So, by the tangled logic of Nazi racism, the hospital was allowed to remain. It was mainly staffed by Jews and other unfortunate people who did not count as properly Aryan: Slavs from Eastern Europe, people of mixed ancestry, and those married to Jews. But there were not enough nurses, so Carla helped out.

The hospital was harassed constantly by the Gestapo, critically short of supplies, especially drugs, understaffed and almost completely without funds.

Carla was breaking the law as she took the temperature of an eleven-year-old boy whose foot had been crushed in an air raid. It was also a crime for her to smuggle medicines out of her everyday hospital and bring them here. But she wanted to prove, if only to herself, that not everyone had given in to the Nazis.

As she finished her ward round she saw Werner outside the door, in his air force uniform.

For several days he and Carla had lived in fear, wondering whether anyone had survived the bombing of the school and lived to condemn Werner; but it was now clear they had all died, and no one else knew of Macke's suspicions. They had got away with it, again.

Werner had recovered quickly from his bullet wound.

And they were lovers. Werner had moved into the von Ulrichs' large, half-empty house, and he slept with Carla every night. Their parents made no objection: everyone felt they could die any day, and people should take what joy they could from a life of hardship and suffering.

But Werner looked more solemn than usual as he waved to Carla through the glass panel in the door to the ward. She beckoned him inside and kissed him. 'I love you,' she said. She never tired of saying it.

He was always happy to say: 'I love you, too.'

'What are you doing here?' she said. 'Did you just want a kiss?'

'I've got bad news. I've been posted to the Eastern Front.'

'Oh, no!' Tears came to her eyes.

'It's really a miracle I've avoided it this long. But General Dorn can't keep me any longer. Half our army consists of old men and schoolboys, and I'm a fit twenty-four-year-old officer.'

She whispered: 'Please don't die.'

'I'll do my best.'

Still whispering, she said: 'But what will happen to the network? You know everything. Who else could run it?'

He looked at her without speaking.

She realized what was in his mind. 'Oh, no – not me!'

'You're the best person. Frieda's a follower, not a leader. You've shown the ability to recruit new people and motivate them. You've never been in trouble with the police and you have no record of political activity. No one knows the role you played in opposing Aktion T4. As far as the authorities are concerned, you are a blameless nurse.'

'But Werner, I'm scared!'

'You don't have to do it. But no one else can.'

Just then they heard a commotion.

The neighbouring ward was for mental patients, and it was not unusual to hear shouting and even screaming; but this seemed different. A cultured voice was raised in anger. Then they heard a second voice, this one with a Berlin accent and the insistent, bullying tone that outsiders said was typical of Berliners.

Carla stepped into the corridor, and Werner followed.

Dr Rothmann, wearing a yellow star on his jacket, was arguing with a man in SS uniform. Behind them, the double doors to the psychiatric ward, normally locked, were wide open. The patients were leaving. Two more policemen and a couple of nurses were herding a ragged line of men and women, most in pyjamas, some walking upright and apparently normal, others shambling and mumbling as they followed one another down the staircase.

Carla was immediately reminded of Ada's son, Kurt, and Werner's brother, Axel, and the so-called hospital in Akelberg. She did not know where these patients were going, but she was quite sure they would be killed there.

Dr Rothmann was saying indignantly: 'These people are sick! They need treatment!'

The SS officer replied: 'They're not sick, they're lunatics, and we're taking them where lunatics belong.'

'To a hospital?'

'You will be informed in due course.'

'That's not good enough.'

Carla knew she should not intervene. If they found out she was not Jewish she would be in deep trouble. She did not look particularly Aryan or otherwise, with dark hair and green eyes. If she kept quiet, probably they would not bother her. But if she protested about what the SS were doing she would be arrested and questioned, and then it would come out that she was working illegally. So she clamped her teeth together.

The officer raised his voice. 'Hurry up – get those cretins in the bus.'

Rothmann persisted. 'I must be informed where they are going. They are my patients.'

They were not really his patients – he was not a psychiatrist.

The SS man said: 'If you're so concerned about them, you can go with them.'

Dr Rothmann paled. He would almost certainly be going to his death.

Carla thought of his wife, Hannelore; his son, Rudi; and his daughter in England, Eva; and she felt sick with fear.

The officer grinned. 'Suddenly not so concerned?' he jeered.

Rothmann straightened up. 'On the contrary,' he said. 'I accept your offer. I swore an oath, many years ago, to do all I can to help sick people. I'm not going to break my oath now. I hope to die at peace with my conscience.' He limped down the stairs.

An old woman went by wearing nothing but a robe open at the front, showing her nakedness.

Carla could not remain silent. 'It's November out there!' she cried. 'They have no outdoor clothing!'

The officer gave her a hard look. 'They'll be all right on the bus.'

'I'll get some warm clothing.' Carla turned to Werner. 'Come and help me. Grab blankets from anywhere.'

The two of them ran around the emptying psychiatric ward, pulling blankets off beds and out of the cupboards. Each carrying a pile, they hurried down the stairs.

The garden of the hospital was frozen earth. Outside the main door was a grey bus, its engine idling, its driver smoking at the wheel. Carla saw that he was wearing a heavy coat plus a hat and gloves, which told her that the bus was not heated.

A small group of Gestapo and SS men stood in a knot, watching the proceedings.

The last few patients were climbing aboard. Carla and Werner boarded the bus and began to distribute the blankets.

Dr Rothmann was standing at the back. 'Carla,' he said. 'You ... you'll tell my Hannelore how it was. I have to go with the patients. I have no choice.'

'Of course.' Her voice was choked.

'I may be able to protect these people.'

Carla nodded, though she did not really believe it.

'In any event, I cannot abandon them.'

'I'll tell her.'

'And say that I love her.'

Carla could no longer stop the tears.

Rothmann said: 'Tell her that was the last thing I said. I love her.'

Carla nodded.

Werner took her arm. 'Let's go.'

They got off the bus.

An SS man said to Werner: 'You, in the air force uniform, what the hell do you think you're doing?'

Werner was so angry that Carla was frightened he would start a fight. But he spoke calmly. 'Giving blankets to old people who are cold,' he said. 'Is that against the law now?'

'You should be fighting on the Eastern Front.'

'I'm going there tomorrow. How about you?'

'Take care what you say.'

'If you would be kind enough to arrest me before I go, you might save my life.'

The man turned away.

The gears of the bus crashed and its engine note rose. Carla and Werner turned to look. At every window was a face, and they were all different: babbling, drooling, laughing hysterically, distracted, or distorted with spiritual distress – all insane. Psychiatric patients being taken away by the SS. The mad leading the mad.

The bus pulled away.

(vi)

'I might have liked Russia, if I'd been allowed to see it,' Woody said to his father.

'I feel the same.'

'I didn't even get any decent photographs.'

They were sitting in the grand lobby of the Hotel Moskva, near the entrance to the subway station. Their bags were packed and they were on their way home.

Woody said: 'I have to tell Greg Peshkov that I met a Volodya Peshkov. Though Volodya was not so pleased about it. I guess anyone with connections in the West might fall under suspicion.'

'You bet your socks.'

'Anyway, we got what we came for – that's the main thing. The allies are committed to the United Nations organization.'

'Yes,' said Gus with satisfaction. 'Stalin took some persuading, but he saw sense in the end. You helped with that, I think, by your straight-talking to Peshkov.'

'You've fought for this all your life, Papa.'

'I don't mind admitting that this is a pretty good moment.'

A worrying thought crossed Woody's mind. 'You're not going to retire now, are you?'

Gus laughed. 'No. We've won agreement in principle, but the job has only just begun.'

Cordell Hull had already left Moscow, but some of his aides were still here, and now one of them approached the Dewars. Woody knew him, a young man called Ray Baker. 'I have a message for you, Senator,' he said. He seemed nervous.

'Well, you just caught me in time – I'm about to leave,' said Gus. 'What is it?'

'It's about your son Charles – Chuck.'

Gus went pale and said: 'What is the message, Ray?'

The young man was having trouble speaking. 'Sir, it's bad news. He's been in a battle in the Solomon Islands.'

'Is he wounded?'

'No, sir, it's worse.'

'Oh, Christ,' said Gus, and he began to cry.

Woody had never seen his father cry.

'I'm sorry, sir,' said Ray. 'The message is that he's dead.'

18

Woody stood in front of the mirror in his bedroom at his parents' Washington apartment. He was wearing the uniform of a second lieutenant in the 510th Parachute Regiment of the United States Army.

He had had the suit made by a good Washington tailor, but it did not look well on him. Khaki made his complexion sallow, and the badges and flashes on the tunic jacket just seemed untidy.

He could probably have avoided the draft, but he had decided not to. Part of him wanted to continue to work with his father, who was helping President Roosevelt plan a new global order that would avoid any more world wars. They had won a triumph in Moscow, but Stalin was inconstant, and seemed to relish creating difficulties. At the Tehran conference in December, the Soviet leader had revived the halfway-house idea of regional councils, and Roosevelt had had to talk him out of it. Clearly the United Nations organization was going to require tireless vigilance.

But Gus could do that without Woody. And Woody was feeling worse and worse about letting other men fight the war for him.

He was looking as good as he ever would in the uniform, so he went into the drawing room to show his mother.

Rosa had a visitor, a young man in navy whites, and after a moment Woody recognized the freckled good looks of Eddie Parry. He was sitting on the couch with Rosa, holding a walking stick. He got to his feet with difficulty to shake Woody's hand.

Mama had a sad face. She said: 'Eddie was telling me about the day Chuck died.'

Eddie sat down again, and Woody sat opposite. 'I'd like to hear about that,' Woody said.

'It doesn't take long to tell,' Eddie began. 'We were on the beach at Bougainville for about five seconds when a machine gun opened up from somewhere in the swamp. We ran for cover, but I got a couple of bullets in my knee. Chuck should have gone on to the tree line. That's the drill – you leave the wounded to be picked up by the medics. Of course, Chuck disobeyed that rule. He stopped and came back for me.'

Eddie paused. There was a cup of coffee on the small table beside him, and he took a gulp.

'He picked me up in his arms,' he went on. 'Darn fool. Made hisself a target. But I guess he wanted to get me back in the landing craft. Those boats have high sides, and they're made of steel. We would have been safe, and I could have got medical attention right away on the ship. But he shouldn't have done it. Soon as he stood upright, he got hit by a spray of bullets – legs, back and head. I think he must have died before he hit the sand. Anyway, by the time I was able to lift my head and look at him, he just wasn't there any more.'

Woody saw that his mother was controlling herself with difficulty. He was afraid that if she cried, he would too.

'I lay on that beach beside his body for an hour,' Eddie said. 'I held his hand all the time. Then they brought a stretcher for me. I didn't want to go. I knew I'd never see him again.' He buried his face in his hands. 'I loved him so much,' he said.

Rosa put her arm around his big shoulders and hugged him. He laid his head on her chest and sobbed like a child. She stroked his hair. 'There, there,' she said. 'There, there.'

Woody realized that his mother knew what Chuck and Eddie were.

After a minute Eddie began to pull himself together. He looked at Woody. 'You know what this is like,' he said.

He was talking about the death of Joanne. 'Yes, I do,' Woody said. 'It's the worst thing in the world – but it hurts a little less every day.'

'I sure hope so.'

'Are you still in Hawaii?'

'Yes. Chuck and I work in the enemy land unit. Used to work.' He swallowed. 'Chuck decided we needed to get a better feel for how our maps were used in action. That's why we went to Bougainville with the marines.'

'You must be doing a good job,' Woody said. 'We seem to be beating the Japs in the Pacific.'

'Inch by inch,' Eddie said. He glanced at Woody's uniform. 'Where are you stationed?'

'I've been at Fort Benning, in Georgia, doing parachute training,' Woody said. 'Now I'm on my way to London. I leave tomorrow.'

He caught his mother's eye. Suddenly she looked older. He realized her face was lined. Her fiftieth birthday had passed with no big fuss. However, he guessed that talking about Chuck's death while her other son stood there in army uniform had struck her a hard blow.

Eddie did not pick that up. 'People say we'll invade France this year,' he said.

'I assume that's why my training was accelerated,' Woody said.

'You should see some action.'

Rosa muffled a sob.

Woody said: 'I hope I'll be as brave as my brother.'

Eddie said: 'I hope you never find out.'

(ii)

Greg Peshkov took dark-eyed Margaret Cowdry to an afternoon symphony concert. Margaret had a wide, generous mouth that loved kissing. But Greg had something else on his mind

He was following Barney McHugh.

So was an FBI agent called Bill Bicks.

Barney McHugh was a brilliant young physicist. He was on leave from the US Army's secret laboratory at Los Alamos, New Mexico, and had brought his British wife to Washington to see the sights.

The FBI had found out in advance that McHugh was coming to the concert, and Special Agent Bicks had managed to get Greg two seats a few rows behind McHugh's. A concert hall, with hundreds of strangers crowding together to come in and go out, was the perfect location for a clandestine rendezvous, and Greg wanted to know what McHugh might be up to.

It was a pity they had met before. Greg had talked to McHugh in Chicago on the day the nuclear pile was tested. It had been a year and a half ago, but McHugh might remember. So Greg had to make sure McHugh did not see him.

When Greg and Margaret arrived, McHugh's seats were empty. Either side were two ordinary-looking couples, a middle-aged man in a cheap grey chalk-stripe suit and his dowdy wife on the left, and two elderly ladies on the right. Greg hoped McHugh was going to show up. If the guy was a spy Greg wanted to nail him.

They were going to hear Tchaikovsky's first symphony. 'So, you like classical music,' said Margaret chattily as the orchestra tuned up. She had no idea of the real reason she had been brought here. She knew that Greg was working in weapons research, which was secret, but like almost all Americans she had no inkling of the nuclear bomb. 'I thought you only listened to jazz,' she said.

'I love Russian composers – they're so dramatic,' Greg told her. 'I expect it's in my blood.'

'I was raised listening to classical. My father likes to have a small orchestra at dinner parties.' Margaret's family were rich enough to make Greg feel a pauper by comparison. But he still had not met her parents, and he suspected they would disapprove of the illegitimate son of a famous Hollywood womanizer. 'What are you looking at?' she said.

'Nothing.' The McHughs had arrived. 'What's your perfume?'

'Chichi by Renoir.'

'I love it.'

The McHughs looked happy, a bright and prosperous young couple on holiday. Greg wondered if they were late because they had been making love in their hotel room.

Barney McHugh sat next to the man in the grey chalk stripe. Greg knew it was a cheap suit by the unnatural stiffness of the padded shoulders. The man did not look at the newcomers. The McHughs started to do a crossword, their heads leaning together intimately as they studied the newspaper Barney was holding. A few minutes later the conductor appeared.

The opening piece was by Saint-Saëns. German and Austrian composers had declined in popularity since war broke out, and concertgoers were discovering alternatives. There was a revival of Sibelius.

McHugh was probably a Communist. Greg knew this because J. Robert Oppenheimer had told him. Oppenheimer, a leading theoretical physicist from the University of California, was director of the Los Alamos laboratory and scientific leader of the entire Manhattan Project. He had strong Communist ties, though he insisted he had never joined the party.

Special Agent Bicks had said to Greg: 'Why does the army have to have all these pinkos? Whatever it is you're trying to achieve out there in the desert, aren't there enough bright young conservative scientists in America to do it?'

'No, there aren't,' Greg had told him. 'If there were, we would have hired them.'

Communists were sometimes more loyal to their cause than to their country, and might think it right to share the secrets of nuclear research with the Soviet Union. This would not be like giving information to the enemy. The Soviets were America's allies against the Nazis – in fact, they had done more of the fighting than all the other allies put together. All the same it was dangerous. Information intended for Moscow might find its way to Berlin. And anyone who thought about the postwar world for more than a minute could guess that the USA and the USSR might not always be friends.

The FBI thought Oppenheimer was a security risk and kept trying to persuade Greg's boss, General Groves, to fire him. But Oppenheimer was the outstanding scientist of his generation, so the General insisted on keeping him.

In an attempt to prove his loyalty, Oppenheimer had named McHugh as a possible Communist, and that was why Greg was tailing him.

The FBI were sceptical. 'Oppenheimer is blowing smoke up your ass,' Bicks had said.

Greg said: 'I can't believe it. I've known him for a year now.'

'He's a fucking Communist, like his wife and his brother and his sister-in-law.'

'He's working nineteen hours a day to build better weapons for American soldiers – what kind of traitor does that?'

Greg hoped McHugh did turn out to be a spy, for that would lift suspicion from Oppenheimer, bolster General Groves's credibility, and boost Greg's own status too.

He watched McHugh constantly throughout the first half of the concert, not wanting to take his eyes off him. The physicist did not look at the people either side of him. He seemed absorbed in the music, and only moved his gaze from the stage to look lovingly at Mrs McHugh, who was a pale English rose. Had Oppenheimer simply been wrong about McHugh? Or, more subtly, was Oppenheimer's accusation a distraction to divert suspicion away from himself?

Bicks was watching, too, Greg knew. He was upstairs in the dress circle. Perhaps he had seen something.

In the interval, Greg followed the McHughs out and stood in the same line for coffee. Neither the dowdy couple nor the two old ladies were anywhere nearby.

Greg felt thwarted. He did not know what to conclude. Were his suspicions unfounded? Or was it simply that this visit by the McHughs was innocent?

As he and Margaret were returning to their seats, Bill Bicks came up beside him. The agent was middle-aged, a little overweight, and losing his hair. He wore a light-grey suit that had sweat stains under the armpits. He said in a low voice: 'You were right.'

'How do you know?'

'That guy sitting next to McHugh.'

'In a grey striped suit?'

'Yeah. He's Nikolai Yenkov, a cultural attaché at the Soviet Embassy.'

Greg said: 'Good God!'

Margaret turned around. 'What?'

'Nothing,' Greg said.

Bicks moved away.

'You've got something on your mind,' she said as they took their seats. 'I don't believe you heard a single bar of the Saint-Saëns.'

'Just thinking about work.'

'Tell me it's not another woman, and I'll forget it.'

'It's not another woman.'

In the second half he began to feel anxious. He had seen no contact between McHugh and Yenkov. They did not speak, and Greg saw nothing pass from one to the other: no file, no envelope, no roll of film.

The symphony came to an end and the conductor took his bows. The audience began to file out. Greg's spy hunt was a washout.

In the lobby, Margaret went to the ladies' room. While Greg was waiting, Bicks approached him.

'Nothing,' Greg said.

'Me neither.'

'Maybe it's a coincidence, McHugh sitting by Yenkov.'

'There are no coincidences.'

'Perhaps there was a snag. A wrong code word, say.'

Bicks shook his head. 'They passed something. We just didn't see it.'

Mrs McHugh also went to the ladies' room and, like Greg, McHugh waited nearby. Greg studied him from behind a pillar. He had no briefcase, no raincoat under which to conceal a package or a file. But all the same, something about him was wrong. What was it?

Then Greg realized. 'The newspaper!' he said.

'What?'

'When Barney came in he was carrying a newspaper. They did the crossword while waiting for the show. Now he doesn't have it!'

'Either he threw it away – or he passed it to Yenkov, with something concealed inside.'

'Yenkov and his wife have left already.'

'They may still be outside.'

Bicks and Greg ran for the door.

Bicks shoved his way through the crowd still filing out of the exits. Greg stayed close behind. They reached the sidewalk outside and looked both ways. Greg could not see Yenkov, but Bicks had sharp eyes. 'Across the street!' he cried.

The attaché and his dowdy wife were standing at the kerb, and a black limousine was approaching them slowly.

Yenkov was holding a folded newspaper.

Greg and Bicks ran across the road.

The limousine stopped.

Greg was faster than Bicks and reached the far sidewalk first.

Yenkov had not noticed them. Unhurriedly, he opened the car door then stepped back to let his wife get in.

Greg threw himself at Yenkov. They both fell to the ground. Mrs Yenkov screamed.

Greg scrambled to his feet. The chauffeur had got out of the car and was coming around it, but Bicks yelled: 'FBI!' and held up his badge.

Yenkov had dropped the newspaper. Now he reached for it. But Greg was faster. He picked it up, stepped back, and opened it.

Inside was a sheaf of papers. The top one was a diagram. Greg recognized it immediately. It showed the working of an implosion trigger for a plutonium bomb. 'Jesus Christ,' he said. 'This is the very latest stuff!'

Yenkov jumped into the car, slammed the door, and locked it from the inside.

The chauffeur got back in and drove away.

(iii)

It was Saturday night, and Daisy's apartment in Piccadilly was heaving. There had to be a hundred people there, she thought, feeling pleased.

She had become the leader of a social group based on the American Red Cross in London. Every Saturday she gave a party for American servicemen, and invited nurses from St Bart's hospital to meet them. RAF pilots came too. They drank her unlimited Scotch and gin, and danced to Glenn Miller records on her gramophone. Conscious that it might be the last party the men ever attended, she did everything she could to make them happy – except kiss them, but the nurses did plenty of that.

Daisy never drank liquor at her own parties. She had too much to think about. Couples were always locking themselves in the toilet, and having to be dragged out because the room was needed for its regular purpose. If a really important general got drunk he had to be seen safely home. She often ran out of ice – she could not make her British staff understand how much ice a party needed.

For a while after she split up with Boy Fitzherbert her only friends had been the Leckwith family. Lloyd's mother, Ethel, had never judged her. Although Ethel was the height of respectability now, she had made mistakes in her past, and that made her more understanding. Daisy still went to Ethel's house in Aldgate every Wednesday evening, and drank cocoa around the radio. It was her favourite night of the week.

She had now been socially rejected twice, once in Buffalo and again in London, and the depressing thought occurred to her that it might be her fault. Perhaps she did not really belong in those prissy high-society groups, with their strict rules of conduct. She was a fool to be attracted to them.

The trouble was that she loved parties and picnics and sporting events and any gathering where people dressed up and had fun.

However, she now knew she did not need British aristocrats or old-money Americans to have fun. She had created her own society, and it was a lot more exciting than theirs. Some of the people who had refused to speak to her after she left Boy now hinted heavily that they would like an invitation to one of her famous Saturday nights. And many guests came to her apartment to let their hair down after an excruciatingly grand dinner in a palatial Mayfair residence.

Tonight was the best party so far, for Lloyd was home on leave.

He was openly living with her at the flat. She did not care what people thought: her reputation in respectable circles was already so bad that no further damage could be done. Anyway, the urgency of wartime love had driven many people to break the rules in similar ways. Domestic staff could sometimes be as rigid as duchesses about such things, but all Daisy's employees adored her, so she and Lloyd did not even pretend to be occupying separate bedrooms.

She loved sleeping with him. He was not as experienced as Boy, but he made up for that in enthusiasm – and he was eager to learn. Every night was a voyage of exploration in a double bed.

As they looked at their guests talking and laughing, drinking and smoking, dancing and smooching, Lloyd smiled at her and said: 'Happy?'

'Almost,' she said.

'Almost?'

She sighed. 'I want to have children, Lloyd. I don't care that we're not married. Well, I do care, of course, but I still want a baby.'

His face darkened. 'You know how I feel about illegitimacy.'

'Yes, you explained it to me. But I want some part of you to cherish if you die.'

'I'll do my best to stay alive.'

'I know.' But if her suspicion was correct, and he was working undercover in occupied territory, he could be executed, as German spies were executed in Britain. He would be gone, and she would have nothing left. 'It's the same for a million women, I realize that, but I can't face the thought of life without you. I think I'll die.'

'If I could make Boy divorce you I would.'

'Well, this is no kind of talk for a party.' She looked across the room. 'What do you know? I believe that's Woody Dewar!'

Woody was wearing a lieutenant's uniform. She went over and greeted him. It was strange to see him again after nine years – though he did not look much different, just older.

'There are thousands of American soldiers here now,' Daisy said as they foxtrotted to 'Pennsylvania Six-Five Thousand'. 'We must be about to invade France. What else?'

'The top brass certainly don't share their plans with greenhorn lieutenants,' Woody said. 'But like you I can't think of any other reason why I'm here. We can't leave the Russians to bear the brunt of the fighting much longer.'

'When do you think it will happen?'

'Offensives always begin in the summer. Late May or early June is everyone's best guess.'

'That soon!'

'But no one knows where.'

'Dover to Calais is the shortest sea crossing,' Daisy said.

'And for that reason the German defences are concentrated around Calais. So maybe we'll try to surprise them – say by landing on the south coast, near Marseilles.'

'Perhaps then it will be over at last.'

'I doubt it. Once we have a bridgehead, we still have to conquer France, then Germany. There's a long road ahead.'

'Oh, dear.' Woody seemed to need cheering up. And Daisy knew just the girl to do it. Isabel Hernandez was a Rhodes Scholar doing a Master's in history at St Hilda's College, Oxford. She was gorgeous, but the boys called her a ball-buster because she was so fiercely intellectual. However, Woody would be oblivious to that. 'Come over here,' she called to Isabel. 'Woody, this is my friend Bella. She's from San Francisco. Bella, meet Woody Dewar from Buffalo.'

They shook hands. Bella was tall, with thick dark hair and olive skin just like Joanne Rouzrokh's. Woody smiled at her and said: 'What are you doing here in London?' Daisy left them.

She served supper at midnight. When she could get American supplies it was ham and eggs; otherwise, cheese sandwiches. It provided a lull when people could talk, a bit like the interval at the theatre. She noticed that Woody Dewar was still with Bella Hernandez, and they seemed to be deep in conversation. She made sure everyone had what they needed then sat in a corner with Lloyd.

'I've decided what I'd like to do after the war, if I'm still alive,' he said. 'As well as marry you, that is.'

'What?'

'I'm going to try for Parliament.'

Daisy was thrilled. 'Lloyd, that's wonderful!' She put her arms around his neck and kissed him.

'It's too early for congratulations. I've put my name down for Hoxton, the constituency next to Mam's. But the local Labour party may not pick me. And if they do I may not win. Hoxton has a strong Liberal MP at the moment.'

'I want to help you,' she said. 'I could be your right-hand woman. I'll write your speeches – I bet I'd be good at that.'

'I'd love you to help me.'

'Then it's settled!'

The older guests left after supper, but the music continued and the drink never ran out, so the party became even more uninhibited. Woody was now slow-dancing with Bella: Daisy wondered if this was his first romance since Joanne.

The petting got heavier, and people began disappearing into the two bedrooms. They could not lock the doors – Daisy had taken the keys out – so there were sometimes several couples in the same room, but no one seemed to mind. Daisy had once found two people in the broom cupboard, fast asleep in each other's arms.

At one o'clock her husband arrived.

She had not invited Boy, but he showed up in the company of a couple of American pilots, and Daisy shrugged and let him in. He was amiably squiffy, and danced with several nurses, then politely asked her.

Was he just drunk, she wondered, or had he softened towards her? And if so, might he reconsider the divorce?

She consented, and they did the jitterbug. Most of the guests had no idea they were a separated husband and wife, but those who knew were amazed.

'I read in the papers that you bought another racehorse,' she said, making small talk.

'Lucky Laddie,' he said. 'Cost me eight thousand guineas – a record price.'

'I hope he's worth it.' She loved horses, and she had thought they would buy and train racehorses together, but he had not wanted to share that enthusiasm with his wife. It had been one of the frustrations of her marriage.

He read her mind. 'I disappointed you, didn't I?' he said.

'Yes.'

'And you disappointed me.'

That was a new thought to her. After a minute's reflection she said: 'By not turning a blind eye to your infidelities?'

'Exactly.' He was drunk enough to be honest.

She saw her opportunity. 'How long do you think we should punish one another?'

'Punish?' he said. 'Who's punishing anyone?'

'We're punishing each other by staying married. We should get divorced, as sensible people do.'

'Perhaps you're right,' he said. 'But this time on a Saturday night is not the best moment to discuss it.'

Her hopes rose. 'Why don't I come and see you?' she said. 'When we're both fresh – and sober.'

He hesitated. 'All right.'

She pressed her advantage eagerly. 'How about tomorrow morning?'

'All right.'

'I'll see you after church. Say twelve noon?'

'All right,' said Boy.

(iv)

As Woody was walking Bella home through Hyde Park, to a friend's flat in South Kensington, she kissed him.

He had not done this since Joanne died. At first he froze. He liked Bella a lot: she was the smartest girl he had met since Joanne. And the way she had clung to him while they were slow-dancing had let him know he could kiss her if he wanted to. All the same he had been holding back. He kept thinking about Joanne.

Then Bella took the initiative.

She opened her mouth and he tasted her tongue, but that only made him think of Joanne kissing him that way. It was only two and a half years since she had died.

His brain was forming words of polite rejection when his body took over. He was suddenly consumed with desire. He began to kiss her back hungrily.

She responded eagerly to his excess of passion. She took both his hands and put them on her breasts, which were large and soft. He groaned helplessly.

It was dark and he could hardly see but he realized, by the half-smothered sounds coming from the surrounding vegetation, that there were numerous couples doing similar things nearby.

She pressed her body against his, and he knew she could feel his erection. He was so excited he felt he would ejaculate any second. She seemed as madly aroused as he was. He felt her unbuttoning his pants with frantic fingers. Her hands were cool on his hot penis. She eased it out of his clothing, then, to his surprise and delight, she knelt down. As soon as her lips closed over the head, he spurted uncontrollably into her mouth. She sucked and licked feverishly as he did so.

When the climax was over she continued to kiss it until it softened. Then she gently put it away and stood up.

'That was exciting,' she whispered. 'Thank you.'

He had been about to thank her. Instead, he put his arms around her and pulled her close. He felt so grateful to her that he could have wept. He had not realized how badly he needed a woman's affection tonight. Some kind of shadow had been lifted from him. 'I can't tell you . . .' he began, but he could not find words to explain how much it meant to him.

'Then don't,' she said. 'I know, anyway. I could feel it.'

They walked to her building. At the door he said: 'Can we—'

She put a finger on his lips to silence him. 'Go and win the war,' she said.

Then she went inside.

(v)

When Daisy went to a Sunday service, which was not often, she now avoided the elite churches of the West End, whose congregations had snubbed her, and instead caught the Tube to Aldgate and attended the Calvary Gospel Hall. The doctrinal differences were wide, but they did not matter to her. The singing was better in the East End.

She and Lloyd arrived separately. People in Aldgate knew who she was, and they liked having a rogue aristocrat sitting on one of their cheap seats; but it would have been pushing their tolerance too far for a married-and-separated woman to walk in on the arm of her paramour. Ethel's brother Billy had said: 'Jesus did not condemn the adulteress, but he did tell her to sin no more.'

During the service she thought about Boy. Had he really meant last night's conciliatory words, or were they just the softness of the drunken moment? Boy had even shaken hands with Lloyd as he left. Surely that meant forgiveness? But she told herself not to let her hopes rise. Boy was the most completely self-absorbed person she had ever known, worse than his father or her brother Greg.

After church Daisy often went to Eth Leckwith's house for Sunday dinner, but today she left Lloyd to his family and hurried away.

She returned to the West End and knocked on the door of her husband's house in Mayfair. The butler showed her into the morning room.

Boy came in shouting. 'What the hell is this?' he roared, and he threw a newspaper at her.

668

She had seen him in this mood plenty of times, and she was not afraid of him. Only once had he raised a hand to strike her. She had seized a heavy candlestick and threatened to bop him.' It had not happened again.

Though not scared, she was disappointed. He had been in such a good mood last night. But perhaps he might still listen to reason.

'What has happened to displease you?' she said calmly.

'Look at that bloody paper.'

She bent and picked it up. It was today's edition of the *Sunday Mirror*, a popular left-wing tabloid. On the front page was a photograph of Boy's new horse, Lucky Laddie, and the headline:

LUCKY LADDIE WORTH
28 COAL MINERS

The story of Boy's record-breaking purchase had appeared in yesterday's press, but today the *Mirror* had an outraged opinion piece, pointing out that the price of the horse, £8,400, was exactly twenty-eight times the £300 standard compensation paid to the widow of a miner who died in a pit accident.

And the Fitzherbert family wealth came from coal mines.

Boy said: 'My father is furious. He was hoping to be Foreign Secretary in the postwar government. This has probably ruined his chances.'

Daisy said in exasperation: 'Boy, kindly explain why this is my fault?'

'Look who wrote the damned thing!'

Daisy looked.

By Billy Williams
Member of Parliament for Aberowen

Boy said: 'Your boyfriend's uncle!'

'Do you imagine he consults me before writing his articles?'

He wagged a finger. 'For some reason, that family hates us!'

'They think it's unfair that you should make so much money from coal, when the miners themselves get such a raw deal. There is a war on, you know.'

'You live on inherited money,' he said. 'And I didn't see much sign of wartime austerity at your Piccadilly apartment last night.'

'You're right,' she said. 'But I gave a party for the troops. You spent a fortune on a horse.'

'It's my money!'

'But you got it from coal.'

'You've spent so much time in bed with that Williams bastard that you've become a bloody Bolshevik.'

'And that's one more thing that's driving us apart. Boy, do you really want to stay married to me? You could find someone who suits you. Half the girls in London would love to be Viscountess Aberowen.'

'I won't do anything for that damned Williams family. Anyway, I heard last night that your boyfriend wants to be a Member of Parliament.'

'He'll make a great one.'

'Not with you in tow. He won't even get elected. He's a bloody socialist. You're an ex-Fascist.'

'I've thought about this. I know it's a bit of a problem—'

'Problem? It's an insuperable barrier. Wait till the papers get that story! You'll be crucified the way I've been today.'

'I suppose you'll give the story to the *Daily Mail.*'

'I won't need to – his opponents will do that. You mark my words. With you by his side, Lloyd Williams doesn't stand a bloody chance.'

(vi)

For the first five days of June, Lieutenant Woody Dewar and his platoon of paratroopers, plus a thousand or so others, were isolated at an airfield somewhere north-west of London. An aircraft hangar had been converted into a giant dormitory with hundreds of cots in long rows. There were movies and jazz records to entertain them while they waited.

Their objective was Normandy. By means of elaborate deception plans, the Allies had tried to convince the German High Command that the target would be two hundred miles north-east at Calais. If the Germans had been fooled, the invasion force would meet relatively light resistance, at least for the first few hours.

The paratroopers were to be the first wave, in the middle of the night. The second wave would be the main force of 130,000 men, aboard a fleet of five thousand vessels, landing on the beaches of Normandy at dawn. By then, the paratroopers should have already destroyed inland strongpoints and taken control of key transport links.

Woody's platoon had to capture a bridge across a river in a small town called Eglise-des-Soeurs, ten miles inland. When they had done so, they had to keep control of the bridge, blocking any German units that might be sent to reinforce the beach, until the main invasion force caught up with them. At all costs they must prevent the Germans from blowing up the bridge.

While they waited for the green light, Ace Webber ran a marathon poker game, winning a thousand dollars and losing it again. Lefty Cameron obsessively cleaned and oiled his lightweight M1 semi-automatic carbine, the paratrooper model with a folding stock. Lonnie Callaghan and Tony Bonanio, who did not like one another, went to mass together every day. Sneaky Pete Schneider sharpened the commando knife he had bought in London until he could have shaved with it. Patrick Timothy, who looked like Clark Gable and had a similar moustache, played a ukulele, the same tune over and over again, driving everybody crazy. Sergeant Defoe wrote long letters to his wife, then tore them up and started again. Mack Trulove and Smoking Joe Morgan cropped and shaved each other's hair, believing that would make it easier for the medics to deal with head injuries.

Most of them had nicknames. Woody had discovered that his own was Scotch.

D-Day was set for Sunday 4 June, then postponed because of bad weather.

On Monday 5 June, in the evening, the colonel made a speech. 'Men!' he shouted. 'Tonight is the night we invade France!'

They roared their approval. Woody thought it was ironic. They were safe and warm here, but they could hardly wait to get over there, jump out of airplanes, and land in the arms of enemy troops who wanted to kill them.

They were given a special meal, all they could eat, steak, pork, chicken, fries, ice cream. Woody did not want any. He had more idea than the men of what was ahead of him, and he did not want to do it on a full stomach. He got coffee and a donut. The coffee was American, fragrant and delicious, unlike the frightful brew served up by the British, when they had any coffee at all.

He took off his boots and lay down on his cot. He thought about Bella Hernandez, her lopsided smile and her soft breasts.

Next thing he knew, a hooter was sounding.

For a moment, Woody thought he was waking from a bad dream in which he was going into battle to kill people. Then he realized it was true.

They all put on their jump suits and assembled their equipment. They had too much. Some of it was essential: a carbine with 150 rounds of .30 ammunition; anti-tank grenades; a small bomb known as a Gammon grenade; K-rations; water purifying tablets; a first-aid kit with morphine. Other things they might have done without: an entrenching tool, shaving kit, a French phrase book. They were so overloaded that the smaller men struggled to walk to the planes lined up on the runway in the dark.

Their transport aircraft were C-47 Skytrains. To Woody's surprise, he saw by the dim lights that they had all been painted with distinctive black and white stripes. The pilot of his aircraft, a bad-tempered Midwesterner called Captain Bonner, said: 'That's to prevent us being shot down by our own goddamn side.'

Before boarding, the men were weighed. Donegan and Bonanio both had disassembled bazookas packed in bags that dangled from their legs, adding eighty pounds to their weight. As the total mounted, Captain Bonner became angry. 'You're overloading me!' he snarled at Woody. 'I won't get this motherfucker off the ground!'

'Not my decision, Captain,' Woody said. 'Talk to the colonel.'

Sergeant Defoe boarded first and went to the front of the plane, taking a seat beside the open arch leading to the flight deck. He would be the last to jump. Any man who developed a last-minute reluctance to leap into the night would be helped along with a good shove from Defoe.

Donegan and Bonanio, carrying the leg bags holding their bazookas as well as everything else, had to be helped up the steps. Woody as platoon commander boarded last. He would be first out, and first on the ground.

The interior was a tube with a row of simple metal seats on either side. The men had trouble fastening seat belts around their equipment, and some did not bother. The door closed and the engines roared into life.

Woody felt excited as well as scared. Against all reason, he felt eager for the battle to come. To his surprise he found himself impatient to get down on the ground, meet the enemy, and fire his weapons. He wanted the waiting to be over.

He wondered if he would ever see Bella Hernandez again.

He thought he could feel the plane straining as it lumbered down the runway. Painfully, it picked up speed. It seemed to rumble along on the ground for ever. Woody found himself wondering how long the damn runway was anyhow. Then at last it lifted. There was little sensation of flying, and he thought the plane must be remaining just a few feet above the ground. Then he looked out. He was sitting by the rearmost of the seven windows, next to the door, and he could see the shrouded lights of the base dropping away. They were airborne.

The sky was overcast, but the clouds were faintly luminous, presumably because the moon had risen beyond them. There was a blue light at the tip of each wing, and Woody could see as his plane moved into formation with others, forming a giant V shape.

The cabin was so noisy that men had to shout into one another's ears to be heard, and conversation soon ceased. They all shifted in their hard seats, trying in vain to get comfortable. Some closed their eyes, but Woody doubted that anyone actually slept.

They were flying low, not much above a thousand feet, and occasionally Woody saw the dull pewter gleam of rivers and lakes. At one point he glimpsed a crowd of people, hundreds of faces all staring up at the planes roaring overhead. Woody knew that more than a thousand aircraft were flying over southern England at the same time, and he realized it must be a remarkable sight. It occurred to him that those people were watching history being made, and he was part of it.

After half an hour they crossed the English beach resorts and were over the sea. For a moment the moon shone through a break in the cloud, and Woody saw the ships. He could hardly believe what he was looking at. It was a floating town, vessels of all sizes sailing in ragged rows like assorted houses in city streets, thousands of them, as far as the eye could see. Before he could call the attention of his comrades to the remarkable sight, the clouds covered the moon again and the vision was gone, like a dream.

The planes headed right in a long curve, aiming to hit France to the west of the drop area and then follow the coastline eastwards, checking position by terrain features to ensure the paratroopers landed where they should.

The Channel Islands, British though closer to France, had been occupied by Germany at the end of the Battle of France in 1940; and now, as the armada overflew the islands, German anti-aircraft guns opened fire. At such a low altitude the Skytrains were terribly vulnerable. Woody realized he could be killed even before he reached the battlefield. He would hate to die pointlessly.

Captain Bonner zigzagged to avoid the flak. Woody was glad he did, but the effect on the men was unfortunate. They all felt airsick, Woody included. Patrick Timothy was the first to succumb, and vomited on the floor. The foul smell made others feel worse. Sneaky Pete threw up next, then several men all at once. They had stuffed themselves with steak and ice cream, all of which now came back up. The stink was appalling and the floor became disgustingly slippery.

The flight path straightened as they left the islands behind. A few minutes later the French coast appeared. The plane banked and turned left. The co-pilot got up from his seat and spoke in the ear of Sergeant Defoe, who turned to the platoon and held up ten fingers. Ten minutes to drop.

The plane slowed from its cruising speed of 160mph to the approximate speed for a parachute jump, about 100mph.

Suddenly they entered fog. It was heavy enough to blot out the blue light at the tip of the wing. Woody's heart raced. For planes flying in close formation this was very dangerous. How tragic it would be to die in a plane crash, not even in combat. But Bonner could do nothing but fly straight and level and hope for the best. Any change of direction would cause a collision.

The plane left the fog bank as suddenly as it had entered it. To either side, the other planes were still miraculously in formation.

Almost immediately, anti-aircraft fire broke out, the flak exploding in deadly blossoms among the serried planes. In these circumstances, Woody knew, the pilot's orders were to maintain speed and fly straight to the target zone. But Bonner defied orders and broke formation. The roar of the engines went to full throttle. He began to zigzag again. The nose of the plane dipped as he tried for more speed. Looking out of the window, Woody saw that many other pilots had been equally undisciplined. They could not control the urge to save their own lives.

The red light went on over the door: four minutes to go.

Woody felt certain the crew had put the light on too soon, desperate to dump their troops and fly to safety. But they had the charts and he could not argue.

He got to his feet. 'Stand up and hook!' he yelled. Most of the men could not hear him, but they knew what he was saying. They got up, and each man clipped his static line to the overhead cable, so that he could not be thrown through the door accidentally. The door opened, and the wind roared in. The plane was still going too fast. Jumping at this speed was unpleasant, but that was not the main problem. They would land farther apart, and it would take Woody much longer to find his men on the ground. His approach to his objective would be delayed. He would begin his mission behind schedule. He cursed Bonner.

The pilot continued to bank one way then the other, dodging flak. The men struggled to keep their footing on a floor that was slimy with vomit.

Woody looked out of the open door. Bonner had lost height while trying to gain speed, and the plane was now at about five hundred feet – too low. There might not be enough time for the parachutes to open fully before the men hit the ground. He hesitated, then beckoned his sergeant forward.

Defoe stood beside him and looked down, then shook his head. He put his mouth to Woody's ear and shouted: 'Half our men will break their ankles if we jump at this height. The bazooka carriers will kill themselves.'

Woody made a decision.

'Make sure no one jumps!' he yelled at Defoe.

Then he unhooked his static line and went forward, pushing through the double row of standing men, to the flight deck. There were three crew. Yelling at the top of his voice, Woody said: 'Climb! Climb!'

Bonner yelled: 'Get back there and jump!'

'No one is going to jump at this altitude!' Woody leaned over and pointed at the altimeter, which showed 480 feet. 'It's suicide!'

'Get off the flight deck, Lieutenant. That's an order.'

Woody was outranked, but he stood his ground. 'Not until you gain height.'

'We'll be past your target zone if you don't jump now!'

Woody lost his temper. 'Climb, you dumb fuck! Climb!'

Bonner looked furious, but Woody did not move. He knew the pilot would not want to return home with a full plane. He would face a military inquiry into what had gone wrong. Bonner had disobeyed too many orders tonight for that. With a curse, he jerked the control lever back. The nose went up immediately, and the aircraft began to gain height and lose speed.

'Satisfied?' Bonner snarled.

'Hell, no.' Woody was not going to go aft now and give Bonner the chance to reverse the manoeuvre. 'We jump at a thousand feet.'

Bonner went to full throttle. Woody kept his eyes on the altimeter.

When it touched 1,000 he went aft. He pushed through his men, reached the door, looked out, gave the men the thumbs-up, and jumped.

His chute opened immediately. He dropped fast through the air while it spread its dome, then his fall was arrested. Seconds later he hit water. He suffered a split-second of panic, fearing that the cowardly Bonner had dropped them all in the sea. Then his feet touched solid ground, or at least soft mud, and he understood that he had come down in a flooded field.

The silk of the parachute fell around him. He struggled out of its folds and unfastened his harness.

Standing in two feet of water, he looked around. This was either a water meadow or, more likely, a field that had been flooded by the Germans to impede an invasion force. He saw no one, enemy or friend, and no animals either, but the light was poor.

He checked his watch – it was 3.40 a.m. – then looked at his compass and oriented himself.

Next he took his M1 carbine out of its case and unfolded the stock. He snapped a 15-round magazine into the slot, then worked the slide to chamber a round. Finally, he rotated the safety lever into the disengaged position.

He reached into a pocket and took out a small tin object like a child's toy. When pressed, it made a distinctive clicking sound. It had been issued to everyone so that they could recognize each other in the dark without resorting to giveaway English passwords.

When he was ready, he looked around again.

Experimentally, he pressed the click twice. After a moment, an answering click came from directly ahead.

He splashed through the water. He smelled vomit. In a low voice he said: 'Who's there?'

'Patrick Timothy.'

'Lieutenant Dewar here. Follow me.'

Timothy had been second to jump, so Woody figured if he continued in the same direction he had a good chance of finding the others.

Fifty yards along he bumped into Mack and Smoking Joe, who had found one another.

They emerged from the water on to a narrow road, and found their first casualties. Lonnie and Tony, with their bazookas in leg bags, had both landed too hard. 'I think Lonnie's dead,' said Tony. Woody checked: he was right. Lonnie was not breathing. He looked as if he had broken his neck. Tony himself could not move, and Woody thought the man's leg was broken. He gave him a shot of morphine, then dragged him off the road into the next field. Tony would have to wait there for the medics.

Woody ordered Mack and Smoking Joe to hide Lonnie's body, for fear it might lead the Germans to Tony.

He tried to see the landscape around him, straining to recognize something that corresponded to his map. The task seemed impossible, especially in the dark. How was he going to lead these men to the objective if he did not know where he was? The only thing of which he could be reasonably sure was that they had not landed where they were supposed to.

He heard a strange noise and, a moment later, he saw a light.

He motioned the others to duck down.

The paratroopers were not supposed to use flashlights, and French people were subject to a curfew, so the person approaching was probably a German soldier.

In the dim light Woody saw a bicycle.

He stood up and aimed his carbine. He thought of shooting the rider immediately, but could not bring himself to do it. Instead he shouted: '*Halt! Arretez!*'

The cycle stopped. 'Hello, Loot,' said the rider, and Woody recognized the voice of Ace Webber.

Woody lowered his weapon. 'Where did you get the bike?' he said incredulously.

'Outside a farmhouse,' Ace said laconically.

Woody led the group the way Ace had come, figuring that the others were more likely to be in that direction than any other. He looked anxiously for terrain features to match his map, but it was too dark. He felt useless and stupid. He was the officer. He had to solve such problems.

He picked up more of his platoon on the road, then they came to a windmill. Woody decided he could not blunder around any longer, so he went to the mill house and hammered on the door.

An upstairs window opened, and a man said in French: 'Who is it?'

'The Americans,' Woody said. '*Vive la France!*'

'What do you want?'

'To set you free,' Woody said in schoolboy French. 'But first I need some help with my map.'

The miller laughed and said: 'I'm coming down.'

A minute later Woody was in the kitchen, spreading his silk map over the table under a bright light. The miller showed him where he was. It was not as bad as Woody had feared. Despite Captain Bonner's panic, they were only four miles north-east of Eglise-des-Soeurs. The miller traced the best route on the map.

A girl of about thirteen crept into the room in a nightdress. 'Maman says you're American,' she said to Woody.

'That's right, mademoiselle,' he said.

'Do you know Gladys Angelus?'

Woody laughed. 'As it happens, I did meet her once, at the apartment of a friend's father.'

'Is she really, really beautiful?'

'Even more beautiful than she looks in the movies.'

'I knew it!'

The miller offered him wine. 'No, thanks,' said Woody. 'Maybe after we've won.' The miller kissed him on both cheeks.

Woody went back outside and led his platoon away, heading in the direction of Eglise-des-Soeurs. Including himself, nine of the original eighteen were now together. They had suffered two casualties, Lonnie dead and Tony wounded, and seven more had not yet appeared. His orders were not to spend too much time trying to find everyone. As soon as he had enough men to do the job, he was to proceed to the target.

One of the missing seven showed up right away. Sneaky Pete emerged from a ditch and joined the group with a casual 'Hi, gang,' as if it was the most natural thing in the world.

'What were you doing in there?' Woody asked him.

'I thought you were German,' Pete said. 'I was hiding.'

Woody had seen the pale gleam of parachute silk in the ditch. Pete must have been hiding there since he landed. He had obviously panicked and curled up in a ball. But Woody pretended to accept his story.

The one Woody really wanted to find was Sergeant Defoe. He was an experienced soldier, and Woody had been planning to rely heavily on him. But he was nowhere to be seen.

They were approaching a crossroads when they heard noises. Woody identified the sound of an engine idling, and two or three voices in conversation. He ordered everyone down on their hands and knees, and the platoon advanced crawling.

Up ahead, he saw that a motorcycle rider had stopped to talk to two men on foot. All three were in uniform. They were speaking German.

There was a building at the crossroads, perhaps a small tavern or a bakery.

He decided to wait. Perhaps they would leave. He wanted his group to move silently and unobserved for as long as possible.

After five minutes he ran out of patience. He turned around. 'Patrick Timothy!' he hissed.

Someone else said: 'Pukey Pat! Scotch wants you.'

Timothy crawled forward. He still smelled of vomit, and now it had become his name.

Woody had seen Timothy play baseball, and knew he could throw hard and accurately. 'Hit that motorcycle with a grenade,' Woody said.

Timothy took a grenade from his pack, pulled the pin, and lobbed it.

There was a clang. One of the men said in German: 'What was that?' Then the grenade detonated.

There were two explosions. The first knocked all three Germans to the ground. The second was the motorcycle's fuel tank blowing up, and it sent a starburst of flame that burned the men, leaving a stink of scorched flesh.

'Stay where you are!' Woody shouted to his platoon. He watched the building. Was there anyone inside? During the next five minutes, no one opened a window or a door. Either the place was empty, or the occupants were hiding under their beds.

Woody got to his feet and waved the platoon on. He felt strange as he stepped over the grisly bodies of the three Germans. He had ordered their deaths – men who had mothers and fathers, wives or girlfriends, perhaps sons and daughters. Now each man was an ugly mess of blood and burned flesh. Woody should have felt triumphant. It was his first encounter with the enemy, and he had vanquished them. But he just felt a bit sick.

Past the crossroads, he set a brisk pace, and ordered no talking or smoking. To keep up his strength he ate a bar of D-ration chocolate, which was a bit like builder's putty with sugar added.

After half an hour he heard a car and ordered everyone to hide in the fields. The vehicle was travelling fast with its headlights on. It was probably German, but the Allies were sending over jeeps by glider, along with anti-tank guns and other artillery, so it was just possible this was a friendly vehicle. He lay under a hedge and watched it go by.

It went too fast for him to identify it. He wondered whether he should have ordered the platoon to shoot it up. No, he thought, on balance they did better to focus on their mission.

They passed through three hamlets that Woody was able to identify on his map. Dogs barked occasionally but no one came to investigate. Doubtless the French had learned to mind their own business under enemy occupation. It was eerie, creeping along foreign roads in the dark, armed to the teeth, passing quiet houses where people slept unconscious of the deadly firepower outside their windows.

At last they came to the outskirts of Eglise-des-Soeurs. Woody ordered a short rest. They entered a little stand of trees and sat on the ground. They drank from their canteens and ate rations. Woody still would not permit smoking: the glow of a cigarette could be seen from surprisingly far.

The road they were on should lead straight to the bridge, he reckoned. There was no hard information about how the bridge was guarded. Since the Allies had decided it was important, he assumed the Germans thought the same, therefore some security was likely; but it might be anything from one man with a rifle to a whole platoon. Woody could not plan the assault until he saw the target.

After ten minutes he moved them on. The men did not have to be nagged about silence now: they sensed the danger. They trod quietly along the street, past houses and churches and shops, keeping to the sides, peering into the gloomy night, jumping at the least sound. A sudden loud cough from an open bedroom window almost caused Woody to fire his carbine.

Eglise-des-Soeurs was a large village rather than a small town, and Woody saw the silver glint of the river sooner than he expected. He raised a hand for them all to halt. The main street led gently downhill at a slight angle to the bridge, so he had a good view. The waterway was about a hundred feet wide, and the bridge had a single curved span. It must be an old structure, he guessed, because it was so narrow that two cars could not have passed.

The bad news was that there was a pillbox at each end, twin concrete domes with horizontal shooting slits. A pair of sentries patrolled the bridge between the pillboxes. They stood one at each end. The nearer one was speaking through a firing slit, presumably chatting to whomever was inside. Then they both walked to the middle, where they looked over the parapet at the black water flowing beneath. They did not appear very tense, so Woody deduced they had not yet learned that the invasion had begun. On the other hand, they were not slacking. They were awake and moving and looking about them with some degree of alertness.

Woody could not guess how many men were inside, nor how they were armed. Were there machine guns behind those slits, or just rifles? It would make a big difference.

Woody wished he had some experience of battle. How was he supposed to deal with this situation? He guessed there must be thousands of men like him, new junior officers who just had to make it up as they went along. If only Sergeant Defoe were here.

The easy way to neutralize a pillbox was to sneak up and put a grenade through one of the slits. A good man could probably crawl to the nearer one unobserved. But Woody needed to take out both at the same time – otherwise the attack on the first would forewarn the occupants of the second.

How could he reach the farther pillbox without being seen by the patrolling sentries?

He sensed his men getting restless. They did not like to think their leader might be unsure as to what to do next.

'Sneaky Pete,' he said. 'You'll crawl up to that nearest pillbox and put a grenade through the slit.'

Pete looked terrified, but he said: 'Yes, sir.'

Next, Woody named the two best shots in the platoon. 'Smoking Joe and Mack,' he said. 'Choose one each of the sentries. As soon as Pete deploys his grenade, take the sentries out.'

The two men nodded and hefted their weapons.

In the absence of Defoe, he decided to make Ace Webber his deputy. He named four others and said: 'Go with Ace. As soon as the shooting starts, run like hell across the bridge and storm the pillbox on the other side. If you're quick enough you'll catch them napping.'

'Yes, sir,' said Ace. 'The bastards won't know what's hit them.' His aggression was masking fear, Woody guessed.

'Everyone not in Ace's group, follow me into the near pillbox.'

Woody felt bad about giving Ace and those with him the more dangerous assignment, and himself the relative safety of the nearer pillbox; but it had been drummed into him that an officer must not risk his life unnecessarily, for then he might leave his men leaderless.

They walked towards the bridge, Pete in the lead. This was a dangerous moment. Ten men going along a street together could not remain unnoticed for long, even at night. Anyone looking carefully in their direction would sense movement.

If the alarm was raised too soon, Sneaky Pete might not get to the pillbox, and then the platoon would lose the advantage of surprise.

It was a long walk.

Pete reached a corner and stopped. Woody guessed he was waiting for the near sentry to leave his post outside the pillbox and walk to the middle.

The two sharpshooters found cover and settled in.

Woody dropped to one knee and signalled the others to do likewise. They all watched the sentry.

The man took a long pull on his cigarette, dropped it, trod on the end to put it out, and blew a long cloud of smoke. Then he eased himself upright, settled his rifle strap on his shoulder, and started walking.

The sentry on the far side did the same.

Pete ran the next block and came to the end of the street. He got down on his hands and knees and crawled rapidly across the road. He reached the pillbox and stood up.

No one had noticed. The two sentries were still approaching one another.

Pete took out a grenade and pulled the pin. Then he waited a few seconds. Woody guessed he did not want the men inside to have time to throw the grenade out again.

Pete reached around the curve of the dome and gently dropped the grenade inside.

Joe and Mack's carbines barked. The nearer sentry fell, but the farther one was unhurt. To his credit he did not turn and run, but courageously went down on one knee and unslung his rifle. He was too slow, though: the carbines spoke again, almost simultaneously, and he fell without firing.

Then Pete's grenade exploded inside the nearer pillbox with a muffled thump.

Woody was already running full pelt, and the men were close behind him. Within seconds he reached the bridge.

The pillbox had a low wooden door. Woody flung it open and stepped inside. Three men in German uniforms were dead on the floor.

He moved to a firing slit and looked out. Ace and his four men were haring across the short bridge, shooting at the farther pillbox as they ran. The bridge was only a hundred feet long, but that proved to be fifty feet too much. As they reached the middle, a machine gun opened up. The Americans were trapped in a narrow corridor with no cover. The machine gun clacked insanely and in seconds all five of them had fallen. The gun continued to rake them for several seconds, to be certain they were dead – and, in the process, making sure of the two German sentries too.

When it stopped, they were all still.

Silence fell.

Beside Woody, Lefty Cameron said: 'Jesus Christ Almighty.'

Woody could have wept. He had sent ten men to their deaths, five Americans and five Germans, yet he had failed to achieve his objective. The enemy still held the far end of the bridge and could stop Allied forces crossing it.

He had four men left. If they tried again, and ran across the bridge together, they would all be killed. He needed a new plan.

He studied the townscape. What could he do? He wished he had a tank.

He had to act fast. There might well be enemy troops elsewhere in the town. They would have been alerted by the gunfire. They would respond soon. He could deal with them if he had both pillboxes. Otherwise he would be in trouble.

If his men could not cross the bridge, he thought desperately, perhaps they could swim the river. He decided to take a quick look at the bank. 'Mack and Smoking Joe,' he said. 'Fire at the other pillbox. See if you can get a bullet through the slit. Keep them busy while I scout around.'

The carbines opened up and he went out through the door.

He was able to shelter behind the near pillbox while he looked over the parapet at the upstream bank. Then he had to scuttle across the road to see the other edge. However, no fire came from the enemy position.

There was no river wall. Instead an earth slope went down to the water. It looked the same on the far bank, he thought, though there was not enough light to be sure. A good swimmer might get across. Under the span of the arch he would not be easy to see from the enemy position. Then he could repeat on the far side what Sneaky Pete had done this side, and grenade the pillbox.

Looking at the structure of the bridge he had a better idea. Below the level of the parapet was a stone ledge a foot wide. A man with steady nerves could crawl across, all the time remaining out of sight.

He returned to the captured pillbox. The smallest man was Lefty Cameron. He was also feisty, not the type to get the shakes. 'Lefty,' said Woody. 'There's a hidden ledge that runs across the outside of the bridge below the parapet. Probably used by workmen doing repairs. I want you to crawl across and grenade the other pillbox.'

'You bet,' said Lefty.

It was a gutsy response from someone who had just seen five comrades killed.

Woody turned to Mack and Smoking Joe and said: 'Give him cover.' They began to shoot.

Lefty said: 'What if I fall in?'

'It's only fifteen or twenty feet above the water at most,' Woody said. 'You'll be fine.'

'Okay,' said Lefty. He went to the door. 'I can't swim, though,' he said. Then he was gone.

Woody saw him dart across the road. He looked over the parapet, then straddled it and eased down the other side until he was lost to view.

'Okay,' he said to the others. 'Hold your fire. He's on his way.'

They all stared out. Nothing moved. It was dawn, Woody realized: the town was coming more clearly into view. But none of the inhabitants showed themselves: they knew better. Perhaps German troops were mobilizing in some neighbouring street, but he could hear nothing. He realized he was listening for a splash, fearful that Lefty would fall in the river.

A dog came trotting across the bridge, a medium-size mongrel with a curled tail that stuck up jauntily. It sniffed the dead bodies with curiosity, then moved on purposefully, as if it had an important rendezvous elsewhere. Woody watched it pass the far pillbox and continue into the other side of the town.

Dawn meant the main force was now landing on the beaches. Someone had said it was the largest amphibious attack in the history of warfare. He wondered what kind of resistance they were meeting. There was no one more vulnerable than an infantryman loaded with gear splashing through the shallows, the flat beach ahead of him offering a clear field of fire to gunners in the dunes. Woody felt grateful for this concrete pillbox.

Lefty was taking a long time. Had he fallen in the water quietly? Could something else have gone wrong?

Then Woody saw him, a slim khaki form bellying over the parapet of the bridge at the far end. Woody held his breath. Lefty dropped to his knees, crawled to the pillbox, and came upright with his back flat against the curved concrete. With his left hand he drew out a grenade. He pulled the pin, waited a couple of seconds, then reached around and threw the grenade through the slit.

Woody heard the boom of the explosion and saw a flash of lurid light from the firing slits. Lefty raised his arms above his head like a champion.

'Get back under cover, asshole,' Woody said, though Lefty could not hear him. There could be a German soldier hiding in a nearby building waiting to avenge the deaths of his friends.

But no shot rang out, and after a brief victory dance Lefty went inside the pillbox, and Woody breathed more easily.

However, he was not yet fully secure. At this point a sudden sally by a couple of dozen Germans could win the bridge back. Then it would all have been in vain.

He forced himself to wait another minute to see if any enemy troops showed themselves. Still nothing moved. It was beginning to look as if there were no Germans in Eglise-des-Soeurs other than those manning the bridge: they were probably relieved every twelve hours from a barracks a few miles away.

'Smoking Joe,' he said. 'Get rid of the dead Germans. Throw them in the river.'

Joe dragged the three bodies out of the pillbox and disposed of them, then did the same with the two sentries.

'Pete and Mack,' Woody said. 'Go over to the other pillbox and join Lefty. Make sure the three of you stay alert. We haven't killed all the Germans in France yet. If you see enemy troops approaching your position, don't hesitate, don't negotiate, just shoot them.'

The two men left the pillbox and walked briskly across the bridge to the far end.

There were now three Americans in the far pillbox. If the Germans tried to retake the bridge they would have a hard time of it, especially in the growing light.

Woody realized that the dead Americans on the bridge would

forewarn any approaching enemy forces that the pillboxes had been captured. Otherwise he might retain an element of surprise.

That meant he had to get rid of the American corpses too.

He told the others what he was going to do, then stepped outside.

The morning air tasted fresh and clean.

He walked to the middle of the bridge. He checked each body for a pulse, but there was no doubt: they were all dead.

One by one, he picked up his comrades and dropped them over the parapet.

The last one was Ace Webber. As he hit the water, Woody said: 'Rest in peace, buddies.' He stood still for a minute with his head bent and his eyes closed.

When he turned around, the sun was coming up.

(vii)

The great fear of Allied planners was that the Germans would rapidly reinforce their troops in Normandy, and mount a powerful counter-attack that would drive the invaders back into the sea, in a repeat of the Dunkirk disaster.

Lloyd Williams was one of the people trying to make sure that did not happen.

His job helping escaped prisoners get home had low priority after the invasion, and he was now working with the French Resistance.

At the end of May the BBC broadcast coded messages that triggered a campaign of sabotage in German-occupied France. During the first few days of June hundreds of telephone lines were cut, usually in hard-to-find places. Fuel depots were set on fire, roads were blocked by trees, and tyres were slashed.

Lloyd was assisting the railwaymen, who were strongly Communist and called themselves *Résistance Fer*. For years they had maddened the Nazis with their sly subversion. German troop trains somehow got diverted down obscure branch lines and sent many miles out of their way. Engines broke down unaccountably and carriages were derailed. It was so bad that the occupiers brought railwaymen from Germany to run the system. But the disruption got worse. In the spring of 1944 the railwaymen began to damage their own network. They blew up tracks and sabotaged the heavy lifting cranes required for moving crashed trains.

The Nazis did not take this lying down. Hundreds of railwaymen were executed, and thousands deported to camps. But the campaign escalated, and by D-Day rail traffic in some parts of France had come to a halt.

Now, on D-Day plus one, Lloyd lay at the summit of an embankment beside the main line to Rouen, capital city of Normandy, at a point where the track entered a tunnel. From his vantage point he could see approaching trains a mile away.

With Lloyd were two others, codenamed Legionnaire and Cigare. Legionnaire was leader of the Resistance in this neighbourhood. Cigare was a railwayman. Lloyd had brought the dynamite. Supplying weaponry was the main role played by the British in the French Resistance.

The three men were half hidden by long grass dotted with wild flowers. It was the kind of place to bring a girl on a fine day such as this, Lloyd thought. Daisy would like it.

A train appeared in the distance. Cigare scrutinized it as it came nearer. He was about sixty, wiry and small, with the lined face of a heavy smoker. When the train was still a quarter of a mile away he shook his head in negation. This was not the one they were waiting for. The engine passed them, puffing smoke, and entered the tunnel. It was hauling four passenger coaches, all full, carrying a mixture of civilians and uniformed men. Lloyd had more important prey in his sights.

Legionnaire looked at his watch. He had dark skin and a black moustache, and Lloyd guessed he might have a North African somewhere in his ancestry. Now he was jumpy. They were exposed here, in the open air and in daylight. The longer they stayed, the higher the chance they would be spotted. 'How much longer?' he said worriedly.

Cigare shrugged. 'We'll see.'

Lloyd said in French: 'You can leave now, if you wish. Everything is set.'

Legionnaire did not reply. He was not going to miss the action. For the sake of his prestige and authority he had to be able to say: 'I was there.'

Cigare tensed, peering into the distance, the skin around his eyes creasing with the effort. 'So,' he said cryptically. He raised himself to his knees.

Lloyd could hardly see the train, let alone identify it, but Cigare was alert. It was moving a lot faster than the previous one, Lloyd could tell. As it came closer he observed that it was longer, too: twenty-four carriages or more, he thought.

'This is it,' said Cigare.

Lloyd's pulse quickened. If Cigare was right, this was a German troop train carrying more than a thousand officers and men to the Normandy battlefield – perhaps the first of many such trains. It was Lloyd's job to make sure neither this train nor any following passed through the tunnel.

Then he saw something else. A plane was tracking the train. As he watched, the aircraft matched course with the train and began to lose height.

The plane was British.

Lloyd recognized it as a Hawker Typhoon, nicknamed a Tiffy, a one-man fighter-bomber. Tiffies were often given the dangerous mission of penetrating deep behind enemy lines to harass communications. There was a brave man at the controls, Lloyd thought.

But this formed no part of Lloyd's plan. He did not want the train to be wrecked before it reached the tunnel.

'Shit,' he said.

The Tiffy fired a machine-gun burst at the carriages.

Legionnaire said: 'But what is this?'

Lloyd replied in English: 'Fucked if I know.'

He could see now that the engine was hauling a mixture of passenger coaches and cattle trucks. However, the cattle trucks probably also contained men.

The plane, travelling faster, strafed the carriages as it overhauled the train. It had four belt-fed 20mm cannon, and they made a fearsome rattling sound that could be heard over the roar of the plane's engine and the energetic puffing of the train. Lloyd could not help feeling sorry for the trapped soldiers, unable to get out of the way of the lethal hail of bullets. He wondered why the pilot did not fire his rockets. They were highly destructive against trains or cars, though difficult to fire accurately. Perhaps they had been used up in an earlier encounter.

Some of the Germans bravely put their heads out of the windows and fired pistols and rifles at the plane, with no effect.

But Lloyd now saw a light anti-aircraft battery emplaced on a flatbed car immediately behind the engine. Two gunners were hastily deploying the big gun. It swivelled on its base and the barrel lifted to aim at the British plane.

The pilot did not appear to have seen it, for he held his course, rounds from his cannon tearing through the roofs of the carriages as he overhauled them.

The big gun fired and missed.

Lloyd wondered if he knew the flyer. There were only about five thousand pilots on active service in the UK at any one time. Quite a lot of them had been to Daisy's parties. Lloyd thought of Hubert St John, a brilliant Cambridge graduate with whom he had been reminiscing about student days a few weeks ago; of Dennis Chaucer, a West Indian from Trinidad who complained bitterly about tasteless English food, especially the mashed potatoes that seemed to be served with every meal; and of Brian Mantel, an amiable Australian he had brought across the Pyrenees on his last trip. The brave man in the Tiffy could easily be someone Lloyd had met.

The anti-aircraft gun fired again, and missed again.

Either the pilot still had not seen the gun, or he felt it could not hit him; for he took no evasive action, but continued to fly dangerously low and wreak carnage on the troop train.

The engine was just a few seconds from the tunnel when the plane was hit.

Flame flared from the plane's engine, and black smoke billowed. Too late, the pilot veered away from the railway track.

The train entered the tunnel, and the carriages flashed past Lloyd's position. He saw that every one was packed full with dozens, hundreds of German soldiers.

The Tiffy flew directly at Lloyd. For a moment he thought it would crash where he lay. He was already flat on the ground, but he stupidly put his hands over his head, as if that could protect him.

The Tiffy roared by a hundred feet above him.

Then Legionnaire pressed the plunger of the detonator.

There was a roar like thunder inside the tunnel as the track blew up, followed by a terrible screeching of tortured steel as the train crashed.

At first the carriages full of soldiers continued to flash by, but a second later their charge was arrested. The ends of two linked carriages rose in the air, forming an inverted V. Lloyd heard the men inside screaming. All the carriages came off the rails and tumbled like dropped matchsticks around the dark O of the tunnel's mouth. Iron crumpled like paper, and broken glass rained on the three saboteurs watching from the top of the embankment. They were in danger of being killed by their own explosion, and without a word they all leaped to their feet and ran.

By the time they had reached a safe distance it was all over. Smoke was billowing out of the tunnel: in the unlikely event that any men in there had survived the crash, they would burn to death.

Lloyd's plan was a success. Not only had he killed hundreds of enemy troops and wrecked a train, he had also blocked a main railway line. Crashes in tunnels took weeks to clear. He had made it much more difficult for the Germans to reinforce their defences in Normandy.

He was horrified.

He had seen death and destruction in Spain, but nothing like this. And he had caused it.

There was another crash, and when he looked in the direction of the sound he saw that the Tiffy had hit the ground. It was burning, but the fuselage had not broken up. The pilot might be alive.

He ran towards the plane, and Cigare and Legionnaire followed.

The downed aircraft lay on its belly. One wing had snapped in half. Smoke came from the single engine. The perspex dome was blackened by soot and Lloyd could not see the pilot.

He stepped on the wing and unfastened the hood catch. Cigare did the same on the other side. Together, they slid the dome back on its rails.

The pilot was unconscious. He wore a helmet and goggles, and an oxygen mask over his nose and mouth. Lloyd could not tell whether it was someone he knew.

He wondered where the oxygen tank was, and whether it had yet burst.

Legionnaire had a similar thought. 'We have to get him out before the plane blows up,' he said.

Lloyd reached inside and unfastened the safety harness. Then he put his hands under the pilot's arms and pulled. The man was completely limp. Lloyd had no way of knowing what his injuries might be. He was not even sure the man was alive.

He dragged the pilot out of the cockpit, then got him over his shoulder in a fireman's lift and carried him a safe distance from the burning wreckage. As gently as he could, he laid the man on the ground face up.

He heard a noise that was a cross between a whoosh and a thump, and looked back to see that the whole plane was ablaze.

He bent over the pilot and carefully removed the goggles and the oxygen mask, revealing a face that was shockingly familiar.

The pilot was Boy Fitzherbert.

And he was breathing.

Lloyd wiped blood from Boy's nose and mouth.

Boy opened his eyes. At first there seemed no intelligence behind them. Then, after a minute, his expression altered and he said: 'You.'

'We blew up the train,' Lloyd said.

Boy seemed unable to move anything but his eyes and mouth. 'Small world,' he said.

'Isn't it?'

Cigare said: 'Who is he?'

Lloyd hesitated, then said: 'My brother.'

'My God.'

Boy's eyes closed.

Lloyd said to Legionnaire: 'We have to bring a doctor.'

Legionnaire shook his head. 'We must get out of here. The Germans will be coming to investigate the train crash within minutes.'

Lloyd knew he was right. 'We'll have to take him with us.'

Boy opened his eyes and said: 'Williams.'

'What is it, Boy?'

Boy seemed to grin. 'You can marry the bitch now,' he said.

Then he died.

(viii)

Daisy cried when she heard. Boy had been a rotter, and treated her badly, but she had loved him once, and he had taught her a lot about sex; and she felt sad that he had been killed.

His brother, Andy, was now a viscount and heir to the earldom; Andy's wife, May, was a viscountess; and Daisy's name, according to the elaborate rules of the aristocracy, was the Dowager Viscountess Aberowen – until she married Lloyd, when she would be relieved to become plain Mrs Williams.

However, that might be a long time coming, even now. Over the summer, hopes of a quick end to the war came to nothing. A plot by German army officers to kill Hitler on 20 July failed. The Germany army was in full retreat on the Eastern Front, and the Allies took Paris in

August, but Hitler was determined to fight on to the terrible end. Daisy had no idea when she would see Lloyd, let alone marry him.

One Wednesday in September, when she went to spend the evening in Aldgate, she was greeted by a jubilant Eth Leckwith. 'Great news!' Ethel said when Daisy walked into the kitchen. 'Lloyd has been selected as Prospective Parliamentary Candidate for Hoxton!'

Lloyd's sister Millie was there with her two children, Lennie and Pammie. 'Isn't it wonderful?' she said. 'He'll be Prime Minister, I bet.'

'Yes,' said Daisy, and she sat down heavily.

'Well, I can see you're not happy about that,' said Ethel. 'As my friend Mildred would say, it went down like a cup of cold sick. What's the matter?'

'It's just that having me as a wife isn't going to help him get elected.' It was because she loved him so much that she felt so bad. How could she blight his prospects? But how could she give him up? When she thought like this her heart felt heavy and life seemed desolate.

'Because you're an heiress?' said Ethel.

'Not just that. Before Boy died he told me Lloyd would never get elected with an ex-Fascist as his wife.' She looked at Ethel, who always told the truth, even when it hurt. 'He was right, wasn't he?'

'Not entirely,' Ethel said. She put the kettle on for tea, then sat opposite Daisy at the kitchen table. 'I'm not going to say it doesn't matter. But I don't think you should despair.'

You're just like me, Daisy thought. You say what you think. No wonder he loves me: I'm a younger version of his mother!

Millie said: 'Love conquers all, doesn't it?' She noticed that four-year-old Lennie was hitting two-year-old Pammie with a wooden soldier. 'Don't bash your sister!' she said. Turning back to Daisy, she went on: 'And my brother loves you to bits. I don't think he's ever loved anyone else, to tell you the truth.'

'I know,' said Daisy. She wanted to cry. 'But he's determined to change the world, and I can't bear the thought that I'm standing in his way.'

Ethel took the crying two-year-old on to her knee, and the toddler calmed down immediately. 'I'll tell you what to do,' she said to Daisy. 'Be prepared for questions, and expect hostility, but don't dodge the issue and don't hide your past.'

'What should I say?'

'You might say you were fooled by Fascism, as millions of others were; but you drove an ambulance in the Blitz, and you hope you've paid your dues. Work out the exact words with Lloyd. Be confident, be your irresistibly charming self, and don't let it get you down.'

'Will it work?'

Ethel hesitated. 'I don't know,' she said after a pause. 'I really don't. But you have to try.'

'It would be awful if he had to give up what he loves most for my sake. Something like that could destroy a marriage.'

Daisy was half hoping Ethel would deny this, but she did not. 'I don't know,' she said again.

19

Woody Dewar got used to the crutches quickly.

He was wounded at the end of 1944, in Belgium, in the Battle of the Bulge. The Allies pushing towards the German border had been surprised by a powerful counter-attack. Woody and others of the 101st Airborne Division had held out at a vital crossroads town called Bastogne. When the Germans sent a formal letter demanding surrender, General McAuliffe sent back a one-word message that became famous: 'Nuts!'

Woody's right leg was smashed up by machine-gun bullets on Christmas Day. It hurt like hell. Even worse, it was a month before he got out of the besieged town and into a real hospital.

His bones would mend, and he might even lose the limp, but his leg would never again be strong enough for parachuting.

The Battle of the Bulge was the last offensive of Hitler's army in the West. After that they would never counter-attack again.

Woody returned to civilian life, which meant he could live at his parents' apartment in Washington and enjoy being fussed over by his mother. When the plaster cast came off he went back to work at his father's office.

On Thursday 12 April 1945, he was in the Capitol building, the home of the Senate and the House of Representatives, hobbling slowly through the basement, talking to his father about refugees. 'We think about twenty-one million people in Europe have been driven from their homes,' said Gus. 'The United Nations Relief and Rehabilitation Administration is ready to help them.'

'I guess that will start any day now,' said Woody. 'The Red Army is almost in Berlin.'

'And the US Army is only fifty miles away.'

'How much longer can Hitler hold out?'

'A sane man would have surrendered by now.'

Woody lowered his voice. 'Somebody told me the Russians found what seems to have been an extermination camp. The Nazis killed hundreds of people a day there. A place called Auschwitz, in Poland.'

Gus nodded grimly. 'It's true. The public don't know yet, but they'll find out sooner or later.'

'Someone should be put on trial for that.'

'The UN War Crimes Commission has been at work for a couple of years now, making lists of war criminals and collecting evidence. Someone will be put on trial, provided we can keep the United Nations going after the war.'

'Of course we can,' Woody said indignantly. 'Roosevelt campaigned on that basis last year, and he won the election. The United Nations conference opens in San Francisco in a couple of weeks.' San Francisco had a special significance for Woody, because Bella Hernandez lived there, but he had not yet told his father about her. 'The American people want to see international co-operation, so that we never have another war like this one. Who could be against that?'

'You'd be surprised. Look, most Republicans are decent men who simply have a view of the world that is different from ours. But there is a hard core of fucking nutcases.'

Woody was startled. His father rarely swore.

'The types who planned an insurrection against Roosevelt in the thirties,' Gus went on. 'Businessmen like Henry Ford, who thought Hitler was a good strong anti-Communist leader. They sign up for right-wing groups such as America First.'

Woody could not remember him speaking this angrily before.

'If these fools have their way, there will be a third world war even worse than the first two,' Gus said. 'I've lost a son to war, and if I ever have a grandson I don't want to lose him too.'

Woody suffered a stab of grief: Joanne would have given Gus grandchildren, if she had lived.

Right now Woody was not even dating, so grandchildren were a distant prospect – unless he could track down Bella in San Francisco . . .

'We can't do anything about complete idiots,' Gus went on. 'But perhaps we can deal with Senator Vandenberg.'

Arthur Vandenberg was a Republican from Michigan, a conservative and an opponent of Roosevelt's New Deal. He was on the Senate Foreign Relations Committee with Gus.

'He's our greatest danger,' Gus said. 'He may be self-important and vain, but he commands respect. The President has been wooing him, and he's come around to our point of view, but he could backslide.'

'Why would he do that?'

'He's strongly anti-Communist.'

'Nothing wrong with that. We are too.'

'Yes, but Arthur is kind of rigid about it. He'll get riled if we do anything he thinks is kowtowing to Moscow.'

'Such as?'

'God knows what kind of compromises we might have to make in San Francisco. We've already agreed to admit Belorussia and the Ukraine as separate states, which is just a way of giving Moscow three votes in the General Assembly. We have to keep the Soviets on board – but if we go too far, Arthur could turn against the whole United Nations project. Then the Senate may refuse to ratify it, exactly the way they rejected the League of Nations in 1919.'

'So our job in San Francisco is to keep the Soviets happy without offending Senator Vandenberg.'

'Exactly.'

They heard running footsteps, an unusual sound in the dignified hallways of the Capitol. They both looked around. Woody was surprised to see the vice-president, Harry Truman, running through the hallway. He was dressed normally, in a grey double-breasted suit and a polka-dot tie, though he had no hat. He seemed to have lost his normal escort of aides and Secret Service guards. He was running steadily, breathing hard, not looking at anyone, going somewhere in a terrific hurry.

Woody and Gus watched in astonishment. So did everyone else.

When Truman disappeared around a corner, Woody said: 'What the heck . . . ?'

Gus said: 'I think the President must have died.'

(ii)

Volodya Peshkov entered Germany in a ten-wheeler Studebaker US6 army truck. Made in South Bend, Indiana, the truck had been carried by rail to Baltimore, shipped across the Atlantic and around the Cape of Good Hope to the Persian Gulf, then sent by train from Persia to central Russia. Volodya knew it was one of two hundred thousand Studebaker trucks given to the Red Army by the American government. The Russians liked them: they were tough and reliable. The men said the letters 'USA' stencilled on the side stood for *Ubit Sukina syna Adolf*, which meant 'Kill that son of a bitch Adolf.'

They also liked the food the Americans were sending, especially the cans of compressed meat called Spam, strangely bright pink in colour but gloriously fatty.

Volodya had been posted to Germany because the intelligence he was getting from spies in Berlin was now not as up-to-date as information that could be gained by interviewing German prisoners of war. His fluent German made him a first-class front-line interrogator.

When he crossed the border he had seen a Soviet government poster that said: 'Red Army soldier: You are now on German soil. The hour of revenge has struck!' It was among the milder pieces of propaganda. The Kremlin had been whipping up hatred of Germans for some time, believing it would make soldiers fight harder. Political commissars had calculated – or said they had – the number of men killed in battle, the number of houses torched, the number of civilians murdered for being Communists or Slavs or Jews, in every village and town overrun by the German army. Many front-line soldiers could quote the figures for their own neighbourhoods, and were eager to do the same kind of damage in Germany.

The Red Army had reached the river Oder, which snaked north–south across Prussia, the last barrier before Berlin. A million Soviet soldiers were within fifty miles of the capital, poised to strike. Volodya was with the Fifth Shock Army. Waiting for the fighting to begin, he was studying the army newspaper, *Red Star*.

What he read horrified him.

The hate propaganda went further than anything he had read before. 'If you have not killed at least one German a day, you have wasted that day,' he read. 'If you are waiting for the fighting, kill a German before combat. If you kill one German, kill another – there is nothing more amusing for us than a heap of German corpses. Kill the German – this is your old mother's prayer. Kill the German – this is what your children beseech you to do. Kill the German – this is the cry of your Russian earth. Do not waver. Do not let up. Kill.'

It was a bit sickening, Volodya thought. But worse was implied. The writer made light of looting: 'German women are only losing fur coats and silver spoons that were stolen in the first place.' And there was a sidelong joke about rape: 'Soviet soldiers do not refuse the compliments of German women.'

Soldiers were not the most civilized of men in the first place. The way the invading Germans had behaved in 1941 had enraged all Russians. The government was fuelling their wrath with talk of revenge. And now the army newspaper was making it clear they could do anything they liked to the defeated Germans.

It was a recipe for Armageddon.

(iii)

Erik von Ulrich was consumed by a yearning that the war should be over.

With his friend Hermann Braun and their boss Dr Weiss, Erik set up a field hospital in a small Protestant church; then they sat in the nave with nothing to do but wait for the horse-drawn ambulances to arrive loaded with horribly torn and burned men.

The German army had reinforced Seelow Heights, overlooking the Oder river where it passed closest to Berlin. Erik's aid station was in a village a mile back from the line.

Dr Weiss, who had a friend in army intelligence, said there were 110,000 Germans defending Berlin against a million Soviets. With his usual sarcasm he said: 'But our morale is high, and Adolf Hitler is the greatest genius in military history, so we are certain to win.'

There was no hope, but German soldiers were still fighting fiercely. Erik believed this was because of the stories filtering back about how the Red Army behaved. Prisoners were killed, homes were looted and wrecked, women were raped and nailed to barn doors. The Germans believed they were defending their own families from Communist brutality. The Kremlin's hate propaganda was backfiring.

Erik was looking forward to defeat. He longed for the killing to stop. He just wanted to go home.

He would have his wish soon – or he would be dead.

Sleeping on a wooden pew, Erik was awakened at three o'clock in the morning on Monday 16 April by the Russian guns. He had heard artillery bombardments before, but this was ten times as loud as anything in his experience. For the men on the front line it must have been literally deafening.

The wounded started to arrive at dawn, and the team went wearily to work, amputating limbs, setting broken bones, extracting bullets, and cleaning and bandaging wounds. They were short of everything from drugs to clean water, and they gave morphine only to those who were screaming in agony.

Men who could still walk and hold a gun were sent back to the line.

The German defenders held out longer than Dr Weiss expected. At the end of the first day they were still in position, and as darkness fell the rush of wounded slowed. The medical unit got some sleep that night.

Early on the next day Werner Franck was brought in, his right wrist horribly crushed.

He was a captain now. He had been in charge of a section of the line with thirty 88mm Flak guns. 'We only had eight shells for each gun,' he said while Dr Weiss's clever fingers worked slowly and meticulously to set his smashed bones. 'Our orders were to fire seven at the Russian tanks, then use the eighth to destroy our own gun so that it could not be used by the Reds.' He had been standing by an 88 when it suffered a direct hit from the Soviet artillery and turned over on him. 'I was lucky it was only my hand,' he said 'It might have been my damn head.'

When his wrist had been taped up, he said to Erik: 'Have you heard from Carla?'

Erik knew that his sister and Werner were now a couple. 'I haven't had any letters for weeks.'

'Nor me. I hear things are pretty grim in Berlin. I hope she's all right.'

'I worry, too,' said Erik.

Surprisingly, the Germans held the Seelow Heights for another day and night.

The dressing station got no warning that the line had collapsed. They were triaging a fresh cartload of wounded when seven or eight Soviet soldiers crashed into the church. One fired a machine-gun burst at the vaulted ceiling and Erik threw himself to the ground, as did everyone else capable of moving.

Seeing that no one was armed, the Russians relaxed. They went around the room taking watches and rings from those who had them. Then they left.

Erik wondered what would happen next. This was the first time he had been trapped behind enemy lines. Should they abandon the field hospital and try to catch up with their retreating army? Or were their patients safer here?

Dr Weiss was decisive. 'Carry on with your work, everyone,' he said.

A few minutes later a Soviet soldier came in with a comrade over his shoulder. Pointing his gun at Weiss, he spoke a rapid stream of Russian. He was in a panic, and his friend was covered in blood.

Weiss replied calmly. In halting Russian he said: 'No need for the gun. Put your friend on this table.'

The soldier did so, and the team went to work. The soldier kept his rifle pointed at the doctor.

Later in the day, the German patients were marched or carried out and put into the back of a truck which drove away east. Erik watched Werner Franck disappear, a prisoner of war. As a boy, Erik had often been told the story of his Uncle Robert, who had been imprisoned by the Russians during the First World War, and had walked home from Siberia, a journey of four thousand miles. Erik wondered now where Werner would end up.

More wounded Russians were brought in, and the Germans took care of them as they would have of their own men.

Later, as Erik fell into an exhausted sleep, he realized that now he, too, was a prisoner of war.

(iv)

As the Allied armies closed in on Berlin, the victorious countries began squabbling among themselves at the United Nations conference in San Francisco. Woody would have found it depressing, except that he was more interested in trying to reconnect with Bella Hernandez.

She had been on his mind all through the D-Day invasion and the fighting in France, his time in hospital and his convalescence. A year ago she had been at the end of her period at Oxford University and planning to do a doctorate at Berkeley, right here in San Francisco. She would probably be living at her parents' home in Pacific Heights, unless she had an apartment near the campus.

Unfortunately, he was having trouble getting a message to her.

His letters were not answered. When he called the number listed in the phone book, a middle-aged woman who he suspected was Bella's mother said with icy courtesy: 'She's not at home right now. May I give her a message?' Bella never called back.

She probably had a serious boyfriend. If so he wanted her to tell him. But perhaps her mother was intercepting her mail and not passing on messages.

He should probably give up. He might be making a fool of himself. But that was not his way. He recalled his long, stubborn courtship of Joanne. There seems to be a pattern here, he thought; is it something about me?

Meanwhile, every morning he went with his father to the penthouse at the top of the Fairmont Hotel, where Secretary of State Edward Stettinius held a briefing for the American team at the conference. Stettinius had taken over from Cordell Hull, who was in hospital. The USA also had a new president, Harry Truman, who had been sworn in on the death of the great Franklin D. Roosevelt. It was a pity, Gus Dewar observed, that at such a crucial moment in world history the United States should be led by two inexperienced newcomers.

Things had begun badly. President Truman had clumsily offended Soviet foreign minister Molotov at a pre-conference meeting at the White House. Consequently Molotov arrived in San Francisco in a foul mood. He announced he was going home unless the conference agreed immediately to admit Belorussia, Ukraine, and Poland.

No one wanted the USSR to pull out. Without the Soviets, the United Nations were not the United Nations. Most of the American delegation were in favour of compromising with the Communists, but the bow-tied Senator Vandenberg prissily insisted that nothing should be done under pressure from Moscow.

One morning when Woody had a couple of hours to spare he went to Bella's parents' house.

The swanky neighbourhood where they lived was not far from the Fairmont Hotel on Nob Hill, but Woody was still walking with a cane, so he took a taxi. Their home was a yellow-painted Victorian mansion on Gough Street. The woman who came to the door was too well dressed to be a maid. She gave him a lopsided smile just like Bella's: she had to be the mother. He said politely: 'Good morning, ma'am. I'm Woody Dewar. I met Bella Hernandez in London last year and I'd sure like to see her again, if I may.'

The smile disappeared. She gave him a long look and said: 'So you're him.'

Woody had no idea what she was talking about.

'I'm Caroline Hernandez, Isabel's mother,' she said. 'You'd better come in.'

'Thank you.'

She did not offer to shake hands, and she was clearly hostile, though there was no clue as to why. However, he was inside the house.

Mrs Hernandez led Woody into a large, pleasant parlour with a breathtaking ocean view. She pointed to a chair, indicating that he should sit down with a gesture that was barely polite. She sat opposite him and gave him another hard look. 'How much time did you spend with Bella in England?' she asked.

'Just a few hours. But I've been thinking about her ever since.'

There was another pregnant pause, then she said: 'When she went to Oxford, Bella was engaged to be married to Victor Rolandson, a splendid young man she has known most of her life. The Rolandsons are old friends of my husband's and mine – or, at least, they were, until Bella came home and broke off the engagement abruptly.'

Woody's heart leaped with hope.

'She would only say she had realized she did not love Victor. I guessed she'd met someone else, and now I know who.'

Woody said: 'I had no idea she was engaged.'

'She was wearing a diamond ring that was pretty hard to miss. Your poor powers of observation have caused a tragedy.'

'I'm very sorry,' Woody said. Then he told himself to stop being a pussy. 'Or rather, I'm not,' he said. 'I'm very glad she's broken off her engagement, because I think she's absolutely wonderful and I want her for myself.'

Mrs Hernandez did not like that. 'You're mighty fresh, young man.'

Woody suddenly felt resentful of her condescension. 'Mrs Hernandez, you used the word "tragedy" just now. My fiancée, Joanne, died in my arms at Pearl Harbor. My brother, Chuck, was killed by machine-gun fire on the beach at Bougainville. On D-Day I sent Ace Webber and four other young Americans to their deaths for the sake of a bridge in a one-horse town called Eglise-des-Soeurs. I know what tragedy is, ma'am, and it's not a broken engagement.'

She was taken aback. He guessed young people did not often stand up to her. She did not reply, but looked a little pale. After a moment she got up and left the room without explanation. Woody was not sure what she expected him to do, but he had not yet seen Bella so he sat tight.

Five minutes later, Bella came in.

Woody stood up, his pulse quickening. Just the sight of her made him smile. She wore a plain pale-yellow dress that set off her lustrous dark hair and coffee skin. She would always look good in dramatically simple clothing, he guessed; just like Joanne. He wanted to put his arms around her and crush her soft body to his own, but he waited for a sign from her.

She looked anxious and uncomfortable. 'What are you doing here?' she said.

'I came looking for you.'

'Why?'

'Because I can't get you out of my mind.'

'We don't even know each other.'

'Let's put that right, starting today. Will you have dinner with me?'

'I don't know.'

He crossed the room to where she stood.

She was startled to see him using a walking stick. 'What happened to you?'

'My knee got shot up in France. It's getting better, slowly.'

'I'm so sorry.'

'Bella. I think you're wonderful. I believe you like me. We're both free of commitments. What's worrying you?'

She gave that lopsided grin that he liked so much. 'I guess I'm embarrassed. About what I did, that night in London.'

'Is that all?'

'It was a lot, for a first date.'

'That kind of thing went on all the time. Not to me, necessarily, but I heard about it. You thought I was going to die.'

She nodded. 'I've never done anything like that, not even with Victor. I don't know what came over me. And in a public park! I feel like a whore.'

'I know exactly what you are,' Woody said. 'You're a smart, beautiful woman with a big heart. So why don't we forget that mad moment in London, and start getting to know one another like the respectable well-brought-up young people that we are?'

She began to soften. 'Can we, really?'

'You bet.'

'Okay.'

'I'll pick you up at seven?'

'Okay.'

That was an exit line, but he hesitated. 'I can't tell you how glad I am that I found you again,' he said.

She looked him in the eye for the first time. 'Oh, Woody, so am I,' she said. 'So glad!' Then she put her arms around his waist and hugged him.

It was what he had been longing for. He embraced her and put his face into her wonderful hair. They stayed like that for a long minute.

At last she pulled away. 'I'll see you at seven,' she said.

'You bet.'

He left the house in a cloud of happiness.

He went from there straight to a meeting of the steering committee in the Veterans Building next to the opera house. There were forty-six members around the long table, with aides such as Gus Dewar sitting behind them. Woody was an aide to an aide, and sat up against the wall.

The Soviet foreign minister, Molotov, made the first speech. He was not impressive to look at, Woody reflected. With his receding hair, neat moustache, and glasses, he looked like a store clerk, which was what his father had been. But he had survived a long time in Bolshevik politics. A friend of Stalin's since before the revolution, he was the architect of the Nazi–Soviet pact of 1939. He was a hard worker, and was nicknamed Stone-Arse because of the long hours he spent at his desk.

He proposed that Belorussia and Ukraine be admitted as original members of the United Nations. These two Soviet republics had borne the brunt of the Nazi invasion, he pointed out, and each had contributed more than a million men to the Red Army. It had been argued that they were not fully independent of Moscow, but the same argument could be applied to Canada and Australia, dominions of the British Empire that had each been given separate membership.

The vote was unanimous. It had all been fixed up in advance, Woody knew. The Latin American countries had threatened to dissent unless Hitler-supporting Argentina was admitted, and that concession had been granted to secure their votes.

Then came a bombshell. The Czech foreign minister, Jan Masaryk, stood up. He was a famous liberal and anti-Nazi who had been on the cover of *Time* magazine in 1944. He proposed that Poland should also be admitted to the UN.

The Americans were refusing to admit Poland until Stalin permitted elections there, and Masaryk as a democrat should have supported that stand, especially as he, too, was trying to create a democracy with Stalin looking over his shoulder. Molotov must have put terrific pressure on Masaryk to get him to betray his ideals in this way. And, indeed, when Masaryk sat down he wore the expression of one who has eaten something disgusting.

Gus Dewar also looked grim. The prearranged compromises over Belorussia, Ukraine and Argentina should have ensured that this session went smoothly. But now Molotov had thrown them a low ball.

Senator Vandenberg, sitting with the American contingent, was outraged. He took out a pen and notepad and began writing furiously. After a minute he tore the sheet off, beckoned Woody, gave him the note, and said: 'Take that to the Secretary of State.'

Woody went to the table, leaned over Stettinius's shoulder, put the note in front of him, and said: 'From Senator Vandenberg, sir.'

'Thank you.'

Woody returned to his chair up against the wall. My part in history, he thought. He had glanced at the note as he handed it over. Vandenberg had drafted a short, passionate speech rejecting the Czech proposal. Would Stettinius follow the senator's lead?

If Molotov got his way over Poland, then Vandenberg might sabotage the United Nations in the Senate. But if Stettinius took Vandenberg's line now, Molotov might walk out and go home, which would kill off the UN just as effectively.

Woody held his breath.

Stettinius stood up with Vandenberg's note in his hand. 'We've just honoured our Yalta engagements on behalf of Russia,' he said. He meant the commitment made by the USA to support Belorussia and Ukraine. 'There are other Yalta obligations which equally require allegiance.' He was using the words Vandenberg had written. 'One calls for a new and representative Polish Provisional Government.'

There was a murmur of shock around the room. Stettinius was going up against Molotov. Woody glanced at Vandenberg. He was purring.

'Until that happens,' Stettinius went on, 'the Conference cannot, in good conscience, recognize the Lublin government.' He looked directly at Molotov and quoted Vandenberg's exact words. 'It would be a sordid exhibition of bad faith.'

Molotov looked incandescent.

The British foreign secretary, Anthony Eden, unfolded his lanky figure and stood up to support Stettinius. His tone was faultlessly courteous, but his words were scathing. 'My government has no way of knowing whether the Polish people support their provisional government,' he said, 'because our Soviet allies refuse to let British observers into Poland.'

Woody sensed the meeting turning against Molotov. The Russian clearly had the same impression. He was conferring with his aides loudly enough for Woody to hear the fury in his voice. But would he walk out?

The Belgian foreign minister, bald and podgy with a double chin, proposed a compromise, a motion expressing the hope that the new Polish government might be organized in time to be represented here in San Francisco before the end of the conference.

Everyone looked at Molotov. He was being offered a face-saver. But would he accept it?

He still looked angry. However, he gave a slight but unmistakable nod of assent.

And the crisis was over.

Well, Woody thought, two victories in one day. Things are looking up.

(v)

Carla went out to queue for water.

There had been no water in the taps for two days. Luckily, Berlin's housewives had discovered that every few blocks there were old-fashioned street pumps, long disused, connected to underground wells. They were rusty and creaky but, amazingly, they still worked. So every morning now the women stood in line, holding their buckets and jugs.

The air raids had stopped, presumably because the enemy was on the point of entering the city. But it was still dangerous to be on the street, because the Red Army's artillery was shelling. Carla was not sure why they bothered. Much of the city had gone. Whole blocks and even larger areas had been completely flattened. All utilities were cut off. No trains or buses ran. Thousands were homeless, perhaps millions. The city was one huge refugee camp. But the shelling went on. Most people spent all day in their cellars or in public air-raid shelters, but they had to come out for water.

On the radio, shortly before the electricity went off permanently, the BBC had announced that the Sachsenhausen concentration camp had been liberated by the Red Army. Sachsenhausen was north of Berlin, so clearly the Soviets, coming from the east, were encircling the city instead of marching straight in. Carla's mother, Maud, deduced that the Russians wanted to keep out the American, British, French and Canadian forces rapidly approaching from the west. She had quoted Lenin: 'Who controls Berlin, controls Germany; and who controls Germany, controls Europe.'

Yet the German army had not given up. Outnumbered, outgunned, short of ammunition and fuel, and half starved, they slogged on. Again and again their leaders hurled them at overwhelming enemy forces, and again and again they obeyed orders, fought with spirit and courage, and died in their hundreds of thousands. Among them were the two men Carla loved: her brother, Erik, and her boyfriend, Werner. She had no idea where they were fighting or even whether they were alive.

Carla had wound up the spy ring. The fighting was deteriorating into chaos. Battle plans meant little. Secret intelligence from Berlin was of small value to the conquering Soviets. It was no longer worth the risk. The spies had burned their code books and hidden their radio transmitters in the rubble of bombed buildings. They had agreed never to speak of their work. They had been brave, they had shortened the war, and they had saved lives; but it was too much to expect the defeated German people to see things that way. Their courage would remain forever secret.

While Carla waited her turn at the tap, a Hitler Youth tank-hunting squad went past, heading east, towards the fighting. There were two men in their fifties and a dozen teenage boys, all on bicycles. Strapped to the front of each bicycle were two of the new one-shot anti-tank weapons called *Panzerfäuste*. The uniforms were too large for the boys, and their oversize helmets would have looked comical if their plight had not been so pathetic. They were off to fight the Red Army.

They were going to die.

Carla looked away as they passed: she did not want to remember their faces.

As she was filling her bucket, the woman behind her in line, Frau Reichs, spoke to her quietly, so that no one else could hear: 'You're a friend of the doctor's wife, aren't you?'

Carla tensed. Frau Reichs was obviously talking about Hannelore Rothmann. The doctor had disappeared along with the mental patients from the Jewish Hospital. Hannelore's son, Rudi, had thrown away his yellow star and joined those Jews living clandestinely, called U-boats in Berlin slang. But Hannelore, not herself Jewish, was still at the old house.

For twelve years a question such as the one just asked – are you a friend of a Jew's wife? – had been an accusation. What was it today? Carla did not know. Frau Reichs was only a nodding acquaintance: she could not be trusted.

Carla turned off the tap. 'Dr Rothmann was our family physician when I was a child,' she said guardedly. 'Why?'

The other woman took her place at the standpipe and began to fill a large can that had once held cooking oil. 'Frau Rothmann has been taken away,' she said. 'I thought you'd like to know.'

It was commonplace. People were 'taken away' all the time. But when it happened to someone close to you it came as a blow to the heart.

There was no point in trying to find out what had happened to them – in fact, it was downright dangerous: people who inquired about disappearances tended to disappear themselves. All the same, Carla had to ask. 'Do you know where they took her?'

This time there was an answer. 'The Schulstrasse transit camp.' Carla felt hopeful. 'It's in the old Jewish Hospital, in Wedding. Do you know it?'

'Yes, I do.' Carla sometimes worked at the hospital, unofficially and illegally, so she knew that the government had taken over one of the hospital buildings, the pathology lab, and surrounded it with barbed wire.

'I hope she's all right,' said the other woman. 'She was good to me when my Steffi was ill.' She turned off the tap and walked away with her can of water.

Carla hurried away in the opposite direction, heading for home.

She had to do something about Hannelore. It had always been nearly impossible to get anyone out of a camp, but now that everything was breaking down perhaps there might be a way.

She took the bucket into the house and gave it to Ada.

Maud had gone to queue for food rations. Carla changed into her nurse's uniform, thinking it might help. She explained to Ada where she was going and left again.

She had to walk to Wedding. It was two or three miles. She

wondered if it was worth it. Even if she found Hannelore, she probably would not be able to help her. But then she thought of Eva in London and Rudi in hiding somewhere here in Berlin: how terrible it would be if they lost their mother in the last hours of the war. She had to try.

The military police were on the streets, stopping people and demanding papers. They worked in threes, forming summary courts, and were mainly interested in men of fighting age. They did not bother Carla in her nurse's uniform.

It was strange that in this blasted cityscape the apple and cherry trees were gorgeous with white and pink blossoms, and that in the quiet moments between explosions she could hear the birds singing as optimistically as they did every spring.

To her horror she saw several men hanged from lamp posts, some in uniform. Most of the bodies had a card hanging around the neck saying 'Coward' or 'Deserter'. These had been found guilty by those three-man street courts, she knew. Was there not already enough killing to satisfy the Nazis? It made her want to weep.

She was forced to take shelter from artillery bombardments three times. On the last occasion, when she was only a few hundred yards from the hospital, the Soviets and the Germans seemed to be fighting only a few streets away. The shooting was so heavy that Carla was tempted to turn back. Hannelore was probably doomed, and might already be dead: why should Carla add her own life to the toll? But she went on anyway.

It was evening when she reached her destination. The hospital was in Iranische Strasse, on the corner of Schul Strasse. The trees lining the streets were in new leaf. The laboratory building, which had been turned into a transit camp, was guarded. Carla considered going up to the guard and explaining her mission, but it seemed an unpromising strategy. She wondered if she might slip inside from the tunnel system.

She went into the main building. The hospital was functioning. All the patients had been moved into the basements and tunnels. The staff were working by the light of oil lamps. Carla could tell by the smell that the toilets were not flushing. Water was being carried in buckets from an old well in the garden.

Surprisingly, soldiers were bringing wounded comrades in for help. Suddenly they did not care that the doctors and nurses might be Jewish.

She followed a tunnel under the garden to the basement of the laboratory. As she expected, the door was guarded. However, the young Gestapo man looked at her uniform and waved her through without questioning her. Perhaps he no longer saw any point in his job.

She was inside the camp, now. She wondered whether it would be as easy to get out.

The smell here was worse, and she soon saw why. The basement was overcrowded. Hundreds of people were packed into four storerooms. They sat or lay on the floor, the lucky ones having a wall to lean against. They were dirty, smelly and exhausted, and they looked at her with dull, uninterested gazes.

She found Hannelore after a few minutes.

The doctor's wife had never been beautiful, but she had once been a statuesque woman with a strong face. Now she was gaunt, like most people, and her hair was grey and lifeless. She was hollow-cheeked and lined with strain.

She was talking to an adolescent who was at the age when a girl can seem too voluptuous for her years, having womanly breasts and hips but the face of a child. The girl was sitting on the floor, crying, and Hannelore was kneeling beside her, holding her hand and speaking in a low, soothing voice.

When Hannelore saw Carla she stood up, saying: 'Good God! Why are you in here?'

'I thought maybe if I tell them you're not Jewish they might let you go.'

'That was brave.'

'Your husband saved many lives. Someone ought to save yours.'

For a moment, Carla thought Hannelore was going to cry. Her face seemed about to crumple. Then she blinked and shook her head. 'This is Rebecca Rosen,' she said in a controlled voice. 'Her parents were killed by a shell today.'

Carla said: 'I'm so sorry, Rebecca.'

The girl did not speak.

Carla said: 'How old are you, Rebecca?'

'Nearly fourteen.'

'You're going to have to be a grown-up now.'

'Why didn't I die too?' Rebecca said. 'I was right beside them. I should have died. Now I'm all alone.'

'You're not alone,' Carla said briskly. 'We're with you.' She turned back to Hannelore. 'Who's in charge here?'

'His name is Walter Dobberke.'

'I'm going to tell him he must let you go.'

'He's left for the day. And his second-in-command is a sergeant with the brains of a warthog. But look, here comes Gisela. She's Dobberke's mistress.'

The young woman walking into the room was pretty, with long fair hair and creamy skin. No one looked at her. She wore a defiant expression.

Hannelore said: 'She has sex with him on the bed in the electrocardiogram room upstairs. She gets extra food in exchange. No one will speak to her except me. I just don't think we can judge people for the compromises they make. We are living in hell, after all.'

Carla was not so sure. She would not befriend a Jewish girl who slept with a Nazi.

Gisela met Hannelore's eye and came over. 'He's had new orders,' she said, speaking so quietly that Carla had to strain to hear her. Then she hesitated.

Hannelore said: 'Well? What are the orders?'

Gisela's voice fell to a whisper. 'To shoot everyone here.'

Carla felt a cold hand grasp her heart. All these people – including Hannelore and young Rebecca.

'Walter doesn't want to do it,' Gisela said. 'He's not a bad man, really.'

Hannelore spoke with fatalistic calm. 'When is he supposed to kill us?'

'Immediately. But he wants to destroy the records first. Hans-Peter and Martin are putting the files into the furnace right now. It's a long job, so we have a few hours left. Maybe the Red Army will get here in time to save us.'

'And maybe they won't,' Hannelore said crisply. 'Is there any way we can persuade him to disobey his orders? For God's sake, the war is almost over!'

'I used to be able to talk him into anything,' Gisela said sadly. 'But he's getting tired of me now. You know what men are like.'

'But he should be thinking of his own future. Any day now the Allies will be in charge here. They will punish Nazi crimes.'

Gisela said: 'If we're all dead, who's going to accuse him?'

'I will,' said Carla.

The other two stared at her, not speaking.

Carla realized that even though she was not Jewish she, too, would be shot, to prevent her bearing witness.

Casting about for ideas, she said: 'Perhaps, if Dobberke spared us, it would help him with the Allies.'

'That's a thought,' said Hannelore. 'We could all sign a declaration saying that he saved our lives.'

Carla looked enquiringly at Gisela. Her expression was dubious, but she said: 'He might do it.'

Hannelore looked around. 'There's Hilde,' she said. 'She acts as a secretary for Dobberke.' She called the woman over and explained the plan.

'I'll type out release documents for everyone,' Hilde said. 'We'll ask him to sign them before we give him the declaration.'

There were no guards within the basement area, just at the ground-floor door and the tunnel, so the prisoners could move around freely inside. Hilde went into the room that served as Dobberke's underground office. She typed the declaration first. Hannelore and Carla went around the basement explaining the plan and getting everyone to sign. Meanwhile Hilde typed the release documents.

By the time they finished it was the middle of the night. There was no more they could do until Dobberke showed up in the morning.

Carla lay on the floor next to Rebecca Rosen. There was nowhere else to sleep.

After a while Rebecca began to cry quietly.

Carla was not sure what to do. She wanted to give comfort, but no words came. What did you say to a child who had just seen both her parents killed? The muffled weeping continued. In the end Carla rolled over and put her arms around Rebecca.

She knew immediately that she had done the right thing. Rebecca cuddled up to her, head on her breast. Carla patted her back as if she were a baby. Slowly the sobs eased and eventually Rebecca fell asleep.

Carla did not sleep. She spent the night making imaginary speeches to the camp commandant. Sometimes she appealed to his better nature, sometimes she threatened him with Allied justice, sometimes she argued from his own self-interest.

She tried not to think about the process of being shot. Erik had explained to her how the Nazis executed people twelve at a time in Russia. She supposed they would have an efficient system here too. It was hard to imagine. Perhaps that was just as well.

She could probably escape shooting if she left the camp right now, or first thing in the morning. She was not an inmate, nor a Jew, and her papers were perfectly in order. She could go out the way she came in, dressed in her nurse's uniform. But that would mean abandoning both Hannelore and Rebecca. She could not bring herself to do that, no matter how badly she longed to get out of here.

The fighting in the streets outside continued until the small hours, then there was a short pause. It began again at dawn. Now it was close enough for her to hear machine-gun fire as well as artillery.

Early in the morning the guards brought an urn of watery soup and a sack of bread, all discarded parts of stale loaves. Carla drank the soup and ate the bread and then, reluctantly, used the toilet, which was unspeakably dirty.

With Hannelore, Gisela and Hilde she went up to the ground floor to wait for Dobberke. The shelling had resumed, and they were in danger every second, but they wanted to confront him the moment he arrived.

He did not appear at his usual hour. He was normally punctual, Hilde said. Perhaps he had been delayed by the fighting in the streets. He might have been killed, of course. Carla hoped not. His second-in-command, Sergeant Ehrenstein, was too stupid to argue with.

When Dobberke was an hour late, Carla began to lose hope.

After another hour, he arrived.

'What's this?' he said when he saw the four women waiting in the hall. 'A mothers' meeting?'

Hannelore replied: 'All the prisoners have signed a declaration saying you saved their lives. It may save *your* life, if you accept our terms.'

'Don't be ridiculous,' he said.

Carla spoke up. 'According to the BBC, the United Nations has a list of the names of Nazi officers who have taken part in mass murders. In a week's time you could be on trial. Wouldn't you like to have a signed declaration that you spared people?'

'Listening to the BBC is a crime,' he said.

'Though not as serious as murder.'

Hilde had a file folder in her hand. She said: 'I have typed release orders for all the prisoners here. If you sign them, you can have the declaration.'

'I could just take it from you.'

'No one will believe in your innocence if we're all dead.'

Dobberke was angered by the situation he found himself in, but not confident enough just to walk away. 'I could shoot the four of you for insolence,' he said.

Carla spoke impatiently. 'This is what defeat is like,' she said. 'Get used to it.'

His face darkened with anger, and she realized she had gone too far. She wished she could take back her words. She stared at Dobberke's furious expression, trying not to let her fear show.

At that moment a shell landed outside the building. The doors rattled and a window smashed. They all ducked instinctively, but no one was hurt.

When they straightened up, Dobberke's face had changed. Rage was replaced by something like disgusted resignation. Carla's heartbeat quickened. Had he given up?

Sergeant Ehrenstein ran in. 'No one hurt, sir,' he reported.

'Very good, Sergeant.'

Ehrenstein was about to go out again when Dobberke called him back. 'This camp is now closed,' Dobberke said.

Carla held her breath.

'Closed, sir?' There was aggression as well as surprise in the sergeant's voice.

'New orders. Tell the men to go . . .' Dobberke hesitated. 'Tell them to report to the railway bunker at Freidrich Strasse Station.'

Carla knew Dobberke was making this up, and Ehrenstein seemed to suspect it too. 'When, sir?'

'Immediately.'

'Immediately.' Ehrenstein paused, as if the word 'immediately' required further elucidation.

Dobberke stared him out.

'Very good, sir,' said the sergeant. 'I'll tell the men.' He went out.

Carla felt a surge of triumph, but told herself she was not yet free.

Dobberke said to Hilde: 'Show me the declaration.'

Hilde opened her folder. There were a dozen sheets, all with the same wording typed at the top, the rest of the space covered with signatures. She handed them over.

Dobberke folded the papers and stuffed them in his pocket.

Hilde placed the release orders in front of him. 'Sign these, please.'

'You don't need release orders,' Dobberke said. 'And I don't have time to sign my name hundreds of times.' He stood up.

Carla said: 'The police are on the streets. They're hanging people from the lamp posts. We need papers.'

He patted his pocket. 'They'll hang me if they find this declaration.' He went to the door.

Gisela cried: 'Take me with you, Walter!'

He turned to her. 'Take you?' he said. 'What would my wife say?' He went out and slammed the door.

Gisela burst into tears.

Carla went to the door, opened it, and watched Dobberke stride away. There were no other Gestapo men in sight: they had already obeyed his orders and abandoned the camp.

The commandant reached the street and broke into a run.

He left the gate open.

Hannelore was standing beside Carla, looking out with incredulity.

'We're free, I think,' said Carla.

'We must tell the others.'

Hilde said: 'I'll tell them.' She went down the basement stairs.

Carla and Hannelore walked fearfully along the path that led from the laboratory entrance to the open gate. There they hesitated and looked at one another.

Hannelore said: 'We're frightened of freedom.'

Behind them a girlish voice said: 'Carla, don't go without me!' It was Rebecca, running down the path, her breasts bouncing under a grubby blouse.

Carla sighed. I've acquired a child, she thought. I don't feel ready to be a mother. But what can I do?

'Come on, then,' she said. 'But be ready to run.' She realized she did not need to worry about Rebecca's agility: the girl could undoubtedly run faster than either Carla or Hannelore.

They crossed the hospital garden to the main gate. There they paused and looked up and down Iranische Strasse. It seemed quiet. They crossed the road and ran to the corner. As Carla looked along Schul Strasse she heard a burst of machine-gun fire and saw that farther up the street there was a firefight. She saw German troops retreating towards her and Red Army soldiers coming after them.

She looked around. There was nowhere to hide except behind trees, and that was hardly any protection at all.

A shell landed in the middle of the road fifty yards away and exploded. Carla felt the blast, but she was not hurt.

Without conferring, all three women ran back inside the hospital grounds.

They returned to the laboratory building. Some of the other prisoners were standing just inside the barbed wire, as if not quite daring to come out.

Carla said to them: 'The basement stinks, but right now it's the safest place.' She went inside the building and down the stairs, and most of the others followed.

She wondered how long she would have to stay here. The German army must give up, but when? Somehow she could not imagine Hitler agreeing to surrender under any circumstances. The man's whole life had been based on arrogantly shouting that he was the boss. How could such a man admit that he had been wrong, stupid and wicked? That he had murdered millions and caused his country to be bombed to ruins? That he would go down in history as the most evil man who had ever lived? He could not. He would go mad, or die of shame, or put a pistol in his mouth and pull the trigger.

But how long would it take? Another day? Another week? Longer?

There was a shout from upstairs. 'They're here! The Russians are here!'

Then Carla heard heavy boots clattering down the steps. Where had the Russians got such good boots? From the Americans?

Then they were in the room, four, six, eight, nine men with dirty faces, carrying submachine guns with drum magazines, ready to kill as quick as look at you. They seemed to take up a lot of room. People shrank away from them, even though they were the liberators.

The soldiers took in their surroundings. They saw that they were in no danger from the emaciated prisoners, mainly female. They lowered their guns. Some moved into the adjoining rooms.

A tall soldier pulled up his left sleeve. He was wearing six or seven wristwatches. He shouted something in Russian, pointing at the watches with the stock of his gun. Carla thought she knew what he was saying, but she could hardly believe it. The man then grabbed an elderly woman, took her hand, and pointed to her wedding ring.

Hannelore said: 'Are they going to rob us of what little the Nazis didn't steal?'

They were. The tall soldier looked frustrated and tried to pull off the woman's ring. When she realized what he wanted, she took it off herself and gave it to him.

The Russian took it, nodded, then pointed all around the room.

Hannelore stepped forward. 'These people are prisoners!' she said in German. 'Jews, and families of Jews, persecuted by the Nazis!'

Whether he understood her or not, he took no notice, but just pointed insistently at the watches on his arm.

Those few who had any valuables that had not been stolen or traded for food handed them over.

Liberation by the Red Army was not going to be the happy event many people had been looking forward to.

But there was worse to come.

The tall soldier pointed at Rebecca.

She cringed away from him and tried to hide behind Carla.

A second man, small with fair hair, grabbed Rebecca and pulled her away. Rebecca screamed, and the small man grinned as if he liked the sound.

Carla had a dreadful feeling she knew what was going to happen next.

The short man held Rebecca firmly while the tall man squeezed her breasts roughly, then said something that made them both laugh.

There were cries of protest from the people all around.

The tall man levelled his gun. Carla was terrified he would fire. He would kill and wound dozens of people if he pulled the trigger of a submachine gun in a crowded room.

Everyone else realized the danger, and they went quiet.

The two soldiers backed towards the door, taking Rebecca with them. She yelled and struggled, but she could not break the small soldier's grip.

When they reached the door, Carla stepped forward and cried: 'Wait!'

Something in her voice made them stop.

'She's too young,' Carla said. 'Only thirteen!' She did not know whether they understood her. She held up two hands, showing ten fingers, then one hand showing three. 'Thirteen!'

The tall soldier seemed to understand her. He grinned and said in German: *'Frau ist frau.'* A woman is a woman.

Carla found herself saying: 'You need a real woman.' She walked slowly forward. 'Take me, instead.' She tried to smile seductively. 'I'm not a child. I know what to do.' She came close, close enough to smell the rank odour of a man who had not bathed for months. Trying to conceal her distaste, she lowered her voice and said: 'I understand what a man wants.' She touched her own breast suggestively. 'Forget the child.'

The tall soldier looked again at Rebecca. Her eyes were red with weeping and her nose was running, which helpfully made her look more like a child, less like a woman.

He looked back at Carla.

She said: 'There's a bed upstairs. Shall I show you where?'

Again she was not sure he understood the words, but she took him by the hand and he followed her up the steps to the ground floor.

The fair one let go of Rebecca and came after.

Now that she had succeeded, Carla regretted her bravado. She wanted to break away from the Russians and run. But they would probably shoot her down then go back to Rebecca. Carla thought of the devastated child who had lost both parents yesterday. To be raped the next day would surely destroy her spirit for ever. Carla had to save her.

I will not be smashed by this, Carla thought. I can live through it. I will be myself again afterwards.

She led them to the electrocardiogram room. She felt cold, as if her heart were freezing and her thoughts becoming sluggish. Next to the bed was a can of the grease used by the doctors to improve the conductivity of the terminals. She pulled off her underpants, then took a large dob of grease and pushed it into her vagina. That might save her from bleeding.

She had to keep her act up. She turned back to the two soldiers. To her horror, three more followed them into the room. She tried to smile, but she could not.

She lay on her back and parted her legs.

The tall one knelt between her knees. He ripped open her uniform blouse to expose her breasts. She could see that he was manipulating himself, making his penis erect. He lay on top and entered her. She told herself this had no connection with what she and Werner had done together.

She turned her head to the side, but the soldier grasped her chin and turned her face back, making her look at him as he thrust inside her. She closed her eyes. She felt him kissing her, trying to force his tongue into her mouth. His breath smelled like rotting meat. When she clamped her mouth shut, he punched her face. She cried out and opened her bruised lips to him. She tried to think how much worse this would have been for a thirteen-year-old virgin.

The soldier grunted and ejaculated inside her. She tried not to let her disgust show on her face.

He climbed off, and the fair-haired one took his place.

Carla tried to close down her mind, to make her body into something detached, a machine, an object that had nothing to do with her. This one did not want to kiss her, but he sucked her breasts and bit her nipples, and when she cried out in pain he seemed pleased and did it harder.

Time passed, and he ejaculated.

Then another one got on top.

She realized that when this was over she would not be able to bathe or shower, for there was no running water in the city. That thought pushed her over the top. Their fluids would be inside her, their smell would be on her skin, their saliva in her mouth, and she would have no effective way to wash. Somehow that was worse than everything else. Her courage failed her, and she started to cry.

The third soldier satisfied himself, then the fourth lay on her.

20

1945 (II)

Adolf Hitler killed himself on Monday 30 April 1945, in his bunker in Berlin. Exactly a week later in London, at twenty to eight in the evening, the Ministry of Information announced that Germany had surrendered. A holiday was declared for the following day, Tuesday 8 May.

Daisy sat at the window of her apartment in Piccadilly, watching the celebrations. The street was thronged with people, making it almost impassable to cars and buses. The girls would kiss any man in uniform, and thousands of lucky servicemen were taking full advantage. By early afternoon many people were drunk. Through the open window Daisy could hear distant singing, and guessed that the crowd outside Buckingham Palace was singing 'Land of Hope and Glory'. She shared their happiness, but Lloyd was somewhere in France or Germany, and he was the only soldier she wanted to kiss. She prayed he had not been killed in the last few hours of the war.

Lloyd's sister, Millie, showed up with her two children. Millie's husband, Abe Avery, was also with the army somewhere. She and the children had come to the West End to join in the celebrations, and they took a break from the crowds at Daisy's place. The Leckwith home in Aldgate had long been a place of refuge for Daisy, and she was glad to have a chance to reciprocate. She made tea for Millie – her staff were out there celebrating – and poured orange juice for the children. Lennie was five now and Pammie three.

Since Abe had been conscripted, Millie had been running his leather wholesaling business. His sister, Naomi Avery, was the bookkeeper, but Millie did the selling. 'It's going to change, now,' Millie said. 'For the past five years the demand has been for tough hides for boots and shoes. Now we're going to need softer leathers, calf and pigskin, for handbags and briefcases. When the luxury market comes back, there'll be decent money to be made at last.'

Daisy recalled that her father had the same way of thinking as Millie. Lev, too, was always looking ahead, searching out the opportunities.

Eva Murray appeared next, with her four children in tow. Jamie, aged eight, organized a game of hide-and-seek, and the apartment became like a kindergarten. Eva's husband, Jimmy, now a colonel, was also somewhere in France or Germany, and Eva was suffering the same agonies of anxiety as Daisy and Millie.

'We'll hear from them, any day now,' Millie said. 'And then it will really be all over.'

Eva was also desperate for news of her family in Berlin. However, she thought it might be weeks or months before anyone could learn the fates of individual Germans in the postwar chaos. 'I wonder whether my children will ever know my parents,' she said sadly.

At five o'clock Daisy made a pitcher of martinis. Millie went into the kitchen and, with characteristic speed and efficiency, produced a plate of sardines on toast to eat with the drinks. Eth and Bernie arrived just as Daisy was making a second round.

Bernie told Daisy that Lennie could read already, and Pammie could sing the National Anthem. Ethel said: 'Typical grandfather, thinks there have never been bright children before,' but Daisy could tell that in her heart she was just as proud of them.

Feeling relaxed and happy halfway down her second martini, she looked around at the disparate group gathered in her home. They had paid her the compliment of coming to her door without an invitation, knowing they would be welcomed. They belonged to her, and she to them. They were, she realized, her family.

She felt very blessed.

(ii)

Woody Dewar sat outside Leo Shapiro's office, looking through a sheaf of photographs. They were the pictures he had taken at Pearl Harbor, in the hour before Joanne died. The film had stayed in his camera for months, but eventually he had developed it and printed the pictures. Looking at them had made him so sad that he had put them in a drawer in his bedroom at the Washington apartment and left them there.

But this was a time for change.

He would never forget Joanne, but he was in love again, at last. He adored Bella and she felt the same. When they parted, at the Oakland train station outside San Francisco, he had told her that he loved her, and she had said: 'I love you, too.' He was going to ask her to marry him. He would have done so already but it seemed too soon – less than three months – and he did not want to give her hostile parents a pretext for objecting.

Also, he needed to make a decision about his future.

He did not want to go into politics.

This was going to shock his parents, he knew. They had always assumed he would follow in his father's footsteps and end up as the third Senator Dewar. He had gone along with this assumption unthinkingly. But in the war, and especially while in hospital, he had asked himself what he *really* wanted to do, if he survived; and the answer was not politics.

This was a good time to leave. His father had achieved his life's ambition. The Senate had debated the United Nations. It was at a similar point in history that the old League of Nations had foundered, a painful memory for Gus Dewar. But Senator Vandenberg had spoken passionately in favour, speaking of 'the dearest dream of mankind', and the UN Charter had been ratified by eighty-nine votes to two. The job was done. Woody would not be letting his father down by quitting now.

He hoped Gus would see it that way too.

Shapiro opened his office door and beckoned. Woody stood up and went in.

Shapiro was younger than Woody had expected, somewhere in his thirties. He was Washington bureau chief for the National Press Agency. He sat behind his desk and said: 'What can I do for Senator Dewar's son?'

'I'd like to show you some photographs, if I may.'

'All right.'

Woody spread his pictures on Shapiro's desk.

'Is this Pearl Harbor?' Shapiro said.

'Yes. December seventh, 1941.'

'My God.'

Woody was looking at them upside-down, but still they brought tears to his eyes. There was Joanne, looking so beautiful; and Chuck, grinning happily to be with his family and Eddie. Then the planes coming over, the bombs and torpedoes dropping from their bellies, the black-smoke explosions on the ships, and the sailors scrambling over the sides, dropping into the sea, swimming for their lives.

'This is your father,' Shapiro said. 'And your mother. I recognize them.'

'And my fiancée, who died a few minutes later. My brother, who was killed at Bougainville. And my brother's best friend.'

'These are fantastic photographs! How much do you want for them?'

'I don't want money,' Woody said.

Shapiro looked up in surprise.

Woody said: 'I want a job.'

(iii)

Fifteen days after VE Day, Winston Churchill called a General Election.

The Leckwith family were taken by surprise. Like most people, Ethel and Bernie had thought Churchill would wait until the Japanese surrendered. The Labour leader, Clement Attlee, had suggested an election in October. Churchill wrong-footed them all.

Major Lloyd Williams was released from the army to stand as Labour candidate for Hoxton, in the East End of London. He was full of eager enthusiasm for the future envisioned by his party. Fascism had been vanquished, and now British people could create a society that combined freedom with welfare. Labour had a well-thought-out plan for avoiding the catastrophes of the last twenty years: universal comprehensive unemployment insurance to help families through hard times, economic planning to prevent another Depression, and a United Nations Organization to keep the peace.

'You don't stand a chance,' said his stepfather, Bernie, in the kitchen of the house in Aldgate on Monday 4 June. Bernie's pessimism was the more convincing for being so uncharacteristic. 'They'll vote Tory because Churchill won the war,' he went on gloomily. 'It was the same with Lloyd George in 1918.'

Lloyd was about to reply, but Daisy got in first. 'The war wasn't won by the free market and capitalist enterprise,' she said indignantly. 'It was people working together and sharing the burdens, everybody doing his bit. That's socialism!'

Lloyd loved her most when she was passionate, but he was more deliberate. 'We already have measures that the old Tories would have condemned as Bolshevism: government control of railways, mines and shipping, for example, all brought in by Churchill. And Ernie Bevin has been in charge of economic planning all through the war.'

Bernie shook his head knowingly, an old-man gesture that irritated Lloyd. 'People vote with their hearts, not brains,' he said. 'They'll want to show their gratitude.'

'Well, no point sitting here arguing with you,' Lloyd said. 'I'm going to argue with voters instead.'

He and Daisy took a bus a few stops north to the Black Lion pub in Shoreditch, where they met up with a canvassing team from the Hoxton Constituency Labour Party. In fact canvassing was not about arguing with voters, Lloyd knew. Its main purpose was to identify supporters, so that on election day the party machine could make sure they all went to the polling station. Firm Labour supporters were noted; firm supporters of other parties were crossed off. Only people who had not yet made up their minds were worth more than a few seconds: they were offered the chance to speak to the candidate.

Lloyd got some negative reactions. 'Major, eh?' one woman said. 'My Alf is a corporal. He says the officers nearly lost us the war.'

There were also accusations of nepotism. 'Aren't you the son of the MP for Aldgate? What is this, a hereditary monarchy?'

He remembered his mother's advice. 'You never win a vote by proving the constituent a fool. Be charming, be modest, and don't lose your temper. If a voter is hostile and rude, thank him for his time and go away. You'll leave him thinking maybe he misjudged you.'

Working-class voters were strongly Labour. A lot of people told Lloyd that Attlee and Bevin had done a good job during the war. The waverers were mostly middle-class. When people said that Churchill had won the war, Lloyd quoted Attlee's gentle put-down: 'It wasn't a one-man government, and it wasn't a one-man war.'

Churchill had described Attlee as a modest man with much to be modest about. Attlee's wit was less brutal, and for that reason more effective; at least, Lloyd thought so.

A couple of constituents mentioned the sitting MP for Hoxton, a Liberal, and said they would vote for him because he had helped them solve some problem. Members of Parliament were often called upon by constituents who felt they were being treated unjustly by the government, an employer or a neighbour. It was time-consuming work but it won votes.

Overall, Lloyd could not tell which way public opinion was leaning.

Only one constituent mentioned Daisy. The man came to the door with his mouth full of food. Lloyd said: 'Good evening, Mr Perkinson, I understand you wanted to ask me something.'

'Your fiancée was a Fascist,' the man said, chewing.

Lloyd guessed he had been reading the *Daily Mail*, which had run a spiteful story about Lloyd and Daisy under the headline THE SOCIALIST AND THE VISCOUNTESS.

Lloyd nodded. 'She was briefly fooled by Fascism, like many others.'

'How can a socialist marry a Fascist?'

Lloyd looked around, spotted Daisy, and beckoned her. 'Mr Perkinson here is asking me about my fiancée being an ex-Fascist.'

'Pleased to meet you, Mr Perkinson.' Daisy shook the man's hand. 'I quite understand your concern. My first husband was a Fascist in the thirties, and I supported him.'

Perkinson nodded. He probably believed a wife should take her views from her husband.

'How foolish we were,' Daisy went on. 'But, when the war came, my first husband joined the RAF and fought against the Nazis as bravely as anyone.'

'Is that a fact?'

'Last year he was flying a Typhoon over France, strafing a German troop train, when he was shot down and killed. So I'm a war widow.'

Perkinson swallowed his food. 'I'm sorry to hear that, of course.'

But Daisy had not finished. 'For myself, I lived in London throughout the war. I drove an ambulance all through the Blitz.'

'Very brave of you, I'm sure.'

'Well, I just hope you think that my late husband and I both paid our dues.'

'I don't know about that,' Perkinson said sulkily.

'We won't take up any more of your time,' said Lloyd. 'Thank you for explaining your views to me. Good evening.'

As they walked away, Daisy said: 'I don't think we won him round.'

'You never do,' Lloyd said. 'But he's seen both sides of the story now, which might make him a bit less vociferous about it, later this evening, when he talks about us in the pub.'

'Hmm.'

Lloyd sensed he had failed to reassure Daisy.

Canvassing finished early, for tonight the first of the radio election broadcasts would be aired on the BBC, and all party workers would be listening. Churchill had the privilege of making the first one.

On the bus home, Daisy said: 'I'm worried. I'm an election liability to you.'

'No candidate is perfect,' Lloyd said. 'It's how you deal with your weaknesses that matters.'

'I don't want to be your weakness. Perhaps I should stay out of the way.'

'On the contrary, I want everyone to know all about you from the start. If you are a liability, I will get out of politics.'

'No, no! I'd hate to think I made you give up your ambitions.'

'It won't come to that,' he said, but once again he could see that he had not succeeded in assuaging her anxiety.

Back in Nutley Street, the Leckwith family sat around the radio in the kitchen. Daisy held Lloyd's hand. 'I came here a lot while you were away,' she said. 'We used to listen to swing music and talk about you.'

The thought made Lloyd feel very lucky.

Churchill came on. The familiar rasp was stirring. For five grim years that voice had given people strength and hope and courage. Lloyd felt despairing: even he was tempted to vote for this man.

'My friends,' the Prime Minister said. 'I must tell you that a socialist policy is abhorrent to the British ideas of freedom.'

Well, that was routine knockabout stuff. All new ideas were condemned as foreign imports. But what would Churchill offer people? Labour had a plan, but what did the Conservatives propose?

'Socialism is inseparably interwoven with totalitarianism,' Churchill said.

Lloyd's mother, Ethel, said: 'Surely he's not going to pretend we're like the Nazis?'

'I think he is, though,' Bernie said. 'He'll say we've defeated the enemy abroad, now we must defeat the enemy in our midst. Standard conservative tactic.'

'People won't believe that,' Ethel said.

Lloyd said: 'Hush!'

Churchill said: 'A socialist state, once thoroughly completed in all its details and its aspects, could not afford to suffer opposition.'

'This is outrageous,' said Ethel.

'But I will go farther,' said Churchill. 'I declare to you, from the bottom of my heart, that no socialist system can be established without a political police.'

'Political police?' Ethel said indignantly. 'Where is he getting this stuff from?'

Bernie said: 'This is good, in a way. He can't find anything to criticize in our manifesto, therefore he's attacking us for things we aren't actually proposing to do. Bloody liar.'

Lloyd shouted: 'Listen!'

Churchill said: 'They would have to fall back on some form of Gestapo.'

Suddenly they were all on their feet, shouting protests. The Prime Minister was drowned out. 'Bastard!' Bernie yelled, shaking his fist at the Marconi radio set. 'Bastard, bastard!'

When they had quietened down, Ethel said: 'Is that going to be their campaign? Just lies about us?'

'It bloody well is,' said Bernie.

Lloyd said: 'But will people believe it?'

(iv)

In southern New Mexico, not far from El Paso, there is a desert called Jornada del Muerto, the Voyage of the Dead. All day long the cruel sun beats down on needlethorn mesquite and sword-leafed yucca plants. The inhabitants are scorpions and rattlesnakes, fire ants and tarantula spiders. Here the men of the Manhattan Project tested the most dreadful weapon the human race had ever devised.

Greg Peshkov was with the scientists watching from ten thousand yards away. He had two hopes: first, that the bomb would work; and second, that ten thousand yards was far enough.

The countdown started at nine minutes past five in the morning, Mountain War Time, on Monday 16 July. It was dawn, and there were streaks of gold in the sky to the east.

The test was codenamed Trinity. When Greg had asked why, the senior scientist, the pointy-eared Jewish New Yorker J. Robert Oppenheimer, had quoted a poem by John Donne: 'Batter my heart, three-person'd God.'

'Oppie' was the cleverest person Greg had ever met. The most brilliant physicist of his generation, he also spoke six languages. He had read Karl Marx's *Capital* in the original German. The kind of thing he did for fun was learn Sanskrit. Greg liked and admired him. Most physicists were geeks but Oppie, like Greg himself, was an exception: tall, handsome, charming, and a real ladykiller.

In the middle of the desert, Oppie had instructed the Army Corps of Engineers to build a one-hundred-foot tower of steel struts in concrete footings. On top was an oak platform. The bomb had been winched up to the platform on Saturday.

The scientists never used the word 'bomb'. They called it 'the gadget'. At its heart was a ball of plutonium, a metal that did not exist in nature but was created as a by-product in nuclear piles. The ball weighed ten pounds and contained all the plutonium in the world. Someone had calculated that it was worth a billion dollars.

Thirty-two detonators on the surface of the ball would go off simultaneously, creating such powerful inward pressure that the plutonium would become more dense and go critical.

No one really knew what would happen next.

The scientists were running a betting pool, dollar a ticket, on the force of the explosion measured in equivalent tons of TNT. Edward Teller bet 45,000 tons. Oppie bet 300 tons. The official forecast was 20,000 tons. The night before, Enrico Fermi had offered to take side bets on whether the blast would wipe out the entire state of New Mexico. General Groves had not found it funny.

The scientists had had a perfectly serious discussion about whether the explosion would ignite the atmosphere of the entire earth, and destroy the planet; but they had come to the conclusion that it would not. If they were wrong, Greg just hoped it would happen fast.

The trial had originally been scheduled for 4 July. However, every time they tested a component, it failed; so the big day had been postponed several times. Back at Los Alamos, on Saturday, a mock-up they called the Chinese Copy had refused to ignite. In the betting pool, Norman Ramsey had picked zero, gambling that the bomb would be a dud.

Today detonation had been scheduled for 2 a.m., but at that time there had been a thunderstorm – in the desert! Rain would bring the radioactive fallout down on the heads of the watching scientists, so the blast was postponed.

The storm had ended at dawn.

Greg was at a bunker called S-10000, which was the control room. Like most of the scientists, he was standing outside for a better view. Hope and fear struggled for mastery of his heart. If the bomb was a dud, the efforts of hundreds of people – plus about two billion dollars – would have gone for nothing. And if the bomb was not a dud, they might all be killed in the next few minutes.

Beside him was Wilhelm Frunze, the young German scientist he had first met in Chicago. 'What would have happened, Will, if lightning had struck the bomb?'

Frunze shrugged. 'No one knows.'

A green Verey rocket shot into the sky, startling Greg.

'Five-minute warning,' Frunze said.

Security had been haphazard. Santa Fe, the nearest town to Los Alamos, was crawling with well-dressed FBI agents. Leaning nonchalantly against walls in their tweed jackets and neckties, they were obvious to local residents, who wore blue jeans and cowboy boots.

The Bureau was also illegally tapping the phones of hundreds of people involved in the Manhattan Project. This bewildered Greg. How could the nation's premier law enforcement agency systematically commit criminal acts?

Nevertheless, army security and the FBI had identified some spies and quietly removed them from the project, including Barney McHugh. But had they found them all? Greg did not know. Groves had been forced to take risks. If he had fired everyone the FBI asked him to, there would not have been enough scientists left to build the bomb.

Unfortunately, most scientists were radicals, socialists and liberals. There was hardly a conservative among them. And they believed that the truths discovered by science were for humankind to share, and should never be kept secret in the service of one regime or country. So, while the American government was keeping this huge project top secret, the scientists held discussion groups about sharing nuclear technology with all the nations of the world. Oppie himself was suspect: the only reason he was not in the Communist Party was that he never joined clubs.

Right now Oppie was lying on the ground next to his kid brother, Frank, also an outstanding physicist, also a Communist. They both held pieces of welding glass through which to observe the explosion. Greg and Frunze had similar pieces of glass. Some of the scientists were wearing sunglasses.

Another rocket went off. 'One minute,' said Frunze.

Greg heard Oppie say: 'Lord, these affairs are hard on the heart.'

He wondered if those would be Oppie's last words.

Greg and Frunze lay on the sandy earth near Oppie and Frank. They all held their visors of welding glass in front of their eyes and gazed towards the test site.

Facing death, Greg thought about his mother, his father, and his sister Daisy in London. He wondered how much they would miss him. He thought, with mild regret, of Margaret Cowdry, who had dumped him for a guy who was willing to marry her. But most of all he thought of Jacky Jakes and his son, Georgy, now nine years old. He passionately wanted to watch Georgy grow up. He realized Georgy was the main reason he was hoping to stay alive. Stealthily, the child had crept into his soul and stolen his love. The strength of this feeling surprised Greg.

A gong chimed, a strangely inappropriate sound in the desert.

'Ten seconds.'

Greg suffered an impulse to get up and run away. Silly though it was – how far could he get in ten seconds? – he had to force himself to lie still.

The bomb went off at five twenty-nine and forty-five seconds.

First there was an awesome flash, impossibly bright, the fiercest glare Greg had ever seen, stronger than the sun.

Then a weird dome of fire seemed to come out of the ground. With terrifying speed it grew monstrously high. It reached the level of the mountains and continued to rise, rapidly dwarfing the peaks.

Greg whispered: 'Jesus . . .'

The dome morphed into a square. The light was still brighter than noonday, and the distant mountains were so vividly illuminated that Greg could see every fold and crevice and rock.

Then the shape changed again. A pillar appeared below, seeming to push miles into the sky, like the fist of God. The cloud of boiling fire above the pillar spread like an umbrella, until the whole thing looked like a mushroom seven miles tall. The colours in the cloud were hellish orange, green and purple.

Greg was hit by a wave of heat, as if the Almighty had opened a giant oven. At the same moment the bang of the explosion reached his ears like the crack of doom. But that was only the beginning. A noise like supernaturally loud thunder rolled over the desert, drowning all other sound.

The blazing cloud began to diminish but the thunder went on and on, impossibly sustained, until Greg wondered if this was the sound of the end of the world.

At last it faded away, and the mushroom cloud began to disperse.

Greg heard Frank Oppenheimer say: 'It worked.'

Oppie said: 'Yes, it worked.'

The two brothers shook hands.

And the world is still here, Greg thought.

But it has been forever changed.

(v)

Lloyd Williams and Daisy went to Hoxton Town Hall on the morning of 26 July to watch the votes being counted.

If Lloyd lost, Daisy was going to break off the engagement.

He fervently denied that she was a political liability, but she knew better. Lloyd's political enemies made a point of calling her 'Lady Aberowen'. Voters reacted to her American accent by looking indignant, as if she had no right to take part in British politics. Even Labour Party members treated her differently, asking if she would prefer coffee when they were all drinking tea.

As Lloyd had forecast, she was often able to overcome people's initial hostility, by being natural and charming, and helping the other women wash up the tea cups. But was that enough? The election results would give the only definite answer.

She was not going to marry him if it meant his giving up his life's work. He said he was willing to do it, but it was a hopeless foundation for marriage. Daisy shuddered with horror as she imagined him doing some other job, working at a bank or in the civil service, miserably unhappy and trying to pretend it was not her fault. It did not bear thinking about.

Unfortunately, everyone thought the Conservatives were going to win the election.

Some things had gone Labour's way in the campaign. Churchill's 'Gestapo' speech had backfired. Even Conservatives had been dismayed. Clement Attlee, broadcasting the following evening for Labour, had been coolly ironic. 'When I listened to the Prime Minister's speech last night, in which he gave such a travesty of the policy of the Labour Party, I realized at once what was his object. He wanted the voters to understand how great was the difference between Winston Churchill, the great leader in war of a united nation, and Mr Churchill, the party leader of the Conservatives. He feared lest those who had accepted his leadership in war might be tempted out of gratitude to follow him further. I thank him for having disillusioned them so thoroughly.' Attlee's magisterial disdain had made Churchill seem a rabble-rouser. People had had too much of blood-red passion, Daisy thought; they would surely prefer temperate common sense in peacetime.

A Gallup poll taken the day before voting showed Labour winning, but no one believed it. The idea that you could forecast the result by asking a small number of electors seemed a bit unlikely. The *News Chronicle*, which had published the poll, was predicting a tie.

All the other papers said the Conservatives would win.

Daisy had never before taken any interest in the mechanics of democracy, but her fate was in the balance now, and she watched, mesmerized, as the voting papers were taken out of the boxes, sorted, counted, bundled, and counted again. The man in charge was called the Returning Officer, as if he had been away for a while. He was, in fact, the Town Clerk. Observers from each of the parties monitored the proceedings to make sure there was no carelessness or dishonesty. The process was long, and Daisy felt tortured by suspense.

At half past ten, they heard the first result from elsewhere. Harold Macmillan, a protégé of Churchill's and a wartime Cabinet Minister, had lost Stockton-on-Tees to Labour. Fifteen minutes later there was news of a huge swing to Labour in Birmingham. No radios were allowed into the hall, so Daisy and Lloyd were relying on rumours filtering in from outside, and Daisy was not sure what to believe.

It was midday when the Returning Officer called the candidates and their agents into a corner of the room, to give them the result before making the announcement publicly. Daisy wanted to go with Lloyd but she was not permitted.

The man spoke quietly to all of them. As well as Lloyd and the sitting MP, there was a Conservative and a Communist. Daisy studied their faces, but could not guess who had won. They all went up on to the platform, and the room fell silent. Daisy felt nauseous.

'I, Michael Charles Davies, being the duly appointed Returning Officer for the Parliamentary Constituency of Hoxton . . .'

Daisy stood with the Labour Party observers and stared at Lloyd. Was she about to lose him? The thought squeezed her heart and made her breathless with fear. In her life she had twice chosen a man who was disastrously wrong. Charlie Farquharson had been the opposite of her father, nice but weak. Boy Fitzherbert had been much like her father, wilful and selfish. Now, at last, she had found Lloyd, who was both strong and kind. She had not picked him for his social status or for what he could do for her, but simply because he was an extraordinarily good man. He was gentle, he was smart, he was trustworthy, and he adored her. It had taken her a long time to realize that he was what she was looking for. How foolish she had been.

The Returning Officer read out the number of votes cast for each candidate. They were listed alphabetically, so Williams came last. Daisy was so anxious that she could not keep the numbers in her head. 'Reginald Sidney Blenkinsop, five thousand four hundred and twenty-seven...'

When Lloyd's vote was read out, the Labour Party people all around Daisy burst out cheering. It took her a moment to realize that meant he had won. Then she saw his solemn expression turn into a broad grin. Daisy began to clap and cheer louder than anyone. He had won! And she did not have to leave him! She felt as if her life had been saved.

'I therefore declare that Lloyd Williams is duly elected Member of Parliament for Hoxton.'

Lloyd was a Member of Parliament. Daisy watched proudly as he stepped forward and made an acceptance speech. There was a formula for such speeches, she realized, and he tediously thanked the Returning Officer and his staff, then thanked his losing opponents for a fair fight. She was impatient to hug him. He finished with a few sentences about the task that lay ahead, of rebuilding war-torn Britain and creating a fairer society. He stood down to more applause.

Coming off the stage, he walked straight to Daisy, put his arms around her, and kissed her.

She said: 'Well done, my darling,' then she found she could no longer speak.

After a while they went outside and caught a bus to Labour Party headquarters at Transport House. There they learned that Labour had already won 106 seats.

It was a landslide.

Every pundit had been wrong, and everyone's expectations were confounded. When all the results were in, Labour had 393 seats, the Conservatives 210. The Liberals had twelve and the Communists one – Stepney. Labour had an overwhelming majority.

At seven o'clock in the evening Winston Churchill, Britain's great war leader, went to Buckingham Palace and resigned as Prime Minister.

Daisy thought of one of Churchill's jibes about Attlee: 'An empty car drew up and Clem got out.' The man he called a nonentity had thrashed him.

At half past seven Clement Attlee went to the palace in his own car, driven by his wife, Violet, and King George VI asked him to become Prime Minister.

In the house in Nutley Street, after they had all listened to the news on the radio, Lloyd turned to Daisy and said: 'Well, that's that. Can we get married now?'

'Yes,' said Daisy. 'As quick as you like.'

(vi)

Volodya and Zoya's wedding reception was held in one of the smaller banqueting halls in the Kremlin.

The war with Germany was over, but the Soviet Union was still battered and impoverished, and a lavish celebration would have been frowned upon. Zoya had a new dress, but Volodya wore his uniform. However, there was plenty to eat, and the vodka flowed freely.

Volodya's nephew and niece were there, the twin children of his sister, Anya, and her unpleasant husband, Ilya Dvorkin. They were not yet six years old. Dimka, the dark-haired boy, sat quietly reading a book, while blue-eyed Tania was running around the room crashing into tables and annoying the guests, in a reversal of the expected behaviour of boys and girls.

Zoya looked so desirable in pink that Volodya would have liked to leave right away and take her to bed. That was out of the question, of course. His father's circle of friends included some of the most senior generals and politicians in the country, and many of them had come to toast the happy couple. Grigori was hinting that one extremely distinguished guest might arrive later: Volodya hoped it was not the depraved NKVD boss Beria.

Volodya's happiness did not quite let him forget the horrors he had seen and the profound misgivings he had developed about Soviet Communism. The unspeakable brutality of the secret police, the blunders of Stalin that had cost millions of lives, and the propaganda that had encouraged the Red Army to behave like crazed beasts in Germany, had all caused him to doubt the most fundamental things he had been brought up to believe. He wondered uneasily what kind of country Dimka and Tania would grow up in. But today was not the day to think about that.

The Soviet elite were in a good mood. They had won the war and defeated Germany. Their old enemy Japan was being crushed by the USA. The insane honour code of Japan's leaders made it difficult for them to surrender, but it was only a matter of time now. Tragically, while they clung to their pride, more Japanese and American troops would die, and more Japanese women and children would be bombed out of their homes; but the end result would be the same. Sadly, it seemed there was nothing the Americans could do to hasten the process and prevent unnecessary deaths.

Volodya's father, drunk and happy, made a speech. 'The Red Army has occupied Poland,' he said. 'Never again will that country be used as a springboard for a German invasion of Russia.'

All the old comrades cheered and thumped the tables.

'In Western Europe Communist parties are being endorsed by the masses as never before. In the Paris municipal elections last March, the Communist party won the largest share of the vote. I congratulate our French comrades.'

They cheered again.

'As I look around the world today, I see that the Russian revolution, in which so many brave men fought and died . . .' He tailed off as drunken tears came to his eyes. A hush descended on the room. He recovered himself. 'I see that the revolution has never been as secure as it is today!'

They raised their glasses. 'The revolution! The revolution!' Everyone drank.

The doors flew open, and Comrade Stalin walked in.

Everyone stood up.

His hair was grey, and he looked tired. He was about sixty-five, and he had been ill: there were rumours that he had suffered a series of strokes or minor heart attacks. But his mood today was ebullient. 'I have come to kiss the bride!' he said.

He walked up to Zoya and put his hands on her shoulders. She was a good three inches taller than he, but she managed to stoop discreetly. He kissed her on both cheeks, allowing his grey-moustached mouth to linger just long enough to make Volodya feel resentful. Then he stepped back and said: 'How about a drink for me?'

Several people hastened to get him a glass of vodka. Grigori insisted on giving Stalin his chair in the centre of the head table. The buzz of conversation resumed, but it was subdued: they were thrilled he was here, but now they had to be careful of every word and every move. This man could have a person killed with a snap of his fingers, and he frequently had.

More vodka was brought, the band began to play Russian folk dances, and slowly people relaxed. Volodya, Zoya, Grigori and Katerina did a four-person dance called a kadril, which was intended to be comic and always made people laugh. After that more couples danced, and the men started to do the barynya, in which they had to squat and kick up their legs, which caused many of them to fall over. Volodya kept checking on Stalin out of the corner of his eye – as did everyone else in the room – and he seemed to be enjoying himself, tapping his glass on the table in time with the balalaikas.

Zoya and Katerina were dancing a troika with Zoya's boss, Vasili, a senior physicist working on the bomb project, and Volodya was sitting out, when the atmosphere changed.

An aide in a civilian suit came in, hurried around the edge of the room, and went right up to Stalin. Without ceremony, he leaned over the leader's shoulder and spoke to him quietly but urgently.

Stalin at first looked puzzled, and asked a sharp question, then another. Then his face changed. He went pale, and seemed to stare at the dancers without seeing them.

Volodya said under his breath: 'What the hell has happened?'

The dancers had not yet noticed, but those sitting at the head table looked frightened.

After a moment Stalin stood up. Those around him deferentially did the same. Volodya saw that his father was still dancing. People had been shot for less.

But Stalin had no eyes for the wedding guests. With the aide at his side he left the table. He walked towards the door, crossing the dance floor. Terrified revellers jumped out of his way. One couple fell over. Stalin did not seem to notice. The band ground to a halt. Saying nothing, looking at nobody, Stalin left the room.

Some of the generals followed him out, looking scared.

Another aide appeared, then two more. They all sought out their bosses and spoke to them. A young man in a tweed jacket went up to Vasili. Zoya seemed to know the man, and listened intently to him. She looked shocked.

Vasili and the aide left the room. Volodya went to Zoya and said: 'For God's sake, what's going on?'

Her voice was shaky. 'The Americans have dropped a nuclear bomb in Japan.' Her beautifully pale face seemed even whiter than normal. 'At first the Japanese government couldn't figure out what had happened. It took them hours to realize what it was.'

'Are we sure?'

'It flattened five square miles of buildings. They estimate that seventy-five thousand people were killed instantly.'

'How many bombs?'

'One.'

'One bomb?'

'Yes.'

'Good God. No wonder Stalin turned pale.'

They both stood silent. The news was spreading around the room visibly. Some people sat stunned; others got up and left, heading for their offices, their telephones, their desks and their staff.

'This changes everything,' Volodya said.

'Including our honeymoon plans,' said Zoya. 'My leave is sure to be cancelled.'

'We thought the Soviet Union was safe.'

'Your father has just made a speech about how the revolution has never been so secure.'

'Now nothing is secure.'

'No,' said Zoya. 'Not until we have a bomb of our own.'

(vii)

Jacky Jakes and Georgy were in Buffalo, staying at Marga's apartment for the first time. Greg and Lev were there too, and on Victory Japan Day – Wednesday 15 August – they all went to Humboldt Park. The paths were crowded with jubilant couples and there were hundreds of children splashing in the pond.

Greg was happy and proud. The bomb had worked. The two devices dropped on Hiroshima and Nagasaki had wreaked sickening devastation, but they had brought the war to a quick end and saved thousands of American lives. Greg had played a role in that. Because of what they had all done, Georgy was going to grow up in a free world.

'He's nine,' Greg said to Jacky. They were sitting on a bench, talking, while Lev and Marga took Georgy to buy ice cream.

'I can hardly believe it.'

'What will he be, I wonder?'

Jacky said fiercely: 'He's not going to do something stupid like acting or playing the goddamn trumpet. He's got brains.'

'Would you like him to be a college professor, like your father?'

'Yes.'

'In that case . . .' Greg had been leading up to this, and was nervous about how Jacky might react – 'he ought to go to a good school.'

'What did you have in mind?'

'How about boarding school? He could go where I went.'

'He'd be the only black pupil.'

'Not necessarily. When I was there we had a coloured guy, an Indian from Delhi called Kamal.'

'Just one.'

'Yes.'

'Was he teased?'

'Sure. We called him Camel. But the boys got used to him, and he made some friends.'

'What happened to him, do you know?'

'He became a pharmacist. I hear he already owns two drugstores in New York.'

Jacky nodded. Greg could tell that she was not opposed to this plan. She came from a cultured family. Although she herself had rebelled and dropped out, she believed in the value of education. 'What about the school fees?'

'I could ask my father.'

'Would he pay?'

'Look at them.' Greg pointed along the path. Lev, Marga and Georgy were returning from the ice-cream vendor's cart. Lev and Georgy were walking side by side, eating ice-cream cones, holding hands. 'My conservative father, holding the hand of a coloured child in a public park. Trust me, he'll pay the school fees.'

'Georgy doesn't really fit anywhere,' Jacky said, looking troubled. 'He's a black boy with a white daddy.'

'I know.'

'People in your mother's apartment building think I'm the maid – did you know that?'

'Yes.'

'I've been careful not to set them straight. If they thought Negroes were in the building as guests, there might be trouble.'

Greg sighed. 'I'm sorry, but you're right.'

'Life is going to be tough for Georgy.'

'I know,' said Greg. 'But he's got us.'

Jacky gave him a rare smile. 'Yeah,' she said. 'That's something.'

Part Three

THE COLD PEACE

21

After the wedding Volodya and Zoya moved into an apartment of their own. Few Russian newlyweds were so lucky. For four years the industrial might of the Soviet Union had been directed to making weapons. Hardly any homes had been built, and many had been destroyed. But Volodya was a major in Red Army Intelligence, as well as the son of a general, and he was able to pull strings.

It was a compact space: a living room with a dining table, a bedroom so small the bed almost filled it; a kitchen that was crowded with two people in it; a cramped toilet with a washbasin and shower, and a tiny hall with a closet for their clothes. When the radio was on in the living room, they could hear it all over the flat.

They quickly made it their own. Zoya bought a bright yellow coverlet for the bed. Volodya's mother produced a set of crockery that she had bought in 1940, in anticipation of his wedding, and saved all through the war. Volodya hung a picture on the wall, a graduation photograph of his class at the Military Intelligence Academy.

They made love more now. Being alone made a difference Volodya had not anticipated. He had never felt particularly inhibited when sleeping with Zoya at his parents' place, or in the apartment she had used to share; but now he realized it had an influence. You had to keep your voice down, you listened in case the bed squeaked, and there was always the possibility, albeit remote, that somebody would walk in on you. Other people's homes were never completely private.

They often woke early, made love, then lay kissing and talking for an hour before getting dressed for work. Lying with his head on her thighs on one such morning, the smell of sex in his nostrils, Volodya said: 'Do you want some tea?'

'Yes, please.' She stretched luxuriously, reclining on the pillows.

Volodya put on a robe and crossed the tiny hallway to the little kitchen, where he lit the gas under the samovar. He was displeased to see the pots and dishes from last night's dinner stacked in the sink. 'Zoya! he said. 'This kitchen's in a mess!'

She could hear him easily in the small apartment. 'I know,' she said.

He went back to the bedroom. 'Why didn't you clean up last night?'

'Why didn't you?'

It had not occurred to him that it might be his responsibility. But he said: 'I had a report to write.'

'And I was tired.'

The suggestion that it was his fault irritated him. 'I hate a filthy kitchen.'

'So do I.'

Why was she being so obtuse? 'If you don't like it, clean it!'

'Let's do it together, right away.' She sprang out of bed. She pushed past him with a sexy smile and went into the kitchen.

Volodya followed.

She said: 'You wash, I'll dry.' She took a clean towel from a drawer.

She was still naked. He could not help but smile. Her body was long and slim, and her skin was white. She had flat breasts and pointed nipples, and the hair of her groin was fine and blonde. One of the joys of being married to her was her habit of moving around the apartment in the nude. He could stare at her body for as long as he liked. She seemed to enjoy it. If she caught his eye she showed no embarrassment, but just smiled.

He rolled up the sleeves of his robe and began to wash the dishes, passing them to Zoya to dry. Washing up was not a very manly activity – Volodya had never seen his father do it – but Zoya seemed to think such chores should be shared. It was an eccentric idea. Did Zoya have a highly developed sense of fairness in marriage? Or was he being emasculated?

He thought he heard something outside. He glanced into the hall: the apartment door was only three or four steps from the kitchen sink. He could see nothing out of the ordinary.

Then the door was smashed open.

Zoya screamed.

Volodya picked up the carving knife he had just washed. He stepped past Zoya and stood in the kitchen doorway. A uniformed policeman holding a sledgehammer was just outside the ruined door.

Volodya was filled with fear and rage. He said: 'What the fuck is this?'

The policeman stepped back, and a small, thin man with a face like a rodent entered the flat. It was Volodya's brother-in-law, Ilya Dvorkin, an agent of the secret police. He was wearing leather gloves.

'Ilya!' said Volodya. 'You stupid weasel.'

'Speak respectfully,' said Ilya.

Volodya was baffled as well as angry. The secret police did not normally arrest the staff of Red Army Intelligence, and vice versa. Otherwise it would have been gang warfare. 'Why the hell have you bust my door? I would have opened it!'

Two more agents stepped into the hall and stood behind Ilya. They wore their trademark leather coats, despite the mild late-summer weather.

Volodya was fearful as well as angry. What was going on?

Ilya said in a shaky voice: 'Put the knife down, Volodya.'

'No need to be afraid,' said Volodya. 'I was just washing up.' He handed the knife to Zoya, standing behind him. 'Please step into the living room. We can talk while Zoya gets dressed.'

'Do you imagine this is a social call?' Ilya said indignantly.

'Whatever kind of call it is, I'm sure you don't want the embarrassment of seeing my wife naked.'

'I am here on official police business!'

'Then why did they send my brother-in-law?'

Ilya lowered his voice. 'Don't you understand that it would be much worse for you if someone else had come?'

This looked like bad trouble. Volodya struggled to keep up the facade of bravado. 'Exactly what do you and these other assholes want?'

'Comrade Beria has taken over the direction of the nuclear physics programme.'

Volodya knew that. Stalin had set up a new committee to direct the work and made Beria chairman. Beria knew nothing about physics and was completely unqualified to organize a scientific research project. But Stalin trusted him. It was the usual problem of Soviet government: incompetent but loyal people were promoted into jobs they could not cope with.

Volodya said: 'And Comrade Beria needs my wife in her laboratory, developing the bomb. Have you come to drive her to work?'

'The Americans created their nuclear bomb before the Soviets.'

'Indeed. Could they perhaps have given research physics higher priority than we did?'

'It is not possible that capitalist science should be superior to Communist science!'

'This is a truism.' Volodya was puzzled. Where was this heading? 'So what do you conclude?'

'There must have been sabotage.'

That was exactly the kind of ludicrous fantasy the secret police would dream up. 'What kind of sabotage?'

'Some of the scientists deliberately delayed the development of the Soviet bomb.'

Volodya began to understand, and he felt afraid. But he continued to respond belligerently: it was always a mistake to show weakness with these people. 'Why the hell would they do that?'

'Because they are traitors – and your wife is one!'

'You'd better not be serious, you piece of shit.'

'I am here to arrest your wife.'

'What?' Volodya was flabbergasted. 'This is insane!'

'It is the view of my organization.'

'There is no evidence.'

'For evidence, go to Hiroshima!'

Zoya spoke for the first time since she had screamed. 'I'll have to go with them, Volodya. Don't get yourself arrested too.'

Volodya pointed a finger at Ilya. 'You are in so much fucking trouble.'

'I'm carrying out my orders.'

'Step out of the way. My wife is going into the bedroom to get dressed.'

'No time for that,' said Ilya. 'She must come as she is.'

'Don't be ridiculous.'

Ilya put his nose in the air. 'A respectable Soviet citizen would not walk around the apartment with no clothes on.'

Volodya wondered briefly how his sister felt being married to this creep. 'You, the secret police, morally disapprove of nudity?'

'Her nakedness is evidence of her degradation. We will take her as she is.'

'No you fucking won't.'

'Stand aside.'

'You stand aside. She's going to get dressed.' Volodya stepped into the hall and stood in front of the three agents, holding his arms out so that Zoya could pass behind him.

As she moved, Ilya reached past Volodya and grabbed her arm.

Volodya punched him in the face, twice. Ilya cried out and staggered back. The two men in leather coats stepped forward. Volodya aimed a punch at one, but the man dodged it. Then each man took one of Volodya's arms. He struggled, but they were strong and seemed to have done this before. They slammed him against the wall.

While they held him, Ilya punched him in the face with leather-gloved fists, twice, three times, four, then in the stomach, again and again until Volodya puked blood. Zoya tried to intervene, but Ilya punched her, too, and she screamed and fell back.

Volodya's bathrobe came open in front. Ilya kicked him in the balls, then kicked his knees. Volodya sagged, unable to stand, but the two men in leather coats held him up, and Ilya punched him some more.

At last Ilya turned away, rubbing his knuckles. The other two released Volodya, and he crumpled to the floor. He could hardly breathe and felt unable to move, but he was conscious. Out of the corner of his eye he saw the two heavies grab Zoya and march her naked out of the apartment. Ilya followed.

As the minutes went by, the pain changed from sharp agony to deep, dull ache, and Volodya's breathing began to return to normal.

Motion eventually returned to his limbs, and he dragged himself upright. He made it to the phone and dialled his father's number, hoping the old man had not yet left for work. He was relieved to hear his father's voice. 'They've arrested Zoya,' he said.

'Fucking bastards,' Grigori said. 'Who was it?'

'It was Ilya.'

'What?'

'Make some calls,' Volodya said. 'See if you can find out what the fuck is going on. I have to wash off the blood.'

'What blood?'

Volodya hung up.

It was only a couple of steps to the bathroom. He dropped his bloodstained robe and got into the shower. The warm water brought some relief to his bruised body. Ilya was mean but not strong, and he had not broken any bones.

Volodya turned off the water. He looked in the bathroom mirror. His face was covered with cuts and bruises.

He did not bother to dry himself. With considerable effort, he got dressed in his Red Army uniform. He wanted the symbol of authority.

His father arrived as he was trying to tie the laces of his boots. 'What the fucking hell happened here?' Grigori roared.

Volodya said: 'They were looking for a fight, and I was foolish enough to give them one.'

His father was unsympathetic at first. 'I'd have expected you to know better.'

'They insisted on taking her away naked.'

'Fucking creeps.'

'Did you find out anything?'

'Not yet. I talked to a couple of people. No one knows anything.' Grigori looked worried. 'Either someone has made a really stupid mistake . . . or for some reason they're very sure of themselves.'

'Drive me to my office. Lemitov is going to be mad as hell. He won't let them get away with this. If they are allowed to do it to me, they'll do it to all of Red Army Intelligence.'

Grigori's car and driver were waiting outside. They drove to the Khodynka airfield. Grigori stayed in the car while Volodya limped into Red Army Intelligence headquarters. He went straight to the office of his boss, Colonel Lemitov.

He tapped on the door, walked in, and said: 'The fucking secret police have arrested my wife.'

'I know,' said Lemitov.

'You know?'

'I okayed it.'

Volodya's jaw dropped. 'What the fuck?'

'Sit down.'

'What is going on?'

'Sit down and shut up, and I'll tell you.'

Volodya eased himself painfully into a chair.

Lemitov said: 'We have to have a nuclear bomb, and fast. At the moment, Stalin is playing it tough with the Americans, because we're fairly sure they don't have a big enough arsenal of nuclear weapons to wipe us out. But they're building a stockpile, and at some point they will use them – unless we are in a position to retaliate.'

This made no sense. 'My wife can't design the bomb while the secret police are punching her in the face. This is insane.'

'Shut the fuck up. Our problem is that there are several possible designs. The Americans took five years to figure out which would work. We don't have that much time. We have to steal their research.'

'We'll still need Russian physicists to copy the design – and for that they have to be in their laboratories, not locked in the basement of the Lubyanka.'

'You know a man called Wilhelm Frunze.'

'I was at school with him. The Berlin Boys' Academy.'

'He gave us valuable information about British nuclear research. Then he moved to the States, where he worked on the nuclear bomb project. The Washington staff of the NKVD contacted him, scared him by their incompetence, and fucked up the relationship. We need to win him back.'

'What has all this got to do with me?'

'He trusts you.'

'I don't know that. I haven't seen him for twelve years.'

'We want you to go to America and talk to him.'

'But why did you arrest Zoya?'

'To make sure you come back.'

(ii)

Volodya told himself he knew how to do this. In Berlin, before the war, he had shaken off Gestapo tails, met with potential spies, recruited them, and made them into reliable sources of secret intelligence. It was never easy – especially the part where he had to talk someone into turning traitor – but he was expert.

However, this was America.

The Western countries he had visited, Germany and Spain in the thirties and forties, were nothing like this.

He was overwhelmed. All his life he had been told that Hollywood movies gave an exaggerated impression of prosperity, and that in reality most Americans lived in poverty. But it was clear to Volodya, from the day he arrived in the USA, that the movies hardly exaggerated at all. And poor people were hard to find.

New York was jammed with cars, many driven by people who clearly were not important government officials: youngsters, men in work clothes, even women out shopping. And everybody was so well dressed! All the men appeared to be wearing their best suits. The women's calves were clad in sheer stockings. Everyone seemed to have new shoes.

He had to keep reminding himself of the bad side of America. There was poverty, somewhere. Negroes were persecuted, and in the South they could not vote. There was a lot of crime – Americans themselves said that it was rampant – although, strangely, Volodya did not actually see any evidence of it, and he felt quite safe walking the streets.

He spent a few days exploring New York. He worked on his English, which was not good, but it hardly mattered: the city was full of people who spoke broken English with heavy accents. He got to know the faces of some of the FBI agents assigned to tail him, and identified several convenient locations where he would be able to lose them.

One sunny morning he left the Soviet consulate in New York, hatless and wearing only grey slacks and a blue shirt, as if he were going to run a few errands. A young man in a dark suit and tie followed him.

He went to the Saks Fifth Avenue department store and bought underwear and a shirt with a small brown checked pattern. Whoever was tailing him had to think he was probably just shopping.

The NKVD chief at the consulate had announced that a Soviet team would shadow Volodya throughout his American visit, to make sure of his good behaviour. He could barely contain his rage at the organization that had imprisoned Zoya, and he had to repress the urge to take the man by the throat and strangle him. But he had remained calm. He had pointed out sarcastically that in order to fulfil his mission he would have to evade FBI surveillance, and in doing so he might inadvertently also lose his NKVD tail; but he wished them luck. Most days he shook them off in five minutes.

So the young man tailing him was almost certainly an FBI agent. His crisply conservative clothes corroborated that.

Carrying his purchases in a paper bag, Volodya left the store by a side entrance and hailed a cab. He left the FBI man at the kerb waving his arm. When the cab had turned two corners Volodya threw the driver a bill and jumped out. He darted into a subway station, left again by a different entrance, and waited in the doorway of an office building for five minutes.

The young man in the dark suit was nowhere to be seen.

Volodya walked to Penn Station.

There he double-checked that he was not being followed, then bought his ticket. With nothing but that and his paper bag he boarded a train.

The journey to Albuquerque took three days.

The train sped through mile after endless mile of rich farmland, mighty factories belching smoke, and great cities with skyscrapers pointing arrogantly at the heavens. The Soviet Union was bigger, but apart from the Ukraine it was mostly pine forests and frozen steppes. He had never imagined wealth on this scale.

And wealth was not all. For several days something had been nagging at the back of Volodya's mind, something strange about life in America. Eventually he realized what it was: no one asked for his papers. After he had passed through immigration control in New York, he had not shown his passport again. In this country, it seemed, anyone could walk into a railway station or a bus terminus and buy a ticket to any place without having to get permission or explain the purpose of the trip to an official. It gave him a dangerously exhilarating sense of freedom. He could go anywhere!

America's wealth also heightened Volodya's sense of the danger his country faced. The Germans had almost destroyed the Soviet Union, and this country was three times as populous and ten times as rich. The thought that Russians might become underlings, frightened into subservience, softened Volodya's doubts about Communism, despite what the NKVD had done to him and his wife. If he had children, he did not want them to grow up in a world tyrannized by America.

He travelled via Pittsburgh and Chicago and attracted no attention en route. His clothes were American, and his accent was not noticed for the simple reason that he spoke to no one. He bought sandwiches and coffee by pointing and paying. He flicked through newspapers and magazines that other travellers left behind, looking at the pictures and trying to work out the meanings of the headlines.

The last part of the journey took him through a desert landscape of desolate beauty, with distant snowy peaks stained red by the sunset, which probably explained why they were called the Blood of Christ Mountains.

He went to the toilet where he changed his underwear and put on the new shirt he had bought in Saks.

He expected the FBI or Army security to be watching the train station in Albuquerque, and sure enough he spotted a young man whose check jacket – too warm for the climate of New Mexico in September – did not quite conceal the bulge of a gun in a shoulder holster. However, the agent was undoubtedly interested in long-distance travellers who might be arriving from New York or Washington. Volodya, with no hat or jacket and no luggage, looked like a local man coming back from a short trip. He was not followed as he walked to the bus station and boarded a Greyhound for Santa Fe.

He reached his destination late in the afternoon. He noted two FBI men at the Santa Fe bus station, and they scrutinized him. However, they could not tail everyone who got off the bus, and once again his casual appearance caused them to dismiss him.

Doing his best to look as if he knew where he was going, he strolled along the streets. The low flat-roofed pueblo-style houses and squat churches baking in the sun reminded him of Spain. The storefront buildings overhung the sidewalks, creating pleasantly shady arcades.

He avoided La Fonda, the big hotel on the town square next to the cathedral, and checked in to the St Francis. He paid cash and gave his name as Robert Pender, which might have been American or one of several European nationalities. 'My suitcase will be delivered later,' he said to the pretty girl behind the reception desk. 'If I'm out when it comes, can you make sure it gets sent up to my room?'

'Oh, sure, that won't be a problem,' she said.

'Thank you,' he said, then he added a phrase he had heard several times on the train: 'I sure appreciate it.'

'If I'm not here, someone else will deal with the bag, so long as it has your name on it.'

'It does.' He had no luggage, but she would never realize that.

She looked at his entry in the book. 'So, Mr Pender, you're from New York.'

There was a touch of scepticism in her voice, no doubt because he did not sound like a New Yorker. 'I'm from Switzerland originally,' he explained, naming a neutral country.

'That accounts for the accent. I haven't met a Switzerland person before. What's it like there?'

Volodya had never been to Switzerland, but he had seen photographs. 'It snows a lot,' he said.

'Well, enjoy our New Mexico weather!'

'I will.'

Five minutes later he went out again.

Some of the scientists lived at the Los Alamos laboratory, he had learned from his colleagues in the Soviet Embassy, but it was a shanty town with few civilized comforts, and they preferred to rent houses and apartments nearby if they could. Willi Frunze could afford it easily: he was married to a successful artist who drew a syndicated strip cartoon called Slack Alice. His wife, also called Alice, could work anywhere, so they had a place in the historic downtown neighbourhood.

The New York office of the NKVD had provided this information. They had researched Frunze carefully, and Volodya had his address and phone number and a description of his car, a pre-war Plymouth convertible with whitewall tyres.

The Frunzes' building had an art gallery on the ground floor. The apartment upstairs had a large north-facing window that would appeal to an artist. A Plymouth convertible was parked outside.

Volodya preferred not to go in: the place might be bugged.

The Frunzes were an affluent, childless couple, and he guessed they would not stay at home listening to the radio on a Friday night. He decided to wait around and see if they came out.

He spent some time in the art gallery, looking at the paintings for sale. He liked clear, vivid pictures and would not have wanted to own any of these messy daubs. He found a coffee shop down the block and got a window seat from which he could just see the Frunzes' door. He left there after an hour, bought a newspaper, and stood at a bus stop pretending to read it.

The long wait permitted him to establish that no one was watching the Frunze apartment. That meant that the FBI and Army security had not tagged Frunze as a high risk. He was a foreigner, but so were many of the scientists, and presumably nothing else was known against him.

This was a downtown commercial district, not a residential neighbourhood, and there were plenty of people on the streets; but all the same after a couple of hours Volodya began to worry that someone might notice him hanging around.

Then the Frunzes came out.

Frunze was heavier than he had been twelve years ago – there was no shortage of food in America. His hair was beginning to recede, although he was only thirty. He still had that solemn look. He wore a sports shirt and khaki pants, a common American combination.

His wife was not so conservatively dressed. Her fair hair was pinned up under a beret, and she wore a shapeless cotton dress in an indistinct brown colour, but she had an assortment of bangles on both wrists, and numerous rings. Artists had dressed like that in Germany before Hitler, Volodya remembered.

The couple set off along the street, and Volodya followed.

He wondered what the wife's politics were, and what difference her presence would make in the difficult conversation he was about to have. Frunze had been a staunch social democrat back in Germany, so it was not likely that his wife would be a conservative; a speculation that was borne out by her appearance. On the other hand, she probably did not know he had given secrets to the Soviets in London. She was an unknown quantity.

He would prefer to deal with Frunze alone, and he considered leaving them and trying again tomorrow. But the hotel receptionist had noticed his foreign accent, so by the morning he might have an FBI tail. He could deal with that, he thought, though not as easily in this small town as in New York or Berlin. And tomorrow was Saturday, so the Frunzes would probably spend the day together. How long might Volodya have to wait before catching Frunze alone?

There was never an easy way to do this. On balance he decided to go ahead tonight.

The Frunzes went into a diner.

Volodya walked past the place and glanced through the window. It was an inexpensive restaurant with booths. He thought of going in and sitting down with them, but he decided to let them eat first. They would be in a good mood when full of food.

He waited half an hour, watching the door from a distance. Then, full of trepidation, he went in.

They were finishing their dinner. As he crossed the restaurant, Frunze glanced up then looked away, not recognizing him.

He slid into the booth next to Alice and spoke quietly in German. 'Hello, Willi, don't you remember me from school?'

Frunze looked hard at him for several seconds, then his face broke into a smile. 'Peshkov? Volodya Peshkov? Is it really you?'

A wave of relief washed over Volodya. Frunze was still friendly. There was no barrier of hostility to overcome. 'It's really me,' Volodya said. He offered his hand and they shook. Turning to Alice, he said in English: 'I am very bad speaking your language, sorry.'

'Don't bother to try,' she replied in fluent German. 'My family were immigrants from Bavaria.'

Frunze said in amazement: 'I've been thinking about you lately, because I know another guy with the same surname – Greg Peshkov.'

'Really? My father had a brother called Lev who came to America in about 1915.'

'No, Lieutenant Peshkov is much younger. Anyway, what are you doing here?'

Volodya smiled. 'I came to see you.' Before Frunze could ask why, he said: 'Last time I saw you, you were secretary of the Neukölln Social Democratic Party.' This was his second step. Having established a friendly footing, he was reminding Frunze of his youthful idealism.

'That experience convinced me that democratic socialism doesn't work,' Frunze said. 'Against the Nazis we were completely impotent. It took the Soviet Union to stop them.'

That was true, and Volodya was pleased Frunze realized it; but, more importantly, the comment showed that Frunze's political ideas had not been softened by life in affluent America.

Alice said: 'We were planning to have a couple of drinks at a bar around the corner. A lot of the scientists go there on a Friday night. Would you like to join us?'

The last thing Volodya wanted was to be seen in public with the Frunzes. 'I don't know,' he said. In fact he had been too long with them in this restaurant. It was time for step three: reminding Frunze of his terrible guilt. He leaned forward and lowered his voice. 'Willi, did you know the Americans were going to drop nuclear bombs on Japan?'

There was a long pause. Volodya held his breath. He was gambling that Frunze would be wracked by remorse.

For a moment he feared he had gone too far. Frunze looked as if he might burst into tears.

Then the scientist took a deep breath and got control of himself. 'No, I didn't know,' he said. 'None of us did.'

Alice interjected angrily: 'We assumed the American military would give *some* demonstration of the power of the bomb, as a threat to make the Japanese surrender earlier.' So she had known about the bomb beforehand, Volodya noted. He was not surprised. Men found it hard to keep such things from their wives. 'So we expected a detonation some time, somewhere,' she went on. 'But we imagined they would destroy an uninhabited island, or maybe a military facility with a lot of weapons and very few people.'

'That might have been justifiable,' Frunze said. 'But ...' His voice fell to a whisper. 'Nobody thought they would drop it on a city and kill eighty thousand men, women and children.'

Volodya nodded. 'I thought you might feel this way.' He had been hoping for it with all his heart.

Frunze said: 'Who wouldn't?'

'Let me ask you an even more important question.' This was step four. 'Will they do it again?'

'I don't know,' Frunze said. 'They might. Christ forgive us all, they might.'

Volodya concealed his satisfaction. He had made Frunze feel responsible for future use of nuclear weapons, as well as past.

Volodya nodded. 'That's what we think.'

Alice said sharply: 'Who's *we*?'

She was shrewd, and probably more worldly-wise than her husband. She would be hard to fool, and Volodya decided not to try. He had to risk levelling with her. 'A fair question,' he said. 'And I didn't come all this way to deceive an old friend. I'm a major in Red Army Intelligence.'

They stared at him. The possibility must have crossed their minds already, but they were surprised by the stark admission.

'I have something I need to say to you,' Volodya went on. 'Something hugely important. Is there somewhere we can go to talk privately?'

They both looked uncertain. Frunze said: 'Our apartment?'

'It has probably been bugged by the FBI.'

Frunze had some experience of clandestine work, but Alice was shocked. 'You think so?' she said incredulously.

'Yes. Could we drive out of town?'

Frunze said: 'There's a place we go sometimes, around this time of the evening, to watch the sunset.'

'Perfect. Go to your car, get in, and wait for me. I'll be a minute behind you.'

Frunze paid the check and left with Alice, and Volodya followed. During the short walk he established that no one was tailing him. He reached the Plymouth and got in. They sat three across the front seat, American style. Frunze drove out of town.

They followed a dirt road to the top of a low hill. Frunze stopped the car. Volodya motioned for them all to get out, and led them a hundred yards away, just in case the car was bugged too.

They looked across the landscape of stony soil and low bushes towards the setting sun, and Volodya took step five. 'We think the next nuclear bomb will be dropped somewhere in the Soviet Union.'

Frunze nodded. 'God forbid, but you're probably right.'

'And there's absolutely nothing we can do about it,' Volodya went on, pressing home his point relentlessly. 'There are no precautions we can take, no barriers we can erect, no way we can protect our people. There is no defence against the nuclear bomb – the bomb that you made, Willi.'

'I know it,' said Frunze miserably. Clearly he felt it would be his fault if the USSR was attacked with nuclear weapons.

Step six. 'The only protection would be our own nuclear bomb.'

Frunze did not want to believe that. 'It's not a defence,' he said.

'But it's a deterrent.'

'It might be,' he conceded.

Alice said: 'We don't want these bombs to spread.'

'Nor do I,' said Volodya. 'But the only sure way to stop the Americans flattening Moscow the way they flattened Hiroshima is for the Soviet Union to have a nuclear bomb of its own, and threaten retaliation.'

Alice said: 'He's right, Willi. Hell, we all know it.'

She was the tough one, Volodya saw.

Volodya made his voice light for step seven. 'How many bombs do the Americans have right now?'

This was a crucial moment. If Frunze answered this question he would have crossed a line. So far the conversation had been general. Now Volodya was requesting secret information.

Frunze hesitated for a long moment. Finally he glanced at Alice.

Volodya saw her give an almost imperceptible nod.

Frunze said: 'Only one.'

Volodya concealed his triumph. Frunze had betrayed trust. It was the difficult first move. A second secret would come more easily.

Frunze added: 'But they'll have more soon.'

'It's a race, and if we lose, we die,' Volodya said urgently. 'We have to build at least one bomb of our own before they have enough to wipe us out.'

'Can you do that?'

That gave Volodya the cue for step eight. 'We need help.'

He saw Frunze's face harden, and guessed he was remembering whatever it was that had made him refuse to co-operate with the NKVD.

Alice said to Volodya: 'What if we say we can't help you? That it's too dangerous?'

Volodya followed his instinct. He held up his hands in a gesture of surrender. 'I go home and report failure,' he said. 'I can't make you do anything you don't want to do. I wouldn't want to pressure you or coerce you in any way.'

Alice said: 'No threats?'

That confirmed Volodya's guess that the NKVD had tried to bully Frunze. They tried to bully everyone: it was all they knew. 'I'm not even trying to persuade you,' Volodya said to Frunze. 'I'm laying out the facts. The rest is up to you. If you want to help, I'm here as your contact. If you see things differently, that's the end of it. You're both smart people. I couldn't fool you even if I wanted to.'

Again they looked at each other. He hoped they were thinking how different he was from the last Soviet agent who had approached them.

The moment stretched out agonizingly.

It was Alice who spoke at last. 'What kind of help do you need?'

That was not a yes, but it was better than rejection, and it led logically to step nine. 'My wife is one of the physicists on the team,' he said, hoping this would humanize him at a moment when they might be in danger of seeing him as manipulative. 'She tells me there are several routes to a nuclear bomb, and we don't have time to try them all. We can save years if we know what worked for you.'

'That makes sense,' Willi said.

Step ten, the big one. 'We have to know what type of bomb was dropped on Japan.'

Frunze's expression was agonized. He looked at his wife. This time she did not give him the nod, but neither did she shake her head. She seemed as torn as he did.

Frunze sighed. 'Two kinds,' he said.

Volodya was thrilled and startled. 'Two different designs?'

Frunze nodded. 'For Hiroshima they used a uranium device with a gun ignition. We called it Little Boy. For Nagasaki, Fat Man, a plutonium bomb with an implosion trigger.'

Volodya could hardly breathe. This was red-hot data. 'Which is better?'

'They both worked, obviously, but Fat Man is easier to make.'

'Why?'

'It takes years to produce enough U-235 for a bomb. Plutonium is quicker, once you have a nuclear pile.'

'So the USSR should copy Fat Man.'

'Definitely.'

'There is one more thing you could do to help save Russia from destruction,' Volodya said.

'What?'

Volodya looked him in the eye. 'Get me the design drawings,' he said.

Willi paled. 'I'm an American citizen,' he said. 'You're asking me to commit treason. The penalty is death. I could go to the electric chair.'

So could your wife, Volodya thought; she's complicit. Thank God you haven't thought of that.

He said: 'I've asked a lot of people to put their lives at risk in the last few years. People like yourselves, Germans who hated the Nazis, men and women who took terrible risks to send us information that helped us win the war. And I have to say to you what I said to them: a lot more people will be killed if you don't do it.' He fell silent. That was his best shot. He had nothing more to offer.

Frunze looked at his wife.

Alice said: 'You made the bomb, Willi.'

Frunze said to Volodya: 'I'll think about it.'

(iii)

Two days later he handed over the plans.

Volodya took them to Moscow.

Zoya was released from jail. She was not as angry about her imprisonment as he was. 'They did it to protect the revolution,' she said. 'And I wasn't hurt. It was like staying in a really bad hotel.'

On her first day at home, after they made love, he said: 'I have something to show you, something I brought back from America.' He rolled off the bed, opened a drawer, and took out a book. 'It's called the *Sears-Roebuck Catalogue*,' he said. He sat beside her on the bed and opened the book. 'Look at this.'

The catalogue fell open at a page of women's dresses. The models were impossibly slender, but the fabrics were bright and cheerful, stripes and checks and solid colours, some with ruffles, pleats, and belts. 'That's attractive,' Zoya said, putting her finger on one. 'Is two dollars ninety-eight a lot of money?'

'Not really,' Volodya said. 'The average wage is about fifty dollars a week, rent is about a third of that.'

'Really?' Zoya was amazed. 'So most people could easily afford these dresses?'

'That's right. Maybe not peasants. On the other hand, these

catalogues were invented for farmers who live a hundred miles from the nearest store.'

'How does it work?'

'You pick what you want from the book and send them the money, then a couple of weeks later the mailman brings you whatever you ordered.'

'It must be like being a tsar.' Zoya took the book from him and turned the page. 'Oh! Here are some more.' The next page showed jacket-and-skirt combinations for four dollars ninety-eight. 'These are elegant too,' she said.

'Keep turning the pages,' Volodya said.

Zoya was astonished to see page after page of women's coats, hats, shoes, underwear, pyjamas, and stockings. 'People can have *any* of these?' she said.

'That's right.'

'But there's more choice on one of these pages than there is in the average Russian shop!'

'Yes.'

She carried on slowly leafing through the book. There was a similar range of clothing for men, and again for children. Zoya put her finger on a heavy woollen winter coat for boys that cost fifteen dollars. 'At that price, I suppose every boy in America has one.'

'They probably do.'

After the clothes came furniture. You could buy a bed for twenty-five dollars. Everything was cheap if you had fifty dollars a week. And it went on and on. There were hundreds of things that could not be bought for any money in the Soviet Union: toys and games, beauty products, guitars, elegant chairs, power tools, novels in colourful jackets, Christmas decorations, and electric toasters.

There was even a tractor. 'Do you think,' Zoya said, 'that any farmer in America who wants a tractor can have one *right away*?'

'Only if he has the money,' said Volodya.

'He doesn't have to put his name down on a list and wait for a few years?'

'No.'

Zoya closed the book and looked at him solemnly. 'If people can have all this,' she said, 'why would they want to be Communist?'

'Good question,' said Volodya.

22

1946

The children of Berlin had a new game called *Komm, Frau* – Come, Woman. It was one of a dozen games in which boys chase girls, but it had a new twist, Carla noticed. The boys would team up and target one of the girls. When they caught her, they would shout: '*Komm, Frau!*' and throw her to the ground. Then they would hold her down while one of their number lay on top of her and simulated sexual intercourse. Children of seven and eight, who ought not to know what rape was, played this game because they had seen what Red Army soldiers did to German women. Every Russian knew that one phrase of the German language: '*Komm, Frau.*'

What was it about the Russians? Carla had never met anyone who had been raped by a French, British, American or Canadian soldier, though she supposed it must happen. By contrast, every woman she knew between fifteen and fifty-five had been raped by at least one Soviet soldier: her mother, Maud; her friend Frieda; Frieda's mother, Monika; Ada, the maid; all of them.

Yet they were lucky, for they were still alive. Some women, abused by dozens of men, hour after hour, had died. Carla had heard of a girl who had been bitten to death.

Only Rebecca Rosen had escaped. After Carla had protected her, the day the Jewish Hospital was liberated, Rebecca moved into the von Ulrich town house. It was in the Soviet zone, but she had nowhere else to go. She hid for months like a criminal in the attic, coming down only late at night when the bestial Russians had fallen into drunken sleep. Carla spent a couple of hours up there with her when she could, and they played card games and told each other their life stories. Carla wanted to be like an older sister, but Rebecca treated her like a mother.

Then Carla found she was going to be a mother for real.

Maud and Monika were in their fifties, and too old to have babies, mercifully; and Ada was lucky; but both Carla and Frieda were pregnant by their rapists.

Frieda had an abortion.

It was illegal, and a Nazi law that threatened the death penalty was still in force. So Frieda went to an elderly 'midwife' who did it for five cigarettes. Frieda contracted a severe infection, and would have died but that Carla was able to steal scarce penicillin from the hospital.

Carla decided to have her baby.

Her feelings about it swung violently from one extreme to another. When suffering morning sickness she raged against the beasts who had violated her body and left her with this burden. At other times she found herself sitting with her hands on her belly staring into space and thinking dreamily about baby clothes. Then she would wonder if the baby's face would remind her of one of the men, and cause her to hate her own child. But surely it would have some von Ulrich features too? She felt anxious and frightened.

She was eight months pregnant in January 1946. Like most Germans she was also cold, hungry and destitute. When her pregnancy became obvious she had to give up nursing and join the millions of unemployed. Food rations were issued every ten days. The daily amount, for those without special privileges, was 1,500 calories. It still had to be paid for, of course. And even for customers with cash and ration cards, sometimes there was simply no food to buy.

Carla had considered asking the Soviets for special treatment because of her wartime work as a spy. But Heinrich had tried that and suffered a frightening experience. Red Army Intelligence had expected him to continue to spy for them, and asked him to infiltrate the US military. When he said he would rather not, they became nasty and threatened to send him to a labour camp. He got out of it by saying he spoke no English, therefore was no use to them. But Carla was well warned, and decided it was safest to keep quiet.

Today Carla and Maud were happy because they had sold a chest of drawers. It was a Jugendstil piece in burled light oak that Walter's parents had bought when they got married in 1889. Carla, Maud and Ada had loaded it on to a borrowed handcart.

There were still no men in their house. Erik and Werner were among millions of German soldiers who had disappeared. Perhaps they were dead. Colonel Beck had told Carla that almost three million Germans had died in battle on the Eastern Front, and more had died as prisoners of the Soviets – killed by hunger, cold and disease. But another two million were still alive and working in labour camps in the Soviet Union.

Some had come back: they had either escaped from their guards or had been released because they were too ill to work, and they had joined the thousands of displaced persons on the tramp all over Europe, trying to find their way home. Carla and Maud had written letters and sent them care of the Red Army, but no replies had ever come.

Carla felt torn about the prospect of Werner's return. She still loved him, and hoped desperately that he was alive and well, but she dreaded meeting him when she was pregnant with a rapist's baby. Although it was not her fault, she felt irrationally ashamed.

So the three women pushed the handcart through the streets. They left Rebecca behind. The Red Army orgy of rape and looting had passed its nightmare peak, and Rebecca no longer lived in the attic, but it was still not safe for a pretty girl to walk the streets.

Huge photographs of Lenin and Stalin now hung over Unter den Linden, once the promenade of Germany's fashionable elite. Most Berlin roads had been cleared, and the rubble of destroyed buildings stood in stacks every few hundred yards, ready to be re-used, perhaps, if ever Germans were able to rebuild their country. Acres of houses had been flattened, often entire city blocks. It would take years to deal with the wreckage. There were thousands of bodies rotting in the ruins, and the sickly sweet smell of decaying human flesh had been in the air all summer. Now it smelled only after rain.

Meanwhile, the city had been divided into four zones: Russian, American, British and French. Many of the buildings still standing had been commandeered by the occupying troops. Berliners lived where they could, often seeking inadequate shelter in the surviving rooms of half-demolished houses. The city had running water again, and electric power came on fitfully, but it was hard to find fuel for heating and cooking. The chest of drawers might be almost as valuable chopped up for firewood.

They took it to Wedding, in the French zone, where they sold it to a charming Parisian colonel for a carton of Gitanes. The occupation currency had become worthless, because the Soviets printed too much of it, so everything was bought and sold for cigarettes.

Now they were returning triumphant, Maud and Ada steering the empty cart while Carla walked alongside. She ached all over from pushing the cart, but they were rich: a whole carton of cigarettes would go a long way.

Night fell and the temperature dropped to freezing. Their route home took them briefly into the British sector. Carla sometimes wondered whether the British might help her mother if they knew the hardship she was suffering. On the other hand, Maud had been a German citizen for twenty-six years. Her brother, Earl Fitzherbert, was wealthy and influential, but he had refused to support her after her marriage to Walter von Ulrich, and he was a stubborn man: it was not likely he would change his attitude.

They came across a small crowd, thirty or forty ragged people, outside a house that had been taken over by the occupying power. Stopping to find out what they were staring at, the three women saw a party going on inside. Through the windows they could observe brightly lit rooms, laughing men and women holding drinks, and waitresses moving through the throng with trays of food. Carla looked around her. The crowd was mostly women and children – there were not many men left in Berlin, or indeed in Germany – and they were all staring longingly at the windows, like rejected sinners outside the gates of paradise. It was a pathetic sight.

'This is obscene,' said Maud, and she marched up the path to the door of the house.

A British sentry stood in her way and said: '*Nein, nein,*' probably the only German he knew.

Maud addressed him in the crisp upper-class English she had spoken as a girl. 'I must see your commanding officer immediately.'

Carla admired her mother's nerve and poise, as always.

The sentry looked doubtfully at Maud's threadbare coat, but after a moment he tapped on the door. It opened, and a face looked out. 'English lady wants the CO,' said the sentry.

A moment later the door opened again and two people looked out. They might have been caricatures of a British officer and his wife: he in his mess kit with a black bow tie, she in a long dress and pearls.

'Good evening,' Maud said. 'I'm frightfully sorry to disturb your party.'

They stared at her, astonished to be spoken to that way by a woman in rags.

Maud went on: 'I just thought you should see what you're doing to these wretched people outside.'

The couple looked at the crowd.

Maud said: 'You might draw the curtains, for pity's sake.'

After a moment the woman said: 'Oh, dear, George, have we been terribly unkind?'

'Unintentionally, perhaps,' the man said gruffly.

'Could we possibly make amends by sending some food out to them?'

'Yes,' Maud said quickly. 'That would be a kindness as well as an apology.'

The officer looked dubious. It was probably against some kind of regulation to give canapés to starving Germans.

The woman pleaded: 'George, darling, may we?'

'Oh, very well,' said her husband.

The woman turned back to Maud. 'Thank you for alerting us. We really didn't mean to do this.'

'You're welcome,' Maud said, and she retreated down the path.

A few minutes later, guests began to emerge from the house with plates of sandwiches and cakes, which they offered to the starving crowd. Carla grinned. Her mother's impudence had paid off. She took a large piece of fruit cake, which she wolfed in a few starved bites. It contained more sugar than she had eaten in the past six months.

The curtains were drawn, the guests returned to the house, and the crowd dispersed. Maud and Ada grasped the handles of the cart and recommenced pushing it home. 'Well done, Mother,' said Carla. 'A carton of Gitanes *and* a free meal, all in one afternoon!'

Apart from the Soviets, few of the occupying soldiers were cruel to Germans, Carla reflected. She found it surprising. American GIs gave out chocolate bars. Even the French, whose own children had gone hungry under German occupation, often showed kindness. After all the misery we Germans have inflicted on our neighbours, Carla thought, it's astonishing they don't hate us more. On the other hand, what with the Nazis, the Red Army and the air raids, perhaps they think we've been punished enough.

It was late when they got home. They left the cart with the neighbours who had loaned it, giving them half a pack of Gitanes as payment. They entered their house, which was luckily still intact. There was no glass in most of the windows, and the stonework was pocked with craters, but the place had not suffered structural damage, and it still kept the weather out.

All the same, the four women now lived in the kitchen, sleeping there on mattresses they dragged in from the hall at night. It was hard enough to warm that one room, and they certainly did not have fuel to heat the rest of the house. The kitchen stove had burned coal in the old days, but that was now virtually unobtainable. However, they had found the stove would burn many other things: books, newspapers, broken furniture, even net curtains.

They slept in pairs, Carla with Rebecca and Maud with Ada. Rebecca often cried herself to sleep in Carla's arms, as she had the night after her parents were killed.

The long walk had exhausted Carla, and she immediately lay down. Ada built up the fire in the stove with old news magazines Rebecca had brought down from the attic. Maud added water to the remains of the lunchtime bean soup and reheated it for their supper.

Sitting up to drink her soup, Carla suffered a sharp abdominal pain. This was not a result of pushing the handcart, she realized. It was something else. She checked the date and counted back to the date of the liberation of the Jewish Hospital.

'Mother,' she said fearfully, 'I think the baby's coming.'

'It's too soon!' Maud said.

'I'm thirty-six weeks pregnant, and I'm getting cramps.'

'Then we'd better get ready.'

Maud went upstairs to fetch towels.

Ada brought a wooden chair from the dining room. She had a useful length of twisted steel from a bomb site that served her as a sledgehammer. She smashed the chair into manageable pieces, then built up the fire in the stove.

Carla put her hands on her distended belly. 'You might have waited for warmer weather, Baby,' she said.

Soon she was in too much pain to notice the cold. She had not known anything could hurt this much.

Nor that it could go on so long. She was in labour all night. Maud and Ada took turns holding her hand while she moaned and cried. Rebecca looked on, white-faced and scared.

The grey light of morning was filtering through the newspaper taped over the glassless kitchen window when at last the baby's head emerged. Carla was overwhelmed by a feeling of relief like nothing she had ever experienced, even though the pain did not immediately cease.

After one more agonizing push, Maud took the baby from between her legs.

'A boy,' she said.

She blew on his face, and he opened his mouth and cried.

She gave the baby to Carla, and propped her upright on the mattress with some cushions from the drawing room.

He had lots of dark hair all over his head.

Maud tied off the cord with a piece of cotton, then cut it. Carla unbuttoned her blouse and put the baby to her breast.

She was worried she might have no milk. Her breasts should have swollen and leaked towards the end of her pregnancy, but they had not, perhaps because the baby was early, perhaps because the mother was undernourished. But, after a few moments of sucking, she felt a strange pain, and the milk began to flow.

Soon he fell asleep.

Ada brought a bowl of warm water and a rag, and gently washed the baby's face and head, then the rest of him.

Rebecca whispered: 'He's so beautiful.'

Carla said: 'Mother, shall we call him Walter?'

She had not intended to be dramatic, but Maud fell apart. Her face crumpled and she bent double, wracked by terrible sobs. She recovered herself sufficiently to say, 'I'm sorry,' then she was convulsed by grief again. 'Oh, Walter, my Walter,' she wept.

Eventually her crying subsided. 'I'm sorry,' she said again. 'I didn't mean to make a fuss.' She wiped her face with her sleeve. 'I just wish your father could see the baby, that's all. It's so unfair.'

Ada surprised them both by quoting the Book of Job: 'The Lord giveth and the Lord taketh away,' she said. 'Blessed be the name of the Lord.'

Carla did not believe in God – no holy being worthy of the name could have allowed the Nazi death camps to happen – but all the same she found comfort in the quotation. It was about accepting everything in human life, including the pain of birth and the sorrow of death. Maud seemed to appreciate it too, and she became calmer.

Carla looked adoringly at baby Walter. She would care for him and feed him and keep him warm, she vowed, no matter what difficulties stood in the way. He was the most wonderful child that had ever been born, and she would love and cherish him for ever.

He woke up, and Carla gave him her nipple again. He sucked contentedly, making small smacking noises with his mouth, while four women watched him. For a little while, in the warm, dim-lit kitchen, there was no other sound.

(ii)

The first speech made by a new Member of Parliament is called a maiden speech, and is usually dull. Certain things have to be said, stock phrases are used, and the convention is that the subject must not be controversial. Colleagues and opponents alike congratulate the newcomer, the traditions are observed and the ice is broken.

Lloyd Williams made his first *real* speech a few months later, during the debate on the National Insurance Bill. That was more scary.

In preparing it he had two orators in mind. His grandfather, Dai Williams, used the language and rhythms of the Bible, not just in chapel but also – perhaps especially – when speaking of the hardship and injustice of the life of a coal miner. He relished short words rich in meaning: toil, sin, greed. He spoke of the hearth and the pit and the grave.

Churchill did the same, but had humour that Dai Williams lacked. His long, majestic sentences often ended with an unexpected image or a reversal of meaning. Having been editor of the government newspaper the *British Gazette* during the General Strike of 1926, he had warned trade unionists: 'Make your minds perfectly clear: if ever you let loose upon us again a general strike, we will loose upon you another *British Gazette.*' A speech needed such surprises, Lloyd believed; they were like the raisins in a bun.

But when he stood up to speak, he found that his carefully wrought sentences suddenly seemed unreal. His audience clearly felt the same, and he could sense that the fifty or sixty MPs in the chamber were only half listening. He suffered a moment of panic: how could he be boring about a subject that mattered so profoundly to the people he represented?

On the government front bench he could see his mother, now Minister for Schools, and his Uncle Billy, Minister for Coal. Billy Williams had started work down the pit at the age of thirteen, Lloyd knew. Ethel had been the same age when she began scrubbing the floors of Tŷ Gwyn. This debate was not about fine phrases, it was about their lives.

After a minute he abandoned his script and spoke extempore. He recalled instead the misery of working-class families made penniless by unemployment or disability, scenes he had witnessed first hand in the East End of London and the South Wales coalfield. His voice betrayed the emotion he felt, somewhat to his embarrassment, but he ploughed on. He sensed his audience beginning to pay attention. He spoke of his grandfather and others who had started the Labour movement with the dream of comprehensive employment insurance to banish forever the fear of destitution. When he sat down there was a roar of approval.

In the visitors' gallery his wife Daisy smiled proudly and gave him a thumbs-up sign.

He listened to the rest of the debate in a glow of satisfaction. He felt he had passed his first real test as an MP.

Afterwards, in the lobby, he was approached by a Labour Whip, one of the people responsible for making sure MPs voted the right way. After congratulating Lloyd on his speech, the Whip said: 'How would you like to be a parliamentary private secretary?'

Lloyd was thrilled. Each minister and secretary of state had at least one PPS. In truth a PPS was often little more than a bag-carrier, but the job was the usual first step on the way to a ministerial appointment. 'I'd be honoured,' Lloyd said. 'Who would I be working for?'

'Ernie Bevin.'

Lloyd could hardly believe his luck. Bevin was Foreign Secretary and the closest colleague of Prime Minister Attlee. The intimate relationship between the two men was a case of the attraction of opposites. Attlee was middle class: the son of a lawyer, an Oxford graduate, an officer in the First World War. Bevin was the illegitimate child of a housemaid, never knew his father, started work at the age of eleven, and founded the mammoth Transport and General Workers Union. They were physical opposites, too: Attlee slim and dapper, quiet, solemn; Bevin a huge man, tall and strong and overweight, with a loud laugh. The Foreign Secretary referred to the Prime Minister as 'little Clem'. All the same they were staunch allies.

Bevin was a hero to Lloyd and to millions of ordinary British people. 'There's nothing I'd like more,' Lloyd said. 'But hasn't Bevin already got a PPS?'

'He needs two,' the Whip said. 'Go to the Foreign Office tomorrow morning at nine and you can get started.'

'Thank you!'

Lloyd hurried along the oak-panelled corridor, heading for his mother's office. He had arranged to meet Daisy there after the debate. 'Mam!' he said as he entered. 'I've been made PPS to Ernie Bevin!'

Then he saw that Ethel was not alone. Earl Fitzherbert was with her.

Fitz stared at Lloyd with a mixture of surprise and distaste.

Even in his shock Lloyd noticed that his father was wearing a perfectly cut light-grey suit with a double-breasted waistcoat.

He looked back at his mother. She was quite calm. This encounter was not a surprise to her. She must have contrived it.

The earl came to the same conclusion. 'What the devil is this, Ethel?'

Lloyd stared at the man whose blood ran in his veins. Even in this embarrassing situation, Fitz was poised and dignified. He was handsome, despite the drooping eyelid that resulted from the Battle of the Somme. He leaned on a walking stick, another consequence of the Somme. A few months short of sixty years old, he was immaculately groomed, his grey hair neatly trimmed, his silver tie tightly knotted, his black shoes shining. Lloyd, too, always liked to look well turned out. That's where I get it from, he thought.

Ethel went and stood close to the earl. Lloyd knew his mother well enough to understand this move. She frequently used her charm when she wanted to persuade a man. All the same, Lloyd did not like to see her being so warm to one who had exploited her then let her down.

'I was so sorry when I heard about the death of Boy,' she said to Fitz. 'Nothing is as precious to us as our children, is it?'

'I must go,' Fitz said.

Until this moment, Lloyd had met Fitz only in passing. He had never before spent this much time with him or heard him speak this number of words. Despite feeling uncomfortable, Lloyd was fascinated. Grumpy though he was right now, Fitz had a kind of allure.

'Please, Fitz,' said Ethel. 'You have a son whom you have never acknowledged – a son you should be proud of.'

'You shouldn't do this, Ethel,' said Fitz. 'A man is entitled to forget the mistakes of his youth.'

Lloyd cringed with embarrassment, but his mother pressed on. 'Why should you want to forget? I know he was a mistake, but look at him now – a Member of Parliament who has just made a thrilling speech and been appointed PPS to the Foreign Secretary.'

Fitz pointedly did not look at Lloyd.

Ethel said: 'You want to pretend that our affair was a meaningless dalliance, but you know the truth. Yes, we were young and foolish, and randy too – me as much as you – but we loved each other. We *really* loved each other, Fitz. You should admit it. Don't you know that if you deny the truth about yourself you lose your soul?'

Fitz's face was no longer merely impassive, Lloyd saw. He was struggling to maintain control. Lloyd understood that his mother had put her finger on the real problem. It was not so much that Fitz was ashamed of having an illegitimate son; he was too proud to accept that he had loved a housemaid. He probably loved Ethel more than his wife, Lloyd guessed. And that upset all his most fundamental beliefs about the social hierarchy.

Lloyd spoke for the first time. 'I was with Boy at the end, sir. He died bravely.'

For the first time, Fitz looked at him. 'My son doesn't require your approval,' he said.

Lloyd felt as if he had been slapped.

Even Ethel was shocked. 'Fitz!' she said. 'How can you be so mean?'

At that point Daisy came in.

'Hello, Fitz!' she said gaily. 'You probably thought you'd got rid of me, but now you're my father-in-law again. Isn't that amusing?'

Ethel said: 'I'm just trying to persuade Fitz to shake Lloyd's hand.'

Fitz said: 'I try to avoid shaking hands with socialists.'

Ethel was fighting a losing battle, but she would not give up. 'See how much of yourself there is in him! He resembles you, dresses like you, shares your interest in politics – he'll probably end up Foreign Secretary, which you always wanted to be!'

Fitz's expression darkened further. 'It is now most unlikely that I shall ever be Foreign Secretary.' He went to the door. 'And it would not please me in the least if that great office of state were to be held by my Bolshevik bastard!' With that he walked out.

Ethel burst into tears.

Daisy put her arm around Lloyd. 'I'm so sorry,' she said.

'Don't worry,' Lloyd said. 'I'm not shocked or disappointed.' This was not true, but he did not want to appear pathetic. 'I was rejected by him a long time ago.' He looked at Daisy with adoration. 'I'm lucky to have plenty of other people who love me.'

Ethel said tearfully: 'It's my fault. I shouldn't have asked him to come here. I might have known it would turn out badly.'

'Never mind,' said Daisy. 'I have some good news.'

Lloyd smiled at her. 'What's that?'

She looked at Ethel. 'Are you ready for this?'

'I think so.'

'Come on,' said Lloyd. 'What is it?'

Daisy said: 'We're going to have a baby.'

(iii)

Carla's brother, Erik, came home that summer, near to death. He had contracted tuberculosis in a Soviet labour camp, and they had released him when he became too ill to work. He had been sleeping rough for weeks, travelling on freight trains and begging lifts on lorries. He arrived at the von Ulrich house barefoot and wearing filthy clothes. His face was like a skull.

However, he did not die. It might have been being with people who loved him; or the warmer weather as winter turned into spring; or perhaps just rest; but he coughed less and regained enough energy to do some work around the house, boarding up smashed windows, repairing roof tiles, unblocking pipes.

Fortunately, at the beginning of the year Frieda Franck had struck gold.

Ludwig Franck had been killed in the air raid that destroyed his factory, and for a while Frieda and her mother had been as destitute as everyone else. But she got a job as a nurse in the American zone, and soon afterwards, she explained to Carla, a little group of American doctors had asked her to sell their surplus food and cigarettes on the black market in exchange for a cut of the proceeds. Thereafter she turned up at Carla's house once a week with a little basket of supplies: warm clothing, candles, flashlight batteries, matches, soap, and food – bacon, chocolate, apples, rice, canned peaches. Maud divided the food into portions and gave Carla double. Carla accepted without hesitation, not for her own sake, but to help her feed baby Walli.

Without Frieda's illicit groceries, Walli might not have made it.

He was changing fast. The dark hair with which he had been born had now gone, and instead he had fine, fair hair. At six months he had Maud's wonderful green eyes. As his face took shape, Carla noticed a fold of flesh in the outer corners of his eyes that gave him a slant-eyed look, and she wondered if his father had been a Siberian. She could not remember all the men who had raped her. Most of the time she had closed her eyes.

She no longer hated them. It was strange, but she was so happy to have Walli that she could hardly bring herself to regret what had happened.

Rebecca was fascinated by Walli. Now just fifteen, she was old enough to have the beginnings of maternal feelings, and she eagerly helped Carla bathe and dress the baby. She played with him constantly, and he gurgled with delight when he saw her.

As soon as Erik felt well enough, he joined the Communist Party.

Carla was baffled. After what he had suffered at the hands of the Soviets, how could he? But she found that he talked about Communism in the same way he had talked about Nazism a decade earlier. She just hoped that this time his disillusionment would not be so long coming.

The Allies were keen for democracy to return to Germany, and city elections were scheduled for Berlin later in 1946.

Carla felt sure the city would not return to normal until its own people took control, so she decided to stand for the Social Democratic Party. But Berliners quickly discovered that the Soviet occupiers had a curious notion of what democracy meant.

The Soviets had been shocked by the results of elections in Austria the previous November. The Austrian Communists had expected to run neck-and-neck with the Socialists, but had won only four seats out of 165. It seemed that voters blamed Communism for the brutality of the Red Army. The Kremlin, unused to genuine elections, had not anticipated that.

To avoid a similar result in Germany, the Soviets proposed a merger between the Communists and the Social Democrats in what they called a united front. The Social Democrats refused, despite heavy pressure. In East Germany the Russians started arresting Social Democrats, just as the Nazis had in 1933. There the merger was forced through. But the Berlin elections were supervised by the four Allies, and the Social Democrats survived.

Once the weather warmed up, Carla was able to take her turn queuing for food. She carried Walli with her wrapped in a pillowcase – she had no baby clothes. Standing in line for potatoes one morning, a few blocks from home, she was surprised to see an American jeep pull up with Frieda in the passenger seat. The balding, middle-aged driver kissed her on the lips, and she jumped out. She was wearing a sleeveless blue dress and new shoes. She walked quickly away, heading for the von Ulrich house, carrying her little basket.

Carla saw everything in a flash. Frieda was not trading on the black market, and there was no syndicate of doctors. She was the paid mistress of an American officer.

It was not unusual. Thousands of pretty German girls had been faced with the choice: see your family starve, or sleep with a generous officer. French women had done the same under German occupation: officers' wives back here in Germany had spoken bitterly about it.

All the same, Carla was horrified. She believed that Frieda loved Heinrich. They were planning to get married as soon as life returned to some semblance of normality. Carla felt sick at heart.

She reached the head of the line and bought her ration of potatoes, then hurried home.

She found Frieda upstairs in the drawing room. Erik had cleaned up the room and put newspaper in the windows, the next best thing to glass. The curtains had long ago been recycled as bed linen, but most of the chairs had survived so far, their upholstery faded and worn. The grand piano was still there, miraculously. A Russian officer had discovered it and announced that he would return next day with a crane to lift it out through the window, but he had never come back.

Frieda immediately took Walli from Carla and began to sing to him. '*A, B, C, die Katze lief im Schnee.*' The women who had not yet had children, Rebecca and Frieda, could hardly get enough of Walli, Carla observed. Those who had had children of their own, Maud and Ada, adored him but dealt with him in a briskly practical way.

Frieda opened the lid of the piano and encouraged Walli to bang on the keys as she sang. The instrument had not been played for years: Maud had not touched it since the death of her last pupil, Joachim Koch.

After a few minutes Frieda said to Carla: 'You're a bit solemn. What is it?'

'I know how you get the food you bring us,' Carla said. 'You're not a black marketeer, are you?'

'Of course I am,' Frieda said. 'What are you talking about?'

'I saw you this morning, getting out of a jeep.'

'Colonel Hicks gave me a lift.'

'He kissed you on the lips.'

Frieda looked away. 'I knew I should have got out earlier. I could have walked from the American zone.'

'Frieda, what about Heinrich?'

'He'll never know! I'll be more careful, I swear.'

'Do you still love him?'

'Of course! We're going to get married.'

'Then why . . . ?'

'I've had enough of hard times! I want to put on pretty clothes and go to nightclubs and dance.'

'No, you don't,' Carla said confidently. 'You can't lie to me, Frieda – we've been friends too long. Tell me the truth.'

'The truth?'

'Yes, please.'

'You're sure?'

'I'm sure.'

'I did it for Walli.'

Carla gasped with shock. That had never occurred to her, but it made sense. She could believe Frieda would make such a sacrifice for her and her baby.

But she felt dreadful. This made her responsible for Frieda's prostituting herself. 'This is terrible!' Carla said. 'You shouldn't have done it – we would have managed somehow.'

Frieda sprang up from the piano stool with the baby still in her arms. 'No, you wouldn't!' she blazed.

Walli was frightened, and cried. Carla took him and rocked him, patting his back.

'You wouldn't have managed,' Frieda said more quietly.

'How do you know?'

'All last winter, babies were brought into the hospital naked, wrapped in newspapers, dead of hunger and cold. I could hardly bear to look at them.'

'Oh, God.' Carla held Walli tight.

'They turn a peculiar bluish colour when they freeze to death.'

'Stop it.'

'I have to tell you, otherwise you won't understand what I did. Walli would have been one of those blue frozen babies.'

'I know,' Carla whispered. 'I know.'

'Percy Hicks is a kind man. He has a frumpy wife back in Boston and I'm the sexiest thing he's ever seen. He's nice and quick about intercourse and always uses a condom.'

'You should stop,' Carla said.

'You don't mean that.'

'No, I don't,' Carla confessed. 'And that's the worst part. I feel so guilty. I am guilty.'

'You're not. It's my choice. German women have to make hard

choices. We're paying for the easy choices German men made fifteen years ago. Men such as my father, who thought Hitler would be good for business; and Heinrich's father, who voted for the Enabling Act. The sins of the fathers are visited on the daughters.'

There was a loud knock at the front door. A moment later they heard scampering steps as Rebecca hurried upstairs to hide, just in case it was the Red Army.

Then Ada's voice said: 'Oh! Sir! Good morning!' She sounded surprised and a bit worried, though not scared. Carla wondered who would induce that particular mixture of reactions in the maid.

There was a heavy masculine tread on the stairs, then Werner walked in.

He was dirty and ragged and thin as a rail, but there was a broad smile on his handsome face. 'It's me!' he said ebulliently. 'I'm back!'

Then he saw the baby. His jaw dropped and the happy smile disappeared. 'Oh,' he said. 'What . . . who . . . whose baby is that?'

'Mine, my darling,' said Carla. 'Let me explain.'

'Explain?' he said angrily. 'What explanation is necessary? You've had someone else's baby!' He turned to go.

Frieda said: 'Werner! In this room are two women who love you. Don't walk out without listening to us. You don't understand.'

'I think I understand everything.'

'Carla was raped.'

He went pale. 'Raped? Who by?'

Carla said: 'I never knew their names.'

'Names?' Werner swallowed. 'There . . . there was more than one?'

'Five Red Army soldiers.'

His voice fell to a whisper. 'Five?'

Carla nodded.

'But . . . couldn't you . . . I mean . . .'

Frieda said: 'I was raped, too, Werner. And so was Mother.'

'Dear God, what has been going on here?'

'Hell,' said Frieda.

Werner sat down heavily in a worn leather chair. 'I thought hell was where I've been,' he said. He buried his face in his hands.

Carla crossed the room, still holding Walli, and stood in front of Werner's chair. 'Look at me, Werner,' she said. 'Please.'

He looked up, his face twisted with emotion.

'Hell is over,' she said.

'Is it?'

'Yes,' she said firmly. 'Life is hard, but the Nazis have gone, the war is finished, Hitler is dead, and the Red Army rapists have been brought under control, more or less. The nightmare has ended. And we're both alive, and together.'

He reached out and took her hand. 'You're right.'

'We've got Walli, and in a minute you'll meet a fifteen-year-old girl called Rebecca who has somehow become my child. We have to make a new family out of what the war has left us, just as we have to build new houses with the rubble in the streets.'

He nodded acceptance.

'I need your love,' she said. 'So do Rebecca and Walli.'

He stood up slowly. She looked at him expectantly. He said nothing but, after a long moment, he put his arms around her and the baby, gently embracing them both.

(iv)

Under wartime regulations still in force, the British government had a right to open a coal mine anywhere, regardless of the wishes of the owner of the land. Compensation was paid only for loss of earnings on farmland or commercial property.

Billy Williams, as Minister for Coal, authorized an open-cast mine in the grounds of Tŷ Gwyn, the palatial residence of Earl Fitzherbert on the outskirts of Aberowen.

No compensation was payable as the land was not commercial.

There was uproar on the Conservative benches in the House of Commons. 'Your slag heap will be right under the bedroom windows of the countess!' said one indignant Tory.

Billy Williams smiled. 'The earl's slag heap has been under my mother's window for fifty years,' he said.

Lloyd Williams and Ethel both travelled to Aberowen with Billy the day before the engineers began to dig the hole. Lloyd was reluctant to leave Daisy, who was due to give birth in two weeks, but it was a historic moment, and he wanted to be there.

Both his grandparents were now in their late seventies. Granda was almost blind despite his pebble-lensed glasses, and Grandmam was bent-backed. 'This is nice,' Grandmam said when they all sat around the old kitchen table. 'Both my children here.' She served stewed beef with mashed turnips and thick slices of home-made bread spread with the butcher's fat called dripping. She poured large mugs of sweetened milky tea to go with it.

Lloyd had eaten like this frequently as a child, but now he found it coarse. He knew that even in hard times French and Spanish women managed to serve up tasty dishes delicately flavoured with garlic and garnished with herbs. He was ashamed of his fastidiousness, and pretended to eat and drink with relish.

'Pity about the gardens at Tŷ Gwyn,' Grandmam said tactlessly.

Billy was stung. 'What do you mean? Britain needs the coal.'

'But people love those gardens. Beautiful, they are. I've been there at least once every year since I was a girl. Shame it is to see them go.'

'There's a perfectly good recreation ground right in the middle of Aberowen!'

'It's not the same,' said Grandmam imperturbably.

Granda said: 'Women will never understand politics.'

'No,' said Grandmam. 'I don't suppose we will.'

Lloyd caught his mother's eye. She smiled and said nothing.

Billy and Lloyd shared the second bedroom, and Ethel made up a bed on the kitchen floor. 'I slept in this room every night of my life until I went in the army,' Billy said as they lay down. 'And I looked out the window every morning at that fucking slag heap.'

'Keep your voice down, Uncle Billy,' Lloyd said. 'You don't want your mother to hear you swear.'

'Aye, you're right,' said Billy.

Next morning after breakfast they all walked up the hill to the big house. It was a mild morning, and for a change there was no rain. The ridge of mountains at the skyline was softened with summer grass. As Tŷ Gwyn came into view, Lloyd could not help seeing it more as a beautiful building than as a symbol of oppression. It was both, of course: nothing was simple in politics.

The great iron gates stood open. The Williams family passed into the grounds. A crowd had gathered already: the contractor's men with their machinery, a hundred or so miners and their families, Earl Fitzherbert with his son Andrew, a handful of reporters with notebooks, and a film crew.

The gardens were breathtaking. The avenue of ancient chestnut trees was in full leaf, there were swans on the lake, and the flower beds blazed with colour. Lloyd guessed the earl had made sure the place looked its best. He wanted to brand the Labour government as wreckers in the eyes of the world.

Lloyd found himself sympathizing with Fitz.

The Mayor of Aberowen was giving an interview. 'The people of this town are against the open-cast mine,' he said. Lloyd was surprised: the town council was Labour, and it must have gone against the grain for them to oppose the government. 'For more than a hundred years, the beauty of these gardens has refreshed the souls of people who live in a grim industrial landscape,' the Mayor went on. Switching from prepared speech to personal reminiscence, he added: 'I proposed to my wife under that cedar tree.'

He was interrupted by a loud clanking sound like the footsteps of an iron giant. Turning to look back along the drive, Lloyd saw a huge machine approaching. It looked like the biggest crane in the world. It had an enormous boom ninety feet long and a bucket into which a lorry could easily fit. Most astonishing of all, it moved along on rotating steel shoes that made the earth shake every time they hit the ground.

Billy said proudly to Lloyd: 'That's a walking monighan dragline excavator. Picks up six tons of earth at a time.'

The film camera rolled as the monstrous machine stomped up the drive.

Lloyd had only one misgiving about the Labour Party. There was a streak of puritan authoritarianism in many socialists. His grandfather had it, and so did Billy. They were not comfortable with sensual pleasures. Sacrifice and self-denial suited them better. They dismissed the ravishing beauty of these gardens as irrelevant. They were wrong.

Ethel was not that way, nor was Lloyd. Perhaps the killjoy strain had been bred out of their line. He hoped so.

Fitz gave an interview on the pink gravel path while the digger driver manoeuvred his machine into position. 'The Minister for Coal has told you that when the mine is exhausted the garden will be subject to what he calls an effective restoration programme,' he said. 'I say to you that that promise is worthless. It has taken more than a century for my grandfather and my father and I to bring the garden to its present pitch of beauty and harmony. It would take another hundred years to restore it.'

The boom of the excavator was lowered until it stood at a forty-five-degree angle over the shrubbery and flower beds of the west garden.

The bucket was positioned over the croquet lawn. There was a long moment of waiting. The crowd fell silent. Billy said loudly: 'Get on with it, for God's sake.'

An engineer in a bowler hat blew a whistle.

The bucket was dropped to the earth with a massive thud. Its steel teeth dug into the flat green lawn. The drag rope tautened, there was a loud creak of straining machinery, then the bucket began to move back. As it was dragged across the ground it dug up a bed of huge yellow sunflowers, the rose garden, a shrubbery of summersweet and bottlebrush buckeye, and a small magnolia tree. At the end of its travel the bucket was full of earth, flowers and plants.

The bucket was then lifted to a height of twenty feet, dribbling loose earth and blossoms.

The boom swung sideways. It was taller than the house, Lloyd saw. He almost thought the bucket would smash the upstairs windows, but the operator was skilled, and stopped it just in time. The drag rope slackened, the bucket tilted, and six tons of garden fell to the ground a few feet from the entrance.

The bucket was returned to its original position, and the process was repeated.

Lloyd looked at Fitz and saw that he was crying.

23

At the beginning of 1947 it seemed possible that all Europe might go Communist.

Volodya Peshkov was not sure whether to hope for that or its opposite.

The Red Army dominated Eastern Europe, and Communists were winning elections in the West. Communists had gained respect for their role in resisting the Nazis. Five million people had voted Communist in the first French postwar election, making the Communists the most popular party. In Italy a Communist-Socialist alliance won 40 per cent of the vote. In Czechoslovakia the Communists on their own won 38 per cent and led the democratically elected government.

It was different in Austria and Germany, where voters had been robbed and raped by the Red Army. In the Berlin city elections, the Social Democrats won 63 of 130 seats, the Communists only 26. However, Germany was ruined and starving, and the Kremlin still hoped that the people might turn to Communism in desperation, just as they had turned to Nazism in the Depression.

Britain was the great disappointment. Only one Communist had been sent to Parliament in the postwar election there. And the Labour government was delivering everything Communism promised: welfare, free health care, education for all, even a five-day week for coal miners.

But in the rest of Europe, capitalism was failing to lift people out of the postwar slump.

And the weather was on Stalin's side, Volodya thought as the layers of snow grew thick on the onion domes. The winter of 1946–7 was the coldest in Europe for more than a century. Snow fell in St Tropez. British roads and railways became impassable, and industry ground to a halt – something that had never happened in the war. In France, food rations fell below wartime levels. The United Nations Organization calculated that 100 million Europeans were living on 1,500 calories a day – the level at which health begins to suffer from the effects of malnutrition. As the engines of production ran slower and slower, people began to feel they had nothing to lose, and revolution came to seem the only way out.

Once the USSR had nuclear weapons, no other country would be able to stand in its way. Volodya's wife, Zoya, and her colleagues had built a nuclear pile, at Laboratory No. 2 of the Academy of Sciences, a deliberately vague name for the powerhouse of Soviet nuclear research. The pile had gone critical on Christmas Day, six months after the birth of Konstantin, who was at the time sleeping in the laboratory's crèche. If the experiment went wrong, Zoya had whispered to Volodya, it would do little Kotya no good to be a mile or two away: all of central Moscow would be flattened.

Volodya's conflicting feelings about the future took on a new intensity with the birth of his son. He wanted Kotya to grow up a citizen of a proud and powerful country. The Soviet Union deserved to dominate Europe, he felt. It was the Red Army that had defeated the Nazis, in four cruel years of total warfare: the other Allies had stood on the sidelines, fighting minor wars, joining in only for the last eleven months. All their casualties put together were only a fraction of those suffered by the Soviet people.

But then he would think of what Communism meant: arbitrary purges, torture in the basements of the secret police, conquering soldiers urged on to excesses of bestiality, the whole vast country forced to obey the wayward decisions of a tyrant more powerful than a tsar. Did Volodya really want to extend that brutal system to the rest of the continent?

He remembered walking into Penn Station in New York and buying a ticket to Albuquerque, without asking anyone's permission or showing any papers; and the exhilarating sense of total freedom that had given him. He had long ago burned the *Sears-Roebuck Catalogue*, but it lived in his memory, with its hundreds of pages of good things available for everyone to have. Russian people believed that stories of Western freedom and prosperity were just propaganda, but Volodya knew better. A part of him longed for Communism to be defeated.

The future of Germany, and therefore of Europe, was to be decided at the Conference of Foreign Ministers held in Moscow in March 1947.

Volodya, now a colonel, was in charge of the intelligence team assigned to the conference. Meetings were held in an ornate room at Aviation Industry House, conveniently close to the Hotel Moskva. As always, the delegates and their interpreters sat around a table, with their aides on several rows of chairs behind them. The Soviet Foreign Minister, Vyacheslav Molotov, Old Stone Arse, demanded that Germany pay ten billion dollars to the USSR in war reparations. The Americans and British protested that this would be a death blow to Germany's sickly economy. That was probably what Stalin wanted.

Volodya renewed his acquaintance with Woody Dewar, who was now a news photographer assigned to cover the conference. He was married, too, and showed Volodya a photo of a striking dark-haired woman holding a baby. Sitting in the back of a ZIS-110B limousine, returning from a formal photo session at the Kremlin, Woody said to Volodya: 'You realize that Germany doesn't have the money to pay your reparations, don't you?'

Volodya's English had improved, and they could manage without an interpreter. He said: 'Then how are they feeding their people and rebuilding their cities?'

'With handouts from us, of course,' said Woody. 'We're spending a fortune in aid. Any reparations the Germans paid you would be, in reality, our money.'

'Is that so wrong? The United States prospered in the war. My country was devastated. Maybe you should pay.'

'American voters don't think so.'

'American voters may be wrong.'

Woody shrugged. 'True – but it's their money.'

There it was again, Volodya thought: the deference to public opinion. He had remarked on it before in Woody's conversation. Americans talked about voters the way Russians talked about Stalin: they had to be obeyed, right or wrong.

Woody wound down the window. 'You don't mind if I take a cityscape, do you? The light is wonderful.' His camera clicked.

He knew he was supposed to take only approved shots. However, there was nothing sensitive on the street, just some women shovelling snow. All the same, Volodya said: 'Please don't.' He leaned past Woody and wound up the window. 'Official photos only.'

He was about to ask for the film out of Woody's camera when Woody said: 'Do you remember me mentioning my friend Greg Peshkov, with the same surname as you?'

Volodya certainly did. Willi Frunze had said something similar. It was probably the same man. 'No, I don't remember,' Volodya lied. He wanted nothing to do with a possible relative in the West. Such connections brought suspicion and trouble to Russians.

'He's on the American delegation. You should talk to him. See if you're related.'

'I will,' said Volodya, resolving to avoid the man at all costs.

He decided not to insist on taking Woody's film. It was not worth the fuss for a harmless street scene.

At the next day's conference the American Secretary of State, George Marshall, proposed that the four Allies should abolish the separate sectors of Germany and unify the country, so that it could once again become the beating economic heart of Europe, mining and manufacturing and buying and selling.

That was the last thing the Soviets wanted.

Molotov refused to discuss unification until the question of reparations had been settled.

The conference was stalemated.

And that, Volodya thought, was exactly where Stalin wanted it.

(ii)

The world of international diplomacy was a small one, Greg Peshkov reflected. One of the young aides in the British delegation at the Moscow conference was Lloyd Williams, the husband of Greg's half-sister, Daisy. At first Greg did not like the look of Lloyd, who was dressed like a prissy English gentleman; but he turned out to be a regular guy. 'Molotov is a prick,' Lloyd said in the bar of the Hotel Moskva over a couple of vodka martinis.

'So what are we going to do about him?'

'I don't know, but Britain can't live with these delays. The occupation of Germany is costing money we can't afford, and the hard winter has turned the problem into a crisis.'

'You know what?' said Greg, thinking aloud. 'If the Soviets won't play ball, we should just go ahead without them.'

'How could we do that?'

'What do we want?' Greg counted points on his fingers. 'We want to unify Germany and hold elections.'

'So do we.'

'We want to scrap the worthless Reichsmark and introduce a new currency, so that Germans can start to do business again.'

'Yes.'

'And we want to save the country from Communism.'

'Also British policy.'

'We can't do it in the east because the Soviets won't come to the party. So fuck them! We control three quarters of Germany – let's do it in our zone, and let the eastern part of the country go to blazes.'

Lloyd looked thoughtful. 'Is this something you've discussed with your boss?'

'Hell, no. I'm just running off at the mouth. But listen, why not?'

'I might suggest it to Ernie Bevin.'

'And I'll put it to George Marshall.' Greg sipped his drink. 'Vodka is the only thing the Russians do well,' he said. 'So, how's my sister?'

'She's expecting our second baby.'

'What is Daisy like as a mother?

Lloyd laughed. 'You think she's probably terrible.'

Greg shrugged. 'I never saw her as the domestic type.'

'She's patient, calm and organized.'

'She didn't hire six nurses to do all the work?'

'Just one, so that she can come out with me in the evenings, usually to political meetings.'

'Wow, she's changed.'

'Not completely. She still loves parties. What about you – still single?'

'There's a girl called Nelly Fordham that I'm pretty serious about. And I guess you know that I have a godson.'

'Yes,' said Lloyd. 'Daisy told me all about him. Georgy.'

Greg felt sure, from the slightly embarrassed look on Lloyd's face, that he knew Georgy was Greg's child. 'I'm very fond of him.'

'That's great.'

A member of the Russian delegation came up to the bar, and Greg caught his eye. There was something very familiar about him. He was in his thirties, handsome apart from a brutally short military haircut, and he had a slightly intimidating blue-eyed gaze. He nodded in a friendly way, and Greg said: 'Have we met before?'

'Perhaps,' the Russian said. 'I was at school in Germany – the Berlin Boys' Academy.'

Greg shook his head. 'Ever been to the States?'

'No.'

Lloyd said: 'This is the guy with the same surname as you, Volodya Peshkov.'

Greg introduced himself. 'We might be related. My father, Lev Peshkov, emigrated in 1914, leaving behind a pregnant girlfriend, who then married his older brother, Grigori Peshkov. Could we be half-brothers?'

Volodya's manner altered immediately. 'Definitely not,' he said. 'Excuse me.' He left the bar without buying a drink.

'That was abrupt,' Greg said to Lloyd.

'It was,' said Lloyd.

'He looked kind of shocked.'

'It must have been something you said.'

(iii)

It could not be true, Volodya told himself.

Greg claimed that Grigori had married a girl who was already pregnant by Lev. If that was the case, the man Volodya had always called father was not his father but his uncle.

Perhaps it was a coincidence. Or the American could just be stirring up trouble.

All the same Volodya was reeling with shock.

He returned home at his usual time. He and Zoya were rising fast and had been given an apartment in Government House, the luxury block where his parents lived. Grigori and Katerina came to the apartment at Kotya's suppertime, as they did most evenings. Katerina bathed her grandson, then Grigori sang to him and told him Russian fairy tales. Kotya was nine months old and not yet talking, but he seemed to like bedtime stories just the same.

Volodya followed the evening routine as if sleepwalking. He tried to behave normally, but he found he could hardly speak to either of his parents. He did not believe Greg's story, but he could not stop thinking about it.

When Kotya was asleep, and the grandparents were about to leave, Grigori said to Volodya: 'Have I got a boil on my nose?'

'No.'

'Then why have you been staring at me all evening?'

Volodya decided to tell the truth. 'I met a man called Greg Peshkov. He's part of the American delegation. He thinks we're related.'

'It's possible.' Grigori's tone was light, as if it did not much matter, but Volodya saw that his neck had reddened, a giveaway sign of suppressed emotion in his father. 'I last saw my brother in 1919. Since then I haven't heard from him.'

'Greg's father is called Lev, and Lev had a brother called Grigori.'

'Then Greg could be your cousin.'

'He said brother.'

Grigori's blush deepened and he said nothing.

Zoya put in: 'How could that be?'

Volodya said: 'According to this American Peshkov, Lev had a pregnant girlfriend in St Petersburg who married his brother.'

Grigori said: 'Ridiculous!'

Volodya looked at Katerina. 'You haven't said anything, Mother.'

There was a long pause. That in itself was significant. What did they have to think about, if there was no truth in Greg's story? A weird coldness descended on Volodya, like a freezing fog.

At last his mother said: 'I was a flighty girl.' She looked at Zoya. 'Not sensible, like your wife.' She sighed deeply. 'Grigori Peshkov fell in love with me, more or less at first sight, poor idiot.' She smiled fondly at her husband. 'But his brother, Lev, had fancy clothes, cigarettes, money for vodka, gangster friends. I liked Lev better. More fool me.'

Volodya said amazedly: 'So it's true?' Part of him still hoped desperately for a denial.

'Lev did what such men always do,' Katerina said. 'He made me pregnant then left me.'

'So Lev is my father.' Volodya looked at Grigori. 'And you're just my uncle!' He felt as if he might fall over. The ground under his feet had shifted. It was like an earthquake.

Zoya stood beside Volodya's chair and put her hand on his shoulder, as if to calm him, or perhaps restrain him.

Katerina went on: 'And Grigori did what men such as Grigori always do: he took care of me. He loved me, he married me, and he provided for me and my children.' Sitting on the couch next to Grigori, she took his hand. 'I didn't want him, and I certainly didn't deserve him, but God gave him to me anyway.'

Grigori said: 'I have dreaded this day. Ever since you were born I have dreaded it.'

Volodya said: 'Then why did you keep the secret? Why didn't you just speak the truth?'

Grigori was choked up, and spoke with difficulty. 'I couldn't bear to tell you that I wasn't your father,' he managed to say. 'I loved you too much.'

Katerina said: 'Let me tell you something, my beloved son. Listen to me, now, and I don't care if you never listen to your mother again, but hear this. Forget the stranger in America who once seduced a foolish girl. Look at the man sitting in front of you with tears in his eyes.'

Volodya looked at Grigori and saw a pleading expression that tugged at his heart.

Katerina went on: 'This man has fed you and clothed you and loved you unfailingly for three decades. If the word *father* means anything at all, this is your father.'

'Yes,' Volodya said. 'I know that.'

(iv)

Lloyd Williams got on well with Ernie Bevin. They had a lot in common, despite the age difference. During the four-day train journey across snowy Europe Lloyd had confided that he, like Bevin, was the illegitimate son of a housemaid. They were both passionate anti-Communists: Lloyd because of his experiences in Spain, Bevin because he had seen Communist tactics in the trade union movement. 'They're slaves to the Kremlin and tyrants over everyone else,' Bevin said, and Lloyd knew exactly what he meant.

Lloyd had not warmed to Greg Peshkov, who always looked as if he had dressed in a rush: shirtsleeves unbuttoned, coat collar twisted, shoelaces untied. Greg was shrewd, and Lloyd tried to like him, but he felt that underneath Greg's casual charm there was a core of ruthlessness. Daisy had said that Lev Peshkov was a gangster, and Lloyd could imagine that Greg had the same instincts.

However, Bevin jumped at Greg's idea for Germany. 'Was he speaking for Marshall, do you suppose?' said the portly Foreign Secretary in his broad West Country accent.

'He said not,' Lloyd replied. 'Do you think it could work?'

'I think it's the best idea I've heard in three bloody weeks in bloody Moscow. If he's serious, arrange an informal lunch, just Marshall and this youngster with you and me.'

'I'll do it right away.'

'But tell nobody. We don't want the Soviets to get a whisper of this. They'll accuse us of conspiring against them, and they'll be right.'

They met the following day at No. 10 Spasopeskovskaya Square, the American Ambassador's residence, an extravagant neoclassical mansion built before the revolution. Marshall was tall and lean, every inch a soldier; Bevin rotund, nearsighted, a cigarette frequently dangling from his lips; but they clicked immediately. Both were plain-speaking men. Bevin had once been accused of ungentlemanly speech by Stalin himself, a distinction of which the Foreign Secretary was very proud. Beneath the painted ceilings and chandeliers they got down to the task of reviving Germany without the help of the USSR.

They agreed rapidly on the principles: the new currency, the unification of the British, American, and – if possible – French zones; the demilitarization of West Germany; elections; and a new transatlantic military alliance. Then Bevin said bluntly: 'None of this will work, you know.'

Marshall was taken aback. 'Then I fail to understand why we're discussing it,' he said sharply.

'Europe's in a slump. This scheme will fail if people are starving. The best protection against Communism is prosperity. Stalin knows that – which is why he wants to keep Germany impoverished.'

'I agree.'

'Which means we've got to rebuild. But we can't do it with our bare hands. We need tractors, lathes, excavators, rolling stock – all of which we can't afford.'

Marshall saw where he was going. 'Americans aren't willing to give Europeans any more handouts.'

'Fair enough. But there must be a way the USA can lend us the money we need to buy equipment from you.'

There was a silence.

Marshall hated to waste words, but this was a long pause even by his standards.

Then at last he spoke. 'It makes sense,' he said. 'I'll see what I can do.'

The conference lasted six weeks and, when they all went home again, nothing had been decided.

(v)

Eva Williams was a year old when she got her back teeth. The others had come fairly easily, but these hurt. There was not much Lloyd and Daisy could do for her. She was miserable, she could not sleep, she would not let them sleep, and they were miserable too.

Daisy had a lot of money, but they lived unostentatiously. They had bought a pleasant row house in Hoxton, where their neighbours were a shopkeeper and a builder. They got a small family car, a new Morris Eight with a top speed of almost sixty miles per hour. Daisy still bought pretty clothes, but Lloyd had just three suits: evening dress, a chalk stripe for the House of Commons, and tweeds for constituency work at the weekends.

Lloyd was in his pyjamas late one evening, trying to rock the grizzling Evie to sleep, and at the same time leafing through *Life* magazine. He noticed a striking photograph taken in Moscow. It showed a Russian woman, wearing a headscarf and a coat tied with string like a parcel, her old face deeply lined, shovelling snow on the street. Something about the way the light struck her gave her a look of timelessness, as if she had been there for a thousand years. He looked for the photographer's name and found it was Woody Dewar, whom he had met at the conference.

The phone rang. He picked it up and heard the voice of Ernie Bevin. 'Turn your wireless on,' Bevin said. 'Marshall's made a speech.' He hung up without waiting for a reply.

Lloyd went downstairs to the living room, still carrying Evie, and switched on the radio. The show was called *American Commentary.* The BBC's Washington correspondent, Leonard Miall, was reporting from Harvard University in Cambridge, Massachusetts. 'The Secretary of State told alumni that the rebuilding of Europe is going to take a longer time, and require a greater effort, than was originally foreseen,' said Miall.

That was promising, Lloyd thought with excitement. 'Hush, Evie, please,' he said, and for once she quietened.

Then Lloyd heard the low, reasonable voice of George C. Marshall. 'Europe's requirements, for the next three or four years, of foreign food and other essential products – principally from America – are so much greater than her present ability to pay that she must have substantial additional help ... or face economic, social and political deterioration of a very grave character.'

Lloyd was electrified. 'Substantial additional help' was what Bevin had asked for.

'The remedy lies in breaking the vicious circle and restoring the confidence of the European people in the economic future,' Marshall said. 'The United States should do whatever it is able to assist in the return of normal economic health in the world.'

'He's done it!' Lloyd said triumphantly to his uncomprehending baby daughter. 'He's told America they have to give us aid! But how much? And how, and when?'

The voice changed, and the reporter said: 'The Secretary of State did not outline a detailed plan for aid to Europe, but said it was up to the Europeans to draft the programme.'

'Does that mean we have carte blanche?' Lloyd eagerly asked Evie.

Marshall's voice returned to say: 'The initiative, I think, must come from Europe.'

The report ended, and the phone rang again. 'Did you hear that?' said Bevin.

'What does it mean?'

'Don't ask!' said Bevin. 'If you ask questions, you'll get answers you don't want.'

'All right,' Lloyd said, baffled.

'Never mind what he meant. The question is what we do. The initiative must come from Europe, he said. That means me and you.'

'What can I do?'

'Pack a bag,' said Bevin. 'We're going to Paris.'

24

1948

Volodya was in Prague as part of a Red Army delegation holding talks with the Czech military. They were staying in art deco splendour at the Imperial Hotel.

It was snowing.

He missed Zoya and little Kotya. His son was two years old and learning new words at bewildering speed. The child was changing so fast that he seemed different every day. And Zoya was pregnant again. Volodya resented having to spend two weeks apart from his family. Most of the men in the group saw the trip as a chance to get away from their wives, drink too much vodka, and maybe fool around with loose women. Volodya just wanted to go home.

The military talks were genuine, but Volodya's part in them was a cover for his real assignment, which was to report on the activities in Prague of the ham-fisted Soviet secret police, perennial rivals of Red Army Intelligence.

Volodya had little enthusiasm for his work nowadays. Everything he had once believed in had been undermined. He no longer had faith in Stalin, Communism, or the essential goodness of the Russian people. Even his father was not his father. He would have defected to the West if he could have found a way of getting Zoya and Kotya out with him.

However, he did have his heart in his mission here in Prague. It was a rare chance to do something he believed in.

Two weeks ago the Czech Communist party had taken full control of the government, ousting their coalition partners. Foreign Minister Jan Masaryk, a war hero and democratic anti-Communist, had become a prisoner on the top floor of his official residence, the Czernin Palace. The Soviet secret police had undoubtedly been behind the coup. In fact Volodya's brother-in-law, Colonel Ilya Dvorkin, was also in Prague, staying at the same hotel, and had almost certainly been involved.

Volodya's boss, General Lemitov, saw the coup as a public relations catastrophe for the USSR. Masaryk had constituted proof, to the world, that East European countries could be free and independent in the shadow of the USSR. He had enabled Czechoslovakia to have a Communist government friendly to the Soviet Union and at the same time wear the costume of bourgeois democracy. This had been the perfect arrangement, for it gave the USSR everything it wanted while reassuring the Americans. But that equilibrium had been upset.

However, Ilya was crowing. 'The bourgeois parties have been smashed!' he said to Volodya in the hotel bar one night.

'Did you see what happened in the American Senate?' Volodya said mildly. 'Vandenberg, the old isolationist, made an eighty-minute speech in favour of the Marshall Plan, and he was cheered to the rafters.'

George Marshall's vague ideas had become a plan. This was mainly thanks to the rat-like cunning of British Foreign Secretary Ernie Bevin. In Volodya's opinion, Bevin was the most dangerous kind of anti-Communist: a working-class social democrat. Despite his bulk he moved fast. With lightning speed he had organized a conference in Paris that had given a resounding collective European welcome to George Marshall's Harvard speech.

Volodya knew, from spies in the British Foreign Office, that Bevin was determined to bring Germany into the Marshall Plan and keep the USSR out. And Stalin had fallen straight into Bevin's trap, by commanding the East European countries to repudiate Marshall Aid.

Now the Soviet secret police seemed to be doing all they could to assist the passage of the bill through Congress. 'The Senate was all set to reject Marshall,' Volodya said to Ilya. 'American taxpayers don't want to foot the bill. But the coup here in Prague has persuaded them that they have to, because European capitalism is in danger of collapse.'

Ilya said indignantly: 'The bourgeois Czech parties wanted to take the American bribe.'

'We should have let them,' said Volodya. 'It might have been the quickest way to sabotage the whole scheme. Congress would then have rejected the Marshall Plan – they don't want to give money to Communists.'

'The Marshall Plan is an imperialist trick!'

'Yes, it is,' said Volodya. 'And I'm afraid it's working. Our wartime allies are forming an anti-Soviet bloc.'

'People who obstruct the forward march of Communism must be dealt with appropriately.'

'Indeed they must.' It was amazing how consistently people such as Ilya made the wrong political judgements.

'And I must go to bed.'

It was only ten, but Volodya went too. He lay awake thinking about Zoya and Kotya and wishing he could kiss them both goodnight.

His thoughts drifted to his mission. He had met Jan Masaryk, the symbol of Czech independence, two days earlier, at a ceremony at the grave of his father, Thomas Masaryk, the founder and first President of Czechoslovakia. Dressed in a coat with a fur collar, head bared to the falling snow, the second Masaryk had seemed beaten and depressed.

If he could be persuaded to stay on as Foreign Minister, some compromise might be possible, Volodya mused. Czechoslovakia could have a thoroughly Communist domestic government, but in its international relations it might be neutral, or at least minimally anti-American. Masaryk had both the diplomatic skills and the international credibility to walk that tightrope.

Volodya decided he would suggest it to Lemitov tomorrow.

He slept fitfully and woke before six o'clock with a mental alarm ringing in his imagination. It was something about last night's conversation with Ilya. Volodya ran over it again in his mind. When Ilya had said *People who obstruct the forward march of Communism* he had been talking about Masaryk; and when a secret policeman said someone had to be *dealt with appropriately* he always meant *killed*.

Then Ilya had gone to bed early, which suggested an early start this morning.

I'm a fool, Volodya thought. The signs were there and it took me all night to read them.

He leaped out of bed. Perhaps he was not too late.

He dressed quickly and put on a heavy overcoat, scarf and hat. There were no taxis outside the hotel – it was too early. He could have called a Red Army car, but by the time a driver was awakened and the car brought it would take the best part of an hour.

He set out to walk. The Czernin Palace was only a mile or two away. He headed west out of Prague's gracious city centre, crossed the St Charles Bridge, and hurried uphill towards the castle.

Masaryk was not expecting him, nor was the Foreign Minister obliged to give audience to a Red Army colonel. But Volodya felt sure Masaryk would be curious enough to see him.

He walked fast through the snow and reached the Czernin Palace at six-forty-five. It was a huge baroque building with a grandiose row of Corinthian half-columns on the three upper storeys. The place was lightly guarded, he found to his surprise. A sentry pointed to the front door. Volodya walked unchallenged through an ornate hall.

He had expected to find the usual secret police moron behind a reception desk, but there was no one. This was a bad sign, and he was filled with foreboding.

The hall led to an inner courtyard. Glancing through a window, he saw what looked like a man sleeping in the snow. Perhaps he had fallen there drunk: if so, he was in danger of freezing to death.

Volodya tried the door and found it open.

He ran across the quadrangle. A man in blue silk pyjamas lay face down on the ground. There was no snow covering him, so he could not have been there many minutes. Volodya knelt beside him. The man was quite still and did not appear to be breathing.

Volodya looked up. Rows of identical windows like soldiers on parade looked into this courtyard. All were closed tightly against the freezing weather – except one, high above the man in pyjamas, that stood wide open.

As if someone had been thrown out of it.

Volodya turned the lifeless head and looked at the man's face.

It was Jan Masaryk.

(ii)

Three days later in Washington, the Joint Chiefs of Staff presented to President Truman an emergency war plan to meet a Soviet invasion of Western Europe.

The danger of a third world war was a hot topic in the press. 'We just *won* the war,' Jacky Jakes said to Greg Peshkov. 'How come we're about to have another?'

'That's what I keep asking myself,' said Greg.

They were sitting on a park bench while Greg took a breather from throwing a football with Georgy.

'I'm glad he's too young to fight,' Jacky said.

'Me, too.'

They both looked at their son, standing talking to a blonde girl about his age. The laces of his Keds were undone and his shirt was untucked. He was twelve years old and growing up. He had a few soft black hairs on his upper lip, and he seemed three inches taller than last week.

'We've been bringing our troops home as fast as we can,' Greg said. 'So have the British and the French. But the Red Army stayed put. Result: they now have three times as many soldiers in Germany as we do.'

'Americans don't want another war.'

'You can say that again. And Truman hopes to win the Presidential election in November, so he's going to do everything he can to avoid war. But it may happen anyway.'

'You're getting out of the army soon. What are you going to do?'

There was a quaver in her voice that made him suspect the question was not as casual as she pretended. He looked at her face, but her expression was unreadable. He answered: 'Assuming America is not at war, I'm going to run for Congress in 1950. My father has agreed to finance my campaign. I'll start as soon as the Presidential election is over.'

She looked away. 'Which party?' She asked the question mechanically.

He wondered if he had said something to upset her. 'Republican, of course.'

'What about marriage?'

Greg was taken aback. 'Why do you ask that?'

She was looking hard at him now. 'Are you getting married?' she persisted.

'As it happens, I am. Her name is Nelly Fordham.'

'I thought so. How old is she?'

'Twenty-two. What do you mean, you thought so?'

'A politician needs a wife.'

'I love her!'

'Sure you do. Is her family in politics?'

'Her father is a Washington lawyer.'

'Good choice.'

Greg felt annoyed. 'You're being very cynical.'

'I know you, Greg. Good Lord, I fucked you when you weren't much older than Georgy is now. You can fool everyone except your mother and me.'

She was perceptive, as always. His mother had also been critical of his engagement. They were right: it was a career move. But Nelly was pretty and charming and she adored Greg, so what was so wrong? 'I'm meeting her for lunch near here in a few minutes,' he said.

Jacky said: 'Does Nelly know about Georgy?'

'No. And we must keep it that way.'

'You're right. Having an illegitimate child is bad enough; a black one could ruin your career.'

'I know.'

'Almost as bad as a black wife.'

Greg was so surprised that he came right out with it. 'Did you think I was going to marry *you*?'

She looked sour. 'Hell, no, Greg. If I was given a choice between you and the Acid Bath Murderer, I'd ask for time to think about it.'

She was lying, he knew. For a moment he contemplated the idea of marrying Jacky. Interracial marriages were unusual, and attracted a good deal of hostility from blacks as well as whites, but some people did it and put up with the consequences. He had never met a girl he liked as much as Jacky; not even Margaret Cowdry, whom he had dated for a couple of years, until she got fed up of waiting for him to propose. Jacky was sharp-tongued, but he liked that, maybe because his mother was the same. There was something deeply attractive about the idea of the three of them being together all the time. Georgy would learn to call him Dad. They could buy a house in a neighbourhood where people were broad-minded, some place that had a lot of students and young professors, maybe Georgetown.

Then he saw Georgy's blonde friend being called away by her parents, a cross white mother wagging a finger in admonition, and he realized that marrying Jacky was the worst idea in the world.

Georgy returned to where Greg and Jacky sat. 'How's school?' Greg asked him.

'I like it better than I used to,' the boy said. 'Math is getting more interesting.'

'I was good at math,' Greg said.

Jacky said: 'Now there's a coincidence.'

Greg stood up. 'I have to go.' He squeezed Georgy's shoulder. 'Keep working on the math, buddy.'

'Sure,' said Georgy.

Greg waved at Jacky and left.

She had been thinking about marriage at the same time as he, no doubt. She knew that coming out of the army was a decision moment for him. It forced him to think about his future. She could not really have thought he would marry her, but all the same she must have harboured a secret fantasy. Now he had shattered it. Well, that was too bad. Even if she had been white he would not have married her. He was fond of her, and he loved the kid, but he had his whole life ahead of him, and he wanted a wife who would bring him connections and support. Nelly's father was a powerful man in Republican politics.

He walked to the Napoli, an Italian restaurant a few blocks from the park. Nelly was already there, her copper-red curls escaping from under a little green hat. 'You look great!' he said. 'I hope I'm not late.' He sat down.

Nelly's face was stony. 'I saw you in the park,' she said.

Greg thought: Oh, shit.

'I was a little early, so I sat for a while,' she said. 'You didn't notice me. Then I started to feel like a snoop, so I left.'

'So you saw my godson?' he said with forced cheerfulness.

'Is that who he is? You're a surprising choice for a godfather. You never even go to church.'

'I'm good to the kid!'

'What's his name?'

'Georgy Jakes.'

'You've never mentioned him before.'

'Haven't I?'

'How old is he?'

'Twelve.'

'So you were sixteen when he was born. That's young to be a godfather.'

'I guess it is.'

'What does his mother do for a living?'

'She's a waitress. Years ago she was an actress. Her stage name was Jacky Jakes. I met her when she was under contract to my father's studio.' That was more or less true, Greg thought uncomfortably.

'And his father?'

Greg shook his head. 'Jacky is single.' A waiter approached. Greg said: 'How about a cocktail?' Perhaps it might ease the tension. 'Two martinis,' he said to the waiter.

'Right away, sir.'

As soon as the waiter had left, Nelly said: 'You're the boy's father, aren't you?'

'Godfather.'

Her voice became contemptuous. 'Oh, stop it.'

'What makes you so sure?'

'He may be black, but he looks like you. He can't keep his shoelaces tied or his shirt tucked in, and nor can you. And he was charming the pants off that little blonde girl he was talking to. Of course he's yours.'

Greg gave in. He sighed and said: 'I was going to tell you.'

'When?'

'I was waiting for the right moment.'

'Before you proposed would have been a good time.'

'I'm sorry.' He was embarrassed, but not really contrite: he thought she was making an unnecessary fuss.

The waiter brought menus and they both looked at them. 'The spaghetti bolognese is great,' said Greg.

'I'm going to get a salad.'

Their martinis arrived. Greg raised his glass and said: 'To forgiveness in marriage.'

Nelly did not pick up her drink. 'I can't marry you,' she said.

'Honey, come on, don't overreact. I've apologized.'

She shook her head. 'You don't get it, do you?'

'What don't I get?'

'That woman sitting on the park bench with you – she loves you.'

'Does she?' Greg would have denied it yesterday, but after today's conversation he was not sure.

'Of course she does. Why hasn't she married? She's pretty enough. By now she could have found a man willing to take on a stepson, if she'd really been trying. But she's in love with you, you rotter.'

'I'm not so sure.'

'And the boy adores you, too.'

'I'm his favourite uncle.'

'Except that you're not.' She pushed her glass across the table. 'You have my drink.'

'Honey, please relax.'

'I'm leaving.' She stood up.

Greg was not used to girls walking out on him. He found it unnerving. Was he losing his allure?

'I want to marry you!' he said. He sounded desperate even to himself.

'You can't marry me, Greg,' she said. She slipped the diamond ring off her finger and put it down on the red checked tablecloth. 'You already have a family.'

She walked out of the restaurant.

(iii)

The world crisis came to a head in June, and Carla and her family were at the centre of it.

The Marshall Plan had been signed into law by President Truman, and the first shipments of aid were arriving in Europe, to the fury of the Kremlin.

On Friday 18 June the Western Allies alerted Germans that they would make an important announcement at eight o'clock that evening. Carla's family gathered around the radio in the kitchen, tuned to Radio Frankfurt, and waited anxiously. The war had been over for three years, yet still they did not know what the future held: capitalism or Communism, unity or fragmentation, freedom or subjugation, prosperity or destitution.

Werner sat beside Carla with Walli, now two and a half, on his knee. They had married quietly a year ago. Carla was working as a nurse again. She was also a Berlin city councillor for the Social Democrats. So was Frieda's husband, Heinrich.

In East Germany the Russians had banned the Social Democratic Party, but Berlin was an oasis in the Soviet sector, ruled by a council of the four main Allies called the Kommandatura, which had vetoed the ban. As a result, the Social Democrats had won, and the Communists had come a poor third after the conservative Christian Democrats. The Russians were incensed and did everything they could to obstruct the elected council. Carla found it frustrating, but she could not give up the hope of independence from the Soviets.

Werner had managed to start a small business. He had searched through the ruins of his father's factory and scavenged a small horde of electrical supplies and radio parts. Germans could not afford to buy new radios, but everyone wanted their old ones repaired. Werner had found some engineers formerly employed at the factory and set them to work fixing broken wireless sets. He was the manager and salesman, going to houses and apartment buildings, knocking on doors, drumming up business.

Maud, also at the kitchen table this evening, worked as an interpreter for the Americans. She was one of the best, and often translated at meetings of the Kommandatura.

Carla's brother Erik was wearing the uniform of a policeman. Having joined the Communist Party – to the dismay of his family – he had got a job as a police officer in the new East German force organized by the Russian occupiers. Erik said the Western Allies were trying to split Germany in two. 'You Social Democrats are secessionists,' he said, quoting the Communist line in the same way he had parroted Nazi propaganda.

'The Western Allies haven't divided anything,' Carla retorted. 'They've opened the borders between their zones. Why don't the Soviets do the same? Then we would be one country again.' He seemed not to hear her.

Rebecca was almost seventeen. Carla and Werner had legally adopted her. She was doing well at school, and good at languages.

Carla was pregnant again, though she had not told Werner. She was thrilled. He had an adopted daughter and a stepson, but now he would have a child of his own as well. She knew he would be delighted when she told him. She was waiting a little longer to be sure.

But she yearned to know in what kind of country her three children were going to live.

An American officer called Robert Lochner came on the air. He had been raised in Germany and spoke the language effortlessly. Beginning at seven o'clock on Monday morning, he explained, West Germany would have a new currency, the Deutsche Mark.

Carla was not surprised. The Reichsmark was worth less every day. Most people were paid in Reichsmarks, if they had a job at all, and the currency could be used for basics such as food rations and bus fares, but everyone preferred to get groceries or cigarettes. Werner in his business charged people in Reichsmarks but offered overnight service for five cigarettes and delivery anywhere in the city for three eggs.

Carla knew from Maud that the new currency had been discussed at the Kommandatura. The Russians had demanded plates so that they could print it. But they had debased the old currency by printing too much, and there was no point in a new currency if the same thing was going to happen. Consequently the West refused and the Soviets sulked.

Now the West had decided to go ahead without the co-operation of the Soviets. Carla was pleased, for the new currency would be good for Germany, but she felt apprehensive about the Soviet reaction.

People in West Germany could exchange sixty inflated old Reichsmarks for three Deutsche Marks and ninety new pennies, said Lochner.

Then he said that none of this would apply in Berlin, at least at first, whereupon there was a collective groan in the kitchen.

Carla went to bed wondering what the Soviets would do. She lay beside Werner, part of her brain listening in case Walli in the next room should cry. The Soviet occupiers had been getting angrier for the last few months. A journalist called Dieter Friede had been kidnapped in the American zone by the Soviet secret police, then held captive: the Soviets at first denied all knowledge, then said they had arrested him as a spy. Three students had been expelled from university for criticizing the Russians in a magazine. Worst of all, a Soviet fighter aircraft buzzed a British European Airways passenger plane landing at Gatow airport and clipped its wing, causing both planes to crash, killing four BEA crew, ten passengers and the Soviet pilot. When the Russians got angry, someone else always suffered.

Next morning the Soviets announced it would be a crime to import Deutsche Marks into East Germany. This included Berlin, the statement said, 'which is part of the Soviet zone'. The Americans immediately denounced this phrase and affirmed that Berlin was an international city, but the temperature was rising, and Carla remained anxious.

On Monday, West Germany got the new currency.

On Tuesday, a Red Army courier came to Carla's house and summoned her to city hall.

She had been summoned this way before, but all the same she was fearful as she left home. There was nothing to stop the Soviets imprisoning her. The Communists had all the same arbitrary powers the Nazis had assumed. They were even using the old concentration camps.

The famous Red City Hall had been damaged by bombing, and the city government was based in the New City Hall in Parochial Strasse. Both buildings were in the Mitte district where Carla lived, which was in the Soviet zone.

When she got there she found that Acting Mayor Louise Schroeder and others had also been called for a meeting with the Soviet liaison officer, Major Otshkin. He informed them that the East German currency was to be reformed, and in future only the new Ostmark would be legal in the Soviet zone.

Acting Mayor Schroeder immediately saw the crucial point. 'Are you telling us that this will apply in all sectors of Berlin?'

'Yes.'

Frau Schroeder was not easily intimidated. 'Under the city constitution, the Soviet occupying power cannot make such a rule for the other sectors,' she said firmly. 'The other Allies must be consulted.'

'They will not object.' He handed over a sheet of paper. 'This is Marshal Sokolovsky's decree. You will bring it before the city council tomorrow.'

Later that evening, as Carla got into bed with Werner, she said: 'You can see what the Soviet tactic is. If the city council were to pass the decree, it would be difficult for the democratically minded Western Allies to overturn it.'

'But the council won't pass it. The Communists are a minority, and no one else will want the Ostmark.'

'No. Which is why I'm wondering what Marshal Sokolovsky has up his sleeve.'

The next morning's newspapers announced that from Friday there would be two competing currencies in Berlin, the Ostmark and the Deutsche Mark. It turned out that the Americans had secretly flown in 250 million in the new currency in wooden boxes marked 'Clay' and 'Bird Dog' which were now stashed all over Berlin.

During the day Carla began to hear rumours from West Germany. The new money had brought about a miracle there. Overnight, more goods had appeared in shop windows: baskets of cherries and neatly tied bundles of carrots from the surrounding countryside, butter and eggs and pastries, and long-hoarded luxuries such as new shoes, handbags, and even stockings at four Deutsche Marks the pair. People had been waiting until they could sell things for real money.

That afternoon Carla set off for City Hall to attend the council meeting scheduled for four o'clock. As she drew near she saw dozens of Red Army trucks parked in the streets around the building, their drivers lounging around, smoking. They were mostly American vehicles that must have been given to the USSR as Lend-Lease aid during the war. She got an inkling of their purpose when she began to hear the sound of an unruly mob. What the Soviet governor had up his sleeve, she suspected, was a truncheon.

In front of City Hall, red flags fluttered above a crowd of several thousand, most of them wearing Communist Party badges. Loudspeaker trucks blared angry speeches, and the crowd chanted: 'Down with the secessionists.'

Carla did not see how she was going to reach the building. A handful of policemen looked on uninterestedly, making no attempt to help councillors get through. It reminded Carla painfully of the attitude of police on the day the Brownshirts had trashed her mother's office, fifteen years ago. She was quite sure the Communist councillors were already inside, and that if Social Democrats did not get into the building the minority would pass the decree and claim it to be valid.

She took a deep breath and began to push through the crowd.

For a few steps she made progress unnoticed. Then someone recognized her. 'American whore!' he yelled, pointing at her. She pressed on determinedly. Someone else spat at her, and a gob of saliva smeared her dress. She kept going, but she felt panicky. She was surrounded by people who hated her, something she had never experienced, and it made her want to run away. She was shoved, but managed to keep her feet. A hand grasped her dress, and she pulled free with a tearing sound. She wanted to scream. What would they do, rip all her clothes off?

Someone else was fighting his way through the crowd behind her, she realized, and she looked back and saw Heinrich von Kessel, Frieda's husband. He drew level with her and they barrelled on together. Heinrich was more aggressive, stamping on toes and vigorously elbowing everyone within range, and together they moved faster, and at last reached the door and went in.

But their ordeal was not over. There were Communist demonstrators inside too, hundreds of them. They had to fight through the corridors. In the meeting hall the demonstrators were everywhere – not just in the visitors' gallery but on the floor of the chamber. Their behaviour here was just as aggressive as outside.

Some Social Democrats were here, and others arrived after Carla. Somehow most of the sixty-three had been able to fight their way through the mob. She was relieved. The enemy had not managed to scare them off.

When the speaker of the assembly called for order, a Communist assemblyman standing on a bench urged the demonstrators to stay. When he saw Carla he yelled: 'Traitors stay outside!'

It was all grimly reminiscent of 1933: bullying, intimidation, and democracy being undermined by rowdyism. Carla was in despair.

Glancing up to the gallery, she was appalled to see her brother, Erik, among the yelling mob. 'You're German!' she screamed at him. 'You lived under the Nazis. Have you learned *nothing*?'

He seemed not to hear her.

Frau Schroeder stood on the platform, calling for calm. She was jeered and booed by the demonstrators. Raising her voice to a shout, she said: 'If the city council cannot hold an orderly debate in this building, I will move the meeting to the American sector.'

There was renewed abuse, but the twenty-six Communist councillors saw that this move would not suit their purpose. If the council met outside the Soviet zone once, it might do so again, and even move permanently out of the range of Communist intimidation. After a short discussion, one of them stood up and told the demonstrators to leave. They filed out, singing the '*Internationale*'.

'It's obvious whose command they're under,' Heinrich said.

At last there was quiet. Frau Schroeder explained the Soviet demand, and said that it could not apply outside the Soviet sector of Berlin unless it was ratified by the other Allies.

A Communist deputy made a speech accusing her of taking orders directly from New York.

Accusations and abuse raged to and fro. Eventually they voted. The Communists unanimously backed the Soviet decree – after accusing others of being controlled from outside. Everyone else voted against, and the motion was defeated. Berlin had refused to be bullied. Carla felt wearily triumphant.

However, it was not yet over.

By the time they left it was seven o'clock in the evening. Most of the mob had disappeared, but there was a thuggish hardcore still hanging around the entrance. An elderly woman councillor was kicked and punched as she left. The police looked on with indifference.

Carla and Heinrich left by a side door with a few friends, hoping to depart unobserved, but a Communist on a bicycle was monitoring the exit. He rode off quickly.

As the councillors hurried away, he returned at the head of a small gang. Someone tripped Carla, and she fell to the ground. She was kicked painfully once, twice, three times. Terrified, she covered her belly with her hands. She was almost three months pregnant – the stage at which most miscarriages occurred, she knew. Will Werner's baby die, she thought desperately, kicked to death on a Berlin street by Communist thugs?

Then they disappeared.

The councillors picked themselves up. No one was badly injured. They moved off together, fearful of a recurrence, but it seemed the Communists had roughed up enough people for one day.

Carla got home at eight o'clock. There was no sign of Erik.

Werner was shocked to see her bruises and torn dress. 'What happened?' he said. 'Are you all right?'

She burst into tears.

'You're hurt,' Werner said. 'Should we go to the hospital?'

She shook her head vigorously. 'It's not that,' she said. 'I'm just bruised. I've had worse.' She slumped in a chair. 'Christ, I'm tired.'

'Who did this?' he asked angrily.

'The usual people,' she said. 'They call themselves Communists instead of Nazis, but they're the same type. It's 1933 all over again.'

Werner put his arms around her.

She could not be consoled. 'The bullies and the thugs have been in power for so long!' she sobbed. 'Will it ever end?'

(iv)

That night the Soviet news agency put out an announcement. From six o'clock in the morning, all passenger and freight transport in and out of West Berlin – trains, cars and canal barges – would be stopped. No supplies of any kind would get through: no food, no milk, no medicines, no coal. Because the electricity generating stations would therefore be shut down, they were switching off the supply of electricity – to Western sectors only.

The city was under siege.

Lloyd Williams was at British military headquarters. There was a short Parliamentary recess, and Ernie Bevin had gone on holiday to Sandbanks, on the south coast of England, but he was worried enough to send Lloyd to Berlin to observe the introduction of the new currency and keep him informed.

Daisy had not accompanied Lloyd. Their new baby, Davey, was only six months old, and anyway Daisy and Eva Murray were organizing a birth control clinic for women in Hoxton that was about to open its doors.

Lloyd was desperately afraid that this crisis would lead to war. He had fought in two wars, and he never wanted to see a third. He had two small children who he hoped would grow up in a peaceful world. He was married to the prettiest, sexiest, most lovable woman on the planet and he wanted to spend many long decades with her.

General Clay, the workaholic American military governor, ordered his staff to plan an armoured convoy that would barrel down the autobahn from Helmstedt, in the west, straight through Soviet territory to Berlin, sweeping all before it.

Lloyd heard about this plan at the same time as the British governor, Sir Brian Robertson, and heard him say in his clipped soldierly tones: 'If Clay does that, it will be war.'

But nothing else made any sense. The Americans came up with other suggestions, Lloyd heard, talking to Clay's younger aides. The Secretary of the Army, Kenneth Royall, wanted to halt the currency reform. Clay told him it had gone too far to be reversed. Next, Royall proposed evacuating all Americans. Clay told him that was exactly what the Soviets wanted.

Sir Brian wanted to supply the city by air. Most people thought that was impossible. Someone calculated that Berlin required 4,000 tons of fuel and food per day. Were there enough airplanes in the *world* to move that much stuff? No one knew. Nevertheless, Sir Brian ordered the Royal Air Force to make a start.

On Friday afternoon Sir Brian went to see Clay, and Lloyd was invited to be part of the entourage. Sir Brian said to Clay: 'The Russians might block the autobahn ahead of your convoy, and wait and see if you have the nerve to attack them; but I don't think they'll shoot planes down.'

'I don't see how we can deliver enough supplies by air,' Clay said again.

'Nor do I,' said Sir Brian. 'But we're going to do it until we think of something better.'

Clay picked up the phone. 'Get me General LeMay in Wiesbaden,' he said. After a minute he said: 'Curtis, have you got any planes there that can carry coal?'

There was a pause.

'Coal,' said Clay more loudly.

Another pause.

'Yes, that is what I said – coal.'

A moment later, Clay looked up at Sir Brian. 'He says the US Air Force can deliver anything.'

The British returned to their headquarters.

On Saturday Lloyd got an army driver and went into the Soviet zone on a personal mission. He drove to the address at which he had visited the von Ulrich family fifteen years ago.

He knew that Maud was still living there. His mother and Maud had resumed correspondence at the end of the war. Maud's letters put a brave face on what was undoubtedly severe hardship. She did not ask for help, and anyway there was nothing Ethel could do for her – rationing was still in force in Britain.

The place looked very different. In 1933 it had been a fine town house, a little run down but still gracious. Now it looked like a dump. Most of the windows had boards or paper instead of glass. There were bullet holes in the stonework, and the garden wall had collapsed. The woodwork had not been painted for many years.

Lloyd sat in the car for a few moments, looking at the house. Last time he came here he had been eighteen, and Hitler had only just become Chancellor of Germany. The young Lloyd had not dreamed of the horrors the world was going to see. Neither he nor anyone else had suspected how close Fascism would come to triumphing over all Europe, and how much they would have to sacrifice to defeat it. He felt a bit like the von Ulrich house looked, battered and bombed and shot at but still standing.

He walked up the path and knocked.

He recognized the maid who opened the door. 'Hello, Ada, do you remember me?' he said in German. 'I'm Lloyd Williams.'

The house was better inside than out. Ada showed him up to the drawing room, where there were flowers in a glass tumbler on the piano. A brightly patterned blanket had been thrown over the sofa, no doubt to hide holes in the upholstery. The newspapers in the windows let in a surprising amount of light.

A two-year-old boy walked into the room and inspected him with frank curiosity. He was dressed in clothes that were evidently home-made, and he had an Oriental look. 'Who are you?' he said.

'My name is Lloyd. Who are you?'

'Walli,' he said. He ran out again, and Lloyd heard him say to someone outside: 'That man talks funny!'

So much for my German accent, Lloyd thought.

Then he heard the voice of a middle-aged woman. 'Don't make such remarks! It's impolite.'

'Sorry, Grandma.'

Next moment Maud walked in.

Her appearance shocked Lloyd. She was in her mid-fifties, but looked seventy. Her hair was grey, her face was gaunt, and her blue silk dress was threadbare. She kissed his cheek with shrunken lips. 'Lloyd Williams, what a joy to see you!'

She's my aunt, Lloyd thought with a rather queer feeling. But she did not know that: Ethel had kept the secret.

Maud was followed by Carla, who was unrecognizable, and her husband. Lloyd had met Carla as a precocious eleven-year-old: now, he calculated, she was twenty-six. Although she looked half-starved – most Germans did – she was pretty, and had a confident air that surprised Lloyd. Something about the way she stood made him think she might be pregnant. He knew from Maud's letters that Carla had married Werner, who had been a handsome charmer back in 1933 and was still the same.

They spent an hour catching up. The family had been through unimaginable horror, and said so frankly, yet Lloyd still had a sense that they were editing out the worst details. He told them about Daisy, Evie and Dave. During the conversation a teenage girl came in and asked Carla if she could go to her friend's house.

'This is our daughter, Rebecca,' Carla said to Lloyd.

She was about sixteen, so Lloyd supposed she must be adopted.

'Have you done your homework?' Carla asked the girl.

'I'll do it tomorrow morning.'

'Do it now, please,' Carla said firmly.

'Oh, Mother!'

'No argument,' said Carla. She turned back to Lloyd, and Rebecca stomped out.

They talked about the crisis. Carla was deeply involved, as a city councillor. She was pessimistic about the future of Berlin. She thought the Russians would simply starve the population until the West gave in and handed the city over to total Soviet control.

'Let me show you something that may make you feel differently,' Lloyd said. 'Will you come with me in the car?'

Maud stayed behind with Walli, but Carla and Werner went with

Lloyd. He told the driver to take them to Tempelhof, the airport in the American zone. When they arrived he led them to a high window from which they could look down on the runway.

There on the tarmac were a dozen C-47 Skytrain aircraft lined up nose to tail, some with the American star, some with the RAF roundel. Their cargo doors were open, and a truck stood at each one. German porters and American airmen were unloading the aircraft. There were sacks of flour, big drums of kerosene, cartons of medical supplies, and wooden crates containing thousands of bottles of milk.

While they watched, empty aircraft were taking off and more were coming in to land.

'This is amazing,' said Carla, her eyes glistening. 'I've never seen anything like it.'

'There has never *been* anything like it,' Lloyd replied.

She said: 'But can the British and Americans keep it up?'

'I think we have to.'

'But for how long?'

'As long as it takes,' said Lloyd firmly.

And they did.

25

Almost halfway through the twentieth century, on 29 August 1949, Volodya Peshkov was on the Ustyurt Plateau, east of the Caspian Sea in Kazakhstan. It was a stony desert in the deep south of the USSR, where nomads herded goats in much the same way as they had in Bible times. Volodya was in a military truck that bounced uncomfortably along a rough track. Dawn was breaking over a landscape of rock, sand, and low thorny bushes. A bony camel, alone beside the road, stared malevolently at the truck as it passed.

In the dim distance, Volodya saw the bomb tower, lit by a battery of spotlights.

Zoya and the other scientists had built their first nuclear bomb according to the design Volodya had got from Willi Frunze in Santa Fe. It was a plutonium device with an implosion trigger. There were other designs, but this one had worked twice before, once in New Mexico and once at Nagasaki.

So it should work today.

The test was codenamed RDS-1, but they called it First Lightning.

Volodya's truck pulled up at the foot of the tower. Looking up, he saw a clutch of scientists on the platform, doing something with a snake's nest of cables that led to detonators on the skin of the bomb. A figure in blue overalls stepped back, and there was a toss of blonde hair: Zoya. Volodya felt a flush of pride. My wife, he thought; top physicist *and* mother of two.

She conferred with two men, the three heads close together, arguing. Volodya hoped nothing was wrong.

This was the bomb that would save Stalin.

Everything else had gone wrong for the Soviet Union. Western Europe had turned decisively democratic, scared off Communism by bully-boy Kremlin tactics and bought off by Marshall Plan bribes. The USSR had not even been able to take control of Berlin: when the airlift had gone on relentlessly day after day for almost a year, the Soviet Union had given up and reopened the roads and railways. In Eastern Europe, Stalin had retained control only by brute force. Truman had been re-elected President, and considered himself leader of the world. The Americans had stockpiled nuclear weapons, and had stationed B-29 bombers in Britain, ready to turn the Soviet Union into a radioactive wasteland.

But everything would change today.

If the bomb exploded as it should, the USSR and the USA would be equals again. When the Soviet Union could threaten America with nuclear devastation, American domination of the world would be over.

Volodya no longer knew whether that would be good or bad.

If it did not explode, both Zoya and Volodya would probably be purged, sent to labour camps in Siberia or just shot. Volodya had already talked to his parents, and they had promised to take care of Kotya and Galina.

As they would if Volodya and Zoya were killed by the test.

In the strengthening light Volodya saw, at various distances around the tower, an odd variety of buildings: houses of brick and wood, a bridge over nothing, and the entrance to some kind of underground structure. Presumably the army wanted to measure the effect of the blast. Looking more carefully he saw trucks, tanks, and obsolete aircraft, placed for the same purpose, he imagined. The scientists were also going to assess the impact of the bomb on living creatures: there were horses, cattle, sheep, and dogs in kennels.

The confab on the platform ended with a decision. The three scientists nodded and resumed their work.

A few minutes later Zoya came down and greeted her husband.

'Is everything all right?' he said.

'We think so,' Zoya replied.

'You *think* so?'

She shrugged. 'We've never done this before, obviously.'

They got into the truck and drove, across country that was already a wasteland, to the distant control bunker.

The other scientists were close behind.

At the bunker they all put on welders' goggles as the countdown ticked away.

At sixty seconds, Zoya held Volodya's hand.

At ten seconds, he smiled at her and said: 'I love you.'

At one second, he held his breath.

Then it was as if the sun had suddenly risen. A light stronger than noon flooded the desert. In the direction of the bomb tower, a ball of fire grew impossibly high, reaching for the moon. Volodya was startled by the lurid colours in the fireball: green, purple, orange and violet.

The ball turned into a mushroom whose umbrella kept rising. At last the sound arrived, a bang as if the largest artillery piece in the Red Army had been fired a foot away, followed by rolling thunder that reminded Volodya of the terrible bombardment of the Seelow Heights.

At last the cloud began to disperse and the noise faded.

There was a long moment of stunned silence.

Someone said: 'My God, I didn't expect *that*.'

Volodya embraced his wife. 'You did it,' he said.

She looked solemn. 'I know,' she said. 'But *what* did we do?'

'You saved Communism,' said Volodya.

(ii)

'The Russian bomb was based on Fat Man, the one we dropped on Nagasaki,' said Special Agent Bill Bicks. 'Someone gave them the plans.'

'How do you know?' Greg asked him.

'From a defector.'

They were sitting in Bicks's carpeted office in the Washington headquarters of the FBI at nine o'clock in the morning. Bicks had his jacket off. His shirt was stained in the armpits with sweat, though the building was comfortably air-conditioned.

'According to this guy,' Bicks went on, 'a Red Army intelligence colonel got the plans from one of the scientists on the Manhattan Project team.'

'Did he say who?'

'He doesn't know which scientist. That's why I called you in. We need to find the traitor.'

'The FBI checked them all out at the time.'

'And most of them were security risks! There was nothing we could do. But you knew them personally.'

'Who was the Red Army colonel?'

'I was coming to that. You know him. His name is Vladimir Peshkov.'

'My half-brother!'

'Yes.'

'If I were you, I'd suspect me.' Greg said it with a laugh, but he was very uneasy.

'Oh, we did, believe me,' Bicks said. 'You've been subjected to the most thorough investigation I have seen in twenty years with the Bureau.'

Greg gave him a sceptical look. 'No kidding.'

'Your kid's doing well in school, isn't he?'

Greg was shocked. Who could have told the FBI about Georgy? 'You mean my godson?' he said.

'Greg, I said *thorough*. We know he's your son.'

Greg was annoyed, but he suppressed the feeling. He had probed the personal secrets of numerous suspects during his time in Army security. He had no right to object.

'You're clean,' Bicks went on.

'I'm relieved to hear it.'

'Anyway, our defector insisted the plans came from a scientist, rather than any of the normal army personnel working on the project.'

Greg said thoughtfully: 'When I met Volodya in Moscow, he told me he had never been to the United States.'

'He lied,' said Bicks. 'He came here in September 1945. He spent a week in New York. Then we lost him for eight days. He resurfaced briefly then went home.'

'Eight days?'

'Yeah. We're embarrassed.'

'It's enough time to go to Santa Fe, stay a couple of days, and come back.'

'Right.' Bicks leaned forward across his desk. 'But think. If the scientist had already been recruited as a spy, why wasn't he contacted by his regular controller? Why bring someone from Moscow to talk to him?'

'You think the traitor was recruited on this two-day visit? It seems too quick.'

'Possibly he had worked for them before but lapsed. Either way, we're guessing the Soviets needed to send *someone whom the scientist already knew*. That means there ought to be a connection between Volodya and one of the scientists.' Bicks gestured at a side table covered with tan file folders. 'The answer is in there somewhere. Those are our files on every one of the scientists who had access to those plans.'

'What do you want me to do?'

'Go through them.'

'Isn't that your job?'

'We've already done it. We didn't find anything. We're hoping you'll spot something we've missed. I'll sit here and keep you company, do some paperwork.'

'It's a long job.'

'You've got all day.'

Greg frowned. Did they know . . . ?

Bicks said confidently: 'You have no plans for the rest of the day.'

Greg shrugged. 'Got any coffee?'

He had coffee and doughnuts, then more coffee, then a sandwich at lunchtime, then a banana mid-afternoon. He read every known detail about the lives of the scientists, their wives and families: childhood, education, career, love and marriage, achievements and eccentricities and sins.

He was eating the last bite of banana when he said: 'Jesus fucking Christ.'

'What?' said Bicks.

'Willi Frunze went to the Berlin Boys' Academy.' Greg slapped the file triumphantly down on the desk.

'And . . . ?'

'So did Volodya – he told me.'

Bicks thumped his desk in excitement. 'School friends! That's it! We've got the bastard!'

'It's not proof,' said Greg.

'Oh, don't worry, he'll confess.'

'How can you be sure?'

'Those scientists believe that knowledge should be shared with everyone, not kept secret. He'll try to justify himself by arguing that he did it for the good of humanity.'

'Maybe he did.'

'He'll go to the electric chair all the same,' said Bicks.

Greg was suddenly chilled. Willi Frunze had seemed a nice guy. 'Will he?'

'You bet your ass. He's going to fry.'

Bicks was right. Willi Frunze was found guilty of treason and sentenced to death, and he died in the electric chair.

So did his wife.

(iii)

Daisy watched her husband tie his white bow tie and slip into the tailcoat of his perfectly fitting dress suit. 'You look like a million dollars,' she said, and she meant it. He should have been a movie star.

She remembered him thirteen years earlier, wearing borrowed clothes at the Trinity Ball, and she felt a pleasant frisson of nostalgia. He had looked pretty good then, she recalled, even though his suit was two sizes too big.

They were staying in her father's permanent suite at the Ritz-Carlton hotel in Washington. Lloyd was now a junior minister in the British Foreign Office, and he had come here on a diplomatic visit. Lloyd's parents, Ethel and Bernie, were thrilled to be looking after two grandchildren for a week.

Tonight Daisy and Lloyd were going to a ball at the White House.

She was wearing a drop-dead dress by Christian Dior, pink satin with a dramatically spreading skirt made of endless folds of flaring tulle. After the years of wartime austerity she was delighted to be able to buy gowns in Paris again.

She thought of the Yacht Club Ball of 1935 in Buffalo, the event that she imagined, at the time, had ruined her life. The White House was obviously a lot more prestigious, but she knew that nothing that happened tonight could ruin her life. She reflected on that while Lloyd helped her put on her mother's necklace of rose-coloured diamonds with matching earrings. At the age of nineteen she had desperately wanted high-status people to accept her. Now she could hardly imagine worrying about such a thing. As long as Lloyd said she looked fabulous, she did not care what anyone else thought. The only other person whose approval she might seek was her mother-in-law, Eth Leckwith, who had little social status and had certainly never worn a Paris gown.

Did every woman look back and think how foolish she had been when young? Daisy thought again about Ethel, who had certainly behaved foolishly – getting pregnant by her married employer – but never spoke regretfully about it. Maybe that was the right attitude. Daisy contemplated her own mistakes: becoming engaged to Charlie Farquharson, rejecting Lloyd, marrying Boy Fitzherbert. She was not quite able to look back and think about the good that had come of those choices. It was really not until she had been decisively rejected by high society, and had found consolation at Ethel's kitchen in Aldgate, that her life had taken a turn for the better. She had stopped yearning for social status and had learned what real friendship was, and she had been happy ever since.

Now that she no longer cared, she enjoyed parties even more.

'Ready?' said Lloyd.

She was ready. She put on the matching evening coat that Dior had made to go with the dress. They went down in the elevator, left the hotel, and stepped into the waiting limousine.

(iv)

Carla persuaded her mother to play the piano on Christmas Eve.

Maud had not played for years. Perhaps it saddened her by bringing back memories of Walter: they had always played and sung together, and she had often told the children how she had tried, and failed, to teach him to play ragtime. But she no longer told that story, and Carla suspected that nowadays the piano made Maud think of Joachim Koch, the young officer who had come to her for piano lessons, whom she had deceived and seduced, and whom Carla and Ada had killed in the kitchen. Carla herself was not able to shut out the recollection of that nightmare evening, especially getting rid of the body. She did not regret it – they had done the right thing – but, all the same, she would have preferred to forget it.

However, Maud at last agreed to play 'Silent Night' for them all to sing along. Werner, Ada, Erik, and the three children, Rebecca, Walli, and the new baby, Lili, gathered around the old Steinway in the drawing room. Carla put a candle on the piano, and studied the faces of her family in its moving shadows as they sang the familiar German carol.

Walli, in Werner's arms, would be four years old in a few weeks' time, and he tried to sing along, alertly guessing the words and the melody. He had the Oriental eyes of his rapist father: Carla had decided that her revenge would be to raise a son who treated women with tenderness and respect.

Erik sang the words of the hymn sincerely. He supported the Soviet regime as blindly as he had supported the Nazis. Carla had at first been baffled and infuriated, but now she saw a sad logic to it. Erik was one of those inadequate people who were so scared by life that they preferred to live under harsh authority, to be told what to do and what to think by a government that allowed no dissent. They were foolish and dangerous, but there were an awful lot of them.

Carla gazed fondly at her husband, still handsome at thirty. She recalled kissing him, and more, in the front of his sexy car, parked in the Grunewald, when she was nineteen. She still liked kissing him.

When she thought over the time that had passed since then, she had a thousand regrets, but the biggest was her father's death. She missed him constantly and still cried when she remembered him lying in the hall, beaten so cruelly that he did not live until the doctor arrived.

But everyone had to die, and Father had given his life for the sake of a better world. If more Germans had had his courage the Nazis would not have triumphed. She wanted to do all the things he had done: to raise her children well, to make a difference to her country's politics, to love and be loved. Most of all, when she died, she wanted her children to be able to say, as she said of her father, that her life had meant something, and that the world was a better place for it.

The carol came to an end; Maud held the final chord; and little Walli leaned forward and blew the candle out.

ACKNOWLEDGEMENTS

My principal history advisor for The Century Trilogy is Richard Overy. I am grateful also to historians Evan Mawdsley, Tim Rees, Matthias Reiss and Richard Toye for reading the typescript of *Winter of the World* and making corrections.

As always I had invaluable help from my editors and agents, especially Amy Berkower, Leslie Gelbman, Phyllis Grann, Neil Nyren, Susan Opie and Jeremy Trevathan.

I met my agent Al Zuckerman in about 1975 and he has been my most critical and inspiring reader ever since.

Several friends made helpful comments. Nigel Dean has an eye for detail like no one else. Chris Manners and Tony McWalter were as sharply perceptive as ever. Angela Spizig and Annemarie Behnke saved me from numerous errors in the German sections.

We always thank our families, and so we should. Barbara Follett, Emanuele Follett, Jann Turner, and Kim Turner read the first draft and made useful criticisms, as well as giving me the matchless gift of their love.

Ocean
Greenland
Denmark
ICELAND
CANADA
UNITED STATES
North Atlantic Ocean
MEXICO
BAHAMAS
CUBA
DOMINICAN REP.
HAITI
BELIZE
JAMAIQUE
HONDURAS
GUATEMALA
EL SALVADOR
NICARAGUA
COSTA RICA
PANAMA
ANTIGUA AND BARBUDA
DOMINICA
SAINT LUCIA
BARBADOS
GRENADA
CAPE VERDE
WESTERN SAHARA
SENEGAL
GAMBIA
GUINEA BISSAU
SIERRA LEONE
VENEZUELA
COLOMBIA
GUYANA
SURINAM
French Guiana
ECUADOR
BRAZIL
PERU
BOLIVIA
PARAGUAY
URUGUAY
ARGENTINA
South Pacific Ocean